Fall Technical Meeting of the Western States Section of the Combustion Institute (WSS/CI 2017 Fall Meeting)

Laramie, Wyoming, USA
2 – 3 October 2017

ISBN: 978-1-5108-5203-7

Printed from e-media with permission by:

Curran Associates, Inc.
57 Morehouse Lane
Red Hook, NY 12571

Some format issues inherent in the e-media version may also appear in this print version.

Printed by Curran Associates, Inc. (2018)

For permission requests, please contact Western States Section/Combustion Institute at the address below.

Western States Section/Combustion Institute
P.O. Box 969 M S 9052
Livermore, California 94551-0969
USA

Phone: 925-294-3840
Fax: 925-294-2276

http://wssci.us/

Additional copies of this publication are available from:

Curran Associates, Inc.
57 Morehouse Lane
Red Hook, NY 12571 USA
Phone: 845-758-0400
Fax: 845-758-2633
Email: curran@proceedings.com
Web: www.proceedings.com

WSS/CI Western States Section/Combustion Institute

2017 FALL TECHNICAL MEETING
WESTERN STATES SECTION OF THE COMBUSTION INSTITUTE
Hosted by University of Wyoming, Laramie, WY

Sunday, 1 October 2017

4:00 pm **WSSCI Board Meeting:** Salon B in the University of Wyoming Conference Center

Monday, 2 October 2017

7:30 – 5:00 **Registration:** University of Wyoming Conference Center Lobby
7:30 – 8:00 **Continental Breakfast:** University of Wyoming Conference Center Lobby
8:00 - 8:15 **Welcome Address:** Dr. Edmund Synakowski, *Vice President for Research and Economic Development, University of Wyoming*
Welcome Remarks: Prof. Fletcher Miller, *WSSCI Chair, San Diego State University*
Prof. Erica Belmont, *University of Wyoming*

8:15 – 9:15 **Plenary Lecture in Salon C: Dr. Timothy Ombrello, *Air Force Research Lab***
Session Chair: Erica Belmont, *University of Wyoming*

9:15 – 9:25	**BREAK**		
	Laminar Flames I **Salon A** **Session Chair:** A.L. Sánchez	**Turbulent Flames I** **Salon B** **Session Chair:** G. Blanquart	**Fire I** **Salon C** **Session Chair:** S. McAllister
9:25 – 9:45	**1A01:** Porous wall-fed liquid fueled miniature tubular flame burner 1 *V.M. Sauer, J.S. Batther, D. Dunn-Rankin*	**1B01:** Flame stability of turbulent premixed jet flames of large hydrocarbon fuels 170 *N.E. Schorn, J. Bonebreak, D.L. Blunck*	**1C01:** Investigation of merging flames in horizontal and vertical geometries 354 *M. Rhamati, C. Harper, M. Butler, T.H. Fletcher*
9:45 – 10:05	**1A02:** Flame propagation in narrow channels at varying Lewis number 7 *X. Ma, S. Shen, J. Wongwiwat, J. Gross, P. Ronney*	**1B02:** Implementing and assessing the importance of multicomponent transport properties using direct numerical simulations of premixed, turbulent flames *A.J. Fillo, J. Schlup, G. Blanquart, K.E. Niemeyer* 177	**1C02:** Modeling the Coanda effect with FDS and STARCCM+ to predict the effect of fires on slopes for implications of wildland firefighter safety 360 *B. Billings, M. Roberts, B. Butler, T.H. Fletcher*
10:05 – 10:25	**1A03:** Laminar flame speeds of associated gas mixtures measured with flat flame method 16 *F. Arafin, E. Belmont*	**1B03:** Dependence of intermittency on turbulence intensity, fuel type, and simulation fidelity in premixed reacting flows 201 *S.H.R. Whitman, C.A.Z. Towery, A.Y. Poludnenko, P.E. Hamlington*	**1C03:** Plume attachment in wildfires on slopes: Temperature measurements in a laboratory-scale analog 370 *T.P. Grumstrup, M.A. Finney*
10:25 – 10:45	**1A04:** Experimental and numerical study of freely propagating propane cool flames 25 *M. Hajilou, E. Belmont*	**1B04:** Lagrangian analysis of enstrophy in turbulent premixed flames 209 *R. Darragh, C.A.Z. Towery, A.Y. Poludnenko, P.E. Hamlington*	**1C04:** The onset of puffing of pool fires *W. Coenen, D. Moreno, A.L. Sánchez, F.A. Williams* 377
10:45 – 11:00	**BREAK**		

	Laminar Flames II Salon A **Session Chair:** Y.-C. Chien	Turbulent Flames II Salon B **Session Chair:** P. Hamlington	Fire II Salon C **Session Chair:** S. Bhattacharjee
11:00 – 11:20	**1A05:** Temperature profiles and extinction limits of a coflow water-laden methane/air diffusion flame 36 *M. Vicariotto, P. Martinez, D. Dunn-Rankin*	**1B05:** PLIF-assisted discrimination of gas and liquid velocities for 5 kHz PIV in a liquid-fueled pressurized combustor 219 *H. Ozogul, H. Ek, I. Chterev, B. Emerson, T. Lieuwen*	**1C05:** Model sensitivities in LES predictions of buoyant methane fire plumes *H. Koo, J.C. Hewson, S.P. Domino, R.C. Knaus* 382
11:20 – 11:40	**1A06:** Experimental and computational investigation of autoignition of jet fuels and surrogates in nonpremixed flows at elevated pressures 46 *G. Mairinger, A. Frassoldati, A. Cuoci, M. Pelucchi, E. Pucher, K. Seshadri*	**1B06:** Laboratory investigation of down-scaling effects on low swirl burners 229 *A. Frank, P. Therkelsen, J.-Y. Chen, V.H. Rapp, R.K. Cheng*	**1C06:** Direct numerical simulation of wildland fires at small scales *N.T. Wimer, A.S. Makoweicki, A.Y. Poludnenko, C.M. Hoffman, J.W. Daily, G.B. Rieker, P.E. Hamlington* 393
11:40 – 12:00	**1A07:** A novel formulation for unsteady counterflow flames using a thermal-conductivity-weighted coordinate 52 *A. Weiss, M. Vera, A. Liñán, A.L. Sánchez, F. Williams*	**1B07:** Evaporation and autoignition studies of liquid n-alkane droplets in lean, high pressure methane/air mixtures in a rapid compression machine 239 *J. Mohr, C. Gould, A. Zdanowicz, A. Marchese*	**1C07:** The effect of pressure and external heating on the Nomex flame spread limits (LOC) 401 *M. Thomsen, X. Huang, C. Fernandez-Pello, A. Alonso, D.L. Urban, G.A. Ruff*
12:00 – 12:20	**1A08:** Global diffusion flame extinction strain rate experiments of single large hydrocarbon fuels 67 *A. Jadhav, R. Alsulami, B.C. Windom*	**1B08:** Preliminary study of the interactions between detonation chemistry and magnetohydrodynamics *M.F. Zaiger, K.E. Niemeyer* 252	**1C08:** Heat transfer optimization in simulated microgravity combustion 408 *S. Hossain, I.S. Wichman*
12:20 – 1:20	**LUNCH** – Box Lunches are available in the University of Wyoming Conference Center Lobby **Women in Combustion Lunch** – Salon B		
	Microcombustion and Transport **Salon A** **Session Chair:** P.D. Ronney	**Kinetics I** **Salon B** **Session Chair:** K.E. Niemeyer	**Fire III** **Salon C** **Session Chair:** T.P. Grumstrup
1:20 – 1:40	**1A09:** Hydrocarbon-fueled electrical power generator with no moving parts 73 *J. Wongwiwat, P. Bhuripanyo, V. McCloyn, P. Ronney*	**1B09:** Uncertainty of transport parameters in flame models: A database from virial coefficient measurements 259 *D.I. Pineda, T.A. Casey, X. Shi, J.-Y. Chen*	**1C09:** Correlating mass burning rate and flame spread rate for thin PMMA: Implications on pyrolysis temperature 414 *L. Carmignani, S. Bhattacharjee*
1:40 - 2:00	**1A10:** Impact of pressure on simulated FREI combustion 82 *P. Sharma, M. Ayoobi, M. McNenly, I. Schoegl*	**1B10:** Influence of blending methyl decanoate with *n*-decane on ignition delay times in a fuel ignition tester 268 *R. Leathers, S. Stuhlman, K. Kumar, S. Beyerlein, C.-J. Sung*	**1C10:** Two-dimensional heat transfer analysis within non-thermally thin poly(methyl methacrylate) that is burned in a narrow channel apparatus 426 *N.A. Lage, F.J. Miller*
2:00 - 2:20	**1A11:** Computational performance of chemistry and transport calculations in NGA 89 *J. Schlup, N. Burali, G. Beardsell, G. Blanquart*	**1B11:** Ignition delay times and Derived Cetane Numbers of canola, corn, and soy derived bio-diesel *S. Stuhlman, C. Dunkel, R. Leathers, K. Kumar, D. Shrestha, S. Beyerlein, C.-J. Sung* 279	**1C11:** Temperature measurement of glowing firebrands with multi-color pyrometry 440 *J.L. Urban, D. Shirazi, M. Vicariotto, D. Murphy, D. Dunn-Rankin, C. Fernandez-Pello*

	Microcombustion and Transport **Salon A** **Session Chair:** P.D. Ronney	**Kinetics I** **Salon B** **Session Chair:** K.E. Niemeyer	**Fire III** **Salon C** **Session Chair:** T.P. Grumstrup
2:20 - 2:40	**1A12:** Modeling a pressurized coal feed system 100 *T. Schroedter, B. Adams*	**1B12:** Missing experimental data and rate parameter inference for $H_2+OH=H_2O+H$ 288 *T.A. Casey, H. Najm*	**1C12:** Pyrolysis of live vegetation at slow heating rates 446 *E. Amini, M.-S. Safdari, M. Rahmati, J. Howarth, J. DeYoung, T.H. Fletcher*
2:40 – 3:00	**1A13:** A new efficient model for multicomponent membrane separation and application to the Argon Power Cycle 111 *F. Chourou, M.S. Aznar, J.-Y. Chen, A. Dreizler*	**1B13:** Probing the pyrolysis chemistry of ethyl propionate in a microreactor 297 *C. Rogers, J. Porterfield, J. Daily, B. Ellison, N. Labbe*	**1C13:** Effect of reduced plume entrainment on the burning rate of porous fuel beds 451 *S. McAllister*
3:00 - 3:15	**BREAK**		
	Innovative Technologies **Salon A** **Session Chair:** J. Cole	**Kinetics II** **Salon B** **Session Chair:** H. Najm	**Coal and Biomass I** **Salon C** **Session Chair:** B. Adams
3:15 - 3:35	**1A14:** Kinetic study of ethane pyrolysis in an internal combustion engine 127 *Z. Taie, D. Wagner, C. Hagen*	**1B14:** A new jet-stirred reactor for chemical kinetics experiments 305 *A.A. Davani, P.D. Ronney*	**1C14:** Comparison of laminar pulverized coal flame experiment and simulation 463 *C. Dunn, E. Beagle, M. Stoellinger, E. Belmont*
3:35 - 3:55	**1A15:** Flame stability and liftoff heights of a surrogate jet fuel burned near MILD combustion conditions 138 *P. Weide, D. Blunck*	**1B15:** Numerical design of a novel microreactor to study short residence time combustion 318 *T. Fan, C. Rogers, J.W. Daily, B. Ellison, N. Labbe*	**1C15:** A fast-running simulation tool for axisymmetric oxy-coal combustors 475 *T. Williams, B. Adams*
3:55 - 4:15	**1A16:** Estimation of multiple model parameters using a data-assimilated CFD algorithm 147 *Y. Wang, X. Gao*	**1B16:** Comparison and analysis of chemical kinetic models for toluene autoignition 325 *M.A. Mayer, K.E. Niemeyer*	**1C16:** Modeling effects of annealing on coal char reactivity to O_2 and CO_2 based on preparation conditions 486 *T. Holland, T.H. Fletcher*
4:15 – 4:35	**1A17:** Investigation of argon oxy combustion with low molecular weight fuels: Improving thermodynamic efficiency 157 *C. Scudiere, J.-Y. Chen, R. Dibble, M. Aznar*	**1B17:** Investigating stiffness detection metrics for chemical kinetics 335 *A. Alferman, K.E. Niemeyer*	**1C17:** Co-gasification of coal and biochar in modified drop tube reactor and TGA 510 *E. Beagle, Y. Wang, D. Bell, E. Belmont*
4:35 – 4:55	**1A18:** Preliminary design and development of an ultra high efficiency Stirling engine heater head 163 *M.S. Aznar, A. Frank, P.L. Therkelesen, V. Rapp, R.K. Cheng*	**1B18:** pyMARS: An open software package for reducing chemical kinetics models 347 *P.O. Mestas, P. Clayton, K.E. Niemeyer*	**1C18:** Comparison of raw and torrefied healthy and beetle kill pine in co-combustion with coal *A. Howell, E. Beagle, E. Belmont* 524
6:00	**Reception** – Laramie Historic Railroad Depot 1st & S 1st St, Laramie		

WSS/Ci Western States Section/Combustion Institute

Tuesday, 3 October 2017

7:30 – 1:00 **Registration:** University of Wyoming Conference Center Lobby
7:30 – 8:00 **Continental Breakfast:** University of Wyoming Conference Center Lobby

8:00 - 8:05 **Opening Remarks and Announcement:** Prof. Erica Belmont, *University of Wyoming*

8:05 – 9:05 **Plenary Lecture: Dr. Daniel Dietrich,** ***NASA***
Session Chair: Dr. Fletcher Miller, *University of California San Diego*

9:05 – 9:15	**BREAK**		
	Diagnostics **Salon A** **Session Chair:** B.C. Windom	**Engines** **Salon B** **Session Chair:** D.B. Olsen	**Coal and Biomass II** **Salon C** **Session Chair:** T.H. Fletcher
9:15 – 9:35	**2A01:** Exploring continuous monitoring methods for SO_3 in flue gas conditions 536 *A. Biasioli, Y.-C. Chien, D. Dunn-Rankin*	**2B01:** Investigation of non-ideal gasoline-ethanol vaporization behavior using an ethanol, cyclohexane, and benzene surrogate 630 *S. Burke, M. Ratcliff, R. McCormick, D. Bartholet, B. Windom*	**2C01:** Sub-model fidelity in pulverized coal combustion simulations 731 *C. Dunn, M. Stoellinger*
9:35 - 9:55	**2A02:** Strategy for background-free measurements of high-speed OH in turbulent flames using SLIPI 545 *B. Zhou, J.H. Frank*	**2B02:** Performance of a compression ignition engine fueled with renewable diesel blends produced from hydrothermal liquefaction, fast pyrolysis, and conversion of ethanol to diesel *J. Tryner, K. Albrecht, J. Billing, R.T. Hallen, A.J. Marchese* 636	**2C02:** Smoldering behavior of compacted sawdust 746 *P.P. Radyjowski, R.H. Bush, J.L. Ellzey*
9:55 - 10:15	**2A03:** Direct comparison of simulated OH fluorescence and experimental results in a non-premixed laminar diffusion coflow flame at high pressure conditions 554 *D. Escofet-Martin, A. Torredemer, Y.-C. Chien, D. Dunn-Rankin*	**2B03:** Crankcase blow-by gas particulate matter characterization and filtration on a John Deere 4.5 l diesel engine 647 *B.D. Rice, J. Tillotson, D.B. Olsen*	**2C03:** Impact of particle properties on radiative heat flux in an oxy-coal reactor 754 *T. Hosler, B. Adams*
10:15 - 10:35	**2A04:** Light extinction based image analysis technique for rotating fluidized beds 571 *Z. Lavrich, Z. Taie, D. Wagner, C. Hagen*	**2B04:** Evaluation of fuel savings for a ship system with variable frequency drives and energy storage using a Cummins QSK50 Tier 4 diesel engine 664 *J. Moothart, D.M. Wise, D.B. Olsen*	**2C04:** CFD simulations of hydropyrolysis *G.D. Baldwin, C.W. Dunn, M. Stoellinger* 766
10:35 – 10:50	**BREAK**		

	Heterogeneous and Soot **Salon A** **Session Chair:** D.O. Lignell	**Turbulent Flames III** **Salon B** **Session Chair:** M. Stoellinger	**Laminar Flames III** **Salon C** **Session Chair:** K. Seshadri
10:50 – 11:10	**2A05:** Opposed-flow heterogeneous reactor for partial oxidation of methane 580 *Y. Lin, X. Li, M.V. Twigg, W.F. Northrop*	**2B05:** Lagrangian analysis of premixed autoignition in compressible turbulence 678 *C.A.Z. Towery, R. Darragh, A.Y. Poludnenko, P.E. Hamlington*	**2C05:** Impact of acoustic-induced pressure fluctuations on the dynamics of laminar premixed flames 776 *G. Beardsell, G. Blanquart*
11:10 – 11:30	**2A06:** Preliminary analysis of smoldering combustion in cellulose and hemicellulose mixtures 593 *T.C. Mulky, K.E. Niemeyer*	**2B06:** The effect of spray burner outlet configuration, nozzle position and nozzle type on the flame behavior and stability of liquid fuels 689 *R.A. Alsulami, B.C. Windom*	**2C06:** Impact of non-ideal corrections to equation of state, thermodynamics, and chemical kinetics on one-dimensional freely propagating flame structure 787 *S. Yellapantula, J.B. Bell, E. Motheau, M.S. Day, R.W. Grout*
11:30 – 11:50	**2A07:** Effects of fuel composition and packing density on horizontal smoldering propagation *B.D. Smucker, D.A. Cowan, D.L. Blunck* 606	**2B07:** Nitrogen oxide emissions from reacting jets in vitiated crossflow 698 *M.D. Sirignano, V. Nair, B. Emerson, J. Seitzman, T.C. Lieuwen*	**2C07:** Numerical simulations of a co-flow methane/air flame including ions and excited species under different gravity conditions 798 *C.-F. Lopez-Camara, D. Dunn-Rankin*
11:50 - 12:10	**2A08:** Flame temperature effect on the transition between soot and graphitic carbon products in premixed stagnation flames 615 *J. Camacho*	**2B08:** Effect of stoichiometric mixture fraction on hydrogen edge-flames in a counter-flow burner 713 *Z. Zhou, P.D. Ronney*	**2C08:** Influences of stoichiometry on steadily propagating triple flames in counterflows 819 *P. Rajamanickam, W. Coenen, A.L. Sánchez, F.A. Williams*
12:10 – 12:30	**2A09:** Modeling soot formation from solid complex fuels 622 *A.J. Josephson, E. Hopkins, R.R. Linn, D.O. Lignell*	**2B09:** Transported and presumed joint PDF modeling of turbulent jet flames 720 *V. Jaganath, M. Stoellinger*	**2C09:** A new reduced thermal diffusion model for H and H_2 830 *J. Schlup, G. Blanquart*
12:30 – 12:50			**2C10:** Thermal explosions in spherical vessels at large Rayleigh numbers 846 *D. Moreno, I. Iglesias, A.L. Sánchez, A. Liñán, F.A. Williams*
12:50	**Adjourn**		

2017 WSSCI Fall Technical Meeting Author List

Western States Section of the Combustion Institute – Fall 2017 Meeting
Hosted by the University of Wyoming
October 2-3, 2017

Porous Wall-Fed Liquid Fueled Miniature Tubular Flame Burner

Vinicius Maron Sauer *Jagdeep Singh Batther* *Derek Dunn-Rankin*

Department of Mechanical and Aerospace Engineering, University of California Irvine, 4200 Engineering Gateway, Irvine, CA 92697, USA

Nonpremixed swirl-type tubular flames were described recently both experimentally and theoretically. In that new configuration liquid fuel flows inwardly through a porous tube and vaporizes becoming the fuel source for the chemical reaction with the swirling air coming from the bottom inlet of the burner. The permeable walls prevent from optical access to the combustion chamber and only the flame structure produced above the burner rim can be evaluated. Observations of the external flame along with measurements of external combustor wall temperatures provide valuable information on the system stability. However, further evaluation of the tubular flame anchored in the combustion chamber is not possible with the diagnostic tools available. In order to allow flame visualization, and consequently, future studies on the structure of nonpremixed swirl-type tubular flames, a new configuration is proposed and evaluated experimentally. The outer permeable cylinder is replaced by an impermeable translucent tube, and a porous cylinder, from where liquid fuel is injected, is positioned at the axis of the burner. The combustion chamber consists of the annular space between the inner (porous) and the outer (impermeable and translucent) tubes instead of a circular cross section from the original nonpremixed swirl-type tubular flame burner. Experimental results show the feasibility of the proposed configuration as well as how the system stability and the characteristics of external flame compare to the conventional nonpremixed swirl-type tubular flame. The transparent outer wall allows the evaluation of the anchoring mechanisms and the overall structure of the tubular flame produced in the combustion chamber.

1 Introduction

A tubular flames is a canonical configuration used in fundamental studies in combustion science. In such systems, introduced by Ishizuka [1], a flame long in the longitudinal direction and circular in cross section is produced. Many of the studies in which tubular flames are employed have been restricted to systems in which fuel and oxidizer are premixed before entering the combustion chamber. More recently, the interest in nonpremixed versions of the counterflow- and swirl-type tubular flames has increased. Experimental counterflow-type tubular flames produced by gaseous fuels have provided the complete description of the temperature and species concentration fields [2, 3]. However, due to its design characteristics, the experimental setup in which nonpremixed swirl-type tubular flames are generated does not allow the detailed characterization of the flame as a consequence of the lack of optical access to the interior of the combustion chamber [4].

The nonpremixed counterflow tubular flame is obtained experimentally using a modified version of the premixed counterflow tubular burner, which is altered by installing a porous cylinder along the center axis [2]. One reactant flows outwardly from the porous cylinder in the radial direction, while the outer contoured nozzle issues inwardly the other reactant. Windows are aligned with the axis of symmetry of the outer nozzle allowing flame visualization as well as the application of diagnostics techniques that require optical access to the combustion chamber.

The nonpremixed swirl-type tubular flame is obtained experimentally by employing a burner consisting of two concentric cylindrical tubes and an aluminum base used for support. Liquid fuel is injected into the small annular gap between the inner and the outer tubes from injectors located close to the base of the tubes. The walls of the inner tube, which are permeable, allow the surface tension to spread the liquid over its volume. The hollow central region of the inner tube acts as the combustion chamber. The outer tube is solid, with the choice of material depending on the goal of the experiment. If the evaluation of the liquid fuel level in the annular region is necessary, a translucent material, such as quartz can be used [5]. On the other hand, if lower overall wall outer temperatures are needed, the outer tube can be made out of a material with higher thermal conductivity (e.g., aluminum), which helps heat to diffuse axially from regions at higher temperatures [4]. Air is injected axially from the bottom of the combustion chamber through a swirl generator. The porous tube consists of a commercial stainless steel filter element, such that it prevents flame visualization along its longitudinal direction, as well as the evaluation of temperature and species concentrations through optical diagnostics techniques.

In order to permit a complete characterization of nonpremixed swirl-type tubular flames, a modified version of the original nonpremixed swirl-type burner is presented. The fundamental characteristics of the system remain the same as the original version, with the injection of fuel through porous walls. However, in this new configuration, instead of injecting fuel though the outer walls, the condensed fuel is delivered via an inner permeable tube with a smaller diameter positioned along the axis of the burner. The condensed fuel vaporizes and becomes the gaseous fuel source for the chemical reaction with the swirling air coming from the bottom inlet in the annular region created by the inner permeable tube and the outer quartz tube. The tubular flame is located in the combustion chamber, namely, the annular zone between the burner inner and outer walls.

2 Experimental Setup

A schematic of the burner is presented in Fig. 1. The system is composed of two concentric cylindrical tubes and an aluminum base used for alignment and support. The liquid fuel flows radially outwards into the combustion chamber through the walls of the inner tube, which are permeable. The porosity of the walls allows the surface tension to spread the liquid over its volume. The outer tube is a quartz solid tube with circular cross section.

The annular region between both tubes acts as the combustion chamber. Fuel is delivered to the porous tube trough one of its ends, which is connected to external fuel lines. A jetted photopolymer rapid prototyping machine is used to produce the swirl generators, represented schematically in Fig. 2, which consist of three blades with a vane angle of $15°$. The porous tube consists of a sintered stainless steel element with internal diameter $3\,\mathrm{mm}$ and outer diameter $6\,\mathrm{mm}$. A tube of $5\,\mu\mathrm{m}$ pore size is considered in the evaluation of the burning conditions. The inner diameter of the quartz tube is $18\,\mathrm{mm}$.

The length the combustion chamber is $50.8\,\mathrm{mm}$. The flow rate of compressed air is controlled a

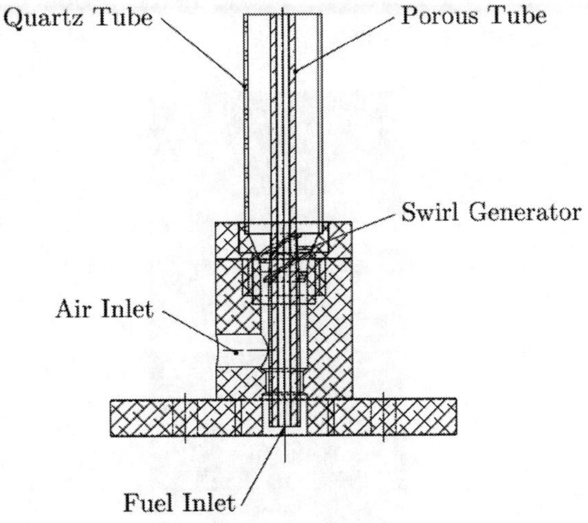

Figure 1: Schematic section of the combustor.

rotameter calibrated with a bubble flow meter in the range of 10 to $40\,\mathrm{L\,min^{-1}}$. Two continuous syringe pumps control the flow of liquid n-heptane. The volumetric flow rates of air and fuel are varied independently such that several inlet equivalence ratios are considered.

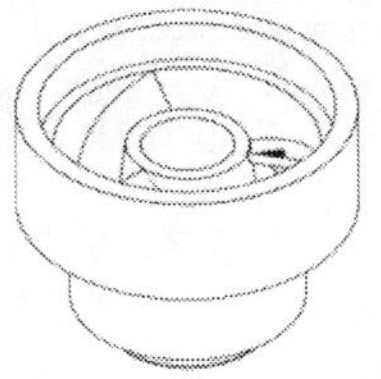

Figure 2: Swirl vane three-dimensional model.

In this study, a description of the burner operation limits is presented. The experimental conditions presented herein can be employed in future evaluations and complete characterization of the flame produced by the burner.

3 Results and Discussion

During the operation of the nonpremixed swirl-type tubular combustor, the stability of the observed flame can vary with the choice of inlet conditions. Unstable flames are characterized by either weak or strong oscillations in its luminous core that can eventually lead to extinction. The overall shape of stable flames does not vary significantly and it is steady over the observed time length. The diameter of the tubular flame obtained in stable operation conditions varies, approximately, between 12 mm and 8 mm, depending on the inlet conditions. The system has not been evaluated over time intervals greater than 90 s. Therefore, flames denoted as unstable, might not be able to sustain themselves for longer periods. A photographic image of a stable flame produced by the system is shown in Fig. 3, which shows that the flame produced by the burner presents the geometric characteristics of a tubular flame.

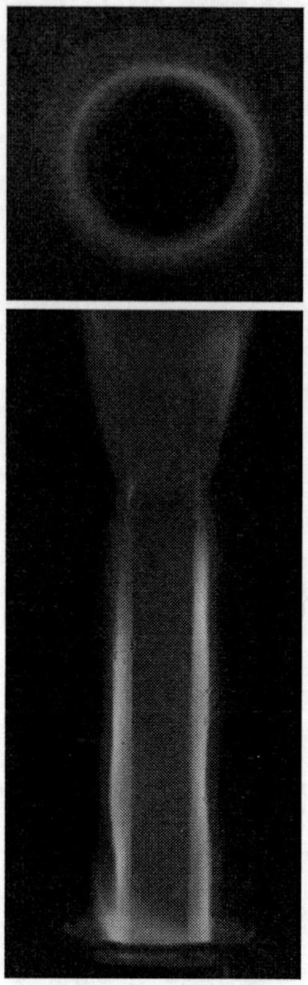

Figure 3: Photographic images of the experiment. Top: Top view; Bottom: Side view.

The observed flame configurations presented in Fig. 3 are dependent on the total flow rate of air and fuel. In order to identify the operational limits, the stable and unstable burning conditions, the working-limit chart of the burner is obtained and discussed.

3.0.1 Stability Limits

In order to identify the operational limits and the stability of burning conditions working-limit charts are presented in Fig. 4, where the abscissa represents the inlet liquid fuel flow rate and the ordinate the inlet air flow rate. The procedure used to obtain the working-limit chart consists in selecting the fuel flow rate on the syringe pumps and varying the air flow rate independently. The black line denotes the overall stoichiometric inlet condition (though the reaction is non-premixed so we presume that any reaction interface is always stoichiometric). The regions above and below the black line separates the map into global lean (upper) and rich operation zones (lower). The map shows that all measured points lie in the region of lean conditions. Depending on the amount of air introduced into the chamber it is possible to change the flame shape and combustion behavior. When the flame is in the stable region, represented in Fig. 4 by circles, oscillations are not observed and the shape of the luminous core is similar to the one from Fig. 3. Torch flames, represented in Fig. 4 by squares, are observed in richer regions of the map, and characterized by a long external structure mainly due to the reduced amount of oxidizer in the combustor, which allows an excess of unburned fuel to leave the burner and react with ambient air. For leaner conditions the flame start to oscillate and becomes unstable, and the flame cannot sustain its shape for long periods of time. In the unstable region, represented by triangles in Fig. 4, a great part of the liquid fuel vaporizes at the lower region of the burner. A zone of high temperatures develops at the top portion of the permeable tube due to the lack of fuel in the liquid phase, which can be noticed by the glowing red color emitted by the walls of the porous tube.

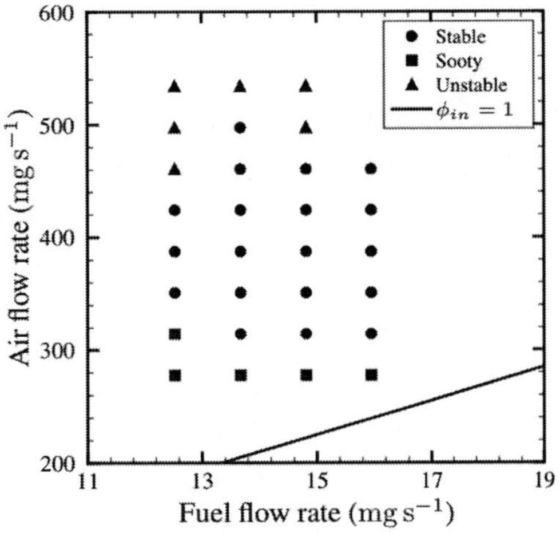

Figure 4: Combustor working-limit map.

4 Conclusions

The modified porous based nonpremixed tubular burner demonstrates very promising combustion behavior as compared to the original nonpremixed swirl-type tubular flame combustor. The new configuration with an inner porous and an outer quartz tube allows optical access to the tubular flame, which permits the application of optical diagnostics techniques to the evaluation of temperature and species concentrations fields. The working-limit map shows that stable flames can be obtained for a wide range of lean inlet conditions. The experimental conditions presented herein can be employed in future evaluations and complete characterization of the nonpremixed swirl-type tubular flame established in the burner.

Acknowledgments

Vinicius M. Sauer would like to acknowledge CNPq-Brazil for the support.

References

[1] Satoru Ishizuka. *Proc. Combust. Inst.*, 20 (1985) 287–294.

[2] Shengteng Hu, Peiyong Wang, Robert W. Pitz, and Mitchell D. Smooke. *Proc. Combust. Inst.*, 31 (2007) 1093–1099.

[3] Shengteng Hu and Robert W. Pitz. *Combust. Flame*, 156 (2009) 51–61.

[4] Vinicius M. Sauer and Derek Dunn-Rankin. Porous Wall Fed Liquid Fuel Nonpremixed Swirl-Type Tubular Flames. In *Proc. 26th ICDERS*, pages 1–4, Boston, MA, August 2017. Paper No. 1114.

[5] Vinicius M. Sauer. *Analysis of Laminar Nonpremixed Stagnation Flow and Swirl-Type Tubular Flames*. Phd, University of California, Irvine, Irvine, CA, August 2017.

Paper # 29LF-0057

Topic: Laminar Flames

Western States Section of the Combustion Institute – Fall 2017 Meeting
Hosted by the University of Wyoming
October 2-3, 2017

Flame Propagation in Narrow Channels at Varying Lewis Number

X. Ma[1] *S. Shen*[2] *J. Wongwiwat, J. Gross, P. Ronney*

Department of Aerospace and Mechanical Engineering
University of Southern California
Los Angeles, California, United States of America
Corresponding Author Email: [1] *xiaotonm@usc.edu*
[2]*sishen@usc.edu*

Abstract: The propagation of quasi-2D premixed-gas flames in H_2-O_2-inert mixtures having various effective Lewis numbers (Le_{eff}) from fuel-lean to fuel-rich conditions were studied using a Hele-Shaw cell. Various instabilities were observed including those attributed to thermal expansion of the burned gas (Darrieus-Landau, DL), buoyancy (Rayleigh-Taylor, RT), viscosity contrast across the flame (Saffman-Taylor, ST), and diffusive-thermal (DT) effects. By varying the concentrations of fuel, oxygen, and insert gas, a range of equivalence ratios and Le_{eff} were obtained at fixed adiabatic flame temperatures. The direction of propagation relative to the gravity vector was also varied. It was found that the flame front shapes and propagation rates were strongly affected by all 4 of the aforementioned instability mechanisms, but in many cases the contributions of each mechanism were indefinable and nearly independent of each other.
Keywords: diffusive-thermal instability, Darrieus-Landau instability, buoyance effect, Saffman-Taylor instability, Hele-Shaw cell

1. Introduction

In spite of the challenges associated with its distribution and storage, hydrogen remains an appealing fuel due to increased concerns about the environmental impacts of hydrocarbon fuels. While the advantages of hydrogen as a fuel (e.g. combustion at flame temperatures lower than those possible with hydrocarbons) could be exploited to reduce NOx and other emissions, the most significant is the absent of carbon, and thus does not contribute any carbon dioxide to the atmosphere. On other words, hydrogen is a great alternative green fuel creating little emissions and not greenhouse gas in its product. However, despite the long history of investigation into hydrogen combustion, factors that influence the propagation rates and extinction conditions of flames in H_2-O_2-inert mixtures are still not well understood.

Lean combustion is especially challenging due to the low molecular weight of hydrogen and thus the low effective Lewis numbers (Le_{eff}) of flames. This distinct property results in a substantial influence of diffusive-thermal instability (DT), which may also couple with other instabilities. The Le_{eff} of lean H_2-O_2-N_2 mixtures is about 0.3, whereas for rich mixtures, where O_2 rather than fuel is the stoichiometrically limiting reactant, Le_{eff} can be around 1.3. Consequently, the objective of this is work is to examine the interactions of Darrius-Landau instability (DL), Rayleigh-Taylor instability (RT), Saffman-Taylor instability (ST), diffusive-thermal instability (DT) and potentially other instabilities on the behavior of flames in H_2-O_2-inert mixtures of varying equivalence ratios, and thus the varying Le_{eff} mixtures, in a very simple, well-defined geometry. The experiment employs a Hele-Shaw cell for its structural simplicity. Moreover, the apparatus enables assessment of DL instabilities [1, 2] in a quasi-2D geometry as well as the combined influences, including the buoyancy and viscosity effects [3] studied by Joulin and Sivashinsky [4].

2. Methods / Experimental

The experiment uses a planar combustion chamber, known as a Hele-Shaw cell, a spark generator, a partial pressure mixing system, a Vision Research Miro eX2 high speed camera, a vacuum pump, and a computer controlling the system through LabVIEW programs. The overall diagram of this experimental setup is illustrated below in figure 1a. The Hele-Shaw cell is made of two acrylic plates separated by a hollow aluminum frame to create a combustion chamber that is 59.5 cm long, 39.5 cm wide and 0.5 inches thick. The chamber has three spark electrodes installed at one end so that after ignition, three circular flames quickly merge into a single roughly planar flame. At the ignition end of the cell an exhaust manifold is connected to the atmosphere through a ball valve. The Hele-Shaw cell and a separate mixing chamber were first evacuated. The mixing chamber was then filled using the partial pressure method. Next, the Hele-Shaw cell was filled with the mixed gas to a pressure slightly above one atmosphere to prevent back flow of air when the ball valve opens. Just before ignition the ball valve was opened so that combustion takes place at a nearly constant pressure (as confirmed by time-resolved chamber pressure measurements). The camera was commanded to begin recording right before ignition. Videos of flame propagating through the cell were collected using the Vision Research Phantom Camera Control software. After processing the video files to adjust flame visibility, a typical position vs. time plot as shown below in figure 1b is produced. The time plot reveal that after the initial transient period, a nearly constant slope (thus steady propagation speed) is observed until the flame approaches the opposite end of the Hele-Shaw cell. This is a quasi-steady state period of flame propagation. The burning rate was determined by counting the number of pixels over which the flame had passed as a measure of the burned fraction vs. time within the quasi-steady state period.

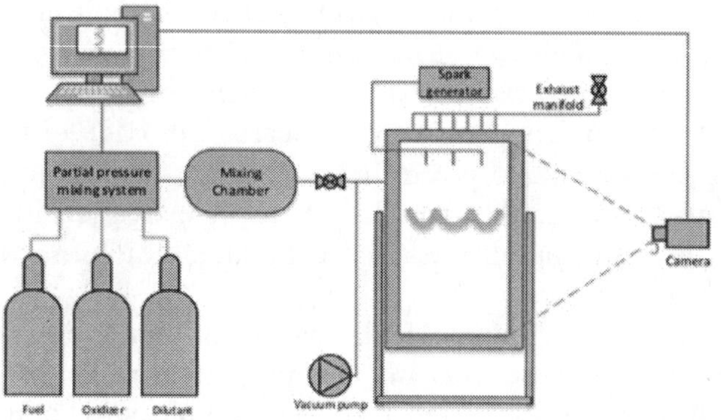

(a)

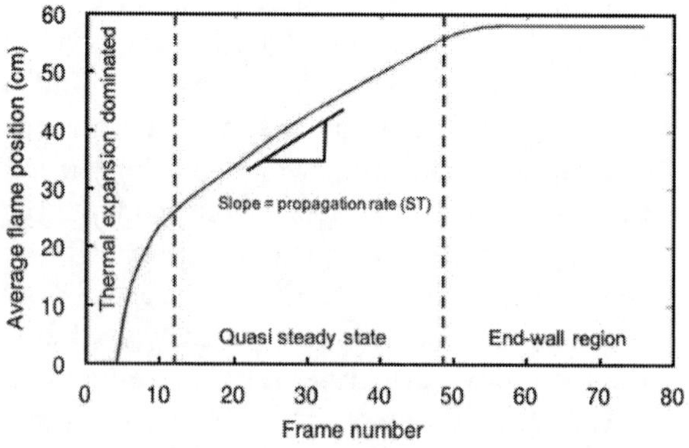

(b)

Figure 1: Left: Experimental apparatus. Right: Typical plot of flame position vs frame number.

3. Results and Discussion

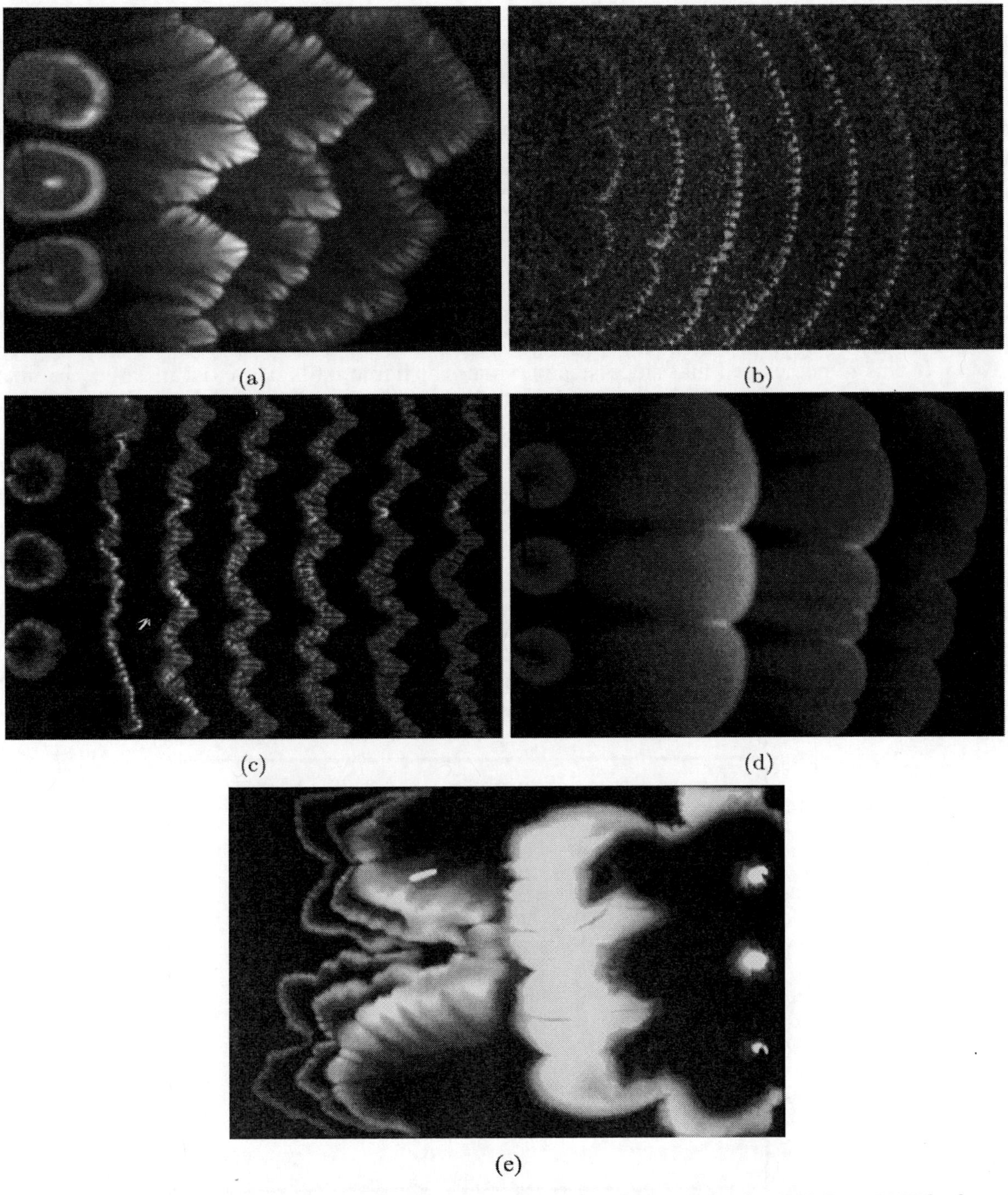

Figure 2: (a) H_2-O_2-N_2, $\phi = 0.8$, $T_{ad} = 1300$ K, horizontal propagation. (b) H_2-O_2-N_2, $\phi = 0.45$, $T_{ad} = 900$K horizontal propagation. (c) H_2-O_2-N_2, $\phi = 0.35$, $T_{ad} = 1050$K downward propagation. (d) H_2-O_2-N_2, $\phi = 2.0$, $T_{ad} = 1300$K horizontal propagation. (e) H_2-O_2-N_2, $\phi = 0.8$, $T_{ad} = 1400$ K, horizontal propagation, 1/8 inch.

Typical time elapsed images of flames are shown below in figure 2. Figure 2a shows a fuel lean mixture flame with low Le_{eff} (the fuel, H_2, is the scarce reactant), where both small-scale DT instabilities and larger-scale DL instabilities can be observed. Figure 2b shows a close extinction flame, where due to substantial heat loss in burned gases, large-scale wrinkling caused by DL and ST effects are suppressed, but the smaller-scale DT cells are retained. Figure 2c shows a downward propagating flame for a low Le_{eff} mixture, where the RT effect is stabilizing the flame, and consequently the large-scale structures are suppressed. This results in a flame that is nearly flat on a large scale, but is still highly wrinkled at smaller scales. Figure 2d shows a high Le_{eff} flame in a rich mixture (the oxygen, O_2, is the scarce reactant). The DT instability, as shown in low Le_{eff} mixtures, is no longer present, but the larger-scale DL and ST instabilities can still be observed. Moreover, unlike flames in low Le_{eff} mixtures, where the combined effects of DT and DL/ST are presented in angular shaped flames (figure 2a), high Le_{eff} mixtures have cusped flames (figure 2d). Figure 2e shows a flame shape with viscous fingering that is characteristic of thinner plate spacing (0.125in) when ST instabilities are prominent.

It was hypothesized that since steady, spherical flame balls can exist in large chambers [5], a 2D analog (flame circles) could be possible in the quasi-cylindrical geometry in a Hele-Shaw cell. As in the case of flame balls, such structures would not be possible under the influence of buoyancy. However, unlike flame balls, for which buoyancy must be eliminated using microgravity experiments, in the case of flame circles, it may be possible to observe them in a horizontally oriented Hele-Shaw cell. While flame circles cannot be perfectly steady because Laplace's equation does not have a steady solution in cylindrical, unconfined geometry due to the boundary condition, which cannot be satisfied as $r \rightarrow \infty$ (unlike the spherical case). The singularity as $r \rightarrow \infty$ is very weak (logarithmic), and thus one might expect such structures to persist on long time scales [6]. In fact, no such behavior was observed. It can be stated that low-Le flames in a Hele-Shaw cell behave more like propagating flames with DT instabilities rather than flame circles with weak effects of propagation (thus convection).

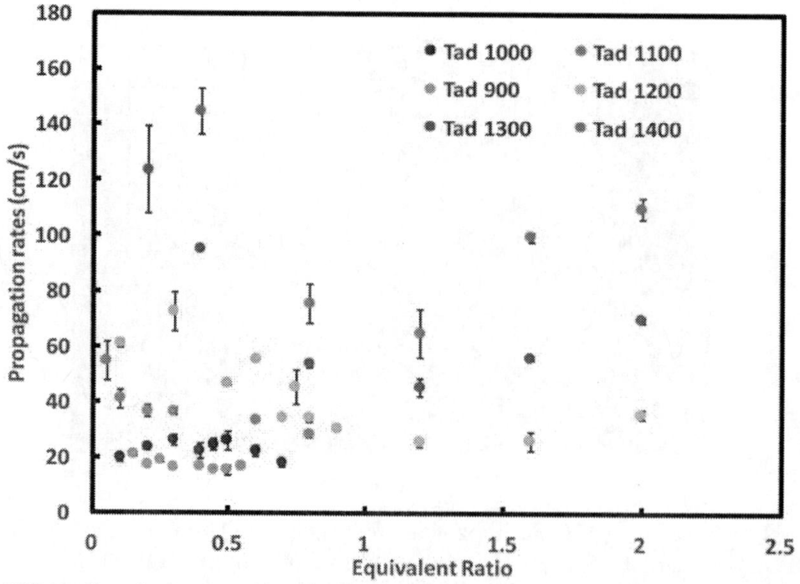

Figure 3 Effect of equivalence ratio and adiabatic flame temperature on flame propagation rates.

The effect of mixture strength (characterized by calculated adiabatic flame temperature) and Le_{eff} number (characterized by equivalence ratio, varying from Le → 0.3 for lean mixtures to Le → 1.3 for rich mixtures) in H_2-O_2-N_2 is shown in figure 3. It can be seen that as equivalence ratio increases, in general, the propagation speed decreases due to the increased Le_{eff}. However, for the stronger mixtures at high equivalence ratio, there is a slight increase in propagation rate. In addition, for the weaker mixtures (lower adiabatic flame temperature) there is very little effect of equivalence ratio on propagation speed for the range of conditions where propagation is possible (though the weaker mixtures will not burn at all at higher equivalence ratios).

It is also interesting to note that there is almost no effect of orientation relative to the gravity vector on the propagation rate for low-Le flames (low equivalence ratio, figure 4). In contrast, for higher Le flames (higher equivalence ratio, but still lean, figure 5) there is a substantial buoyance effect which, as expected, diminishes for faster flames (higher adiabatic flame temperature) because for the faster flames buoyancy-induced convection velocities, i.e. RT effects, becomes slower relative to flame propagation speeds. Furthermore, different Hele-Shaw cell thickness present little effect on flame propagation rate in these low-Le flames (lean mixtures, figure 4 and figure 5). In high Le flames (rich mixtures, figure 6), there is big difference between propagation rates (S_T) of flames propagating through different Hele-Shaw cell thickness. There is increasing heat loss in the thinner Hele-Shaw cell, which slowed down the flames. Moreover, buoyance effect is stronger in larger cell than in thinner cell. As shown in figure 6, the flame propagation velocity ratios show much more difference between flames in separate orientations in 0.5in thick cell than in 0.125in thick cell. Comparing figure 5 and 6, ratio between flame propagation rates (S_T) and laminar burning velocities (S_L) is much lower in high Le_{eff} (figure 6) than in low Le_{eff} mixture (figure 5) due to the lack of cellular structure (figure 2d).

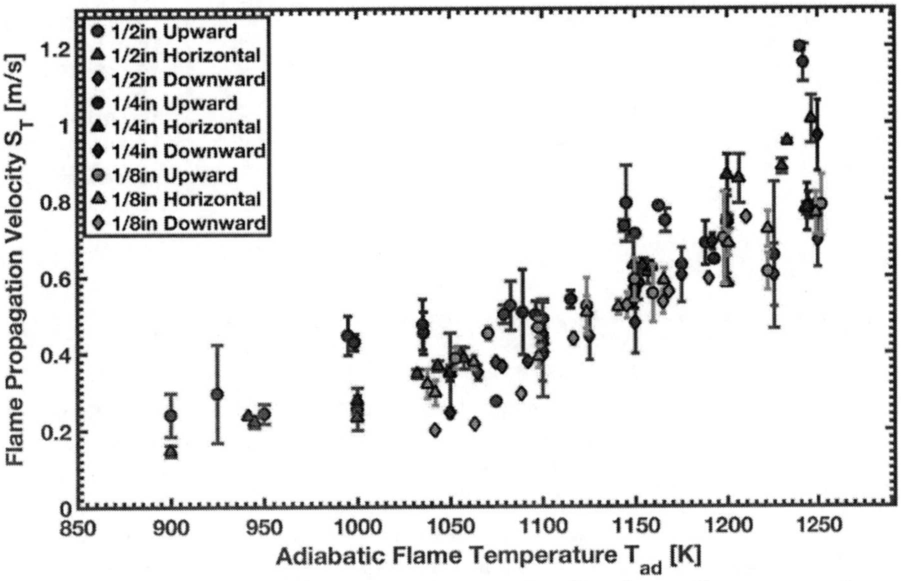

Figure 4 Effect of adiabatic flame temperature and propagation direction on flame propagation rates(S_T) at equivalence ratio 0.2.

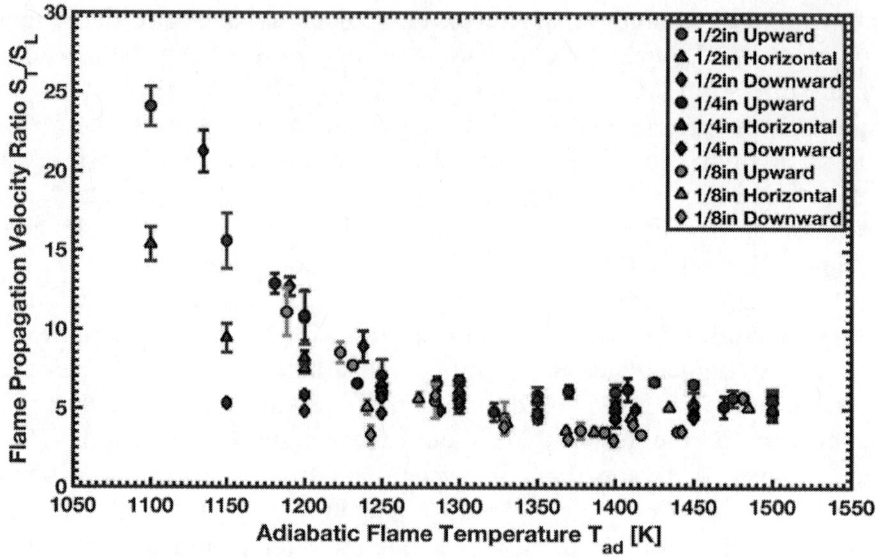

Figure 5 Effect of adiabatic flame temperature and propagation direction on ratio between flame propagation rates (S_T) and laminar burning velocity (S_L) at equivalence ratio 0.8.

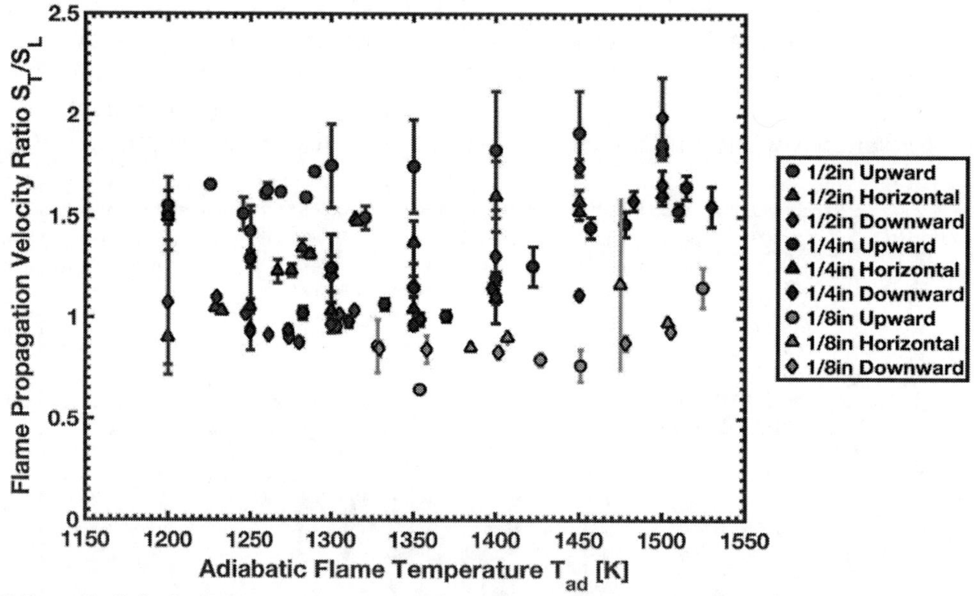

Figure 6 Effect of adiabatic flame temperature and propagation direction on ratio between flame propagation rates (S_T) and laminar burning velocity (S_L) at equivalence at equivalence ratio 2.

4. Scaling Parameters

The effects of all instabilities caused by thermal expansion, viscosity changes across the flame front, and buoyancy can be modeled with a single parameter, JS parameter, derived by Joulin & Sivashinsky [4]. The analysis process involves a linear stability analysis of the propagating flame assuming a unity Le, thus omitting the DT effect due to the absence of damping at small wavelengths. The hypothesis is that there will be a linear relationship between the growth rate of JS analysis divided by wavenumber, i.e. phase velocity, and ratio of ST and SL. The result is a dispersion relation between the effects of DL, ST, and RT instabilities on the relationship between scaled wavelength and growth rate. The derivation for this relation are as follows; refer to appendix A for the definitions of the variables.

$$\Omega^2 + (1+\Lambda)\Omega - \left\{\frac{1-\varepsilon^2}{4\varepsilon} + \frac{1+\varepsilon}{4}(F+G)\Lambda\right\} = 0;$$

$$\Omega \equiv \frac{\sigma(1+\varepsilon)}{2kU};\ \Lambda \equiv \frac{f_{av}}{\rho_u Uk};\ F \equiv \frac{f_b - \varepsilon f_u}{\varepsilon f_{av}};$$

$$G \equiv \frac{\rho_u(1-\varepsilon)g}{f_{av}U};\ \varepsilon \equiv \frac{\rho_b}{\rho_u};\ \mathrm{f}_{\mathrm{av}} \equiv \frac{f_u + f_b}{2}$$

$$\lambda_{ST} = \frac{\pi}{6}\frac{\rho_u S_L w^2}{\mu_{av}}$$

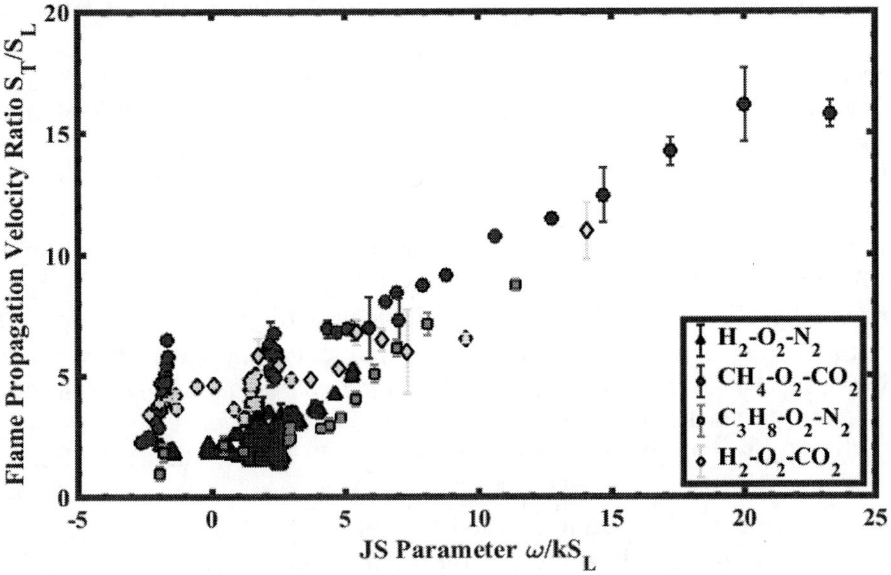

Figure 7 JS Parameter Fitting for gas mixtures with Le_{eff} near unity.

In premixed gas mixtures where the Le_{eff} is close to unity, the data falls into the predicted linearity. The JS scaling analysis is done on data obtained from hydrogen, propane, and methane as well as two different inert gases for hydrogen gas – nitrogen and carbon dioxide. As shown in figure 7, the fitted data for all four mixtures align to a roughly linear slope.

5. Conclusions

Flames in H_2-O_2-inert mixtures propagating in narrow channels exhibit a wide range of behaviors including the combined effects of Darrieus-Landau (DL), Rayleigh-Taylor (RT), Saffman-Taylor (ST) and diffusive-thermal (DT) effects, all in terms of flame shapes and propagation rates. These, especially thermal expansion (DL) effects, are always present, except for very weak mixtures. Buoyancy (RT) effects are destabilizing/stabilizing at long wavelengths for upward/downward propagation, and presenting stronger influence in thicker cell than in thinner cell. Heat loss has little effect except where the effect is so strong as to suppress thermal expansion. There is clear evidence of both DL and DT instabilities. Moreover, some new phenomena have been observed, e.g. angular large-scale structures (figure 1a vs. cusped for Le larger than 1 (figure 1d) as well as very stable saw-tooth patterns for downward propagation (figure 1c)). For Le_{eff} near unity, DL, RT, and ST instabilities in flames of H_2-O_2-inert as well as CH_4- O_2-inert and C_3H_8-O_2-inert can be linearly correlated through the JS-parameter. However, this correlation does not appear linear for Le_{eff} further from unity and further research is required in order to develop a successful correlation for the data. Finally, we remark that most experiments are conducted in open flames (Bunsen flames, counter flowing jets, etc.) where gas expansion (DL effects) are at least partially relaxed in the 3rd dimension, whereas most practical applications are in confined geometries, where unavoidable thermal expansion (DL) instabilities cause propagation to be higher than the laminar burning velocity even when heat loss, Lewis number and buoyancy effects are negligible.

5. Acknowledgements

This work was supported by the National Science Foundation under Grant CBET-1236892.

6. References

[1] G. Darrieus, "Propagation d'un front de flame", unpublished work presented at La Technique Moderne, Paris., 1938.

[2] L. D. Landau, "On the theory of slow combustion," *Acta Physico-Chem.* (USSR), vol. 19, p. 77, 1944

[3] P. G. Saffman and G. I. Taylor, "The penetration of a fluid into a porous medium of Hele-Shaw cell containing a more viscous fluid," *Proceedings of the Royal Society A: Mathematical, Physical and Engineering Sciences*, vol. 245, no. 1242, pp. 312-329, 1958.

[4] G. Joulin and G. Sivashinsky, "Influence of momentum and heat losses on the large-scale stability of quasi-2d premixed flames," *Combustion Science and Technology*, vol. 98, no. 1-3, pp. 11-23, April 1994.

[5] G. I. Sivashinsky, "Diffusional-thermal theory of cellular flames," *Combustion Science and Technology*, vol. 15, no. 3-4, pp. 137-145, 1977.

[6] M. Wu, P. Ronney, R. Colantonio, and D. Vanzandt, "Detailed numerical simulation of flame ball structure and dynamics," *Combustion and Flame*, vol. 16, no. 3, pp. 387-397, 199

Appendix A

Table 1. Variables for JS-parameter Derivation

ε	Effects of thermal expansion
σ	Linear growth rate of JS parameter
Λ	Scaled wavelength
λ_{ST}	Wavelength
ρ_b	Density of burned gas
ρ_u	Density of unburned gas
Ω	Scaled growth rate
μ_{av}	Average viscosity
U	Flame propagation velocity
k	Wavenumber
w	Cell thickness
G	Buoyancy effect
F	Viscosity change across the flame front
f_{av}	Average viscosity factor
S_L	Laminar flame speed
f_b	Burned gas viscosity factor
f_u	Unburned gas viscosity factor

Western States Section of the Combustion Institute – Fall 2017 Meeting
Hosted by the University of Wyoming
October 2-3, 2017

Laminar Flame Speeds of Associated Gas Mixtures Measured with Flat Flame Method

F. Arafin[a], E. Belmont[a]*

[a]Department of Mechanical Engineering, University of Wyoming, Laramie, WY, USA
**Corresponding author, email: ebelmont@uwyo.edu*

Abstract: Associated gas worth of tens of billions of dollars is flared annually during oil and gas production, which leads to resource waste and environmental issues. The primary objective of this work is to determine the laminar flame speeds of an associated gas mixture at lean to mildly rich equivalence ratios (ϕ). The component gases (methane, ethane, propane, and butane) and an associated mixture from the Bakken gas field in North Dakota are investigated for this purpose. The study uses a McKenna burner facility for stabilization of premixed, laminar flat flames with measured heat loss from the flames. Equivalence ratios of ϕ=0.7 to ϕ=1.1 are investigated at local pressure over a range of reactant flow rates. Adiabatic laminar flame speed at each equivalence ratio is obtained using linear extrapolation to zero heat loss. Numerical simulations are also performed in Cantera using the USC-Mech II mechanism to predict the laminar flame speed of component gases and associated gas mixture. Experimental data obtained for the component gases are compared with simulation results to validate the experimental setup. Finally, the experimental and numerical results for the associated gas mixture are presented.

1. Introduction

Natural gas (NG) is one of the most extensively used sources of energy for heating, cooking, transportation, and electricity generation [1]. However, the composition of natural gas can vary significantly with location and season. A previous study [2] shows that the major constituent of natural gas, methane, can vary from 55.8% to 98.1% in composition, while the natural gas also contains ethane (0.5% to 13.3%), propane (0% to 23.7%) and traces of other hydrocarbons depending upon the source of extraction and refining process. Among these mixtures of natural gas constituents, those which have lower concentrations of methane than the pipeline standard, are frequently known as associated gas. They are usually not processed and transported as fuel due to the isolation of wells and low value of the gas. Rather, a common practice is to burn off this gas via flaring during oil production when it comes to the surface in places having no infrastructure for its utilization [3]. Estimates by the EIA [4] and the World Bank [3] show that 5 trillion cubic feet of natural gas, which is 4% of total production, was flared worldwide in 2011. Natural gas worth 55 billion USD was flared worldwide and gas worth 2.3 billion USD was flared in the US in 2011. In the Bakken formation in North Dakota, storage and transportation infrastructure have been developed in recent years, which reduced the fraction of flared gas; however, production capacity also increased substantially. Consequently, the total quantity of gas flared in 2015 remained almost the same as 2010 [5]. Given this ongoing waste of fuel and source of emissions, it is important to investigate the utilization of associated gas as an alternate fuel source and understand its combustion characteristics.

Laminar premixed flame speed is a fundamental physical property of a combustible mixture related to the thermal diffusivity, reaction rate, and temperature through the flame zone. A substantial number of investigations [5-11] have been done to determine the laminar flame speed of the pure species components, including methane, ethane, propane, and butane. In addition, some studies [12-14] also focused on the laminar flame speeds of fuel blends for the validation of chemical kinetics mechanisms. Khan et al. [13] suggested that the presence of higher hydrocarbons increases the laminar flame speed. Bourque et al. [14] performed atmospheric and high pressure laminar flame speed experiments with two natural gas blends and captured the basic trends of laminar flame speed with composition and pressure using a developed chemistry mechanism.

The present work measured the laminar flame speed of a mixture of associated gas with high levels of higher hydrocarbons, including ethane, propane and butane, using the flat flame burner technique. The flame speeds of hydrocarbon blends are assessed using additive analysis, with the objectives of assessing whether additive or non-additive flame speed behaviors are observed and understanding the kinetic pathways underlying non-additive behavior. This paper presents flame speeds of component gases for validation of the experimental setup and calculation of additive flame speeds, as well as measured flame speeds for one associated gas mixture. Experimental and numerical results are presented and compared, and ongoing and future work is discussed.

2. Methods

2.1 Experimental techniques

The measurement of laminar flame speeds was conducted using a McKenna burner at local pressure, which is 75 kPa in Laramie, WY. A schematic of the experimental setup is shown in Figure 1. The burner is constructed with a bronze porous plug that can produce a flat laminar flame. A cooling water jacket embedded inside the porous plug facilitates extraction of heat through the water jacket. The inlet and outlet temperatures of the coolant were monitored by a pair of K-type thermocouples throughout the experiments. The McKenna burner is equipped with a shroud flow to isolate the flame from the ambient, for which nitrogen (N_2) was used in the current work. The shroud gas was heated to the coolant outlet temperature in order to avoid heat loss to the ambient. The flow rate of N_2 was varied such that the velocity of the shroud flow was equal to the reactant flow rate.

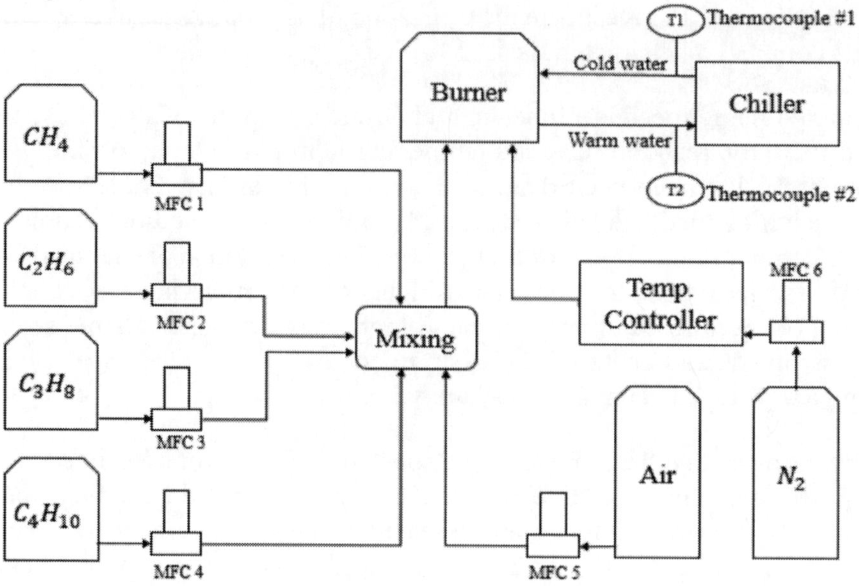

Figure 1. Schematic of the experimental setup

The premixed reactants were prepared by adding the required amount of fuel to a metered quantity of air through a mixing manifold. The flow rates of the component gases and air were controlled by Teledyne Hastings mass flow controllers. Dried and filtered air was used, and all the component gases, including methane (CH_4), ethane (C_2H_6), propane (C_3H_8) and butane (C_4H_{10}), had purity equal to or greater than 99.5%. While multiple associated gas compositions are being investigated, the composition that is presented in this paper is given in Table 1. All studied compositions are derived from measured and representative compositions of associated gas for the Bakken formation, as presented in [15]. The four highest concentration component gases (methane, ethane, propane, and butane) were considered and their concentrations were normalized as presented in Table 1. Lower amounts of pentane, carbon dioxide and N_2 were not considered. Component gases and the associated gas mixture were assessed over a range of ϕ from 0.7 to 1.1.

Table 1. Assumed composition (normalized) of associated gas in the Bakken gas field

Component	Composition (mol%)
Methane	62.2
Ethane	21.5
Propane	12.2
Butane	4.1

A technique similar to that described in the study of Botha and Spalding [16] was followed to determine the adiabatic laminar flame speed free from the effect of stretch. Following this technique, the flame was established with a flow rate that gives a flat, stabilized flame above the burner surface.

Flow rates were increased after each measurement, and the coolant temperature increased or decreased with decreased or increased heat loss from the flame to the porous plug. When steady state was achieved at each reactant flow rate, inlet and outlet coolant temperatures were recorded and the flow rate of water through the cooling jacket was measured in order to calculate the rate of heat extraction from the flame. There was a region of linear variation between the unburned mixture velocity and the heat extracted per milliliter of fuel consumed. The adiabatic laminar flame speed was then obtained through linear extrapolation to zero heat loss. An example of the laminar flame speed extrapolation for C_3H_8/air mixture at a stoichiometric ratio is shown in Figure 2.

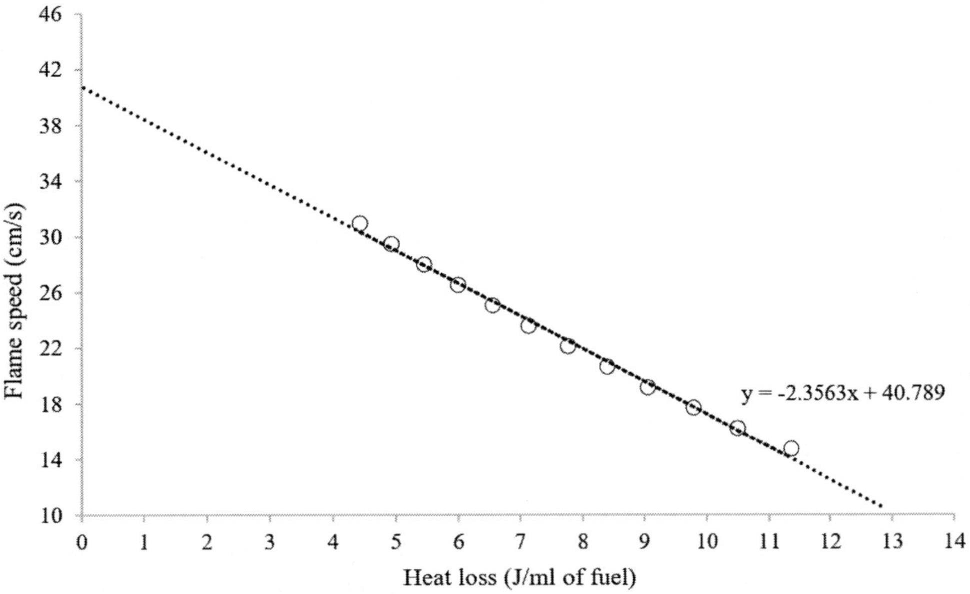

Figure 2. Flame speeds of C_3H_8/air at a stoichiometric ratio over a range of tested flow rates, which produce varying amounts of heat loss from the McKenna burner.

In general, the data points used for extrapolation fell within a range of 20-60% of the flame speed with some exceptions for the leanest mixtures used in the experiment. Higher flow rates lead to wrinkling of the flame and were not utilized for flame speed determination. This range produced a maximum of approximately 5% in the variation of the linearity, depending upon the quantity of data points. The primary sources of uncertainty are the accuracy of the flow controllers, the measurement of heat loss, and the quality of the extrapolation.

2.2 Numerical simulations

One-dimensional simulations of freely propagating, premixed flames were performed for a constant pressure, adiabatic system using *Cantera.* In doing so, USC-Mech II [17] was selected as the chemical kinetic model. Inputs required for the simulation were reactant temperature, pressure, and reactant composition. All of the simulations were performed at a reactant temperature of 293 K and 75 kPa pressure. A range of equivalence ratios was selected from ϕ=0.7 to 1.1 in 0.1 increments to compare the results with experimental data.

2.3 Additive analysis

Additive analysis for the laminar flame speed was performed on a mole fraction basis using Eqn. 1. Mole fractions of the component gases, X_i, were multiplied by their respective laminar flame speeds, $S_{L,i}$ using both numerical and experimental flame speeds, and summed to get the additive $S_{L,mixture}$ for a particular equivalence ratio.

$$S_{L,mixture} = X_{CH4}S_{L,CH4} + X_{C2H6}\,S_{L,C2H6} + X_{C3H8}\,S_{L,C3H8} + X_{C4H10}\,S_{L,C4H10} \quad (1)$$

The additive values for $S_{L,mixture}$ were then compared with the simulation and experimental values.

3. Results and Discussion

3.1 Component gases

As the investigated mixture of associated gas was composed of four component gases, the experimental setup was first validated for these gases by comparing the experimental data with simulation results. Fig. 3a shows the measured and simulated laminar flame speeds of CH_4/air over a range of equivalence ratios. It can be seen that the maximum laminar flame speed obtained from experiments was 38.3 cm/sec at ϕ=1, above and below which lower flame speeds were measured, as expected. It can be seen from comparison with numerical data that experimental and simulated flame speeds are in good agreement, with a maximum deviation of 7.9% at ϕ=0.8.

Fig. 3b shows the measured and simulated laminar flame speeds of C_2H_6/air mixture over a range of equivalence ratios. Similar trends are observed between experimental and numerical laminar flame speeds, with flame speeds increasing with increase in ϕ and the highest flame speed measured at the highest tested ϕ of 1.1. However, simulation results show higher values than experimental measurements throughout the range of equivalence ratios, with deviation ranging from 4.8% to 7.3%. To investigate this, further simulations were performed for C_2H_6/air mixtures at standard temperature and pressure, and the values were compared with the available experimental values from [7][10][12]. It was found that the USC-Mech II mechanism consistently overpredicted the laminar flame speed for C_2H_6/air mixture. Thus, the discrepancy between experimental and numerical values was not attributed to the experimental setup.

The comparison plot for experimental and numerical laminar flame speeds of C_3H_8/air mixture over a range of ϕ is shown in Fig. 3c, and the numerical and experimental results are found to be in good agreement. Relatively higher deviation is found for leanest and richest condition, with the highest deviation between experimental and numerical flame speeds found to be 2.8% at ϕ=0.7.

Finally, Fig. 3d shows experimental and numerical laminar flame speeds of C_4H_{10}/air mixtures over a range of equivalence ratios. The highest experimental flame speed was measured to be 38.6 cm/s at ϕ=1.1, and flame speed decreased with decrease in ϕ. The largest discrepancies between experimental and numerical values is observed at the leanest tested C_4H_{10} equivalence ratios of ϕ=0.7 and ϕ=0.8. At these equivalence ratios, C_4H_{10} had narrower flow rate ranges over which a stable and flat flame was attained as compared to other component gases and equivalence ratios, which may affect the accuracy of the extrapolated flame speed and contribute to the discrepancy between experimental and numerical flame speeds.

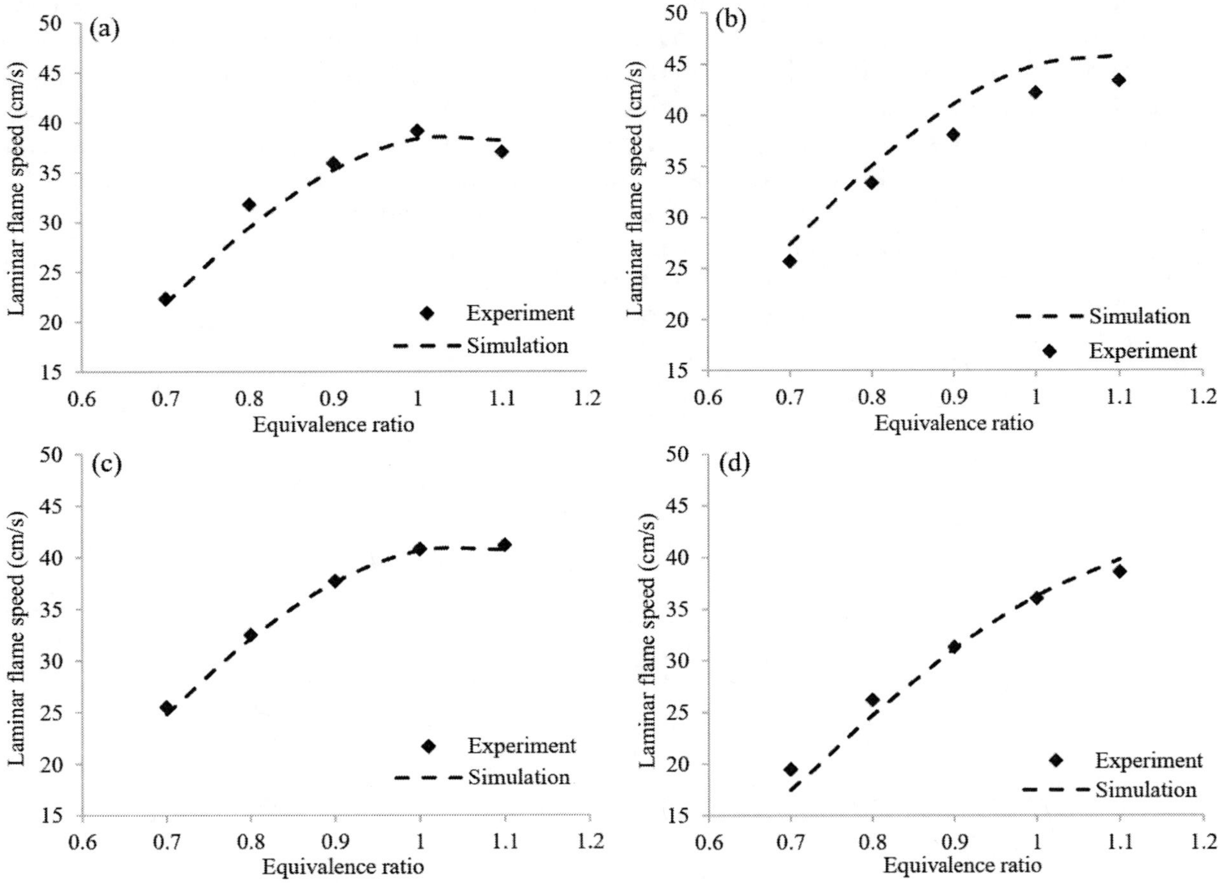

Figure 3. Experimental and simulated adiabatic laminar flame speeds for (a) methane/air, (b) ethane/air, (c) propane/air, (d) butane/air at 293 K and 75 kPa pressure.

The comparison plots in Fig. 3 show reasonably good agreement for all of the component gases except C_2H_6, although the deviation for C_2H_6 is attributed to a consistent overprediction of flame speed by the chemical mechanism used in the simulations. Therefore, the experimental setup used in this work is considered validated and is expected to produce accurate flame speed measurements for the associated gas mixture.

3.2 Associated gas experiments

Fig. 4 summarizes the measured and numerically simulated flame speeds for the associated gas composition described in Table 1. Similar to the component gases, adiabatic laminar flame speeds increase as equivalence ratio is increased from lean to stoichiometry. The peak value for the adiabatic laminar flame speed was found to be 41.6 cm/s at ϕ=1.0. Comparing the simulated and experimental flame speed values it is found that the data are in reasonably good agreement near stoichiometry and in the mildly rich mixture. However, deviation increases for the leaner mixtures with the highest deviation of 10.2% observed at ϕ=0.7. According to the flame instability limit associated with the flat flame burner [18], increased instability of the flame is expected at very lean equivalence ratios and may increase the uncertainty in zero heat extrapolation. It is also to be noted that the chemical mechanism used in the simulation does not predict laminar flame speeds for C_2H_6 well. As the mixture contains around 21.5% ethane, the uncertainty due to the chemical mechanism is cumulative in the deviation between experimental and simulation results. In addition, along with chemical kinetics, flame speeds depend on diffusive properties of reactive mixtures. The multi-component transport properties used in simulation can also be responsible for the deviation between numerical and experimental values. Overall, relatively low quantitative differences between numerical and experimental values confirm that the basic trends for laminar flame speed with equivalence ratio have been captured well. Experimental S_L values suggest that the reactivity and thermal diffusivity of the associated mixture are comparable to the component gases. For lean mixtures, methane gives the closest S_L value to associated gas, while for ϕ=1.0 and 1.1, S_L for propane are comparable to associated gas S_L.

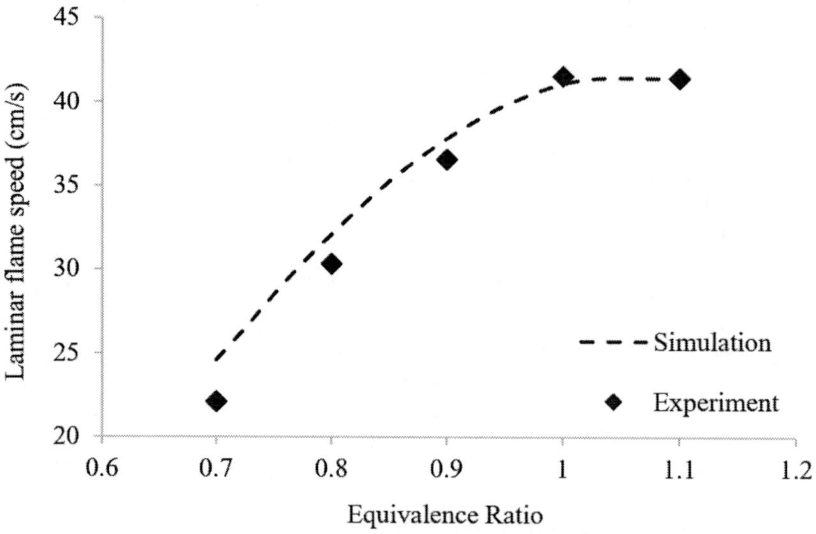

Figure 4. Comparison of experimental and numerical laminar flame speeds for the associated gas mixture

3.3 Additive analysis

Fig. 5 shows additive laminar flame speed values in comparison to numerical and experimental values. As shown in Fig. 5a, simulated associated gas flame speeds are greater than additive values based on simulated component gases across all tested equivalence ratios, with the deviation ranging from 3.0% to 6.0%. These results suggest that the chemical kinetic model accounts for non-additive behavior in gas mixtures. Similarly, data are not in agreement in Fig. 5b, in which experimentally measured flame speeds of the associated gas mixture are compared with additive flame speeds based on experimentally measured flame speeds of the component gases. Deviations of 0.2-7.0% are observed between experimental and additive S_L values across all equivalence ratios, except at ϕ=0.9 where excellent agreement is observed. This additive analysis warrants for further comprehensive investigation for the sources of these discrepancies, with the objective of forming a basis for predictive capabilities for flame speeds based upon associated gas composition.

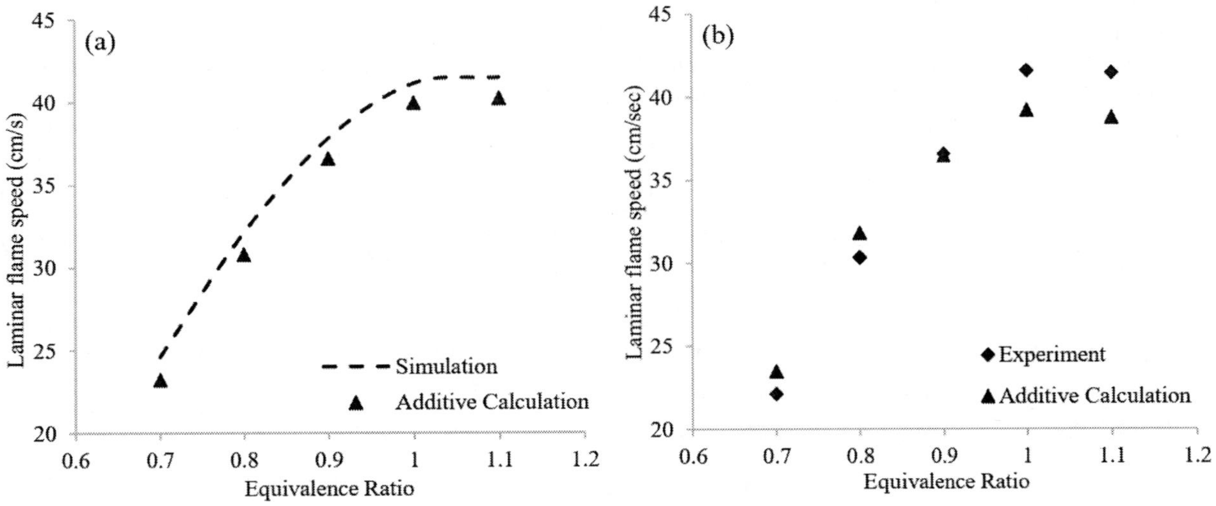

Figure 5. Comparison of adiabatic laminar flame speed profiles between (a) simulation result and calculated additive values, (b) experimental data and calculated additive values.

4. Conclusions

A laminar flame speed measurement technique based on zero heat extrapolation, using a flat flame McKenna burner, has been employed in this study for the investigation of component gas and associated gas mixture flame speeds. After the validity of the experimental setup was checked with satisfactory comparison with numerical data for component gas flame speeds, this study measured adiabatic laminar flame speeds for an associated gas mixture from the Bakken formation over a range of equivalence ratios from 0.7 to 1.1. General agreement between numerical and experimental results near stoichiometric and mildly rich equivalence ratios confirms that the basic trend for S_L with composition has been captured satisfactorily. In general, the laminar flame speed profile of the mixture suggests that the diffusivity and the reactivity of the investigated associated mixture are comparable to that of methane and propane depending on the equivalence ratio. Additive analysis of the mixture warrants future investigations on reaction path in order to identify how the species affect laminar flame speeds. Sensitivity analysis on the component gases can also be done to identify the effects of their concentrations on laminar flame speed. Additionally, ongoing investigations are examining other associated gas mixtures in order to evaluate the expected range of laminar flame speeds for Bakken associated gas.

References

[1] U.S. Energy Information Agency. International Energy Outlook 2013. 2013. doi:EIA-0484(2013).

[2] Liss W. E., Thrasher W. H., Steinmetz G. F., Chowdiah P., and Attari A., 1992, Variability of Natural Gas Composition in Select Major Metropolitan Areas of the United States, GRI Report.

[3] World Bank 2012 Global Gas Flaring Reduction program World Bank.

[4] US Energy Information Administration (www.eia.gov).

[5] Pederstad A., Gallardo M., Saunier S., Improving utilization of associated gas in US tight oil fields, Carbon Limits, 2015.

[6] Tahtouh T., Halter F., Mounaïm-Rousselle C., Measurement of laminar burning speeds and Markstein lengths using a novel methodology, Combust. Flame 2009, 156, 1735–1743.

[7] Konnov A. A., Dyakov I. V., De Ruyck J., Measurement of adiabatic burning velocity in ethane–oxygen–nitrogen and in ethane–oxygen–argon mixtures, Exp. Thermal Fluid Sci. 2003, 27, 379–384.

[8] Bosschaart K. J., De Goey L. P. H, The laminar burning velocity of flames propagating in mixtures of hydrocarbons and air measured with the heat flux method, Combust. Flame 2004, 136, 261–269.

[9] Tang Z., Huang J. H. C., Jin X., Wang, Miao H., Effects of N2 dilution on laminar burning characteristics of propane-air premixed flames, Energy & Fuels 23 (2009) 151–156.

[10] Egolfopolous F.N., Vagelopoulos C.M., Direct experimental determination of laminar flame speeds, Twenty-Seventh Symposium (International) on Combustion (1998) 513–519.

[11] Arkan F.S. (2000). Effect of initial temperature on flame propagation through tube for (methane, propane, LPG and butane)-air mixtures, MSc Thesis, University of Technology, 2000.

[12] Dirrenberger P., Gall H.L., Bounaceur R., Herbinet O., Glaude P.A., Konnov A., Battin-Leclerc F., Measurements of Laminar Flame Velocity for Components of Natural Gas, Energy & Fuels 2011, 25, 3875-3884

[13] Khan A.R., Ravi M.R, Ray A., Effect of Higher Hydrocarbon Content on Laminar Burning Velocity and Flame Stability of Natural Gas, 10th U.S National Combustion Meeting.

[14] Bourque G., Healy D., Vries J. D., Petersen E., Ignition and Flame Speed Kinetics of Two Natural Gas Blends with High Levels of Heavier hydrocarbons. J Eng. Gas Turbines Power 2010; 132. doi:10.1115/1.3124665.

[15] Wocken C., Stevens B.G., Almlie J.C., Schlasner S.M., End-Use Technology Study – An Assessment of Alternative Uses For Associated Gas, Energy and Environmental Research Center, 2012.

[16] Botha J. P., Spalding D. B., The laminar flame speed of propane/air mixture with heat extraction from the flame. Proc. R. Soc. London, Ser. A 1954, 225, 71−96.

[17] Wang H., You X., Joshi A.V., Davis S.G., Laskin A., Egolfopoulos F., Law C.K., USC Mech Version II. High-Temperature Combustion Reaction Model of H2/CO/C1-C4 Compounds, 2007.

[18] Kurdyumov V. N., Matalon M., The porous-plug burner: Flame stabilization, onset of oscillation, and restabilization. Combust. Flame 2008, 153, 105−118.

Western States Section of the Combustion Institute – Fall 2017 Meeting
Hosted by the University of Wyoming
October 2-3, 2017

Experimental and numerical study of freely propagating propane cool flames

*Mohammadhadi Hajilou, Erica Belmont**

Department of Mechanical Engineering, The University of Wyoming, Laramie, WY, USA

Corresponding author, email: ebelmont@uwyo.edu

Abstract:

This paper employs a recently developed experimental platform to investigate low temperature oxidation of premixed propane cool flames at sub-atmospheric pressures. Propane cool flames are studied over a range of equivalence ratios (ϕ) from 0.17 to 1.0 in 0.17 increments. Due to the sensitivity of propane cool flame stability to pressure, a pressure stability map is presented for the cool flames over a range of equivalence ratios and different stability regimes are defined. The height of the flame above the burner surface is measured at different equivalence ratios and pressures for a range of flow rates, and the cool flame propagation speed is estimated at the intersection of burner-stabilized and freely propagating flame operating regimes. Based on the pressure stability map, a cool flame of ϕ=0.17 at 17.3 kPa is chosen for detailed study. A temperature profile of the freely propagating propane cool flame of ϕ=0.17 at 17.3 kPa is presented and used in fixed-temperature numerical simulations. In order to perform non-fixed temperature simulations, the method of directed relation graph (DRG) is employed to reduce chemical kinetic models and isolate cool flame chemistry. Experimental and numerical simulation results are compared.

Keywords:

Cool flame; Low temperature oxidation; Propane; Reduced mechanism

1. Introduction

Both high and low temperature oxidation kinetics have a significant role in the performance of hydrocarbon fuels, and a detailed understanding of chemical kinetics can aid in yielding higher thermal efficiency in engines and other combustors and lower amounts of pollutants [1]. Several works have reported on the importance of low temperature fuel oxidation to engine knock in diesel engines [2-5]. Although various studies have been performed to investigate chemical kinetics of high temperature oxidation of fuels [6-9], low temperature fuel oxidation, where negative temperature coefficient (NTC) behavior is observed, is still lesser understood [10].

In this regard, there exists a need for an experimental platform which can be employed to investigate low temperature oxidation of fuels under strongly coupled chemistry-transport conditions [11]. In a recent work by Hajilou et al. [11], an experimental setup was developed and utilized for the study of freely propagating dimethyl ether (DME) cool flames, in which a flat flame laminar Hencken burner facility was used as a platform [12, 13] and ozone (O_3) enhancement was employed [14-16] to produce a stabilized freely propagating DME cool flame for the study of flame lift-off heights, propagation speed, stability, and temperature and species measurements.

In studying combustion chemical kinetics, whether for implementation in fixed or non-fixed temperature simulations, many methods have been developed to reduce chemical kinetic models, such as the method of directed relation graph (DRG) [17-20]. Generally, mechanism reduction is performed to improve the usability of these models in large scale simulations. Lu and Law [17] reported the use of auto-ignition simulations for low temperature to mid-high temperature oxidation and perfectly stirred reactor (PSR) simulations for high temperature oxidation in the implementation of DRG. In cases where there was a need to include the NTC regime in the reduced mechanism as well, the starting temperature was simply chosen in the NTC regime [18].

The experimental platform utilized in [11] was employed in the present study to investigate the combustion characteristics of ozone-enhanced freely propagating premixed propane cool flames at sub-atmospheric pressures. Due to the sensitivity of propane cool flame stability to pressure, a pressure stability map was first developed for a range of equivalence ratios to identify stable operating points, and different pressure stability regimes were defined. Lift-off height measurements of the flame above the burner surface were taken to identify burner-stabilized and freely propagating operating regimes, and the cool flame propagation speed was identified at the intersection of these regimes. Temperature measurements of a lean propane cool flame as a function of height above the burner (HAB) were taken and utilized in fixed temperature numerical simulations, and the numerical propagation speed was compared with the value attained from the experiments. In order to facilitate a non-fixed temperature simulation of the freely propagating flame, a chemical kinetics model was reduced employing the DRG technique to isolate the cool flame chemistry without the transition to hot flame usually seen in non-fixed temperature numerical simulations [21], and the results were compared with the experiments.

2. Materials and Methods

2.1. Experimental Techniques

The experimental setup in this study is what was recently used by Hajilou et al. [11]. As was explained in their work, it consisted of a stainless steel pressure chamber, 87 cm in height and 60 cm in diameter, inside which a Hencken burner (Technologies for Research, Model RT1X1) was mounted on a translation stage, facilitating temperature and species measurements from upstream to downstream of the flame. In order to maintain the pressure in the experiments at a desired level, the chamber was connected to a vacuum pump (Edwards XDS 10). Two Borosilicate windows mounted on both sides of the chamber permitted optical access. Pictures of the flame were taken through the chamber windows with a Nikon D90 camera utilizing a NIKKOR 50mm/f1.8 lens, and the pictures were analyzed using ImageJ software [22]. Full details of the experimental setup can be found in [11].

In order to meter the flow rates of propane (C_3H_8), oxidizer (O_2) and inert (Ar), Teledyne mass flow controllers were used to produce mixtures with equivalence ratios (ϕ) ranging from 0.17 to 1.0 in 0.17 increment. To stabilize the cool flame, ozone (O_3) was produced in the oxidizer flow using an ozone generator (Ozone Solutions TG-40), and the concentration of O_3 was monitored using an InUSA O_3 analyzer (Model H1). For this study, an O_3 concentration of 6.2 mol% in O_2/O_3 was chosen based on a number of preliminary experiments.

To develop a pressure stability map, the cool flame was ignited at a pressure and equivalence ratio based on a number of preliminary experiments to find a condition at which the flame was stable. Then, using the regulation valve of the vacuum pump, the pressure was varied to find unstable and stable flame conditions. Increasing the pressure in increments of 5.3 kPa from the minimum stable value, pictures of the flame were taken at each step to measure the lift-off height of propane cool flames for a range of mixture flow rates.

Temperature measurements were conducted using two butt-welded K-type thermocouples of diameters 130 μm and 380 μm. The thermocouples were positioned at the center of the Hencken burner surface and mounted parallel to each other, and measurements from upstream to downstream were performed using the translation stage. To account for conductive and radiative heat losses in the measurements, the two-thermocouple correction method [23] was employed.

In each set of experiments, the chamber was first pumped down to the desired pressure level and then C_3H_8, O_2 and O_3 were introduced to the central surface of the Hencken burner, while Ar flow was introduced to the surrounding area of the center burner surface as a curtain flow. $C_3H_8/O_2/O_3$ cool flames were ignited using a glow plug downstream of the burner, to which the power was gradually provided until the cool flame was ignited.

2.2. Numerical Simulations

The Directed Relation Graph method (DRG) was applied to a full propane mechanism in order to isolate cool flame chemistry. The full chemical kinetic model reduced in this study was Aramco 1.3 (254 species and 1568 reactions) [24], and the model was altered to incorporate an O_3 sub-model [14, 21]. An auto-ignition simulation was performed to capture the low temperature chemistry, and the starting temperature of the simulations was chosen as 600 K in order to begin the simulation within the NTC regime. The auto-ignition simulation was stopped when the temperature of the mixture rose to approximate 850 K in order to prevent the simulation from entering the double flame and hot flame regimes and abbreviate the chemistry accordingly. Time steps of 0.1 milliseconds, and 153 time steps, were chosen to capture sensitive reactions, and the starting species for DRG were chosen to be C_3H_8, O_2 and O_3. The cut-off value (ε) in DRG was varied in small increments, and a single value of ε was chosen for further simulations.

3. Results and Discussions

3.1. Pressure stability map

Figure 1 shows a pressure stability map for propane cool flames of ϕ=0.17 to 1.0 in ϕ=0.17 increments. Flow rates of the mixtures were varied from 1.0 SLPM to 1.8 SLPM, except for ϕ=0.17 at 14.7 kPa where the flow rate was varied from 0.6 SLPM to 1.5 SLPM due limited flame stability at higher flow rates. Two flow regimes of burner-stabilized and freely propagating were identified for cool flames, where in the burner-stabilized regime no significant change in the lift-off height, or height of the flame above the burner surface, was observed with an increase in the flow rate of the mixture, and in the freely propagating regime the cool flame lift-off height increased approximately linearly with an increase in the mixture flow rate [11-13]. This methodology was applied to the propane cool flames in this study at the different tested equivalence ratios and pressures. Then, different pressure regimes were defined and are identified in Fig. 1 based on the cool flame lift-off data at each pressure and equivalence ratio. The first pressure regime was defined as stable without lift-off, indicating that a stable flame was achieved but no increase in lift-off height of greater than of 1.5 mm was seen with an increase in the mixture flow rate within the aforementioned flow rate ranges, and the cool flame was observed only in the burner-stabilized regime. It should be noted that a transition from burner-stabilized regime to freely propagating regime might be observed at higher flow rates than those tested in this study, but testing of higher flow rates than the reported range was not possible due to limitations of the experimental setup. The second pressure regime was defined as stable with lift-off, where the cool flames at these pressures transitioned from burner-stabilized to freely propagating regimes within the tested flow rate range. The last pressure regime in Fig. 1 was defined as unstable, where the cool flames at these pressures were either unstable and near extinction or blow-off, or could not be ignited, and therefore were not suitable for a prolonged study.

As can be observed in Fig. 1, at ϕ=0.84 and ϕ=1.0, the cool flame only existed in the regime of stable without lift-off and therefore did not lift off at any of the tested flow rates. For equivalence ratios ϕ=0.34, ϕ=0.5 and ϕ=0.67, the cool flame did not lift off at the highest tested pressure of 36.0 kPa, 41.3 kPa and 46.7 kPa, respectively, but the cool flame was categorized as stable with lift-off at the two lower operating pressures of 25.3 kPa and 30.7 kPa for ϕ=0.34, 30.7 kPa and 36.0 kPa for ϕ=0.5, and 36.0 kPa and 41.3 kPa for ϕ=0.67. Lastly, the cool flame of ϕ=0.17 was stable without lift-off at pressures of 30.7 kPa and 25.3 kPa, but was stable with lift-off at pressures of 14.7 kPa and 20.0 kPa. Additionally, at all the tested equivalence ratios, the lowest pressure reported on the stability map for each equivalence ratio was an unstable point for that equivalence ratio. Based on the pressure stability map, a propane cool flame of ϕ=0.17 at 17.3 kPa was chosen for further investigation.

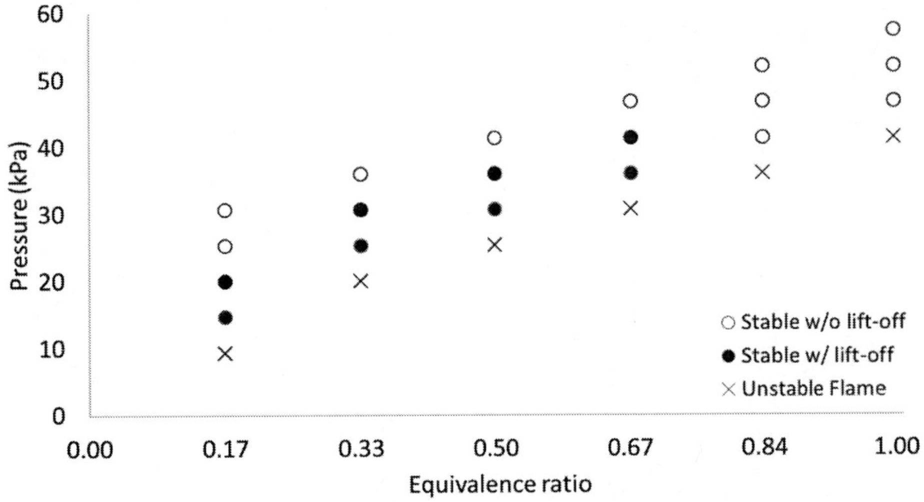

Figure 1. Pressure stability map of propane cool flames

3.2. Lift-off height

Fig. 2 shows the variation of lift-off height, or the flame height above the burner surface (HAB), in terms of reactant flow rate for the $C_3H_8/O_2/O_3$ cool flame of ϕ=0.17 at 17.3 kPa. As can be seen in Fig. 2, there was no significant change in the height of the flame above the burner with an increase in the reactant flow rate from 0.78 SLPM to 1.46 SLPM; hence, this regime is labeled as burner-stabilized. At higher flow rates, a linear increase in the lift-off height of the cool flame was observed with an increase in the mixture flow rate from 1.55 SLPM to 1.75 SLPM; this regime is labeled as freely propagating. As has been discussed in [11-13], the intersection of these regimes can be utilized to estimate adiabatic propagation speed of the flame. The intersection of these two regimes occurs at 1.50 SLPM and the corresponding adiabatic propagation speed for the cool flame is calculated to be 24.7 cm/sec. The uncertainties associated with the lift-off height and the propagation speed are calculated to be ± 0.81 mm and ± 0.5 cm/sec, respectively.

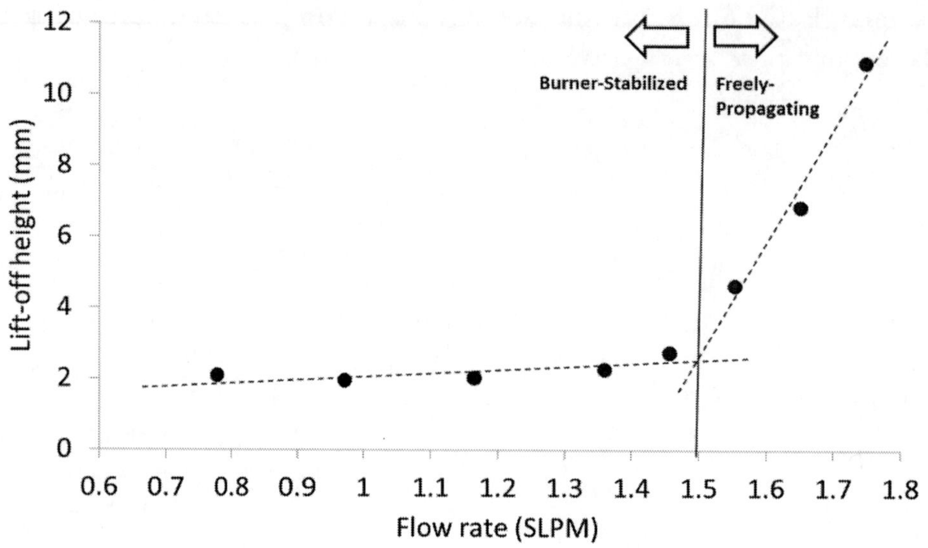

Figure 2. Lift-off height in terms of reactant flow rate for a propane cool flame of ϕ=0.17 at 17.3 kPa

3.3. Temperature profile

Based on the lift-off height results (Fig. 2), 1.55 SLPM was chosen as the point in the freely propagating regime for temperature measurements. Figure 3 shows the temperature profile of the freely propagating $C_3H_8/O_2/O_3$ cool flame of ϕ=0.17 at 17.3 kPa. The thermocouples were moved from upstream to downstream of the flame and vice versa for temperature measurements. It can be seen in Fig. 3 that the downstream gradients are zero within the calculated uncertainty, suggesting no significant exothermic reactions are progressing in the downstream.

As was discussed in [11], due to the unique flow field imposed in the experimental setup used in this study, a trailing hot flame after the cool flame which is predicted by non-fixed temperature numerical simulations is not observed in the experiments; therefore, fixed temperature numerical simulations are used to reproduce the flame observed in the experiments. In order to perform numerical simulations for comparison with the experimentally determined propagation speed, the experimentally measured temperature profile was utilized and the PREMIX code in CHEMKIN was employed using Aramco 1.3 [24], along with an O_3 sub-mechanism [14, 21].

As can be seen in Fig. 3, no temperatures are reported between 3.6 mm and 9.3 mm above the burner surface. This is due to the fact that, at those locations, the flame was distorted by the thermocouples which resulted in disrupted temperature measurements. Thus, to use the experimental temperature profile in numerical simulations, two limiting scenarios for the change of temperature between upstream and downstream were considered: the first is denoted by the solid line and the second by the dashed line. The reason for choosing these two limiting cases in particular is based on the temperature profile observed in freely propagating DME cool flame in [11] utilizing the same experimental setup, where the trend of temperature change from upstream to downstream followed an approximately linear trend. Using the kinetic model of Aramco 1.3 [24] with the O3 sub-mechanism added in the model [14, 21], and employing both temperature profile scenarios in the simulations using CHEMKIN, the numerical value of the flame propagation speed was calculated to be 20.0 cm/sec, which deviates from the experimental adiabatic propagation speed of 24.7 cm/sec by 18.8%. This result shows that, regardless of which temperature profile was chosen, the flame propagation speed was the same in both cases. The downstream temperature of the $C_3H_8/O_2/O_3$ cool flame of ϕ=0.17 at 17.3 kPa was measured to be 800 K and the average uncertainty for the temperature measurements was calculated to be ± 34.4 K.

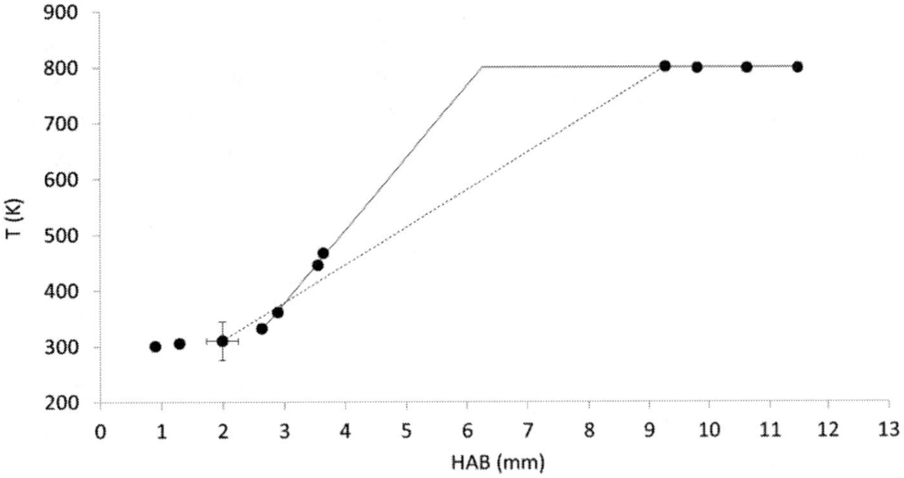

Figure 3. Temperature profile of the freely propagating propane cool flame of ϕ=0.17 at 17.3 kPa in terms of HAB

3.4. Kinetic model reduction

While the fixed temperature simulations provide valuable insight into the flame characteristics, such as flame speed and species evolution, the structure of the flame is somewhat imposed by the prescribed temperature profile. There is additional value in the ability to simulate a non-fixed temperature, freely propagating cool flame in which the cool flame chemistry is allowed to evolve without a prescribed temperature, as the chemistry evolves in the experimental flames. However, as noted in this study and [11], non-fixed temperature simulations performed using a full chemistry mechanism results in hot flame transition, thereby obscuring the cool flame chemistry. In order to focus only on cool flame chemistry and permit non-fixed temperature cool flame simulations, the DRG chemistry model reduction technique [17-20] was implemented in Cantera [25] to reduce chemical kinetic models for cool flames and isolate the cool flame chemistry. After performing the DRG technique for a range of cut-off values and, considering all of the reduced mechanisms, a reduced mechanism with $\varepsilon = 0.15$ was chosen, which most closely approximated the experimental cool flame temperature was chosen. The reduced chemical kinetic model consisted of 30 species and 97 reactions for $C_3H_8/O_2/O_3$ cool flame of ϕ=0.17 at 17.3 kPa and was utilized for further simulations and comparison with experimental results.

Non-fixed temperature, freely propagating cool flame simulations were conducted using the reduced mechanism. Figure 4 shows the numerical temperature profile attained from these simulations, utilizing the reduced kinetic model for the $C_3H_8/O_2/O_3$ cool flame of ϕ=0.17 at P=17.3 kPa. It is noteworthy that the numerical temperature profile predicts a nearly linear trend between upstream and downstream, which validates the assumption of a linear temperature trend between upstream and downstream temperatures that was utilized in fixed temperature simulations in this study. However, downstream temperature predicted by the numerical simulations was 999 K from the reduced Aramco 1.3 model [24], which is higher than the experimentally measured peak temperature of 800 K and this discrepancy warrants further investigation.

Moreover, propagation speed from non-fixed temperature simulations of the $C_3H_8/O_2/O_3$ cool flame of ϕ=0.17 at 17.3 kPa predicted by the reduced model of Aramco 1.3 [24] was calculated to be 10.5 cm/sec. This value deviates from the experimental propagation speed of 24.7 cm/sec by 58%, which is greater than the difference observed in fixed temperature simulations and likewise warrants further investigation.

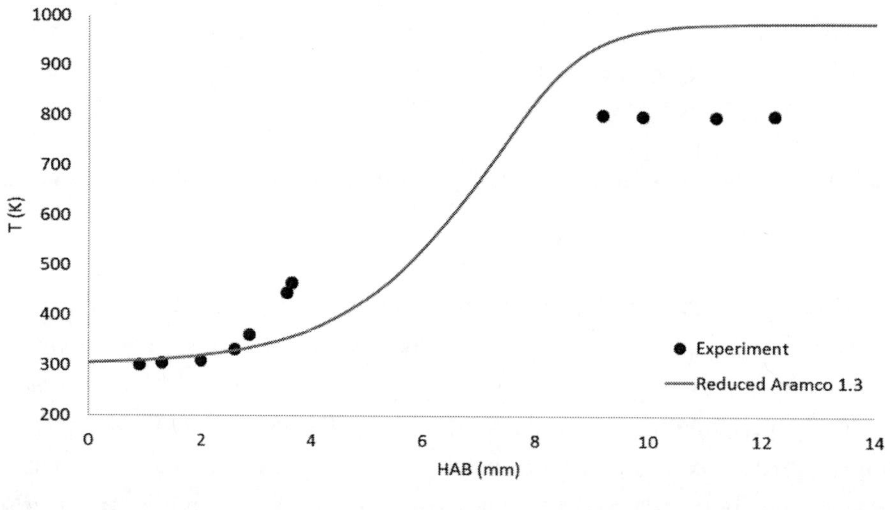

Figure 4. Comparison of the numerical temperature profile attained from non-fixed temperature simulations using a reduced kinetic model with experimental data for a $C_3H_8/O_2/O_3$ cool flame of ϕ=0.17 at 17.3 kPa

4. Conclusions

Propane cool flames of ϕ=0.17 to ϕ=1.0 in 0.17 increments were investigated experimentally using a previously developed experimental facility for the study of cool flames at sub-atmospheric pressures. Significant sensitivity of propane cool flame stability to pressure was observed, and a pressure stability map was presented which introduced three different pressure regimes: I) stable without lift-off, where the propane cool flame only operated in a burner-stabilized flame regime; II) stable with lift-off, where the propane cool flame was measured in both burner-stabilized and freely propagating regimes; and III) unstable.

Based upon the pressure stability map, a propane cool flame of ϕ=0.17 at 17.3 kPa was chosen for further investigation. Flame lift-off height in terms of mixture flow rate showed existence of the two flame regimes: burner-stabilized and freely propagating. The intersection of these two regimes was utilized to estimate the adiabatic propagation speed of the cool flame, which was calculated to be 24.7 cm/sec. Temperature measurements of the flame were performed upstream and downstream of the freely propagating propane cool flame. Fixed temperature numerical simulations were performed which showed a numerical propagation speed of 20.0 cm/sec, which deviates from the experimental value by 18.8%.

In order to facilitate non-fixed temperature simulations of freely propagating cool flames, reduction of the Aramco 1.3 chemical kinetic model was performed using the DRG method. Auto-ignition simulations were conducted and abbreviated to isolate the cool flame chemistry and eliminate the transition of cool flame to hot flame that is observed in non-fixed temperature numerical simulations using full kinetic models. Results from the reduced model produced a numerical downstream temperature of 999 K , which is substantially higher than the maximum temperature of 800 K that was measured experimentally. The propagation speed from the non-fixed temperature simulation was calculated to be 10.5 cm/sec using the reduced Aramco 1.3 mechanism, which was deviant from the experimental value by 58%.

Acknowledgements

This research was supported by NASA under award No. NNX16AQ95A.

References

[1] F. L. Dryer, "Chemical kinetic and combustion charactersitics of transportation fuels," *Proc. Combust. Inst.*, vol. 35, no. 1, pp. 117-144, 2015.

[2] T.A. Litzinger, "A review of experimental studies of knock chemistry in engines," *Prog. in Energy and Combust. Sci.*, vol. 16, no. 3, pp. 155-167, 1990.

[3] J.E. Dec, "Advanced compression-ignition engines—understanding the in-cylinder processes," *Proc. Combust. Inst.*, vol. 32, no. 2, pp. 2727-2742, 2009.

[4] R.D Reitz, G. Duraisamy, "Review of high efficiency and clean reactivity controlled compression ignition

(RCCI) combustion in internal combustion engines," *Prog. in Energy and Combust. Sci.*, vol. 46, pp. 12-71, 2015.

[5] O. Welz, M P. Burke, I. O. Antonov, C. F. Goldsmith, J. D. Savee, D. L. Osborn, C. A. Taatjes, S. J. Klippenstein, L. Sheps, "New Insights into Low-Temperature Oxidation of Propane from Synchrotron Photoionization Mass Spectrometry and Multiscale Informatics Modeling," *J. Phys. Chem. A*, vol. 119, p. 7116–7129, 2015.

[6] D. F. Davidson R.K. Hanson, "Recent advances in laser absorption and shock tube methods for studies of combustion chemistry," *Prog. in Energy and Combust. Sci.*, vol. 44, pp. 103-114, 2014.

[7] F.N. Egolfopoulos, N. Hansen, Y. Ju, K. Kohse-Höinghaus ,C.K. Law, F. Qi, "Advances and challenges in laminar flame experiments and implications for combustion chemistry," *Prog. in Energy and Combust. Sci*, vol. 43, pp. 36-67, 2014.

[8] N. Hansen, T. A. Cool, P. R. Westmoreland, K. Kohse-Höinghaus, "Recent contributions of flame-sampling molecular-beam mass spectrometry to a fundamental understanding of combustion chemistry," *Prog.in Energy and Combust. Sci.*, vol. 35, no. 2, pp. 168-191, 2009.

[9] J. K. Lefkowitz, J. S. Heyne, S. H. Won, S. Dooley, H. H. Kim, F. M. Haas, S. Jahangirian, F. L. Dryer, Y. Ju, "A chemical kinetic study of tertiary-butanol in a flow reactor and a counterflow diffusion flame," *Combust. Flame*, vol. 159, no. 3, pp. 968-978, 2012.

[10] F. Battin-Leclerc, "Detailed chemical kinetic models for the low-temperature combustion of hydrocarbons with application to gasoline and diesel fuel surrogates," *Prog. in Energy and Combust. Sci.*, vol. 34, no. 4, pp. 440-498, 2008.

[11] M. Hajilou, T. Ombrello, S.H. Won, E. Belmont, "Experimental and numerical characterization of freely propagating ozone-activated dimethyl ether cool flames," *Combust. Flame*, vol. 176, pp. 326-333, 2017.

[12] T. Ombrello, C. Carter, V. Katta, "Burner platform for sub-atmospheric pressure flame studies," *Combust. Flame*, vol. 159, no. 1, pp. 2363-2773, 2012.

[13] E. Belmont, T. Ombrello, M. Brown, C. Carter, J. Ellzey, "Experimental and numerical investigation of freely propagating flames stabilized on a Hencken Burner," *Combust. Flame*, vol. 162, no. 6, pp. 2679–2685, 2015.

[14] T. Ombrello, S.H. Won, Y. Ju, S. Williams, "Flame propagation enhancement by plasma excitation of oxygen. Part I: Effects of O_3," *Combust. Flame*, vol. 157, no. 10, pp. 1906-1915, 2010.

[15] S.H. Won, B. Jiang, P. Dievart, C.H. Sohn, Y. Ju, "Self-sustaining n-heptane cool diffusion flames activated by ozone," *Proc. Combust. Inst.*, vol. 35, pp. 881-888, 2015.

[16] C.H. Sohn, J.W. Son, S.H. Won, Y. Ju, "Computational studies of diffusion cool flame structures of n-heptane with/without ozone sensitization with a reduced chemistry," *J. Mech. Sci. Tech.*, vol. 39, no. 3, pp. 1297-1305, 2015.

[17] T. Lu, C. K. Law, "A directed relation graph method for mechanism reduction," *Proc. Combust. Inst.*, vol. 30, pp. 1333-1341, 2005.

[18] T. Lu, C.K. Law, "Linear time reduction of large kinetic mechanisms with directed relation graph: n-Heptane and iso-octane," *Combust. Flame*, vol. 144, pp. 24-36, 2006.

[19] T. Lu, C.K. Law, "On the applicability of directed relation graphs to the reduction of reaction mechanisms,"

Combust. Flame, vol. 144, pp. 472-483, 2006.

[20] T. Lu, C.K. Law, "Toward accommodating realistic fuel chemistry in large-scale computations," *Prog. in Energy and Combust. Sci.*, vol. 35, pp. 192-215, 2009.

[21] Y. Ju, C.B. Reuter, S. H. Won, "Numerical simulations of premixed cool flames of dimethyl ether/oxygen mixtures," *Combust. Flame*, vol. 162, no. 10, pp. 3580-3588, 2015.

[22] ImageJ, U. S. National Institutes of Health, Bethesda, Maryland, USA W.S. Rasband. (1997-2014) http://imagej.nih.gov/ij/.

[23] J. Stein, "Uncertainties in radiation correction for thermocouples in high temperature gases", 2007, B.Sc. Thesis, Princeton University, New Jersey, USA.

[24] W. K. Metcalfe, S. M. Burke, S. S. Ahmed, H. J. Curran , "A Hierarchical and Comparative Kinetic Modeling Study of C_1-C_2 Hydrocarbon and Oxygenated Fuels," *Intl. J. Chem. Kin.*, vol. 45, pp. 638-675, 2013.

[25] D. G. Goodwin, H. K. Moffat, R. L. Speth, "Cantera: An object- oriented software toolkit for chemical kinetics, thermodynamics, and transport processes," 2017, Version 2.3.0. [Online]. http://www.cantera.org

Western States Section of the Combustion Institute - Fall 2017 Meeting
Hosted by the University of Wyoming
October 2-3, 2017

Temperature profiles and extinction limits of a coflow water-laden methane/air diffusion flame

Michela Vicariotto *Patricia Martinez* *Derek Dunn-Rankin*

Department of Mechanical and Aerospace Engineering, University of California, Irvine

Temperature profiles and extinction limits of a non-premixed coflow methane flame were studied when water vapor was added on the fuel side of the flame. The understanding of thermal and chemical effects of water addition on the fuel side of a diffusion flame is relevant for processes such as the combustion of methane hydrates, emulsified fuels and wet biomass. In this work, temperature profiles were measured experimentally by applying thin-filament pyrometry (TFP) to the flame, at different conditions of inlet fuel velocity, and increasing water amounts up to extinction. A SiC 14 μm fiber and a commercial DSLR camera were used for the TFP measurements. Results show that increasing amounts of water can be sustained by the flame for decreasing methane/water mixture inlet velocities. The TFP temperature profiles were compared to numerical CFD simulations, showing good agreement in the trend of temperature as increasing molar fractions of water were added, and in the maximum amount of water the flame can sustain before extinction. The good comparison suggests that the simulations capture accurately the role of water in these flames.

1 Introduction

Water diluted combustion systems have been widely studied in applications such as NO_x emission reduction [1, 2], soot reduction [3, 4], and fire suppression [5, 6, 7]. Many of the studies found in literature regarding water diluted combustion, investigate the role of liquid and vapor water addition in premixed flames [8, 9, 10]. However, nonpremixed combustion is important in fires and power applications. Most of the works regarding water addition in diffusion flames utilize either a counterflow configuration with dilution on the air [11, 12, 13] or fuel side [14, 15, 16], or a coflow configuration with water added to the oxidizer stream [17].

The present study is focused on water dilution in nonpremixed coflow flames where the diluent is added on the fuel side. This configuration is relevant for a number of combustion applications. Some examples include: combustion of watery fuels such as methane hydrates [18], emulsified fuels [19, 20], gel fuels [21] and biomass, or the behavior of pool fires developing from LNG or oil on water and steam-assisted flares for the reduction of emissions.

This work reports the temperature profiles and extinction limits of a water-laden fuel coflow diffusion flame. Temperature measurements are obtained by applying thin-filament pyrometry (TFP) to the flame. TFP was first introduced for combustion by Vilimpoc and Goss [22] and has then been widely used in several reacting flow environments. Uncertainties related to the use of thin-filament pyrometry can be found in the literature. Maun at al. [23] estimated uncertainties up to $\pm$30 K in the temperature measurements of a coflowing methane/air diffusion flame. Bin at al. [24] reported temperature uncertainties to be between 36 K and 45 K, depending on the experimental approach employed. The study also investigated fiber aging and spectral emissivities. The current work employs the best practices recommended by these earlier comprehensive works in TFP.

2 Methods

2.1 *Experimental setup*

The laminar diffusion flame is established on a co-annular burner with a fuel tube (inner diameter of 1.8 mm) surrounded by a 2.54 cm inner diameter air nozzle. The oxidizer tube is filled with brass beads, and a honeycomb mesh is placed at its exit to ensure a top-hat air velocity profile. The fuel tube extends above the honeycomb mesh by 5.5 mm. Before reaching the burner, the fuel gas flows through a water vapor addition system. A 40 mL glass bubbler, filled with distilled water, is placed into a water bath. The temperature of the bath, and and thus of the water in the bubbler, is controlled by a heater with a temperature accuracy of 0.1 °C. The heater also provides recirculation in the bath. Methane saturates with water vapor as it bubbles through the heated distilled water. The molar fraction of water vapor in the flow, at the exit of the bubbler, is derived from the water saturation pressure p_{H_2O} at the temperature of the bath T_w:

$$X_{H_2O} = \frac{p_{H_2O}(T_w)}{p}$$

where p is the total pressure, assumed to be 1 atm. The water addition system was tested at different bath temperature conditions with an air flow of 40 mL/min as carrier gas. At each test temperature air flowed through the system for 30 minutes, and the mass of water lost was then measured. Water mole fraction was calculated assuming it to be uniform during the test time. Each condition was run three times, the averages of the three measured mole fractions are shown in Figure 1, and compared to the theoretical predictions. Up to a bath temperature of 87 °C (corresponding to $X_{H_2O} = 0.61$), the difference between the predicted and the measured mole fraction is below 1%. Above 87 °C, this difference increases as temperature is raised and it is maximum (2.8%) at the last tested condition of 93 °C. Each tested condition showed high repeatability with a standard deviation in the measured X_{H_2O}, for the 3 tests, always below 0.0015.

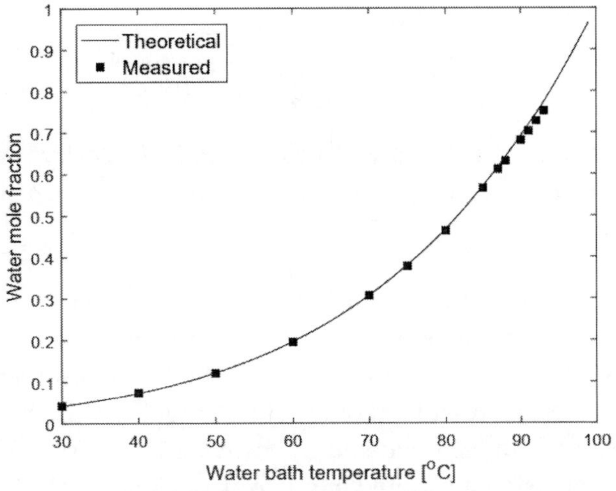

Figure 1: Theoretical vs measured water mole fraction

The fuel line between the water addition system and the burner was wrapped in heating tape to prevent condensation. The oxidizer line and the burner are also heated, respectively with a heating band and heat tape to avoid local condensation or water droplet effects in the burner or at the fuel exit. The temperature of fuel and air at the burner exit were measured with a type K thermocouple to be 510 ± 10 K.

2.2 *Thin-filament pyrometry*

The incandescence of a 14 μm SiC fiber placed in the flame is imaged, in this work, with a Nikon D90 DSLR camera. Planck's law for graybodies can be used to correlate the signal of the glowing fiber I to its temperature T:

$$I(\lambda, T) = \frac{2\epsilon\pi hc^2}{\lambda^5 e^{hc/\lambda kT} - 1}$$

where ϵ is the fiber emissivity, λ the wavelength and h,c and k are the Planck's constant, the speed of light and the Stefan-Boltzmann constant, respectively. The camera detected signal $D(T)$ can then be calculated integrating I over the wavelength range of response for the camera (λ_1 = 400 nm and λ_2 = 720 nm):

$$D(T) = \eta \int_{\lambda_1}^{\lambda_2} I(\lambda, T)\tau(\lambda)S(\lambda)d\lambda$$

where $\tau(\lambda)$ accounts for the transmission characteristics of the optics, $S(\lambda)$ is the spectral response function of the camera and η is a constant that accounts for the efficiency of the light collection system. In this study a BG40 color filter, manufactured by SCHOTT Inc., is used to prevent saturation of the red channel, and $\tau(\lambda)$ represents its spectral transmittance. The spectral response $S(\lambda)$ was obtained through calibration of the camera; details can be found in [25, 26]. An intensity based approach needs a calibration point in order to determine the constant η. An alternative is the color-ratio approach, where any color ratio C_l between the three RGB channels can be calculated by integrating the Planck's law:

$$C_l = \frac{\int_{\lambda_1}^{\lambda_2} I(\lambda, T)\tau_1(\lambda)S_1(\lambda)d\lambda}{\int_{\lambda_1}^{\lambda_2} I(\lambda, T)\tau_2(\lambda)S_2(\lambda)d\lambda}$$

In this study, a single color-ratio point is used to calibrate the intensity based approach. However, Figure 2 shows good agreement in the temperature profiles obtained using three different color ratios and the intensity ratio.

Pictures of the glowing fiber were taken starting at 0.25 mm from the burner rim and moving upwards by 0.7 mm at each step with the fiber positioned horizontally. A Nikkor 50 mm f/1.2 lens was used, the camera was set with an exposure time of 1/1000 s and ISO sensitivity of 320. Pictures were taken in raw ".NEF " format to minimize the internal built-in processing of the camera. The pictures were then processed with the open-source software DCRAW [27] to apply linear gamma correction and demosaicking using a variable number of gradients (VNG) interpolation.

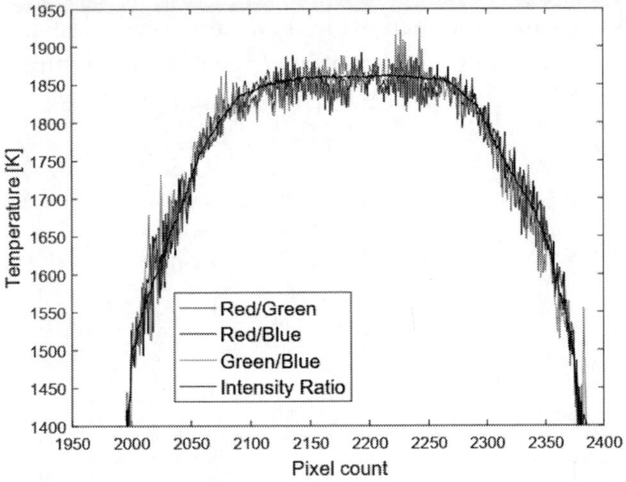

Figure 2: Temperature profile for color-ratio and intensity-ratio approaches

2.3 *Radiation corrections*

Thin-filament pyrometry allows to determine the temperature of the fiber, and not directly the flame temperature. A radiation correction is necessary to retrieve the temperature difference between the fiber (T_f) and the surrounding gas (T_g). The fiber is heated by the hot gas and cooled by radiation losses, this can be described by the following heat balance where conduction along the fiber can be neglected due to the low fiber thermal conductivity and diameter [23, 24, 22, 28]:

$$h\left(T_g - T_f\right) = \epsilon\sigma\left(T_f^4 - T_\infty^4\right)$$

where ϵ is the total emissivity of the fiber (0.88), σ is the Stefan-Boltzman constant and T_∞ is the ambient temperature. h is the convective heat transfer coefficient calculated as:

$$h = \frac{Nu\,k}{d_f}$$

where k is the gas thermal conductivity, d_f is the fiber diameter (14 μm) and Nu is the Nusselt number. In this work, the Nusselt number is calculated with the following correlation, as in [23]:

$$Nu = (0.8237 - 0.5\ln(Pe))^{-1}$$

where

$$Pe = \frac{U\,d_f}{\alpha}$$

is the Peclet number. U is the velocity and α is the thermal diffusivity of the gas. The correlation used for Nu is valid for $Pe < 0.2$, the maximum Pe number for the conditions considered in this work is 0.149. The properties of the gas phase needed to perform the corrections (k, U and α) were determined by running a single-step chemistry CFD simulation in OpenFOAM, for each dilution condition. More details on the simulation setup can be found in [29]. Figure 3 shows a comparison between a temperature profile corrected with data obtained from the single-step chemistry simulation, and a profile obtained using data from a GRI-Mech 3.0 [30] simulation. The small difference between the two corrected profiles (3.5 K at maximum), justifies the use of the single-step chemistry simulations to correct the temperature data.

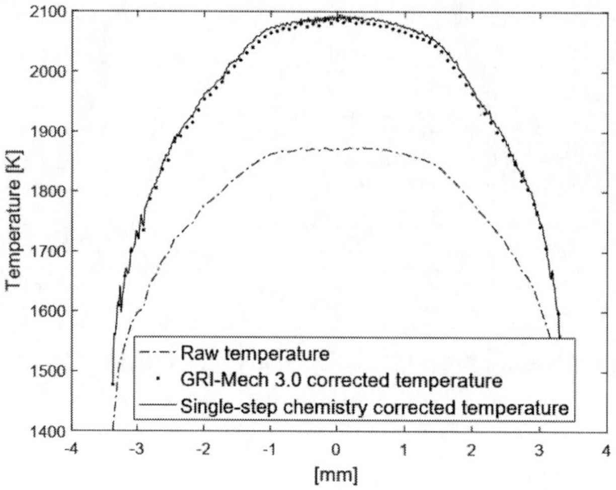

Figure 3: Temperature corrections: Single-step vs GRI-Mech 3.0 chemistry

3 Results and Discussion

Figures 4 and 5 show contour plots of the corrected temperature profiles obtained at different levels of water dilution in two inlet fuel flow rate conditions. Figure 4 results are obtained with a constant methane flow rate of 31 mL/min, which translates into a velocity of 35.3 cm/s at 510 K with no water added to the flow. As water mole fraction is increased, the velocity of the methane/water mixture rises. Figure 5 refers to a constant total mass flow rate on the fuel side. Both experiments were run with an air coflow at constant velocity of 46.2 cm/s. In both cases, the flame was able to sustain water dilution up to 0.705 mole fraction. A further minor increases in the diluent mole fraction of about 0.01 resulted in flame extinction via a blowoff process in which the lifted flame starts oscillating before extinguishing. In the case of increasing total mass flow rate, as the velocity increases, the lateral stoichiometric region of the flame does not expand so that the flame becomes taller. At zero dilution, the lower part of the reaction zone is positioned on the side of the burner fuel tube tip (the black line at zero vertical position in the Figures show the fuel tube exit). As the dilution level increases the flame rises and becomes lifted from the tip at about 0.5 water mole fraction. Above 0.6 water mole fraction, small increases in dilution result in larger lifting of the flame as the extinction limit is approached. A similar behavior in terms of lifting is shown by the constant total mass flow rate experiment. However, in this case, as dilution is increased a smaller amount of methane is introduced in the flame. The fuel does not diffuse to the air as fast and the reaction zone shrinks. Heat release to the surroundings is enhanced by the smaller size of the flame, and temperature is lower. It appears that in this condition of constant total mass flow rate and lower velocity, the flame can be sustained at a lower temperature. This result is similar to counterflow flame results where more water can be added to lower strained flames. However a finer vertical spatial resolution should be sought in order to properly map the temperature profile when the hot region of the flame becomes only a few millimeters tall at high water dilution levels.

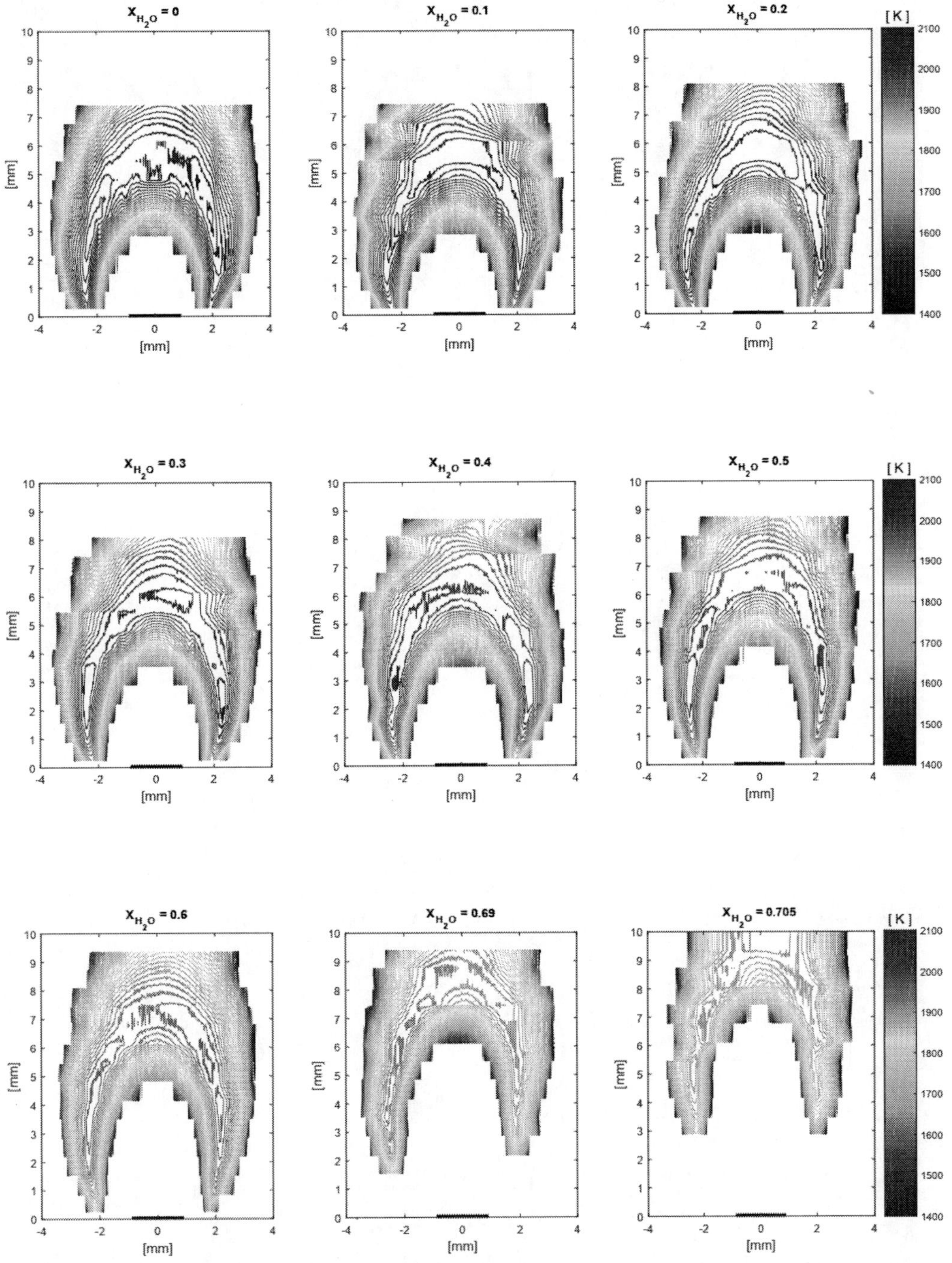

Figure 4: Temperature contours for the constant methane flow rate condition diluted flame

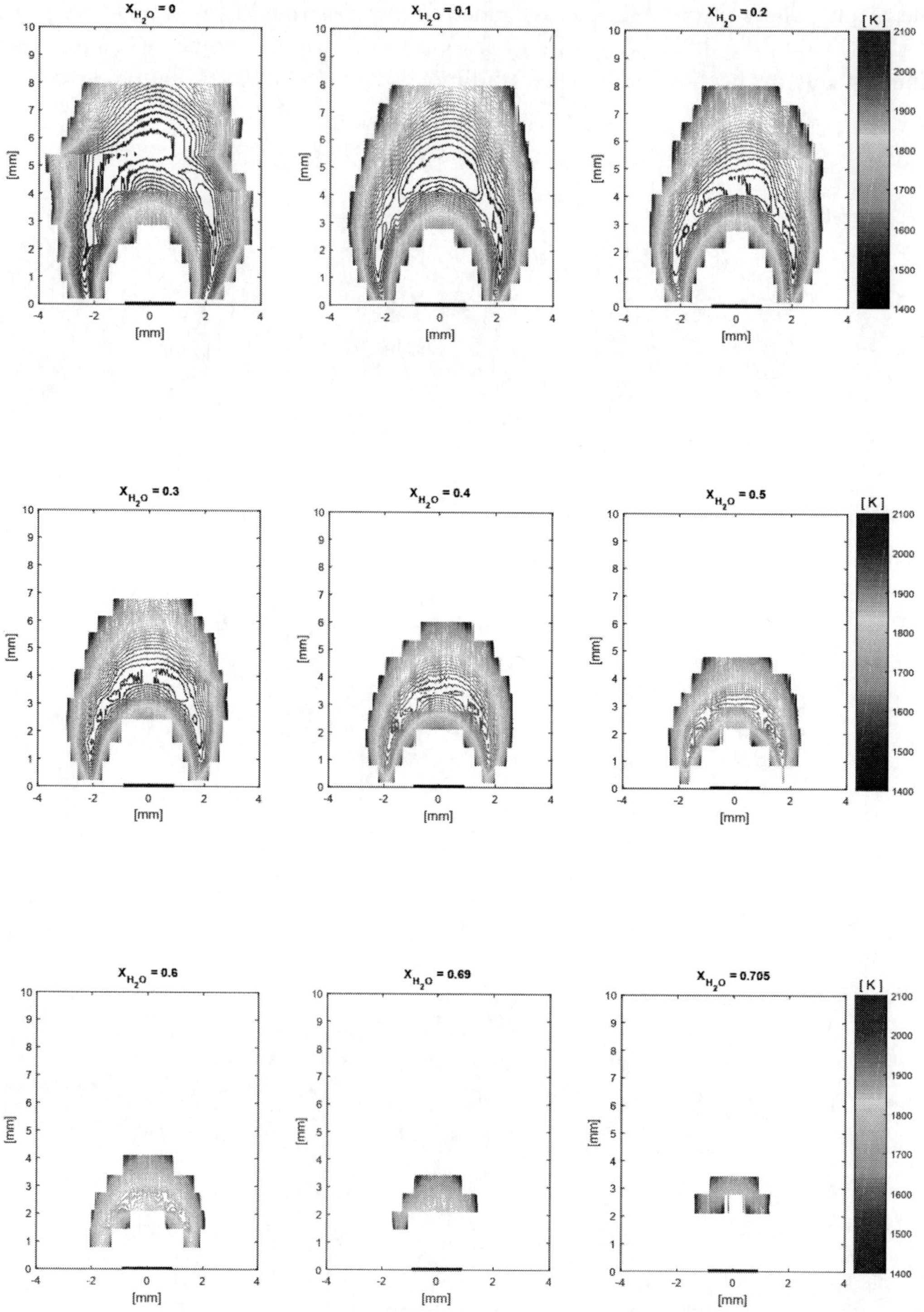

Figure 5: Temperature contours for the constant total mass flow rate condition diluted flame

Figure 6 shows a comparison in the peak temperatures between TFP measurements and preliminary results from GRI-Mech 3.0 chemistry OpenFOAM simulations. The decreasing trend of temperature is well matched by the simulation which also captures the lower temperature in the condition of constant total mass flow rate with respect to the constant methane flow rate one. At 0.6 water mole fraction, this temperature difference is 68.1 K in the experimental results and 62.9 K according to the simulations. Similar to what is seen in the experiments, the numerical results predict extinction between 0.67 and 0.68 water mole fraction. The simulations overpredict the flame temperature however and do so more as the water mole fraction increases. The source of this discrepancy is not yet clear and is the subject of future work, but the measured results are of high enough fidelity to provide an experimental benchmark for the simulations.

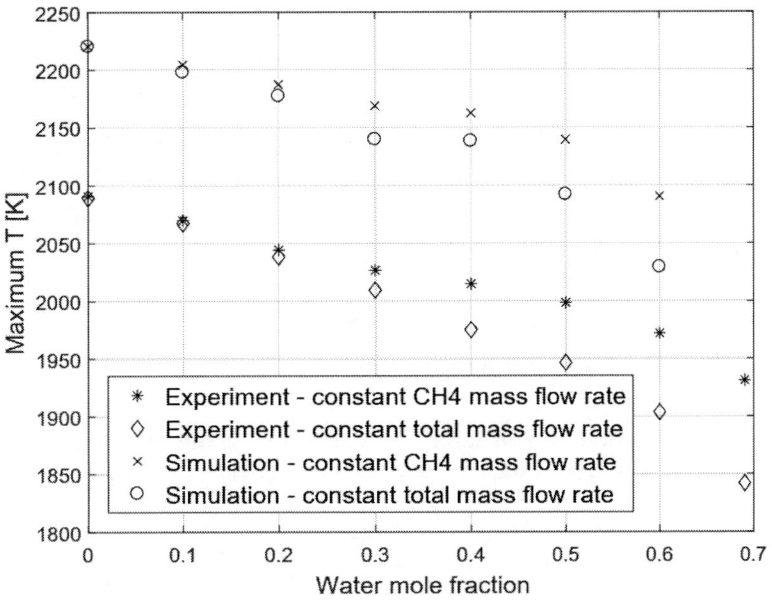

Figure 6: Experimental vs numerical peak temperatures

4 Conclusions

This study focused on measuring the temperature profiles and the extinction limits of a water diluted coflow diffusion flame. Water is added on the fuel side of the flame up to water mole fractions of 0.716 before reaching extinction through blowoff. A 14 μm SiC fiber was used to perform thin-filament pyrometry. A calibrated commercial DSLR Nikon D90 camera captured the images of the glowing fiber, from which the fiber temperature was determined. Radiation corrections allowed to calculate the gas temperature. Peak temperatures are compared to CFD numerical results showing good agreements in the trends and suggesting the simulation can capture the overall role of water in the flame but absolute differences suggest that water chemistry on the fuel side may require refinement. Future work will focus on understanding the absolute difference in these temperature results.

Acknowledgments

This work was supported with funds from the National Science Foundation under grant CBET 1605533, with Song-Charng Kong as program manager. The authors appreciate the useful suggestions and comments from colleagues David Escofet-Martin, Alex Torredemer Serra and Dr. Alice Chien.

References

[1] D. Zhao, H. Yamashita, K. Kitagawa, N. Arai, and T. Furuhata. Behavior and effect on NOx formation of OH radical in methane-air diffusion flame with steam addition. *Combustion and Flame*, 130:352–360, 2002.

[2] F. Bedford, C. Rutland, P. Dittrich, A. Raab, and F. Wirbeleit. Effects of direct water injection on di diesel engine combustion. In *SAE Technical Paper*, 2000.

[3] N. Samec, R.W. Dibble, J.Y. Chen, and A. Pagon. Reduction of NOx and soot emission by water injection during combustion in a diesel engine. In *Seoul 2000 FISITA World Automotive Congress*, 2000.

[4] C. Arnal, M.U. Alzueta, A. Millera, and R. Bilbao. Influence of water vapor addition on soot oxidation at high temperature. *Combustion and Flame*, 43:55–63, 2012.

[5] A.M. Lentati and H.K. Chelliah. Dynamics of water droplets in a counterflow field and their effect on flame extinction. *Combustion and Flame*, 115:158–179, 1998.

[6] R. Ananth and R.C. Mowrey. Ultra-fine water mist extinction dynamics of a co-flow diffusion flame. *Combustion Science and Technology*, 180:1659–1692, 2008.

[7] G.O. Thomas. The quenching of laminar methane-air flames by water mists. *Combustion and Flame*, 130:147–160, 2002.

[8] W. Yang and R.J. Kee. The effect of monodispersed water mists on the structure, burning velocity, and extinction behavior of freely propagating, stoichiometric, premixed, methane-air flames. *Combustion and Flame*, 130(4):322 – 335, 2002.

[9] A.N. Mazas, B. Fiorina, D.A. Lacoste, and T. Schuller. Effects of water vapor addition on the laminar burning velocity of oxygen-enriched methane flames. *Combustion and Flame*, 158(12):2428 – 2440, 2011.

[10] S.P. Fuss, E.F. Chen, W. Yang, R.J. Kee, B.A. Williams, and J.W. Fleming. Inhibition of premixed methane/air flames by water mist. *Proceedings of the Combustion Institute*, 29(1):361 – 368, 2002.

[11] R. Seiser and K. Seshadri. The influence of water on extinction and ignition of hydrogen and methane flames. *Proceedings of the Combustion Institute*, 30(1):407 – 414, 2005.

[12] J. Suh and A. Atreya. The effect of water vapor on counterflow diffusion flames. 1995.

[13] J. Park, S. Kim, K.M. Lee, and T.K. Kim. Chemical effect of diluents on flame structure and no emission characteristic in methane-air counterflow diffusion flame. *International Journal of Energy Research*, 26:1141–1160, 2002.

[14] S. Lee, S. Padilla, D. Dunn-Rankin, T. Pham, and O.C. Kwon. Extinction limits and structure of counterflow nonpremixed H2O-laden CH4/air flames. *Energy*, 93:442–450, 2015.

[15] R.E. Padilla. Structure and behavior of nonpremixed water-laden methane/air flames.University of California, Ivine, 2015.

[16] S. Lee, H. Haa, D. Dunn-Rankin, and O.C. Kwon. Effects of pressure on structure and extinction limits of counterflow nonpremixed water-laden methane/air flames. *Energy*, 134:545–533, 2017.

[17] F. Liu, J. Consalvi, and A. Fuentes. Effects of water vapor addition to the air stream on soot formation and flame properties in a laminar coflow ethylene/air diffusion flame. *Combustion and Flame*, 161:1724–1734, 2014.

[18] M. Roshandell, J. Santacana-Vall, S. Karnani, J. Botimer, P. Taborek, and D. Dunn-Rankin. Burning ice Direct combustion of methane clathrates. *Combustion Science and Technology*, 188:2137–2148, 2016.

[19] T. Kadota and H. Yamasaki. Recent advances in the combustion of water fuel emulsion. *Progress in Energy and Combustion Science*, 28:385 – 404, 2002.

[20] J.M. Ballester, N. Fueyo, and C. Dopazo. Combustion characteristics of heavy oil-water emulsions. *Fuel*, 75:695 – 705, 1996.

[21] Y. Solomon, B. Natan, and Y. Cohen. Combustion of gel fuels based on organic gellants. *Combustion and Flame*, 156:261 – 268, 2009.

[22] V. Vilimpoc and L.P. Goss. SiC-based thin-filament pyrometry: Theory and thermal properties. *Symposium (International) on Combustion*, 22:1907–1914, 1989.

[23] J.D. Maun, P.B. Sunderland, and D.L. Urban. Thin-filament pyrometry with a digital still camera. *Applied Optics*, 46:51483–488, 2007.

[24] B. Ma, G. Wang, G. Magnotti, R. Barlow, and M. Long. Intensity-ratio and color-ratio thin-filament pyrometry: Uncertainties and accuracy. *Combustion and Flame*, 161:908–916, 2014.

[25] A. Llado Gambin. Thin filament pyrometry in a water-laden methane/air counterflow flame. University of California, Ivine, 2015.

[26] M. Minniti. Thin filament pyrometry for temperature measurements in fuel hydrate flames and non-premixed water-laden methane-air flames. Politecnico di Milano, 2013.

[27] D. Coffin. http://www.cybercom.net/d̃coffin/dcraw/. (July 2017).

[28] L.P. Goss, V. Vilimpoc, B. Sarka, and W F. Lynn. Thin-filament pyrometry: A novel thermometry technique for combusting flows. *Journal of Engineering for Gas Turbines and Power*, 111:46–52, 1989.

[29] A. Torredemer Serra. OpenFOAM simulations for a high-pressure nonpremixed coflow laminar flame. Universitat Politecnica de Catalunya, 2017.

[30] G.P. Smith, D.M. Golden, M. Frenklach, N.W. Moriarty, B. Eiteneer, M. Goldenberg, C.T. Bowman, R.K. Hanson amd S. Song, V.V. Lissianski W.C. Gardiner, , and Z. Qin. http://www.me.berkeley.edu/gri-mech/.

Topic: Laminar Flames

2017 WSSCI Fall Technical Meeting
Organized by the Western States Section Combustion Institute
October 2–3, 2017
University of Wyoming, Laramie , WY

Experimental and Computational Investigation of Autoignition of Jet Fuels and Surrogates in Nonpremixed Flows at Elevated Pressures

Gerald Mairinger[(a)] *Alessio Frassoldati*[(b)] *Alberto Cuoci*[(b)]
Matteo Pelucchi[(b)] *Ernst Pucher*[(c)] *Kalyanasundaram Seshadri*[(a)]

[(a)]*Department of Mechanical and Aerospace Engineering, University of California at San Diego, La Jolla, California 92093-0411, USA*
[(b)]*Dipartimento di Chimica Materiali, e Ingegneria Chimica "G. Natta", Politecnico di Milano, Piazza L. da Vinci 32, 20133 Milano, Italy*
[(c)]*Institute for Powertrains and Automotive Technology, Vienna University of Technology, Gusshausstrasse 27-29/315, A-1040 Vienna, Austria [6pt]*

Experimental and computational investigation is carried out to elucidate the mechanism of autoignition of Jet Fuels and Surrogates in nonpremixed flows at elevated pressures up to 6 bar. The counterflow configuration is employed. In this configuration a axisymmetric flow of a gaseous oxidizer stream is directed over the surface of an evaporating pool of a liquid fuel. The oxidizer stream is a mixture of oxygen and nitrogen. The jet fuels tested are JP5, JetA, and JP8, and the surrogates tested are the Aachen surrogate, consisting of 80% n-decane and 20% 1,3,5-trimethylbenzene by mass, Surrogate C made up of 60% dodecane, 20% methylcyclohexane, and 20% o-xylene by volume, and the Princeton Surrogate made up of 49.6% n-dodecane, 24.3% iso-octane, 19.8% n-propylbenzene, and 6.3% 1,3,5-trimethylbenzene by mass. The experiments are conducted at a fixed value of mass fraction of oxygen and at a fixed low value of the strain rate. The temperature of the oxidizer stream at autoignition, Tig, is measured as a function of pressure, p. Computations were carried out using a detailed mechanism and critical conditions of autoignition for surrogates were predicted and compared with measurements.

Improved understanding of the combustion of jet-fuels at elevated pressures is essential for accurate prediction of chemical processes taking place in air-breathing propulsion systems. Jet fuels are a mixture of numerous aliphatic and aromatic compounds. The major components of this fuel are straight chain paraffins, branched chain paraffins, cycloparaffins, aromatics, and alkenes [1, 2]. It has been established that an useful approach to obtaining fundamental understanding of combustion of jet-fuels is to first develop surrogates that reproduce selected aspects of combustion of these fuels [3–6]. The present investigation is concerned with developing surrogates for describing autoignition of jet fuels at elevated pressures.

In a previous study [4], several batches of jet fuels, and fifteen possible surrogates of jet fuels were tested in non-premixed systems at pressure of 1 atm. In this study [4] and other studies at 1 atm [3, 7] a surrogate made up of n-decane (80 %) and trimethylbenzene (20 %) by weight, called the

Aachen surrogate, and a surrogate made up of *n*-dodecane (57 %), methylcyclohexane (21 %), and *o*-xylene (22 %), by weight, called Surrogate C were found to reproduce key aspects of nonpremixed combustion of JP-8. Modelling studies, using detailed chemical kinetic mechanism were also carried out [7]. The various aspects of nonpremixed combustion of these surrogates calculated using the detailed mechanism agreed well with experimental data.

Experimental and computational investigation were carried out recently [8] to elucidate the fundamental mechanism of autoignition of *n*-heptane, *n*-decane, and *n*-dodecane in nonpremixed flows at elevated pressures up to 6 bar [8]. The counterflow configuration was employed. In this configuration a axisymmetric flow of a gaseous oxidizer stream was directed over the surface of an evaporating pool of a liquid fuel. The oxidizer stream was a mixture of oxygen and nitrogen. The experiments were conducted at a fixed value of mass fraction of oxygen and at a fixed low value of the strain rate. The temperature of the oxidizer stream at autoignition, T_{ig}, was measured as a function of pressure, p. Computations were carried out using skeletal mechanisms constructed from a detailed mechanism and critical conditions of autoignition were predicted. The experimental data and predictions show that for all fuels tested T_{ig} decreases with increasing p. At a fixed value of p, T_{ig} for *n*-dodecane was the lowest followed by *n*-decane and *n*-heptane. This indicates that *n*-dodecane is most easy to ignite followed by *n*-decane and *n*-heptane. This was in agreement with previous experimental and computational studies at 1 atm where a similar order of reactivities for these fuels were observed at low strain rates. Flame structures at conditions before and at conditions immediately after autoignition were calculated. A noteworthy finding was that low temperature chemistry was found to play a dominant role in promoting autoignition. The influence of low temperature chemistry was found to increase with increasing pressure. The present study extends this to jet fuels and surrogates.

1 Experiment

Figure 1 is a schematic illustration of the "condensed-fuel" counterflow configuration employed

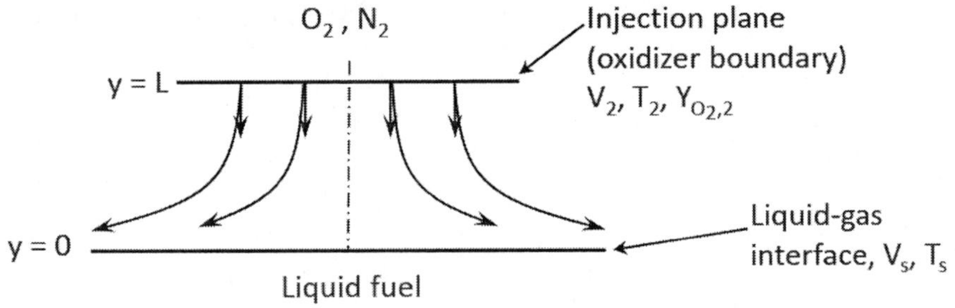

Figure 1: Schematic illustration of the counterflow configuration. V_2 and V_s are the velocities at the oxidizer-boundary and on the gas side of the liquid-gas interface, respectively. T_2 and T_s are the temperatures at the oxidizer-boundary and the liquid-gas interface, respectively.

here in the experimental and computational study. In this configuration an axisymmetric flow of a gaseous oxidizer stream is directed over the surface of an evaporating pool of a liquid fuel. The oxidizer stream is a mixture of oxygen (O_2) and nitrogen (N_2). It is injected from the oxidizer-duct, the exit of which is the oxidizer-boundary. The origin is placed on the axis of symmetry at the surface of the liquid pool, and y is the axial co-ordinate and r the radial co-ordinate. The distance between the liquid-gas interface and the oxidizer-boundary is L. At the oxidizer-boundary, the magnitude of the injection velocity is V_2, the temperature T_2, the density ρ_2, and the mass fraction of oxygen $Y_{\mathrm{O}_2,2}$. Here subscript 2 represents conditions at the oxidizer-boundary. The radial component of the flow velocity at the oxidizer-boundary is presumed to be equal to zero. The temperature at the liquid-gas interface is T_{s}, and the mass averaged velocity on the gas side of the liquid-gas interface is V_{s}. Here subscript s represents conditions on the gas side of the liquid-gas interface. It has been shown previously [9] that the radial component of the flow velocity at the liquid-gas interface is small and can be presumed to be equal to zero. It has been shown that in the asymptotic limit of large Reynolds number the stagnation plane formed between the oxidizer stream and the fuel vapors is close to the liquid-gas interface and a thin boundary layer is established there. The inviscid flow outside the boundary layer is rotational. The local strain rate, a_2, at the stagnation plane, is given by [9, 10].

$$a_2 = 2V_2/L. \tag{1}$$

Figure 2 shows a typical autoignition event recorded by the high speed camera. The figure shows the progression of the ignition event at successive values of the time, t. The first image shows the onset of autoignition where the flame is seen as a small circular disc above the liquid pool and close to the the axis of symmetry. In the subsequent images the diameter of the flame increases. The repeatability of the measurements of autoignition temperatures is $\pm 10\,\mathrm{K}$. The accuracy is expected to be 30 K and is attributed to uncertainties in evaluating the correction that must be applied to the measured temperature due to heat losses from radiation from the thermocouple.

Critical conditions of autoignition were measured for JP-8, Jet-A, JP-5, Aachen Surrogate, Surrogate C and the Princeton Surrogate made up of 49.6 % n-dodecane, 24.3 % *iso*-octane, 19.8 % n-propylbenzene and 6.3% 1,3,5 trimethylbenzene by mass. The experimental data were obtained at a fixed strain rate $a_2 = 150\,\mathrm{s}^{-1}$, $Y_{\mathrm{O}_2,2} = 0.15$. Kinetic modeling is carried out using skeletal chemical-kinetic mechanisms constructed from the POLIMI-1412 detailed chemical-kinetic mechanism [11].

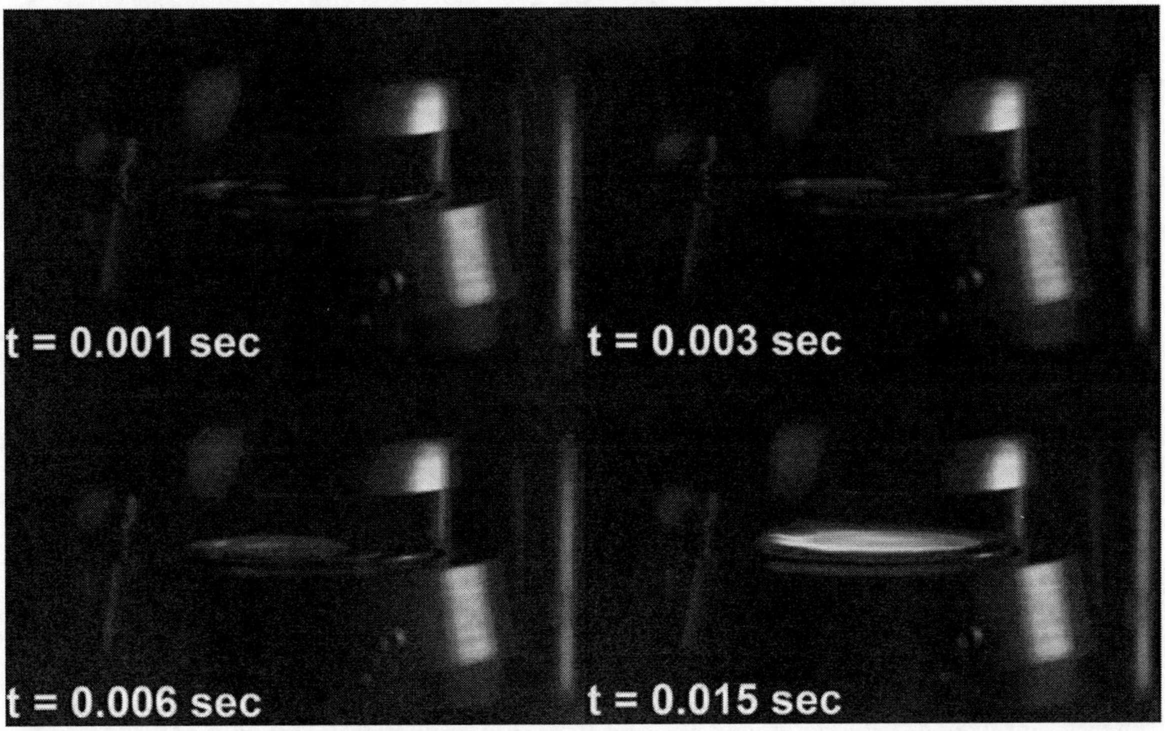

Figure 2: High speed photograph of a typical autoignition event at $p = 3.0\,\text{bar}$ $Y_{O_2,2} = 0.15$, strain rate $a_2 = 150\,\text{s}^{-1}$. The fuel is n-heptane

2 Results and Discussion

Figure 3 shows the measured temperature of the oxidizer stream at autoignition, T_{ig} as a function

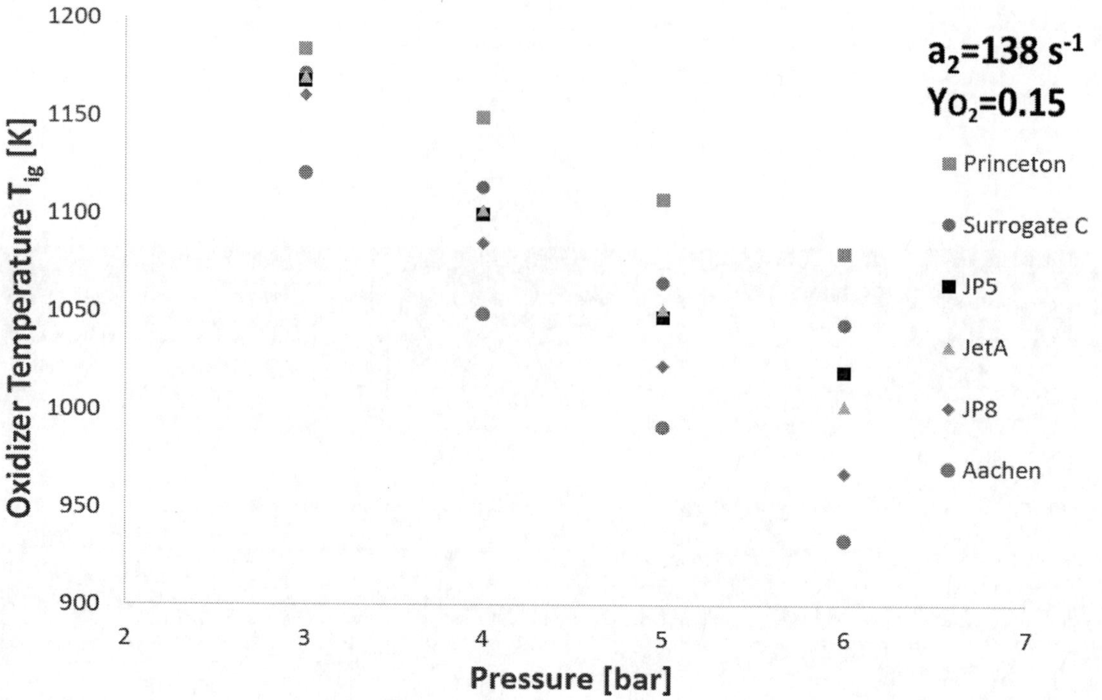

Figure 3: The measured temperature of the oxidizer stream at autoignition, T_{ig} as a function of pressure at fixed $a_2 = 138\,\mathrm{s}^{-1}$, and $Y_{O_2,2} = 0.15$

of pressure at fixed $a_2 = 138\,\mathrm{s}^{-1}$, and $Y_{O_2,2} = 0.15$. Experimental data is shown for the jet fuels JP-8, Jet-A, JP-5, Princeton Surrogate, the Aachen Surrogate, and Surrogate C. For all fuels the temperature of the oxidizer stream at autoignition, T_{ig} decreases with increasing pressure. The measured temperatures of autoignition of the Aachen surrogate and Surrogate C agrees best with those of JP-8.

Figure 4 compares the measured and predicted values of critical conditions of autoignition of Princeton Surrogate, the Aachen Surrogate, and Surrogate C with experimental data. The predictions agree well with the measurements.

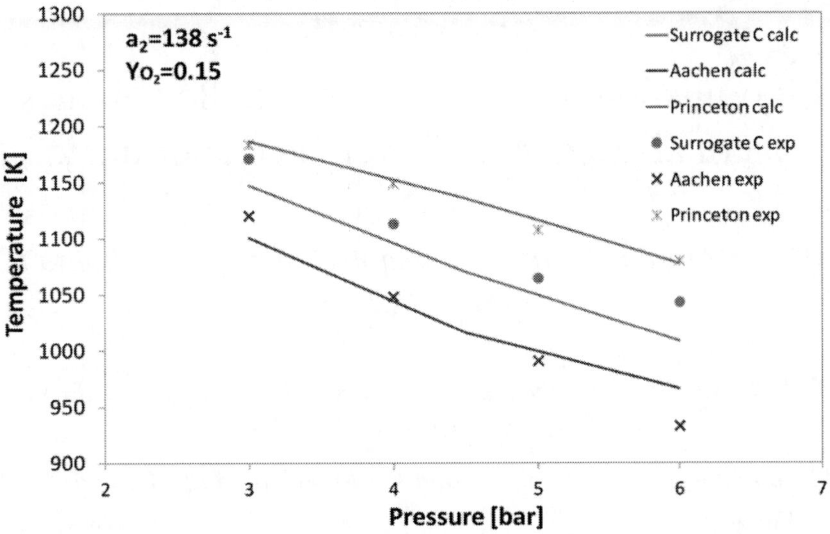

Figure 4: The temperature of the oxidizer stream at autoignition, $T_{\rm ig}$ as a function of pressure at fixed $a_2 = 150\,\mathrm{s}^{-1}$, and $Y_{\rm O_2,2} = 0.15$. The symbols represent experimental data and the lines are predictions.

Acknowledgments

The authors would like to acknowledge Professor F. A. Williams for his contributions and helpful discussions. The research at UCSD is supported by the U. S. Army Research Office, Grant # W911NF1610054 (Program Manager Dr. Ralph A. Anthenien Jr

References

[1] T. Edwards and L. Q. Maurice. *Journal of Propulsion and Power*, 17 (2001) 461–466.

[2] T. Edwards. *2001 Fall Technical Meeting, Eastern States Section of the Combustion Institute*, (2001) journal.

[3] S. Honnet, K. Seshadri, U. Niemann, and N. Peters. *Proceedings of the Combustion Institute*, 32 (2008) 485–492.

[4] S. Humer, R. Seiser, and K. Seshadri. *Journal of Propulsion and Power*, 27 (2011) 847–855.

[5] R. Gehmlich, A. Kuo, and K. Seshadri. *Proceedings of the Combustion Institute*, 35 (2014) 937–943.

[6] S. Dooley, S. H. Won, J. Heyne, T. I. Farouk, Y. Ju, F. L. Dryer, K. Kumar, X. Hui, C.-J. Sung, H. Wang, M. A. Oehlschlaeger, V. Iyer, S. Iyer, T. A. Litzinger, R. J. Santoro, T. Malewicki, and K. Brezinsky. *Combustion and Flame*, 159 (2012) 1444–1466.

[7] S. Humer, A. Frassoldati, S. Granata, T. Faravelli, E. Ranzi, R. Seiser, and K. Seshadri. *Proceedings of the Combustion Institute*, 31 (2007) 393–400.

[8] Gerald Mairinger Alessio Frassoldati Ryan Gehmlich Ulrich Niemann Alessandro Stagni Eliseo Ranzi Kalyanasundaram Seshadri. *Combustion Theory and Modelling*, 20 (2016) 995–1009.

[9] K. Seshadri, S. Humer, and R. Seiser. *Combustion Theory and Modelling*, 12 (2008) 831–855.

[10] K. Seshadri and F. A. Williams. *International Journal of Heat and Mass Transfer*, 21 (1978) 251–253.

[11] E. Ranzi, A. Frassoldati, R. Grana, A. Cuoci, T. Faravelli, A.P. Kelley, and C.K. Law. *Progress in Energy and Combustion Science*, 38 (2012) 468–501.

Western States Section of the Combustion Institute - Fall 2017 Meeting
Hosted by University of Wyoming
October 2-3, 2017.

A novel formulation for unsteady counterflow flames using a thermal-conductivity-weighted coordinate

Adam Weiss[1] *Marcos Vera*[2] *Amable Liñán*[3] *Antonio L. Sánchez*[1]
Forman Williams[1]

[1]*Department of Mechanical and Aerospace Engineering, University of California San Diego, La Jolla, CA 92093-0411, USA*

[2]*Grupo de Mecánica de Fluidos, Universidad Carlos III de Madrid, Leganés 28911, Spain*

[3]*ETSI Aeronáuticos, Pl. Cardenal Cisneros 3, Madrid 28040, Spain*

A general formulation is given for the description of reacting mixing layers in stagnation-type flows subject to both time-varying strain and pressure. The salient feature of the formulation is the introduction of a thermal-conductivity-weighted transverse coordinate that leads to a compact transport operator that facilitates numerical integration and theoretical analysis. For steady counterflow mixing layers the associated transverse mass flux is shown to be effectively linear in terms of the new coordinate, so that the conservation equations for energy and chemical species uncouple from the mass and momentum conservation equations, thereby greatly simplifying the solution. Comparisons are shown with computations of diffusion flames with infinitely fast reaction using both the classic Howarth-Dorodnitzyn density-weighted coordinate and the new thermal-conductivity-weighted coordinate, illustrating the advantages of the latter.

1 Introduction

Transformation methods aimed at facilitating the analysis of laminar boundary-layer type problems involving variable density were first introduced in the seminal work of Howarth and Dorodnitzyn [1, 2, 3]. These methods employ a density-weighted transverse coordinate that, when combined with assumptions regarding the variation of the transport properties (i.e. constant Prandtl and Lewis numbers and a constant value of the product of the density and viscosity), reduce the molecular–transport operators to second-order derivatives with respect to the transverse coordinate. In time-dependent flows, use of the Howarth-Dorodnitzyn transformation along with the introduction of an appropriately defined stream function simplifies the analysis in a similar fashion in what is often referred to as the Moore-Stewartson transformation [4, 5].

The above-mentioned transformations find application in the analysis of reactive boundary-layer flows (see [6] for a general formulation of the problem), characterized by large variations of the temperature and composition, which in turn produce significant relative changes of the density and transport coefficients. In the Howarth-Dorodnitzyn transformation these changes are embodied in the so-called Chapman-Rubesin parameter (i.e. the product of the density and viscosity expressed in dimensionless form) appearing in the reduced viscous term of the momentum equation, and also in the heat-conduction and diffusion terms of the energy and species conservation equations,

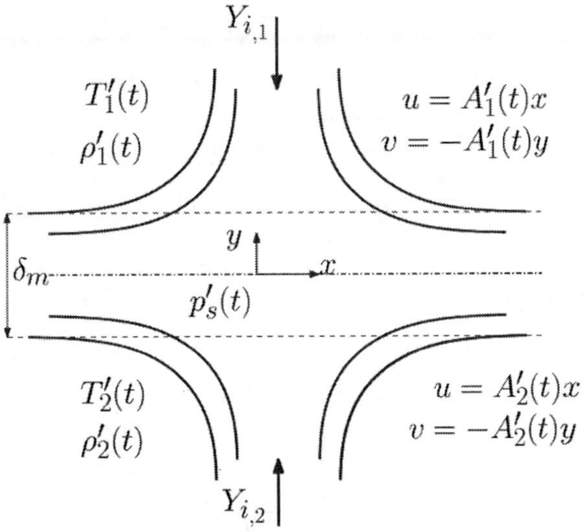

Figure 1: Schematic of stagnation-point mixing layer of characteristic thickness δ_m subject to variable strain $A_1'(t)$ and variable pressure $p_s'(t)$.

when these are written with constant Prandtl and Lewis numbers. In the presence of a flame, the widely used simplifying assumption of unity Chapman-Rubesin parameter is in general not justified, thereby complicating the analysis.

As is apparent from observation of the conservation equations, an alternative to the Howarth-Dorodnitzyn transformation that readily simplifies the molecular–transport operator involves consideration of a transverse coordinate weighted with the inverse of the viscosity or thermal conductivity. A review of the literature reveals that such a transformation has been largely overlooked, perhaps because the simplification that it offers is not immediately obvious. Only recently has this transformed variable been used in an investigation of steady diffusion flames [7], shedding light on the utility of this alternative variable for describing situations characterized by large variation in transport properties, that being the case in reactive flows of interest in combustion.

The purpose of this short communication is to explore further the applicability of the thermal-conductivity-weighted coordinate, extending the treatment to the case of unsteady flows by introduction of an appropriately defined apparent transverse mass flux. The flow selected for purposes of comparison will be the mixing layer separating planar counterflowing streams, depicted in Figure 1, which has been widely employed as a canonical problem to analyze effects of flow strain on premixed, partially premixed, and nonpremixed flames [8]. The formulation given includes unsteady effects through variable strain and pressure as well as a general chemistry description, thereby enabling future analyses involving time-dependent finite-rate systems. Results for steady diffusion flames will be compared to those generated with the classical Howarth-Dorodnitzyn variable. It will be shown that the new formulation is particularly advantageous in this case, because the transverse mass flux is linearly proportional to the thermal-conductivity-weighted coordinate, so that the solution effectively reduces to that corresponding to constant-density flow.

2 Unsteady mixing layers subject to strain and variable pressure

Let us consider the counterflow mixing layer separating two opposed nearly inviscid planar streams. A reference system moving with the dividing streamline will be employed in the description, with x and y denoting the streamwise and transverse coordinates, as indicated in Figure 1. Following [9], attention is directed to the self-similar stagnation-point region, described in terms of the streamwise and transverse velocity components $u = A'(y,t)x$ and $v(y,t)$, temperature $T'(y,t)$, density $\rho'(y,t)$, and mass fractions $Y_i(y,t)$, with $A'(y,t)$ being the strain rate. A prime $'$ is introduced in the notation to differentiate between equivalent non-dimensional variables that will appear later. Subscripts 1 and 2 will be used to denote conditions in each feed stream.

In the general time-dependent formulation given below the pressure, although uniform in the first approximation because of the prevailing low-Mach-number conditions, is allowed to vary with time. The stagnation-point pressure $p'_s(t)$ will be used as a convenient reference value to describe these temporal pressure changes, which must be accounted for in the equation of state

$$p'_s(t) = \rho' R T'/M, \tag{1}$$

where R is the universal gas constant and $M(y,t) = [\sum_i Y_i/M_i]^{-1}$ is the mean molecular mass, with M_i denoting the molecular mass of species i. As a result of the compression work, the feed-stream values of the temperature and density vary in time and satisfy the relationships

$$\rho'_1 c_{p1} \frac{\mathrm{d}T'_1}{\mathrm{d}t} = \rho'_2 c_{p2} \frac{\mathrm{d}T'_2}{\mathrm{d}t} = \frac{\mathrm{d}p'_s}{\mathrm{d}t}, \tag{2}$$

obtained by writing the energy equation in the outer chemically frozen uniform streams, with c_p denoting the specific heat at constant pressure. Using $p'_s = \rho'_1 T'_1 R/M_1 = \rho'_2 T'_2 R/M_2$ in (2) gives the familiar equation

$$\frac{\gamma_1}{\gamma_1 - 1}\frac{1}{T'_1}\frac{\mathrm{d}T'_1}{\mathrm{d}t} = \frac{\gamma_1}{\rho'_1}\frac{\mathrm{d}\rho'_1}{\mathrm{d}t} = \frac{\gamma_2}{\gamma_2 - 1}\frac{1}{T'_2}\frac{\mathrm{d}T'_2}{\mathrm{d}t} = \frac{\gamma_2}{\rho'_2}\frac{\mathrm{d}\rho'_2}{\mathrm{d}t} = \frac{1}{p'_s}\frac{\mathrm{d}p'_s}{\mathrm{d}t} \tag{3}$$

involving the specific-heat ratios $\gamma_1 = c_{p_1}/(c_{p_1} - R/M_1)$ and $\gamma_2 = c_{p_2}/(c_{p_2} - R/M_2)$. For systems such that the approximation $\gamma_1 = \gamma_2$ is sufficiently accurate, the ratio of boundary temperatures and boundary densities at all times takes constant values according to

$$T'_2/T'_1 = (M_2/M_1)/(\rho'_2/\rho'_1) = \Theta \tag{4}$$

as can be obtained from integration of (3). This approximation is to be adopted below with $\gamma = \gamma_1 = \gamma_2$ representing in the following the constant ratio of specific heats.

The spatial pressure differences from the stagnation-point value $\tilde{p} = p' - p'_s(t)$ are smaller than p'_s by a factor of the order of the Mach number squared, and they correspondingly have been neglected in writing (1) and (2). They are, however, fundamental in establishing the motion along the mixing layer. At leading order in the boundary-layer approximation $\tilde{p}(x,t)$ does not vary with the transverse coordinate. This condition of negligible pressure jump across the mixing layer can be used in evaluating the streamwise component of the momentum equation in the outer streams to yield

$$\rho'_1\left(\frac{\mathrm{d}A'_1}{\mathrm{d}t} + A'^2_1\right)x = \rho'_2\left(\frac{\mathrm{d}A'_2}{\mathrm{d}t} + A'^2_2\right)x = -\frac{\partial \tilde{p}}{\partial x}, \tag{5}$$

as an expression relating A'_1 and A'_2. For given values of $p'_s(t)$ and $A'_1(t)$, the first equation in (5) together with (4) provide a Ricatti equation to determine $A'_2(t)$, while (3) determines ρ'_1, ρ'_2, T'_1 and T'_2 as functions of time. These serve as boundary conditions for the mixing–layer problem, to be formulated below, following [9].

The solution for the self-similar reactive counterflow mixing layer involves the integration of

$$\frac{\partial \rho'}{\partial t} + \frac{\partial}{\partial y}(\rho' v) + \rho' A' = 0 \tag{6}$$

$$\rho'\left(\frac{\partial A'}{\partial t} + v\frac{\partial A'}{\partial y} + A'^2\right) = \rho'_1\left(\frac{\mathrm{d}A'_1}{\mathrm{d}t} + A'^2_1\right) + \frac{\partial}{\partial y}\left(\mu\frac{\partial A'}{\partial y}\right) \tag{7}$$

$$\rho' c_p\left(\frac{\partial T'}{\partial t} + v\frac{\partial T'}{\partial y}\right) = \frac{\partial}{\partial y}\left(\lambda'\frac{\partial T'}{\partial y}\right) + \frac{\mathrm{d}p'_s}{\mathrm{d}t} - \sum_i h_i^o \dot{m}_i \tag{8}$$

$$\rho'\left(\frac{\partial Y_i}{\partial t} + v\frac{\partial Y_i}{\partial y}\right) = \frac{\partial}{\partial y}\left(\rho' D_i \frac{\partial Y_i}{\partial y}\right) + \dot{m}_i \tag{9}$$

with $v = 0$ at $y = 0$ and boundary conditions obtained by matching to outer inviscid regions

$$A' - A'_1(t) = T' - T'_1(t) = Y_i - Y_{i,1} = 0 \quad \text{as} \quad y \to \infty \tag{10}$$

$$A' - A'_2(t) = T' - T'_2(t) = Y_i - Y_{i,2} = 0 \quad \text{as} \quad y \to -\infty. \tag{11}$$

A Fickian description with Soret diffusion neglected has been adopted for the diffusion velocities, with D_i representing the diffusion coefficient of species i in the mixture. Differences of specific heat of each species from the mean value c_p have been neglected along with Dufour effects in writing the conductive heat flux, with λ' representing the thermal conductivity. In the chemical reaction terms h_i^o and $\dot{m}_i$ are the enthalpy of formation and chemical rate of production of species i. The above equations (6)–(9) must be supplemented by the equation of state (1) and by expressions for the variation with T' and composition of the specific heat c_p, viscosity μ, and the transport coefficients λ' and D_i.

3 Simplified non-dimensional formulation

The formulation given above can be used for the analysis of time-dependent combustion problems, including transient ignition events in Diesel-engine applications [9] and periodic solutions, such as those arising in studies of flame acoustic response [10, 11]. The nondimensionalization of the problem begins by writing the stagnation-point pressure and the strain rate in the dimensionless form $p_s = p'_s/p'_c$ and $A = A'/A'_c$ with use made of appropriately selected scales p'_c and A'_c, the latter used also in the definition of the dimensionless time $\tau = A'_c t$. For steady problems the constant values of p'_s and A'_1 are used as scales, whereas for time-dependent systems the specific selection of p'_c and A'_c depends on the problem at hand. For instance, for transient ignition or extinction problems the natural scales are the initial values $p'_s(0)$ and $A'_1(0)$ whereas for studies of flame response to periodic oscillations the time-average values of $p'_s(t)$ and $A'_1(t)$ would be preferred. For a given value of p'_c, the accompanying characteristic values of the temperature and density T'_c and ρ'_c for the variables $T = T'/T'_c$ and $\rho = \rho'/\rho'_c$ are to be selected from the properties of stream 1 to reduce (3) to

$$T_1 = T_2/\Theta = p_s^{(\gamma-1)/\gamma} \quad \text{and} \quad \rho_1 = \rho_2\Theta/(M_2/M_1) = p_s^{1/\gamma}. \tag{12}$$

These dimensionless variables can be used to write (5) in the form

$$\frac{\mathrm{d}A_2}{\mathrm{d}\tau} + A_2^2 = \frac{\Theta}{M_2/M_1}\left(\frac{\mathrm{d}A_1}{\mathrm{d}\tau} + A_1^2\right), \tag{13}$$

whereas the equation of state (1) becomes

$$p_s(t) = \rho T \sum_i (M_1/M_i) Y_i. \tag{14}$$

The key component of the formulation is the introduction of the dimensionless inverse-thermal-conductivity-weighted coordinate

$$\eta(y,t) = (\lambda'_c/\delta_m)\int_0^y \frac{\mathrm{d}y}{\lambda'(y,t)}, \tag{15}$$

involving the characteristic thermal conductivity $\lambda'_c = \lambda'(T'_c, Y_{i_1})$ of stream 1 and the characteristic mixing layer thickness $\delta_m = (\lambda'_c/\rho'_c c_p A'_c)^{1/2}$. Introduction of the transformed coordinate η along with the modified nondimensional transverse mass flux

$$F = -\frac{\rho' v + \rho'\lambda' \int_0^y \frac{\partial}{\partial t}\left(\frac{1}{\lambda'}\right)\mathrm{d}y}{\rho'_c \delta_m A'_c} \tag{16}$$

reduce (6)–(9) to

$$\frac{\partial}{\partial \tau}(\rho\lambda) = \frac{\partial F}{\partial \eta} - \rho\lambda A \tag{17}$$

$$\rho\lambda\frac{\partial A}{\partial \tau} = F\frac{\partial A}{\partial \eta} + Pr\frac{\partial^2 A}{\partial \eta^2} + \lambda\left[p_s^{1/\gamma}\left(\frac{\mathrm{d}A_1}{\mathrm{d}\tau} + A_1^2\right) - \rho A^2\right] \tag{18}$$

$$\rho\lambda\frac{\partial T}{\partial \tau} = F\frac{\partial T}{\partial \eta} + \frac{\partial^2 T}{\partial \eta^2} + \frac{\gamma-1}{\gamma}\lambda\frac{\mathrm{d}p_s}{\mathrm{d}\tau} - \lambda\sum_i\left(\frac{h_i^o}{c_p T'_c}\right)\left(\frac{\dot{m}_i}{\rho'_c A'_c}\right) \tag{19}$$

$$\rho\lambda\frac{\partial Y_i}{\partial \tau} = F\frac{\partial Y_i}{\partial \eta} + \frac{1}{L_i}\frac{\partial^2 Y_i}{\partial \eta^2} + \lambda\frac{\dot{m}_i}{\rho'_c A'_c}. \tag{20}$$

where

$$\lambda(T, Y_i) = \lambda'/\lambda'_c \tag{21}$$

is the nondimensional thermal conductivity, a function of the temperature and composition. A constant specific heat c_p has been assumed in writing (19) while constant values of the Prandtl number $Pr = \mu c_p/\lambda'$ and Lewis numbers $L_i = \lambda'/(\rho' c_p D_i)$ have been used in (18) and (20), respectively. These approximations are introduced to exhibit most clearly the advantage of the transformation being investigated, since variable specific heats and Prandtl or Lewis numbers introduce distracting complications with any transformation. Comparison of (17)–(20) with (6)–(9) reveals that, in this transformed formulation, the product $\rho\lambda$, which corresponds to the Chapman-Rubesin parameter of the Howarth-Dorodnitzyn transformation, curiously plays essentially the same role as the gas density in the original dimensional formulation.

For given functions $p_s(\tau)$ and $A_1(\tau)$, the problem reduces to that of integrating (17)–(20) supplemented with (14) and (21) subject to the initial conditions $A(\eta,0) = A_{\text{init}}(\eta)$, $T(\eta,0) = T_{\text{init}}(\eta)$, and $Y_i(\eta,0) = Y_{i_{\text{init}}}(\eta)$ and the boundary conditions $F = 0$ at $\eta = 0$ and

$$\begin{cases} A - A_1(\tau) = T - p_s^{(\gamma-1)/\gamma} = Y_i - Y_{i,1} = 0 & \text{as} \quad \eta \to \infty \\ A - A_2(\tau) = T - \Theta p_s^{(\gamma-1)/\gamma} = Y_i - Y_{i,2} = 0 & \text{as} \quad \eta \to -\infty, \end{cases} \tag{22}$$

with the fuel-side strain rate $A_2(\tau)$ determined from integration of (13). Inspection of (17)–(20) reveals that introduction of the transverse coordinate η and the accompanying mass flux F leads to a compact convective-diffusive operator that facilitates numerical integration. As shown below, an additional advantage of the current formulation lies in the remarkable fact that in steady systems the transverse mass flux across the mixing layer can be approximated by the linear distribution $F = \eta$, so that the conservation equations for energy and species reduce to those obtained when employing the thermo-diffusive approximation (constant density and transport properties), resulting in the mixture-fraction profile being expressible in explicit form in terms of the complementary error function.

4 An illustrative application of the formulation

4.1 Steady counterflow diffusion flames

As an illustrative example the counterflow formulation is applied to steady nonpremixed methane-air flames. Stream 1 will be taken to be the air stream, with corresponding air-side oxygen mass fraction $Y_{O_{2,1}} = 0.232$. For generality, dilution of the fuel stream with nitrogen is permitted, and results will be presented for different values $Y_{F,2} \leq 1$ of the methane mass fraction in its feed stream. Since the focus is on the flow structure, the oxidation of the fuel will be modeled with a one-step irreversible reaction CH_4 + $2O_2$ $\rightarrow$ CO_2 + $2H_2O$ with infinitely fast rate. As is well known, if a unity Lewis number is assumed for both reactants, a good approximation for methane-air flames, then the associated Burke-Schumann problem [12] can be described in terms of the mixture fraction

$$Z = \frac{S\hat{Y}_F - \hat{Y}_O + 1}{S+1} = \frac{T - 1 + q(\hat{Y}_O - 1)}{\Theta - 1 - q}, \tag{23}$$

written for the normalized reactant mass fractions $\hat{Y}_F = Y_F/Y_{F,2}$ and $\hat{Y}_O = Y_{O_2}/Y_{O_{2,1}}$. Two thermochemical parameters appear in the description, namely, the mass of air needed to burn the unit mass of fuel stream $S = 4Y_{F,2}/Y_{O_{2,1}}$ and the dimensionless heat of reaction $q = (q'Y_{O_{2,1}})/(c_pT'_c)$, where $q' = (h^o_F - 2h^o_{H_2O} - h^o_{CO_2})/(2M_{O_2}) \simeq 12,537$ kJ/g is the amount of heat released per unit mass of oxygen consumed in the reaction. The flame appears as a surface $\eta = \eta_f$ separating a fuel region with $\hat{Y}_O = 0$ from an oxidizer region with $\hat{Y}_F = 0$, with the reactants reaching the flame from opposite sides by diffusion and convection. Both reactant mass fractions vanish at the flame sheet η_f, where

$$Z = Z_s = \frac{1}{S+1} = \frac{1}{4Y_{F,2}/0.232 + 1} \tag{24}$$

and

$$T = T_f = 1 + (\Theta - 1)Z_s + q(1 - Z_s). \tag{25}$$

Use of (23) allows for the temperature and composition to be expressed as piecewise linear functions of Z according to

$$\begin{cases} Z > Z_s : \hat{Y}_{\rm O} = 0, \quad \hat{Y}_{\rm F} = (Z - Z_s)/(1 - Z_s), \quad T = (\Theta - 1 - q)Z + 1 + q, \\ Z < Z_s : \hat{Y}_{\rm O} = 1 - Z/Z_s, \quad \hat{Y}_{\rm F} = 0 \quad T = (\Theta - 1 - q)Z + 1 + qZ/Z_s. \end{cases} \tag{26}$$

The problem reduces to the integration of the continuity and momentum equations coupled with a transport equation for the mixture fraction. In terms of the thermal-conductivity-weighted coordinate η defined in (15) and the nondimensional transverse mass flux $F = -(\rho' v)/(\rho'_1 \delta_m A'_1)$, with

$$\delta_m = \left(\frac{\lambda'_1}{\rho'_1 c_p A'_1} \right)^{1/2} \tag{27}$$

representing the characteristic mixing-layer thickness, the needed conservation equations become

$$\begin{aligned} \frac{\mathrm{d}F}{\mathrm{d}\eta} - \rho\lambda A = 0 \\ Pr\frac{\mathrm{d}^2 A}{\mathrm{d}\eta^2} + F\frac{\mathrm{d}A}{\mathrm{d}\eta} + \lambda(1 - \rho A^2) = 0 \\ \frac{\mathrm{d}^2 Z}{\mathrm{d}\eta^2} + F\frac{\mathrm{d}Z}{\mathrm{d}\eta} = 0 \end{aligned} \tag{28}$$

to be integrated with the boundary conditions $F(0) = 0$ and

$$\begin{cases} \eta \to +\infty : \quad A - 1 = Z = 0 \qquad \text{(oxidizer side)} \\ \eta \to -\infty : \quad A - A_2 = Z - 1 = 0. \qquad \text{(fuel side)} \end{cases} \tag{29}$$

As follows from (13), for steady counterflows the fuel-side strain rate is given by $A_2 = [\Theta(M_1/M_2)]^{1/2}$, where the molecular–mass ratio varies with the fuel-side dilution according to

$$\frac{M_1}{M_2} = \frac{1 + (M_{\rm N_2}/M_{\rm CH_4} - 1)Y_{\rm F,2}}{1 + (M_{\rm N_2}/M_{\rm O_2} - 1)Y_{\rm O_{2},1}}, \tag{30}$$

with $M_{\rm N_2}/M_{\rm CH_4} = 28/16$ and $M_{\rm N_2}/M_{\rm O_2} = 28/32$. In the integration, equations (28) are supplemented with the presumed power-law expression $\lambda = T^{0.7}$ for the thermal conductivity and the equation of state

$$1 = \rho T \left\{ \frac{M_1}{M_{\rm N_2}} + Y_{\rm F,2}\frac{M_1}{M_{\rm N_2}}\left(\frac{M_{\rm N_2}}{M_{\rm CH_4}} - 1\right)\hat{Y}_{\rm F} + \left(1 - \frac{M_1}{M_{\rm N_2}}\right)\hat{Y}_{\rm O} \right\}, \tag{31}$$

which is written by assuming that the mean molecular mass of the combustion products $M_{\rm P}$ is equal to that of nitrogen, an excellent approximation for methane-air combustion, for which $M_{\rm P} = (M_{\rm CO_2} + 2M_{\rm H_2O})/3 = 27 \approx M_{\rm N_2} = 28$ g/mol.

Results obtained with (28) will be compared to integrations employing the alternative conservation equations

$$\begin{aligned} \frac{\mathrm{d}F}{\mathrm{d}\zeta} - A = 0 \\ Pr\frac{\mathrm{d}}{\mathrm{d}\zeta}\left(C\frac{\mathrm{d}A}{\mathrm{d}\zeta}\right) + F\frac{\mathrm{d}A}{\mathrm{d}\zeta} + \frac{1}{\rho} - A^2 = 0 \\ \frac{\mathrm{d}}{\mathrm{d}\zeta}\left(C\frac{\mathrm{d}Z}{\mathrm{d}\zeta}\right) + F\frac{\mathrm{d}Z}{\mathrm{d}\zeta} = 0 \end{aligned} \quad (32)$$

employing the standard Howarth-Dorodnitzyn coordinate

$$\zeta = (\rho_1'\delta_m)^{-1}\int_0^y \rho'(y)\mathrm{d}y. \quad (33)$$

The corresponding boundary conditions are identical to those written above for (28). Appearing as a factor in the viscous stress and in the diffusive flux of (32) is the Chapman-Rubesin function $C = \rho\lambda$ which describes the relation between ζ and η according to $\mathrm{d}\zeta/\mathrm{d}\eta = C$ as follows from the respective definition of each coordinate. In general, $C \neq 1$ and the resulting transport operators in (32) are more complicated than those obtained with the coordinate η, as can be seen by comparing (28) and (32).

4.2 Sample numerical results

Integrations were performed for $Pr = 0.7$, $\Theta = 1$, and $q = 7$, as corresponds to methane-air combustion with equal feed-stream temperatures. Solutions are given for different extents of dilution of the fuel stream, measured by $Y_{\mathrm{F},2}$, which enters in the equation of state (31), in the boundary value of the strain rate $A_2 = [\Theta(M_1/M_2)]^{1/2}$ through the molecular-mass-ratio (30), and in the definition of the stoichiometric mixture fraction Z_s defined in (24). In particular, Figure 2 shows the mixing-layer structure in terms of the density-weighted coordinate ζ for methane-air ($Y_{\mathrm{F},2} = 1$) and diluted methane-air ($Y_{\mathrm{F},2} = 0.1$), with corresponding results plotted in terms of the thermal-conductivity-weighted coordinate η in Figure 3. Besides profiles of mixture fraction Z and transverse momentum flux F, the plots include profiles of reactant mass fractions and normalized temperature increment evaluated using the piecewise relations (26).

The right plots in Figure 2 display also the variation of the Chapman-Rubesin parameter $C = \rho\lambda$ across the mixing layer, which shows a discontinuous slope at the flame, with an associated value there given by

$$C_f = T_f^{-0.3} M_{\mathrm{N}_2}/M_1, \quad (34)$$

with $T_f = 1 + q(1 - Z_s)$, as follows from (25). In addition, the graphs show the results obtained by writing (32) with the commonly used approximation $C = 1$, with associated profiles represented by the light curves. The departures from unity of C are seen to be substantial, especially for undiluted fuel feed. Correspondingly, the mixing-layer structures computed with $C = 1$, including the resulting flame location in the transformed coordinate, differ significantly from the exact results.

For completeness, dashed curves are included in Figure 2 to illustrate the applicability of the approximation $F = \zeta$, which when combined with the assumption $C = 1$ leads to the analytic result

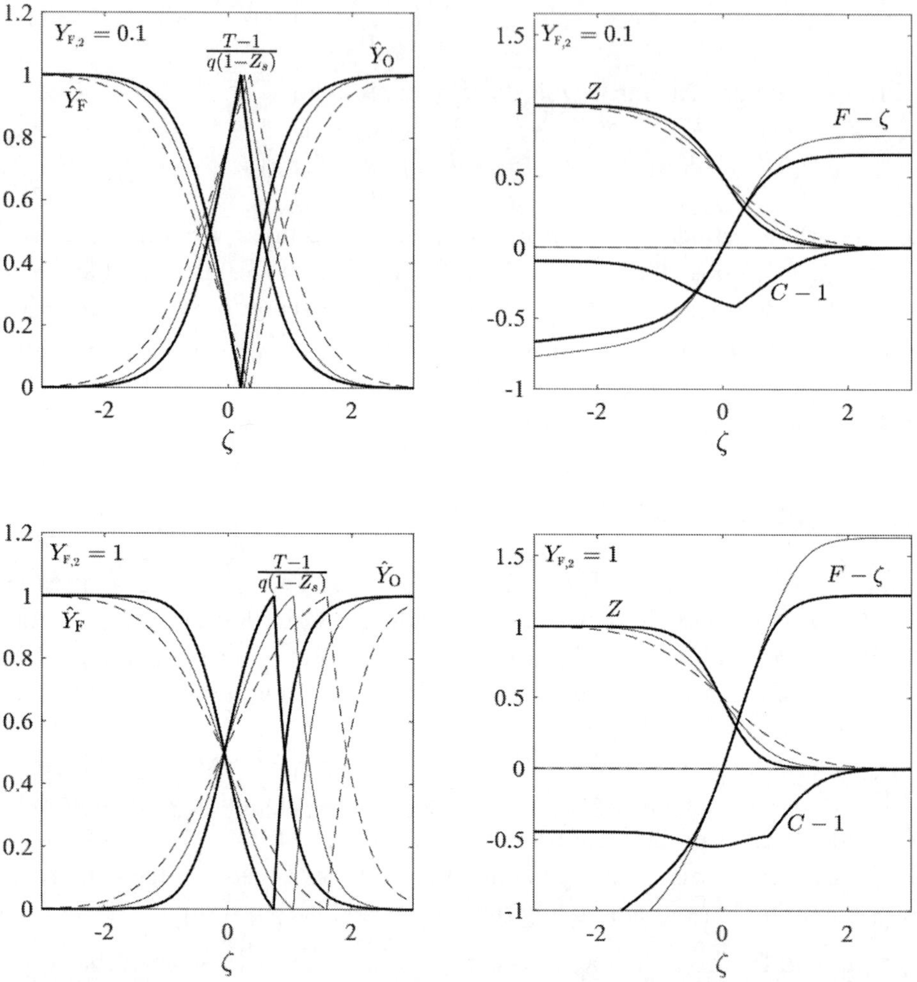

Figure 2: Results of integrations (dark solid curves) using the density-weighted Howarth-Dorodnitzyn variable ζ for diluted $Y_{F,2} = 0.1$ (top) and pure $Y_{F,2} = 1$ (bottom) methane-air counterflow respectively. Shown in light curves are integrations employing $C = 1$ in (32), with the dashed curves corresponding to results with the additional approximation $F = \zeta$, for which $Z = (1/2)\text{erfc}(\zeta/\sqrt{2})$.

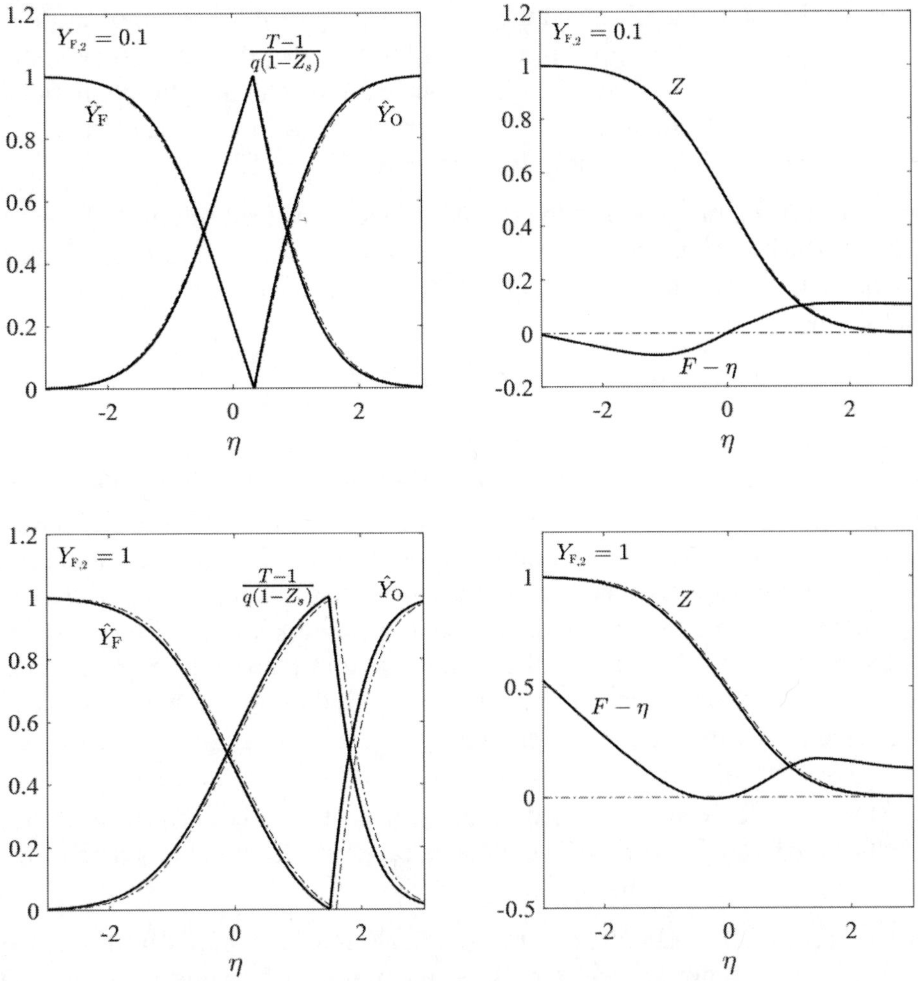

Figure 3: Integrations (solid lines) using the thermal-conductivity variable η for diluted $Y_{F,2} = 0.1$ (top) and pure $Y_{F,2} = 1$ (bottom) methane-air counterflow respectively. Shown in dot-dashed lines are analytic results obtained using the approximation $F = \eta$, including $Z = (1/2)\mathrm{erfc}(\eta/\sqrt{2})$ from which T, $\hat{Y}_F$ and $\hat{Y}_O$ may be found using the piecewise relation (26).

$Z = (1/2)\mathrm{erfc}(\zeta/\sqrt{2})$, quoted in early flamelet studies [13], where erfc is the complementary error function. As can be seen, the resulting accuracy is far from satisfactory, as would be expected, in view of the profiles of $F - \zeta$ shown in the right-hand-side plots of Figure 2.

By way of contrast, the profiles shown on the right-hand-side plots of Figure 3 indicate that the departures of the transverse mass flux from the linear distribution are moderately small when expressed in terms of the thermal-conductivity-weighted coordinate η. Correspondingly, the solution for the mixture fraction $Z = (1/2)\mathrm{erfc}(\eta/\sqrt{2})$, obtained with the approximation $F = \eta$ by integrating the third equation in (28), and the accompanying profiles of temperature and reactant mass fraction, evaluated from (26), give excellent predictions for the mixing-layer structure, as can be seen in the comparisons of Figure 3. The differences of the analytic predictions, represented by the dot-dashed curves, from the results of numerical integrations of (28) are almost imperceptible for $Y_{\mathrm{F},2} = 0.1$ and remain remarkably small even for $Y_{\mathrm{F},2} = 1$.

For completeness, Figure 4 shows the variation with fuel-feed dilution of the flame location y_f, defined in the Burke-Schumann limit by the condition $Z = Z_s$. Naturally, when no simplifications are introduced, both formulations (28) and (32) provide the same value, which can be expressed in the dimensionless form

$$\frac{y_f}{\delta_m} = \int_0^{\eta_f} \lambda \mathrm{d}\eta = \int_0^{\zeta_f} \frac{\mathrm{d}\eta}{\rho} \tag{35}$$

in terms of the dimensionless flame locations η_f and ζ_f, with δ_m given for the steady counterflow in (27). For the range of $Y_{\mathrm{F},2}$ investigated, the resulting flame always lies on the air side of the mixing layer. Consequently, since $\hat{Y}_{\mathrm{O}} = 0$ between the stagnation plane and the flame, the first expression in (26) must be used in evaluating the integrals in (35). The exact numerical results, shown as a dark solid curve, are compared with those obtained by employing in (32) the approximation $C = 1$ (light solid curve) and also with those obtained with $Z = (1/2)\mathrm{erfc}(\zeta/\sqrt{2})$ (dashed curve). Clearly, the predictions of flame location resulting from these approximations are far from satisfactory, with both approximations significantly overpredicting the flame distance to the stagnation plane, in agreement with the profiles shown in Figure 2. The figure also tests the approximation $F = \eta$ and associated prediction $y_f/\delta_m = \int_0^{\eta_f}[1+q(1-Z)]^{0.7}\mathrm{d}\eta$, where $\eta_f = \sqrt{2}\mathrm{erfc}^{-1}(2Z_s)$ and $Z = (1/2)\mathrm{erfc}(\eta/\sqrt{2})$, with erfc^{-1} denoting the inverse (not the reciprocal) of the complementary error function. As can be seen, over the entire range of dilution considered the accuracy of this approximation is remarkably good.

In summary, the results in Figures 2 and 4 indicate that departures from unity of the Chapman-Rubesin parameter cannot be neglected in computations of counterflow flames using a density-weighted coordinate, and that the analytic solutions stemming from using the approximations $C = 1$ and $F = \zeta$ are not capable of giving acceptable accuracy under most conditions. In contrast, the alternative coordinate η results in a simpler transport operator and, for the steady flames shown in Figure 3, effectively reduces the solution to one involving a linearly varying transverse mass flux. Therefore, somewhat unexpectedly, the resulting solution is analogous to that obtained in the thermo-diffusive approximation of constant density and constant transport properties, a drastic simplification worthy of being considered in future work.

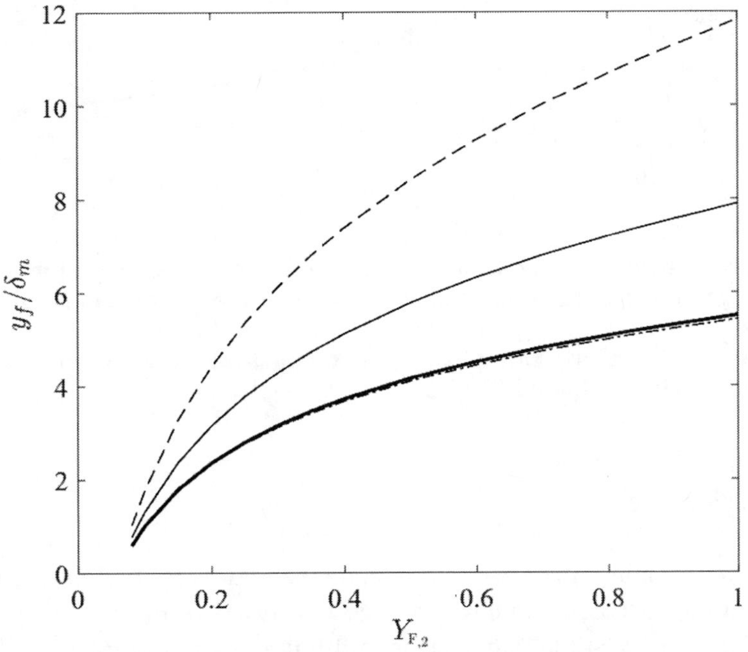

Figure 4: The variation with the fuel-stream dilution of the flame location (35) **obtained from numerical integration of** (28) **or** (32) **(dark solid curve), with the approximation** $C = 1$ **used in integrating** (32) **(light solid curve), with the presumed profile** $Z = (1/2)\text{erfc}(\zeta/\sqrt{2})$ **used to evaluate the density in the last expression of** (35) **(dashed curve), and with the approximation** $y_f/\delta_m = \int_0^{\eta_f}[1+q(1-Z)]^{0.7}\text{d}\eta$ **with** $\eta_f = \sqrt{2}\text{erfc}^{-1}(2Z_s)$ **and** $Z = (1/2)\text{erfc}(\eta/\sqrt{2})$ **(dot-dashed curve).**

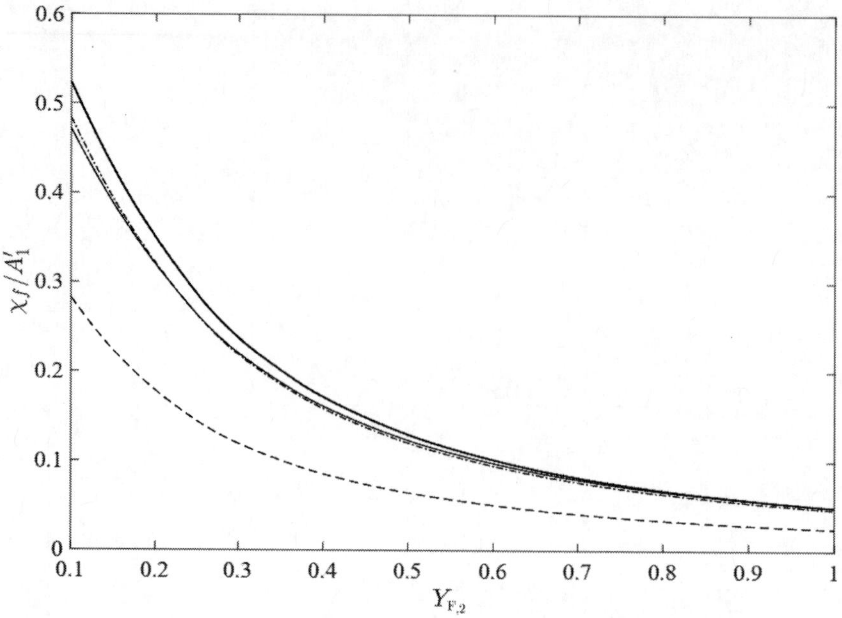

Figure 5: The variation with the fuel-stream dilution of the scalar dissipation rate at the flame (36) with the mixture-fraction gradient computed by numerical integration of (28) or (32). The dark solid curve represents exact numerical results; the light curve is the prediction obtained by employing $C = 1$ in computing $\mathrm{d}Z/\mathrm{d}\zeta|_f$ along with $C_f = 1$; the dashed curve is obtained with $C_f = 1$ and (37); and the dot-dashed curve is obtained from (38).

4.3 Scalar dissipation rate

The above counterflow results can be used to evaluate the scalar dissipation rate $\chi = 2[\lambda'/(\rho' c_p)]|\nabla Z|^2$, a quantity of utmost importance in nonpremixed combustion [8]. For the two formulations considered above the value χ_f of χ at the flame, scaled with the air-side strain rate A_1', can be expressed as

$$\frac{\chi_f}{A_1'} = \frac{2}{C_f}\left(\left.\frac{\mathrm{d}Z}{\mathrm{d}\eta}\right|_f\right)^2 = 2C_f\left(\left.\frac{\mathrm{d}Z}{\mathrm{d}\zeta}\right|_f\right)^2, \tag{36}$$

where the mixture-fraction derivatives are to be evaluated at $\eta = \eta_f$ or $\zeta = \zeta_f$, with C_f correspondingly representing the flame value of the Chapman-Rubesin parameter, given in (34) in terms of the flame temperature (25). The corresponding value of χ_f/A_1' is represented by a dark solid curve in Figure 5 as a function of the fuel-feed dilution.

The different approximations discussed above are also tested in Figure 5. The curve styles have been selected to match those employed in the corresponding curves of Figures 2–4. Thus, consistent with the results in Figures 2 and 4, the light curve represents results obtained from the last expression in (36) with $C_f = 1$ and with $\mathrm{d}Z/\mathrm{d}\zeta|_f$ computed from (32) with $C = 1$, whereas the dashed curve is the prediction obtained by using simultaneously $C_f = 1$ together with

$$\left(\left.\frac{\mathrm{d}Z}{\mathrm{d}\zeta}\right|_f\right)^2 = \frac{1}{2\pi}\exp[-\{\sqrt{2}\mathrm{erfc}^{-1}(2Z_s)\}^2] \tag{37}$$

the latter stemming from the commonly used approximation $Z = (1/2)\mathrm{erfc}(\zeta/\sqrt{2})$ [8]. Also tested in the figure is the approximation $F = \eta$, leading to $Z = (1/2)\mathrm{erfc}(\eta/\sqrt{2})$ and to

$$\frac{\chi_f}{A_1'} = \frac{1}{\pi C_f} \exp[-\{\sqrt{2}\mathrm{erfc}^{-1}(2Z_s)\}^2], \tag{38}$$

with corresponding predictions, obtained with use made of (24), (25), and (34), shown as a dot-dashed curve, that being the curve style used for this approximation in Figures 3 and 4.

Consistent with the degree of accuracy exhibited by the dot-dashed curves in Figures 3 and 4, the comparisons indicate that the analytical prediction (38) is remarkably good, with relative errors that remain smaller than 5% over the whole range of dilutions considered in the figure. Somewhat unexpectedly, the prediction obtained by using the Howarth-Dorodnitzyn transformation together with the approximation $C = 1$ yields similar accuracy. This is the result of a fortuitous cancellation of errors in (36), as can be inferred from the profiles shown in Figure 3. Thus, the approximation $C = 1$ yields flames that are too far from the stagnation place, thereby leading to severe under-predictions of $\mathrm{d}Z/\mathrm{d}\zeta$, but these are almost exactly corrected when the value of $C_f \simeq 0.5 - 0.6$ is replaced by $C_f = 1$ in (36). As can be seen in Figure 5, this fortuitous accuracy degrades significantly when the gradient of mixture fraction is evaluated with use of (37).

5 Concluding remarks

A compact formulation has been presented based on introduction of a new inverse-thermal-conductivity-weighted variable and associated transformed mass flux which may be applied towards the description of reacting counterflow mixing layers subject to variable strain rates and pressures. When compared to classical methods for dealing with boundary-layer-type problems, namely that of Howarth and Dorodnitzyn, the resulting equations posses a simpler convective-diffusive operator stemming from the fact that the variation with temperature of the transport coefficients is removed from the diffusive terms, thereby facilitating numerical integration and theoretical analyses of these flows, of direct interest in flamelet modeling. The advantage of the new transformation for combustion problems must be related to the effect of an increasing thermal conductivity tending to increase the distances over which variables change, thereby approaching constant-density conditions in the transformed coordinate. The formulation given can be extended readily, with minor modifications, to other stagnation-point flows of practical interest, including axisymmetric counterflow mixing layers separating opposed round jets, as well as stagnation-point boundary-layer flows generated by a jet impinging normally on a wall.

Numerical integrations reveal that, for steady situations, the transformed transverse mass flux F varies linearly with the rescaled coordinate η, a simplification that does not arise with the classical density-weighted coordinate ζ. This remarkable fact may be used to uncouple species and energy conservation from the continuity and momentum equations which then no longer need to be considered in the analysis. This simplification, akin to the thermo-diffusive approximation of constant density and constant transport properties, illustrates an additional advantage of the inverse-thermal-conductivity-weighted coordinate η in the description of stagnation-type flows.

Acknowledgements

This research was funded by the US AFOSR grant no. FA9550-16-1-0321.

References

[1] Leslie Howarth. *Proceedings of the Royal Society of London A: Mathematical, Physical and Engineering Sciences*, 164 (1938) 547–579.

[2] Lo Howarth. Concerning the effect of compressibility on laminar boundary layers and their separation. In *Proceedings of the Royal Society of London A: Mathematical, Physical and Engineering Sciences*, Vol. 194, pages 16–42. The Royal Society, 1948.

[3] A Dorodnitsyn. Laminar boundary layer in compressible fluid. In *Dokl. Akad. Nauk SSSR*, Vol. 34, pages 213–219, 1942.

[4] Franklin K Moore. Unsteady laminar boundary-layer flow. Technical report, DTIC Document, 1951.

[5] K t Stewartson. *The Quarterly Journal of Mechanics and Applied Mathematics*, 4 (1951) 182–198.

[6] Forman A Williams. *Cummings Publ. Co*, (1985) journal.

[7] Amable Liñán, Daniel Martínez-Ruiz, Marcos Vera, and Antonio L Sánchez. *Combustion and Flame*, 175 (2017) 91–106.

[8] Norbert Peters. *Turbulent combustion*. Cambridge university press, 2000.

[9] A Liñán and F A Williams. *Combustion and flame*, 95 (1993) 31–46.

[10] JS Kim and F A Williams. *Combustion and flame*, 98 (1994) 279–299.

[11] HG Im, CK Law, JS Kim, and F A Williams. *Combustion and Flame*, 100 (1995) 21–30.

[12] Amable Liñán, Marcos Vera, and Antonio L Sánchez. *Annual Review of Fluid Mechanics*, 47 (2015) 293–314.

[13] Norbert Peters. *Progress in energy and combustion science*, 10 (1984) 319–339.

Western States Section of the Combustion Institute – Fall 2017 Meeting
Hosted by the University of Wyoming
October 2-3, 2017

Global Diffusion Flame Extinction Strain Rate Experiments of Single Large Hydrocarbon Fuels

Anish Jadhav, Radi Alsulami, and Bret C. Windom
Department of Mechanical Engineering, Colorado State University, Fort Collins, CO 80523

Abstract
An experimental validation study of the flame extinction characteristics of large hydrocarbon fuels is carried out. A counterflow flame burner combined with a liquid fuel vaporization system is utilized at a reduced pressure of 0.84 atm to measure the diffusion flame global extinction strain rate of liquid fuels as a function of fuel dilution. A numerical study is also employed using CHEMKIN to calculate the global extinction strain rate of hydrocarbon fuels, including n-heptane, n-decane and toluene. To date, the global extinction strain rate calculations correlate well with the experimental data. The flame stability behavior of carefully selected hydrocarbon fuel blends is also analyzed using the gas phase counterflow flame burner experiment. Flame extinction measurements of the blended fuels are compared to global flame extinction predictions using a previously published radical index methodology. The gas phase extinction strain rates of the selected fuel blends provide a baseline comparison of gas phase flame behavior between a comparative set of fuels that will be useful to extract the impact of evaporation phenomenon on flame stability in future multi-phase flame stability experiments.
Keywords: Counterflow burner, n-heptane, n-decane, Toluene, Radical index, Flame stability

1. Introduction

Fossil fuels such as coal and petroleum have been the key drivers for the industrialization of developing countries. This increased demand of energy due to large scale industrialization has led to overdependence on fossil fuels and raised worldwide concern with regards to global warming. Nevertheless, fossil fuels provide high energy density especially useful for transportation, most significantly in the aviation sector. As such, the development of new technologies capable of replacing petroleum fueled aviation devices will prove to be highly difficult[1]. Making the chemical energy conversion more efficient and the combustion cleaner through the development of biofuels and advanced high efficiency/fuel flexible engines is the primary solution for reducing fossil fuel dependency. The development of new engine strategies will rely on the ability to predict and simulate the chemical energy conversion process, including the coupling between the complex chemistry and the fluid dynamics. However, performing tests with research fuels is expensive and simulating the chemistry of these fuels is very complicated as they often involve many species. For this reason, fuel surrogates are used which can reproduce the chemical kinetics, transport, and physical properties of the real fuels[2].

A fuel surrogate needs to be validated over a broad range of temperature and pressures. Common combustion platforms used in the development of surrogate fuels include counter flow flames to measure flame stability behavior and flame speeds, flow reactors to quantify reaction pathways and key species generation, and shock tubes/rapid compression machines to quantify ignition delays and fuel pyrolysis[4]. Flame stability/flame robustness represents a critical behavior that is needed to develop a surrogate fuel. Flame extinction is not only affected by the chemical kinetics of the fuel but also by thermal and mass transport properties and is important in ensuring the surrogate fuel can capture combustion/flame limitations, for example blowout limits, as well as, turbulent flame let dynamics and their localized extinction behaviors. Using a counterflow flame burner flame robustness can be quantified by measuring the flame extinction strain rate. Previous work has shown that the extinction behavior of jet fuels can be mimicked properly by a mixture of large alkanes and aromatic fuels[6]. There is a need to be able to model the extinction behavior of a fuel to allow for easier identification of potential surrogate blends. However, predicting extinction strain rates, depending on the chemical kinetic model, can be tedious and time consuming often requiring model reduction/combination steps which can contribute to unexpected error. Thus, there is a need to employ blending strategies that can predict the extinction behaviors of a mixture if the extinction of the neat components is known. The objective of this study is to calculate the extinction strain rate of diffusion flames for single large hydrocarbon species and their mixtures. Extinction limits for n-heptane, n-decane, and toluene are studied. The radical index approach is applied to predict the extinction behaviors of a blended fuel comprised of the three-species noted above which are compared to experimental measurements.

2. Methods

Figure 1 shows a schematic illustration of experimental setup, which consists of a counterflow burner, vaporization system and flow controllers. The counter flow burner consists of two opposing converging nozzles with an inner exit diameter of 10 mm that have been aerodynamically designed to generate a uniform velocity at the nozzle exit. The separation distance between the two nozzles is set at 9 mm. The fuel, which is delivered from the top burner, is pre-vaporized and blended with nitrogen to control the fuel mass fraction. Air is used as the oxidizer and is delivered from the bottom nozzle. To avoid fuel condensation and reduce heat loss, the upper chamber is heated with the help of band heater. The fuel exit temperature used for all the experiments was maintained at 500K (±10). The experiment temperatures were monitored using K-type thermocouples at multiple locations along the path of the fuel flow. A syringe pump was used to control the liquid fuel flow rate, and the flow rates of the nitrogen and air were controlled by calibrated sonic nozzles. The temperature of the oxidizer was kept at 298K and all the experiments were carried out at 0.84 atm, the local ambient pressure of the laboratory located in Fort Collins, CO. A nitrogen co-flow was applied to shield the flame from the ambient air and was applied with flow rates to match the velocity of the reactant streams thus minimizing shear effects.

To calculate the extinction strain rates, the fuel and air flow rates are increased simultaneously while maintaining a momentum balance, creating a stagnation plane located halfway between the two nozzles, until the flame extinguishes. Once the flame extinguishes, the fuel and oxidizer nozzle exit velocities are noted and the global strain rate is calculated with the following formula:

$$a = \frac{2Uo}{L}(1 + \frac{Uf}{Uo}(\sqrt{\frac{\rho f}{\rho o}}))$$

where, Uo is the velocity of oxidizer, Uf is the velocity of fuel, L is the axial distance, ρf is the density of fuel, and ρo is the density of oxidizer. The fuel mole fraction is adjusted and the experiment is repeated measuring the extinction strain rates for varying levels of fuel dilution.

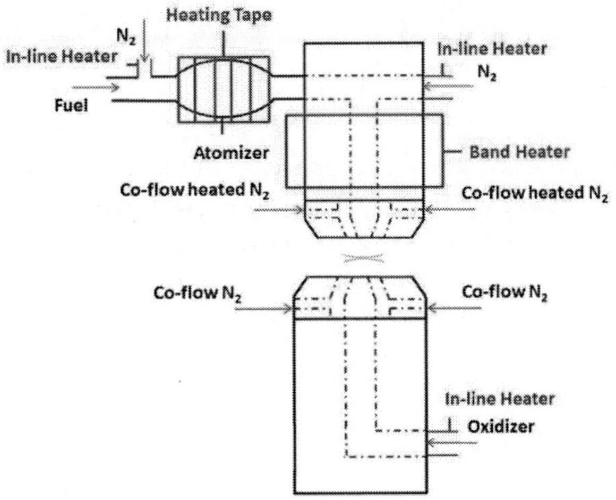

Figure 1: Counterflame burner with the configuration.

Numerical computations were performed using OPPDIF module of CHEMKIN software. The separation distance used in the computations was 9 mm and consistent extinction strain rates were obtained. The mole fraction of the fuel side was increased from 6% to 25% to recreate the experimental conditions. All the computations were done at reduced pressure of 0.84 atm. For n-heptane, the reduced kinetic mechanism from LLNL was used[8], for n-decane and toluene kinetic mechanism reduction was performed in the CHEMKIN Reaction Workbench using the closed homogeneous reactor. The detailed chemical kinetic for the n-decane was obtained from the LLNL high temperature alkane mechanism[7] and for toluene from Metcalfe et al.[3].

3. Results and Discussion

Figure 2 shows a plot of experimental and numerical global extinction strain rates of the three fuels plotted against the fuel mole fraction. The fuels were selected to differentiate the kinetic effect as well as the role of transport on the extinction flame behavior. Clearly, the extinction limits vary according to the molecular structure of the fuel, with n-decane producing the most robust flame and toluene the weakest. This trend follows previously reported data and can be attributed to the differences in the chemical kinetics and the molar specific heat of combustion between the three fuels. Though the trends match previous data, the absolute experimental results when compared with the previously obtained results (measured at sea level and a P = 1 atm) are reduced by ~135 s^{-1} [5]. This difference can be attributed to the difference of pressure at which the experiments were performed. The extinction simulations over predicted the measured global strain rates for the n-alkane fuels but provided a good agreement for the toluene flames.

The concept of radical index was introduced to predict the extinction strain rates for single and blended fuels based on the formation of radical pool. OH concentration plays an important role in determining the radical index as well as extinction limit. Radical index for n-alkanes and toluene were used from the previous literature.[5] The comparison between the radical index predicted flame extinction curves to the measured values can be seen in Figure 3 for n-heptane, n-decane, and toluene. The radical index predicted strain rates over predict the experimental measurements for both of the n-alkane fuels but closely match closely the toluene extinction measured strain rates. The obvious disagreement between the radical index predicted and the measured extinction strain rates for the n-alkane fuels most likely stems from the inability of the kinetic model to predict the flame's maximum

OH radical concentration at P = 0.84 atm, a key input for the radical index extinction strain rate calculation.

To explore the ability of the radical index to determine the extinction of blended fuels, the extinction behavior of a blend comprised of equal parts (by volume) of n-heptane, n-decane, and toluene was predicted and measured, the comparison of which can be seen in **Figure 3**. For the blended fuel, the radical index was calculated by the linear summation of component fuel radical indices. Overall, the radical index does a good job at predicting the extinction behavior of the blend. A slight over prediction is observed for the blended fuel which most likely stems from the over prediction of the radical index of the n-alkane fuels.

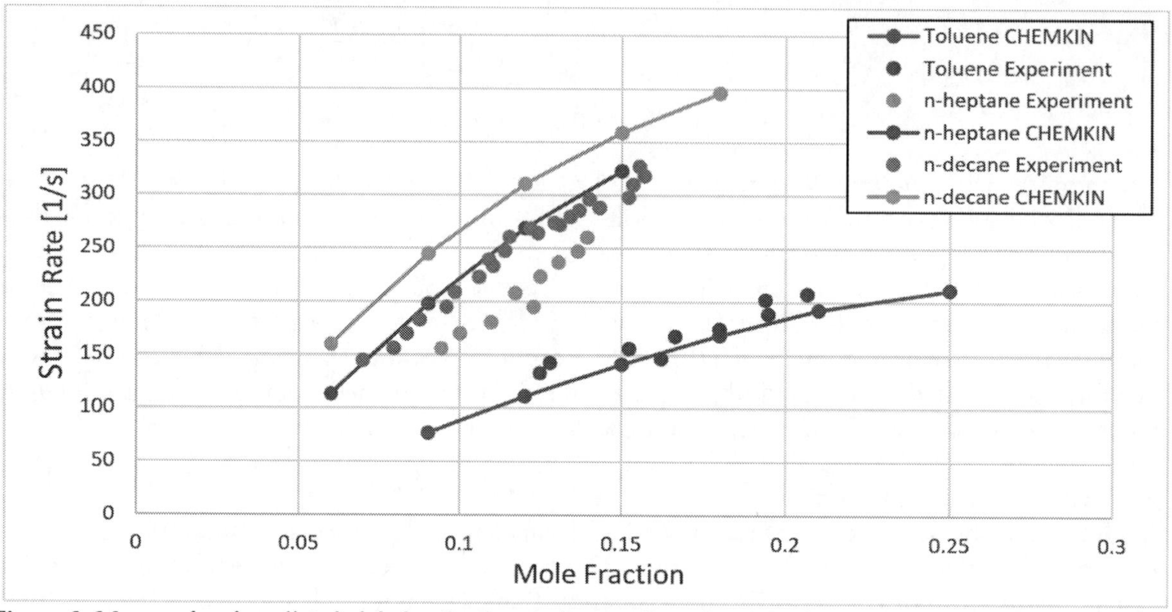

Figure 2. Measured and predicted global extinction strain rates for n-decane, n-heptane, and toluene as a function of fuel dilution at 500 K and 0.84 atm.

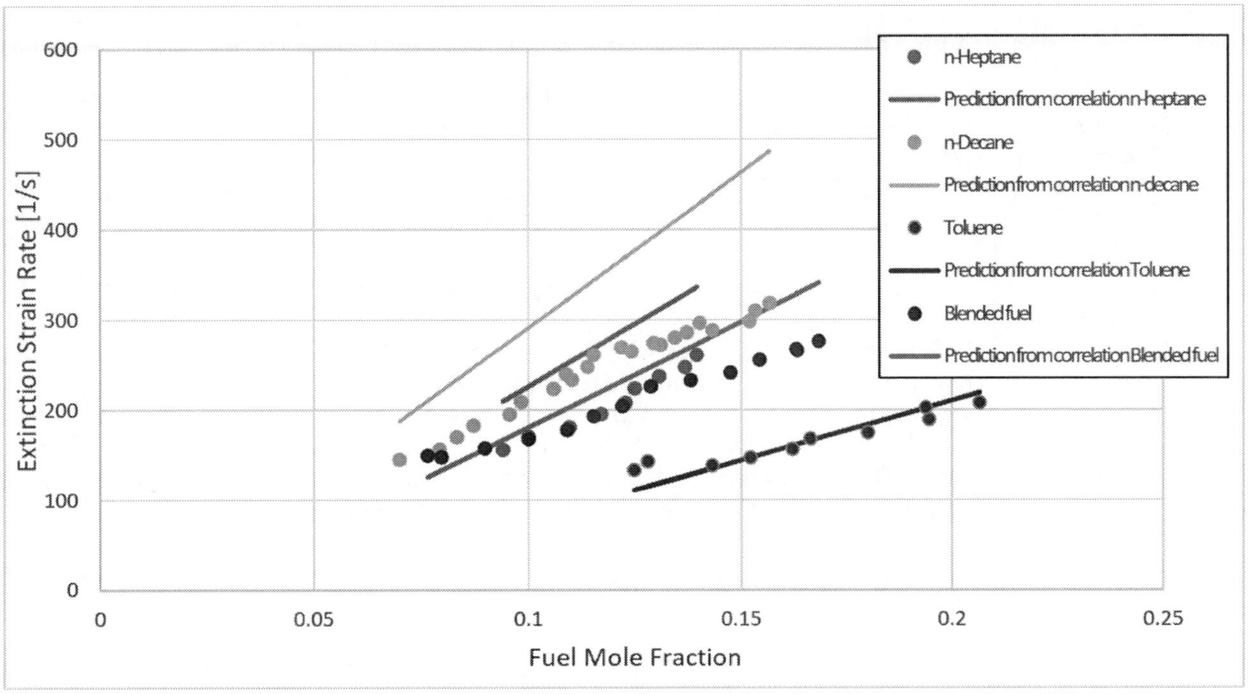

Figure 3. Measured and predicted data from the radical index correlation for n-decane, n-heptane, toluene, and a blended fuel comprised of equal parts (by volume) of n-heptane, n-decane, and toluene.

4. Conclusions

The global extinction limits for n-heptane, n-decane and toluene using the counterflow diffusion flame burner have been measured. Numerical computations of these fuels have been performed, which shows a reasonable agreement with the experimental data. The experimental data is also compared with the data obtained from a previous study[5]. There was a noticeable difference in the extinction limits between the experimental measurements and the predictions for the n-alkane fuels possibly stemming from the inability of the mechanisms to capture the kinetics at a reduced pressure of 0.84 atm. Future work will explore the utilization of alternative chemical mechanisms for the prediction of the n-alkane extinction strain rates and the applicability of the radical index for the predictions of blended fuels extinction limits at reduced pressures. The extinction limits of additional blends will be measured/calculated to provide a baseline comparison of gas phase flame behaviors between a comparative set of fuels that will be used in future multi-phase spray flame experiments to better understand the role of evaporation phenomenon on flame stability.

References

[1] Coppola, Gianfilippo & Coriton, Bruno & Gomez, Alessandro. (2009). Highly turbulent counterflow flames: A laboratory scale benchmark for practical systems. Combustion and Flame. 156. 1834-1843. 10.1016/j.combustflame.2009.03.017.

[2] Deepti Singh, Takayuki Nishiie & Li Qiao (2011): Experimental and Kinetic Modeling Study of the Combustion of n-Decane, Jet-A, and S-8 in Laminar Premixed Flames, Combustion Science and Technology, 183:10, 1002-1026

[3] Comprehensive Detailed Chemical Kinetic Modeling Study of Toluene Oxidation W. K. Metcalfe, S. Dooley, and F. L. Dryer *Energy & Fuels* **2011** *25* (11), 4915-4936 DOI: 10.1021/ef200900q

[4] M. Colket, T. Edwards, S. Williams, N.P. Cernansky, D.L. Miller, F. Egolfopoulos, P. Lindstedt, K. Seshadri, F.L. Dryer, C.K. Law, D. Friend, D.B. Lenhert, H. Pitsch, A. Sarofim, M. Smooke, W. Tsang, Development of an experimental database and kinetic models for surrogate jet fuels, in: 45th AIAA Aerospace Science Meeting and Exhibit, Reno, Nevada, AIAA 2007-770, 2007.

[5] Hee Won, Sang & Dooley, Stephen & Dryer, Frederick & Ju, Yiguang. (2012). A radical index for the determination of the chemical kinetic contribution to diffusion flame extinction of large hydrocarbon fuels. Combustion and Flame - COMBUST FLAME. 159. . 10.1016/j.combustflame.2011.08.020.

[6] Hee Won, Sang & Sun, Wenting & Ju, Yiguang. (2010). Kinetic Effects of Toluene Blending on n-Decane Diffusion Flame Extinction Limit. Combustion and Flame - COMBUST FLAME. 157. 411-420. 10.1016/J.COMBUSTFLAME.2009.11.016.

[7] Westbrook, C. K., W. J. Pitz, O. Herbinet, H. J. Curran, and E. J. Silke, "A Detailed Chemical Kinetic Reaction Mechanism for n-Alkane Hydrocarbons from n-Octane to n-Hexadecane," Combust. Flame **156** (1) (2009) 181-199, http://dx.doi.org/10.1016/j.combustflame.2008.07.014 LLNL-JRNL-401196.

[8] Seiser, H., H. Pitsch, K. Seshadri, W. J. Pitz, and H. J. Curran, "Extinction and Autoignition of n-Heptane in Counterflow Configuration," Proceedings of the Combustion Institute **28**, p. 2029-2037, 2000; Lawrence Livermore National Laboratory, Livermore, CA, UCRL-JC-137080.

Western States Section of the Combustion Institute – Fall 2017 Meeting
Hosted by the University of Wyoming
October 2-3, 2017

Hydrocarbon-Fueled Electrical Power Generator with no Moving Parts

J.Wongwiwat[1] *P.Bhuripanyo*[1] *V.McCloyn*[2] *P.Ronney*[1]

[1]*Department of Aerospace and Mechanical Engineering,*
University of Southern California, Los Angeles, California, United States of America
[2]*Charles Evans Hughes Middle School,*
Long Beach, California United States of America

Hydrocarbon fuels possess roughly 50 times higher energy density compared to even the very best batteries. However, the developments of scaled-down electrical power generators for portable applications employing internal combustion engines have been far from successful owing to the difficulties in minimizing heat and friction losses. Hence, an alternate design solution without moving parts incorporating single-chamber solid oxide fuel cells (SCFCs), thermal transpiration (gas pumping in a nanoporous medium relying on temperature gradient) and catalytic combustion is proposed. The concept involves having thermal transpiration membrane utilizes the temperature gradient created by SCFCs and catalytic combustion to supply reactants at high temperature to SCFC for its operation.

With this concept, catalytic combustion modeling and thermal transpiration membrane modeling were studied individually and integrated together for an airbreathing cylindrical chamber. The chamber design involves an aluminum cylinder wrapped with glass microfiber filter as transpiration membrane. Aluminum structure provides practically uniform temperature on inside surface of the membranes. Platinum mesh was used as a catalyst for rich butane-air combustion, the products of which supply the SCFCs. It was found that the temperature and flow rates are appropriate for SCFC operation and equivalence ratio can be varied from 1.5 to 18.9. The integrated system demonstrates capability in self-sustaining power generation using hydrocarbon fuel without moving parts.

Keywords: fuel cell, catalytic combustion, thermal transpiration membrane

1. Introduction

Portable electronic devices have ever-growing needs for electrical power source and/or energy storage unit. Batteries hold tremendously less specific energy compared to hydrocarbon fuels, which typically contain up to 12,000 Wh/kg whereas commercial batteries such as lithium ion contain only 200 Wh/kg, just only 1.6% of hydrocarbon fuels. Even though energy density of batteries has significantly increased for more than twice in the last decade, the numbers remain distant from being able to support various mobile applications such as wheel chairs, mobile robots and portable military devices. In addition, such batteries have to be occasionally recharged and/or replaced unlike hydrocarbon fuels that can be stored in a container that can be replenished easily.

Scaling down of internal combustion engines and electrical generators to *Micro-Electro-Mechanical Systems (*MEMS) scales has been unsuccessful due to issues with heat losses, friction losses and the difficulty of precise manufacturing parts at such small scales. A small aircraft engine is an example of small scale power generator using hydrocarbon fuel. Menon et al. [1], Walther et al. [2] and Dessornes et al. [3] attempted to study small-scale engine e.g. a reciprocating engine, Wankel rotary engine and a microscale gas turbine; however, all of them were confronted with various issues including, unacceptable noise, manufacturing tolerance concerns, friction losses and heat losses, resulting in the efficiency of their engines to be too low to be function as power generators.

Small-scale power generators with no moving parts have been studied throughout a lot of research groups due to friction loss concerns. Hsu et al. [4] proposed an idea of using thermoelectric to generate electricity from burning hydrocarbon fuel without any moving parts, but the efficiency of thermoelectric itself was proved too low. Another solution is a proton exchange membrane fuel cell (PEMFC), a portable power generator with no moving parts, has difficulty involving hydrogen storage and fuel impurity as described in Shan et al.'s work [5]. Solid Oxide Fuel Cell (SOFC) is a robust fuel cell that can operate with various hydrocarbon fuels but it has to operate under high temperature at least 600°C. SOFC can also operate in a single chamber [6,7,8] and no energy storage issue for SOFC if either propane or butane is used as fuel. SOFC could be a feasible solution for a small portable electrical power generator.

To operate this fuel cell, oxygen has to be continuously supplied to maintain the reaction. Using small fan requires energy input and friction loss would be dominant at small scale. Using thermal transpiration pump from nanoporous membranes can eliminate loss due to friction from moving parts. The pumping power is generated from temperature gradient across the membrane such that ambient air will be driven from the cold side to the hot side of the membrane. The temperature gradient can be provided from catalytic combustion on platinum foil placed inside a combustion chamber. One solution for small-scale power generator is thermal transpiration membrane and catalytic combustion from butane combining with single-chamber solid oxide fuel cells (SCFCs) for electrical power generation. This self-sustaining power generation device without moving parts was previously introduced by Wang et.al. [9] as in *Figure 1.* It was designed without optimization of heat transfer and energy consumption; therefore, the power generator provided only 0.092% of efficiency due to heat losses and unconsumed fuel. To achieve the best flow and power management, parameters for thermal transpiration membrane, catalytic combustion and SCFC have to be investigated separately. The system will be designed and optimized based on maximum efficiency.

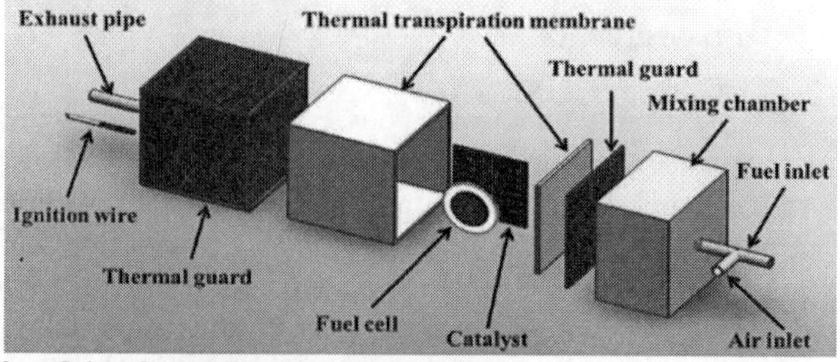

Figure 1 - Schematic of the thermal transpiration device
(Left) assembled view and (Right) exploded view

2. Methods

2.1 Thermal Transpiration Membrane

Thermal transpiration is a rarefied gas dynamics phenomenon describing the gas flow through a narrow channel with an imposed temperature gradient. While the term "transpiration" is often invoked in two-phase liquid/gas processes, thermal transpiration as discussed in this study is refers to a gas interacting via elastic and/or inelastic collisions with a solid, stationary surface without vaporization or condensation. As gas flow from cold end to the hot end, a pressure gradient is created across the channel. The concept of thermal transpiration has been studied since Osborne Reynolds since 1879 [10] with high thermal conductivity porous media causing high thermal energy input for pumping power. Later on, Sone et al. [11], Vargo et al. [12] and Muntz et al. [13] have studied about the mass flow rate ($\dot{M}$) induced by thermal transpiration membrane. The relation of all related parameters is shown as:

$$\dot{M} = P_{avg}\sqrt{\frac{m}{2kT_{avg}}}A\frac{L_r}{L_x}\left[\frac{\Delta T}{T_{avg}}Q_T - \frac{\Delta P}{P_{avg}}Q_P\right] \quad (1)$$

and can be rearranged as

$$\Delta P = \left[\frac{\Delta T}{T_{avg}}Q_T - \frac{\dot{M}}{P_{avg}\sqrt{\frac{m}{2kT_{avg}}}A\frac{L_r}{L_x}}\right]\frac{P_{avg}}{Q_P} \quad (2)$$

When ΔP is the pressure difference between both side of the membrane, P_{avg} is the average pressure across the membrane, ΔT is the temperature difference across the membrane, T_{avg} is the membrane average temperature, Q_T is the thermally driven flow coefficient, Q_P is the pressure driven return flow coefficient, A is the channel cross-section area, L_r is the pore size radius, L_x is the capillary channel length, m is the molecular mass of the gas and k is Boltzmann's constant. The relation of Q_T and Q_P as a function of Knudsen number, $K_n \equiv \lambda/L_r$ (Knudsen number is the ratio of gas mean free path to channel radius) is shown in *Figure 2* (Left) and the thermal transpiration diagram and parameters are shown in *Figure 2* (Right)

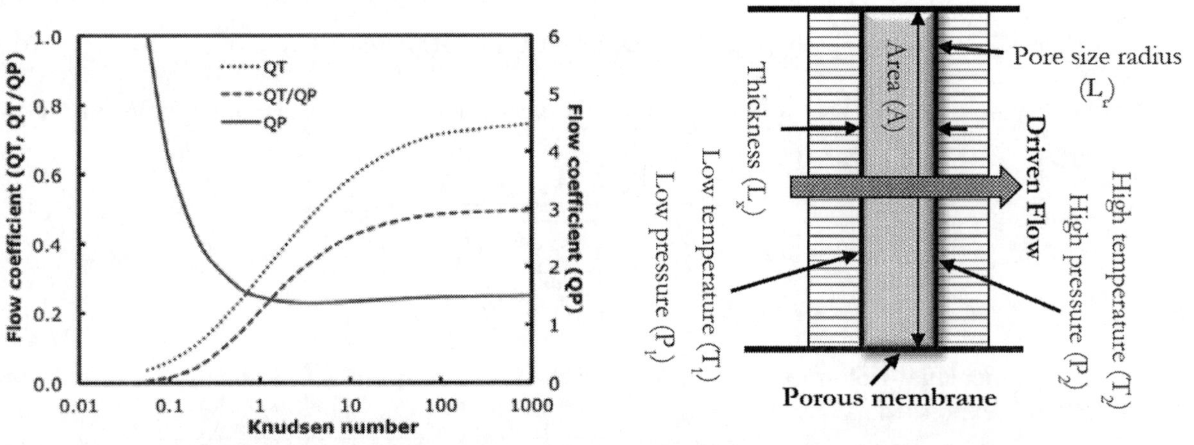

Figure 2 - (Left) Flow coefficient vs. Knudsen number for circle capillary channel [12]
(Right) Thermal transpiration membrane diagram and parameters

2.2 Catalytic Combustion

Catalytic combustion is chemical reactions between oxidizer and fuel occurring only on catalytic surface. The catalyst helps decrease activation energy of the reaction so that the surface combustion can sustain at temperature lower than that of gas phase combustion. Well-known catalyst is a noble metal i.e. platinum which has been used in this study because platinum is one of the most reactive catalysts. Langmuir-Hinshelwood (LH) proposed adsorption rate model that is commonly used to express reaction rate for catalytic combustion as

$$r_{ads} = k^{ads} C_s \theta(s) - k^{des}\theta \tag{3}$$

where k^{ads} is rate constant for adsorption, k^{des} is rate constant for desorption, $\theta(s)$ is empty space on the surface, θ is fraction molecule adsorbed on the catalyst surface and C_s is gas phase concentration adjacent to the catalyst surface. The rate constant of adsorption and the rate constant of desorption for each molecule can be distinctively described by modified Arrhenius expression as

$$k^{ads} = \frac{s}{\Gamma}\sqrt{\frac{RT}{2\pi M}}\, e^{-\frac{Ea^{ads}}{RT}} \left(\frac{T}{T_{ref}}\right)^{\beta_{ads}} \tag{4}$$

and

$$k^{des} = A e^{-\frac{Ea^{des}}{RT}} \left(\frac{T}{T_{ref}}\right)^{\beta_{des}} \tag{5}$$

when s is sticking coefficient, Γ is site density (2.5×10^{-9} $\frac{mol}{cm^2}$ for platinum), R is universal gas constant, T is catalyst surface temperature, T_{ref} is reference temperature (300 K), Ea is activation energy of adsorption and desorption, β is temperature exponent and A is desorption coefficient. Generally, high temperature increases the rate constant of adsorption and desorption which tend to increase the global reaction rate of catalytic combustion.

Butane (C_4H_{10}) is used as fuel but its surface reaction contains a lot of detail reaction mechanism that requires a lot of computational time; therefore, Deshmukh et al. [14] and Kaisare et.al. [15] suggested simpler simulation model involving one-step mechanism to estimate catalytic combustion of butane on platinum surface with the following global reaction;

$$C_4H_{10} + 6.5O_2 \rightarrow 4CO_2 + 5H_2O \tag{6}$$

The reaction rate for fuel-lean mixture when C_4H_{10} is deficient molecule can be expressed as

$$r = \frac{\Gamma k^{ads}_{C_4H_{10}} C_{s,C_4H_{10}}}{\left(1 + \sqrt{\frac{k^{des}_{O}}{k^{ads}_{O} C_{s,O_2}}}\right)^2} \tag{7}$$

However, under fuel-rich mixture condition when O_2 is deficient molecule, reaction rate was modified and expressed as

$$r = \Gamma k^{ads}_{O_2} C_{s,O_2} \left[\frac{-\left[\left(\frac{k^{ads}_{CO_2} C_{s,CO_2}}{k^{des}_{CO_2}}\right) + \left(\frac{k^{ads}_{H_2O} C_{s,H_2O}}{k^{des}_{H_2O}}\right) + 1\right] \pm \sqrt{\left[\left(\frac{k^{ads}_{CO_2} C_{s,CO_2}}{k^{des}_{CO_2}}\right) + \left(\frac{k^{ads}_{H_2O} C_{s,H_2O}}{k^{des}_{H_2O}}\right) + 1\right]^2 - 4\left(\frac{k^{ads}_{C_4H_{10}} C_{s,C_4H_{10}}}{k^{des}_{C_4H_{10}}}\right)(-1)}}{2\left(\frac{k^{ads}_{C_4H_{10}} C_{s,C_4H_{10}}}{k^{des}_{C_4H_{10}}}\right)}\right]^2 \tag{8}$$

2.3 Single-Chamber Solid Oxide Fuel Cell (SCFC)

A solid oxide fuel cell (SOFC) is an electrochemical conversion device that can generate electricity from fuel and oxidizer which are C_4H_{10} and O_2. The process of SOFC starts from oxygen molecules passing through cathode and recharge electrons to generate oxygen ions. When these ions go through electrolyte and reach the anode side, they react with fuel and release electrons. The electrolyte works as one-way gate to prevent these electrons from coming back to the cathode side. If the anode and the cathode are connected to a load, electrons will go through the load from anode side to cathode side to generate electricity.

Ideally, electrical power generation from a SOFC can be estimate from the change in Gibbs free energy between reactants and products which can be estimated as

$$\Delta G = \Delta H - T\Delta S \tag{9}$$

which leads to ideal efficiency as

$$\eta = \frac{\Delta G}{HV_{fuel}} \tag{10}$$

in which HV_{fuel} is the fuel heating value.

Typically, an SOFC requires the separation of fuel and oxidizer to operate but the disadvantage of the separated chamber design includes the presence of thermal stress and thermal expansion that must be accounted for, which can potentially lead to design difficulties. Under such small scale, high temperature and sealing become major concerns for the chamber design. However, Hibino et.al. [16] introduced a single-chamber SOFC (SCFC) that can avoid sealing problem in a small-scale device. Furthermore, yttria-stabilized zirconia (YSZ)-samaria-doped ceria (SDC) has been investigated under various sintering temperature and operating temperature by R.J. Milcarek et.al. [17] and it was found that the power generation of SCFC increased significantly when the SCFC was operated with H_2-air mixture. YSZ-SDC SCFCs may provide a promising solution for power generation devices.

3. Results and Discussion

3.1 Thermal Transpiration Membrane

Even though the pore size radius of glass microfiber membrane has been provided by manufacturer, Muntz et al. [18] only described the flow in a circular and zero thickness capillary channel. An effective pore radius and an effective capillary's cross-sectional area have to be determined from the experimental data comparing with Equation (2) and ANSYS, comercial simulation software. Temperature gradient across the membrane was controlled by an electrical power supply to a heating wire for three cases of temperature difference. Volume flow rate was controlled by varying outlet size. Smaller outlet generates higher pressure on the outlet side of the membrane that makes less volume flow rate. The varying outlet size is similar to a valve in a piping system. After various flow conditions, the effective pore radius and the capillary's cross-sectional area were found as 2.92 μm and 21.7% respectively by fitting Equation (2) to the experimental data. Simulation was considered based on experiemental results and theory curve fitting. User-defined function (UDF) had been modified by the effective pore radius and the effective capillary's cross-sectional area depending on temperature difference across the membrane. The simulation results and performance curves from the experiment, theory and simulation results are shown in *Figure 3.*

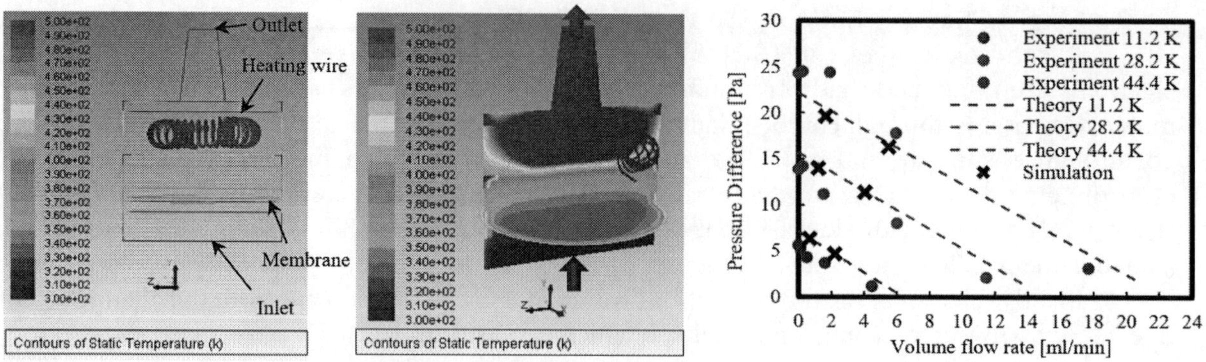

Figure 3 – (Left and Middle) Temperature gradient cross-section along the flow path (Right) Performance curve of thermal transpiration membrane in various temperature gradients across the membrane from experiment, theory and simulation results

3.2 Catalytic Combustion

To determine performance of surface reaction on platinum, catalytic combustion was setup in a 7-mm diameter quartz glass tube with 1-cm long plain platinum foil as shown in *Figure 4*. C_4H_{10}-air mixture with various equivalence ratios (ϕ) from 0.8 to 4 and various flow speed from 4 cm/s to 20 cm/s were supplied through the glass tube controlling by flow controllers. A thermocouple was used to measure gas temperature after combustion and combustion product compositions were analyzed by gas chromatography.

The results from all conditions show that the reaction on the catalyst surface in the simulation was slightly higher than the experiments with approximately 2% off as shown in *Figure 5* and they were following the same trend. When equivalence ratio increased at 12 cm/s constant flow speed, C_4H_{10} mole fraction at outlet increased due to the excess of C_4H_{10}. O_2 was deficient reactant and amount of O_2 at outlet was roughly constant. CO_2 and H_2O concentrations also indicated the lower reaction rate at higher equivalence ratio. When flow speed was varied and the equivalence ratio was fixed at 1.5, the concentration of C_4H_{10} and O_2 in exhaust increased because of insufficient residual time. The experimental results were slightly less than simulation results about 2%. Even these results do not provide perfect fit trend but they followed the same behavior and this model can be used to design the power generator.

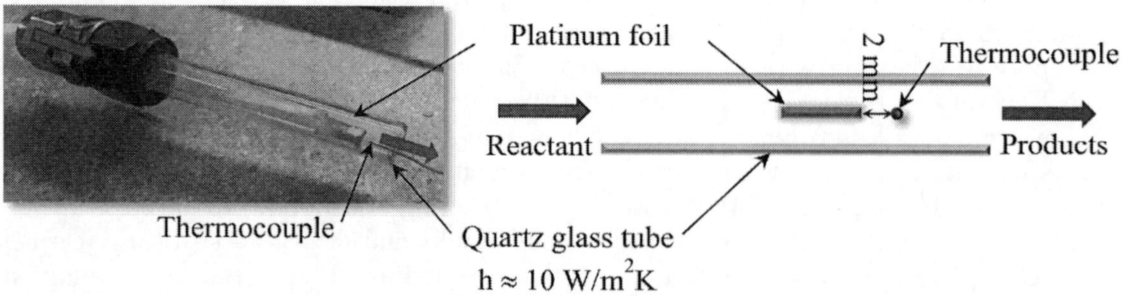

Figure 4 - Experimental setup and the diagram of catalytic combsution on plantinum foil

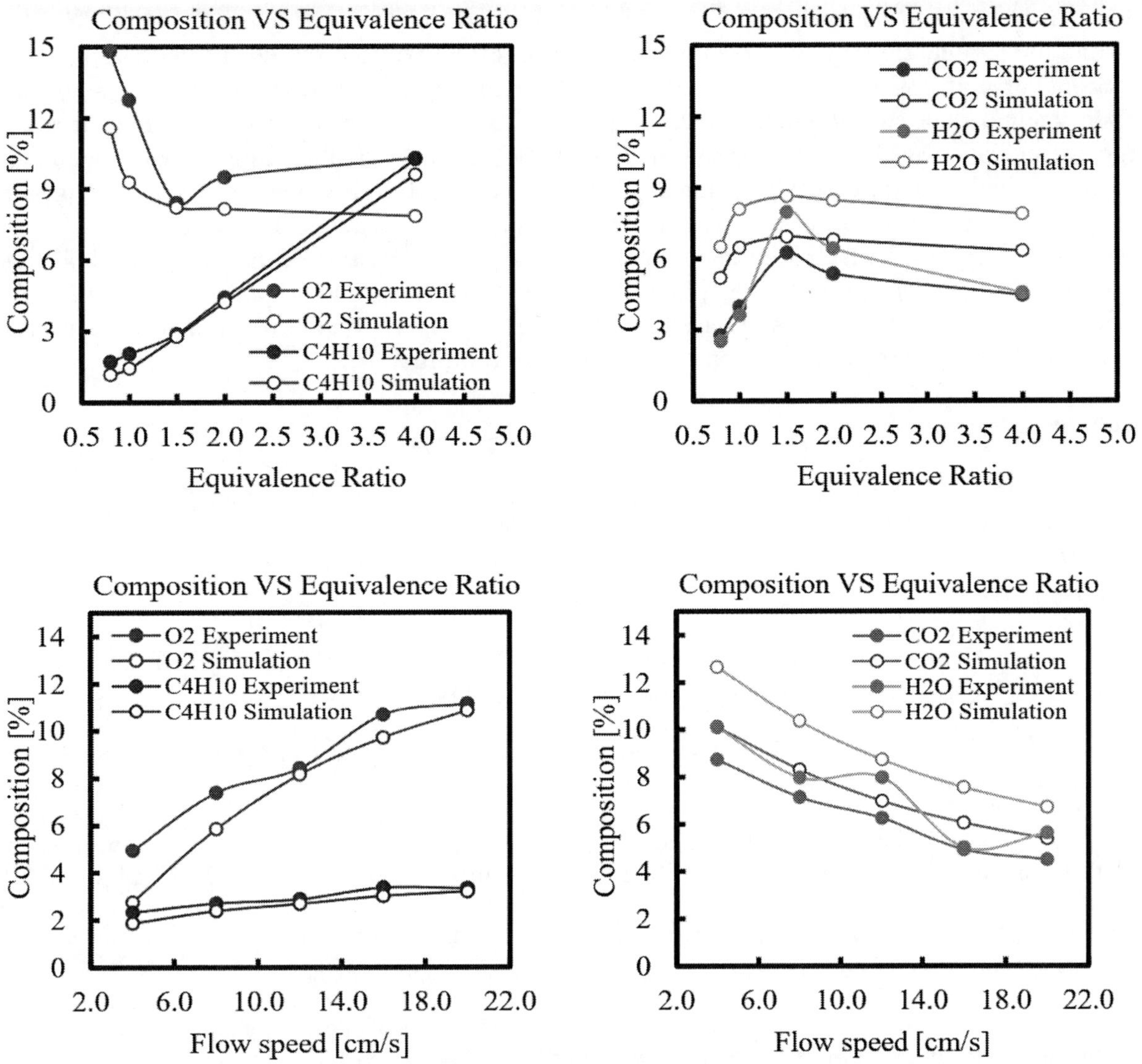

Figure 5 The exhaust compositions of experimental results and simulation results (Top) constant flow speed at 12 cm/s and (Bottom) constant equivalence ratio at 1.5

4. Conclusions

Thermal transpiration membranes integrating with catalytic combustion and SCFCs can be combined as power generator without any moving parts. The membrane provides air pumping power depending on temperature gradient across the membrane itself. C_4H_{10} which can be stored in liquid phase is used as fuel mixing with the air from the membrane. The platinum foil is used as a catalyst for catalytic combustion in fuel rich-mixture conditions. The left-over C_4H_{10} and air are supplied to the SCFCs for power generation purpose.

Both thermal transpiration membrane and catalytic combustion provided reasonable results that can be used to design the power generator without any moving parts. Thermal transpiration membrane simulation modeling was made based on effective pore radius and the capillary's cross-sectional area at 2.92 μm and 21.7%, respectively. Catalytic combustion modeling was also studied. Experimental data and simulation data from one-step reduced mechanism provided a reasonable agreement covering the SCFC operating region. SCFCs provided maximum power generation at about equivalence ratio 2.0 and higher reactant flow rate provides higher power generation. However, SCFC efficiency at higher flow rate decreases because of residue time of reactants. Thermal transpiration membrane integrated with catalytic combustion can provide equivalence ratio from 1.5 up to 18.9 by adjusting only fuel flow rate from 0.4 cm/s to 0.8 cm/s. The result shows that temperatures and compositions are appropriate for SCFC operation and thus provide self-sustaining power generation with no moving parts using hydrocarbon fuel.

Acknowledgements

The authors would like to gratefully acknowledge the National Science Foundation (NSF) for the financial support through grant number CBET-1402142.

References

[1] S. Menon, N. Moulton, *C. Cadou, Development of a Dynamometer for Measuring Small Internal-Combustion Engine Performance*, J. of Propulsion Power, 23 (2007) 194-202.

[2] D.C. Walther, A.P. Pisano, MEMS rotary engine power system: project overview and recent research results, Proc. 4th int'l symposium on MEMS and nanotechnology (2003) 227–234.

[3] O. Dessornes, S. Landais, R. Valle, A. Fourmaux, S. Burguguru, C. Zwyssig, Z. Kozanecki, Advances in the development of a Microturbine Engine, J. of Eng. Gas Turbines Power 136 (2014) 071201-1 – 071201-9.

[4] Y. Hsu, I. Sapir, P. Ronney, G.J. Snyder, Micro-combustion power system with dual path counter-flow system, US 8,614,392 B1, 2013.

[5] A. Shan, P. Sui, G. Kim, S. Ye., A transient PEMFC model with CO poisoning and mitigation by O_2 bleeding and Ru-containing catalyst, J. Power Sources, 166 (2007) 1-21.

[6] R.S. Rajaura, S. Srivastava, V. Sharma, P.K. Sharma, C. Lal, M. Singh, H.S. Palsania, Y.K. Vijay, Role of interlayer spacing and functional group on the hydrogen storage properties of graphene oxide and reduced graphene oxide, Int. J. of Hydrogen Ener., 41 (2016) 9454-9461.

[7] S.C. Li, F.C. Wang, The development of a sodium borohydride hydrogen generation system for proton exchange membrane fuel cell, Int. J. of Hydrogen Ener., 41 (2016) 3038-3051.

[8] X. Li, A. Faghri, Review and advance of direct methanol fuel cells (DMFCs) part I: Design, fabrication, and testing with high concentration methanol solutions, J. of Power Sources, 226 (2013) 223-240.

[9] R. Osborne, On certain dimensional properties of matter in the gaseous state, Phil. Trans. R. Soc. Lond. 1879 170, 727-845.

[10] Y. Sone, T. Kataoka, T. Ohwada, H. Sugimoto, K. Aoki, Numerical Examination of Applicability of the Linearized Boltzmann Equation, Eur. J. Mech. Fluids. Vol. 13(5), 573-589, (1994).

[11] S.E. Vargo, E.P. Muntz, G.R. Shiflett, Knudsen compressor as a micro- and macroscale vacuum pump without moving parts or fluids, J. Vac. Sci. Technol. A17(4), 2308-2313, Jul/Aug 1999

[12] E.P. Muntz, Y. Sone, K. Aoki, S. Vargo, M. Young, Performance analysis and optimization considerations for a Knudsen compressor in transitional flow, J. Vac. Sci. Technol. A20(1), 214-224, Jan/Feb 2002.

[13] T. Hibino, A. Hashimoto, T. Inoue, J. Tokuno, S. Yoshida, M. Sano, A solid oxide fuel cell using an exothermic reaction as the heat source, J of the Elec. Chem Society 148(6), (2001), A544-A549

[14] S.R. Deshmukh, D.G. Vlachos, A reduced mechanism for methane and one-step rate expressions for fuel-lean catalytic combustion of small alkanes on noble metals, Combustion and Flame 149 (2007) 366-383

[15] N.S. Kaisare, S.R. Deshmukh, D.G. Vlachos, Stability and performance of catalytic microreactors: Simulations of propane catalytic combustion on Pt, Chemical Engineering Science 63, (2008), 1098-1116.

[16] J. Ahn, P.D. Ronney, Z. Shao, S.M. Haile, A thermally self-sustaining miniature solid oxide fuel cell, J. Fuel Cell Sci. and Tech, 6, (2009), 041004-1 – 041004-4

[17] R.J. Milcarek, K. Wang, M.J. Garrett, J. Ahn, Performance Investigation of Dual Layer Yttria-Stabilized Zirconia–Samaria-Doped Ceria Electrolyte for Intermediate Temperature Solid Oxide Fuel Cells, J. of Electrochemical Ener. Conversion and Storage, Vol.13 (2016) 011002-1 – 011002-7.

Western States Section of the Combustion Institute – Fall 2017 Meeting
Hosted by the University of Wyoming
October 2&3, 2017

Impact of Pressure on Simulated FREI Combustion

Pawan Sharma[1,*], *Mohsen Ayoobi*[2], *Matthew McNenly*[3], *and Ingmar Schoegl*[1]

[1]*Mechanical Engineering, Louisiana State University, Baton Rouge, LA 70803, USA*
[2]*Wayne State University, Detroit, MI 48202, USA*
[3]*Lawrence Livermore National Laboratory, Livermore, CA 94550, USA*
[*]*Corresponding author:* pshar15@lsu.edu

Abstract: Flames with repetitive extinction and ignition (FREI) are self-excited instabilities that are observed in externally heated tubular micro-combustors with a controlled wall temperature profile. In FREI, flames ignite in the downstream hot region and move upstream while losing heat, and eventually quench in the upstream cold region. This periodic phenomenon repeats itself at several hundred Hz. The present numerical study seeks to investigate effects of pressure on the behavior of FREI inside a 1mm tubular microchannel. In this study, the impact of pressure is investigated for stoichiometric ($\Phi = 1$) premixed propane/air mixtures in an axisymmetric 2D configuration. Transient numerical simulations are performed using detailed chemistry. Externally imposed wall temperatures are kept the same for all four cases. Results show that temperature levels characteristic for extinction and ignition decrease as the pressure increases. The decrease is more pronounced for extinction temperatures, which is consistent with a known decrease of the quenching diameter for increasing pressures.
***Keywords:* ignition, Extinction, FREI, Micro-Combustion**

1. Introduction

Understanding micro-scale combustion is necessary to make it useful for fundamental combustion studies. Unlike meso-scale combustion, micro-scale combustion is difficult to sustain due to the increase of surface to volume ratio and associated wall quenching effects. The minimum diameter of a tube that supports flame propagation is known as quenching diameter d_T, and is in the order of a few millimeters for gaseous alkane hydrocarbon fuels [1]. The quenching diameter decreases as the tube wall temperature increases, so as a result, external heating enables combustion inside a micro-combustor. There are three combustion regimes in an externally heated quartz tube micro-combustors with a controlled wall temperature profile: stable flames at high flow rates, flames with repetitive extinction and ignition (FREI) at intermediate flow rates, and weak flames at low flow rates [2]. FREI are self-excited instabilities that are observed at intermediate flow rates and caused by flame-wall thermal interactions [2–4]. In FREI, flames ignite in the downstream hot region and move upstream while losing heat, and eventually quench in the upstream cold region [2–4]. FREI results from the loss stability via a Hopf bifurcation [5–8]. FREI in externally heated micro-combustor are caused by a negative phase in heat transfer through the wall [9]. Previous studies with weak flames confirmed that externally heated micro-combustor with controlled temperature profile could be used for fundamental combustion studies [10, 11]. The present numerical study seeks to extend fundamental combustion studies for weak flames towards the FREI regime.

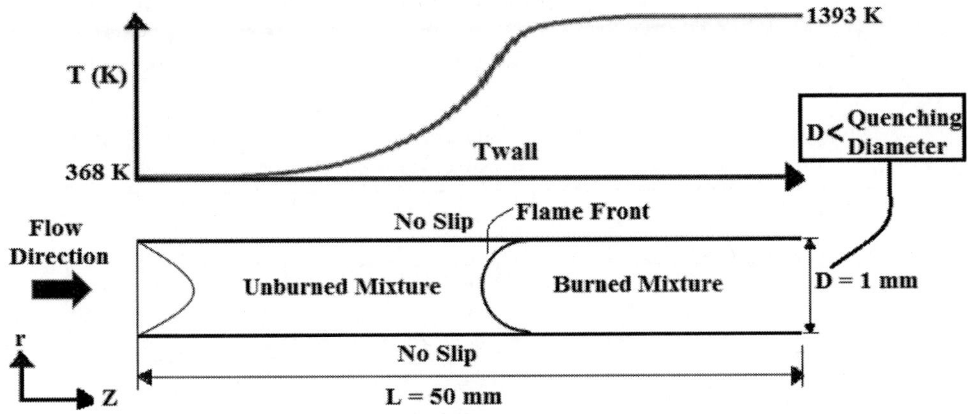

Figure 1: 1mm tubular microchannel with externally imposed temperature profile.

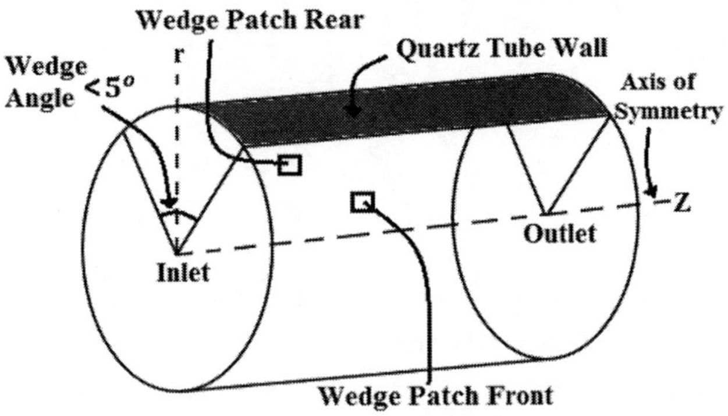

Figure 2: 2D axisymmetric geometry.

2. Geometry of Computational Domain and Numerical Methods

A cylindrical tube of length L = 50 mm and diameter D = 1 mm, shown in Figure 1, was used in numerical simulations. The cylindrical tube was simulated based on the 2D axisymmetric geometry shown in Figure 2. Hydrodynamic and chemical phenomena were approximated as axi-symmetric because the diameter of the circular tube is narrow enough in comparison of the length [12], which results in circumferential gradients being much smaller than axial gradients. A uniform grid was used across the two-dimensional geometry. Figure 1 illustrates an experimental temperature profile that was applied along the wall of the computational domain. The velocity profile was assumed to be fully developed at the inlet of the geometry, and no-slip boundary conditions were applied on the cylinder geometry walls. Velocity-derived and pressure-derived boundary conditions for the 2D axisymmetric computational domain are summarized in Table 1.

3. Cases

Simulations are performed using the open-source CFD package OpenFOAM [13]. In this work, numerical methods use discretized equations for continuity, momentum, energy, and species. For discretization, a standard Gaussian finite volume integration scheme is used with a second-order scheme for terms including gradients and Laplacians and first order upwind for terms including

Table 1: OpenFoam boundary conditions.

Boundary	Type	Velocity-Derived Type/Value	Pressure-Derived Type/Value
Inlet	Patch	Fixed Value/Non-Uniform (Fully Developed Flow)	Zero Gradient
Outlet	Patch	Zero Gradient	Fixed Value/Unifrom (Atmospheric Pressure)
Wall	Wall	Fixed Value/Unifrom (No Slip)	Zero Gradient
Axi-Symmetric-Front	Wedge	Wedge	Wedge
Axi-Symmetric-Rear	Wedge	Wedge	Wedge

Table 2: Simulation parameters.

Pressure (atm)	u_{in} (cm/s)	T_{in}(K)	Diameter (mm)	Φ	Grid
1	30	298	1	1	10×525
2	15	298	1	1	10×525
5	6	298	1	1	10×525
10	3	298	1	1	10×1025

divergence. A first order implicit Euler scheme is used for time integration. The preconditioned conjugate gradient (PCG) is used to solve pressure and density, whereas preconditioned biconjugate gradient (PBiCG) is used to solve velocity, species, and temperature. Cholesky (DIC) and diagonal incomplete lower-upper (DILU) techniques are used to preconditioned PCG and PBiCG, respectively. The laminarSMOKE solver is used to resolve the detailed chemistry of laminar reacting flows [14]. The laminarSMOKE solver is developed and maintained by the CRECK group in Milan and built on the top of open-source CFD package OpenFOAM. The San Diego mechanism with 53 species and 235 reactions is used to model combustion of propane/air mixture at the stoichiometric ratio for four different pressures [15]. Parallel computing is utilized to solve the computational domain. The computational domain is divided into 64 equal size blocks and solved on 64 processors (4 nodes and 16 processors on each node). A total of four 2D transient cases of pressures 1 atm, 2 atm, 5 atm, and 10 atm was investigated. Case details are summarized in Table 2. All cases are investigated for otherwise identical conditions, i.e. burning mass flux, stoichiometric fuel/air ratio, and externally employed experimental temperature profile. At the inlet, the wall temperature is 353 K and at the outlet temperature is 1371 K. The burning mass flux is kept constant by reducing the velocity to compensate for increased density at elevated pressures. Stoichiometric propane/air mixtures enter the channel with uniform inlet temperature T_{in} and a fully developed velocity profile with mean velocity u_{in}. A grid resolution of 10×525 is used for pressures of 1 atm, 2 atm, and 5 atm, i.e. 10 grid points in the radial direction and 525 grid points in the axial direction. Since the flame thickness decreases with increasing pressure, the grid resolution is increased to 10×1025 for the 10 atm case to better resolve the flame thickness. . Data are saved every 500 µs, and simulations are continued until two ignition-extinction cycles are observed.

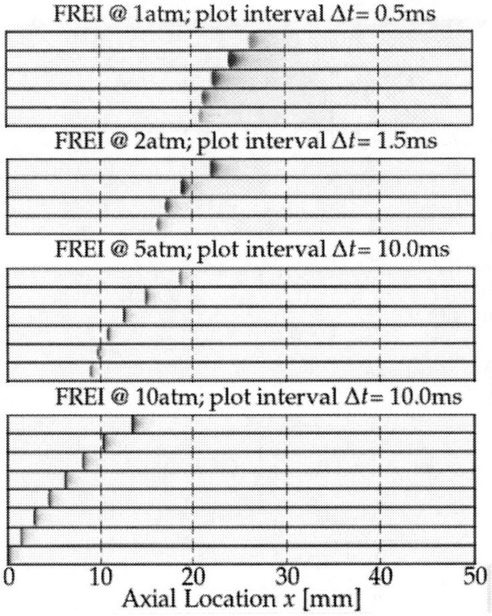

Figure 3: Instantaneous FREI snapshots.

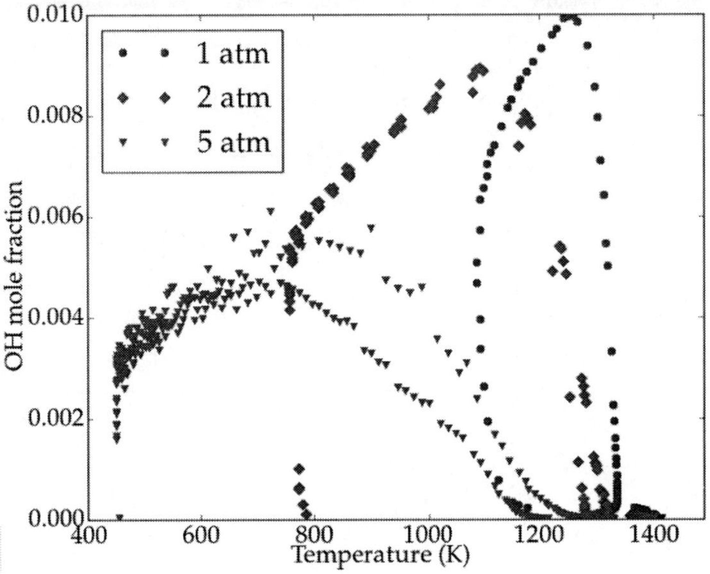

Figure 4: Temporal OH mole fraction as a function of the wall temperature for different pressures.

4. Results

Figure 3 shows instantaneous FREI snapshots for 1 atm, 2 atm, 5 atm and 10 atm. For all four cases, the unburned premixed stoichiometric propane/air mixture enters from the left and moves towards the right slowly gaining heat from the wall. Eventually auto-ignition occurs far downstream when sufficiently high gas temperature are reached due to heat transfer from the walls. After ignition, flame fronts form and propagate upstream where they eventually extinguish. This extinction is attributed to excessive heat losses to relatively cold walls. After a significant delay, a new unburned premixed mixture re-ignites downstream and quickly moves upstream, where it extinguishes again. Figure 3 illustrates that an increase in pressure affects the FREI behavior drastically, as both flames ignition and extinction shift far upstream, where wall temperatures are lower. For 10 atm, the flame does not quench and moves to the tube inlet.

Figure 4 shows the temporal OH mole fraction as a function of wall temperature for different pressures, where the periodic variation of the OH mole fraction is plotted with respect to the wall temperature during the FREI cycle. Figure 4 illustrates that during ignition the OH mole fraction increases sharply, and subsequently remains at an elevated level while the flame propagates upstream. At the upstream extinction location, the OH mole fraction decreases and eventually vanishes. Thus, this plot can be used to obtain ignition and extinction temperatures by finding wall temperatures at which the OH mole fraction sharply increases and decreases, respectively. Table 3 shows results for ignition and extinction temperatures of FREI for all four cases. Both ignition and extinction temperatures decrease with increasing pressure, although the decrease is more pronounced for extinction temperatures, which is consistent with a known decrease of the quenching diameter for increasing pressure ($d_T \propto p^{-1}$) [1]. For 10 atm, the flame ignites and propagates to the inlet where it stabilizes, i.e. no extinction is observed. It is noted that first and subsequent ignition/extinction cyclces show slightly different temporal OH mole fraction contours,

Table 3: Ignition temperature, Extinction temperature, FREI frequency, and FWHM reaction zone thickness for different pressures.

Pressure (atm)	Ignition Temperature (K)	Extinction Temperature (K)	Frequency (Hertz)	FWHM Reaction Zone Thickness (mm)
1	1333	1087	142	1.11
2	1309	744	41	0.63
5	1235	433	7	0.55
10	999	N/A	N/A	0.33

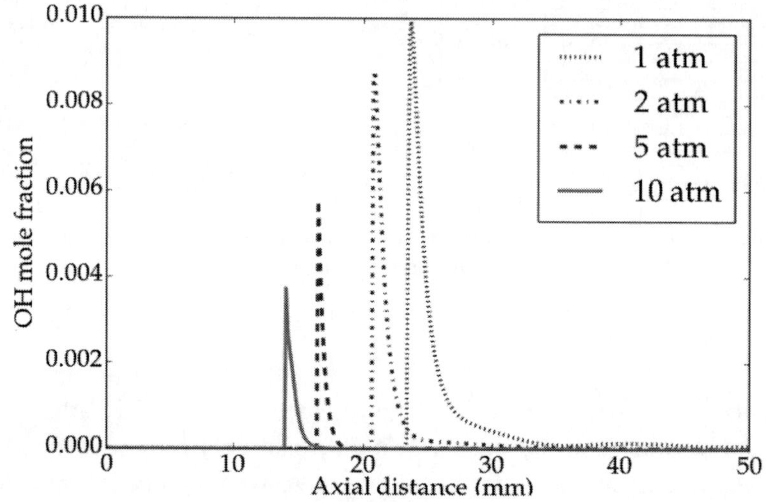

Figure 5: Impact of pressure on peak OH mole fraction on the FREI cycle.

which is visible for both 1 atm and 2 atm cases. In the case of 5 atm, data show some oscillations which are attributed to imperfect grid spacing, which no longer adequately resolves the flame thickness at the given pressure.

Figure 5 shows the impact of pressure on peak OH mole fraction during the FREI cycle. Results illustrate that with increasing pressure there is a significant upstream shift of the location for peak OH mole fraction, which corresponds to results shown in Figure 4. The decrease of peak OH mole concentrations with a decrease in pressure is attributed to a reduction of OH creation rates due to lower temperatures. Table 3 further illustrates the Full-Width Half Maximum (FWHM) calculated for FREI reaction zone thickness at different pressures. Here, the reaction zone thickness is calculated by taking the difference between two axial positions at which OH mole fraction is equal to half of its maximum value. From Table 3 it is evident that the reaction zone thickness decreases with an increase in pressure which is in direct correlation with the literature that laminar flame thickness (δ) decreases with increasing pressure ($\delta \propto p^{-1}$) [1]. The same results are evident in Figure 3, which illustrates that FREI reaction zones narrow as the pressure increases. While results are consistent with observations from adiabatic flames, it is noted that FREI are strongly non-adiabatic due to flame-wall thermal interactions, where further details will be investigated in future work.

Table 3 shows the frequency of FREI propagation for different pressures. The FREI frequency is calculated by taking the inverse of the time difference in two ignition periods. Simulations of FREI result in frequencies of 142 Hz for 1 atm, 41 Hz for 2 atm, and 7 Hz for 5 atm. The decrease in FREI frequency as pressure increase is attributed to the larger drop of extinction temperatures in comparison to ignition temperatures as well as a reduction of flame propagation velocities, i.e., flames will travel a larger distance before quenching. Overall, results clearly show that FREI relies on flame quenching, where a decrease of the quenching diameter at high pressures will require a reduction of diameters of suitable micro-combustors.

5. Conclusions

This numerical study presented in this paper investigates the impact of pressure on FREI behavior at four pressures (1 atm, 2 atm, 5 atm, and 10 atm) while holding the mass flux and other external conditions constant. It is found that while both ignition and extinction temperatures decrease with increasing pressure, the pressure effect is more pronounced for extinction, which is consistent with a known reduction of quenching diameters at elevated pressures. The FREI reaction zone thickness was found to decrease with pressure which corresponds to the behavior of laminar flames. Peak OH mole fractions decrease with increasing pressure due to a reduction of OH production rates at lower temperatures. Further, frequencies of FREI decrease with an increase of pressure due to of an increase of the traveled distance as well as a reduction of flame speeds. Overall there are strong indicators for high-pressure FREI requiring reduced micro-combustor diameters ($D < 1$ mm). Further, trends for auto-ignition, extinction, and reaction zone thickness observed in FREI closely resemble known trends for corresponding adiabatic phenomena, i.e. FREI constitute a promising approach for fundamental combustion studies.

Acknowledgements

The research reported in this paper was sponsored by the U.S. Department of Energy (DOE) Bioenergy Technologies Office (BETO) and Vehicles Technologies Office (VTO) under the DOE Co-- Optimization of Fuels and Engines Initiative. The authors also would like to acknowledge the support provided by the Economic Development Assistantship Program and High Performance Computing at Louisiana State University.

References

[1] I. Glassman, Combustion, Academic Press, San Diego, 1996.

[2] K. Maruta, T. Kataoka, N. I. Kim, S. Minaev, and R. Fursenko, Characteristics of combustion in a narrow channel with a temperature gradient, Proc. Combust. Inst. 30 (2005) 2429–2436.

[3] Y. Fan, Y. Suzuki, and N. Kasagi, Experimental study of micro-scale premixed flame in quartz channels, Proc. Combust. Inst. 32 (2009) 3083–3090.

[4] Y. Ju and K. Maruta, Microscale combustion: Technology development and fundamental research, Progress in Energy and Combustion Science 37 (2011) 669–715.

[5] H. Nakamura, A. Fan, S. Minaev, E. Sereshchenko, R. Fursenko, Y. Tsuboi, and K. Maruta, Bifurcations and negative propagation speeds of methane/air premixed flames with repetitive extinction and ignition in a heated microchannel, Combust. Flame 159 (2012) 1631–1643.

[6] G. Pizza, C. E. Frouzakis, J. Mantzaras, A. G. Tomboulides, and K. Boulouchos, Dynamics of premixed hydrogen/air flames in microchannels, Combust. Flame 152 (2008) 433–450.

[7] S. Minaev, K. Maruta, and R. Fursenko, Nonlinear dynamics of flame in a narrow channel with a temperature gradient, Combust. Theory Model. 11 (2007) 187–203.

[8] A. Fan, S. S. Minaev, E. V. Sereshchenko, Y. Tsuboi, H. Oshibe, H. Nakamura, and K. Maruta, Dynamic Behavior of Splitting Flames in a Heated Channel, Combust. Explos. Shock Waves 45 (2009) 245–250.

[9] E. Miyata, N. Fukushima, Y. Naka, M. Shimura, M. Tanahashi, and T. Miyauchi, Direct numerical simulation of micro combustion in a narrow circular channel with a detailed kinetic mechanism, Proc. Combust. Inst. 35 (2015) 3421–3427. DOI: 10.1016/j.proci.2014.07.057.

[10] H. Oshibe, H. Nakamura, T. Tezuka, S. Hasegawa, and K. Maruta, Stabilized three-stage oxidation of DME/air mixture in a micro flow reactor with a controlled temperature profile, Combust. Flame 157 (2010) 1572–1580.

[11] T. Kamada, H. Nakamura, T. Tezuka, S. Hasegawa, and K. Maruta, Study on combustion and ignition characteristics of natural gas components in a micro flow reactor with a controlled temperature profile, Combust. Flame 161 (2015) 37–48. DOI: 10.1016/j.combustflame.2013.08.013.

[12] G. Pizza, C. E. Frouzakis, J. Mantzaras, A. G. Tomboulides, and K. Boulouchos, Three-dimensional simulations of premixed hydrogen/air flames in microtubes, J. Fluid Mech 658 (2010) 463–491.

[13] OpenFoam, *OpenFoam - The Open CFD Toolbox*, http://www.openfoam.com/, 2012.

[14] A. Cuoci, A. Frassoldati, T. Faravelli, and E. Ranzi, A computational tool for the detailed kinetic modeling of laminar flames: application to C_2H_4/CH_4 coflow flames, Combust. Flame 160 (2013) 870–886.

[15] Chemical-Kinetic Mechanisms for Combustion Applications, *Mechanical and Aerospace Engineering (Combustion Research), University of California at San Diego*, 2009, URL: http://combustion.ucsd.edu.

Paper # 29OT-0044 **Topic: Other**

Western States Section of the Combustion Institute – Fall 2017 Meeting
Hosted by the University of Wyoming
October 2-3, 2017

Computational performance of chemistry and transport calculations in NGA

J. Schlup[1,*], N. Burali[2], G. Beardsell[2], and G. Blanquart[2]

[1] *Graduate Aerospace Laboratories*
[2] *Mechanical Engineering*
California Institute of Technology, Pasadena, USA
[*] *Corresponding author, e-mail: jschlup@caltech.edu*

For many computational fluid dynamics (CFD) codes, it is critical to understand and quantify the time spent calculating various quantities appearing in the governing equations. Evaluation of fluid properties, such as viscosity, conductivity, and diffusivity, and (for reacting flows) chemical source terms, can lead to large computational overheads that severely limit the size of numerical simulations. In this work, a detailed analysis of the computational time for the evaluation of chemistry source terms and transport properties is done for the reacting flow code NGA (Desjardins *et al.*, J. Comput. Phys., 2008). Computational scalings for the viscosity, conductivity, mass and thermal diffusion coefficients, and chemical source terms are investigated. A cost comparison between single versus double precision calculations for the transport properties is also considered. These scalings are performed using a range of chemical models and different levels of approximation for the transport properties. Results are obtained using a three-dimensional domain, and the computational scalings based on the number of species in the chemical model are presented.

1 Introduction

For reacting flow Navier-Stokes solvers, a careful balance between the accuracy and computational cost of each part of the code must be considered. High physical fidelity is often associated with prohibitive computational costs, which limit the complexity of the configurations that can be investigated [1].

The NGA code [2] has been developed as a versatile tool in combustion research, and has been extensively used to investigate constant and variable density laminar and turbulent flows, with chemical models of varying complexity [2, 3]. The governing equations are solved either in a low-Mach limit [2] or in the compressible formulation [4]. Problems of interest are often investigated using Direct Numerical Simulations (DNS) with detailed chemistry and Large Eddy Simulations (LES) with tabulated chemistry.

It is the goal of the current work to present a detailed discussion of the computational efficiency of the NGA code. More precisely, the goals are (*i*) to analyze and compare the computational cost of different chemistry integration techniques and transport models, and (*ii*) to evaluate the scalability with the number of scalars being transported.

This paper is organized as follows. Section 2 presents an overview of NGA. In Section 3, the chosen test cases are described. A detailed analysis of the cost of the chemistry integration techniques and the transport models is presented in Section 4, and Section 5 discusses the scalings against the number of species considered. Finally, the conclusions are discussed in Section 6.

2 NGA

2.1 Governing equations

In this work, the low-Mach number Navier-Stokes equations [5, 6] are solved

$$\frac{\partial \rho}{\partial t} + \nabla \cdot (\rho u) = 0, \tag{1}$$

$$\frac{\partial \rho \boldsymbol{u}}{\partial t} + \nabla \cdot (\rho \boldsymbol{u}\boldsymbol{u}) = -\nabla p + \nabla \cdot \boldsymbol{\tau}, \tag{2}$$

$$\frac{\partial \rho Y_i}{\partial t} + \nabla \cdot (\rho \boldsymbol{u} Y_i) = -\nabla \cdot \boldsymbol{j_i} + \rho \dot{\omega}_i, \tag{3}$$

$$\frac{\partial \rho T}{\partial t} + \nabla \cdot (\rho \boldsymbol{u} T) = \nabla \cdot (\rho \alpha \nabla T) + \rho \dot{\omega}_T - \frac{1}{c_p} \sum_i c_{p,i} \boldsymbol{j_i} \cdot \nabla T + \frac{\rho \alpha}{c_p} \nabla c_p \cdot \nabla T. \tag{4}$$

In the above equations, the subscript "i" refers to the i-th species, ρ is the density, $\boldsymbol{u}$ is the velocity vector, $\boldsymbol{\tau}$ is the viscous stress tensor, Y_i is the species mass fraction, $\boldsymbol{j_i}$ is the species mass diffusion flux, $\dot{\omega}_i$ is the species source term, T is the temperature, $c_{p,i}$ and c_p are the species and the mixture specific heats, respectively, $\alpha = \lambda/(\rho c_p)$ is the mixture heat diffusivity, λ is the mixture heat conductivity, and $\dot{\omega}_T = -\frac{1}{c_p}\sum_i h_i \dot{\omega}_i$ is the heat release.

The diffusion fluxes $\boldsymbol{j_i}$ are computed as [7]

$$\boldsymbol{j_i} = -\rho D_{i,m} \frac{Y_i}{X_i} \nabla X_i - D_i^T \frac{\nabla T}{T} - \rho Y_i \boldsymbol{u_c}, \tag{5}$$

where D_i^T are the thermal diffusion coefficients (discussed in Section 2.2) and the correction velocity

$$\boldsymbol{u_c} = -\sum_i \left[D_{i,m} \frac{Y_i}{X_i} \nabla X_i + \frac{1}{\rho} D_i^T \frac{\nabla T}{T} \right] \tag{6}$$

ensures zero net mass diffusion flux, i.e. $\sum_i \boldsymbol{j_i} = 0$. The species diffusivities, $D_{i,m}$, are computed either by supplying a set of constant Lewis numbers

$$D_{i,m} = \frac{\alpha}{\mathrm{Le}_i}, \tag{7}$$

or through a mixture-averaged formulation

$$D_{i,m} = \frac{1 - Y_i}{\sum_{k \neq i} \frac{X_k}{\mathcal{D}_{k,i}}}. \tag{8}$$

In Eq. (8), X_k is the k-th species mole fraction and $\mathcal{D}_{k,i}$ is the binary diffusion coefficient of the k-th species relative to the i-th species. The binary diffusion coefficients are computed using classical molecular gas theory [7].

The mixture-averaged thermal conductivity, λ, is computed following Mathur *et al.* [8]

$$\lambda = \frac{1}{2}\left[\sum_i X_i\lambda_i + \left(\sum_i \frac{X_i}{\lambda_i}\right)^{-1}\right]. \tag{9}$$

The species thermal diffusivities, λ_i, are evaluated using a modified version of Eucken's formula [9]. The stress tensor is given by

$$\boldsymbol{\tau} = \mu\left[\nabla\boldsymbol{u} + (\nabla\boldsymbol{u})^T\right] - \frac{2}{3}\mu(\nabla\cdot\boldsymbol{u})\boldsymbol{I}, \tag{10}$$

where $\boldsymbol{I}$ is the identity matrix and μ is the fluid viscosity, which is calculated using a modified form of Wilke's formula, similar to the one used for viscosity [10, 11]

$$\mu = \frac{1}{2}\left[\sum_i X_i\mu_i + \left(\sum_i \frac{X_i}{\mu_i}\right)^{-1}\right]. \tag{11}$$

The species viscosites, μ_i, are obtained using standard gas kinetic theory [12]. For case F-W (see Table1), the fluid viscosity is computed using Wilke's formula [10]

$$\mu = \sum_i \left(\frac{X_i\mu_i}{\sum_j X_j\Phi_{ij}}\right) \tag{12}$$

where

$$\Phi_{ij} = \frac{\left[1 + \left(\frac{\mu_i}{\mu_j}\right)^{1/2}\left(\frac{W_j}{W_i}\right)^{1/4}\right]^2}{\left[1 + \left(\frac{W_i}{W_j}\right)\right]^{1/2}}. \tag{13}$$

2.2 Thermal diffusion coefficients

Thermal diffusion is used in case F-MATD, described in Section 3. For these flames, the thermal diffusion coefficients are computed using a mixture-averaged formulation based on the work of Chapman and Cowling [13]. This formulation of thermal diffusion coefficients takes the form [14]

$$D_i^T = \rho\frac{W_i}{W}D_{i,m}k_{Ti}, \tag{14}$$

where k_{Ti} is the thermal diffusion ratio of species i defined as [13]

$$k_{Ti} = \frac{W^2}{R\rho}\sum_j \frac{1.2C_{ij}^* - 1}{\mathcal{D}_{i,j}}\frac{Y_i a_j - Y_j a_i}{W_i + W_j}. \tag{15}$$

In Eq. (15), C^*_{ij} is a ratio of collision integrals [15]

$$C^*_{ij} = \frac{1}{3}\frac{\Omega^{(1,2)}_{ij}}{\Omega^{(1,1)}_{ij}}, \tag{16}$$

a_i is given by [16]

$$a_i = \lambda_{i,mon}\left[1 + \frac{1.065}{2\sqrt{2}X_i}\sum_{j\neq i} X_j \Phi_{ij}\right]^{-1}, \tag{17}$$

and the monatomic species conductivities, $\lambda_{i,mon}$, are defined as [13]

$$\lambda_{i,mon} = \frac{15}{4}\frac{R\mu_i}{W_i}. \tag{18}$$

2.3 Numerical methods

The governing equations are solved using the finite differences, discretely energy-conserving code NGA, designed for the simulation of variable density laminar and turbulent flows [2]. A fractional step approach is used, whereby the momentum, scalar, and density fields are staggered in time. A semi-implicit Crank-Nicolson scheme is used for time marching; for the integration of the chemical source terms, the preconditioned iterative method of Savard *et al.* [3] is used. Mass conservation is enforced through a pressure Poisson equation. For the present work, the Navier-Stokes equations are solved using a second-order accurate scheme in both space and time. The third-order accurate Bounded QUICK (BQUICK) scheme [17] is used for scalar transport, which ensures boundedness of the transported scalars. All simulations were run on TACC's Stampede cluster using the Extreme Science and Engineering Discovery Environment (XSEDE) [18].

3 Test cases

The chosen test cases are summarized in Table 1. For each test case, four premixed, three-dimensional flames are considered: an H_2/air flame with an equivalence ratio, $\phi = 0.4$; a CH_4/air case with $\phi = 0.9$, and an n-C_7H_{16}/air flame with $\phi = 0.9$ (run with two chemical models, detailed in Section 3.2). These configurations are selected as they have extensively been investigated in previous studies from the authors; further, the flames require chemical models with increasing number of species (n_s) and reactions (n_r), thus presenting an opportunity to investigate the scalability of different parts of the NGA code with the number of transported species. These chemical models are described in Section 3.2. All flames are laminar. It should be noted that for all cases except F-MADP, several variables in the calculation of diffusion coefficients are treated as single-precision floating-point variables. The single-precision variables are detailed in Section 4.1.

3.1 Geometry

A longitudinal slice of the computational domain is shown in Fig. 1. The domain is discretized using $1024 \times 32 \times 32$ uniformly spaced grid points in the longitudinal, transverse, and spanwise

	Properties	Case					
		F-BL	F-W	F-Ex	F-MA	F-MADP	F-MATD
μ	modified Wilke	✓	✗	✓	✓	✓	✓
	Wilke	✗	✓	✗	✗	✗	✗
$\dot{\omega}_i$	semi-implicit	✓	✓	✗	✓	✓	✓
	explicit	✗	✗	✓	✗	✗	✗
D_i	Le_i =const	✓	✓	✓	✗	✗	✗
	mix.-av.	✗	✗	✗	✓	✓	✓
	D_i^T	✗	✗	✗	✗	✗	✓

Table 1: Test cases considered in this work. Acronyms, specifying which transport or chemistry model is being timed: F-, fuel, which can be H_2, CH_4 or n-C_7H_{16}; BL, baseline; W, Wilke's formula; Ex, explicit chemistry integration; Im, implicit chemistry integration; CL, constant Lewis numbers; MA, mixture-averaged species diffusion coefficients; MADP, mixture-averaged species diffusion coefficients with double-precision floating-point variables; MATD, mixture-averaged species diffusion coefficients and thermo-diffusion (Soret and Dufour effects).

directions, respectively. For each flame, the grid spacing Δ is such that there are at least 20 grid points through the laminar flame thickness, l_F. Table 2 reports the grid resolution used for each flame.

	H_2/air	CH_4/air	n-C_7H_{16}
l_F [μm]	650	474	379
Δ [μm]	32.5	23.7	19.0

Table 2: Spatial resolution used for each flame.

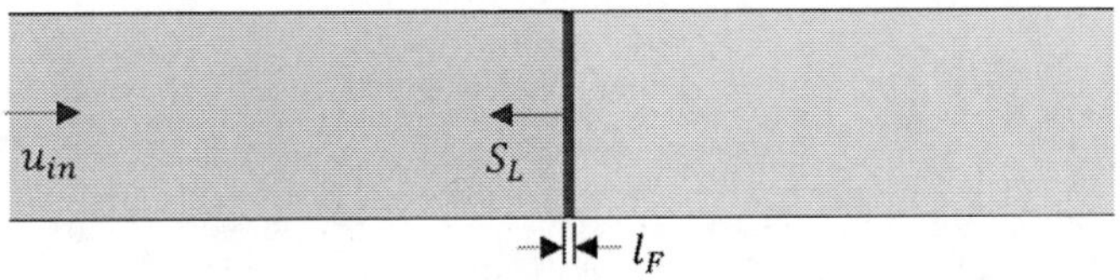

Figure 1: Sketch of the computational domain.

3.2 Chemistry models

Four chemical models are used in the present simulations. The nine species chemical model of Hong *et al.* [19] is used for the combustion of H_2, which has 54 reactions[1]. The mechanism for the CH_4 flame is the GRIMech3.0 [20] chemical model, which includes 53 species and 634

[1]For all mechanisms discussed in this section, forward and backward reactions are counted separately.

reactions. Finally, for the heptane flames, two chemistry models are considered: the full Caltech-Mech2.4 [21], which contains 171 species and 1895 reactions, and the 35-species, 217 reaction reduced version of the mechanism of Bisetti *et al.* [22] (aromatic species have been removed).

4 Computational cost

The cost of running all cases presented in Table 1 are shown in Fig. 2. This figure shows the overall computational cost of each case, in terms of the CPU time (in microseconds) per simulation time step (iteration), multiplied by the number of CPUs and divided by the number of computational points (ghost points are not considered since their impact is small for the current domain partition). Because each case has an identical number of points and was run on the same number of CPUs (and identical domain decomposition), the comparison of these timings is direct.

As can be seen, case F-Ex has the lowest computational cost per iteration of any of the considered cases. However, the time step required for F-Ex is often orders of magnitude lower than the baseline case with semi-implicit time integration (see Savard *et al.* [3] for a complete discussion). Thus, the semi-implicit method can reach the desired simulation time significantly faster than the explicit method. Further, one can see that the overall cost for F-BL and F-Ex increases nearly linearly with the number of species, with scalings between linear and quadratic for the remaining cases. In the subsequent sections, the specific transport models and chemistry integration techniques are investigated, showing the computational cost and scaling with n_s.

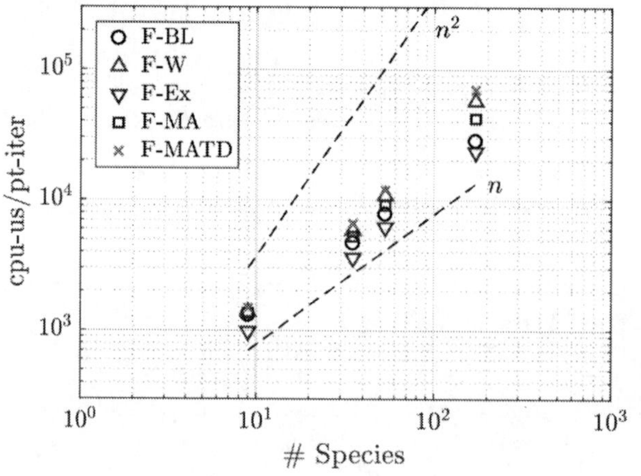

Figure 2: Overall computational cost of NGA for F-BL (black circles), F-W (green triangles), F-Ex (red inverted triangles), F-MA (blue squares), and F-MATD (orange crosses).

4.1 Single- versus double-precision

For nearly all of the cases investigated in Table 1, a small portion of the diffusion coefficient calculations are performed using a mixture of single- and double-precision floating point variables to assess the impact of double-precision on the computational time. For this comparison, the species molecular weights, ratio of Boltzmann constant to species potential well depth, k_B/ϵ_i, collision diameter, σ_i, and the coefficients used in calculating species viscosity,

$$\mu_{\text{coeff}} = \frac{5}{16}\frac{\sqrt{\pi k_B m_i}}{\pi \sigma_i^2}, \tag{19}$$

are converted from single- to double-precision floating-point variables for case F-MADP. Each other variable in the computation of $D_{i,m}$ is a double-precision floating-point variable, and the rest of NGA remains double-precision.

For the computation of $D_{i,m}$, the use of several single-precision floating-point variables decreases the computational cost. The cost savings found using single-precision increases slightly as n_s increases, simply due to the increased number of floating-point operations for a larger chemical mechanism. For case H_2-MADP, the computational cost of computing $D_{i,m}$ (with the double-precision variables listed above) increased by 14.5%, while for n-C_7H_{16}-MADP with Caltech-Mech2.4, the cost increased by 17.5%.

5 Scalings

Figure 3 shows the cost of each portion of the F-BL case. The scalar, chemistry, viscosity, conductivity, and diffusivity timings will be investigated in more detail in the subsequent sections. Timings for the velocity and pressure solvers are included for completeness, though a detailed analysis of the velocity and pressure solvers is outside the scope of this work. It is clear that the majority of the computational expense lies in the scalar transport and chemistry calculations, where the cost is orders of magnitude above the remainder of the code. An approximately equal amount of time is spent in the viscosity, conductivity, and diffusivity calculations.

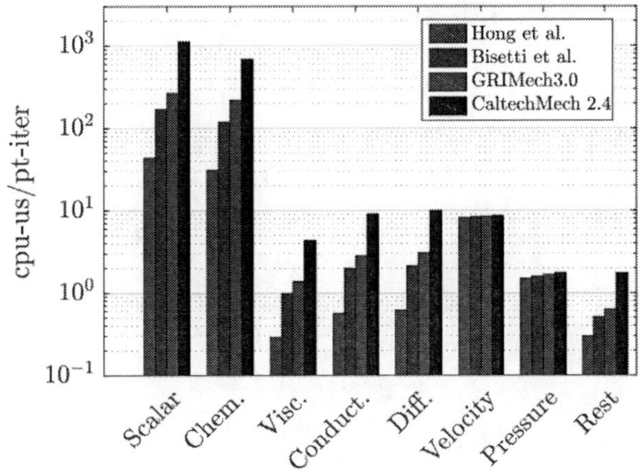

Figure 3: Overall computational cost of the F-BL case, showing scalar (transport and time integration), chemistry, viscosity, conductivity, diffusion, velocity, pressure, and the remainder of the code.

5.1 Viscosity

Figure 4 compares the cost associated with the computation of viscosity, using a modified form of Wilke's formula (F-BL, Eq. (11)) and Wilke's original formula (F-W, Eq. (12)). The double sum in

Eq. (12) explains why F-W scales quadratically with the number of species, while Eq. (11) is linear with respect to the number of species. It is worth noting that even for the nine species hydrogen mechanism, evaluating viscosity with Eq. (11) is more than an order of magnitude less expensive than using Eq. (12).

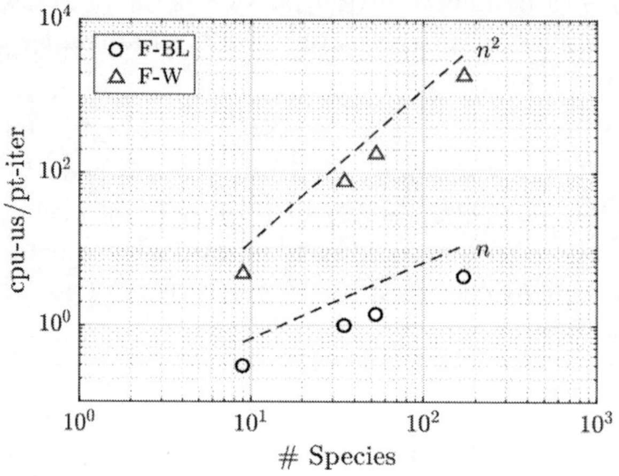

Figure 4: Viscosity computational cost using a modified form of Wilke's formulation [9] (black circles) and Wilke's original formulation [10] (green triangles).

5.2 Conductivity

Figure 5 shows the cost associated with the evaluation of conductivity using Eq. (9). As expected, the cost is linear with the number of species.

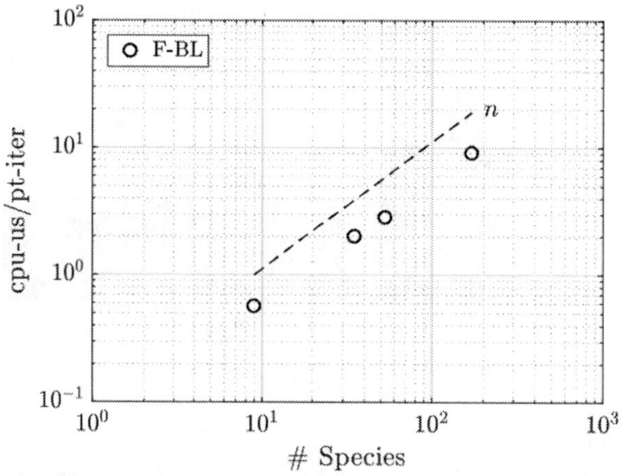

Figure 5: Conductivity computational cost using Eq. (9) [8].

5.3 Time integration of scalar fields

Figure 6 shows a comparison of the computational cost associated with the scalar transport for cases F-BL and F-Ex. For both cases four sub-iterations were used. As can be seen, the cost per time step is almost the same in both cases, and scales almost linearly with the number of species. However, it should be noted that the maximum time step in the explicit case remains constrained by the smallest chemistry time scales, and is typically orders of magnitude smaller than with the semi-implicit approach (see Fig. 21 in Savard *et al.* [3]).

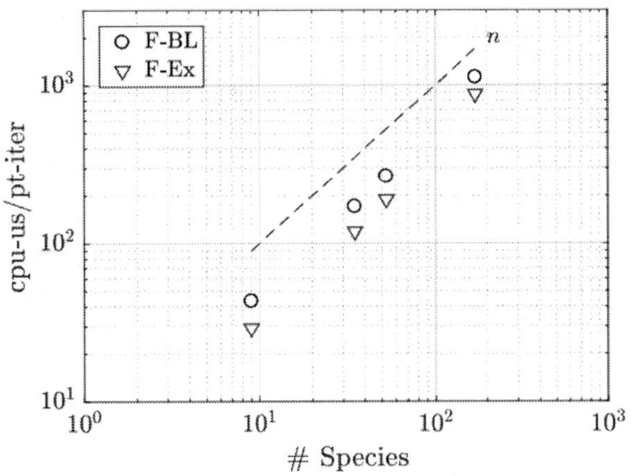

Figure 6: Scalar computational cost.

5.4 Diffusion coefficients

Figure 7a compares the computational costs associated with constant Lewis numbers (Eq. (7)) and mixture-averaged diffusion coefficients (Eq. (8)). The cost of computing diffusion coefficients using a constant Lewis number is typically over an order of magnitude lower than the mixture-averaged diffusion coefficient cost. Additionally, as expected from Eq. (7), the cost of the constant Lewis number approach increases linearly with the number of species. The mixture-averaged diffusion model, however, scales quadratically with the number of species. This can be explained through Eq. (8), where it is shown that for each species diffusion coefficient, a summation over all species (except itself) is required.

When incorporating thermal diffusion, one can first compute a_i from Eq. (17), which requires $\mathcal{O}(n_s^2)$ operations, similar to $D_{i,m}$. Then, with this value stored, an additional set of $\mathcal{O}(n_s^2)$ operations are used to compute k_{Ti} and, subsequently, D_i^T. This methodology yields a quadratic scaling with the number of species, as shown in Fig. 7b. Further, the timings shown in Fig. 7b can be compared with Fig. 7a, showing that the D_i^T coefficients are two to three times more costly to compute than the $D_{i,m}$ coefficients.

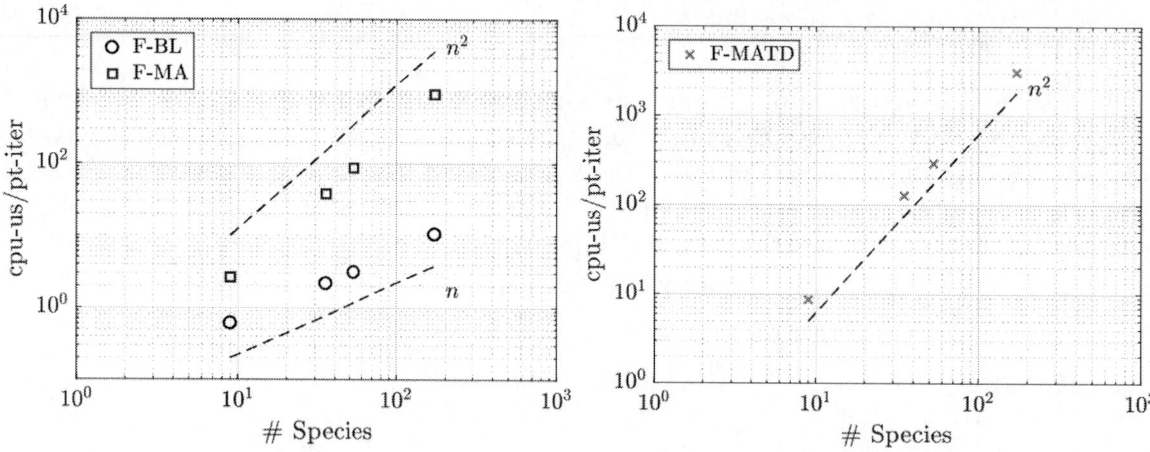

(a) Molecular diffusion coefficient cost.

(b) Thermal diffusion coefficient cost.

Figure 7: Computational cost for (a) computing diffusion coefficients using the baseline case with constant Lewis numbers (black circles) and mixture-averaged model (blue squares), and (b) computing thermal diffusion coefficients using the mixture-averaged thermal diffusion model (orange crosses).

6 Conclusions

A detailed set of computational scalings for multiple transport property and chemistry models have been presented, covering chemical models containing nine to 171 species. Overall, NGA has a computational cost nearly linear with the number of species present in the mechanism. This is due to the large number of implemented transport models that have a linear cost, including conductivity, viscosity, constant Lewis numbers, and the semi-implicit time integration of the chemistry. More complex models, such as mixture-averaged diffusion, thermal diffusion, and the full Wilke model for viscosity, scale quadratically with the number of species. Yet, even with these more costly models, NGA still has an almost linear scaling. Further, a brief investigation of using single-precision floating-point variables in the calculation of $D_{i,m}$ indicates additional cost reductions are possible.

Acknowledgments

The authors gratefully acknowledge funding from the U.S. Department of Energy-Basic Energy Sciences (DE-SC0006591) under the supervision of Dr. Wade Sisk. This work used the Extreme Science and Engineering Discovery Environment (XSEDE), which is supported by National Science Foundation grant number ACI-1053575. N.B. received Government support under contract FA9550-11-C-0028 and awarded by the Department of Defense, Air Force Office of Scientific Research, National Defense Science and Engineering Graduate (NDSEG) Fellowship, 32 CFR 168a.

References

[1] C. K. Westbrook, Y. Mizobuchi, T. J. Poinsot, P. J. Smith, and J. Warnatz. Proc. Combust. Inst., 30 (2005) 125–157.

[2] O. Desjardins, G. Blanquart, G. Balarac, and H. Pitsch. J. Comput. Phys., 227 (2008) 7125–7159.

[3] B. Savard, Y. Xuan, B. Bobbitt, and G. Blanquart. J. Comput. Phys., 295 (2015) 740–769.

[4] L. R. Boeck, S. Lapointe, J. Melguizo-Gavilanes, and G. Ciccarelli. Proc. Combust. Inst., 36 (2017) 2799–2806.

[5] A. Majda. Compressible Fluid Flow and Systems of Conservation Laws in Several Space Variables. Springer-Verlag Berlin Heidelberg, 1984.

[6] A. Majda and J. Sethian. Combust. Sci. Technol., 42 (1985) 185–205.

[7] J. O. Hirschfelder, C. F. Curtiss, R. B. Bird, and M. G. Mayer. Molecular theory of gases and liquids, Vol. 26. Wiley New York, 1954.

[8] S. Mathur, P. K. Tondon, and S. C. Saxena. Molecular physics, 12 (1967) 569–579.

[9] J. H. Ferziger and H. G. Kaper. Mathematical Theory of Transport Processes in Gases. North-Holland, Amsterdam, 1972.

[10] C. R. Wilke. J. Chem. Phys., 18 (1950) 517–519.

[11] S. Lapointe, B. Savard, and G. Blanquart. Combust. Flame, 162 (2015) 3341–3355.

[12] J. O. Hirschfelder, C. F. Curtiss, and R. B. Bird. Molecular theory of gases and liquids. John Wiley, 1954.

[13] S. Chapman and T. G. Cowling. The Mathematical Theory of Non–Uniform Gases. Cambridge University Press, Cambridge, 3 edition, 1970.

[14] P. Paul and J. Warnatz. Symp. (Int.) Combust., 27 (1998) 495 – 504.

[15] G. Dixon-Lewis. Flame structure and flame reaction kinetics. II. Transport phenomena in multicomponent systems. In Proc. R. Soc. London, Ser. A, Vol. 307, pages 111–135. The Royal Society, 1968.

[16] E. S. Oran and J. P. Boris. Prog. Energy Combust. Sci., 7 (1981) 1–72.

[17] M. Herrmann, G. Blanquart, and V. Raman. AIAA J., 44 (2006) 2879–2886.

[18] J. Towns, T. Cockerill, M. Dahan, I. Foster, K. Gaither, A. Grimshaw, V. Hazlewood, S. Lathrop, D. Lifka, G. D. Peterson, R. Roskies, J. R. Scott, and N. Wilkins-Diehr. Comput. Sci. Eng., 16 (2014) 62–74.

[19] Z. Hong, D. F. Davidson, and R. K. Hanson. Combust. Flame, 158 (2011) 633–644.

[20] G. P. Smith, D. M. Golden, M. Frenklach, N. W. Moriarty, B. Eiteneer, M. Goldenberg, C. T. Bowman, R. K. Hanson, S. Song, Jr. Gardiner, W. C., V. V. Lissianski, and Z. Qin. GRI-Mech 3.0. http://combustion.berkeley.edu/gri-mech/, 1999.

[21] G. Blanquart. CaltechMech, version 2.4. http://www.theforce.caltech.edu/CaltechMech, 2017.

[22] F. Bisetti, G. Blanquart, M. E. Mueller, and H. Pitsch. Combust. Flame, 159 (2012) 317–335.

Western States Section of the Combustion Institute – Fall 2017 Meeting
Hosted by the University of Wyoming
October 2-3, 2017

Modeling a Pressurized Coal Feed System

Taylor Schroedter[1] *Bradley Adams*[1]

[1]*Mechanical Engineering, Brigham Young University, Provo, Utah*

A gas-solid model was developed for a feed system to a pressurized oxy-coal (POC) reactor utilizing the multiphase particle-in-cell (MP-PIC) method. The model was created using CPFD Software's Barracuda Virtual Reactor 17.2.0. The geometry of the feed system was simple and contained a vertical standpipe of five inches feeding into a horizontal pipe of twelve inches. The coal particles from the standpipe were transported into the reactor with gas. To ensure appropriate combustion in the reactor, the feed system needed to maintain both a steady flow rate of gas and solids in the pipe and uniform distribution of solids through the cross-section of the channel. Steadiness was determined by the mass flow rate of the coal leaving the feed system. Uniformity was assessed by plotting gas and particle velocities at the end of the channel. The uniformity and steadiness of the flow was assessed for different grid sizes, starting pressure and boundary conditions, and other parameters within Barracuda.

1. Introduction

Coal combustion has been widely used since the 1800s to generate power. Even with the dawning of inexpensive natural gas, coal continues to be a major source of electricity in the United States [1]. One of the challenges with coal combustion is the greenhouse gases created and released into the environment, primarily carbon dioxide (CO_2). A potential solution to this involves the use of oxy-coal combustion, where coal is burned with pure oxygen, rather than air as is commonly done [2]. This process creates a higher purity CO_2 flue gas stream which is readily captured for storage or industrial use. Combustion of oxygen and coal at high pressures can provide higher efficiency power cycles due to enhanced thermal energy recovery [2].

A pressurized oxy-coal (POC) reactor will be built with funding from the U.S. Department of Energy (DOE) to study design and performance issues related to pressurized oxy-coal combustion. The first step of the reactor process, a feed system, is required to produce a steady and uniform flow of coal into the reactor. To produce a steady flow, a bed of coal particles in a pressurized pipe must be fluidized and pneumatically transported. The objective of this research is to identify an accurate and functional CPFD model of the feed system.

2. Methods

Barracuda Virtual Reactor 17.2.0 is a software package from CPFD Software that specializes in modeling gas-solid flows, particularly dense phase flows. Barracuda was utilized in the running of all simulations.

The geometry of the model consists of an L-shaped pipe. The initial vertical section is five inches in height. Moving down the vertical standpipe it comes to a T-bend. In the right-direction (See Figure 1), towards the reactor, it is twelve inches and in the opposite direction to the CO_2 inlet is 0.3 inches. The interior diameter of the pipe is a quarter inch.

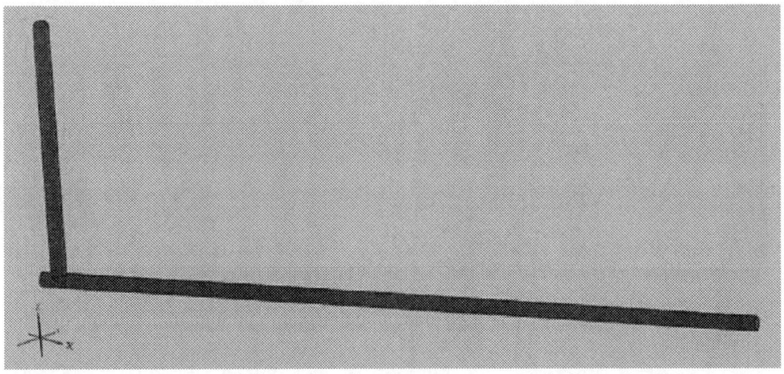

Figure 1. CAD model of the feed system in SolidWorks 2016 x64 Edition

Barracuda creates a computational grid based on a distribution of uniform square cells or prisms. The spacing and sizing of the grid can be changed, but it remains rectangular prisms. Because a cylindrical geometry is being broken up into cubes, it was important to have enough cells to accurately represent the round duct shape. Barracuda creates both real and null cells around the geometry, and this creates a 12" x 5" x 0.25" rectangular "box" to encapsulate the entire feed system. Cells within the pipe flow are considered real cells. Those outside the pipe are null cells. Three different meshes were tested, maintaining the same mass of particles in the system.

The simulations were run at a constant temperature of 300 K with a bituminous Pittsburgh coal and carbon dioxide. The coal particles had a diameter of 75 μm and a density of 1300 kg/m^3. The carbon dioxide had a density of 1.782 kg/m^3 at 101 kPa and 35.29 kg/m^3 at 2 MPa. Regarding the particle-to-particle interaction, the close pack volume fraction was equal to 0.6 and the initial packing fraction was equal to 0.4. The maximum momentum redirection from collision was 40%. For particle-to-wall interaction, the normal-to-wall momentum retention as well as the tangent-to-wall momentum retention was 0.8 with a diffuse bounce of zero.

A drag model was selected based on its accuracy in prediction of gas-particle behavior as noted in previous experiments and simulations. There are many choices for a drag model in Barracuda and the Wen-Yu-Ergun is the most fitting for these purposes [3, 4, 5, 6]. This drag model assumes a particle sphericity equal to one. While sphericity can play an important role in particle behavior, it was not a focus in this study. An important number to consider in two phase flow is the Archimedes number, Ar, defined as:

$$Ar = \frac{g d_p^3 \rho_f (\rho_p - \rho_f)}{\mu_f^2}$$

where g is the gravitational constant, d_p is the particle diameter, ρ_p and ρ_f are the particle density and fluid density respectively, and μ_f is the fluid viscosity. This number represents the ratio between external (gravity and buoyancy) forces and internal viscous forces and it is a representation of the physical properties of the fluid-particle system. The Reynolds number, Re, is defined as:

$$\mathrm{Re} = \frac{\rho_f v d_p}{\mu_f}$$

where ρ_f is the fluid density, v is the velocity of the fluid, d_p is the particle diameter, and μ_f is the fluid viscosity. From the simple relationships between the Archimedes and Reynolds numbers, three zones are presented to determine the superficial velocity in a particle-fluid system and the corresponding type of flow [7]. These three zones are: 1) where the pickup velocity increases as the particles diameter increased due to an increase of the gravity force, 2) the pickup velocity increases as the particle diameter decreases due to increase in cohesive forces, and 3) where powders that are cohesive that are so high that the particles move as agglomerates rather than single particles [8]. The zones are not clearly labeled, but an overview of dual-phase horizontal flow is described in Figure 2.

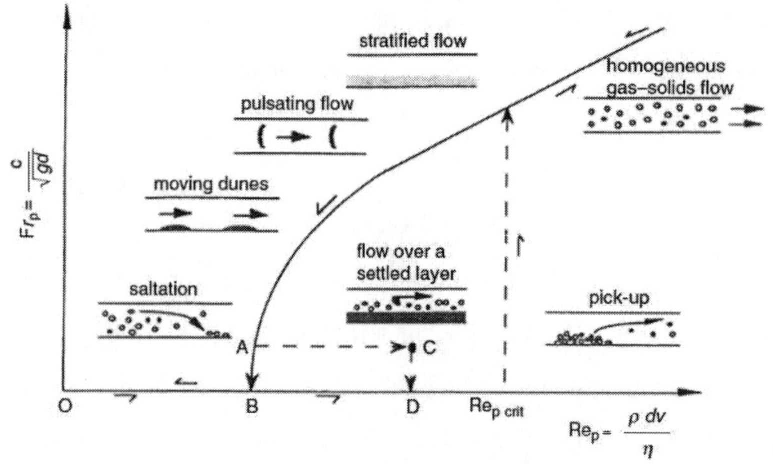

Figure 2. Model of flow regimes observed for horizontal pneumatic conveying including pick-up and saltation.

For a given flow rate equal to 0.004 kg/s (based on calculations to maintain optimal combustion) three different grid sizes were tested at 2 MPa and three different pressures were tested with the most suitable grid. The coal particles initially filled up the vertical standpipe and were pulled down by gravity to the horizontal section. The CO_2 then entrained the dropping coal and transported it towards the end of the feed system. The initial pressure was constant throughout the piping. A pressure boundary condition was set at the outlet of the feed system maintaining the given initial pressure. A special boundary condition was set up to maintain constant flow of coal while also reducing the need for a taller or wider standpipe. When the coal left the end horizontal section, it was recycled back into the top of the standpipe. Each case ran for ten seconds and depending on the grid size, took anywhere from a few hours to a few days to run.

3. Results and Discussion

3.1 Gridding

Three different grids were tested to observe the differences in steadiness and uniformity. In each case, all other parameters were kept the same except for the number of grid cells. The pressure was kept at 2 MPa for Cases 1-3. Case 1 was considered the base case. Case 2 kept the same volume of the pipe occupied by particles, but the grid was refined to double the amount of Case 1 cells. Case 3 quadrupled the amount of cells in the base case. Table 1 summarizes each of the first three cases:

Table 1. Summary of the three cases varying grid size and number of cells all at 2 MPa

						Average Flow Rate (kg/s)		
Case	Total Number of Cells	Real Cells	Null Cells	Numerical Particles	Total Mass (g)	CO2	Coal	Pressure Drop (Pa)
1	2E+06	115205	1790707	7.59E+05	2.15	0.004	0.00360	535
2	4E+06	222993	3558087	1.53E+06	2.19	0.004	0.00360	550
3	8E+06	431684	7173028	3.05E+06	2.63	0.004	0.00635	1006

Figure 3 shows the computational grids at the junction of the standpipe and horizontal duct for each of the three cases. For Case 1, each cell size was equal to 0.50 mm and along the diameter of the pipe there were 13 cells; vertically there were 257 cells and horizontally there were 618 cells. For Case 2, each cell was equal to 0.40 mm and along the diameter of the pipe there were 16 cells; vertically there were 324 cells and horizontally there were 778 cells. For Case 3, each cell was equal to 0.32 mm and along the diameter of the pipe there were 20 cells; vertically there were 408 cells and horizontally there were 981 cells. The cell size is important to know because it defines the limits of how many particles are able to fit inside one cell. The cell size does not have to be significantly larger than the largest particle size, but it should allow for four to five particles across the cell. Case 3 has the smallest cell size and pushes the limit of the cell size to particle size ratio.

All other things being equal, it is unclear how for Case 3, there is 22% more mass than Case 1 when the close pack volume fraction, geometry, and the initial and boundary conditions are all the same. Coal was specified to fill the entirety of the standpipe and each case should have the same amount of mass. The flow rate of coal for Case 3, 0.00635 kg/s is almost double the flow rate for Cases 1 and 2, 0.00360 kg/s. The CO_2-to-coal ratio is equal to 1.11 for Case 1 and 2, and this falls within the one to two range that will be acceptable for combustion in the reactor. The CO_2-to-coal ratio is equal to 0.630 for Case 3. This will lead to a fuel rich coal stream entering the reactor and will cause problems further down the reactor system.

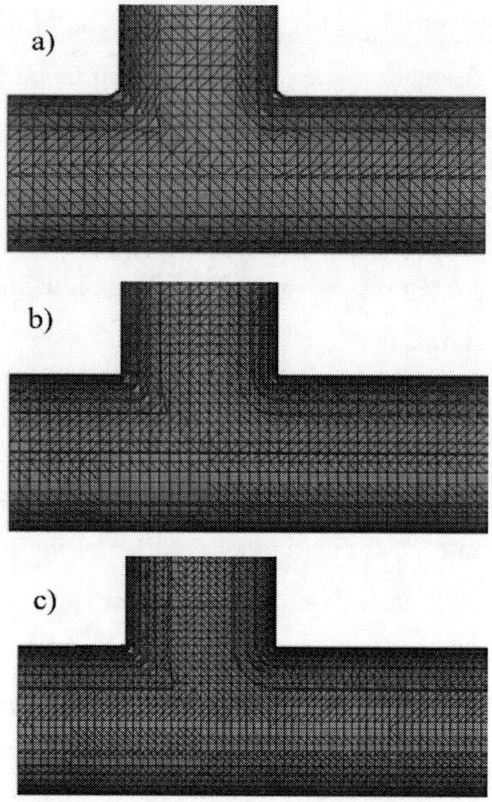

Figure 3. Snapshots of the gridding for each case: a) Case 1, b) Case 2, c) Case 3

For a grid, it is important to capture all the important geometry. Luckily, this geometry was simple and no special grid lines or precautions needed to be taken. Another important grid characteristic is for it to have enough resolution to accurately calculate the gas-particle dynamics. A grid should not have more cells than necessary to produce accurate results. But a grid must also avoid being too coarse in the number of cells it contains.

Figure 4 helps summarize the results from running all three grid cases.

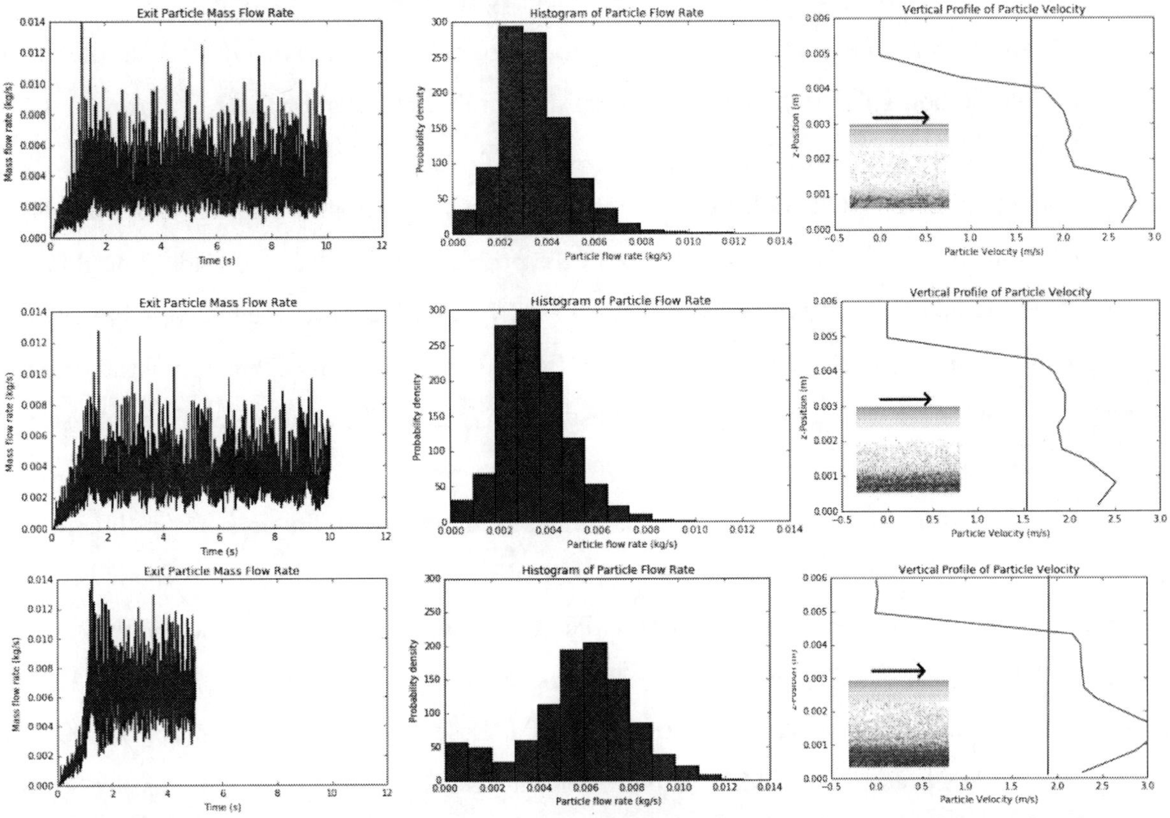

Figure 4. Three plots from each case showing the variation of the mass flow rate of the coal with respect to time (first column), a histogram of the probability density of the coal mass flow rate with respect to coal mass flow rate over entire time period through a flux plane at the end of the system (second column), and the velocity of the coal along the vertical diameter of the pipe at the end of the feed system (third column); the overall average velocity is shown in red. Top row: Case 1. Middle row: Case 2. Bottom row: Case 3.

For each of the three cases the average flow rate of the coal is approximately the same (See Table 1). The pressure drop recorded in the table is the difference of the average pressure over the course of the ten simulated seconds. There is no coal initially in the horizontal section and for each case it takes about 1.5 to 2 seconds to reach a steady state (See Figure 4). Pulsing will always occur with gas-particle flow, but based on the histogram plots, it is safe to assume that the flow is steady state. The flow seems to fluctuate quite a bit, but the second column of plots helps break down the first column of plots.

From the histogram plots, well over the majority of the flow rates recorded fall within the range of 0.002 to 0.005 kg/s. A flow rate above 0.005 kg/s is caused by minor slug formations and dunes. The coal particles will build up at the beginning of the duct and then suddenly be pushed forward, causing for a short time more coal flowing than CO_2. Again, Case 3 is the exception and most of the flow falls within the range from 0.004 to 0.008 kg/s.

Regarding run times, Case 1 and Case 2 both took a little less than a day to run the full 10 seconds, about 20 hours for each one. Because of the number of cells for Case 3, to run out to only 5 seconds, it took 2.5 days. With regards to other simulations from the literature, Case 3 is a very refined grid [4, 5, 6]. Others have been able to match simulations with their experiments with much coarser grids. This may be the root of the problem and too fine of grids should be avoided for skewed results as well as long run times. Typically the finer the grid, the better the results. This seems to not be the case because the flow turns into a dense phase flow rather than a dilute phase flow in Cases 1 and 2. Further investigation is needed in this grid refinement study.

3.2 Pressure

After choosing to use the coarser grid, the base case (Case 1), the pressure in the pipe was varied to see the effect pressure had on the flow rate of the gas and particles. A realistic pressure was chosen to be atmospheric pressure of 100 kPa. The other pressure was at the halfway point between the base case, 2 MPa, and atmospheric pressure, 1 kPa. Case 1 is the same Case 1 from the gridding cases and has the same pressure of 2 MPa. Case 4 was run at a pressure of 100 kPa. Case 5 was run at a pressure of 1 MPa. Table 2 summarizes the results from the pressure-varying simulations.

Table 2. Summary of the different pressure simulations.

		Average Flow Rate Leaving System (kg/s)				
Case	Pressure (Pa)	CO2	Coal	Pressure Drop (Pa)	Density (kg/m^3)	Average Velocity (m/s)
4	100000	0.004	0.00601	16000	1.7644	14.78
5	1000000	0.004	0.00441	13000	17.644	3.9445
1	2000000	0.004	0.0036	535	35.288	1.56

Based on the pressure drop, for Case 1 from the beginning of the horizontal section to the end of the feed system, it has a 0.03 percent pressure decrease dropping from 2,000,535 to 2,000,000 Pa. Case 4 is quite different; its pressure drops from 115,000 to 100,000 Pa. This is a 15 percent decrease in pressure. Case 5 is similar to Case 4, but the pressure drops from 1,013,000 to 1,000,000 Pa. This is a 1.3 percent decrease. The large pressure drop may be caused from too many particles in the standpipe. When the CO_2 and coal are recycled back to the top of the standpipe, they have nowhere to flow and this could have caused a large increase in pressure all along the standpipe.

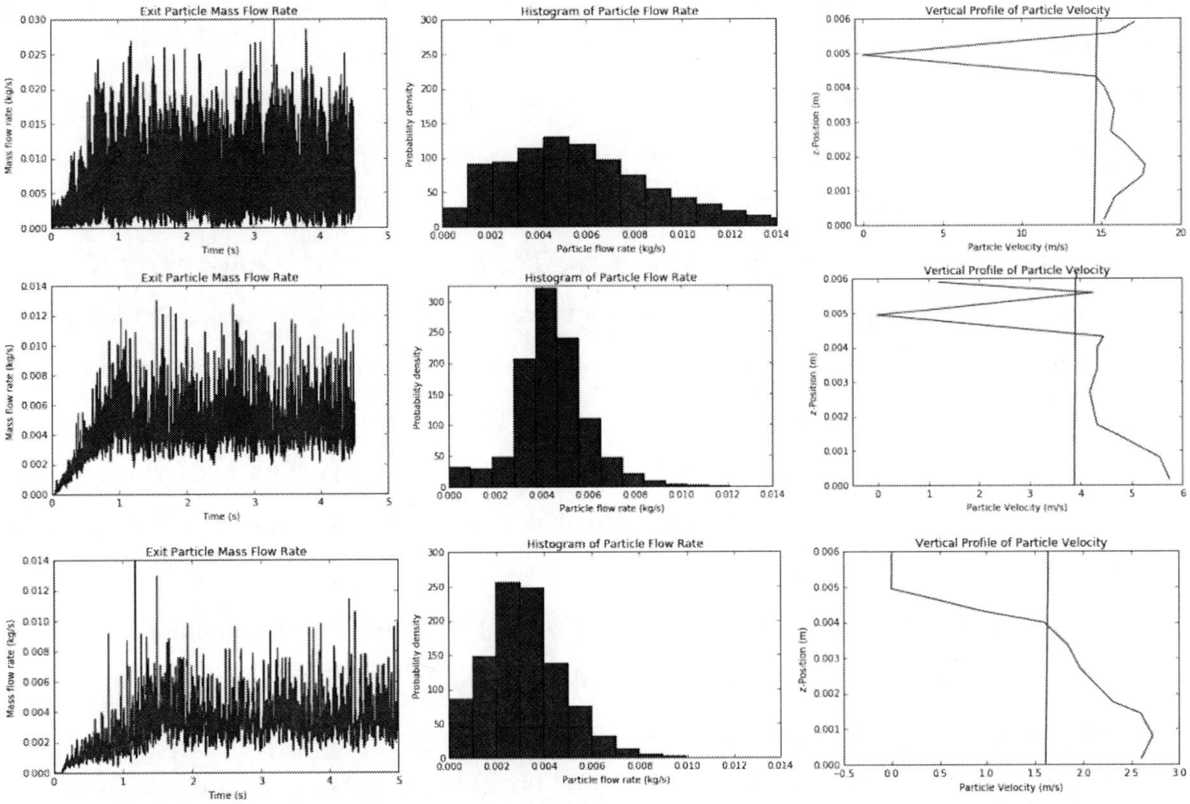

Figure 5. Three plots from each case (different pressures) showing the variation of the mass flow rate of the coal with respect to time (first column), a histogram of the probability density of the coal mass flow rate with respect to coal mass flow rate over entire time period through a flux plane at the end of the system (second column), and the velocity of the coal along the vertical diameter of the pipe at the end of the feed system (third column); the overall average velocity is shown in red. Top row: Case 4, 100 kPa. Middle row: Case 5, 1 MPa. Bottom row: Case 1, 2 MPa. Note the scaling is different for the first and last plot of the top row.

Due to the limit in time, Cases 4 and 5 were run out to only 4.5 seconds. From the plots it appears to reach steady state in about two seconds. But based on the velocity profile plots in the third column of Figure 5, it still seems there is a transient period of coal particles settling to the bottom of the duct. If the cases were able to run the full 10 seconds, perhaps the upper two plots would resemble the bottom plot. Case 1 was able to run out to the full 10 seconds, but the plot in Figure 5 is only based on the first five seconds.

Looking at the right column of graphs, this shows the velocity profile of the particles leaving the feed system. Case 4, except for one data point, approaches homogenous flow (See Figure 2). As the pressure is increased from 100 kPa to 1 MPa and then 2 MPa it is apparent from the velocity profile that the particles start falling to the bottom of the duct and become more of a stratified flow.

4. Conclusions

After analyzing the three different grid cases, Cases 1 (100,000 real cells) and 2 (200,000 real cells) yield similar results, while Case 3 (400,000 real cells), which has a finer grid is different from the other two. For reasons unknown, Case 3 is the outlier, because as a grid is more refined the results should converge. A more thorough grid study is needed to determine if Case 3 is accurate in its predictions or not. Previous work has been done with smaller gridding to make sure the gas-particle dynamics were consistent. The geometry was the same, but other modeling parameters were slightly different and they are not included in this study. For future work, both coarser and finer grids will be investigated to see if they maintain the results found in Case 1.

The three different pressure cases (Case 4, 5, and 1) each produced different results. It is clear that as the pressure increases the coal mass flow rate and velocity decrease. This is largely due to the increasing density as the flow rate of the incoming CO_2 remains constant. At atmospheric pressure, the flow is relatively uniform, but at higher pressures uniformity is not maintained.

The feed system will operate at a high pressure of 2 MPa and the feed system seems capable of delivering a steady amount of coal at this pressure. Given the dimensions and current flow rate parameter of CO_2, by the time the coal reaches the end of the pipe, the majority of it is along the bottom half.

The default values for hard particles were chosen for the particle-particle and particle-wall interactions. A sensitivity study will take place altering slightly the maximum momentum redirection from collisions, as well as the normal-to-wall and tangent-to-wall momentum retention values. Another sensitivity study that could be done is looking at different drag models that take into account a sphericity value that is less than one.

Future planned experiments geared specifically to this geometry will develop confidence in this Barracuda model.

5. Acknowledgements

This material is based upon work supported by the Department of Energy under Award Number DE-FE0029157. The authors are grateful to Sam and Rosemary Clark and others with CPFD Software for their support and assistance in using Barracuda Virtual Reactor.

Disclaimer: This report was prepared as an account of work sponsored by an agency of the United States Government. Neither the United States Government nor any agency thereof, nor any of their employees, makes any warranty, express or implied, or assumes any legal liability or responsibility for the accuracy, completeness, or usefulness of any information, apparatus, product, or process disclosed, or represents that its use would not infringe privately owned rights. Reference herein to any specific commercial product, process, or service by trade name, trademark, manufacturer, or otherwise does not necessarily constitute or imply its endorsement, recommendation, or favoring by the United States Government or any agency thereof. The views and opinions of authors expressed herein do not necessarily state or reflect those of the United States Government or any agency thereof.

6. References

[1] Tyler Hodge. "Coal May Surpass Natural Gas as Most Common Electricity Generation Fuel This Winter." *U.S. Energy Information Administration* – EIA – Independent Statistics and Analysis (2016).

[2] Jongsup Hong, Gunaranjan Chaudhry et al. "Analysis of oxy-fuel combustion power cycle utilizing a pressurized coal combustor." *Energy*, 34 (2009) 1332–1340.

[3] W. K. Hiromi Ariyaratne, Chandana Ratnayake, and Morten C. Melaaen. "Application of the MP-PIC method for predicting pneumatic conveying characteristics of dilute phase flows." *Powder Technology*, 310 (2017) 318-328.

[4] Qinggong Wan, Hairui Yang, et al. "Application of CPFD method in the simulation of circulating fluidized bed with a loop seal, part I-Determination of modeling parameters." *Powder Technology*, 253 (2014) 814-821.

[5] Yongshi Liang, Yongmin Zhang et al. "A critical validation study on CPFD model in simulating gas-solid bubbling fluidized beds." *Powder Technology*, 263 (2014) 121-134.

[6] Alireza Abbasi, Paul E. Ege, Hugo I. de Lasa. "CPFD simulation of a fast fluidized bed steam coal gasifier feeding section." *Chemical Engineering Journal*, 174 (2011) 341-350.

[7] Haim Kalman, Andrei Satran, Dikla Meir, Evgeny Rabinovich. "Pickup (critical) velocity of particles" *Powder Technology*, 160 (2005) 103–113.

[8] G.E. Klinzing, F. Rizk, R. Marcus and L.S. Leung. *Pneumatic Conveying of Solids: A theoretical and practical approach*, 3E (2009) 187-188

Western States Section of the Combustion Institute – Fall 2017 Meeting
Hosted by the University of Wyoming
October 2-3, 2017

A new efficient model for multicomponent membrane separation and application to the Argon Power Cycle

Farouk Chourou[1,2,3], *Miguel Sierra Aznar*[1,3], *Jyh-Yuan Chen*[1] and *Andreas Dreizler*[2]

[1]*Department of Mechanical Engineering, University of California at Berkeley, United States*
[2]*Fachgebiet Reaktive Strömungen und Messtechnik (RSM), Center of Smart Interfaces (CSI), Technische Universität Darmstadt, Germany*
[3]*Noble Thermodynamic Systems Inc., United States*

Carbon capture has been deemed crucial by the Intergovernmental Panel on Climate Change if the world is to achieve the ambitious goals stated in the Paris agreement. Emerging power technologies that integrate carbon capture such as the Allam Cycle and the Argon Power Cycle are gaining momentum. In the case of the later, polymeric membrane separation is a promising technology for cost-effective integration of carbon capture onto large size internal combustion engines used today to compensate for intermittent renewable sources. In an effort to support the development and integration of this technology, this paper presents and discusses a newly developed 1D-model for hollow fiber membrane gas separation that provides the user with the crucial process parameter for further integration into process optimization software. Unlike the majority of published models, this new model accounts for the pressure drop across the retentate and pressure build-up on the permeate side of the membrane. Moreover, momentum and concentration transport phenomena as well as energy are modeled. The Finite Difference Method in conjunction with the Cantera toolkit was chosen to solve the system of differential equations. The model is validated against publicly available experimental data and agreement falls within 3% error margins. Finally, the model has been used to evaluate the performance of the membrane units on the closed loop Argon Power Cycle.

1. Introduction

Carbon capture has been deemed crucial by the Intergovernmental Panel on Climate Change if the world is to achieve the ambitious goals stated in the Paris agreement. Amine-based carbon separation is a well developed carbon capture technology with large scale pilot projects already operational. However, this technology is capital intensive and adds substantially to the operational cost of the power generation plants as described by Yang et al. in [1]. Polymeric membrane separation, on the other hand, is a promising technology for cost-effective integration of carbon capture but it is yet to prove its effectiveness to gain the favors of the power generation sector [2]. The separation efficacy of membrane systems are strongly dependent on the gas stream pressure and the CO_2 concentration. As a result, membranes are not optimal for natural gas power systems whose exhaust streams are often at little above atmospheric pressure and with CO_2 concentrations well bellow 10%. Yet, polymeric gas membrane separation is widely used in the power and industrial sectors, from refinery hydrogen recovery

(H_2-HC), syn-gas upgrading (H_2-CO), sour gas treating (H_2S-HC), natural gas dehydration (H_2O-HC) to enhanced oil recovery (EOR) and landfill gas upgrading (CO_2-CH_4) proving their efficacy if the conditions are right. Polymeric gas membrane separation presents the advantage, compared to adsorption, absorption or cryogenics, of low capital investment, ease of operation and low energy consumption [3, 4]. The Argon Power Cycle (APC), an emerging power cycle technology developed at UC Berkeley, makes use of membrane separation carbon capture as an intrinsic component of the system in an effort to render a system in which carbon capture and power generation processes work synergistically to deliver a cost effective emissions free power system. A simplified scheme of the power system with emphasis on the gas separation process is depicted in fig. 1 .

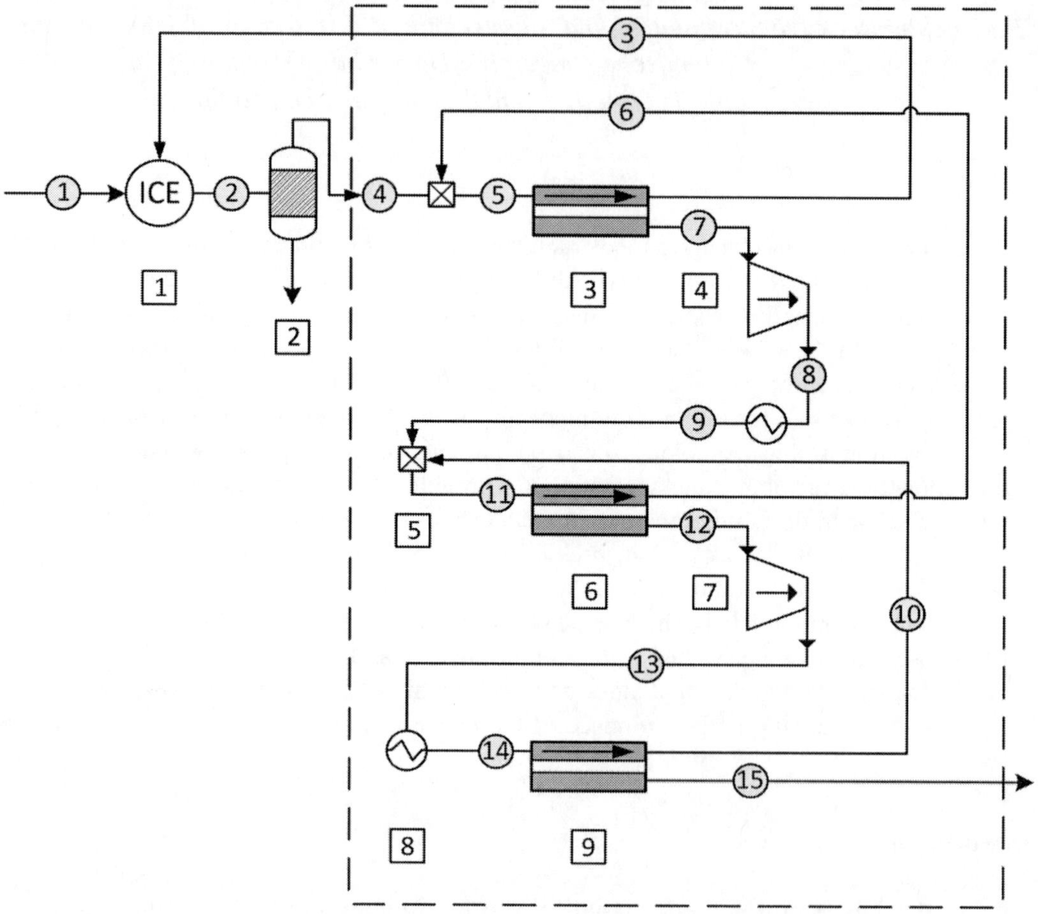

Figure 1: Simplified Argon Power Cycle process flow diagram.

The scientific foundation behind the APC increase is the influence the specific heat ratio (κ) has on the thermodynamic efficiency of internal combustion engines. Since engines typically use air as a the oxidizer, the specific heat ratio of the working fluid is dominated by that of nitrogen. Ideal working fluids with the highest possible specific heat ratio are monoatomic (noble) gases with only transnational modes of molecular motion. Gases such as helium, neon and argon have a specific heat ratio of 1.667, 20% higher than that of air (1.4). This difference in κ translates into higher temperatures and pressures conditions reached within the engine cylinder, ultimately increasing the boundary work exerted over the piston. Previous research by Aznar et al. [5] demonstrated a substantial increase in thermodynamic efficiency with CH_4-O_2-Ar in the order of 30%-37% improvement for port fuel injection (PFI), further extended to high compression ratio regimes through gaseous direct injection (DI). At this conditions higher efficiency gains (50%) were achieved. Research by Killingsworth et al. [6] and Kuroki et al. [7] focused on H_2-O_2-Ar combustion, and demonstrated large (30%-35%) relative efficiency gains for a given compression ratio under PFI operation. Unlike air, monoatomic gases are not freely available forcing the cycle to recirculate the working fluid in a closed loop configuration, also understood as 100% exhaust gas recirculation (EGR). This enables the APC system to freely increase its nominal pressure at no cost, while conserving its valuable working fluid. Although hydrogen is the perfect fuel for the APC, hydrocarbon fuels such as natural gas or biogas are economically more appealing. In the case of carbonaceous fuels, a membrane gas separation unit is required. This unit separates the resulting CO_2 from the exhaust gas and delivers the argon back to the engine intake. The high pressure of the exhaust stream and the closed-loop nature of the cycle makes this separation possible by providing a high-pressure gradient across the membrane unit and enabling an increased carbon dioxide concentration on the exhaust stream through its combined recirculation with argon.

2. Hollow fiber membranes

In an effort to prove the feasibility of the CO_2 gas separation under APC conditions, this paper presents a validated numerical model of a polymeric membrane unit, applied to APC operating conditions. In this work we have restricted ourselves to hollow fiber type membranes because they have proven effective at industrial scale gas separation applications [8]. Nonetheless, many different membrane materials and configuration have been investigated by the art, but to the best of the authors knowledge, no radically new membrane system have been recently introduced. Thus, we deem our scope of work appropriate for our specific purposes.

Hollow fiber membrane systems resemble tube and shell heat exchangers as can be seen in fig. 2. The working fluid or feed, enters on one end of the membrane unit, flows through the tube or shell side, and permeates through the tube walls. After permeation, the remaining flow, the retentate, leaves the unit through the other end. The fluid that has permeated, the permeate, leaves the unit through an outlet passage at either end of the membrane depending on whether it has been designed for counter current or co-current operation.

The total flux of gas across the membrane J, in kg m^{-2} s^{-1}, results from the sum of each individual species flux J_i, present in the feed and permeate streams. Equation 1 shows how the species permeation phenomena is driven by the partial pressure gradient of each species across membrane.

$$J = \sum_{i=1}^{N_c} J_i, \qquad J_i = P_i\left(X_i^r p^r - X_i^p p^p\right), \qquad i = 1, 2, ...N_c \tag{1}$$

where P_i is the species i permeability in mol m^{-2} s^{-1} Pa^{-1}, X_i^r and X_i^p are the mole fraction of species i at the retentate and permeate side respectively, and p^r and p^p are the pressures in Pa at the permeate and retentate side respectively. It can be easily observed that for efficient permeation, a combination of high pressure and concentration gradient operating conditions are desired.Moreover, regarding the membrane itself, two major performance indicators need to be taken into account to fulfill the gas separation requirements. The first one, the permeability, indicates the magnitude of a species flow permeating the membrane and thus its magnitude will determine the permeating area requirements for a membrane unit. The second one, the selectivity (γ), shown in eq. (2), is a non dimensional term

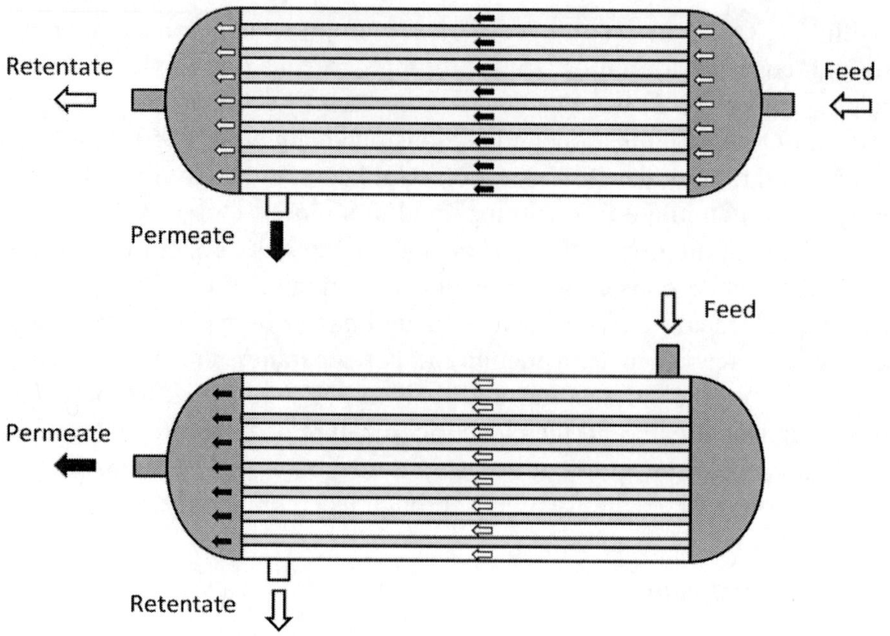

Figure 2: Representation of a hollow fiber membrane setup. Upper: Bore feed type; Lower: Shell feed type; White arrows indicate retentate flow, and black arrows indicate permeate flow.

indicative of how well the membrane permeates a specific species versus the rest of species present on the feed. In general, for cost-effective permeation, a combination of a high selectivity membrane, with substantial permeability for the species of interest and feed stream conditions of high pressure and species concentration is required.

$$\gamma_i = \frac{P_i}{P_{min}}, \qquad i = 1, 2, ...N_c \tag{2}$$

where N_c is the total amount of species present in the stream and P_{min} is the minimum within the group of all species permeabilities. Another important parameter regarding membrane operation is the stage cut ϕ, shown in eq. (3) for shell feed membranes. This parameter indicates the amount of permeating flow with respect to the feed stream volumetric flow rate entering the system.

$$\phi = \frac{Q^p}{Q^f} = \frac{u_z^p \pi R^{p2}}{u_{z,in}^r \pi \left[R^{o2} - R^{r2}\right]} \tag{3}$$

where Q^P and Q^f are the permeate and feed volumetric flow rates in $m^3\,s^{-1}$, u_z^p and u_r^p are the axial velocity in $m\,s^{-1}$ of the gas streams at the permeate and retentate side, and R^0, R^r and R^p are characteristic radiuses in m of the hollow fiber geometry as shown in fig. 3.
For more in depth description of the working principles of membranes, we invite the reader to consult [3, 8–10] among others, which provide in depth fundamentals of mass transfer through membranes and discuss the advantages and disadvantages of the different membrane material and system types.

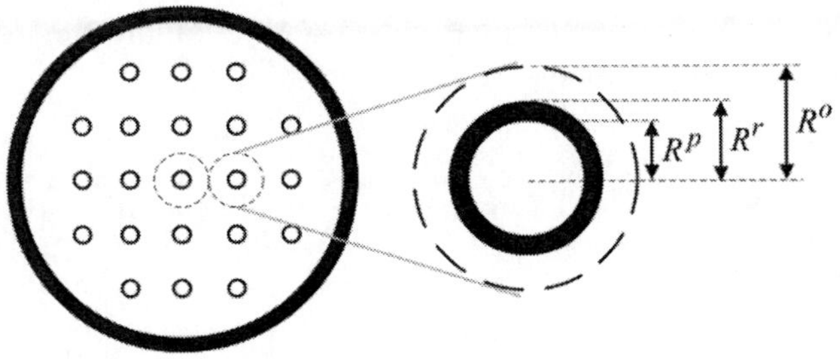

Figure 3: Cross section view of the fiber bundle in a shell feed hollow fiber membrane setup.

2.1. Numerical Model

In this model we reduce the entire membrane unit to a single hollow fiber equivalent, comprised of a hollow fiber with wall thickness $R^r - R^p$ and a cylindrical control volume of radius R^0. The model solves for a system of conservation equations, eqs. (4) to (8), at both sides of the membrane.

$$\frac{\partial(\rho)}{\partial t} + \frac{\partial(\rho u_z)}{\partial z} = \sum_{i=1}^{N_c} \dot{\omega}_i \tag{4}$$

$$\frac{\partial(\rho u_z)}{\partial t} + \frac{\partial(\rho u_z u_z)}{\partial z} = -f - \frac{\partial p}{\partial z} \tag{5}$$

$$\frac{\partial(\rho C_{p,mix} T)}{\partial t} + \frac{\partial(\rho C_{p,mix} T u_z)}{\partial z} = \frac{\partial}{\partial z}\left[k_{mix}\frac{\partial T}{\partial z}\right] + \sum_{i=1}^{N_c} \dot{\omega}_i h_i \tag{6}$$

$$\frac{\partial(\rho Y_i)}{\partial t} + \frac{\partial(\rho Y_i u_z)}{\partial z} = \frac{\partial}{\partial z}\left[D_i \rho \frac{\partial Y_i}{\partial z}\right] + \dot{\omega}_i, \qquad i = 1, 2, \ldots N_c - 1. \tag{7}$$

$$\sum_{i=1}^{N_c} Y_i = 1 \tag{8}$$

With the density ρ in $\mathrm{kg\,m^{-3}}$, momentum dissipation through friction f in $\mathrm{kg\,m^{-2}\,s^{-2}}$, explained in the next section, heat capacities at constant pressure $C_{p,mix}$ and $C_{p,i}$ in $\mathrm{J\,kg^{-1}\,K^{-1}}$ of mixture and single species, respectively, temperature T in K, thermal conductivity k_{mix} in $\mathrm{W\,m^{-1}\,K^{-1}}$, the specific enthalpy h_i in $\mathrm{J\,kg^{-1}}$, diffusion coefficient of specie i D_i in $\mathrm{m^2\,s^{-1}}$. Furthermore, we exploit the symmetries of the problem and reduce the system to a 1D problem along the z coordinate, in which the cylindrical geometry is accounted for by a geometric factor β in $\mathrm{m^{-1}}$. This factor is contained within the source term $\dot{\omega}_i$ in $\mathrm{kg\,m^{-3}\,s^{-1}}$, and acknowledges the difference in surface area at the interface between the retentate and permeate streams.

$$\dot{\omega}_i = \beta J_i \qquad \text{with} \qquad \beta = \begin{cases} \beta^r = -\frac{2R^r}{R^{o2} - R^{r2}} \\ \beta^p = \frac{2R^r}{R^{p2}} \end{cases} \tag{9}$$

The source term in this system of equations accomplishes the numerical coupling of energy and species conservation between the permeate and retentate. Notice, β takes a negative sign for the retentate, as flow is leaving, and positive for the permeate, as the flow is entering. Thermodynamic and transport property libraries are used to update the fluid properties changes along the membrane. Equally, a equation of state could be used, such as ideal gas law, to evaluate the density as function of pressure and temperatures of the gaseous stream.

$$p = \rho \left(\frac{R_u}{M_{mix}} \right) T \quad \text{with} \quad M_{mix} = \left(\sum_{i=1}^{n} \frac{Y_i}{M_i} \right)^{-1} \tag{10}$$

where R_u is the universal gas constant in J kg^{-1} K^{-1} and M_{mix} is the average molecular weight of the gas stream in kg mol^{-1}. fig. 4 illustrates the cross section of the hollow fiber across the length of the membrane, L in m. The superscripts r indicate retentate properties whereas the superscripts p indicates permeate properties. The control volume is indicated by dashed lines and arrows are used to indicate the direction of the flow fields. Boundary conditions are indicated at each end and side of the hollow fiber.

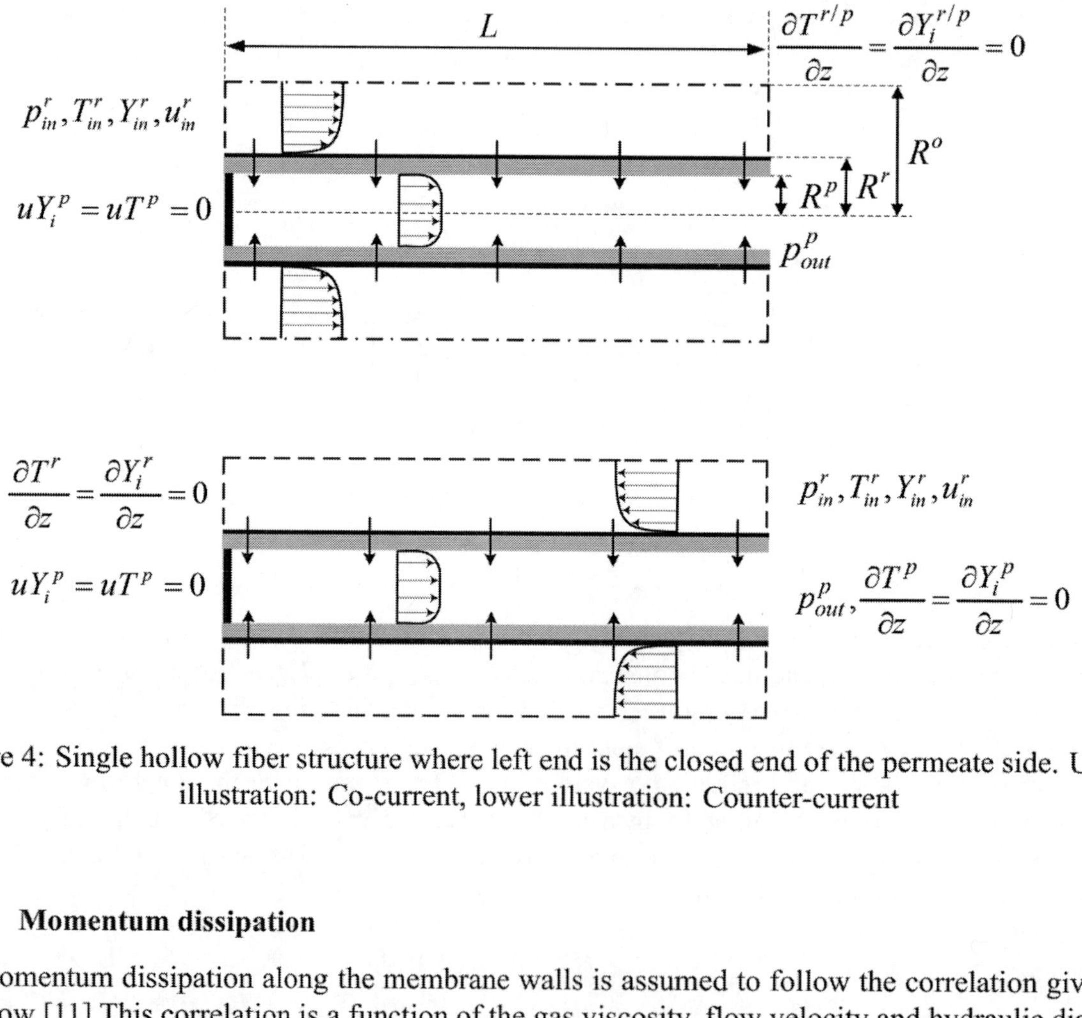

Figure 4: Single hollow fiber structure where left end is the closed end of the permeate side. Upper illustration: Co-current, lower illustration: Counter-current

2.1.1. Momentum dissipation

The momentum dissipation along the membrane walls is assumed to follow the correlation given for pipe flow [11].This correlation is a function of the gas viscosity, flow velocity and hydraulic diameter D_h in m of the flow passage.

$$f = \frac{32 u_z \nu_{mix} \rho}{D_h^2} = \begin{cases} D_h^r = 4[R^o - R^r] & \text{for retentate} \\ D_h^p = 2R^p & \text{for permeate} \end{cases}, \quad Re_{D_h} = \frac{D_h u_z}{\nu_{mix}} \tag{11}$$

Notice this correlation hold only for low Reynolds number, typically encountered in gas separation (with very small hydraulic diameters), i.e. $Re_{D_h} < 2100$ for pipe flow and $Re_{D_h} < 2000$ for annulus flow, [11]. The friction factor f, on the retentate side is multiplied by a factor (0.5), as only half of the parabolic velocity field is within our control volume. A slight discrepancy arises from the pipe flow correlation as it assumes the density is constant and that the flow to be developed. This assumption does not hold at the inlet region, estimated to be $0.035 D_h Re_{D_h}$ [11], due to the permeating phenomena along the membrane walls.

2.1.2. Joule-Thomson expansion

As the gas permeates through the membrane it undergoes an isenthalpic expansion from the feed pressure to that of the permeate. This effect induces a temperature change in the gas, which may drive temperature across the membrane down enough to the point of water condensation. This has a negative effect on the membrane performance, as water droplets quickly clog the membrane pores.

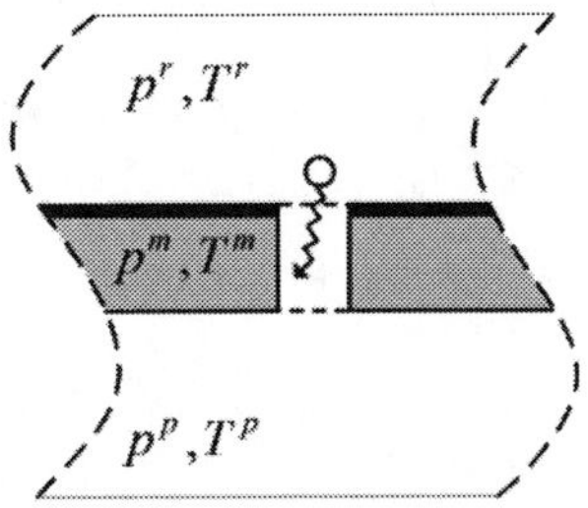

Figure 5: Illustration of molecule permeating from the retentate to the permeate through the membrane interface.

Although the pore clogging phenomena is not accounted for in our model, knowledge on whether condensation may occur is of critical relevance. Therefore, we evaluate the Joule-Thomson effect via the following assumptions: The pressure and temperature of a molecule right at the retentate interface is that of the retentate bulk stream, i.e p^r, T^r). Within the membrane, the isenthalpic expansion occurs, and the molecule is assumed to reach the permeate side at a pressure p^m equal to that of the permeate stream p^p and at a temperature T^m that results from the integration of eq. (12).

$$dh_i = C_{p,i}\left[dT - \mu_i dp\right] \tag{12}$$

where μ_i is the Joule-Thomson coefficient for species i in K Pa^{-1}. When eq. (12) is integrated between the retentate and permeate of the membrane, as shown in fig. 5, it yields:

$$h_{p,i} = C^m_{p,i} T^m = C^r_{p,i}\left[T^r - \mu_i p^r\right] + C^m_{p,i}\mu_i p^m \tag{13}$$

Finally, the overall temperature at a point z within the permeate $T^p(z)$ results from the solution of the energy equation considering the source term entering the permeate at T^m.

2.1.3. Simplified model

Several assumptions are made to effectively simplify the model: 1. The flow is steady; 2. Joule-Thomson effect is negligible on the axial direction; 3. The membrane geometry is fix and axisymmetric; it does not deform as demonstrated by Pan, [12]; 4.Permeabilities P_i are constant, and based on pure species permeation data. It may be expanded to be a function of pressure, temperature and gas composition if higher accuracy is required [13, 14];

With the aforementioned assumptions, and the different models for mass transfer eq. (1), momentum dissipation eq. (11) and isenthalpic expansion eq. (12), the system of eqs. (4) to (7) reduces to:

$$\frac{\partial(\rho u_z)}{\partial z} = \sum_{i=1}^{n} \beta J_i \tag{14}$$

$$\frac{\partial(\rho Y_i u_z)}{\partial z} = \frac{\partial}{\partial z}\left[D_i \rho \frac{\partial Y_i}{\partial z}\right] + \beta J_i \qquad \text{with} \qquad i = 1 \cdots N_c - 1 \tag{15}$$

$$\frac{\partial(\rho u_z u_z)}{\partial z} = -\frac{\partial p}{\partial z} - f \tag{16}$$

$$\frac{\partial(\rho C_{p,mix} T u_z)}{\partial z} = \frac{\partial}{\partial z}\left[k_{mix}\frac{\partial T}{\partial z}\right] + \sum_{i=1}^{n}(\beta J_i)h_i \tag{17}$$

with the following boundary conditions for the retentate side:

$$(u_z T)^r (z = 0) = (u_z T)^r_{in} \quad ; \quad \left.\frac{\partial (u_z T)^r}{\partial z}\right|_{z=L} = 0 \tag{18}$$

$$(u_z Y_i)^r (z = 0) = (u_z Y_i)^r_{in} \quad ; \quad \left.\frac{\partial (u_z Y_i)^r}{\partial z}\right|_{z=L} = 0 \tag{19}$$

$$p^r (z = 0) = p^r_{in} \tag{20}$$

and the following boundary conditions for the permeate side:

$$(u_z T)^p (z = 0) = 0 \quad ; \quad \left.\frac{\partial (u_z T)^p}{\partial z}\right|_{z=L} = 0 \tag{21}$$

$$(u_z Y_i)^p (z = 0) = 0 \quad ; \quad \left.\frac{\partial (u_z Y_i)^p}{\partial z}\right|_{z=L} = 0 \tag{22}$$

$$p^p (z = L) = p^p_{out} \tag{23}$$

Notice, the set eqs. (14) to (17) requires boundary conditions at both ends for temperature and mass fraction, whereas one boundary condition is sufficient for pressure and velocity. The following boundary conditions apply for the co-current configuration, though, by reversing the z-axis on the retentate side, they may be extended to counter-current configurations. As a result we obtain a non linear boundary value problem.

2.2. Numerical solution

The numerical solution of the boundary value problem is carried using Finite-Difference Method (FDM). The convective and diffusive term are discretized using 2nd order central difference scheme; eqs. (24) and (25), and using the trapezoidal rule to interpolate the source/sink term $\dot{\omega}$, eq. (26) at cell center $\tilde{i}$. First derivative boundary condition is enforced using ghost points, fig. 6. The advantage of this method compared to the classical FDM is its conservativeness [15].

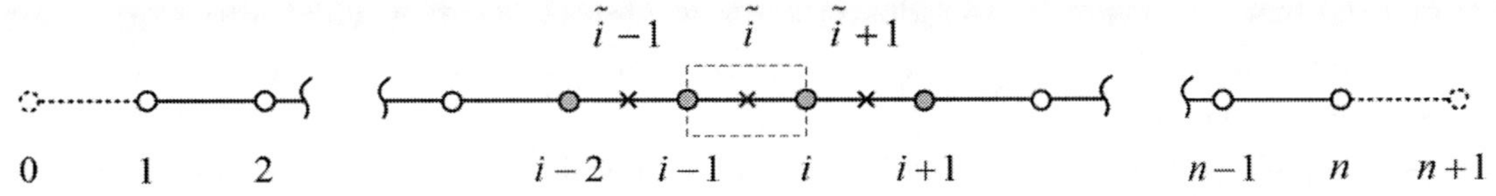

Figure 6: Numerical grid

$$\left.\frac{\partial u_z \phi}{\partial z}\right|_{\tilde{i}} = \frac{u_z^i \phi^i - u_z^{i-1} \phi^{i-1}}{z^i - z^i - 1} + \mathcal{O}(\Delta z^2) \tag{24}$$

$$\left.\frac{\partial}{\partial z}\left[\kappa \frac{\partial \phi}{\partial z}\right]\right|_{\tilde{i}} = \frac{1}{z^i - z^{i-1}}\left[\kappa^i \left[\frac{\phi^{i+1} - \phi^{i-1}}{z^{i+1} - z^{i-1}}\right] - \kappa^{i-1}\left[\frac{\phi^i - \phi^{i-2}}{z^i - z^{i-2}}\right]\right] + \mathcal{O}(\Delta z^2) \tag{25}$$

$$\left.\dot{\omega}\right|_{\tilde{i}} = \frac{\dot{\omega}^i + \dot{\omega}^{i-1}}{2} \tag{26}$$

Note that the used discretization is truly second order only in case of isotropic grid. The set of non linear algebraic equations obtained after discretization takes the following form:

$$\underline{\underline{A}}\left(\underline{x}\right) \underline{x} = \underline{b}\left(\underline{x}\right) \tag{27}$$

Where the matrix $\underline{\underline{A}}$ contains the discrete convective and diffusive term, the vector $\underline{x}$ the set of unknowns $\{u, y_i, T, p\}$ in the nodes noted i and the vector $\underline{b}$ the source/sink terms along with the boundary conditions. The solution of eq. (27) is carried using the *Picard-Iteration* using the following algorithm:

$$\underline{\underline{A}}\left({}^t\underline{x}\right) {}^{t+1}\underline{x}^* = \underline{b}\left({}^t\underline{x}\right) \tag{28}$$

eq. (28) is solved for ${}^{t+1}\underline{x}^*$ using direct matrix inversion using the LU decomposition as recommended by Schäfer [16] for low computational effort and memory requirement, which is then used to update ${}^{t+1}\underline{x}$ as following:

$${}^{t+1}\underline{x} = \delta \; {}^{t+1}\underline{x}^* + (1 - \delta) \; {}^t\underline{x} \tag{29}$$

With δ the relaxation factor, used to stabilize the solver for the unknown quantities $\{u, y_i, T, p\}$, in next denoted with the subscript u, y, T, p and the super script t specifies the iteration. The solution is computed using MATLAB and Cantera [17] module for MATLAB, combined with the GRI 3.0 mechanism [18] to compute the transport parameters. The following algorithm is used:

1. Solve the set of coupled eqs. (14) to (17) for the retentate and permeate side along with the boundary conditions eqs. (18) to (23) with the parameter ν_{mix}, k_{mix}, D_i, k_{mix}, $C_{p,mix}$, $C_{p,i}$ computed under retentate inlet boundary conditions, eqs. (18) to (20).

2. Compute the updated parameter ν_{mix}, D_i, k_{mix}, $C_{p,mix}$, $C_{p,i}$ for certain number of nodes and interpolate over the rest.

3. Solve the equations eqs. (14) to (17) using the parameters computed in Step 2.

4. reiterate over Step 2 and Step 3 until relative change in the transport properties in under a certain threshold.

This method has the advantage to reduce the computing time relied with calling the Cantera module at every iteration. No difference has been noticed between calculating the transport data for all nodes and the actual procedure.
Relaxation factors for the pressure, velocity, mass fraction and temperature were used and varied depending on the operating conditions to stabilize the solver, beginning with a value of 0.1 and ramped up to 1. Note that for certain cases, the use of a frozen solver for the energy equation, $\delta_T = 0$, for the first iterations was mandatory to achieve convergence. Convergence is ensured through monitoring the outlet velocity on the permeate side and the relative change in the solution for the set of unknowns between two consecutive Iterations ϵ, given by:

$$^{t+1}\epsilon_y = \left\| ^{t+1}\underline{y}_i - {}^{t}\underline{y}_i \right\| \tag{30}$$

with $||.||$ the euclidean norm.

2.3. Validation

All the considered membranes in this section are Shell side feed, and the shown results were investigated of grid convergence using a isotropic mesh refinement.

2.3.1. Air separation

Giglia et al. [4] considered synthetic air and Helium-Nitrogen separation using hollow membrane fibers in co-current and counter-current configuration. Experimental setup data along with uncertainty in the boundary conditions are shown in table 1. Feed inlet flow rate measurement accuracy was of $\pm 1\%$. Different stage cuts were achieved through changing the feed flow rate, which resulted in a different pressure profile. Species mole fraction and pressure at permeate closed end were measured using a gas chromatograph and a pressure transducer. In this configuration, an interesting phenomena in accentuated: the pressure build-up on the permeate side, which is due to the high fibers surface and permeability The results obtained through the model show excellent agreement with experimental data for high stage cuts. For low stage cuts, the deviation in the permeate velocity is in the order of 3%, in the concentration of 2% and in the pressure of 2%, outside of the error range (figs. 7 to 9). The model under predicts the pressure at permeate side in co-current configuration for low stage cuts for the following reason: At low stage cuts, the velocity profile over the membrane length is steeper than at high stage cuts, which leads to a higher velocity gradients in axial direction, in contradiction with the used correlation for momentum loss which assumes developed flow (section 2.1.1). An argument for that is also the low velocity on the outlet of the permeate side for the counter-current configuration, which leads to lower velocity gradients in the axial direction, and a better agreement in the pressure at the permeate closed end for the counter-current configuration. Giglia et al. [4] state that the diameter measurements uncertainty of the fiber may cause such discrepancies in the pressure; a relative change of 3.4% of membrane radius R^p at 5% stage cut may lead to a pressure decrease at permeate closed end of 7.6 kPa, as the momentum loss scales as $\propto 1/{R^p}^2$, eq. (11).

2.3.2. Carbon dioxide gas separation

Tranchino et al. [19] investigated CO_2 separation from methane using Polysulfone hollow fiber membranes in co-current and counter-current configuration. The results obtained, shown in fig. 10 are in

$X^r_{O_2,in}$	$0.2078 \pm 0.01\%$
$X^r_{N_2,in}$	$0.7922 \pm 0.01\%$
P_{O_2}	7.155×10^{-9} mol s^{-1} m^{-2} Pa^{-1}
P_{N_2}	1.255×10^{-9} mol s^{-1} m^{-2} Pa^{-1}
p^r	691 ± 7 kPa
p^p	103 ± 0.7 kPa
R^r	$186.5 \pm 5\%$ µm
R^p	$72.5 \pm 5\%$µm
R^o	≈ 268.3 µm
L	175 cm
N	80
T^r	298 K

Table 1: Synthetic air experimental setup [4]

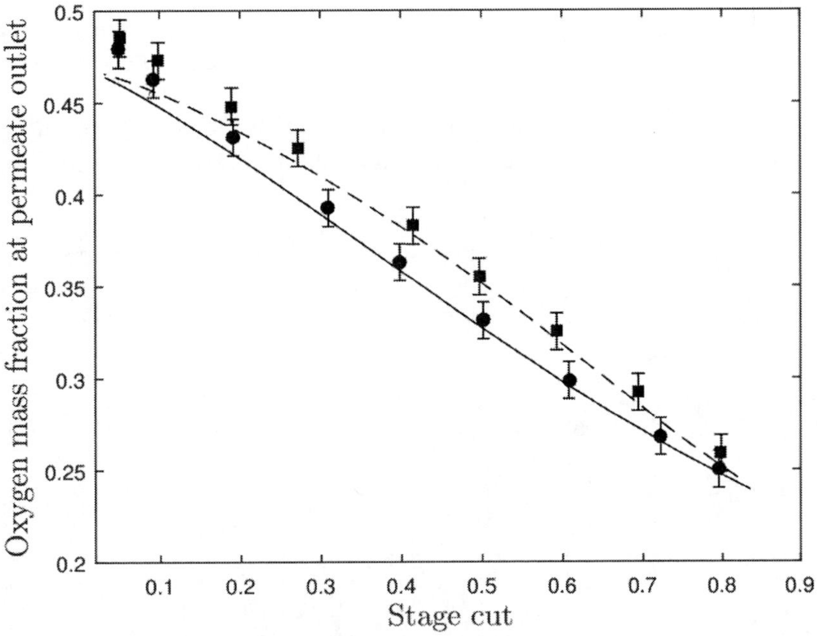

Figure 7: O_2 mole fraction at permeate outlet as function of stage cut for O_2-N_2 separation. (——,- - - -): Model, (•, ■): Experimental data for co-current and counter-current configuration, respectively

excellent agreement with the experimental data. Pressure build-up in the permeate side is negligible because of the low permeability of the membrane and the short length of the membrane, which resulted in the low velocity and hence low pressure build-up.

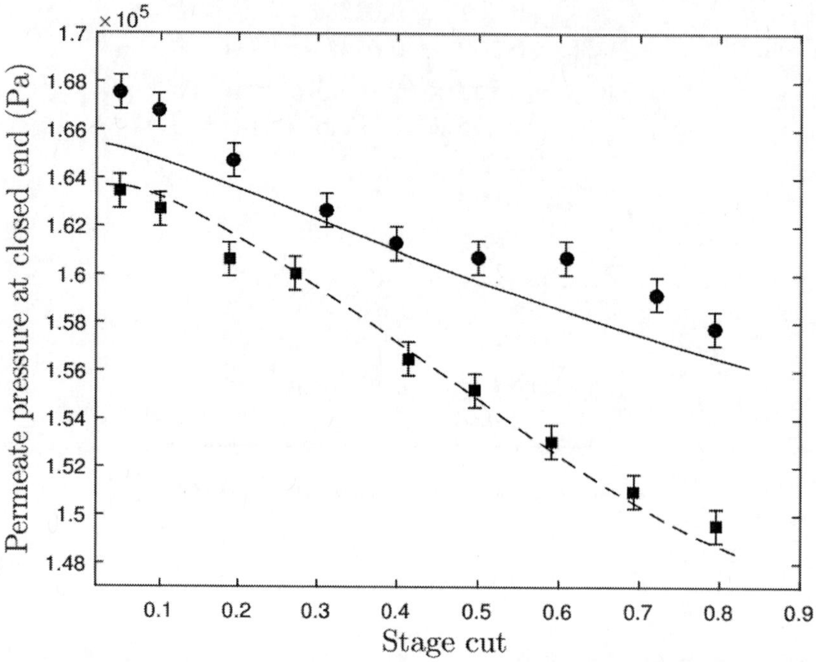

Figure 8: Pressure at fiber closed as function of stage cut for O_2-N_2 separation. (——,- - - -): Model, (•, ■): Experimental data for co-current and counter-current configuration, respectively

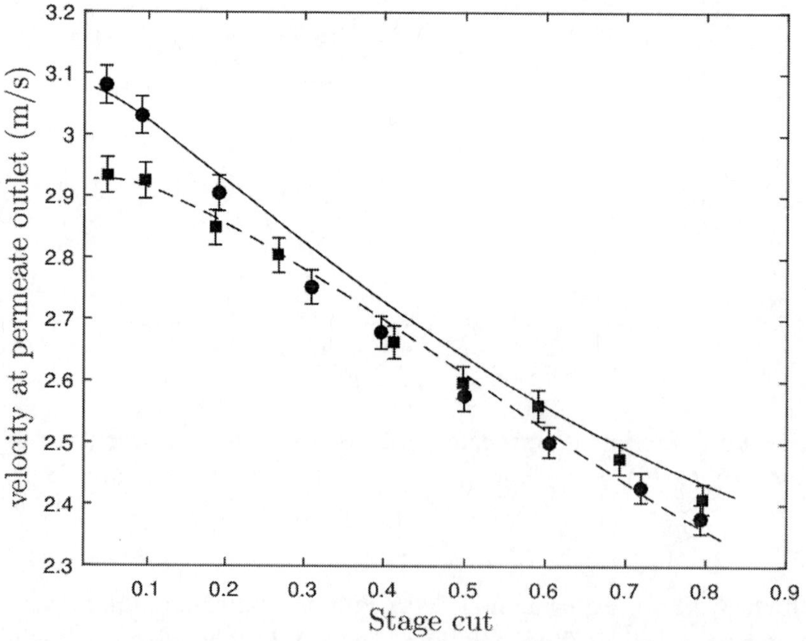

Figure 9: Velocity at permeate outlet as function of stage cut for O_2-N_2 separation. (——,- - - -): Model, (•, ■): Experimental data for co-current and counter-current configuration, respectively

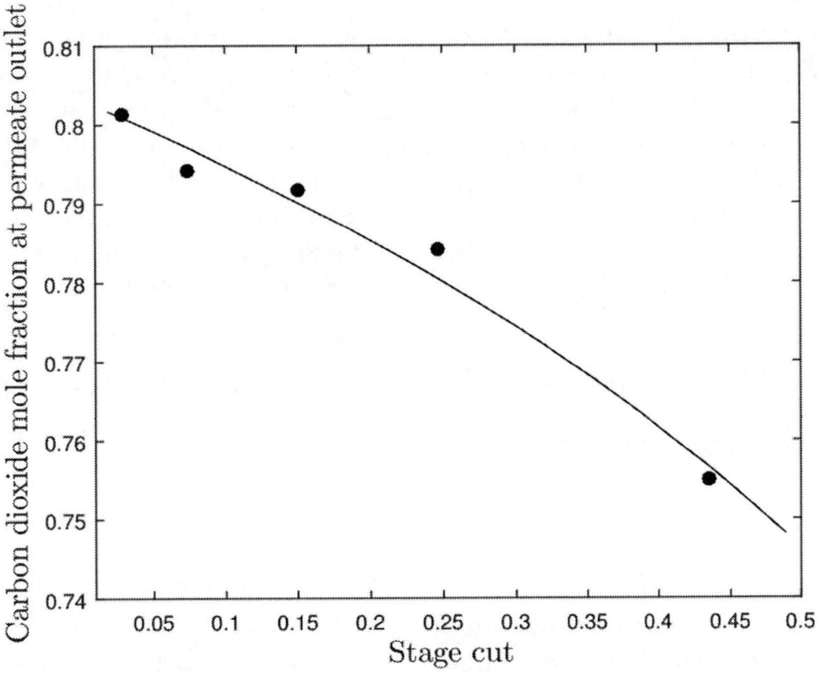

Figure 10: CO_2 at permeate outlet as function of stage cut for CO_2-CH_4 separation. (——): Model, (•): Experimental data

$X^r_{CO_2,in}$	0.6
$X^r_{CH_4,in}$	0.4
P_{CO_2}	$31.6 \times 10^{-10}\,\mathrm{mol\,s^{-1}\,m^{-2}\,Pa^{-1}}$
P_{CH_4}	$8.81 \times 10^{-10}\,\mathrm{mol\,s^{-1}\,m^{-2}\,Pa^{-1}}$
p^r	405.3 kPa
p^p	101.3 kPa
R^r	367.5 ± 4.5 µm
R^p	194.5 ± 6 µm
L	15 cm
N	6
T^r	301 K

Table 2: CO_2 Separation [19]

3. Application to the Argon Power Cycle:

Following figure 1, we evaluate the performance of the membrane system under expected APC operating conditions. The effect of cycle pressure, ζ and CO_2 concentration in the recirculating stream $X_{CO2,r}$ are studied in order to asses their influence on the energy consumption of the entire membrane gas separation system. For the purpose of this work we assume the conditions at node 4 in fig. 1 to be dry, at the engine boost pressure, and with an amount of CO_2 result from the sum of the recirculated stream and the products of complete stoichiometric combustion. Note, as indicated by Aznar et al. [5], that the optimum oxygen dilution is 15% and thus we maintain that parameter fix. In future work we will evaluate the influence of recirculated CO_2 on the optimum oxygen dilution. In general, we use an artificial working fluid with different amount of recirculated CO_2 such that at any given moment in the engine the following reaction occurs:

$$a\mathrm{CH_4} + \mathrm{bO_2} + \mathrm{cAr} + \mathrm{dCO_2} \longrightarrow \mathrm{eCO_2} + \mathrm{fH_2O} + \mathrm{gAr} \tag{31}$$

We continue by defining the pressure ratio ζ as follows

$$\zeta = \frac{p^r_{in}}{p^p_{out}} \tag{32}$$

and we restrict our work to a constant ζ over the whole separation process. Figure 11 shows the resulting energy consumption of the membrane setup for of a 1 MW engine operating at different conditions. The different conditions consist of three different pressure ratios: 4, 5 and 6; three different percentages of recirculated $X_{\mathrm{CO_2},r}$: 5%, 10% and 15% and three different engine operating efficiency η: 50%, 60% and 70%. Note that the compressors efficiency is kept constant at 90% over all the process conditions. The membrane characteristics used on this study are summarized in Table 3.

$\gamma_{\mathrm{H_2O}}$	9000
$\gamma_{\mathrm{CO_2}}$	1000
$\gamma_{\mathrm{O_2}}$	80
γ_{Ar}	60
$\gamma_{\mathrm{CH_4}}$	1
p^r	6,5,4 kPa
p^p	1 kPa

Table 3: Membrane characteristics.

Note, compressor (4) has, independently of the used parameters, a higher energetic consumption compared to compressor (7). This is due to the high volumetric flow rate at node 5. At low percentages of recirculated CO_2, the energy consumption E of the separation process reaches its peak for low pressure ratios while the opposite occurs at high recirculated CO_2, hence at low recirculation of CO_2, high pressure ratios are the key to a more efficient cycle. This is due to a balance between higher permeation efficiencies at higher compression ratios. At lower compression ratios, the stage cut increases and so the amount of argon to be compressed. This increase in argon flow rate offset the extra cost of few extra bars of compressions. In general, the separation process penalty on the APC is, at worse, less than 5% absolute efficiency for low CO_2 recirculation, and lower than 4% for high percentage of CO_2 recirculation. It is obvious, yet worth noticing, that the more efficient the engine performs the more cost effective is the separation process. This, in turn, means that a better optimization of the combustion process may yield indirect savings downstream, during the separation process.

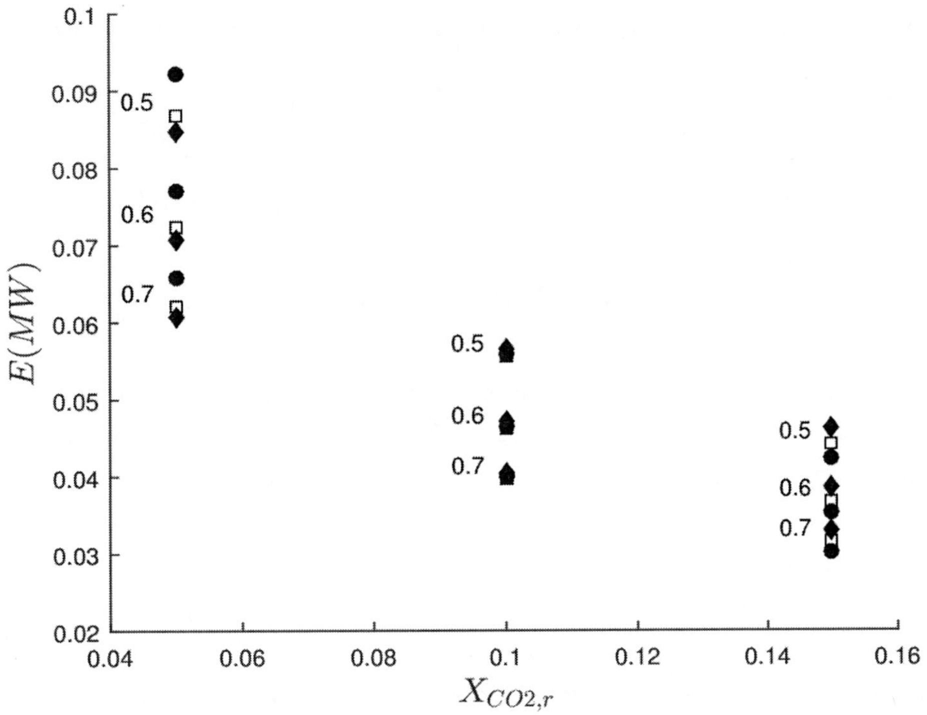

Figure 11: Energy consumption for the separation process. (•, ■, ♦): $\zeta = 4$, 5 and 6, respectively

4. Conclusions

A new efficient membrane gas separation model was developed and integrates models for all the relevant phenomena at stake under the context of the closed loop Argon Power Cycle. The model was first validated against experimental data and later used to model a a three stage membrane separation process that would achieve CO_2 separation from Ar at outlet CO_2 concentration sufficient for industrial standards ($> 95\%$). Results show a promising future for this technology, as the energy penalty is less than 5% for the investigated cases. With the immense gain in efficiency granted by the use of argon gas, this penalty represent to added cost to a system that can deliver electric power to the grid at current and higher efficiencies, yet at no cost to the environment or extra cost to the consumers.

Acknowledgements

We would like to acknowledge Membrane Technology and Research Inc. (MTR) for helping our understanding of polymeric membranes and their consideration to provide advice when it was needed.

References

[1] H. Yang *et al. Journal of Environmental Sciences,* 20 (2008) 14–27.

[2] E. Favre. *Journal of Membrane Science,* 294 (2007) 50–59.

[3] S. P. Nunes and K. V. Peinemann. *Membrane Technology in the Chemical Industry* (2016).

[4] S. Giglia, B. Bikson, J. E. Perrin and A. A. Donatelli. *Industrial and Engineering Chemistry Research,* 30 (1991) 1239–1248.

[5] M. S. Aznar *et al. Proceedings of the European Combustion Meeting 2017* (2017) 1–6.

[6] N. J. Killingsworth *et al. Proceedings of the Combustion Institute,* 33 (2011) 3141–3149.

[7] R. Kuroki, A. Kato, E. Kamiyama and D. Sawada. *SAE Technical Paper* (2010).

[8] J. Marriott. *Detailed modelling and optimal design of membrane separation systems* PhD thesis (2001).

[9] R. W. Baker. *Membrane Technology and Applications* (2012).

[10] W. L. McCabe, J. C. Smith and P. Harriott. *Unit Operations Of Chemical Engineering* (1993) 1–1113.

[11] R. B. Bird, W. E. Stewart and E. N. Lightfoot. *Transport Phenomena* (2007) 1–905.

[12] C. Y. Pan. *AIChE Journal,* 29 (1983) 545–552.

[13] S. P. Kaldis, G. C. Kapantaidakis, T. I. Papadopoulos and G. P. Sakellaropoulos. *Journal of Membrane Science,* 142 (1998) 43–59.

[14] S. P. Kaldis, G. C. Kapantaidakis and G. P. Sakellaropoulos. *Journal of Membrane Science,* 173 (2000) 61–71.

[15] C. Hirsch. *Numerical Computation of Internal and External Flows* (2007) 1–412.

[16] M. Schäfer. *Computational Engineering — Introduction to Numerical Methods* (2006) 1–321.

[17] D. Goodwin, H. Moffat and R. Speth. *Cantera: An object-oriented software toolkit for chemical kinetics, thermodynamics, and transport processes* 2016.

[18] G. P. Smith *et al. Gas Research Institute (http://www. me. berkeley. edu/gri_mech)* (1999).

[19] L. Tranchino, R. Santarossa, F. Carta, C. Fabiani and L. Bimbi. *Separation Science and Technology,* 24 (1989) 1207–1226.

Sub Topic: Innovative Technologies

Western States Section Combustion Institute 2017 Fall Meeting
October 2-3, 2017 • Laramie, Wyoming

Kinetic Study of Ethane Pyrolysis in an Internal Combustion Engine

Z. Taie, D. Wagner, C. Hagen*

School of Mechanical, Industrial and Manufacturing Engineering,
Oregon State University, Bend, OR, USA
Corresponding Author Email: taiez@oregonstate.edu

Abstract: Oregon State University is investigating the use of Internal Combustion (IC) engines in conjunction with catalysts as intensified, distributed, scalable reactors for ethylene production. In this work, we investigated whether using an unmodified IC engine could replicate the yield seen in the more complex catalytic system. The engine reactor was modeled using Chemkin routines and appropriate chemical kinetic models. Investigations included admitting pure ethane, air-ethane, methane-ethane, and steam-ethane mixtures, all at various preheat temperatures. The diluent gasses were necessary to increase the specific heat ratio of the mixture, such that an engine with a reasonable compression ratio (<40) could raise the mixture to pyrolysis temperatures. Modeling revealed that large dilution percentages (<1% ethane) and high preheat temperature (>400 °C) were necessary for substantial ethylene yield. However, high dilution and preheat would result in low ethylene throughput and increased operating costs, making the technology unattractive relative to large commercial cracking plants.
Keywords: Ethylene Production, Ethane Pyrolysis, Internal Combustion Engine

1. Introduction

The consumption of ethane in the United States is projected to double between 2016 and 2018, primarily driven by an increase in demand for plastic, of which ethylene is a major feedstock. Today, there exists a need for more ethylene production capacity in the United States. The current technology fleet is primarily composed of large ethane cracking plants which are expensive, and have long construction lead times. Building more of these plants cannot satisfy ethylene demand in the near term, and is an expensive bet that ethylene demand will remain high. If possible, small, modular, scalable reactors that have low capital expense and short construction times would help to alleviate the current capacity issue. There are many ways to envision how these small modular plants could be utilized. They could be operated in large banks as a stand-alone reactor plant, they could be incorporated into existing chemical processing plants that have extra balance of plant capacity, or they could be located near economically stranded natural gas well sites. However, in order for these reactors to be possible, new highly intensified chemical processes and reactor technology must be developed.

By design, engines act as variable volume chemical reactors that efficiently control reactant delivery and product exhaust at speeds higher than traditional reactor technology. Potentially, this increase in gas handling capability could allow for increased chemical throughput compared to larger chemical cracking plants [1]. This, combined with high temperatures and pressures already created inside of a reciprocating internal combustion engine could provide an environment that is favorable for pyrolysis reactions, especially when aided by a catalyst.

Previous work, conducted during this same project, investigated placing a solid phase catalyst, designed to assist in ethylene selectivity, inside of the engine. Two catalyst locations are currently under investigation: 1) inside of the engine cylinder, and 2) inside of the engine head, directly after the cylinder exhaust valves. Neither of these catalyst locations, nor the performance of the catalyst itself has been experimentally tested. For these reasons, numerical investigations into the possibility of operating a reciprocating internal combustion engine (ICE) without catalytic assistance for the purpose of producing ethylene was conducted. If possible, traditional ICEs could be retrofitted to provide a cheap and robust intensified chemical reactor that can be easily scaled to meet varying demand.

2. Initial Investigation

A literature investigation showed the ignition delay of ethane to be sufficiently short at about 1000 K [2]. Sufficiently short was assumed to mean that the ethane cracking reaction would be able to happen within a few crank angle degrees (~ 10 ms) in an engine that was running at 1000 RPM. This determination was selected so that the pyrolysis reaction could occur in the time frame where the pressure and temperature within the engine cylinder were maximized near top dead center. According to the literature, the pressure had little to no effect on the reaction rates [3]. The major driver of the pyrolysis was the gas temperature. For this reason, it was determined that the engine reactor must reach 1000 K in the cylinder in order to initiate ethane conversion.

Initially, the hope was to intake pure ethane (in order to maximize the possible ethylene yield) at standard conditions (to minimize amount of input energy). However, initial Chemkin simulations revealed that an ICE with a typical compression ratio was not able to raise the ethane to 1000 K. Upon investigation, the ratio of specific heats for ethane at 300 K was discovered to be 1.185, which is substantially lower than that of air (1.4 at 300 K) [4]. For an ideal gas undergoing isentropic compression, the temperature after compression is given by the following equation

$$T_2 = T_1(CR)^{\gamma-1}$$

where CR is the compression ratio of the engine, T1 and T2 are the temperatures of the gas at BDC and TDC respectively, and gamma is the ratio of specific heats of the gas being compressed (in this form, typically evaluated at the average temperature being investigated). From this equation, it can be seen that for an engine with a fixed CR the ratio of specific heats and the inlet temperature are what determine the peak in-cylinder temperature.

An isentropic compression calculation was conducted on the pure ethane with an initial temperature of 300 K to determine what compression ratio was needed to reach a final temperature of 1000 K at TDC (evaluating the ratio of specific heats at 300 K). This resulted in a necessary CR of 661 which is not feasible in an ICE. Compounding the problem, the ratio of specific heats for ethane decreases with increasing temperature, requiring an even larger CR to reach the desired reaction temperature. Thus, the need to dilute the ethane with a gas that has a larger ratio of specific heats was necessary.

3. Air and Ethane Mixture

The most abundant and inexpensive gas is air. Thus a parametric study was run to determine the CR needed to compress a given mixture of air and ethane to 1000 K. The results of this study are shown in Figure 1. The x-axis shows the mixture composition in terms of the equivalence ratio for ethane. The black line assumes that the ratio of specific heats during the entire compression was equal to the ratio at 1000 K, which is the most conservative case. The blue line (with circles) used the ratio of specific heats calculated at 300 K, which would be the best case scenario. Neither of these cases are correct, but they provide bounding limits.

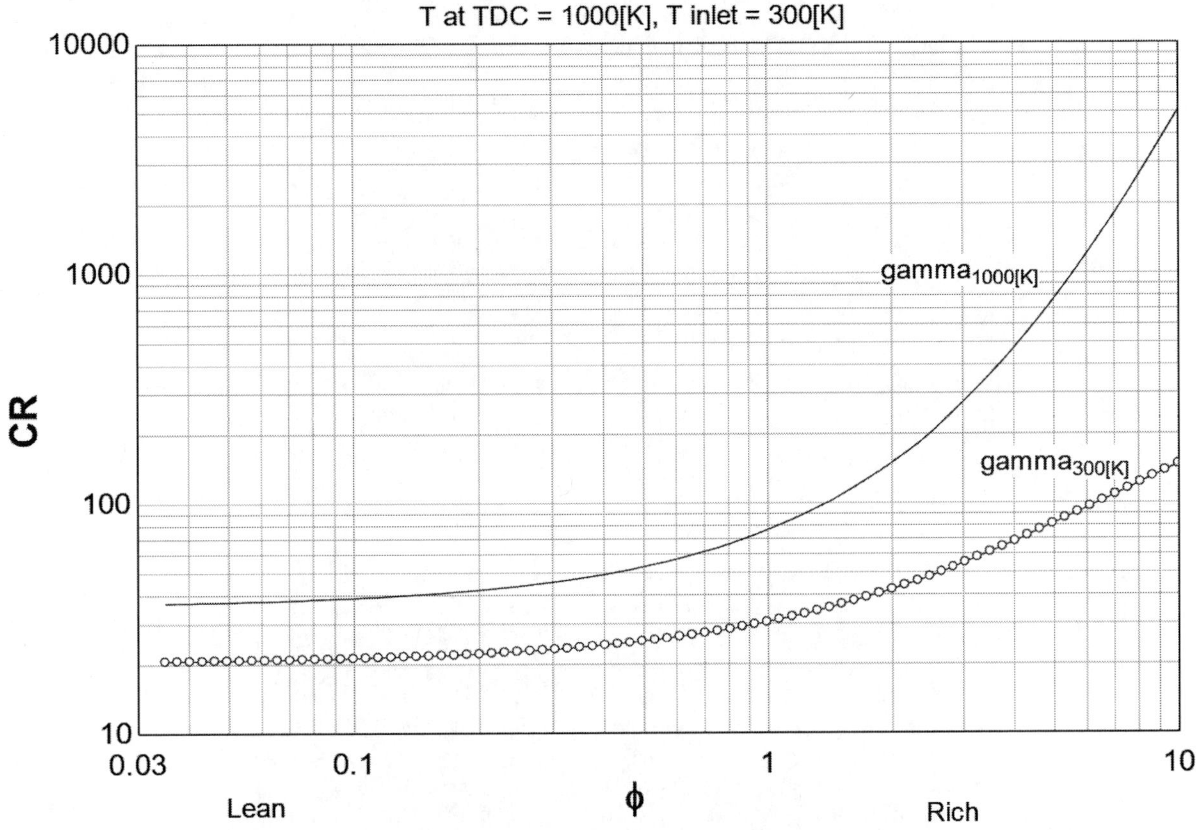

Figure 1. Compression ratio needed to reach 1000 K for various mixtures of air and ethane.

The results show that for mixtures with an equivalence ratio less than 0.15 the maximum compression ratio needed to reach 1000 K is about 40. This provides a mixture range for which an ICE could be used (albeit a high compression engine). It is not feasible to produce ethylene in an air-ethane mixture with an equivalence ratio higher than 0.15 in an ICE.

A model was constructed using the closed HCCI IC engine module in CHEMKIN 17.2 and the GRI Mech 3.0 chemical kinetics mechanism to investigate the amount of ethylene that could be produced using the approach outlined above. The initial temperature was set to 300 K, the equivalence ratio was varied from 0.01 to 2, and the CR was set to 40. The results are shown in Figure 2.

As was expected, there is no ethylene production at the higher equivalence ratios. There is some ethylene production at the lower equivalence ratios, but due to the high dilution ratio of the air to ethane at intake, the mole fraction of the produced ethylene does not surpass 0.017 %. This would not be a large enough ethylene production to warrant the energy cost required for compression and product separation.

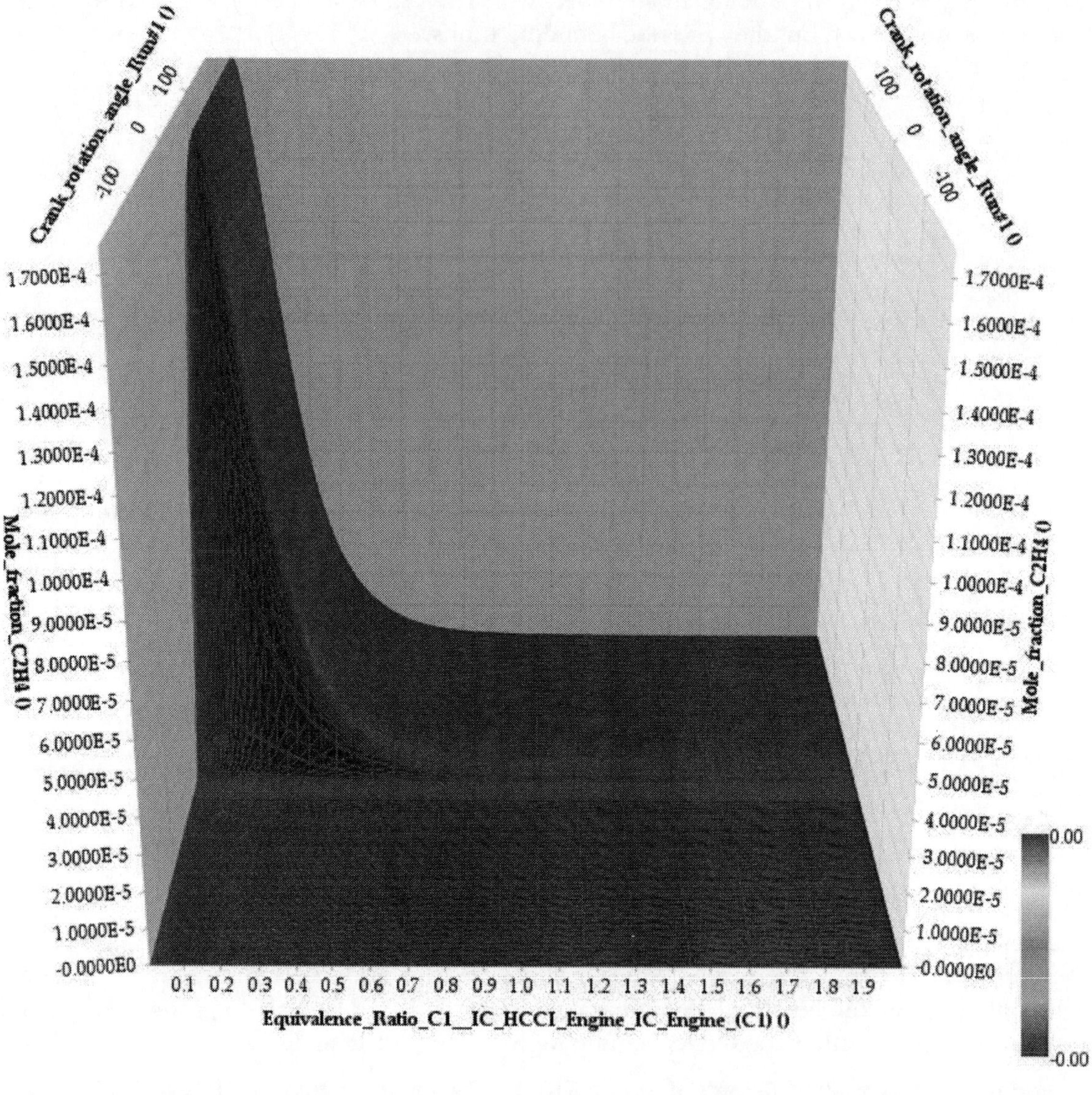

Figure 2. Mole fraction of ethylene vs. equivalence ratio. Inlet 300 K, TDC is at 0 crank angle degrees.

This model was built using physical dimensions corresponding to a 28.5 cc two-stroke 3w-28i model plane engine that is present in the Energy Systems Laboratory. The dimensions of the engine that were used in the simulation are: bore 36 mm, stroke 28 mm, connecting rod length 28 mm, and compression ratio 40. The connecting rod length was not from the 3w-28i model; it was added to the model to allow the simulation to run. Using these dimensions, the mass of ethylene produced was found for each mixture. The yield was also calculated by assuming that the best theoretical production of ethylene was 1 mol of ethylene per 1 mol of ethane. The yield was calculated from the mass of ethylene produced, and the mass of ethane input into the cylinder. These results are shown in Figure 3.

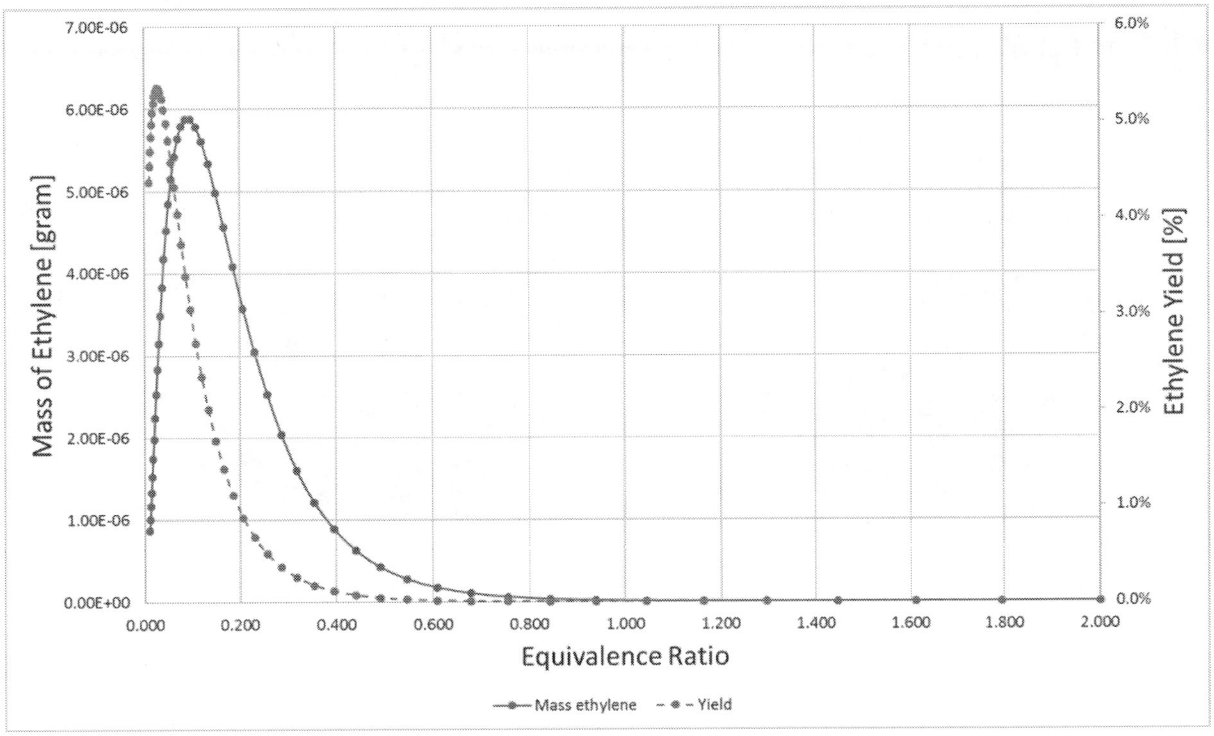

Figure 3. Ethylene yield and mass production vs. equivalence ratio for an air-ethane mixture at standard temperture.

The highest ethylene yield was observed at the lower equivalence ratios. This is most likely due to the higher temperatures that were achieved from the higher ratio of specific heats. The 1000 K temperature threshold (the earlier target) was reached in all of the mixtures below an equivalence ratio of 1, as can be seen in Figure 4. However, the higher temperatures and increased length of time above 1000 K, seem to produce a higher yield. The ethylene yield peaks around an equivalence ratio of 0.028; however, this does not correspond to the highest mass production of ethylene, which occurs closer to an equivalence ratio of 0.092. This is believed to be because the mole fraction of ethane admitted into the cylinder decreases as the equivalence ratio decreases, meaning the maximum amount of ethylene able to be produced decreases as well.

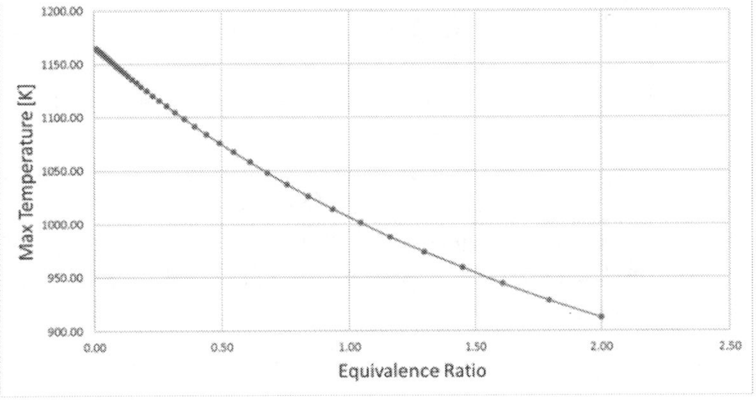

Figure 4. Max cylinder temperature for inlet temp of 300 K and a CR of 40.

Since the ethylene production seems to depend strongly on temperature, preheating the reactants in order to obtain a higher peak temperature was investigated. Two cases were run: a preheat of 50 °C, and a preheat of 100 °C. Even a moderate preheat of 50 °C (to 350 °C) caused combustion to initiate. This can be seen in Figure 5. In these cases, the ethylene is an intermediate combustion species and is nearly immediately consumed, except for a single run in either case (which is likely an outlier or simulation error).

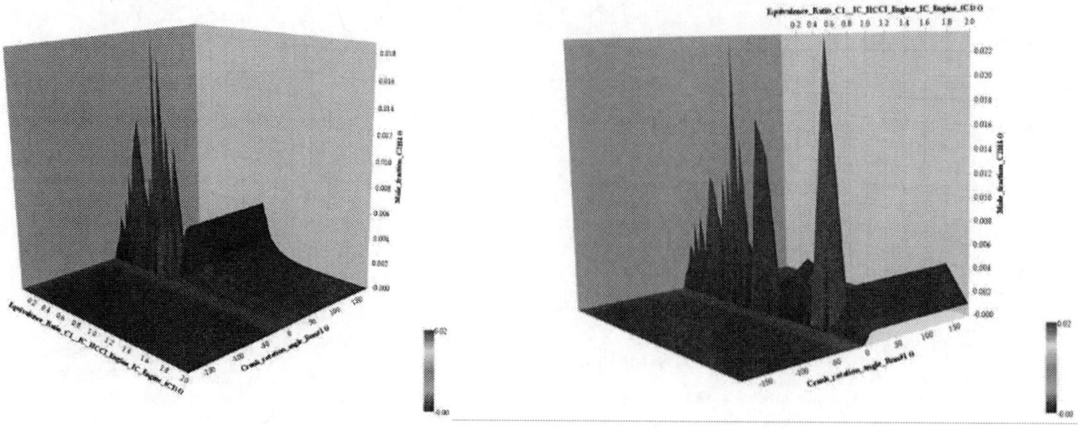

Figure 5. Ethylene production for an air-ethane mixture preheated to 350(left) and 400 °C (right).

4. Non-Oxidative Diluent Gasses

From the prior study, when using a cold air-ethane mixture, ethylene production is limited to low yield and low mole fractions. When pre-heating the mixture, combustion occurs and rapidly consumes any ethylene that was produced from cracking. Neither of these cases are a viable option for ethylene production. In order to reach the necessary temperatures for the cracking reactions to take place, while simultaneously prohibiting the possibility of combustion, another non-oxidative diluent gas is necessary.

Nitrogen alone has a ratio of specific heats of 1.4 and would be mostly inert, however, it is not as economically available as air is; but it is widely available in the United States and is included in this investigation. Two other diluent gasses were identified and investigated: steam and methane. The ratio of specific heats was calculated as a function of temperature for each of these two gasses and was compared to that of ethane in Figure 6. At 300 K the steam and methane have similar ratios of specific heat (1.3 vs 1.328). However, methane's ratio of specific heats decreases much more quickly than steam's as the temperature is increased. For this reason, steam would be a preferable choice for the diluent since less compression would be necessary to reach the same peak temperature.

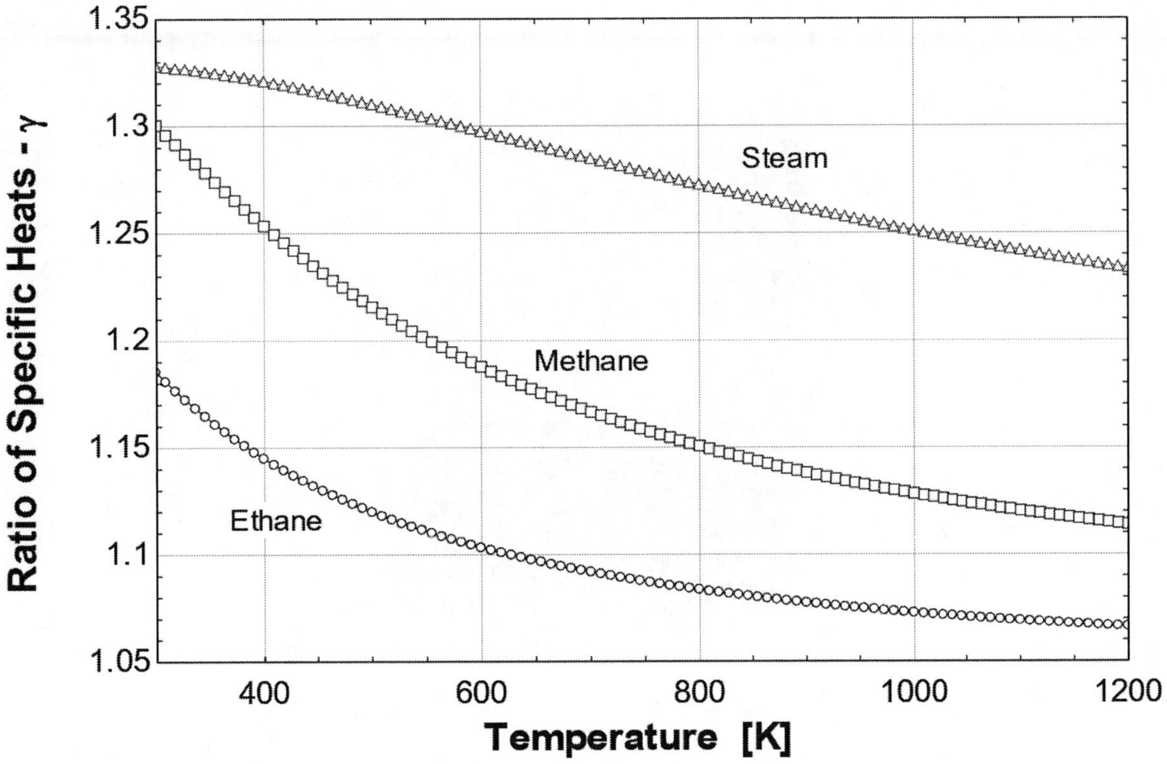

Figure 6. Ratio of specific heats for methane and steam compared to ethane.

5. Ethane and Steam Mixture

Water should be readily available at the reactor engine site, which could be vaporized into steam at 100 °C. The lack of free oxygen means much higher compression is possible since little to no combustion will occur at the increased temperatures. A potential downside to using steam is that the energy requirements for generating the steam could potentially make the process economically unattractive. This determination would require an in depth techno-economic analysis that is outside the scope of this work.

Ethane and steam mixtures between 1 and 100 % ethane by mole were evaluated. No substantial ethylene production was seen until a reactant preheat temperature of approximately 400 °C was reached. At a preheat temperature of 475 °C there was ethylene production for each mixture as can be seen in Figure 7. Clearly, there is a much larger mole fraction of ethylene (approaching 2.6 %) at the lower ethane concentration mixtures. However, it is not readily apparent from Figure 7 which case has the highest yield or mass flow out of ethylene. These data were calculated and are shown in Figure 8. The ethylene yield is not notably higher than the air-ethane mixture, but the mass of ethylene generated is higher at the lower ethane inlet mixtures. The maximum ethylene production was 0.4 mg/cycle, compared to 0.006 mg/cycle in the air-ethane case.

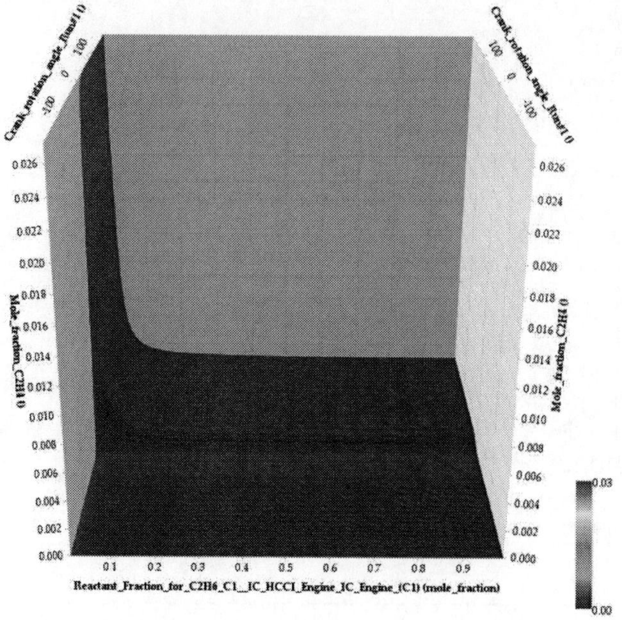

Figure 7. Ethylene production for steam-ethane mixtures entering at 475 °C.

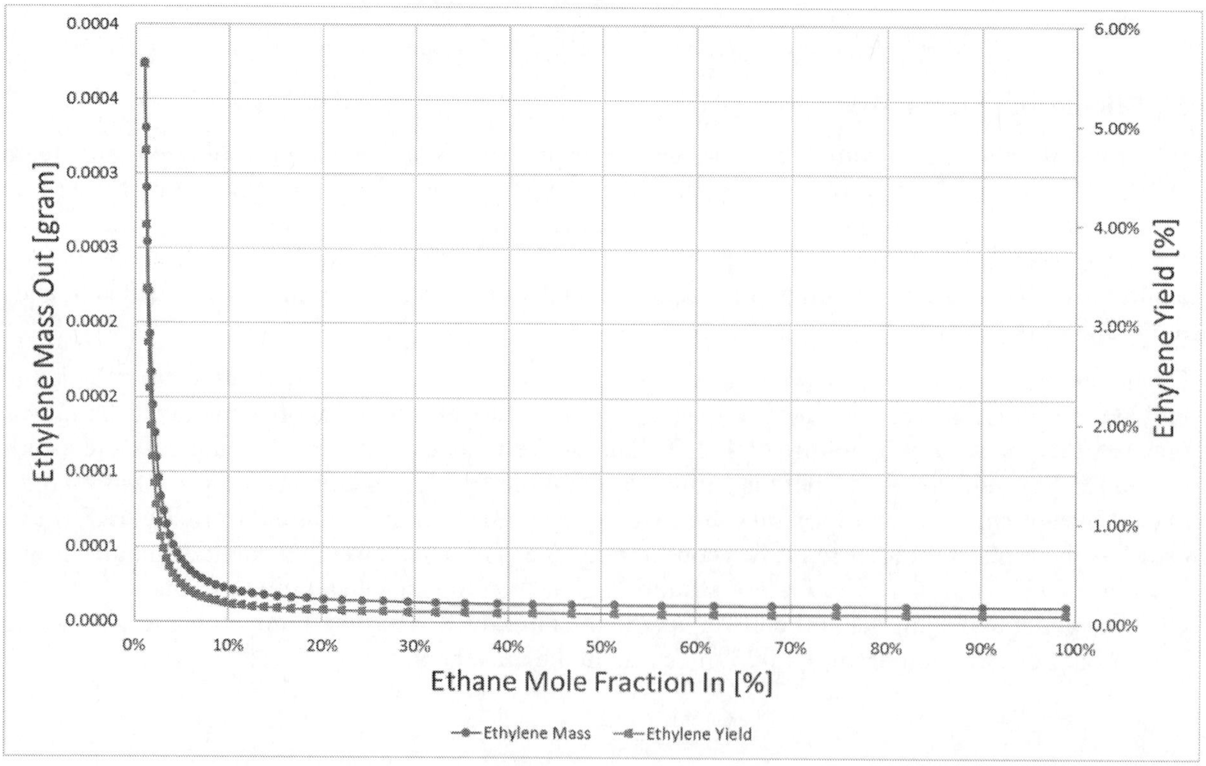

Figure 8. Ethylene mass and yield for the Steam-Ethane mix incoming at 475 °C.

6. Ethane and Methane Mixture

Locating the reactor engine at an economically stranded natural gas well site would provide access to an abundant supply of cheap methane. As was seen in the analysis of the ethane-steam mixtures, a substantial preheat was needed in order to reach the required maximum temperature. Since the ratio of specific heats is lower for methane than it is for steam, it is expected that the methane preheat would need to be even higher than that of the steam. In fact, to obtain approximately the same final mole fraction of ethylene as the ethane and steam case, a preheat temperature of 550 °C was necessary. The results for this preheat level are shown in Figure 9. These results show that the methane adds little to no value to the ethylene production. However, increasing the preheat level to these temperatures does have a substantial effect on the ethylene yield, most likely due to the rise in peak temperature. The yield and mass of ethylene produced can be seen in Figure 10.

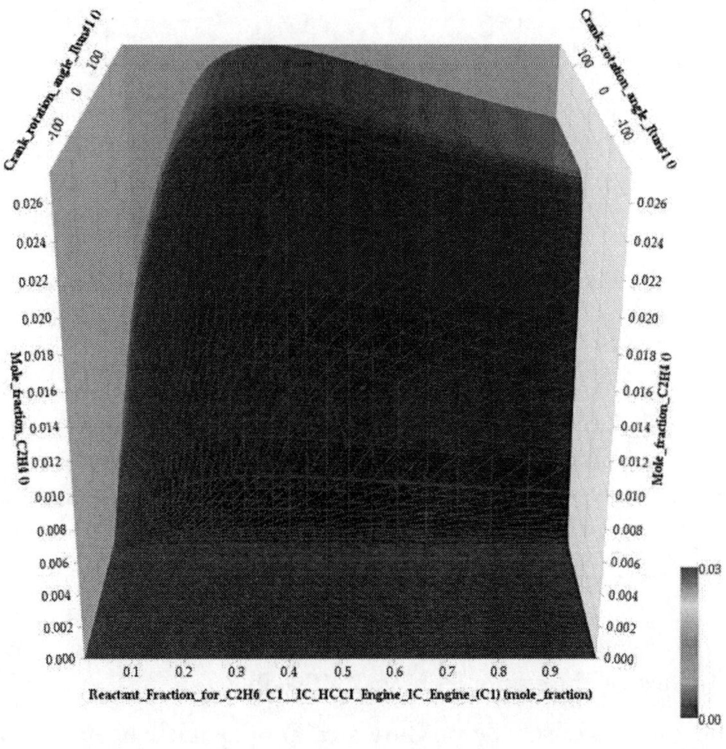

Figure 9. Ethylene production from ethane-methane mixtures preheated to 550 °C.

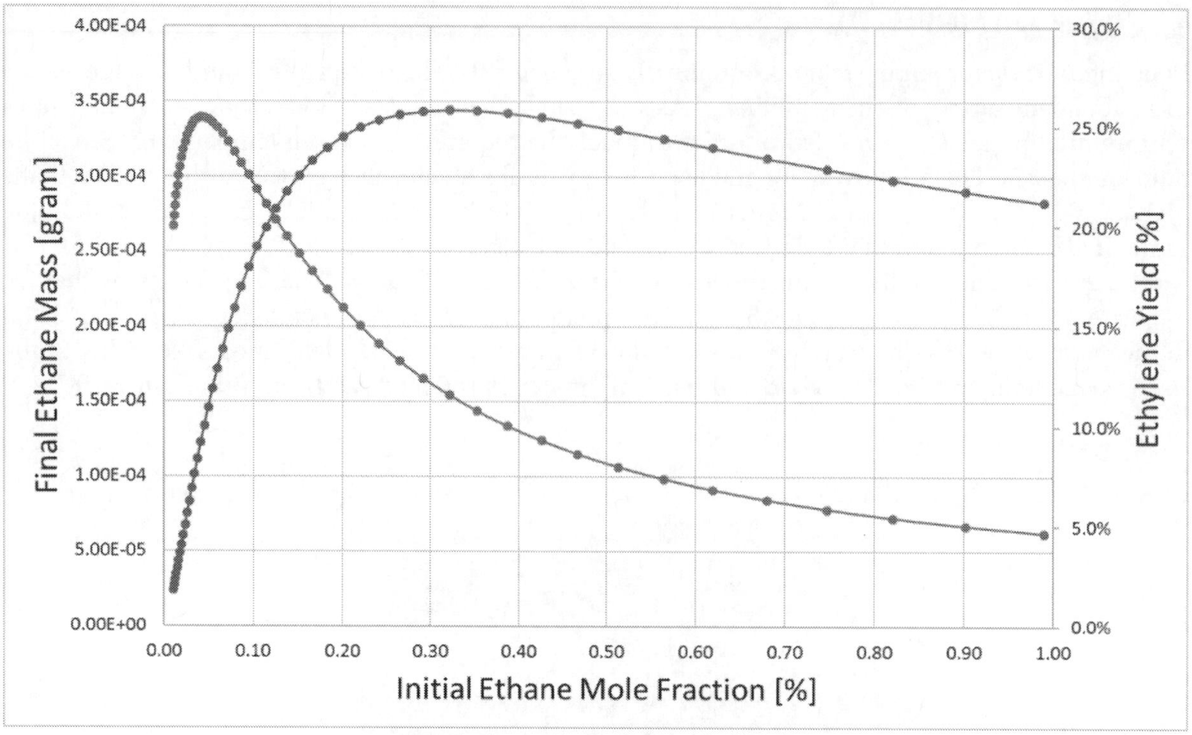

Figure 10. Ethylene yield for an ethane-methane mixture preheated to 550 °C.

7. Conclusion

The closed HCCI model in Chemkin was used to simulate running a modified internal combustion engine as a modular, scalable reactor to produce ethylene through pyrolysis. Four gas mixtures were evaluated including pure ethane, ethane-air, ethane-steam, and ethane-methane to determine the operation conditions that would produce the highest ethylene yield and mass output. Parametric studies of each mixture were conducted over inlet temperature and initial mixture concentration.

Initial investigations revealed that ethane must be heated to at least 1000 K for pyrolysis to occur within a few crank angle degrees of an ICE operating at 1000 RPM. Using only compression to reach this temperature would require an ICE to have a compression ratio of 661, due to ethane's low ratio of specific heats. This is not technically feasible.

Diluting the ethane with air raises the mixture's ratio of specific heats. However, for an ICE with a compression ratio of 40, the mixture would have to be at a stoichiometric ratio of less than 0.15 to reach a peak temperature of 1000 K. This low percentage of ethane admitted into the cylinder yielded a miniscule amount of ethylene. Attempting to preheat the mixture did not yield more ethylene. Instead, it initiated combustion which destroyed any ethylene that was produced.

Without substantial preheating, an ethane-methane or ethane-steam mixture will not produce a substantial amount of ethylene, either. If a large preheat is present the ethylene yield is more favorable, with the methane-ethane mixture having the highest yield. However, a techno-economic analysis is necessary to determine if the produced ethylene yield outweighs the separation and preheat energy requirements.

Other variables such as the engine speed and heat transfer were not investigated. The mass of ethylene produced should scale linearly with the size of the engine being used since the peak temperature is only a function of compression ratio. Heat transfer would adversely affect the ethylene production by decreasing the peak temperature of the mixture. These are variables that should be investigated in future studies.

8. Acknowledgements

This information, data or work presented herein was funded in part by the Advanced Research Projects Agency-Energy (ARPA-E), U.S. Department of Energy, under Award Number DE-AR0000681. The views and opinions of authors expressed herein do not necessarily state or reflect those of the United States Government or any agency thereof.

9. References

[1] C. H. S. Menon, Z. Taie, "Internal combustion engines as chemical reactors: Issues and challenges," in *Western states section of the combustion institute*, 2016.

[2] E. Hu, Y. Chen, Z. Zhang, X. Li, Y. Cheng, and Z. Huang, "Experimental Study on Ethane Ignition Delay Times and Evaluation of Chemical Kinetic Models," *Energy and Fuels*, vol. 29, pp. 4557–4566, 2015.

[3] M. K. Sabbe, K. M. Van Geem, M.-F. Reyniers, and G. B. Marin, "First principle-based simulation of ethane steam cracking," *AIChE J.*, vol. 57, no. 2, pp. 482–496, Feb. 2011.

[4] f-Chart, "Engineering Equation Solver." 2016.

Sub Topic: Turbulent Flames

Western States Section Combustion Institue Meeting
Organized by the Western States Section of the Combustion Institute
October 02–03, 2017
Laramie, Wyoming

Flame Stability and Liftoff Heights of a Surrogate Jet Fuel Burned near Mild Combustion Conditions

Parker Weide[1] *and David Blunck*[1,*]

[1]*School of Mechanical, Industrial, and Manufacturing Engineering, Oregon State University, 204 Covell Hall Corvallis, OR 97331-2409, USA*
[*]*Corresponding author:* David.Blunck@oregonstate.edu

Abstract: This research aims to identify how differences in temperature and oxygen concentration in the reaction zone affects liftoff heights for moderate or intense low-oxygen dilution (MILD) combustion of a jet fuel and a surrogate. MILD combustion offers a method of burning fuels with reduced CO and NO_x production. Jet fuels, and large hydrocarbon fuels in general, are significant because of their relatively high energy density and use in the transportation field. Surrogate fuels are of particular interest due to the need to replicate the combustion characteristics of jet fuels (which can have 100's of constituents). In this study, MILD conditions are created above the burner by injecting preheated and vaporized fuel into a high temperature, low oxygen concentration coflow. The temperature of the coflow was varied between 1300K to 1400K, and the oxygen concentration was varied between 3% to 9% by volume. Liftoff height was measured using the chemiluminescence of OH*. It was found that in coflows with lowest temperature and lowest O_2 concentration, the liftoff height of both jet-A and surrogate fuel are highest. As the coflow temperature increases, or O_2 concentration increases, the jet fuel's liftoff height decreases. Both fuels displayed similar characteristics at higher temperature coflows, as well as at higher O_2 concentrations. However, at a coflow temperature of 1300K and O_2 concentration of 3%, larger liftoff heights were measured for the surrogate fuel. It is hypothesized that this discrepancy is caused by differences in fuel chemistry and changes in the formation of CH_2O.
Keywords:* *MILD Combustion, Surrogate Fuels, Low Temperature Chemistry

1. Introduction

Emphasis is being placed on finding ecologically friendlier technologies for energy production as global demand for energy increases. One approach is to operate combustion devices at higher temperatures to increase the overall efficiency of the system. However, this often increases NO_x production by thermal mechanisms. This can be countered by using the heat from the cycle's combustion products to preheat and dilute the reactants which in turn reduces the production of NO_x by lowering the maximum flame temperature [1, 2]. If the reactants are preheated to a temperature above the autoignition temperature, and the temperature increase in the reaction zone is less than the autoignition temperature, MILD, or Moderate or Intense Low-Oxygen Dilution combustion can occur [3]. The lowering of the temperature increase is generally established by reducing O_2 concentration [4]. The low O_2 concentration causes the temperature increase within the reaction zone to be less than conventional combustion. This reduces the formation of NO_x [5].

MILD combustion offers a range of advantages when compared to conventional combustion. In MILD combustion, the reaction zone is more distributed, in part because of the high level of mixing of products with reactants [4, 6]. This also leads to a lower peak temperature than normally observed in conventional diffusion flames [5]. The reduction in peak temperature can lessen thermal loading on burners and turbines, potentially reducing cooling requirements and maintenance requirements. By using heat from combustion products to preheat reactants or by using the products themselves within the combustion chamber, MILD combustion has the potential to allow for higher levels of thermodynamic efficiencies for systems.

MILD combustion of large hydrocarbon fuels is of interest because real jet fuels are comprised of complex mixtures (e.g. of 100's of constituents) of large molecules, with additives blended to meet requirements for physical properties. Surrogates fuels, however, are a simple blend (e.g., 3) of large hydrocarbons designed to reproduce the more important characteristics of real fuels [7]. The use of surrogates in simulations allows for tractable models of the physical and chemical properties of real fuels. It is noted that most surrogate fuels have been evaluated for laminar conditions, although most practical devices are turbulent. This is significant because different transport processes may be important under turbulent conditions.

While the characteristics of gaseous fuels [1, 8] and some large hydrocarbon fuels [9] have been researched in MILD or MILD-like conditions, significantly less information is available on the MILD combustion of surrogate fuels [10, 11]. One method of evaluating the stability of fuels is to measure the liftoff height of the flames for laminar and turbulent conditions.

Motivated by the advantages of MILD combustion and the interest in jet and surrogate fuels, the objectives of this study are as follows:

- Determine the influence of fuel preheat temperature on the liftoff height of a surrogate jet fuel in MILD combustion.
- Ascertain the effects of O_2 concentration on liftoff heights of a surrogate jet fuel.
- Identify similarities and differences between combustion characteristics for surrogate and jet fuels when combusted at MILD or MILD-like laminar and turbulent conditions.

2. Experimental Approach

A JHC burner was used to study the MILD combustion of the jet and surrogate fuels. This burner has been used previously when studying MILD combustion of gaseous fuels [6, 12, 13]. A model of the experimental arrangement is shown in Figure 1. The large hydrocarbon fuels are introduced through a central jet, which is surrounded by a high temperature and low oxygen coflow. For this research, the burner arrangement has been modified with a vaporizer to allow for liquid fuels to be studied. To measure liftoff heights of flames, an Andor Solis iStar 334t ICCD camera was used with a bandpass filter to view OH chemiluminescence. The

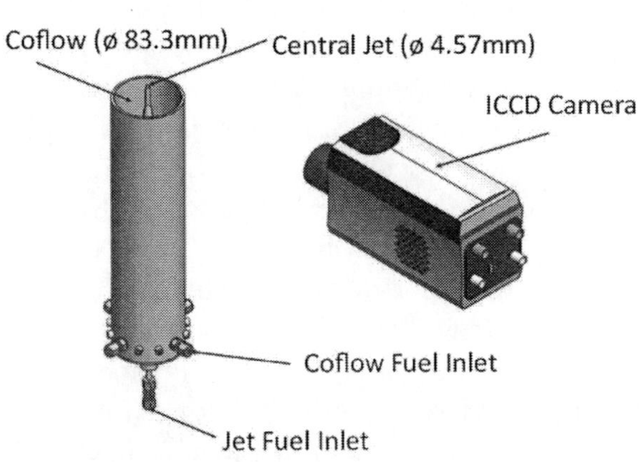

Figure 1: Diagram of the experimental setup used to measure liftoff heights.

bandpass filter is cnetered on a 310 nm wavelength, with a FHWM of 11 nm. The OH* radical's location and intensity is representative of the reaction zone [14].

The JHC burner works by creating the temperature and O_2 concentration required for MILD combustion in a coflow and having the fuel of interest flow through the central jet. The coflow is created by burning a mixture of CH_4, N_2, H_2, and air. The flame anchors 150 mm below the central fuel jet in a series of ceramic balls. This porous media helps to reduce hot-spots, ensure proper mixing, and stabilize the coflow flame. The operating conditions of the coflow can be seen in Table 1. Flow rates of the gases are controlled by setting the pressure upstream of calibrated orifices. The flow rates are monitored using two pressure transducers measuring the pressure drop across a calibrated orifice. Gas temperatures of the reactants are measured with a type-K Omega thermocouple and accounted for in a Labview program. The program uses these temperatures, along with the appropriate physical properties to display the volumetric flow rate. At each operating condition, the velocity of the coflow is roughly 2.3 meters per second at the burner exit, based on continuity calculations.

Table 1: Coflow conditions for JHC burner near MILD combustion conditions

Coflow Temperature [K]	O_2 [vol %]	Air [LPM]	Nitrogen [LPM]	Methane [LPM]	Hydrogen [LPM]	Equivalence Ratio
1300	3	81.6	48.6	5.00	6.36	0.77
1350	3	79.1	33.5	5.00	6.36	0.79
1400	3	77.2	21.5	5.00	6.36	0.81
1300	6	101	31.7	5.00	6.36	0.62
1350	6	95.6	18.7	5.00	6.36	0.66
1400	9	120	14.6	5.00	6.36	0.52

The fuel of interest exits from the burner through a 4.6 mm jet after passing through a series of electric heaters and a vaporizer. The fuel is stored in a pneumatic cylinder. The fuel flow rate is controlled by adjusting the pressure on the air side of the piston via a high precision regulator. The flow rate of the fuel is measured using a differential pressure transducer and a calibrated orifice. Fuel is dispersed into small droplets at the vaporizer inlet using an atomizing nozzle. The fuel is then vaporized in the high temperature environment of the vaporizer. Fuel then travels through four stages of electronic heaters. The first two heaters step the temperature up to 300°C, while the second two maintain this temperature as the fuel flows through the central tube. The heaters are managed by PID controllers to automatically adjust to changes in mass flow rate. The Reynolds number of the central jet was adjusted by changing the mass flow rate of the fuel. These mass flow rates can be seen in Table 2. The viscosity of the vaporized jet fuel was determined using the approach reported by Pederson et. all [15]. The approach consists of predicting the viscosity of large hydrocarbons at high temperatures using critical constants and molecular weights of each component. For jet-A, the eight components with the largest volume percentage were used to predict the viscosity. For the surrogate fuel, the three components were n-dodecane, n-hexadecane, and 1,3,5 Trimethyl Benzene. Results for these analyses are shown in Table 3.

Table 2: Mass flow rates and exit velocities of fuels used for testing

Reynolds Number	Jet-A Mass Flow Rate (g/s)	Jet-A Exit Velocity (m/s)	Surrogate Mass Flow Rate (g/s)	Surrogate Exit Velocity (m/s)
3750	0.142	2.566	0.141	2.568
5000	0.190	3.421	0.188	3.424
6250	0.237	4.277	0.235	4.280
7500	0.285	5.132	0.282	5.136
8750	0.332	5.987	0.329	5.992
10 000	0.380	6.843	0.376	6.848

Two fuels were studied in this experiment. The first fuel, jet-A (POSF-10325), is composed of varying concentrations of n-alkanes, iso-alkanes, aromatics, and cycloalkanes. This fuel has been extensively studied and its characteristics documented. The surrogate fuel used in this experiment (POSF-12765) was designed to replicate the combustion characteristics of jet-A. The method described by Dooley et. all was used in formulating the surrogate fuel [16]. This technique uses a chemical group additivity method to match key combustion characteristics of a fuel. The surrogate fuel is comprised of three main components; n-dodecane, iso-octane, and 1,3,5 Trimethyl Benzene [17]. Physical properties of these fuels are shown in Table 3.

Table 3: Physical properties of the two fuels studied.

	jet-A (POSF-10325)	Surrogate fuel (POSF-12765)
Flash Point (°C)	48	24
Density (g/cm^3)	0.780	0.777
Aromatics (vol. %)	11.2	23
Heat of Combustion (mJ/kg)	43.24	42.92
Viscosity at 300°C (kg/m-s)	1.051E-05	1.040E-05

The liftoff height of the flame was determined from chemiluminescence measurements. The camera captured a series of 120 images at 1.3 Hz for each operating condition. The camera is positioned 20 inches from the central jet. To determine the liftoff height, the average peak intensity was determined for the 3% 1300K condition. This condition had the lowest peak intensity. The liftoff heights were then identified by the location of 10% of the threshold for each operating condition.

3. Results and Discussion

The flame liftoff height for both the surrogate and jet-A fuels decreased as the O_2 concentration of the coflow increased, as shown in Figure 2. For the 6% and 9% O_2 concentration conditions, both fuels exhibited lower liftoff heights than the 3% condition. This trend was observed for all Reynolds numbers. This is expected, as the higher O_2 concentration tends to create more conventional combustion. Hence, increases in the O_2 concentration increases the laminar flame speed which tends to decrease the flame height. Note, however, that at the 3% O_2 condition, the flame liftoff height increased for both fuels as their Reynolds number increased until a Reynolds number near 7500, after which the liftoff height tended to decrease. The decrease is attributed to an increase in CH_2O production, which increases with flame stretch (i.e., greater turbulence) [9]. The increase in CH_2O production as the Reynolds number increases has been observed with small hydrocarbon fuels using PLIF measurements. As the oxygen concentration increases, the influence of CH_2O on liftoff height is reduced, and therefore the trend of decreasing liftoff height at higher Reynolds numbers is not replicated in the coflow conditions with reduced O_2 concentration [9]. It is noted that the liftoff height for the surrogate and jet-A fuel are similar for the conditions near 6% and 9% O_2 concentration. This is expected, considering that these conditions are more like conventional combustion; the conditions at which surrogate fuels are evaluated (although at laminar conditions). In contrast, at 3% O_2, the liftoff height of the surrogate fuel flame is higher than the jet-A fueled flame. It is plausible that the discrepancy is caused by changes in CH_2O production because of the fuel chemistry.

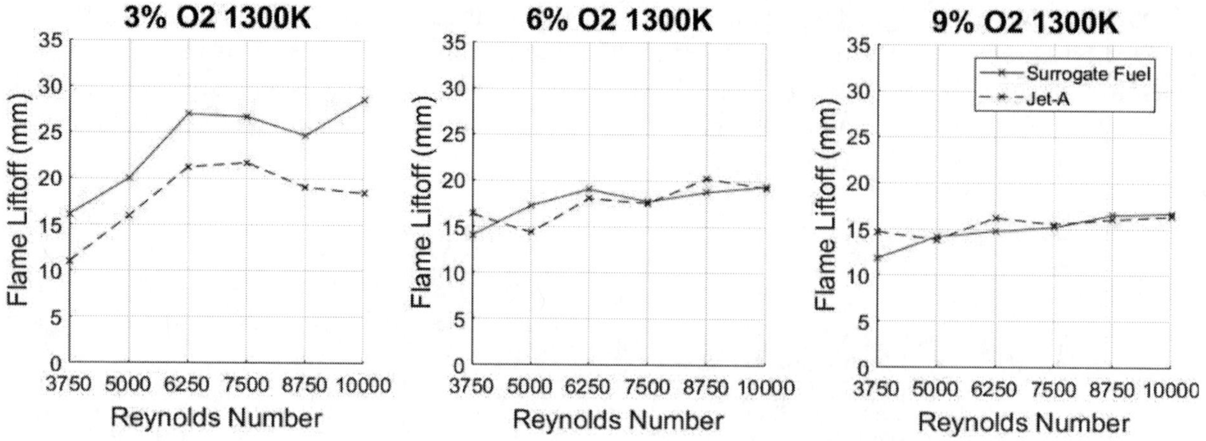

Figure 2: Liftoff heights for a surrogate fuel and jet-A with a constant coflow temperature and varying O_2 concentration and Reynolds number.

Significant changes in average liftoff height and fuel sensitivity occurred when changing the temperature of the coflow. The liftoff height of both fuels tends to decrease as the temperature of the coflow increases. Figure 3 shows the liftoff heights of the fuels when burned in a 1300K, 1350K, and 1400K coflow. At the lowest temperature, the liftoff heights are the highest, with the highest liftoffs occurring near Reynolds numbers of 6250 or 7500. As the flames become more turbulent and the turbulence intensity increases, flame stretch also increases, which can increase CH_2O production [18]. The higher turbulence intensity is also known to increase the production of CH_2O [19]. As the temperature of the coflow increases, the liftoff heights of both decrease. The higher temperature gas in the coflow increases the reactivity of the fuel, thereby lowering liftoff heights. The increased temperature also decreases the effectiveness of CH_2O in decreasing ignition delay times [9].

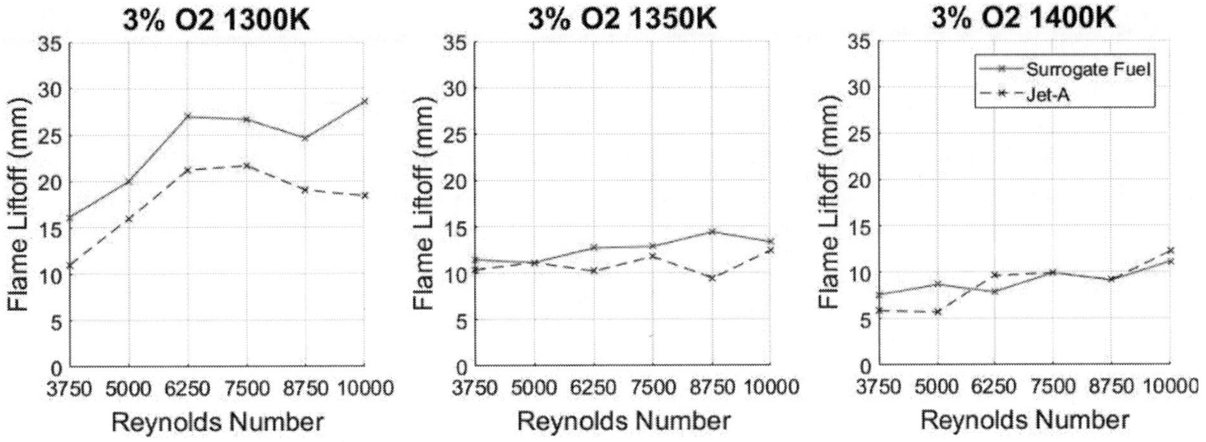

Figure 3: The liftoff heights of jet-A and a surrogate jet fuel flames relative to changes in the Reynolds number and coflow temperature while maintaining a constant coflow O_2 concentration.

4. Summary

In summary, the physical processes affecting the liftoff heights for two types of jet fuels when burned in MILD or near MILD conditions were studied. This work was motivated by a lack of research on surrogate fuels in turbulent flow and MILD conditions. To study these effects, a burner with a high temperature and reduced O_2 concentration coflow encircling a central jet was used. The Reynolds number of the central jet, as well as the temperature and O_2 concentration of the coflow was altered to investigate their affects on the liftoff height of the jet fuels. The major observations of this work are as follows.

- Decreasing the O_2 concentration of the coflow in the JHC burner caused an increase in liftoff heights for both fuels tested. When the coflow is at 3% O_2, the surrogate fuel was consistently measured at a higher liftoff height than jet-A. This is attributed to different rates of CH_2O production for the surrogate fuel in this environment, although this requires further investigation.

- Decreasing the temperature of the coflow resulted in increased liftoff heights for both fuels. At a higher temperatures, both fuels displayed similar liftoff heights. Deviation of liftoff heights between Reynolds numbers also decreased as temperature of the coflow increased.

- When differences in liftoff height were observed, jet-A displayed a lower liftoff height more often than the surrogate. This trend is clearly shown at a low temperature coflow with 3% O_2 concentration. It is plausible that this difference is caused by differences in CH_2O production rates.

Sub Topic: Turbulent Flames

5. Acknowledgements

The authors thank Scott Boreham and Heather Miller for helping to operate the burner. Tim Edwards from Air Force Research Laboratory provided the fuels. Acknowledgment is made to the Donors of the American Chemical Society Petroleum Research Fund (55535-DN16) for support of this research as part of a Doctoral New Investigator Award.

References

[1] J. Wünning and J. Wünning, Flameless oxidation to reduce thermal no-formation, Progress in Energy and Combustion Science 23 (1997) 81–94.

[2] P. Li, J. Mi, B. B. Dally, F. Wang, L. Wang, Z. Liu, S. Chen, and C. Zheng, Progress and recent trend in MILD combustion, Science China Technological Sciences 54 (2011) 255–269.

[3] M. De Joannon, A. Cavaliere, R. Donnarumma, and R. Ragucci, Dependence of autoignition delay on oxygen concentration in mild combustion of high molecular weight paraffin, Proceedings of the Combustion Institute 29 (2002) 1139–1146.

[4] A. Cavaliere and M. de Joannon, Mild Combustion, Progress in Energy and Combustion Science 30 (2004) 329–366.

[5] M. Derudi and R. Rota, Experimental study of the mild combustion of liquid hydrocarbons, Proceedings of the Combustion Institute 33 (2011) 3325–3332.

[6] İ. B. Özdemir and N. Peters, Characteristics of the reaction zone in a combustor operating at mild combustion, Experiments in Fluids 30 (2001) 683–695.

[7] E. Ranzi, *A Wide-Range Kinetic Modeling Study of Oxidation and Combustion of Transportation Fuels and Surrogate Mixtures*, CMIC Department, 2006, URL: `http://pubs.acs.org/doi/pdf/10.1021/ef060028h`.

[8] P. Li, F. Wang, Y. Tu, Z. Mei, J. Zhang, Y. Zheng, H. Liu, Z. Liu, J. Mi, and C. Zheng, Moderate or Intense Low-Oxygen Dilution Oxy-combustion Characteristics of Light Oil and Pulverized Coal in a Pilot-Scale Furnace, Energy & Fuels 28 (2014) 1524–1535, eprint: `http://dx.doi.org/10.1021/ef402240w`.

[9] E. M. Walters, Stability and Liftoff of Non-premixed Large Hydrocarbon Flames in MILD Conditions, PhD thesis, OSU Libraries, 2016, URL: `https://ir.library.oregonstate.edu/xmlui/bitstream/handle/1957/59597/WaltersEricM2017.pdf?sequence=1`.

[10] R. Weber, J. P. Smart, and W. vd Kamp, On the (MILD) combustion of gaseous, liquid, and solid fuels in high temperature preheated air, Proceedings of the Combustion Institute 30 (2005) 2623–2629.

[11] S.-R. Wu, W.-C. Chang, and J. Chiao, Low NOx heavy fuel oil combustion with high temperature air, Fuel 86 (2007) 820–828.

[12] P. R. Medwell, P. A. Kalt, and B. B. Dally, Imaging of diluted turbulent ethylene flames stabilized on a Jet in Hot Coflow (JHC) burner, Combustion and Flame 152 (2008) 100–113.

[13] A. Mardani and S. Tabejamaat, Effect of hydrogen on hydrogen–methane turbulent non-premixed flame under MILD condition, International Journal of Hydrogen Energy 35 (2010) 11324–11331.

[14] J. Kojima, Y. Ikeda, and T. Nakajima, Spatially resolved measurement of OH*, CH*, and C2* chemiluminescence in the reaction zone of laminar methane/air premixed flames, Proceedings of the Combustion Institute 28 (2000) 1757–1764.

[15] K. S. Pedersen, A. Fredenslund, P. L. Christensen, and P. Thomassen, Viscosity of crude oils, Chemical Engineering Science 39 (1984) 1011–1016.

[16] S. Dooley, S. H. Won, M. Chaos, J. Heyne, Y. Ju, F. L. Dryer, K. Kumar, C.-J. Sung, H. Wang, M. A. Oehlschlaeger, R. J. Santoro, and T. A. Litzinger, A jet fuel surrogate formulated by real fuel properties, Combustion and Flame 157 (2010) 2333–2339.

[17] J. T. Edwards, Reference Jet Fuels for Combustion Testing, 55th AIAA Aerospace Sciences Meeting (2017), DOI: 10.2514/6.2017-0146.

[18] S. A. Filatyev, J. F. Driscoll, C. D. Carter, and J. M. Donbar, Measured properties of turbulent premixed flames for model assessment, including burning velocities, stretch rates, and surface densities, Combustion and Flame 141 (2005) 1–21.

[19] J. Sjöholm, J. Rosell, B. Li, M. Richter, Z. Li, X.-S. Bai, and M. Aldén, Simultaneous visualization of OH, CH, CH2O and toluene PLIF in a methane jet flame with varying degrees of turbulence, Proceedings of the Combustion Institute 34 (2013) 1475–1482.

Western States Section of the Combustion Institute - Fall 2017 Meeting
Hosted by University of Wyoming
October 2-3, 2017.

Estimation of Multiple Model Parameters Using a Data-Assimilated CFD Algorithm

Yijun Wang and Xinfeng Gao

Computational Fluid Dynamics and Propulsion Laboratory, Department of Mechanical Engineering, Colorado State University, Fort Collins, CO, 80523, USA

Physical models are often used for modeling the unresolved scales for combustion simulations. Uncertainties in model parameters can lead the simulation to deviate significantly from the real combustion behavior. Therefore, understanding the model parameters and using their proper values are important for the simulation. In our previous study, we developed a data-assimilation CFD modeling framework to improve the estimate of model parameter, which has demonstrated that data assimilation has a great impact on the improvement of single model parameter estimation. In the present study, we will focus on using the ensemble Kalman filter (EnKF) algorithm to investigate the performance of multiple model parameter estimation in the convection-diffusion-reaction (CDR) equation. The CFD algorithm is a fourth-order finite-volume scheme. For the present investigation, we neglect numerical errors in the CFD algorithm by simply focusing on uncertainty in the model parameters. By restricting our study to one-dimensional space, we hope to develop a greater understanding on the performance of the data assimilation technique on the estimation of multiple parameters. This study provides us with insight on where, when, and how the data have significant impact on the estimation of multiple parameters, and the limitations of EnKF for the CDR problem. In future, we will experiment with the maximum likelihood ensemble filter.

1 Introduction

When error/uncertainty exists in physical models or the model parameters used in the equations governing combustion, computational fluid dynamics (CFD) predictions of the state will not be accurate. In previous studies [1, 2, 3], we have applied data assimilation (DA) to improve estimation of a model parameter and the results were promising. In this study, we focus on applying ensemble Kalman filter (EnKF) algorithm to estimate multiple parameters in the convection-diffusion-reaction (CDR) model. The primary goal of this work is to investigate the performance of EnKF on the estimation of multiple parameters. Specifically, this study examines how the data assimilation process for the estimation of multiple parameters is impacted by the spatial and temporal distribution of the data.

2 Problem Statement

The one-dimensional transient in-homogeneous CDR equation from our previous study [2] is again used for this study, since it contains three different fluid dynamics mechanism phenomena, representing the typical physical processes in Navier-Stokes equation and has an analytical solution for performance assessment. The space is restricted to one dimension for the computational efficiency and a thorough understanding on the correlation of uncertainty in multiple parameters.

The governing equation for the one-dimensional transient in-homogeneous CDR equation can be written as:

$$\frac{\partial \phi}{\partial t} + a\frac{\partial \phi}{\partial x} - \mu\frac{\partial^2 \phi}{\partial x^2} + c\phi = f\,, \quad 0 \leq x \leq 1 \tag{1}$$

where ϕ is the solution variable. The true values for the convection speed, a, the diffusivity, μ, and source term coefficient, c, are all unity. The other source term, f, is modeled as $f = 2(x-1)e^{-t}$. The exact solution [4] for CDR problem is in form of

$$\phi_{\text{exact}}(x,t) = x^2 e^{-t}\,. \tag{2}$$

Figure 1 plots the exact solution in space at different times, showing its spatial distribution and change in time.

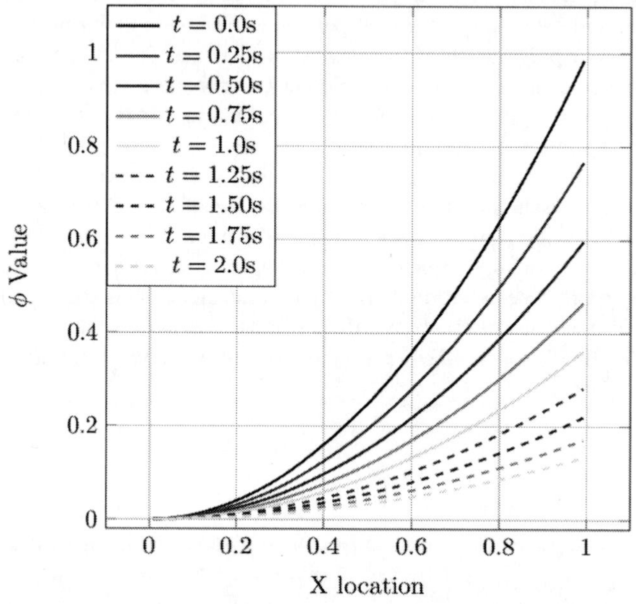

Figure 1: The exact solution for CDR equation.

2.1 Model Parameters Under Estimation

Although the true values for a, μ, and c are known, we purposely assume that there is uncertainty in each of the model parameters, then the parameters will be estimated dynamically by assimilating available data during the CFD solution process. The estimation will be compared to the true value for the evaluation of the assimilation performance.

2.2 Data Assimilation Setups

The EnKF algorithm as described in Gao et al. [2] is employed and its implementation is also described in that reference. To briefly repeat, there are two major reasons for using EnKF in this work. It does not require a closure term and it is simple to implement and efficient on parallel computers. However, to keep in mind, it is a linear operator and there maybe challenges in coping with nonlinear problems.

As the first step, an initial ensemble must be created based on a certain probability distribution function. From our previous work [5], a 16-member ensemble is each created for a, μ, and c, using Gaussian distribution:

$$f(a;\bar{a},\sigma_a^2) = \frac{1}{\sqrt{2\pi\sigma_a^2}} e^{-\frac{(a_i-\bar{a})^2}{2\sigma_a^2}}$$
$$f(\mu;\bar{\mu},\sigma_\mu^2) = \frac{1}{\sqrt{2\pi\sigma_\mu^2}} e^{-\frac{(\mu_i-\bar{\mu})^2}{2\sigma_\mu^2}}$$
$$f(c;\bar{c},\sigma_c^2) = \frac{1}{\sqrt{2\pi\sigma_c^2}} e^{-\frac{(c_i-\bar{c})^2}{2\sigma_c^2}} .$$

The ensemble mean is $\bar{a} = 8$, $\bar{\mu} = 4$, and $\bar{c} = 8$. The variance is $\sigma_a^2 = 9$, $\sigma_\mu^2 = 4$ and $\sigma_c^2 = 9$. The ensembles are shown in Fig. 2.

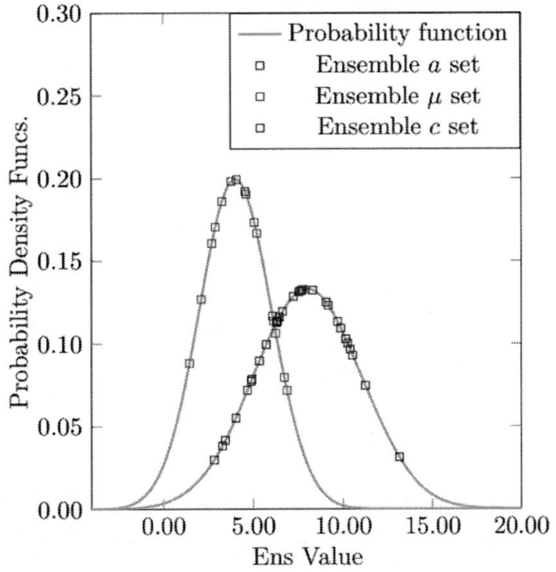

Figure 2: The initial ensemble guess

As for observations, they are created from the perfect model, i.e., the CFD solution generated by using the exact model parameters. There are totally 64 observations available; that is that every cell in the mesh has an observation value. When all of them are used for assimilation, we call it a "Full-Observation" case. The data-assimilation frequency is set to be every 250 time steps and the time step is $\Delta t = 1e^{-5}s$.

2.3 Objectives

The objective of this study is to investigate how the 3 model parameters simultaneously respond when data are assimilated from different spatial and temporal zones. Based on the rate of change of the solution variable (ϕ) in spatial and temporal as shown in Fig. 1, the domain is separated into 3 spatial zones (See Fig. 3) and 3 temporal zones which are (a) time zone 1 (T_1): $0 \sim 0.5$ sec, (b) time zone 2 (T_2): $0.5 \sim 1.5$ sec, and (c) time zone 3 (T_3): $1.5 \sim 2.5$ sec.

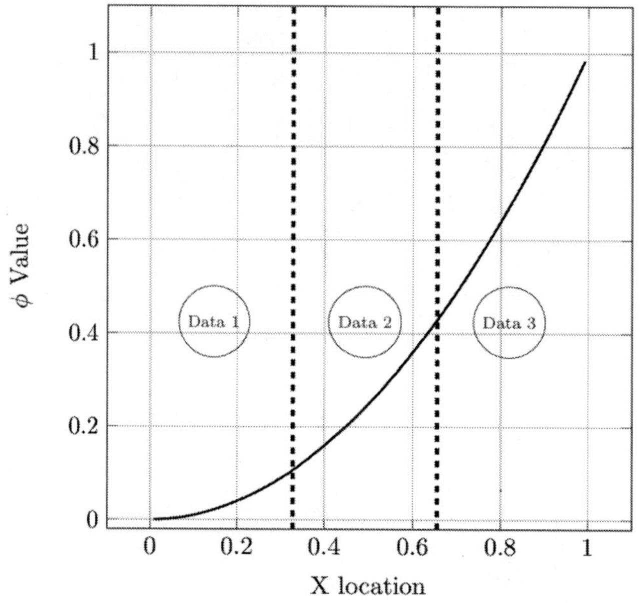

Figure 3: The illustration of separated observation region in computational domain

3 Results and Discussion

3.1 The Impact of Spatial Distributions

As shown in Fig. 3, data are assimilated for 3 individual zones, and their results are compared with the base case where the assimilation is performed by using the "Full-Observation" case.

The convergence histories for the ensemble mean $\bar{a}$, $\bar{\mu}$ and $\bar{c}$ are plotted in Figs. 4 - 6. From the convergence trajectories, each model parameter converges toward its true value, although the assimilation process for the use of "Data 3" is oscillatory and requires more iterations to finally converge to the true value. From our previous study [5], we found that the error modeling in observations is critical to the data assimilation process. Recall the Kalman gain matrix (**K**), the important term in the EnKF algorithm,

$$\mathbf{K} = \mathbf{P}_h^f \mathbf{H}_h^T \left(\mathbf{H}_h \mathbf{P}_h^f \mathbf{H}_h^T + \mathbf{R} \right)^{-1}, \tag{3}$$

where, **R** is the observation error covariance matrix. In this study, the exact solution are used as the observations, which leads the observation error covariance matrix become nearly zero. Naturally, the performance of the Kalman gain matrix during the data assimilation process is maximized. By examining the Kalman gain matrix at various times, we found that the magnitude of the matrix entries is much larger at early times, indicating that the propagated solutions are pulled more strongly toward the truth at that time. This also explains that the cases always take $50 \sim 150$ iterations to converge their proper values. Interestingly, the base case where the assimilation is performed using the "Full-Observation" data has the largest error in its final estimate in comparison to its true value. Next, we explain the oscillatory phenomenon and the large error in the base case by examining the error covariance matrix.

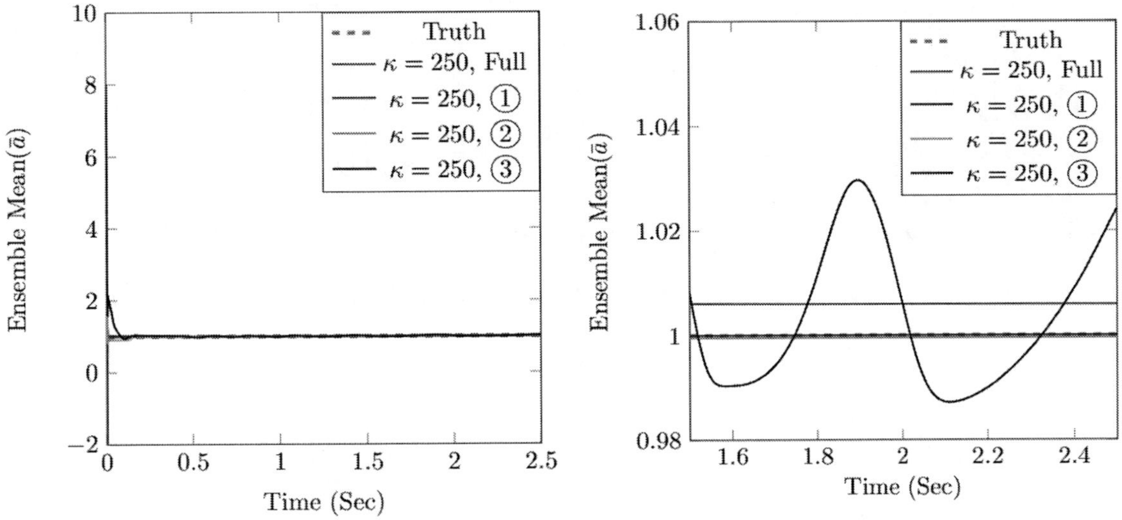

(a) Overall View

(b) Zoom-in View

Figure 4: The convergence history of model parameter a.

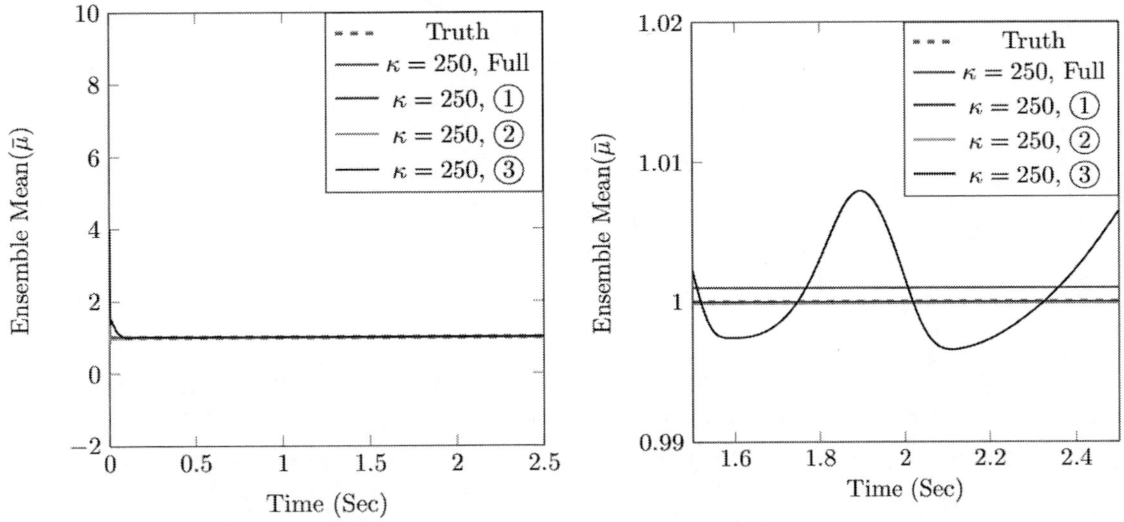

(a) Overall View

(b) Zoom-in View

Figure 5: The convergence history of model parameter μ.

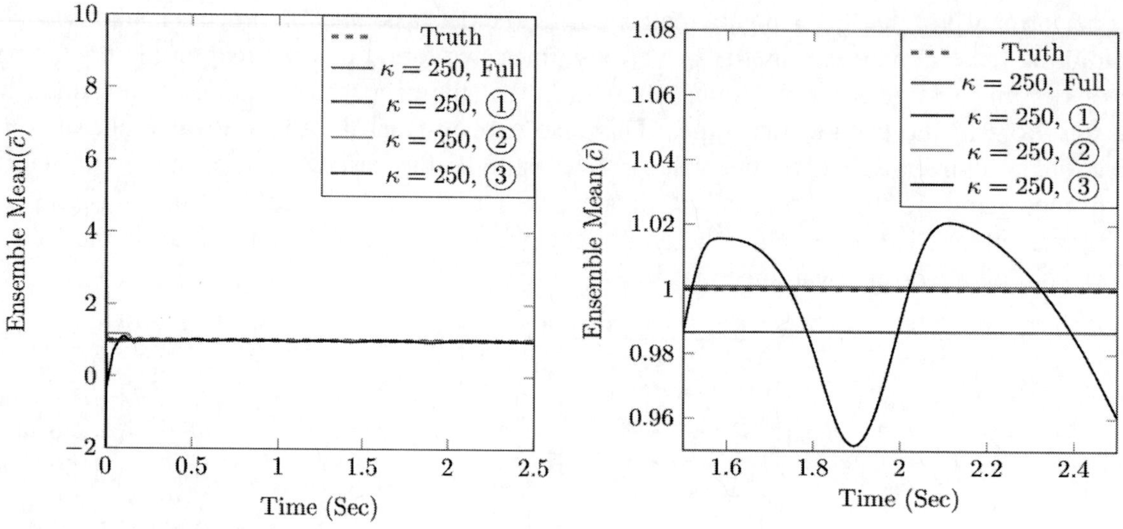

(a) Overall View (b) Zoom-in View

Figure 6: The convergence history of model parameter c.

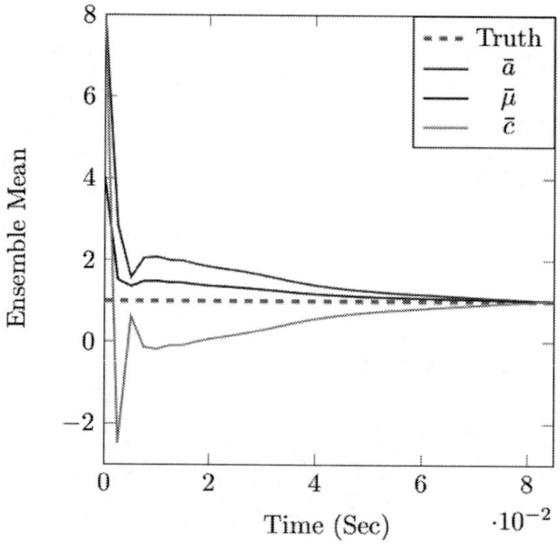

Figure 7: The convergence slop comparison

By comparing the convergence slop of each parameters in Fig. 7, we observed that the source term coefficient, c, is more difficult to converge than others. The covariance correlation between any two model parameters is output for the base case and the three types of spatial distribution (Data

1, Data 2 and Data 3) at an arbitrary time step:

$$\mathbf{P}^f_{full} = \begin{bmatrix} 4.31912e^{-11} & 6.71273e^{-12} & -1.0659e^{-10} \\ 6.71273e^{-12} & 1.04417e^{-12} & -1.65589e^{-11} \\ -1.0659e^{-10} & -1.65589e^{-11} & 2.63148e^{-10} \end{bmatrix} \begin{matrix} a \\ \mu \\ c \end{matrix}, \quad \text{columns: } a, \mu, c$$

$$\mathbf{P}^f_{D1} = \begin{bmatrix} 2.97266e^{-11} & 3.44601e^{-12} & -8.29026e^{-11} \\ 3.44601e^{-12} & 4.05621e^{-13} & -9.54489e^{-12} \\ -8.29026e^{-11} & -9.54489e^{-12} & 2.32069e^{-10} \end{bmatrix} \begin{matrix} a \\ \mu \\ c \end{matrix}, \quad \text{columns: } a, \mu, c$$

$$\mathbf{P}^f_{D2} = \begin{bmatrix} 9.59858e^{-12} & 1.80177e^{-12} & -2.11352e^{-11} \\ 1.80177e^{-12} & 3.67814e^{-13} & -3.65798e^{-12} \\ -2.11352e^{-11} & -3.65798e^{-12} & 4.98198e^{-11} \end{bmatrix} \begin{matrix} a \\ \mu \\ c \end{matrix}, \quad \text{columns: } a, \mu, c$$

$$\mathbf{P}^f_{D3} = \begin{bmatrix} 5.11749e^{-8} & 1.36722e^{-8} & -8.33682e^{-8} \\ 1.36722e^{-8} & 3.65278e^{-9} & -2.22731e^{-8} \\ -8.33682e^{-8} & -2.22731e^{-8} & 1.35814e^{-7} \end{bmatrix} \begin{matrix} a \\ \mu \\ c \end{matrix}, \quad \text{columns: } a, \mu, c$$

From the matrix, we can observe that covariance correlation value is larger in the base case than that in the Data 1 and Data 2 cases. However, the largest covariance correlation value is obtained in the Data 3 case. Essentially, the source term coefficient is always showing negative covariance correlation to the other two parameters, which implies that it can be considered as the most sensitive parameter among the multiple parameters. And, the **P** matrix directly impacts the **K** matrix, which illustrates that the observation location is critical to the evaluation of covariance correlation during the data assimilation process. In this case, Data 3 seems critical. It is consistent with the convergence histories plots. In addition, this is consistent with the fact that the slope in the solution change in space increases as the location increases from the origin.

3.2 The Impact of Temporal Distribution

Assume data is available in each cell and the data assimilation frequency κ is again 250. The impact of the temporal distribution of data on the assimilation process is investigated. However, the cases where the assimilation is performed for the time zone 2 and the time zone 3 failed. Fig. 8 plots the convergence history for the ensemble mean $\bar{a}$, $\bar{\mu}$ and $\bar{c}$ for the case where data is assimilated for time zone 1 and the case where data is assimilated for full time zone ($\mathrm{T_f} = 0 \sim 2.5$sec).

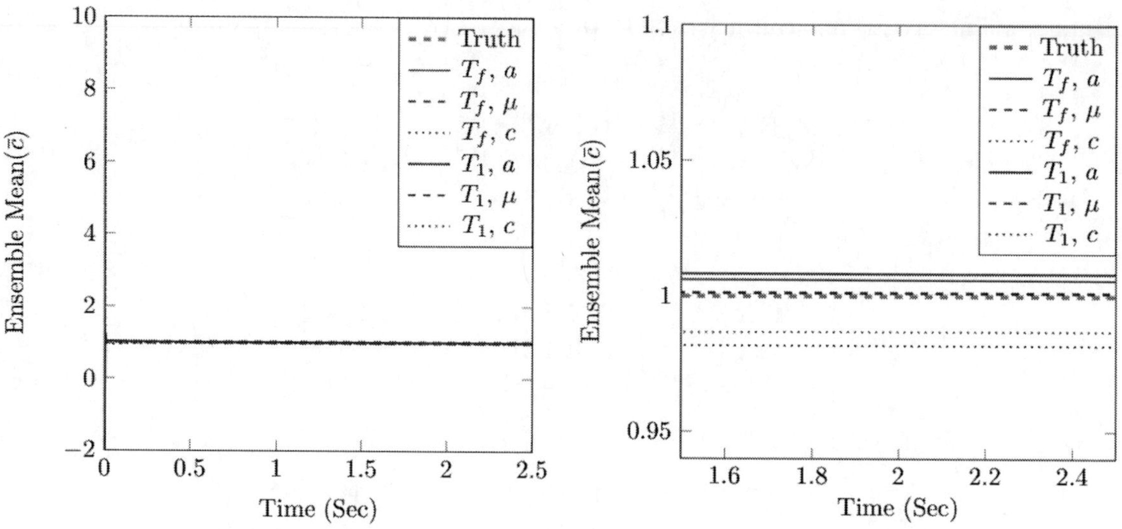

(a) Overall View (b) Zoom-in View

Figure 8: The convergence history of model parameters.

The failed cases are interesting, because they simply indicate that it is always better to assimilate data as early as possible if data is available. The error covariance in these two cases are large, and the strong nonlinear covariance correlation between the model parameters in the Kalman gain matrix results in an excessive correction, leading to a nonphysical diffusion coefficient and thus failing the simulation. This is potentially coincident with the fact that EnKF can have a difficulty in coping with nonlinear problems. To further verify this, we run a set of cases, using the same temporal distribution of the data by only estimating a single model parameter. In Figs. 9 - 11, the convergence history is plotted for each individual ensemble mean, and clearly each model parameter can converge toward the true value with a smooth convergence trajectory. Therefore, the nonlinear interaction of the model parameters plays an importance role on the estimate of the multiple model parameters using EnKF. For general EnKF's application, if the model and observation operators are linear, and the errors are normally distributed (i.e., Gaussian), the Kalman gain matrix would lead an exact, unique, stationary point of the objective function in the DA problem at the analysis stage [6]. For nonlinear problem, the Kalman gain matrix can only give a linear/first-order approximation to an optimal solution at the analysis stage. While some constraints can be enforced in the CFD model, for example, flooring/ceiling the diffusion coefficient, it will be worthwhile to investigate other data assimilation method, such as maximum likelihood ensemble filter (MLEF), for nonlinear problem.

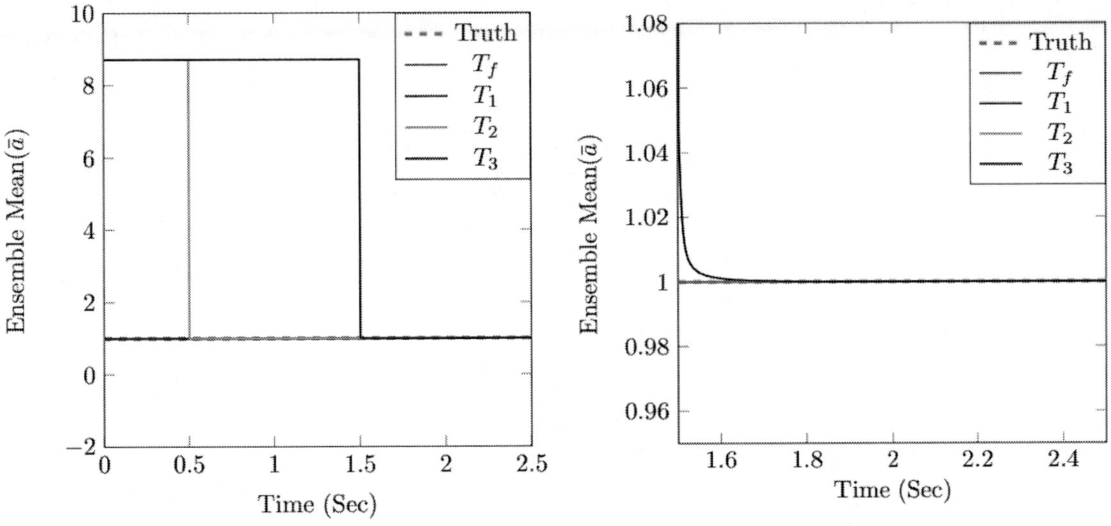

(a) Overall View (b) Zoom-in View

Figure 9: The convergence history of model parameter a.

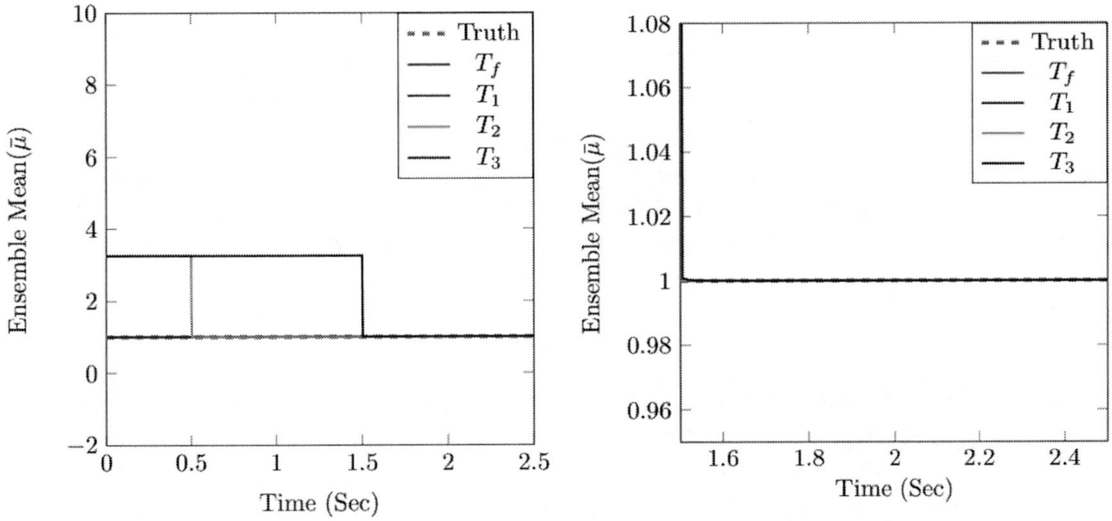

(a) Overall View (b) Zoom-in View

Figure 10: The convergence history of model parameter μ.

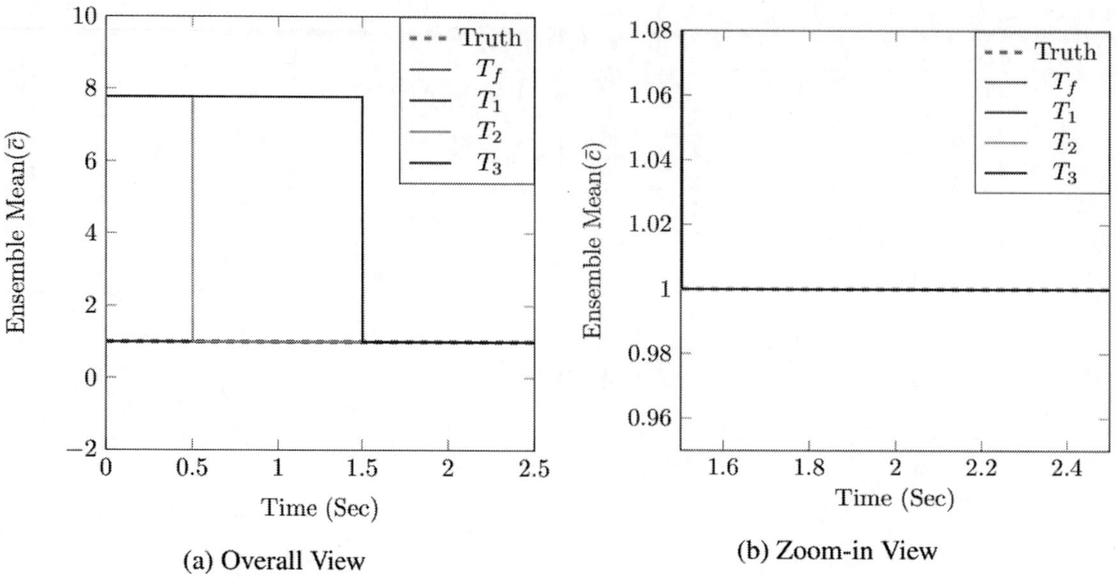

Figure 11: The convergence history of model parameter c.

4 Future Work

Through this study, we examined the impact of data spatial distribution and temporal distribution on estimating the multiple parameters simultaneously by EnKF algorithm. The result of this work provide insight into the effect of spatial and temporal distributions of how the data have significant impact on the estimation of multiple model parameters, show the most sensitive model parameter, and demonstrate the limitation of EnKF for the CDR problem. An immediate follow-up study will be focused on the investigation of other DA method (e.g. MLEF) for nonlinear problems.

References

[1] X. Gao, Y. Wang, N. Overton, I. May, , and X. Tu. Estimation of flame speed model parameter using ensemble kalman filter algorithm. In *Fall Meeting of the Western States Section of the Combustion Institute*, October 2015. WSSCI 2015-134IE-0013.

[2] X. Gao, Y. Wang, N. Overton, I. May, and X. Tu. Data assimilated computational fluid dynamics algorithm for combustion. AIAA 2016-1810, 54th AIAA Aerospace Sciences Meeting, 2016.

[3] X. Gao, Y. Wang, N. Overton, M. Zupanski, and X. Tu. *Journal of Computational Science*, 21 (2017) 38–59.

[4] Tony W. H. Sheu, S. K. Wang, and R. K. Lin. *Journal of Computational Physics*, 164 (2000) 123–142.

[5] Y. Wang and X. Gao. Covariance correction to the ensemble Kalman filter in a data-assimilated CFD algorithm for the convection-diffusion-reaction equation. In *23rd AIAA Computational Fluid Dynamics Conference*, number AIAA 2017-3630. AIAA AVIATION Forum, 2017. https://doi.org/10.2514/6.2017-3630.

[6] W. Lahoz, B. Khattatov, and R. Ménard. *Data Assimilation: Making Sense of Observations*. Springer, 2007.

Western States Section of the Combustion Institute - Fall 2017 Technical Meeting
Hosted by University of Wyoming
October 2-3, 2017.

INVESTIGATION OF ARGON OXY COMBUSTION WITH LOW MOLECULAR WEIGHT FUELS: IMPROVING THERMODYNAMIC EFFICIENCY

Charles Scudiere[1] *Jyh-Yuan Chen*[1] *Robert Dibble*[1,2]
Miguel Aznar[1,3]

[1]*Department of Mechanical Engineering, University of California, Berkeley, USA*
[2]*King Abdullah University of Science and Technology, Thuwal, Makkah Province, Saudi Arabia*
[3]*Lawrence Berkeley National Laboratory, Berkeley, USA*

Utilizing fuel and oxidizing mixtures of oxygen and argon in spark ignition (SI) engines have been shown to increase thermal efficiency relative to standard fuel and air oxidizing mixtures due to the increased ratio of specific heats [1], [2] et. al. Motivated by these results, stoichiometric methane mixtures with various oxidizing volumetric ratios of argon and oxygen (90/10, 85/15, 80/20) are explored and studied in a simulated research engine in a fuel port injection configuration. Computational fluid dynamics (CFD) and combustion simulations using reduced chemical kinetic mechanisms were performed to investigate overall combustion performance and thermodynamic efficiencies. A successive approach of increasingly higher dimensional models (zero dimensional, quasi one dimensional, and three dimensional CFD) with adiabatic, Angelberger, and Woschni heat loss models are compared in the context of thermodynamic theory. Significantly higher efficiency gains on the order of 10% relative to standard fuel and air oxidizing mixtures are predicted.

1 Introduction

Temperature is an important state variable to all gas dynamics pertinent to combustion and is the most significant factor to theoretical maximum thermodynamic efficiency, among other key combustion characteristics. In terms of cycle energy efficiency, exhaust temperature leaving the engine places an upper bound on the efficiency as exhaust gases carry away a portion of the cycle's energy. The ideal Carnot Cycle gives the maximal theoretical efficiency, limited by the ratio of the inlet and outlet temperature.

When thermal energy is added to a system operating at typical atmospheric and engine operating conditions, energy is absorbed and stored at the molecular level. The shape and molecular structure of each of the species in the mixture dictates the different modes of energy storage available. The ways energy can be stored at the molecule level include translational, rotational, and vibrational energy modes.

With more energy storage modes able to hold energy, there is a corresponding increase in the number of microstates available to the system. More concretely, the quantity of microstates (Ω) of a given system increase with increasing energy storage modes (μ). The entropy change for a given hypothetical change in the number of energy storage modes is therefore,

$$\frac{\partial S}{\partial \mu} = \frac{k_B}{\Omega}\frac{\partial \Omega}{\partial \mu} \tag{1}$$

This is directly tied to temperature, where temperature is defined in statistical thermodynamics as:

$$T \equiv \frac{\partial U}{\partial S}|_{N,V}, \tag{2}$$

where U is the internal energy of the system, S is the system entropy, V is the system volume held constant, and N is the number of moles in system also held constant.

Given equations 1 and 2 and comparing two hypothetical ideal isentropic cycles with a equal system internal energy, a lower number of energy storage modes results in higher system temperatures, while a higher numbers of energy storage modes results in lower temperatures. Therefore, a higher ratio of inlet and exhaust temperatures is possible with species which have fewer molecular energy storage modes in a thermodynamically adiabatic perspective.

Systems which have fewer energy storage modes carry less energy at a given temperature, commonly referred to as heat capacity. Therefore, comparing monoatomic dilution species with diatomic dilution species at equal exhaust temperatures, systems with monoatomic species will have higher temperatures to utilize more of the available energy in the cycle and exhaust less thermal energy to the surroundings. This energy is directly available and utilized for conversion to additional mechanical work in an engine. At the same time, higher temperatures increase the heat transfer rate.

Exploring the use of argon in engines to realize the theoretical efficiency gains has been shown to work with hydrogen engines [1], [2] et. al. Less has been shown with gaseous oxycombustion of methane and other hydrocarbons mixed with high argon dilutions in engine applications. The competing effects of the increase in theoretical potential efficiency and heat losses is further explored in this paper.

2 Results

2.1 Quasi 1-D

An initial quasi one dimensional in space parameter study using Steady premixed CH_4-O_2-Ar adiabatic flames at various temperatures and pressures using CHEMKIN PREMIX with multi-component transport with a reduced chemical kinetic mechanism was performed to determine the key combustion characteristics at various temperatures and pressures. The ideal flame characteristics of burning methane (CH4) in the following various oxidizing mixture concentrations of argon and oxygen: argon % / oxygen %: air (01/21), 75/25, 80/20, 85/15, 90/10. These simulations demonstrated a relative 30% higher flame speeds with argon as a diluent compared to a nitrogen diluent (air) for the same oxygen concentrations.

2.2 Two Zone Model

A Two Zone model study of Port Injected (PI) engine mode over a range of oxidizing mixture concentrations (argon % / oxygen %: air (01/21), 75/25, 80/20, 85/15, 90/10) highlighted promising oxidizing mixture concentrations. With the higher flame speeds and temperatures achieved in the argon mixtures, the empirical Woschni heat loss model[3] modeled the heat losses to the engine walls. Spark timing, configuration, and dilution concentration was varied to eliminate knock and maximize thermal efficiency of the cycle. The simulated pressure traces at the maximally efficient oxidizer mixture, with an engine compression ratio of 10, are presented in Figure 1.

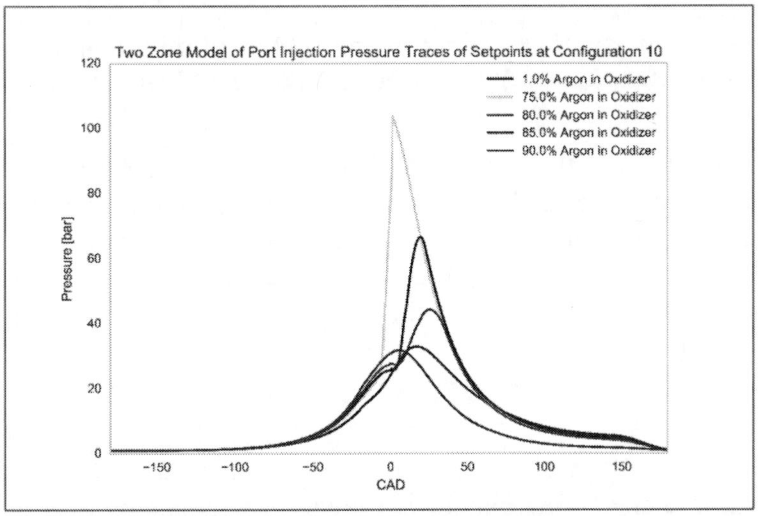

Figure 1: Two Zone model pressure traces of PI mode methane combustion with various oxidizer mixtures at the engine configuration 10 using the Woschni heat loss model vs. CAD at optimized compression ratio and spark timing.

Despite the increased heat losses from the increased gas temperature, higher efficiencies on the order of 10% overall and 30% relative were shown in the model study with oxidizing mixtures of 85% Argon and 15% Oxygen in the mid-pressure regions compared with air among all of the spark timing (0-20bTDC) and engine configurations (Compression Ratios of 3.0-17.5) considered.

2.3 Three Dimensional Port Injection Simulations

Upon analysis of the Two Zone model predictions, three dimensional simulations were performed in a Port Injection engine mode utilizing a commercial CFD solver with a coupled chemistry solver. Two reduced chemical kinetic mechanisms were used with mutli component transport to simulate the chemical kinetics and thermal transport properties. Comparisons were made to the Two Zone model with good agreement. See Figure 2 for the PI mode temperature plots for a visualization of the three dimensional temperatures following ignition. Graphs were cut along the y-axis showing vertical and horizontal cross-section as the flame propagated through the chamber.

With constant temperature wall boundary conditions utilizing the Angelberger heat loss model [4], the ignition and flame spread of the Port Injection engine modes were observed to propagate

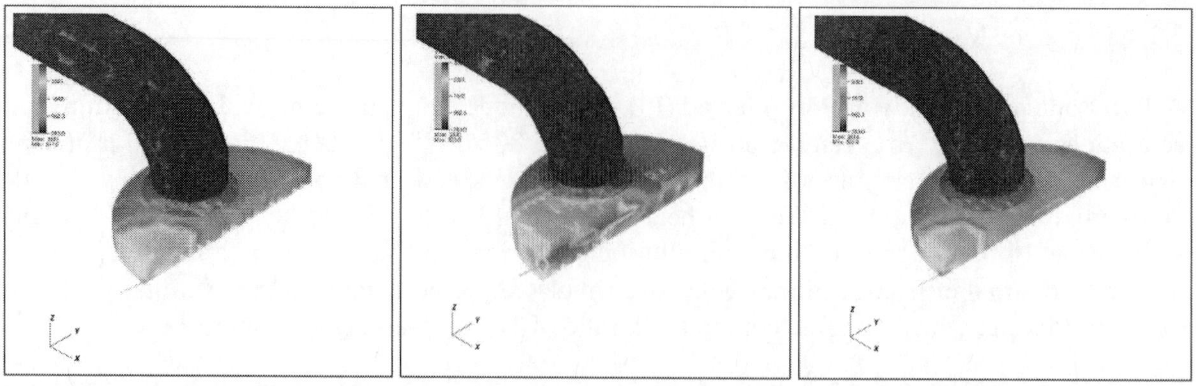

Figure 2: 3-D PI engine mode after 2.3 ms. Combustion of methane and air (left), methane and 80% Ar/20%O_2 oxidizer (center), methane and 85% Ar/15%O_2 oxidizer (right).

well for the argon mixtures of 80% argon and 20% oxygen and the 85% argon and 15% oxygen mixtures. In all of the cases considered for this system, the flame speeds of the air mixtures were lower and difficult to sustain both a high temperature reaction zone and subsequent complete combustion within the engine cycle. On the other hand, argon rich oxidizing mixtures had higher mean temperatures and flame speeds, facilitating higher combustion and energy efficiencies.

3 Discussion

Homonuclear diatomic species, such as nitrogen, have three translational, two vibrational, and two rotational independent energy storage modes. Conversely, monoatomic species, such as argon, only have three translational independent energy storage modes. The difference of 4 independent molecular energy storage modes available is tied to the system entropy at a given state. For each of the models, higher gas temperatures was achieved by the fewer energy states.

Heat losses from the higher mean gas temperatures were not modeled in the adiabatic quazi 1D cases and showed a higher temperature and flame characteristics than the higher order counterparts.

The two-zone model better approximated the temperature of the bulk gas with the Woschni heat loss model but models the thermal boundary layer, limiting the accuracy in determining an accurate temperature distribution in the chamber for the gas mixtures of interest. It was able to utilize the bulk gas and wall temperatures to estimate the losses empirically. The model also predicted large heat release rates for the PI mode with high oxygen concentration leading to undesirable pressure rise rates and peak pressures in the cylinder. In particular, reviewing figure 1, the 75% argon oxidizing mixture has a significantly higher and undesirable rapid pressure rise rate following ignition.

Following recent experimental results of running a CFR engine in a PI engine mode operation, analysis shows excellent agreement of the Two Zone model with the measured experimental values. These experimental results also confirmed the superior oxidizing mixture of 85% argon and 15% oxygen in lowering specific fuel consumption predicted with the Two Zone model.

In agreement with the Two Zone model, the three dimensional CFD model 85% argon, 15% oxygen out performed the higher and lower oxygen oxidizer mixtures. This higher order model was also able to more accurately resolve the thermodynamic states of the system and resolve the thermal boundary layer. Although the mean gas temperatures were at a higher temperature for the argon dilution gases, the thermal boundary layer did not penetrate far enough into the bulk gas to significantly out-weight the increased temperature and pressures. This thermal penetration time was too low compared to the engine cycle time. The mixtures never reached a steady state value before the expansion from the piston movement cooled the gas. The overall cycle had a net positive boost in thermal efficiency and was in fair agreement with the previous models with gains on the order of 10% overall and 30% relative to a efficiency from a cycle with air.

4 Conclusion

Significantly higher combustion efficiencies were noted in the argon mixtures on the order of 10% overall and 30% relative than simulated with the air mixture, with subsequent higher thermal efficiencies. The increased combustion stability and the faster temperature rise rate as compared with the heat losses was a competing effect with the decrease in the thermodynamic microstates available to the engine system. With successively higher order models, higher resolution of the thermodynamic and heat transfer, the effects of fewer molecular storage modes became increasingly apparent. The temperatures were achieved and subsequent higher efficiencies were realized. Comparisons with recent experimental results are ongoing.

With further incorporating additional insight from our recent experimental results, the author is optimistic that higher efficiencies approaching the theoretical limitations can be achieved in other configurations such as HCCI, DI, and other modes.

Acknowledgments

Many thanks to the students and the PIs in the Combustion Modeling Lab, the Combustion Analysis Laboratory, and the Combustion and Fire Processes Laboratory as well as my advisor for all of their support, guidance, advice, and feedback.

This investigation was conducted at the University of California, Berkeley, USA and was supported in part by the CEC EISG PIER grant # 58074A/14-07G. Additionally, this research used Chemkin and an in-house FORTRAN Engine Analysis code of which there are multiple authors and contributors. In addition, CONVERGE[5] was used to perform the three-dimensional CFD computations. Countless open source and GNU licensed software packages were utilized through out the progress of this project and report including VisIt[6] and Gnuplot for post processing.

References

[1] Vi Hai Rapp. *Investigation of Alternative Fuels and Advanced Engine Technology: Improving Engine Efficiency and Reducing Emissions.* (PhD Doctoral Dissertation). UMI: 3616245, Mechanical Engineering Department, University of California, Berkeley, 2011.

[2] Rentaro Kuroki, Akira Kato, Eiichi Kamiyama, and Daisaku Sawada. (2010) journal.

[3] G. Woschni. *Technical Report 670931, SAE Technical Paper*, (1967) journal.

[4] C. Angelberger, T. Poinsot, and B. Delhaye. *SAE paper 972881*, (1997) journal.

[5] Convergent Science. (2016) journal.

[6] Hank Childs, Eric Brugger, Brad Whitlock, Jeremy Meredith, Sean Ahern, David Pugmire, Kathleen Biagas, Mark Miller, Cyrus Harrison, Gunther H. Weber, Hari Krishnan, Thomas Fogal, Allen Sanderson, Christoph Garth, E. Wes Bethel, David Camp, Oliver Rübel, Marc Durant, Jean M. Favre, and Paul Navrátil. VisIt: An End-User Tool For Visualizing and Analyzing Very Large Data. In *High Performance Visualization–Enabling Extreme-Scale Scientific Insight*, pages 357–372. Oct 2012.

Western States Section of the Combustion Institute – Fall 2017 Meeting
Hosted by the University of Wyoming
October 2-3, 2017

Preliminary Design and Development of an Ultra High Efficiency Stirling Engine Heater Head

Miguel Sierra Aznar[1*], Alex Frank[1,2], Peter L. Therkelesen[1], Vi Rapp[1] and Robert K. Cheng[1]

[1]*Enviromental Technology Area, Lawrence Berkeley National Laboratory*
Berkeley, CA 94720 United States
[2]*Mechanical Engineering Department, University of California-Berkeley*
Berkeley, CA 94720, United States
[*]*Corresponding author: m.sierra.aznar@lbl.gov*

Abstract: Stirling engines have long offered the promise of ultra-high efficiency operation with near zero pollutant emissions if coupled with the right burner and heat transfer technologies. However, modern Stirling engines lack an ultra-high efficiency heater head that combines low emission combustion with an efficient heat transfer system capable of transferring the vast majority of heat from combustion products to the working fluid in the engine. This paper presents the preliminary efforts by Lawrence Berkeley National Laboratory (LBNL) to develop and optimize an integrated burner and heat exchanger system based upon the Low Swirl Burner for small scale (1kW$_e$) Stirling engine units. This text presents a short review of the state of the art on combustion system for Stirling engines and an analysis of the potential key concepts open for improvement.
Keywords: Stirling Engines, Low Swirl Burner, heater head, distributed power.

1. Introduction.

Distributed power generation (DG) is becoming a development priority for a diverse set of factors REF. One factor that hinders wider integration of renewable energy sources (RES) is the impact they have on the power system infrastructure. Renewable farms are subject to weather and climate conditions: they are built where natural resources are abundant and they operate when nature dictates. As a result, power is generated far away from the consumer, limiting the net power that gets delivered. Transmission losses are estimated to be 5% of the total power generated as indicated by the U.S Energy Information Agency (IEA) and the world bank [1][2]. Moreover, when the sun is shining and wind is blowing, large power plants must reduce their power output, operating far from their optimal baseload conditions. This, in turn, causes a substantial decrease in their efficiency, cost effectiveness and ultimately, an increase in their emissions (3). DGs can provide the grid with a generous population of small power systems, operating in an ON/OFF mode, effectively generating power when needed at peak efficiency and at optimized operating conditions. The U.S. Department of Energy (DOE) estimates that the combination of small scale DGs fueled by natural gas, readable available in the U.S., could yield a 10% reduction in CO_2 emissions from the power sector, while supporting its RES portfolio [3].

Seeking to address this need, through its advanced research projects agency – energy (*arpa-e*), DOE has implemented a program to aid in the development of 1 kW$_e$ (electrical output) scale generators for small electrical and thermal systems (GENSETS) in the residential sector. The scale of the setup has been dictated by a combination of factors and assumptions, primarily the average heat and power needs of the average U.S. household and an assumed electric efficiency of the system of 40%. The GENSETS program expected outcome is the realization of a combine heat and power (CHP) system that meets or exceed the specifications listed in Table 1. A more extensive list of specifications is targeted by *arpa-e* but only the items listed here are of technical nature relating to the realization of this project.

Table 1. GENSETS program primary target specifications[3].

Property	Specification
Electric Power Output	$1kW_e$
Fuel to Electricity efficiency	40%
Heat power output	>1 kWt/1kW_e
Heat output quality	>80°C
NOx	$\leq$ 0.07 lb/MWh / 0.031 g/kWh
CO	$\leq$ 0.1 lb/MWh / 0.045 g/kWh
VOC	$\leq$ 0.02 lb/MWh / 0.009 g/kWh
Particulate Matter	$\leq$ 0.4 g/MWh / 0.0004 g/kWh

It is clear from the above that the targets are ambitious and the constrains are challenging. To aid in the development of a Stirling Engine (SE) CHP system that accomplishes the above targets, this paper presents the preliminary efforts by Lawrence Berkeley National Laboratory (LBNL) in the optimization of an integrated burner and heat exchanger system based upon the Low Swirl Burner (LSB) as the combustion system for small scale ($1kW_e$) SE units.

2. Stirling Engine and Heat Transfer.

SE have existed since the 1800's [4]. Much work has been done to optimize its thermodynamic efficiency, but it is only recently that interest has been rekindled due to the energy transition the world is undertaking [5]. Considerable innovation is still required to reach a level of cost effectiveness such that it becomes competitive with current technologies in the market. The main allures of SEs are their fuel flexibility and wide temperature operational range. Because of the external nature of their heat input, they can use almost any fuel and can operate at substantially low temperatures, as long as a lower temperature heat sink is available [6]. However, as the external nature of the heat source affords the aforementioned capabilities, it also comes at the cost of limited heat transfer rates. The optimum tradeoff between heat transfer and thermal efficiency needs to be found for optimum performance.
The SE maximum theoretical efficiency is dependent upon the magnitude of the temperature difference between heat source and heat sink as described by the Carnot cycle.

$$\eta_c = 1 - \frac{T_c}{T_h} \tag{1}$$

When the temperature of the cycle $T_h[K]$ is increased to reach higher thermodynamic efficiencies, the temperature gradient across the external heat source $T_s[K]$ and the engine T_h $[K]$ is reduced.

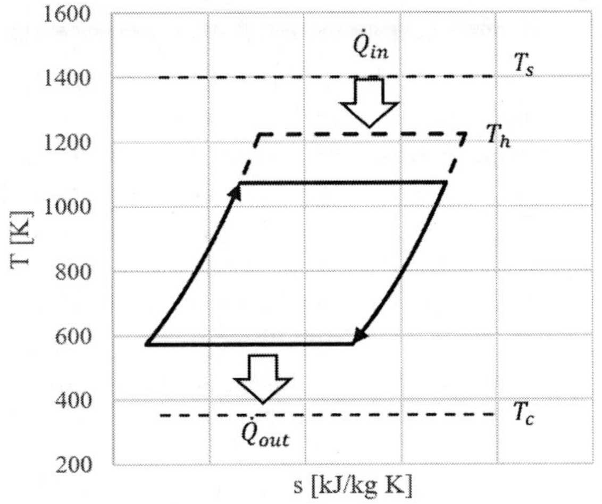

Consequently, the overall achievable heat transfer rate $\dot{Q}\,[\frac{kW}{m^2}]$ is proportionally reduced.

$$\dot{Q} = UA(T_s - T_h) \tag{2}$$

Where $U[\frac{W}{m^2K}]$ is the overall heat transfer coefficient of the heat exchanger system between the engine head and the heat source. The engine's power output could be maintained by an increase in heat transfer area $A\ [m^2]$ but this would defeat the purpose of distributed scale system as the volume of the system would be too large for residential use. An obvious approach to overcome the resulting heat transfer limitation is to raise the source temperature $T_s[K]$ further. At this point, heat transfer is no longer the performance limiting factor, but the mechanical properties of the interfacing material become the limitation. Pressure difference across the engine wall are often larger than 2MPa [7][8]. At elevated temperatures (500-1000 °C) the yield strength of a common material such as stainless steel is substantially reduced and thus, the longevity of the SE is jeopardized. For this purpose, materials with excellent mechanical properties are used, such as Inconel 625 or Mar-M-247. These allows are nickel based and are resistant to oxidizing and corrosive environments[9][10]. These superalloys, although resistant to substantially severe operating conditions are limited to temperatures below 982 °C. It is clear that on the SE side, an increase in cycle temperature combined with superalloys material may afford an increase on the heat source temperature while maintaining the robustness of the system. Unfortunately, superalloys are not well suited for heat transfer and its thermal conductivity and diffusivity is one order of magnitude below top thermal conductors such as copper or silver. For that reason, a properly sized interface wall is necessary to reduce the impact of low conductivities. Furthermore, and favored by the close loop configuration of SEs, an improvement on their heat transfer rate is achieved by using tailored working fluids. Many studies have shown H_2 to be the ideal working fluid for its mass and heat transfer qualities, mainly low viscosity $\mu\ [m/s^2]$ and high thermal conductivity $\kappa\ [W/mK]$, but is often disregarded as it raises safety concern for industrial or residential applications [11][12][13]. On the other hand, He, which presents no safety concerns, possess similar thermal properties to that of H_2 with slightly lower thermal conductivity and higher dynamic viscosities. As a result, there is a slight reduction in the potential heat transfer rate achievable.

With all the above (Sterling cycle temperature, specific working fluid and material properties) the remaining factor influencing heat transfer is the heat source. For the application targeted in this work, combustion of methane in air is purposed as the heat source. Methane has the potential to provide flame temperatures (at normal pressure and temperature conditions) of 1900 °C. However, extreme high temperatures, above 1500 °C will trigger the generation of thermal NOx preventing the system from fulfilling the emission targets stated in Table 1. Flame temperature can be reduced by leaning the fuel air mixture. At sufficiently low equivalence ratios $\phi < 0.7$, generation of NOx is substantially reduced and fully diminished at $\phi < 0.5$. However, leaning the fuel air mixture reduces the flame stability and may approach the lean blow off limit depending on the burner configuration. The burner used for this work is the LSB (fig. 1). The scale of the burner is chosen to deliver 3.3 kW of thermal power, assuming a SE minimum efficiency of 30%. At the recommended modified swirl number (0.7-0.8) and velocities (10-20 m/s), a burner outer diameter of 14mm is obtained for a nominal $\phi = 0.65$. Following the previous studies from this group, the optimum design for this burner was found to have 16-constant radius, at 37 degree, thin-walled vanes and an outer to inner channel diameter of 0.66. The center plate has 34 strategically placed 0.84 mm diameter holes, creating a geometric open area of 28.2% [14][15].

Figure 1. Low Swirl Burner and center plate (14 mm OD, 37-degree vanes, 0.66 ID/OD).

The blow off limit for this burner was previously found to be at $\phi = 0.58$ under ambient temperature and pressure conditions. Consequently, operating this burner at the nominal conditions should be sufficient to provide a heat source temperature of 1500 °C while preventing NOx formation. This needs to be further downgraded to temperatures the interface wall can withstand and thus it needs to be diluted with air or recirculated exhaust gas (approx. +5% mass) down to average mixture temperatures of 982 °C.

Table 2. Heat transfer problem parameters.

Wall Material	Inconel 625
Working Fluid	He
Max Cycle Temperature	800-982 °C
Heat source temperature	982 °C
Power input	3 kW thermal
Fuel	CH_4
Mixture Equivalence ratio	0.65
Mass recirculation	5%

Another important factor regarding the effective heat transfer rate is the flow regime within the heat exchange system. Forced flow will decrease the thickness of the velocity and thermal boundary layers thus increasing the heat transfer rate. The velocities found in our system has been measure through particle image velocimetry and results are shown in Figure 2. At inlet velocities of 8m/s, ambient air conditions and with the 14mm pipe inner diameter, the incoming flow Reynolds number falls right below 8000. After the flame, a decrease in gas density due to the elevated temperatures, an enlargement of the burner diameter (from 1 to 3) and the radical increase in dynamic viscosity cause the flow to slow down and transition to laminar regime. Mean velocities can be observed to be around 0.7 times lower than that of the bulk flow. At this stage, the Reynolds numbers for the exhaust stream yields numbers below 1000, well into the laminar regime. Turbulent intensity, a measure of the magnitude is also low, in the order of 1m/s. At higher velocities, little improvement is achieved as shown in the far right of Figure 2.

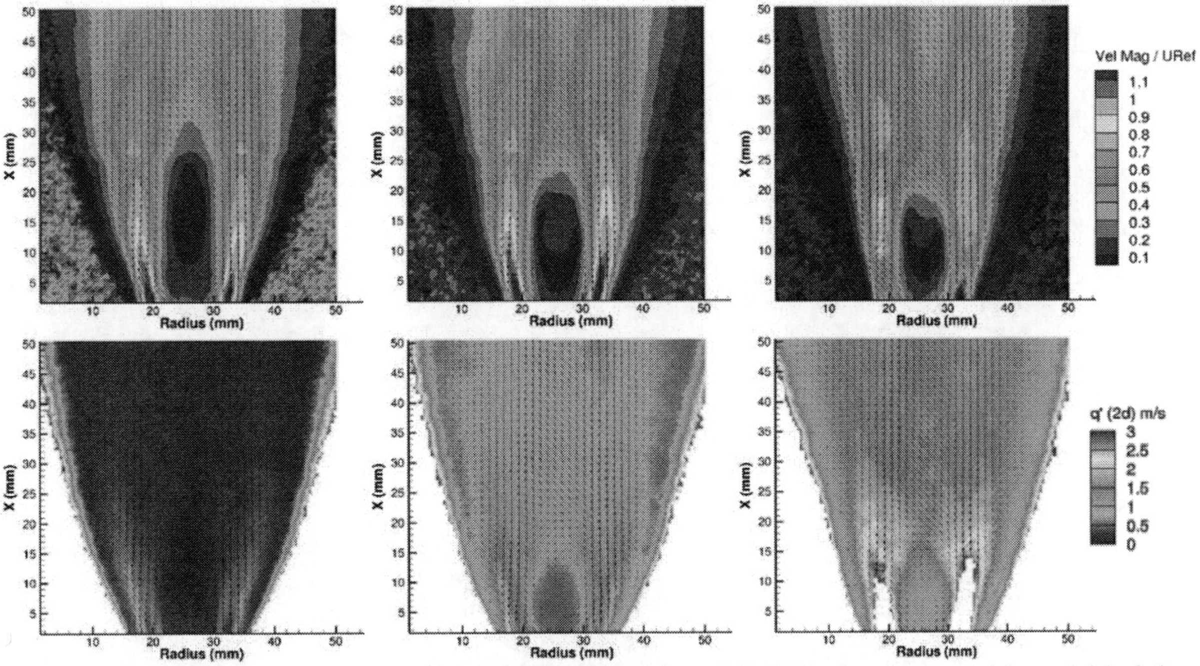

Figure 2. Flow field pass a 14mm LSB for a reacting $\phi = 0.7$ CH4 air mixture with variable inlet flow velocity from left to right: 4, 8, and 13m/s. Top: Normalized velocity field; Bottom: turbulent intensity q' in m/s.

With all the boundary conditions properly defined, we ambition to perform a parametric study of three different heat exchangers optimized for the use of LSB in SEs. We intent to explore the effect of obstructing element ahead of the flame. Preliminary experimental work has shown little to no negative effect on the flame stability. If this indeed applied to a broad set of shapes, this feature could be exploited to increase the hydrodynamic characteristics of the flow, ultimatey increasing the heat transfer rates within the heat exchanger chamber.

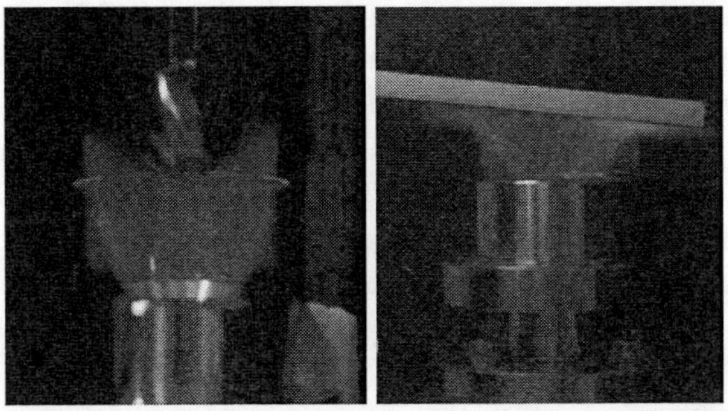

Figure 3. Preliminary evaluation of object intrusion downstream the LSB flame with no visible effect on the stability of the flame. Left: spherical object, Right: flat rectangular object.

In the best-case scenario, the exhaust gas will leave the heat exchanger at the SE temperature. This means that nearly two thirds of the energy released by the fuel will not reach the working fluid and will leave as waste heat. Heat recuperation is required if the fuel to electricity efficiency is to be achieved.

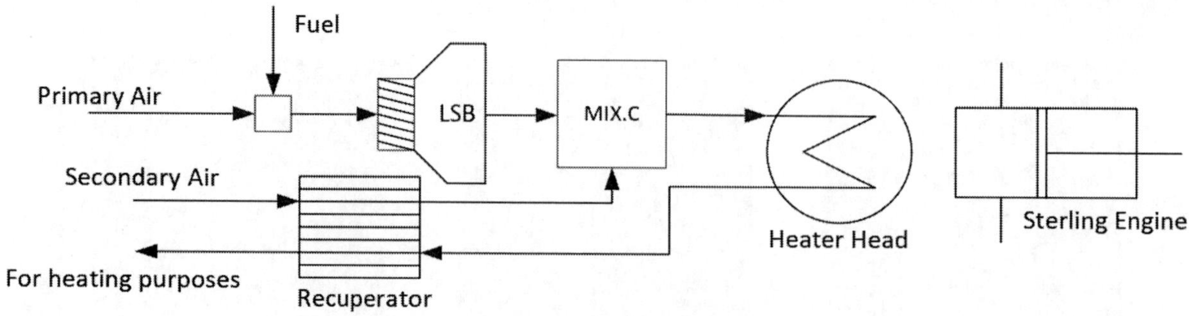

Figure 4. Heater head with regenerator setup.

To evaluate the performance a SE+LSB we will construct a setup as the one depicted in Figure 4. With it we intent to study the influence of intrusion object into the LSB flame and deliver a cost effective and clean solution for the effective integration of SE. Figure 6 shows the initial test rig at LBNL.

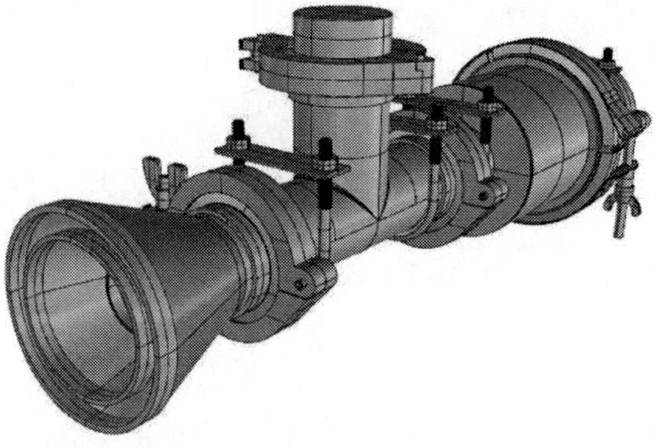

Figure 5. Sterling engine heater head test setup design at LBNL.

3. Conclusions

Heat transfer is a well-known phenomenon to the power generation community as much of their operational constrains are due to heat losses across the system boundaries. Ironically, in the case of the SE, lack of enough heat transfer is one of the major reasons for SEs poor performance in real applications. The features of the LSB may prove to be a game changer on the development of SE for distributed generation applications. So far, our preliminary evaluations have shown the LSB to be a perfect candidate to serve as heat source, both for its stability at lean combustion regimes as the stability at low Reynold flow fields. At the end of this project that begins we hope to report the outstanding performance of the LSB+SE system.

Acknowledgements
This work is supported by the U.S. Department of Energy, Advanced Research Projects Agency – Energy under contract No. DE-AE0000607.

References

[1] U.S. Energy information Administration, "State Electricity Profiles - Energy Information Administration," 2017. [Online]. Available: https://www.eia.gov/electricity/state/. [Accessed: 31-Aug-2017].

[2] The World Bank, "Electric power transmission and distribution losses (% of output) | Data," 2014. [Online]. Available: https://data.worldbank.org/indicator/EG.ELC.LOSS.ZS?end=2014&start=1960&view=chart. [Accessed: 31-Aug-2017].

[3] U.S. Department of Energy, "GENSETS Program Overview.," 2017. [Online]. Available: https://arpae.energy.gov/sites/default/files/documents/files/GENSETS_ProgramOverview.pdf.

[4] D. G. Thombare and S. K. Verma, "Technological development in the Stirling cycle engines," *Renew. Sustain. Energy Rev.*, vol. 12, no. 1, pp. 1–38, Jan. 2008.

[5] K. Wang, S. R. Sanders, S. Dubey, F. H. Choo, and F. Duan, "Stirling cycle engines for recovering low and moderate temperature heat: A review," *Renew. Sustain. Energy Rev.*, vol. 62, pp. 89–108, Sep. 2016.

[6] M. H. Ahmadi, M.-A. Ahmadi, and F. Pourfayaz, "Thermal models for analysis of performance of Stirling engine: A review," *Renew. Sustain. Energy Rev.*, vol. 68, pp. 168–184, Feb. 2017.

[7] F. A. Farret and M. G. Simões, Eds., *Integration of Alternative Sources of Energy*. Hoboken, NJ, USA: John Wiley & Sons, Inc., 2005.

[8] A. J. Organ, *The regenerator and the Stirling engine*. Mechanical Engineering Publications, 1997.

[9] J. Gary Wood, A. Buffalino, E. Holliday, B. Penswick Consultant, and W. David Gedeon, "Free-Piston Stirling Power Conversion Unit for Fission Surface Power, Phase I Final Report NASA STI Program . . . in Profi le," 2010.

[10] Coulson M. Scheuermann *et al.*, "Materials Technology Assessment for a 1050 K Stirling Space Engine Design," 1988.

[11] H. Karabulut, H. S. Yücesu, and C. Çinar, "Nodal analysis of a Stirling engine with concentric piston and displacer," *Renew. Energy*, vol. 31, no. 13, pp. 2188–2197, Oct. 2006.

[12] H. Bankaitis, "System Safety in Stirling Engine Development," 1981.

[13] C. M. Invernizzi, "Stirling engines using working fluids with strong real gas effects," *Appl. Therm. Eng.*, vol. 30, no. 13, pp. 1703–1710, Sep. 2010.

[14] A. Frank, P. Therkelsen, J. Chen, and R. K. Cheng, "kW Scale Combustor for Power Generation," in *10th U. S. National Combustion Meeting*, 2017, no. 2SC–0248.

[15] P. L. Therkelsen, D. Littlejohn, and R. K. Cheng, "Parametric Study of Low-Swirl Injector Geometry on its Operability," in *Volume 2: Combustion, Fuels and Emissions, Parts A and B*, 2012, p. 309.

Sub Topic: Turbulent Flame

Western States Section of the Combustion Institute–Fall 2017 Meeting
Hosted by the University of Wyoming
October 2–3, 2017

Flame Stability of Turbulent Premixed Jet Flames of Large Hydrocarbon Fuels

Nathan E. Schorn[1], *Jonathan Bonebreak*[1], *and David L. Blunck*[1,*]

[1]*Mechanical, Industrial and Manufacturing Engineering, Oregon State University, 104 Rogers Hall, Corvallis OR, United States*
[*]*Corresponding author:* `David.Blunck@oregonstate.edu`

Abstract: In this study, flame stability limits of various jet and jet-like fuels are examined by measuring turbulent consumption speeds and visual imaging of flame tip quenching at fuel lean operating conditions. Tip quenching is of interest because of its indication of the progression of the flame toward instabilities or blow-off. The experiments were performed on a piloted turbulent Bunsen burner coupled with a fuel vaporizer. Flame speed data was collected at atmospheric pressure, at Reynolds number of 10,000, turbulence intensities near 20%, and burner exit temperatures near 475 K. During the experiments, the equivalence ratio was dropped from stoichiometric to values near 0.7, at which quenching of the flame tip was observed and fuel exited through the flame sheet. Turbulent flame speeds were used to quantify the equivalence ratios at which the flames were stable. Six large hydrocarbon fuels were studied with varying chemical compositions (e.g., aromatics, n-paraffin's). Jet-A had the greatest stability (i.e., least prone to tip quenching) along with a test fuel composed of mostly iso-paraffins. The variations observed in flame stability between the fuels are attributed to differences in flame stretch sensitivities.
Keywords: Turbulent Consumption Speed, Tip Quenching, Flame Stretch

1. Introduction

Operating combustion devices near lean stability limits is advantageous for some conditions. However, running at lean operating conditions can cause combustion instabilities. There has been strides to improve lean operating conditions, especially for combustors [1, 2]. Several studies have considered stability limits of large hydrocarbon fuels [3, 4]. However, few studies have considered at turbulent conditions. This is of importance because combustors are turbulent in nature. By observing flame stability limits of turbulent jet flames of large hydrocarbon fuels, further insight can be gained on extinguish conditions of flames.

Two methods of flame extinguishing (i.e., stability) are often considered: extinguishing near the base of the flame and tip quenching [5]. Tip quenching was studied by Law and colleagues [6] for methane/air mixtures at laminar conditions. Law showed that tip quenching can often occur due to flame stretch on a Bunsen burner flame [6]. Law also did a study on flame blow off of symmetrical counter flow flames. Law was able to find critical laminar flame speeds and stretch rates at which the flame would extinguish by using laser Doppler velocimetry [7].

More recent to this study, Carbone et al. [8] did a comprehensive study of eight different fuel types ranging from gaseous to liquid (but vaporized) fuels. Instantaneous flame images revealed that intense reactivity fluctuations occur at the tip of the jet flames even when they are on average ignited [8]. The fluctuations became more severe for tip-quenched flames whereas they were dampened for tip-ignited flames [8].

Prior work from this research group [9] determined the turbulent consumption speeds of three jet and jet-like fuels for varying turbulence conditions. The study was performed on an turbulent Bunsen burner, the same burner used in this study. UV chemiluminescence measurements were used to determine turbulent consumption speeds of the fuels. The fuels that were tested were conventional jet-A fuel, and 2 experimental blends containing different paraffins and aromatics. It was shown that the flow conditions effected the flame speeds and stability of the fuels. This was attributed to the sensitivity of the fuel chemistry, specifically the aromatic content of each of the fuels [9]. The greater the aromatic content in the fuel, the less sensitive the fuel was to flame stretch [9].

With this background, the objective of this study is to identify the conditions when tip quenching occurs for Bunsen burner flames burning jet and jet-like fuels. A greater range of fuels are evaluated than was tested previously. Visual and analytical techniques are used to determine the lean stability limits of the large hydrocarbon fuels.

2. Experimental Approach

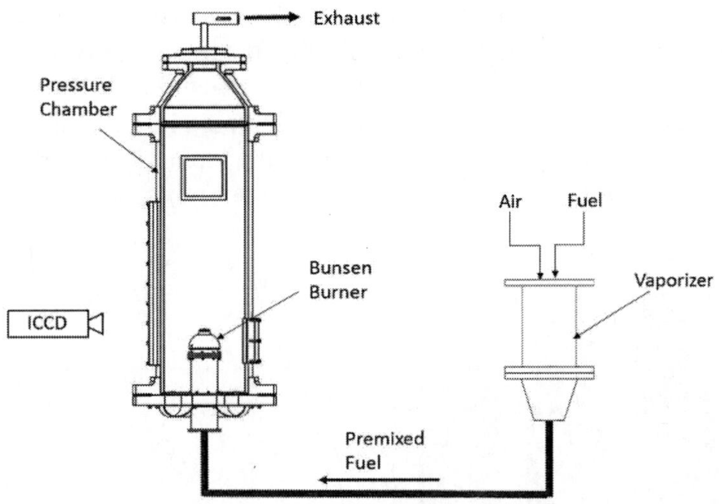

Figure 1: Schematic of the turbulent Bunsen burner and vaporizer.

Turbulent consumption speeds are measured using a turbulent piloted Bunsen burner, and used to identify the equivalence ratio at which tip quenching occurred. The burner apparatus includes a pressure chamber, Bunsen burner and a fuel vaporizer, as seen in Figure 1. The Bunsen burner was based on the work reported by Venkateswaran et al. [10]. Thermal mass flow meters are used to meter the air flow to desired rates. The fuel is pumped through an atomizer nozzle in the vaporizer by a syringe pump. The fuel and air are mixed within the vaporizer at temperatures near 475 K. The premixed fuel/air flow through the piloted Bunsen burner at temperatures near 475 K after being heated. A pilot flame at the burner tip burns premixed methane/air and anchors the main flame to the burner. The heat release from the pilot is less than 11.5 % of the primary flame. A turbulence generator within the burner produces estimated turbulence intensities (TI) near 10 to 20% [9], although only the latter was evaluated in this study.

Chemiluminescence measurements are used to determine the turbulent flame speeds. The measurements were collected using an AndorSolis iStar 334 CCD camera that is sensitive to wavelengths between 200 and 1100 nm. OH* and CH* radicals have been shown to emit brightly at wave lengths between 300-500 nm [11] and are markers of a flame brush for visual imaging techniques. A quartz window allows visual access inside the pressure vessel. The camera captures 1 image every 0.7 seconds for a total of 360 images. Using the images captured by the camera a turbulent consumption speed is determined using the relationship, [12]:

$$S_T = \frac{\dot{m}}{A_{\langle c \rangle}\rho_u}. \quad (1)$$

Where $\dot{m}$ is the mass flow rate, $A_{\langle c \rangle}$ is the area of flame brush, and ρ_u is the unburned density. ρ_u was determined from the air and fuel densities and the exit temperature of the burner. $\langle c \rangle$ is the progress variable for the unburned and burned faces of the flame. A $\langle c \rangle = 0.05$ corresponds to the unburned inner surface of the flame brush and a $\langle c \rangle = 1$ corresponds to the fully burned inner surface of the flame brush [13]. Kobayashi [13] determined that a $\langle c \rangle = 0.5$ to be the mean inner surface of the flame brush that can be used when determining the mean cone area. The mean cone area of the flame was determined using the steps illustrated in Figure 2. First, the image is cropped from easier processing. Second the image is mirrored along an axis of symmetry. Third, the image is filtered of any bad pixels that might be present. Lastly, the image is processed using a three point inverse Abel convolution scheme [14]. This step returns an estimated spatial field represented by the two-dimensional images collected and a centerline outlined in red in figure 2a. This allows for the location of the $\langle c \rangle = 0.5$ to be found in the center of the flame. A cone area is formed from this point to the base of burner exit.

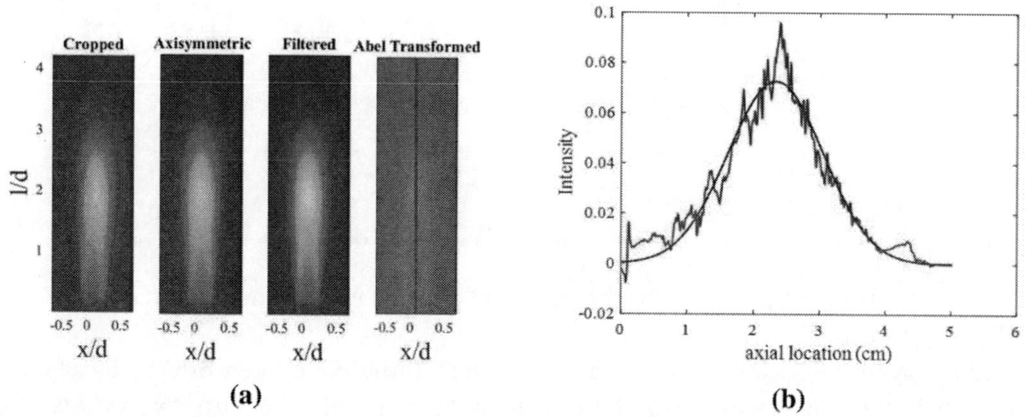

Figure 2: Figure 2a shows the time average flame images as it is being process from left to right. Figure 2b is the curve generated by the Abel transformation of the flame image, Gaussian fit (blue) to the Abel deconvoluted centerline intensity (red).

The stability of six fuels were evaluated. The chemical composition of each fuel varied greatly, as summarized in Table 1 [15]. A1, A2 and A3 are jet fuels found in industry corresponding to their more common names, JP-8, jet-A, and JP-5, respectively. These fuels have a wide range of constituents. C1 and C4 are large hydrocarbon fuels composed primarily of iso-paraffins. n-Dodecane is among the major straight chain paraffins found in diesel and jet fuels [16].

Table 1: Compositions of fuel evaluated [15].

Fuel Designation	Aromatics (% vol)	Iso-Paraffins (% vol)	n-Paraffins (% vol)
A1	11.2	39.07	27.23
A2	17	31.46	21.23
A3	18	20.42	14.47
C1	1	near 99	0
C4	2.3	near 98	0
n-Dodecane	0	0	near 100

Each fuel was studied at a Reynolds number of 10,000 and an estimated turbulence intensity near 20%. The air-to-fuel ratio of the premixed fuel was decreased from stoichiometric to the lowest possible ratio until visible confirmation that the flame tip was fully extinguished. Figure 3 provides an example of the progression toward tip extinguishing. Equivalence ratios ranging from 0.6 to stoichiometric were evaluated while identifying tip quenching.

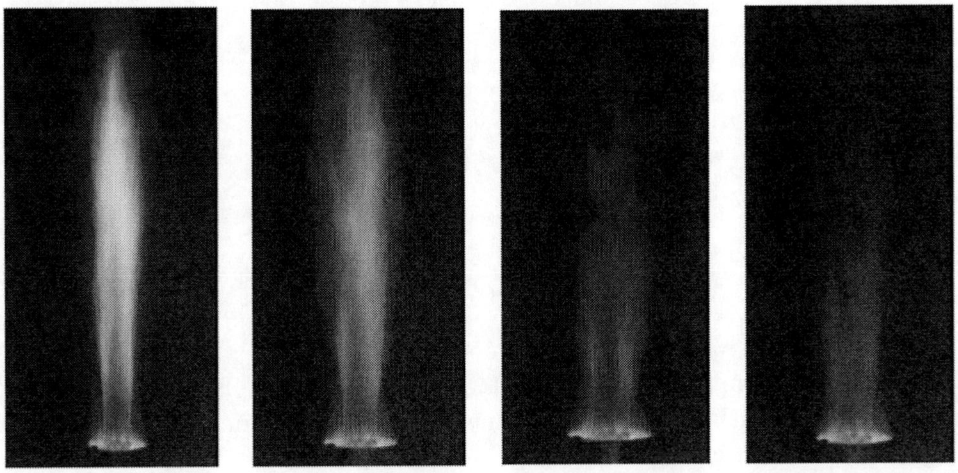

(a) ϕ value of 0.9. **(b)** ϕ value of 0.8. **(c)** ϕ value of 0.7. **(d)** ϕ value of 0.6.

Figure 3: Visual inspection of full flame tip ignition (i.e panel a), quasi ignition/quenching, and full tip quench (i.e., panel d).

The turbulent flame speeds were used to identify the conditions when tip quenching occurred. Flame speeds decreased linearly as the equivalence ratio decreased [3–5, 9]. The flame speed increased however when local tip quenching occurred. This unphysical response occurred because of fuel breaking through the flame front, biasing the assumptions used for equation 1. The lowest equivalence ratio at which a flame speed is reported for each fuel corresponds to the conditions just prior to local quenching and the increase in the flame speeds.

3. Results and Discussion

Figure 4 reports the turbulent flame speeds for each fuel. Each fuel follows a linear, increasing trend with increasing equivalence ratio. Peak flame speeds near 2.25 m/s are observed at ϕ value of 0.85. The turbulent flame speedsfor the different fuels are generally statistically similar to each other. This is significant when considering the variability of the chemical classes within the fuel. For example, n-dodecane (which is a paraffin) has flame speeds similar to the jet fuels (which have aromatics, paraffins and cycloparafgins). At a ϕ value of 0.8, the cluster of average flame speed values are tighter, while at a ϕ value of 0.7 the values are more spread. C1 is the exception to the trends just described. It has a non-linear increasing trend and the flame speed is statistically lower then the rest of the fuels. Note that C1 is composed primarily of iso-parrafin whereas the jet fuels have multiple constituents (as noted earlier). The reason why C4, which also is primarily iso-paraffins, has flame speed similar to the jet fuels is not clear. It is possible that difference in the distribution of the carbon number for the two fuels caused the sensitivity.

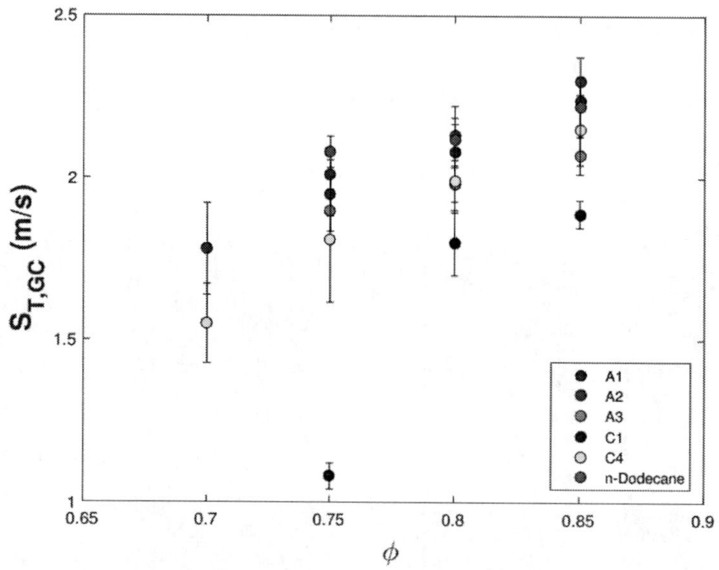

Figure 4: Turbulent consumption speed for six large hydrocarbon fuels. Error bars includes instrumentation error and a t-distribution of mean values. Note that the flame speeds when tip quenching occurs was omitted for clarity.

Table 2 lists the equivalence ratios at which local tip quenching occurs. Comparing Tables 1 and 2, there is no apparent correlation between the chemical composition of the fuels and their stability. It is noteworthy that A2 is noted as a more conventional jet fuel while A1 and A3 were designed to be extremes in the fuel specifications; however, A2 tends to be the most stable. C4 and C1 are both iso-paraffins but C4 exhibits a greater stability. Prior work has suggested that difference in flame stability are attributed to changes in stretch sensitivity [3, 4, 7, 9, 13]. Evaluation of stretch sensitivity for these fuels is needed to verify that this is the correct phenomenon.

Table 2: Equivalence ratios at which each fuel experienced local tip quenching.

Fuel Types	ϕ
A1	0.7
A2	0.65
A3	0.7
C1	0.7
C4	0.65
n-Dodecane	0.7

4. Summary and Future Work

The sensitivity of tip quenching to the fuel composition of large hydrocarbon fuels has been shown qualitatively and quantitatively. Chemiluminescence of OH* and CH* radicals allowed for the determination of turbulent consumption speeds. By taking visual images of the flame, a correlation between the unphysical trend of flame speed data and tip quenching was formed. Jet-A and a test jet fuel composing of mostly iso-paraffins showed the greatest stability. Comparing the stability results to the chemical compositions of the large hydrocarbon fuels, no apparent correlation can be made. Future work will include evaluating flame stability at different flow conditions, turbulence intensities and pressures. Evaluating more fuels with differing compositions, with empathises on the aromatic and paraffin content, can provide more insight on the effects of chemical kinetics and stability.

5. Acknowledgments

This work was funded by the US Federal Aviation Administration (FAA) Office of Environment and Energy as a part of ASCENT Project 27B (13-C-AJFE-OSU-02) as well as the National Science Foundation (1314109-DGE). Opinions, findings, and conclusions or recommendations expressed in this material are those of the authors and do not necessarily reflect the views of the FAA or other ASCENT Sponsors.

The authors would like to thank: Dr. Tim Lieuwen and colleagues for burner specifications, Dr. Scott Stouffer (University of Dayton Research Institute) and the Air Force Research Laboratory (AFRL) for the vaporizer design, Dr. Tim Edwards (AFRL) for supplying the fuel. Arron Fillo for burner development and data analysis. Ankit Agarwal, Mila Bakker, Delphine Le Brun Colon and Brandon Pendergrass for assistance with data collection.

References

[1] Y. Huang and V. Yang, Dynamics and stability of lean-premixed swirl-stabilized combustion, Progress in Energy and Combustion Science 35 (2009) 293–364.

[2] T. Lieuwen, V. McDonell, E. Petersen, and D. Santavicca, Fuel flexibility influences on premixed combustor blowout, flaskback, autoignition and stability, Combustion and Fuels, Education 1 (2006) 601–615.

[3] X. Hui and C. J. Sung, Laminar flame speeds of transportation-relevant hydrocarbons and jet fuels at elevated temperatures and pressures, Fuel 109 (2012) 191–200.

[4] K. Kumar, C. J. Sung, and X. Hui, Laminar flame speeds and extinction limits of conventional and alternative jet fuels, Fuel 90 (2011) 1004–1011.

[5] C. K. Law, Combustion Physics 1st ed. Cambridge University Press, 2006.

[6] C. K. Law, S. Ishizuka, and P. Cho, On the Opening of Premixed Bunsen Flame Tips, Combustion Science and Technology 28 (1981) 89–96.

[7] C. K. Law, D. L. Zhu, and G. Yu, Propagation and Extinction of Stretched Premixed Flames, Symposium (International) on Combustion 21 (1986) 1419–1426.

[8] F. Carbone, J. L. Smolke, A. M. Fincham, and F. N. Egolfopoulos, Comparative behavior of piloted turbulent premixed jet flames of C1-C8 hydrocarbons, Combustion and Flame 180 (2017) 88–101.

[9] A. J. Fillo, J. M. Bonebrake, and D. Blunck, Impact of fuel chemistry and stretch rate on global consumption speed of large hydrocarbon fuel/air flames, National Combustion Meeting (2017) 1–12.

[10] P. Venkateswaran, A. Marshall, D. H. Shin, and D. Noble, Measurements and analysis of turbulent consumption speeds of H2/CO mixtures, Combustion and Flame 158 (2010) 1602–1614.

[11] D. Guyot, F. Guethe, B. Schuermans, A. Lacarelle, and C. O. Pascheret, CH*/OH* Chemiluminescence Response of an Atmospheric Premixed Flame Under Varying Operating Conditions, ASME Tubro Expo 2010: Power of Land, Sea, and Air 2 (2010) 933–944.

[12] S. R. Turns, An Introduction to Combustion: Concepts and Applications 3rd ed, McGraw-Hill Education, 2011.

[13] H. Kobayashi, T. Tamura, K. Maruta, and T. Niioka, Burning velocity of turbulent premixed flames in a high-pressured environment, Symposium (International) on Combustion 26 (1996) 389–396.

[14] C. J. Dasch, One-dimensional tomography: a comparison of Abel, onion-peeling, and filtered backprojection methods, Applied Optics 31 (1992) 1146–1152.

[15] T. Edwards, Reference jet fuels for combustion testing, 55th AIAA Aerospace Science Meeting (2017) 1–58.

[16] K. Kumar and C. J. Sung, Laminar flame speeds and extiniction limits of preheated n-decane/O2/N2 and n-dodecane/O2/N2 mixtures, Combustion and Flame 151 (2007) 209–224.

10th Meeting of the Western States Section of The Combustion Institute
Organized by the University of Wyoming
October 2–3, 2017
Laramie, Wyoming

Implementing and assessing the importance of multicomponent transport properties using direct numerical simulations of premixed, turbulent flames

Aaron J. Fillo[1], *Jason Schlup*[2], *Guillaume Blanquart*[3], *and Kyle E. Niemeyer*[1,*]

[1]*School of Mechanical, Industrial, and Manufacturing Engineering, Oregon State University, Corvallis, OR 97331, USA*
[2]*Graduate Aerospace Laboratories, California Institute of Technology, Pasadena, USA*
[3]*Mechanical Engineering, California Institute of Technology, Pasadena, USA*
[*]*Corresponding author:* Kyle.Niemeyer@oregonstate.edu

Abstract: Implementing multicomponent diffusion models in numerical combustion studies is computationally expensive due to the challenges involved in computing diffusion coefficients. As a result, mixture-averaged diffusion treatments are used to avoid these costs. However, to the authors' knowledge, the accuracy and appropriateness of the mixture-averaged diffusion models has not been verified for three-dimensional turbulent premixed flames. This study will evaluate the role of multicomponent mass diffusion in premixed, high-Karlovitz hydrogen flames, neglecting secondary Soret and Dufour effects. Direct numerical simulation (DNS) of these flames is performed by implementing the Stefan–Maxwell equations in NGA. A semi-implicit algorithm decreases computational expense of inverting the full multicomponent ordinary diffusion array while maintaining simulation accuracy and fidelity. The algorithm is unconditionally stable provided the components of the mixture-averaged and multicomponent diffusion coefficient matrices are of order one over the number of species or smaller, and performance scales approximately with the number of species squared. One-dimensional simulations of premixed hydrogen flames are performed and compared with matching cases in Cantera to verify this method. Premixed two-dimensional, unstable and three-dimensional, turbulent hydrogen flames are simulated and compared with previous mixture-averaged DNS results. Simulation conditions are carefully selected to match previously published results and ensure valid comparison. A priori analysis shows similar relative angles between the species diffusion flux vectors and the species gradient vectors between mixture-averaged and multicomponent results. Further, a posteriori analysis demonstrates negligible differences in conditional means of the fuel mass fraction and its diffusion source term against temperature for mixture-average and multicomponent transport, respectively, between the two cases.
Keywords: ***Turbulent flames, Direct numerical simulation, multicomponent diffusion, mixture-averaged diffusion***

1. Introduction

Mixture-averaged diffusion models are commonly used to reduce the high computational expense associated with numerical combustion studies [1]. This approach reduces the complexity of the system of equations by approximating the full multicomponent diffusion coefficient matrix as a single diagonal [2]. Several approaches, such as those used by Warnatz [3] and Coltrin et al. [4], reduce the system's complexity even further by approximating multicomponent diffusion processes in terms of equivalent Fickian processes. However, to our knowledge, the accuracy and appropriateness of mixture-averaged approximations hast not been evaluated for three-dimensional turbulent flame simulations, despite common usage.

As further motivation for this study, Lapointe and Blanquart [5] recently investigated the impact of the unity-Lewis number approximation and reported that n-heptane, isooctane, toluene, and methane flames have similar normalized turbulent flame speeds and fuel burning rates. They observed similarities in the turbulent flame speeds and burning rates for both the unity and non-unity Lewis number assumption models, although the normalized flame speeds decreased below unity when including differential diffusion effects. Conversely, Fillo et al. [6, 7] experimentally observed significant differences between the normalized turbulent flame speeds of three jet fuel surrogates, with similar chemistry to the neat n-heptane and isooctane flames studied by Lapointe and Blanquart [5]. The observed discrepancies between similar fuels warrant a detailed investigation of the fundamental transport phenomena involved.

Several studies have investigated the impact of multicomponent transport for a wide range of simulations. For example, Dworkin et al. [8] investigated the fidelity of the Fickian diffusion model for soot formation in ethylene/air flames. They showed that for low strain-rate flames, detailed multicomponent transport and thermal diffusion models reduce error in soot volume fractions in counter-flow diffusion flames. Similarly, Giovangigli [9] demonstrated that multicomponent Soret effects significantly impact a wide range of laminar hydrogen/air flames. Specifically, they noted that multicomponent Soret effects influenced laminar flame speeds and extinction stretch rates for flat and strained premixed flames, respectively. For high-pressure systems, Borchesi et al. [10] developed and analyzed a multi-species turbulent mixing model for large eddy simulations. They focused on turbulent crossflow mixing of a five-species combustion-relevant mixture which includes n-heptane, O_2, CO_2, N_2, and H_2O. This analysis showed that the multi-species transport model significantly improved the accuracy and fidelity of the solution throughout the mixing layer; however, this study was restricted to non-reacting flows and, as a result, did not assess the impact of multicomponent transport on the stiff chemistry inherent in turbulent combustion.

These studies provide compelling evidence that multicomponent transport is important and can affect the accuracy of combustion models. However, the studies failed to assess the impact of multicomponent transport on three-dimensional systems with detailed chemistry. Motivated by the dearth of affordable three-dimensional multicomponent transport models, Ambikasaran and Narayanaswamy [11] proposed an efficient algorithm to compute multicomponent diffusion velocities which scales as $\mathcal{O}(nSpecies)$. This offers a significant improvement over previous methods that directly invert the Stephan–Maxwell equations and scale as $\mathcal{O}(nSpecies^3)$.

In this article, we present an a priori and a posteriori assessment of the importance of multicomponent transport using DNS of turbulent hydrogen/air flames. In addition, we demonstrate and analyze an efficient, dynamic algorithm to reduce the computational expense of calculating the multicomponent diffusion fluxes. Freely propagating one-dimensional, two-dimensional unsteady, and three-dimensional, turbulent, premixed flames are simulated. a priori analysis of the relative direction and magnitude of species flux vectors is performed as a means of assessing the relative differences between mixture-averaged and multicomponent mass diffusion transport. Finally, we provide an a posteriori assessment of the impact of mixture-averaged and multicomponent mass diffusion transport on the turbulent statistics of the three-dimensional hydrogen simulation.

2. Governing equations

This section presents the governing equations used in this study for variable-density, low-Mach number reacting flows. In addition, this section outlines the method used to determine the ordinary multicomponent mass diffusion fluxes based on the Boltzmann's equation of kinetic theory [12–16].

The variable-density, low-Mach number, reacting flow equations are solved using the finite-difference code NGA [17, 18]. The conservation equations are

$$\frac{\partial \rho}{\partial t} + \nabla \cdot \rho \mathbf{u} = 0\,, \tag{1}$$

$$\frac{\partial \rho \mathbf{u}}{\partial t} + \nabla \cdot (\rho\ \mathbf{u}\mathbf{u}) = -\nabla p + \nabla \cdot \tau\,, \tag{2}$$

$$\frac{\partial \rho T}{\partial t} + \nabla \cdot (\rho \mathbf{u} T) = \nabla \cdot (\rho \alpha \nabla T) + \rho \dot{\omega}_T - \frac{1}{c_p} \sum_i c_{p,i} \mathbf{j}_i \cdot \nabla T + \frac{\rho \alpha}{c_p} \nabla c_p \cdot \nabla T\,, \tag{3}$$

$$\frac{\partial \rho Y_i}{\partial t} + \nabla \cdot (\rho \mathbf{u} Y_i) = -\nabla \cdot \mathbf{j}_i + \dot{\omega}_i\,, \tag{4}$$

where ρ is the mixture density, $\mathbf{u}$ is the velocity, p is the hydrodynamic pressure, τ is the viscous stress tensor, T is the temperature, α is the mixture thermal diffusivity, $c_{p,i}$ is the constant-pressure specific heat of species i, c_p constant-pressure specific heat of the mixture, $\mathbf{j}_i$ is the diffusion flux of species i, Y_i is the mass fraction of species i, and $\dot{\omega}_i$ is the production rate of species i. In equation (3), the temperature source term is given by

$$\dot{\omega}_T = -c_p^{-1} \sum_i h_i(T) \dot{\omega}_i\,, \tag{5}$$

where $h_i(T)$ is the specific enthalpy of species i as a function of temperature. Similarly, the density is determined from the ideal gas equation of state

$$\rho = \frac{p_o W}{RT}\,, \tag{6}$$

where p_o is the thermodynamic pressure, R is the universal gas constant, and W is the mixture molecular weight.

2.1 Mixture-averaged species diffusion flux

The diffusion flux is calculated with both the mixture-averaged and multicomponent models based on Boltzmann's equation for the kinetic theory of gases [12–16]. Both models assume a constant-pressure environment and neglect body forces acting on the mixture along with Soret and Dufour effects. The species diffusion flux for the mixture-averaged diffusion model is related to the species gradients by a Fickian formulation and is expressed as

$$\mathbf{j}_i = -\rho D_{im} \frac{Y_i}{X_i} \nabla X_i - \rho Y_i \mathbf{u}'_c\,, \tag{7}$$

where $D_{i,m}$ is the ith species mixture-averaged diffusion coefficient as expressed by Bird et al. [2],

$$D_{im} = \frac{1-Y_i}{\sum_{i \neq j}^{N} X_j / \mathscr{D}_{ji}}, \tag{8}$$

and u_c is the correction velocity used to ensure mass continuity:

$$\mathbf{u}_c = -\sum_i \rho D_{im} \frac{Y_i}{X_i} \nabla X_i \,. \tag{9}$$

2.2 Multicomponent species diffusion flux

The multicomponent diffusion model for the species diffusion flux is

$$\mathbf{j}_i = \frac{\rho Y_i}{X_i W} \sum_{i \neq k}^{N} W_j D_{i,j} \nabla X_j \,, \tag{10}$$

where W_j is the molar mass of the jth species and $D_{i,j}$ is the ordinary multicomponent diffusion coefficient computed using the `MCMDIF` subroutine of CHEMKIN II [19] with the method outlined by Dixon-Lewis [20]. In this approach the multicomponent diffusion coefficients, thermal conductivities, and thermal diffusion coefficients are computed from the solution of a system of equations defined by the L matrix and composed of nine sub-matrices:

$$\begin{bmatrix} L^{00,00} & L^{00,10} & 0 \\ L^{10,00} & L^{10,10} & L^{10,01} \\ 0 & L^{01,10} & L^{01,01} \end{bmatrix} \begin{bmatrix} a_1^{00} \\ a_1^{10} \\ a_1^{01} \end{bmatrix} = \begin{bmatrix} 0 \\ X \\ X \end{bmatrix}, \tag{11}$$

where the right-hand side is composed of the one dimensional mole fraction arrays X_i. From this system of equations, the inverse of the $L^{00,00}$ block provides the multicomponent diffusion coefficients:

$$D_{i,j} = X_i \frac{16T}{25P} \frac{W}{W_j} (p_{i,j} - p_{i,i}) \,, \tag{12}$$

where

$$p = (L^{00,00})^{-1} \,. \tag{13}$$

The $L^{00,00}$ sub-matrix block is given by

$$L_{ij}^{00,00} = \frac{16T}{25p} \sum_{k=1}^{N} \frac{X_k}{m_i [\mathscr{D}_{ik}]_1} \left\{ m_j X_j (1-\delta_{ik}) - m_i X_i (\delta_{ij} - \delta_{jk}) \right\} \,, \tag{14}$$

where $\mathscr{D}_{ik}$ is the binary diffusion coefficient, m_j is the molecular mass of the jth species, and δ_{ij} is the reduced dipole moment of the jth species.

The mixture-averaged and multicomponent mass diffusion models will be abbreviated here as MA and MC, respectively. These models are compared using one-dimensional steady, two-dimensional unsteady, and three-dimensional, turbulent, premixed DNS of hydrogen/air flames to determine the accuracy and appropriateness of the mixture-averaged diffusion model as an approximation of full multicomponent diffusion.

3. Methods

This section presents the discretized equations, numerical algorithm, and preconditioner proposed in this paper. The method is modeled on semi-implicit time-marching scheme for species mass fraction fields proposed by Savard et al. [18].

3.1 Multicomponent model implementation

As discussed previously, multicomponent mass diffusion has not yet been demonstrated with three-dimensional turbulent flames due to its high computational expense. To reduce this expense while maintaining the accuracy and fidelity of the ordinary multicomponent diffusion coefficient matrix, our approach calculates the diffusion coefficients once per grid-point and uses a simple dynamic algorithm to reduce memory requirements. The approach capitalizes on the knowledge that it is not necessary to store all intermediate flux values through the entire computation; only the (i) and $(i+1)$ fluxes in each direction are stored at any time during the calculation of the diffusion source terms. This is possible because the discretized forms of equations (3) and (4) are additive and commutative.

To ensure a stable solution, the scalar and momentum fields are advanced using a modified form of the semi-implicit preconditioning strategy outlined by Savard et al. [18]. The preconditioner corresponds to an approximation of the diagonal of the chemical Jacobian and diffusion coefficient matrices, and uses an iterative approach to maintain stability and reduce error. This approach is second-order accurate in time, free of lagging errors, and the cost per iteration compares with that of an explicit time integration [18]. This performance makes the semi-implicit strategy ideal for the simulation of turbulent flames with stiff chemistry and multicomponent diffusion transport.

3.1.1 Overview of scalar field solver

An iterative procedure is applied to fully cover the nonlinearities in the Navier–Stokes equations and the species diffusion terms. This iterative process has been demonstrated to be critically important for stability and accuracy [17, 18, 21, 22]. Savard et al. [18] fully detail the numerical algorithm sequence; we describe all modifications to the preconditioning strategy employed in this method next.

A uniform time step Δt is used. The density, pressure, and scalar fields are advanced from time level $t^{n+1/2}$ to $t^{n+2/3}$, and the velocity fields are advanced from time t^n to t^{n+1}, where t^n is the current time. A total number of iterations, Q, are assumed. For the sub-iterations $k = 1, \ldots, Q$, the scalar fields are advanced in time using the semi-implicit Crank–Nicolson method [22, 23]:

$$\begin{aligned}\rho_k^{n+3/2}\mathbf{Y}_{k+1}^{n+3/2} = \rho^{n+1/2}\mathbf{Y}^{n+1/2} + \Delta t\left[(\mathbf{C}_k^{n+1} + \mathbf{Diff}_k^{n+1})\cdot\mathbf{Y}_k^* + \Omega_k^*\right] \\ + \frac{\Delta t}{2}\left(\frac{\partial\mathbf{C}}{\partial\mathbf{Y}} + \frac{\partial\mathbf{Diff}}{\partial\mathbf{Y}}\right)_k^{n+1}\cdot(\mathbf{Y}_{k+1}^{n+3/2} - \mathbf{Y}_k^{n+3/2}),\end{aligned} \tag{15}$$

where

$$\mathbf{Y}_k^* = \frac{\mathbf{Y}^{n+1/2} + \mathbf{Y}_k^{n+3/2}}{2} \tag{16}$$

$$\Omega_k^* = \frac{\Omega^{n+1/2} + \Omega_k^{n+3/2}}{2} \tag{17}$$

and

$$\mathbf{Diff} = -\nabla \cdot \mathbf{j}_i \,. \tag{18}$$

To simplify the discrete notations for spatial differential operators, the operators corresponding to the convective and diffusive terms in equation (3) are written as $\mathbf{C}$ and $\mathbf{Diff}$, respectively. $\frac{\partial \mathbf{C}}{\partial \mathbf{Y}}$ and $\frac{\partial \mathbf{Diff}}{\partial \mathbf{Y}}$ are the Jacobian matrices corresponding to the convective and diffusive terms, respectively. $\mathbf{C}$ and $\frac{\partial \mathbf{C}}{\partial \mathbf{Y}}$ are functions of the density and the velocity, while $\mathbf{Diff}$ and $\frac{\partial \mathbf{D}}{\partial \mathbf{Y}}$ are functions of the density and the kinematic viscosity. They are consistently updated at each sub-iteration [18]. Convergence of the sub-iterations recovers the fully implicit, second-order time-accurate, Crank–Nicolson formulation.

3.1.2 Preconditioning

Expansion of the above numerical procedure is based only on modifying the time-marching step for species mass fraction fields in the method by Savard et al. [18]. Specifically, this method modifies the treatment of the mass-diffusion source term in the species mass fraction fields. All other intermediate steps, as described by by Savard et al. [18], are unchanged.

3.1.3 Preconditioning iterative method

For simpler implementation, the set of equations (equation (15)) is solved in its residual form

$$\begin{aligned} &\left[\rho_k^{n+3/2}\mathbf{I} - \left(\frac{\partial \mathbf{C}}{\partial \mathbf{Y}} + \frac{\partial \mathbf{Diff}}{\partial \mathbf{Y}}\right)_k^{n+1}\right] \cdot \left(\mathbf{Y}_{k+1}^{n+3/2} - \mathbf{Y}_k^{n+3/2}\right) \\ &= \rho^{n+1/2}\mathbf{Y}^{n+1/2} - \rho_k^{n+3/2}\mathbf{Y}_k^{n+3/2} + \Delta t\left[\left(\mathbf{C}_k^{n+1} + \mathbf{Diff}_k^{n+1}\right)\cdot \mathbf{Y}_k^* + \Omega_k^*\right] . \end{aligned} \tag{19}$$

This equation can be restated as

$$\mathbf{Y}_{k+1}^{n+3/2} = \mathbf{Y}_k^{n+3/2} - \Delta t \mathbf{J}^{-1} \cdot \Theta_k \,, \tag{20}$$

where the matrix $\mathbf{J}$ is

$$\mathbf{J} = \rho_k^{n+3/2}\mathbf{I} - \frac{\Delta t}{2}\left(\frac{\partial \mathbf{C}}{\partial \mathbf{Y}} + \frac{\partial \mathbf{Diff}}{\partial \mathbf{Y}}\right)_k^{n+1} , \tag{21}$$

and the vector

$$\Theta_k = \frac{\rho_k^{n+3/2}\mathbf{Y}_k^{n+3/2} - \rho_k^{n+1/2}\mathbf{Y}_k^{n+1/2}}{\Delta t} - \left[\left(\mathbf{C}_k^{n+1} + \mathbf{Diff}_k^{n+1}\right)\cdot \mathbf{Y}_k^* + \Omega_k^*\right] \tag{22}$$

is the residual of the species transport equation at the previous sub-iterations, and limits to zero as the sub-iterations fully converge.

Written in its residual form, the time advancement of the species transport equations described above resembles the standard preconditioned Richardson-type iterative method [18, 24], where the matrix **J** acts as a preconditioner. The choice of **J** as a preconditioner can be arbitrary, and does not modify the discrete form the of the equations to solve (i.e., $\Theta_k = 0$). It only changes the convergence characteristics of the iterative method [18]. For example,

$$\mathbf{J} = \rho_k^{n+3/2}\mathbf{I} \tag{23}$$

is equivalent to the fully-explicit integration of the convective, diffusive, and chemical source terms in the species transport equations, while setting

$$\mathbf{J} = \rho_k^{n+3/2}\mathbf{I} - \frac{\Delta t}{2}\left(\frac{\partial \mathbf{C}}{\partial \mathbf{Y}} + \frac{\partial \mathbf{Diff}}{\partial \mathbf{Y}} + \frac{\partial \Omega}{\partial \mathbf{Y}}\right)_k^{n+1} \tag{24}$$

is equivalent to the fully-implicit integration of the convective, diffusive, and chemical source terms in the species transport equations [18].

There is a clear trade-off in the selection of the preconditioner. Since preconditioning is applied to each step of the iterative methods, the form of matrix **J** should be optimized for low computational and inversion cost while maintaining strong convergence. The fully-explicit preconditioner provides the cheapest option but results in poor convergence performance, requiring extremely small time steps. On the other end of the spectrum, the fully-implicit preconditioner would provide excellent convergence criteria and be unconditionally stable, but would be unfeasible due to high computation cost. Unfortunately, both the chemical and diffusion source terms are generally dependent on a large number of species and, as a result, their Jacobian matrices are typically not sparse [2, 19, 25]. Thus, the construction of the preconditioner may become prohibitively expensive when considering a large number of chemical species.

3.1.4 Semi-implicit preconditioning for multicomponent mass diffusion

In an attempt to achieve strong convergence while maintaining a low cost form for the preconditioner, an approximation of the chemical and diffusion Jacobians that lies between the fully-implicit and fully-explicit extremes is proposed. The proposed preconditioner is of the form

$$\mathbf{J} = \rho_k^{n+3/2}\mathbf{I} - \frac{\Delta t}{2}\left(\frac{\partial \mathbf{C}}{\partial \mathbf{Y}} + \Lambda\right)_k^{n+1} \tag{25}$$

where Λ is a diagonal matrix defined as

$$\Lambda_{i,i} = D_{MA,i,i} - \frac{\dot{\omega}_i^-}{Y_i} \,. \tag{26}$$

The matrix $\frac{\dot{\omega}_i^-}{Y_i}$ is a good approximation of the diagonal of the chemical Jacobian [18]. In addition, the mixture-averaged diffusion coefficient matrix, an approximation of the multicomponent diffusion coefficient matrix, approximates the Jacobian of the species diffusion source terms. The species diffusion source term equals the divergence of the diffusion mass flux (equation (15)) and depends on the diffusion coefficient (equations (7) and (10)). As a result, since the diffusion source term and the diffusion coefficient are of similar order, and both depend on the underlying species diffusivities, the mixture-averaged diffusion coefficient matrix is a reasonable approximation for the Jacobian of the diffusion source term. Moreover, the diagonal nature of the the mixture-averaged diffusion coefficient matrix allows inversion with a low computational cost as evidenced by the ubiquity of the mixture-averaged mass diffusion assumption in the current computational literature. Thus, the mixture-averaged diffusion coefficient matrix provides a reasonable and low-cost estimate of the Jacobian matrix for the species diffusion source terms.

The diagonal preconditioning is applied within an iterative procedure for each time step in this algorithm. This allows for further reduction of the residual with each iteration making this method suitable for unsteady reacting flow simulations. Finally, Savard et al. [18] thoroughly demonstrated the convergence and accuracy of a similar semi-implicit diagonal preconditioner matrix.

3.2 Dynamic memory algorithm

In practice, the portions of the enthalpy (equation (3)) and species diffusion source terms (equation (4)) that can be computed from the (i) and $(i+1)$ flux vectors for a given i,j,k grid-point are calculated and stored. At the next grid-point $(i+1,j,k)$ the process is repeated and the remaining information for the previous grid-point (i,j,k) is calculated and stored along with half the solution of the next grid-point position. Figure 1 summarizes this process; fluxes are located at cell faces while source terms are at cell centers.

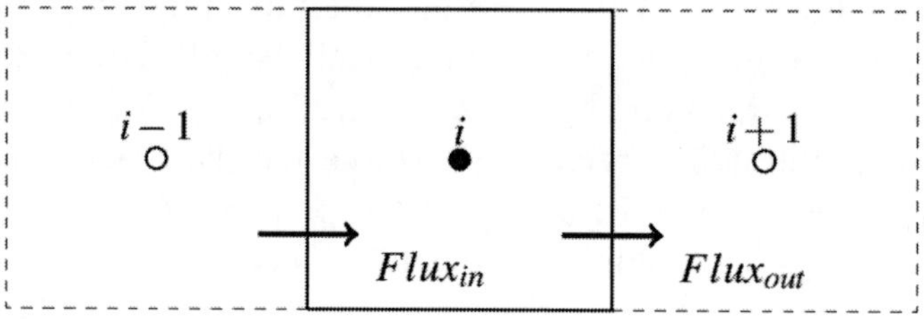

for *i=1:X* **do**
- Calculate diffusion coefficient matrix;
- **for** *isc=1:nSpec* **do**
  - $Flux(i-1/2)+=Flux_{in,isc}$;
  - $Flux(i+1/2)+=Flux_{out,isc}$;
- **end**
- $Source(i)=Source(i)+$ influence from $Flux(i-1/2)$ and $Flux(i+1/2)$;
- $Source(i-1)=Source(i-1)+$ influence from $Flux(i-1/2)$;
- $Source(i+1)=Source(i+1)+$ influence from $Flux(i+1/2)$;

end

Figure 1: Dynamic algorithm for calculating multicomponent enthalpy and species diffusion source term. Fluxes are located at cell faces while source terms are at cell centers.

3.3 Theoretical stability analysis

To simplify the stability analysis, we rewrote equation (19) in a point-wise form that neglects both the chemical source term—demonstrated to be stable by Savard et al. [18]—and the convective transport term, which is assumed to be integrated explicitly (i.e., not modified by sub-iterations). In addition, both the density and diffusion coefficients are assumed to be constant in space. Although in reality both density and the diffusion coefficients vary across the domain, the local variation in these values within the discretized stencil is negligible compared with other terms. Thus, assuming constant density and constant diffusion-coefficients should be valid for a point-wise stability analysis. Finally, the resulting form of equation (19) is transformed into modified wavenumber space by assuming a solution of the form

$$\mathbf{Y}(x,y,z,t) = f(t)e^{i\kappa x + i\kappa y + 8i\kappa z}\,, \tag{27}$$

such that equation (19) reduces to the form

$$\left[\mathbf{I} + \frac{\Delta t}{2}\mathbf{D}_{MA}\kappa'^2\right]\left(f_{k+1}^{n+1} - f_k^{n+1}\right) = f_0^n - f_k^{n+1} - \frac{\Delta t}{2}\mathbf{D}_{MC}\kappa'^2\left(f_k^{n+1} + f_0^n\right)\,, \tag{28}$$

where D_{MA} and D_{MC} are the mixture-averaged and multicomponent diffusion coefficient matrices calculated from equations (8) and (12) respectively, f_0 is the initial value of the sub-iteration, and κ' is the modified wavenumber given by

$$\kappa'^2 = \frac{1}{\Delta x^2}\left[1 - cos(\kappa\Delta x)\,,\right. \tag{29}$$

where κ is the wavenumber. Dropping the superscripts n and $n+1$ for clarity, equation (28) reduces to

$$f_{k+1} = \mathbf{A}f_0 + \mathbf{B}f_k\,, \tag{30}$$

where

$$A = \left[\mathbf{I} + \frac{\Delta t}{2}\kappa'^2\mathbf{D}_{MA}\right]^{-1}\left[\mathbf{I} - \frac{\Delta t}{2}\kappa'^2\mathbf{D}_{MA}\right]\,, \tag{31}$$

and

$$B = \left[\mathbf{I} + \frac{\Delta t}{2}\kappa'^2\mathbf{D}_{MA}\right]^{-1}\left[\frac{\Delta t}{2}\kappa'^2\left(\mathbf{D}_{MA} - \mathbf{D}_{MC}\right)\right]\,. \tag{32}$$

Inspecting equation (30), the coefficient $\mathbf{A}$ in front of f_0 is strictly less than one for all values of $\mathbf{D}_{MA,i,j}$. Therefore, convergence of the sub-iterations is assured as long as all elements of $\mathbf{B} \leq 1$.

Expanding matrix $\mathbf{B}$ yields

$$\mathbf{B} = \begin{bmatrix} \frac{\frac{\Delta t}{2}\kappa'^2 D_{MA,1,1}}{1+\frac{\Delta t}{2}\kappa'^2 D_{MA,1,1}} & \frac{-\frac{\Delta t}{2}\kappa'^2 D_{MC,1,2}}{1+\frac{\Delta t}{2}\kappa'^2 D_{MA,2,2}} & \frac{-\frac{\Delta t}{2}\kappa'^2 D_{MC,1,3}}{1+\frac{\Delta t}{2}\kappa'^2 D_{MA,3,3}} & \cdots & \frac{-\frac{\Delta t}{2}\kappa'^2 D_{MC,1,j}}{1+\frac{\Delta t}{2}\kappa'^2 D_{MA,i,j}} \\ \frac{-\frac{\Delta t}{2}\kappa'^2 D_{MC,2,1}}{1+\frac{\Delta t}{2}\kappa'^2 D_{MA,1,1}} & \frac{\frac{\Delta t}{2}\kappa'^2 D_{MA,2,2}}{1+\frac{\Delta t}{2}\kappa'^2 D_{MA,2,2}} & \frac{-\frac{\Delta t}{2}\kappa'^2 D_{MC,2,3}}{1+\frac{\Delta t}{2}\kappa'^2 D_{MA,3,3}} & \cdots & \frac{-\frac{\Delta t}{2}\kappa'^2 D_{MC,2,j}}{1+\frac{\Delta t}{2}\kappa'^2 D_{MA,i,j}} \\ \frac{-\frac{\Delta t}{2}\kappa'^2 D_{MC,3,1}}{1+\frac{\Delta t}{2}\kappa'^2 D_{MA,1,1}} & \frac{-\frac{\Delta t}{2}\kappa'^2 D_{MC,3,2}}{1+\frac{\Delta t}{2}\kappa'^2 D_{MA,2,2}} & \ddots & \cdots & \vdots \\ \vdots & \vdots & \vdots & \ddots & \vdots \\ \frac{-\frac{\Delta t}{2}\kappa'^2 D_{MC,i,1}}{1+\frac{\Delta t}{2}\kappa'^2 D_{MA,1,1}} & \frac{-\frac{\Delta t}{2}\kappa'^2 D_{MC,i,2}}{1+\frac{\Delta t}{2}\kappa'^2 D_{MA,2,2}} & \cdots & \cdots & \frac{\frac{\Delta t}{2}\kappa'^2 D_{MA,i,j}}{1+\frac{\Delta t}{2}\kappa'^2 D_{MA,i,j}} \end{bmatrix} \tag{33}$$

The eigenvalues of this matrix control the stability of the overall method. In the present case, the absolute values of these eigenvalues are all smaller than unity, making the method stable.

3.4 Method verification

To verify the method implementation, an artificial species profile was generated such that the direction and relative magnitudes of the flux vectors could be predicted intuitively. Specifically, a two-dimensional V-shaped species profile with a central angle of 45° was created and projected into three dimensions as shown in Figure 2.

Figure 2: Normalized artificial input species profile for method verification; slice from *x*-*z*-plane.

For such a profile, the flux vectors should be constant in the *y*-direction, of equal magnitude and opposite sign in the *z*-direction reflected over the *x*-*y*-plane, and have varying magnitude but constant sign matching the initial input profile in the *x*-direction. These predictions should be consistent independent of chemical species or other scalar value for the artificial input profile. We ran the algorithm for one "complete" set of subiterations to convergence and normalized the resulting diffusion flux vectors to ensure the relative magnitudes and direction were consistent with expectation.

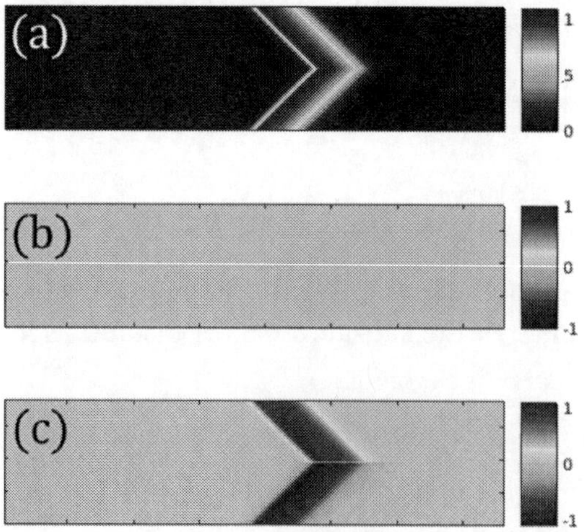

Figure 3: Normalized flux vectors resulting form artificial species profile after one full iteration of semi-implicit multicomponent diffusion calculation; (a) *x*-component of flux, (b) *y*-component of flux, (b) *z*-component of flux.

Figure 3 shows the results of this artificial test case. The resulting normalized flux vectors agree with expectation and have equal magnitudes in the *x*- and *z*-direction corresponding to the 45° artificial flame angle. This result indicates proper functionality of the proposed method.

4. Multicomponent diffusion verification

To ensure accuracy and verify the multicomponent mass diffusion model, one-dimensional unstretched (flat) laminar flame simulations were performed and results compared with similar mixture-averaged and multicomponent results from Cantera [26]. We selected the one-dimensional flat flame configuration for this purpose, since it restricts all transport to the streamwise direction. Thus, for mass continuity to be maintained, the stream-wise species diffusion fluxes and source terms for the mixture-averaged and multicomponent cases should be similar in magnitude and sum to zero. This is because the spanwise fluxes are zero by definition for this geometry and cannot contribute to mass conservation. This simplified geometry allows for a direct comparison of the multicomponent mass diffusion model to the accepted mixture-averaged diffusion model.

A one-dimensional unstretched (flat) laminar hydrogen/air flame with an equivalence ratio of $\phi = 0.4$ was simulated in NGA using both the mixture-averaged and multicomponent diffusion models. The inlet mixture has an unburnt temperature of 298 K and pressure of 1 atm. The inlet velocity boundary condition, u_{BC}, matches the laminar flame speed. The computational domain is 1.2 cm in length, with the flame located near the center of the domain. Each NGA case, regardless of diffusion model, uses an identical grid. The grid consists of 720 points, with at least 20 points through the laminar flame thickness defined using the maximum temperature gradient, $l_F = (T_{\max} - T_{\min})/|\nabla T|_{\max}$. Schlup and Blanquart [27] used an identical configuration.

The Cantera simulations were similarly run using both mixture-averaged and multicomponent diffusion models with matching inlet conditions, equivalence ratio, and domain size. The freely-propagating adiabatic flat flame solver was used with grid refinement criterion for slope and curve set to 0.1 and a refinement ratio of 2. Finally, the Cantera simulations were solved to steady-state with the energy equation enabled. All simulations used the same nine-species hydrogen mechanism of Hong et al. [28] with updated rate constants from the same group [29, 30].

4.1 One-dimensional a priori flux comparison

To verify the multicomponent diffusion model, we calculated the species flux vectors while solving the governing equations. The fluxes are determined using the same input parameters as stated above for a single time step. This a priori comparison is intended to provide a one-to-one comparison of the mixture-averaged and multicomponent models as the input parameters are identical. It is important that this comparison be performed on the first time step as the input parameters to the diffusion flux calculation may evolve over time.

Figure 4 presents the a priori flux results for H_2, H_2O, and OH relative to the local mixture temperature. As expected, the flux profiles for the mixture-averaged and multicomponent cases have a similar shape and magnitude. The multicomponent case has an approximately 18 % higher maximum magnitude for H_2 and OH species fluxes compared with the mixture-averaged case; however, this is attributed to the inclusion of the correction velocity to maintain mass continuity in the mixture-averaged model (equation (9)). The correction velocity is weighted based on the mole and mass fractions of the species; as a result, a large portion of the correction for mass continuity is lumped into the N_2 mass flux. This weighting of the correction velocity accounts for the observed differences in the flux magnitude for the other species, as presented in Figure 5. In Figure 5 the maximum magnitude of the mixture-averaged model is approximately 40 % higher than the multicomponent model; this difference corresponds to the observed differences in the H_2

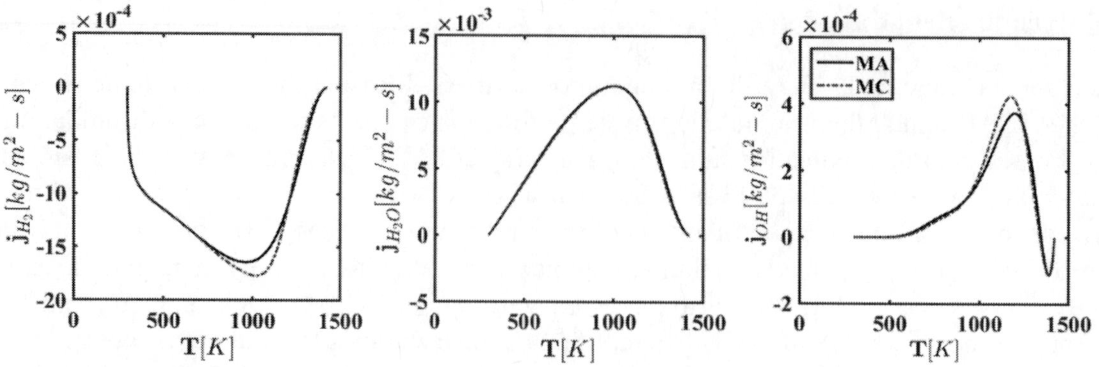

Figure 4: A priori comparison of mass diffusion flux for H_2, H_2O, and OH relative to mixture local temperature in hydrogen/air flame with $\phi = 0.4$.

and OH species fluxes and fulfills mass-continuity requirements. As a secondary check, the species mass fluxes for both models sum to zero, confirming that mass is conserved.

4.2 One-dimensional a posteriori comparison

To further verify the multicomponent model, we present an a posteriori comparison of the species profiles and unstretched laminar flame speeds. The unstretched laminar flame speed is defined as

$$S_L^o = -\frac{\int \rho \dot{\omega}_{\mathrm{H}_2} dx}{\rho_u Y_{\mathrm{H}_2,u}}, \tag{34}$$

where ρ_u is the unburnt mixture density and $Y_{\mathrm{H}_2,u}$ is the unburnt fuel mass fraction. The laminar flame speeds (S_L^o) are approximately $23.0\,\mathrm{cm\,s^{-1}}$ and $22.3\,\mathrm{cm\,s^{-1}}$ for the mixture-averaged and multicomponent diffusion cases, respectively. Figure 6 compares several species profiles for the steady-state one-dimensional flat flame solutions relative to local mixture temperature; the profiles are within 1 % at all points. Similarly, profiles agree closely for the remaining six species. This strong agreement between the one-dimensional NGA and Cantera results for both the mixture-averaged and multicomponent cases verifies that the multicomponent model is functional and accurate.

5. Results and discussion

This section presents a priori results for a two-dimensional, unsteady premixed hydrogen flame simulation as well as a priori and a posteriori results for a three-dimensional, fully turbulent premixed hydrogen flame. Additionally, stability results based on the previously discussed theoretical stability criterion are presented. The section will conclude with a brief discussion of one-dimensional and two-dimensional scaling for the proposed method.

5.1 Two-dimensional a priori flux comparison

This section provides an a priori assessment of the species mass diffusion fluxes for a two-dimensional, freely propagating, unsteady hydrogen/air flame. Previous numerical studies used configurations

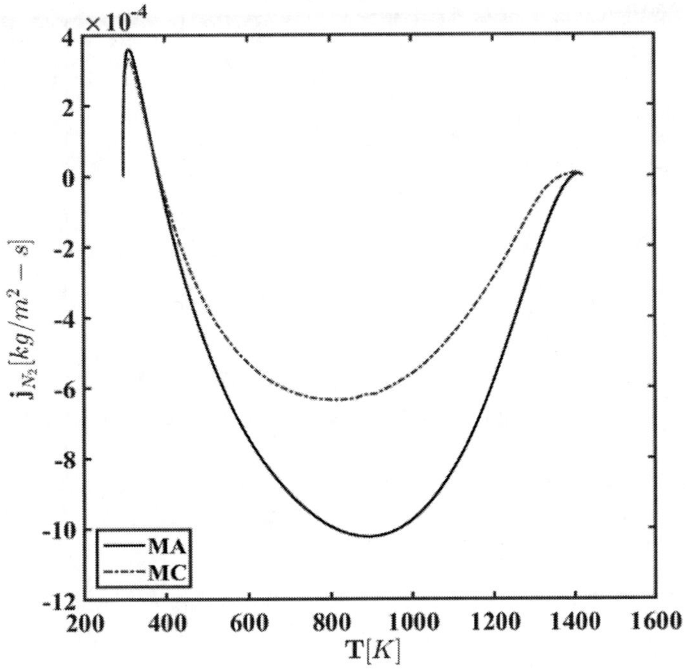

Figure 5: A priori comparison of mass diffusion flux for N_2 relative to mixture local temperature in hydrogen/air flame with $\phi = 0.4$.

similar to this two-dimensional freely-propagating flame [1, 27, 31–33]. The flame is initialized with an array of solutions corresponding to the one-dimensional flat flame simulations run previously, with an equivalence ratio of $\phi = 0.4$ and an inlet temperature and pressure of 298 K and 1 atm, respectively. The flame profile is initially perturbed in the span-wise direction by two sinusoidal modes.

The initial velocity boundary condition equals the unstretched laminar flame speed; however, once instabilities commence, the unstable flame propagates faster than the laminar flame speed. To avoid this, the velocity boundary condition changes to match the mean effective burning velocity S_{eff}^{2D}, defined as:

$$S_{\text{eff}}^{2D} = -\frac{\int_A \rho \dot{\omega}_{H_2} dA}{\rho_u Y_{H_2,u} L}, \tag{35}$$

where L is the spanwise dimension of the computational domain. This matched-velocity boundary condition allows the simulation to run for an arbitrary time, while keeping the flame within the domain.

The physical size of the domain is approximately $120\, l_F$ in the stream-wise direction by $30\, l_F$ in the spanwise direction. The grid is a structured, uniform mesh with 1888×472 cells, with a cell size corresponding to $\Delta x = \Delta y = l_F/16$. The top and bottom boundaries are periodic boundary conditions, with the inflow at the left boundary and outflow at the right boundary. Figure 7 shows the initial flame profile and grid parameters. Burali et al. [1] and Schlup et al. [27] used identical configurations. Figure 8 shows the temperature contour at an instant in time for the two-dimensional, unsteady configuration as reference for the flame structure.

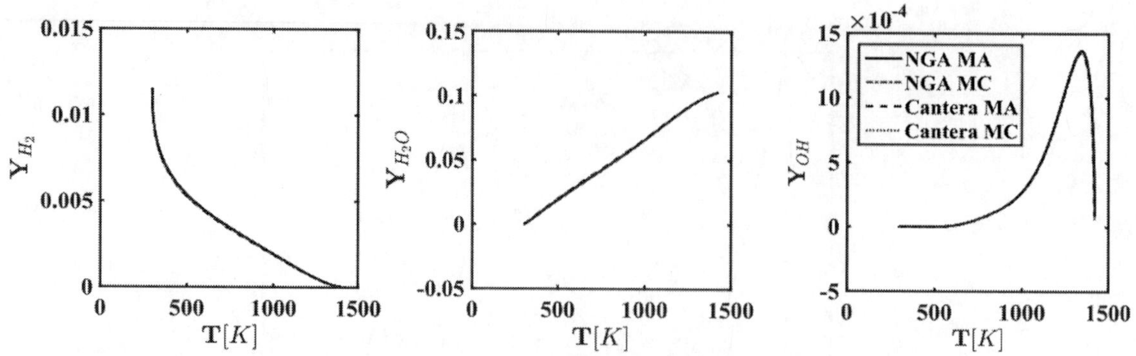

Figure 6: A posteriori comparison of species mass fraction for H_2, H_2O, and OH relative to mixture local temperature in a hydrogen/air flame with $\phi = 0.4$ using NGA and Cantera.

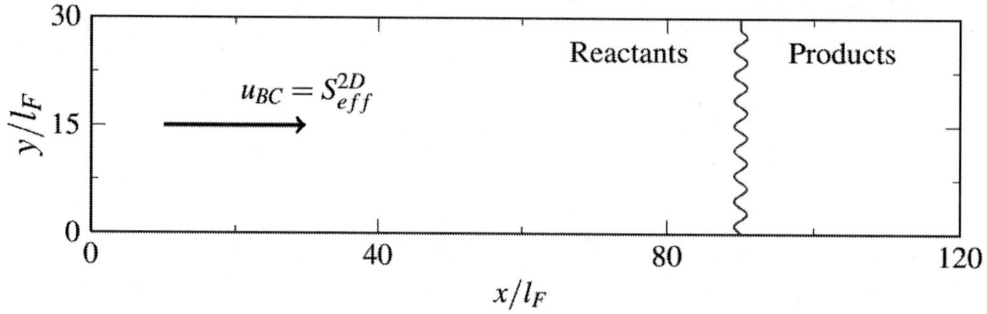

Figure 7: Initial flame profile for two-dimensional freely propagating unsteady hydrogen/air flame with $\phi = 0.4$.

As with the one-dimensional flames, we performed an a priori assessment of the species mass diffusion fluxes. However, for the two-dimensional unsteady flames, both the flux vector's magnitude and direction must be considered. To assess the impact of the relative diffusion direction, Figure 9 presents a probability density function (PDF) of the angle between the species flux vector and species gradient vector for both the mixture-averaged and multicomponent models through the flame front. The angles between vectors are computed using the geometric definition of the dot product of two vectors and are only presented for points across the flame sheet. To ensure a one-to-one comparison, the results presented correspond to identical initial data files for both the mixture-averaged and multicomponent models and have been run for a single time step. As before, the results for H_2, H_2O, and OH are presented.

No observable difference is evident between the mixture-averaged and multicomponent models with respect to the PDF of the angle between species flux vectors and species gradient vectors in Figure 9. Similar trends are observed comparing the PDF of the angle between the species flux vectors and temperature gradient vectors. In addition, the two diffusion models exhibit no differences between the magnitude of the flux vectors. Finally, the PDFs peak at π for all three species; these peaks are expected and indicate that mass diffuses in the direction of the negative gradient vector (i.e., from high concentration to low concentration).

Expanding the PDF analysis to include the flame wake, there are similarly few observable

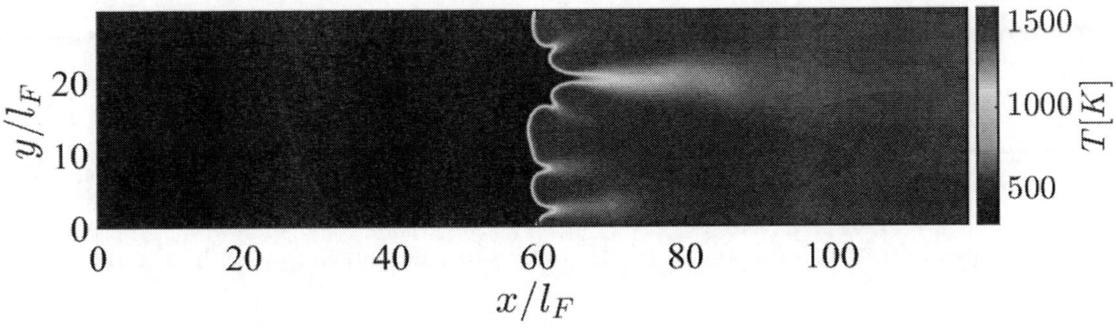

Figure 8: Temperature contour of multicomponent mass diffusion model for two-dimensional freely propagating unsteady hydrogen/air flame with $\phi = 0.4$.

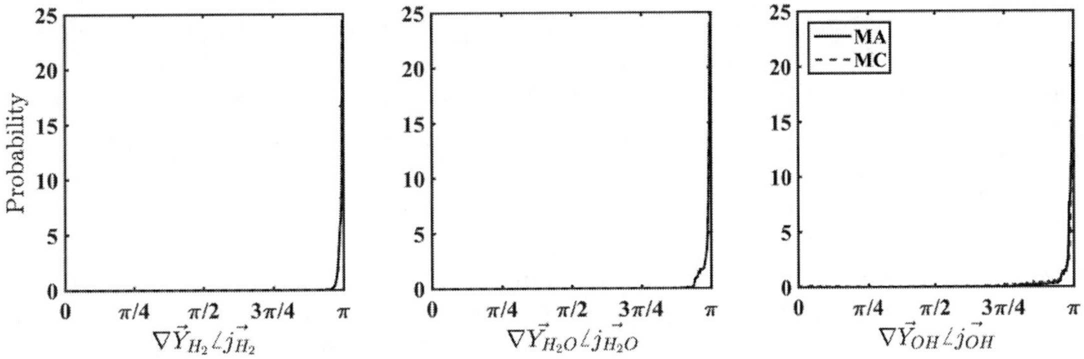

Figure 9: A priori assessment of the mixture-averaged and multicomponent models comparing the probability density function of the angle between species flux vectors and species gradient vectors for H_2, H_2O, and OH in a two-dimensional freely propagating unsteady hydrogen/air flame with $\phi = 0.4$.

differences between the relative flux angles and magnitudes for the mixture average and multicomponent cases.

5.2 Three-dimensional turbulent flux comparison

As a final test of the multicomponent mass diffusion model, and an assessment of its impact on mass transport, we simulated a three-dimensional, turbulent, premixed, freely propagating flame. Again, the computational setup for this case is similar to those of Burali et al. and Schlup et at. [1, 27]. For complete configuration details, the reader is referred to those papers. The inlet equivalence ratio is $\phi = 0.4$, with an unburnt Karlovitz number, $\mathrm{Ka}_u = \tau_F/\tau_\eta = 149$, where $\tau_\eta = (\nu/\varepsilon)^{1/2}$ is the Kolmogorov time scale of the incoming turbulence with unburnt kinematic viscosity, ν, and turbulent energy dissipation, ε. The unburnt turbulent Reynolds number is $\mathrm{Re}_t = u'l/\nu = 289$, where u' is the fluctuation of the mean velocity and l is the integral length scale. The inlet boundary condition has a mean inflow velocity approximately matching the turbulent flame speed such that the flame remains relatively center in the domain, and arbitrarily long simulations can

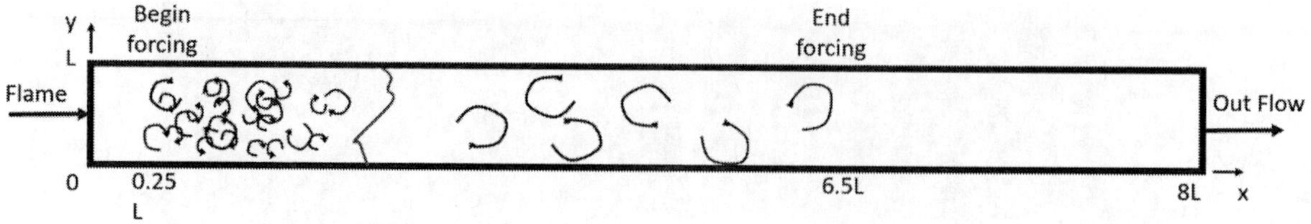

Figure 10: Two-dimensional schematic of the three-dimensional flame configuration. Adapted from [1, 27].

be performed. Once the turbulence has fully developed, the simulations are each run for 25 eddy turnover times, $\tau = k/\varepsilon \approx 500\,\mu s$.

The domain has 1520 points in the streamwise direction and 190 points in both spanwise directions, with a uniform grid size of $\Delta x = l_F/16$. This domain is about $100 l_F$ in length and $12 l_F$ in the spanwise directions; thus, the spanwise dimensions of this case are approximately 40% of the two-dimensional freely propagating dimensions. The two spanwise directions have periodic boundary conditions, with inflow and outflow boundary conditions in the streamwise direction. The flame is kept near the center of the domain to prevent effects from the inlet and outlet. Given the prescribed turbulence intensity, this mesh has a grid spacing equivalent to $\Delta x \approx 2\eta$, where η is the Kolmogorov length scale. Because no mean shear exists in this configuration, the production of turbulent kinetic energy is maintained using a linear forcing method [34, 35]. Figure 10 shows a two-dimensional schematic of the domain, including the locations of the flame and the forcing region. Figure 11 shows a three-dimensional view of the iso-surface of T_{peak} defining the flame front. The flame surface shows the complex behavior of the flame in the turbulent field.

Figure 11: Iso-surface of peak temperature colored by OH mass fraction for a three-dimensional turbulent hydrogen/air flame with multicomponent mass diffusion.

5.2.1 A priori comparison of species fluxes

To assess the impact of the relative diffusion direction for the three-dimensional turbulent simulation, Figure 12 presents a PDF of the angles between the species flux vector and species gradient vector for both the mixture-averaged and multicomponent models through the flame front. As with the two-dimensional unsteady simulation, the angles between vectors are computed using the geometric definition of the dot product of two vectors and are only presented for points across flame sheet. To ensure a one-to-one comparison, the results presented correspond to identical initial data files for both the mixture-averaged and multicomponent models and have been run for a single time step. As before, the results for H_2, H_2O, and OH are presented.

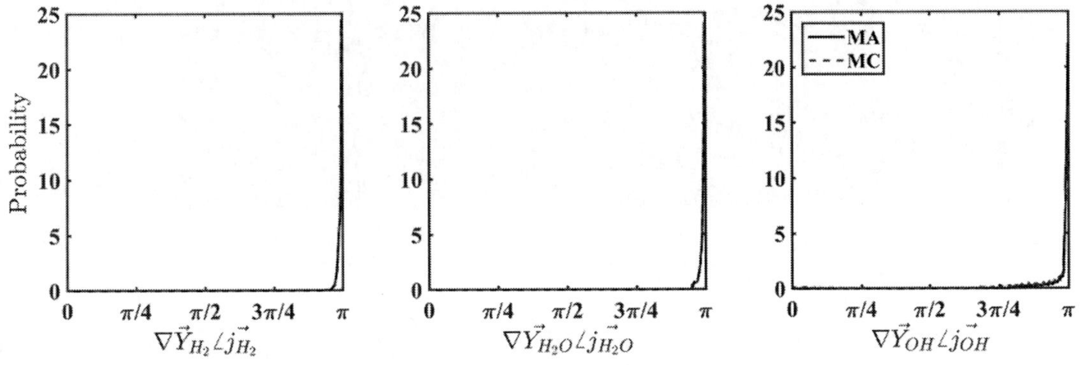

Figure 12: A priori assessment of the mixture-averaged and multicomponent models comparing the PDF of the angle between species flux vectors and species gradient vectors for H_2, H_2O, and OH in a three-dimensional, freely propagating, premixed, hydrogen/air flame with $\phi = 0.4$.

As with the two-dimensional simulation, no differences are observed between the mixture-averaged and multicomponent models with respect to the PDF of the angle between species flux vectors and species gradient vectors in Figure 9. Similar trends were observed when comparing the PDFs of the angle between the species flux vectors and temperature gradient vectors. In addition, no differences exist between the magnitude of the flux vectors for the two diffusion models. Finally, as with the two-dimensional simulation, the PDFs peak at π for all three species; these peak are expected and indicate that mass (mostly) diffuses in the direction of the negative gradient vector.

Figure 13 presents contour slices of the three-dimensional domain showing absolute flux angles relative to the species gradient for H_2. Qualitative inspection of these contours does suggest minor differences in the relative direction of H_2 flux, primarily in the wake. Similar differences are observed in all of the species flux angle contours; however, it is unlikely that these observed differences will significantly impact the flame. The magnitudes of diffusion flux vectors for the mixture-averaged and multicomponent models remain equivalent within the wake, and throughout the domain. Thus, the average transport is likely equivalent between the mixture averaged and multicomponent models.

5.2.2 A posteriori comparison of turbulent statistics

To assess the impact of the observed differences in the relative direction of the species diffusion between the mixture-averaged and multicomponent mass diffusion models, we present an a posteriori comparison of the turbulent statistics. The flames were allowed to develop in a turbulent flow field, and the statistics computed after the initial transients of the initial flow and scalar fields have advected through the domain. As an initial assessment, the effective flame propagation speeds for

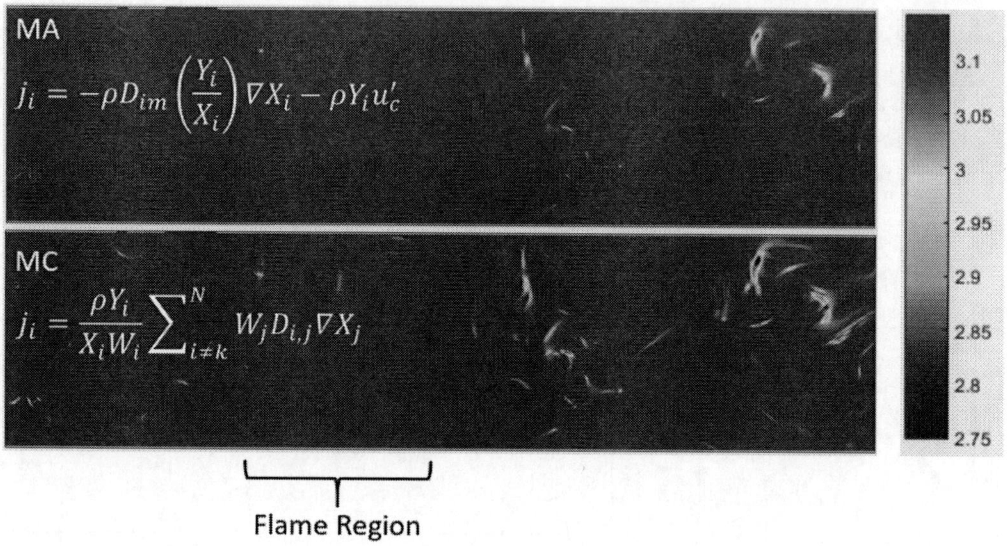

Figure 13: Contours of a section of the three-dimensional turbulent simulation showing absolute flux angles relative to the species gradient for H_2 for mixture-averaged (top) and multicomonent (bottom).

the two cases were calculated in a similar manner to the two-dimensional freely propagating flames

$$S_{\text{eff}} = -\frac{\int_V \rho\,\dot{\omega}_{\text{H}_2} dV}{\rho_u Y_{\text{H}_2,u} L}\,. \tag{36}$$

Figure 14 shows a time history of the turbulent flame speed over eight eddy turn-over times. The average normalized flames speeds from the mixture-averaged and multicomponent models are within 5% each other and are $S_T^{MA}/S_l^0 = 31.6$ and $S_T^{MC}/S_l^0 = 30.0$, respectively. Although the normalized turbulent flame speed for the multicomponent diffusion model is smaller than the mixture averaged model, this difference is negligible and supports the a priori assessment that the two models are equivalent.

To further assess any differences between the mixture-averaged and multicomponent mass diffusion models, Figure 15 presents the conditional means of the fuel mass fraction and its source term. As with the turbulent flame speeds, the differences in the calculated conditional means are negligible. This further demonstrates that although there are minor differences in the relative direction and magnitude of the diffusion mass fluxes between the mixture averaged and multicomponent models, these differences are negligible and do not meaningfully impact the turbulent statistics of the flame.

5.3 Stability results

Using the theoretical stability criterion determined previously, the numerical stability of the scheme is now analyzed numerically. While the theoretical analysis was performed assuming explicit transport of the convective terms and constant density and diffusion coefficient, this test case was performed with semi-implicit transport and variable density and diffusion coefficients. It is important to demonstrate the stability of the proposed preconditioner for the semi-implicit multicomponent

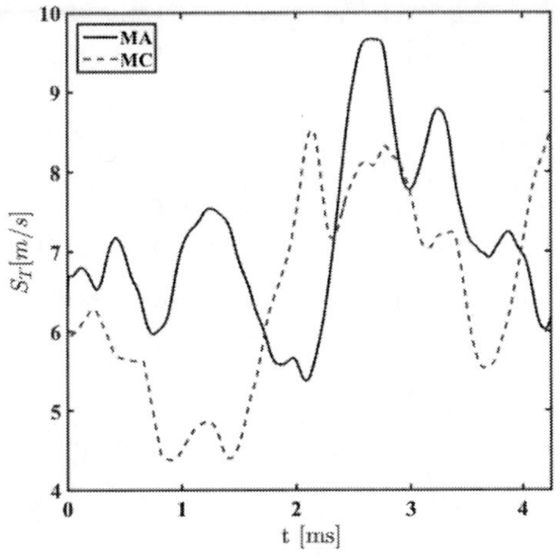

Figure 14: Flame speed history for three-dimensional, freely propagating, premixed, turbulent hydrogen/air flame with $\phi = 0.4$, for eight eddy turnover times.

species diffusion transport in a practical numerical simulation. The proposed stability criterion was found to be much less than unity for all test cases.

Figure 16 presents the numerical stability results for the one-dimensional test case. For the time step size tested, converging (as opposed to converged) sub-iterations implies a stable simulation, which agrees with similar results by Savard et al. [18]. In other words, unless the sub-iterations diverge, the simulation remains stable. In fact, the spectral radius of the proposed stability criterion was $\ll 1$ for all times steps examined, supporting the theoretical analysis that the semi-implicit diffusion calculation is unconditionally stable provided $D_{MA,i,i} \leq 1$ and $D_{MC,i,j} \leq 1$ at all points. These results suggest that stability can be well approximated by the theory. This should also correspond to the numerical stability limit for the three-dimensional turbulent premixed flame which will be investigates in a later study.

5.4 Computational cost

The implemented multicomponent mass diffusion model is computationally efficient. While we do not present a detailed analysis of the computational cost or scaling of the algorithms, a brief discussion of the increased cost to include full multicomponent mass diffusion is necessary to provide context for its use. The presented timing comparisons examine how the method scales with both species and spatial dimension.

Three chemical models (containing 9, 35, and 172 species) were tested on a one-dimensional flat flame to determine the cost of multicomponent mass diffusion in a wide range of mechanism sizes. Figure 17 shows the computational time per grid point for computing the diffusion mass fluxes on a desktop work-station. The time to compute coefficients for the mixture-averaged model increases linearly with the number of species. Alternatively, the time to compute diffusion mass fluxes increases by approximately the square of the number of species and is higher than the mixture-averaged scheme for all chemical models, with a cost more than an order of magnitude larger than the mixture-averaged model in simulations with large numbers of species. From these

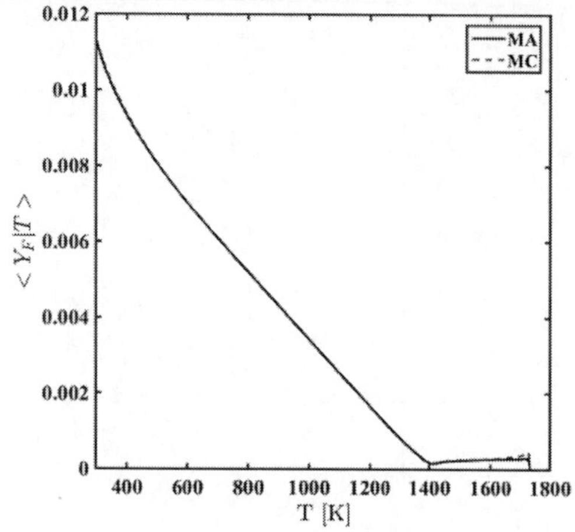

(a) Conditional means of Y_F

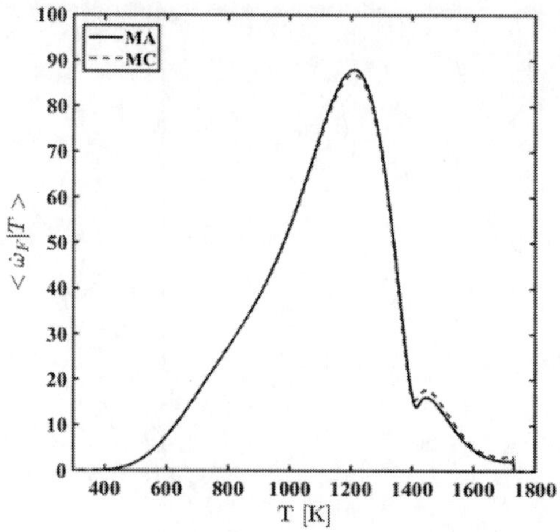

(b) Conditional means of $\dot{\omega}_F$

Figure 15: Turbulence statistics for the three-dimensional, freely propagating, premixed, turbulent hydrogen/air flame with $\phi = 0.4$, with conditional means of fuel mass fraction and source term presented against flame temperature.

results, it is clear that the use of the multicomponent mass diffusion model results in a considerable increase in time over the mixture-averaged model.

Timings were then acquired for one-dimensional and two-dimensional configurations covering the cases presented in this work. Figure 17 presents the computational time for each part of the code for both diffusion models. These timings were acquired on the California Institute of Technology cluster. It is interesting to note that the cost of calculating the multicomponent diffusion mass fluxes does not increase between the one-dimensional and two-dimensional freely propagating flame cases. This consistency in timing is likely due to the high efficiency of the dynamic memory allocation algorithm used to implement this model. However, the computational time increases for calculating the mass diffusion coefficients in Chemkin, as demonstrated by the increase in time taken for "Scalar" in Figure 17. This slow down could be resolved by implementation of a more efficient package for calculating the mass diffusion coefficients.

6. Summary and future work

This article proposes an efficient and stable scheme for implementation of full multicomponent mass diffusion in reacting flow DNS. It presents an a priori and a posteriori assessment of DNS mixture-averaged and multicomponent species diffusion models for one-dimensional steady; two-dimensional unsteady; and three-dimensional, turbulent, premixed, freely propagating hydrogen/air flames. No notable differences were observed between the two models when the relative direction and magnitude of the flux vectors, or turbulent statistics, were compared. Moreover, although the proposed scheme is efficient and stable, the mixture-averaged model is demonstrated to have a lower computational expense for chemical kinetic models with large numbers of species. For the presented hydrogen studies, this result suggests the mixture-averaged mass diffusion model is sufficient for DNS of three-dimensional, premixed turbulent flames. However, despite these results,

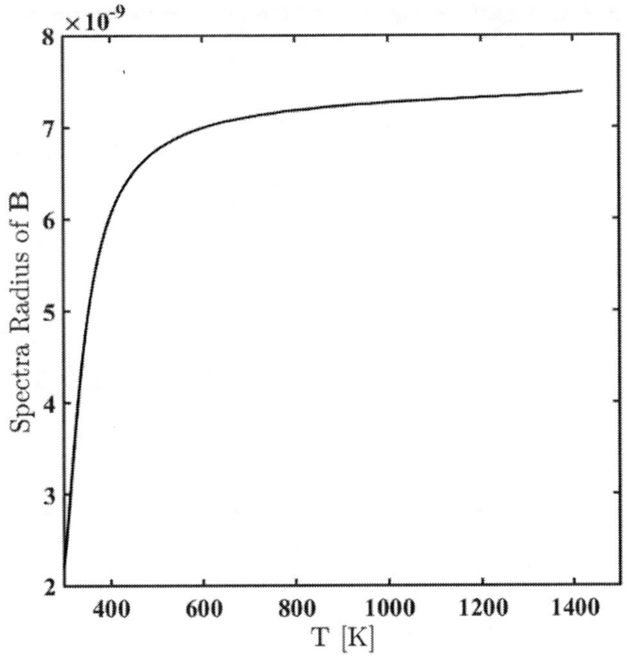

Figure 16: Spectral radius of **B** as a function of temperature in the one-dimensional premixed flame for $\Delta t = 2 \times 10^{-6}$.

there is insufficient data to draw firm conclusions on the accuracy and appropriateness of mixture-averaged assumptions for all flames. Additional data for different fuels, namely hydrocarbon fuels, and mechanisms with a greater number of species are needed.

In support of this effort, we are currently performing several three-dimensional turbulent flame DNS for heptane/air and toluene/air flames. The models for these species include 35 and 47 species, respectively. The three-dimensional turbulent simulations include the full multicomponent diffusion model and will provide a detailed assessment of the mixture-averaged and multicomponent diffusion models.

7. Acknowledgements

This material is based upon work supported by the National Science Foundation under Grant No. 1314109-DGE. This research used resources of the National Energy Research Scientific Computing Center, a DOE Office of Science User Facility supported by the Office of Science of the U.S. Department of Energy under Contract No. DE-AC02-05CH11231.

References

[1] N. Burali, S. Lapointe, B. Bobbitt, G. Blanquart, and Y. Xuan, Assessment of the constant non-unity Lewis number assumption in chemically-reacting flows, Combustion Theory and Modelling 20 (2016) 632–657.

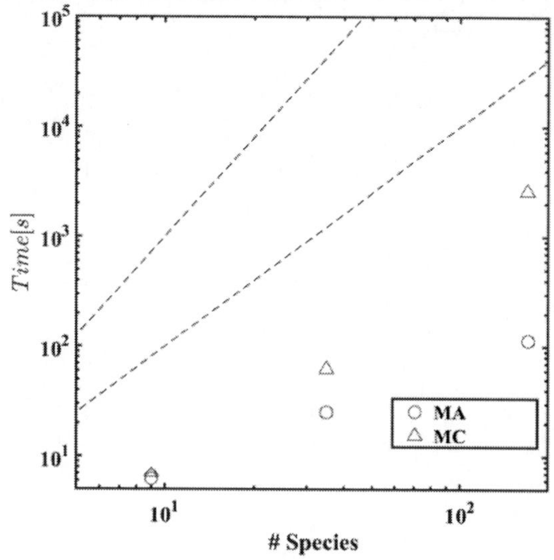

(a) Species scaling

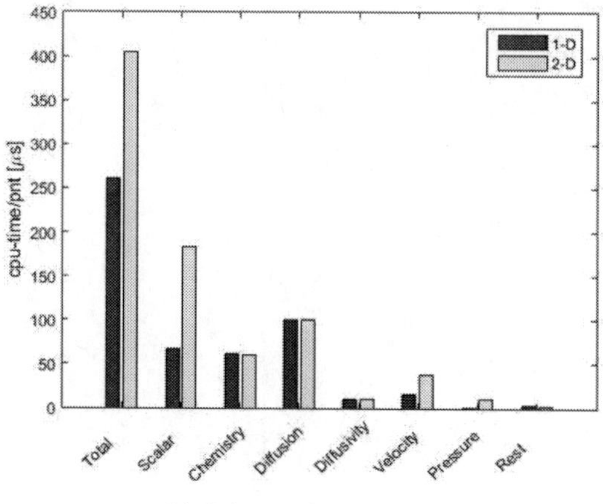

(b) Dimensional scaling

Figure 17: Computational time per grid point of computing diffusion coefficients and diffusion mass fluxes with (a) three different chemical models and (b) increasing spatial dimension.

[2] R. B. Bird, W. E. Stewart, and E. N. Lightfoot, Transport Phenomena, John Wiley & Sons, Inc., New York, 1960.

[3] J. Warnatz, Calculation of the Structure of Laminar Flat Flames I: Flame Velocity of Freely Propagating Ozone Decomposition Flames, Berichte der Bunsengesellschaft für physikalische Chemie 82 (1978) 193–200.

[4] M. E. Coltrin, R. J. Kee, and J. A. Miller, A Mathematical Model of Silicon Chemical Vapor Deposition, Journal of The Electrochemical Society 133 (1986) 1206. DOI: 10.1149/1.2108820.

[5] S. Lapointe and G. Blanquart, Fuel and chemistry effects in high Karlovitz premixed turbulent flames, Combustion and Flame 167 (2016) 294–307. DOI: 10.1016/j.combustflame.2016.01.035.

[6] A. Fillo and D. Blunck, Effects of Fuel Chemistry and Turbulence Intensity on Turbulent Consumption Speed for Large Hydrocarbon Fuels, Western States Section of the Combustion Institute (2015).

[7] A. J. Fillo, J. M. Bonebrake, and D. L. Blunck, Impact of fuel chemistry and stretch rate on the global consumption speed of large hydrocarbon fuel/air flames, 10th U. S. National Combustion Meeting The Combustion Institute, College Park, MD, (2017), DOI: AbstractID:2TF-0268.

[8] S. Dworkin, M. Smooke, and V. Giovangigli, The impact of detailed multicomponent transport and thermal diffusion effects on soot formation in ethylene/air flames, Proceedings of the Combustion Institute 32 (2009) 1165–1172. DOI: 10.1016/j.proci.2008.05.061.

[9] V. Giovangigli, Multicomponent transport in laminar flames, Proceedings of the Combustion Institute 35 (2015) 625–637. DOI: 10.1016/j.proci.2014.08.011.

[10] G. Borghesi and J. Bellan, *A priori* and *a posteriori* investigations for developing large eddy simulations of multi-species turbulent mixing under high-pressure conditions, Physics of Fluids 27 (2015) 035117.

[11] S. Ambikasaran and K. Narayanaswamy, An accurate, fast, mathematically robust, universal, non-iterative algorithm for computing multi-component diffusion velocities, Proceedings of the Combustion Institute 36 (2017) 507–515. DOI: 10.1016/j.proci.2016.05.055.

[12] C. F. Curtiss and J. O. Hirschfelder, Transport Properties of Multicomponent Gas Mixtures, The Journal of Chemical Physics 17 (1949) 550–555. DOI: 10.1063/1.1747319.

[13] J. O. Hirschfelder, C. F. Curtiss, and R. B. Bird, Molecular Theory of Gases and Liquids, Wiley, New York, 1954.

[14] S. Chapman and T. G. Cowling, The Mathematical Theory of NonUniform Gases, 3rd ed., Cambridge University Press, Cambridge, 1970.

[15] K. H. Ferziger and H. G. Kaper, Mathematical Theory of Transport Processes in Gases, North-Holland, Amsterdam, 1972.

[16] S. H. Lam, Multicomponent diffusion revisited, Physics of Fluids 18 (2006) 073101. DOI: 10.1063/1.2221312.

[17] O. Desjardins, G. Blanquart, G. Balarac, and H. Pitsch, High order conservative finite difference scheme for variable density low Mach number turbulent flows, Journal of Computational Physics 227 (2008) 7125–7159. DOI: 10.1016/j.jcp.2008.03.027.

[18] B. Savard, Y. Xuan, B. Bobbitt, and G. Blanquart, A computationally-efficient, semi-implicit, iterative method for the time-integration of reacting flows with stiff chemistry, Journal of Computational Physics 295 (2015) 740–769. DOI: 10.1016/j.jcp.2015.04.018.

[19] R. Kee, F. Rupley, and J. Miller, *Chemkin-II: A Fortran chemical kinetics package for the analysis of gas-phase chemical kinetics*, Sandia National Laboratories Report SAND89-8009, 1989.

[20] G. Dixon-Lewis, Flame Structure and Flame Reaction Kinetics. II. Transport Phenomena in Multicomponent Systems, Proceedings of the Royal Society of London A: Mathematical, Physical and Engineering Sciences 307 (1968).

[21] L. Shunn, F. Ham, and P. Moin, Verification of variable-density flow solvers using manufactured solutions, Journal of Computational Physics 231 (2012) 3801–3827.

[22] C. Pierce, Progress-variable approach for large-eddy simulation of turbulent combustion, PhD thesis, Stanford University, 2001.

[23] O. Desjardins, G. Blanquart, G. Balarac, and H. Pitsch, High order conservative finite difference scheme for variable density low Mach number turbulent flows, Journal of Computational Physics 227 (2008) 7125–7159.

[24] L. F. Richardson, The Approximate Arithmetical Solution by Finite Differences of Physical Problems Involving Differential Equations, with an Application to the Stresses in a Masonry Dam, Philosophical Transactions of the Royal Society of London A: Mathematical, Physical and Engineering Sciences 210 (1911), DOI: 10.1098/rsta.1911.0009.

[25] F. Perini, E. Galligani, and R. D. Reitz, A study of direct and Krylov iterative sparse solver techniques to approach linear scaling of the integration of chemical kinetics with detailed combustion mechanisms, Combustion and Flame 161 (2014) 1180–1195.

[26] D. G. Goodwin, H. Moffat K., and R. L. Speth, *Cantera: An Object-oriented Software Toolkit for Chemical Kinetics, Thermodynamics, and Transport Processes*, 2017.

[27] J. Schlup and G. Blanquart, *Validation of a mixture-averaged thermal diffusion model for premixed lean hydrogen flames*, Submitted, 2016.

[28] Z. Hong, D. F. Davidson, and R. K. Hanson, An improved H2/O2 mechanism based on recent shock tube/laser absorption measurements, Combustion and Flame 158 (2011) 633–644. DOI: 10.1016/j.combustflame.2010.10.002.

[29] K.-Y. Lam, D. F. Davidson, and R. K. Hanson, A Shock Tube Study of H_2 + OH $\rightarrow$ H_2O + H Using OH Laser Absorption, International Journal of Chemical Kinetics 45 (2013) 363–373. DOI: 10.1002/kin.20771.

[30] Z. Hong, K.-Y. Lam, R. Sur, S. Wang, D. F. Davidson, and R. K. Hanson, On the rate constants of OH + HO_2 and HO_2 + HO_2: A comprehensive study of H_2O_2 thermal decomposition using multi-species laser absorption, Proceedings of the Combustion Institute 34 (2013) 565–571. DOI: 10.1016/j.proci.2012.06.108.

[31] J. F. Grcar, J. B. Bell, and M. S. Day, The Soret effect in naturally propagating, premixed, lean, hydrogen–air flames, Proceedings of the Combustion Institute 32 (2009) 1173–1180. DOI: 10.1016/j.proci.2008.06.075.

[32] R. Bastiaans and A. Vreman, Numerical simulation of instabilities in lean premixed hydrogen combustion, International Journal of Numerical Methods for Heat & Fluid Flow 22 (2012) 112–128.

[33] J. D. Regele, E. Knudsen, H. Pitsch, and G. Blanquart, A two-equation model for non-unity Lewis number differential diffusion in lean premixed laminar flames, Combustion and Flame 160 (2013) 240–250. DOI: 10.1016/j.combustflame.2012.10.004.

[34] C. Rosales and C. Meneveau, Linear forcing in numerical simulations of isotropic turbulence: Physical space implementations and convergence properties, Physics of Fluids 17 (2005) 095106.

[35] P. L. Carroll and G. Blanquart, A proposed modification to Lundgren's physical space velocity forcing method for isotropic turbulence, Physics of Fluids 25 (2013) 105114.

Western States Section of the Combustion Institute – Fall 2017 Meeting
Hosted by the University of Wyoming
October 2-3, 2017

Dependence of Intermittency on Turbulence Intensity, Fuel Type, and Simulation Fidelity in Premixed Reacting Flows

Samuel H.R. Whitman[1,*], *Colin A.Z. Towery*[1], *Alexei Y. Poludnenko*[2], *and Peter E. Hamlington*[1]

[1]*Department of Mechanical Engineering, University of Colorado, Boulder, CO, USA*
[2]*Department of Aerospace Engineering, Texas A&M University, College Station, TX, USA*
[*]*Corresponding author:* `samuel.whitman@colorado.edu`

Abstract: Intermittency is an important metric for determining the likelihood of extreme fluctuating quantities in turbulent premixed flames. Such extreme quantities may lead to flow-altering events including extinction, auto- and re-ignition, and deflagration to detonation transitions. Here we analyze intermittency of enstrophy (i.e., vorticity magnitude) and temperature gradient magnitude fields based on data from direct numerical simulations (DNS) of stoichiometric premixed flames in unconfined domains. The DNS are performed for different turbulence intensities and fuels, for single- and multi-step chemistry models, for varying spatial resolution, and for temperature-dependent and constant viscosities. These simulations thus enable the study of physical effects on intermittency, as well as the study of effects based on simulation fidelity. We show that intermittency in the temperature gradient magnitude varies with the chemistry model but has little dependence on viscosity, while enstrophy intermittency varies with the viscosity model, but has little dependence on the fidelity of the chemistry.
Keywords: Intermittency, Turbulence, Premixed

1. Introduction

In turbulent premixed combustion, the statistical distributions of velocity gradient (e.g., vorticity and strain rate) and scalar (e.g., temperature) gradient quantities are of considerable practical interest, since they are connected to ignition, extinction, and other transient flame phenomena. A number of studies have shown that such distributions are highly intermittent in non-reacting and reacting turbulence using both experimental (e.g., [1, 2]) and numerical (e.g., [3, 4]) data. From a statistical standpoint, such intermittency manifests itself as extended tails of probability density functions (PDFs) and represents a higher-than-Gaussian probability of extreme events. As a result, transient flame phenomena may be much more likely than would be expected based on the assumption of Gaussian statistics.

In this study, we examine the dependence of intermittency in premixed reacting flows on turbulence intensity, on the temperature dependence of the viscosity, and on the level of fidelity with which the chemical kinetics are modeled (i.e., single- versus multi-step kinetics), including varying fuel type. Intermittency is studied for these conditions using conditional PDFs of temperature gradient magnitude and enstrophy. Intermittency in the temperature gradient is of particular interest since this quantity is closely connected to the scalar dissipation rate, which is included in some reacting flow models [5, 6].

The present study is based on computational data from direct numerical simulations (DNS) of premixed stoichiometric hydrogen-air and methane-air flames at atmospheric conditions. A set of thirteen total simulations are performed, corresponding to three different turbulence intensities, both single- and multi-step chemical mechanisms, two different fuel types and varying temperature dependence of viscosities in the single-step simulations.

2. Numerical Simulations

All simulations are performed using Athena-RFX [7] to solve the fully compressible, reactive flow equations. These equations are solved for premixed stoichiometric hydrogen-air and methane-air turbulent flames in an unconfined domain and at atmospheric conditions. Both single- and multi-step models of hydrogen-air reaction kinetics are used to model source terms in the species equations, while the methane-air case is only modeled with multi-step kinetics.

The single-step hydrogen-air combustion model is identical to that used by Towery et al. [8]. Viscosity, mass diffusivity, and thermal diffusivity all have a power-law dependence on temperature with an exponent $n = 0.7$. Simulations are also performed using exponents of $n = 0$ and $n = -1$ for the viscosity, such that the dynamic viscosity μ is a temperature-independent constant in the first case, and the kinematic viscosity ν is temperature-independent in the second case. This is intended to provide insight into the need for a physical viscosity in numerical simulation codes where such a choice exists. For all three temperature dependencies in the single-step chemistry simulations, the Schmidt, Prandtl, and Lewis numbers are all unity in the reactants. Thermodynamic quantities are related to each other using the equations of state for a calorically-perfect ideal gas. The resulting laminar flame speed is $S_L = 302$ cm/s and the laminar flame thermal width based on the maximum temperature gradient is $\delta_L = 0.032$ cm. The unburnt reactants temperature is $T_u = 290$ K and the burnt products temperature is $T_b = 2135$ K.

Multi-step Hydrogen kinetics are modeled using the 2014 San Diego Mechanism [9], comprised of 21 reactions between 8 reacting species: H, H_2, O, O_2, OH, H_2O, HO_2, H_2O_2, and inert N_2. The NASA seven-coefficient format is used to compute the thermodynamic functions for pure chemical species in conjunction with the equations of state for a thermally-perfect ideal gas. Mixture averaging is used to calculate molecular transport coefficients as prescribed by [10]. Mixture-averaged diffusion coefficients are calculated following the TRANSPORT library [11]. Thermal conduction coefficients are evaluated for pure species following Warnatz [12]. Binary diffusion coefficients and pure species viscosity are evaluated using standard kinetic methods. The resulting laminar flame speed is $S_L = 250$ cm/s and the laminar flame thermal width based on the maximum temperature gradient is $\delta_L = 0.034$ cm. The unburnt reactants temperature is $T_u = 300$ K and the burnt products temperature is $T_b = 2135$ K.

Multi-step methane kinetics are modeled using the reduced Stanford & University of Connecticut mechanism with 19 species and 54 reactions. The rest of the model is the same as the multi-step hydrogen kinetics. The resulting laminar flame speed is $S_L = 37.3$ cm/s and the laminar flame thermal width based on the maximum temperature gradient is $\delta_L = 0.044$ cm. The unburnt reactant temperature is $T_u = 300$ K and the burnt product temperature is $T_b = 2232$ K.

Three-dimensional simulations were performed for premixed flames in unconfined domains, with periodic boundary conditions in spanwise directions and open boundaries normal to the direction of flame propagation. For all simulations, the domain width was set to $L = 8\delta_L$, resulting in a width of $L = 0.350$ cm for CH_4, $L = 0.272$ cm for multi-step hydrogen and $L = 0.259$ cm for the single-step hydrogen cases. Single-step chemistry simulations were run for two resolutions, the lower at $256 \times 256 \times 8192$ with a physical domain length of $32L$ and the higher-resolution cases at $512 \times 512 \times 8192$ with a physical domain length of $16L$. Because of the higher computational cost associated with multi-step chemisty, both multi-step simulations were run only at the lower resolution, and in the case of the higher-intensity H_2 runs the lengthwise dimension was truncated to 2048 grid cells and physical domain length of $8L$ such that each grid cell has the same physical dimensions in all cases.

Hydrogen simulations were performed for three different turbulence intensities. Low-intensity cases (I1) had Karlovitz numbers of 52 (multi-step chemistry) and 58 (single-step), and Damköhler numbers of 0.376 and 0.377 respectively. Medium intensity cases (I2) had Karlovitz numbers of 103 (multi-step) and 116 (single-step), and Damköhler numbers of 0.188 and 0.186 respectively. High-intensity cases (I3) had Karlovitz numbers of 294 (multi-step) and 325 (single-step), and Damköhler numbers of 0.094 in both cases. The Karlovitz and Damköhler numbers are defined according to the definitions in Towery et al. [8]. The Methane simulation was run only at the lowest intensity (I1), corresponding to a Karlovitz number of 25.5 and a Damköhler number of 0.377.

3. Results

Using data generated by the DNS, PDFs of temperature gradient magnitude and enstrophy are conditioned on a temperature-based progress variable $c = (T - T_u)/(T_b - T_u)$. This progress-variable

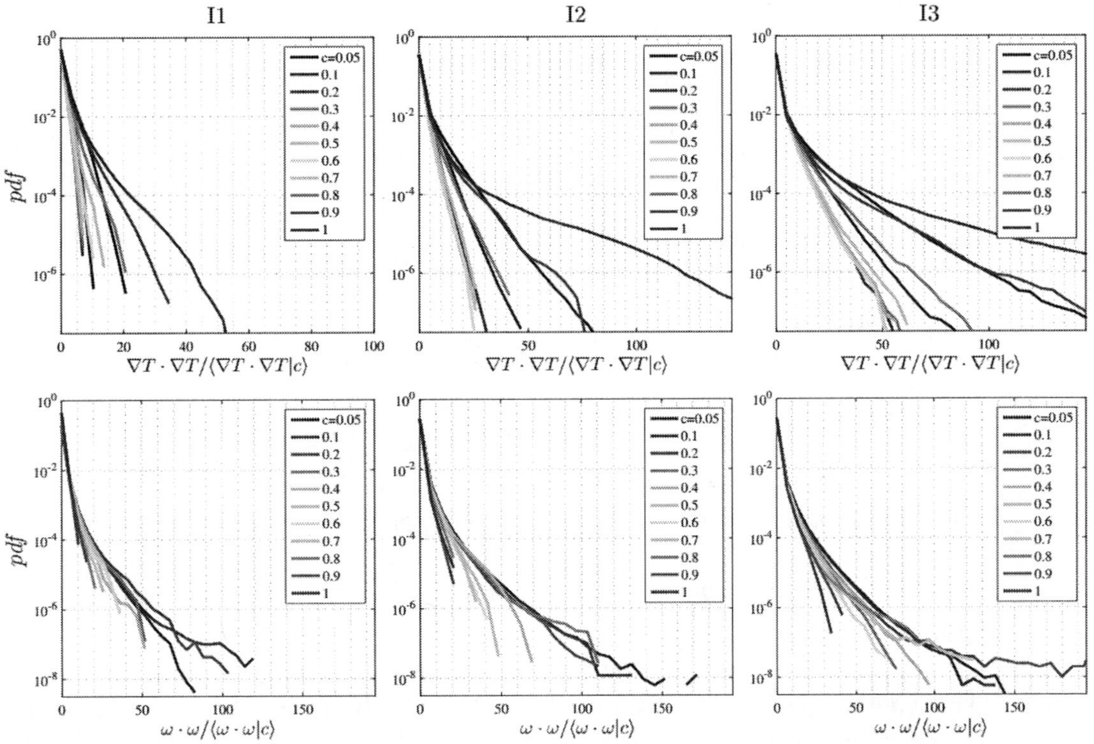

Figure 1: PDFs of temperature gradient magnitude (top) and enstrophy (bottom). Panels show multi-step cases I1, I2 and I3 from left to right in order of increasing turbulence intensity.

is used to track location within the flame, defined from $c = 0.05$ (reactants) to $c = 1$ (products). Statistics are collected over spatial regions spanning the turbulent flame brush and over many temporal snapshots. In general, intermittency in the resulting PDFs can be determined from the prominence of the tails in the PDFs, corresponding to the probabilities of extreme values of the temperature gradient magnitude and enstrophy.

The effect of turbulence intensity is shown in Fig. 1 on the previous page, where we see PDFs for the three prescribed intensities I1, I2 and I3. The multi-step hydrogen case is specifically presented to ensure that all simulation parameters are as realistic as possible. It is seen that intermittency in the temperature gradient magnitude increases steadily with the intensity. The enstrophy intermittency also increases steadily, but to a lower degree when compared relative to the temperature gradient magnitude. The qualitative shape of each PDF is very similar; the tails are simply lengthened at the higher intensities.

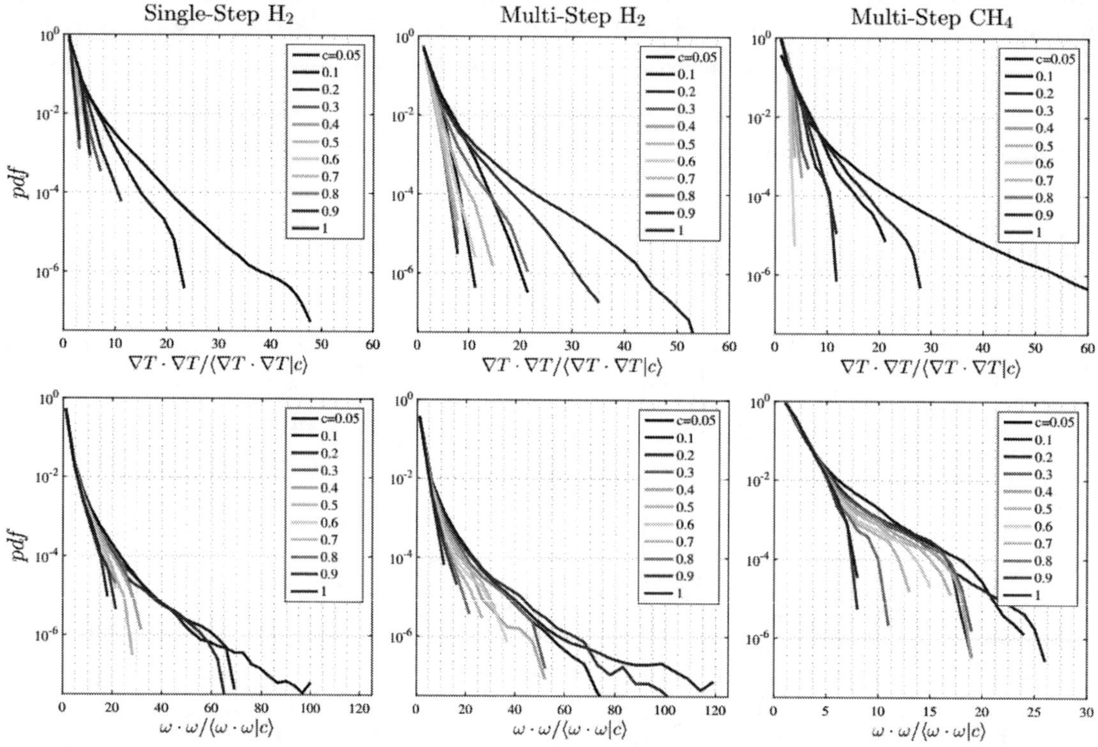

Figure 2: PDFs of temperature gradient magnitude (top) and enstrophy (bottom). Panels show multi-step hydrogen cases I1, I2 and I3 from left to right in order of increasing turbulence intensity.

The effect of chemistry is shown in Fig. 2, where we see that intermittency in the temperature gradient magnitude (shown in the top row) varies distinctly across each case. The single-step hydrogen chemistry shows a monotonic decrease in temperature gradient intermittency across the flame brush since higher c values have shorter tails, while multi-step hydrogen and methane both show a recovery of intermittency in the products region of the flame. This progressive recovery of intermittency throughout the flame brush and into the products is not observed in any of the single-step simulations. While the chemistry is seen to strongly affect the intermittency of the temperature gradient magnitude, the qualitative behavior in the enstrophy PDFs is consistent, showing a steady

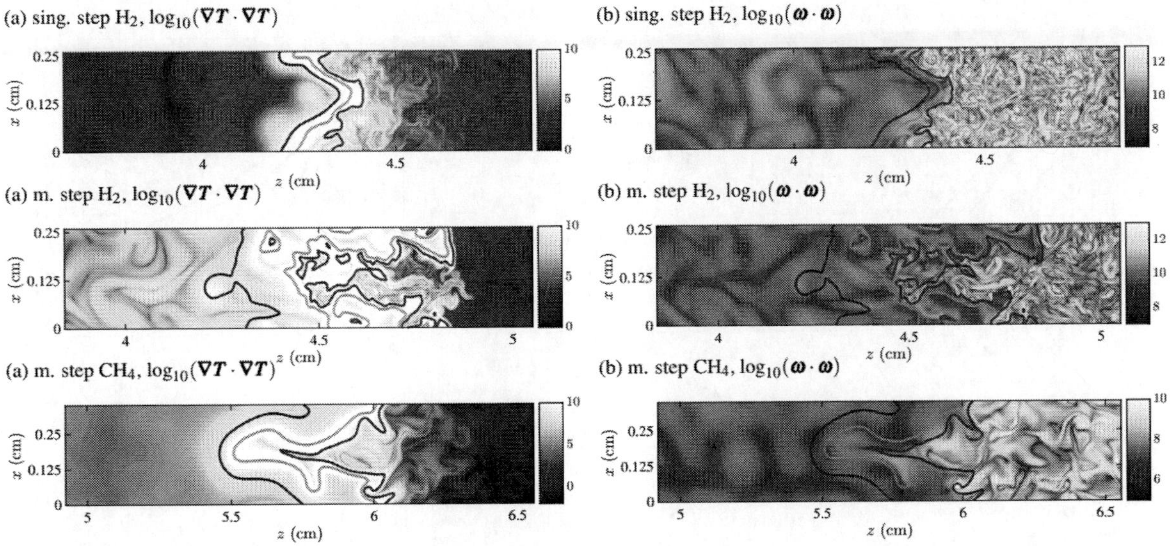

Figure 3: Two-dimensional fields of temperature gradient magnitude and enstrophy are shown for varying chemistry models. We have zoomed in to show the near-flame region more clearly. In descending order are single-step chemistry with $n = 0.7$, multi-step H_2 and multi-step CH_4. Note that the color axes for CH_4 are slightly modified for clarity. Each figure shows blue, green, and black line contours of $T = 400$, 1200, and 2000 K, respectively.

suppression of intermittency as the flame progresses from the reactants to the products. While the enstrophy PDF for methane is less intermittent, as evidenced by the shorter range on the x-axis due to the different species present in the premixed fluid, the suppression of intermittency is consistently observed across all chemistries. Two-dimensional fields showing representative slices of the temperature gradient magnitude and enstrophy for varying chemistry are shown in Fig. 3.

The relationship between temperature and viscosity has a noticeable effect on enstrophy in the single-step chemistry cases, as can be seen on the following page in Figs. 4 and 5. As the temperature dependence of the viscosity is decreased to $n = 0$ and then $n = -1$, the tails in enstrophy PDFs are elongated (by an order of magnitude for the combustion products), enough so in the case of $n = -1$ that the expanded binning range is still insufficient to capture the greatest values (resulting in their collection in the final bin). Meanwhile, varying n has no noticeable effect on the qualitative behavior of the temperature gradient magnitude intermittency. In all cases the pattern is the same as already noted for single-step chemistry: the temperature gradient magnitude intermittency decreases monotonically throughout the flame into the products, unaffected by choice of viscosity temperature dependence. Figure 5 shows temperature gradient magnitude and enstrophy fields of the flame brush regions for these cases. It can be observed that the enstrophy fields are noticeably different with increasing n. For $n = 0.7$ the heat released by the flame suppresses the enstrophy intermittency throughout the flame, but the effect is lessened at $n = 0$ and at $n = -1$ the enstrophy is similar in appearance throughout the domain while its intermittency is actually observed to increase in Fig. 4.

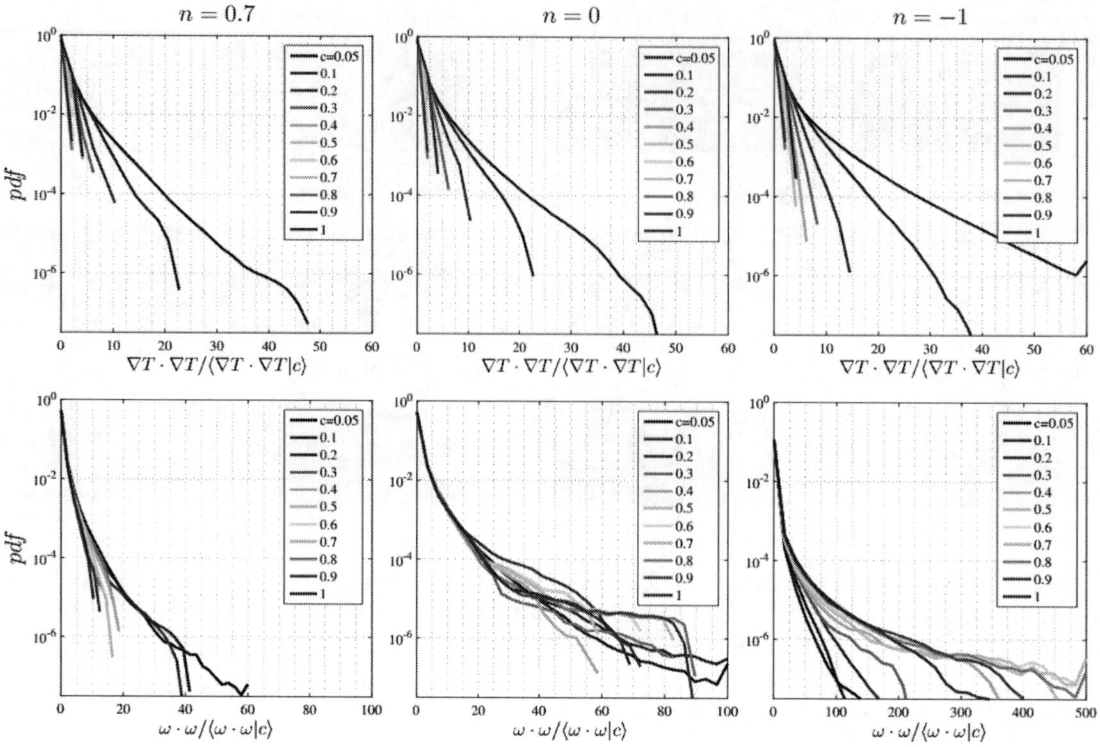

Figure 4: PDFs of temperature gradient magnitude and enstrophy. Panels show single-step hydrogen simulations with viscosity temperature exponent $n = 0.7$, $n = 0$ and $n = -1$.

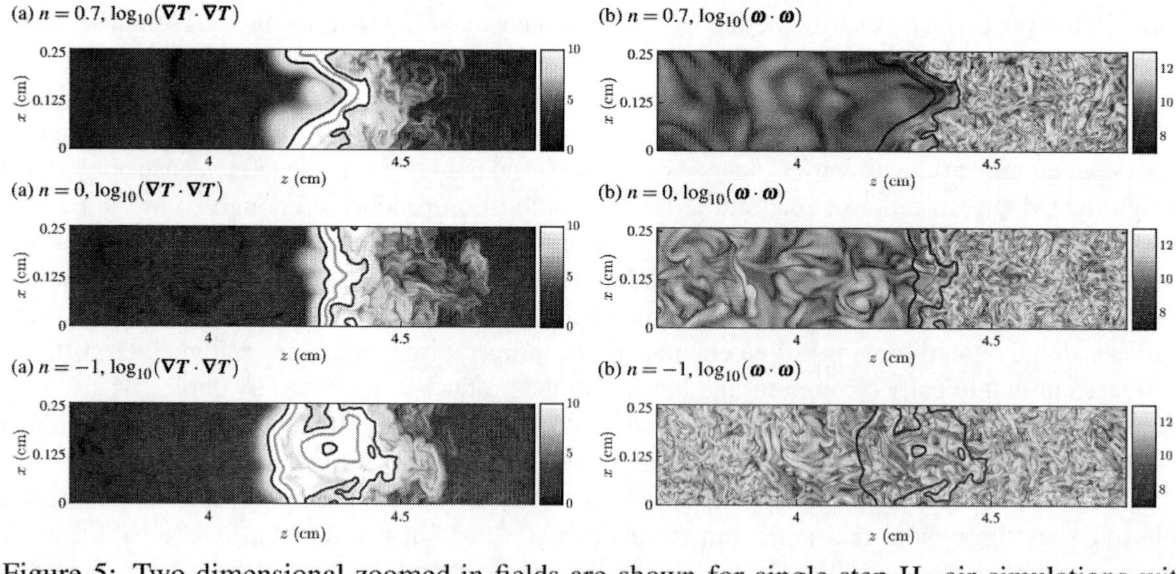

Figure 5: Two-dimensional zoomed-in fields are shown for single-step H_2-air simulations with varying viscosity temperature dependence. In descending order are $n = 0.7$ (realistic), $n = 0$ (constant dynamic viscosity μ) and $n = -1$ (constant kinematic viscosity ν). Each figure shows blue, green, and black line contours of $T = 400$, 1200, and 2000 K, respectively.

4. Conclusions

The results presented show a clear dependence of intermittency in the temperature gradient magnitude on the fidelity of the chemical kinetic model. In particular, a minimum in the intermittency develops for intermediate values of the progress variable in the multi-step cases, but intermittency decreases monotonically through the flame in the single-step cases. Conversely, enstrophy intermittency is shown to depend on whether the viscosity is temperature-dependent or constant, but not on the fidelity of the chemical kinetic model. In particular, enstrophy intermittency decreases substantially through the flame for temperature-dependent viscosity, but is essentially constant through the flame when the viscosity is constant. This is consistent with the results of Bobbitt et al. [13] who found no significant impact of the combustion model on conditional mean enstrophy throughout the flame brush in n-heptane/air combustion. Moreover, these results extend the earlier study by Hamlington et al. [4] by showing the impacts of complex chemistry and temperature dependent viscosity on intermittency in premixed reacting flows.

5. Acknowledgements

SHRW, CAZT and PEH were supported, in part, by AFOSR award FA9550-17-1-0144 and by DoD HPCMP. AYP was supported in part by AFOSR award F4FGA05260G001. Computing resources were provided by DoD HPCMP under a Frontier project award.

References

[1] L. K. Su and W. J. A. Dahm, Scalar imaging velocimetry measurements of the velocity gradient tensor field in turbulent flows. II. Experimental results, Physics of Fluids 8 (1996) 1883–1906.

[2] B. W. Zeff, D. D. Lanterman, R. Mcallister, R. Roy, E. J. Kostelich, and D. P. Lathrop, Measuring intense rotation and dissipation in turbulent flows, Nature 421 (2003) 146–149.

[3] M. Overholt and S. Pope, Direct numerical simulation of a statistically stationary, turbulent reacting flow, Combustion Theory and Modelling 3 (1999) 371–408.

[4] P. E. Hamlington, A. Y. Poludnenko, and E. S. Oran, Intermittency in premixed turbulent reacting flows, Physics of Fluids 24 (2012) 075111.

[5] K. Sreenivasan, Possible Effects of Small-Scale Intermittency in Turbulent Reacting Flows, Flow, Turbulence and Combustion 72 (2004) 115–131.

[6] A. Y. Klimenko and R. W. Bilger, Conditional moment closure for turbulent combustion, publisher not identified, 1998.

[7] A. Poludnenko and E. Oran, The interaction of high-speed turbulence with flames: Global properties and internal flame structure, Combustion and Flame 157 (2010) 995–1011.

[8] C. A. Z. Towery, A. Y. Poludnenko, J. Urzay, J. O'brien, M. Ihme, and P. E. Hamlington, Spectral kinetic energy transfer in turbulent premixed reacting flows, Physical Review E 93 (2016).

[9] A. L. Sánchez and F. A. Williams, Recent advances in understanding of flammability characteristics of hydrogen, Progress in Energy and Combustion Science 41 (2014) 1–55.

[10] A. Ern and V. Giovangigli, Multicomponent transport algorithms, Springer, 1994.

[11] R. J. Kee, A FORTRAN computer code package for the evaluation of gas-phase, multicomponent transport properties, Sandia National Laboratories, 1988.

[12] J. Warnatz, Influence of Transport Models and Boundary Conditions on Flame Structure, Numerical Methods in Laminar Flame Propagation (1982) 87–111.

[13] B. Bobbitt, S. Lapointe, and G. Blanquart, Vorticity transformation in high Karlovitz number premixed flames, Physics of Fluids 28 (2016) 015101.

Western States Section of the Combustion Institute – Fall 2017 Meeting
Hosted by the University of Wyoming
October 2-3, 2017

Lagrangian Analysis of Enstrophy in Turbulent Premixed Flames

Ryan Darragh[1,*], *Colin A.Z. Towery*[2], *Alexei Y. Poludnenko*[3], *and Peter E. Hamlington*[2]

[1]*Ann and H.J. Smead Aerospace Engineering Sciences, University of Colorado, Boulder, CO 80309, USA*
[2]*Department of Mechanical Engineering, University of Colorado, Boulder, CO 80309, USA*
[3]*Department of Aerospace Engineering, Texas A&M University, College Station, TX 77843, USA*
**Corresponding author:* ryan.darragh@colorado.edu

Abstract: Turbulent combustion is a multi-scale and multi-physics problem depending upon both chemical and fluid dynamic processes. These processes are often examined using an Eulerian framework, but recently the Lagrangian framework, a long-time tool in non-reacting flow research, has become increasingly common for the study of turbulent combustion. The two analysis frameworks are in fact equivalent, with the only difference being a change in reference frame. In this study, a Lagrangian fluid parcel tracking algorithm is used to analyze the enstrophy (i.e., vorticity magnitude) dynamics in turbulent premixed reacting flows. The analysis of vorticity dynamics in the premixed flame case is based on data from a three dimensional direct numerical simulation of a premixed stoichiometric hydrogen-air flame in an unconfined domain. Vorticity budget terms are tracked along Lagrangian trajectories as fluid parcels travel through the flame, with particular focus on understanding the dynamical causes of turbulence variations through the flame preheat and reaction zones.
Keywords: ***premixed, turbulent, Lagrangian, enstrophy***

1. Introduction

Simulating turbulent reacting flows is typically cheaper and safer than performing experiments, especially with high-speed flows. However, simulations often do not fully capture all of the physics in a problem and instead employ approximations due to either a lack of physics included in the model, or a lack of resolution in the simulation. A lack of resolution can typically be solved by using additional and/or more powerful computers, while a defect in the model is more difficult to rectify. Chief among these difficulties is the challenge of accurately understanding the physics that a model fails to represent. From a computational standpoint, the best way to examine this physics is through direct numerical simulation (DNS), a type of simulation that resolves all scales within a flow. By taking what is learned at the smallest scales, a new model can be developed that captures more of the physics. Typically these models are used in cheaper Large Eddy Simulations (LES) where only the largest scales are resolved, and the subgrid scales (SGS) are modeled. Recently, there has been reason to believe that enstrophy production near the flame is not properly accounted for in SGS models and, as such, there has been a push to better understand the enstrophy at the smallest scales. This is particularly true for the baroclinic torque term, which arises from a misalignment in density and pressure gradients, and is expected to be larger at the edge of a flame than away from the flame brush. In an effort to aid in the development of more accurate SGS models, this paper investigates all of the budget terms in the enstrophy balance equation along Lagrangian trajectories.

It has long been known that flames and turbulence interact with each other. Combustion is accompanied by energy release which increases the turbulence kinetic energy and pressure, thus altering the flow [1, 2] and in some cases generating turbulence [3]. Similarly, turbulence also influences a flame by broadening the preheat zone [4] and wrinkling the flamelet [5] as vorticity increases. Additional studies into how a flame and vorticity interact have been performed from an Eulerian perspective, finding that viscous effects redistribute enstrophy within a flamelet and that the baroclinic torque term is large near a flamelet [6]. Kolmogorov's first hypothesis has also been tested by using the vorticity in a premixed flame [7].

This paper will focus on the enstrophy budget equation which can be found by contracting vorticity with the curl of the momentum transport equation, resulting in

$$\frac{1}{2}\frac{D\Omega}{Dt} = \underbrace{\omega_i\omega_j\frac{\partial u_i}{\partial x_j}}_{\text{Vortex Stretching}} - \underbrace{\Omega\frac{\partial u_j}{\partial x_j}}_{\text{Dilatation}} + \underbrace{\frac{1}{\rho^2}\omega_i\left(\varepsilon_{ijk}\frac{\partial \rho}{\partial x_j}\frac{\partial p}{\partial x_k}\right)}_{\text{Baroclinic Torque}} + \underbrace{\omega_i\varepsilon_{ijk}\frac{\partial}{\partial x_j}\left(\frac{1}{\rho}\frac{\partial \tau_{kl}}{\partial x_l}\right)}_{\text{Diffusion}}, \tag{1}$$

where u_i is the velocity, ε_{ijk} the Levi-Cevita tensor, $\omega_i = \varepsilon_{ijk}\partial u_k/\partial x_j$ is the vorticity, enstrophy is defined as $\Omega = \omega_j\omega_j$, D/Dt is the material derivative, ρ is the density, $\tau_{ij} = \mu\left(2S_{ij} + \frac{1}{3}S_{mm}\delta_{ij}\right)$ is the viscous stress tensor (including the bulk viscosity term with the bulk viscosity assumed to be equal to μ), and p is the pressure. The first term of the budget equation represents vortex stretching, the second dilitation, the third baroclinic torque, and the fourth accounts for diffusion. The stretching term changes vortex line orientation and allows for these lines to be stretched. Dilatation causes the size of a vortex to change due to flow compressibility. The baroclinic torque produces vorticity from the misalignment of pressure and density gradients and the diffusion term dissipates vorticity thereby reducing enstrophy.

Analyses of the vorticity near a flame usually take the Eulerian, or laboratory, frame of reference. However, this study is performed in the Lagrangian reference frame by following fluid parcels that are small enough to not be deformed by shear or strain, but still large enough to allow fluid properties, such as species mass fraction, to be defined. If enough fluid parcels are followed, the analysis can provide the same information as an Eulerian analysis with the added benefit of providing a time history through specific regions (in the Eulerian sense) of the flow. That is, if a fluid parcel is found to experience high baroclinic torque, this analysis allows for the possibility of looking back in the history of the fluid parcel and seeing what led up to that event. By taking this approach, additional information can be learned about the influence of the flame on turbulence.

The Lagrangian frame of reference is commonly used in research on nonreacting flows (see [8, 9] for a review on the topic), but has only recently gained attention in the reacting flow community. Hamlington *et al.* used this approach to investigate the thermochemical properties of fluid parcels as they passed through a flame, finding non-monotonicities in the temperature and fuel-mass fraction to be common occurrences [10]. Steinberg *et al.* used an experimental approach to determine flame residence times [11, 12], and Day *et al.* used a joint experimental-computational study of a low-swirl burner to investigate diffusion along Lagrangian path lines. A related method has also been explored by Uranakara *et al.* [13] who tracks flame surfaces rather than fluid parcels.

2. Description of Numerical Simulations

All simulations are performed using `Athena-RFX` [14–16], which solves the compressible, reactive-flow Navier-Stokes equations [17] on a fixed, equi-spaced, three dimensional mesh using an unsplit corner transport upwind scheme [18, 19] with a nonlinear HLLC Riemann solver and piecewise parabolic spatial reconstruction. The governing equations solved by Athena-RFX are given by

$$\frac{\partial \rho}{\partial t}+\frac{\partial(\rho u_j)}{\partial x_j}=0, \tag{2}$$

$$\frac{\partial(\rho u_i)}{\partial t}+\frac{\partial}{\partial x_j}\left[\rho u_i u_j+p\delta_{ij}-\mu\left(2S_{ij}+\frac{1}{3}S_{kk}\delta_{ij}\right)\right]=\mathscr{F}_i, \tag{3}$$

$$\frac{\partial(\rho e_0)}{\partial t}+\frac{\partial}{\partial x_j}\left[\rho e_0 u_j+p u_j-k\frac{\partial\Theta}{\partial x_j}-\mu u_i\left(2S_{ij}+\frac{1}{3}S_{kk}\delta_{ij}\right)\right]=-\rho q\dot{\omega}+u_i\mathscr{F}_i, \tag{4}$$

$$\frac{\partial(\rho Y)}{\partial t}+\frac{\partial}{\partial x_j}\left(\rho Y u_j-\lambda\frac{\partial Y}{\partial x_j}\right)=\rho\dot{\omega}, \tag{5}$$

where μ the temperature-dependent shear viscosity, $S_{ij}=\frac{1}{2}\left(\frac{\partial u_i}{\partial x_j}+\frac{\partial u_j}{\partial x_i}\right)$ the strain rate, δ_{ij} the Kronecker delta tensor, $\mathscr{F}_i$ the large-scale energy injection term, e_0 the specific total energy, k the thermal conductivity, Θ the temperature, q the chemical energy release, $\dot{\omega}$ the reaction rate, Y the reactant mass fraction, and λ the mass diffusivity. The bulk viscosity is set to μ and an ideal gas equation of state is used. A power-law dependence on Θ is used for

$$\mu=\rho_r\nu_r\left(\frac{\Theta}{\Theta_r}\right)^{0.7}, \qquad k=\rho_r\alpha_r c_p\left(\frac{\Theta}{\Theta_r}\right)^{0.7}, \qquad \lambda=\rho_r D_r\left(\frac{\Theta}{\Theta_r}\right)^{0.7}, \tag{6}$$

where ν_r is the kinematic viscosity, α_r the thermal diffusivity, D_r the mass diffusivity, and $c_p=\frac{\gamma R}{M(\gamma-1)}$ the specific heat capacity at constant pressure where the r subscript denotes that the value is taken in the reactants. Periodic forcing is performed throughout the entire domain at large scales and has been confirmed to be capable of producing a statistically stationary turbulent flame [16]. Overall, the simulation has third order spatial and second order temporal accuracy.

Single-step chemistry is used for the H_2-air reaction represented through an Arrhenius reaction [16],

$$\dot{\omega}=-\rho Y_\alpha B\exp\left(-\frac{Q}{R\Theta}\right), \tag{7}$$

where B is a pre-exponential factor, Q the activation energy, and R the gas constant. The initial fuel temperature is set to 293 K and the adiabatic flame temperature is 2135 K. All model parameters are provided in Table 1.

The simulation used for this study is performed in an unconfined domain, as shown in Figure 1, which has an aspect ratio of $1\times1\times16$ with periodic boundary conditions in the x and y directions. Prior to flame ignition, periodic boundaries are used in the z direction, and zeroth order extrapolation afterwards. A uniform mesh of $256\times256\times4096$ grid cells with a physical size of $L=0.259$ cm in the x and y directions, and $L_z=16L$ in the z direction, providing a grid resolution of $\Delta\approx1\times10^{-3}$ cm. The laminar flame is resolved with 32 grid cells. The Kolmogorov scale,

Θ_r	293 K	Initial temperature
p_r	1.01×10^6 erg/cm^3	Initial pressure
ρ_r	8.73×10^{-4} g/cm^3	Initial density
γ	1.17	Adiabatic index
M	21 g/mol	Molecular weight
Θ_p	2135 K	Post-flame temperature
ρ_p	1.2×10^{-4} g/cm^3	Post-flame density
B	6.85×10^{12} cm^3/(g s)	Pre-exponential factor
Q	$46.37\ RT_r$	Activation energy
q	$43.28\ RT_r/M$	Chemical energy release
ν_r	0.189 cm^2/s	Initial kinematic viscosity
α_r	1.77 cm^2/s	Initial thermal diffusivity
D_r	1.77 cm^2/s	Initial mass diffusivity

Table 1: Parameters used to define the single-step H_2-air chemical model.

$\eta_K = \left(\frac{\nu^3}{\varepsilon}\right)^{1/4}$ where ε is the kinetic energy dissipation rate, is resolved to 8.8 grid cells in the product and 0.7 grid cells in the fuel. The flow is less-well resolved in the reactants, but resolution improves as products are created. Because the primary interest is inside and near the flame, a lower resolution in the reactants is not expected to influence the results.

The simulation is initialized as a uniform, cold reactive mixture and turbulence is allowed to develop for two domain-scale eddy turnover times $\tau_{ed} = L/U$, where U is the turbulent velocity at the domain scale. Once the turbulence is developed, a laminar flame is initialized near the center of the domain and allowed to develop for an additional τ_{ed} before data collection begins. This initialization procedure has been verified by Poludnenko and Oran [16].

The Lagrangian fluid parcel tracking algorithm is implemented as a post-processing step. It

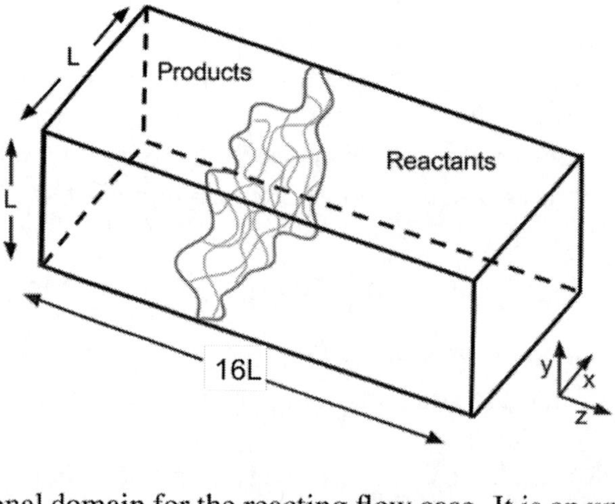

Figure 1: The computational domain for the reacting flow case. It is an unconfined domain with an aspect ratio of $1 \times 1 \times 16$ with periodic boundary conditions in the x and y directions, and zeroth order extrapolation in the z direction. The flame, represented in orange, is initialized near the center of the domain. Reactants are located at high z values and products at low z values. The flame slowly propagates in the direction of larger z into the reactants.

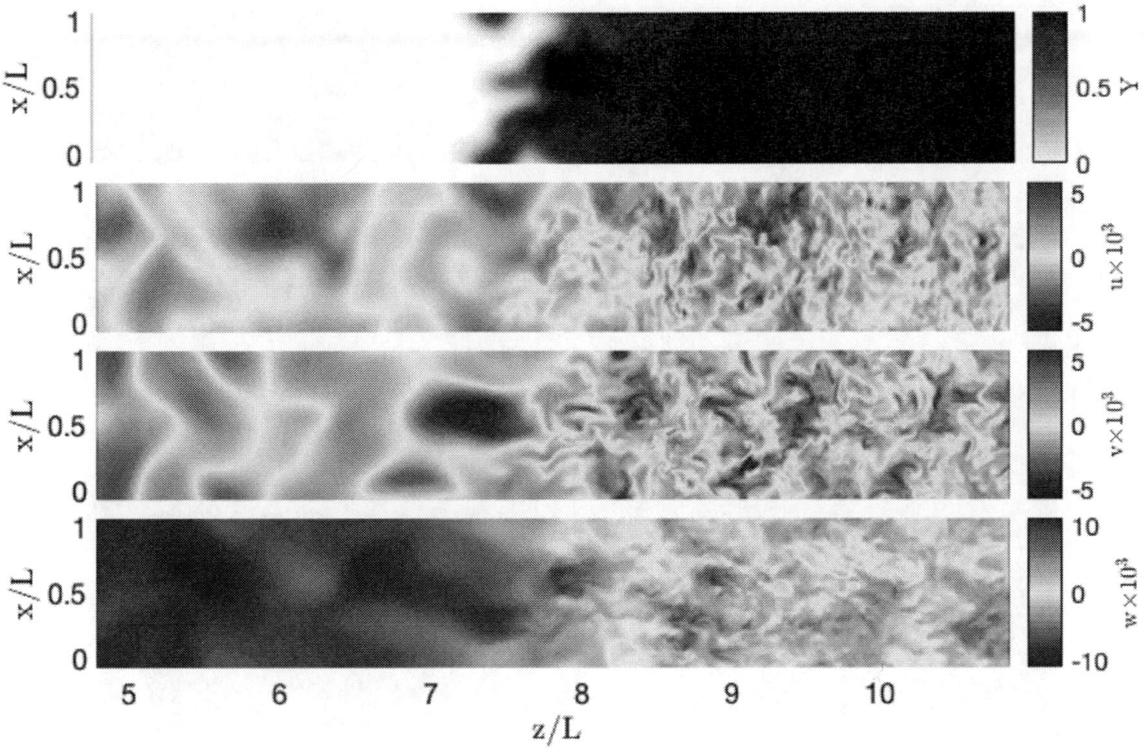

Figure 2: Visualization of the fuel-mass fraction and all three velocity fields in a constant y plane at the time fluid parcels are released. The velocities shown are in units of cm/s, reactants are on the right and products on the left.

uses Akima splines for spatial and temporal interpolation and independently evolves fluid parcels initialized from the same location using fourth and fifth order Runge-Kutta as an error-checking mechanism. A full description of the algorithm can be found in Darragh *et al.* [20].

3. Enstrophy Budget Results

Figure 2 shows the fuel-mass fraction, Y, and each of the velocity components near the flame brush at the time fluid parcels are released. As fluid begins to combust, the small-scale flow features are destroyed leaving only the large-scale variations. This can be seen as the changes in velocity lose most of their finer details after passing through the flame. Eulerian temperature and enstrophy fields are shown in Figure 3, also at the time fluid parcels are released. The temperature matches up well with Y, as is to be expected, while all four budget terms are fairly constant in the products. Near the flame the baroclinic torque term has the greatest variation with a noticeable increase in magnitude.

While these Eulerian fields are not the focus of this study, they do orient us within the flow and provide motivation for the Lagrangian study that follows. Of most interest is the baroclinic torque component to the vorticity budget equation because of its variations within the flame brush. However, these variations are not planar and thus not all fluid that passes from reactants to products will experience these variations. By tracking Lagrangian fluid parcels that move through these

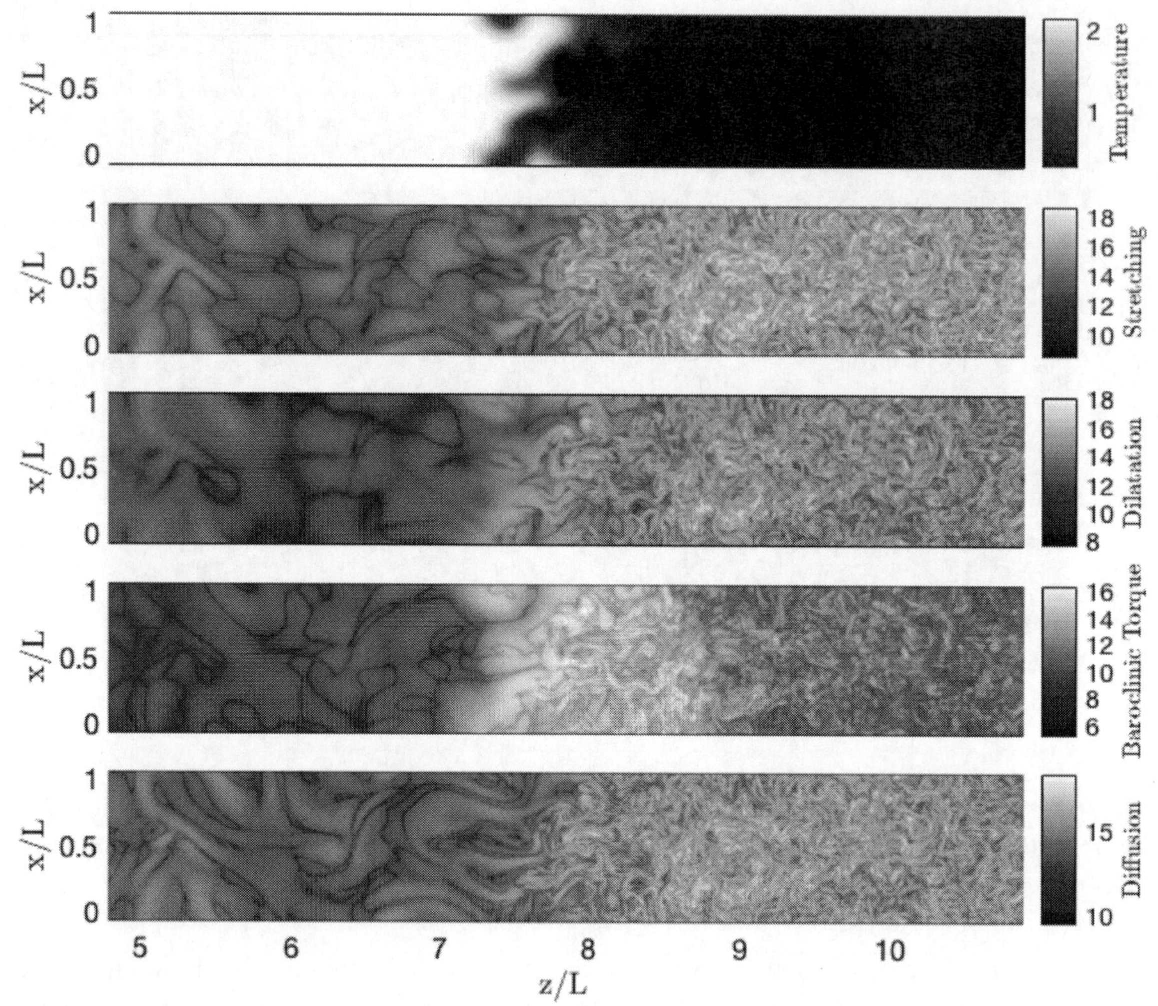

Figure 3: Temperature and enstrophy fields at constant y at the time fluid parcels are released. The temperature is scaled by 10^3 and the enstrophy budget terms are shown as the base 10 logarithm of the absolute value of the quantity. The baroclinic torque term has a noticeable increase in the flame.

regions, we are able to isolate the fluid that experiences a large variation in baroclinic torque, and also characterize in a statistical sense what portion of the flow experiences increased baroclinic torque.

Fluid parcels are seeded on the 500 K isosurface over a uniform grid in the x-y plane. They are then evolved through the flame. Figure 4 shows a selection of several fluid parcel trajectories. Note that the curvature of the trajectories decreases in the products once the small scale flow features have been destroyed.

The enstrophy budget for single trajectories is shown in Figure 5. Three representative cases are shown: no increase in baroclinic torque, a spike in baroclinic torque near the interface between preheat and reaction zones, and a constant increase in baroclinic torque through the preheat and reaction zone interface. In all three cases there is a lot of variation in the unresolved reactants

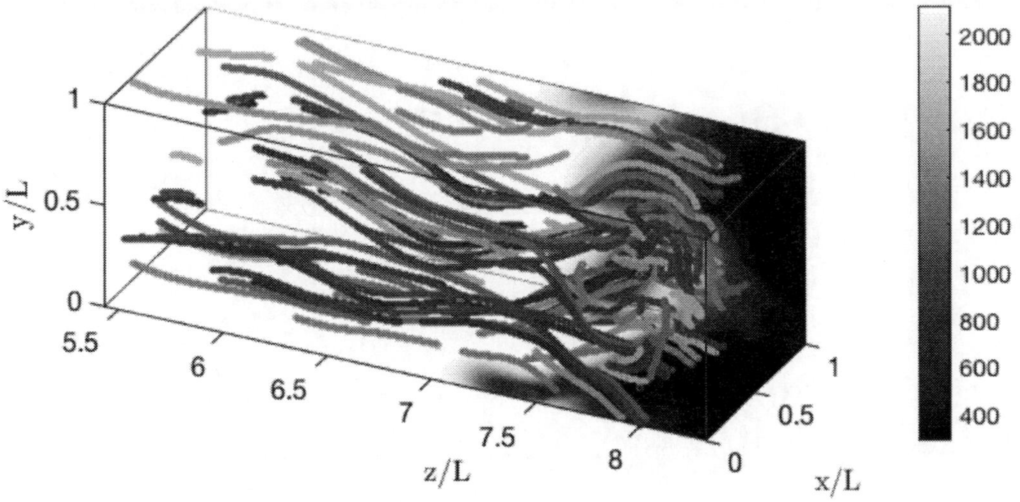

Figure 4: 44 sample trajectories initialized on the $T = 500$ K isosurface and then moving through the flame.

which subsides as the fluid is burnt. Here the preheat zone is considered the region of fluid between 600 K and 1000 K and the reaction zone the region between 1000 K and 1400 K (see Hamlington *et al.* [10] for justification on the zone boundaries). The peaks in enstrophy implies the mismatch in pressure and density gradients dominates the enstrophy evolution for some of the fluid near the interface. However the other two examples also show that the baroclinic torque is not always the dominate term in the balance equaiton. An accurate SGS model will need to account for all three cases to accurately describe the flow within and near the flame.

4. Conclusions and Future Work

A Lagrangian analysis of a turbulent premixed flame has found increased variation in the baroclinic torque at the interface of the preheat and reaction zones, confirming the findings of previous Eulerian studies. Additional work needs to be done to determine what history leads up to these increased variation events. Higher turbulence intensities should also be investigated to see if the baroclinic variation persists.

Acknowledgements

RD, CAZT and PEH were supported in part by AFOSR awards FA9550-14-1-0273 and FA9550-17-1-0144, and by a DoD HPCMP Frontier Project award. RD was also supported by a National Defense Science and Engineering Graduate Fellowship. AYP was supported in part by AFOSR award F4FGA05260G001. Computing resources were provided by DoD HPCMP under the Frontier Project award.

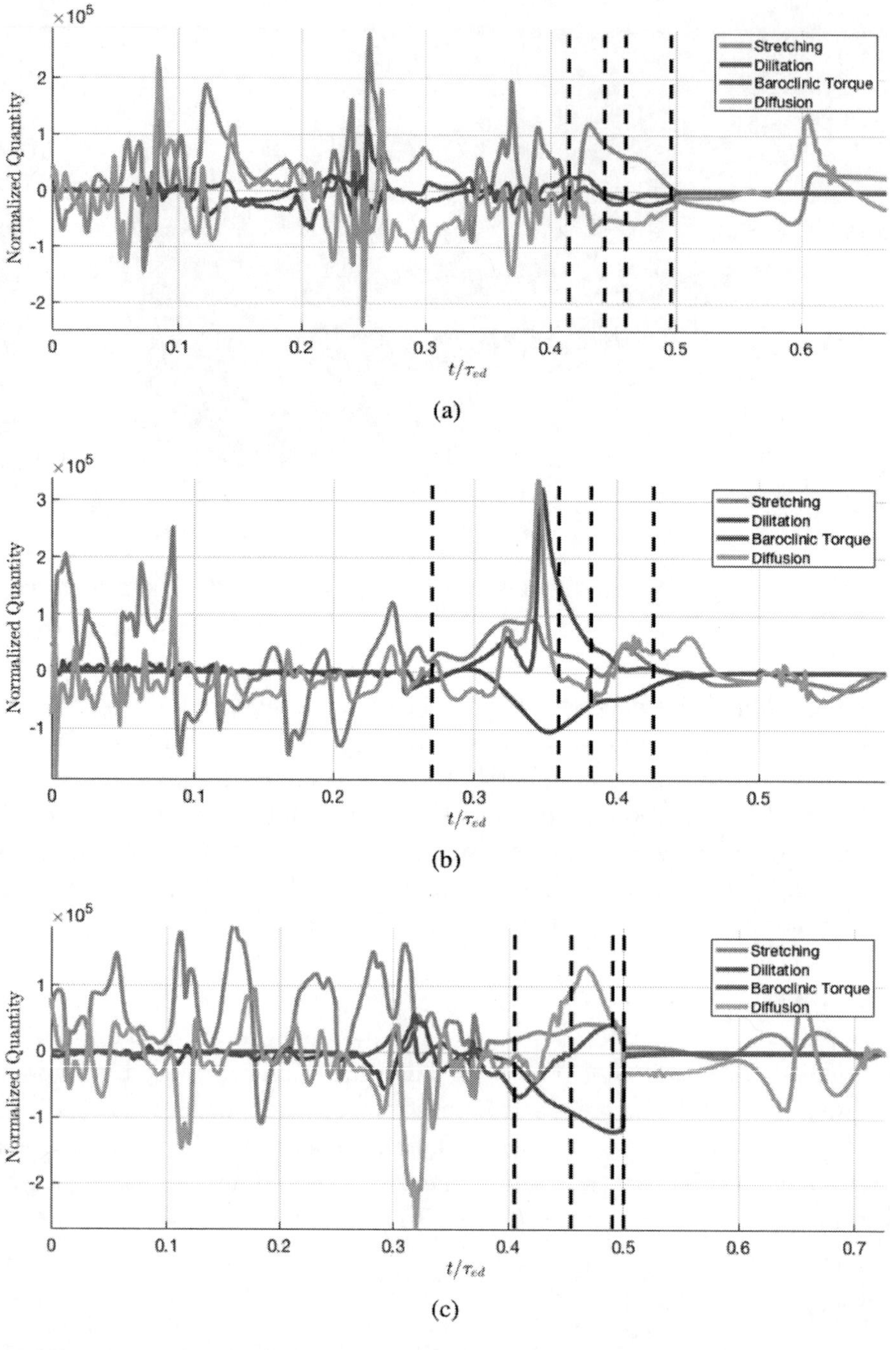

Figure 5: Enstrophy budget terms for three single fluid parcels. The four dashed black lines denote the time the fluid parcel first reaches a temperature of 600 K, 1000 K, 1400 K, and 2000 K. Panel (a) shows a fluid parcel that does not experience a dominating spike in baroclinic torque, panel (b) shows a fluid parcel that experiences a spike near the preheat and reaction zone interface, and panel (c) shows a fluid parcel that experiences a constant increase in baroclinic torque over the preheat and reaction zone interface.

References

[1] S. Zhang and C. J. Rutland, Premixed flame effects on turbulence and pressure-related terms, Combustion and Flame 102 (1995) 447–461.

[2] K. Luo, Combustion effects on turbulence in a partially premixed supersonic diffusion flame, Combustion and Flame 119 (1999) 417–435.

[3] Y. Liu and B. Lenze, Investigation of flame-generated turbulence in premixed flames at low and high burning velocities, Experimental Thermal and Fluid Science 5 (1992) 410–415.

[4] P. E. Hamlington, A. Y. Poludnenko, and E. S. Oran, Interactions between turbulence and flames in premixed reacting flows, Physics of Fluids (1994-present) 23 (2011) 125111.

[5] V. Sankaran and S. Menon, Subgrid combustion modeling of 3-D premixed flames in the thin-reaction-zone regime, Proceedings of the Combustion institute 30 (2005) 575–582.

[6] A. Lipatnikov, S. Nishiki, and T. Hasegawa, A direct numerical simulation study of vorticity transformation in weakly turbulent premixed flames, Physics of Fluids (1994-present) 26 (2014) 105104.

[7] B. Bobbitt, S. Lapointe, and G. Blanquart, Vorticity transformation in high Karlovitz number premixed flames, Physics of Fluids (1994-present) 28 (2016) 015101.

[8] P. Yeung, Lagrangian investigations of turbulence, Annual review of fluid mechanics 34 (2002) 115–142.

[9] F. Toschi and E. Bodenschatz, Lagrangian properties of particles in turbulence, Annual Review of Fluid Mechanics 41 (2009) 375–404.

[10] P. E. Hamlington, R. Darragh, C. A. Briner, C. A. Towery, B. D. Taylor, and A. Y. Poludnenko, Lagrangian analysis of high-speed turbulent premixed reacting flows: Thermochemical trajectories in hydrogen-air flames, In press at Combustion and Flame (2017).

[11] A. Steinberg, B. Coriton, and J. Frank, Influence of combustion on principal strain-rate transport in turbulent premixed flames, Proc. Comb. Inst. 35 (2015) 1287–1294.

[12] J.R. Osborne, S.A. Ramji, C.D. Carter, and A.M. Steinberg, Relationship between local reaction rate and flame structure in turbulent premixed flames from simultaneous 10 kHz TPIV, OH PLIF, and CH2O PLIF, Proc. Combust. Inst. 36 (2017) 1835–1841.

[13] H. A. Uranakara, S. Chaudhuri, H. L. Dave, P. G. Arias, and H. G. Im, A flame particle tracking analysis of turbulence–chemistry interaction in hydrogen–air premixed flames, Combust. Flame 163 (2016).

[14] J. M. Stone, T. A. Gardiner, P. Teuben, J. F. Hawley, and J. B. Simon, Athena: a new code for astrophysical MHD, The Astrophysical Journal Supplement Series 178 (2008) 137.

[15] A. Poludnenko and E. Oran, The Interaction of High-Speed Turbulence with Flames, APS Division of Fluid Dynamics Meeting Abstracts (2009).

[16] A. Y. Poludnenko and E. S. Oran, The interaction of high-speed turbulence with flames: Global properties and internal flame structure, Combustion and Flame 157 (2010) 995–1011.

[17] C. Towery, A. Poludnenko, J. Urzay, J. O'Brien, M. Ihme, and P. Hamlington, Spectral kinetic energy transfer in turbulent premixed reacting flows, Physical Review E 93 (2016) 053115.

[18] P. Colella, Multidimensional upwind methods for hyperbolic conservation laws, Journal of Computational Physics 87 (1990) 171–200.

[19] J. Saltzman, An unsplit 3D upwind method for hyperbolic conservation laws, Journal of Computational Physics 115 (1994) 153–168.

[20] R. Darragh, C. Towery, P. Hamlington, and A. Poludnenko, Lagrangian Analysis of Vorticity Dynamics in Turbulent Premixed Flames, (AIAA Paper 2017-3466, June 2017).

Western States Section of the Combustion Institute – Fall 2017 Meeting
Hosted by the University of Wyoming
October 2-3, 2017

PLIF-assisted discrimination of gas and liquid velocities for 5 kHz PIV in a liquid-fueled pressurized combustor

Hamdullah Ozogul[1] *Hanna Ek*[1] *Ianko Chterev*[1] *Ben Emerson*[1] *Tim Lieuwen*[1]

[1]*Ben T. Zinn Combustion Laboratory, Georgia Institute of Technology, Atlanta, GA 30332*

Particle image velocimetry (PIV) is an increasingly popular diagnostic for combustor flow field measurements. However, in liquid-fueled combustors the participation of liquid droplets in the PIV algorithm may bias the PIV measurement if the liquid droplets are poor flow tracers. This motivates an analysis technique to produce separate liquid and gaseous flow fields. This study has developed a filtering method for particle image velocimetry (PIV) data. The method is demonstrated on a stereoscopic PIV dataset from a high-pressure, reacting, liquid fueled, swirl-stabilized combustor. The method uses fuel planar laser induced fluorescence (PLIF) to spatio-temporally classify vectors as gas-phased or liquid phased. This information is then used to produce flow statistics conditioned on gas or liquid phase velocities. For the particular flow field studied here, we note that the time-averaged flow of the droplets follows the gas velocity field, but this is not the case for instantaneous flow field.

1. Introduction

Liquid-fueled gas turbine combustors are the most widespread combustor type in aircraft engines. One of the major challenges that faces combustion system design is to understand the flow dynamics inside of the combustor. This is particularly challenging due to the many dynamical flow features, three-dimensional features (such as swirl), two-phased flow associated with fuel-air mixing, and the harsh conditions of elevated temperature and pressure [1,2]. Recent advances in laser-based diagnostics and associated analysis techniques have revolutionized the measurement opportunities available to the combustion community [3–6]. In particular, high-speed particle image velocimetry (PIV) is a promising technique to measure the flow field in such an environment without intrusion and has been a breakthrough for the understanding of gas turbine combustor flow dynamics [7–10]. The PIV technique seeds the flow with flow-tracer material, images these flow tracers, and calculates a velocity field from the resulting spatio-temporal flow tracer information. An important consideration for PIV measurements is that their accuracy depends on the performance of the seeding material as a flow tracer.

A significant body of literature has been published for high-pressure, liquid fueled, reacting PIV [11–16]. In the case of multiphase reacting flows, the seeding material is not the only flow tracer, since the imaging system also detects the larger liquid droplets [13,15,14]. In order to qualify as a good flow tracer, seeding particles should follow the flow dynamics in the range of accelerations of interest without disturbing the flow [15]. However, liquid droplets also participate in the PIV velocity field calculation algorithm and may bias the PIV measurement if the liquid droplets are poor flow tracers. In this case, the liquid droplets represent the liquid-phase velocity field, but not the gas-phase velocity field. Therefore, this paper presents a technique which provides the possibility to separate liquid phase velocity field and gaseous phase velocity field by filtering the PIV data. In order to filter the data, fuel planar laser induced fluorescence (PLIF) is used to spatio-temporally locate the liquid fuel droplets [17–19]. The fuel-PLIF data is obtained by synchronizing (in both space and time) a second laser sheet with the PIV laser illumination sheet. This second laser sheet is produced with a high speed tunable dye laser and is tuned to roughly 283 nm, which is readily absorbed by the fuel. The resulting stimulated fluorescence of the fuel is imaged with an intensified high speed camera to locate the fuel droplets. These spatio-temporal fuel droplet locations are used to filter the velocity vector fields, and thus condition the velocity vectors as liquid phase or gas phase.

Hence, this paper investigates the separation of the gas and liquid phase velocities for a liquid-fueled, high pressure, reacting combustor through a method based on a simultaneous PIV and fuel-PLIF measurements. The paper demonstrates this technique on an existing dataset, and sheds light on the fidelity of fuel droplets as flow-tracers in a typical combustor environment.

2. Methods

A high-pressure liquid-fueled swirl stabilized combustor rig is used to demonstrate the method of filtering the PIV data by phase. Details of the facility are provided in Chterev et al. [19], so only a brief summary is provided here. The rig consists of both air and fuel supplies, a swirler, a quartz cylindrical test section and an exhaust. It houses several acoustic sensors, pressure, and temperature probes. The operating conditions are summarized in Table 1. The fuel, referred to as "C1," consists of 99.6% iso-paraffins. Compared to Jet-A, it has a lower boiling range, lower density and a higher viscosity. A great difference is noticed in the Cetane number with 17.1, compared with 48.3 for Jet-A [20,21].

Table 1

P_{comb} [kPa]	T [K]	ϕ	Fuel
345	450	0.45	C1

The optical system for the PIV measurement is composed of a high speed Nd:YLF 527 nm laser, two 12 bit Photron SA5 cameras to acquire the stereo PIV data at 5kHz, and interference filters centered at 528 nm to reject light from combustion and other diagnostics. The cameras were operated with a resolution of 896×848 pixels. The optical system for the fuel PLIF measurement consists of a pulsed Nd:YAG-pumped dye laser and two 12 bit intensified cameras are used to excite the fuel and the OH species and then collect the PLIF data. The two laser sheets are aligned and the fours cameras are synchronized in a thoughtful way. The OH PLIF is not treated in this paper. Figure 1 gives an overview of the optical system setup.

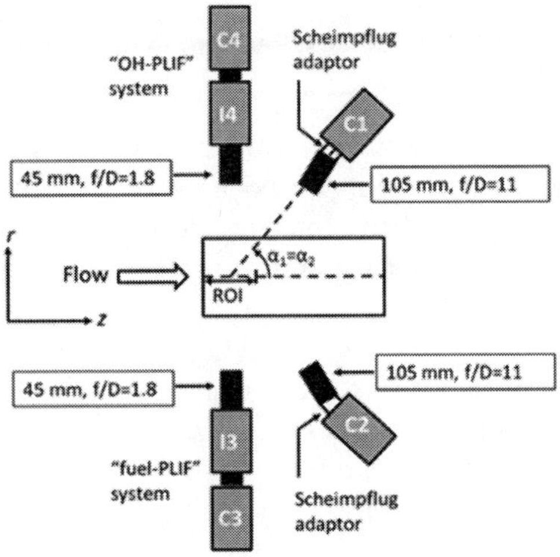

Figure 1: Optical systems setup (from [19])

Titanium dioxide (TiO_2) particles were the chosen seeding material due to their resistance to burning in the combustor. The choice of the seed particle size is based on the comparison of the time scale of the flow τ_f with the time scale of the particle response τ_p [22,23]. The Stokes number defined by equation (1), can be used to quantify the degree to which particles follow the gaseous flow.

$$Stk = \tau_p/\tau_f \quad (1)$$

A common formula for the particle response is the one derived from the Stokes drag coefficient:

$$\tau_p = \rho_p {d_p}^2/(18\mu_g) \quad (2)$$

with ρ_p and d_p respectively the density and the diameter of the seed particles and μ_g the air dynamic viscosity. The formula for τ_f is defined as the ratio of a characteristic length scale D (swirler diameter) over the maximum of the velocity fluctuation $max(\tilde{u}_{RMS})$ as $\tau_f = D/max(\tilde{u}_{RMS})$. We used 0.5-1 µm TiO_2 particles and obtained $Stk \approx 0.01$ in the worst case which means that the seed particles are good tracers.

Gas-Liquid Phase Filter

To filter and separate the liquid fuel droplets from the seed particles, the authors used the following steps:

- perform the PIV calculation on the pre-processed unfiltered PIV frames
- compute fuel PLIF histogram to identify fuel signal threshold
- identify and store spatio-temporal fuel droplet locations
- apply filter based on these localizations to tag velocity vectors as gas or liquid phase
- post-process each phase independently

The rest of this section details each of these steps.

PIV calculation

The PIV frames are processed with LaVision DaVis 8.3. The PIV processing steps are detailed in Chterev et al. [19] and outlined here. Dual frames are imaged with 8 µs delay and a velocity field is calculated every 200 µs. Among the processing steps several image pre-processing methods have been applied to obtain better results, such as a sliding minimum subtraction to erase any static reflections. Self-calibration is performed with DaVis 8.3. Finally, the velocity field calculation is performed with multiple passes of differently-sized interrogation windows. The interrogation windows have a decreasing size from 96×96 pixels to 32×32 pixels to cover a great velocity range. An overlap of 50% is applied to reach a better resolution.

Fuel PLIF pre-processing

The RAW fuel PLIF data underwent several image corrections before processing: (1) a dark field subtraction is applied to take in account the no light offset of the camera, (2) a fixed pattern noise correction for the spatial variation of the intensity response, (3) a laser intensity correction to adjust the spatial variation of the laser intensity across the LIF laser sheet, (4) image synchronization and calibration on DaVis 8.3.

Fuel threshold determination

The emission spectrum of the fuel is broad and covers the emission of the OH, but as the liquid fuel intensity is orders of magnitude higher than the OH intensity, there is no need to reject the OH emission. Firstly, a histogram [24] of the intensity is computed for the full data set by interrogating exclusively the zone close to the fuel entrance to the combustion chamber.

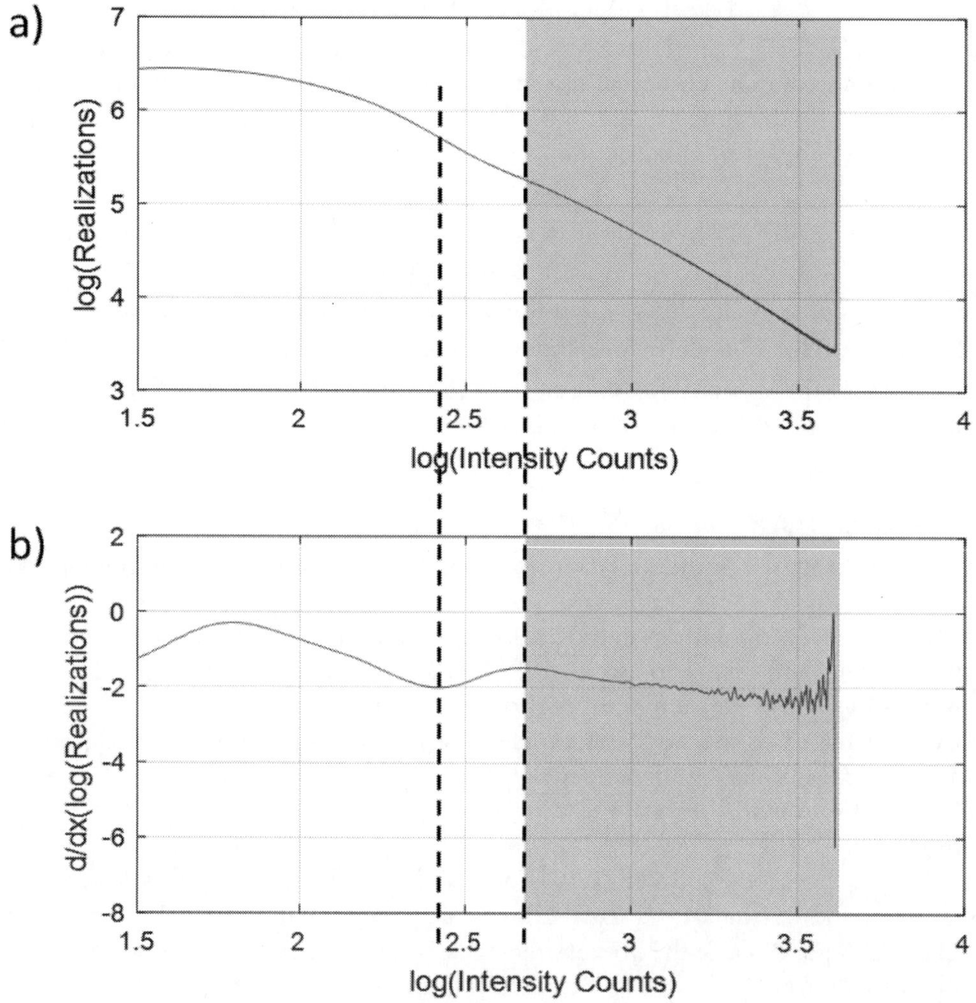

Figure 2: Fuel PLIF intensity histogram and derivative (Fuel signal in green)

Figure 2a presents fuel PLIF intensity histogram, and Figure 2b presents the derivative of the histogram (to help identify points of inflection in the histogram). Two populations are evident in this histogram. The one with highest intensities corresponds to fuel signal from droplets large enough to be properly resolved. The second one corresponds to a weaker signal from noise and very small droplets. The derivative of the histogram has been computed to get accurate thresholds to define these populations, since its local minima and maxima represent the points of inflection that separate the two populations. The fuel-containing regions are defined as regions containing fuel PLIF intensities greater than the higher of these two points of inflection; this is represented graphically as the green zone in Figure 2.

Phase separated PIV

The phase separation algorithm considers each location in space and time in the fuel PLIF data, and defines liquid phase regions as those regions with fuel PLIF intensity above-threshold, and gas-phase regions as those regions with fuel PLIF intensity below-threshold. This information is then used to tag each of the velocity vectors as either gas-phase or liquid phase.

3. Results and Discussion

The filter described above has been applied to the data set of Table 1 on 1000 instantaneous velocity maps, which corresponds to a total time of 0.2 ms. Figure 3 presents the resulting time-averaged velocity fields for the gas phase (Figure 3a) and the liquid phase (Figure 3b). Note that each vector field represents only a portion of the total measurement domain, since there are regions which are dominated by liquid fuel and rarely see gas phase, and likewise there are regions that the liquid phase never or rarely occupies. A minimum fraction of realizations of 5% has been applied to avoid poorly converged averages. Hence, a vector appears on the Figure 3 only if more than 50 vectors exist for that phase at that location.

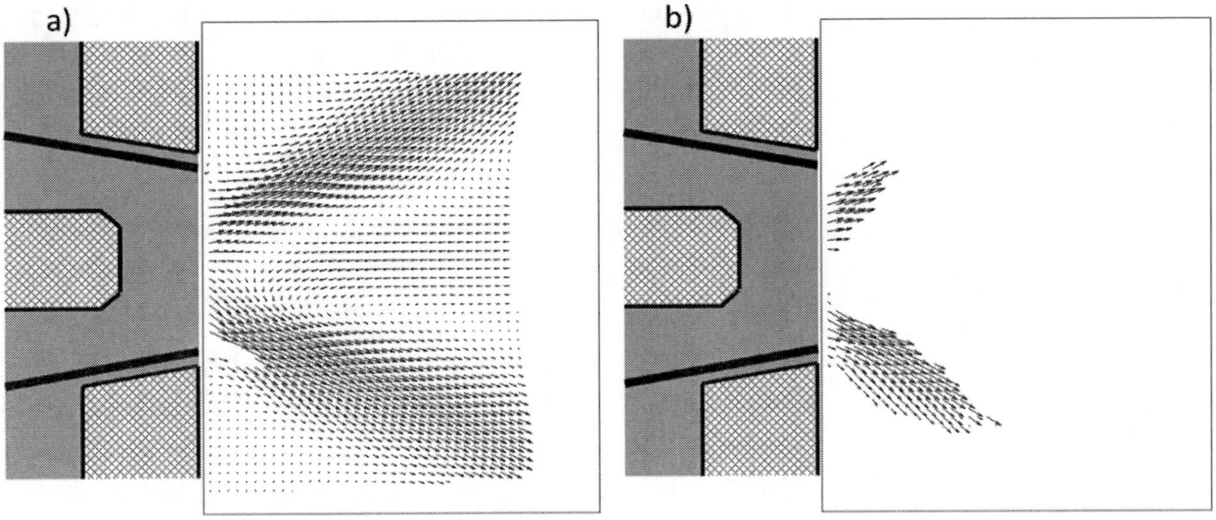

Figure 3: (a) Gas phase and (b) liquid phase average velocity maps

Figure 3 shows that liquid fuel is present close to the dump plane, with more fuel present near the bottom half of the studied plane. We hypothesize that this is due to slight asymmetry in the setup, which could introduce more fuel to the lower filming surface. It should also be noticed that a region without vectors is present in the bottom half of the gas phase map. This means liquid fuel is absent from this region less than 5% of the time.

In order to compare the gaseous and liquid velocity fields from Figure 3 more quantitatively, profiles have been plotted in Figure 4 in the radial direction at a distance of 0.1×D and D/2 from the dump plane. The average velocity profiles of both phases are close to the non-filtered data. Hence, from a time-averaged point of view, the gas phase measurement is not considerably affected by the presence of the fuel droplets. This suggests that the fuel droplets follow the flow well enough to perform as good flow tracers for time averaged flow.

A Stokes number is defined in order to provide some insight into the time averaged flow-tracing fidelity of the liquid droplets. Since the droplets sizes are not known, the Stokes number can be evaluated to determine a "cutoff" droplet size, below which the droplets should be good flow tracers. The droplets should be sized to satisfy the inequality $Stk < 0.1$ in order to be good flow tracers. Hence, the equation for the cutoff droplet size is:

$$\frac{\rho_{fuel} d_{fuel}^{2}}{(18\mu_g) \times \tau_{f-fuel}} < 0.1 \Leftrightarrow d_{fuel} < \left(1.8 \times \mu_g \times \tau_{f-fuel}/\rho_{fuel}\right)^{0.5} \quad (3)$$

where d_{fuel} is the maximum fuel droplet diameter for a good flow tracer and τ_{f-fuel} a characteristic flow time of the droplets.

To be able to compute the equation (3), τ_{f-fuel} has to be defined. As Figure 4 shows profiles close to each other, this characteristic time could be defined from the mean velocity magnitude $\bar{V}_{fuel}$ and an average acceleration magnitude $\bar{a}$ as $\tau_{f-fuel} = \bar{V}_{fuel}/\bar{a}$. However, each instantaneous velocity map is spaced by 200 μs which prevents the calculation of time averaged acceleration fields in an Eulerian framework. Thus, the time averaged acceleration field is calculated in a Lagrangian framework as $\vec{a} = \vec{V} \cdot \nabla\vec{V}$. The stereo PIV measurement provides a planar measurement of three-component velocity vectors, such that in-plane velocity gradients are available but out of plane velocity gradients are not. Hence, the in-plane acceleration is used to define τ_{f-fuel}. The acceleration is computed for the non-filtered data and τ_{f-fuel} has been computed for each position. The minimum of those values could be chosen as the worst case to define τ_{f-fuel}. Finally the condition on the diameter is $d_{fuel} < 60\ \mu m$.

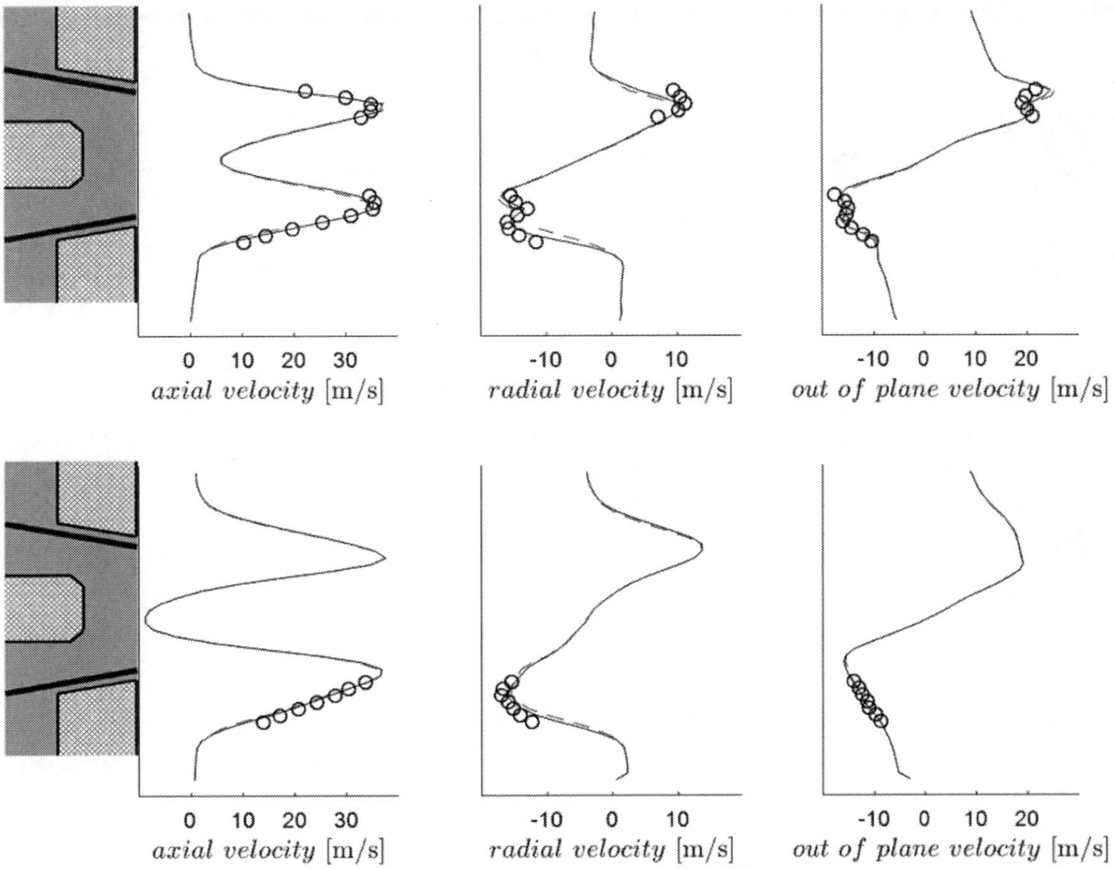

Figure 4: Average velocity profiles of each phase: non-filtered data in full line; gas phase in dashed line; liquid phase with circle symbols. The top plots and the bottom plots are at 0.15×D and D/2 from the dump plane respectively.

The above results have explored the time averaged flow tracing fidelity of the liquid droplets. Another important consideration for the liquid droplets is their dynamical flow tracing fidelity. Therefore, the RMS velocity fluctuations are presented on Figure 5 at the same locations as Figure 4. Here the gas phase profiles are close to the non-filtered data profiles. However, the liquid phase profiles have lower values close to the dump plane and these differences become smaller at D/2 from the dump plane. Hence, from a dynamic point of view, the larger droplets do not follow well the flow at the dump plane. This highlights a shortcoming of traditional PIV measurements of liquid-fueled combustors, since liquid droplets may follow the flow well enough to provide a convincing time-averaged picture but may significantly deviate from and potentially bias the velocities and/or accelerations of the dynamical flow features. Therefore, this study demonstrates the importance of an experimental methodology to isolate and conditionally analyze the gas and liquid phases when studying unsteady flow fields from PIV measurements in liquid fueled combustors.

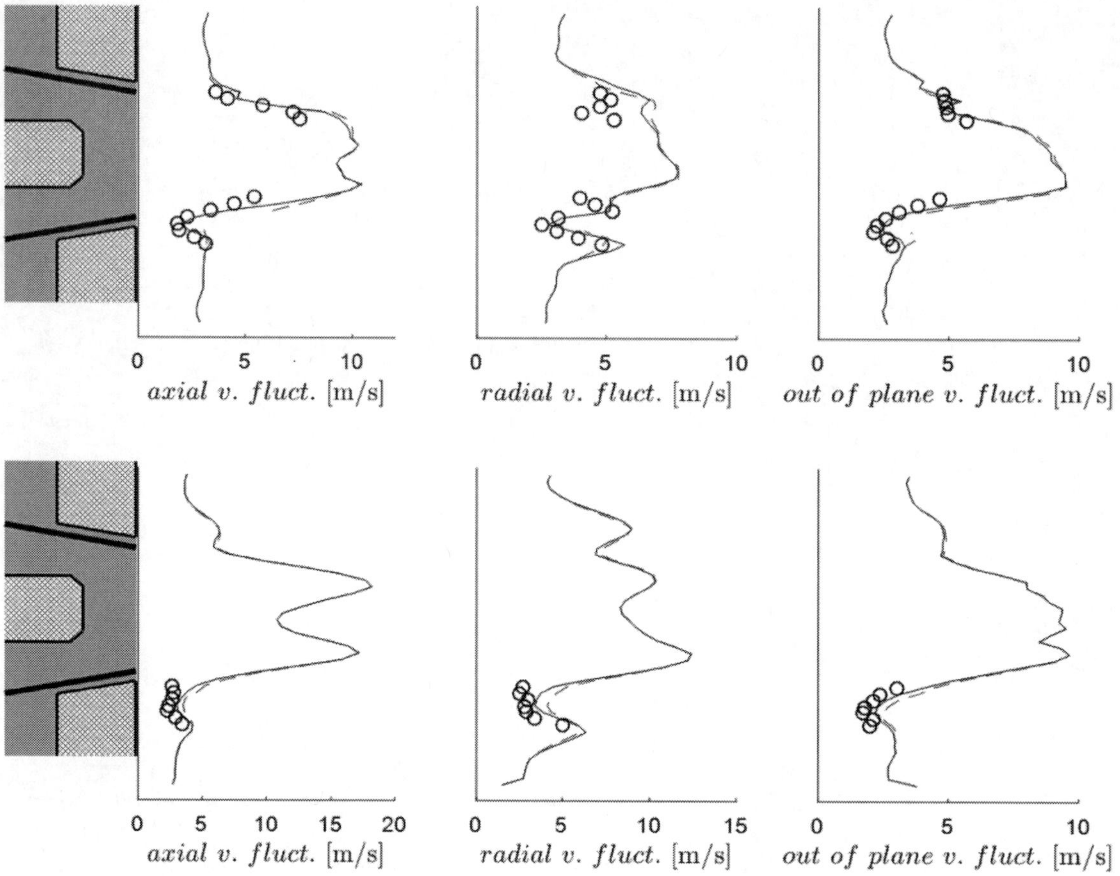

Figure 5: RMS velocity fluctuation profiles of each phase: non-filtered data in full line; gas phase in dashed line; liquid phase with circle symbols. The top plots and the bottom plots are at 0.15×D and D/2 from the dump plane respectively.

4. Conclusions

This paper presents a methodology to separate and conditionally analyze PIV-based vector fields by gas or liquid phase. This separation is enabled by a coincident fuel PLIF measurement. The method is demonstrated on a reacting, high-pressure, swirl-stabilized, liquid-fueled combustor dataset. The results show that practically sized liquid droplets may be poor flow tracers that could bias PIV measurements, but that this bias may be predominantly present in the dynamical flow field rather than the time-averaged flow field.

Acknowledgements

This research was partially supported by the Air Force Office of Scientific Research (contract #FA9550-16-1-0442), contract monitor Dr. Chiping Li, and by the US Federal Aviation Administration (FAA) Office of Environment and Energy as a part of ASCENT Project 13-C-AJFE-

GIT-008 under FAA Award Number: 27a. Any opinions, findings, and conclusions or recommendations expressed in this material are those of the authors and do not necessarily reflect the views of the FAA or other ASCENT Sponsors.

References

[1] T.C. Lieuwen, *Unsteady Combustor Physics*, Cambridge University Press, (2012).

[2] A.H. Lefebvre, D.R. Ballal, *Gas Turbine Combustion: Alternative Fuels and Emissions*, (2010).

[3] B.R. Halls, N. Jiang, T.R. Meyer, S. Roy, M.N. Slipchenko, J.R. Gord, *Opt. Lett.*, 42 (2017) 2830–2833.

[4] T.R. Meyer, B.R. Halls, N. Jiang, M.N. Slipchenko, S. Roy, J.R. Gord, *Opt. Express.*, 24 (2016) 29547–29555.

[5] J.D. Miller, T.R. Meyer, M.N. Slipchenko, J.G. Mance, S. Roy, *AIAA Pap.*, (2014).

[6] S. Roy, J.-C. Hua, W. Barnhill, G.H. Gunaratne, J.R. Gord, *Phys. Rev.*, 91 (2015) 013001.

[7] A. Upatnieks, J.F. Driscoll, S.L. Ceccio, *Proc. Combust. Inst.*, 29 (2002) 1897–1903.

[8] A.M. Steinberg, J.F. Driscoll, S.L. Ceccio, *Exp. Fluids.*, 44 (2008) 985–999.

[9] J. Westerweel, G.E. Elsinga, R.J. Adrian, *Annu. Rev. Fluid Mech.*, 45 (2013) 409–436.

[10] V. Sick, *Proc. Combust. Inst.*, 34 (2013) 3509–3530.

[11] C.T. Chong, S. Hochgreb, *Appl. Mech. Mater.*, 316–317 (2013) 17–22.

[12] C.T. Chong, S. Hochgreb, *Fuel*, 115 (2014) 551–558.

[13] M. Schroll, J. Klinner, L. Lange, C. Willert, *10th Int. Symp. Part. Image Velocim. PIV*, (2013).

[14] U. Meier, J. Heinze, E. Magens, M. Schroll, C. Hassa, S. Bake, T. Doerr, *Proceedings of ASME Turbo Expo 2015*, (2015).

[15] C.D. Slabaugh, A.C. Pratt, R.P. Lucht, *Appl. Phys. B*, 118 (2015) 109–130.

[16] C.D. Slabaugh, I. Boxx, S. Werner, R.P. Lucht, W. Meier, *AIAA J.*, 54 (2016) 946–961.

[17] B.D. Ritchie, J.M. Seitzman, AIAA Pap. (2001).

[18] S. Shani, T. Tran, F. Genin, J. Matlach, S. Menon, J.M. Seitzman, *12th AIAA Intl Space Planes Hypersonic Syst. Technol.*, (2003).

[19] I. Chterev, N. Rock, H. Ek, B.L. Emerson, J.M. Seitzman, T. Lee, J. Gord, T.C. Lieuwen, *55th AIAA Aerospace Sciences Meeting*, (2017).

[20] M. Colket, J. Heyne, M. Rumizen, M. Gupta, T. Edwards, W.M. Roquemore, G. Andac, R. Boehm, J. Lovett, R. Williams, J. Condevaux, D. Turner, N. Rizk, J. Tishkoff, C. Li, J. Moder, D. Friend, V. Sankaran, *AIAA J.*, 55 (2017) 1087–1104.

[21] J.T. Edwards, *55th AIAA Aerospace Sciences Meeting*, (2017).

[22] M. Raffel, C.E. Willert, Steven Wereley, J. Kompenhans, *Particle Image Velocimetry: A Practical Guide*, Springer Science & Business Media, (2007).

[23] F. Picano, F. Battista, G. Troiani, C.M. Casciola, *Exp. Fluids.*, 50 (2011) 75–88.

[24] H. Ek, N. Rock, H. Ozogul, B.L. Emerson, T.C. Lieuwen, *ILASS-Am. 29th Annu. Conf.*, (2017).

Western States Section of the Combustion Institute – Fall 2017 Meeting
Hosted by the University of Wyoming
October 2-3, 2017

Laboratory Investigation of Down-scaling Effects on Low Swirl Burners

Alex Frank[1,2*], *Peter Therkelsen*[2], *Jyh-Yuan Chen*[1], *Vi H. Rapp*[2], *Robert K. Cheng*[2]

[1]*Mechanical Engineering Department, University of California-Berkeley, Berkeley, CA 94720, United States*
[2]*Energy Storage and Distributed Resources Department, Lawrence Berkeley National Laboratory, 1 Cyclotron Road, Berkeley, CA 94720, United States*
[*]*Corresponding Author Email: alex.frank@berkeley.edu*

Abstract: This paper reports a preliminary experimental investigation by the use of particle image velocimetry (PIV) on the effects of burner size and bulk flow velocity on the flowfield characteristics of a low swirl burner (LSB). The impetus for our study is a recent development in which the LSB was found to be scalable to small sizes by the successful demonstration of an Advanced Research Projects Agency-Energy (ARPA-E) Generators for Small Electrical and Thermal Systems (GENSETS) 3 kW natural gas combustor for use in a recuperated Brayton Cycle. The study presented in this paper utilizes three LSBs that have diameters ranging from 12 mm to 25.4 mm representing an output range of less than 1 kW to 23 kW. Inlet bulk flow velocities are varied from 3 m/s to 13 m/s representing a wide operational range and flow regime. Preliminary results have shown that the small-scale burners tested exhibit similar flowfield characteristics to their larger-scale counterparts in the non-reacting cases studied. Further investigation is required to understand the reacting flowfield information and is the subject of future work.

Keywords: Gas Turbines, Combustor, Small-Scale, Particle Image Velocimetry

1. Introduction

The technology for small-scale power generation has long been the near exclusive domain of reciprocating internal combustion engines due to their durability, affordability, low emissions, and ease of maintenance. To broaden the technology base for residential combined heat and power (CHP) systems, the U.S. Department of Energy's Advanced Research Projects Agency - Energy (ARPA-E) has launched the Generators for Small Electrical and Thermal Systems (GENSETS) program which aims to provide 1 kW of electricity from natural gas while also supplying heat for use within the household such as hot water and heat [1]. Towards this goal, Lawrence Berkeley National Lab (LBNL) partnered with Metis Design Corporation to develop a microturbine with the low swirl burner (LSB) as the combustion system.

Recently, we reported on the adoption of the LSB to a miniaturized scale of 14 mm in diameter for the microturbine [2]. This represents a 3.3:1 reduction from our previous smallest LSB developed for domestic air-heater application. Both reduced-scale LSBs, i.e. the 25.4 mm LSB for air heaters and the 14 mm LSB for the microturbine have been optimized to operate within a small range of conditions at fixed heat outputs. This is because air heater and microturbine systems operate at single optimized output levels and do not need turndown capability. From previous experimental investigations, it has been demonstrated that for LSBs in diameter > 31.8 mm, turndown ratios of 10 or greater can be achieved [3].

The objective of this study is to investigate the operating regime of the reduced-scale LSBs. During the course of our experimentation, we also found that the reduced scale LSBs are capable of operating at bulk inlet velocities (U_0), that are lower than those of their larger scale counterparts. At these lower bulk inlet velocities, the flame stabilization mechanism established for LSB operation, based on a self-similar divergent flowfield coupled with a linear correlation of the turbulent displacement flame speed with turbulence intensity, may not be applicable. This is theorized to be because the foundation of the LSB operating concept is incumbent on a fully developed turbulent flow regime with a U_0 typically greater than 10 m/s [4]. The capability of the reduced-scale LSBs to operate at lower U_0 indicates that the LSBs flame stabilization mechanism extends to the transitional and under-developed turbulent flow regimes. To obtain a bettering understanding of the evolution of the LSB flowfields within unstable transitional regimes, we conducted a study using three reduced-size LSBs.

2. Apparatus, Diagnostics, and Experimental Setup

The effects of down-scaling on the LSB flowfield, were studied through the use of three geometrically similar swirlers. The smallest LSB is 12 mm in diameter, the GENSETS LSB is 14 mm in diameter, and the largest LSB, equivalent to the one for air-heater. is 25.4 mm in diameter (Figure 1). The swirlers and center plates are all linear scales of each other and feature 16 vanes at a 37 degree angle. Center plates all have 34 strategically placed holes and have an open area of 28.2%. By keeping the geometry constant between all three LSBs, it is possible to study the flowfield evolution of each burner in isolation.

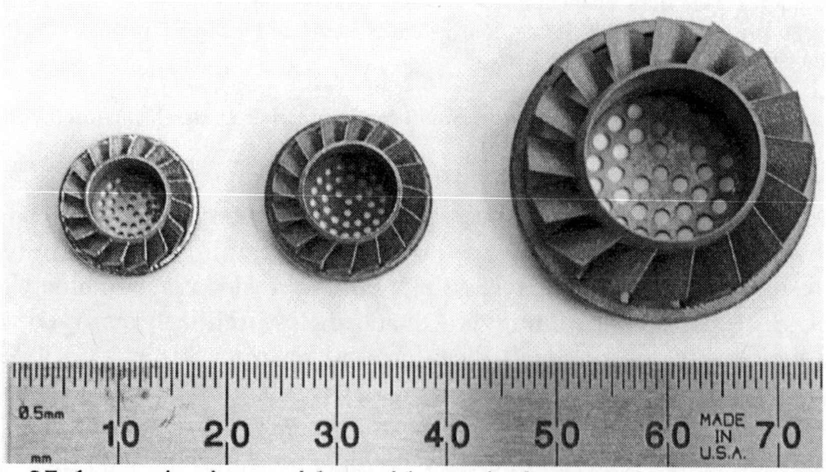

Figure 1. 16 vane, 37-degree titanium swirlers with attached center plates. From left to right, 12 mm, 14 mm, and 25.4 mm diameter swirlers.

To interrogate the flowfield, particle image velocimetry (PIV) was utilized with the LSB flame burning in the open without an enclosure. The LSB was mounted in a stainless-steel holder which is in turn mounted into an aerodynamically contoured plenum in that supplies a near top-hat flowfield relatively free of disturbances. Fuel (CH_4) and air are premixed far upstream of the swirler and both are controlled by Brooks mass flow controllers (MFCs). In order to ensure accuracy at a wide range of fuel flow measurements, multiple Brooks model GF40 with varying flow ranges were used. In low (< 8 kW), medium (> 8 kW, <15 kW) and high > 15 kW fuel ranges, 15 SLPM, 25 SLPM, and 44 SLPM methane MFC's were used. For airflow measurement and control, a Brooks model 5853E 500 SLPM MFC was used for all conditions. Details of the experimental setup can be seen in Figure 2.

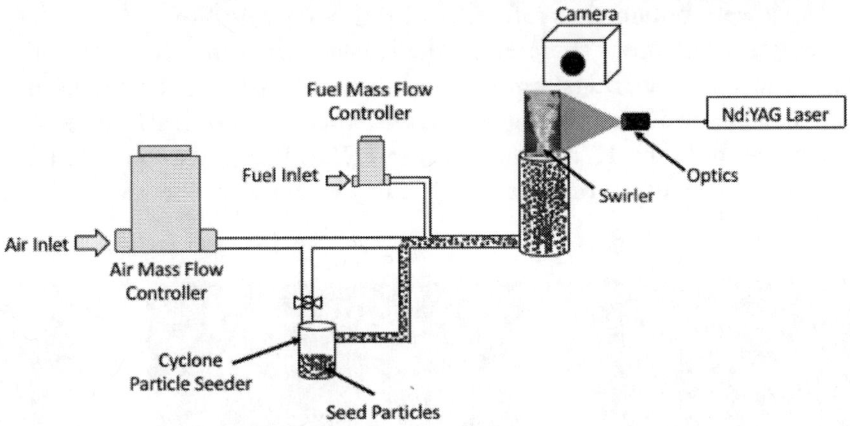

Figure 2. Schematics of LSB PIV setup at LBNL.

In this apparatus, a portion of the airflow is diverted into a cyclone particle seeder which entrains Aerosil 200 particles that have an approximate diameter of approximately 12 nm [5]. The PIV system is comprised of a New Wavo Solo PIV laser with double 120 mJ pulses at 532 nm and a Kodak/Redlake ES 4.0 digital camera. The camera has a 2048 x 2048 pixel resolution which captures a field of view of approximately 5 cm x 5 cm with a pixel resolution of 0.025 mm/pixel.

Data acquisition and analysis was performed by software provided by Wernet [6]. In order to minimize out-of-plane motion of the swirling particles, a short interframe time of 25 μs was utilized for all conditions measured. Furthermore, the PIV data was processed using 64 x 64 pixel regions with 25% overlap between regions with 32 x 32 region sub processing. 224 image pairs were taken in order to obtain an ensemble average for each of the points analyzed.

3. Results and Discussion

Table 1
Laboratory Test Conditions at STP with CH_4

Run	ϕ	Reactant Flow (g/s)	U_0 (m/s)	Heat Release (kW)
LSB12-0	0.0	0.407-1.764	3-13	0
LSB12-8	0.8	0.393-1.704	3-13	0.96-4.19
LSB12-9	0.9	0.391-1.696	3-13	1.07-4.67
LSB12-1	1.0	0.389-1.690	3-13	1.18-5.14
LSB14-0	0.0	0.554-2.400	3-13	0
LSB14-7	0.7	0.537-2.327	4-13	1.16-5.04
LSB14-8	0.8	0.535-2.317	3-13	1.31-5.70
LSB14-9	0.9	0.532-2.308	3-13	1.46-6.35
LSB14-1	1.0	0.530-2.298	3-13	1.61-6.99
LSB254-0	0.0	1.825-7.905	3-13	0
LSB254-7	0.7	1.762-7.664	3-13	3.83-16.60
LSB254-8	0.8	1.764-7.633	3-13	4.33-18.79
LSB254-9	0.9	1.754-7.601	3-13	4.83-20.94
LSB254-1	1.0	1.747-7.570	3-13	5.31-23.05

To map the operational regime of the LSB, a wide range of U_0 and equivalence ratios (ϕ) were studied (Table 1). The reactants flow rate and U_0 are based on combined fuel and air flow rates. Heat release is calculated using the lower heating value of methane. For all swirlers, non-reacting flowfields at $3 < U_o < 13$ m/s were obtained as references for the reacting flows.

Figure 3 depicts the flames generated by the LSBs at the same velocity and equivalence ratio. For all three LSBs, the flame shape is similar, with a highly stable, flat bowl-shaped flame. The scale difference between the three flames is apparent with the 25.4 mm LSB producing a flame significantly wider and taller than both the 12 mm and 14 mm LSBs. It can also be noted that the leading edge of the flame brush for the three burners is appears to exist in the same location, this could be a consequence of linearly scaling.

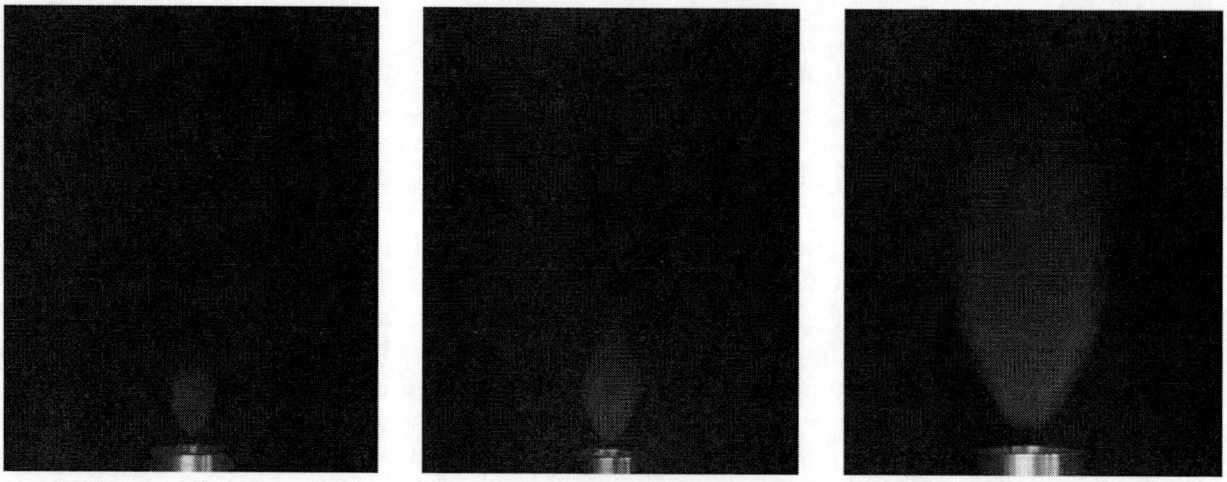

Figure 3. Images of 12 mm, 14 mm, 25.4 mm (from left to right) LSB operating with a U_o of 10 m/s at an equivalence ratio of 0.8 with CH_4.

Non-Reacting PIV Results

PIV data were collected for all conditions listed in Table 1. Compared in Figure 4 are the velocity vectors and 2D turbulent kinetic energy (q' shown in colored contours) obtained from the three LSBs at the same inlet velocity. All three LSBs show a central zone of low q' caused by the low velocity downstream of the center plate. The length in the axial direction and width in the radial direction is proportional to the burner diameter. Conversely, the highest q' is shown in the swirling regime, the region with the highest velocity. At high (> 10 m/s) U_0, there is no recirculation zone in any of the non-reacting cases which is consistent with all previous research on the LSB.

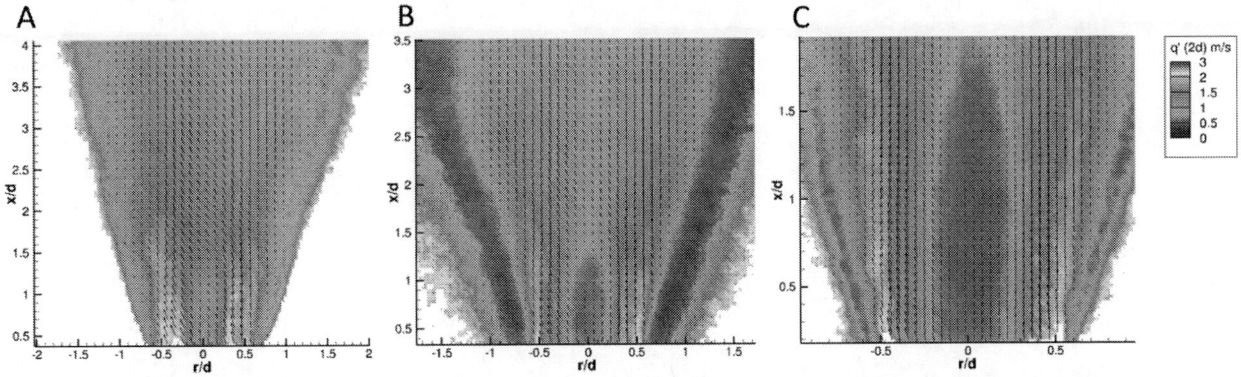

Figure 4. Mean velocity vectors superimposed on contours of 2D turbulent kinetic energy for 12 mm (A), 14 mm (B), and 25.4 mm (C) diameter LSBs. For all cases, $U_o = 10$ m/s, $\phi = 0$.

In order to gain an understanding of the burner flowfield along the centerline, an analysis was performed to determine the normalized rate of divergence ($dU/dX/U_0$) and virtual origin position (X_0) for all U_0 on all three LSBs (Figure 5). The axial divergence rate was calculated for each velocity by analyzing the slope of the normalized axial velocity vs. axial distance. The slope was calculated based on a fitted line for the portion of the curve that had a constant slope which typically occurs upstream of the leading edge of the flame brush (X_f). The virtual origin was calculated using the normalized axial divergence and extrapolating the linear curve to the point where $U/U_0 = 1$.

The data presented in Figure 5 (left) shows that the 25.4 mm diameter LSB has a relatively constant $dU/dX/U_0$ across the velocity range with a small scatter. The data for the 12 mm and 14 mm LSBs follow a similar trend but show significantly more scatter. Furthermore, the order of magnitude in the normalized axial divergence is significantly smaller for the larger scale burner. In other words, the rate of velocity deceleration along the burner centerline is much slower on the 25.4 mm diameter burner compared to the 12 mm and 14 mm LSBs. This indicates that scale does play a major role in the development of the burner flowfield.

The virtual origin position plot (Figure 5 right) shows that for the 25.4 mm LSB, the virtual origin position increases axially towards the burner exit with increasing velocity. The 12 mm and 14 mm LSBs both have a more constant virtual origin position across the velocity range and also have a large scatter. As expected, the magnitude of X_0 scales with the burner diameter.

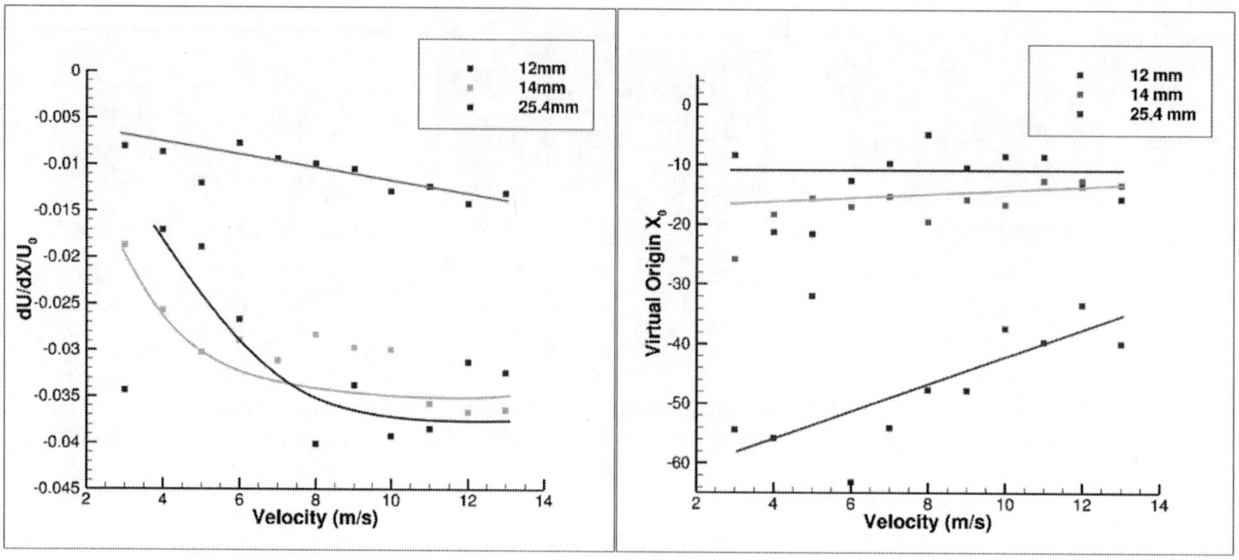

Figure 5. Processed data for the normalized axial divergence rate, $dU/dX/U_0$, vs velocity (left), and virtual origin (X_0) position vs. velocity for $U_0 = 10$ m/s, $\phi = 0.8$.

Reacting Flows

In an effort to understand the reacting flowfield of the small-scale LSB, in addition to PIV data, high speed video of the 14 mm LSB was taken using a Photron Mini AX camera. Two still frames from the video are shown in Figure 6 that depict the flame having both a varied flame shape as well as highly varied X_f. The large movement of the flame suggests that at this scale, X_f has a large amount of axial movement relative to the burner diameter.

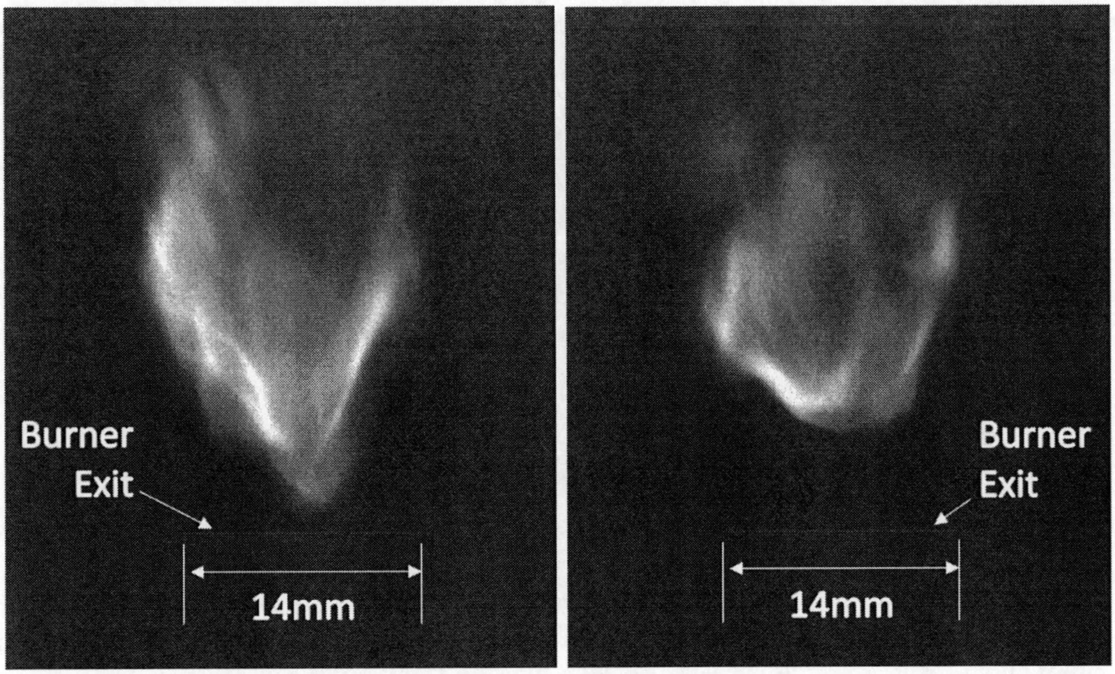

Figure 6. Still frames from a high-speed camera video of a 14 mm diameter LSB. $U_o = 10$ m/s, $\phi = 0.80$.

Additionally, there is evidence heat release rate may play a role in the LSB stability of the 12 and 14 mm LSBs. Results from lean blow off (LBO) experiments indicated that while the 25.4 mm LSB had a nearly constant LBO across the velocity range. Below $\phi = 0.60$, the 14 mm 12 mm LSB both had LBO values associated with richer fuel mixtures as velocity decreased. Furthermore, all values of LBO tested across the velocity range for the 12 mm LSB were richer than the 14 mm LSB meaning that as the burner scales down, a higher heat release rate is required to sustain a flame. This indicates that the heat release rate and scale may both play an important role in the burner flowfield evolution and stability.

PIV data for the reacting flow case are shown in Figure 7. The plots appear similar to the non-reacting case but have noticeable differences. First, the radial width (r/d) of all the flowfields are noticeably wider due to the expansion of the burned gas. Additionally, a recirculation zone exists on all cases and shifts far downstream in the 25.4 mm LSB. The central region of low q' also enlarges with burner size.

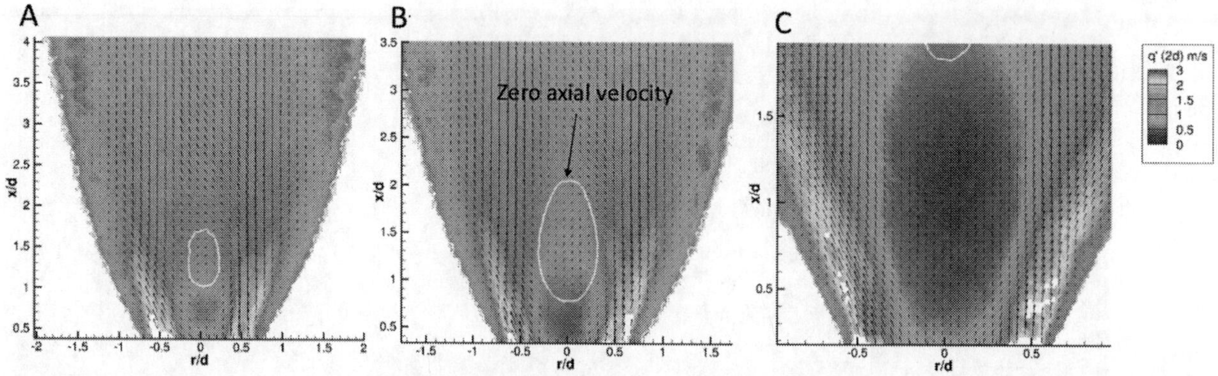

Figure 7. Mean velocity vectors superimposed on contours of 2D turbulent kinetic energy for 12 mm (A), 14 mm (B), and 25.4 mm (C) diameter LSBs. White lines represent zero velocity in the x – direction. For all cases, $U_o = 10$ m/s, $\phi = 0.80$.

A comparison of the centerline velocity profiles is presented in Figure 8 for both the non-reacting and reacting flowfields. First analyzing the non-reacting flowfields, the linear deceleration from the LSB exit is clearly visible in the nearfield for all three LSBs. The location of the inflection point in axial velocity shifts downstream with the 25.4 mm LSB but appears to be in the same vicinity on the 12 mm and 14 mm LSBs. It can also be seen that the flowfield never reverses for the non-reacting cases. These data are consistent with previous studies that have been performed on larger scale LSBs.

An analysis of the reacting flowfield is less conclusive. The 12 mm and 14 mm LSBs both show the linearly decreasing axial velocity in the nearfield but the 25.4 mm LSB has an accelerating flow in this region which is caused by the flame. From data obtained for larger LSBs, it is typical to see a linearly decelerating region before the acceleration. However, for these reduced-scale LSBs, it is inferred that the flame position is very close to the burner exit and thus the PIV sampled the velocities exclusively in the products of the flames. The low flame lift-off distance seems to be a consequence of down-scaling of the LSB geometry. Since the flame position is relatively constant and very close (if not inside) the burner exit tube, to capture and analyze the flow velocities in the reactants regions of the turbulent flame brush would require a conditional sampling approach (i.e. the use of non-refractory seed materials such as silicone oil).

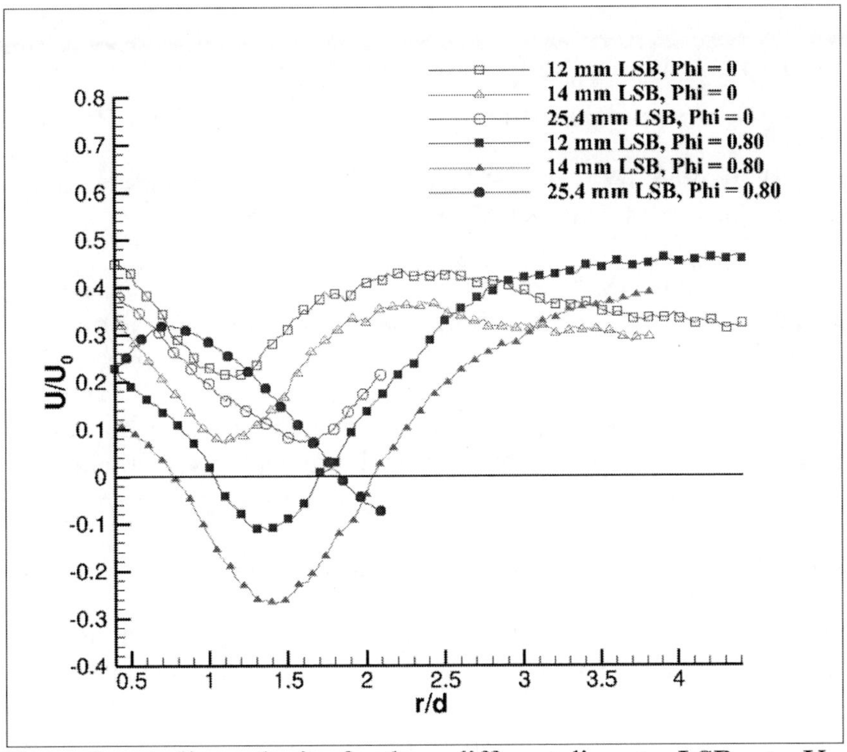

Figure 8. Plots of centerline velocity for three different diameter LSBs at a U_0 of 10 m/s.

4. Conclusions

Experiments have been performed on three small-scale LSBs with diameters of 25.4 mm, 14 mm, and 12 mm. These burners are smaller than any previous by the authors. The burners were investigated at a range of bulk inlet velocities from 3 m/s to 13 m/s with methane flames at equivalence ratios ranging from 0.7 to 1.0. These ranges are consistent with the larger scale burners tested in past investigations. Examination of the non-reacting flowfields using PIV indicates that the burner at this small scale has the same divergence characteristics of their larger scale counterparts. The results obtained from the reacting flow however are less illuminating because the flame position is not yet known and is very close to the burner exit. The next focus of this work will be to measure the flame position in order to provide more insights to this data. This will be accomplished by seeding the flow with non-refractory oil which vaporizes in the flame front and with high-speed imaging. The oil seed will also provide the scattering sites for conditional PIV sampling of the velocity vectors in the reactants region of the fluctuating flames.

Acknowledgements

This work is supported by the U.S. Department of Energy, Advanced Research Projects Agency – Energy under contract No. DE-AE0000607. The authors of this paper would like to thank Mr. Gary Hubbard for creating the control software and data acquisition systems for this project. Additionally, the authors would like to thank Ashley Frank for her photographs of the burner and hardware.

References

[1] *GENSETS Program Overview*. (2014). Retrieved from https://arpae.energy.gov/sites/default/files/documents/files/GENSETS_ProgramOverview.pdf

[2] A. Frank, P. Therkelsen, J.Y. Chen, R.K. Cheng, kW scale combustor for power generation, 10th U. S. National Combustion Meeting (2017), paper 2SC-0248.

[3] R.K. Cheng, D.T. Yegian, M.M. Miyasato, G.S. Samuelsen, C.E. Benson, R. Pellizzari, Scaling and Development of Low-Swirl Burners for Low-Emission Furnaces and Boilers, Proc. Combust. Inst. 28 (2000) 1305-1313.

[4] M. Day, S. Tachibana, J. Bell, M. Lijewski, V. Beckner, R.K. Cheng, A combined computational and experimental characterization of lean premixed turbulent low swirl labatory flames, Combust. Flame 159 (2012) 275-290.

[5] R. Eyjolfsson, Design and Manufacture of Pharmaceutical Tablets, Elsevier, U.K., (2015), pp 4.

[6] M.P. Wernet, Fuzzy logic enhanced digital PIV processing software, International Congress on Instrumentation for Aerospace Simulation Facilities, Toulouse, France (1999).

Western States Section of the Combustion Institute – Fall 2017 Meeting
Hosted by the University of Wyoming
October 2-3, 2017

Evaporation and Autoignition Studies of Liquid n-Alkane Droplets in Lean, High Pressure Methane/Air Mixtures in a Rapid Compression Machine

Jeffrey Mohr, Colin Gould, Andrew Zdanowicz and Anthony Marchese
Department of Mechanical Engineering
Colorado State University
Fort Collins, CO 80525

The combustion of two fuels with disparate reactivity (e.g. natural gas/diesel) in conventional diesel engines has been demonstrated as a means to increase fuel efficiency, reduce fuel costs and reduce pollutant formation compared to standard diesel operation. To maximize the economic benefits, it is desirable to maximize the natural gas substitution rate. However, the natural gas substation rate is constrained by onset of uncontrolled fast combustion (i.e. engine knock) and/or incomplete combustion, which results in high unburned hydrocarbon emissions. Previous detailed computational engine modeling with reduced chemical kinetics and simplified spray models have captured these effects but little data are available to validate chemistry and spray models at these engine-like conditions. In this study, a rapid compression machine (RCM) was used as a platform to provide a high-temperature/high-pressure environment to better understand the thermodynamic, transport and chemical kinetic phenomena of dual-fuel combustion. The RCM was modified to perform evaporation and combustion experiments on single n-alkane fuel droplets in gaseous inert, O_2/inert and O_2/CH_4/inert environments. Droplet evaporation experiments were performed on C_5 to C_{12} n-alkane droplets in inert gas to measure droplet evaporation rates at near supercritical and supercritical conditions (18 bar $< P <$ 35 bar; 450 K $< T <$ 850 K). Droplet ignition experiments were conducted to measure ignition delay as a function of temperature, pressure and ambient CH_4 concentration

1. Introduction

Natural gas (NG)/diesel dual-fuel combustion has shown potential to be a cleaner and more cost effective strategy than burning diesel fuel or NG alone in internal combustion (IC) engines [1][2]. Dual-fuel engines have been shown to have better load response than those fueled only by NG and have the potential to meet EPA emission standards without the need for costly diesel exhaust aftertreatment systems [3]. The common practice in dual-fuel internal combustion engines is to deliver NG to the intake manifold through fumigation, thereby forming a lean homogenous air/NG mixture that is drawn into the cylinder during the intake stroke. Upon compression, this lean mixture is sufficiently nonreactive such that it will not autoignite. Rather, a relatively small amount of diesel fuel, injected near top dead center (TDC), is required to initiate combustion. Delivering the two fuels to the combustion chamber in this way results in a unique chemically reacting system, wherein the premixed autoignition event bifurcates into two separate flames: a stationary, non-premixed stationary flame that remains in the vicinity of the spray and a propagating premixed NG/air flame. After the premixed autoignition phase (which consumes both diesel fuel and entrained natural gas), the total rate of heat release during second phase is the sum of that released by the two distinct flames.

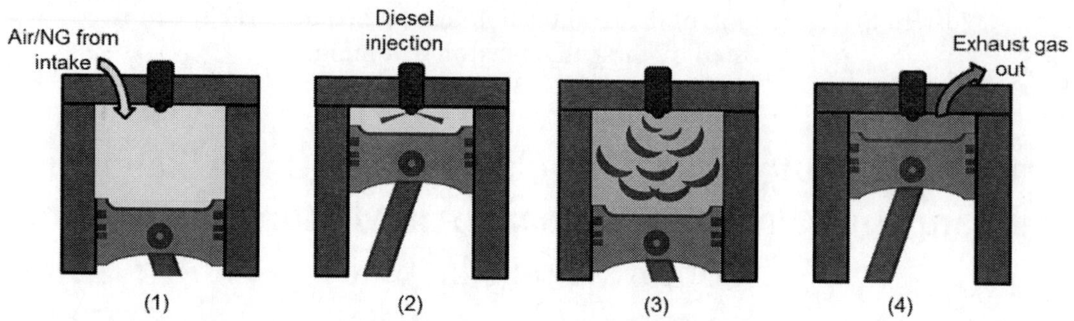

Figure 1. Illustration of standard 4-stroke dual-fuel cycle: (1) intake stroke, (2) compression stroke, (3) power stroke, (4) exhaust stroke.

Since NG is historically less expensive and produces less greenhouse CO_2 emissions on a per unit energy basis, it is of interest to maximize the substitution percentage of NG for dual-fuel operation. However, the NG substitution percentage has been shown to be limited by the onset of engine knock and/or incomplete combustion [4]. Premature autoignition, often taking the form of engine knock, is characterized by a volumetric heat release of the air/fuel mixture that results in extreme in-cylinder pressure rise rates. This phenomenon is still not well understood in dual-fuel engine operation and it can have a significant impact on peak-pressure phasing and can cause excessive component wear over time and mechanical damage to the engine. Furthermore, increasing substitution percentages can lead to decreasing combustion efficiency, meaning that less of the available air/fuel mixture actually burns. The resulting unburned hydrocarbon emissions, which consist predominantly of methane (CH_4), negate the greenhouse gas benefits of natural gas substation since CH_4 is a thirty times more powerful greenhouse gas (GHG) compared to CO_2 on a 100 year basis [5].

The potential economic and environmental benefits of dual-fuel combustion strategies are potentially valuable but a thorough understanding of the combustion phenomena are required to develop efficient, reliable systems that can be adopted on a larger scale. Researchers at the Colorado State University (CSU) Engine and Energy Conversion Laboratory (EECL) have developed computational dual-fuel engine models with reduced chemical kinetics and simplified injection spray models that have successfully characterized the apparent heat release rates (AHRR), in-cylinder pressure data, and autoignition limitations of NG/diesel dual-fuel engines with reasonable accuracy[4,6]. In fact, the previous work has identified a new form of engine knock unique to dual-fuel engines [4]. While the modeling performed to date has been show to adequately reproduce experimentally measured AHRR for multiple operating conditions, it has not been possible to independently validate the individual sub-models that govern the overall combustion phenomena. Specifically, the simplified injection spray and fuel vaporization models and the reduced chemical kinetic mechanisms need to be independently verified under the elevated temperature and pressure conditions that are commensurate with engine operation.

To better understand the vaporization process of liquid diesel fuel under engine-like conditions and to ultimately improve vaporization and spray models, the evaporation characteristics of single n-alkane droplets, C_5 to C_{12}, were observed in the CSU rapid compression machine (RCM). The RCM is an instrument capable of producing high-temperature, high-pressure transient conditions meant to replicate the compression stroke of an internal combustion engine. This system allows for the isolated observation of the fuel evaporation and dual-fuel combustion phenomena, without the need to account for cycle-to-cycle variations and other challenges that are present in conventional engine testing.

In a previous study, CFD simulations of dual-fuel droplet combustion in the RCM were performed by researchers at the EECL in an effort to better understand the dual-fuel combustion process [6]. The simulation configuration consisted of a single n-heptane droplet surrounded by a lean homogenous mixture of CH_4 and air. The simulations used previously developed reduced chemical kinetic mechanism called CSU86, which consisted of 86 species and 393 reactions [4]. The results of these simulations characterized dual-fuel droplet combustion phenomena such as ignition delay, non-premixed flame position and premixed flame propagation speed, which can be directly compared against experimental data acquired in the RCM. In this paper, we present experiments on droplet vaporization at near critical and supercritical conditions along with preliminary droplet dual-fuel droplet ignition results. The objectives of this research are twofold:

(1) Characterize the evaporation characteristics of n-alkane fuel droplets under subcritical, near critical and supercritical temperature and pressure conditions produced by the RCM.
(2) Validate dual-fuel autoignition CFD simulation results with corresponding RCM experiments.

2. Experimental Approach

The CSU RCM was employed in this study to create a stagnant, homogeneous, high pressure/temperature environment for droplet evaporation and dual-fuel droplet combustion studies. Other methods such as using falling droplets that vaporize and/or ignite in a furnace [8] are inadequate for this research, as there is a high probability of premature ignition of the NG due to heating. Performing falling droplet experiments at elevated pressures represents another challenge. To create an accurate representation of dual-fuel combustion, the diesel fuel surrogate droplet must autoignite first, which then ignites a propagating flame in the NG/air mixture. Performing the dual-fuel experiments in the RCM not only results in the highest probability of a successful combustion event as it has been defined, but also allows researchers to visually observe the combustion phenomena of interest including ignition delay, droplet evaporation rate, non-premixed flame position and premixed flame speed.

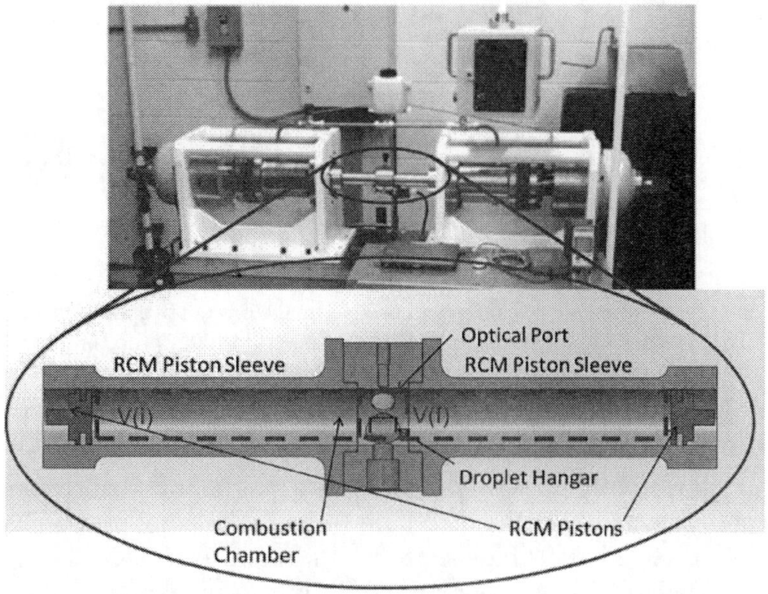

Figure 2. A photograph of the rapid compression machine (top) and a cross section of the combustion chamber (bottom) with the initial and final volumes indicated.

The CSU RCM is a pneumatically-fired horizontally-opposed dual-piston machine. As represented in Figure 2, the opposing pistons compress the chamber volume at a compression ratio of 11.6 in approximately 20 ms. The dual piston design allows the system to compress symmetrically and minimizes the impact of fluid motion resulting from high compression speeds. Additionally, this RCM utilizes crevice volumes around the circumference of the pistons that help reduce roll up vortices and create a near homologous environment [7]. The data measurement and acquisition system fitted to the RCM consists of a high speed pressure transducer and LVDT transducers that record piston position with respect to time. The RCM has also been outfitted with two sapphire windows through which a Schlieren imaging system records visual data of the combustion events as described in the following section.

2.1 Schlieren Imaging System

A Schlieren imaging system [6] was used to make visual observations of the combustion chamber interior during the droplet evaporation and dual-fuel combustion experiments. Schlieren imaging techniques utilize collimated light to detect refraction differences induced by the density gradients in the observed fluid. The more severe the density gradient, the more light refracted and the darker that portion of the image appears. For the evaporation experiments conducted, density gradients are used to quantify the alkane droplet size a function of time. Conversely, in the dual-fuel combustion experiments, high density gradients are indicative of flame fronts, as high temperatures result in lower local gas densities, thus creating a visible gradient.

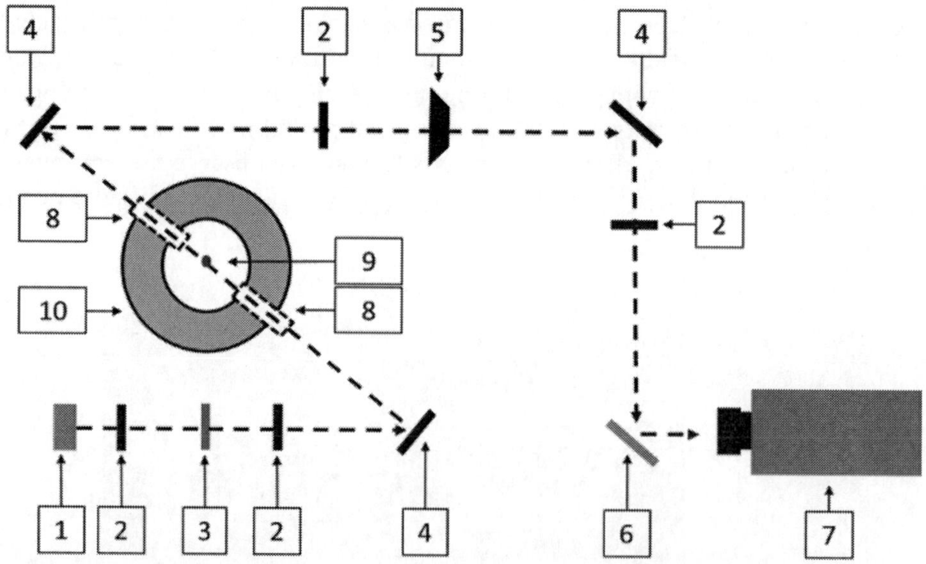

Figure 3. Illustration of the Schlieren Optical System used to image droplet evaporation and dual-fuel combustion inside the RCM chamber. The system components are as follows: 1) 405nm LED, 2) focusing lens, 3) iris, 4) silver mirror, 5) knife edge, 6) dielectric mirror, 7) high-speed camera, 8) sapphire window, 9) liquid droplet.

A schematic diagram of the Schlieren imaging set-up is presented in Figure 3. A 405 nm light emitting diode (LED) is used as the primary light source for this system. Using a mechanical iris and focusing lens, the light from the LED is reduced to a single point source and then passed through a series of lenses and reflected as collimated light through the combustion chamber. Upon exiting the chamber, the light is directed to the high speed CCD camera as shown. The final dielectric mirror just before the camera (labeled 6 in the schematic) reflects only a narrow band of light (375-450 nm) into the camera. Imaging within this range of wavelengths reduces the image saturation and increases contrast by eliminating a large portion of the broadband black body radiation emitted by soot created during the combustion process.

2.2 Combustion Chamber Design

As illustrated in Figure 4(a), a glass droplet hanger was fabricated to suspend fuel droplets in the center of the combustion chamber and field of view. The droplet suspension device consists of a single 3mm glass rod necked down to a minimum diameter of 330 μm and shaped into a hook. The tip of the rod is covered by a bead of high temperature epoxy that promotes droplet adhesion. This droplet suspension design was shown to minimize fluid flow disturbance and to survive extreme environments within the RCM while not impeding the optical view of the droplet. The suspension device is secured to a threaded plug and inserted into the combustion chamber opposite a septum/valve assembly through which droplets are grown and deposited with a syringe (see Figure 4(b)).

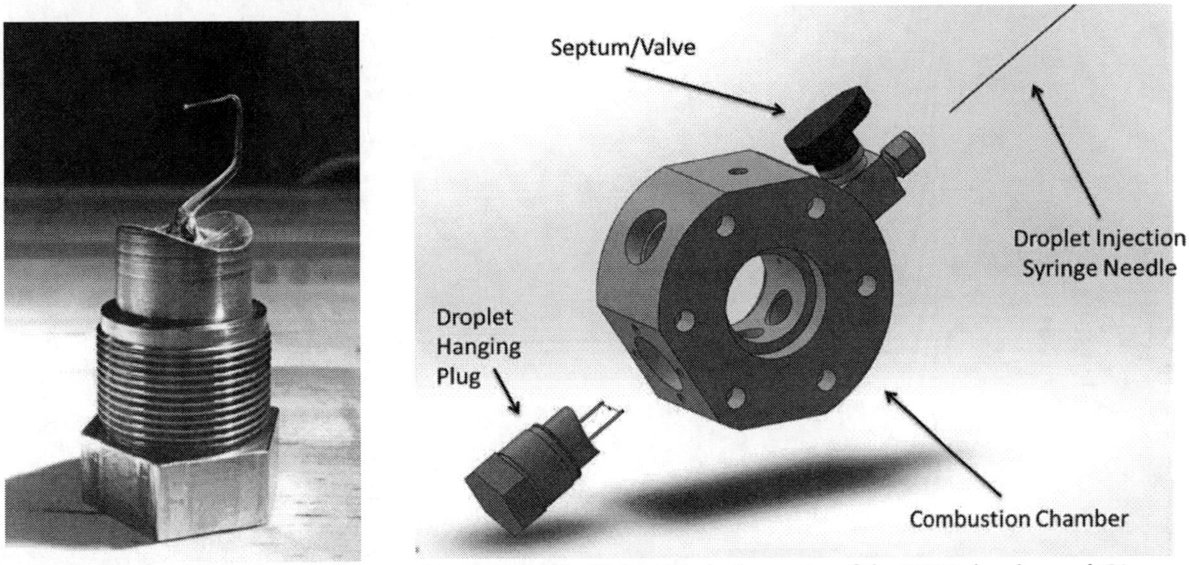

Figure 4. (a) The droplet hanger plug used to suspend liquid droplets in the center of the RCM chamber and **(b)** an exploded assembly of the RCM combustion chamber and droplet hanging plug demonstrating droplet delivery

For the evaporation experiments, the combustion chamber was filled with nitrogen and evaporation experiments were performed with a variety of n-alkane fuels (C_5, C_7, C_{10}, and C_{12}) during which droplet diameter was measured as a function of time after compression of the pistons to TDC. For the dual-fuel combustion experiments, n-heptane droplets were suspended in gaseous mixtures of CH_4, nitrogen, argon, and oxygen of the desired equivalence ratio.

2.3 Image Analysis

The images collected during the experiments were analyzed using Image J2. For the evaporation experiments, the raw images were smoothed and converted to binary files using the Image J stock programs. The droplet was then fit to an ellipse allowing for the major and minor diameters of the droplet to be averaged and recorded, serving as the droplet diameter measurement. The transient diameter measurements were then smoothed using a 7pt average. Combustion data were converted into image sequences and analyzed to primarily be a proof of concept of the experimental setup. In future studies, the Image J analysis method will be adapted to measure flame propagation rates as well.

3. Results and Discussion

3.1 Droplet Evaporation

Evaporation experiments were performed in inert environments of dry nitrogen. Measured pressures were consistent between tests and largely unaffected by the presence of the fuel droplets. To calculate the compressed temperature at TDC, the compression process was assumed to be isentropic, which is a reasonable assumption since little heat transfer from the gas to the chamber walls occurs during the rapid compression event. Using isentropic compression relationships and the ideal gas law, the chamber temperature with respect to time can be determined from the pressure data. The measured pressure data and calculated chamber temperature are shown in Figure 5 for a representative experiment with initial pressure of 1.25 bar, initial temperature of 313 K and a compression ratio 11.6.

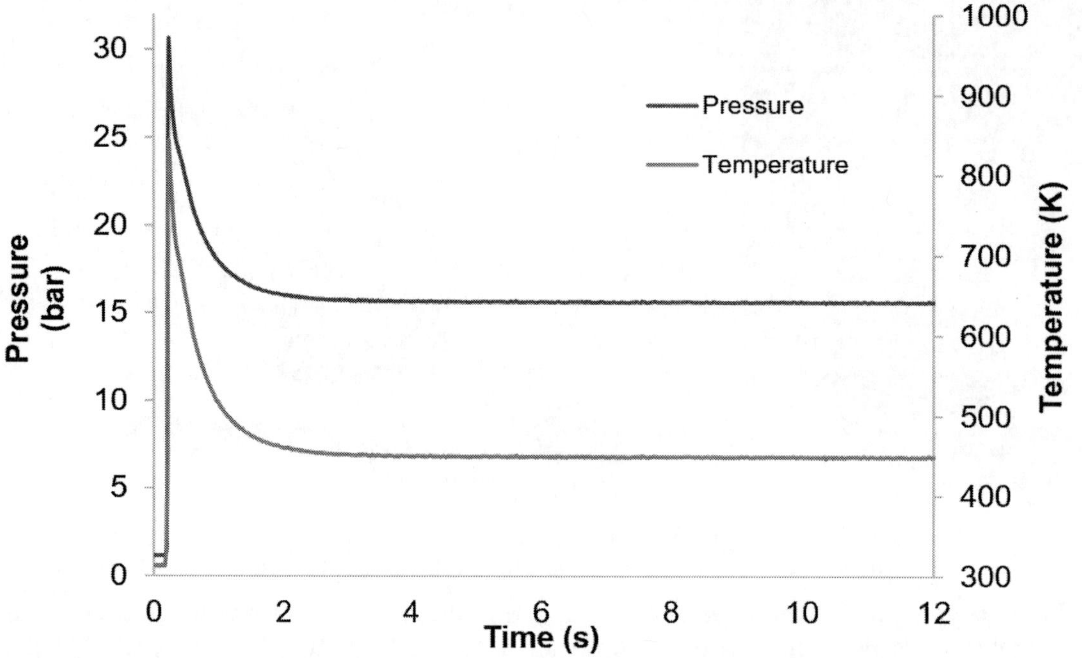

Figure 5. Measured instantaneous pressure and calculated temperature in the RCM nitrogen gas at an initial pressure of 1.25 bar, initial temperature of 313 K and a compression ratio 11.6, resulting in compressed pressure and temperature of 35 bar and 845 K, respectively.

As shown in Fig. 5, the rapid compression event is indicated by the 1.25 to 35 bar pressure increase at the start of the experiment. Compression takes place in 20 ms and results in an equally rapid increase in temperature. Immediately after TDC, heat loss to the chamber walls results in a decrease in pressure and temperature back down until they reach a relatively steady state for the remainder of the data collection period. Critical and supercritical environments are achieved in the first 400 ms, with transient temperature and pressure conditions maintained for approximately 3 s, and evaporation rates are recorded for the entirety of the test.

3.1.1 Subcritical Droplet Evaporation Results

Droplet diameters ranged from 1.5 mm to 2.25 mm with an average diameter of 2 mm for all tests. The instantaneous droplet diameter measurements were normalized by the initial diameter and the square of the normalized droplet diameter was plotted as a function of normalized time (Figure 6). In The initial high frequency oscillations in the droplet diameter data were attributed to initial disturbances imparted on the liquid droplet by an initial shock wave and subsequent convective flows, which deformed the droplet after compression. The normalized diameter squared results suggest a variation in the droplet evaporation rate with time (as signified by the variation in slope of the diameter squared plots), which is a consequence of the variation in pressure and far field temperature as a function of time due to heat loss from the RCM combustion chamber (see Fig. 5). As the pressure and far field temperature approach a nearly constant value, the slope of the normalized diameter squared plots approach a constant value as expected from the classical d^2 law for droplet vaporization.

Figure 7 contains plots of instantaneous vaporization rate, which were calculated by taking the derivative of the diameter squared with respect to time. The instantaneous vaporization rate plots further demonstrate that, after an initial transient period, the vaporization rates approaches a constant value. It is notable that the n-pentane and n-heptane exhibit high vaporization rates during the transient period and do not exhibit any droplet diameter increase during the transient period whereas the n-decane and n-dodecane droplets appear to exhibit a period of droplet diameter increase, presumably due to density decrease caused by droplet heating. Figures 6 and 7 also show that the droplet evaporation rate decreases with increasing carbon number of the fuel as expected due to the decrease in volatility.

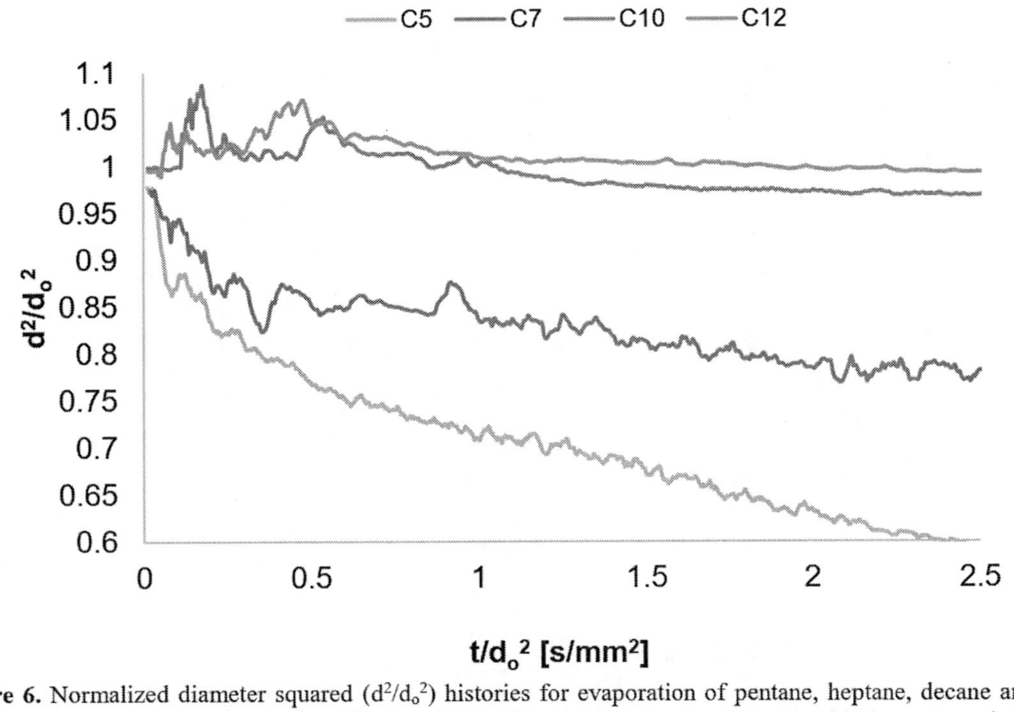

Figure 6. Normalized diameter squared (d^2/d_o^2) histories for evaporation of pentane, heptane, decane and dodecane in nitrogen gas in the RCM at an initial compressed temperature and pressure of 845 K and 35 bar, respectively.

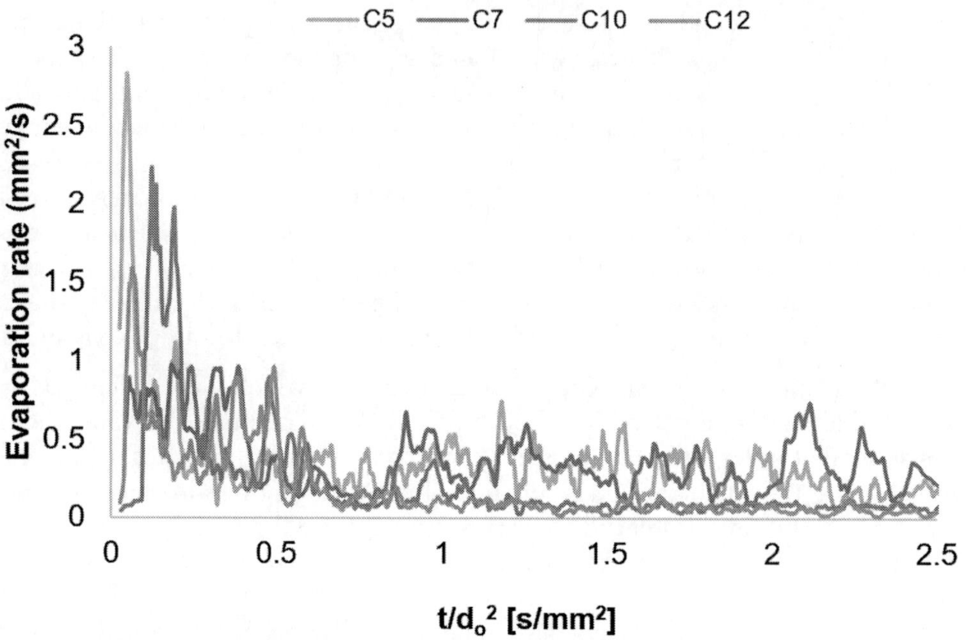

Figure 7. Instantaneous evaporation rate for n-pentane, n-heptane, n-decane and n-dodecane in nitrogen gas in the RCM at an initial compressed temperature and pressure of 845 K and 35 bar, respectively.

3.1.2 Supercritical Evaporation Results

The critical pressures and temperatures of the n-alkanes testing in this study are shown below in Table 1. From the table, it can be seen that the critical pressure and temperature of the n-alkane fuels vary inversely with increasing carbon number. Specifically, as carbon number increases, critical pressure decreases while critical temperature increases. It should also be noted that these pressure and temperatures are relatively low compared to the operating pressures and temperatures inside the cylinder of a modern diesel engine. Furthermore, the critical property data suggest that the RCM is capable of creating a supercritical conditions for n-dodecane in terms of the compressed pressure and compressed far field temperature. It should be noted, however, that the temperature at the droplet surface is initially subcritical since its initial temperature is approximately 300 K.

Table 1. Critical properties (pressure and temperature) for n-alkane fuels tested in this study.

Fuel	Critical Pressure (Bar)	Critical Temperature (K)
n-pentane	33.6	469.8
n-heptane	27.4	540.0
n-decane	21.1	617.8
n-dodecane	18.0	658.2

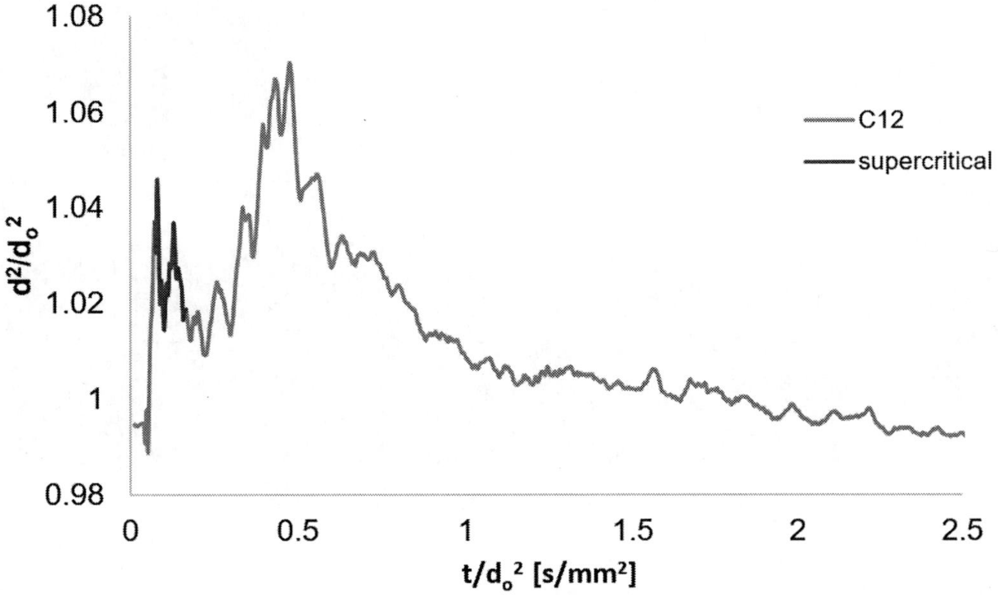

Figure 8. Normalized diameter squared (d^2/d_o^2) histories for evaporation of n-dodecane in nitrogen gas in the RCM at an initial compressed temperature and pressure of 845 K and 35 bar, respectively. Note the highlighted red portion where the far field environment was above the critical point of n-dodecane

Figure 8 is a plot of the normalized droplet diameter squared for an n-dodecane test with respect to normalized time. To aid comparison, the portion of the plot where the chamber far field is above the critical pressure and temperature is highlighted in red. For n-dodecane, supercritical far field conditions last for approximately 400 ms. During this period, the droplet diameter increases in the initial stages due to thermal expansion. Two peaks in the d^2 plot are observed – one corresponding to the supercritical state and the other after it has subsided. It should be noted that, although the far field reaches supercritical conditions, more work needs to be done to quantify the conditions in the vicinity of the droplet surface, which is initially well below the critical temperature and steadily heats up toward the boiling point. These experiments demonstrate the potential for the RCM to produce the pressures and temperatures necessary for transient supercritical droplet studies.

3.2 Combustion Results

As described in Ref. [6], dual-fuel droplet combustion has been modeled computationally and preliminary dual-fuel droplet combustion experiments have recently been performed in the CSU RCM. However, measurement of the relevant combustion parameters (ignition delay, droplet burning rate, non-premixed flame diameter and premixed flame speed) have not yet been quantified experimentally due to difficulties introduced by gas convection during the compression stroke of the RCM. Figure 9 is a sequence of images for an n-heptane droplet surrounded by a lean ($\phi = 0.7$) mixture of methane, argon, nitrogen, and oxygen. The premixed gas composition (47% N_2, 24% Ar, 29% O_2 by mole fraction) was adjusted to achieve desired peak temperatures under compression.

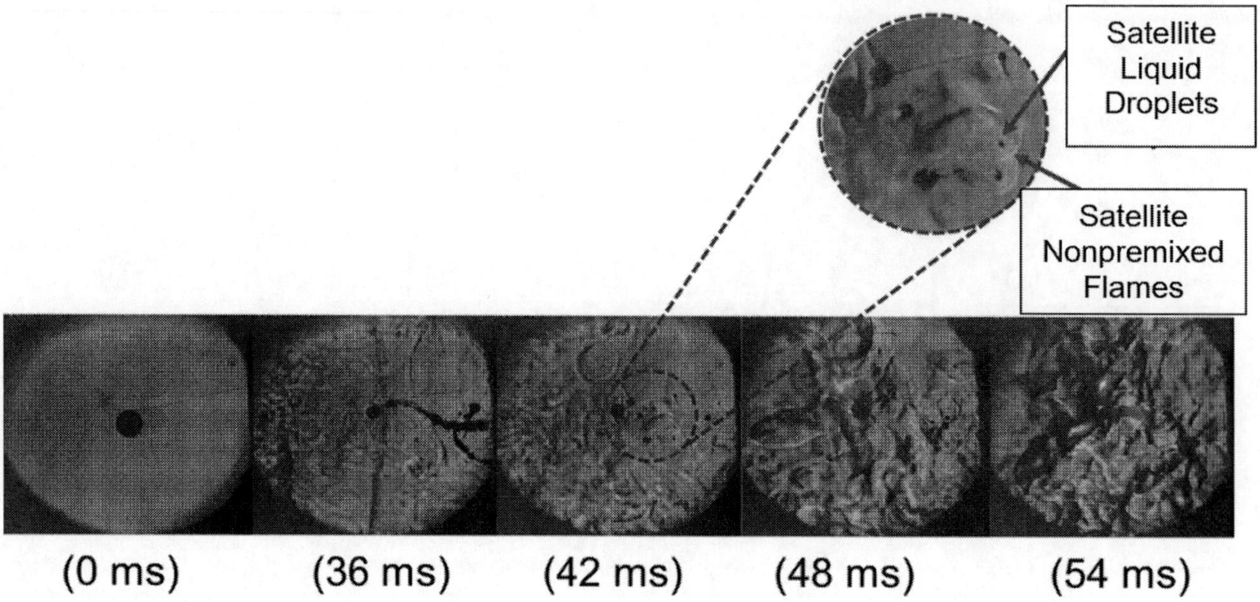

Figure 9. Image sequence of n-heptane droplet combustion in a lean ($\phi = 0.7$) methane/argon/nitrogen/oxygen mixture before (0 ms), during (36 ms), and after compression (42-54 ms).

Figure 9 presents the image sequence of a dual-fuel combustion experiment performed in the RCM. These images are taken perpendicular to the chamber axis, with the pistons out of camera view to the left and right. In the first frame, the droplet can be seen on the suspension fiber prior to compression. The RCM is fired and at 36 ms the pistons reach TDC. However, in this experiment, the pistons arrive at TDC at slightly different times, with the left piston arriving 4.82 ms before the right piston. The gas convection produced by this piston offset causes the droplet to dislodge from the suspension fiber and impedes quantitative comparisons of combustion phenomena with computational results [6]. However, dual-fuel combustion did occur before the droplet left the image frame. At 42 ms, the region that has been circled in red indicates the presence small satellite liquid droplets traveling through the chamber and enveloped with a nonpremixed flame. Soon after this image (48-54 ms), the turbulence of the system overwhelmed the image and distorted the view of the chamber. The flame fronts enveloping the individual droplets and the resulting turbulence thereafter are clear indicators of dual-fuel combustion. A diffusion flame and premixed flame were not simultaneously observed in this sequence. This is thought to be because of the 250 fps frame rate limitation on the camera or the turbulence resulting from the piston offset obscuring the flame front. Future tests with an increased frame rate and a smaller piston offset will allow the diffusion and premixed flames to be captured simultaneously.

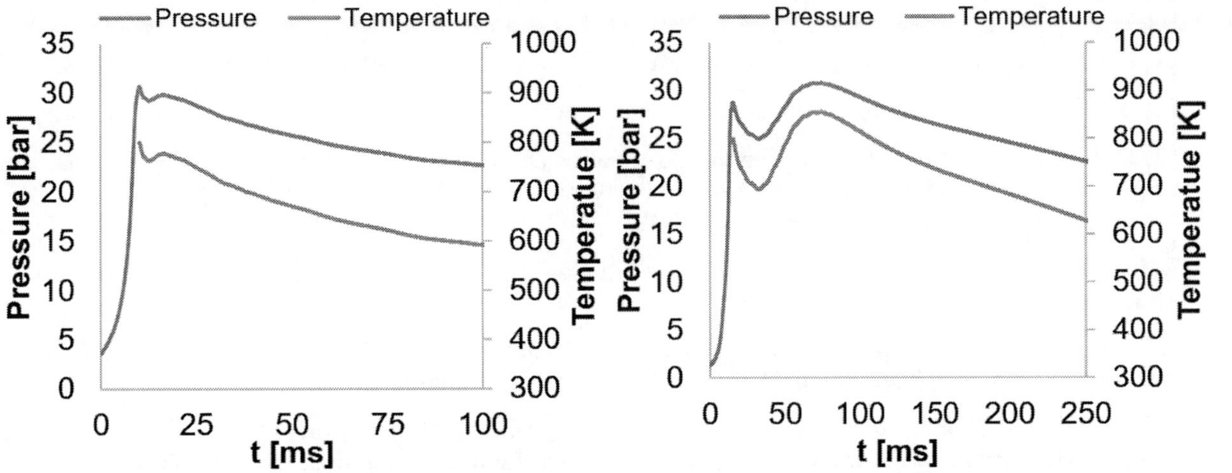

Figure 5. (a) Measured pressure and calculated temperature for combustion of a single n-heptane droplet in an $O_2/N_2/Ar$ oxidizing environment in the RCM and **(b)** measured pressure and calculated temperature for combustion of a single n-heptane droplet in lean ($\phi = 0.7$) $CH_4/ O_2/N_2/Ar$ environment in the RCM as shown in the image sequence in Figure 9.

Although the images were not conclusive, dual-fuel combustion was verified by observing the instantaneous pressure data from the high speed pressure transducer. As shown in Fig. 10(a), when an n-heptane droplet was compressed and ignited in simulated air, the heat release from the droplet combustion resulted in 2 to 3 bar pressure increase followed by the pressure decrease due to heat loss. As shown in Fig. 10(b), however, the n-heptane droplet combustion in a lean $CH_4/Ar/N_2/O_2$ mixture produced a much high pressure rise rate, which is indicative of ignition and propagation of a premixed CH_4 flame as indicated in Fig. 9. Therefore, the measured pressure and calculated temperature data suggest that the heptane droplet(s) ignited and created a premixed methane/air flame that propagated through the chamber. It can be concluded, then, that dual-fuel combustion was achieved in this experiment. However, additional tests will be necessary to make the visual observations necessary for comparison against the CFD simulation results.

4. Conclusions

The results of the subcritical n-alkane fuel evaporation rate tests performed in this study were consistent to conventional droplet evaporation theories, with a strong dependency of the far field temperature on the droplet vaporization rate. An inverse relationship between n-alkane carbon number and evaporation rate was also observed. Far field chamber conditions momentarily reached supercritical conditions during the n-dodecane evaporation experiment, resulting in two distinct peaks of its droplet diameter plotted against time corresponding with the super- and subcritical vaporization rates. Although the surface temperature of the droplet during this time interval is believed to be well below the critical point, these experiments demonstrate ability to produce supercritical conditions for droplet vaporization studies in the future. Successful dual-fuel combustion experiments were performed in the RCM, but have not yet yielded quantitative results. Further work is ongoing to quantitatively measure the combustion parameters (ignition delay, droplet burning rate, nonpremixed flame position and premixed flame speed) for comparison to the dual-fuel CFD simulation results [6].

Acknowledgements

This research was funded by the National Science Foundation Grant #1510550

References

[1] U.S. Energy Information Administration, Annual Energy Outlook 2016 with projections to 2040; 2016

[2] U.S. Energy Information Administration - EIA - Independent Statistics and Analysis. ID 73; 2016

[3] D. Serrano, L. Bertrand, Exploring the Potential of Dual Fuel Diesel-CNG combustion for Passenger Car Engine, in Proc. FISITA 2012 World Automotive Congress (2012)

[4]A. Hockett, G. Hampson, A. J. Marchese, Development and Validation of a Reduced Chemical Kinetic Mechanism for Computational Fluid Dynamics Simulations of Natural Gas/Diesel Dual-Fuel Engines. *Energy and Fuels*, 30 (3), (2016). pp. 2414-2427.

[5] Environmental Protection Agency, "Understanding Global Warming Potentials"; 2016 https://www.epa.gov/ghgemissions/understanding-global-warming-potentials

[6] C. Gould, S. Bhoite, J. Mohr, M. Baumgardner, A. Marchese, Autoignition of Liquid Hydrocarbon Droplets in Lean, High Pressure Natural Gas Mixtures Using in a Rapid Compression Machine, The 10th US National Meeting of the Combustion Institute (2017), paper 3J04.

[7] G. Mittal and S. Sung, Aerodynamics inside a rapid compression machine, Combustion and Flame, 2006

[8] T. Grumstrump, A. J. Marchese, A. Yalin, NOx formation in methyl ester, alcohol, and alkane droplet autoignition and combustion: PLIF measurements and detailed kinetic modeling. Doctoral Dissertation, Colorado State University, Fort Collins, CO 2014

[9] S. Turns, An Introduction to Combustion: Concepts and Applications. New York: McGraw-Hill, 2012.

[10] H. Kim, S. Baek and D. Chang, A single n-heptane droplet behavior in rapid compression machine, International Journal of Heat and Mass Transfer, 2014.

[11] M. Baumgartner, *Characterizing fuel reactivity in advanced internal combustion engines.* Doctoral Dissertation, Colorado State University, Fort Collins, CO. 2014

[12] C. Sung and H. Curran, Using Rapid Compression Machines for Chemical Kinetic Studies, Progress in Energy and Combustion Science, 2014

[13] G. Kukkadapu, K. Kumar, C. Sung, M. Mehl and W. Pitz, Autoignition of Gasoline and its Surrogates in Rapid Compression Machine, Proceedings of the Combustion Institute, 2003

[14] G. Godsave, Studies of the Combustion of Drops in a Fuel Spray- The Burning of Single Drops of Fuel, Symposium on Combustion, 1953

[15] G. Zhu, S. Aggarwal, Fuel Droplet Evaporation in a Super Critical Environment, Journal of Engineering for Gas Turbines and Power, 2002

[16] H. Ghassemi, S. Baek, Q. Khan, Experimental Study on Evaporation of Kerosene Droplets at Elevated Pressures and Temperatures, Combustion Science and Technology, 2006

[17] L. Kapusta, I. Pielecha, K. Wislocki, Autoignition and Combustion of n-hexane in Subcritical and Supercritical Environments, Journal of Thermal Analysis and Calorimetry, 2016

[18] M. Johnson, G. Zhu, S. Aggarwal, S. Golsborough, Droplet Evaporation Characteristic due to Wet Compression under RCM Conditions, International Journal of Heat and Mass Transfer, 2010

[19] R. Matlosz, S. Leipziger, T. Torda, Investigation of Liquid Drop Evaporation in a High Temperature and High Pressure Environment, International Journal of Heat and Mass Transfer, 1972

[20] G. Faeth, Current Status of Droplet and Liquid Combustion, Progress in Energy and Combustion Science, 1977

[21] E. Curtis, P. Farrell, A Numerical Study of High-Pressure Droplet Vaporization, Combustion and Flame, 1992

[22] C.K. Law, Combustion Physics, Cambridge University Press, New York, NY, USA, 2006

Sub Topic: Detonations, Explosions, and Supersonic Combustion

10th Meeting of the Western States Section of The Combustion Institute
Organized by the University of Wyoming
October 2–3, 2017
Laramie, Wyoming

Preliminary study of the interactions between detonation chemistry and magnetohydrodynamics

Matthew F. Zaiger[1] *and Kyle E. Niemeyer*[1,*]

[1]*Mechanical, Industrial and Manufacturing Engineering, Oregon State University, Corvallis OR, United States*

[*]*Corresponding author:* kyle.niemeyer@oregonstate.edu

Abstract: Oxy-fuel combustion can be realized with detonation engines as a strategy for reducing the challenges of carbon sequestration by eliminating NOx emissions from combustion-based power generation. A magnetohydrodynamic (MHD) generator can directly extract power from ionized gases behind the detonation front via the Lorentz force. This eliminates moving components found in gas turbine engines, which are unable to withstand high temperatures encountered in oxy-fuel combustion. Modeling detonations from a pulse detonation engine (PDE) coupled with MHD requires the simultaneous consideration of interactions between compressible flow, chemical reactions, and an electromagnetic field. In this work, we study a PDE-MHD system by coupling a Riemann solver, implicit ordinary differential equation integrators, and a Galerkin finite-element solver, which respectively solve ideal-MHD Euler, stiff chemical kinetics, and decoupled Maxwell equations. First we validate preliminary, two-dimensional simulations used in this study with experimental measurements from a PDE fueled by propane and nitrous oxide. Then, we analyze the effects of N_2/CO_2 dilution to better understand the effects on detonation speed, cell size, and interaction with a magnetic field. These results provide insights into the complex, potentially parasitic interactions between detonation and MHD effects.
***Keywords:* Detonation, Magnetohydrodynamics, Oxyfuel combustion**

1. Introduction

Oxy-fuel combustion has potential to eliminate NOx emissions from combustion-based power generation [1]. However, the high temperatures resulting from oxy-fuel combustion can have negative impacts on the components of standard gas turbine engines [2]. Detonation engines can withstand high temperatures [3], providing a strategy to use oxy-fuel combustion by removing the complexity associated with fragile moving components. Unlike gas turbine engines, detonation engines experience pressure gain during combustion, and their performance can be modeled by a Humphrey cycle [4]. Studies have shown that a detonation engine paired with a magnetohydrodynamic (MHD) system could be feasible as a topping cycle, potentially increasing combustion-based power plant efficiency through direct power extraction [5].

Two common types of detonation engines are pulse detonation engines (PDEs) and rotating detonation engines. For a PDE, ignition of the fuel-oxidizer mixture occurs at one end of the engine. A flame front sets up within the engine and begins to deflagrate down the length of the tube. This deflagrating flame front interacts with the walls of the tube and accelerates. Once the flame front accelerates to supersonic speeds it transitions to detonation where the flame front is coupled with a preceding shock. Temperatures and pressures within detonations ionize the gases behind the detonation front [3], resulting in an electrically conductive fluid. Power can be directly extracted from this working fluid using an MHD generator.

Numerical simulations can provide insight into the complex interactions between a detonation front and magnetic field. Modeling of these coupled phenomena requires a multifaceted approach that incorporates Euler's equations, stiff chemical kinetics, and Galerkin finite-element solvers into a single simulation. For this paper a two-dimensional model is implemented to better understand this coupling and the dilution effects on detonation cell size. During evaluation of the two-dimensional solution, discontinuities arise requiring a Riemann solver–which we use from Conservation LAW (CLAWpack) package [6, 7]. The time-splitting and integration required in solving the stiff chemical kinetics and thermodynamic properties are managed through Cantera [8]. Lastly the determination of the electrical potential field are solved by decoupling Maxwell's equations and use of a Galerkin solver. At each point in time, the velocity field from the fluid solver can be input into the Galerkin solver and then updated to determine the electrical potential. This can be used for increasing understanding of how power extraction is influenced by a detonation front and dilution of the flow.

Previous experimental work has been performed to observe the impacts of diluents CO_2 and N_2 on detonation speeds [9]. This study aims to simulate two–dimensional detonations representative of the experimental work to better understand the observed impacts of diluents and MHD on detonation cell size and speeds. First, we will model detonation of an H_2/O_2 system, and later progress to more complex fuels such as C_3H_8/N_2O with N_2 and CO_2 as diluents.

2. Methods

This section reviews the governing equations and boundary conditions solved and satisfied by the two-dimensional model. The numerical solvers use a finite-volume approach with fractional-step splitting methods for coupling reactions and fluid flow influenced by MHD.

2.1 Governing equations

The inviscid Euler equations paired with reactive source terms are solved in this simulation [10]. The two-dimensional conservation equations used in CLAWpack are modeled by

$$\frac{\partial Q}{\partial t} + \frac{\partial F(Q)}{\partial x} + \frac{\partial G(Q)}{\partial y} = S\,, \tag{1}$$

where Q is state of the system, F and G are the flux functions in x and y, and S is the source term. Expanding the state, flux, and source terms with the conservation of mass, momentum, and energy of the system gives

$$Q = \begin{bmatrix} \rho_i \\ \rho u \\ \rho v \\ \rho E \end{bmatrix}, \quad F = \begin{bmatrix} \rho_i u \\ \rho uu + p \\ \rho vu \\ (\rho E + p)u \end{bmatrix}, \quad G = \begin{bmatrix} \rho_i u \\ \rho uv \\ \rho vv + p \\ (\rho E + p)v \end{bmatrix}, \quad \text{and } S = \begin{bmatrix} W_i \dot{\omega}_i \\ 0 \\ 0 \\ 0 \end{bmatrix},$$

where ρ_i is the density of species i, $i = 1, \ldots, n$, and n represents the number of species, u is the velocity in space x, and E is the specific total energy. In addition, p represents the static pressure found in the equation of state

$$E = \frac{p}{\gamma - 1} + \frac{1}{2}\rho(u^2 + v^2)\,, \tag{2}$$

where γ is the ratio of specific heats. The source term variables W_i and $\dot{\omega}_i$ are the molecular weight and production rate for species i. The chemical species considered in the H_2–O_2 model, taken from GRI-Mech [11], are

$$H_2,\ O_2,\ H,\ O,\ OH,\ HO_2,\ H_2O_2,\ H_2O\,.$$

The reaction rates are given from the Arrhenius law

$$k(T) = AT^{\alpha}e^{-E_a/R^{\circ}T} \tag{3}$$

where A is the pre-exponential constant, α is the temperature exponent, E_a is the activation energy, and R° is the universal gas constant. The ideal MHD equation can be represented as

$$Q = \begin{bmatrix} \rho \\ \rho u \\ B \\ E \end{bmatrix}, \quad F = \begin{bmatrix} \rho u \\ \rho uu + I((p + \frac{1}{2}B^2) - BB) \\ uB - Bu \\ (E + p + \frac{1}{2}B^2)u - \mathbf{B}(u \cdot B) \end{bmatrix}, \quad S = 0\,, \tag{4}$$

where B is the applied magnetic field, I is the identity matrix, and total energy E is defined as

$$E = \frac{p}{\gamma - 1} + \frac{1}{2}\rho(u \cdot u) + \frac{1}{8\pi}B^2\,. \tag{5}$$

A Dirichlet boundary condition is used for the inflow, a Neumann boundary condition for the outflow, and symmetry conditions along with wall interactions will be applied. A requirement to satisfy the coupling of Maxwell's equations is have a magnetic flux of zero and is achieved by enforcing Gauss's Law.

2.2 Modeling

Modern shock approximations use a Riemann similarity solution to replicate the behavior across discontinuities. CLAWpack [7], first developed by Leveque in 1998 [12], contains a variety of Riemann solvers for different applications. CLAWpack uses a Godunov-type finite volume method where solutions are averaged across cells, and the fluxes are solved at the interfaces with the Riemann solver and limiter. A simple one-dimensional Cartesian coordinate system given by

$$Q_i^{n+1} = Q_i^n - \frac{\Delta t}{\Delta x}(F_{i+1/2}^n - F_{i-1/2}^n) \tag{6}$$

where Q_i^{n+1} is the next cell-average time-step approximation. When extending the model to two dimensions, higher-order schemes are necessary for proper resolution of the interactions between the reactive Euler and Maxwell's equations. Modification of Eq. 6

$$Q_i^{n+1} = Q_i^n - \frac{\Delta t}{\Delta x}(A^- Q_{i+1/2}^n + A^+ Q_{i+1/2}^n) - \frac{\Delta t}{\Delta x}(F_{i+1/2}^n - F_{i-1/2}^n) \tag{7}$$

increases the scheme to a second order approach. This adaptation (and others) can be selected using the PyClaw extension of CLAWpack, which includes higher-order WENO methods that use the same Riemann solvers, have adaptive mesh refinement, and parallel processing capabilities [7].

The reaction rates and species concentrations are solved with Cantera [8]. Cantera pulls the required information from the GRI-Mech 3.0 kinetic model [11]. To solve for the reaction rates, an adiabatic constant-volume reactor class was used with initial properties and conditions from the states in the flow. This includes the density, temperature, and volume of each cell. The input temperature to solve the source term comes from an approximation at the next step.

At each time step the integration of the reaction rates is considered adiabatic, meaning the total density and specific energy of the system is held constant. To perform integrations and step forward in time, the `Reactor` class in Cantera uses the `step()` function. This progresses the state of the reaction system to a desired amount of time for each cell. The integration of the system of ordinary differential equations is handled by the SUNDIALS package CVODE solver [13]. The integration process can require splitting of the time step to ensure fast reactions occurring in detonations are not overlooked. After the desired time has been reached, the net production rate for each species at each cell is determined from comparing the species creation and destruction rates:

$$\dot{\omega}_i = \dot{C}_i - \dot{D}_i \tag{8}$$

where $\dot{C}_i$ and $\dot{D}_i$ are the creation and destruction rates for each species.

Electrical potential can be determined from the system by the decoupling of Maxwell's equations and through integration of a Galerkin solver. The assumptions required for this method states that a the magnetic Reynolds number is less than one. At each point in time, the velocity field from the solver can be input into the Galerkin solver and then updated. This can be used for better understanding how the power extraction is influenced by the detonation front and ionization characteristics of the flow.

3. Results and discussion

The current state of the research has a completed hyperbolic flow shock solver (with CLAWpack) coupled to a reactive solver (with Cantera). The one-dimensional Sod shock tube case is a verification tool for compressible-flow solvers [14]. The analytical solution was solved with a shock tube calculation package from [15] and the system was verified within 6% error. After the fluid solver was verified, the system was coupled and one-dimensional detonations have been verified against the shock and detonation toolbox (SDtoolbox) [16] for H_2/O_2 within 6% error of the Chapman–Jouguet (CJ) velocity. This established the foundation to extend the solver into two dimensions. The two-dimensional reactive solver was initialized with a grid size of 0.03 m Δx and 0.0006 m Δy.

Figure 1 shows the preliminary results of the two dimensional detonation solver using schlieren-like density contours for analysis. In Figure (1a) we have the initial condition set as a high pressure and density region mimicking a shock wave. At time = 0 the shock is released and after a period period of time, changes in the density gradient can be seen in Figure (1b), indicating reactions and then followed by a transition to a detonation. The detonation begins to proceed down the length of the tube as seen in Figures (1c) and (1d). The detonation represented by Figure 1 is traveling at 2170 m/s which is slower than the CJ velocity of 2810 m/s by 22%. Grid refinement is critical for capturing reaction chemistry. If the grid size is too large reactions can be limited, which in

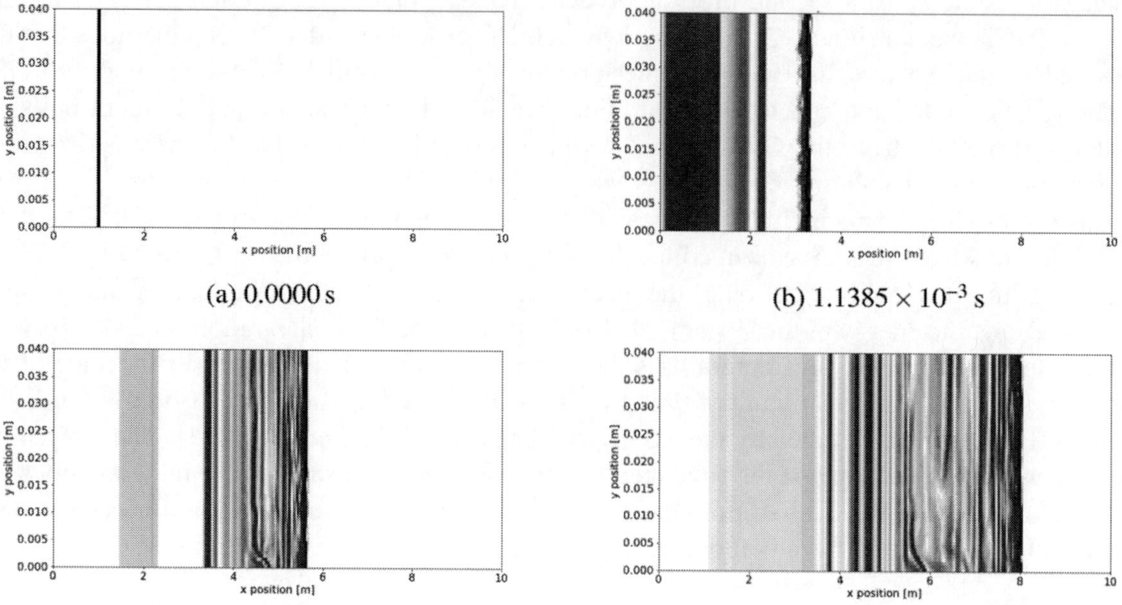

(a) 0.0000 s

(b) 1.1385×10^{-3} s

(c) 2.2770×10^{-3} s

(d) 3.4500×10^{-3} s

Figure 1: Initiation and progression of a H_2/O_2 detonation represented by schlieren-like density contours solved with a two-dimensional simulation.

turn impacts the detonation velocity [17]. As the mesh density is increased the simulation velocity is expected to approach the CJ velocity. For one–dimensional detonations it was found that a Δx of 0.0005 m was required to attain error less than 6% of the CJ velocty. It is expected that the two-dimensional solver will follow a similar trend.

4. Conclusions and future work

The two-dimensional fluid dynamic and chemical reaction solvers are coupled and will soon be updated to incorporate MHD. The detonation solver will be verified against the SDtoolbox and validated using the modified PPD model developed by [18] against Oregon State University's experimental data [9]. Once validated the MHD solver will be coupled with the reactive Euler equations and the electrical potential will be verified against an analytical solution.

With a verified solver for coupled MHD–detonation problems, we will perform parametric studies of properties including: the physical dimensions of the tube, equivalence ratios, and diluent ratios. A detailed understanding will be gleaned concerning the implications on detonation cell sizes and velocities. From [9] it was shown that decreasing linear trends in detonation speeds were measured with increasing dilution of both CO_2 and N_2. It is expected that the simulations will agree with the experimental data. In addition the interaction of the detonation with MHD is expected to have parasitic effects on the detonation chemistry as well as the velocity of the fluid. Results will be compared with experimental data, giving a picture of MHD and dilution impacts on the detonation front, emissions, and chemical kinetics.

Sub Topic: Detonations, Explosions, and Supersonic Combustion

Acknowledgements

This research was funded by the Department of Energy/National Energy Technology Laboratory (NETL) under award number DE-FE0025822.

References

[1] B. J. P. Buhre, L. K. Elliott, C. D. Sheng, R. P. Gupta, and T. F. Wall, Oxy-fuel combustion technology for coal-fired power generation, Progress in Energy and Combustion Science 31 (2005) 283–307. DOI: 10.1016/j.pecs.2005.07.001.

[2] R. E. Anderson, S. MacAdam, F. Viteri, D. O. Davies, J. P. Downs, and A. Paliszewski, Adapting Gas Turbines to Zero Emission Oxy–Fuel Power Plants, Proceedings of ASME Turbo Expo 2008: Power for Land, Sea and Air, American Society of Mechanical Engineers, (2008).

[3] G. D. Roy, S. M. Frolov, A. A. Borisov, and D. W. Netzer, Pulse detonation propulsion: challenges, current status, and future perspective, (2004), DOI: 10.1016/j.pecs.2004.05.001.

[4] T. Bussings and G. Pappas, Pulse Detonation Engine Theory and Concepts, Developments in High-Speed-Vehicle Propulsion Systems, Progress in Aeronautics and Astronautics 165 (1996) 421–472.

[5] N. Kayukawa, Open-cycle magnetohydrodynamic electrical power generation: A review and future perspectives, Progress in Energy and Combustion Science 30 (2004) 33–60. DOI: 10.1016/j.pecs.2003.08.003.

[6] Clawpack Development Team, Clawpack software, Version 5.4.0, 2017, DOI: 10.5281/zenodo.262111, URL: http://www.clawpack.org.

[7] K. T. Mandli, A. J. Ahmadia, M. Berger, D. Calhoun, D. L. George, Y. Hadjimichael, D. I. Ketcheson, G. I. Lemoine, and R. J. LeVeque, Clawpack: building an open source ecosystem for solving hyperbolic PDEs, PeerJ Computer Science 2 (2016) e68. DOI: 10.7717/peerj-cs.68.

[8] D. G. Goodwin, H. K. Moffat, and R. L. Speth, *Cantera: An Object-oriented Software Toolkit for Chemical Kinetics, Thermodynamics, and Transport Processes*, Version 2.3.0, 2017.

[9] D. R. Bean, M. F. Zaiger, R. Nelke, P. Beck, I. Walters, and D. L. Blunck, Effect of Combustion Products on Detonation Velocities, 10th U.S. National Combustion Meeting (2017).

[10] J. P. B. Elaine S. Oran, Numerical Simulation of Reactive Flow 2nd ed. Cambridge Univeristy Press, 2000.

[11] G. P. Smith, D. M. Golden, M. Frenklach, N. W. Moriarty, B. Eiteneer, M. Goldenberg, C. T. Bowman, R. K. Hanson, S. Song, W. C. Gardiner, V. V. Lissianski, and Z. Qin, *GRI-Mech 3.0*, http://www.me.berkeley.edu/gri_mech/.

[12] R. J. LeVeque, D. Mihalas, E. A. Dorfi, and E. Muller, Nonlinear Conservation Laws and Finite Volume Methods, in: Computational Methods for Astrophysical Fluid Flow, vol. 27, Springer-Verlag, Berlin/Heidelberg, 1998, pp. 1–159.

[13] A. C. Hindmarsh, P. N. Brown, K. E. Grant, S. L. Lee, R. Serban, D. E. Shumaker, and C. S. Woodward, SUNDIALS: Suite of nonlinear and differential/algebraic equation solvers, ACM Transactions on Mathematical Software (TOMS) 31 (2005) 363–396.

[14] G. A. Sod, A survey of several finite difference methods for systems of nonlinear hyperbolic conservation laws, Journal of Computational Physics 27 (1978) 1–31.

[15] J. Škifić, shocktubecalc, version 0.12, URL: https://pypi.python.org/pypi/shocktubecalc.

[16] J. E. Shepherd, Shock and Detonation Toolbox, version 2.7, URL: http://shepherd.caltech.edu/EDL/public/cantera/html/SD_Toolbox/.

[17] D. Schwer and K. Kailasanath, Blast mitigation by water mist, (1) simulation of confined blast waves, Report No. NRL/MR/6410 – 02 – 8636, 2002.

[18] O. A. Powell, C. Dreyer, and P. Papas, Laminar flame speeds of propane–nitrous oxide mixtures, 47th AIAA Aerospace Sciences Meeting including The New Horizons Forum and Aerospace Exposition The American Institute of Aeronautics and Astronautics, (2009).

Paper # 29KI-0026 **Topic: Kinetics**

Western States Section of the Combustion Institute - Fall 2017 Meeting
Hosted by the University of Wyoming
October 2-3, 2017.

Uncertainty of transport parameters in flame models: a database from virial coefficient measurements

Daniel I. Pineda[1] *Tiernan A. Casey*[2] *Xian Shi*[3] *Jyh-Yuan Chen*[3]

[1]*Department of Mechanical & Aerospace Engineering, University of California, Los Angeles, CA*

[2]*Combustion Research Facility, Sandia National Laboratories, Livermore, CA*

[3]*Department of Mechanical Engineering, University of California, Berkeley, CA*

In combustion models, the uncertainty in transport data is often either not considered or not quantified when validating chemical kinetic mechanisms. The molecular parameters used to determine the coefficients of diffusivity, viscosity, and thermal conductivity are subject to experimental uncertainty and thus can influence model predictions. In this study, we build on previous efforts establishing a framework for quantifying the uncertainty in flame models due to transport. We examine decades of experimental data on the coefficients of the virial equation of state for several key combustion molecules, and use Markov Chain Monte Carlo methods to generate a distribution of possible fits consistent with the data. With these distributions, we establish a new publicly-available transport database and combine it with a relatively untuned mechanism in the literature, propagating the distributions through transport models and flame models to obtain uncertainties on species diffusivities and flame speed. Results show significant differences from established parameter values for some species, and that the uncertainty in flame behavior from transport uncertainty is largely governed by the species that dominate the fresh gas mixture, along with species involved in rate-limiting reactions. This work establishes a protocol for future comparison with experimental flame data.

1 Introduction

Concerns have been raised in the community regarding uncertainty in combustion models [1], and many efforts have been made to quantify the uncertainty in chemical kinetic mechanisms [2, 3, 4]. Notable examples with specific regard to chemistry include investigations in correlations among Arrhenius parameters in ignition [5, 6] and in flame structure and speed [7, 8].

Although the uncertainty in chemistry is large (especially for large fuels), sensitivities to transport parameters can be on the same order of magnitude as reaction rates [9]. Researchers have investigated sensitivity analyses of diffusion coefficients in flames with significantly off-unity Lewis numbers [10, 11] and in flame extinction [12, 13]. Most relevant to this work, some studies have investigated the sensitivities of molecular transport parameters (i.e. collision diameter σ, well depth ϵ, dipole moment μ, and polarizability α, etc) [14, 9]. Macroscale transport coefficients such as diffusivity D, conductivity λ, and viscosity η are calculated from these molecular parameters in the transport models of various combustion simulation codes, using either empirical fits or kinetic theory to do so [15]. The uncertainties in these macroscale coefficients influence the uncertainty of flame models, and so the underlying uncertainties in the molecular transport parameters must be quantified. Expanding on previous research [16], this paper aims to provide the combustion community with an active and accessible database of the best estimates and uncertainties of molecular transport parameters.

2 Methods

2.1 Molecular interaction model and Markov Chain Monte Carlo procedure

To quantify the uncertainties in flame models due to transport, the uncertainties in the transport must first be quantified. Molecular transport parameters such as collision diameter (σ) and well depth (ϵ) are used in models that describe the binary intermolecular interactions between molecules [17]. The uncertainties in these parameters stem from the uncertainties in the experiments used to determine them. For relatively stable species, various experimental methods exist to measure macroscale properties of these molecular species (viscosity, diffusivity, the second coefficient of the virial equation of state, etc.) and subsequently infer the molecular parameters from statistical mechanical or kinetic theory expressions for the measured macroscale quantities [18]. Viscosity data is often considered the most reliable [19], but many measurements from prior to 1970 are considered unreliable due to the suspected presence of catalyzing reactions in the experimental measurement apparatus at high temperatures [15, 18]. As in our initial study [16], we continue to determine the best estimates and uncertainties of molecular transport parameters based on virial coefficient measurements, for which thorough compilations of experimental data are available spanning the years 1897-1998 [20, 21].

For each experiment that measures the second coefficient of the virial equation of state $B_{2,j,exp}(T)$ for molecular species j, a relationship exists that models $B_{2,j}(T)$ as a function of σ_j, ϵ_j, and other molecular parameters, though we don't consider those in this study. Thus, $B_{2,j,model}(T, \sigma_j, \epsilon_j)$ attempts to replicate $B_{2,j,exp}(T)$ via best-estimates for σ_j and ϵ_j. The expression for $B_{2,j,model}(T, \sigma_j, \epsilon_j)$ we use in this study is given in Equation 1 [22]:

$$B_{2,j,model}(T, \sigma_j, \epsilon_j) = -2\pi \int_0^\infty r^2 \left[e^{-U(r)/k_B T} - 1\right] dr \tag{1}$$

where, for a Lennard-Jones fluid, $U(r)$ is given by Equation 2:

$$U(r) = 4\epsilon \left[\left(\frac{\sigma}{r}\right)^{12} - \left(\frac{\sigma}{r}\right)^{6}\right] \tag{2}$$

Though technically derived for non-polar monatomic species, this and similar expressions have been shown to reasonably reproduce the data of polyatomic [18] and slightly polar [19] molecules.

For each experiment measuring $B_{2,j,exp}(T)$, there is an associated uncertainty $\Delta B_{2,j,exp}(T)$. Because of this, there is an uncertainty in $B_{2,j,model}(T, \sigma_j, \epsilon_j)$, ultimately corresponding to the uncertainties $\Delta\sigma_j$ and $\Delta\epsilon_j$. As in our previous study, for each species j, a joint posterior PDF of (σ_j, ϵ_j) in Equation 1 is generated using a Bayesian inference procedure comparing $B_{2,j,model}(T, \sigma_j, \epsilon_j)$ to all available $B_{2,j,exp}(T) \pm \Delta B_{2,j,exp}(T)$. Experimental uncertainties are interpreted as one standard deviation in an additive Gaussian error model. The inference is conducted using a Markov Chain Monte Carlo (MCMC) technique employing adaptive Metropolis sampling [23]. The MCMC chain is run for 50,000 samples with 50,000 samples for burn-in. A multivariate normal distribution is fitted to the resulting joint posterior PDF for each molecule j, with a peak corresponding to $\sigma_{j,opt} = \sigma_{j,mean}$ and $\epsilon_{j,opt} = \epsilon_{j,mean}$. We randomly sample 30 combinations of σ_j and ϵ_j from this multivariate normal distribution for subsequent use in flame simulations. Other parameters, such as permanent dipole moment μ_j, polarizability α_j, and rotational relaxation $Z_{rot,j}$ are much less influential in flame models [9] than σ_j and ϵ_j and are not considered here.

2.2 Flame simulations

We run 1-D adiabatic free flame simulations for a range of ϕ using $\sigma_{j,opt}$ and $\epsilon_{j,opt}$ for all species investigated using CANTERA [24] with multicomponent transport and the Soret effect enabled. In many cases, $\sigma_{j,opt}$ and $\epsilon_{j,opt}$ deviate significantly from the values reported in the original CHEMKIN database [25]. Thus, to reduce the influence of biases associated with chemical kinetic mechanisms that have been tuned (with potentially inaccurate transport data) to match flame experiments, we use a relatively "un-tuned" mechanism in the literature [26] for CH_4-air combustion. For a single species j, we use the 30 aforementioned samples to change σ_j and ϵ_j while holding the properties of the other species constant, and re-run the simulations for the same range of ϕ. We estimate the associated uncertainty in unburned flame speed, $\Delta s^0_{u,j}$, by equating it to the overall range of $s^0_{u,j}$ reported by all of the simulations for a given ϕ. Finally, we sum the $\Delta s^0_{u,j}$ values in quadrature for all species to get the total uncertainty in flame speed, $\Delta s^0_{u,tot}$. This approach assumes that the uncertainties due to each species are independent of one another [27]. We use this same approach to determine the self-diffusion coefficients D_{jj} of the selected species j at flame temperatures.

In this paper, we only consider the uncertainties for the molecules CH_4, O_2, N_2, H_2, CO, Ar, CO_2, C_2H_6, C_2H_2, and C_2H_4. These are molecules that are simultaneously in the mechanism we use [26] and for which experimental virial coefficient data exist [20, 21]. For the significant but unstable species H, H_2O_2, HO_2, O, and OH—for which no experimental virial coefficient data exist—the transport parameter values recommended by Paul et al. [14, 28] are assumed. These values have been shown [15] to better reproduce some experimental diffusivity data in the literature than those in the original CHEMKIN database. The transport properties of all other species are unchanged from that original database, however.

3 Results and Discussion

3.1 Optimized molecular transport parameters and virial coefficient data agreement

The results of the MCMC optimization procedure, including the optimized molecular transport parameters and their corresponding variances and covariances, are listed in Table 1. For some species, the optimized values are significantly different from those listed in the original CHEMKIN database; the optimized parameters often show much better agreement with the experimental virial coefficient data, and some notable examples are shown in Figure 1. While the values for some well-known species, such as N_2, are in good agreement, others, such as C_2H_2, are far off. Similar plots of the agreements with experiment for all of the species in Table 1 are available in our online GitHub repository [29], along with all of the original experimental data points and uncertainties

Table 1: Summary of optimized molecular transport parameters with covariance parameters.

Species	σ_{opt} [A]	ϵ_{opt} [K]	var_σ [A^2]	var_ϵ [K^2]	$covar_{\sigma\epsilon}$ [A K]
CH_4	3.861	146.2	9.726E-7	1.980E-3	-4.273E-5
O_2	3.523	117.2	1.421E-5	2.968E-2	-6.298E-4
N_2	3.769	95.34	1.423E-6	9.254E-4	-2.542E-5
H_2	2.898	30.63	3.515E-5	1.432E-2	-5.455E-4
CO	3.791	99.99	4.213E-5	4.583E-2	-1.259E-3
Ar	3.462	118.7	1.463E-6	4.456E-3	-7.815E-5
CO_2	4.316	200.2	9.890E-7	2.947E-3	-4.309E-5
C_2H_6	4.854	205.7	5.197E-6	1.848E-2	-3.058E-4
C_2H_2	8.088	110.2	1.126E-2	1.248E+0	-1.141E-1
C_2H_4	4.770	185.3	1.060E-5	3.230E-2	-5.771E-4

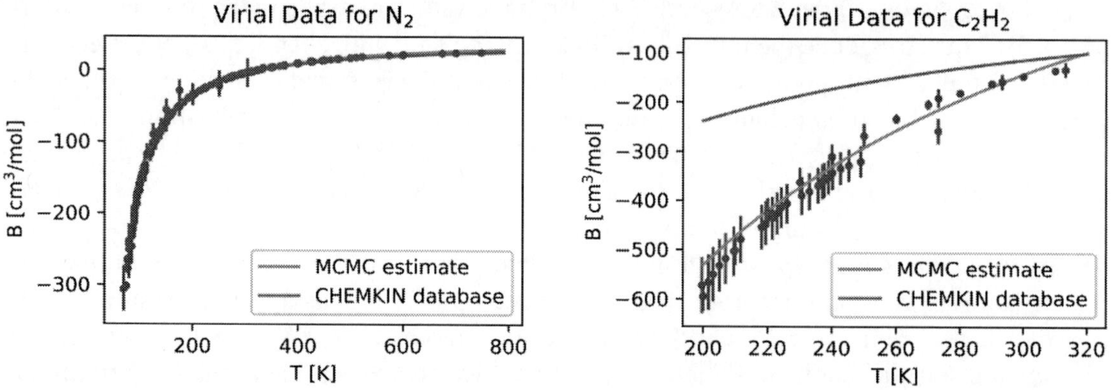

Figure 1: Comparison of Equation 1 with experimental data using both the optimized parameters in Table 1 and those in the original CHEMKIN database for nitrogen (left) and acetylene (right).

for other researchers to implement in their own models. An example of our multivariate normal distribution fitting results is shown in Figure 2 for CH_4. The means, variances, and covariances of these multivariate normal fits are also available in our repository, and they will be regularly updated to reflect the latest additions of experimental data. Notably, the model in Equation 1 fails to satisfactorily reproduce experimental virial coefficient data for highly polar molecules, such as H_2O. This is not surprising; we did not account for permanent dipole moments in our model. These molecules have historically presented problems for transport parameter fitting even when more appropriate intermolecular interaction models (i.e., Stockmayer potential) are used [19].

3.2 Effects on flame speed and self-diffusion coefficients

The uncertainty results for three different atmospheric pressure CH_4-air flames are shown in Table 2. Although Ar was included in the list of species whose parameters were modified, it had no effect in these simulations because there was no Ar in the air and so is not listed in Table 2. In all cases, the inclusion of the Soret effect reduced flame speeds slightly. Though not shown here, it is noteworthy that the predicted flame speeds using our optimized transport parameters with the "relatively untuned" mechanism [26] are lower than those predicted using the transport parameters

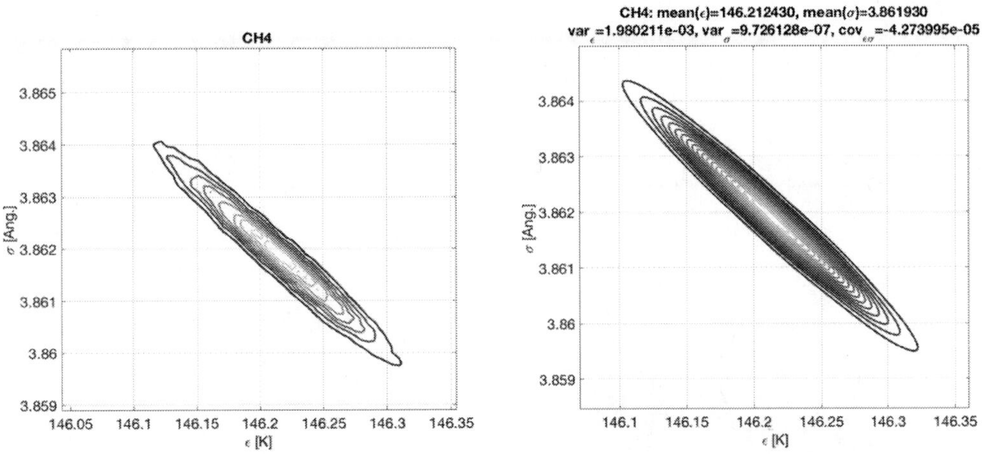

Figure 2: Example of joint posterior PDF generated from MCMC sampling (left) and a multivariate normal distribution fit (right) applied to the joint posterior PDF for CH_4, the parameters of which are listed in Table 1.

Table 2: Influence of the parameters of each species on 1 atm s_u^0 for different ϕ in [cm/s].

ϕ	s_u^0	$\Delta s_{u,tot}^0$	Δ_{CH_4}	Δ_{O_2}	Δ_{N_2}	Δ_{H_2}	Δ_{CO}	Δ_{CO_2}	$\Delta_{C_2H_6}$	$\Delta_{C_2H_2}$	$\Delta_{C_2H_4}$
0.60	8.6242	0.017	0.001	0.005	0.008	0.006	0.012	0.001	0.000	0.001	0.000
1.00	33.996	0.030	0.001	0.021	0.020	0.006	0.002	0.000	0.001	0.001	0.001
1.40	18.027	0.058	0.004	0.053	0.010	0.021	0.001	0.001	0.001	0.005	0.001

in the original CHEMKIN database [25] and much lower than those predicted using the GRI-MECH 3.0 [30] mechanism, especially for lean flames. The reasons for this are presently unclear, but highlight that chemistry and transport are intrinsically coupled in flame speed calculations. Example histograms of laminar flame speeds for lean, stoichiometric, and rich atmospheric flames resulting from the random transport parameter samples are shown in Figure 3 and are grouped by color according to the species whose parameters were varied. Consistent with previous observations [16], we found that the larger relative uncertainties in laminar flame speed occur more frequently for rich values of ϕ, and that, for all ϕ, the largest source of uncertainty is due to the uncertainty in the transport parameters of N_2 and/or O_2. In lean conditions, the uncertainty in the transport parameters of CO and H_2 play a more significant role in the overall flame speed uncertainty than at stoichiometric conditions, which was also known previously. The additions of the species C_2H_4, C_2H_6, and particularly C_2H_2 to our analysis increases the uncertainty over previous estimates at rich conditions, since these species predominantly show up in rich flames. Additionally, the inclusion of the Soret effect in the flame models in this paper increases the contribution to the overall uncertainty from H_2 for stoichiometric-to-rich ϕ over previous estimates.

Example histograms of the self-diffusion coefficients D_{jj} of H_2 and O_2 are shown in Figure 4. The effect of modifying transport parameters on diffusivities is larger in an absolute sense for light species like H_2, and smaller for heavier species like O_2. The histograms in Figure 4 appear to be normal distributions, but we cannot be certain with only thirty samples.

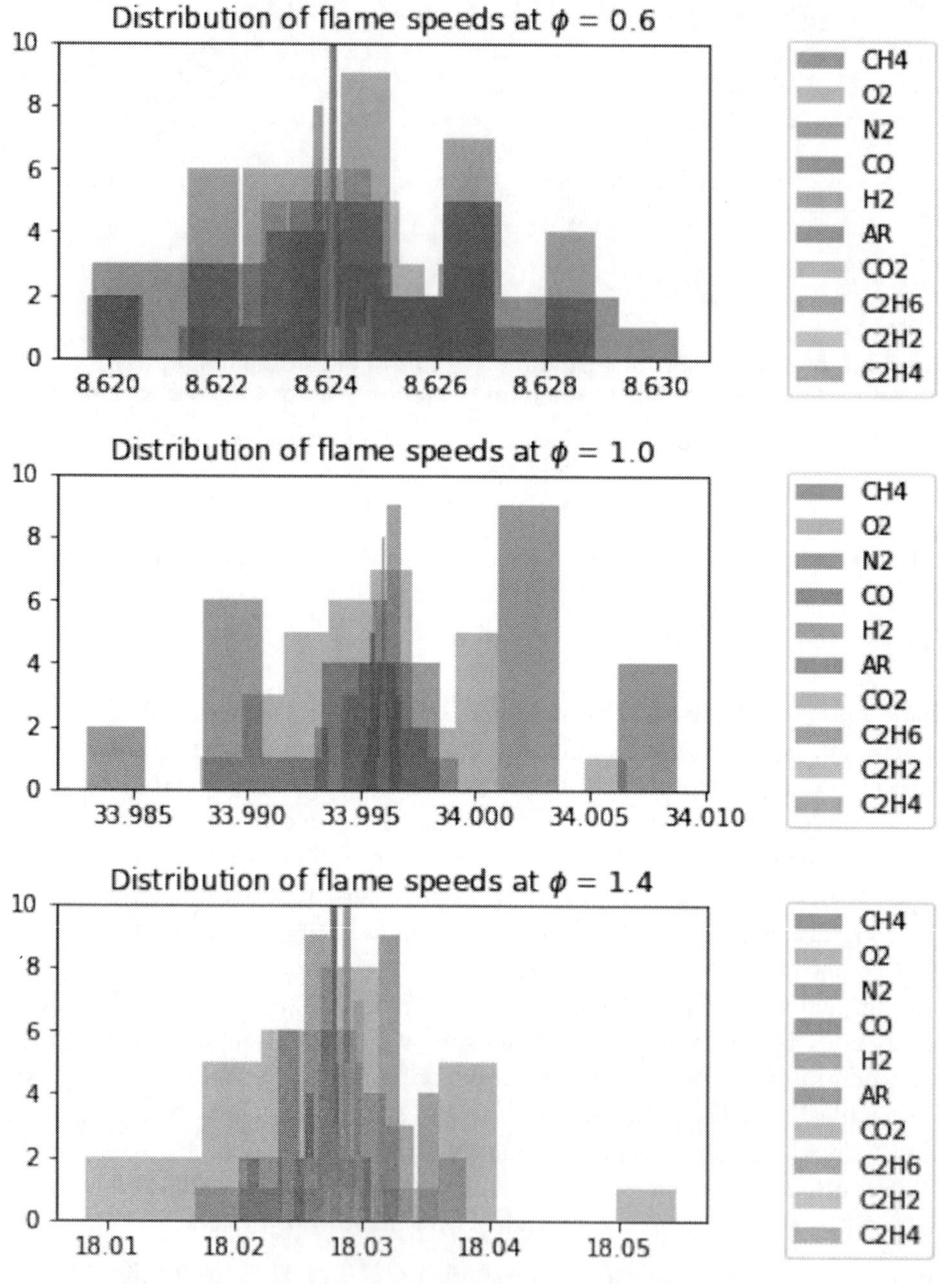

Figure 3: Visual representation of the effect of modifying transport parameters on flame speeds, grouped by species. The x-axis is flame speed in [cm/s].

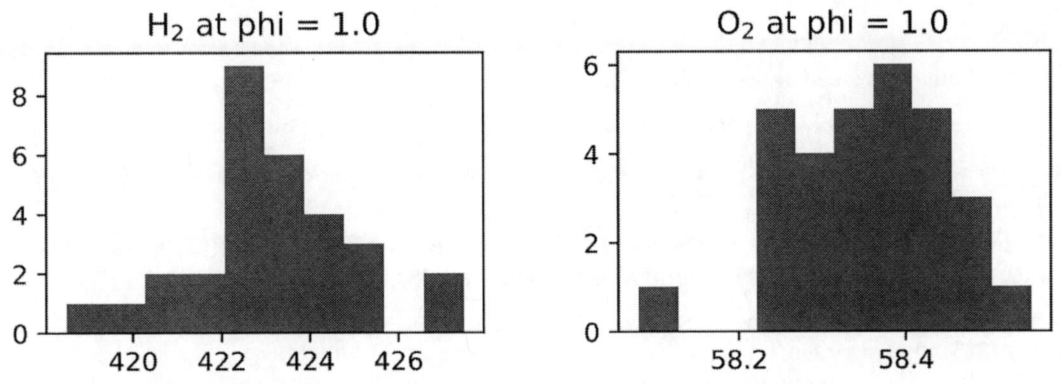

Figure 4: Visual representation of the effect of modifying transport parameters on species diffusivities for H_2 (left) and O_2 (right) at a flame temperature of 2216 K. The x-axis is diffusivity [cm^2/s]

4 Conclusions

We are certainly not the first to notice that the properties listed in the original CHEMKIN database do not satisfactorily reproduce experimental results [28, 15] and do not make any claims of the sort. However, this work adds to the existing evidence that transport is not a solved problem in flame models. We noticed that flame speeds were overall lower after modifying the transport parameters in the mechanism, and we encourage the community to consider *both* the transport and the chemistry uncertainties in tandem when refining mechanisms for 1-D flames. Otherwise, chemical rate parameters might incidentally be adjusted such that fortuitous error cancellations mask underlying inaccuracies in the transport data. We note that we are not the first to express this word of caution [9]. Transport parameters derived from other physical measurements by others (e.g. from viscosity measurements by Mouritis & Rummens [19]) have slightly different values. For non-polar molecules, our MCMC-optimized $\sigma_{j,opt}$ are in decent agreement, but our $\epsilon_{j,opt}$ are slightly larger. This could be a consequence of experimental virial coefficient data being heavily weighted towards low temperatures, where stronger intermolecular interactions are more apparent [18]. For slightly polar molecules, our MCMC-optimized $\sigma_{j,opt}$ are in not-as-good agreement (very poor agreement for extremely polar molecules), and our $\epsilon_{j,opt}$ are in worse agreement. Luckily, first-order macroscopic properties are not that sensitive to ϵ_j [9]. Future studies will incorporate experimental viscosity data and models into our estimates for $\sigma_{j,opt}$ and $\epsilon_{j,opt}$.

The larger flame speed uncertainty for rich flames with the inclusion of more species that are often present in rich flames lends more support to the notion that rich combustion chemistry is not well characterized. Indeed, many mechanisms (particularly for larger fuels) deviate from experimental flame speed data [31], speciation data [32], or blow-off data [33] for very rich conditions. Inadequate transport parameter data/models for species that predominantly exist in these regimes may provide an explanation, and this will be the subject of future investigation.

Acknowledgments

DIP wishes to thank Rebecca L. Freund, Lianne Ippel, and Dr. Joris Mulder for helpful discussions regarding statistics.

References

[1] Alison S. Tomlin. *Proceedings of the Combustion Institute*, 34 (2013) 159–176.

[2] Michael Frenklach. *Proceedings of the Combustion Institute*, 31 I (2007) 125–140.

[3] Trent Russi, Andrew Packard, Ryan Feeley, and Michael Frenklach. *The Journal of Physical Chemistry A*, 112 (2008) 2579–2588.

[4] T. Turányi, T. Nagy, I. Gy. Zsély, M. Cserháti, T. Varga, B. T. Szabó, I. Sedyó, P. T. Kiss, A. Zempléni, and H. J. Curran. *International Journal of Chemical Kinetics*, 44 (2012) 284–302.

[5] Jens Prager, Habib N. Najm, Khachik Sargsyan, Cosmin Safta, and William J. Pitz. *Combustion and Flame*, 160 (2013) 1583–1593.

[6] Tiernan A Casey, Mohammad Khalil, and Habib Najm. Inference of H2O2 thermal decomposition rate parameters from experimental statistics. In *10th U.S. National Combustion Meeting*, pages 1–6, College Park, MD, 2017. The Combustion Institute.

[7] M. T. Reagan, H. N. Najm, P. P. Pébay, O. M. Knio, and R. G. Ghanem. *International Journal of Chemical Kinetics*, 37 (2005) 368–382.

[8] David A. Sheen, Xiaoqing You, Hai Wang, and Terese Løvås. *Proceedings of the Combustion Institute*, 32 (2009) 535–542.

[9] Nancy J. Brown and Kenneth L. Revzan. *International Journal of Chemical Kinetics*, 37 (2005) 538–553.

[10] Manoj K. Mishra, Richard Yetter, Yakir Reuven, Herschel Rabitz, and Mitchell D. Smooke. *International Journal of Chemical Kinetics*, 26 (1994) 437–453.

[11] Yufei Dong, Adam T. Holley, Mustafa G. Andac, Fokion N. Egolfopoulos, Scott G. Davis, Prankul Middha, and Hai Wang. *Combustion and Flame*, 142 (2005) 374–387.

[12] G. Esposito and H. K. Chelliah. *Combustion and Flame*, 159 (2012) 3522–3529.

[13] G. Esposito, B. G. Sarnacki, and H. K. Chelliah. *Combustion Theory and Modelling*, 16 (2012) 1029–1052.

[14] Phillip H. Paul. DRFM: A New Package for the Evaluation of Gas-Phase-Transport Properties. Technical Report SAND98-8203, Sandia National Laboratories, Livermore, CA, USA, 1997.

[15] Nancy J. Brown, Lucas A.J. Bastien, and Phillip N. Price. *Progress in Energy and Combustion Science*, 37 (2011) 565–582.

[16] Daniel I. Pineda, Xian Shi, Tiernan A. Casey, and Jyh-Yuan Chen. Analysis of the errors associated with molecular transport parameters in combustion modeling and their effects on one-dimensional flame simulations. In *10th U.S. National Combustion Meeting*, pages 1–6, College Park, MD, 2017. The Combustion Institute.

[17] Joseph O. Hirschfelder, Charles F. Curtiss, and R. Byron Bird. *Molecular Theory of Gases and Liquids*. John Wiley & Sons, Inc., New York, 2nd edition, 1964.

[18] Geoffrey C. Maitland, Maurice Rigby, E. Brian Smith, and William A. Wakeham. *Intermolecular Forces: Their Origin and Determination*. Oxford University Press, Oxford, UK, 1981.

[19] Frank M. Mourits and Frans H. A. Rummens. *Canadian Journal of Chemistry*, 55 (1977) 3007–3020.

[20] J.H. Dymond and E. Brian Smith. *The Virial Coefficients of Pure Gases and Mixtures: A Critical Compilation*. Oxford University Press, New York, NY, 1980.

[21] J. H. Dymond, K. N. Marsh, R. C. Wilhoit, and K. C. Wong. *Virial Coefficients of Pure Gases*, Vol. 21A of *Landolt-Börnstein - Group IV Physical Chemistry*. Springer-Verlag, Berlin/Heidelberg, 2002.

[22] Brian Cowan. Non-Ideal Gases. In *Topics in Statistical Mechanics*, chapter 3. Imperial College Press, London, 2005.

[23] Stefano Andreon and Brian Weaver. *Bayesian Methods for The Physical Sciences*. 2015.

[24] David G. Goodwin, Harry K. Moffat, and Raymond L. Speth. Cantera: An object-oriented software toolkit for chemical kinetics, thermodynamics, and transport processes, 2014.

[25] R. J. Kee, G. Dixon-Lewis, J. Warnatz, M. E. Coltrin, and J. A. Miller. A Fortran Computer Code Package for the Evaluation of Gas-Phase Multicomponent Transport Properties. Technical Report SAND86-8246, Sandia National Laboratories, Livermore, CA, USA, 1986.

[26] Michael Frenklach, Hai Wang, and Martin J Rabinowitz. *Progress in Energy and Combustion Science*, 18 (1992) 47–73.

[27] Hugh W. Coleman and W. Glenn Steele. *Experimentation, Validation, and Uncertainty Analysis for Engineers*. John Wiley & Sons, Inc., Hoboken, NJ, USA, 3rd edition, 2009.

[28] P. Paul and J. Warnatz. *Symposium (International) on Combustion*, 27 (1998) 495–504.

[29] Daniel I. Pineda and Tiernan A. Casey. Virial Coefficient Data for Combustion Species. https://github.com/Combustion-Transport-Analysis/Virial-Coefficient-Data-for-Combustion-Species, 2017. GitHub.

[30] Gregory P. Smith, David M. Golden, Michael Frenklach, Nigel W. Moriarty, Boris Eiteneer, Mikhail Goldenberg, C. Thomas Bowman, Ronald K. Hanson, Soonho Song, William C. Gardiner, Vitali V. Lissianski, and Zhiwei Qin. GRI-MECH 3.0, 1999.

[31] Liming Cai and Heinz Pitsch. *Combustion and Flame*, 162 (2015) 1623–1637.

[32] Markus Köhler, Patrick Oßwald, Hongbin Xu, Trupti Kathrotia, Christian Hasse, and Uwe Riedel. *Chemical Engineering Science*, 139 (2016) 249–260.

[33] Colin H. Smith, Daniel I. Pineda, and Janet L. Ellzey. *Combustion and Flame*, 160 (2013) 557–564.

Western States Section of the Combustion Institute – Fall 2017 Meeting
Hosted by the University of Wyoming
October 2-3, 2017

Influence of Blending Methyl Decanoate with *n*-Decane on Ignition Delay Times in a Fuel Ignition Tester

Richard Leathers[1], Samuel Stuhlman[1], Kamal Kumar[1], Steven Beyerlein[1], Chih-Jen Sung[2]

[1]Department of Mechanical Engineering, University of Idaho, Moscow, ID 83844
[2]Department of Mechanical Engineering, University of Connecticut, Storrs, CT 06269

Abstract

This work reports an experimental investigation on the autoignition of methyl decanoate and *n*-decane, as well as their binary blends in a fuel ignition tester. Ignition delay times were obtained by injecting fuel sprays into pressurized and preheated air in a cylindrical combustion chamber. The pre-injection initial conditions for the oxidizer correspond to pressure and temperature of 24.0 bar and 812.7 K, respectively. The tests were carried out at a fixed fuel injection duration of 5.0 milliseconds. The post injection pressure-time history was used to infer the dependence of derived cetane number on the volumetric fuel blending ratios. The influence of oxidizer temperature on the ignition delay times for neat components of *n*-decane and methyl decanoate was also obtained. A weak first-stage ignition was observed for methyl decanoate at an initial air temperature of 691 K, while no evidence of two-stage ignition behavior was found for *n*-decane over the temperature range investigated. Exhaust gas composition measurements were carried out to deduce the global equivalence ratios. The experimental results were compared to numerical simulations using chemical kinetic models reported in the literature. This work provides valuable data on the autoignition of neat biofuel and hydrocarbon components for diesel like fuels, as well as their blends.

1 Introduction

Biodiesel production has nearly quadrupled over the past decade [1]. The ASTM standard [2] defines biodiesel as, "a fuel comprised of mono-alkyl esters of long chain fatty acids derived from vegetable oils or animal fats, designated B100". Biodiesel is most commonly produced with a lipid feedstock via transesterification with an alcohol, typically methanol [3]. The availability of feedstock determines the choice of raw materials that goes into producing biodiesel. The choice of feedstock in turn determines the fatty-acid composition of the biodiesel and it is the compositional profile that is influential in combustion and ignition performance [4, 5]. Pure biodiesel can be used as a fuel or it can be blended with petroleum derived fuels [3]. Methyl decanoate is one of the fatty acid methyl esters that occurs in biodiesel.

While biodiesel can be used as a drop-in alternative in conventional diesel combustion systems, its usage as a blend with conventional petro-diesel is also likely. Consequently, it is important to develop a fundamental understanding of the autoignition characteristics of both petro and bio derived diesel fuels. Given that methyl decanoate is a constituent of biodiesel, it is essential to acquire fundamental ignition data on blends of methyl decanoate and neat diesel constituent under conditions similar to compression ignition engines. The fuel, *n*-decane, is representative of a typical conventional diesel fuel constituent and is therefore chosen for this study. This study focuses on broadening the experimental understanding of conventional and alternative diesel fuel constituent blends.

A Fuel Ignition Tester (FIT) apparatus is used to obtain the ignition delay times resulting from fuel injection at elevated pressure and low-to-intermediate temperature conditions. The thermodynamic conditions within the FIT are similar to direct-injection compression ignition engines. The objective of this work is to examine the influence of blending methyl decanoate to *n*-decane on the derived cetane number (DCN). In addition, the first stage ignition delay and the total "hot" ignition delay events were obtained using the first and second maximums of the rate of combustion pressure rise as the delay time definitions respectively [6]. The influence of oxidizer temperature on the ignition delay times was also investigated. This work also compares the experimental results to numerical calculations. Homogenous constant volume simulations were carried out using chemical kinetic mechanisms reported in the literature. The global equivalence ratios used in the simulations were deduced from measurements of exhaust products from the FIT. The validity of the zero dimensional simulations to model a constant volume combustion chamber autoignition, under certain conditions, with spray has been previously demonstrated by Allen et al. [7].

2 Experimental Methods

2.1 Fuel Ignition Tester (FIT)

The Fuel Ignition Tester (FIT) is a constant volume combustion chamber manufactured by Waukesha Engine Dresser, Inc. This device is used to experimentally determine the derived cetane number of liquid fuels according to the ASTM D7170 standard [8]. Fuel is injected into the preheated and pressurized air resulting in an autoignition event. The FIT combustion chamber is a cylindrical block with a volume of 0.60±0.03 L. The injector is a standard one-hole nozzle injector conforming to DIN 73372 standard [8]. The nominal combustion chamber pressure is set to 24 bar. The pre-combustion air temperatures are in the range of 695–860 K. Plots of ignition events for pure methyl decanoate at initial air temperatures of 812.8 K and 690.8 K are shown in Fig. 1(a) and Fig. 1(b) respectively. Figures 1(a) and 1(b) also serve to illustrate the definitions of various ignition delays used in this work. The ASTM ignition delay criterion is demonstrated in Fig. 1(a) while the first stage and overall ignition delay criteria are shown in Fig. 1(b).

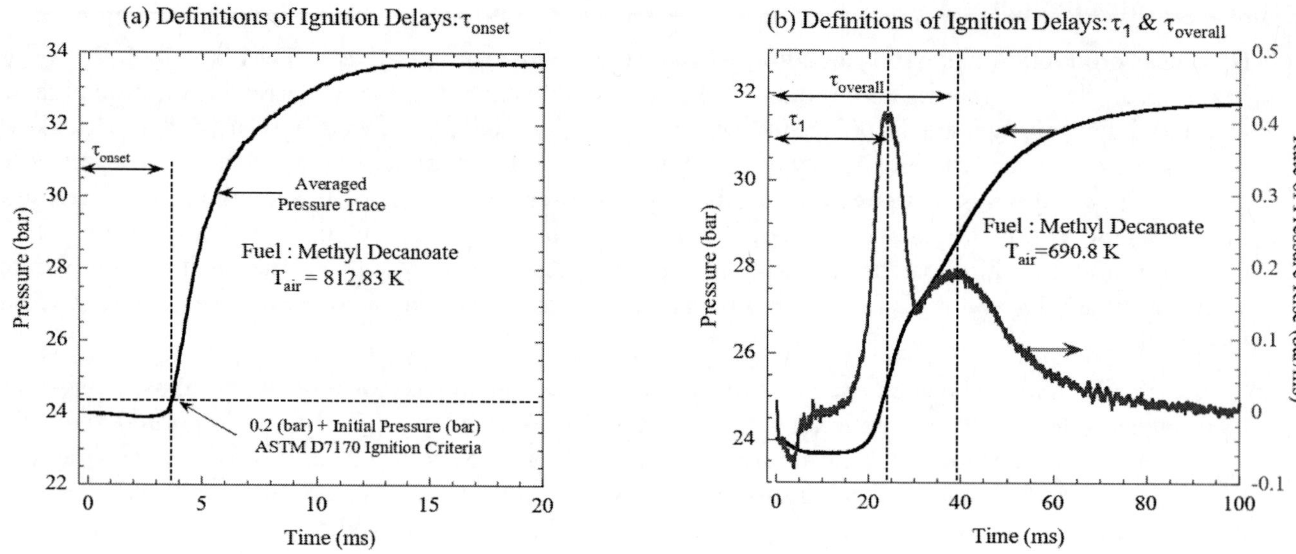

Fig. 1. (a) Definitions of ignition delays based on absolute pressure rise. (b) Definitions of ignition delays based on the rate of pressure rise.

n-Decane used in the experiments was obtained from Fisher Scientific (99.95% purity) and methyl decanoate was obtained from Tokyo Chemical Industry (98.0% purity). The volumetric blending of the fuels was performed using pipettes. The largest uncertainty in the blending was estimated at ±0.05 ml for a 100 ml batch. For the results reported, the overall mean chamber air temperature was 812.7±0.63 K, while the overall averaged charge pressure of the reported results was 24.0±0.012 bar. An oxygen analyzer with a ±3% full scale accuracy was used to determine the oxygen content in the combustion air, which was found to be 20.9% by volume.

The FIT test consists of a series of 25 injections separated by approximately a minute each. The fuel injection mimics a single compression ignition event. There is an induction time between the start of injection and the onset of combustion. The time interval between start of injection and the onset of combustion is the ignition delay time. For the FIT, the onset of ignition, or Ignition Delay (ID), in millisecond, is defined as the time from the start of injection to a point where the chamber pressure is 0.2 bar above the initial charge pressure. The ignition delay times obtained from the set of 25 injections were averaged and correlated to a DCN value [8, 9] using Equation 1:

$$DCN_{FIT} = \frac{171}{ID} \quad (1)$$

This procedure is similar to the ASTM D7170 method [8] for determination of DCN. Additionally, to enable comparison with other work, the first and second maximum rates of pressure rise are also used to define the first stage ignition delay (τ_1) and the overall ignition delay ($\tau_{overall}$), respectively, as demonstrated in Fig. 1(b).

The ignition delay times obtained using the FIT can be adjusted by either the duration of fuel injection or the chamber wall temperature. The injection duration is set by the fuel rack-position which needs to be adjusted to account for the varying physical characteristics of liquid fuels. The position of the fuel pump rack is adjusted such that the duration of the injection remains constant (5 *ms*) for each fuel and blend. The wall calibration temperature is set such that it results in an ignition delay time of

3.15 *ms* for *n*-heptane at 24 bar for an injection duration of 5 *ms*. Daily checks are made to verify the validity of this calibration temperature.

The DCN and other ignition delays were studied for the pure fuels (*n*-decane and methyl decanoate) and a number of blends: R_{MD}=10, 25, 40, 50, 60, 75, and 90 in *n*-decane. R_{MD} denotes the liquid volumetric blending ratio of methyl decanoate in percentage volume; for example, a R_{MD} =60 denotes a binary liquid fuel blend of 60% methyl decanoate and 40% *n*-decane by volume.

2.2 Exhaust Gas Analysis

Exhaust gas composition was measured using the ZRE analyzer manufactured by California Analytical Instruments. The analyzer was used to measure CO, CO_2, and hydrocarbons (HC) in the sample using non-dispersive infrared analyzers. Additionally, the NO concentration in the FIT exhaust was also measured. The NO measurements were carried out with a model CLA-600 chemiluminescence detector. The emissions measurements were used to calculate the global equivalence ratio present in the FIT. Additional air-to-fuel ratio measurements have also been carried out using a Horiba MEXA-584L automotive emissions analyzer as a cross-check. The air-to-fuel equivalence ratio, λ, was derived from the exhaust gas composition using Equation 2 and converted to a fuel-to-air equivalence ratio value using $\phi = 1/\lambda$ [10].

$$\lambda = \frac{[CO_2] + \frac{[CO]}{2} + [O_2] + \left\{\left(\frac{H_{CV}}{4}\times\frac{3.5}{3.5+\frac{[CO]}{[CO_2]}} - \frac{O_{CV}}{2}\right)\times([CO_2]+[CO])\right\}}{\left(1+\frac{H_{CV}}{4}-\frac{O_{CV}}{2}\right)\times\{([CO_2]+[CO])+(K_1\times[HC])\}} \quad (2)$$

The equivalence ratio is dependent on the fuel blend's hydrogen to carbon ratio (H_{CV}) as well as the oxygen to carbon ratio (O_{CV}). All concentration values in Equation 2 are given in % volume except for HC which is given in ppm-vol. The constant K_1 (6×10^{-4}) is used to convert hydrocarbons in ppm to an *n*-hexane, % volume, equivalent. Three conditions were tested with the emissions analyzers: pure *n*-decane, pure methyl decanoate, and R_{MD}=60. The emissions values reported in this work are the averages of the injections that had an injection duration of 5 *ms*.

2.3 Computational Methods

The experimental results in this study were simulated using detailed chemical kinetic schemes for the neat fuels and their blends. Pure fuels were simulated using the LLNL C_8-C_{16} *n*-alkane [11] and LLNL methyl decanoate [12] mechanisms, respectively. The aforementioned mechanisms for the pure fuels include both low and high temperature sub-mechanisms. Mechanisms that can model binary mixtures of *n*-decane and methyl decanoate were scarce, and only one was found in the literature. This blend mechanism developed by Ranzi et al. [13] was used to model the R_{MD}=60 conditions.

Numerical Simulations based on zero-dimensional homogeneous kinetics model were carried out using CHEMKIN® PRO [14]. The initial conditions used for the simulations corresponded to the average pressure and temperature from the FIT experiments, namely 24.0 bar and 812.7 K, respectively. The equivalence ratios were deduced using the exhaust gas composition measurements, discussed earlier. In addition, the effect of oxidizer temperature was modeled using the same averaged pressure and the corresponding experimental air temperature.

3 Results and Discussion

3.1 Ignition Delay Times

Experimental pressure traces for *n*-decane and methyl decanoate are shown in Fig. 2. This figure also shows the rate of pressure rise which was obtained from the average of the raw pressure traces. As seen from the figure, the ignition delay (both τ_{onset} and $\tau_{overall}$) of methyl decanoate is slightly longer than *n*-decane. The peak post-combustion pressure is higher for *n*-decane when compared to methyl decanoate. This is consistent with the lower energy density of methyl decanoate as compared to *n*-decane. However, methyl decanoate exhibits a greater rate of combustion pressure rise (1.36 bar/*ms* larger) than *n*-decane.

Both fuels show a tight grouping of individual test pressure traces, with the scatter in methyl decanoate traces being somewhat lesser after combustion. A representative pressure trace is derived based on the averaging of the 25 individual traces. The averaged pressure trace is an effective approach to characterize the shape, slopes, and locations of peak pressures for the fuels under test. Note, that the averaged trace is difficult to discern in Fig. 2 due to the overlap with the other 25 traces. A clearer representation of the averaged pressure traces can be seen in Figs. 1 and 3.

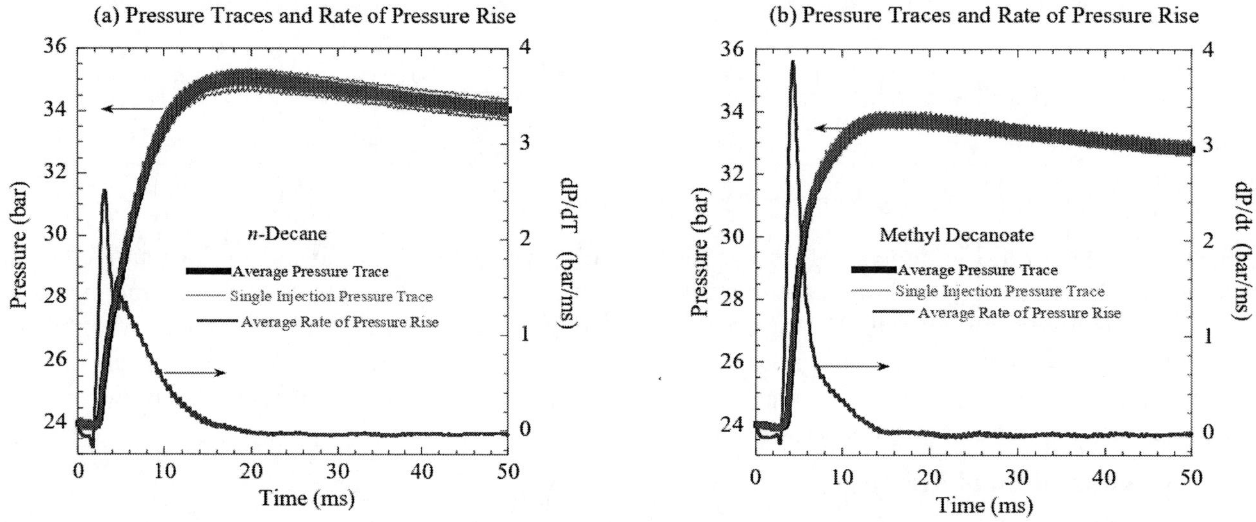

Fig. 2. *Experimental pressure trace data for the pure fuels along with the averaged pressure traces and rates of pressure rise for (a) n-decane and (b) methyl decanoate*

The influence of blending methyl decanoate in *n*-decane on the autoignition characteristics is shown in Figs. 3(a) and 3(b) on different time scales. Figure 3(a) shows that there is a decrease in the peak combustion pressure with increasing methyl decanoate content, consistent with its lower energy density. Figure 3(b) shows that the addition of methyl decanoate into *n*-decane monotonically slows down its reactivity, although it is over the span of a millisecond. Figure 3(b) also shows another clear trend; the rate of pressure rise increases with R_{MD} causing crossover to occur among the averaged pressure traces around 3.5→5ms. This crossover occurs after the τ_{onset} and $\tau_{overall}$ definitions of ignition delay.

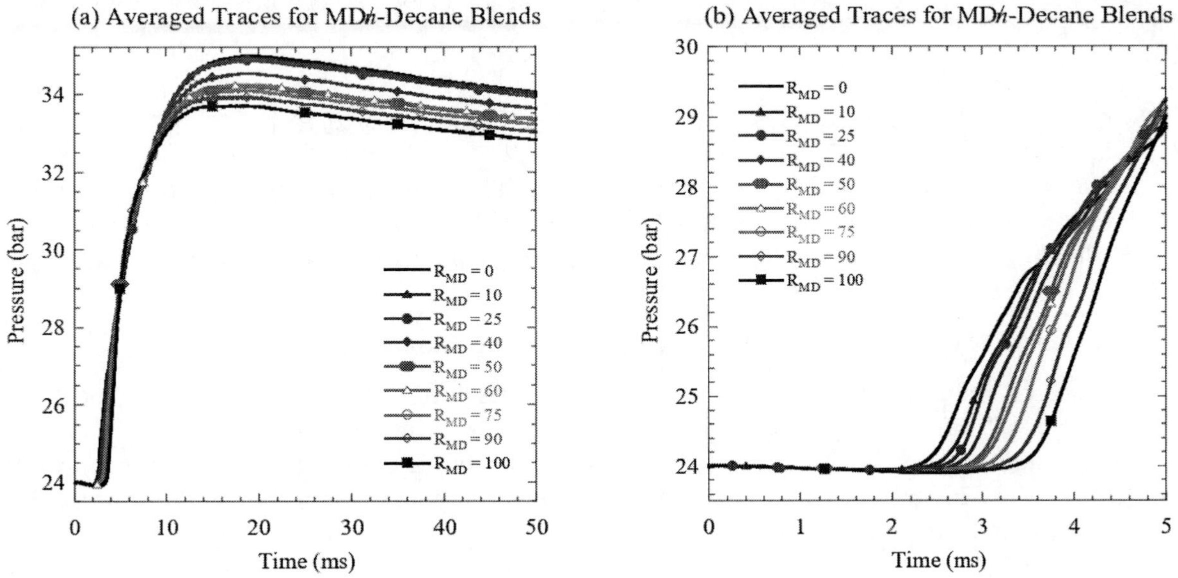

Fig. 3. (a) Averaged pressure traces for methyl decanoate / n-decane blends. (b) First five milliseconds of averaged pressure traces for methyl decanoate / n-decane blends.

The overall ignition delay times based on the maximum rate of pressure rise for methyl decanoate / *n*-decane blends are shown in Fig. 4. The ignition delay time of the blend ($\tau_{overall}$) increases as the methyl decanoate proportion (R_{MD}) is increased. A near linear relationship between the overall ignition delay and the proportion of methyl decanoate appears to hold, as seen from Fig. 4(a).

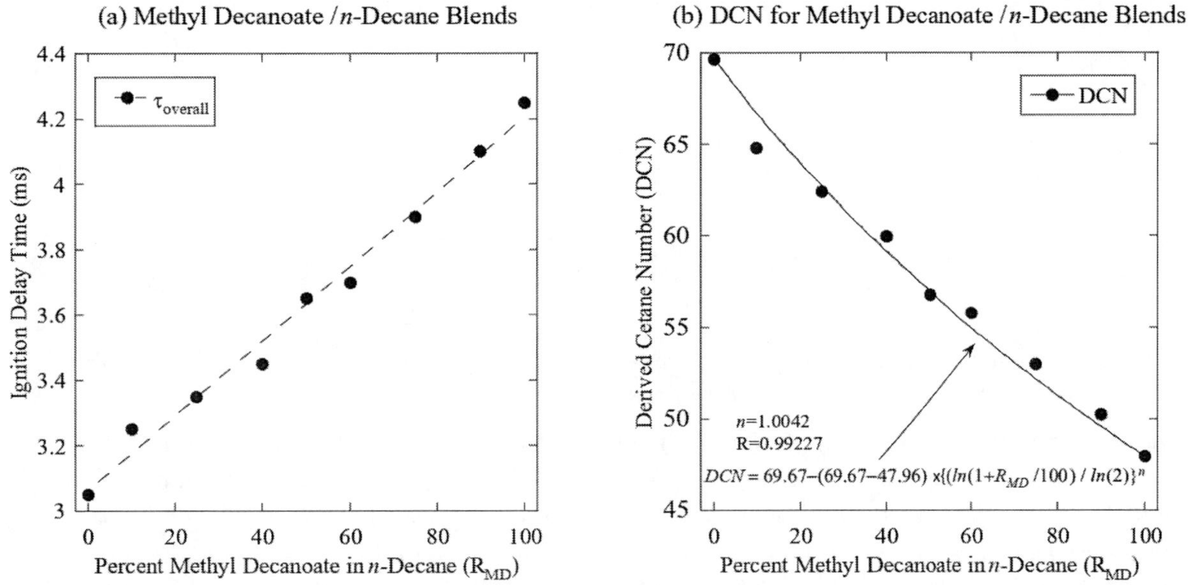

Fig. 4. (a) Ignition delay and (b) DCN as a function of volumetric blending ratio for methyl decanoate / n-decane blends

The derived cetane numbers, using Equation 1 for the binary blends, as a function of the volumetric blending ratio(R_{MD}) are shown in Fig 4(b). As shown, the DCN decreases monotonically as a function of R_{MD}. Furthermore, the DCN variation trend can adequately be captured by a fitting function of the form:

$$DCN\ (R_{MD}/100) = DCN_{base,fuel} - \left(DCN_{base,fuel} - DCN_{methyl\ decanoate}\right) \times \left(\frac{\ln(1 + R_{MD}/100)}{\ln(2)}\right)^n \quad 3$$

Where the base fuel is n-decane and n is a fitting parameter. The form of the fitting function ensures that the end points correspond to the experimentally observed DCN of the neat components. This property makes this functional form preferable over a simple inverse relationship that would appear to be valid based on Fig. 4(a).

3.2 Influence of Oxidizer Temperature on Ignition Delay Times

The dependence of ignition delay time on the oxidizer temperature has also been obtained for the two pure fuels. Figures 5 (a) and 5(b) show the pressure traces and ignition delay times of n-decane at different temperatures, respectively. As shown in Fig. 5 the n-decane ignition delay time exhibits a monotonically decreasing trend as a function of the oxidizer temperature, for the conditions tested. No evidence of a prominent two-stage ignition is found in the pressure traces for the FIT tests for n-decane.

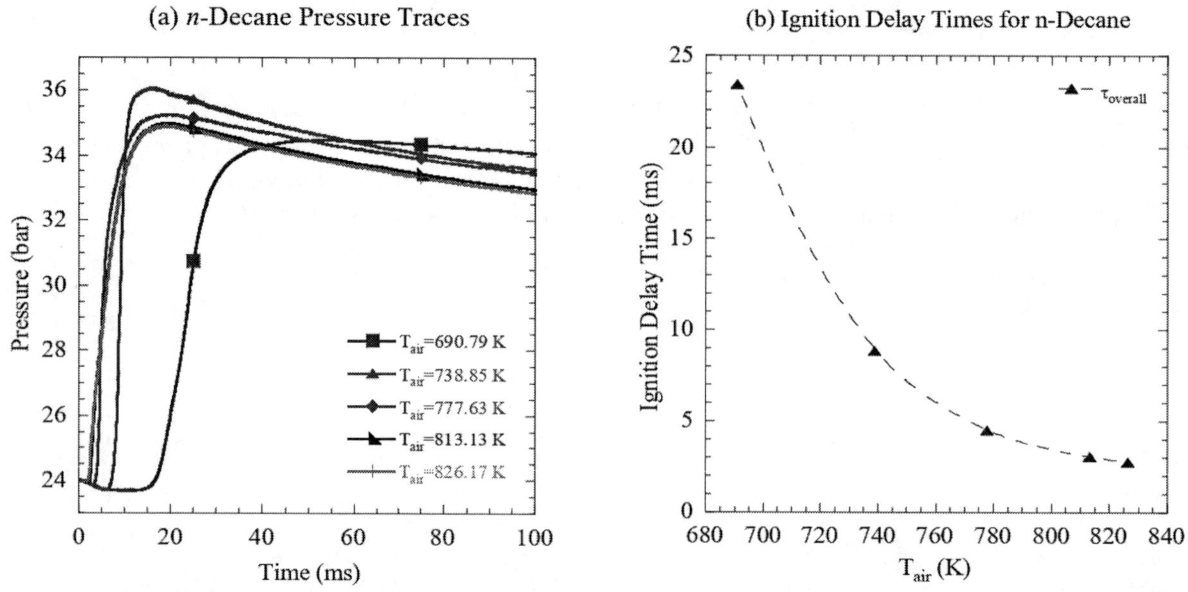

Fig. 5. (a) Averaged pressure traces for n-decane with varying initial air temperatures and (b) ignition delay times for n-decane as a function of initial air temperature.

However, it can be seen from Fig. 5(a) that the post combustion peak pressure of the four tests with the highest oxidizer temperature show a decreasing trend with an increase in temperature. The minimal difference in the duration of fuel injection alone cannot account for the observed trend. An increase in air temperature for a fixed volume and pressure should lead to a slightly larger global equivalence ratio since the number of moles of oxidizer is reduced. Consequently, an increase in the post combustion pressure is expected since the system is globally lean. The contrary peak pressure trend may be related to the reduced reactivity on account of the low-temperature chemistry observed for n-alkanes in a similar temperature window. Additionally, it is seen from Fig. 5(a) that the post-combustion peak pressure drops off for the pressure trace with the lowest initial oxidizer temperature; most likely on account of incomplete combustion.

The dependence of ignition delay time on the oxidizer temperature for the pure fuel, methyl decanoate, is shown in Figs. 6 (a) and 6(b). Figure 6 (a) shows that the ignition event of methyl decanoate exhibits a monotonic decrease in ignition delay with increasing temperature. Negative temperature coefficient (NTC) behavior was not observed for the tested temperature range. Figure 6(b) also shows a sharp decrease in ignition delay with increasing temperature. Similar to *n*-decane, methyl decanoate shows an increasing trend in combustion peak pressure with decreasing temperature, for the three hottest tests. The combustion peak pressure falls off rapidly for the two pressure traces with the lowest initial oxidizer temperature.

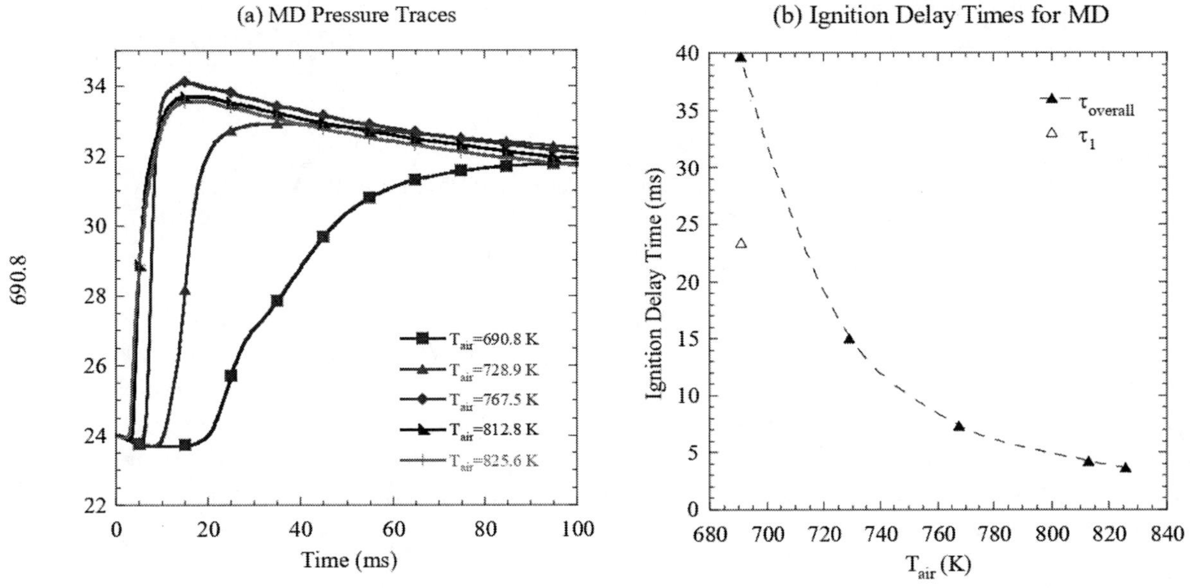

Fig. 6. (a) Averaged pressure traces for methyl decanoate with varying initial air temperatures and (b) ignition delay times for methyl decanoate as a function of initial air temperature.

3.3 Exhaust Emissions

The results of the exhaust gas composition measurements are shown in Figs. 7(a) and 7(b). The graphs represent the averaged volume % emissions results and the averaged volume ppm emissions results, respectively, for the two neat fuels and their blend. Regarding the neat *n*-decane, Figs. 7(a) and 7(b) show that it undergoes near complete combustion with no production of CO. *n*-Decane combustion also produced the most nitric oxide (548 ppm, about 130 ppm more than pure methyl decanoate). For methyl decanoate, CO_2 production was less than that of *n*-decane but very similar to the $R_{MD} = 60$ blend. For pure methyl decanoate, a very small amount of CO was produced; this is consistent with complete combustion. Additionally, methyl decanoate showed the largest residual O_2 after combustion (around 0.5% more than pure *n*-decane and 0.1 % more than the $R_{MD} = 60$ blend, likely due to its fuel bound oxygen atoms. The estimated global equivalence ratios based on the exhaust gas composition measurements and calculated from Equation 2 are: $\phi_{n-decane} = 0.136$, $\phi_{R_{MD}=60} = 0.120$, and $\phi_{\mathrm{MD}} = 0.115$. As expected, the pure methyl decanoate was found to have the lowest equivalence ratio likely because it has oxygen in its molecular makeup.

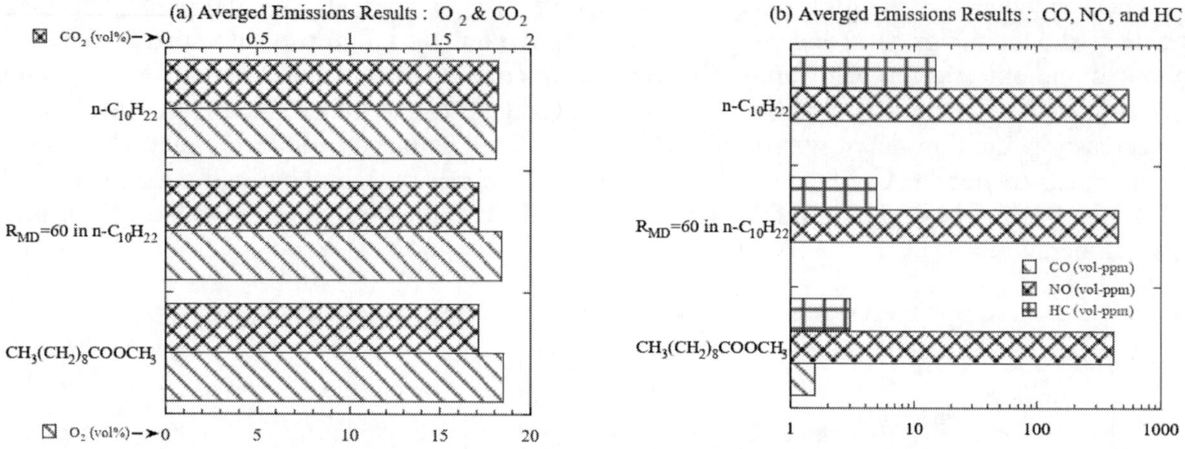

Fig. 7. (a) Oxygen and carbon dioxide concentrations in the raw FIT exhaust (b) Nitric oxide, hydrocarbons, and carbon monoxide concentrations in the raw FIT exhaust. All at $T_{air} \approx 812.7$ K and 24 bar.

3.4 Ignition Delay Simulation Comparisons

Figures 8(a) and 8(b) compare the experimental and simulated pressure traces for neat fuels and binary fuel blends of $R_B = 60$, respectively. For the pure fuels and the LLNL mechanisms[11, 12], the agreement between the experimental and simulated pressure traces is reasonable considering that physical effects related to the spray, vaporization, and non-uniform fuel distributions are not included in the simulations. The mechanism of Ranzi et al. [13] is used for the blend simulations. As such, Fig. 8(b), shows an over prediction of ignition delay. It must be mentioned that this mechanism has been validated for high temperature conditions only.

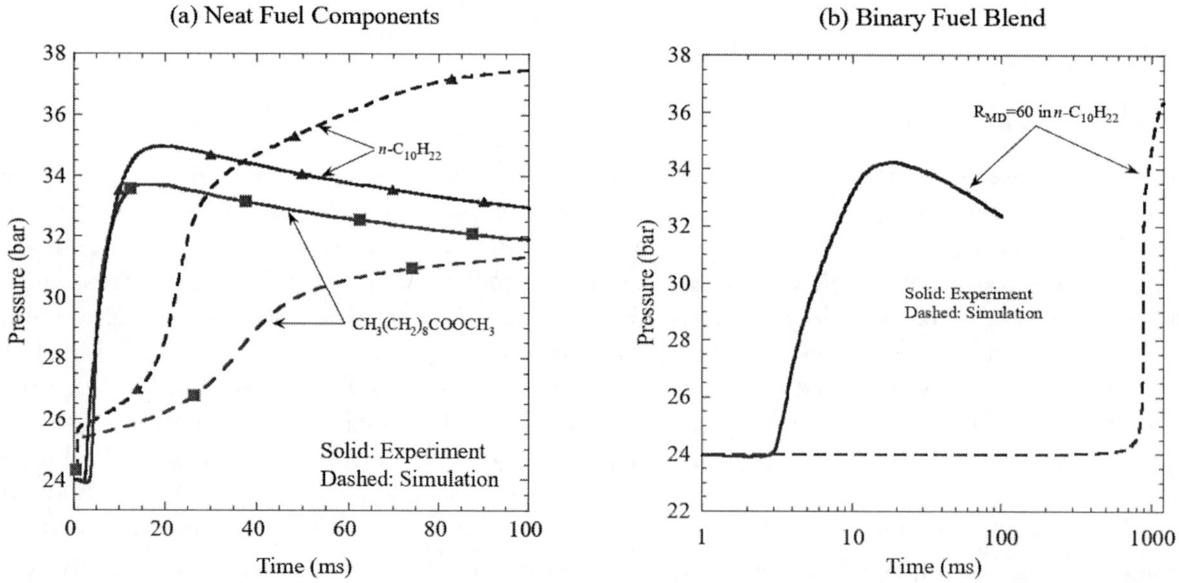

Fig. 8. (a) Comparison of experimental and simulated pressure traces for n-decane ($\phi = 0.136$) [11] and methyl decanoate ($\phi = 0.115$) [12] all at $T \approx 812.7$ K. (b) Comparison of experimental and simulated pressure trace for n-decane/methyl decanoate ($\phi = 0.120$) with R_{MD}=60 and all at $T \approx 812.7$ K [13].

Figures 9(a) and 9(b) compare the experimental and simulated overall and first-stage ignition delay times as a function of air temperature for *n*-decane and methyl decanoate respectively. The simulations exhibit both first and overall ignition delays. With exception of the lowest temperature test for methyl decanoate, only single stage ignition was observed for the experiments. For the *n*-decane fuel the overall ignition delay predictions using the LLNL mechanism [11] were generally higher a by a factor of eight. This could be a consequence of the non-homogeneity within the FIT, with the actual ignition occurring early in fuel rich pockets in the vicinity of the fuel spray. A similar trend holds for the methyl decanoate fuel for the overall ignition delay and also the single data point for the first-stage delay when compared to the simulations [12]. The slope of the Arrhenius plots of the overall ignition delays for both the fuels were however quite similar.

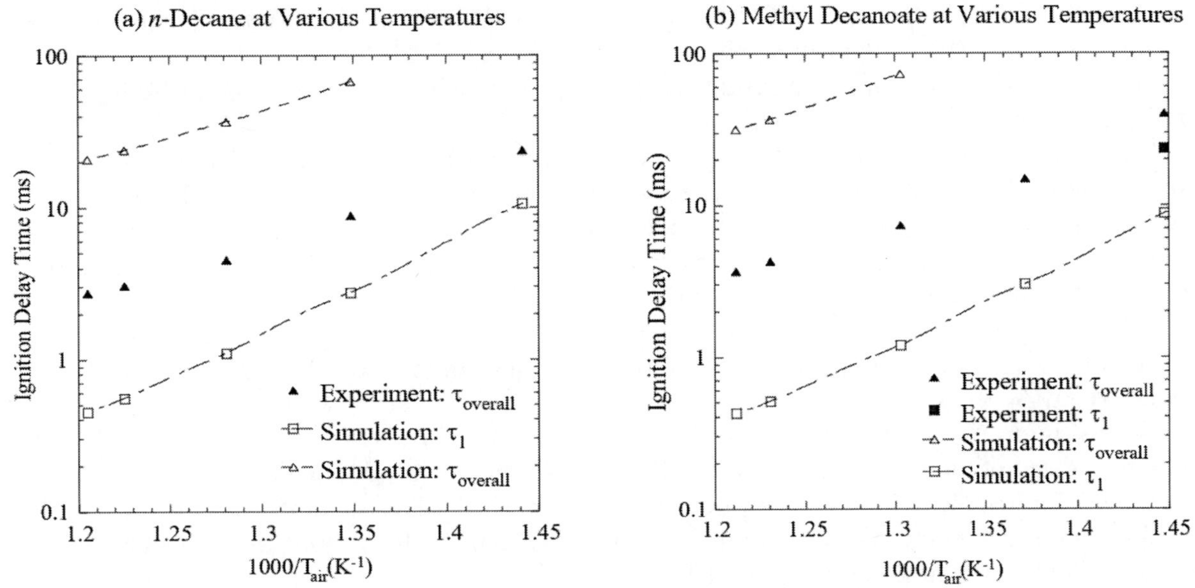

Fig. 9. *Overall ignition delay comparison between experimental and simulated results for (a) n-decane and (b) methyl decanoate.*

4 Conclusions

Autoignition characteristics of *n*-decane, methyl decanoate, and their binary blends have been obtained in a constant volume combustion chamber at a fixed fuel injection duration. Addition of methyl decanoate to *n*-decane was found to slightly reduce reactivity of the blend. A slight inhibition effect of methyl decanoate was observed for the overall ignition delay times. The experimental results were also compared to homogeneous constant volume simulations. In addition, the dependence of ignition delay on the oxidizer temperature was obtained for the two pure fuels. The simulated results over-predicted the experimental overall delays for both the fuels, but the slopes with respect to temperature variation were quite similar. Products of combustion were analyzed for the pure fuels as well as their blends. Nitric oxide production was found to be extremely sensitive to the extent of completion of combustion. The fuel blend tests helped establish a working correlation for DCN as a function of volumetric blending ratio for methyl decanoate / *n*-decane blends.

5 Acknowledgements

R.L. and S.S. acknowledge the assistantship support from the Department of Mechanical Engineering and the College of Graduate Studies at the University of Idaho.

6 References

[1] Energy Information Administration Monthly Energy Review September 2016, Biodiesel and Other Renewable Fuels Overview, 2016.
[2] ASTM D6751-07a, Standard Specification for Biodiesel Fuel Blend Stock (B100) for Middle Distillate Fuels, West Conshohocken, PA, 2007, www.astm.org
[3] T. Issariyakul, A. K. Dalai, Biodiesel from vegetable oils, Renewable & Sustainable Energy Reviews 31 (2014) 446-471.
[4] A. C. Hansen, D. C. Kyristis, C.-f. F. Lee, in: Biomass to Biofuels, A. A. Vertes; N. Qureshi; H. P. Blaschek; H. Yukawa, (Eds.) John Wiley & Sons, Ltd: 2010; pp 27-50.
[5] M. Mittelbach, C. Remschmidt, in: Biodiesel the comprehensive handbook, 1 ed.; Mittelbach, Martin (Publisher): Graz, 2004; pp 109-183.
[6] K. Kumar, Y. Zhang, C.-J. Sung, W. J. Pitz, Autoignition response of n-butanol and its blends with primary reference fuel constituents of gasoline, Combust. Flame 162 (6) (2015) 2466-2479.
[7] J. C. Allen, W. J. Pitz, B. T. Fisher, Experimental and computational study of n-heptane autoignition in a direct-injection constant-volume combustion chamber, Journal of Engineering for Gas Turbines and Power-Transactions of the ASME 136 (9) (2014).
[8] ASTM D7170-09, Standard test method for determination of derived cetane number (DCN) of diesel fuel oils – fixed range injection period, constant volume combustion chamber method, West Conshohocken, PA, 2008,
[9] Waukesha Engine Dresser: Fuel Ignition Tester User Manual, Waukesha, Wisconsin 53188, 2006
[10] Horiba, Automotive Emissions Analyzer MEXA-584L Instruction Manual, 2008
[11] C. K. Westbrook, W. J. Pitz, O. Herbinet, H. J. Curran, E. J. Silke, A comprehensive detailed chemical kinetic reaction mechanism for combustion of n-alkane hydrocarbons from n-octane to n-hexadecane, Combust. Flame 156 (1) (2009) 181-199.
[12] O. Herbinet, W. J. Pitz, C. K. Westbrook, Detailed chemical kinetic oxidation mechanism for a biodiesel surrogate, Combust. Flame 154 (3) (2008) 507-528.
[13] E. Ranzi, A. Frassoldati, R. Grana, A. Cuoci, T. Faravelli, A. P. Kelley, C. K. Law, Hierarchical and comparative kinetic modeling of laminar flame speeds of hydrocarbon and oxygenated fuels, Prog. Energy Combust. Sci. 38 (4) (2012) 468-501.
[14] ANSYS® Academic Research, Release 18.1, CHEMKIN PRO, ANSYS, Inc.

WSSCI 2017 Fall Technical Meeting
Organized by the Western States Section of the Combustion Institute
October 2-3, 2017
University of Wyoming, Laramie

Ignition Delay Times and Derived Cetane Numbers of Canola, Corn, and Soy Derived Bio-Diesel

Samuel Stuhlman[1], Chad Dunkel[2], Richard Leathers[1], Kamal Kumar[1], Dev Shrestha[2], Steven Beyerlein[1], Chih-Jen Sung[3]*

[1]Department of Mechanical Engineering, University of Idaho, Moscow, ID 83844
[2]Department of Biological Engineering, University of Idaho, Moscow, ID 83844
[3]Department of Mechanical Engineering, University of Connecticut, Storrs, CT 06269

Abstract

Experimental results on the ignition delay trends of three in-house produced biofuels made from canola, corn, and soybean are presented. Transesterification was accomplished in a batch reactor using methanol and sodium methoxide as alcohol and catalyst, respectively. Alcohol and soap impurities were removed via water washing. The biodiesel samples produced consist of fatty acid methyl-esters. The fuel makeup for the three biodiesels differ both in terms of the chain length and the degree of unsaturation. These two properties have a strong influence on the autoignition response. A longer chain length leads to a greater autoignition propensity, while a higher degree of unsaturation leads to a reduced tendency to autoignite. The fatty acid profiles for the fuels studied here show that palmitic acid (C16:0), oleic acid (C18:1), linoleic acid (C18:2), and α-linolenic acid (C18:3) are the major constituents. Fuel sprays were injected into pressurized, preheated air in a constant volume combustion chamber. The averaged initial oxidizer pressure and temperature were 24 bar and 812 K, respectively. The injection duration was held constant at 5.0 milliseconds. The ignition delay times were inferred from the combustion pressure traces and were used to obtain the Derived Cetane Number (DCN). The DCN of the soy, corn, and canola biofuels were found to be 43.9, 47.5, and 54.3 respectively. The DCN ranking is consistent with the compositional differences of the fuels. While the chain lengths of the constituents are relatively similar, the degree of unsaturation of the major constituents differs. For example, the canola biodiesel contains the largest amount of methyl oleate (C18:1) and a lesser amount of methyl linoleate (C18:2) than the other two biofuels; this difference in the degree of unsaturation is believed to cause the canola biodiesel to exhibit a higher autoignition propensity, and thereby a higher DCN. At the pressure and temperature conditions of this study, only single stage ignition events were observed. This study provides insights into the influence of fatty acid compositional profile on the autoignition characteristics of plant-derived biodiesel.

1 Introduction

The autoignition phenomenon is central to the compression ignition process occurring in diesel engines. Conventional diesel fuel consists of long chain petroleum derived hydrocarbons but there is a developing trend that favors the adoption of renewable alternatives. Plant-derived biodiesel fuels provide an attractive alternative to petro-diesel. They also provide a cleaner burning alternative on account of the fuel bound oxygen [1, 2]. The substitution of conventional diesel fuel with biodiesel requires us to develop a detailed understanding of the physical and chemical properties of biodiesels originating from varying plant and animal feedstock. This work provides a comparative study related to the autoignition response of canola, corn and soybean derived biodiesel. An important parameter characterizing diesel fuel combustion is the cetane number. It describes how easily a diesel fuel will autoignite under specified operating conditions. There are several approaches to obtaining the cetane number [3, 4]. This work uses a Fuel Ignition Tester (FIT) for the purpose of comparing the autoignition characteristics of the three biofuels mentioned earlier using the Derived Cetane Number (DCN). The FIT is a constant volume combustion chamber which contains preheated and pressurized air into which a fuel spray is injected. The time between the start of injection and the onset of combustion is used to infer the DCN [4, 5].

Three biodiesel fuels were produced in-house for the experimentation of this study; canola, corn, and soybean. These biofuels were made through transesterification. Sodium methoxide was used as catalyst for the reaction. After transesterification the biofuels were water washed to remove remaining impurities like alcohol contaminates and residual catalyst from the reaction. The constituents differ in chain length and unsaturation, both molecular traits affect the ignition delays and DCNs of the fuels. The three biofuels produced for this study have similar chain lengths, but different degrees of unsaturation. This work provides insights into autoignition of plant derived biodiesel.

2 Methods

2.1 Production Methods

Biodiesel was produced through a chemical process known as transesterification. Transesterification of biodiesel was accomplished by chemically reacting a vegetable oil or animal fat with alcohol such as ethanol or methanol [6]. In order for the reaction to take place, a strong base catalyst was used in addition to some form of energy input [7]. The chemical compounds that are produced from this process are called fatty acid esters, or biodiesel, and a glycerol by-product.

Reactor design can vary, with smaller plants using batch reactors and larger facilities using continuous flow processes including continuous stirred-tank reactors (CSTR) or plug flow reactors (PFR) [6]. In a batch reaction configuration, the oil, alcohol, and catalyst are combined in a reactor and mixed for approximately 1 hour at a reaction temperature of 60° C. The reaction process is commonly performed in two steps, and an 80% and 20% reaction will be run consecutively. Performing this procedure in a two-step process ensures that a very complete conversion reaction takes place and as much of the feedstock is converted as possible [6]. Following each reaction step, due to the differences in density, the heavier component containing glycerol, methanol, and catalyst is separated and removed from the lower region of the reactor. The liquid phase remaining in the reactor is primarily biodiesel containing residual alcohol and soaps. Downstream processes include methods of purifying the biodiesel in order to reduce soap concentration, methanol removal and recovery, and glycerin purification [6]. Figure 1 shows the general process used to produce biodiesel and processing of the resulting by-products.

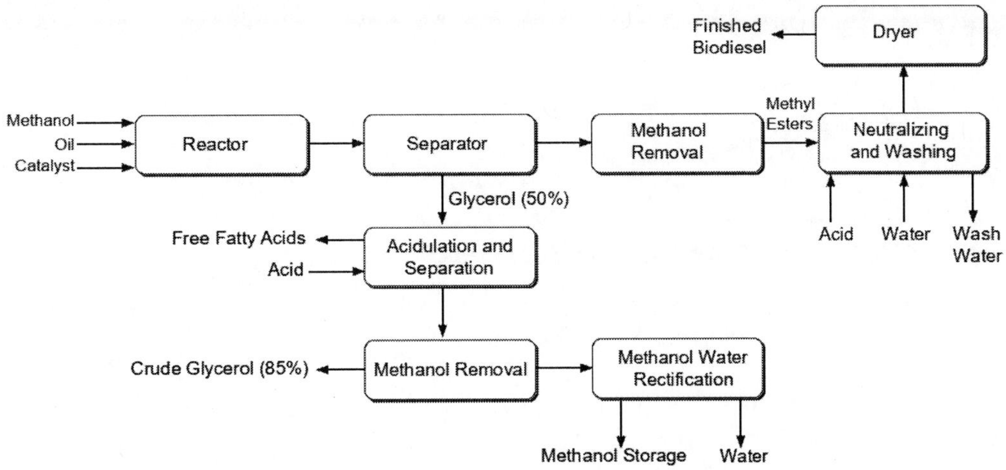

Figure 1: *Process flow schematic for biodiesel production [7].*

In order to remove residual methanol, catalyst, and soap, a water washing process was performed by allowing water to fall through the biodiesel. Due to the non-polar nature of biodiesel and higher density of water, the wash water falls through the fuel and carries away alcohol, catalyst, and soap to the bottom of the reactor. Again, wash water containing contaminants was removed from the lower region of the reactor and purified biodiesel remained. The final step in the production process was to remove moisture content from the fuel via a drying process in which the biodiesel was heated to 100 degrees Celsius until the moisture content of the fuel was less than 0.05% (500 ppm). A schematic for the transesterification process used to produce the biodiesel samples can be seen in Fig. 2.

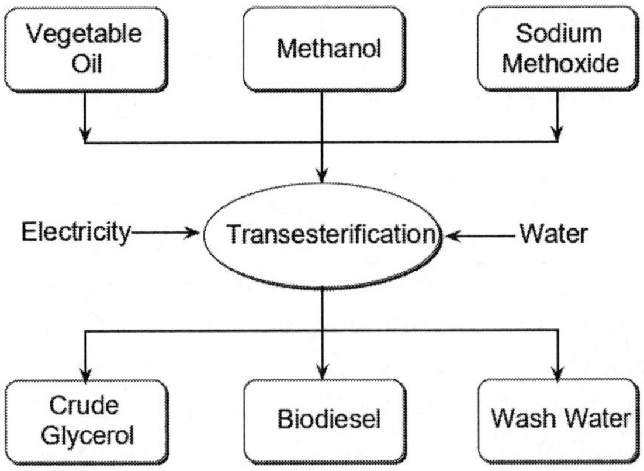

Figure 2: *Transesterification process for producing biodiesel fuel samples.*

The five major Fatty Acid Methyl Esters (FAME's) present in biodiesels tested in this work are shown in Fig. 3.

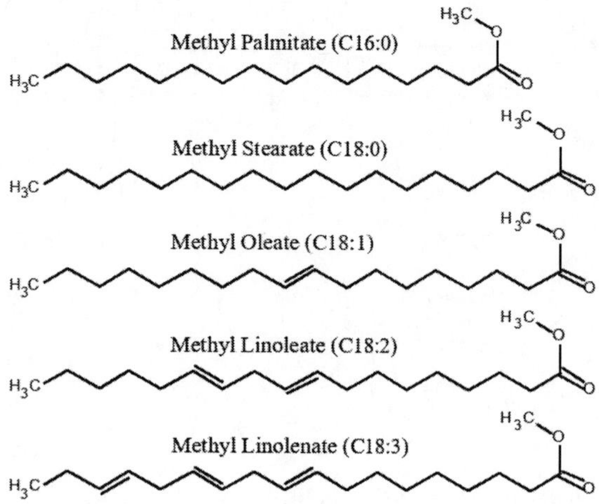

Fatty Acid	*Canola*	*Corn*	*Soy*
C16:0	4.1	11.7	11.1
C18:0	1.8	1.5	4.2
C18:1	65.4	30.9	22.1
C18:2	18.8	54.4	54.2
C18:3	7.6	1	7.9
C20:1	1		

***Fig 3**: Chemical structure of major neat components and percent of total methyl esters for the three plant-derived biofuels tested.*

The compositional profile shows that soy derived biodiesel has the largest proportion of C18:2 and C18:3 (62.1%) followed by corn (55.4%) and canola (26.4%). Additionally, canola derived biodiesel was found to have the largest amount of methyl oleate C18:1 (65.4%).

2.2 Experimental Methods

A Fuel Ignition Tester, manufactured by Waukesha Engine Dresser, Inc., was used to study the autoignition characteristics of the three biodiesels. It was used to obtain the DCN of a diesel fuel in conformance with the ASTM D7170 standard [4]. The FIT combustion chamber is a cylindrical block with a volume of 0.60±0.03 L, with external heating elements, heat shield, and electrically actuated intake and exhaust valves. It is equipped with a fuel injection nozzle assembly with a standard one-hole nozzle injector conforming to DIN 73372 standard [4]. The FIT is computer controlled and instrumented with a dynamic pressure sensor to acquire combustion pressure traces at a high sampling rate.

Combustion air was introduced into the chamber at 24 bar and heated to the desired temperature. Typical operating temperatures were in the range of 695–860 K. Biodiesel was injected into the quiescent oxidizer, leading to an autoignition event. A test comprises of 25 such injections. The fuel autoignites when injected, and the time from the start of injection to the beginning of combustion defines the ignition delay time. For the FIT, the onset of ignition, or Ignition Delay (ID), in millisecond, is defined as the time from the start of injection to a point where the chamber pressure is 0.2 bar above the initial charge pressure. The recorded IDs from the 25 injections were averaged, and then correlated to a DCN value [4, 5] using Equation 1:

$$DCN_{FIT} = \frac{171}{ID} \qquad 1$$

This method of determining the DCN is similar to the ASTM D7170 method [4]. Additionally, to enable comparison with other work, the peak in the rate of pressure rise was used to define an overall ignition delay ($\tau_{overall}$), as demonstrated in Fig. 4. A test set of 25 pressure traces resulting from the autoignition of the soybean biodiesel is shown in Fig. 4. The plot shows the individual pressure traces from the FIT experiment as well as the average pressure trace and a calculated average rate of pressure rise.

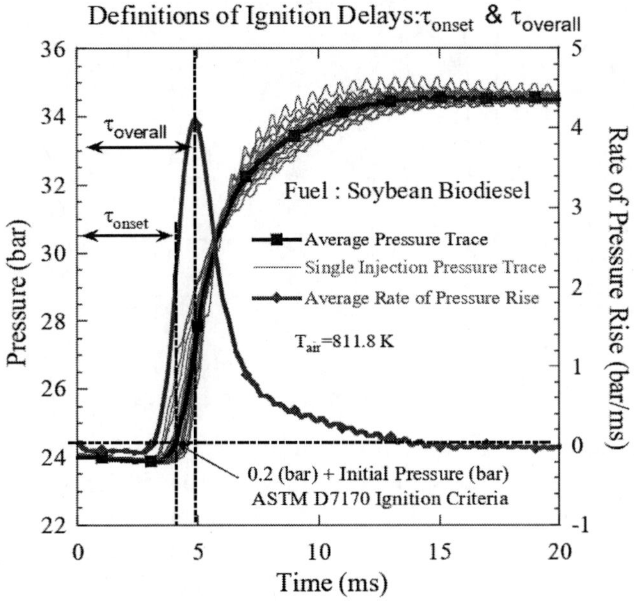

Fig. 4. Definitions of ignition delays based on absolute pressure rise and rate of pressure rise with the soybean biodiesel at an air temperature of 811.8 K.

As previously mentioned, Fig. 4 serves to illustrate the two ignition delay definitions used in this work. The first (τ_{onset}) is based on absolute pressure rise and is used to represent the ASTM ignition delay criteria of ignition. The second ($\tau_{overall}$) is based on the maximum rate of pressure rise and is used for ignition delay comparisons between other works and simulations. Note that the individual pressure traces for soybean derived biodiesel exhibit ringing soon after the occurrence of time corresponding to the highest rates of pressure rise $(dp/dt)_{max}$. As expected, the averaging process is unable to capture the oscillations but nonetheless is very useful for comparison purposes before this time.

3 Results and Discussion

The raw experimental pressure traces for the three fuels tested are shown in Fig. 5 along with the rates of pressure rise on similar scales. A comparison of the plots for the three fuels reveals that the post combustion pressure for soy, corn, and canola are quite similar indicating a similar energy density for the three biodiesels. An interesting feature that distinguishes the set of soy biodiesel pressure traces is the relatively less scatter in the post combustion pressure traces, reasons for which need to be further examined. Additionally, all the three fuels exhibit a slight oscillatory behavior (ringing) post-ignition and during the extended burnout phase ($> 8ms$) after the maximum rates of pressure rise have been achieved. The peak value for the rate of pressure rise is highest for soy followed by corn and canola derived biodiesel. Interestingly, this ranking also follows the amount of saturated methyl esters contained in the respective biodiesels.

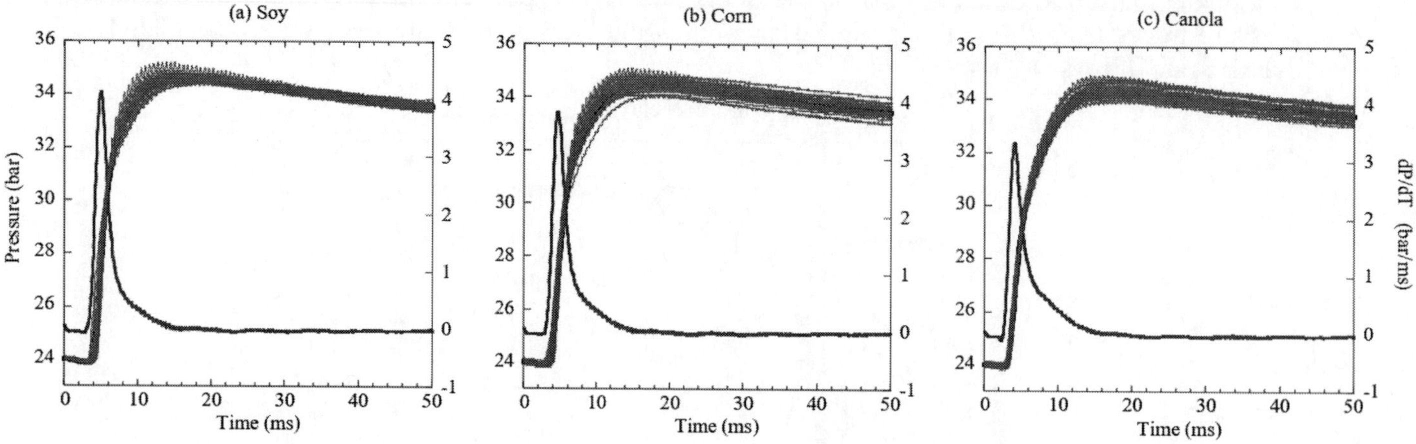

Fig. 5. Raw pressure traces and rate of pressure rise for the biodiesels at an air temperature of 811.8 K.

The averaged pressure traces for the three different biodiesels are shown in Fig. 6. All of the fuels exhibited a single stage ignition behavior. The three biodiesel fuels behave somewhat similarly but a few differences are also observed. The canola biodiesel ignites first followed by the corn and soy biodiesel. Interestingly, even though the canola biodiesel ignites first, the rate of pressure rise does not remain the highest; this results in the other biodiesels showing a crossover in chamber pressures at approximately 30.5 bar and higher. This feature may be attributable to the low temperature fuel chemistry and fatty acid constituents. The post combustion peak pressures are all very similar but show a slight difference; the difference is most likely due to the small change in the averaged injection duration between the tests.

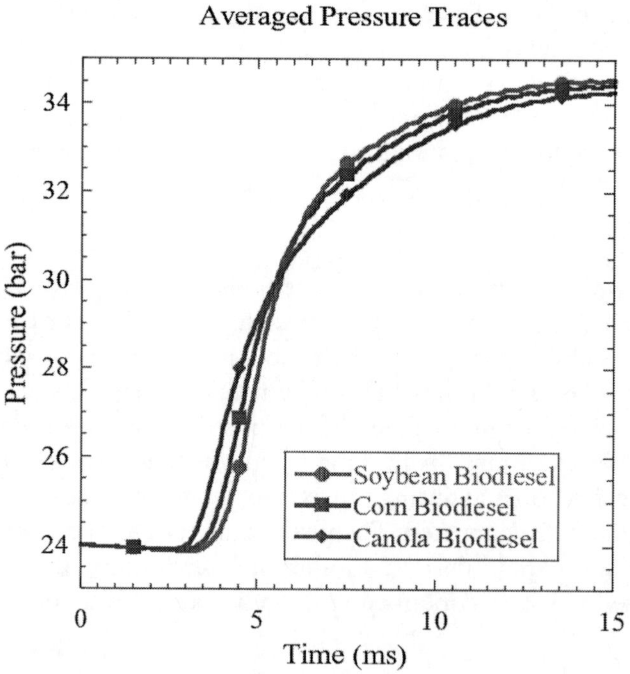

Fig. 6. Averaged pressure traces for soybean, corn, and canola biodiesel.

Figures 7 (a) and 7(b) show the overall ignition delay times and DCNs of the biodiesels, respectively. Note that the ignition delay times in Fig. 7(a) are the overall ignition delay times, not the onset of ignition delay times used for the DCN calculation. The calculated DCNs based on the 0.2 bar pressure rise criterion are shown in Fig. 7(b).

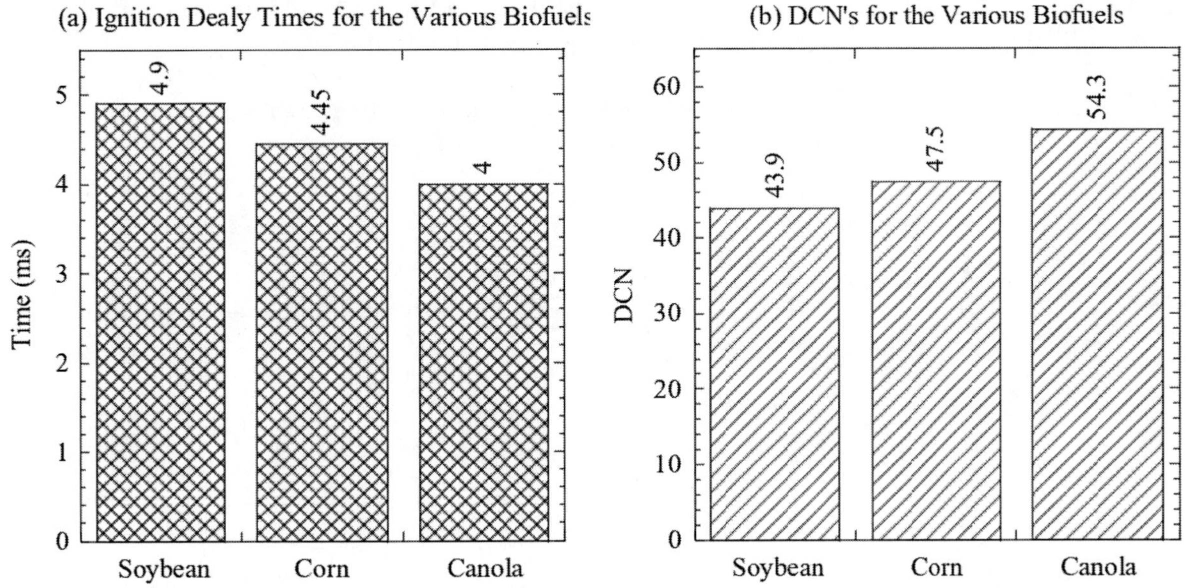

Fig. 7. (a) Overall ignition delay times of the soybean, corn, and canola biodiesel. (b) Derived cetane numbers of the soybean, corn, and canola biodiesel.

The DCN trend in Fig. 7(b) can be explained by examining the makeup of the three biodiesels shown in Table 1. Note that the cetane numbers for pure constituents were obtained from work of Yanowitz et al. [8] and represent the average for the ASTM D613 (CFR) values reported in the compendium. The fuel with the highest amount of unsaturation, in terms of the double bonds, exhibits the lowest DCN. Based on the composition profile in Table 1 the soy biodiesel has the highest C18:2 and C18:3 content, followed by corn and canola. It is the inhibiting effect of these multiple double bonds that leads to reduced reactivity and a lower DCN. Even if one were to include the C18:1 in the analysis, the ranking remains the same in terms of the weighted average of the number of 'double bonds' present in each biodiesel.

(a)

Component	CN	Percent Methyl Ester in Biodiesel		
		Canola	Corn	Soy
Methyl Palmitate (C16:0)	74.4	4.1	11.7	11.1
Methyl Stearate (C18:0)	81.3	1.8	1.5	4.2
Methyl Oleate (C18:1)	56.0	65.4	30.9	22.1
Methyl Linoleate (C18:2)	41.7	18.8	54.4	54.2
Methyl Linolenate (C18:3)	45.9	7.6	1.0	7.9

(b)

Experimental DCN (Current Work)	
Canola	54.3
Corn	47.5
Soy	43.9

***Table 1:** (a) Fatty acid profile and CN's for neat fuels [8] and (b) DCN for biodiesels tested.*

Finally, it is valuable to have an idea of the uncertainty associated with the estimation of DCN using the 25 values that comprise a test set for each fuel. Figure 8 provides a comparative overall view of the distribution and skew for the 25 data points used to derive the DCN for the three fuels. The median value is shown by the line within the box which extends from the 25th to the 75th percentile, and the whiskers extend to 1.5 times the interquartile distance from the closest quartile. The outliers have been represented as individual points in the figure.

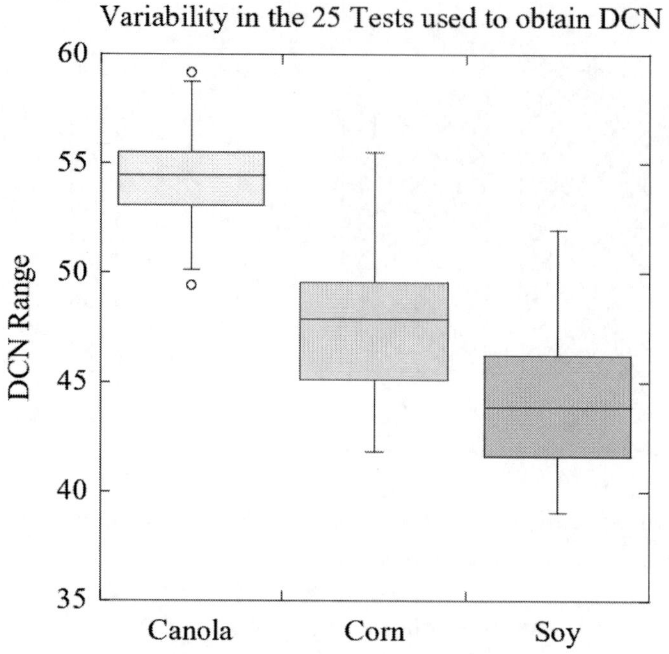

Fig. 8. (a) Variability in the 25 injections used to calculate the averaged DCN value for each fuel.

4 Conclusions

The autoignition characteristics of three biodiesels produced from three plant derived feedstock have been obtained in a constant volume combustion chamber at a fixed fuel injection duration and oxidizer temperature. The autoignition/DCN trend was found to be consistent with the amount and degree of unsaturated components present in a particular biodiesel. The biodiesel derived from canola was the most reactive followed by corn and soy. Similar post combustion pressure rises were observed for the three biodiesel. However, there was a discernible difference in the burn rate for the three fuels with soy-derived biodiesel exhibiting the highest peak value for the rate of pressure rise, followed by corn and canola.

5 Acknowledgements

S.S. and R.L. acknowledge the assistantship support from the Department of Mechanical Engineering and the College of Graduate Studies at the University of Idaho.

6 References

[1] C. Arcoumanis, C. Bae, R. Crookes, E. Kinoshita, The potential of di-methyl ether (DME) as an alternative fuel for compression-ignition engines: A review, Fuel 87 (7) (2008) 1014-1030.
[2] S. C. Sorenson, Dimethyl ether in diesel engines: Progress and perspectives, Journal of Engineering for Gas Turbines and Power-Transactions of the Asme 123 (3) (2001) 652-658.
[3] ASTM D613-08, Standard Test Method for Cetane Number of Diesel Fuel Oil, West Conshohocken, PA, 2008, www.astm.org
[4] ASTM D7170-09, Standard Test Method for Determination of Derived Cetane Number (DCN) of Diesel Fuel Oils—Fixed Range Injection Period, Constant Volume Combustion Chamber Method, West Conshohocken, PA, 2009, www.astm.org
[5] Waukesha Engine Dresser: Fuel Ignition Tester User Manual, Waukesha, Wisconsin 53188, 2006
[6] M. Balat, H. Balat, Progress in biodiesel processing, Appl. Energy 87 (6) (2010) 1815-1835.
[7] J. V. Gerpen, Biodiesel processing and production, Fuel Processing Technology 86 (10) (2005) 1097-1107.
[8] J. Yanowitz, M. A. Ratcliff, R. L. McCormick, J. D. Taylor, M. J. Murphy, Compendium of Experimental Cetane Numbers, NREL/TP-5400-61693, 2014.

Western States Section of the Combustion Institute - Fall 2017 Meeting
Hosted by the University of Wyoming
October 2-3, 2017

Missing experimental data and rate parameter inference for H_2+OH=H_2O+H

Tiernan A. Casey, Habib Najm

Combustion Research Facility, Sandia National Laboratories, Livermore CA 94551, USA

The reaction of OH with H_2 is a crucial chain-propagating step in the H_2-O_2 system thus making the specification of its rate, and its uncertainty, important for predicting the high-temperature combustion of hydrocarbons. In order to obtain an uncertain representation of this reaction rate in the absence of actual experimental data, we perform an inference procedure employing maximum entropy and approximate Bayesian computation methods to discover hypothetical data from a target shock-tube experiment designed to measure the reverse reaction rate. This method attempts to invert the fitting procedure from noisy measurement data to parameters, with associated uncertainty specifications, to arrive at candidate noisy data sets consistent with these reported parameters and their uncertainties. The uncertainty structure of the Arrhenius parameters is obtained by fitting each hypothetical data set in a Bayesian framework and pooling the resulting joint parameter posterior densities to arrive at a consensus density. We highlight the advantages of working with a data-centric representation of the experimental uncertainty with regards to model choice and consistency, and the ability for combining experimental evidence from multiple sources. Finally, we demonstrate the utility of knowledge of the joint Arrhenius parameter density for performing predictive modeling of combustion systems of interest.
Keywords: Arrhenius parameters, Uncertainty quantification, Bayesian inference, Maximum entropy

1. Introduction

The Arrhenius parameters of elementary reaction rate specifications are inherently uncertain due to the underlying uncertainty of the rate constant measurements employed in their determination. Strictly speaking these rate constants are not measurements, rather they are fitting parameters for noisy data in the context of a fitting model, typically a rate law. The effort to understand these data sources and their consequence on combustion modeling from an uncertainty quantification (UQ) perspective has gained increased recent attention[1]. Investigations of elementary reaction rates are often practically limited to specific temperature ranges, meaning that the global expression published in a chemical mechanism is typically a complicated aggregation of multiple individual expressions from different experiments covering different temperature ranges or other varied conditions. The effort of assigning uncertainty to the parameters of such complicated aggregated expressions then often results in the specification of somewhat qualitative 'recommended' uncertainties. From the point of view of UQ, an appropriate first-principles treatment of parametric uncertainty would involve knowledge of the underlying noisy data and the model error in the procedure used to generate these rate constants. This underlying data is typically unavailable or unreported in investigations of elementary reactions, with only the nominal fitting parameters of the data and perhaps some statistics on these parameters given as a measure of the uncertainty of the investigation. A Data-Free Inference (DFI) algorithm was proposed by Berry et al. [2] as a means to infer missing data sets subject to constraints, where a space of consistent data sets is explored in a Bayesian framework with a likelihood designed to enforce statistics on parameters that fit the data, resulting in a maximum entropy-based (MaxEnt [3]) solution for the joint posterior distribution of the data and parameters. This method has been demonstrated for model problems with a variety of constraints on processed data [4], applied to a single-step combustion chemistry [5] by comparing the reduced model to data generated from a detailed model, and applied to the chain-branching reaction $H+O_2 \longrightarrow OH+O$ by inferring the real data from a shock-tube experiment subject to the reported statistics (i.e. an error measure on the evaluated rates) [6]. In this work we apply the method to infer missing data from a shock-tube experiment measuring the rate of an important chain-propagating step in the H_2-O_2 system. This hypothetical data is then used to reveal the structure of the posterior PDF of the Arrhenius rate parameters that fit the data for any choice of chemical model, with samples from this distribution subsequently propagated through models of combustion systems to arrive at uncertain representations of quantities of interest (QoIs).

2. Target experiment

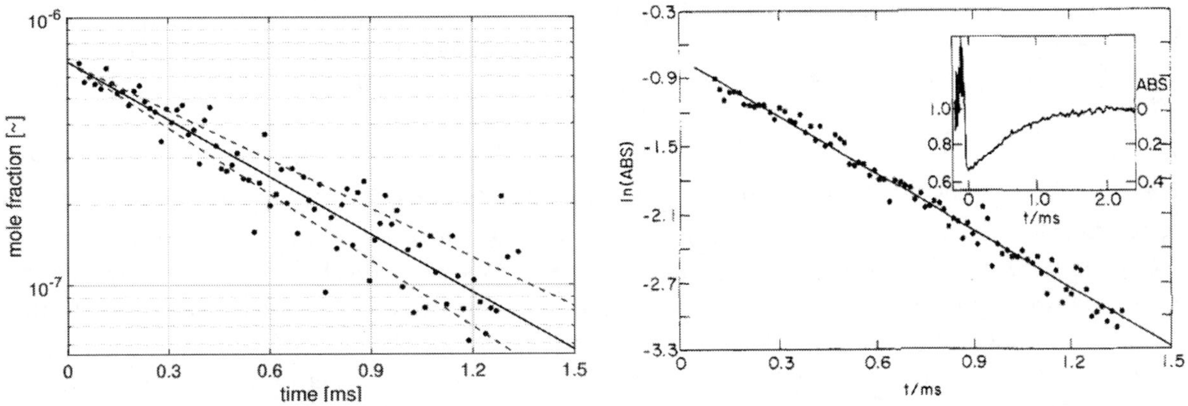

(a) (b)

Figure 1: (a) Noisy H mole fraction vs. time data inferred consistent with the reported constraints for a shock-tube experiment conducted at 1496K, (b) true noisy data reported by the experimentalists under the same conditions.

We investigate the uncertainty of the rate parameters of the reaction

$$H_2+OH \longrightarrow H_2O+H \quad (1)$$

in the context of the shock-tube experiments conducted by Michael & Sutherland [7] which were performed over the temperature range 1246-2297K and measured the rate of the reverse form of eq. 1. This experiment employed flash photolysis to produce H atoms from the photodissociation of H_2O in shocked Ar-H_2O mixtures. The subsequent reaction of H with the plentiful H_2O results in the production of H_2 and O_2. The decay of the H atom concentration is used as the target measured signal (see Fig. 1b) for constraining the rate constant. The experimentalists argue that since no other reactions of significance occur that remove H atoms, the decay is controlled by pseudo-first-order kinetics governed by the reverse of eq. 1 only. The shock-tube campaign encompassed 95 individual experiments spanning temperature, pressure, and H_2O concentration. In this work we choose an initial subset of 20 experiments (see Fig. 2a) to perform data inference.

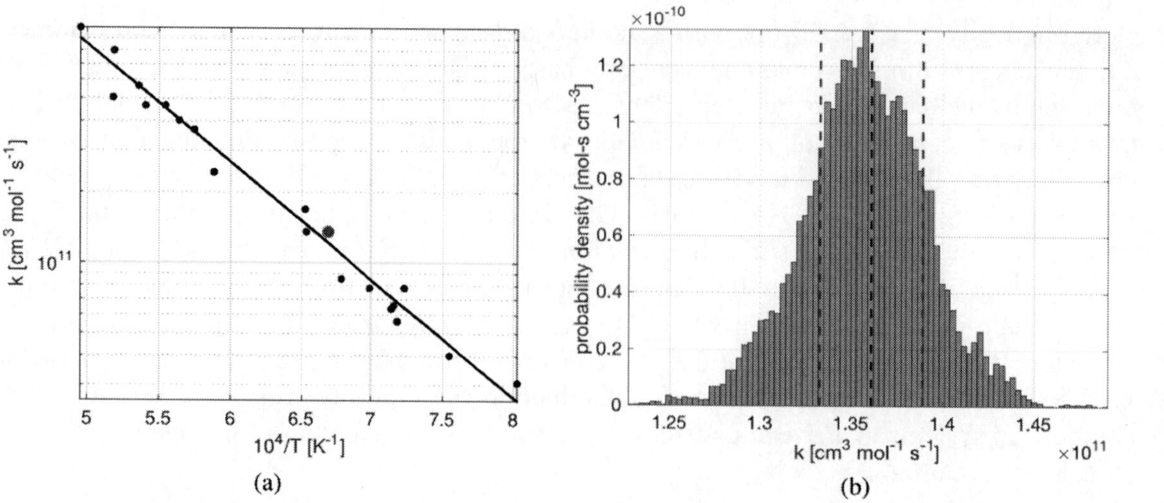

Figure 2: (a) 20 nominal rate constant values determined at various temperatures selected from the 95 reported by the target experiment and employed in the data inference in this work. The red data point corresponds to the data in Fig. 1a and the solid black line is the Arrhenius fit determined using the nominal Arrhenius parameters reported by the experimentalists. (b) Example of adherence to the reported constraint. The dashed lines represent the reported mean (black line) and symmetric error bars (red lines) reported in the experiment for this rate constant. The histogram is the result of a parameter inference constructed to test a hypothetical data set for consistency by comparing the mean and standard deviation of the inferred samples to the reported mean and error bars.

The experiment reports nominal values of the reverse rate constant for eq. 1 at the relevant experimental temperatures, with a summary statement of the uncertainty in determining the H atom decay rates assigning error of 1-3% at the one standard deviation level. We choose a conservative interpretation of this error statement and assign a symmetric error measure of 3% of the nominal value as a standard deviation for observations of the reported rates.

For the purposes of performing parameter estimation using Bayesian inference we assume an experimental data noise model in the form of standard normally distributed additive noise, ε, with a constant nominal strength σ:

$$[\mathrm{H_{data}}]_\mathrm{i} = [\mathrm{H_{model}}]_\mathrm{i} + \sigma\varepsilon_i \tag{2}$$

where the index $\{\mathrm{i}\}$ spans the time series data points that sample the H atom decay.

3. Bayesian data and parameter inference

The inference algorithm proceeds as follows. An exploration of the space of consistent data sets is constructed in a Bayesian framework (eq. 3) which propose hypothetical noisy data, **D** (see Fig. 1a) for a single specific experimental condition (i.e. those in Fig. 2a), and measures its consistency (see Fig. 2b) with the experimentally reported statistics, S,

$$p(\mathbf{D}|S) = \frac{p(S|\mathbf{D})\pi(\mathbf{D})}{p(S)} \tag{3}$$

by employing a likelihood function, $p(S|\mathbf{D})$, constructed using Gaussian error kernels (eq. 4) which penalizes proposed data whose statistics ($S_\mathbf{D}$) do not match those of the experiment

($S_{\text{experiment}}$). This application of approximate Bayesian computation (ABC)[8] is useful in our context where there is no clear means to specify an exact likelihood function.

$$p(S|\mathbf{D}) = \prod_{i=1}^{2} \exp\left[-\delta_i \left(\frac{S_{i,\mathbf{D}} - S_{i,\text{experiment}}}{S_{i,\text{experiment}}}\right)\right] \tag{4}$$

The δ_i values are chosen heuristically to enforce a desired degree of consistency for each statistic. For example, a unique δ_i value can be chosen which will assign a desired threshold acceptance probability, e.g. 20%, to inferred statistics that depart from the reported statistics by some arbitrary level, e.g. 10%. Lower values of δ_i will result in more data being accepted as consistent in the DFI data chain. Practically, these parameters are chosen to deliver reasonable data acceptance rates with as little inconsistency as possible, and can be tuned as required. The approximate likelihood requires the determination of statistics on the uncertain rate constant of the fit model for the proposed data. These statistics are determined from an inner Bayesian parameter inference (eq. 5) which infers the rate constant at the relevant experimental temperature (T_{exp}) and data noise strength as defined in eq. 2,

$$p(\lambda|\mathbf{D}) = \frac{p(\mathbf{D}|\lambda)\pi(\lambda)}{p(\mathbf{D})} \tag{5}$$

with $\lambda := \left\{\mathrm{k}\left(\mathrm{T}_{\mathrm{exp}}\right), \sigma\right\}$. Equation 6 shows the parameter likelihood construction:

$$p(\mathbf{D}|\lambda) = \frac{1}{\sqrt{2\pi}\sigma} \exp\left(-\frac{\left([\mathrm{H}_{\mathrm{data}}]_i - [\mathrm{H}_{\mathrm{model}}]_i\right)^2}{2\sigma^2}\right) \tag{6}$$

The data inference algorithm thus results in a set of nested inferences, which are performed using Markov Chain Monte Carlo (MCMC) techniques employing single-site Metropolis sampling for the outer data chain and adaptive Metropolis sampling for the inner parameter chain. For simplicity and computational efficiency in this work the data vector $\mathbf{D}$ for each experiment is limited to 4 time points along the decay. $\pi(\mathbf{D})$ and $\pi(\lambda)$ are prior PDFs on the data and parameters respectively.

Figure 1a shows an example of an inferred hypothetical data set for a single experimental condition using a large number of data points (82 in total) for demonstration purposes, with Fig. 1b being the corresponding true data set fortuitously reported in the experiment as example data, also using 82 points. It is notable that the hypothetical inferred data appears to have a considerably higher noise amplitude than the real data, despite being selected as consistent with the reported uncertainty. This can be attributed to the fact that the inference approach in this work considers the reported uncertainty to be associated with data noise alone. In reality, uncertainty exists in the initial conditions of the experiment, e.g. in initial temperature and species concentrations. In particular, the uncertainty in the post-shock temperature in the shock-tube is estimated by the experimentalists to be on the order 1-2% of the reported nominal temperature. The solid black line in Fig. 1a represents the noiseless H atom decay curve computed using the experimental best-fit Arrhenius parameters at the nominal experimental temperature, while the dashed red lines are computed using the same Arrhenius parameters but using temperatures $\pm 2\%$ of the nominal value (1494K in this case). It is evident that uncertainty in initial temperature alone is sufficient to span the spread in the noisy hypothetical data and that an inference treatment which accounted for this temperature uncertainty should result in an inferred noise amplitude more comparable to the true data. While accounting for the different sources of uncertainty (both in fitting and in initial conditions) has implications on the form of the inferred hypothetical noisy data, it is not clear what the effect would be on the subsequent combined learning and resulting consensus pooled PDF (see section 4). This is a direction of current work.

4. Combined learning

Once the space of consistent data has been explored - where a typical exploration of the data space in this work results in a data chain with on the order of 1000 samples - for each experimental data point (i.e. each of the selected 20 $k(T_{exp})$ experiments considered) we proceed to construct an uncertain representation of model parameters that fit all the data at each temperature simultaneously. We choose to use a classical Arrhenius fit for simplicity, and construct an inference for the log of the pre-exponential factor, logA (to enforce positivity of the pre-exponential A), and activation temperature, E, for the inferred data across all the experiments. A combined learning approach is taken which performs a parameter inference using data samples from each experiment's data chain simultaneously as evidence. For example, if each individual data inference explored a 4 dimensional data space (i.e. 4 data points in time series along the H decay) and discovered 1000 consistent data sets, and we perform data inference for 20 experiments, then the combined learning approach constructs 1000 super data-set instances of dimension 80. Each of these 1000 super data-sets are used as evidence for parameter inference of the classical Arrhenius model and informs a single posterior parameter PDF. A consensus pooled posterior PDF across all 1000 consistent super data-sets is then calculated by pooling the 1000 posteriors using straight averaging, implicitly assigning an equal weight (or belief) to the posterior densities of each the 1000 super data-sets. The process of assembling the super data-sets can be naturally extended to include the data from more experimental points as evidence, or even data from completely different experiments including real experimental data if available.

An important point to note when constructing the parameter inferences given the consistent hypothetical data is that we are in no way constrained to employing the chemical modeling approach and assumptions used by the experimentalists in their original fitting of their true experimental data. In section 2 we noted that the experimentalists argued that under their conditions eq. 1 was the only active reaction. Seeing as we have full access to a stochastic representation of the true data we are free to abandon this assumption and use any arbitrary chemical model of our choosing to fit the data. For this work we choose to use a full H_2-O_2 chemical mechanism, in particular that of Li et al. [9], chosen subjectively as it is used extensively in simulations of the H_2-O_2 system, with a view of using this model in conjunction with the uncertain representation of the target reaction in later forward modeling of combustion systems. Using this approach, we learn about the parameters of the target reaction of interest in the presence of the other H_2-O_2 reactions which may or may not be active. This approach maximizes the consistency between the uncertain representation of the target reaction, which is learned from data, and the full chemical model we will eventually employ when performing forward modeling.

The parameter inference for each super data-set returns samples from the posterior PDF, which we in turn use to generate an approximation of the PDF function using a kernel density estimation (KDE) approach [10]. The KDE approximation of each posterior is generated on the same grid in logA-E space for each super-data set, enabling a simple point-wise averaging operation to deliver

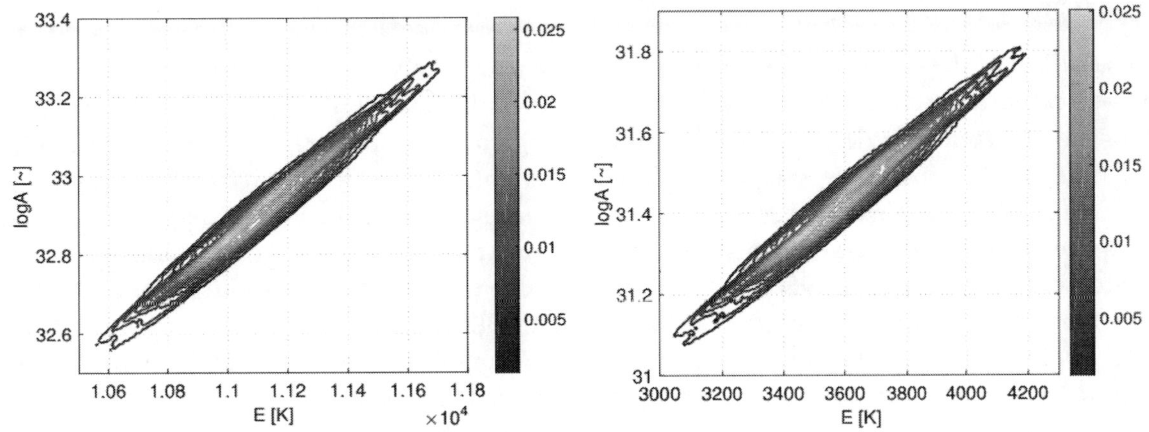

(a) (b)

Figure 3: (a) Pooled parameter posterior PDF resulting from combined learning of the Arrhenius parameters of the reverse reaction, $H_2O+H \longrightarrow H_2+OH$ across hypothetical data for the 20 experiments. (b) Pooled parameter posterior PDF resulting from combined learning of the Arrhenius parameters of $H_2+OH \longrightarrow H_2O+H$ in conjunction with an assumed functional form of the equilibrium constant.

the consensus pooled posterior across the data samples. The result of parameter inference using the Arrhenius model to determine the parameters of the reverse rate of reaction eq. 1 and subsequently pooling (averaging) over the posteriors PDFs of the reverse rate is shown in Fig. 3a, indicating a highly correlated consensus density. This PDF is our summary statement of Arrhenius parameter uncertainty of the reverse reaction given the reported experimental constraints, and the model we choose to re-interpret the hypothetical data. However we are typically interested in the parameters of the forward rate as this is the form of the reaction that appears in many chemical mechanisms. Using our representation of the experimental uncertainty with hypothetical data, it is straightforward to determine the uncertainty in the forward rate by simply changing the fit model for the data. Rather than modeling the rate constant using the reverse rate explicitly, we replace this expression with the ratio of the forward rate to the equilibrium constant of the reaction. Invoking an expression for the equilibrium constant (eq. 7),

$$\mathrm{K} = 4.38\exp\left(-\frac{7513}{T}\right) \tag{7}$$

which is the same expression used by the experimentalists to map their nominal Arrhenius fitting of their data to determine the forward rate of reaction, we can construct an inference for the forward rate parameters directly. The expression in eq. 7 was determined from a least squares approximation to JANAF thermochemistry data [11]. The result of the inference of the forward rate parameters by invoking this functional form of the equilibration constant with subsequent pooling over the posteriors PDFs shown in Fig. 3b, similarly indicating a highly correlated consensus density. Using either approach demonstrates the versatility and endless number of possible strategies that can be employed when operating in a framework based on uncertainty expressed in the form of data.

5. Application

The utility of the representation of the parametric uncertainty of the Arrhenius fit model in the form of a posterior probability density function, in particular having information about the correlation of the parameters, is demonstrated by propagating uncertainty through homogeneous ignition delay time and laminar flame speed calculations. As noted in the sensitivity analysis performed by Li et al. [9], flame speed is particularly sensitive to the reaction under consideration. Ignition delay time calculations are performed using the TCHEM software [12] for stoichiometric H_2-air mixtures at initial temperature 950K, while laminar flame speed calculations utilize the CANTERA steady flame solver for H_2-air mixtures at 1 and 0.1 atm, 300K, and equivalence ratios varying from 0.5-4.5. For the ignition delay time calculations 10,000 samples are drawn from the pooled parameter posterior for the forward rate Arrhenius parameters and propagated through a homogeneous reactor calculation to determine the time for the gas temperature to rise to 1600K. For comparison, 10,000 samples of the parameters logA and E are drawn separately from the univariate marginal distributions for each parameter, i.e. ignoring any correlation between the parameters, to emulate a case where the correlation information is unavailable. Figure 4a shows the comparison between propagating samples from the correlated PDF computed from the DFI data vs. the uncorrelated case, with the latter predicting significantly larger variance in the ignition time. Fig. 4b shows 95% confidence intervals of the laminar flame speed across the interval of equivalence ratio and at two pressures, again indicating the over-prediction of the flame speed uncertainty when one does not have access to the correlation between the parameters.

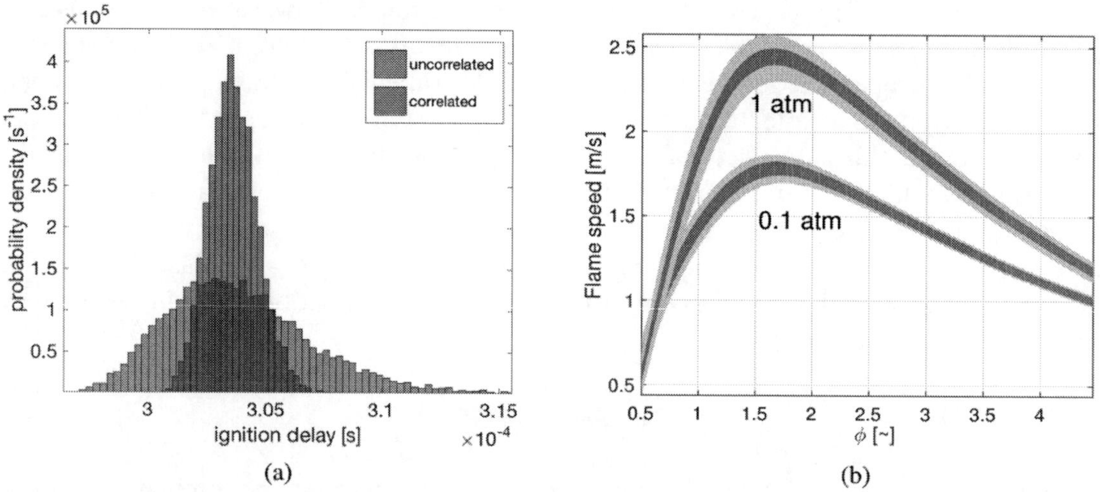

Figure 4: (a) Homogeneous ignition delay and (b) laminar flame speed calculations performed using the correlated DFI pooled posterior parameter density compared to sampling from an uncorrelated joint distribution with the same marginal parameter densities. The light and dark gray regions shaded in (b) indicate 95% confidence intervals of flame speed for uncorrelated and correlated sampling respectively.

6. Conclusion

We present the results of a data inference procedure designed to access missing data from a target shock-tube experiment as a means to construct a representation of the uncertain Arrhenius parameters that fit the true data. The data inference procedure employs a Bayesian framework to explore data space to discover hypothetical noisy data sets consistent with reported constraints in the form of statistics on rate constant values for each experimental condition considered. Having access to a representation of the experimental data allows for a reinterpretation of the Arrhenius parameter fitting procedure by enabling the use of any chemical mechanism of our choosing as a fit model for the noisy data. As such, our representation of the parameter uncertainty is not tied explicitly to the context of the experiment and the fitting strategy and assumptions made by the experimentalists, and naturally accommodates any consistent modeling approach. Uncertain representations of the Arrhenius parameters of both the forward and reverse rates of the target reaction are constructed in the form of posterior probability density functions using combined learning and pooling, highlighting the versatility of our data inference approach. Finally, the uncertain representation of the forward rate is employed to propagate uncertainty through forward models of ignition delay time and laminar flame speed to highlight the necessity of having access to the correlated parameter PDF for making useful predictions of QoIs in combustion systems.

7. Acknowledgements

This work was supported by the US Department of Energy (DOE), Office of Basic Energy Sciences (BES) Division of Chemical Sciences, Geosciences, and Biosciences. Sandia National Laboratories is a multimission laboratory managed and operated by National Technology and Engineering Solutions of Sandia, LLC., a wholly owned subsidiary of Honeywell International, Inc., for the U.S. Department of Energy's National Nuclear Security Administration under contract DE-NA-0003525.

References

[1] H. Wang, D. A. Sheen, Progress in Energy and Combustion Science 47 (2015) 1–31.

[2] R. Berry, H. Najm, B. Debusschere, H. Adalsteinsson, Y. Marzouk, Journal of Computational Physics 231 (2012) 2180–2198.

[3] E. Jaynes, Probability Theory: The Logic of Science, G.L. Bretthorst, Ed, Cambridge University Press, Cambridge, UK, 2003.

[4] K. Chowdhary, H. N. Najm, Statistics and Computing 26 (2016) 149–169.

[5] H. N. Najm, R. D. Berry, C. Safta, K. Sargsyan, B. J. Debusschere, Int. J. Uncertainty Quantification 4 (2014) 111–132.

[6] M. Khalil, K. Chowdhary, C. Safta, K. Sargsyan, H. Najm, Proceedings of the Combustion Institute 36 (2017) 699 – 708.

[7] V. Michael, J. W. Sutherland, J. Phys. Chem 92 (1988) 3853–3857.

[8] M. A. Beaumont, W. Zhang, D. J. Balding, Genetics 162 (2002) 2025–2035.

[9] J. Li, Z. Zhao, A. Kazakov, F. L. Dryer, International journal of chemical kinetics 36 (2004) 566–575.

[10] Z. I. Botev, J. F. Grotowski, D. P. Kroese, Annals of Statistics 38 (2010) 2916–2957.

[11] M. Chase Jr, J. Curnutt, J. Downey Jr, R. McDonald, A. Syverud, E. Valenzuela, Journal of Physical and Chemical Reference Data 11 (1982) 695–940.

[12] C. Safta, H. N. Najm, O. Knio, Sandia Report, SAND2011-3282 (2011).

Western States Section of the Combustion Institute – Fall 2017 Meeting
Hosted by the University of Wyoming
October 2-3, 2017

Probing the Pyrolysis Chemistry of Ethyl Propionate in a Microreactor

Cory Rogers,[1] *Jessica Porterfield,*[2,3] *John Daily,*[1] *Barney Ellison,*[2] *Nicole Labbe*[1,*]

[1]*Department of Mechanical Engineering, University of Colorado Boulder, Boulder, CO 80309*
[2]*Department of Chemistry and Biochemistry, University of Colorado Boulder, Boulder, CO 80309*
[3]*Harvard-Smithsonian Center for Astrophysics, Cambridge, MA 02138*
[*]*Corresponding Author: Nicole Labbe (nicole.labbe@colorado.edu)*

Methyl and ethyl esters are among the primary components of biodiesel, however, the pyrolysis and oxidation pathways of these large esters are not well understood. In this study, a prototypical ethyl ester, ethyl propionate ($CH_3CH_2C(=O)OCH_2CH_3$), was pyrolyzed in a hot microreactor under dilute conditions to elucidate the unimolecular decomposition routes of this biofuel. The microreactor wall temperature was varied over 300-1500 K and the pressure was ~20 Torr. The products were analyzed using photoionization mass spectrometry (PIMS) and Fourier transform infrared spectroscopy (FTIR). The resulting spectra suggest the primary pyrolysis pathway for ethyl propionate is the 6-centered pericyclic reaction to form propionic acid ($CH_3CH_2C(=O)OH$) and ethylene (C_2H_4). Additional products observed suggest the possibility of minor secondary unimolecular pathways, which are further discussed in this work. The pyrolytic experiments were repeated for propionic acid as it is the primary product from ethyl propionate. The primary decomposition pathway for propionic acid is the formation of methyl ketene (CH_3CHCO) and water. Other possible pathways include the loss of hydrogen from the CH_2 group to form the $CH_3CHC(=O)OH$ radical. The results of this work are being used to directly inform the development of high accuracy methyl and ethyl ester combustion reaction models.

1. Introduction

With decreasing availability of fossil fuels and increasing concern with climate change, there is heavy interest in renewable energy. Biofuels are promising renewable fuels and come in many forms including biodiesel. Biodiesel is attractive because it can be used in current diesel engines and can be made from vegetable oils, animal fats, and biological membranes. Biodiesel contains long-chain methyl and ethyl esters, however, modeling the pyrolysis and combustion of these large ester molecules has proven to be difficult. Therefore, it is crucial to understand the combustion and pyrolysis of smaller esters with similar structure and functional groups to inform models of more practical biodiesel components. This paper is a study of the pyrolysis of a simple ethyl ester, ethyl propionate ($CH_3CH_2C(=O)OCH_3CH_2$), in a heated tubular microreactor to elucidate its thermal decomposition pathways.

Ethyl propionate pyrolysis was first studied experimentally in a toluene carrier gas system by Blades et al. [1] which yielded the Arrhenius equation $k = 5.2 \times 10^{12} \exp(-48500 \pm 350 \text{ cal/mol} / RT)$ s^{-1} for the overall decomposition rate of ethyl propionate. Ethyl propionate pyrolysis was then studied in two single-pulse shock tubes by Barnard et al. [2] The authors agreed with the previously reported Arrhenius equation [1] and concluded that the only significant pyrolytic reaction of ethyl propionate is the cyclic reaction to form ethylene (C_2H_4) and propionic acid ($CH_3CH_2C(=O)OH$).

Schwartz et al. [3] studied ethyl propionate in atmospheric coflowing methane/air non-premixed flame experiments and measured the increase in intermediate hydrocarbon concentrations using electron impact mass spectrometry. Schwartz et al. [3] agreed with the earlier conclusion [2] that the formation of ethylene and propionic acid is the dominant pyrolytic pathway. Also, the authors concluded that propionic acid decomposes to form predominantly methyl ketene (CH_3CHCO) and water.

Metcalfe et al. [4, 5] studied ethyl propionate oxidation in a jet stirred reactor and in a shock tube. The authors found that ethyl propionate had significantly faster ignition delay times than other $C_5H_{10}O_2$ isomers. Metcalfe et al. [4,5] attributed this to the low energy pathway to propionic acid and ethylene that methyl esters do not have. The reaction itself is not responsible for the ignition event per se, but instead ignition was sensitive to the faster reactivity of propionic acid and ethylene compared to the other isomers. These data were used to develop a kinetic model of ethyl propionate.

Walton et al. [6] studied ethyl propionate and methyl butanoate in a rapid compression machine. The authors agree with the earlier studies [4,5] that ethyl propionate showed faster ignition delay times. They contributed this to faster kinetics of the route to propionic acid and ethylene compared to slower H abstraction which dominates methyl butanoate decomposition. This work also refined a kinetic model of both fuels.

Yang et al. [7] measured the composition of reaction intermediates of ethyl propionate oxidation in low-pressure premixed flat flames. They concluded that only 10-15% of ethyl propionate goes through the propionic acid and ethylene route. The vast majority of the fuel, in combustion, is consumed via H abstraction. This is contrary to Metcalfe et al. [4, 5] who attributed upwards of 50% of fuel consumption to the propionic acid and ethylene route in a jet-stirred reactor. The authors then used this data, along with the Metcalfe et al. [4,5] and Walton et al. [6] models, to validate an improved ethyl propionate combustion model.

Most recently, Farooq et al. [8] studied ethyl propionate pyrolysis in a shock tube and confirmed that nearly all ethyl propionate dissociated to propionic acid and ethylene immediately. The authors concluded that the most important pyrolysis reactions are the H abstractions from propionic acid and the thermal decomposition of ethylene.

What is clear from the previous work on ethyl propionate is that the thermal decomposition pathways, especially the minor routes outside of the main pathway to propionic acid, remain unclear. There also is uncertainty as to the kinetics of propionic acid, which has been suggested by some of the above mentioned studies to be the dominant product of ethyl propionate combustion, and therefore is important to characterize for accurate combustion models This study attempts to address these two points.

2. Methods

The pyrolysis of ethyl propionate was studied in a tubular microreactor using photoionization mass spectrometry (PIMS) and matrix Fourier transform infrared spectroscopy (FTIR). The microreactor (shown in Fig. 1) is a SiC tube with a 1 mm inner diameter and roughly 3 cm length. SiC was chosen due to its inert chemical behavior and high melting point. The microreactor is resistively heated to fixed temperatures ranging 300 K – 1600 K. SiC is a great thermal conductor and the microreactor is small such that the centerline temperature of the gas is estimated

Figure 1. Tubular microreactor compared for scale to linguini and a quarter.

to be roughly 200 K less than the wall temperatures reported in this study. Ethyl propionate is diluted in a carrier gas (He or Ar) to roughly 0.1% to ensure negligible effects of bimolecular reactions. The mixture is then sent through the heated microreactor using a Parker General valve at 10 Hz. The system at CU Boulder is pulsed due to pumping capacity restrictions, which complicates the pressure and velocity profiles in the microreactor. The pressure inside is roughly 20 Torr and a more detailed CFD analysis on the microreactor flows was performed previously by Guan et al. [9] Products exiting the microreactor are detected using PIMS and FTIR.

In the PIMS setup, the microreactor chamber is held at 10^{-6} Torr using a Varian VHS-6 diffusion pump. The products pass through a skimmer to create a molecular beam in the ionization chamber. The ionization chamber is held at 10^{-7} Torr with a Pfeiffer TMU-521 turbomolecular pump. The molecular beam is intersected and ionized by the 9th harmonic of an Nd:YAG laser (118.2 nm, 10.487 eV) pulsed at 10 Hz. The resulting ions are detected using a reflectron time of flight mass spectrometer (TOFMS) from Jordan Time of Flight.

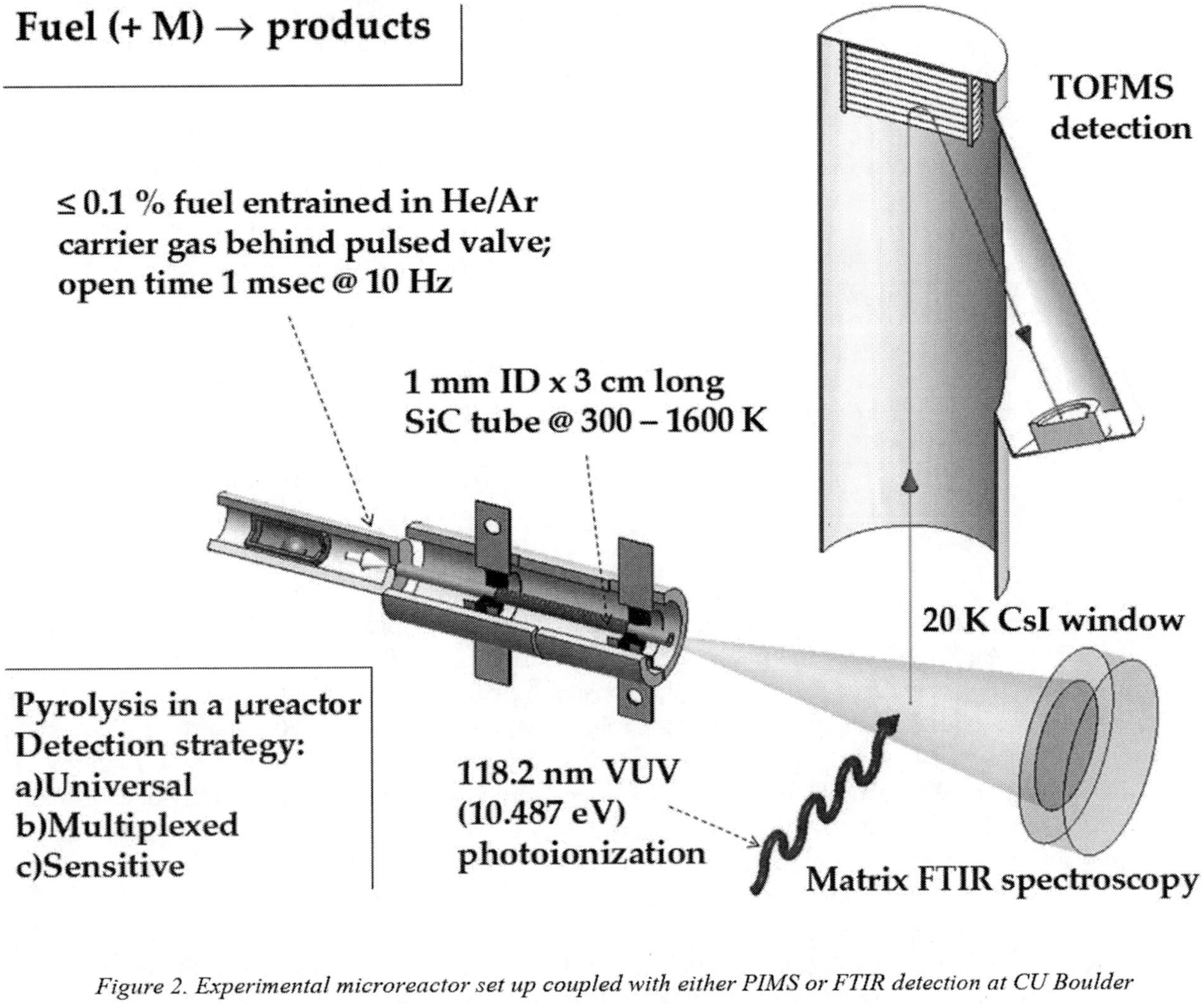

Figure 2. Experimental microreactor set up coupled with either PIMS or FTIR detection at CU Boulder

In the IR setup, the pyrolyzed products from the microreactor and the carrier gas (Ar) are deposited on a CsI window at 20 K. The window chamber is held at 10^{-6} Torr by an Agilent TV 81 M turbomolecular pump. The low pressure and temperature traps products, including radicals, in an Ar matrix. The infrared spectra of the products are taken by a Nicolet 6700 FT-IR. This provides complimentary data to the PIMS to fully show the pyrolysis pathways.

3. Results and Discussion

Table 1. Reaction pathways

Reactants	Label	Products
$CH_3CH_2C(=O)OCH_2CH_3 + (M) \rightleftharpoons$	R1	$C_2H_4 + CH_3CH_2C(=O)OH$
	R2	$H + CH_2CH_2C(=O)OCH_2CH_3$
	R3	$H + CH_3CHC(=O)OCH_2CH_3$
	R4	$H + CH_3CH_2C(=O)OCHCH_3$
	R5	$H + CH_3CH_2C(=O)OCH_2CH_2$
	R6	$CH_3 + CH_2C(=O)OCH_2CH_3$
	R7	$CH_3CH_2 + C(=O)OCH_2CH_3$
	R8	$CH_3CH_2C{=}O + OCH_2CH_3$
	R9	$CH_3CH_2C(=O)O + CH_2CH_3$
	R10	$CH_3CH_2C(=O)OCH_2 + CH_3$
	R11	$CH_3CH{=}C(OH)OCH_2CH_3$
	R12	$CH_3CHO + CH_3CH_2CHO$
	R13	$CH_3CHCO + CH_3CH_2OH$
	R14	$C_2H_4 + HCOOCH_2CH_3$
$CH_3CH_2C(=O)OH + (M) \rightleftharpoons$	R15	$CH_3CHCO + H_2O$
	R16	$H + CH_3CHC(=O)OH$
$CH_3CHCO + (M) \rightleftharpoons$	R17	$2H + CH_2{=}C{=}C{=}O$
$CH_2{=}C{=}C{=}O + (M) \rightleftharpoons$	R18	$CO + HCCH$
$CH_3CHC(=O)OH + (M) \rightleftharpoons$	R19	$CH_3CHCO + OH$
$CH_3CH{=}C(OH)OCH_2CH_3 + (M) \rightleftharpoons$	R20	$H_2O + CH_3CHCO + C_2H_4$

Fig. 3 shows the PIMS spectra when 0.13% ethyl propionate in He is heated in the microreactor. At 300 K, only the parent peak is present at m/z 102. As the microreactor heats to 900 K many more peaks emerge. First we will discuss the peaks that survive to 1300 K. The peak at m/z 74 is propionic acid. R1 is agreed upon as the major pyrolytic pathway in the literature. Thus, the peak at m/z 28 must be ethylene. While we expect to see propionic acid as the ionization energy (IE) is 10.44 eV [10], the 9th harmonic of a Nd:YAG laser should not be able to ionize ethylene with an IE of 10.51 eV[11]. The peak at m/z 28 must be chemically activated ethylene. A more in depth discussion on this can be seen elsewhere [12]. The peak at m/z 56 is methyl ketene (IE 8.95 eV [13]). Both R13 and R15 produce methyl ketene. R13 is unlikely because the barrier to reaction is higher than other thermal decomposition routes that are not seen [14]. Also, R13 should produce ethanol (CH_3CH_2OH) but there is no peak at m/z 46 and ethanol has an IE of 10.48 [15] which is just below the energy of the laser. The peak at m/z 73 is likely a radical of propionic acid. $CH_3CHC(=O)OH$ radical is most probable because the CH_2 group has the weakest C-H bond of the three possibilities in propionic acid. This radical would then perform β-scission, R19, to produce more methyl ketene and OH radicals.

Note that at 1300 K, the parent peak at m/z 102 is gone so all ethyl propionate (IE 10 eV [15]) is consumed.

The picture at 900 K is less clear. Peaks at m/z 29, 43, 45, 57, 75, 77, and 84 are present that are not at 1300 K. We are unable to conclusively say what the source of these peaks is without further experimentation. One possibility is photodissociation of the keto-enol tautomer of ethyl propionate. The keto-enol tautomer is produced via R11. It has been shown [16-18] that the enol IE is roughly 1 eV lower than the keto tautomer. This would put the IE of the keto-enol tautomer of ethyl propionate at 9 eV. With an IE almost 1.5 eV lower than the laser, photodissociation is probable [19]. The excess energy would fragment the tautomer. Possible assignments are $CH_3CH{=}C{=}O{-}CH_2CH_3$ (m/z 84), $CH_3CH_2C(OH)_2$ (m/z 75), CH_3CH_2CO (m/z 57), $CH_3CH{=}OH$ (m/z 45), CH_3CO (m/z 43), and HCO (m/z 29). The peak at m/z 77 is non-authentic and may be due to an impurity in the experiment.

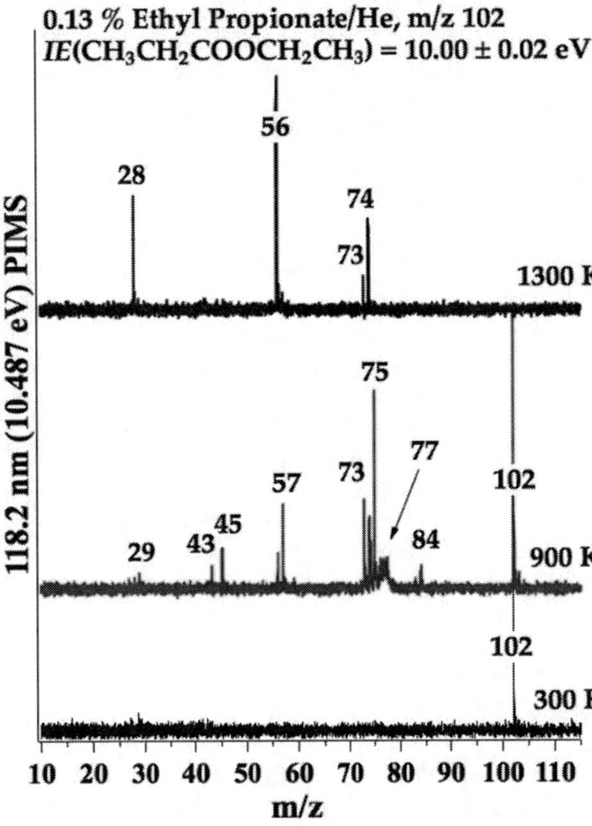

Figure 3. PIMS data of ethyl propionate in He at 300 K, 900 K, and 1300 K

Thermodynamically, R1 and R12 are favored over the formation of the tautomer, R11 [14]. R1 is confirmed but there is no evidence in the PIMS spectra of the products of R12, acetaldehyde (CH_3CHO, m/z 44, IE 10.23 eV [20]) or propanal (CH_3CH_2CHO, m/z 58, IE 9.96 eV [21]). Without seeing the thermodynamically favored R12 reaction it is difficult to be certain that R11 is occurring without direct evidence. Due to this uncertainty, our assignment of the m/z peaks at 29, 43, 45, 57, 75, and 77 to keto tautomer fragmentation is inconclusive at this time.

Notable species that are not seen are methyl radical (CH_3, m/z 15, IE 9.84 eV [22]) and any radical species of ethyl propionate losing a hydrogen at m/z 101. This would be expected as the microreactor experiment probes unimolecular reactions and the barrier to reaction to R1 is 67 kJ/mol lower than the next unimolecular route [14]. Furthermore, a suspected unimolecular pathway of propionic acid produces methyl radical and further study is propionic acid is needed.

FTIR was used to confirm the conclusions made with the PIMS data. Fig. 5-7 show the vibrational signatures of propionic acid, methyl ketene, ethylene, and acetylene (HCCH, m/z 26, IE 11.4 eV [23]). Acetylene is produced, along with carbon monoxide and H radicals, by the thermal decomposition of methyl ketene as seen in R17 and R18. These species were not detected in the PIMS spectra because the IE's are too high. IR spectra for carbon monoxide and carbon dioxide are also included as they cannot be seen in the PIMS but are thermal decomposition products.

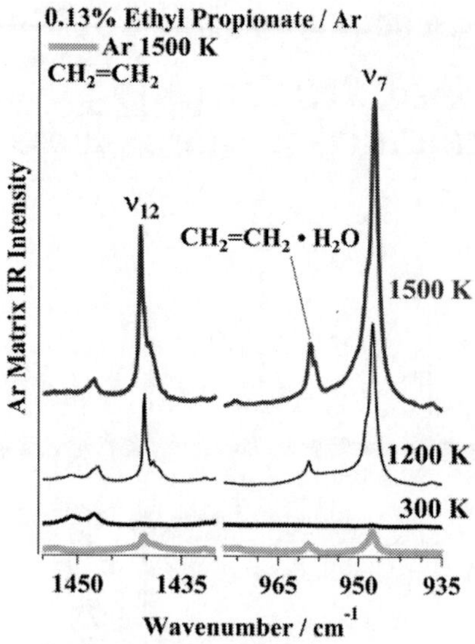

Figure 5. Matrix infrared spectra of ethyl propionate in Ar confirming ethylene as a product. The green trace is a scan of Ar heated to 1400 K.

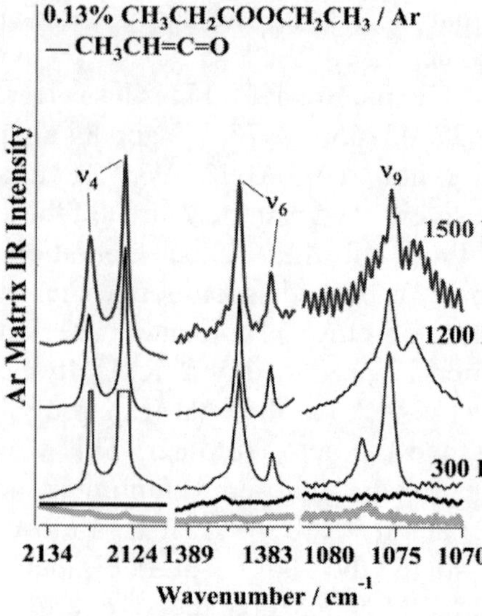

Figure 4. Matrix infrared spectra of ethyl propionate in Ar confirming methyl ketene as a product. The green trace is a scan of Ar heated to 1400 K. The blue trace is authentic methyl ketene.

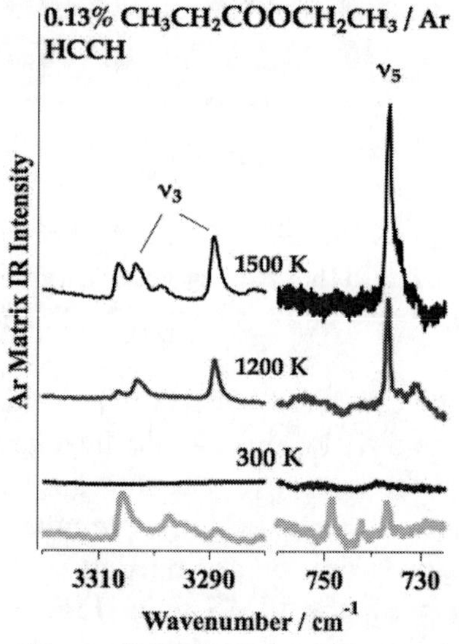

Figure 6. Matrix infrared sprectra of ethyl propionate in Ar confirming acetylene as a product. The green trace is a scan of Ar heated to 1400 K.

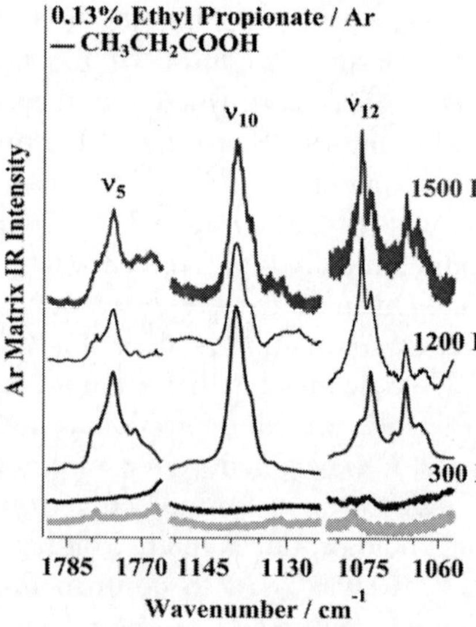

Figure 7. Matrix infrared spectra of ethyl propionate in Ar confirming propionic acid as a product. The green trace is a scan of Ar heated to 1400 K. The blue trace is authentic propionic acid.

4. Conclusions

The pyrolysis of ethyl propionate was performed in a tubular microreactor. The main pathway is the concerted ring decomposition to propionic acid and ethylene. This agrees with earlier conclusions from the literature. Further experiments are needed to form definitive conclusions about the minor pyrolysis pathways. The presence of R11 could easily be confirmed using PIMS with the microreactor and a tunable light source.

Since the majority of ethyl propionate in pyrolysis goes to propionic acid and ethylene, this leads to a question, how well is the pyrolysis of propionic acid understood? Methyl ketene is clearly seen as a product of propionic acid pyrolysis confirming R15 but the presence of the radical at m/z 73 requires further study. Errors in ethyl propionate combustion models in the literature may be solved by through a thorough kinetic study of propionic acid decomposition. This opens the door to future work with the tubular microreactor.

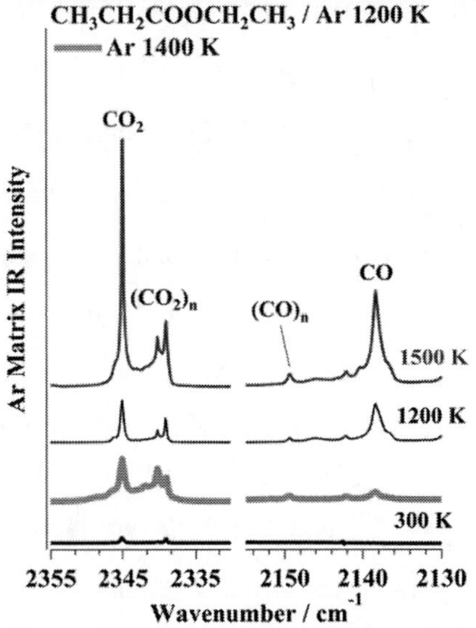

Figure 8. Matrix infrared spectra of ethyl propionate in Ar confirming the presence of carbon monoxide and carbon dioxide. The green trace is a scan of Ar heated to 1400 K.

Acknowledgements

This research was funded through University of Colorado Boulder via the startup funds of Prof. Nicole Labbe and support for J.W.D and G.B.E as well as partial support for C.O.R was provided through NSF Grant CBET-1403979. Support was also provided by J. A. Sewall Chemistry Scholarship and the CU Boulder Graduate School Dissertation Completion Fellowship.

References

[1] A. T. Blades and P. W. Gilderson. *Can. J. Chem.*, 38 (1960) 1412-1415.

[2] J. A. Barnard, A. T. Cocks, and T. K. Parrott. *J. Chem. Soc. Faraday Transactions 1*, 72 (1976) 1456-1463.

[3] W. R. Schwartz, C. S. McEnally, and L. D. Pfefferle. *J. Phys. Chem. A*, 110 (2006) 6643-6648.

[4] W. K. Metcalfe, S. Dooley, H. J. Curran, J. M. Simmie, A. M. El-Nahas, and M. V. Navarro. *J. Phys. Chem. A*, 111 (2007) 4001-4014.

[5] W. K. Metcalfe, C. Togbé, P. Dagaut, H. J. Curran, and J. M. Simmie. *Combust. and Flame*, 156 (2009) 250-260.

[6] S. M. Walton, M. S. Wooldridge, C. K. Westbrook. *Proc. of the Combus. Inst.*, 32 (2009) 255-262.

[7] B. Yang, C. K. Westbrook, T. A. Cool, N. Hansen, and K. Kohse-Höinghaus. *Phys. Chem. Chem. Phys.*, 13 (2011) 6901-6913.

[8] A. Farooq, D. F. Davidson, R. K. Hanson, and C. K. Westbrook. *Fuel*, 134 (2014) 26-38.

[9] Q. Guan, K. N. Urness, T. K. Ormond, D. E. David, G. B. Ellison, J. W. Daily. *Int. Rev. Phys. Chem.*, 33 (2014) 447-487)

[10] R. K. Thomas. *Proc. R. Soc. London A*, 331 (1972) 139-283

[11] S. Willitsch, U. Hollenstein. F. Merkt. *J. Chem. Phys.*, 120 (2004) 1761-1774.

[12] Discussion of chemically activated ethylene

[13] H. Bock, T. Hirabayashi, and S. Mohmand. *Chem. Ber.*, 114 (1981) 2595.

[14] A. M. El-Nahas et al. *J. Phys. Chem. A*, 111 (2007) 3727-3739.

[15] K. Watanabe, T. Nakayama, and J. Mottl. *J. Quant. Spectrosc. Radiat. Transfer*, 2 (1962) 369-382.

[16] G. Y. Matti, O. I. Osman, J. E. Upham, R. J. Suffolk, and H. W. Kroto. *J. Electron Spectrosc. Rel. Phenom.*, 49 (1989) 195-201.

[17] C. A. Taatjes et al. *Science* 308 (2005) 1887-1889

[18] A. Vasiliou et al. *J. Chem. Phys.*, 135 (2011) 014306 5pp.

[19] J. P. Porterfield et al. *J. Phys. Chem. A*, 119 (2015) 12635-12647.

[20] D. J. Knowles and A. J. C. Nicholson. *J. Chem. Phys.*, 60 (1974) 1181-1181.

[21] J. C. Traeger. *Int. J. Mass Spectrom Ion Processes*, 66 (1985) 271-282.

[22] A. M. Schulenburg, C. Alcaraz, G. Grassi, and F. Merkt. *J. Chem. Phys.*, 125 (2006) 104310 10 pp.

[23] S. J. Tang, Y. C. Chou, J. J. M. Lin, and Y. C. Hsu. *J. Chem. Phys.*, 125 (2006) 133201 15 pp.

A New Jet-Stirred Reactor for Chemical Kinetics Experiments

A. Ashkan Davani, Paul D. Ronney
University of Southern California

ABSTRACT

A novel jet-stirred reactor (JSR) is proposed to study combustion processes at low Damköhler number (Da, ratio of residence time within the reactor to the time scale of the chemical reaction) for the purpose of measuring rates of chemical reactions relevant to combustion processes. The goal of this work is to identify an optimal configuration of multiple sets of impinging jets and outlet ports for JSRs.

With this motivation, ANSYS-FLUENT computational fluid dynamics computations using the RANS - Reynolds Stress Model were performed to simulate mixing and reaction in such geometries. The performance of the new JSRs was compared to the classical JSR introduced by Matras and Villermaux (1973) and Dagaut et al. (1986), which we have termed "4 Jets In Plus (+) Pattern" (4JIPP). Initial computations were conducted using a simple one-step reaction of the form A + B --> C + D with zero heat release and zero activation energy. Results showed that a configuration of 8 jets, each surrounded by a concentric annular outlet, termed the Concentric Inlet And Outlet (CIAO) design, at the corners of an imaginary cube circumscribed by a spherical chamber provided far more uniform composition and temperature and thereby more nearly match the idealizations assumed in Perfectly-Stirred Reactor (PSR) theory, even at values of Da higher than those accessible to other JSRs. Moreover, the CIAO design yielded inferred reaction rate constants that were much closer to the ideal values (i.e., PSR predictions) than the classical 4JIPP design.

Subsequent simulations were performed in CIAO and 4JIPP reactors using detailed H_2-O_2 chemistry. Results were compared with PSR model predictions. Again it was found that CIAO more nearly matches PSR predictions than does the 4JIPP reactor. Moreover, in the 4JIPP reactor the predicted compositions were quite different between local and volume-averaged values whereas in CIAO the composition was much more uniform over the chamber volume.

To assess the viability of the CIAO design experimentally, a 3D printed chamber was built and tested in cold-flow mixing experiments using fluorescein + water solutions and Laser Induced Fluorescence imaging to quantify the mixing levels. Consistent with the computational predictions, the initial experimental results indicate much more uniform mixing than 4JIPP type chambers.

1. Motivation

The Jet-Stirred Reactor (JSR) [1] is a classical apparatus for studying the chemical kinetics of combustion reactions. In JSRs, several jets of combustible reactants are fed into a mixing chamber maintained at elevated temperatures and reacted. Analysis of the products of reaction as a function of reactant composition and residence time is used to infer the rates and pathways of reaction. The key assumption required to interpret the data is that mixing of reactants and products is complete and instantaneous, i.e. that the mixing time scale τ_M is much smaller than either the residence time scale τ_R or the chemical time scale τ_C so that no gradients of composition or temperature occur. The JSR has one compelling advantage for chemical kinetics experiments over shock tubes and laminar or turbulent tubular flow reactors which greatly simplifies data interpretation: the JSR is ideally zero-dimensional in <u>both</u> space and time whereas shock tubes are ideally zero-dimensional <u>only</u> in space and one-dimensional time and flow reactors are ideally zero-dimensional <u>only</u> in time and one-dimensional in space. Additionally, JSRs have an inherent advantage over tubular flow reactors: the residence time τ_R is almost completely decoupled from the mixing time τ_M, the latter potentially being far smaller whereas in tubular flow reactors the mixing time and residence time both scale inversely with the mean velocity and thus cannot be decoupled. Another problem with tubular flow reactors is axial dispersion; that is, since the velocity profile is not purely plug-flow, reactants along the centerline will be convected downstream more rapidly than material near the tube wall, which renders the apparatus two-dimensional in space to some extent. Both issues can potentially be avoided with a well-designed JSR.

Of course there are challenges associated with JSR design and operation, the foremost of which are (1) ensuring rapid mixing of incoming reactants with the material already in the JSR, specifically to ensure $\tau_M << \tau_R$ and $\tau_M << \tau_C$ and (2) avoiding pre-reaction before the reactants enter the JSR. While many studies, e.g. [2], have addressed the <u>consequences</u> of an assumed level of imperfect mixing, the <u>prediction</u> and <u>characterization</u> of unmixedness has received very little attention. In fact, in a recent review paper Herbinet and Battin-Leclerc [3] state that, "...no work on this topic has been performed since 1986, and new work with up-to-date experimental techniques could bring valuable information to the subject." We shall show that existing reactors used for obtaining chemical kinetic data for combustion reactions may lead to significant discrepancies between inferred reaction rate constants and the actual values even for an extremely simple test problem. We propose a new reactor design that promises to provide much smaller discrepancies and also addresses the pre-reaction issue.

2. Current approach

To overcome the limitations of current JSRs, we propose to employ a set of impinging jets of reactants with multiple outlet ports, all optimized for rapid mixing times compared to residence times. While we have investigated many geometries based on different icosahedral arrangements of inlet and outlet ports, for brevity here we will only discuss one which computations have shown is <u>far</u> superior to all others, namely a spherical chamber with a novel arrangement of 8 jets at the corners of an imaginary cube (equivalently, a double-tetrahedral configuration), each with a concentric outlet port, that we have termed the Concentric Inlet And Outlet (CIAO) reactor (Fig. 1). The inspiration for this type of geometry comes from the work of Hwang and Eaton [4] who used a similar arrangement of acoustically-forced "synthetic jets" to study particle settling in homogeneous, isotropic, zero-mean-flow turbulence. Our concentric jet arrangement of paired inlets and outlets is analogous to the synthetic jets but provides much higher turbulence intensities over a much larger fraction of the chamber volume.

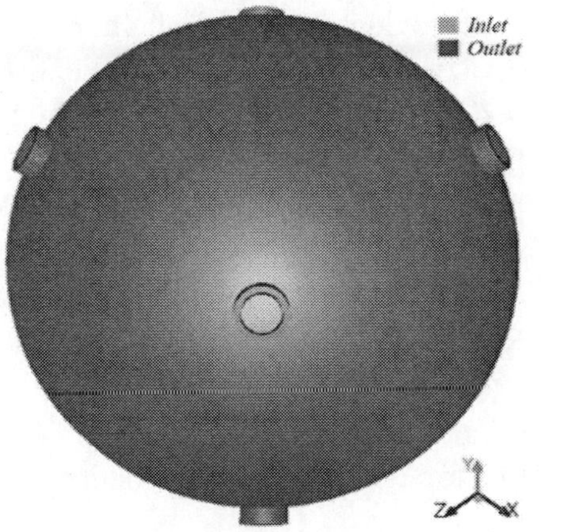

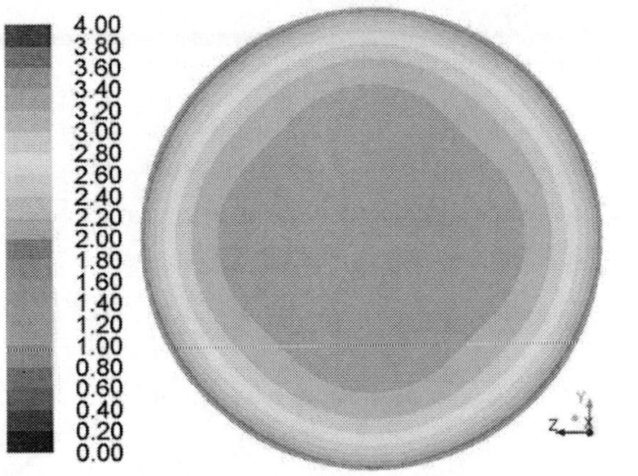

Figure 1. Proposed Concentric Inlet And Outlet (CIAO) reactor.

Figure 2. Computed contours of turbulence intensity (u', units m/s) in the CIAO reactor for inlet jet velocities of 20 m/s.

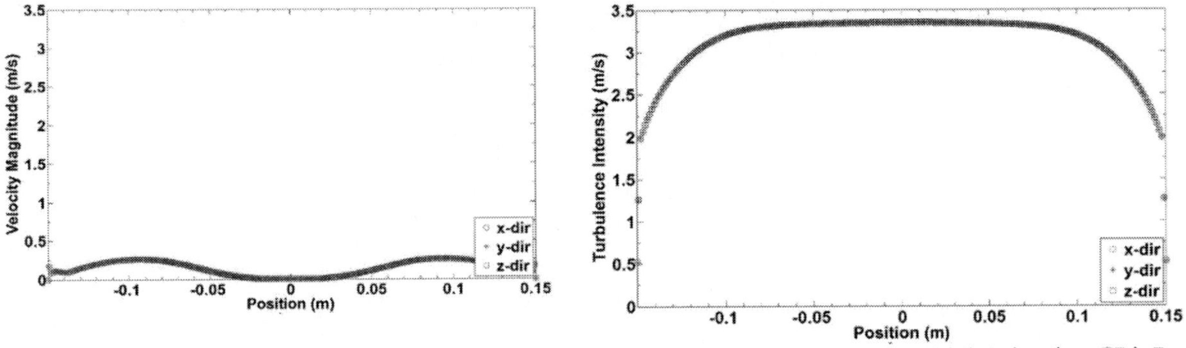

Figure 3. Plots of computed mean velocity (left) and turbulence intensity (right) in the CIAO reactor.

The CIAO apparatus is also very relevant to spherically expanding turbulent premixed flame experiments. The ideal apparatus for such experiments produces that is homogeneous, isotropic and has zero mean flow (more specifically, a very high ratio of turbulence intensity to mean velocity) throughout the entire chamber. Of course, no real apparatus can produce such a flow but a comparison of these properties for the CIAO reactor and two widely-used fan-stirred reactors show that the CIAO reactor flows much more nearly approach the idealized scenario.

3. Results

The aforementioned reactors have been simulated using ANSYS-FLUENT and the Reynolds-Averaged Navier Stokes (RANS) model for simulating turbulent transport. While both the FLUENT flow solver and RANS are often considered inaccurate and unreliable by the academic community, by comparison with experiments we have shown (see Appendix) that it is very accurate for the class of problems of interest here and has the advantages of (1) greatly shortening the design cycle time; (2) interfacing directly with CAD software and 3D printers; and (3) simplifying the sharing of input and output files with other research groups (as opposed to the use of one-of-a-kind in-house codes.) A summary of the testing of the model to demonstrate its accuracy for the current purposes is described in other work [5].

For non-reacting turbulence, the CIAO configuration provides exceptionally homogeneous, isotropic and near-zero mean flow (Figs. 2 and 3), far closer to this idealized situation than conventional fan-stirred reactors (see Appendix). To assess the viability of the CIAO configuration for JSR problems, a very simple test problem was employed, consisting of a one-step or two-step reaction between reactants A and B, in the 2-step case with an intermediate species C and final product D, no heat release and no activation energy of the form

$$A+B\rightarrow C+D \quad \frac{d[A]}{dt}=\frac{d[B]}{dt}=-\frac{1}{2}\frac{d[C]}{dt}=-\frac{1}{2}\frac{d[D]}{dt}=-k_1[A][B] \quad \text{(for 1-step reaction)}$$

$$A+B\rightarrow 2C \quad \frac{d[A]}{dt}=\frac{d[B]}{dt}=-\frac{1}{2}\frac{d[C]}{dt}=-k_1[A][B] \quad \text{(Step 1 for 2-step sequential reaction)} \quad \text{(Eq. 1)}$$

$$C\rightarrow D \quad \frac{d[C]}{dt}=-\frac{d[D]}{dt}=-k_2[C] \quad \text{(Step 2 for 2-step sequential reaction)}$$

If the mixture at the inlets is an equimolar mixture of A and B (i.e. their mole fractions X_i are both 0.5) with no inert, the molar concentrations of A and B ([A] and [B] respectively, units moles/m^3) are the same for all time since any reaction of A also causes a reaction of B. For this case, assuming perfect mixing, some algebra shows that the mole fractions X_i in the reactor are

$$X_A=X_B=\frac{\sqrt{1+Da}-1}{Da};X_C=X_D=\frac{1}{2}-\frac{\sqrt{1+Da_1}-1}{Da_1} \quad \text{(one-step)}$$

$$X_A=X_B=\frac{\sqrt{1+Da}-1}{Da};\; X_C=\frac{2+Da-2\sqrt{1+Da}}{Da(1+\Delta Da)};\; X_D=\Delta\frac{2+Da-2\sqrt{1+Da}}{(1+\Delta Da)} \quad \text{(two-step)} \quad \text{(Eq. 2)}$$

$$\text{where } Da\equiv 4k_1[A]_o\frac{V}{\dot{V}};\; \Delta\equiv\frac{k_2}{4k_1[A]_o}$$

where Da is a Damköhler number for step (1), Δ is the ratio of the Damköhler numbers for reaction step (2) to step (1), V and $\dot{V}$ are the reactor volume and volume flow rate respectively (note $V/\dot{V}$ is the residence time τ_R) and $[A]_o$ is the concentration of reactant A at the inlet. FLUENT / RANS modeling with this simple reaction scheme has been conducted for the 4JIPP reactor and the CIAO reactor (Figs. 4 - 6) with the same V and τ_R. In these computations, by intent no model of turbulence-chemistry interaction is used so that a direct comparison of the performance of the reactors can be made. Moreover, if a model of turbulence-chemistry interaction is needed the key assumption of JSRs – uniform composition, thus no variance in reaction – is already violated.

Figure 5 shows that the CIAO reactor provides much more uniform composition than 4JIPP arrangement. In particular, the 4JIPP geometry results in a central region of nearly solid-body rotation (as evidenced by the particle pathlines, Fig. 6, left) with limited mixing and thus long residence times which in turn lead to high product concentrations in this region. By comparison, in the CIAO reactor (Fig. 6, right) there is no central region of nearly stagnant fluid. This uniformity (or lack thereof) is reflected in Fig. 7 (left) which shows that for the single-step reaction (Δ = 0) the volume-averaged product mole fraction X_C in the CIAO reactor is much closer to the exact solution than in the 4JIPP reactor. Perhaps more significantly is the inverse way of evaluating performance, *i.e.*, the way JSR data are used to obtain kinetic data: when the volume-averaged product mole fraction X_C is inserted into Eq. 2, Da and thus the kinetic constant k_1 is inferred. Figure 7 (right) shows that the simulated volume-averaged values of X_C in the CIAO reactor result in inferred values of k_1 that are much closer to the actual values than those obtained in the 4JIPP reactor. These results suggest that **even in this very simple chemical scenario, the non-uniformity of mixing in widely-used JSRs results in inferred reaction rate constants that are significantly different from the true values.**

In order to facilitate comparisons between the 4JIPP and CIAO reactors, the aforementioned simulations have of necessity employed premixed reactants with the presumption of no reaction before entering the chamber. This was necessary because the 4JIPP reactor has one high mass flow inlet path for air (or other oxidant) and a different low-mass-flow, low-residence-time inlet path for fuel to minimize potential fuel pyrolysis. This is not directly compatible with the symmetrical inlets of the CIAO reactor, hence the use of premixed reactants for comparison. Of course, premixing of reactants is not a practical technique for real experiments because heated premixed reactants will begin to react before entering the "reactor." Consequently, simulations of the CIAO reactor were performed using nonpremixed reactants (4 jets with reactant A, 4 jets with reactant B). Figure 7 shows that **the results with non-premixed reactants are nearly the same as those of premixed reactants up to Da ≈ 10.** Fundamentally this is possible because the CIAO reactor has effectively decoupled the residence time τ_R from the mixing time τ_M, the latter being much smaller. Figure 8 shows that for the two-step model ($\Delta > 0$) the CIAO reactor predicts volume-averaged values of the intermediate [C] and product [D] mole fractions that are closer to the exact solutions than the 4JIPP reactor extends to the two-step model. It may be noted in particular that the concentrations of [C] are predicted reasonably well for conditions of Da and Δ resulting in a nearly 4-decade range of [C] whereas for the 4JIPP configuration there is substantially more difference between the computed and exact solutions.

It is frequently claimed [1, 3] that a sufficient criterion for adequate mixing in a JSR is agreement with the exact solution for the decay of a tracer in a perfectly-stirred reactor when the supply of tracer is suddenly stopped, i.e. $C(t)/C(0) = 1 - \exp(t/\tau_R)$, where C(t) is the tracer concentration. With this motivation, test of a small concentration of CO_2 tracer in N_2 was conducted for 4JIPP, modified 4JIPP [3], and CIAO reactors with the same mean residence time τ_R. Results are shown in Fig. 9 for both volume-averaged tracer concentrations and (to simulate the experimental measurements) tracer concentrations averaged only at the exit planes of the reactors. All 6 of these cases show nearly ideal behavior, yet we have demonstrated that the behavior of the 4JIPP configuration under reacting-flow conditions is far from that of a PSR. Consequently we conclude that **exponential decay of a non-reactive tracer that closely follows the exact solution is not a sufficient criterion to conclude that the apparatus will behave as a PSR in the presence of chemical reactions.**

Following the aforementioned results that employed very simple model chemistry having exact solutions for all properties, simulations were performed in CIAO and 4JIPP reactors using detailed H_2-O_2 chemistry (USC Mech II). Results, examples of which are shown in Fig. 10, were compared with PSR model predictions. Another presentation of the results (Fig. 11) is in terms of the fractional error in mole fraction = $X(x,y,x)/X_{PSR} - 1$ where X is the local mole fraction and X_{PSR} is the mole fraction predicted from a Perfectly-Stirred Reactor computation. As with the results for simple model chemistry, it was found that results in the CIAO reactor more nearly matches PSR predictions than does the 4JIPP reactor. In particular, again the 4JIPP reactor exhibits long residence times and thus more reaction progress near the reactor center where samples are typically extracted, and thus may lead to overestimates of overall reaction rates.

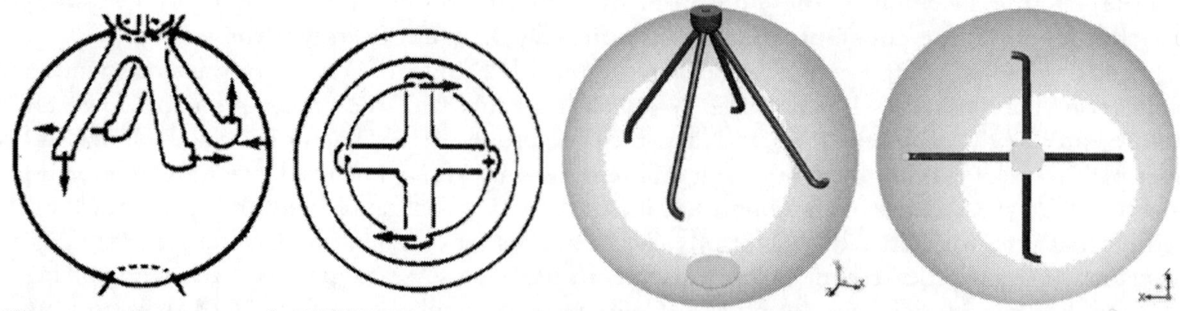

Figure 4: Dagaut *et al.* (4JIPP) [1] reactor. Left 2 images: from the original paper; right 2 images: CAD model used in simulations. Overall diameter is 4 cm and the exit diameter of the 4 fuel jets is 0.1 cm.

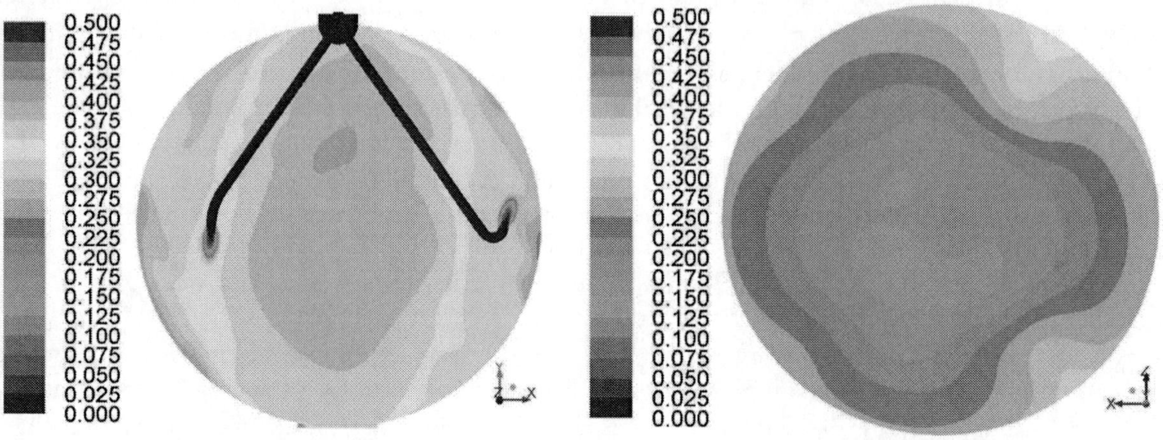

Figure 5. Computed product C mole fraction (X_C) fields for the test problem (Eq. 1) with $\Delta = 0$ (i.e., single-step reaction) at Da = 2.5. Left: 4JIPP reactor; right: proposed CIAO reactor. The exact solution for these conditions is $X_C = 0.268$.

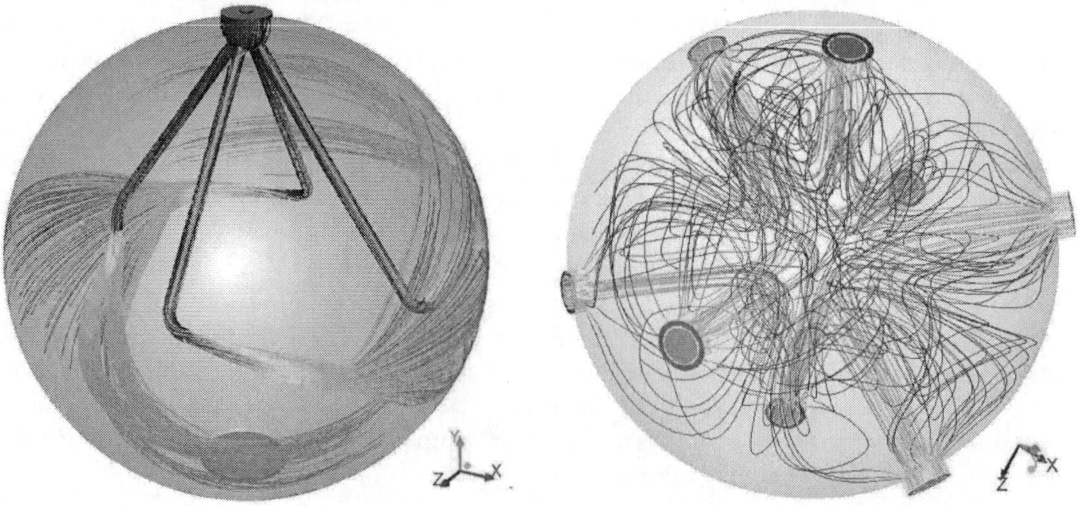

Figure 6. Computed pathlines in JSRs. Left: 4JIPP reactor; right: CIAO reactor.

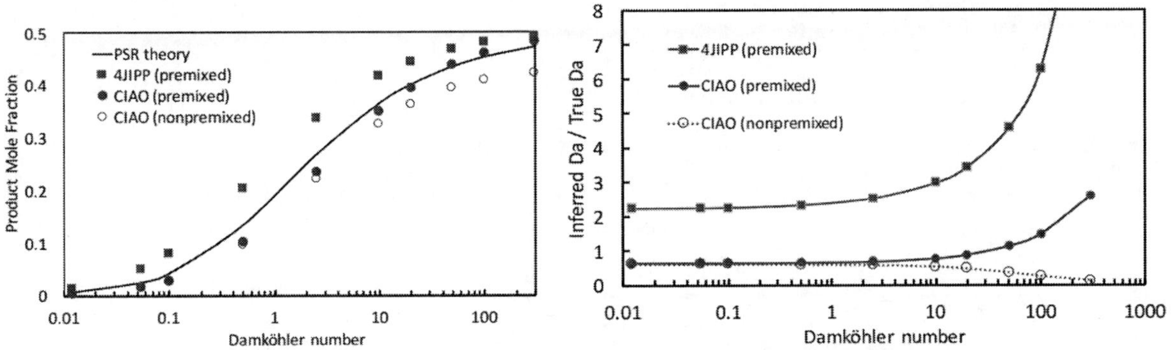

Figure 7. Left: Volume-averaged mole fraction of product C (X_C) as a function of Da for the test problem (Eq. 1) with $\Delta = 0$ (i.e., single-step reaction) predicted by the FLUENT/RANS simulations for the 4JIPP reactor and the concentric jet (CIAO) reactor along with comparison to the exact theory (Eq. 2) assuming perfect mixing. Right: comparison of inferred (from Eq. 2) rate constants using the volume-averaged product mole fraction for the 4JIPP and CIAO reactors (relative to the actual prescribed k_1) as a function of Da.

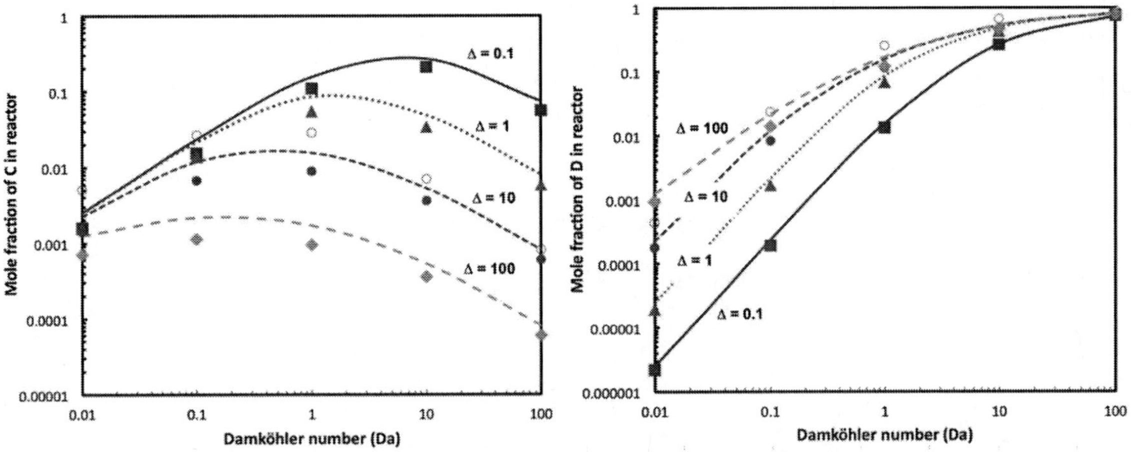

Figure 8. Comparison of computed values (solid symbols: CIAO; open symbols: 4JIPP reactor) to the exact solution (curves) for the two-step reaction (Eq. 1) as a function of the Damköhler number of reaction 1 (Da) and the ratio of Damköhler numbers of reaction 2 to 1 (Δ). Left: mole fraction of intermediate species (X_C); right: mole fraction of final product species (X_D).

It was also found (Fig. 12) that in the 4JIPP reactor, the predicted compositions were quite different between local and volume-averaged values whereas in CIAO the composition was much more uniform over the chamber volume. Somewhat surprisingly, the volume averaged 4JIPP-RANS predictions of mole fractions are close to PSR predictions even though the local mole fractions vary widely. The presence of sample probe in the 4JIPP reactor was found to have a significant effect on the results. 4JIPP experiments for these conditions [6] show large difference from PSR prediction and in fact agree fairly well with predictions at a sample location half-way between chamber center and outlet.

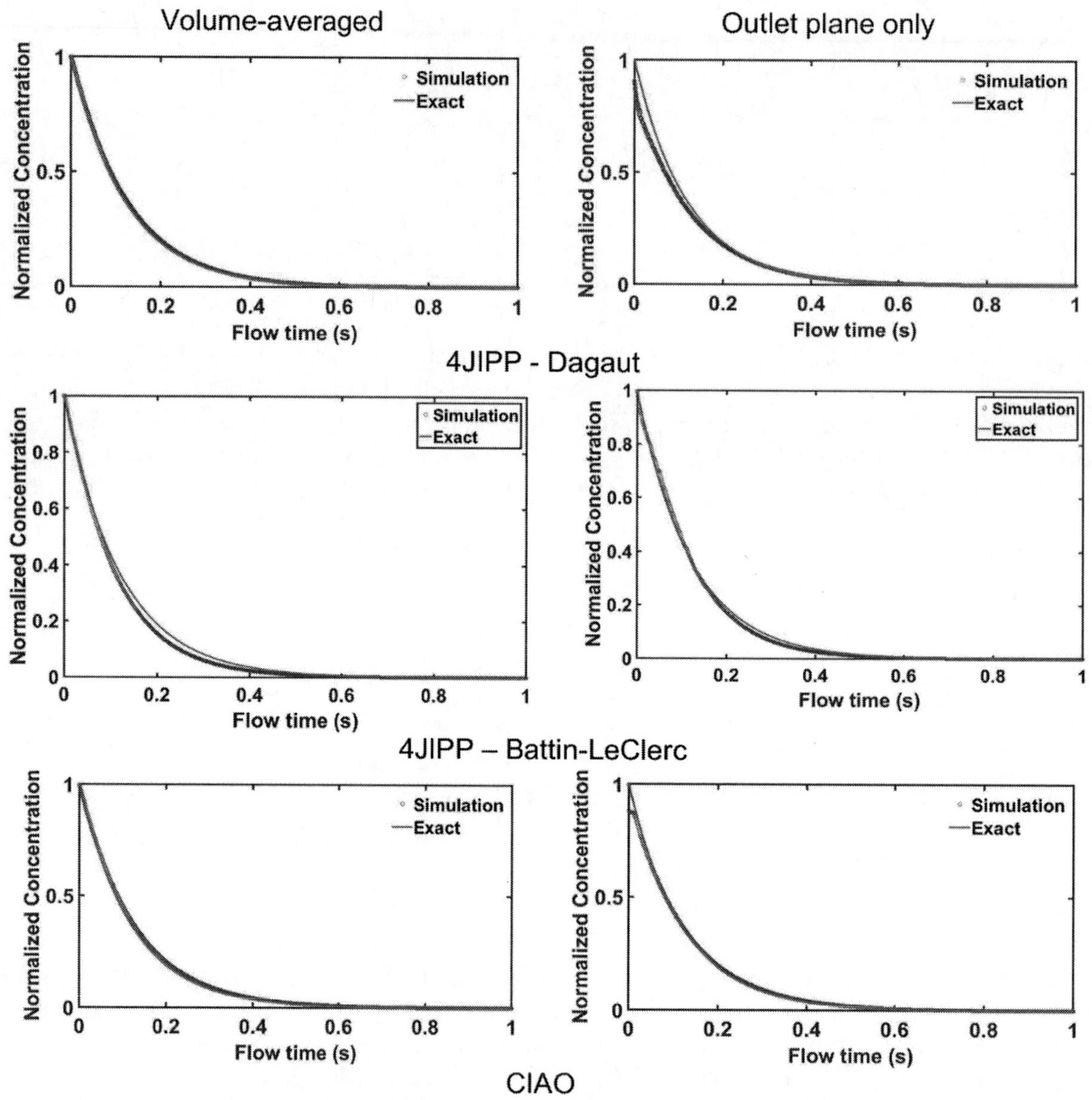

Figure 9. Comparison of computed rates of decay of a passive tracer in the Dagaut-type 4JIPP, Battin-Leclerc 4JIPP and CIAO reactors and comparison with the exact solution for a perfectly-stirred reactor.

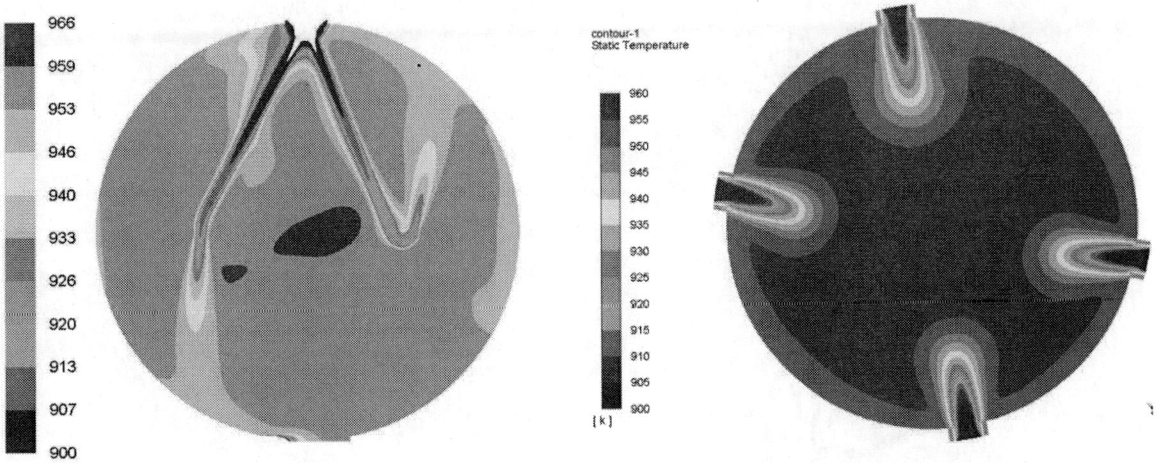

Temperature

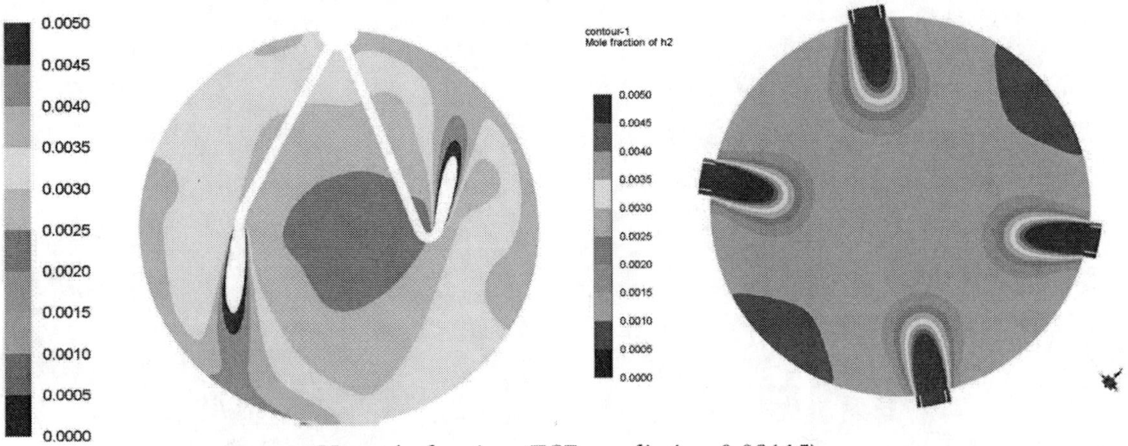

H_2 mole fraction (PSR prediction 0.00115)

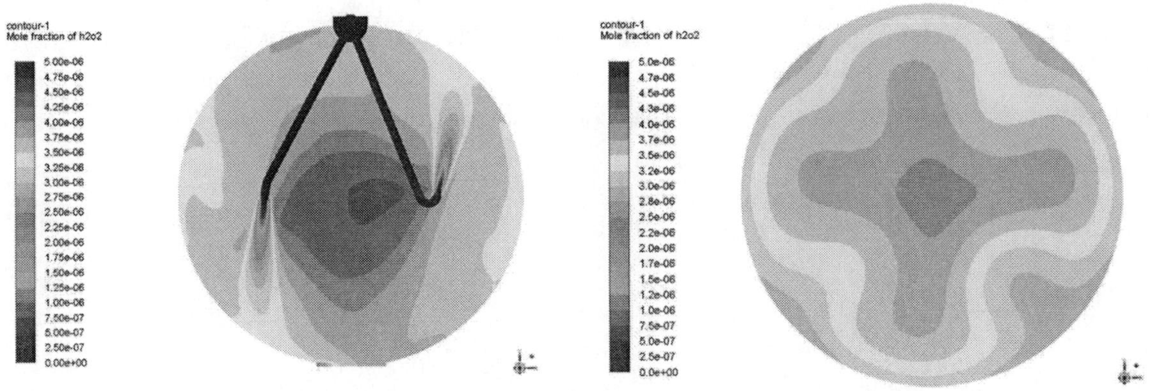

H_2O_2 mole fraction

Figure 10. Computed temperature and species maps for 4JIPP (left column) and CIAO (right column) jet stirred reactors. Mixture $H_2 : O_2 : N_2 = 1 : 2.5 : 96.5$ (equivalence ratio $\phi = 0.2$), inlet temperature 900K, inlet pressure 1 atm, residence time $\tau_R = 0.12$ s.

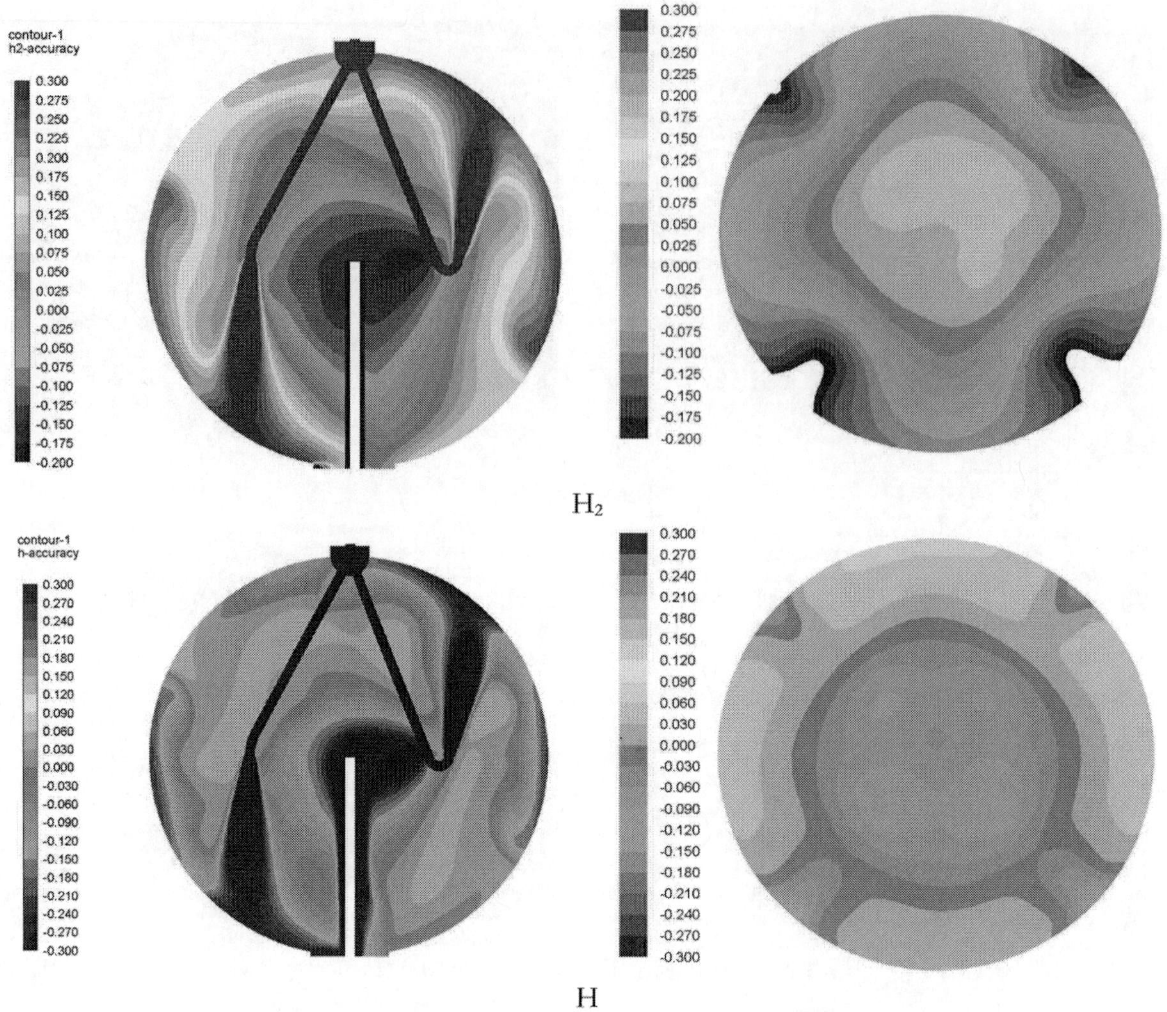

Figure 11. Computed mole fraction fractional error maps for 4JIPP (left column) and CIAO (right column) jet stirred reactors. Mixture $H_2 : O_2 : N_2 = 1 : 5 : 94$ (equivalence ratio $\phi = 0.1$), inlet temperature 900K, inlet pressure 10 atm, residence time $\tau_R = 1.0$ s.

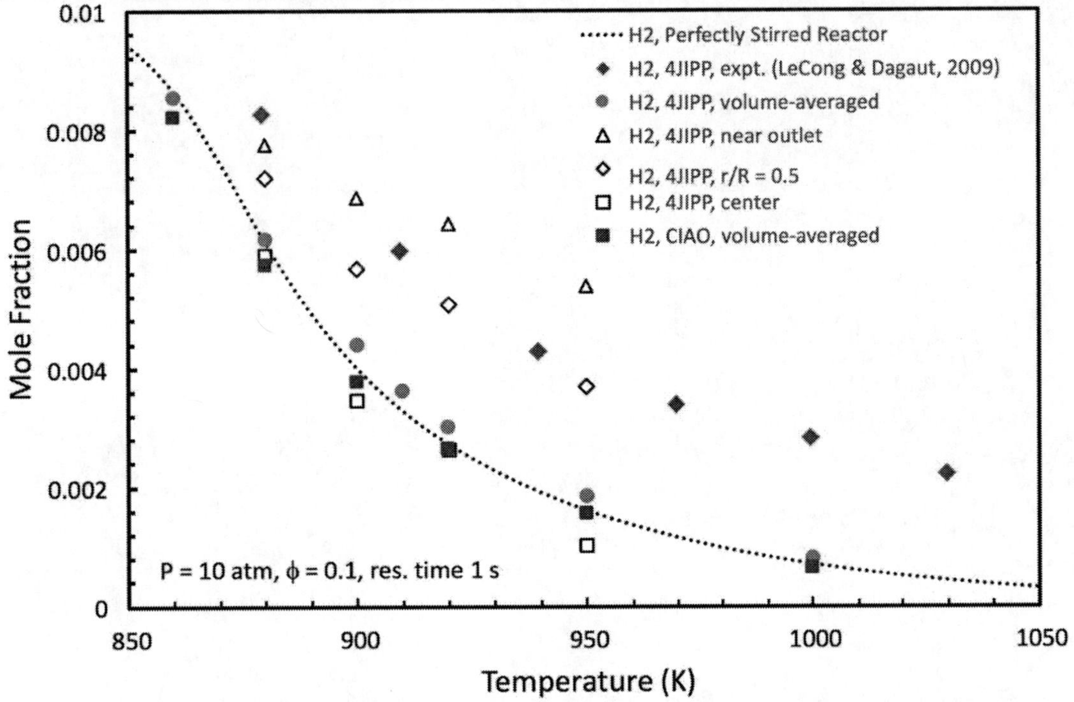

Figure 12. Comparison of H_2 mole fraction as a function of reactor inlet temperature for 4JIPP and CIAO jet-stirred reactors for an ideal Perfectly Stirred Reactor, the 4JIPP reactor (volume-averaged and at specific locations averaged over a 1 mm radius spherical volume: center, half-way toward the outlet and at the outlet plane), CIAO reactor (volume-averaged; location-specific predictions are nearly identical) and experimental results by LeCong and Dagaut [6].

To assess the viability of the CIAO design experimentally, a 3D printed chamber was built and tested in cold-flow mixing experiments using fluorescein + water solutions and Laser Induced Fluorescence imaging to quantify the mixing levels. Consistent with the computational predictions, the initial experimental results indicate much more uniform mixing than 4JIPP type chambers. For this cases the residence time $\tau_R = 0.83$ s and the time between successive frames is 0.00833 s. It can be see that the images are not well correlated, thus even over time period 1/100 of the residence time there is substantial mixing, indicating that near-PSR conditions exist.

Figure 13. Apparatus for cold-flow testing of mixing uniformity in CIAO-type reactors. Left: overall view showing gravity-fed tanks of water (left) and water-fluorescein solution (right) with associated plumbing. Right: two closeup views of the CIAO chamber with optical access.

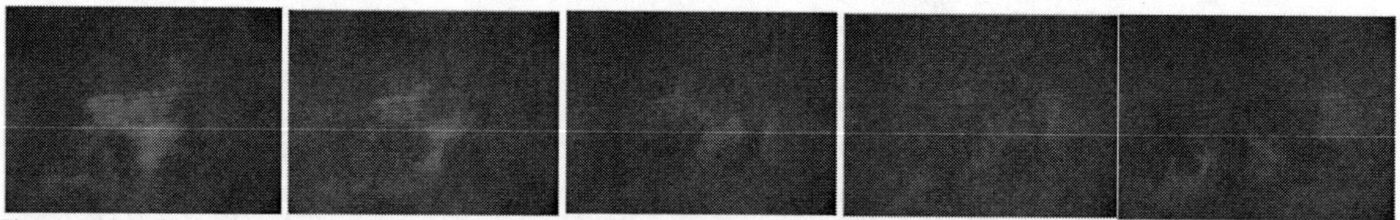

Figure 14. Successive PLIF images (interval 0.0083 s) of fluorescein concentration in a CIAO JSR. Chamber diameter 15 cm, field of view of images 7.62 x 5.71 cm.

4. Conclusion

The proposed CIAO apparatus shows promising results in study of chemical kinetics. CFD computations for single step, two step and detailed reaction mechanisms demonstrate that the apparatus provides an improved performance in study of chemical kinetics since even in very simple chemical scenarios, the non-uniformity of mixing in widely-used JSRs results in inferred reaction rate constants that are significantly different from the true values.

Acknowledgements

Authors would like to appreciate Dr. Philippe Dagaut for providing useful information of their reactor and also for technical discussions. This research was sponsored by National Science Foundation (Proposal No. 1236892)

REFERENCES

[1] P. Dagaut, M. Cathonnet, J. P. Rouan, R. Foulatier, A. Quilgars, J. C. Boettner, F. Gaillard, H. James, "A jet-stirred reactor for kinetic studies of homogeneous gas-phase reactions at pressures up to ten atmospheres (≈1 MPa)," J. *Phys. E: Sci. Instrum.*, 1986, Vol. 19, pp. 207 – 209.

[2] A. Nicolle, P. Dagaut, "Occurrence of NO-reburning in MILD combustion evidenced via chemical kinetic modeling," *Fuel*, Vol.85 (2006) pp. 2469–2478

[3] O. Herbinet, F. Battin-Leclerc, "Progress in Understanding Low-Temperature Organic Compound Oxidation Using a Jet-Stirred Reactor," *International Journal of Chemical Kinetics*, DOI 10.1002/kin.20871 (2014).

[4] W. Hwang, J. K. Eaton, "Creating homogeneous and isotropic turbulence without a mean flow," *Experiments in Fluids*, Vol. 36 (2004) p. 444–454.

[5] Davani A., Ronney P. "A jet-stirred chamber for turbulent combustion experiments" *Comb. Flame*, 185 (2017) pp. 117-128.

[6] T. Le Cong, P. Dagaut, "Experimental and Detailed Modeling Study of the Effect of Water Vapor on the Kinetics of Combustion of Hydrogen and Natural Gas, Impact on NO_X," *Energy and Fuels*, Vol. 23, pp. 725–734 (2009).

Western States Section of the Combustion Institute – Fall 2017 Meeting
Hosted by the University of Wyoming
October 2-3, 2017

Numerical Design of a Novel Microreactor to Study Short Residence Time Combustion

Tianzhu Fan[1], Cory Rogers[1], John W. Daily[1], Barney Ellison[2] and Nicole Labbe[1,]*
[1]Department of Mechanical Engineering, University of Colorado, Boulder, Colorado, 80309, USA
[2]Department of Chemistry and Biochemistry, University of Colorado, Boulder, Colorado, 80309, USA
[*]Corresponding Author: Nicole Labbe (nicole.labbe@colorado.edu)

Abstract
Microreactors have been used as tools to help understand the early time thermal decomposition of biomass polymers and fuel molecules. To date, our studies have involved using very small tubes that are open to vacuum. The exhaust forms a supersonic expansion which is either skimmed into a molecular beam time-of-flight mass spectrometer, or directed onto a cold matrix for IR spectroscopy. While much has been learned from this configuration, qualitative analysis is hampered by the fact that the pressure and velocity vary greatly within the reactor. Therefore, we have explored modifying the geometry to achieve better control over these properties. In particular, we have added a nozzle at the downstream end of the reactor, so that most pressure drop takes place across the nozzle. Here we explore the changes in geometry using CFD simulations.

1. Introduction

At the University of Colorado Boulder, microreactors (Fig. 1) have been employed for over a decade to study the thermolysis of biomass and fuels derived molecules. The reactors are short (~2 cm), small diameter (~1 mm) tubes, which to date have been fabricated from silicon carbide. These reactors are electrically heated and exhaust into a vacuum forming a supersonic expansion. The exhaust is either skimmed into a time-of-flight photoionization mass spectrometer, or directed onto a cold matrix for matrix/IR spectroscopic analysis.

The advantage of these small reactors is their very short residence times, typically on the order of 100 microseconds. The disadvantage is that the thermodynamic properties within the reactor vary considerably in the axial direction, making analysis of the results difficult. While the current design has proven to be extremely useful for understanding thermal decomposition mechanisms for a large number of molecules, it would be beneficial to have better control over the flow properties, making analysis of results more tractable. In particular, extracting kinetic data requires a thorough understanding of the variation of flow parameters in the reactor.

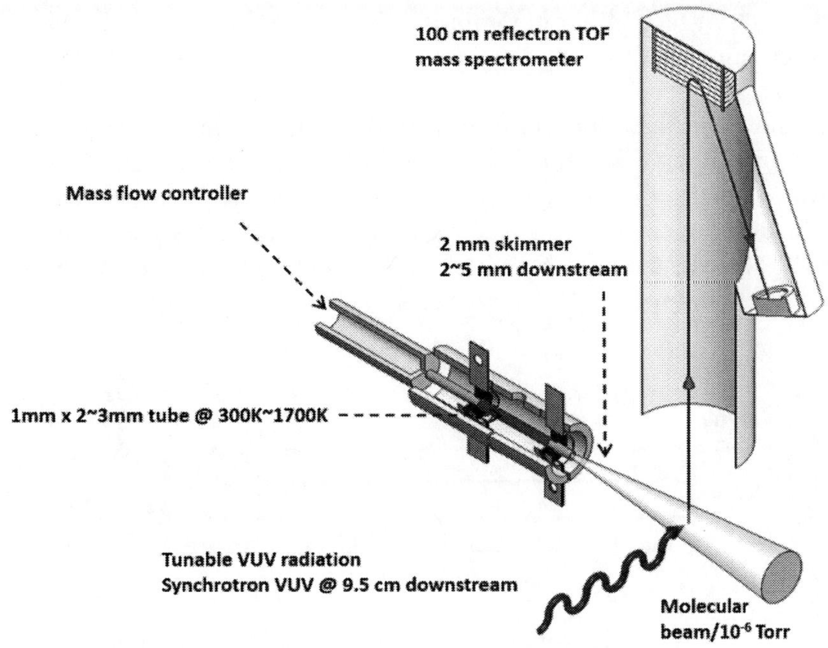

Figure 1: Microreactor shown in the photoionization mass spectrometry configuration when utilized at the Advanced Light Source in Berkeley

To resolve this problem, we have explored adding a flow restriction, or nozzle, at the downstream end of the reactor, with the goal of having most of the pressure drop between inlet and exhaust take place at the nozzle. This should have the effect of generating a more axially uniform flow, and thus a simpler flow to analyze. In the following, we explore the role of design changes to accomplish this goal using computational fluid dynamics (CFD).

2. Methods

The current microreactor geometry is that of a simple cylindrical tube with a length and internal diameter of about 28 mm and 1.0 mm respectively. The flow within the tube is laminar, and assuming for simplicity incompressible flow, the pressure drop ΔP is

$$\Delta P = \frac{\overline{f_{fd,h}} \cdot L \cdot \rho \cdot |\vec{U}|^2}{2D} \qquad \text{Eq. 1}$$

where

$$\overline{f_{fd,h}} = \frac{64}{Re_D} \qquad \text{Eq. 2}$$

is the friction factor, L and D are the length and internal diameter of the cylindrical tube, and Re_D is the Reynolds number for internal flow. Eq. 1 predicts a linear change in pressure with distance, and since our open-ended reactor exhausts to a vacuum, the pressure drops from the inlet value to about zero. The flow is not, in fact, either incompressible or constant density, but detailed CFD simulations show that the pressure does drop to a very small value by the end of the tube.

To achieve a more uniform pressure profile within the microreactor, we propose to add a converging-diverging nozzle structure at the outlet of the reactor main body (see Fig. 2). This will impose a downstream flow resistance such that most of the pressure drop will occur across the nozzle, rather than in the reactor. Fig. 2 shows the geometry. Parameters to study include the length of the cylindrical tube, L, the exit diameter of the converging nozzle, d, and the lengths of both the converging and diverging nozzles sections, G1 and G2.

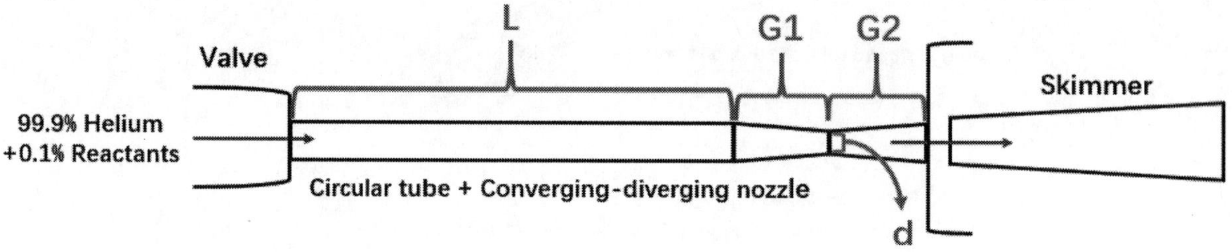

Figure 2: Novel microreactor structure design diagram

To explore the influence of the design parameters, we carried out CFD simulations. The continuity equation, conservation of linear momentum equation, and conservation of energy equation of steady state compressible ideal gas flow are described as follows [1],

$$\nabla \cdot (\rho \vec{U}) = 0 \qquad \text{Eq. 4}$$

$$\rho(\vec{U} \cdot \nabla)\vec{U} = \vec{f} - \nabla p + \mu \nabla^2 \vec{U} - \tfrac{2}{3}\mu \nabla(\nabla \cdot \vec{U}) \qquad \text{Eq. 5}$$

$$c_p \nabla \cdot (\rho \vec{U} T) = \nabla \cdot (k \nabla T) + \lambda \Psi + \mu \Phi + \dot{Q} \qquad \text{Eq. 6}$$

$$\Psi = \left(\frac{\partial u}{\partial x} + \frac{\partial v}{\partial y} + \frac{\partial w}{\partial z}\right)^2 \qquad \text{Eq. 7}$$

$$\Phi = 2\left[\left(\frac{\partial u}{\partial x}\right)^2 + \left(\frac{\partial v}{\partial y}\right)^2 + \left(\frac{\partial w}{\partial z}\right)^2\right] + \left(\frac{\partial u}{\partial y} + \frac{\partial v}{\partial x}\right)^2 + \left(\frac{\partial v}{\partial z} + \frac{\partial w}{\partial y}\right)^2 + \left(\frac{\partial w}{\partial x} + \frac{\partial u}{\partial z}\right)^2 \qquad \text{Eq. 8}$$

where $\vec{f}$ is the body force, k is the thermal conductivity, c_p is the specific heat of gas, μ dynamics viscosity, λ is bulk viscosity, $\vec{U}$ is the velocity vector, and u, v, w represent the velocity magnitude in each of the three Cartesian dimensions.

During experiments, the microreactor wall is heated up to 1300-1700 K, which results in the carrier gas experiencing large temperature changes from room temperature at the inlet within the short residence time of the reactor. Thus, the temperature dependence of thermal conductivity and dynamic viscosity of the gases used must be accounted for. For the simulations reported here, the carrier gas is helium, and we use property results from Petersen [2]. The fitted relationships are

$$k = 2.682 \cdot 10^{-3}(1 + 1.123 \cdot 10^{-3}P)T^{\left(0.71(1-2\cdot10^{-4}P)\right)}\ [W/(m * K)] \qquad \text{Eq. 9}$$

$$\mu = 3.67\times10^{-7} \cdot T^{0.7}\ [Pa{\cdot}s] \qquad \text{Eq. 10}$$

where T and P are in the units of Kelvin and bar respectively. (At our pressures, there is essentially no pressure dependence.) This system of equations has been implemented using ANSYS Fluent and MATLAB. Three dimensional models with different structure parameters were built in Fluent using a tetrahedral mesh. The pressure-based solver was applied with the SIMPLE scheme using a second-order-upwind discretization method.

3. Results and Discussion

For this study, helium was used as the carrier gas, and the reactor diameter held constant.

3.1 Influence of the choke diameter, d

The influence of the nozzle diameter, d, on the reactor center line pressure and velocity distributions was determined for several inlet mass flow rates. The nozzle diameter was varied from 0.30 to 0.10 mm, while holding all other structure variables constant (L = 21mm, G1 = 3mm, and G2 = 2mm). As shown in Fig. 3 , without the nozzle structure, the pressure profile continuously decreases from the inlet pressure to zero along the length of the reactor. Upon addition of the nozzle, the pressure drop within the main body of the reactor decreases. Likewise (see Fig. 4), the Mach number becomes more uniform. Therefore, addition of the nozzle allows considerable control over the pressure and velocity axial distributions.

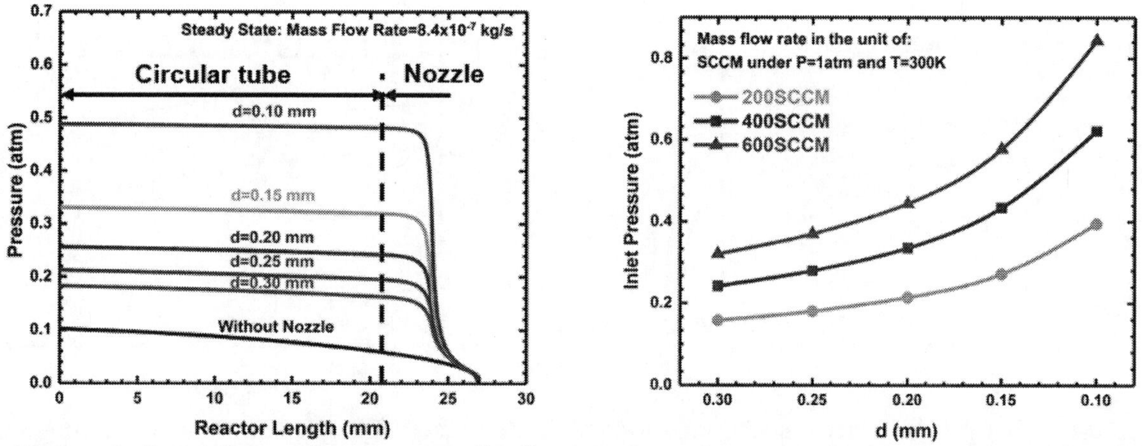

Figure 3: Left: Center line pressure distribution along the reactor length changing with choke diameter, d. **Right:** Microreactor inlet presure under different flow rates changing with choke diameter, d.

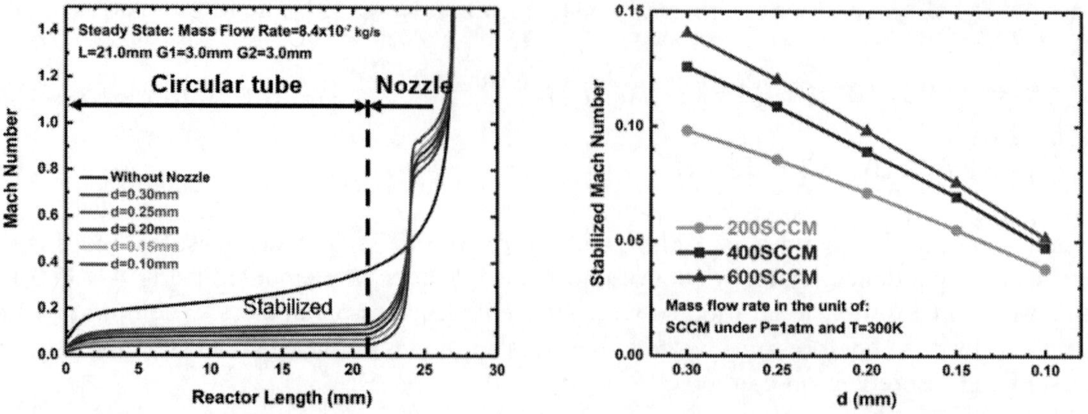

Figure 4: Left: Center line Mach number distribution along the reactor length changing with choke diameter, d. **Right:** Microreactor stabilized Mach number under different flow rates changing with choke diameter, d.

3.2 Influence of G1 & G2 and L

Further calculations were also carried out to determine the influence of the lengths of both the converging and diverging nozzles, G1 and G2 respectively. The simulation results for the center line pressure profiles as they change with nozzle length are shown in the Fig. 5. When the converging nozzle length, G1, increases from 3.0 mm to 5.0 mm, the inlet pressure only changes by ~0.02 atm, which is negligible. The converging nozzle length, G1, also negligibly impacts predicted velocity and residence time, demonstrating that the ideal converging nozzle length for our microreactor is simply that which is easiest to fabricate. The same conclusions may be drawn for adjusting the diverging nozzle length, G2. As such, we have decided to remove the diverging nozzle structure as it does little to affect the properties of the new microreactor other than to increase fabrication difficulty.

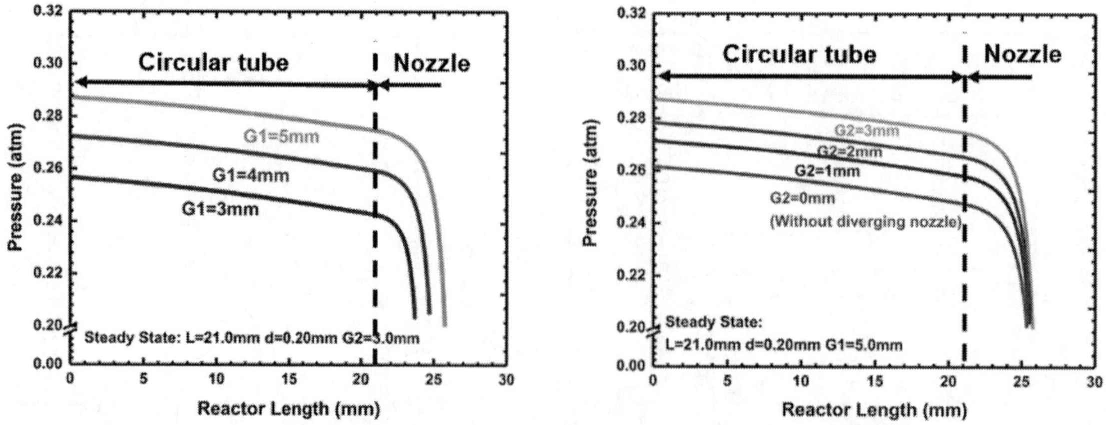

Figure 5: Left: Pressure distribution along along the reactor length changing converging nozzle length G1. **Right:** Pressure distribution along along the reactor length changing diverging nozzle length G2.

Finally, simulations were conducted where the microreactor main body length, L, was adjusted from 20 to 30 mm. These results showed that by increasing the reactor length, the only significant property change observed was an increase in residence time.

3.3 Temperature, Velocity and Knudsen Number Profiles

To make a direct comparison between reactors with and without nozzle, we fixed L = 20mm, G1 = 5mm, and d = 0.15mm. The inlet flow rate was set at 300 SCCM. The simulation results are shown in Fig. 6. The difference in pressure and velocity profiles is readily apparent. In the open-ended reactor, pressure falls continuously to a value close to zero, while the velocity rapidly climbs to the sonic point. With the nozzle in place, the velocity, while not constant, does not vary greatly, and the pressure change is slight, due only to friction. Another significant difference is that the Knudsen number remains at lower values when the nozzle is in place. Guan et al. [3] showed that for a helium carrier gas, free molecular effects begin to occur near the reactor exit when no nozzle is present. This led to slip boundary conditions. With the nozzle, the flow remains in the continuum domain until downstream of the nozzle.

The effect of the parameters d, G1 and L on the flow profiles are described in Table 1. Nozzle diameter, d, has the greatest effect on the fluid properties, while changing the nozzle lengths G1 and G2 have negligible effects. The reactor length, L, had little effect on the flow profile, but directly affected residence time

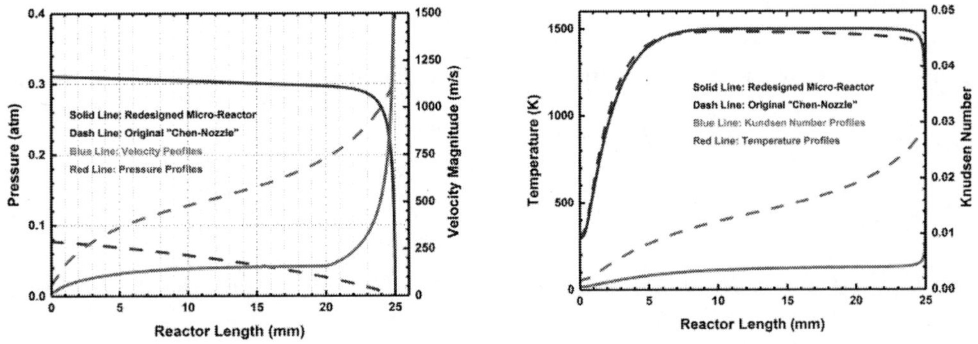

Figure 6: Center line profiles for both the new microreactor design and the current design showing **Left:** pressure and velocity distribution **Right:** temperature and Knudsen number distribution

	Pressure	Velocity	Temperature	Residence time
d	●	●	○	●
G1	◐	◐	○	◐
L	○	○	○	●

● **Distinct Influence** ◐ **Slight Influence** ○ **Almost No Influence**

Table 1: The influence of reactor structure parameters on thermodynamics profiles

4. Conclusions and future work

In this work, the influence of adding a nozzle structure to the existing CU Boulder microreactor system was explored using CFD calculations. Adding a downstream nozzle was demonstrated to cause most of the reactor pressure drop to take place across the nozzle, resulting in more uniform pressure and velocity profiles. The length of the nozzle, G1, was found to have negligible effects. The effect of adjusting the length of the existing reactor, L, was shown mainly to influence the overall microreactor residence time. With a thorough understanding of the influence of the geometric parameters of the new microreactor structure on the thermodynamic properties of the flow within the reactor, a new geometry has been identified and is in the process of being fabricated for testing.

Acknowledgements

This research was funded through University of Colorado Boulder via the startup funds of Prof. Nicole Labbe and support for J.W.D and G.B.E as well as partial support for C.O.R was provided through NSF Grant CBET-1403979

References

[1]. Moukalled, F., L. Mangani, and M. Darwish. "The finite volume method in computational fluid dynamics." (2016).

[2]. Petersen, Helge. *The properties of helium: density, specific heats, viscosity, and thermal conductivity at pressures from 1 to 100 bar and from room temperature to about 1800 K.* 1970.

[3] Guan, Qi, et al. "The properties of a micro-reactor for the study of the unimolecular decomposition of large molecules." *International Reviews in Physical Chemistry* 33.4 (2014): 447-487.

Sub Topic: Reaction Kinetics

10th Meeting of the Western States Section of The Combustion Institute
Organized by the University of Wyoming
October 2–3, 2017
Laramie, Wyoming

Comparison and analysis of chemical kinetic models for toluene autoignition

Morgan A. Mayer[1] and Kyle E. Niemeyer[1,]*

[1] *School of Mechanical, Industrial and Manufacturing Engineering, Oregon State University, Corvallis OR, United States*

[*] *Corresponding author:* kyle.niemeyer@oregonstate.edu

Abstract: Real transportation fuels contain a wide range of hydrocarbon species; however, their complexity poses challenges to modeling. Surrogate models represent real fuels with a blend of hydrocarbons, and can accurately approximate these fuels and enable practical simulation of the reacting flows that occur in engines and combustors. Surrogates for gasoline, diesel, and jet fuels frequently include toluene as a representative for aromatic components. The literature contains several chemical kinetic models for toluene oxidation, but an inclusive and comprehensive comparison has not yet been performed. This study objectively compares several kinetic models for toluene autoignition and identifies the most accurate existing model. Experimental data was collected from the literature for toluene autoignition in shock tubes and rapid compression machines, and converted into the ChemKED standard data format. Then, this dataset was used to perform a validation of the literature kinetic models using the PyTeCK software. Sensitivity analysis will be performed on the best-performing model to identify candidate reactions for improvement. All experimental data encoded in ChemKED for this study are available openly via the Prometheus database.
Keywords: ***Chemical kinetics, Toluene, Model validation, Aromatics, ChemKED***

1. Introduction

Surrogate models for gasolines predict the chemical kinetic behavior of real fuels, which have too many intermediate species and reactions to computationally model. The primary reference fuels (PRFs)—*n*-heptane and isooctane—represent alkanes in surrogate models for gasoline; likewise, the addition of toluene can represent the aromatic hydrocarbons in these surrogate models. It is important to account for aromatic concentration as practical fuels can be comprised of up to 50% aromatic hydrocarbons [1, 2]. Using PRFs allows surrogates to have similar autoignition resistance as the fuels they represent by adjusting the *n*-heptane and isooctane ratio, which determines the octane rating values—research octane number (RON) and motor octane number (MON) [2, 3]. Autoignition resistance, reflected in the octane rating, can be closely matched with that of a PRF mixture only at specified operating conditions. The real fuel sensitivity (RON − MON) cannot be replicated with the two PRF components alone since PRFs have zero octane sensitivity by definition. Other components need to be added to reflect the real fuel sensitivity since RON and MON do not account for reaction kinetics. Toluene reference fuels (TRFs) have a sensitivity similar to gasoline for the same RON, so the surrogate will have a comparable octane rating for a larger range of operating conditions [3]. Toluene is the most common aromatic component in practical fuels, with mole compositions as high as 35%, so toluene can serve as the aromatic in a surrogate [4]. Ethanol has also been of interest in surrogate fuels because of its recent popularity as a (potentially sustainable) fuel additive [3, 5]. Ethanol has an octane rating above the current range defined by the PRFs, so it can represent higher-octane fuels as a surrogate component [2, 3].

Consequently, surrogate fuels with alkane and aromatic representation offer more accuracy in computational engine modeling. The chemical kinetics literature contains many detailed models for toluene oxidation in ranges of engine-relevant conditions [4, 6–10]. However, there is no comprehensive validation that compares the performance of these kinetic models for toluene.

Intermediate reaction rates are difficult to standardize due to the complex nature of combustion reactions and current data collection technologies. Reaction rates are presently based on experimental data, reaction rate estimations, quantum chemical theory calculations, or transition state theory estimations [7, 11, 12]. Models are typically validated against experimental data from several oxidation experiments, then followed by a "brute force" sensitivity analysis for model updates [7, 8, 13]. Model development methods have improved in the combustion community, and are moving towards detailing the kinetics of more complex hydrocarbon mixtures.

Although several models have been published for the same fuels, few performance studies quantify and validate multiple models against a large share of available ignition delay data. Olm et al. [14, 15] comprehensively evaluated many hydrogen and syngas models with data from various fundamental combustion experiments. The hydrogen study compared 19 models with 770 data points, while the syngas study compared 16 models with 732 data points [14, 15]. Niemeyer [16] quantified nine PRF and TRF model performances by applying the same methods to *n*-heptane models using shock tube and rapid compression machine data from 14 autoignition studies. Similarly, Sirumalla et al. [17] employed this method for kinetic models of butanol and its isomers. These are the initial comprehensive quantitative model comparisons, however many earlier studies do not compare several models together in this way. Sakai et al. [9] began with the model from Pitz et al. [18] and modified rate constants based on recent experimental studies and theoretical transition state studies [12]. Davidson et al. [13] found that the models by Pitz et al. [18] and Dagaut et al. [19] significantly overpredict ignition delay times at $\phi = 1$.

This study aims to quantitatively evaluate a selection of published toluene kinetic models in the chemical kinetics literature. The evaluation method validates the performance of proposed models against available experimental data. This study is a next step in describing surrogate component autoignition behavior. The following sections include computational methods, published models, experimental data, simulation methods, results, and conclusions..

2. Methods

We used the Python-based model validation tool PyTeCK [20] to automatically simulate experiments based on kinetic models and experimental conditions. PyTeCK performs Cantera-based [21] simulations using the conditions of each experimental data set. PyTeCK uses the method of Olm et al. [14, 15] to assess the performance of the chemical kinetic models. The agreement between experimental and simulated ignition delay values is given by the average error E over N data sets in the following:

$$E = \frac{1}{N} \sum_{i=1}^{N} E_i \tag{1}$$

where E_i is the average error for the ith dataset

$$E_i = \frac{1}{N_i}\sum_{j=1}^{N_i}\left(\frac{\log\tau_{ij}^{\text{exp}} - \log\tau_{ij}^{\text{sim}}}{\sigma\left(\log\tau_{ij}^{\text{exp}}\right)}\right)^2, \tag{2}$$

N_i is the number of data points in set i, τ_{ij}^{exp} and τ_{ij}^{sim} are the experimental and simulated ignition delay times for the jth data point in set i, respectively, and σ is the standard deviation. The log function represents the natural logarithm. Scatter in experimental results is proportional to the ignition delay, so the logarithm of ignition delay is used in the calculation, as done by Olm et al. [14, 15].

The approach to calculating standard deviation of a dataset, $\sigma\left(\log\tau_{ij}^{\text{exp}}\right)$, follows that of Olm et al [14, 15]. A spline is fit to the experimental ignition delay values with respect to the changing variable, usually temperature, using the `SciPy` function `interpolation.UnivariateSpline`. A minimum standard deviation of $\sigma = 0.10$ is used. It would be preferable to use associated published uncertainties rather than estimations of the deviations, however these values are often not reported.

3. Models

Table 1 includes the detailed chemical kinetic models for toluene used in this study. Seven models are evaluated and compared against experimental data [6, 7, 9, 10, 22–24]. These models are not all of the available detailed models, but a selection of common models that were openly published or obtained from the author. Models were converted to a Cantera [21] format from the original Chemkin model file.

Table 1: Summary of the detailed chemical kinetic models used for this study. "Coverage" describes the range of fuels included in the model. Ar/He describes if the model includes interactions with the inert gases argon or helium in the model in addition to nitrogen, N_2.

Model	Coverage	No. Species	No. Reactions	Ar/He	Reference
CNRS-2009	PRF + toluene+NOx	536	3000	x/x	[6]
Sakai-2009	PRF + toluene	783	2883	x/x	[9]
LLNL-2012	PRF + 1-hexene + toluene	1550	8000	x/x	[10, 25]
Andrae-2013	TRF	138	641	x/–	[7]
Dalian-2013	TRF	56	191	–/–	[22]
Tsinghua-2014	TRF + ethanol	91	411	x/–	[24]
CRECK-2014	TRF + ethanol	317	12353	x/x	[23, 26, 27]

Anderlohr et al. [6] proposed the CNRS-2009 detailed chemical kinetic model for PRF and toluene oxidation. It combines the PRF autoignition model from Buda et al. [28] and the toluene oxidation model from Bounaceur et al. [29]. Additionally, reactions with NO_x are included. The model includes the interactions of NO_x on HC oxidation, and was validated against HCCI and perfectly stirred reactor experiments for neat fuels and fuel blends [6].

The model published by Sakai et al. [9] includes low temperature toluene kinetics relevant to engine conditions. The starting model was by Pitz et al. [18], and rate coefficients for low temperature toluene and benzyl reactions were modified based on theoretical estimations. The model was validated against existing shock tube and flow tube data. The toluene submechanism was merged into the PRF model by Ogura et al. [30], and cross reactions between PRFs and toluene were added.

For the LLNL-2012 model, Mehl et al. [10, 25] updated a previous LLNL mechanism to predict ignition times for PRFs, 1-hexene, toluene, and their mixtures in a rapid compression machine and jet stirred reactor. The cyclopentadiene mechanism was revised for the toluene mechanism, and was validated against the work of Butler et al. [31]. Other interactions occurring between the components identified the effect of different classes of compounds on NTC (negative temperature coefficient) behavior. Radical scavenging effects by toluene and 1-hexene can explain the NTC behavior. In addition, they observed that activation of HO_2 radicals by allyl and benzyl radicals are dependent on the temperature range.

Andrae et al. [7] updated the reaction parameters for neat toluene and its mixtures in the previous model [32]. Updates to a previous toluene submechanism [33] were based on quantum kinetic calculations by Murakami et al. [34], the Sakai-2013 model [9], and individual rates from the literature. The simulations from the updated, less detailed model shows better agreement than the previous detailed model.

The Dalian-2013 model was developed by Liu et al. [22] by combining a previous skeletal PRF model with a reduced model for toluene. Ignition chemistry was described using semi-decoupling method that reduces large molecule mechanisms while leaving the small hydrocarbon mechanisms in its previous form. Liu et al. [22] modified toluene pathways from the models of Andrae et al. [33], Mehl et al. [35], and Sakai et al. [9] based on shock-tube ignition delays and jet-stirred reactor species profiles.

Zhong and Zheng [24] developed the five-component Tsinghua-2014 model by combining their earlier three-component model with reduced models to predict ignition behavior of the PRFs, ethanol, toluene, and diisobutadiene. The toluene submechanism in the current five-component model was revised from Andrae et al. [33].

The CRECK-2014 model version 1412 [23, 26, 27] combines ethanol pathways with a prior model for hydrocarbon oxidation [26]. The model includes chemistry for PRFs, alcohols, and ethers.

4. Experimental data

Ignition delay data was used from a selection of published shock tube studies. The data was taken directly from publications or from direct communication with the authors. Each data set was arranged in the YAML-based ChemKED format. Weber and Niemeyer [36] describe the ChemKED (**Chem**ical **K**inetics **E**xperimental **D**ata) format in detail. In brief, ChemKED is a human- and machine-readible, YAML-based file format for describing fundamental combustion measurements with sufficient information to simulate experimental data points [36].

5. Shock tube simulations

Cantera [21] performed all simulations, driven by PyTeCK [20]. The experimental conditions for each simulation include the reflected shock temperature and pressure, in addition to the initial

Table 2: Summary of the experimental shock tube conditions for toluene oxidation, where $\chi_{O_2}(\%)$ is the mole percent of oxygen gas as reactant.

Study	P (atm)	T (K)	ϕ	$\chi_{O_2}(\%)$	No. data points	Reference
Burcat et al.	1.95–8.85	1339–1797	0.331–1.0	4.5–13.5	54	[37]
Davidson et al.	14–59	855–1269	0.5, 1.0	20.5	25	[13]
Vasudevan et al.	1.5–5.0	1400–2000	0.5–1.875	0.2–4.5	24	[38]
Shen et al.	10–61	1021–1400	0.25, 0.5, 1.0	20.5	58	[11]
Hartmann et al.	39.5	700–1200	0.5, 1.0	20.5	14	[39]

reactant compositions. Shock tube simulations in this study were modeled using the assumption that there are adiabatic constant-volume conditions based on the reflected shock temperature and pressure, and initial reactant composition [20]. It is worth noting that longer ignition delay times above 1 ms observe polytropic pressure increases, and this method does not account for that since there is no pressure history information [40, 41]. Neglecting this could result in greater predicted ignition delay values [42].

6. Results and discussion

This section compares and shows the performance of the seven kinetic models listed in Table 1 using the experimental data described in Table 2. There are a total of 175 data points spread across 21 datasets. Ignition delays in most of the experiments were defined as location of maximum derivative or half the maximum derivative of OH· radical concentration, or the maximum or half-maximum derivative of pressure. Vasudevan et al. [38] specified ignition delay to be half the maximum OH concentration.

Figure 1 describes the performance of the models in predicting ignition delay in toluene oxidation based on experimental data. From the models included in this study, the Sakai-2009 model best predicts ignition delay based on the experimental data included in this study, with the lowest error function value of 28.841. The CNRS-2009 model performed the worst with an error function value of 188.07.

7. Conclusions

This study compared the performance of kinetic models in predicting autoignition delay for toluene. The Sakai-2009 model performs best and the CNRS-2009 model performs worse for the experimental data considered. Six of the seven models considered had relatively similar performances, as their error metrics are in the range of 28.841 to 47.238, while the CNRS-2009 model had an overall error metric over five times larger than the average of the other six models. In addition to quantifying the ability of the models considered to predict ignition delay, a useful data set from shock tube experiments were collected and standardized for such mixtures. These data and the supporting software will be made openly available to support continuing model evaluation. The next steps for this project include collecting more available models for toluene oxidation, and standardizing a more diverse set of autoignition data.

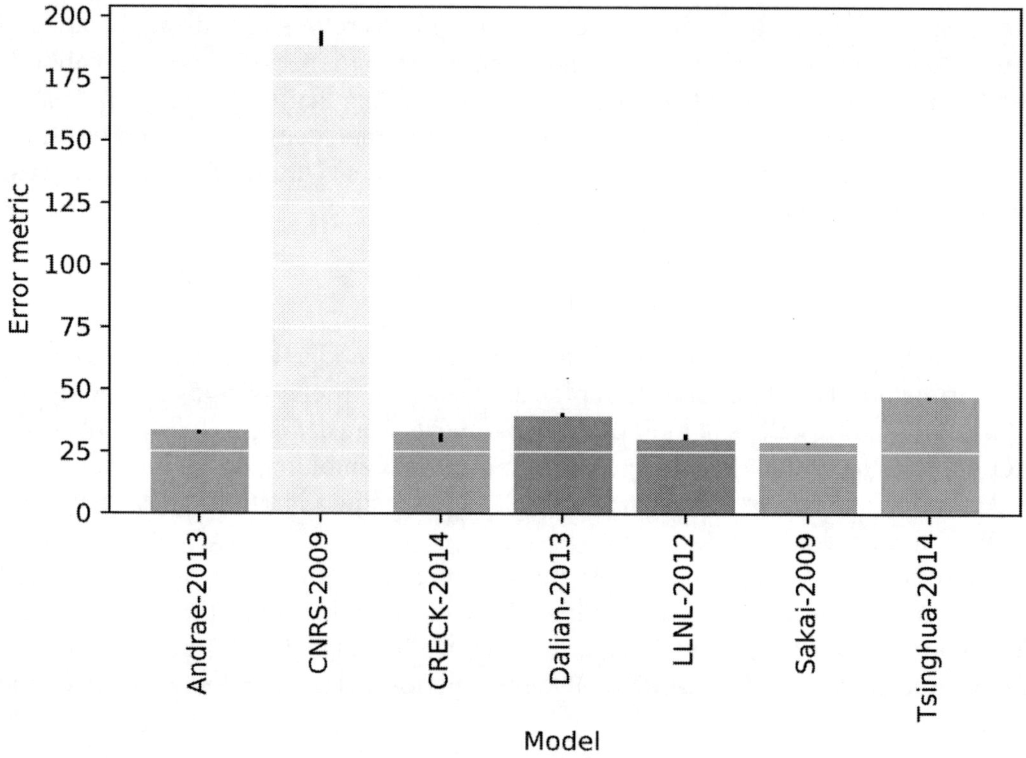

Figure 1: Mean error function for the chemical kinetic models in toluene oxidation simulations of ignition delay times, with error bars indicating the standard deviation among the datasets. Errors were sampled using 175 data points grouped into 21 datasets. Lower values indicate better agreement between a model and the experimental data.

8. Acknowledgements

This material is based upon work supported by the National Science Foundation under grant ACI-1535065 and the Women and Minorities in Engineering program at Oregon State University. The authors also thank Bryan Weber for helpful discussions on experimental uncertainty, as well as numerous colleagues for sharing their kinetic models and experimental data.

References

[1] G. Mittal and C. J. Sung, Autoignition of toluene and benzene at elevated pressures in a rapid compression machine, Combustion and Flame 150 (2007) 355–368. DOI: 10.1016/j.combustflame.2007.04.014.

[2] S. Turns, An Introduction to Combustion: Concepts and Applications, McGraw-Hill Series in Mechanical Engineering, McGraw-Hill, 2012.

[3] G. Kalghatgi, R. Head, J. Chang, Y. Viollet, H. Babiker, and A. Amer, An Alternative Method Based on Toluene/n-Heptane Surrogate Fuels for Rating the Anti-Knock Quality of Practical Gasolines, SAE International Journal of Fuels and Lubricants (2014), DOI: 10.4271/2014-01-2609, arXiv: 1011.1669.

[4] K. R.Sivaramakrishnan R.S.Tranter, A high pressure model for the oxidation of toluene, Proceedings of the Combustion Institute 30 (2005) 1165–1173. DOI: 10.1016/j.proci.2004.08.128.

[5] K. E. Niemeyer and C. J. Sung, Mechanism reduction for multicomponent surrogates: A case study using toluene reference fuels, Combustion and Flame 161 (2014) 2752–2764. DOI: 10.1016/j.combustflame.2014.05.001.

[6] J. M. Anderlohr, R. Bounaceur, A. Pires Da Cruz, and F. Battin-Leclerc, Modeling of autoignition and NO sensitization for the oxidation of IC engine surrogate fuels, Combustion and Flame 156 (2009) 505–521. DOI: 10.1016/j.combustflame.2008.09.009, arXiv: 0903.3809.

[7] J. C. G. Andrae, Comprehensive chemical kinetic modeling of toluene reference fuels oxidation, Fuel 107 (2013) 740–748. DOI: 10.1016/j.fuel.2013.01.070.

[8] W. K. Metcalfe, W. K. Metcalfe, S. Dooley, S. Dooley, F. L. Dryer, and F. L. Dryer, Comprehensive Detailed Chemical Kinetic Modeling Study of Toluene Oxidation, Energy Fuels 25 (2011) 4915–4936. DOI: 10.1021/ef200900q.

[9] Y. Sakai, A. Miyoshi, M. Koshi, and W. J. Pitz, A kinetic modeling study on the oxidation of primary reference fuel-toluene mixtures including cross reactions between aromatics and aliphatics, Proceedings of the Combustion Institute 32 (2009) 411–418. DOI: 10.1016/j.proci.2008.06.154.

[10] M. Mehl, W. J. Pitz, C. K. Westbrook, and H. J. Curran, Kinetic modeling of gasoline surrogate components and mixtures under engine conditions, Proceedings of the Combustion Institute 33 (2011) 193–200. DOI: 10.1016/j.proci.2010.05.027.

[11] H. P. S. Shen, J. Vanderover, and M. A. Oehlschlaeger, A shock tube study of the autoignition of toluene/air mixtures at high pressures, Proceedings of the Combustion Institute 32 (2009) 165–172. DOI: 10.1016/j.proci.2008.05.004.

[12] S. J. Klippenstein, L. B. Harding, and Y. Georgievskii, On the formation and decomposition of C7H8, Proceedings of the Combustion Institute 31 (2007) 221–229. DOI: 10.1016/j.proci.2006.08.045.

[13] D. F. Davidson, B. M. Gauthier, and R. K. Hanson, Shock tube ignition measurements of iso-octane/air and toluene/air at high pressures, Proceedings of the Combustion Institute 30 (2005) 175–1182. DOI: 10.1016/j.proci.2004.08.004.

[14] C. Olm, I. G. Zsély, T. Varga, H. J. Curran, and T. Turányi, Comparison of the performance of several recent syngas combustion mechanisms, Combustion and Flame 162 (2015) 1793–1812. DOI: 10.1016/j.combustflame.2014.12.001.

[15] C. Olm, I. G. Zsély, R. Pálvölgyi, T. Varga, T. Nagy, H. J. Curran, and T. Turányi, Comparison of the performance of several recent hydrogen combustion mechanisms, Combust. Flame 161 (2014) 2219–2234. DOI: 10.1016/j.combustflame.2014.03.006.

[16] K. E. Niemeyer, An autoignition performance comparison of chemical kinetics models for *n*-heptane, Spring 2016 Meeting of the Western States Section of the Combustion Institute Seattle, WA, (2016), DOI: 10.6084/m9.figshare.3120724.

[17] S. K. Sirumalla, M. A. Mayer, K. E. Niemeyer, and R. West, Assessing impacts of discrepancies in model parameters on autoignition model performance: a case study using butanol, (2017), arXiv: arXiv:1708.02232v1.

[18] W. J. Pitz, R. Seiser, J. W. Bozzelli, I. Da Costa, R. Fournet, F. Billaud, F. Battin-Leclerc, K. Seshadri, and C. K. Westbrook, Chemical Kinetic Characterization of Combustion of Toluene, 2nd Joint Meeting of the U.S. Section of the Combustion Institute Oakland, CA, (2001), pp. 1–19.

[19] P. Dagaut, A. Ristori, A. El Bakali, and M. Cathonnet, Experimental and kinetic modeling study of the oxidation of n-propylbenzene, Fuel 181 (2002) 173–184. DOI: 10.1016/S0016-2361(01)00139-9.

[20] K. E. Niemeyer, PyTeCK: a Python-based automatic testing package for chemical kinetic models, Proc. of the 15th Python in Science Conference (2016).

[21] D. G. Goodwin, H. K. Moffat, and R. L. Speth, *Cantera: An Object-oriented Software Toolkit for Chemical Kinetics, Thermodynamics, and Transport Processes*, http://www.cantera.org, Version 2.3.0, 2017, DOI: 10.5281/zenodo.170284.

[22] Y. D. Liu, M. Jia, M. Z. Xie, and B. Pang, Development of a new skeletal chemical kinetic model of toluene reference fuel with application to gasoline surrogate fuels for computational fluid dynamics engine simulation, Energy and Fuels (2013), pp. 4899–4909, DOI: 10.1021/ef4009955.

[23] CRECK Modeling Group, *Primary Reference Fuels (PRF) + PAH + alcohols + ethers (version 1412)*, 2014, URL: http://creckmodeling.chem.polimi.it/.

[24] B. J. Zhong and D. Zheng, A chemical mechanism for ignition and oxidation of multicomponent gasoline surrogate fuels, Fuel (2014), DOI: 10.1016/j.fuel.2014.03.044.

[25] M. Mehl, W. Pitz, M. Sarathy, Y. Yang, and J. E. Dec, Detailed Kinetic Modeling of Conventional Gasoline at Highly Boosted Conditions and the Associated Intermediate Temperature Heat Release, (2012), DOI: 10.4271/2012-01-1109.

[26] E. Ranzi, A. Frassoldati, S. Granata, and T. Faravelli, Wide-range kinetic modeling study of the pyrolysis, partial oxidation, and combustion of heavy n-alkanes, Industrial and Engineering Chemistry Research (2005), DOI: 10.1021/ie049318g.

[27] T. Bieleveld, A. Frassoldati, A. Cuoci, T. Faravelli, E. Ranzi, U. Niemann, and K. Seshadri, Experimental and kinetic modeling study of combustion of gasoline, its surrogates and components in laminar non-premixed flows, Proceedings of the Combustion Institute (2009), DOI: 10.1016/j.proci.2008.06.214.

[28] F. Buda, R. Bounaceur, V. Warth, P. A. Glaude, R. Fournet, and F. Battin-Leclerc, Progress toward a unified detailed kinetic model for the autoignition of alkanes from C4 to C10 between 600 and 1200 K, Combustion and Flame (2005), DOI: 10.1016/j.combustflame.2005.03.005.

[29] R. Bounaceur, I. Da Costa, R. Fournet, F. Billaud, and F. Battin-Leclerc, Experimental and modeling study of the oxidation of toluene, International Journal of Chemical Kinetics (2005), DOI: 10.1002/kin.20047.

[30] T. Ogura, Y. Sakai, A. Miyoshi, M. Koshi, and P. Dagaut, Modeling of the oxidation of primary reference fuel in the presence of oxygenated octane improvers: Ethyl tert-butyl ether and ethanol, Energy and Fuels (2007), DOI: 10.1021/ef700321e.

[31] R. G. Butler and I. Glassman, Cyclopentadiene combustion in a plug flow reactor near 1150 K, Proceedings of the Combustion Institute (2009), DOI: 10.1016/j.proci.2008.05.010.

[32] J. C. G. Andrae, P. Björnbom, R. F. Cracknell, and G. T. Kalghatgi, Autoignition of toluene reference fuels at high pressures modeled with detailed chemical kinetics, Combustion and Flame (2007), DOI: 10.1016/j.combustflame.2006.12.014.

[33] J. C. G. Andrae, T. Brinck, and G. T. Kalghatgi, HCCI experiments with toluene reference fuels modeled by a semidetailed chemical kinetic model, Combustion and Flame (2008), DOI: 10.1016/j.combustflame.2008.05.010.

[34] Y. Murakami, T. Oguchi, K. Hashimoto, and Y. Nosaka, Theoretical study of the benzyl + O2 reaction: Kinetics, mechanism, and product branching ratios, Journal of Physical Chemistry A (2007), DOI: 10.1021/jp075369q.

[35] M. Mehl, T. Faravelli, E. Ranzi, D. Miller, and N. Cernansky, Experimental and kinetic modeling study of the effect of fuel composition in HCCI engines, Proceedings of the Combustion Institute 32 (2009) 2843–2850. DOI: 10.1016/j.proci.2008.08.001.

[36] B. W. Weber and K. E. Niemeyer, ChemKED: a human-and machine-readable data standard for chemical kinetics experiments, (2017), arXiv: arXiv:1706.01987v2.

[37] A. Burcat, C. Snyder, and T. Brabbs, Ignition Delay Times of Benzene and Toluene With Oxygen in Argon Mixtures, (1986).

[38] V. Vasudevan, D. F. Davidson, and R. K. Hanson, Shock tube measurements of toluene ignition times and OH concentration time histories, Proceedings of the Combustion Institute (2005), DOI: 10.1016/j.proci.2004.07.039.

[39] Auto-ignition of toluene-doped n-heptane and iso-octane/air mixtures: High-pressure shock-tube experiments and kinetics modeling, Combustion and Flame (2011).

[40] D. F. Davidson and R. K. Hanson, Interpreting shock tube ignition data, International Journal of Chemical Kinetics 36 (2004) 510–523. DOI: 10.1002/kin.20024.

[41] M. Chaos and F. L. Dryer, Chemical-kinetic modeling of ignition delay: Considerations in interpreting shock tube data, International Journal of Chemical Kinetics (2010), DOI: 10.1002/kin.20471.

[42] F. M. Haas, M. Chaos, and F. L. Dryer, Low and intermediate temperature oxidation of ethanol and ethanol-PRF blends: An experimental and modeling study, Combustion and Flame 156 (2009) 2346–2350. DOI: 10.1016/j.combustflame.2009.08.012.

Sub Topic: Reaction Kinetics

10th Meeting of the Western States Section of The Combustion Institute
Organized by the University of Wyoming
October 2–3, 2017
Laramie, Wyoming

Investigating stiffness detection metrics for chemical kinetics

Andrew Alferman[1] *and Kyle E. Niemeyer*[1,*]

[1]*School of Mechanical, Industrial, and Manufacturing Engineering, Oregon State University, Corvallis, OR 97331, USA*
[*]*Corresponding author:* Kyle.Niemeyer@oregonstate.edu

Abstract: Simulations of combustion and reacting flows often encounter stiffness in the equations governing chemical kinetics. Explicit solvers for these ordinary differential equations offer low computational expense, but typically cannot efficiently handle stiff systems. In contrast, implicit methods demand greater expense but offer unconditional stability–as a result, most reactive flow solvers rely on these methods by default (other than explicit direct numerical simulation solvers). However, explicit or stabilized explicit methods can instead be used to reduce the computational expense while remaining stable and accurate if the chemical kinetics systems exhibit low-to-moderate stiffness. This study investigates metrics for quantifying stiffness, with the goal of identifying one capable of efficiently and robustly determining the appropriate category of integrator required. Methods of measuring the stiffness of chemical kinetics states will be investigated, including stiffness ratio, stiffness index, stiffness indicator, and chemical explosive mode. These will be applied to simulations of hydrogen/carbon monoxide and methane autoignition using initial conditions representative of realistic turbulent combustion, obtained from partially stirred reactor simulations. The stiffness quantification metrics will be compared with the time required to integrate using implicit and explicit methods. We will conclude by analyzing preliminary performance analysis of an integrator scheduler using these metrics.
***Keywords:* Stiffness quantification, Ordinary differential equations, Chemical kinetics, Computational cost reduction**

1. Introduction

An increasing demand for greater efficiency and lower CO_2 emissions in the US energy supply has driven the development of next-generation combustion technologies that use alternative fuels [1]. Computational modeling is one important tool that increasingly drives design and development of these new technologies. Shortening the design cycle and time-to-market of novel and efficient energy devices optimized for alternative fuels requires fundamental advances in combustion science and fuel chemistry; both depend on high-fidelity computational tools. The need to develop new, high-performance computational methods to support combustion research has been recognized by multiple federal agencies as a key objective towards implementing predictive models [2–5].

Species time scales can range from the order of nanoseconds to seconds, requiring greater computational expense for integration by conventional methods (i.e., numerical stiffness) [6]. Most combustion modeling approaches rely on a single, implicit ODE integration method to handle the chemistry in all spatial locations, but these methods are computationally expensive, especially for larger mechanisms. The computational cost of chemistry is either a quadratic or cubic function of the number of species [6]; reaction mechanisms of sizes comparable to those shown above pose computational difficulties even in zero-dimensional (homogeneous) simulations.

Depending on the local conditions, computational flow time-step size, and chemical mechanism being used, it is possible to encounter a wide range of stiffness within a single simulation. We can exploit this situation to reduce the overall simulation expense by selecting the most computationally efficient ODE solver on-the-fly based on local conditions. For example, a low-cost explicit method may be used far away from the reaction zone, while an implicit—or otherwise "stiff"—solver is more economical inside a flame. Making this selection requires a method to detect and classify stiffness.

Although the concept of stiffness has been identified for over 60 years, the term has not been precisely defined despite repeated efforts. The diverse set of problems considered to be "stiff" and the large variety of characterizations used to describe stiffness are amongst the difficulties that have been encountered in developing a precise definition [7]. Nonetheless, a variety of stiffness quantification methods have been developed with the goal of providing a practical means of evaluating a systems ODEs [7–11]. We look to these methods of stiffness quantification to determine their usefulness with respect to the equations governing combustion, and to determine if a reliable and efficient means of switching methods can be developed from it. In doing so, we hope to reduce the computational expense of combustion simulations.

2. Methods

We considered two different chemical kinetic models in this study: the relatively small hydrogen/carbon monoxide (H_2/CO) model of Burke et al. [12] and GRI Mech 3.0 [13], which uses CH_4 as fuel. The H_2/CO model uses 13 species while the GRI Mech 3.0 is larger and uses 53 species, allowing for Jacobian matrices of system of ODEs of 13×13 and 53×53, respectively. These relatively small Jacobian matrices allow for much faster evaluations of stiffness than would be possible with larger, more complicated kinetic systems, which can have hundreds to hundreds of thousands of chemical species [14].

The code used to run autoignition simulations and calculate stiffness metrics was written in Python and made use of open-source software, including the `ode` integrator function of SciPy [15]. Two different integrators of the `ode` integrator function were used to compare explicit to implicit methods. The first integrator was `vode`, which uses an implicit Adams method for non-stiff problems and a method based on the backwards differentiation formulae for stiff problems [16]. The explicit integrator used was `dopri5`, which uses a Runge-Kutta method of order (4)5 developed by Dormand and Prince with automatic step size control [17]. The `vode` integrator was set to use the default behavior of using first-order finite differencing to provide an estimate of the Jacobian matrix. Both integrators use internal time-steps in between the user defined time-steps for accuracy control and make use of `FORTRAN` libraries to perform the integrations. `NumPy` [18] was also extensively used in the code to determine eigenvalues and perform numerous basic mathematical functions. The integrator used default values of 10^{-6} and 10^{-12} for the error control parameters `rtol` and `atol`, respectively.

Computation of the right hand side (RHS) function and the Jacobian matrix of the systems of ODEs were readily achieved using pyJac [19], a Python-based open-source program that generates analytical Jacobian matrices for use in chemical kinetics modeling and analysis [20]. pyJac uses a chemical mechanism developed using Cantera [21] software and a set of initial conditions to generate functions that return the required values.

Our analysis was conducted using information gathered from partially stirred reactor (PaSR) simulations. As described by Niemeyer et al. [20], the PaSR model consists of a number of particles, each with a time-varying composition. At discrete time-steps, events including inflow, outflow, and pairing cause particles to change composition; between these time-steps, mixing and reaction fractional steps evolve the composition of all particles. By using data from a single particle of the PaSR model, the evaluation could be simplified to a zero-dimensional analysis. The H_2/CO simulation used a different set of PaSR data than the GRI Mech 3.0 simulation. Both sets of PaSR data included nine different configurations of temperature and pressure; the initial temperature values were 400, 600, and 800 K, while the pressure values were 1, 10, and 25 atm. Pressure remained constant throughout the simulations.

The time required by `vode` and `dopri5` to advance one user specified time step was measured for each particle of the PaSR simulation. The number of RHS function calls used by the integrator to advance one time-step was measured by adding a global variable, and the number of internal time-steps used by `vode` was measured. Measurement of the time-steps used by `vode` was accomplished without manipulation of the original `FORTRAN` code by limiting the integrator to one time step each time the integrator was called. `dopri5` is explicit and uses six function calls to advance each internal time step, therefore the number of internal time-steps for this integrator was calculated by dividing the number of RHS function calls by six.

The four stiffness quantification methods discussed within this paper are the stiffness ratio, the stiffness index, the stiffness indicator, and the chemical explosive mode. Variable names have been changed in the equations below to maintain consistency throughout this paper.

2.1 Stiffness Ratio

One commonly referenced measure of the stiffness of a system of ODEs is the "stiffness ratio." The stiffness ratio of the system of ODEs is defined as

$$\text{Stiffness Ratio} = \frac{\max|\lambda_p|}{\min|\lambda_p|} \tag{1}$$

in which λ_p is an eigenvalue of the Jacobian matrix of the system of ODEs [22]. A large stiffness ratio is an indication of a large range of time scales in the problem, which is a necessary component for stiffness to arise. This method is readily implemented and carries little computational cost beyond determining the eigenvalues of the Jacobian matrix.

2.2 Stiffness Index

One method of interest regarding stiffness quantification was the "IA-Stiffness Index" proposed by Shampine [23]. This method enables comparison of values of the stiffness index between different equations and different methods. Additionally, the method takes into account the impact of the order of the method selected when evaluating the stiffness. The method is relatively straightforward to implement, requiring computation of a vector of the derivatives of the system of equations, as well as either a weighted norm or the spectral radius of the Jacobian matrix.

The IA-stiffness index of a method of order p introduced by Shampine [23] is

$$\frac{h_{acc}}{h_{iter}} \doteq \tau^{1/(p+1)} \rho[A] \| y^{(p+1)}(x_n) \|^{-1/(p+1)} \left(\frac{|\xi|^{-1/(p+1)}}{|\gamma|} \right) \tag{2}$$

or alternatively,

$$\frac{h_{acc}}{h_{iter}} \doteq \tau^{1/(p+1)} \|A\| \|y^{(p+1)}(x_n)\|^{-1/(p+1)} \left(\frac{|\xi|^{-1/(p+1)}}{|\gamma|} \right) \tag{3}$$

where h_{acc} represents the largest step size which would result in a local accuracy test being passed, h_{iter} represents the minimum step size that will lead to divergence of simple iteration, τ represents the specified tolerance, $\rho[A]$ represents the spectral radius of the Jacobian matrix A, γ is a constant characteristic of the formula, and ξ represents a constant characterizing the accuracy of a reference method. Note that these equations require the $p+1$ derivative of the function $y(x_n)$. Additionally, the matrix norm $\|M\|$ of matrix M with i columns and j rows is given by

$$\|M\| = \max_i \frac{1}{w_i} \sum_j |M_{ij}| w_j \tag{4}$$

in which w_i and w_j are positive weights of the matrix [8].

Although the IA-stiffness index is useful in determining the stiffness of a method for a given system of equations, we are interested in quantifying the stiffness inherent to the system of equations itself. Such a quantification is necessary for investigating method switching mechanisms. Shampine notes that the scalar quantity

$$\text{Stiffness Index} = \rho[A] \|y^{(p+1)}(x_n)\|^{-1/(p+1)} \tag{5}$$

provides a fair "stiffness index" to a given problem [8]. For the remainder of this document, all references to the stiffness index will refer to the value obtained using equation (5). A large value of the stiffness index at a given point is an indication that the system of ODEs is locally stiff at that point.

As previously noted, use of the above equations requires calculation of a vector of the derivatives of a system of equations. In the case of this investigation, this vector is comprised of the derivatives of the thermochemical composition vector with respect to time, with the the vector defined as

$$\Phi = \{T, Y_1, Y_2, \ldots, Y_{N_{sp}}\}^\mathsf{T}, \tag{6}$$

where T is the temperature, Y_i are the species mass fractions, and N_{sp} is the number of species in the mechanism [20]. The vector needed to compute the stiffness index is therefore

$$\frac{\partial \Phi}{\partial t} = \left\{ \frac{\partial Y_1}{\partial t}, \frac{\partial Y_2}{\partial t}, \ldots, \frac{\partial Y_{N_{sp}}}{\partial t} \right\}^\mathsf{T}, \tag{7}$$

where time is denoted by t [20]. In addition to this vector of derivatives, computation of the stiffness index requires either the weighted norm or the spectral radius of the Jacobian matrix for the thermochemical composition vector.

After generating this data using pyJac, the values of the second derivative of the thermochemical composition vector were calculated numerically using a fourth-order central differencing formula. Numerical approximations of higher-level derivatives may also be generated using the same approach. Use of the central differencing formula was possible because the stiffness index was calculated after solving for the thermochemical composition vector, and the reaction was modeled using user defined time-steps of a constant size. Forward and backward difference formulae were used at the boundaries where central differencing could not be used. Variable time-step methods for calculating the second derivative or higher-level derivatives may also be used, however these methods were not necessary at the current stage of the investigation.

To facilitate comparison of the results obtained by Shampine [8], the order of the method was assumed to be 1 for all computations performed in this paper.

2.3 Stiffness Indicator

The "stiffness indicator" of a system of ODEs was proposed by Söderlind et. al. as a mathematically rigorous approach to characterize stiffness that works independent of the integration method used or operational criteria [7]. Söderlind notes that "large" negative values of the stiffness indicator are a necessary condition for stiffness. Söderlind further quantifies how "large" the negative values must be for an equation to be considered locally stiff for a given problem, however we are primarily interested in the values of the stiffness indicator itself to facilitate comparison between different simulations using different models and different timescales.

In the chemical kinetic models of interest, the stiffness indicator is calculated for a given Jacobian matrix $A \in \mathbb{R}^{n \times n}$ as:

$$\text{Stiffness Indicator} = \frac{m[A] + M[A]}{2} \tag{8}$$

where

$$m[A] = \min \lambda [\text{He}(A)]; \qquad M[A] = \max \lambda [\text{He}(A)], \tag{9}$$

with $\lambda[A]$ denoting the eigenvalues of matrix A and $\text{He}(A)$ denoting the Hermitian part of the matrix A, which is defined as

$$\text{He}(A) = \frac{A + A^{\mathsf{T}}}{2}. \tag{10}$$

While determination of the stiffness indicator does require calculation of eigenvalues following a transpose and addition of the Jacobian matrix, it is readily implemented and unlike the stiffness index it does not require storage of prior values of the solution.

2.4 Chemical Explosive Mode Analysis

The "chemical explosive mode analysis" (CEMA) was developed as a diagnostic to identify flame and ignition structure in complex flows [24]. Though this metric was not originally intended as a device to quantify the numerical stiffness of a system of ODEs, we are interested in determining if correlations exist between identification of the flame structure and the computational work required to advance the simulation for either explicit or implicit methods. The chemical explosive modes are associated with positive eigenvalues of the Jacobian of the system of ODEs:

$$\text{Re}(\lambda) > 0. \tag{11}$$

Similar to the stiffness ratio, the CEMA is readily implemented and carries little computational cost beyond determining the eigenvalues of the Jacobian matrix.

3. Results and Discussion

An analysis of each stiffness metric was performed using every particle in the PaSR simulation. In this analysis, a simulation at each particle of the PaSR was advanced only as long as needed to obtain an accurate numerical value of each stiffness metric. Because a fourth order central difference formula was used to numerically approximate the second derivatives in the calculation of the stiffness index, the simulation was advanced until five solution values existed at each particle. The only information needed to calculate the stiffness ratio, stiffness indicator, and chemical explosive modes is the Jacobian matrix, therefore these values could be calculated at every time-step as soon as the solution was known. Constant step sizes of 1×10^{-7} s and 1×10^{-8} s were used for computations of the H_2/CO model and GRI Mech 3.0 models, respectively, however the built in methods of the `ode` package have automatic step size control to ensure that values remain within tolerance, and the integrator may take many steps in between the specified time-step. Using state data obtained during the simulation, the Jacobian matrix was calculated at each time-step using `pyJac`, which allowed the numerical values of each stiffness matrix to be calculated. As previously discussed, the time per integration, number of RHS function calls, and internal time-steps were measured once at each particle, however we are most interested in time per integration for the purposes of this study. The simulation was performed on an early 2015 MacBook Pro with a 2.9 GHz Intel Core i5 processor.

3.1 Results

The H_2/CO model was examined first due to its relative simplicity. The PaSR data for the this model consisted of 900900 different particles. It was found that `vode` used a less time per integration than `dopri5` for 729384 particles or 81.0% of the PaSR data for this model, while `dopri5` had a smaller clock time for 165001 particles or 18.3% of the PaSR data. `vode` and `dopri5` had the same wall clock time per integration within the numerical tolerance of the Python code for 6515 particles, and `dopri5` did not fail when integrating at any particle. In general, `vode` was likely to outperform `dopri5` when pressure was 1 atm, and also when pressure was 10 atm at 400K. At higher pressures and temperatures, the explicit `dopri5` was more likely to outperform the implicit `vode` when using this model. This may be because the H_2/CO model is generally more limited by accuracy requirements as opposed to stability requirements. `vode` uses backward differentiation formulae, which require multiple data points for higher order approximations. At the boundary conditions where few or no adjacent data points exist, lower order approximations are used. The use of lower order methods may be advantageous when few internal time-steps are needed because fewer RHS function calls are required to provide an approximation. Higher order methods may be advantageous when an intermediate number of time-steps are needed to provide the required accuracy because higher order methods are often able to make use of larger time-steps. Additionally, unlike implicit methods, explicit methods do not require calculation of Jacobian matrices to advance time.

As seen in Figure 1a, the stiffness ratio was not highly correlated with time per integration for either `vode` or `dopri5`, nor did it provide a particularly efficient indicator of the relative performance of either method. `dopri5` appears to outperform `vode` at values of the stiffness ratio greater than 10^{19}, however neither method had a clear advantage when values of the stiffness ratio were less than this. The stiffness index did not appear to perform any better than the stiffness ratio in

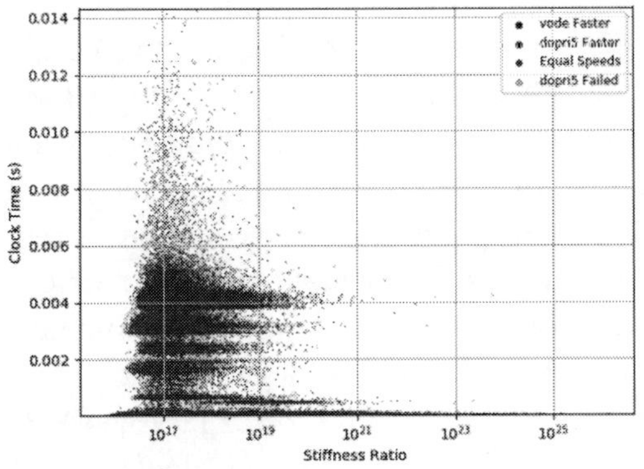

(a) Stiffness ratio versus time per integration at each particle of the PaSR model.

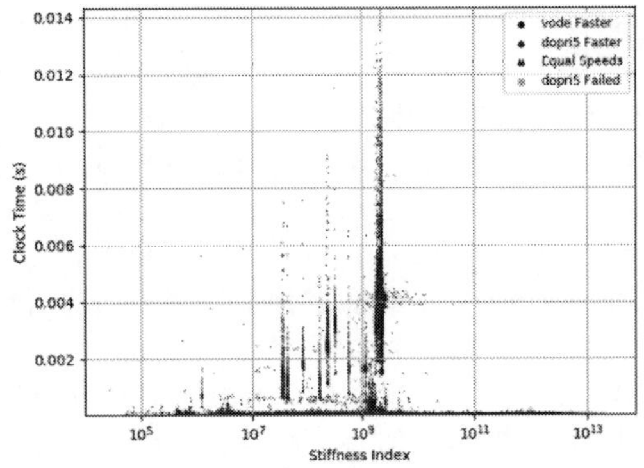

(b) Stiffness index versus time per integration at each particle at each particle of the PaSR model.

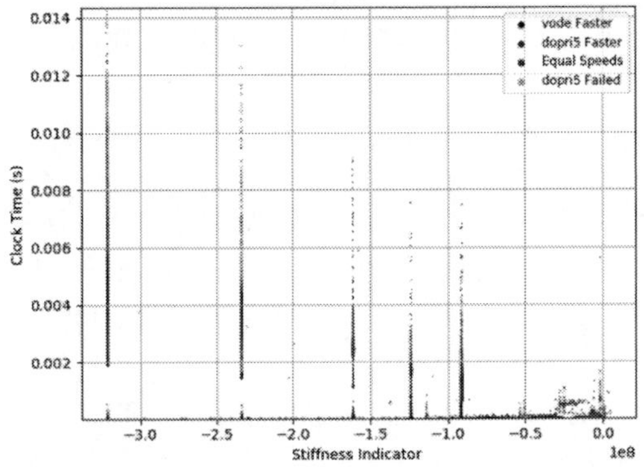

(c) Stiffness indicator versus time per integration at each particle at each particle of the PaSR model.

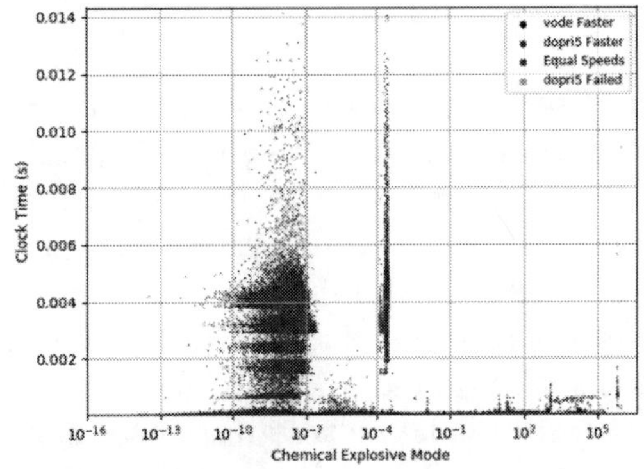

(d) Chemical explosive mode versus time per integration at each particle of the PaSR model.

Figure 1: Stiffness metrics versus time per integration of the H_2/CO model using either `vode` or `dopri5`. Calculated with a user defined constant step size of 10^{-7} s for every initial condition represented in the PaSR model.

the current set of data from this simulation, as seen in Figure 1b. Neither method had a clear advantage at any range of the stiffness index, nor was the stiffness index a good indicator of time per integration. As is the case in all of the plots obtained from this simulation, many of the integrations used very little clock time, making it difficult to draw correlations. Better correlations may become apparent if the user defined step size were increased to a value larger than 1×10^{-7} s. Similar to the stiffness ratio and the stiffness index, the chemical explosive mode was not highly correlated with the time per integraton. When the chemical explosive mode values were greater than 10^2, `dopri5` appears to generally outperform `vode`.

The stiffness indicator had a better correlation with the time per integration and the number of function calls used by either `vode` or `dopri5` than the other stiffness metrics, with a larger negative stiffness indicator correlating to a longer time per integration as seen in Figure 1c. Further investigation will determine if this trend becomes more apparent when the step size is increased to a value larger than 1×10^{-7} s. There did not appear to be a clear correlation between the value of the stiffness indicator and the relative performance of either method in the current set of data.

We looked to the GRI Mech 3.0 model next to determine how the stiffness metrics are impacted by using a stiffer model than the H_2/CO model. The PaSR data for GRI Mech 3.0 consisted of 450900 particles. It was found that `vode` had a smaller clock time than `dopri5` for 408232 particles or 90.5% of the PaSR data for this model, while `dopri5` had a smaller clock time for 165001 particles or 9.4% of the PaSR data. `vode` and `dopri5` had the same clock time within the numerical tolerance of the Python code for only six particles, and `dopri5` did not fail when integrating at any particle. The conditions in which `vode` was likely to outperform `dopri5` when using the GRI Mech 3.0 model ranged from low pressure and low temperature to high temperature and high pressure initial conditions of the PaSR data. Superior performance of `vode` in the high pressure and high temperature conditions is a potential indication that the internal step size of the integrator is limited by stability rather than accuracy requirements.

As seen in Figure 2a, the stiffness ratio was not correlated with the time per integration, however `vode` appeared to generally outperform `dopri5` at values of the stiffness ratio greater than 10^{22}. At values of the stiffness ratio less than 10^{22}, it is not clear which method has better performance. The stiffness index behaved similarly to the stiffness ratio, as seen in Figure 2b While the stiffness ratio was not correlated with the time per integration, `vode` appeared to outperform `dopri5` at index values greater than approximately 5×10^{10}.

The usefulness of the stiffness indicator in predicting the time per integration was more clear in the set of data from the GRI Mech 3.0 model, as seen in Figure 2c. Larger negative values of the stiffness indicator correspond to a longer time per integration for the `dopri5` integrator. Further investigation will determine if additional trends may be made clear for the `vode` integrator if the time step were increased to a value larger than 10^{-8} s. `vode` generally outperformed `dopri5` at values of the stiffness indicator value less than -2×10^9 in the current set of data. Additional investigation will determine if the stiffness indicator can reliably be used to predict relative performance of the two methods.

The usefulness of chemical explosive mode analysis in the GRI Mech 3.0 simulation was similar to the usefulness of the chemical explosive mode in the H_2/CO simulation. While the chemical explosive mode was not useful in predicting values of the time per integration, `dopri5` outperformed `vode` when values of the chemical explosive mode were greater than 1.5×10^2.

We are interested in determining the potential speedup offered by an integrator using these metrics. Because implicit methods such as `vode` are used by default for most simulations of com-

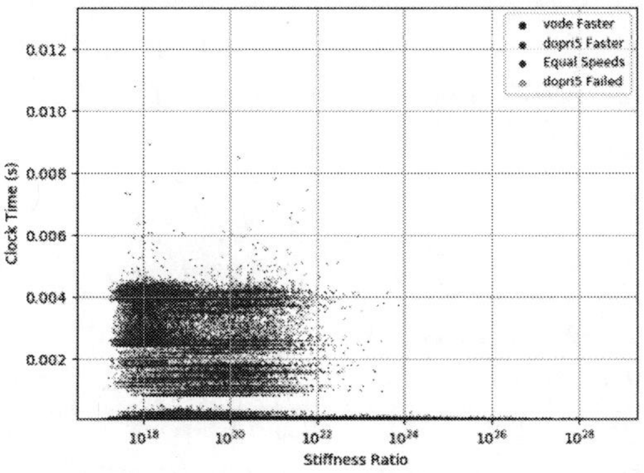

(a) Stiffness ratio versus the time per integarion at each particle of the PaSR model.

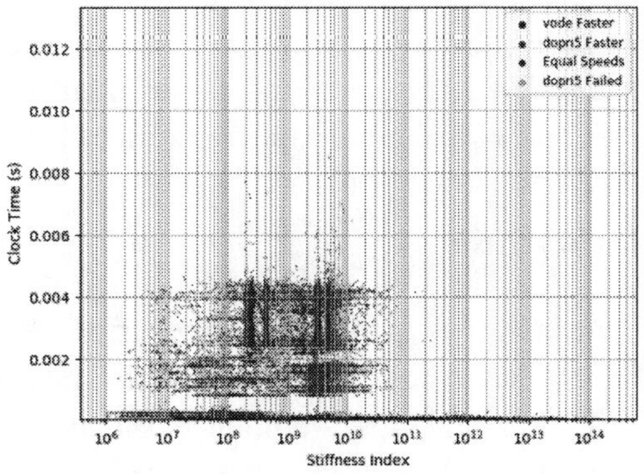

(b) Stiffness index versus the time per integration at each particle of the PaSR model.

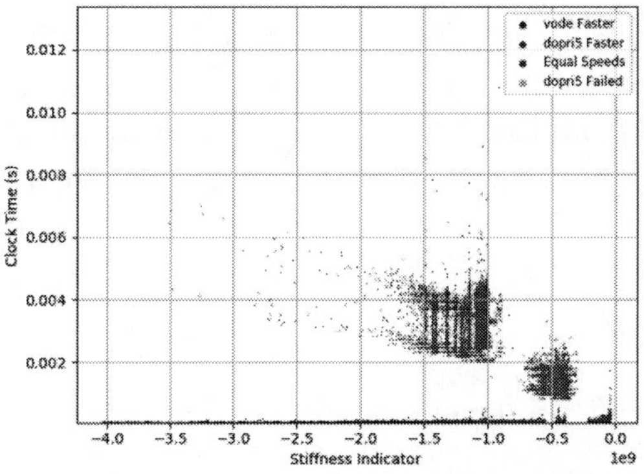

(c) Stiffness indicator versus the time per integration at each particle of the PaSR model.

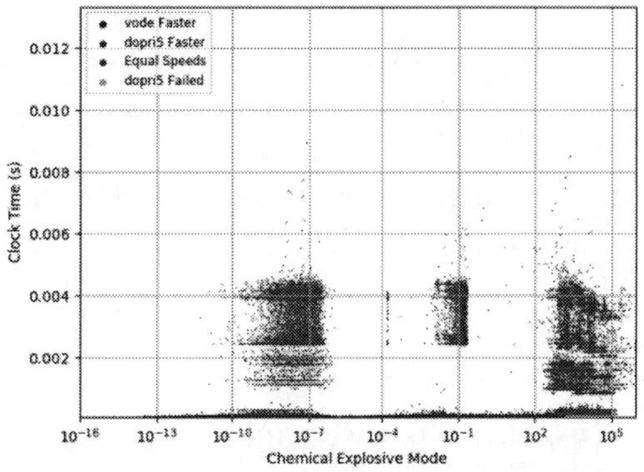

(d) Chemical explosive mode versus the time per integration at each particle of the PaSR model.

Figure 2: Stiffness metrics versus time per integration of the GRI Mech 3.0 model using either `vode` or `dopri5`. Calculated with a constant step size of 10^{-8} s for every initial condition represented in the PaSR model.

bustion, we are most interested in methods that predict when explicit methods such as `dopri5` have superior performance to explicit methods. Using the data obtained from the simulations above, we are able to select integration times from either integrator based on stiffness metric values at those points. For the set of data obtained from the GRI Mech 3.0 simulation, we selected the timing data from `vode` from particles that had a chemical explosive mode value of less than 1.5×10^2, and timing data from `dopri5` for all particles with a chemical explosive mode value of greater than 1.5×10^2. The total clock time `vode` needed to integrate one time step at each particle of the PaSR simulation was 287.3 s, while the "scheduler" had a total wall clock time of 238.5 s to complete all integrations, a speedup of approximately 1.2× compared to `vode`. For comparison, a perfect scheduler that selects the faster method at every particle would have a total integration time of 136.1 s, a speedup of approximately 2.1× compared to `vode`.

4. Conclusions

Four stiffness quantification metrics were investigated using partially stirred reactor simulations for two chemical kinetic models. These metrics were the stiffness ratio, stiffness index, stiffness indicator, and the chemical explosive mode. Our analysis demonstrated that the four stiffness quantification metrics investigated have the potential to be used by a scheduler that automatically selects the most efficient integration method at each point. As an example, it was found that the wall clock time needed to integrate 450900 time steps of 10^{-8} s in the GRI Mech 3.0 model could have been reduced by a factor of 1.2× if a scheduler had selected an explicit method when the chemical explosive mode exceeded a value of 1.5×10^2. Further analysis is needed to determine which stiffness quantification metric may be used to provide the largest speedup, as well as the sensitivity of these metrics to the user defined step size. Additionally, the cost of determining the value of each stiffness quantification metric will be considered when implementing a scheduler in further research.

5. Acknowledgements

This material is based upon work supported by the National Science Foundation under grant ACI-1535065.

References

[1] A. H. Epstein, Aircraft engines' needs from combustion science and engineering, Combustion and Flame 159 (2012) 1791–1792. DOI: 10.1016/j.combustflame.2012.02.022.

[2] A. Trouvé, D. C. Haworth, J. H. Miller, L. K. Su, and A. Violi, Cyber-Based Combustion Science: Report on the NSF Workshop on Cyber-Based Combustion Science, Report No., Available at http://www-personal.umich.edu/~avioli/pdf/nsf-report.pdf, 2006.

[3] U.S. Department of Energy, Basic Research Needs for Clean and Efficient Combustion of 21st Century Transportation Fuels, Report No., Available at http://science.energy.gov/~/media/bes/pdf/reports/files/ctf_rpt.pdf, 2007.

[4] National Research Council, Transforming Combustion Research through Cyberinfrastructure, Available at http://www.nap.edu/catalog.php?record_id=13049, The National Academies Press, Washington, DC, 2011.

[5] National Research Council, Future Directions for NSF Advanced Computing Infrastructure to Support U.S. Science in 2017–2020, Available at http://www.nap.edu/catalog.php?record_id=18972, The National Academies Press, Washington, DC, 2014.

[6] T. Lu and C. K. Law, Toward accommodating realistic fuel chemistry in large-scale computations, Progress in Energy and Combustion Science 35 (2009) 192–215. DOI: 10.1016/j.pecs.2008.10.002.

[7] G. Söderlind, L. Jay, and M. Calvo, Stiffness 1952-2012: Sixty years in search of a definition, BIT Numerical Mathematics 55 (2014) 531–558.

[8] L. F. Shampine, Measuring Stiffness, Applied Numerical Mathematics 1 (1985) 107–119. DOI: 10.1016/0168-9274(85)90020-0.

[9] L. Brugnano, F. Mazzia, and D. Trigiante, Recent Advances in Computational and Applied Mathematics, Recent Advances in Computational and Applied Mathematics (2011) 1–21, arXiv: 0910.3780.

[10] J. D. Lambert, Computational methods in ordinary differential equations, Wiley, 1973.

[11] E. Hairer and G. Wanner, Solving Ordinary Differential Equations II, vol. 14, Springer Series in Computational Mathematics, Springer Berlin Heidelberg, Berlin, Heidelberg, 1996.

[12] M. P. Burke, M. Chaos, Y. Ju, F. L. Dryer, and S. J. Klippenstein, Comprehensive H2/O2 kinetic model for high-pressure combustion, International Journal of Chemical Kinetics 44 (2012) 444–474. DOI: 10.1002/kin.20603.

[13] G. P. Smith, D. M. Golden, M. Frenklach, N. W. Moriarty, B. Eiteneer, M. Goldenberg, C. T. Bowman, R. K. Hanson, S. Song, W. C. Gardiner, V. V. Lissianski, and Z. Qin, *GRI-Mech 3.0*, http://www.me.berkeley.edu/gri_mech/, 1999.

[14] K. E. Niemeyer, Reducing the Cost of Chemistry in Reactive-Flow Simulations: Novel Mechanism Reduction Strategies and Acceleration via Graphics Processing Units, PhD thesis, Cleveland, OH: Case Western Reserve University, 2013.

[15] E. Jones, T. Oliphant, P. Peterson, et al., *SciPy: Open source scientific tools for Python*, [Online; accessed 2017-02-28], 2001–, URL: http://www.scipy.org/.

[16] P. N. Brown, G. D. Byrne, and A. C. Hindmarsh, VODE: A Variable-Coefficient ODE Solver, SIAM Journal on Scientific and Statistical Computing 10 (1989) 1038–1051.

[17] E. Hairer SP Norsett and G. Wanner, Solving ordinary differential equations I: Nonstiff problems, SIAM Review 32 (1990) 485–486. DOI: 10.1016/0378-4754(87)90083-8, arXiv: arXiv:1011.1669v3.

[18] S. Van Der Walt, S. C. Colbert, and G. Varoquaux, The NumPy array: A structure for efficient numerical computation, Computing in Science and Engineering 13 (2011) 22–30. DOI: 10.1109/MCSE.2011.37, arXiv: 1102.1523.

[19] K. E. Niemeyer and N. J. Curtis, *pyJac v1.0.2*, 2017, DOI: 10.5281/zenodo.251144.

[20] K. E. Niemeyer, N. J. Curtis, and C.-J. Sung, pyJac: analytical Jacobian generator for chemical kinetics, Computer Physics Communications (2017), DOI: 10.1016/j.cpc.2017.02.004.

[21] D. G. Goodwin, H. K. Moffat, and R. L. Speth, Cantera: An Object-oriented Software Toolkit for Chemical Kinetics, Thermodynamics, and Transport Processes, http://www.cantera.org, Version 2.3.0, 2017, DOI: 10.5281/zenodo.170284.

[22] R. J. LeVeque, Finite Difference Methods for Ordinary and Partial Differential Equations, 2007, pp. 1–356.

[23] L. F. Shampine, Type-insensitive ODE codes based on implicit A-stable formulas, Mathematics of Computation 39 (1982) 109–123. DOI: 10.1090/S0025-5718-1982-0658216-2.

[24] C. S. Yoo, R. Sankaran, and J. H. Chen, Three-dimensional direct numerical simulation of a turbulent lifted hydrogen jet flame in heated coflow: flame stabilization and structure, Journal of Fluid Mechanics 640 (2009) 453–481.

10th Meeting of the Western States Section of The Combustion Institute
Organized by the University of Wyoming
October 2–3, 2017
Laramie, Wyoming

pyMARS: an Open Software Package for Reducing Chemical Kinetics Models

Phillip O. Mestas[1], *Parker Clayton*[1], *and Kyle E. Niemeyer*[1,*]

[1]*Mechanical, Industrial and Manufacturing Engineering, Oregon State University, 104 Rogers Hall, Corvallis OR, United States*
[*]*Corresponding author:* kyle.niemeyer@oregonstate.edu

Abstract: Chemical kinetic models play a crucial role in reacting-flow simulations. Detailed models can contain hundreds to thousands of independent species, and thousands to tens of thousands of reactions. Performing high-fidelity, multidimensional simulations using such models poses prohibitive computational costs, so kinetic model reduction plays an active role in enabling future research. This study introduces pyMARS (Python-based Model Automated Reduction Software), an open-source software package that performs automatic model reduction using a variety of methods, based on the Cantera suite for handling chemical kinetics. These reduction method include the directed relation graph (DRG), DRG with error propagation (DRGEP), DRG-aided sensitivity analysis (DRGASA), and DRGEP with sensitivity analysis (DRGEPSA) methods. pyMARS (currently) uses autoignition simulations to generate thermochemical data for the reduction process and evaluate the error of potential skeletal models. Future version of pyMARS will include additional skeletal reduction methods, and serve both as a tool for reduction of large kinetic models and a testbed for determining optimal reduction strategies.
***Keywords:* Chemical kinetic models, Model reduction, Reaction kinetics**

1. Introduction

It is critical to have accurate and descriptive chemical kinetic models when performing reaction-flow simulations. Detailed models that describe systems such as gasoline and jet fuel surrogates, like the ones developed by Battin-Leclerc [1], tend to be very large. Due to the complexities of these systems, the size of a detailed model used for simulation can contain hundreds to thousands of independent species and thousands to tens of thousands of individual reactions. Westbrook et al. [2] and Herbinet et al. [3] have both developed reasonable models that are of this magnitude. However, the size of these models comes with a heavy computational cost. Running simulations becomes unreasonable when the chemical kinetic models are this large; thus, it is imperative to have an alternate method for simulating models that alleviates the problem [4, 5].

The solution is chemical kinetic model reduction. Reduced mechanisms relieve the burden of computational costs while still maintaining accuracy if developed correctly [6]. This study introduces pyMARS (Python-based Model Automated Reduction Software), an open-source software package that supports various model reduction techniques documented in literature by making use of the Cantera [7] suite for handling chemical kinetics. pyMARS is written in Python and released openly under the MIT license. The source code can be found at `github.com/Niemeyer-Research-Group/pyMARS`. These automatic reduction techniques can be performed on models, taking out as many details as possible while still maintaining a specified amount of accuracy.

pyMARS currently supports the DRG (Directed Relation Graph) and DRGEP (Directed Relation Graph with Error Propagation) methods for model reduction [8, 9]. Both techniques use ideas from graph theory to determine which species should be eliminated from the model. They both represent the relationships between the species in the model graphically with the edge weights representing the dependency of each species on another. The DRG method only accepts edge weights that fall above a certain threshold and determines what should stay in the model based on what is connected to a few target species on the graph. The DRGEP method accepts all connections and looks at the greatest path from the target species to determine if a species should stay in the model or not. pyMARS iterates through different threshold values until it determines which one provides the desired level of error. The program then produces a new model without including any species that were cut out, or any reactions that involve them. The error is calculated by running an autoignition simulation with the original and reduced models under specified conditions, and comparing the difference between the autoignition times [4].

2. Methods

pyMARS currently has implementations of the previously developed DRGEP and DRG methods for model reduction [8, 9]. Both of the currently supported methods are iterative processes that repeatedly cut out species, and then compare the autoignition delay with that of the original mechanism to determine if the new model is still within a specified accuracy range. Although they have many similarities, the methods differ most greatly by how they use the graph to determine what species should be removed from the model. Before running the program, the user will have to provide the file that holds the detailed model and a file that contains the initial conditions of the system. The user must also provide a list of target species as well as a desired error level.

Once the program has started with the DRG method selected, an autoignition simulation is run to determine thermochemical data used to calculate direct interaction coefficients. These coefficients represent the dependency of one species on another species. The direct interaction coefficient of species A on species B is defined by Lu and Law [8] as r_{AB} where:

$$r_{AB} \equiv \frac{\sum_{i=1,I} |\nu_{A,i}\omega_i\delta_{Bi}|}{\sum_{i=1,I} |\nu_{A,i}\omega_i|} \tag{1}$$

$$\delta_{Bi} = \begin{cases} 1, & \text{if the ith elementary reaction involves species B,} \\ 0, & \text{otherwise.} \end{cases} \tag{2}$$

I represents the number of reactions in the system, $\nu_{A,i}$ the stoichiometric coefficient of species A in the ith reaction, and ω_i the overall reaction rate of the ith reaction. The direct interaction coefficients are used as the weights on a directed graph of the species in the system. One graph is created for each set of initial conditions presented using the tools available in the NetworkX library [10]. An edge is only included on the graph if its weight is greater than a specified threshold value. Once the graph is created, a depth first search is done from each of the target species. If any species is not reached through the search, it is cut out of the mechanism for that threshold value. The program uses an iterative process to determine the maximum amount of species that can be removed before reaching the allowed error level. Iteration occurs by altering the threshold value. The program starts with an arbitrary threshold value that is repeatedly divided by 10 until the error is calculated to be zero. The error percentage is calculated based off of autoignition delay by creating a reduced mechanism and running the autoignition simulation with that mechanism. The autoignition delay from that simulation is compared to the autoignition delay from the original mechanism to determine the error percentage. When a starting threshold value has been determined, the threshold value will be slightly increased and the error will be calculated again. Once the error has gone beyond the allowed level, the iteration will stop. All species not connected to the target species on the final graph are removed from the model, and all reactions involving any of those species are removed as well. Lu and Law [11] showed this method to be efficient, especially on large models. Target species selection for the DRG method is important, but not vital because all of the species that should be maintained will be connected. Since the method uses a depth first search to determine what will be used in the reduced model, all of the species that should be kept will be as long as one of the target species is in that group.

In addition, pyMARS also supports the DRGEP method of model reduction. Although its implementation is similar to that of DRG, the two methods vary in the way the graph is created and searched. Instead of making a new graph to search for each threshold level, the DRGEP method requires only one graph per initial condition before iteration and uses them to make a dictionary to determine what should be cut out. Direct interaction coefficients are calculated similarly to DRG. Details of the calculations were recorded by Niemeyer and Sung [12] as well as Pepiot-Desjardins and Pitsch [9] as:

$$r_{AB} = \frac{\left|\sum_{i=1}^{n_R} \nu_{A,i}\omega_i\delta_{Bi}\right|}{\max(P_A, C_A)} \tag{3}$$

where

$$\delta_{Bi} = \begin{cases} 1, & \text{if the ith elementary reaction involves species B,} \\ 0, & \text{otherwise.} \end{cases} \tag{4}$$

$$P_A = \sum_{i=1}^{n_R} \max(0, \nu_{A,i}\omega_i) \tag{5}$$

$$C_A = \sum_{i=1}^{n_R} \max(0, -\nu_{A,i}\omega_i) \tag{6}$$

n_R represents the number of reactions in the system. To determine the overall interaction coefficient (OIC) for each species, a search is preformed from each target species on each graph to determine the greatest possible path to each other species. This search is done using a modified version of Dijkstra's algorithm [13] that finds the greatest possible path to each node when taking the products of the edge weights. Because all direct interaction coefficients are less than one, this algorithm can be used. Wagner [14] showed many of the optimizations that can be made to Dijkstra's algorithm and Niemeyer and Sung [12] established that this implementation of Dijkstra's algorithm is the best graph searching algorithm for DRGEP specifically. The NetworkX implementation of Dijkstra's algorithm was modified to fit the algorithm needed for pyMARS. The greatest path to each species from any target species on any of the graphs is stored in a dictionary. These values represent each species OIC. Species that are cut out of the mechanism are those with an OIC below the threshold value. The threshold value is iterated in the same way as it is in the DRG method. Each species

OIC is dependent on what target species are used. Because of this, target species selection is vital for the DRGEP method. The distance a species has from the target species will impact its OIC.

The DRG and DRGEP methods for model reduction are both currently supported by pyMARS. The methods use the same iterative process and error calculations. Although they both use ideas in graph theory to determine what should be cut from the mechanism, their implementations vary. DRG limits the connections that are placed on the graph, and only keeps species that are connected to the target species. DRGEP connects all species, and uses distance from the target species to determine what should be cut from the model. Both implementations return an accurate and smaller version of a detailed model without the user being involved in the details of the reduction.

3. Results

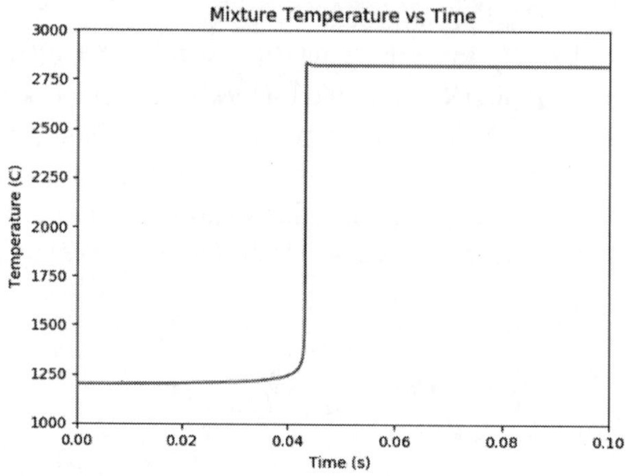

Figure 1: Data from an autoignition simulation done with the original GRI 3.0 mechanism. The initial conditions are a pressure of 1 atm, a temperature of 1200 K, and an equivalence ratio of 1.0.

So far, we have tested pyMARS on smaller sample mechanisms. One of these models is GRI-Mech 3.0 [15]. This model is relatively small, containing only 56 species and 325 reactions, but pyMARS was still able to reduce its size significantly. Figure 1 shows the temperature vs. time for a simulation using the original GRI mechanism. The conditions for this simulation are an initial temperature of 1200 K, a pressure of 1 atm, and an equivalence ratio of 1.0. The graph serves as a good visualization of autoignition delay.

When GRI-Mech 3.0 was used in pyMARS, both the DRG and DRGEP methods were able to reduce the model to nearly half of its original size while still maintaining less than five percent error in autoignition delay. Figure 2 shows a simulation ran on the reduced mechanism output by the DRGEP method with the same initial conditions. It is nearly identical to the plot produced from the original mechanism, with an autoignition delay that is almost the same as it was initially. pyMARS was able to produce a mechanism that is nearly half the size of its original version, but still can be used to accurately simulate the system. This result is significant because it will allow increased performance on necessary simulations with heavy computational costs. The methods supported by pyMARS can now be implemented on models quickly without much work from the user. This has potential to significantly speed up research for those using larger models.

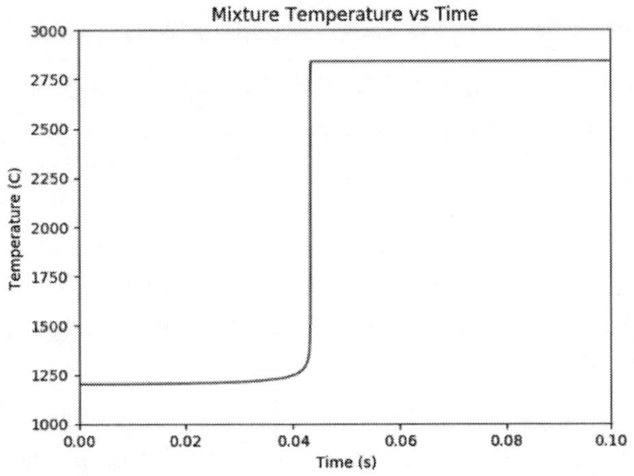

Figure 2: Data from an autoignition simulation done with the reduced GRI 3.0 mechanism with the same initial conditions as before. It is very close to the plot of the data from the original simulation. This shows that the reduced model can accurately describe the system.

4. Conclusions and future work

pyMARS is a valuable toolkit that is able to greatly reduce chemical kinetic models without sacrificing accuracy. It currently includes implementations for two different methods of model reduction. Both methods have so far been effective in reducing model size. Having a smaller model that maintains the accuracy of the original reduces the computational cost associated with detailed models. This result will have a positive impact on future research because it will greatly speed up long simulation times.

Although pyMARS is able to reduce the size of a model efficiently, there is room for improvement by adding additional methods that may prove more effective in some situations. One of these methods is sensitivity analysis as described by Rabitz et al. [16]. This method involves ordering the species based on their overall interaction coefficients. Once the species are ordered, they are removed one by one from the model and the error is calculated. This method suffers from long execution times because each species must be removed one by one. Therefore, it is efficient to use in conjunction with the DRGEP method as shown by Niemeyer et al. [17]. There are also plans to include an isomer lumping reduction method similar to the method described by Huang et al. [18]. With multiple methods for model reduction implemented, pyMARS will be more useful in diverse situations where some methods may be preferred over others. There are also plans to include automated testing for crucial functions to ensure that future versions don't harm the already existing code. More features will be added to pyMARS in the future such as the ability to produce plots from simulations, or write csv files that contain data about the reduction. With increased capabilities, pyMARS will be a more robust toolkit that will allow for models to be reduced even further.

Sub Topic: Reaction Kinetics

Acknowledgements

This material is based upon work supported by the National Science Foundation under grant ACI-1535065, and the URSA Engage program at Oregon State University.

References

[1] F. Battin-Leclerc, Detailed chemical kinetic models for the low temperature combustion of hydrocarbons with application to gasoline and diesel fuel surrogates, (2008), DOI: 10.1016/j.pecs.2007.10.002.

[2] C. K. Westbrook, W. J. Pitz, O. Herbinet, H. J. Curran, and E. J. Silke, A comprehensive detailed chemical kinetic reaction mechanism for combustion of n-alkane hydrocarbons from n-octane to n-hexadecane, Combustion and Flame (2009), DOI: 10.1016/j.combustflame.2008.07.014.

[3] O. Herbinet, W. J. Pitz, and C. K. Westbrook, Detailed chemical kinetic oxidation mechanism for a biodiesel surrogate, Combustion and Flame (2008), DOI: 10.1016/j.combustflame.2008.03.003.

[4] K. E. Niemeyer, C. J. Sung, and M. P. Raju, Skeletal mechanism generation for surrogate fuels using directed relation graph with error propagation and sensitivity analysis, Combustion and Flame (2010), DOI: 10.1016/j.combustflame.2009.12.022, arXiv: arXiv:1607.05079v1.

[5] T. Lu and C. K. Law, Toward accommodating realistic fuel chemistry in large-scale computations, (2009), DOI: 10.1016/j.pecs.2008.10.002.

[6] Y. Ra and R. D. Reitz, A reduced chemical kinetic model for IC engine combustion simulations with primary reference fuels, Combustion and Flame (2008), DOI: 10.1016/j.combustflame.2008.05.002.

[7] D. G. Goodwin, H. K. Moffat, and R. L. Speth, *Cantera: An Object-oriented Software Toolkit for Chemical Kinetics, Thermodynamics, and Transport Processes*, http://www.cantera.org, Version 2.3.0, 2017.

[8] T. Lu and C. K. Law, A directed relation graph method for mechanism reduction, Proceedings of the Combustion Institute (2005), DOI: 10.1016/j.proci.2004.08.145.

[9] P. Pepiot-Desjardins and H. Pitsch, An efficient error-propagation-based reduction method for large chemical kinetic mechanisms, Combustion and Flame (2008), DOI: 10.1016/j.combustflame.2007.10.020.

[10] A. A. Hagberg, D. A. Schult, and P. J. Swart, Exploring Network Structure, Dynamics, and Function using NetworkX, Proceedings of the 7th Python in Science Conference Pasadena, CA USA, (2008), pp. 11–15.

[11] T. Lu and C. K. Law, Linear time reduction of large kinetic mechanisms with directed relation graph: N-Heptane and iso-octane, Combustion and Flame (2006), DOI: 10.1016/j.combustflame.2005.02.015.

[12] K. E. Niemeyer and C.-J. Sung, On the importance of graph search algorithms for DRGEP-based mechanism reduction methods, (2011), DOI: 10.1016/j.combustflame.2010.12.010.

[13] E. Dijkstra, A Note on Two Problems in Connexion with Graphs, Numerische Mathematik (1959).

[14] D. Wagner and T. Willhalm, LNCS 4393 - Speed-Up Techniques for Shortest-Path Computations, (2007), URL: http://i11www.informatik.uni-karlsruhe.de/.

[15] G. P. Smith, D. M. Golden, M. Frenklach, N. W. Moriarty, B. Eiteneer, M. Goldenberg, C. T. Bowman, R. K. Hanson, S. Song, W. C. Gardiner, V. V. Lissianski, and Z. Qin, GRI-Mech 3.0, http://www.me.berkeley.edu/gri_mech/.

[16] H. Rabitz, M. Kramer, and D. Dacol, Sensitivity Analysis in Chemical Kinetics, Ann. Rev. Phys. Chern 34 (1983) 4–9.

[17] K. E. Niemeyer, C. J. Sung, and M. P. Raju, Skeletal mechanism generation for surrogate fuels using directed relation graph with error propagation and sensitivity analysis, Combustion and Flame (2010), DOI: 10.1016/j.combustflame.2009.12.022, arXiv: arXiv:1607.05079v1.

[18] H. Huang, M. Fairweather, J. F. Griffiths, A. S. Tomlin, and R. B. Brad, A systematic lumping approach for the reduction of comprehensive kinetic models, Proceedings of the Combustion Institute (2005), DOI: 10.1016/j.proci.2004.08.001.

Sub Topic: Fire

2017 WSSCI Fall Technical Meeting
Organized by the Western States Section of the Combustion Institute
October 02-03, 2017
Laramie, Wyoming

Investigation of merging flames in Horizontal and vertical Geometries

*Mahmood Rahmati, Caleb Harper, Max Butler, Mohammad-Saeed Safdari, Elham Amini, and Thomas H. Fletcher**

Chemical Engineering Department, Brigham Young University, Provo, UT, USA 84602
Corresponding Author Email: tom_fletcher@byu.edu

Abstract: One of the main needs for accurate shrub combustion modeling is the complicated flame growth pattern due to merging of flames from multiple leaves and small branches. A careful study of flame interactions from fuels in three-dimensional arrangements is a necessary step in strengthening operational field models in complex fuel structures. The purpose of this research is to imitate the flame merging of the solid fuels in different geometries in order to describe fire behavior in complex fuel structures, such as trees or shrubs. Ceramic felts soaked in n-heptane were used as a fuel source to conduct flame-merging experiments. Flame characteristics such as flame length and flame width were measured for both single and merged flames. Our data shows that by altering the separation distance between felts in both vertical and horizontal directions, flame area, height and width vary to a large extent. The region of flame merging is mapped in both the horizontal and vertical directions. These data are being used to develop flame-merging correlations as a function of flame spacing, similar to correlations that have been developed for multiple flames in the horizontal plane.
Keywords: Flame – Merging, Fires

1. Introduction

Understanding the flame propagation behavior in wildland fires is a crucial factor for predicting the fire spread behavior in the burning areas. Investigation of flame propagation behavior among single fuel elements is a preliminary and essential step for recognition of large-scale fire spread behavior. Most of the research in the literature is focused on the interactions between single flames spaced horizontally [1, 2]. Although these researches are important, the next necessary step is to expand our knowledge to 3D spaced flames. The aim of this paper is to include the vertical placement of fuel sources besides horizontally separated fuel sources.

Flame characterization is done by parameters like heat release rate, flame height, flame width or flame area. Among these parameters, dimensionless flame height and dimensionless heat release are two commonly used parameters in the literature.
Dimensionless heat release rate (Q*) is defined as:

$$Q^* = \frac{\dot{Q}}{\rho_\infty C_p T_\infty D^2 \sqrt{gD}} \quad (1)$$

In this equation Q° is the total heat release rate; $\rho_\infty C_p T_\infty$ are the flame environment characteristics; D is the flame characteristic diameter and g is gravitational acceleration constant[3]. McCaffrey [4] suggested a relation to relate non-dimensionalized flame height is to the flame Froude number, where u is the fuel flow velocity.

$$\frac{L}{D} = 31.8\,.Fr^{\frac{2}{5}} \quad (2a)$$

$$Fr = \frac{u}{\sqrt{gD}} \quad (2b)$$

Calculation of heat release is commonly challenging. Zukoski et al correlated the dimensionless heat release to dimensionless flame length by suggesting the relation bellow

$$\frac{L_f}{D} = \begin{cases} 40\,Q^{*2},\; 0.03 < Q^* < 0.15 \\ 3.3\,Q^{*2/3},\; 0.15 < Q^* < 1 \\ 3.3\,Q^{*2/5}\; ,1 < Q^* < 40 \end{cases} \quad (3)$$

Quintiere and Grove [5], also found the correlation (shown as Equation 4) for calculation of heat release by having the laminar flame height

$$Q^* = C_f\left(\frac{L_f}{D}\right)^{\frac{1}{2}}\left(1 + 2C_1\left(\frac{L_f}{D}\right)\right)^2 \quad (4)$$

where C_f and C_1 are coefficients determined by experimental data and are dependent on the geometries of the fuel sources.

Studies also indicated that the non- dimension merged flame heights and heat release are function of separation distance [1, 6]. Baldwin et al. [1] showed that the fully merged flame height reaches up to 41% higher than unmerged flames.

Kamikawa et al. [7] studied the effects of heat release rate and separation distance on merged flame height for arrays of porous propane burners and wood cribs, finding that within a certain separation distance, the merged flame height is the same as if there was no separation. Sugawa and Takahashi [8] conducted similar work with circular pool fires and showed that a fire placed next to an adiabatic wall responds as if it was next to a self-identical fire. Thomas [9] provided scaling relations and a dimensional analysis to determine conditions for merged mass fires. Prince[10] placed individual leaves in realistic geometry and developed a semi-empirical model to predict the fire spread. His results clearly indicated that simple addition of individual leaves flame height is in adequate in predicting the flame height behavior in bushes.

Weng et al. even developed a flame height correlation that relates the flame height (L_f) to the merged flame height (L_m) :

$$\frac{L_f}{L_m} = \left(\frac{2\sqrt{N}(\sqrt{N}-1)S^2+\lambda D^2}{\lambda(2\sqrt{N}(\sqrt{N}-1)S^2+D^2)}\right)^B \quad (5)$$

$$\lambda^{B} = \frac{L_m}{L_1} \tag{6}$$

Where N is the number of fuels in the array and S is the separation distance. λ is the ratio of fully merged flame height to individual flame height. B also can be calculated by having no separation distance between the flames.
All above studies well address the need for models for prediction of flame interactions. However, all of these efforts were for flame merging effects in the horizontal plane, and mostly with flames from jets or liquid pools, and not over solid fuels.

2. Methods / Experimental

Solid fuels shape changes during the combustion. To investigate the flame merging in solid fuels a fuel source was needed with similar flame characteristics as a solid fuel source. For this purpose, ceramic felts soaked in n-heptane were chosen. Ceramic felt structure does not change during the combustion period thus does not affects the flame height and flame shape. In addition to have a steady state flame for enough long time to be able to analyze the flame characteristics two ceramic felts on the top of each other were used. To prevent the heat loss due to conduction thin ceramic rods with very low thermal conductivity (21W/m2 .K) were used to hold the ceramic felts. Test tube racks were used as holder for ceramic rods. A stainless steel sheet was used as a base for doing the experiments. Ruler stickers are used to make measuring the distances between the flames and fuels easier. Figure 2-1 shows the flame merging experimental setup.

Figure 2-1. Flame merging experimental setup

Recorded videos were then analyzed using a Matlab code and the flame height and width were calculated. The code works based on the color contrast in the way that it recognizes the difference the flame brightness from background. Experimental setup background is painted black to prevent any mistake in data analysis by software.

3. Results and Discussion

To analyze the single flame one double layer ceramic felt was soaked in n-heptane for 3 minutes and was ignited. Flame characteristics like length, flame width and flame area was calculated. Flame duration is about 81 seconds. During this period Matlab code recorded the coordination of the flame tip and thickness. Flame area was also calculated by summation of the area of the pictures that had flame in them. Since the flame is, non-continuous histograms were used to analyze the raw data recorded by software. At each separation distance, weighted average flame high, width and area were calculated based on frequency values.
Mean confidence bars at 90 % level were also calculated as a measure for confidence in the observed trend between merged-flame characteristics and separation distance. Figure 1 shows variation in merged-flame height, width and area in two vertical separation distances.

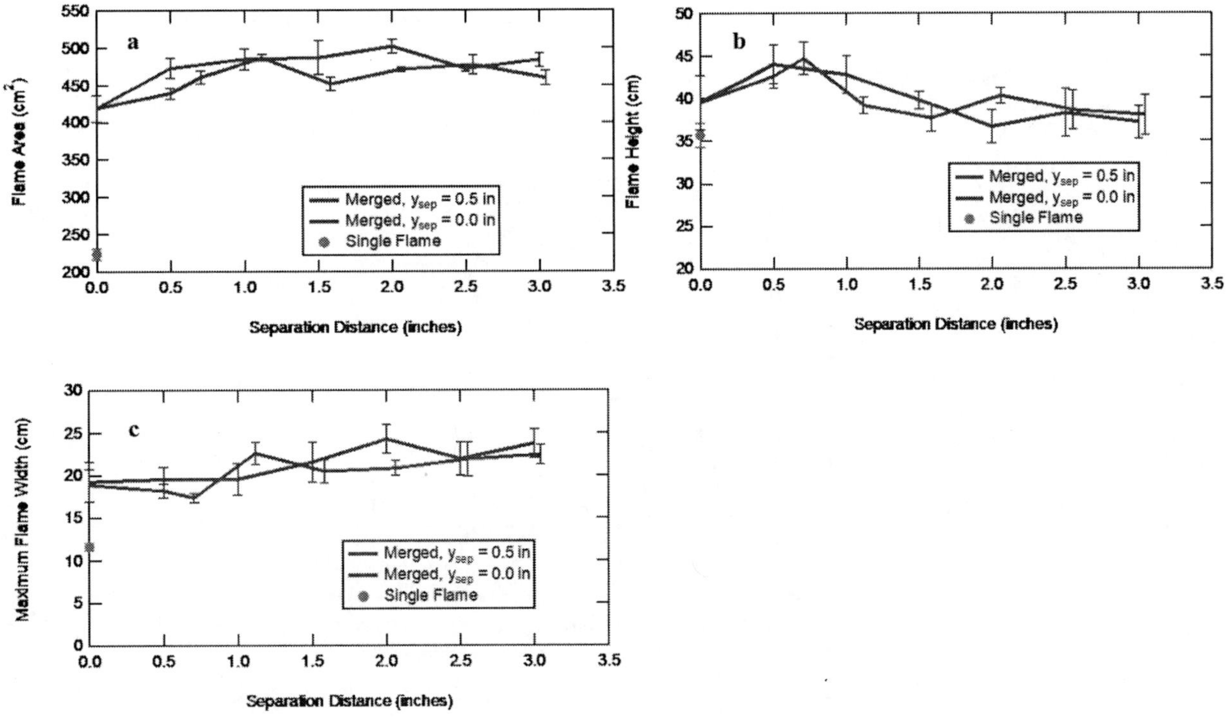

Figure 3-1. Merged-Flame (a) Area (b) Height (c) Width, Blue curve indicates merged-flame characteristics with no vertical separation distance, Red curve indicates merged-Flame characteristics at 0.5 inch vertical separation distance. Green point shows single flame characteristics

Data observed in Figure 3-1 clearly shows that fuel coordination plays a key role in determining the merged-flame characteristics. It can be concluded that, net distance regardless of fuel position cannot be solely used as an accurate measure for prediction of merged-flames height, width and areas.

Better understanding of merged-flame regime was achieved by calculation of Froude number. Figure 3-2 indicates that flames specific to our experiments possess Froude numbers ranging between 0.008 to 0.014

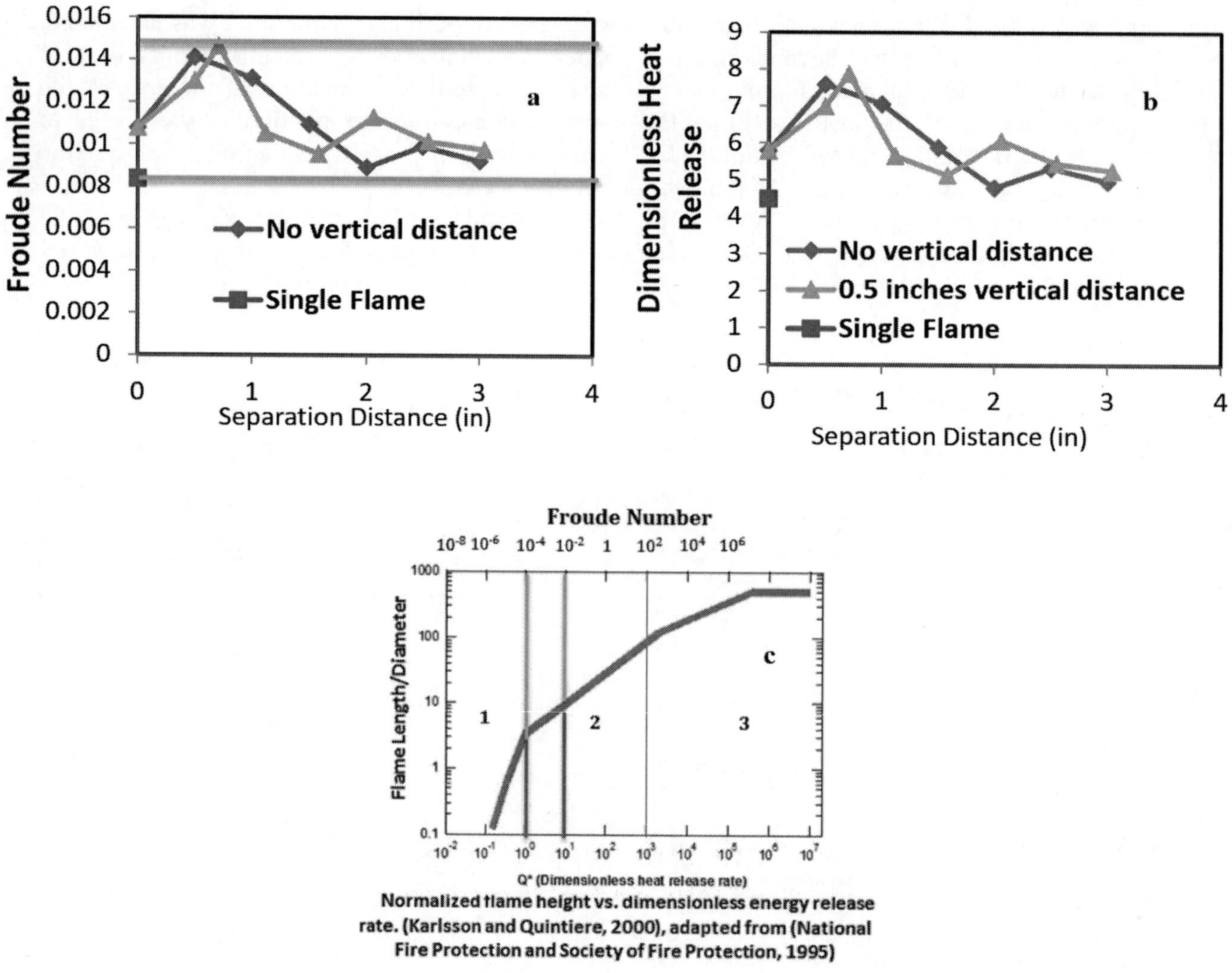

Figure 3-2. (a) Froude number vs. ceramic felt edges separation distance, (b) Dimensionless heat release rate vs. ceramic felt edges separation distance (c) Normalized flame height vs. dimension energy release [11]

Results shown in Figure 3-2. indicate that flames studied in this paper can be categorized as buoyancy driven flames. Furthermore, dimensionless heat release versus separation distance reaches a maximum at different locations depending on the vertical or horizontal spacing between the flames.

5. Acknowledgements

This research was funded by National Science Foundation (NSF)

References

1. Baldwin, R., *Flame merging in multiple fires* Combustion and Flame, 1968. **12**(4): p. 318-324
2. Weng, W.G., et al., *Study on flame height of merged flame from multiple fire sources.* Combustion Science and Technology, 2004. **176**(12): p. 2105-2123.
3. Drysdale, D., *An Introduction to Fire Dynamics.* Second ed. 1999, Chichester, England: John Wiley & Sons. 401.
4. McCaffrey, B., *Momentum implications for buoyant diffusion flames.* Combust. Flame, 1983. **52**: p. 149-167.
5. Quintiere, J. and B. Grove. *A unified analysis for fire plumes.* in *Symposium (International) on Combustion.* 1998. Elsevier.
6. Liu, N., et al., *Global burning rate of square fire arrays: Experimental correlation and interpretation.* Proceedings of the Combustion Institute, 2009. **32**(2): p. 2519-2526.
7. Kamikawa, D., et al., *Experimental study of merged flames from multifire sources in propane and wood crib burners.* Combustion and Flame, 2005.
8. Sugawa, O. and W. Takahashi, *Flame Height Behavior from Multi-fire Sources.* Fire and Materials, 1993. **17**: p. 111-117.
9. Thomas, P.H. *The size of flames from natural fires.* in *Ninth Symposium (International) on Combustion.* 1963. Pittsburgh, PA: Academic Press.
10. Prince, D., et al., *Modeling a Burning Shrub with and without Wind using a Semi-empirical Model*, in *International Association of Wildland Fire 3rd Fire Behavior and Fuels Conference.* 2010, Brigham Young University: Spokane, Washington.
11. Karlsson, B.r. and J.G. Quintiere, *Enclosure fire dynamics.* 2000, Boca Raton, FL: CRC Press.

Western States Section of the Combustion Institute – Fall 2017 Meeting
Hosted by the University of Wyoming
October 2-3, 2017

Modeling the Coanda Effect with FDS and STARCCM+ to Predict the Effect of Fires on Slopes for Implications of Wildland Firefighter Safety

Blake Billings[1], Michael Roberts[1], Bret Butler[2] and Thomas H. Fletcher[1,]*

[1]Chemical Engineering Department, Brigham Young University, Provo, UT 84602, USA
[2]RMRS Missoula Fire Sciences Laboratory, USDA Forest Service, Missoula, MT 59808, USA

**Corresponding Author*

Abstract: The Coanda effect is the behavior of a fluid flowing next to a solid surface which creates a change in pressure, thereby causing the fluid to attach to the surface. Little is known concerning the Coanda effect as it pertains to flame attachment in relation to wildfires. There is a need for improvement in the correlations between fire behavior on or near a slope and fire safety zones. An understanding of this effect could lead to more reliable guidelines for the establishment of firefighter safety zones, which are regions in wildfire areas that have been determined as safe for the survival of the firefighter. This paper discusses the ability of two commercial computational fluid dynamics (CFD) codes (Fire Dynamics Simulator (FDS) and STAR-CCM+) to model the results of actual flame experiments near barren slopes. Simulation results were compared to measured values of flame attachment length and radiative and convective heat fluxes. Pressure and temperature gradient measurements along the slope are reported. Proper modeling of small-scale experiments with such software packages can lead to more confidence in large scale simulations and eventual development of firefighter safety zones in wildfires.

1. Introduction

Over the previous 100 years, wildfires have kill or injured thousands of firefighters in the line of duty [1]. Many of these deaths are due to a phenomenon called firefighter entrapment, which occurs when a fire jumps and passes over a firefighter's location, thereby entrapping the firefighter in the flames and leaving him without a safe escape route. In an effort to decrease the number of entrapment fatalities, improved safety protocols have been developed in the past century. However, fires such as the 2001 Thirty-Mile fire, the 2013 Arizona Blaze fire, and the 2013 Yarnell Hill fire display that the risk of entrapment is still a serious threat. Research has looked towards sophisticated computer modeling programs in an attempt to better predict and describe flame behavior. This research will eventually be used to help establish firefighter safety zones [2]. Safety zones are defined as "an area where a firefighter can survive without a fire shelter" [3].

Previous research has been performed to find flame characteristics over changing and difficult terrain. In Australia, research was performed to help understand the effect of landscape and specifically, downward slopes on burn rate and fire spread [4]. Recent methods to determine dangerous wildfire conditions and behavior have been conducted over the past ten years to more effectively establish safety zones by the USDA Forest Service [1]. Other work has attempted to show the correlation between flame height and radiation heat for establishment of separation distances of safety zones from fires [5]. Research has also been conducted to improve the understanding of the effects of wind and slope on fire spread and radiation [6, 7].

More recently, attention has turned towards the Coanda effect, a property of fluid dynamics that causes flames to attach to nearby slopes regardless of the combustibility of the slope. According to a publication by Butler of the USDA Forest Service summarizing the future needs of establishing the safety zones, one of the needs defined is “improved knowledge of travel rates over various terrains and slopes” [2]. Study of the Coanda effect could help to establish that understanding. Fox and Stewart [8] studied the heat transfer behavior of combusting gases close to slopes. Extensive work has been done to understand a fire that occurred in King’s Cross tube station (i.e., subway) in London in 1987. The Coanda effect was deducted to be one of the causes of the dangerous fire that caused many fatalities on the escalator [9].

Fire Dynamics Simulator, or FDS [10], a robust modeling software that specializes in combustion simulations, has been used with some success in accurately describing the behavior of flames on a slope as well as being able to predict the movements of flames under certain conditions [11]. The FDS Validation Guide has little information on the validation of slope behavior in the program. The closest that is cited is the use of a sloped ceiling above a fire [12].

While FDS shows promise, a means of confirming and validating simulation is needed, especially for this application. An additional CFD code, STARCCM+ [13], is being used to compare with simulations using the FDS code.

2. Experimental Methods

The experimental data was taken from slope attachment experiments performed by Gallacher [14]. The experiments were performed using a pool fire of n-heptane contained in a stainless-steel pan. Adjacent to the pan was a 30 cm long metal sheet that was changed to test different angles of attachment. Each angle was then repeated with 1.25 cm thick fireboard insulation attached to the back side of the sheet. Figure 1 shows a sketch of the experimental setup.

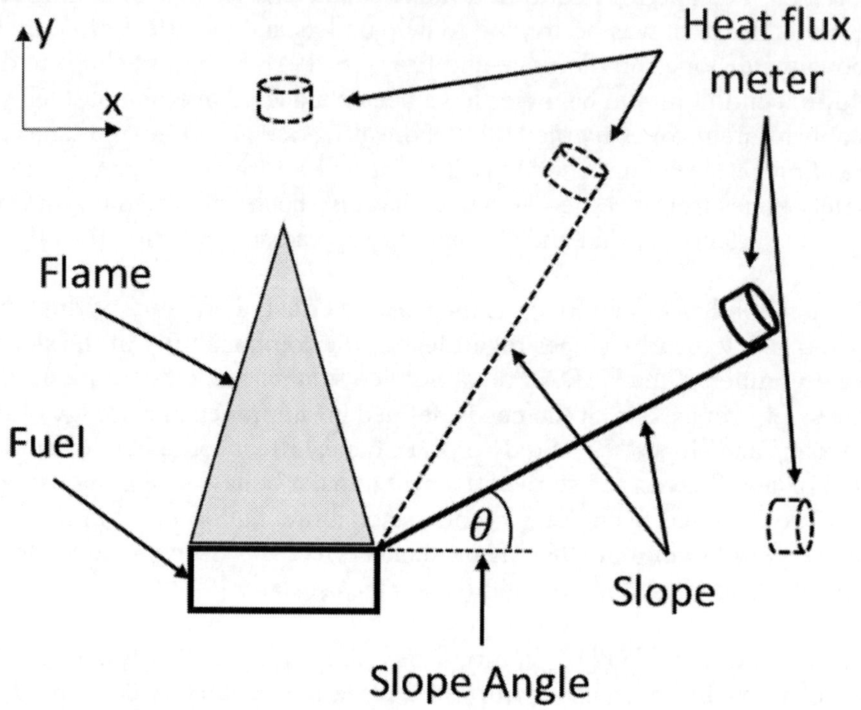

Figure 1. The setup of the flame attachment experiments by Gallacher [15].

The burns at ten different angles ranging from 0 to 63° were recorded by a heat flux meter placed at the top of the slope, and by a video camera to produce images that could be used to calculate the flame attachment length. The video images were then processed using image processing software in Matlab to give an average measurement for the flame area, flame height, and flame attachment length, as seen in Figure 2.

3. Computational Methods

The Fire Dynamics Simulator (FDS) code uses Large Eddy Simulation (LES) as its hydrodynamic model to model the behavior of turbulent flame. The combustion model includes a single step, mixing-controlled chemical reaction of three lumped species; air, fuel, and products [15]. FDS uses a rectilinear mesh to perform calculations. Geometry of objects created in the simulation is restricted to conform to the size of the constructed calculation mesh. Because of this restriction, two different methods were used to create the sloped incline.

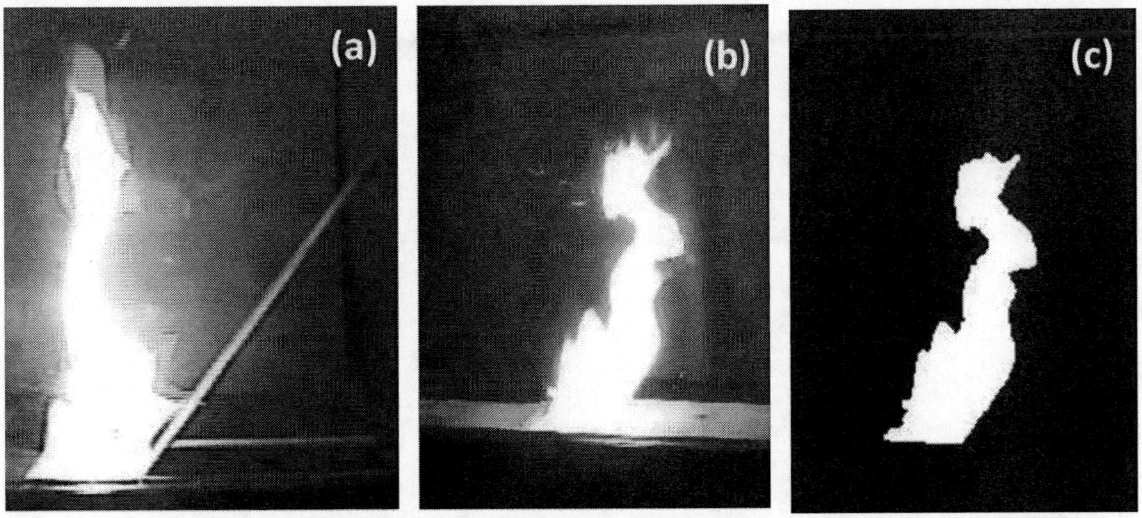

Figure 2. A frame from an experimental burn (a), a frame from a control burn (b), and the resulting binary image from the control burn ready to be processed (c) (taken from Gallacher [10]).

The first method used a function in FDS that allows the gravity to be shifted in a direction other than the default 9.866 m/s^2 in the negative z-direction. Basic vector multiplication was used to create the different angle shifts needed. Figure 3 below shows the initial set up for 36° slope with a gravity shift. This method assures that the slope being used is a flat plane. In contrast, without shifting the gravity vector, a rectilinear mesh must use stair steps to model an inclined geometry. As seen in Figure 3, the major problem with the gravity shift method is the shift occurring across the entire apparatus. The experiment being modeled has a basin at the bottom of the slope that is held horizontally. With shifted gravity, the basin at the bottom of the slope that holds the fuel is shifted as well, which creates enough problems to make this gravity shift approach unusable.

The second method for creating the slope is the use of very small stair step intervals up the slope. Trigonometric calculations were made to create repeating stair steps along the specified angle with 100 steps. The experiment was performed for multiple angles on the slope. Each angle therefore had a different length for the change of the stair step height and width, but the resulting length of the slope was the same. Figure 4 below shows the initial set up for a 36° slope using stair steps.

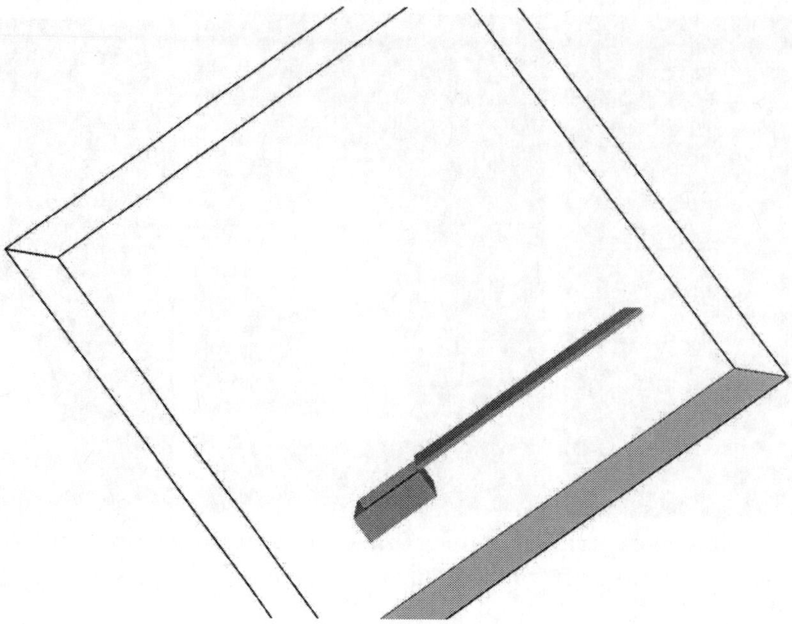

Figure 3. Initial set up of a 36° slope using the gravity shift function with a shift of 5.7642 in the negative x-direction and 7.93375 in the negative z-direction, giving a total gravity vector of 9.8664 in the downward direction as seen by the user.

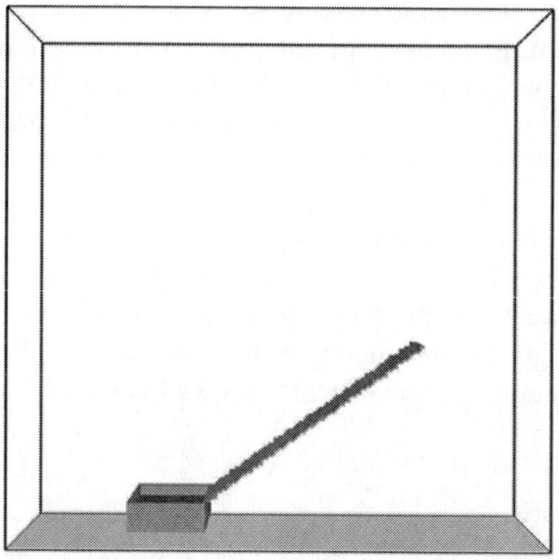

Figure 4. Initial set up of a 36° slope using the stair step method. 100 steps were used. At 36°, each step change was 0.002055m in the x-direction and 0.001493m in the z-direction.

Computation mesh size was a restricting factor in the simulations. Most reasonable cell size that was practical for the computation time was 2.5 mm × 2.5 mm × 2.5 mm in the x,y,z directions respectively. This gives a cell volume of 1.56×10^{-8} m^3. The total calculation domain was split into two different domains to reduce calculation time, each being 0.5 m × 0.2 m × 0.25 m in the x,y,z directions respectively. The cell size used in this set of simulations is about 8 times smaller than the computational cell size used by Ripa et al. [15]. Ripa's previous simulations used a cell size of 5 mm × 5 mm × 5 mm, giving a cell volume of 1.25×10^{-4} m^3. Based on the computational power available and the time needed to run the new simulations, a simulation time of 20 seconds was used.

Three different values were measured in the simulations to make comparisons to the experimental data;

1. mean radiative heat flux ($\overline{\dot{q}''_{rad}}$),
2. mean convective heat flux ($\overline{\dot{q}''_{conv}}$), and
3. mean flame attachment length ($\overline{l_{att}}$).

Predictions of $\overline{\dot{q}''_{rad}}$ and $\overline{\dot{q}''_{conv}}$ were obtained by adding a "device" into the FDS simulation at the top of the slope to measure both values. The values were then time-averaged.

Within FDS there is currently no feature in the software to measure values such as flame height, flame area, or flame attachment length [11]. A MATLAB code was created to calculate the length of attachment along the slope incline from the images created in the FDS code. The video of the mass fraction of the simulation was first rotated in the FDS software and then split up into individual frames rendered as JPEG files. Figure 5 shows an example of a frame from the 36° simulation. In these simulations, each frame was 0.02 seconds. The MATLAB code first allowed a scale to be set based on the length of the incline. Then the code converted the frames to binary images only identifying the area where the oxygen mass fraction is 0 to 0.04. The code uses a chosen position 2 to 5 pixels above the slope and measures the number of pixels along the slope that are inside the low-oxygen boundary. This value is then taken and converted into meters using the previously set scale. The same length is measured for every frame. The entire simulation is then averaged to give the predicted $\overline{l_{att}}$.

4. Results

The obtained predictions were compared to experimental data from Gallacher [5]. Figures 6-8 give comparisons of $\overline{\dot{q}''_{rad}}$, $\overline{\dot{q}''_{conv}}$, and $\overline{l_{att}}$ respectively. In Figure 4, the predicted $\overline{\dot{q}''_{rad}}$ is seen to match the experimental values at lower angles until about 36°. At higher angles, the predictions tend to have greater deviation as the angle increases. This discrepancy may be due to the need for a smaller computation mesh.

In Figure 7, predictions of $\overline{\dot{q}''_{conv}}$ are shown to match the experimental data very well. The predictions tend to pass directly through the same area as both the bare and insulated slopes, as given by Gallacher [15]. Figure 8 shows the comparisons between measurements and predictions of $\overline{l_{att}}$. The predictions and the experimental values of $\overline{l_{att}}$ tend to follow the same shape but the predictions are consistently lower than the experimental values. This discrepancy may be due to the definition that Gallacher used for determining $\overline{l_{att}}$Error! Bookmark not defined.. The discrepancy could also be due in part to the use of the oxygen mass fraction as means of measuring flame position.

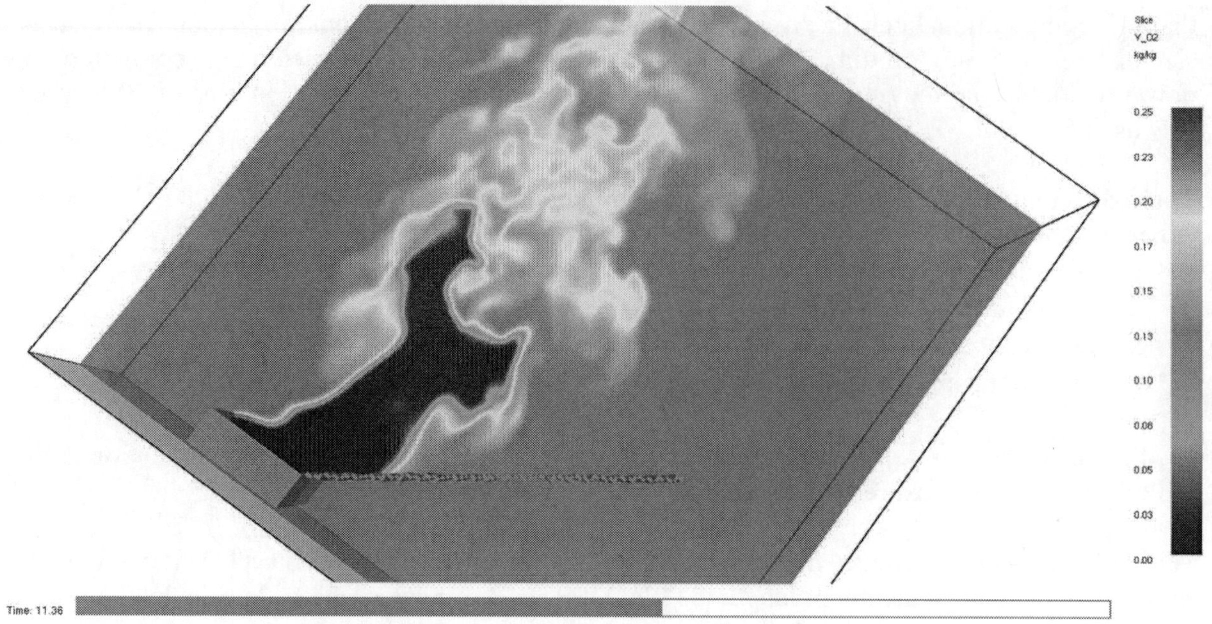

Figure 5. The oxygen mass fraction at frame 568 during the 36° simulation.

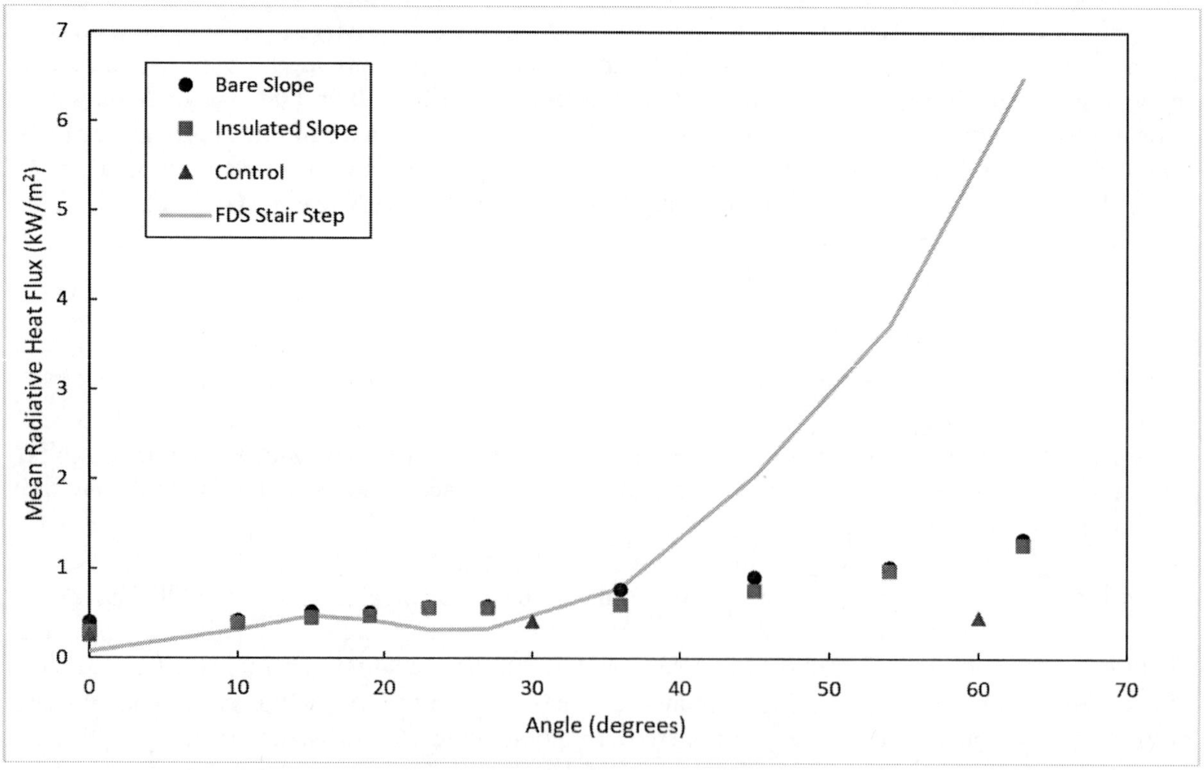

Figure 6. Mean radiative heat flux ($\overline{\dot{q}''_{rad}}$) vs. slope angle.

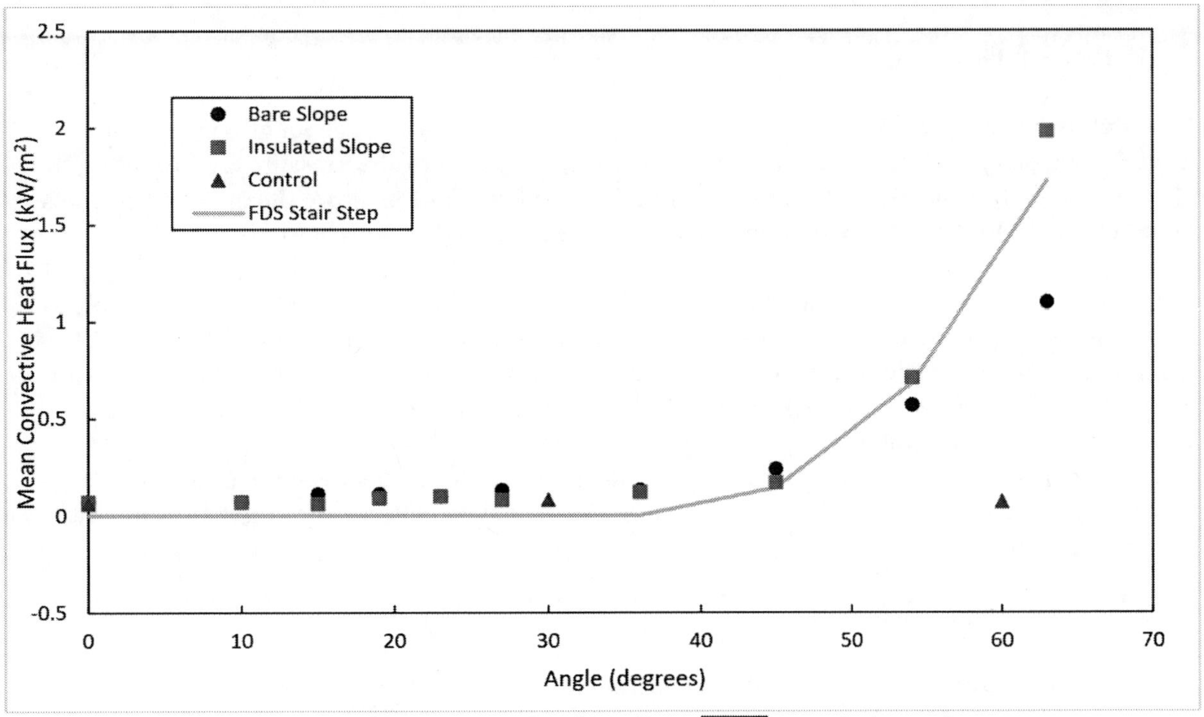

Figure 7. Mean convective heat flux ($\overline{\dot{q}''_{conv}}$) vs. slope angle.

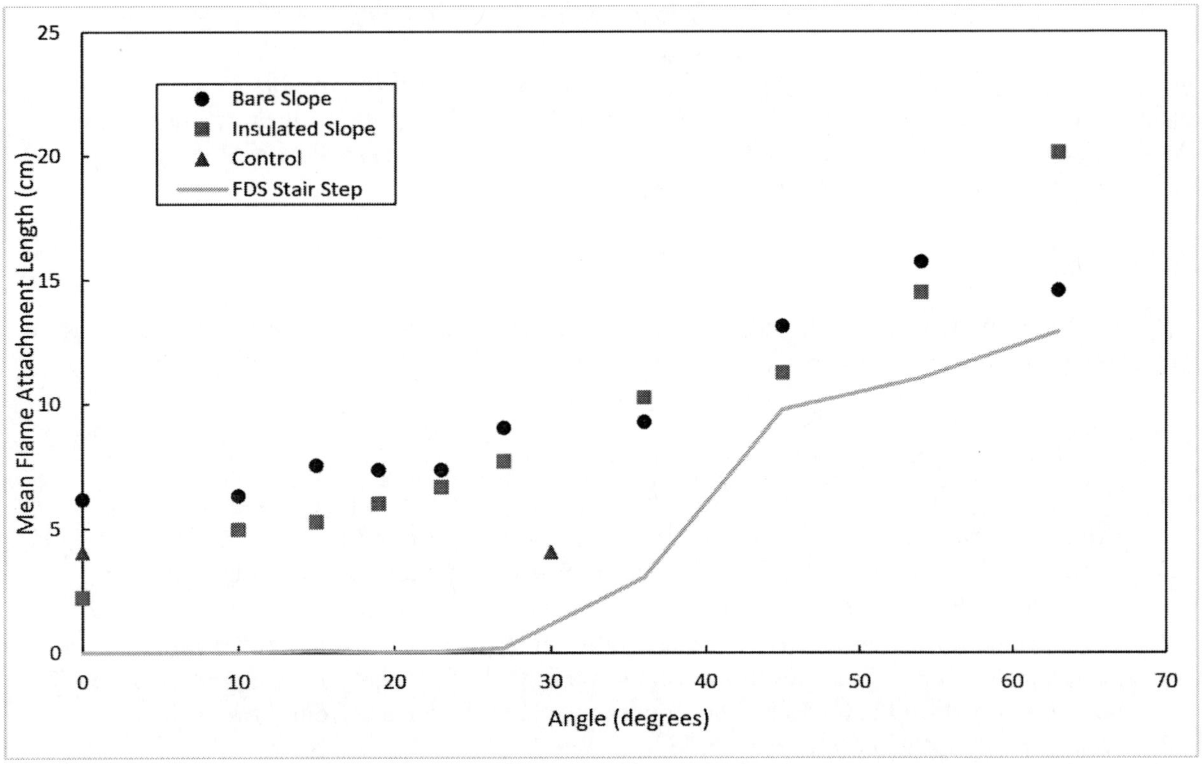

Figure 8. Mean flame attachment length ($\overline{l_{att}}$) vs. slope angle.

5. Future Work

Research is currently being conducted on the pressure change along the length of the slope. There is currently no experimental data to show a comparison of the pressure change along the slope to the predicted values. Future experiments are planned to measure the pressure changes along the slope to facilitate comparison with FDS predictions.

Predictions are currently underway with STARCCM+ using the same conditions. A comparison will be made to see the differences of predictions between STARCCM+ and FDS. STARCCM+ is not constrained to a rectilinear mesh, which will eliminate the stair-step boundary condition along the slope. The two programs will eventually be used to compare large scale predictions with large scale experiments conducted by the United States Forest Service Research and National Fire Decision Support Center. Correlations and patterns between the small and large-scale experiments will then be used and changed to make correlations for larger fires in wildland fire application.

6. Summary and Conclusion

FDS predictions were made of the flame attachment length, mean radiative flux, and mean convective flux for an n-heptane flame near a slope. Slope angle was varied from 0 to 63°. The size of the grid was decreased substantially from earlier modeling efforts. Predicted mean heat flux matched the measured values for all slope angles. Predicted radiative flux increased substantially over measured values above slope angles of about 35°. and STARCCM+ show promise in being able to reliably predict flame behavior on a slope as well as specific flame properties such as temperature and pressure gradients, heat flux, and flame attachment length. As the research progresses, the results from FDS and STARCMM+ will be scaled up and compared with live data from past and current wildfires.

Acknowledgements

Funding was provided by the United States Forest Service Research and National Fire Decision Support Center and by Brigham Young University.

References

[1] Butler, B., 2014, A Study of the Impact of Slope and Wind on Firefighter Safety Zone Effectiveness. Missoula : Joint Fire Science Program, Report No.

[2] Butler, B. W., "Wildland Firefighter Safety Zones: A Review of Past Science and Summary of Future Needs," *International Journal of Wildland Fire*, 23(3) (2014) 295.

[3] 2014, *Incident Response Pocket Guide*. National Wildfire Coordinating Group.

[4] Sullivan, A. L., J. J. Sharples, S. Matthews, and M. P. Plucinski, "A Downslope Fire Spread Correction Factor Based on Landscape-Scale Fire Behaviour," *Environmental Modelling & Software*, 62 (2014) 153-163.

[5] Zárate, L., J. Arnaldos, and J. Casal, "Establishing Safety Distances for Wildland Fires," *Fire Safety Journal*, 43(8) (2008) 565-575.

[6] Page, W. G. and B. W. Butler, "An Empirically Based Approach to Defining Wildland Firefighter Safety and Survival Zone Separation Distances," *International Journal of Wildland Fire*, 26(8) (2017) 655-667.

[7] Viegas, D. X., "On the Existence of a Steady State Regime for Slope and Wind Driven Fires," *International Journal of Wildland Fire*, 13(1) (2004) 101-117.

[8] Fox, J. S. and J. A. Stewart, "The Influence of Coanda Effect on Heat Transfer from Combusting Gases," *Combustion Science and Technology*, 19(1-2) (1978) 73-75.

[9] Moodie, K., "The King's Cross Fire: Damage Assessment and Overview of the Technical Investigation," *Fire Safety Journal*, 18(1) (1992) 13-33.

[10] McGrattan, K., S. Hostikka, R. McDermott, J. Floyd, C. Weinschenk, and K. Overholt, 2017, *Fire Dynamics Simulator User's Guide, 6th Ed.* National Institute of Standards and Technology.

[11] Ripa, B., J. R. Gallacher, D. Kimball, B. Clark, B. Butler, and T. H. Fletcher, " Modeling the Coanda Effect for Fires on Slopes Using Fds, with Implications for Wildland Firefighter Safety," Spring Technical Meeting of the Western States Section of the Combustion Institute, U. of Washington, Seattle, WA (2016).

[12] McGrattan, K., R. McDermott, S. Hostikka, J. Floyd, C. Weinschenk, and K. Overholt, 2017, *Fire Dynamics Simulator Technical Reference Guide Volume 3: Validation. 6th Ed.* National Institute of Standards and Technology.

[13] Star-Ccm+, Last cited: Aug. 23, 2017, https://mdx.plm.automation.siemens.com/star-ccm-plus

[14] *Star-Ccm+*, in *Siemens.com*. Siemens.

[15] Gallacher, J. R., "The Influence of Season, Heating Mode and Slope," PhD Dissertation, Chemical Engineering Department, Brigham Young University, Provo, UT (2016).

Western States Section of the Combustion Institute — Fall 2017 Meeting
Hosted by the University of Wyoming
October 2–3, 2017

Plume Attachment in Wildfires on Slopes: Temperature Measurements in a Laboratory-Scale Analog

Torben P. Grumstrup, Mark A. Finney

Missoula Fire Sciences Laboratory, U. S. Forest Service
Missoula, Montana, 59808, U. S. A.

Fire plume attachment was explored with a propane-fueled flat flame burner fitted flush to a tilting table. The burner was fueled with a fixed flow of propane premixed with a range of air flowrates. A linear array of fine gauge thermocouples recorded gas temperature fluctuations in the plume attachment zone. Temperatures within the attachment zone were found to decrease with increasing air flow to the burner. The trend was attributed to decreasing proximity of flames to the thermocouples as flames transitioned from finger-like to flat geometries. The velocity of fuel and air at the burner exit was found to have no effect on the extent of plume attachment, with one exception. The exceptional case suggested the possibility of a threshold rather than a continuous relationship between the flow of those gases and attachment.

1. Introduction

Among the principal goals of the authors is to understand the physical processes that cause wildfires to spread. In the case of fires on slopes, buoyancy acts in non-intuitive ways to enhance the rate of spread. It is important to study the spread of fire on slopes because many wildfires occur in such terrain. Ideally, spread behavior could be deduced from measurements made in field-scale fires (wildfires and prescribed fire). While field measurement campaigns are planned, they are nonetheless expensive to carry out and occur infrequently. As an alternative (or augmentation) to field-scale experiments, a simple laboratory-scale apparatus is in use at the Missoula Fire Science Lab to examine the relevant physical processes in a controlled and repeatable manner.

Using a variable-angle tilting apparatus with a fixed gas-fired burner for understanding fire behavior is not a new concept. Smith [1] constructed two different sized tilting trenches to examine flame properties during the investigation of the King's Cross Station fire disaster in 1987 [2]. Atkinson et al. [3] used a fixed propane burner in a tilting trench to investigate attachment with temperature and pressure measurements. Wu et al. [4] used an inclined surface to examine the attachment region of a fixed burner with a grid-type schlieren system. Grumstrup et al. 2017a [5] presented a working conceptual model of the flow field of a fire propagating up a slope. Briefly, the tilting of a slope fire plume in the uphill direction is

Figure 1: Tilting table with embedded burner (left); closeup view of 25 μm thermocouple and support wires protruding from table surface (right). Photos by Ian Grob, USFS.

caused by imbalance in the rate of entrainment from the uphill and downhill sides. [5] This explanation is in agreement with findings of experimental and computational studies carried out by a number of workers [2, 3, 4] associated with the Kings Cross fire investigation noted above. Grumstrup et al. 2017a also postulates the attachment zone—a region of elevated temperature and flow velocity immediately uphill of the combustion zone—originates from a combination of the streamwise buoyancy component and a favorable pressure gradient caused by variation in the thermal boundary layer thickness [5]. The former explanation is supported by the pressure measurements of Atkinson et al. [3], while the latter is backed by theoretical work of Umemura and Law on inclined heated plates [6]. Grumstrup et al. 2017b [7] presents observations of dramatic threshold behavior in the transition from unattached to attached plume configurations of a fire on a slope. This paper and Refs. [5, 7] report on an ongoing campaign at the Missoula Fire Sciences Laboratory to understand the physical processes responsible for fire behavior on slopes. Of particular interest is the origin and nature of the plume attachment zone; and how its velocity field, temperature field, and physical extent effect heating of fresh fuel ahead of a fire front. Moreover, it is also important to understand how these three characteristics are influenced by the nature of combustion heat release.

Wildfire is generally thought of as being comprised of non-premixed (or diffusion) flames (e.g. candle, matchstick), wherein gaseous fuel and oxidizer meet only at the thin reaction zone of the flame. Contrast this with a premixed flame (e.g. bunsen burner, gas barbecue grill) where fuel and oxidizer are intermingled prior to burning. [8] In the case of wildfire, the fuel is gas-phase products of pyrolysis and oxidizer is the oxygen fraction of air. Non-premixed flames are undoubtedly important in wildfire. However, premixed combustion is likely a factor as well, given the mixing effects of these highly turbulent phenomena. As such, it is worthwhile to understand how stoichiometry of premixed flames effects attachment. The preliminary results presented here show temperature measurements of a plume from a burner fueled by mixtures of various stoichiometries.

Table 1: Parameters are shown for the series of experiments conducted at two slope angles.

Case	α	$\dot{V}_f$ [lpm]	$\dot{V}_a$ [lpm]	ϕ	v_{exit} [mm/s]
A	20°	4.0	0	-	1.4
B	20°	4.0	20	4.8	8.7
C	20°	4.0	60	1.6	23
D	20°	4.0	80	1.2	30
E	25°	4.0	0	-	1.4
F	25°	4.0	20	4.8	8.7
G	25°	4.0	40	2.4	16
H	25°	4.0	60	1.6	23
J	25°	4.0	80	1.2	30

2. Experimental Methods

Fire plume attachment was explored with a 9.2 × 50 cm propane-fueled flat flame burner [9] fitted flush to a 122 × 122 cm table (Fig. 1) that tilted to arbitrary angles ($0° \leq \alpha \leq 45°$), similar to that of Wu et al. [4]. Flow of propane gas and air were controlled using Omega Engineering FL-2000 series rotometers. The table surface was constructed using non-flammable ceramic fiberboard. Gas temperature of the attached plume was measured by a linear array of twenty-one thermocouples (type K, 25 μm diameter) mounted approximately 8 mm above the table surface. Temperature data were recorded by a National Instruments DAQ system at 500 Hz per channel. The experiments were carried out at the Missoula Fire Sciences Lab where the elevation is 975 m.

Experiments were conducted in an enclosed room where the ambient air (~ 25°C) was made as quiescent as practical. Nonetheless, these experiments were observed to be very sensitive to room currents. We discovered that a very slow room-scale swirl was introduced by running the exhaust hood. The swirl had the effect of lifting one side of the plume off the table and making acquisition of plume attachment data futile. We worked to minimize the influence of such room currents on these data by limiting the number of experiments conducted on a sigle day.

Table 1 shows the parameters of the nine cases tested. Experiments were conducted for two slope angles, α, for a fixed volumetric fuel flowrate, $\dot{V}_f$, and varying volumetric air flowrate, $\dot{V}_a$. Equivalence ratio, ϕ, appears in the next column, followed by the velocity of the fuel-air mixture flowing out of the burner exit prior to combustion, v_{exit}. Note that a case in which $\alpha = 20°$ and $\dot{V}_a = 40$ lpm does not appear because the data were lost.

When acquiring data, the flow of fuel and air was set and the flame allowed to reach steady state with the table in the zero-slope ($\alpha = 0°$) configuration. The table was then raised to the specified slope and the flow field allowed to equilibrate for approximately 15 seconds. The data acquisition system was then triggered and temperature data acquired for approximately 10 seconds. (Ten seconds was much longer than the slowest thermal fluctuation [~1 s] observed in the data.) The table was then returned to the zero-slope configuration and the next case was tested. At the conclusion of experiments, the acquired data were processed and plotted using MATLAB scripts written by the authors.

Uncertainty in the fuel flowrate, air flowrate, slope angle, and thermocouple position were 0.1 slpm, 1.0 slpm, 0.1°, and 1 mm respectively. Uncertainty of K-type thermocouples is typically reported by manufacturers according to ASTM standard as the larger of ±2.2° or 0.75% of the reading in centigrade. [10] Nonetheless, temperature measurement uncertainty was most likely higher because of the extreme nature of the application (namely, large and rapid temperature changes). [10] Moreover, the thermocouples were neither calibrated after installation nor compensated for radiation loss. The combination of these factors mean that the temperature data should be interpreted as being only somewhat accurate. However, the resulting trends are nonetheless revealing of attachment behavior.

3. Results and Discussion

Changes in appearance of the burner flame with varying amounts of premixed air were apparent simply by visual inspection. When no air flowed (Cases A and E), yellow, flickering, finger-like flames were produced similar to those in Fig. 1. This behavior was expected since non-premixed flames such as these will lengthen as needed to access sufficient oxygen for combustion. [8] On the other hand, flowing air at $\dot{V}_a = 80$ lpm (Cases D and J) produced a flat, blue, immobile flame that hovered a few millimeters above the ceramic foam exit of the burner. Likewise, it is well known that a nearly stoichiometric (see Table 1) flame like this one will stabilize very near to the burner exit. [8] For the intermediate air flow cases, a smooth transition from the flickering-yellow to flat-blue flames was observed as $\dot{V}_a$ increased from 0 to 80 lpm. Therefore, the overall effect of mixing increasing amounts of air with propane was to transition from an elongated region of heat release to one that was highly localized. Such behavior has important implications for gas temperatures at the table surface as discussed below.

The mean of the instantaneous temperature measurements, $T_{\mathrm{m}}(d)$, were found and plotted as a function of the distance from the uphill edge of the burner to the respective thermocouple, d. Average temperature measurements for the $\alpha = 20°$ cases are shown in Fig. 3. The attachment zone is the region of elevated temperature for (roughly) $d < 15$. The thermocouple nearest the burner at $d = 1$ cm indicates relatively high temperatures for Cases A, B, and C. The highest temperature is for Case B at nearly 1000°C while both Cases A and C have maximums of around 850°C. Moving further from the burner, Cases A, B, and C show temperatures dropping rapidly with distance before converging to a common trend near $d = 15$ cm. These three cases then slowly decline towards ambient temperature from $d = 15$ cm to the furthest thermocouple at $d = 36.4$ cm. A dramatic difference is seen in Case D where only very slightly elevated temperatures are seen at $d = 1$ cm and essentially ambient temperature everywhere else. The results of the $\alpha = 25°$ slope angle experiments are shown in Fig. 3 where all five cases (E through J) exhibit an overall trend similar to that of Cases A, B, and C. Temperatures at $d = 1$ cm are roughly comparable between the $\alpha = 20°$ and $\alpha = 25°$ experiments, although the maximum, G, slightly exceeds 1100°C. Cases E through J also rapidly fall off in temperature with distance from the burner and converge at around $d = 15$ cm.

In all cases, some scatter in temperature is apparent for roughly $d < 15$ cm. Increasing temperatures correspond to decreasing air flowrate in Fig. 2, while the behavior in Fig. 3

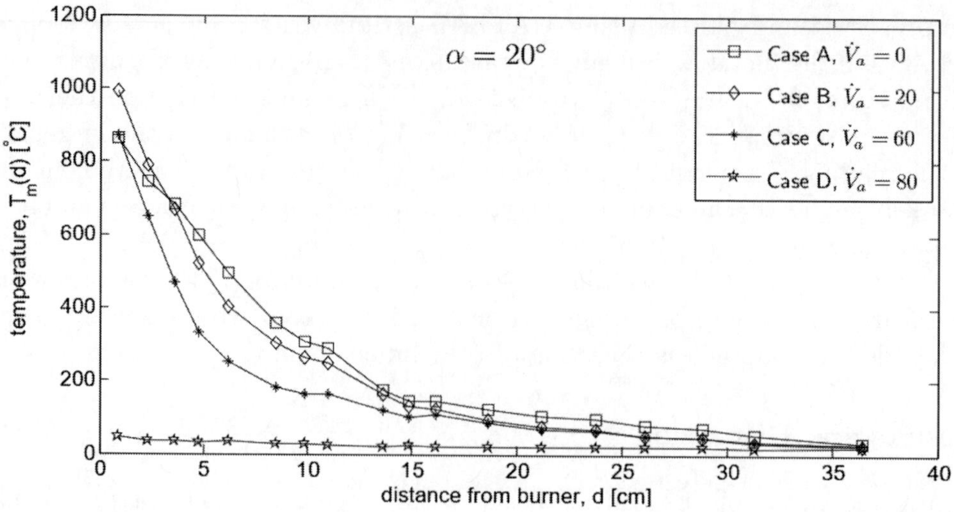

Figure 2: Average gas temperature upslope from the burner for the specified angle.

is less clear. However, consider that Cases E and F (less premixed air) exhibit higher temperatures than Cases G, H, and J (more premixed air). Case H is a counterexample to this argument since it has lower temperatures than Case J. It is impossible to be certain at this point, but the behavior of Case H could easily be explained by the influence of room currents. Discussion of Case D, where temperatures remain essentially at ambient everywhere in the thermocouple array, follows below.

The correlation between increasing temperatures and decreasing air flowrate can be explained by considering the nature of the flame produced by the burner. The elongated flames produced by low to no premixed air flow mean that combustion—and therefore heat release—occur over a region starting at the burner exit and extending to the flame tips. This puts the high temperature products of combustion in close proximity to the thermocouples, causing especially high temperatures to be detected. In contrast, flames with high air flow produce combustion that is localized to the burner surface, giving the hot combustion products an opportunity to mix with cool entrainment air prior to reaching any given thermocouple. Yet this logic seems to contradict the expectation that adiabatic flame temperature should decrease with decreasing air flow as the fuel-air mixture becomes more rich (see, e.g., Section 1.4 of Ref. [8]). Moreover, radiant heat loss from luminous yellow soot will also reduce flame temperature as air flowrate is decreased and soot formation increases. There is no doubt that these flame temperature arguments are valid, but we believe that they are overwhelmed by the effects of proximity of heat release to the thermocouples and mixing.

Another important factor to consider is the exit flow of the mixture leaving the burner, which one would expect to resist attachment since it flows orthogonally to attachment zone flow. However, were that true, the extent of attachment would be expected to decrease with increasing burner exit flow velocity, v_{exit} (see Table 1). Instead, it is apparent in both Figs. 2 and 3 that all but Case D converge near $d = 15$ cm, meaning that attachment length appears to be unaffected by v_{exit}. Case D is unique because it lacks an attachment zone altogether. A cursory examination with shadowgraphy confirmed the plume simply rose vertically without making contact with the table. Here we suspect the buoyancy forces that

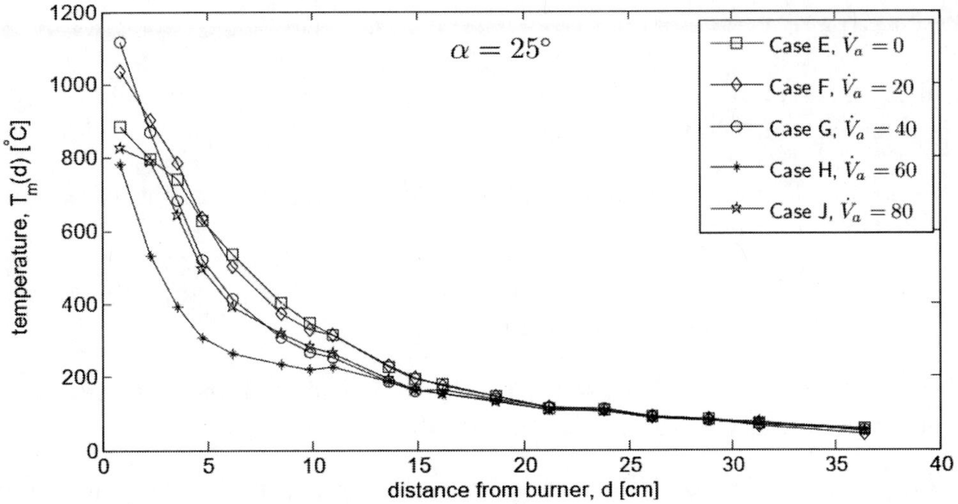

Figure 3: Average gas temperature upslope from the burner for the specified angle.

cause attachment were weaker at a smaller slope angle, enabling the burner exit flow inertia to dominate buoyancy effects and prevent attachment. This theory could be strengthened by attempting to prevent attached flow for $\alpha = 25°$ by increasing $\dot{V}_a$ beyond 80 lpm. Were this possible, it would suggest that the effect of burner exit flow on attachment exhibited threshold rather than smooth, continuous behavior.

4. Conclusions

Fire plume attachment was explored with a propane-fueled flat flame burner fitted flush to a table that tilted to 20° and 25°. The burner was fueled with 4.0 lpm of propane premixed with a range of air flowrates. A linear array of fine gauge thermocouples recorded gas temperature fluctuations in the plume attachment zone approximately 8 mm above the table surface. Temperatures within the attachment zone were found to decrease with increasing air flow. The trend was attributed to decreasing proximity of flames to the thermocouples as flames transitioned from finger-like to flat geometries. The velocity of fuel and air at the burner exit was found to have no effect on the extent of plume attachment, with one exception. The exceptional case suggested the possibility of a threshold rather than a continuous relationship between the flow of those gases and attachment.

Acknowledgements

Significant support during experiments was provided by: Fabricators Randy Pryhorocki and Josh Deering, and Technicians Andrew Gorris and Jon Bergroos. TPG is grateful to colleague Jason D. Forthofer for technical discussions on plume attachment. This research was supported by U. S. Forest Service Research and the National Fire Decision Support Center.

References

[1] D. Smith, "Measurements of flame length and flame angle in an inclined trench," *Fire Safety Journal*, vol. 18, no. 3, pp. 231–244, 1992.

[2] B. Crossland, "The King's Cross Underground fire and the setting up of the investigation," *Fire Safety Journal*, vol. 18, no. 1, pp. 3–11, 1992.

[3] G. Atkinson, D. Drysdale, and Y. Wu, "Fire driven flow in an inclined trench," *Fire safety journal*, vol. 25, no. 2, pp. 141147–145158, 1995.

[4] Y. Wu, H. Xing, and G. Atkinson, "Interaction of fire plume with inclined surface," *Fire Safety Journal*, vol. 35, no. 4, pp. 391–403, 2000.

[5] T. P. Grumstrup, S. S. McAllister, and M. A. Finney, "Qualitative flow visualization of flame attachment on slopes," in *10th U. S. National Combustion Meeting*, 2017.

[6] A. Umemura and C. K. Law, "Natural-convection boundary-layer flow over a heated plate with arbitrary inclination," *Journal of Fluid Mechanics*, vol. 219, pp. 571–584, 1990.

[7] T. P. Grumstrup, J. D. Forthofer, and M. A. Finney, "Temperature measurements of attached fire plumes on a slope," in *Eighth International Symposium on Scale Modeling*.

[8] C. K. Law, *Combustion physics*. Cambridge University Press, 2010.

[9] D. Jimenez, M. A. Finney, and J. Cohen, "Design and construction of gas-fed burners for laboratory studies of flame structure," in *5th International Conference on Forest Fire Research*, 2010.

[10] J. T. Nakos, "Uncertainty analysis of thermocouple measurements used in normal and abnormal thermal environment experiments at Sandia's Radiant Heat Facility and Lurance Canyon Burn Site," Report SAND204-1023, Sandia National Laboratories, 2004.

Paper # 29FI-0020 **Topic: Fire**

Western States Section of the Combustion Institute - Fall 2017 Meeting
Hosted by the University of Wyoming
October 2-3, 2017.

The onset of puffing of pool fires

W. Coenen[1] *D. Moreno*[1] *A.L. Sánchez*[1] *F.A. Williams*[1]

[1]*University of California, San Diego. La Jolla, CA 92093-0411, USA.*

Jet diffusion flames and pool fires are known to undergo a bifurcation to a globally unstable flickering or puffing behavior caused by baroclinic and gravitational vorticity production. Although for pool fires the supercritical puffing regime has been studied extensively in the literature, no detailed account has been given of the critical conditions for the onset of puffing. Apart from the inherent thermochemical parameters, this onset is governed by a single dimensionless number, the Rayleigh number. Equivalently, for a fixed fuel and under fixed ambient conditions, there is a critical fuel pool diameter above which the fire starts puffing. In the present work, we have determined this diameter with laboratory experiments for three fuels: methanol, ethanol, and heptane.

1 Introduction

Pool fires are known to exhibit a self-sustained oscillatory behavior, shedding large toroidal coherent structures at a well established frequency, a phenomenon referred to in the literature as "puffing". This behavior influences the rate of air entrainment, the radiated heat output, the flame height, and also the spreading of the flame. Pool fire puffing has been studied extensively in the literature (e.g. [1, 2, 3, 4, 5, 6, 7, 8, 9]), and it is well known that under normal conditions of temperature and pressure typical hydrocarbon fuel pools of a few centimeters in diameter puff with a frequency on the order of 10 Hz. Nevertheless, no detailed account has been given of the critical conditions for the *onset* of puffing. In that regard, for the physically similar problem of flickering jet diffusion flames, the authors have recently shown that this oscillatory behavior corresponds to a hydrodynamic global instability of the flow, the onset of which can only be predicted with a stability analysis that is also global in nature, taking into account the 2D nonparallel character of the flow [10].

Contrary to a jet diffusion flame, where momentum is injected with the fuel, and consequently the jet Reynolds number plays a role along with the Froude number, in a pool fire the momentum that is transferred to the flow by the evaporation of fuel is negligible. Therefore, apart from the inherent thermochemical parameters, a single dimensionless parameter, expressed as a Rayleigh number $Ra = ga^3/(\nu_A \kappa_A)$, characterizes the flow. Here a is the radius of the fuel pool, ν_A and κ_A are the kinematic viscosity and thermal diffusivity of the ambient air, and g is the gravitational acceleration. Puffing occurs when this Rayleigh number exceeds a critical value. For given burning conditions, the latter is directly equivalent to a critical pool diameter. The objective of the present work is to determine the critical Rayleigh number with laboratory experiments. We present results for three fuels: methanol, ethanol and heptane.

In the next section, we describe the experimental procedure that was followed. The results are presented in section 3. Finally, concluding remarks are given in section 4.

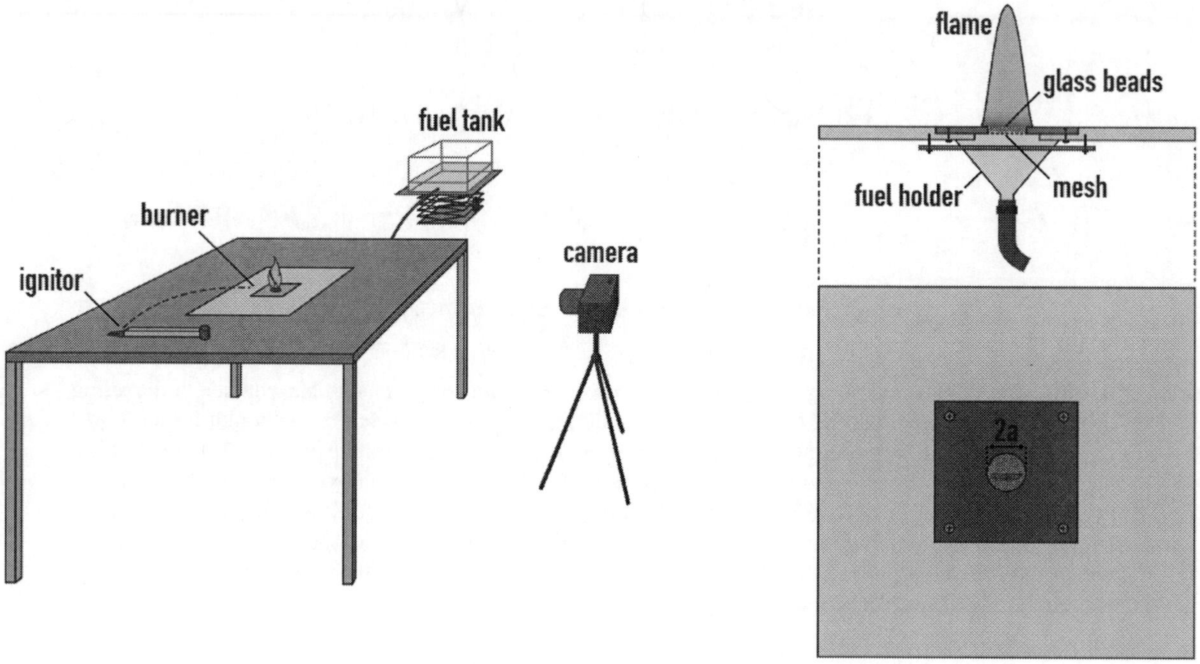

Figure 1: Schematic diagram of the experimental setup. The left hand side shows the general setup, whereas the right hand side shows a more detailed side and top view of the liquid fuel burner.

2 Experimental procedure

To determine the critical conditions for the onset of pool fire puffing, controlled experiments were carried out, in which, for a fixed fuel and under fixed ambient conditions, the flame dynamics were characterized for different fuel pool sizes. Note that the fuel pool radius a is related to the Rayleigh number Ra as $Ra = ga^3/(\nu_A \kappa_A)$ with $g = 9.807\,\mathrm{m/s^2}$, $\nu_A = 1.597 \times 10^{-5}\,\mathrm{m^2/s}$, and $\kappa_A = 2.256 \times 10^{-5}\mathrm{m^2/s}$.

The experimental setup is depicted in figure 1. It consists of a $90\,\mathrm{cm} \times 90\,\mathrm{cm} \times 12.7\,\mathrm{mm}$ PVC table that in its center holds a removable liquid fuel burner. This burner is composed of a $20\,\mathrm{cm} \times 20\,\mathrm{cm} \times 6.35\,\mathrm{mm}$ aluminum plate, a 6 cm-diameter conical polypropylene fuel holder mounted underneath, and a small interchangeable $7.5\,\mathrm{cm} \times 7.5\,\mathrm{cm} \times 3.175\,\mathrm{mm}$ titanium plate placed within a recess in the upper side of the former. The interchangeable plate has a circular hole of a certain radius a in its center, filled up to the rim with a layer of glass beads (0.5 mm in diameter) that rests on a thin stainless steel mesh. This porous layer prevents convective currents in the fuel from affecting the observations. When assembled, all elements form a large flush surface around the circular hole. The fuel holder is connected to an external fuel tank with a cross-sectional area of approximately $350\,\mathrm{cm^2} \gg \pi a^2$. In this manner, once the height of the tank is adjusted so that the burner is filled up to the rim with fuel, this fuel level remains quasi-constant during the course of an experiment (typically a few minutes).

To prevent external perturbations to the flame, the complete experiment is enclosed by a $3\,\mathrm{m} \times 2\,\mathrm{m} \times 2\,\mathrm{m}$ chamber. After closing the chamber, we waited approximately 10 minutes for all air currents to die out before starting the experiment. To ignite the fuel remotely, a surface ignitor

fuel	$2a_c$ [mm]	Ra_c	f [s^{-1}]	St
methanol	17.75–19.25	19030–24270	10.16	0.00344
ethanol	16.00–17.75	13940–19030	10.08	0.00370
heptane	14.25–16.00	9845–13940	13.12	0.00534

Table 1: The critical conditions for the onset of pool fire puffing for different fuels. The diameters and associated Rayleigh numbers are given as an interval, the lower and upper limit of which correspond respectively to the largest value resulting in a stable flame, and the lowest value resulting in an unstable flame. The puffing frequencies and associated Strouhal numbers correspond to the upper limit of the interval.

mounted on a motorized arm was slowly brought close to the fuel and retracted afterwards. The flame dynamics were recorded with a Panasonic Lumix FZ300 camera at a frame rate of 120 fps. For unstable flames, the puffing frequency can be determined from the image sequence by selecting a region of the image near the flame tip and computing the power spectral density of the temporal evolution of the average intensity in that region.

3 Results

For all three fuels, we found a clear transition between a steady and a periodically oscillating flame when the pool diameter (i.e. Rayleigh number) was increased beyond a certain critical value. Figure 2(a) shows a marginally subcritical steady flame, in particular for heptane and for a pool diameter of $2a = 14.25$ mm ($Ra = 9845$). When the diameter is slightly increased to $2a = 16.00$ mm ($Ra = 13940$), the tip of the flame starts oscillating. Further increasing the pool size into the supercritical regime leads to oscillations of the flame that become increasingly violent, such as for the $2a = 17.75$ mm diameter ($Ra = 19030$) flame shown in figure 2(b). For all unstable cases, the computed frequency spectra show a single peak, corresponding to the frequency of puffing. All these findings confirm the underlying nature of the onset of puffing as a Hopf bifurcation to a globally unstable state.

Table 1 summarizes the critical conditions for the onset of pool fire puffing for the three fuels that were analyzed. Note that in the experiments the diameter is varied in discrete steps, so that the actual critical value is contained within the interval declared in the table. For the puffing conditions corresponding to the upper limits of the intervals, the puffing frequencies and corresponding Strouhal numbers are also given. It can be observed that the fuel type has a strong influence on the stability properties of a pool fire. In this context, it must be noted that the flame height also differs greatly between the different fuels, with heptane flames being considerably taller than ethanol or methanol flames.

(a)

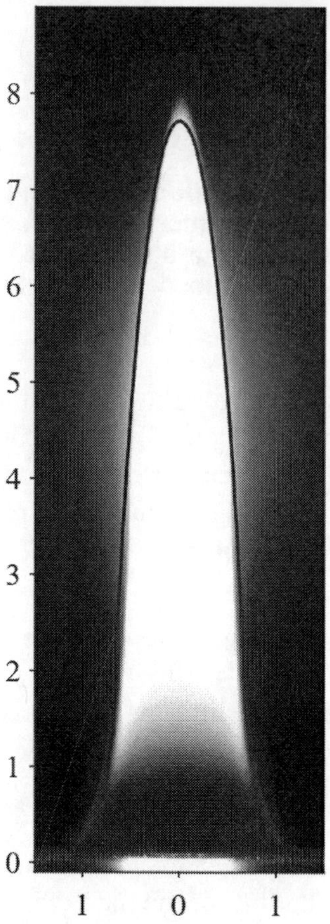

(b)

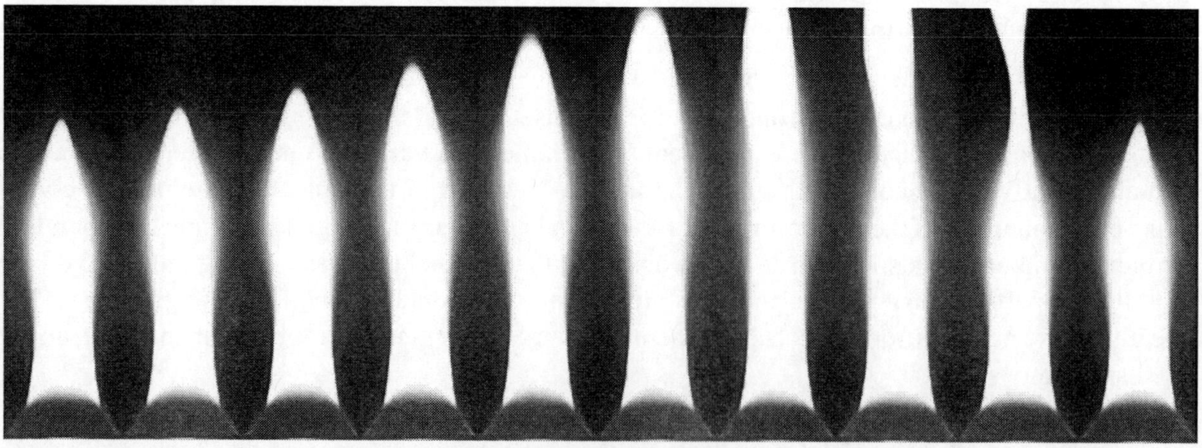

Figure 2: (a) Steady subcritical heptane flame with diameter 2*a*=14.25 mm (*Ra*=9845). The solid line correspond to a preliminary theoretical prediction of the flame front. (b) One puffing cycle for a supercritical heptane pool fire with diameter 2*a*=17.75 mm (*Ra*=19030). The time interval between the images is 1/120 s. The measured puffing frequency for this case is 12.66 Hz (*St* = 0.00464).

4 Conclusions and future work

We have presented an experimental study of the onset of puffing in pool fires. Our results confirm that the onset of puffing is a manifestation of a Hopf bifurcation to a globally unstable state characterized by a single frequency. We have successfully determined the critical conditions in terms of the Rayleigh number for which methanol, ethanol, and heptane fuels become globally unstable (see table 1).

Currently, our efforts are dedicated to the theoretical and numerical description of this problem. In that regard, we have derived a detailed formulation to describe the dynamics of small laminar pool fires, taking advantage of the simplifications associated with the limit of infinitely fast reaction [11], but accounting for the nonunity Lewis number and vaporization characteristics of typical liquid hydrocarbon fuels. The solid curve in figure 2(a) is a preliminary result for the flame front, showing good agreement with the experimental observation. The theoretical description will allow for a global stability analysis to be carried out, using the methodology that has recently proven successful in the prediction of the onset of global instability in jet diffusion flames [10] and in low-density jets [12].

References

[1] V.I. Blinov and G.N. Khudyakov. Diffusion burning of liquids. Technical Report T-1490a-c, Izdatel'stvo Akademii Nauk SSSR, Moscow, 1961.

[2] G.M. Byram and R.M. Nelson. *Fire Tech.*, 6 (1970) 102–110.

[3] M. Hertzberg, K. Cashdollar, C. Litton, and D. Burgess. The diffusion flame in free convection: buoyancy-induced flows, oscillations, radiative balance, and large-scale limiting rates. Technical Report 8263, Dept. of the Interior, Bureau of Mines, 1978.

[4] E. Zukoski, B. Cetegen, and T. Kubota. *Symp. (Int.) Combust.*, 20 (1984) 361–366.

[5] A. Schönbucher, B. Arnold, V. Banhardt, V. Bieller, H. Kasper, M. Kaufmann, R. Lucas, and N. Schiess. *Proc. Combust. Inst.*, 21 (1986) 83–92.

[6] E.J. Weckman and A. Sobiesiak. *Symp. (Int.) Combust.*, 22 (1988) 1299–1310.

[7] B.M. Cetegen and T.A. Ahmed. *Combust. Flame*, 93 (1993) 157–184.

[8] A. Hamins, J.C. Yang, and T. Kashiwagi. *Symp. (Int.) Combust.*, 24 (1992) 1695–1702.

[9] W.M.G. Malalasekera, H.K. Versteeg, and K. Gilchrist. *Fire and materials*, 20 (1996) 261–271.

[10] D. Moreno-Boza, W. Coenen, A. Sevilla, J. Carpio, A.L. Sánchez, and A. Liñan. *J. Fluid Mech.*, 798 (2016) 997–1014.

[11] A. Liñán, M. Vera, and A. L. Sánchez. *Annu. Rev. Fluid. Mech.*, 47 (2015) 293–314.

[12] W. Coenen, L. Lesshafft, X. Garnaud, and A. Sevilla. *J. Fluid Mech.*, 820 (2017) 187–207.

Western States Section of the Combustion Institute – Fall 2017 Meeting
Hosted by the University of Wyoming
October 2–3, 2017

Model sensitivities in LES predictions of buoyant methane fire plumes

Heeseok Koo[1,**], *John C. Hewson*[1], *Stefan P. Domino*[2], *and Robert C. Knaus*[2]

[1] *Fire Science and Technology, Sandia National Laboratories, PO Box 5800, MS 1135, Albuquerque, NM 87185-1135, USA* *

[2] *Compuational Thermodynamics and Fluid Mechanics, Sandia National Laboratories, PO Box 5800, MS 0828, Albuquerque, NM 87185-0828, USA*

[**] *Corresponding author:* `hkoo@sandia.gov`

Abstract: A 1-m diameter methane fire plume has been studied using a large eddy simulation (LES) methodology. Eddy dissipation concept (EDC) and steady flamelet combustion models were used to describe interactions between buoyancy-induced turbulence and gas-phase combustion. Detailed comparisons with experimental data showed that the simulation is sensitive to the combustion model and mesh resolution. In particular, any excessive mixing results in a wider and more diffusive plume. As mesh resolution increases, the current simulations demonstrate a tendency toward excessive mixing.

Keywords: ***Fire plume, buoyant plume, LES, turbulent combustion, SIERRA Fuego***

1. Introduction

Fire plumes have been actively studied in recent years due to the associated impact and danger of industrial accidents involving leaking fuel [1–3]. A number of experiments have been performed to understand the physics [4–8] from which empirical models were obtained [9–11]. Simulating fire plumes, however, has several modeling challenges such as flame ignition/extinction, turbulence-combustion-radiation interaction, laminar-turbulence transition, soot formation, and the fact that they are dominated by a relatively slow, buoyancy-driven flow. Combusting plumes have been a target for many fire modelers [3, 12–16] and efforts by several researchers were compared and discussed on the 2017 Measurement and Computation of Fire Phenomena (MaCFP) workshop, which offered a central forum for fire model validation [17].

LES is becoming the state-of-the-art tool for modeling turbulent combustion and is also widely used for fire calculations [17], though the level of resolution varies and also places simulations in a regime where most of the turbulent kinetic energy is modeled. The EDC approach with fast chemistry and the steady flamelet models are frequently used turbulent combustion models in the fire community [18–20]. These models were used in this study, including assumptions that local extinction/ignition and flame liftoff are not critical in predicting pool fires.

FLAME (the Fire Laboratory for Accreditation of Models by Experimentation) is the high-fidelity, large-scale, controlled, indoor facility for fire validation experiments that is constructed

*Sandia National Laboratories is a multimission laboratory managed and operated by National Technology and Engineering Solutions of Sandia, LLC., a wholly owned subsidiary of Honeywell International, Inc., for the U.S. Department of Energy's National Nuclear Security Administration under contract DE-NA0003525.ă

and maintained by Sandia National Laboratories for pool fire and related experimental studies [2]. A number of data sets are available from experiments performed using different gaseous or liquid fuels, varying pool sizes, under changing operating conditions for various types of fires [4, 5, 21–23]. Across two different versions of the facility, FLAME allows laser diagnostics such as PIV and PLIF at scales larger than typically possible [4] along with heat flux gauges [23], and more recently, coherent anti-Stokes Raman scattering (CARS) [22]. The methane pool fire case that current study focuses on were performed in an older version of the facility from which mean and rms velocities are available [4].

The focus of this paper is to understand sensitivity of the plume predictions to various modeling and mesh choices. In particular, the effects of domain shape, mesh resolution, combustion models and boundary conditions were verified. SIERRA/Fuego [20], an in-house low-Mach turbulent combustion code developed by Sandia National Laboratories, was used for the simulations. The next two sections describe target problem and numerical models, before providing simulation results from Sec. 4.

2. Experimental Configuration

Figure 1 shows an outer view of the facility as well a diagram of the aforementioned FLAME facility. The internal test section of the facility has a 5.8m×5.8m square area with a converging hood 4.5 m above the fuel inlet. A 1m-diameter fuel inlet is located at the middle of the 2m-diameter pedestal-like structure that simulates a ground plane. Air is injected at an outer-ring which lies at 3m below the fuel-inlet height. A cylindrical radiation shield, colored in orange in Fig. 1 in the middle, is installed from the bottom up to 0.9m above the fuel-inlet height to minimize any asymmetric behavior.

PIV measurements were performed on the center plane on a 1m×1m area right above the fuel inlet. Table 1 summarizes three experimental conditions published and available. Here, test #24 is simulated which is in the middle in terms of its fuel-air flow rate ratio.

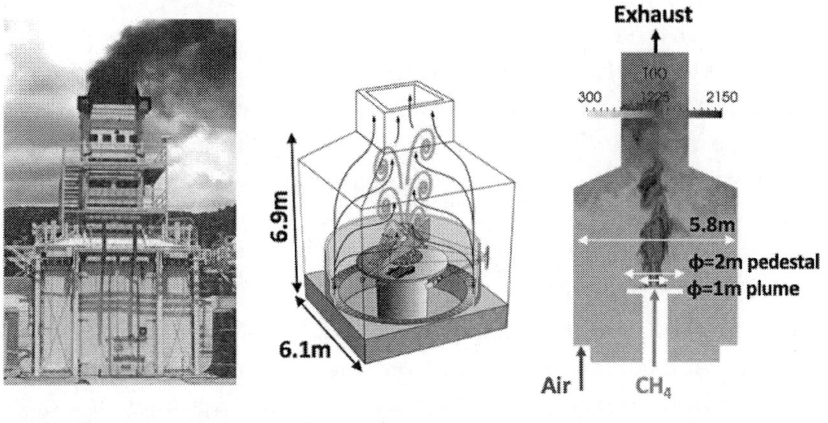

Figure 1: Picture and figures of the FLAME experimental facility

Test No.	Fuel inlet velocity (m/s)	Heat release rate (MW)	Inlet air velocity (m/s)	Equivalence ratio (global)
14	0.074	1.59	0.323	0.19
24	0.097	2.07	0.327	0.25
17	0.117	2.61	0.299	0.34

Table 1: Available test cases

3. Model Details

Simulations are conducted using an unstructured control-volume finite-element code [20, 24, 25], a component of Sandia National Laboratories SIERRA thermal-fluids code suite. The package has a capability of integrating different physics in multiple phases. SIERRA/Fuego, used for the current study, is a low-Mach flow solver suited for non-reacting and reacting flows. Various RANS (Reynolds-averaged Navier-Stokes) and LES turbulence models are available in the code. For LES, constant-coefficient and dynamic versions of the Smagorinsky model and subgrid-scale kinetic energy one-equation models are available. In this study, both Smagorinsky and one-equation models are used to describe the subfilter process with pre-defined coefficients.

Two approaches are used for the gas-phase combustion. In the first approach, the gas-phase combustion is closed with the eddy-dissipation concept (EDC) with fast chemistry [26]. An arbitrary hydrocarbon fuel is decomposed into CO and H_2 at the first step, while the intermediate species, CO and H_2, are further oxidized in the second step provided oxygen is still sufficient. This approach models CO, and potentially soot, in the fuel-rich case. Enthalpy and species mass fraction equations are Favre-filtered along with momentum. Radiative heat transfer is incorporated by solving a radiation transport equation.

In the second approach, the gas-phase combustion is described using a steady flamelet model where a set of one-dimensional flamelets are tabulated for look-up in the computation [27]. The transport equation for a conserved scalar, namely fuel-oxidizer mixture fraction, is solved along with momentum equations. Mixture fraction replaces many variables in EDC, such as enthalpy, species mass fraction, and even radiation, which allows a cost effective computation. A mechanism that contains detailed methane reaction (GRI-Mech 2.11 [28]) and the Flamemaster program [29] were used in the tabulation. A subfilter closure using an assumed, clipped Gaussian PDF was used, with the variance of the mixture fraction being estimated using a gradient model. Tabulated file size was reduced using an adaptive coarsening technique.

4. Domain Sensitivity

The effect of the computational domain shape is discussed in this section. Two configurations were used to assess the effect, marked as 'Full geometry' and 'Cylindrical'. The first mesh includes as many details of the facility as possible (Fig. 2, left). Air and fuel inlets are colored by blue and red in the figure. The oxidizer stream is expected to exhibit recirculation in the unoccupied space below fuel injection pedestal. Moreover, flame interaction with downstream constriction imposed by the upper hood and exhaust chamber is fully considered in this configuration. For meshing convenience, the cylindrical radiation shield is assumed to extend to the top. All elements are hexahedra, except for a small region around the hood where tetrahedral and pyramid elements are

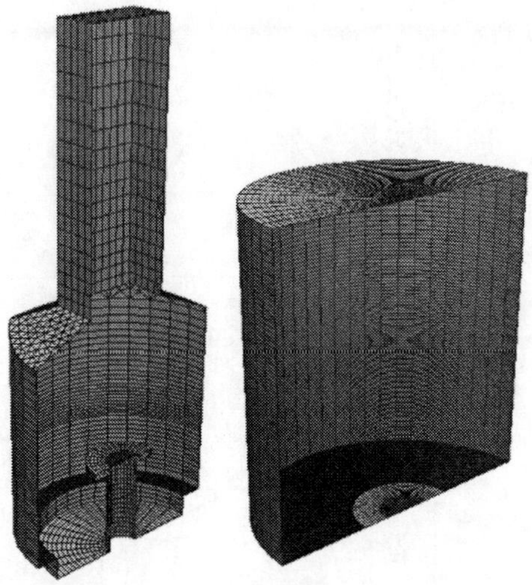

Figure 2: Cut views of full geometry (left) and simplified cylindrical (right) computational domain. Surfaces are colored by fuel inlet (red), air inlet (blue), and wall (grey).

used. A uniform bulk velocity is assumed at both fuel and air inlets; there are no perturbations used to represent inflowing air fluctuations. Mesh and boundary condition information is listed in Table 2. An adiabatic condition is used for all wall surfaces.

A simplified cylindrical domain is also tested (Fig. 2, right) that preserves the same facility diameter of 5.8m and similar height (5m rather than 4.5m). Uniform air velocity is assumed at the coflow section so that total oxidizer flow rate is preserved with the full geometry simulation; this limits the existance of . The top exit boundary, however, is confined to the center part with a diameter of 2m to avoid asymmetric plume development. All elements are hexahedra. Different resolutions are tested using the cylindrical domain. Table 2 lists the meshes used in this study. Around the base of the plume is a nearly uniform mesh with the characteristic cell size listed in the table. Away from the base of the plume, the mesh is progressively coarsened.

Domain shape	Total elements	Mesh size near plume	Air velocity	Fuel velocity
Full geometry	1.7M	2.5cm	0.33	0.097
Cylindrical	0.23M	4cm	0.14	0.097
Cylindrical	1.6M	2cm	0.14	0.097
Cylindrical	7.0M	1cm	0.14	0.097

Table 2: List of the simulated meshes

Figure 3 compares mean velocity profiles of the two domains at three different heights, 0.3m, 0.5m, and 0.9m, from the fuel inlet location. The coarsest results are compared here using the flamelet model—the 1.7M and 0.23M meshes. Peak predicted velocities at the core are comparable to experimental velocities suggesting some general prediction of the buoyant forcing. The full geometry prediction appears to show larger velocities in the tail with larger radial entrainment velocities, indicating stronger external turbulent fluctuations and entrainment or a broader buoyant forcing.

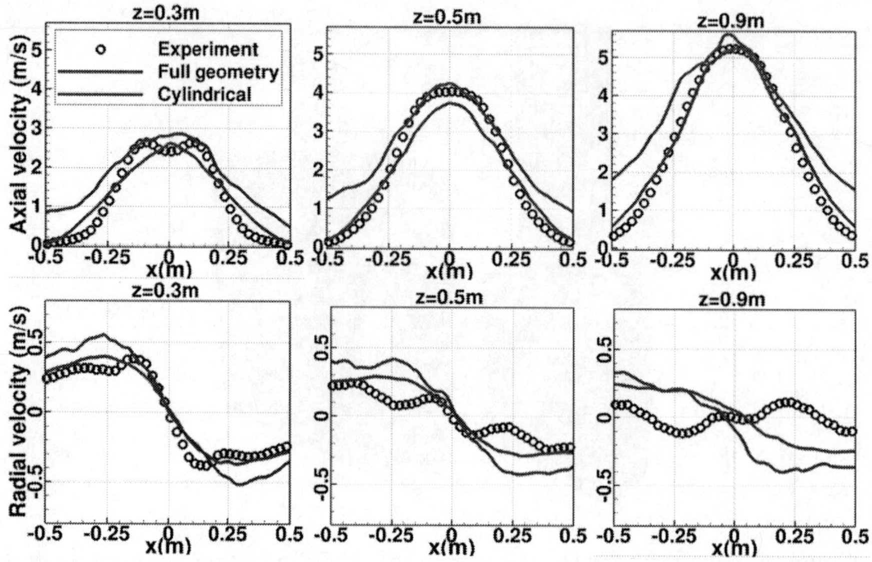

Figure 3: Effect of the configuration geometry for predictions compared to the experimental data [4]. Averaged axial (top) and radial (bottom) velocities at different distances from the fuel inlet.

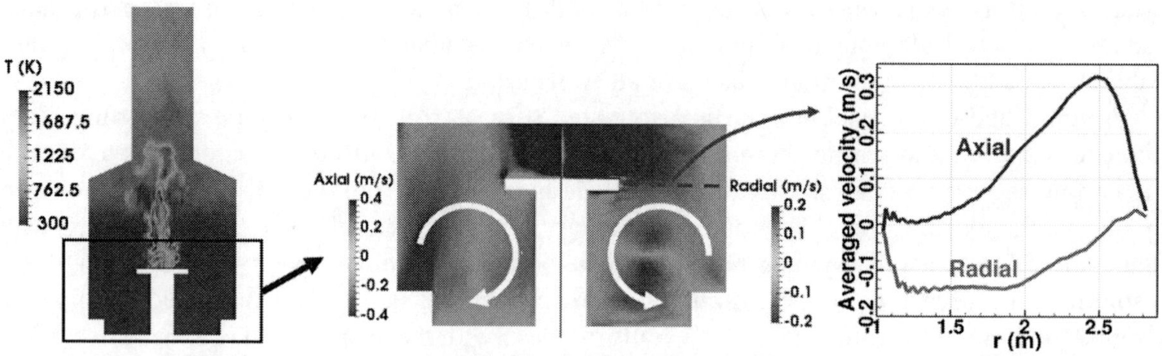

Figure 4: A temperature snapshot (left) and averaged axial and radial velocity contours (middle). On the right is averaged velocity profiles along the red dashed line.

The full geometry simulation is expected to offer a better description of the coflowing oxidizer stream at the fuel injection level. Figures 4 provide a snapshot and averaged velocity contours below the plume source at the middle plane. Also provided is the averaged axial and radial velocity profiles at the fuel plume height. Recirculation develops below the pool (white circular arrows in Fig. 4) in this simulation, and this weakens the vertical flow, tending to convert it to a radial inflow just below the fuel source. This large scale motion appear to contribute to the diffuse plume profile in this simulation (Fig. 3). We note that the outer region is only coarsely resolved, and further assessment of the upstream coflow effect is possible. Meanwhile, it was found that the convergence associated with the top hood affects on the statistics only minimally for this simulation.

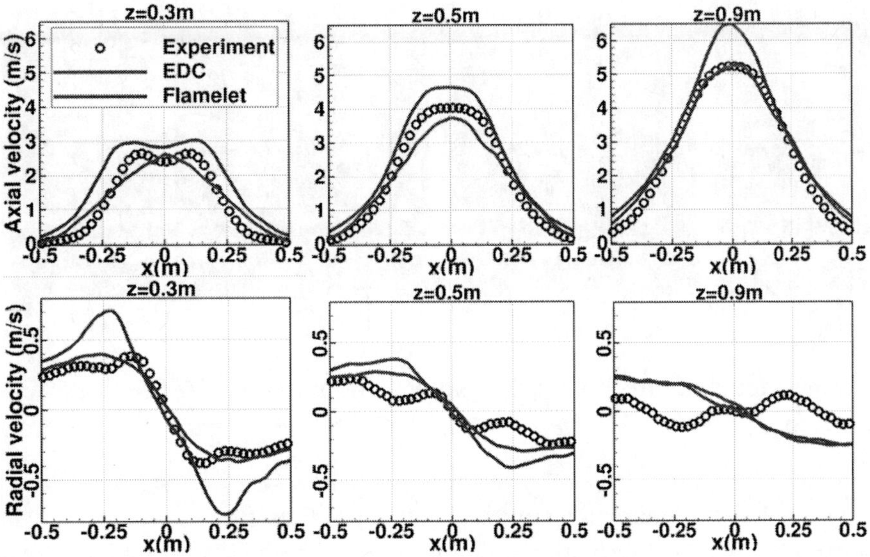

Figure 5: Effect of combustion model compared to the experimental data, for axial (top) and radial (bottom) velocities.

5. Combustion Model Sensitivity

Using the cylindrical domain, two combustion models—the EDC and flamelet models—were tested at three different mesh resolutions. In Fig. 5, the two models were compared on the coarsest 0.23M mesh. The EDC results in a higher centerline velocity which suggests that the reaction rate is over-predicted at this resolution leading to greater buoyant forcing. The plume width is overall similar regardless of the model choice. At the same time, the EDC model successfully captures two peaks and a local minima in the core velocity profile at 0.3 m location, which is missed with flamelet approach at this resolution.

Figure 6 compares averaged temperature contours near the plume for different models and resolutions. The EDC prediction suggests more intense fuel-air mixing in the near-field than the flamelet prediction. This EDC prediction suggests a continuous reaction zone from the plume edge as seen in the experimental pictures [4]. The continuous high-temperature bands are associated with the two peaks in the velocity profile (Fig. 5). Meanwhile, high scalar variance near the plume exit leads to a lower prediction of the mean flame temperature for this coarse mesh. Simulations using better grid resolution better resolve the near-fuel-source temperature field as shown in the same figure; this will be further discussed in the following section.

6. Mesh Resolution Sensitivity

The two combustion models were tested on different meshes with varying levels of resolution. Resolution was varied by doubling the resolution in the near plume region, but the mesh stretching was more aggressive for these cases to manage the overall simulation cost so that the actual mesh size does not increase as the near-plume resolution cubed. The flamelet model is applied at all three resolutions, 0.23M, 1.6M, and 7.0M meshes, and EDC is tested on the 0.23M and 1.6M meshes.

Figure 7 compares mean velocity profiles for different resolutions when the flamelet model

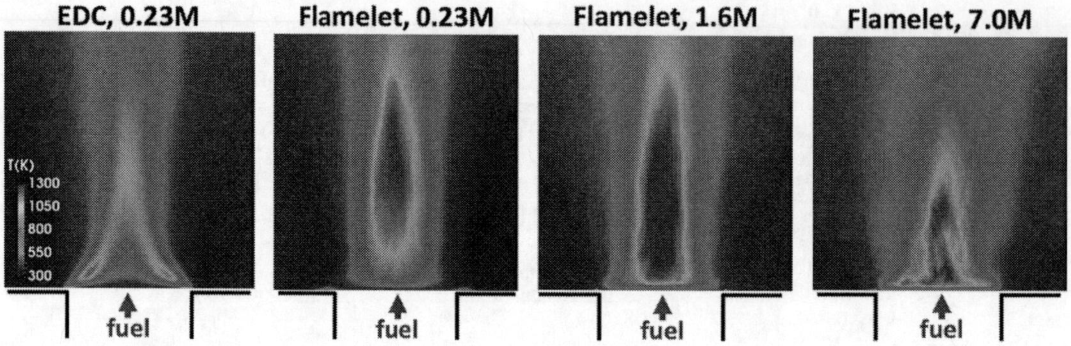

Figure 6: Averaged temperature contours. Note that statistics of 7.0M mesh case are not yet statistically converged.

was used. It shows that the coarser resolution better predicts flow statistics for these cases. As the mesh resolution increases, the plume appears to become weaker and wider. This is associated with stronger lateral entrainment. Figure 6 and 8 show averaged temperature and velocity fields for the three resolutions. The size of the high temperature area in Fig. 6 indicates that the middle resolution (1.6M) results in a broader high temperature region than the other cases. However, a more prominent tendency is an increased resolution of high temperatures nearer the pool surface. This is likely indicative of greater fual-oxidizer mixing right above the fuel exit plane, which is linked to the larger radial entrainment velocity (Fig. 8). This larger radial velocity increases turbulent intensity by inducing lateral motion of the plume, which diffuses the plume (lower velocity peak at Fig. 7). The enhanced plume motion increases mixing efficiency, and in turn, brings the high temperature regions even closer to the plume surface. EDC, with the higher resolution mesh, also predicts higher temperature regions nearer the plume compared to the lower resolution (not shown here). While statistics are not ready yet, this tendency with respect to the mesh resolution is expected to follow that of the flamelet model.

Figure 9 shows that plume puffing occurs at a 1Hz frequency when the 7.0M mesh and flamelet model is used. The motion is slower than the experiment of 1.5Hz. This repeated motion might be affected by the strong crossflow, and the relation between the puffing frequency and entrainment fluctuations is currently under investigation.

Three causes can be imagined regarding predictions of apparently stronger mixing and entrainment with increased mesh refinement than observed in the experiment. First, buoyancy-driven mixing may be incorrectly predicted, and this might influence the statistics. Second, laminar-turbulence transition is often difficult to predict with LES, and this might alter mixing levels. The plume injection is in low velocity ($\sim$0.1m/s) and low Reynolds number ($\sim$6000). For these results, there are no stochastic perturbations introduced into the inflow to represent fluctuations. With the size of the device, it is unlikely that the fuel inlet is spatially uniform with a constant mass flux at all locations throughout the duration of the test. In other simulations not presented here, inlet fluctuations can have a strong effect. Third, the stretched mesh moving away from the fuel inlet may significantly affect the predictions; this stretching is most severe at the highest resolutions.

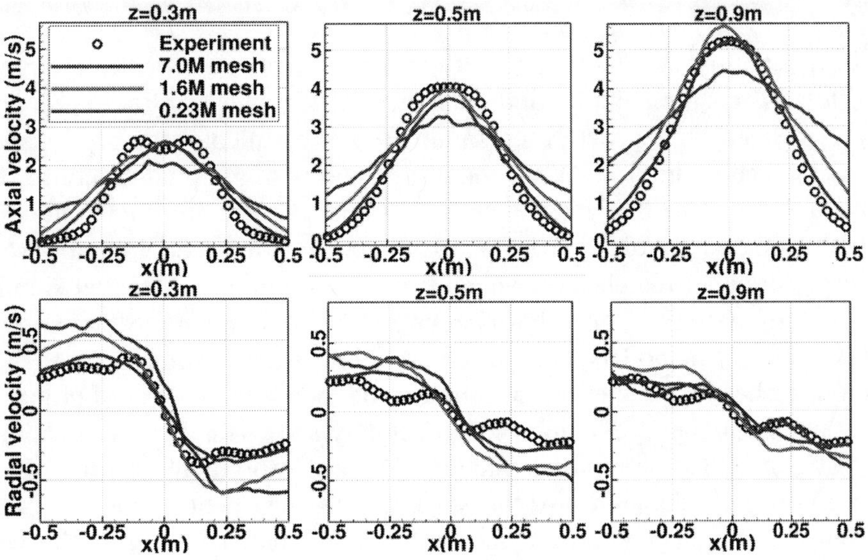

Figure 7: Effect of mesh resolution compared to the experimental data, for axial (top) and radial (bottom) velocities.

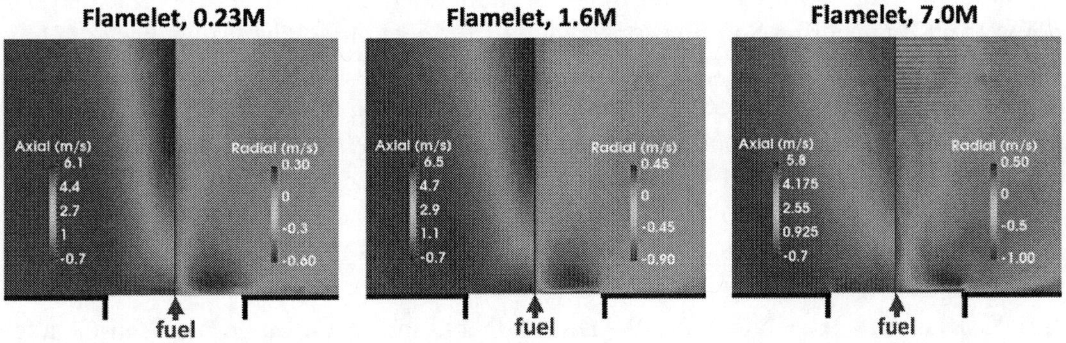

Figure 8: Averaged axial and radial velocity contours of three resolutions.

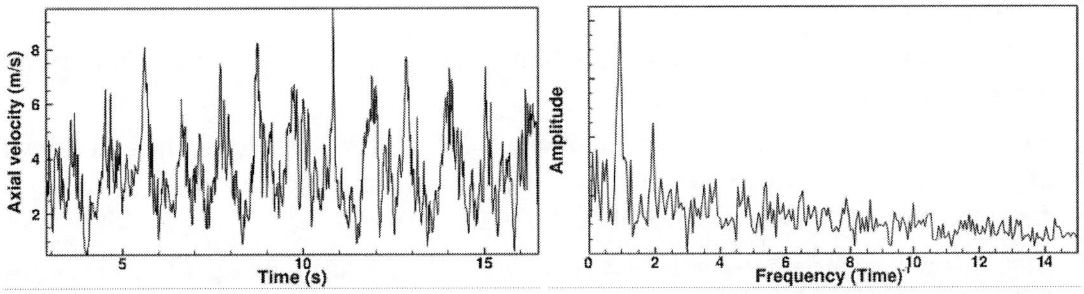

Figure 9: Axial velocity profile in time (left) and in frequency domain (right) at centerline, 0.5m height, from 7.0M flamelet simulation.

7. Conclusion

A series of LES predictions of a 1-m diameter methane fire plume scenario are preformed to understand the effect of mesh resolution and geometric fidelity have on the predictions. First, the full geometry associated with structure containing the experimental facility is compared with a simplified cylindrical configuration. While the full geometry might predict more a realistic coflow velocity profile at the fuel inlet height, the velocity statistics were not improved compared to that of the simplified geometry simulation. Secondly, the combustion model comparison showed that the flamelet approach better matches the velocity statistics while EDC model simulations predict temperature contours that are intuitively appealing with more apparent flame attachment at the fuel inlet edges. Lastly, the models were tested at various mesh resolutions. At the moment, the higher resolution meshes result in excessive mixing and therefore, a diffused plume. However, the coarsest mesh flamelet model prediction, which actually shows the best results among the simulations, shows evidence that the mixture fraction variance is too large, leading to lower averaged flame temperatures near the fuel inlet. At this stage it is not possible to evaluate whether the coarse mesh 'predictivity' is coincidental or not, and further assessment of uncertainties will be continue to clarify these results.

Acknowledgement

Sandia National Laboratories is a multimission laboratory managed and operated by National Technology and Engineering Solutions of Sandia, LLC., a wholly owned subsidiary of Honeywell International, Inc., for the U.S. Department of Energy's National Nuclear Security Administration under contract DE-NA0003525. This work was supported by Sandia National Laboratories' Advanced Simulation and Computing program through the Physics and Engineering Models subprogram and the Verification and Validation subprogram elements.

References

[1] T. K. Blanchat, V. F. Nicolette, W. D. Sundberg, and V. G. Figueroa, Well-Characterized Open Pool Experiment Data and Analysis for Model Validation and Development, SAND 2006-7508 (2006).

[2] T. K. Blanchat, Characterization of the Air Source and the Plume Source at FLAME, SAND 2001-2227 (2001).

[3] G. Maragkos and B. Merci, Large Eddy Simulations of CH4 Fire Plumes, Flow Turbulence and Combustion 99 (2017) 239–278.

[4] S. Tieszen, T. O'Hern, R. Schefer, E. Weckman, and R. Schefer, Experimental study of the effect of fuel mass flux on a 1-m-diameter methane fire and comparison with a hydrogen fire, Combustion and Flame 139 (2004) 126–141.

[5] T. K. Blanchat and J. Suo-Anttila, Hydrocarbon Characterization Experiments in Fully Turbulent Fires - Results and Data Analysis, SAND 2010-6377 (2010).

[6] B. J. McCaffrey, PurelyBuoyantDiffusionFlames:Someexperimentalresults, National Bureau of Standards, NBSIR 79-1910, 1979.

[7] B. M. Cetegen, E. E. Zukoski, and T. Kubota, Entrainment and Flame Geometry of Fire Plumes, National Bureau of Standards, NBS-GCR 80-402, 1980.

[8] E. J. Weckman and A. B. Strong, Experimental investigation of the turbulence structure of medium-scale methanol pool fires, Combustion and Flame 105 (1996) 245–266.

[9] H. Koseki, Large Scale Pool Fires: Results of Recent Experiments, Fire Safety Science 6 (2000) 115–132.

[10] L. Hu, A review of physics and correlations of pool fire behavior in wind and future challenges, Fire Safety Journal 91 (2017) 41–55.

[11] A. Hamins, J. C. Yang, and T. Kashiwagi, An experimental investigation of the pulsation frequency of flames, Symposium (International) on Combustion 24 (1992) 1695–1702.

[12] A. Luketa, Assessment of Simulation Predictions of Hydrocarbon Pool Fire Tests, SAND 2010-2511 (2010).

[13] A. L. Brown, K. J. Dowding, V. F. Nicolette, and T. K. Blanchat, Fire Model Validation for Gas Temperatures and Radiative/Convective Partitioned Heat Flux, Fire Safety Science 9 (2008) 93–104.

[14] A. R. Black, Numerical Predictions and Experiment Results for a 1m Diameter Methane Fire, ASME 2005 International Mechanical Engineering Congress and Exposition (2005).

[15] A. Trouve, CFD Modeling of Large-Scale Pool Fire, The Second International Energy 2030 Conference (2008).

[16] W. M. Eldredge, J. N. Thornock, and P. J. Smith, Development, Verification, and Validation of the Responsive Boundary Model for Pool Fire Simulations, 20th AIAA Computational Fluid Dynamics Conference (2011).

[17] *The 1st Measurement and Computation of Fire Phenomena (MaCFP) Workshop*, 2017, URL: `http://www.iafss.org/macfp/`.

[18] K. McGrattan, S. Hostikka, R. McDermott, J. Floyd, C. Weinschenk, and K. Overholt, Fire Dynamics Simulator Technical Reference Guide Volume 1: Mathematical model, NIST Special Publication 1018 Sixth Edition (2014).

[19] Z. Chen, J. Wen, B. Xu, and S. Dembele, Large Eddy Simulation of Fire Dynamics with the Improved Eddy Dissipation Concept, Fire Safety Science 10 (2011) 795–808.

[20] *SIERRA Low Mach Module: Fuego User Manual – Version 4.42, SAND2016-10164*, Sandia National Laboratories, 2016.

[21] P. DesJardin, T. O'Hern, and S. Tieszen, Large eddy simulation and experimental measurements of the near-field of a large turbulent helium plume, Physics of Fluids 16 (2004) 1866–1883.

[22] K. Frederickson, S. P. Kearney, A. Luketa, J. C. Hewson, and T. W. Grasser, Dual-Pump CARS Measurements of Temperature and Oxygen in a Turbulent Methanol-Fueled Pool Fire, Combustion Science and Technology 182 (2010) 941–959.

[23] T. Blanchat, T. O'Hern, S. Kearney, A. Ricks, and D. Jernigan, Validation experiments to determine radiation partitioning of heat flux to an object in a fully turbulent fire, Proceedings of the Combustion Institute 32 (2009) 2511–2518.

[24] S. R. Tieszen, S. P. Domino, and A. R. Black, Validation of a Simple Turbulence Model Suitable for Closure of Temporally-Filtered Navier-Stokes Equations Using a Helium Plume, Sandia National Laboratories, Albuquerque, NM, USA, (2005).

[25] A. Luketa, V. Romero, S. Domino, D. Glaze, M. Sherman, and V. Figueroa, Validation and uncertainty quantification of Fuego simulations of calorimeter heating in a wind-driven hydrocarbon pool fire, Sandia National Laboratories, Albuquerque, NM, USA, (2009).

[26] B. F. Magnussen, On the Structure of Turbulence and a Generalised Eddy Dissipation Concept for Chemical Reactions in Turbulent Flow, 9th AIAA Aerospace Science Meeting (1981).

[27] N. Peters, Turbulent Combustion, Cambridge University Press, 2000.

[28] C. Bowman, R. Hanson, D. Davidson, J. W.C. Gardiner, V. Lissianski, G. Smith, D. Golden, M. Frenklach, and M. Goldenberg, URL: `http://www.me.berkeley.edu/gri_mech/`.

[29] H. Pitsch, URL: `https://www.itv.rwth-aachen.de/index.php?id=flamemaster`.

Western States Section of the Combustion Institute – Fall 2017 Meeting
Hosted by the University of Wyoming
October 2-3, 2017

Direct Numerical Simulation of Wildland Fires at Small Scales

Nicholas T. Wimer[1,], Amanda S. Makoweicki[1], Alexei Y. Poludnenko[2], Chad M. Hoffman[3], John W. Daily[1], Gregory B. Rieker[1], and Peter E. Hamlington[1]*

[1]*Mechanical Engineering, University of Colorado Boulder, Boulder CO 80309, USA*
[2]*Aerospace Engineering, Texas A&M University, College Station TX 77843, USA*
[3]*Warner College of Natural Resources, Colorado State University, Fort Collins CO 80523, USA*
**Corresponding author:* `Nicholas.Wimer@Colorado.edu`

Abstract: Preliminary results are presented from a new research effort focused on understanding and characterizing wildland fire spread at small scales (roughly 1 m – 1 mm) using direct numerical simulations (DNS). The simulations are intended to directly resolve, with high physical accuracy, all small-scale fluid dynamic and chemical processes relevant to wildland fire spread. Simulation of wildland fires is an incredibly complex and challenging problem due to the vast difference in scales associated with the problem. An understanding is needed not just of the burning of fuel, but also of the atmospheric conditions, weather patterns, topography, and turbulence-flame dynamics. This work is focused on the sub-meter scales associated with wildland fire; in particular, the dynamics of small-scale diffusion flames. Here, preliminary results are presented for DNS of gaseous pool fires coupled with an infinitely-fast chemical reaction mechanism. The results are connected to the fundamental structure and spread of wildland fires, and an outlook is provided for the future expansion of these DNS studies.
***Keywords:* Wildfire, DNS, wildland, pool**

1. Introduction

Wildland fires occur both naturally as the result of lightning strikes or volcanic eruptions, as well as artificially due to accidental (or intentional) human ignitions and "slash-and-burn" farming techniques. In many such cases, the consequences for human life and property can be disastrous. Even prescribed burns intended to mitigate wildland fire risk can have detrimental impacts on air quality and may also become uncontrolled. For both catastrophic and prescribed burns, computational models have the potential to significantly improve our ability to predict wildland fire behavior and health impacts, yet nearly all current models are empirical in nature and do not accurately capture physical and chemical processes occurring at small scales (i.e., less than 1 m). However, with recent advances in computing power and the advent of increasingly sophisticated physics-based direct numerical simulations (DNS) of complex combustion processes, we now have the potential to capture all relevant physical and chemical processes with unprecedented fidelity.

In this study, we seek to understand the small scale dynamics of wildland fires using DNS of a simplified system. We consider a two-dimensional infinite reservoir methane pool fire under the influence of an air co-flow. Additionally, we simplify the complex reaction mechanism to a single-step, infinitely-fast chemical model assuming the reaction timescale is much smaller than the advective and diffusive timescales. This simplified model will allow us to cleanly identify fundamental features of the methane-air diffusion flame that will grant insight when examining the larger, more complex system.

2. Methods

In a real-world wildland fire, many different factors contribute to the spread and structure of the fire, including wind, weather, terrain, and variable fuel types. This is an incredibly complex problem that is very difficult to model accurately given the large range of scales. Therefore, for this initial study, a much simpler system – a gaseous pool fire – is analyzed in an effort to understand some of the underlying dynamics and structure of wildland fires at small scales. The gaseous pool fire is created with a very low inlet velocity of gaseous methane which simulates an infinite reservoir of fuel with a low velocity air coflow that helps to encourage the mixing of the fuel and air for the reaction to take place. Further simplifying the problem is the use of an infinitely fast chemical reaction mechanism instead of a more realistic and complex multi-step chemical reaction mechanism.

2.1 Infinitely-Fast Chemistry

The use of infinitely-fast chemistry assumes the chemical reaction timescale is much smaller than that of the advective and diffusive timescales. In low turbulence diffusion flames, this assumption is much less egregious than in premixed combustion or in fluid flows with high levels of turbulent mixing and has been used by others in DNS of reacting flows [1, 2]. This means that the reaction takes place essentially instantaneously as the reactant species (i.e., CH_4 and O_2) come into contact with each other, immediately yielding products of combustion (i.e., CO_2, H_2O, and heat). This global single step reaction is represented for stoichiometric conditions as

$$CH_4 + 2O_2 + N_2 \rightarrow CO_2 + 2H_2O + N_2 + \text{heat}. \quad (1)$$

In Athena-RFX, this reaction is implemented as follows:

1. At every time step, we check every grid cell in the domain to determine whether CH_4 and O_2 are both present.
2. If, in any given cell, there exists both CH_4 and O_2, the infinitely fast reaction is activated.
3. We determine which reactant (i.e., CH_4 or O_2) is the limiting reactant based on mole fraction and the stoichiometric reaction above.
4. The limiting reactant is completely eliminated along with the stoichiometric equivalent of the non-limiting reactant, and products are added in their place according to the above equation.
5. The amount of heat released is always determined by the amount of CH_4 and O_2 that is depleted, regardless of whether or not it is the limiting reactant.

It should be noted that, since the assumption of infinitely fast chemical reactions is not entirely physical, the heat released is often too large for it to be representative of a real-world reaction, so we have imposed a damping coefficient to prevent the temperatures in the simulation from exceeding that of the adiabatic flame temperature, which for a methane-air reaction is roughly 2200 K.

The idea of a damping coefficient for an infinitely-fast chemical reaction mechanism is not a new one, and the Fire Dynamics Simulator (FDS) from the National Institute of Standards and Technology (NIST) also enforces a damping that is based on the grid resolution of the simulation [3]. Currently, the damping coefficient for Athena-RFX is a constant of 0.25, though a variable damping coefficient more akin to FDS is being considered and the effects of a variable coefficient will be studied in future work.

2.2 Equations Solved

Numerical simulations are performed using Athena-RFX [4, 5], a reacting flow version of the fully-compressible, multi-physics Riemann solver Athena [6, 7]. We perform reacting, fully-compressible simulations of methane-air combustion, which requires the solution of the compressible Navier-Stokes equations written in cell-centered conservative form as

$$\boldsymbol{U}_t + \boldsymbol{F}(\boldsymbol{U})_x + \boldsymbol{G}(\boldsymbol{U})_y = \boldsymbol{S}, \tag{2}$$

where

$$\boldsymbol{U} = \begin{bmatrix} \rho \\ \rho u \\ \rho v \\ E \end{bmatrix}, \qquad \boldsymbol{F} = \begin{bmatrix} \rho u \\ \rho u^2 + p \\ \rho uv \\ u(E+p) \end{bmatrix}, \qquad \boldsymbol{G} = \begin{bmatrix} \rho v \\ \rho uv \\ \rho v^2 + p \\ v(E+p) \end{bmatrix}, \tag{3}$$

and $\boldsymbol{S}$ is a source term that encapsulates the gravitational potential, viscous stress tensor, and heat transfer [7, 8]. Accounting for each of these physical effects in the source term and including the scalar advection, the resulting equations solved can be written as

$$\frac{\partial \rho}{\partial t} + \nabla \cdot [\rho \boldsymbol{v}] = 0 \tag{4}$$

$$\frac{\partial \rho \boldsymbol{v}}{\partial t} + \nabla \cdot [\rho \boldsymbol{v}\boldsymbol{v} + p\boldsymbol{I}] = -\rho \nabla \phi - \nabla \cdot \Pi \tag{5}$$

$$\frac{\partial E}{\partial t} + \nabla \cdot [(E+p)\boldsymbol{v}] = -\rho \boldsymbol{v} \cdot \nabla \phi - \nabla \cdot (-\kappa \nabla T) - \nabla \cdot (\Pi \cdot \boldsymbol{v}) \tag{6}$$

$$\frac{\partial (C_i \rho)}{\partial t} + \nabla \cdot [C_i \rho \boldsymbol{v}] = 0 \tag{7}$$

$$E = \frac{p}{\gamma - 1} + \frac{\rho (\boldsymbol{v} \cdot \boldsymbol{v})}{2} \tag{8}$$

where ρ is the density, $\boldsymbol{v}$ is the velocity vector, E is the total energy density, P is the gas pressure, ϕ is the gravitational vector, γ is the adiabatic index, Π is the viscous stress tensor, κ is the coefficient of heat transfer for the fluid, T is the temperature, and C_i ($i = 1, \ldots, N$) are the mass fractions of N scalars.

For the following simulations, the above equations are solved with an external gravitational field, viscous stresses, thermal conduction and diffusion consistent with that of dry air and methane at a pressure of one atmosphere. The viscous stress tensor, Π, is a temperature dependent quantity, where the coefficient of viscosity obeys a power-law dependence on temperature. Numerically, the Riemann fluxes were calculated using an HLLC scheme [9] and the equations were integrated in time using the directionally unsplit CTU method of Colella [10]. Additionally, all time advancing was done dynamically with a maximum CFL number of 0.3 to ensure stability of the numerical method [11].

2.3 Boundary Conditions

For the simulations in this paper, there are two different types of boundary conditions: inflow and outflow boundary conditions. The inflow boundary condition is a straightforward, standard boundary condition that requires no additional explanation. The treatment of the outflow boundary conditions, however, are not straightforward and since they are of utmost importance to the accuracy of the simulations, will be discussed in full.

For subsonic fluid flow, pressure waves can travel in both the direction of the bulk fluid flow as well as against this direction. This means that, at the location where the fluid should exit the computational domain, pressure waves should be allowed to travel both into and out of the domain. Therefore, the numerical boundary condition for outflow must mimic the effect that the atmospheric ambient conditions have on the fluid flow. To add to the problem, it has been shown by Poinsot *et al.* [12] that even physically well-posed boundary conditions do not always produce physical results in the subsonic outflow regime and that added numerical stability is often required.

In Athena-RFX, we use a version of the Navier-Stokes Characteristic Boundary Conditions (NSCBC) proposed in Poinsot *et al.* [12] with an added pseudo perfectly matched layer (PPML) [13] to deal with the aforementioned numerical stability problems. The NSCBC boundary conditions are written as

$$\rho_{i-1} = \rho_i\,, \tag{9}$$

$$u_{i-1} = u_i\,, \tag{10}$$

$$v_{i-1} = v_i\,, \tag{11}$$

$$p_{i-1} = p_0 - \rho_i g h + \rho(u_i^2 + v_i^2)\,, \tag{12}$$

where ρ is the fluid density, u and v are the components of the velocity, p is the total pressure, p_0 is atmospheric reference pressure, g is the magnitude of gravitational force, and h is the height above the ground.

Although the NSCBCs work well on their own, there are some minor numerical artifacts capable of influencing the solution in an unphysical manner. The addition of the PPML (sometimes called a 'sponge' damping region) on top of the computational domain some distance away from the actual outflow boundary takes care of these issues by damping out spurious pressure waves generate by the computation and gently relaxing the flow to ambient conditions. This is formulated in Athena-RFX as

$$\rho_i = \sigma(\rho_i - \bar{\rho})\,, \tag{13}$$

$$u_i = \sigma(u_i - \bar{u})\,, \tag{14}$$

$$v_i = \sigma(v_i - \bar{v})\,, \tag{15}$$

$$p_i = \sigma(p_i - \bar{p})\,, \tag{16}$$

where σ is the parameter (typical values ranging from 1 to 50) that relaxes the flow to its ambient states $\bar{\rho}, \bar{u}, \bar{v}$, and $\bar{p}$.

These boundary conditions along with the pseudo-perfectly matched layer have been tested and validated using a two-dimensional turbulent buoyant jet and are reported in [11].

2.4 Physical Setup

The physical problem under investigation is a two-dimensional methane pool fire with an air co-flow under the influence of gravity and exhausting into an ambient, hydrostatic atmosphere. The computational domain is 0.5 m $\times$ 2.0 m and the width of the methane pool is 0.1 m and is located in the center of the domain at the bottom boundary. The co-flow region is treated as an inflow boundary condition on the bottom boundary completely surrounding the methane pool. The methane has a small inlet velocity of 1 cm/s to simulate a seeping, infinite reservoir of fuel while the air co-flow region has a slightly larger 4 cm/s inlet velocity; both methane and air enter the domain at a temperature of 300 K. The sides and top of the domain are outflow boundary conditions that relax to a co-flow or a hydrostatic state respectively. A grid resolution for this simulation was chosen to be identical to the previous non-reacting study [11] of 1024 $\times$ 4096, which results in a resolution of slightly smaller than 0.5 mm. This very fine grid resolution leads to roughly 4 grid cells within the reaction zone of our methane-air infinitely-fast combustion mechanism.

3. Results

Figure 1 shows the time evolution of the temperature field from the methane-air pool fire up to a time of six seconds. The white dashed lines denotes the start of the PPML as described in the previous section. As the CH_4 enters the domain, it immediately reacts with the air creating thin puffs of hot products that are visible in the first frame of Figure 1. As the fuel burns, the local temperature increases, leading to an increase in the buoyant force, thereby accelerating and

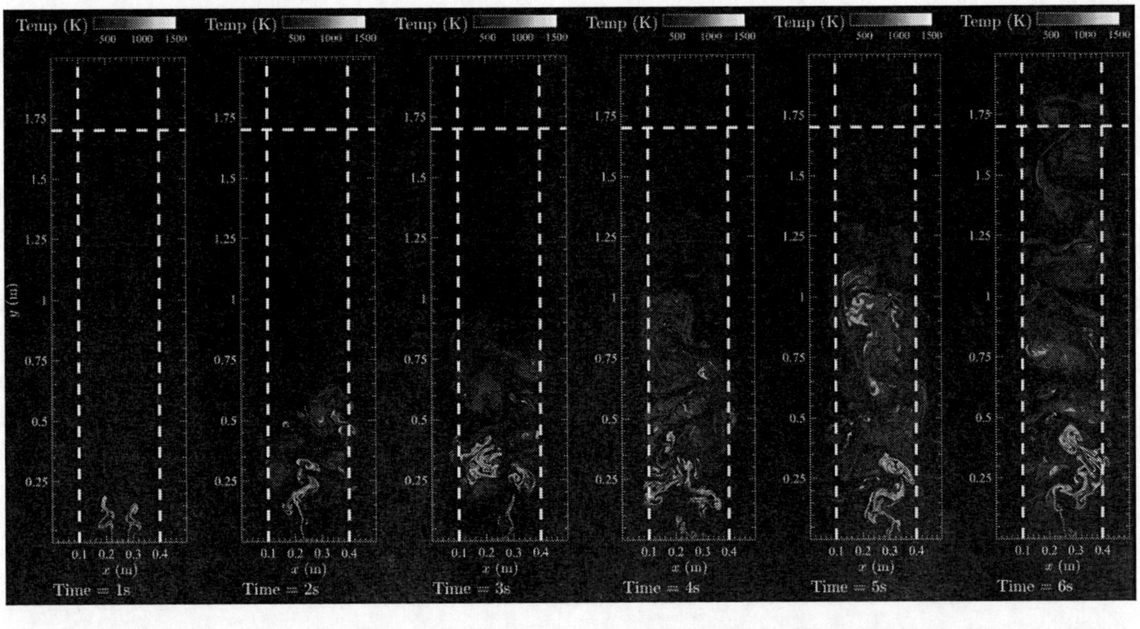

Figure 1: Progression of the temperature field for a methane pool fire with air co-flow using infinitely fast chemistry up to 6 s. The methane inflow occurs between 0.2 and 0.3 m, and the computational domain size is 1024 $\times$ 4096, giving a grid resolution of less than 0.5 mm. White dashed lines show the start of the PPML regions.

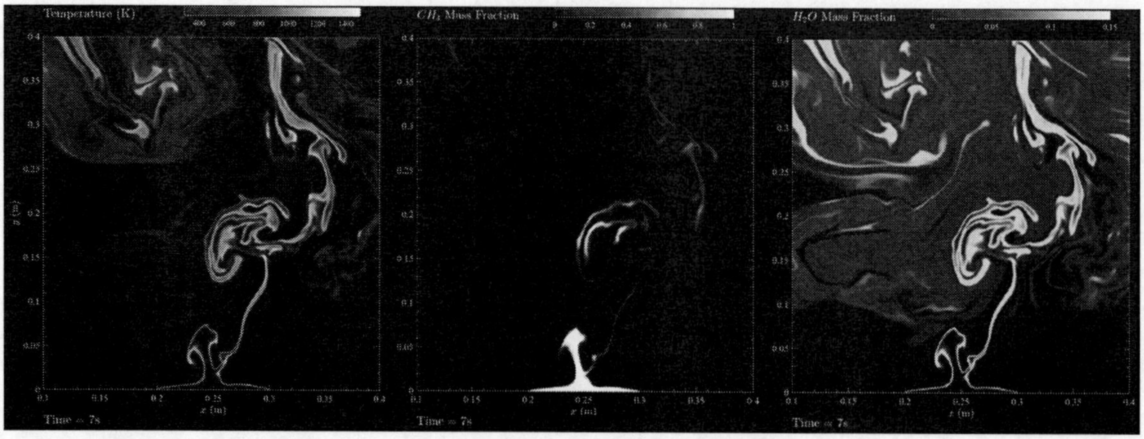

Figure 2: Temperature (left), CH_4 (middle), and H_2O (right) fields near the methane inflow for a methane pool fire with air co-flow using infinitely fast chemistry up to 6 s. The methane inflow occurs between 0.2 and 0.3 m, and the computational domain size is 1024 $\times$ 4096, giving a grid resolution of less than 0.5 mm.

thinning the flame into a meandering ‘finger’. As the simulation progresses, the domain begins to be filled with thin structures that are a result of the vortex shedding from the thin flame front.

The thin, high temperature regions shown in Figure 1 are hot products advecting upwards rather than a thinned-out flame front. This can more clearly be seen in Figure 2, which shows a zoomed view of the temperature field as well as the CH_4 and H_2O mass fractions. Here we can clearly see that the vast majority of the fuel is consumed in the first few centimeters of the domain, turning into H_2O and CO_2. Although the combustion looks to be fairly complete low in the domain, there are clearly patches of methane that go unburned and are carried further up in the domain, leading to other patches of high temperature and H_2O mass fraction, as seen in Figure 2. These pockets seem to be formed from the long ‘finger’ of flame that is stretched until it pinches off of the main reservoir. In the process of pinching into a round pocket, the unburnt fuel on the inside is able to persist for some time as only the edges of the pocket can come into contact with O_2 and react. In a real wildland fire, the advection of pockets of fuel away from the ground can lead to flame ‘jumping’ as an area of previously un-ignited fuel can catch fire due to the sudden spark of high temperature caused by the burning of floating pockets of fuel.

The natural question to answer is then: How far away from the fuel source do these ‘pockets’ of CH_4 travel? Figure 3 shows small pockets of methane traveling up to 1 m from the infinite pool of methane at the bottom of the domain, which equates to roughly 0.5 s to 1.0 s of lifetime in the domain. When these small pockets of methane come into contact with oxygen, they immediately produce high temperature products that are then accelerated upwards and eventually exit the top of the domain. However, if this were a real world wildland fire, instead of a 2D simulation, those pockets of unburned fuel from the ground could be blown horizontally and easily ignite another reservoir of fuel some distance away.

Despite this simulation being a substantial simplification of the true wildland fire problem, the preliminary insights obtained here into the dynamics of flame thinning and pinching can be applied to more complex problems. Future work is needed to expand the complexity of the problem by adding a third dimension and more complex chemistry, as discussed in the next section.

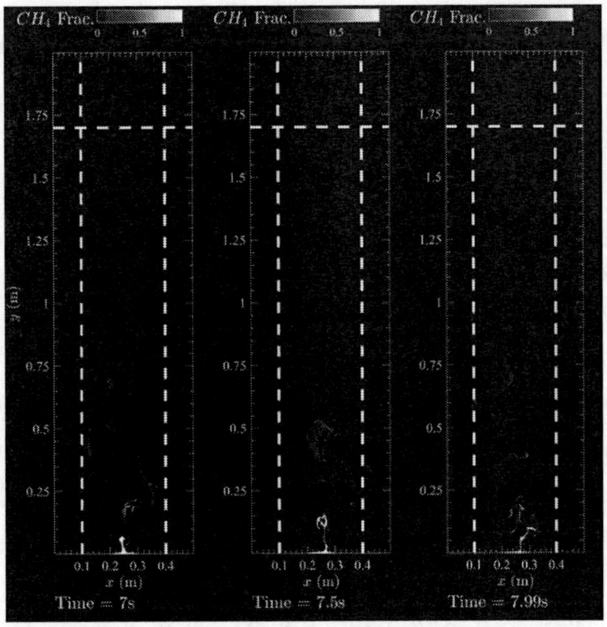

Figure 3: Progression of the CH_4 field for a methane pool fire for one second. Pockets of CH_4 can travel up to 1 m from the infinite pool at the bottom of the domain. Although these are small amounts of methane, upon burning they create pockets of hot products that could, in a real wildfire scenario, ignite a previously unburned reservoir of fuel.

4. Conclusions and Future Work

A high resolution simulation of a gaseous methane pool fire in the presence of an air co-flow was conducted using an infinitely-fast chemical model in Athena-RFX. The resulting temperature and species fields were presented in a preliminary effort to understand the small-scale dynamics of buoyant, diffusion flames and how they might be applied to a real wildland fire. Despite being a two-dimensional simulation with an infinitely-fast chemical model, the simulation sheds some light on a method of wildfire 'jumping' that is observed in nature. Hot embers, or pockets of fuel, being advected some distance before burning completely. As the methane pool fire burns, it is accelerated by an increase in the buoyant force, stretching and thinning the flame until it pinches off into a swirl or pocket of unburnt CH_4. The pocket is then able to persist for some time (0.5 s to 1.0 s) in the domain as only the outside of the pocket can come into contact with air and burn. This simulation shows the pockets of CH_4 persisting up to 1 m from the bottom of the domain.

More work needs to be done examining these dynamics with a multi-step chemical reaction mechanism to see if these results persist, although with an increase in the reaction time for a multi-step model, it is likely that the pockets of fuel will be able to persist for even longer in the domain. This decrease in rate of reaction may be mitigated by a move to three dimensions, which would increase the mixing in the domain and may decrease the life of unburnt fuel pockets. The next steps in this study are therefore to increase the fidelity of the reaction mechanism and to move from two dimensions to three dimensions.

5. Acknowledgments

This research was supported by SERDP under grant W912HQ-16-C-0026.

References

[1] C. Pantano, S. Sarkar, and F. A. Williams, Mixing of a conserved scalar in a turbulent reacting shear layer, Journal of Fluid Mechanics 481 (2003) 291–328.

[2] J. Doom, Y. Hou, and K. Mahesh, A numerical method for DNS/LES of turbulent reacting flows, Journal of Computational Physics 226 (2007) 1136–1151.

[3] K. McGrattan, S. Hostikka, R. McDermott, J. Floyd, C. Weinschenk, and K. Overholt, *Fire Dynamics Simulator User's Guide*, Sixth Edition, National Institute of Standards and Technology.

[4] A. Y. Poludnenko and E. S. Oran, The interaction of high-speed turbulence with flames: Global properties and internal flame structure. Combust. Flame 157 (2010) 995–1011.

[5] A. Y. Poludnenko and E. S. Oran, The interaction of high-speed turbulence with flames: Turbulent flame speed. Combust. Flame 158 (2011) 301–326.

[6] J. M. Stone, T. A. Gardiner, P. Teuben, J. F. Hawley, and J. B. Simon, Athena: A New Code for Astrophysical MHD, The Astrophysical Journal Supplement Series (2008).

[7] *Welcome to the Athena MHD Code Project.* https://trac.princeton.edu/Athena/wiki/AthenaEqns.

[8] E. F. Toro, Riemann Solvers and Numerical Methods for Fluid Dynamics, Springer, 2009.

[9] T. A. Gardiner and J. M. Stone, An unsplit Godunov method for ideal MHD via constrained transport in three dimensions. J. Comp. Phys. 227 (2008) 4123–4141.

[10] P. Colella, Multidimensional upwind methods for hyperbolic conservation laws, Journal of Computational Physics 87 (1990) 171–200.

[11] N. T. Wimer, C. Lapointe, T. R. Hayden, J. D. Christopher, A. Y. Poludnenko, G. B. Rieker, and P. E. Hamlington, Near-and Far-Field Properties of High-Temperature Turbulent Buoyant Jets, AIAA Paper 2017-4423 (2017).

[12] T. J. Poinsot and S. K. Lelef, Boundary conditions for direct simulations of compressible viscous flows, Journal of Computational Physics 101 (1992) 104–129.

[13] D. Appelö and T. Colonius, A high-order super-grid-scale absorbing layer and its application to linear hyperbolic systems, Journal of Computational Physics 228 (2009) 4200–4217.

Western States Section of the Combustion Institute – Fall 2017 Meeting
Hosted by the University of Wyoming
October 2-3, 2017

The effect of pressure and External Heating on the Nomex flame spread limits (LOC)

Maria Thomsen[1], *Xinyan Huang*[1], *Carlos Fernandez-Pello*[1], *Alain Alonso*[2], *David L. Urban*[3], *Gary A. Ruff*[3]

[1]*Department of Mechanical Engineering, University of California, Berkeley, CA 94720, USA*
[2]*GIDAI, University of Cantabria, Avda. Los Castros, s/n, Santander, Spain, 39005*
[3]*NASA John H. Glenn Research Center, Cleveland, OH 44135, USA*

Nomex is used in multiple applications (clothing, curtains, tents, etc.) were protection from a potential fire is needed. Furthermore, its fire-resistant capacity is often tested under very specific conditions that might not represent the real fire situations. For example, when a material is exposed to a near fire or different environmental conditions like reduced pressure, enriched oxygen concentration, its flammability and fire behavior can be altered. In this work, an experimental study was performed to investigate the effect of ambient pressure and oxygen concentration on the upward flame spread over a typical fire-resistant fabric (Nomex HT90-40) exposed to two different external heat sources. One is the radiation from infrared lamps and the other is the flame from a burning polymethyl methacrylate (PMMA) sheet placed below the fabric. The limiting oxygen concentration (LOC) was first quantified under different external heating, and then the upward flame-spread rate above LOC was measured. Experiments show that the flame from nearby burning object not only can ignite the fire-resistant fabric, but also extend the LOC of the material to lower oxygen concentrations. Moreover, the heating from the attached flame is different from an external radiant flux. The results of this work also provide important information about the fire interactions of different materials, and guide the future fire safety design in space exploration.

1. Introduction

The ignition and propagation of fire under the influence of an external heat source have been the subject of several studies evaluating material flammability [1]. However, there is still a need for further understanding of how an external heat source affects the flammability of materials, particularly under varied environmental conditions. These environmental variations become extremely relevant in applications where fire safety is a critical factor as in spacecraft or aircraft applications, where the entire atmospheric condition needs to be carefully controlled to support human habitation [2].

Several researchers have studied the flammability properties of materials in different atmospheres. Hirsch et al. [3,4] studied the fire risk of different materials under a wide range of oxygen concentrations and ambient pressures. Their work showed that the maximum oxygen concentration (MOC) resulting in the flame self-extinguishment decreased with the total pressure; however, the limiting partial pressure of oxygen increased linearly with the increased in total pressure. Kleinhenz and T'ien [5] studied the burning of Nomex III in an upward configuration under different combinations of oxygen concentration and ambient pressure. Their results showed that the limiting oxygen concentration for flame spread (LOC) at 1 atm was 24% and its value increased with decreasing pressure. Nakamura and Aoki [6] found that the flammability of thin cellulosic paper had the same linear trend for oxygen partial pressure and total pressure found in Hirsch et al. [4]

External radiation also plays an important role in the flammability of materials [7–12]. In actual fires, an external radiant heat source may be an adjacent fire, heated walls or an overheated electronic system. Such heat sources can change the material fire behavior as well as its flammability limit. For example, Osorio et al. [11] studied the forced-flow flame spread behavior and the LOC for Nomex HT90-40 under external radiation, and showed that for given oxygen concentration there is a minimum external heat flux that allows flame spread to occur. Thomsen et al. [12] studied the effect of external radiation on the LOC for Nomex HT90-40 under different ambient pressures, showing that as the amount of external heating is increased a decreased in the LOC was observed. One question that arises from these studies is whether an applied external radiation from a heater has the same effect on the flammability of a material as an actual adjacent fire. This the objective of the present work.

In this work, the upward flame spread rate and the flammability of Nomex HT90-40 fabric were investigated under several external heating, ambient pressure, and oxygen concentration configurations. Two different external heat sources were used, a radiant flux from halogen lamps, and a flame from burning PMMA sheet. The latter was considered more representative of the heating from an adjacent fire although less controllable.

2. Experimental setup and Procedures

Experiments were conducted in an apparatus to study the flammability of solid combustible materials under varied ambient conditions [12]. The apparatus consists of a combustion tunnel that is incorporated in a pressure chamber permitting the simultaneous control of oxygen concentration, ambient pressure, flow velocity and external radiant heat flux (see Fig. 1). The first section of the duct serves as a flow straightener, where inlet gases from bottom pass and mix sequentially before entering the test section.

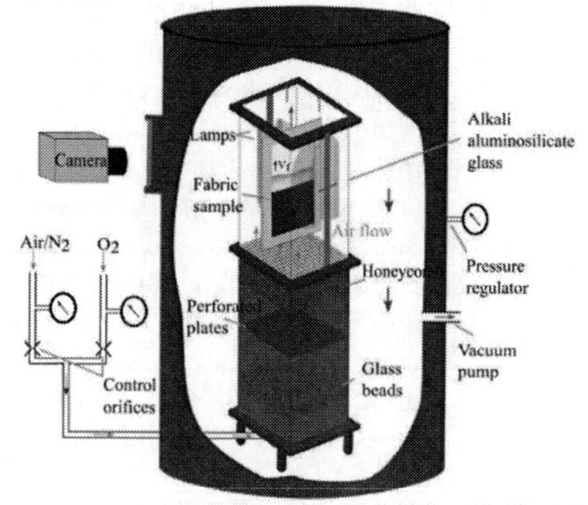

Figure 1. Schematic of the experimental apparatus.

The top segment of the duct is used as the test section. A single layer of a fire resistant material, Nomex HT90-40 (Stern and Stern Industries Inc., Hornell, NY, USA) [13], was placed vertically at the midplane of the duct with both sides exposed to the flow. The area density of Nomex HT90-40 is 24.41 mg/cm^2. In this experiment, 0.33 mm thick Nomex sample was cut into pieces of 140 mm long by 76 mm wide.

A set of experiments were conducted with an external heat source provided by four quartz near-infrared halogen lamps (Ushio QIH120-500T/S), i.e. two in each side of the duct. Each lamp had a lit length of 127 mm and a total length of 220 mm, and it was half surrounded and supported by parabolic reflectors (Research Inc. 5236-05-A). The lamp radiation through a side wall of 0.56 mm alkali-aluminosilicate glass was measured using a Schmidt-Boelter radiometer (Medthermal) under variable power of the lamps. Three constant radiation levels, 5, 10, and 15 kW/m^2 were chosen. The external radiation was applied 120 s ahead of ignition and continued until the end of experiment. The ignition was achieved by a nichrome wire braided to the Nomex fabric. The igniter was energized using a controlled current power supply (BK Precision 1785) set to deliver 40 W for 10 s.

Another set of experiments were conducted with a composite sample consisting of a 0.74 mm thick clear casted PMMA sheet 13 mm long and 76 mm wide placed below the Nomex sample. The same nichrome wire was braided to the PMMA piece for ignition. Once ignited, the PMMA flame provided a source of both heating and pilot to the Nomex. The composite sample was selected to match one of the samples tested in the microgravity spacecraft fire experiment conducted in the Orbital Corporation Cygnus spacecraft (Saffire II) [14].

Two ambient pressures were considered, 100 and 60 kPa. The composition and pressure of concurrent (upward) gas flow were controlled by adding oxidizer gases through critical nozzles (O'Keefe Controls) while constantly evacuating exhaust gases to maintain constant a pressure inside the chamber. The pressure was primarily controlled by a high-capacity vacuum generator (Vaccon JS-300) and a mechanical vacuum regulator connected to the test chamber. Correct operation of this configuration was ensured by monitoring constantly the chamber pressure with an electronic pressure transducer (Omega Engineering, Inc. PX303-015A5V). The flow was fixed to 10 cm/s for every test.

The experimental procedure followed the Oxygen Index tests described in NASA-STD-6001B [15]. Once the sample was in position, the chamber was sealed to adjust the ambient pressure and oxygen concentration. Once the prescribed pressure and oxygen concentration were achieved, and the flow was stabilized, the igniter was activated. The ignition and subsequent concurrent (upward) flame spread were recorded with a resolution of 1280 by 720 at 59 frames per second using a Nikon D3200 camera to track the pyrolysis front. In addition, the mass of the sample was recorded before and after each test using a Sartorius scale (CPA225D) with a 0.01 mg resolution. For each test condition, between three and five replicate experiments were conducted to address the experimental uncertainty.

3. Results

The concurrent flame spread under the lamp radiation was first investigated. A successful flame propagation test was defined as when the flame propagated longer than 10 mm away from the igniter. The LOC for flame spread was determined as the boundary condition between no flame propagation in all repeating tests and at least one flame propagation at the following higher oxygen concentration. Figure 2(a-b) presents the data collected for the concurrent (upward) flame spread over Nomex fabric at 100 and 60 kPa under varying external radiation and oxygen concentration. The flammability boundary is presented by a dashed line that fits the LOC for each external radiant heating. The probability of burning was evaluated as the ratio between the number of cases that result in flame propagation and the total number of repeating tests.

Clearly, the external radiation extends the material flammability by lowering the LOC. For example, with an external radiation of 15 kW/m^2, the LOC at 100 kPa is reduced from 24 to 21% of O_2 and from 26 to 22% of O_2 at 60 kPa.

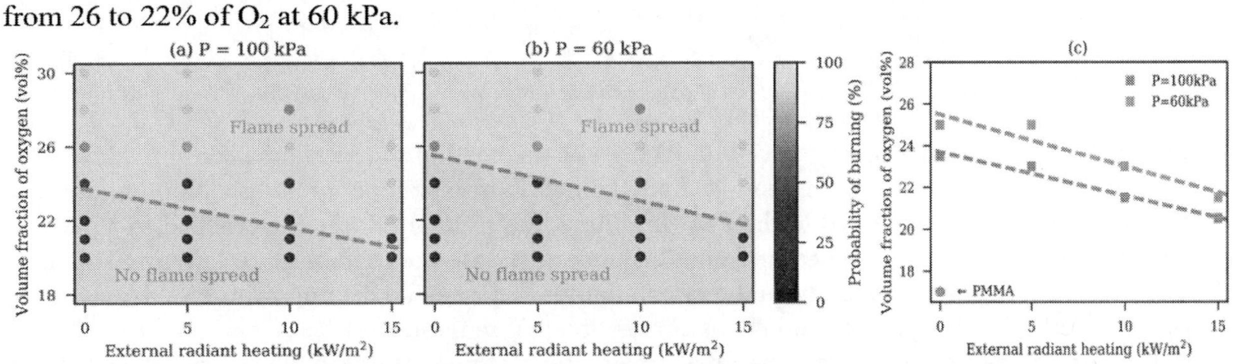

Figure 2. Flame spread (or flammability) boundary for Nomex HT90-40 as a function of oxygen concentration and external radiation for two ambient pressures of (a) 100 kPa and (b) 60 kPa; and (c) the comparison.

Figure 2(c) further compares the pressure effect on the flammability boundary, including as well the LOC for PMMA. Reductions in pressure require a higher oxygen concentration or a larger external radiation to sustain the flame spread.

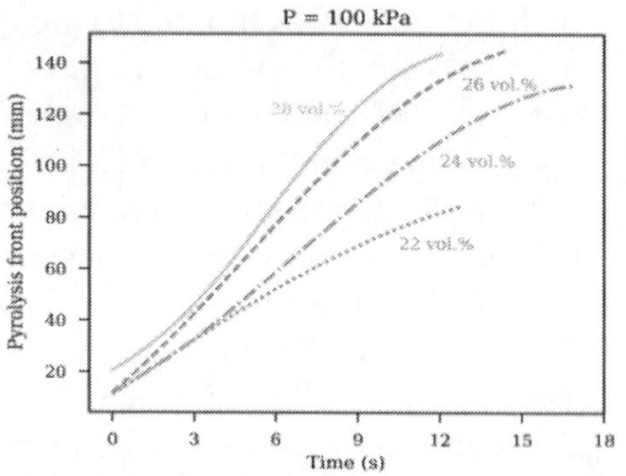

Figure 3. Position of the pyrolysis front in Nomex HT90-40 as a function time with PMMA at 100 kPa.

Average flame spread rates were determined by tracking the pyrolysis front over the fabric. Figure 3 shows the position of the pyrolysis front in Nomex as a function of time for the Nomex HT90-40 plus PMMA configuration at an ambient pressure of 100 kPa and different oxygen concentrations. The flame spread clearly becomes faster as the oxygen concentration increases. Notice that at oxygen concentrations lower than the LOC of Nomex, the PMMA flame allows the burning of fabric until the burnout of the PMMA, so the total propagation distance is shorter (about 80 mm). As the oxygen concentration gets higher than the LOC, the pyrolysis of fabric can keep sustaining the flame spread on its own and therefore propagate over the entire 140-mm sample. Also, two stages of spread were observed (1) before the burnout of PMMA, flame spread in Nomex accelerated, and (2) afterwards, flame spread slowed down.

Figure 4 shows the average concurrent (upward) flame spread rate over the Nomex as a function of oxygen concentration under two different heat sources and two different ambient pressures. The mean values are reported, and error bars show the standard deviation. As expected, the flame spread rate increases with oxygen concentration, independent of changes in the other environmental variables, primarily because of the increasing flame temperature. Also, a higher flame spread rate was found for higher pressures, although the pressure effect is much weaker than the oxygen effect [12]. Moreover, the LOC decreased from 24 to 22% as the external radiation increased from 0 to 10 kW/m^2.

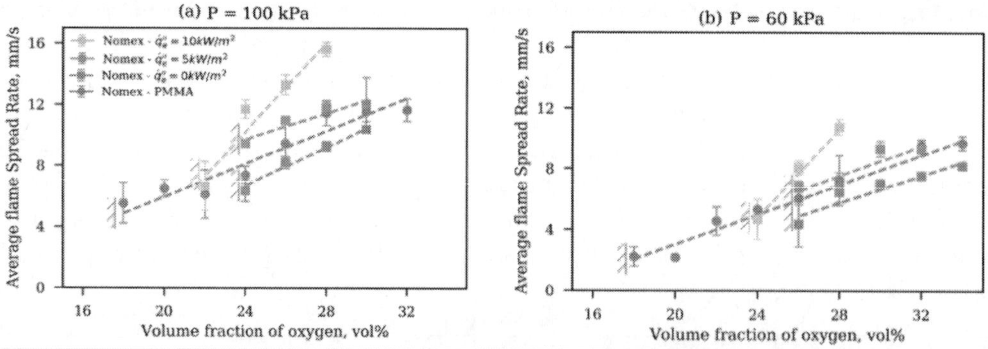

Figure 4. Average flame spread rate as a function of oxygen concentration for two different ambient pressures: (a) 100 kPa and (b) 60 kPa. Hatched line shows the LOC for each heating source.

Figure 4 also shows that for the tests with PMMA as the heating source, flame can spread over Nomex at a lower oxygen concentration than with the Nomex burning alone under the lamp radiation. When the Nomex fabric is subjected to the PMMA flame, the LOC of Nomex is reduced to 17% which is the LOC of PMMA, and it is lower than 22% under the external radiant of 10 kW/m^2. This indicates a different heating mechanism from the radiant lamps, because the heating from PMMA can be higher (about 15 kW/m^2) [16] and include both convection and radiation.

Once ignited, the PMMA flame generated covers the bottom part of the fabric providing enough heat to pyrolyze the material. At the same time, the pyrolysate from Nomex also contributed to the flame, altering both the flame characteristics and the heterogenous reactions in Nomex. It should be noted, that during the Saffire II experiment no flame spread over Nomex under the PMMA flame was observed at 21%, however this limited single data does not allow us to draw further conclusions. Above the LOC, the overall effect of PMMA flame on Nomex flame spread is equivalent to an external radiant heat flux less than 5 kW/m^2, probably because of the early burnout of PMMA and the two-stage flame spread as shown in Figure 3. Although the results presented imply that the heating from the PMMA helps the spread of flames over Nomex and extends the LOC of the Nomex, it is difficult to separate the contributions from the PMMA and the Nomex in the burning process. However, the data can be used to compare flame spread behavior under two heating sources.

To better understand the burning process of the fabric under the different heating sources, we measured mass loss of the sample. The mass loss fraction or the burning fraction (ΔY_b) was defined as

$$\Delta Y_b = \frac{m_b}{m_0} = \frac{m_0 - m_r}{m_0} \tag{1}$$

where m_0, m_b, and m_r are the mass of the original, burnt and residual sample, respectively. Figure 5 shows ΔY_b as a function of oxygen concentration for all experiments. It can be seen that the mass loss fractions of the Nomex burning with the PMMA and with the external radiant heater follow a trend similar to the flame spread rates in Figure 4. Also, the two heating sources have different effects on the overall mass loss (or the overall heat release). When heated by the PMMA flame, the mass loss continuously increases with the oxygen concentration. However, when heated by the radiant lamps, there are two different regions separated by the LOC. Below the LOC, the mass loss fraction is almost constant and is a direct result of the ignition heating (see hollow symbols). Above the LOC, the mass loss fraction has a step increment, showing the clear boundary between no-spread and spread (see filled symbols). Increasing the oxygen concentration further, the mass loss continues to increase, similar to that with PMMA flame.

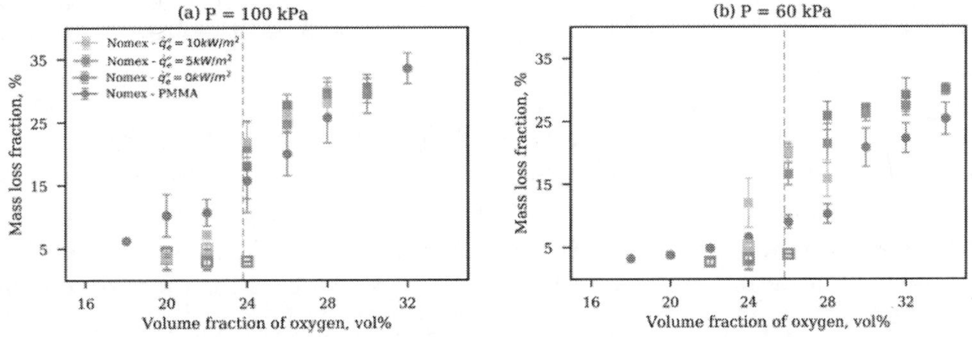

Figure 5. Mass loss fraction of Nomex HT90-40 as a function of oxygen concentration for two different ambient pressures: (a) 100 kPa and (b) 60 kPa. Hollow points represent the no-spread cases.

This difference was also noticed in the flame behavior, specifically, when using the external radiation, the flame spread was more uniform and extended almost through the entire sample. However, when the fabric was exposed to the PMMA flame, the overall flame spread was more irregular.

4. Discussion

The flame spread process over a thin solid fuel can be influenced by the different interactions between the studied environmental variables. When the flame is spreading over the fuel surface different physical, thermal and chemical mechanisms interact at the same time, complicating the flame behavior. A simplified analysis is followed here, with the flame spread rate being expressed as [17]

$$V_f = \left(l_f - l_p\right)\left[\frac{\rho_s c_s s\left(T_p - T_0\right)}{\dot{q}_f'' + \dot{q}_e'' - \dot{q}_{rs}''} - \frac{Cx}{U_\infty}\frac{[F]}{\hat{r}}\right]^{-1} \tag{2}$$

where l_f and l_p are the flame and pyrolysis length, $\dot{q}_f''$ represents the heat transferred by the flame to the solid by radiation and convection, $\dot{q}_e''$ is the external radiant heat flux and $\dot{q}_{rs}''$ is the surface re-radiation, U_∞ is the flow velocity, $\hat{r}$ the reaction rate, x is the coordinate parallel to the fuel surface, *[F]* the fuel concentration, ρ_s and c_s are the solid density and specific heat and s is the solid thickness, T_p and T_o represent the pyrolysis and initial temperatures of the fuel and C is a numerical constant.
Equation 2 focuses on two important mechanisms, the heat transfer to the unburnt solid (left term inside the bracket) and the chemical kinetics in the gas phase (right term inside the bracket). Given Eq. 2, increasing the external radiant heat flux accelerates the flame spread as it mainly affects the heat transfer mechanism. At oxygen concentrations higher than the LOC, the increase in flame temperature promotes the heat transfer from the flame to the solid, and then helps the flame to propagate faster. Furthermore, the higher flame temperatures increase the rate of reactions making the right-side term on the bracket negligible. On the opposite side, for oxygen concentrations below the LOC, the external radiant flux must compensate the reduction in flame heating to allow the flame to keep propagating. As oxygen is further reduced, the lower temperatures would result in slower reactions that would prevent the spread of the flame even in the presence of an external heating source.
Changes in ambient pressure change the heat transfer process by affecting the boundary layer thickness. A reduction in pressure will reduce the gas density and the buoyancy, resulting in a thicker boundary layer. Therefore, moving the flame further away from the solid and reducing the convective heat transferred to the solid. Additionally, as the pressure decreases to lower values, the flame sheet (reaction zone) may become thicker and the flame temperature decreases [18], which could also contributes to the decrease in the flame spread rate.
For the Nomex burning with the PMMA, no flame can be sustained below the LOC of PMMA (~17%). At oxygen concentrations higher than the LOC of PMMA, the PMMA flame can aid the pyrolysis of the fabric. Thus, the resulting flame could then be viewed as a PMMA flame assisted by some Nomex pyrolysates, affecting the in $\dot{q}_f''$, rather than adding $\dot{q}_e''$. It is important to notice that the entire effect of the PMMA flame might not only be affecting the heat transfer mechanism but also the chemical reactions involved, resulting in a total effect much more complex than just adding an external heat flux.
A similar analysis can also be done to understand the changes experienced by the mass loss. Following a similar analysis as presented by Osorio et al.[11], mass loss rate can be defined as

$$m'' = \frac{\left(\dot{q}_f'' + \dot{q}_e'' - \dot{q}_{rs}''\right)}{\Delta H_p} \tag{3}$$

where ΔHp is the heat of pyrolysis of the material. An increase in oxygen concentration and pressure will result in an increase in the heat transfer from the flame to the solid and consequently the mass loss rate as found in the experimental results (see Figure 5).

5. Conclusions

Concurrent flame spread rate, limiting oxygen concentrations (LOC), and mass loss fraction of Nomex HT90-40 fabric burning in an upward configuration were measured as a function of oxygen concentration and ambient pressure. The effects of two external heat sources were also considered: an external radiant heat flux generated by halogen lamps and a flame from a burning sheet of PMMA placed below the fabric. Experiment shows that increases in oxygen concentration and ambient pressure will increase the flammability of the material and promote flame spread. Comparatively, the pressure effect is weaker than the oxygen effect. The two different external heat sources considered resulted in different flame spread behaviors. On one hand, the external radiant flux provided by the lamps increased the flame spread rate and extended its flammability by lowering LOC from 24% to 21% at 15 kW/m^2. On the other hand, the attached PMMA flame helped the fabric to burn at an oxygen concentration of 18%, lower than the LOC of 24%, although there is some uncertainty on how much is the contribution from the PMMA itself in this value. When assisted by the PMMA flame, the flame spread rate continuously increased with oxygen concentration. A simplified analysis of flame spread process allows understanding how some characteristics of spreading flame change in response to variations in environmental conditions. Understanding these relations will provide a better picture of flammability of materials and what to be expected under specific conditions such as a nearby fire.

Acknowledgements
This work was supported by a NASA Grant NNX12AN67A and BecasChile (#72150022).

References

[1] J.G. Quintiere, Fundamentals of Fire Phenomena, Wiley, 2006.
[2] O. Fujita, Proc. Combust. Inst. 35 (2015) 2487–2502.
[3] D.B. Hirsch, J.H. Williams, S.A. Harper, H. Beeson, M.D. Pedley, (2007) 3–6.
[4] D. Hirsch, J. Williams, H. Beeson, J. Test. Eval. (2008) 3–8.
[5] J.E. Kleinhenz, J.S. T'ien, Combust. Sci. Technol. 179 (2007) 2153–2169.
[6] Y. Nakamura, A. Aoki, Adv. Sp. Res. 41 (2008) 777–782.
[7] J.G. Quintiere, M. Harkleroad, D. Walton, Combust. Sci. Technol. 32 (1983) 67–89.
[8] ASTM International, ASTM E1321, Standard Test Method for Determining Material Ignition and Flame Spread Properties, 2013.
[9] J. Quintiere, Fire Mater. 5 (1981) 52–60.
[10] K. Miyamoto, X. Huang, N. Hashimoto, O. Fujita, C. Fernandez-Pello, Fire Saf. J. 86 (2016) 32–40.
[11] A.F. Osorio, C. Fernandez-Pello, D.L. Urban, G.A. Ruff, Proc. Combust. Inst. 34 (2013) 2691–2697.
[12] M. Thomsen, D.C. Murphy, C. Fernandez-pello, D.L. Urban, G.A. Ruff, Fire Saf. J. (2017).
[13] DuPont de Nemours & Company, Technical Guide for Nomex Brand Fiber, Technical Report H-52720, 2001.
[14] G. Jomaas, J.L. Torero, C. Eigenbrod, J. Niehaus, S.L. Olson, P. V Ferkul, G. Legros, A.C. Fernandez-Pello, A.J. Cowlard, S. Rouvreau, N. Smirnov, O. Fujita, J.S. T 'ien, G.A. Ruff, D.L. Urban, Acta Astronaut. 109 (2015) 208–216.
[15] National Aeronautics and Space Administration, NASA-STD-6001B Flammability, Offgassing, and Compatibility Requirements and Test Procedures, Washington, DC, 2011.
[16] M.J. Gollner, X. Huang, J. Cobian, A.S. Rangwala, F.A. Williams, Proc. Combust. Inst. 34 (2013) 2531–2538.
[17] C.A. Fernandez-pello, Fire Explos. Hazards Proc. Fourth Int. Semin. (2004) 13–26.
[18] I. Glassman, R.A. Yetter, Combustion, 4th ed., Academic Press, New York, 2008.

Western States Section of the Combustion Institute – Fall 2017 Meeting
Hosted by the University of Wyoming
October 2-3, 2017

Heat Transfer Optimization in Simulated Microgravity Combustion

Sarzina Hossain[1, 2] and *Indrek S Wichman*[1*]

[1]*Department of Mechanical Engineering, Michigan State University, East Lansing, MI 48824-1226*
[2]*Graduate Student*
[*]*Corresponding Author: wichman@egr.msu.edu*

Heat transfer in simulated microgravity exhibits a competition between buoyancy and convection, which must achieve an optimal balance in order for an on-earth simulation of microgravity to be credible. A procedure has been developed to optimize the gap width in the microgravity simulation device known as the Narrow Channel Apparatus (NCA). The characteristic feature of the NCA is its small gap height, which suppresses buoyancy but enhances convective heat transfer losses. The approach is to determine an optimum (in this case, a *minimum*) of the heat losses. The theoretical predictions are compared with data gathered in the NCA for flame spread over Whatman 44 filter paper for gap heights ranging from 6 - 20 mm (half-gap below and above sample) and average opposed flow velocities ranging from 1 - 40 cm/s. For thin fuels, a full gap of 10 mm (5 mm half-gap) provided a compromise between buoyancy suppression and heat loss. The current theoretical estimates agree well with current and previous experiments. .

1. Introduction

NASA-STD-6001 Test 1 protocol is currently used by NASA to evaluate the material flammability for spacecraft applications [1]. The ground based facility which has been under development as a replacement of this test is the Narrow Channel Apparatus (NCA). Fundamental flame spread research has been carried out with this NCA to simulate microgravity for thermally thin fuels [2]. Different channel configurations have been studied to analyze flame front propagation using analytical, numerical and experimental techniques for thin solid fuels [3]. The most crucial parameter for a NCA is the gap height, h. The gap height should be small enough to suppress buoyancy. But a small gap height leads to high heat losses and possible flame quenching in addition to reduced rates of flame spread. The role of gap height has been evaluated for flame spread over this fuels with different fundamental characteristic scaling parameters [4].

This work focuses on determining optimum gap height of the NCA in the context of heat losses. The evidence of a trade-off between buoyancy suppression and flame quenching through heat losses is of interest for the further development and refinement of a standard test method for spacecraft fire-testing applications. In that light, the first goal is to determine the different types of heat losses involved during combustion in the NCA. The second goal is to calculate optimum gap height from a comparison between buoyant convection and forced convection heat losses.

2. Methods

Theoretical Analysis and Correlations: Heat Losses

2.1 Dimensionless Heat Loss

The first step is to calculate the total heat generation by combustion. This is given by $\dot{Q}_{flame} = \dot{m}c_{p,g}(T_f - T_\infty)$, where $\dot{m}$ is the fuel mass consumption rate and $\Delta h_c = c_{p,g}(T_f - T_\infty)$ is the heat of combustion. This can be evaluated as follows by writing the chemical energy release rate in terms of oxygen consumption,

$$\rho_f(t_f w)\, V_f\, \Delta h_c = \rho_f(t_f w)\, V_f\, c_{p,g}(T_f - T_\infty) = \dot{m}_{ox}\Delta h_{ox}. \quad (1)$$

Here ρ_f is the fuel density, t_f is the fuel thickness, w is the fuel width, V_f is the flame spread rate, $\dot{m}_{ox}$ is the oxygen mass consumption rate and Δh_{ox}=*13.1 MJ/kg* is the lower heating value when the products of combustion are in the gas, or vapor, state.

The second step is to calculate the heat losses during combustion. The first of these heat losses is convective loss to the NCA walls. To evaluate this loss term, the NCA walls are described as flat plates having temperature of $\bar{T} = (T_f + T_\infty)/2$ and boundary layer leading edge located above (and below, on the bottom side) the flame leading edge. In this steady state analysis the origin of the thermal boundary layer is placed at the horizontal middle of the sample (in order to avoid considering the motion of the flame in the NCA). The heated wall length l is then the distance from the center of the sample to the end of the duct, $l = (l_1 + l_2)/2$ is shown in Figure 1.

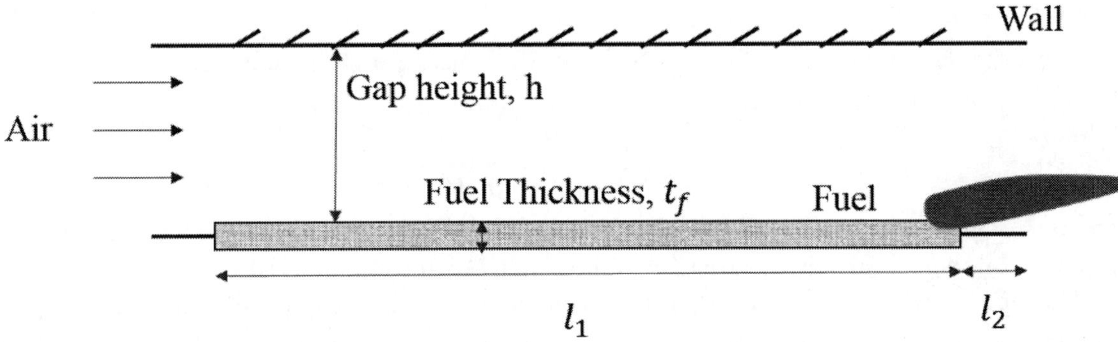

Figure 1: Schematic of spreading flames in narrow channel. Because of symmetry only half of the NCA is shown.

For laminar flow, the Nusselt number $\overline{Nu_l} = h_c l/k$ is $\overline{Nu_l} = 0.664\, Re_l^{\frac{1}{2}} Pr^{\frac{1}{3}}$. Thus, the heat losses $\dot{Q}_w$ from the flame to one NCA wall are given by

$$\dot{Q}_w = h_c l w(\bar{T} - T_\infty) = \overline{Nu_l}\, k w(\bar{T} - T_\infty). \quad (2)$$

Now the heat losses by conduction are calculated. The net heat lost $\dot{Q}_f$ is evaluated by subtracting from the conduction to the gas $\dot{Q}_c$ the heat conducted to the upstream gas and the upstream solid since these are not "lost" but are circulated back to the flame, $\dot{Q}_f = \dot{Q}_c - \dot{Q}_g - \dot{Q}_s$. Heat conduction from the flame is written as $\dot{Q}_c = k_g(T_f - T_\infty)h_* w/\delta_g$, where the characteristic preheat length is δ_g and the characteristic flame length is h_*. The enthalpy rise of the gas is $\dot{Q}_g = \rho_g c_{p,g} V_f (T_f - T_\infty) h_* w$, and the enthalpy rise of the solid fuel is $\dot{Q}_s = \rho_f c_{p,f} V_f t_f (T_s - T_\infty) w$. The heat lost by conduction is then

$$\dot{Q}_f = \frac{k_g(T_f - T_\infty)h_* w}{\delta_g} - \rho_g c_{p,g} V_f (T_f - T_\infty) h_* w - \rho_f c_{p,f} V_f t_f (T_s - T_\infty) w \quad (3)$$

The total heat loss by the flame is the sum of convective losses to the walls and the conductive loss by combining Eq. (2) and (3)

$$\dot{Q}_{loss} = \dot{Q}_w + \dot{Q}_f \;=\; \overline{Nu}_l\, kw(\bar{T} - T_\infty) + \frac{k_g(T_f - T_\infty)hw}{\delta_g} - \rho_g c_{p,g} V_f (T_f - T_\infty) hw - \rho_f c_{p,f} V_f t_f (T_s - T_\infty) w$$

The ratio of heat losses to flame heat generation is given by $\hat{Q}_{loss} = \dot{Q}_{loss}/\dot{Q}_{flame}$:

$$\hat{Q}_{loss} = \frac{2[\overline{Nu}_l\, kw(\bar{T} - T_\infty)] + [\frac{2k_g(T_f - T_\infty)h_* w}{\delta_g} - \rho_g c_{p,g} V_f (T_f - T_\infty) h_* w - \rho_f c_{p,f} V_f t_f (T_s - }{\rho_f (t_f w)\, V_f\, c_{p,g} (T_f - T_\infty)} \quad (4)$$

2.2 Optimum Channel Height for Minimum Heat Loss

In the calculations that follow the heat losses are considered as a combination of forced convection and buoyant convection. In this formulation the heat losses arising from the buoyant flow induced by the flame in a channel of height h are added to heat losses by forced convection to both the upper and lower channel walls. Once these heat losses are written, their sum, the total losses, are minimized by calculating an optimal channel height. The following calculation determines the optimum channel height for minimizing the heat losses.

Consider the heat losses by buoyant convection. The characteristic buoyant velocity is $V \sim \sqrt{gh} = C\sqrt{gh}$ [5] and the buoyant loss rate is [6] $\dot{Q}_{buoy} = \rho C\sqrt{gh}(hw)c_p(\bar{T} - T_w) = \rho C\sqrt{g}(w)c_p(\bar{T} - T_w)h^{3/2}$, or

$$\dot{Q}_{buoy} = Ah^{3/2}, \quad (5)$$

where, $A = \rho C\sqrt{g}(w)c_p(\bar{T} - T_w)$. The understanding herein is that heat lost from the spreading flame impinges on the ceiling surface area above it, which has the characteristic heat-loss area hw.

Consider now the heat losses by forced convection. The slowly propagating flame generates two laminar thermal boundary layers at both inner NCA surfaces (top and bottom), thus $\dot{Q}_{conv} = 2\overline{Nu}_l\, kw(\bar{T} - T_\infty)$, where $\overline{Nu}_l = 0.664\, Re_l^{1/2} Pr^{1/3}$ and the Reynolds number is written for the NCA as $Re_l = \rho \bar{V}_g l/\mu = \dot{m}l/hw\mu$. Here the average opposed gas flow velocity was written in terms of the mass flow rate (which for this analysis is held constant): $\bar{V}_g = \dot{m}/\rho hw$. Thus, one can write

$$\dot{Q}_{conv} = B/h^{1/2} \quad (6)$$

where, $B = 2(0.664)(\dot{m}l/w\mu)^{1/2} Pr^{1/3} kw(\bar{T} - T_\infty)$.

Adding together the buoyant losses (given by Eq. (5)) and convective losses (given by Eq. (6)) yields

$$\dot{Q}_{tot} = Ah^{3/2} + 2B/h^{1/2}, \quad (7)$$

which shows that the buoyant loss increases as h increases whereas the convective loss decreases as h increases. Note that the buoyant heat losses from the flame to the top surface of the NCA cannot continue to increase indefinitely with h but are expected to level off at large h values. These are assumed to lie beyond the range of applicability of our theoretical formulation. Differentiation gives$\partial \dot{Q}_{tot}/\partial h = (3/2)Ah^{1/2} - Bh^{-3/2}$, which vanishes when $h_{opt} = \sqrt{(2B/3A)}$. This value of h minimizes $\dot{Q}_{tot}$ with respect to the channel height h.

After substituting for A, B the result for the optimum channel height which minimizes heat losses as a balance between buoyancy and convection is given by

$$h_{opt} = \left[\frac{2}{3C}(0.664)\alpha Pr^{\frac{1}{3}}\sqrt{\frac{\dot{m}l}{gw\mu}}\right]^{\frac{1}{2}} \quad (8)$$

3. Results and Discussion

From Eq. (4) the dimensionless heat loss is calculated. For this purpose the flame spread rate data for the NCA [4] is used. The dimensionless heat loss is plotted as a function of the dimensionless opposed velocity in Figure 2. In this plot Eq. (4) is the ordinate. The abscissa is the dimensionless opposed flow velocity, $\hat{V}_g = C \cdot Le \cdot Pe^{1/2}$ [7], where we use $C = 1$. In addition, $h_* = 1\ cm$ is used as a characteristic flame length. As seen, the heat losses are largest for the smallest gaps and as the velocity is either increased or decreased, the smallest-gap flames extinguish (low velocity range) or they blow off (higher velocity range) the earliest in both directions of the opposed flow velocity.

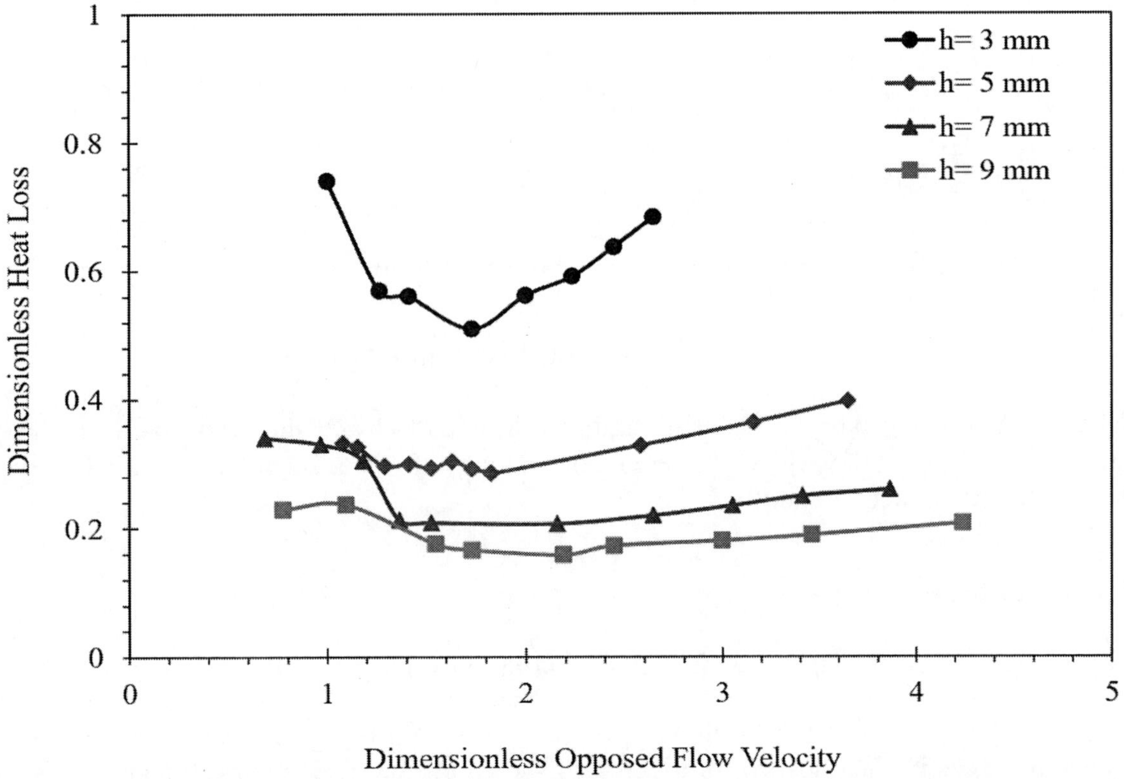

Figure 2: Dimensionless heat loss versus dimensionless opposed flow velocity for gap height, $h = 3$, 5, 7 and 9 mm.

In order to determine a representative range of values, we refer to Figure 2, which shows that the heat loss calculated there is a minimum for a range of dimensionless opposed flow velocities between approximately 1.5-2. The dimensionless opposed flow velocity (1.5-2) corresponds in the NCA to average flow velocities between 7-10 cm/s (which values we use in our calculations). In Figure 3, these inlet velocities (7-10 cm/s) are varied to calculate an optimum height from Eq. (8). The values used in Eq. (8) are $\alpha = 435 \times 10^{-6} \frac{m^2}{s}$, $Pr = 0.685$, $l = (l_1 + l_2)/2 = (35 + 10)/2 = 0.225\, m$, $w = 0.38\, m$, $\mu = 611 \times 10^{-7} Ns/m^2$ and $C \sim O(1)$. The mass flow rates are approximately varied between $\dot{m} = 2.72 \times 10^{-5}$ kg/s to 3.8×10^{-5} kg/s.

The calculated result lies between $h_{opt} \approx (5.2 - 5.8)mm$. This range of values is close to the value of *5 mm* that several previous works [8] have obtained by systematic laboratory experimentation.

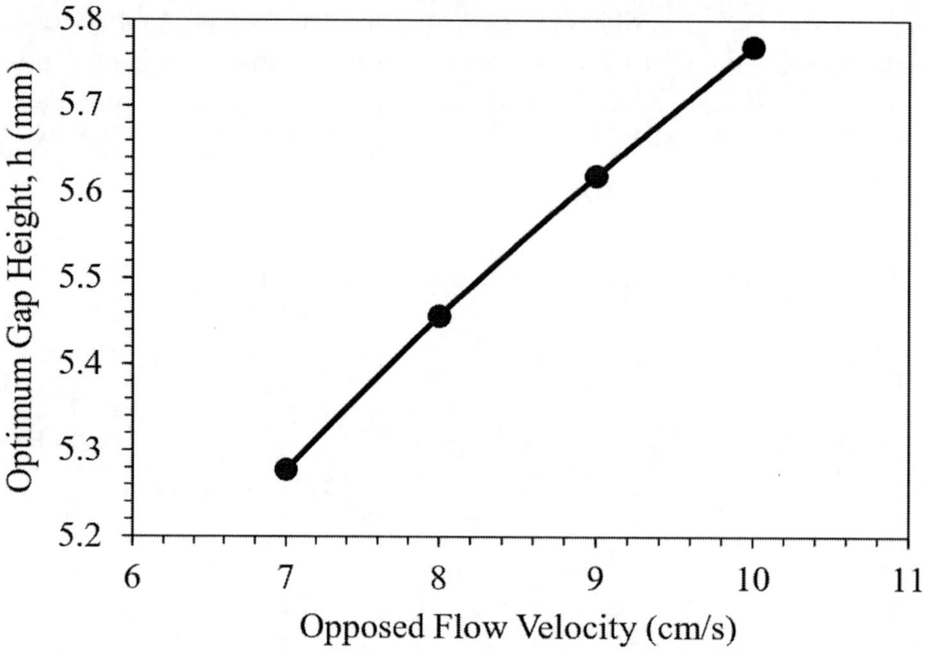

Figure 3: Optimum gap height h_{opt} that minimizes heat losses by channel convection and buoyancy as a function of opposed flow velocity. The calculated range of values is consistent with the experimental value of approximately 5 mm [9].

4. Conclusions

An optimum channel height for simulated microgravity experimentation was produced, and that theoretical optimum agrees well with empirical observation and prior experience. Five millimeters of channel height produces results that appear to closely mirror actual microgravity tests without producing excessive heat losses [8]. It was also shown in Eq. (7) that there is a greater sensitivity to the forced convection heat losses ($\propto h^{3/2}$) than to the buoyant convective heat losses ($\propto h^{-1/2}$).

Acknowledgements

Graduate student Sarzina Hossain and Professor Indrek S Wichman were supported by NASA Cooperative agreement NNX16AC24A S02. We appreciate the discussions and suggestions by Dr. S. L. Olson (NASA-Glenn), Prof. F. J. Miller (SDSU) and Prof. G. Sidebotham (Cooper Institute).

References

[1] D. R. Mulville, NASA-STD-6001, (1998).
[2] S.L. Olson, F.J. Miller, S. Jahangirian and I.S. Wichman, *Combust. Flame,* 156 (2009) 1214-1226.
[3] B. Comas, A. Carmona, T. Pujol, *Fire Safety Journal,* 71 (2015) 162-273.
[4] S. Hossain, G. Sidebotham, S.L. Olson, F.J. Miller and I.S. Wichman, *10th U. S. National Combustion Meeting, College Park, Maryland,* (2017).
[5] I. S. Wichman, S. L. Olson, F. J. Miller and S. A. Tanaya, *Fire Materials*, 37 (7) (2013) 503-519.
[6] J. De Ris, *Symp. (Int.) Combust.,* 12 (1969) 241-252.
[7] G.W. Sidebotham, S.L. Olson, P. Rachow, F. J. Miller and I. S. Wichman, *7th US National Combustion Meeting, Atlanta, GA,* (2011).
[8] I. S. Wichman, *Combust. Flame,* 117 (1999) 384 - 393.
[9] G.W. Sidebotham, I. Glassman, *Combust. Flame,* 90 (1992) 273-283.

Correlating Mass Burning Rate and Flame Spread Rate for Thin PMMA: Implications on Pyrolysis Temperature

Luca Carmignani, Subrata Bhattacharjee

Mechanical Engineering, SDSU, 5500 Campanile Blvd, San Diego, CA 92182, USA

Abstract. Mass burning rate and flame spread rate have been widely studied for their importance in combustion of solid and liquid propellants, as well as fire safety. A number of available models to calculate flame spread rate and mass burning rate are based respectively on the work of de Ris and Emmons, whose closed-form solutions established the baseline in this research field. However, flame spread rate and mass burning rate are experimentally studied separately, with the burning rate work focusing on local burning rate of thick fuels, and numerous work on flame spread rate over thermally thin fuels. In this work, we present experimental data on overall mass burning rate and its dependence on fuel thickness. Given the difficulty of measuring local burning rate for thin fuels, we use extrapolation of data to deduce the thin fuel limit of mass burning rate, consistent with both experimental data and mass conservation through the use of flame spread formula. Moreover, correlating the mass burning rate with flame spread rate, the average pyrolysis temperature of the solid fuel shows a dependence on the fuel thickness. As the fuel thickness decreases, the pyrolysis temperature is found to increase. An existing numerical model is used to predict the pyrolysis temperature as a function of thickness, which shows the same trend as deduced from the experiments.

Nomenclature

ρ	Density, $\mathrm{kg/m^3}$	T_∞	Ambient temperature, K
c	Specific heat, $\mathrm{kJ/(kg\cdot K)}$	V_f	Flame spread rate, $\mathrm{m/s}$
λ	Thermal conductivity, $\mathrm{kW/m\cdot K}$	V_g	Flow velocity, $\mathrm{m/s}$
τ	Semi-thickness of fuel, µm	m	Mass flow rate, $\mathrm{kg/s}$
T_v	Vaporization temperature, K	β	Burn angle, deg
T_f	Flame temperature, K	L_p	Pyrolysis length, m

1. Introduction

Flame spread rate and mass burning rate are crucial characteristics in combustion research. In particular, flames propagation has been largely studied in the last decades for its importance in fire safety and to understand small scale flame phenomena, while mass burning rate is a critical aspect in flame growth, and more generally in propulsion and chemical processes of solid and liquid fuels.

Most of the models for flame spread over solid fuel are based on the pioneering work of de Ris, who obtained a closed-form solution for thermally thin and thick solid fuels [1] (improved by the exact solution of Delichatsios [2]):

$$V_{f,\text{thn}} = \frac{\pi}{4} \frac{\lambda_g}{\rho_s c_s \tau} F$$
$$V_{f,\text{thk}} = V_g \frac{\lambda_g \rho_g c_g}{\lambda_s \rho_s c_s} F^2 \tag{1}$$

Where $F = (T_f - T_v)/(T_v - T_\infty)$ is the so called de Ris coefficient. In previous literature, the value of F varies from about 4.4 to 5.5 for thin PMMA burning in ambient conditions, and in this work we refer to the value of 4.76 according to Laue [3].

The two expressions of eq. 1 were derived using an Oseen flow, infinite kinetics and a constant vaporization temperature of the fuel. Even though these assumptions are not always accurate and have been criticized by many authors, as described in the review of Wichman [4], eq. 1 are still widely used for their ease of use.

The main difference between thin and thick limit formulae is the dependence on the flow velocity V_g; thin fuel spread rate does not depend on the flow velocity using de Ris' assumptions. Another interesting difference is that the spread rate for thick fuels does not depend on thickness (since it was derived from semi-infinite fuel bed).

The starting point for the mass burning rate models is instead Emmons' work. Assuming Prandtl and Schmidt numbers equal to unity, Emmons relates shear stress at the wall and flow velocity to the mass burning rate using the B number [5], obtaining a closed-form solution for burning rate over a liquid fuel (or flat solid fuel). The B number, introduced by Spalding for droplet combustion (it is indeed also called Spalding mass transfer number) [6], depends on fuel properties, and governs the mass transfer. Due to the complexity of the B number, there is some discrepancy in the calculated values also for well-known fuels such as PMMA, going from 1.3 to 1.8, as reported by Rangwala [7]. Singh et al. calculated the local mass burning rate of vertical slabs of PMMA with natural convection measuring temperature profiles and calculating the temperature gradient at the surface [8]. All the mentioned works on mass burning rate, however, consider flames covering the entire fuel surface, and the burning rate is calculated using the B number. Also, most of the methods used to measure mass burning rate in previous works, such as scales, temperature gradients, etc. are not directly applicable to thin fuels, since they burn much faster than thick fuels and they require very high accuracy in the measurements.

In this work, we gather results on spread rate, burning rate and temperature for thin PMMA, with the intent of finding a relation between these important combustion parameters without using the B number, starting from the experimental results obtained with no forced flow. Sibulkin et al. came up with a correlation between spread rate and burning rate for PMMA cylinders with diameters in the range of about 1.5 to 12mm [9], but they focused their research on the influence of the angle of orientation more than the thickness (1.5mm could be considered a transitional thickness between thin and thick limits). They also measured the temperature of the fuel using thermocouples, and they concluded that is almost constant. However, in the last section of this paper we will see how the temperature may actually vary depending on the thickness of thermally thin fuels, as suggested also by the work of Sirignano [10] [11], who criticized the assumption of a fixed value for the vaporization temperature.

2. Mass balance and burn angle

In order to simplify the problem of a burning solid fuel, we can assume the flame in steady-state, and stationary in the chosen reference frame. In other words, the flame spread rate V_f is constant and the fuel is moving toward the flame with velocity V_f. The problem is two-dimensional, so width does not affect the results [1] [12]. Thermally thin fuels are supposed to have uniform temperature in the cross section, therefore in a 2D problem the temperature of the solid will be constant along the vertical direction of the fuel.

Referring to a general case represented in Fig. 1a, there will be a region of the fuel that is not affected by the flame (virgin fuel), a part of fuel heated up by the flame leading edge (preheated layer) and finally the vaporizing region (pyrolyzing region). We can imagine the fuel vaporizing along the semi-perimeter (the problem is symmetric with respect to the x-axis), given by $L_p = L_{px} + L_{py} = L_{px} + \tau$ (where τ is the half thickness of the fuel). The pyrolysis length projection L_{px} can be assumed much larger than the half thickness τ, $(L_{px}/\tau \quad 1)$, as also confirmed by the experimental results presented in section 3. Therefore, $L_p = L_{px} + \tau \quad L_{px}$.

For a better representation of the burning fuel, we can consider Fig. 1b, where the rectangle of the pyrolyzing region in Fig. 1a is substituted by a triangle. The pyrolysis length projection along y then becomes $L_{py} = L_p \sin\beta = \tau$, according to Fig. 1b, and β is called burn angle.

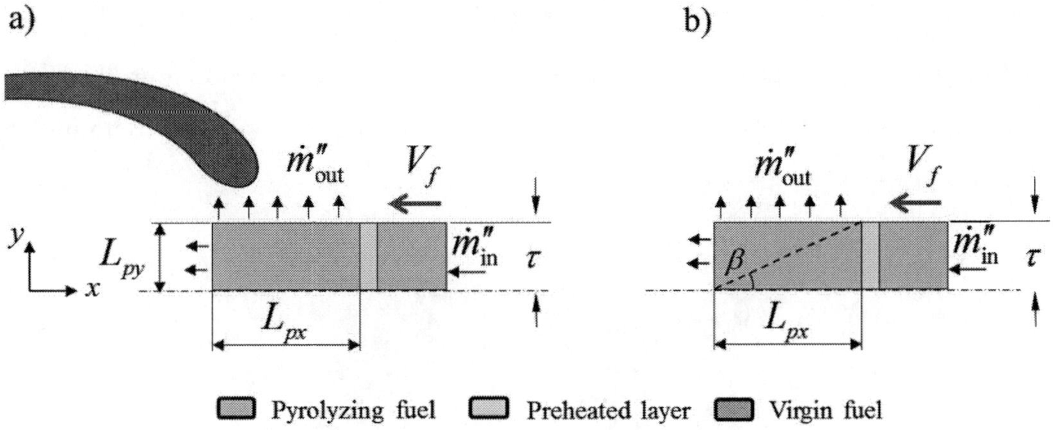

Figure 1. Schematic representation of the problem; the burning semi-perimeter is given ideally by a) the thickness and the length of the pyrolysis region, or more realistically b) by the hypotenuse of a triangle with base L_{px} and height τ .

To satisfy the mass balance in the solid phase, the mass coming in the control volume must be equal to the mass going out. Referring again to Fig. 1, the mass flow rate per unit width is:

$$m'_{in} = \rho_s V_f \tau \tag{2}$$

While the mass going out from the burning region can be expressed as (referring to Fig. 1b):

$$m'_{out} = \bar{m}''_{out} L_p = \bar{m}''_{out} \frac{\tau}{\sin\beta} \tag{3}$$

Where $\bar{m}''_{out}$ is the average mass burning rate. The mass coming in is equal to the mass going out, so we can obtain an expression for the average burning rate combining eq. 2 and eq. 3, that is:

$$\bar{m}''_{out} = \rho_s V_f \sin\beta \tag{4}$$

However, from the assumption of thin fuels, we have:

$$\sin\beta = \frac{\tau}{L_p} \quad 1 \Rightarrow \sin\beta \quad \beta$$

So we can rewrite eq. 4 as:

$$\bar{m}''_{thn} \quad \rho_s V_f \beta \quad \text{or} \quad \bar{m}''_{thn} \quad \rho_s V_f \frac{\tau}{L_p} \tag{5}$$

Where V_f and L_p can be found experimentally.

The burn angle is not only a geometrical result, but a characteristic of the burning phenomena that can relate mass burning rate with flame spread rate. The triangular shape of the solid fuels is based on the shape of the burnt samples observed for $\tau > 750$ μm (as shown in Fig. 2), also obtained by Ayani et al. [13]. The samples in fact presented a constant angle depending on thickness. Even though the burn angle was not directly measurable for very thin fuels, we use its definition to relate burning rate and spread rate.

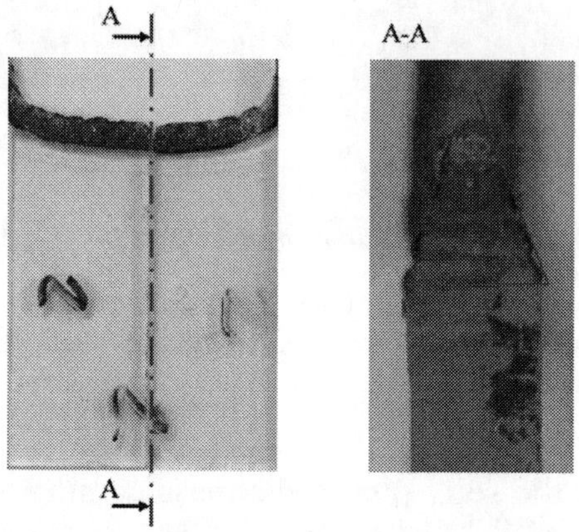

Figure 2. Sample of PMMA with thickness of 2 mm after the experiment (*left*). The sample was then cut along the centerline A-A, in order to measure the burn angle. An enlargement of the pyrolysis section (*right*) shows the typical triangular shape of the burnt surface.

3. Experimental results

Experiments with thin sheets of PMMA were carried on with a vertical sample holder, without forced flow. The thin fuel was pinched between two steel plates held together by magnets, and it was ignited from the top to the bottom with a pilot flame. Flames were suddenly blown-off after the desired amount of burning in steady state. Each experiment was repeated at least three times.

The flame spread rate is calculated using a MATLAB-based code we developed, called Flame Image Analyzer (FIA). It can track the flame in the video of the recorded experiments using a threshold of luminance intensity in the YCbCr color space [14]. Values of pyrolysis length and burn angle were obtained cutting the sample along the centerline (see Fig. 2), and measuring them with ImageJ [15].

In Table 1 the measured values are reported (averaged between the different runs). The ratio τ/L_p 1, assumed for thin fuels, is experimentally justified. The pyrolysis length does not change significantly for $\tau < 400\mu\text{m}$, but for higher values L_p becomes comparable to the half thickness τ, so the thin fuel assumption would not hold for thicknesses over 3mm.

The last column on the right in Table 1 shows the values of the burning rates for thin fuels calculated with eq. 5. Even though the values oscillate a little, it is clear that the burning rate tends to decrease with the thickness.

Table 1. Experimental averaged values of flame spread rate and pyrolysis length, used to determine the burning rate in the thin fuel approximation.

τ (μm)	L_p (mm)	τ/L_p	V_f (mm/s)	$\bar{m}''_{thn}$ $(g/(m^2 s))$
25	1.058	0.024	4.50	126.5
37.5	1.387	0.027	4.30	138.4
100	1.451	0.069	0.85	69.7
200	1.365	0.146	0.45	78.4
400	1.881	0.213	0.19	48.1
750	2.574	0.291	0.12	41.6
1000	3.341	0.299	0.08	30.3
1500	4.756	0.315	0.06	22.5

The burn angle can be obtained in two different ways: (i) direct measurements, (ii) deduced from eq. 5. As previously mentioned, the burn angle was directly measurable only for $\tau > 750$ μm, and the comparison between measured and deduced angles is shown in Fig. 3.

From the plot of Fig. 3 it can be noticed that the burn angle increases almost linearly with τ for very thin fuels, and then the slope starts decreasing. This behavior highlights the transition from the thin fuels regime to the intermediate regime with thick fuels. At first, the angle is very small since the entire region delimited by L_p is burning completely, and an increase in the fuel mass (due to an increase of τ) does not really affect this situation. However, increasing the thickness, the solid fuel at $y<0$ starts acting like a heat sink, decreasing the flame spread rate substantially and causing the burn angle β to increase only slightly.

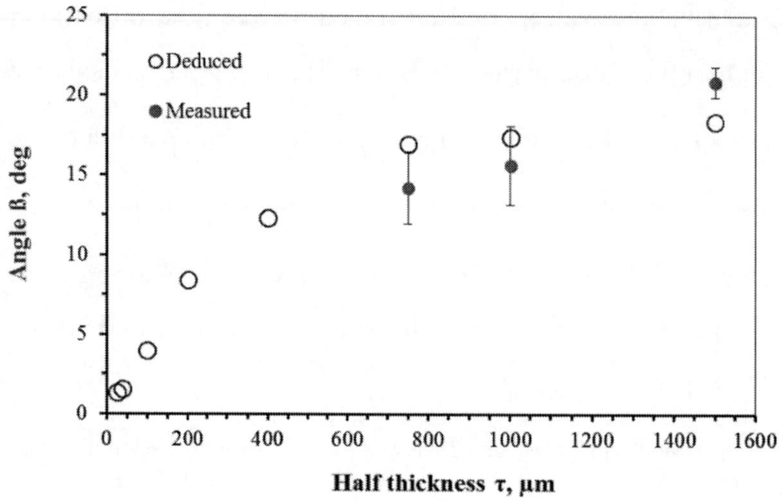

Figure 3. Comparison of the burn angles obtained from eq. 5 and the angles directly measured.

4. Average pyrolysis temperature

Assuming that the pyrolysis kinetics of the solid fuel is described by a first-order Arrhenius equation, we can define the mass flow rate per unit volume as:

$$m'''_{\text{thn}} = \rho_s A \exp\left(-\frac{E_a}{RT_s}\right) = \rho_s A \exp\left(-\frac{T_a}{T_s}\right) \tag{6}$$

Where the ratio E_a/R is substituted with the activation temperature T_a. Since we are considering a two-dimensional problem, we can integrate eq. 6 along x and y, and obtain the mass flow rate per unit width as:

$$m'_{\text{thn}} = \tau L_p \cos\beta \cdot \rho_s A \exp\left(-\frac{T_a}{T_s}\right) \quad \tau L_p \rho_s A \exp\left(-\frac{T_a}{T_s}\right) \tag{7}$$

The last relation follows from the thin fuel assumption $\cos\beta \quad 1$, since $\beta \quad 1$. Also, since $L_p \quad \tau/\beta$:

$$m'_{\text{thn}} \quad \frac{\tau^2}{\beta} \rho_s A \exp\left(-\frac{T_a}{T_s}\right) \tag{8}$$

The mass flow rate per unit width can also be directly obtained plugging de Ris' formula for thin fuels spread rate (eq. 1) in eq. 2:

$$m'_{\text{in}} = \frac{\pi}{4}\frac{\lambda_g}{c_s}F \tag{9}$$

Eq. 9 shows that the mass flow rate is not dependent on thickness for thin fuels, and substituting the known variables in eq. 9 ($\lambda_{g@700K} = 5.2\cdot10^{-5}\ \text{kW}/(\text{m}\cdot\text{K})$; $c_s = 1.4652\ \text{kJ}/(\text{kg}\cdot\text{K})$), and multiplying by 2 because of symmetry, we get m'_{in} $0.238\ \text{g}/(\text{m}\cdot\text{s})$, in good agreement with the experimental data (using experimental V_f), as shown in Fig. 4.

If the mass flow rate per unit width is constant, its variation with τ in eq. 8 must be compensated by another variable. However, all the terms in eq. 8 are constant except for the solid temperature T_s on the pyrolysis surface, therefore this relationship is further investigated.

In order to find T_s, we can match eq. 8 and 9, resulting in:

$$T_s = \frac{T_a}{2\ln\left(\dfrac{\tau}{L_r}\right) - \ln(\beta F)} \tag{10}$$

Where $L_r^2 = (\pi\lambda_g)/(4c_s\rho_s A)$ is a reference length to compare with the semi-thickness τ.

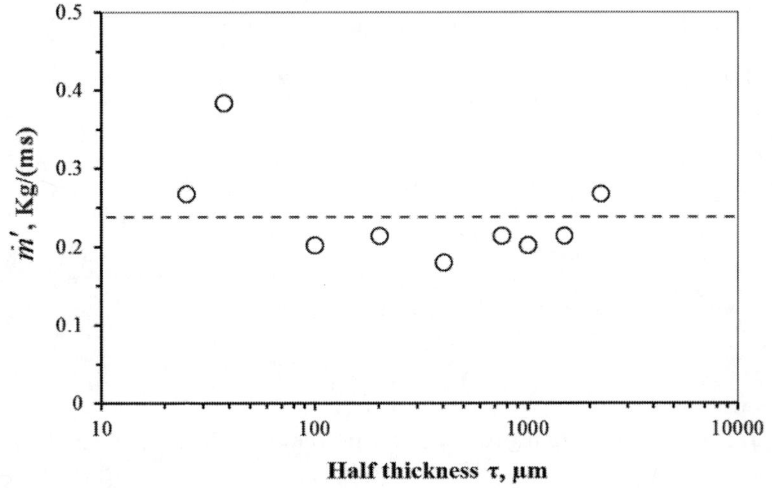

Figure 4. The mass flow rate per unit width is not dependent on thickness according to de Ris, and this is confirmed by experiments.

The computational model previously developed by Bhattacharjee – based on a CFD and a radiation models – was used to simulate a flame in the experimental conditions. The governing equations are separated between solid and gas phase. To model gas phase reaction and fuel pyrolysis, a single-step overall reaction Arrhenius equation is used, while gas and surface radiation are modeled with the global-balance radiation model. The gas-phase equations are solved by iteration of the SIMPLER algorithm [16], generating multiple fields such as temperature, velocity, mass fraction of oxygen and fuel, that can be imported and plotted in MATLAB. A detailed description of the model can be found elsewhere [17] [18].

For the solid phase, temperature profiles obtained numerically are similar to the one shown in Fig. 5. The flame is located at $x = 0$, and temperature rises from the ambient value of 298 K to the vaporization temperature of the pyrolysis region, with the peak indicated by the red dashed box in Fig. 5, and once the flame is passed, it decreases again. The temperature profile after the burning region is not important since the fuel is not actually there anymore, so a better criterion to individuate the pyrolysis front is the use of mass flux. Indeed, isolating the fuel region where mass evaporation was larger than zero (and physically acceptable), it was possible to calculate the average temperature of the burning solid in the simulations. The comparison between pyrolysis temperatures obtained from the converged solutions, and the ones deduced from eq. 10 is shown in Fig. 6.

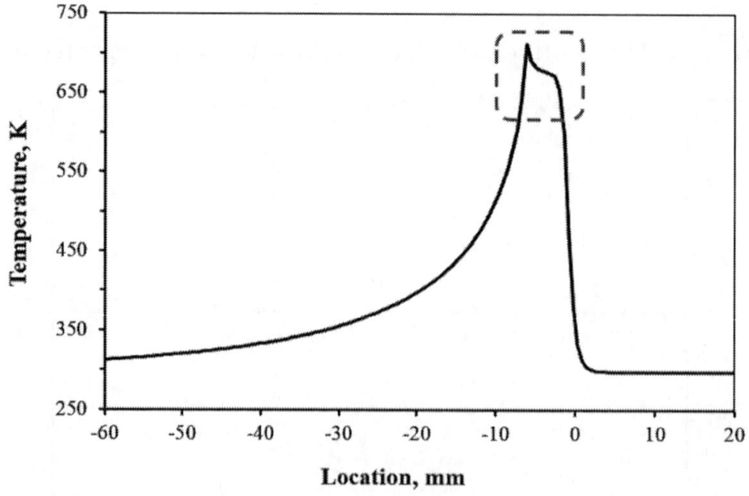

Figure 5. Temperature profile in the solid fuel; the flame is located at $x = 0$.

5. Conclusions

Starting from the mass balance in the solid fuel, we introduced the burn angle, which can be seen as a linking parameter between mass burning rate and flame spread rate over solid fuels. In particular, we assumed that for thin fuels the burn angle is very small, since the pyrolysis length is

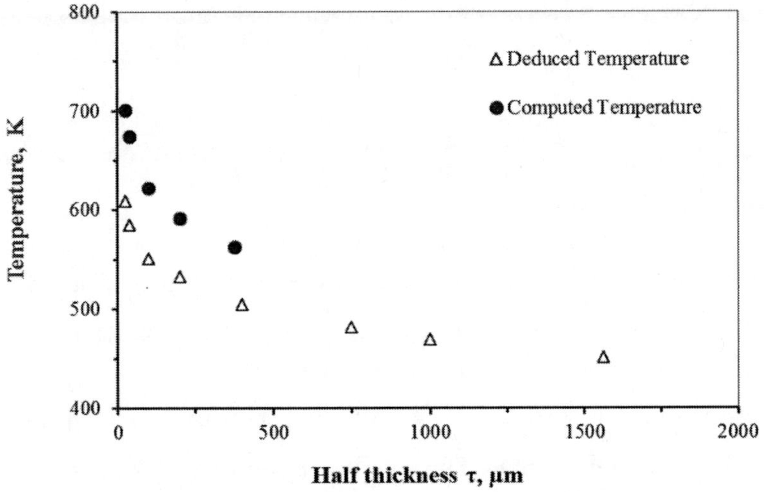

Figure 6. Comparison between temperatures obtained from eq. 10 and computational results.

much larger than fuel thickness. The angle measured for fuels thicker than 1.5 mm are in good agreement with the ones deduced from eq. 3.

According to de Ris, the mass flow rate per unit width is not dependent on thickness, as confirmed by the experimental data. However, eq. 10 shows that following another approach there is a dependence on thickness, which has to be compensated by a variation in the average burning solid temperature. Values following from experimental results are compared with the temperature obtained in simulations in similar conditions with the code developed by Bhattacharjee. Even though the absolute values are different, they both show a decreasing trend of temperature against thickness of the burning solid. This result can be justified thinking about the increase in thickness as a larger heat sink that would lower down the surface temperature of the fuel.

6. Acknowledgments

This work was funded by the NASA ISS Research Project Office with Dr. David Urban serving as the contract monitor.

References

[1] J. de Ris, “Spread of a laminar diffusion flame,” *Proceedings of the Combustion Institute,* vol. 12, no. 1, pp. 241-252, 1969.

[2] M. Delichatsios, “Exact Solution for the Rate of Creeping Flame Spread over Thermally Thin Materials,” *Combust. Sci. and Tech,* vol. 44, pp. 257-267, 1986.

[3] M. Laue, "Experimental Study of the Effect of Fuel Thickness on Opposed Flow Flame Spread over PMMA," San Diego State University, San Diego, 2015.

[4] I. Wichman, "Theory of Opposed-Flow Flame Spread," *Prog. Energy Combust. Sci.,* pp. 646-651 , 1992.

[5] H. Emmons, "The film combustion of liquid fuel," *Z. A. Math. Mech.,* vol. 36, pp. 60-71, 1956.

[6] D. B. Spalding, "Combustion of liquid fuel in gas stream," *Fuel,* vol. 29, pp. 2-7, 1950.

[7] A. Rangwala, "Flame spread analysis using a variable B-number," *Fire Safety Science,* vol. 9, pp. 243-254, 2009.

[8] A. V. Singh and M. J. Gollner, "Estimation of local mass burning rates for steady laminar boundary layer diffusion flames," *Proceedings of the Combustion Institute,* vol. 35, no. 3, pp. 2527-2534, 2015.

[9] M. Sibulkin and C. K. Lee, "Flame Propagation Measurements and Energy Feedback Analysis for Burning Cylinders," *Combustion Science and Technology,* vol. 9, pp. 137-147, 1974.

[10] W. Sirignano, "A critical discussion of theories of flame spread across solid and liquid fuels," *Combust. Sci. and Tech,* p. 95, 1972.

[11] W. Sirignano, "Theory of Flame Spread over Solids," *Acta Astronautica,* pp. 1285-1299, 1974.

[12] R. F. McAlevy III and R. S. Magee, "The mechanism of flame spreading over the surface of igniting condensed-phase materials," *Proc. Combust. Inst,* pp. 215-227, 1969.

[13] M. B. Ayani, J. A. Esfahani and R. Mehrabian, "Downward flame spread over PMMA sheets in quiescent air: Experimental and theoretical studies," *Fire Safety Journal,* vol. 41, pp. 164-169, 2006.

[14] S. Bhattacharjee, L. Carmignani, G. Celniker and B. Rhoades, "Measurement of Instantaneous Flame Spread Rate Over Solid Fuels Using Image Analysis," in *IAFSS* , Lund, Sweden, 2017.

[15] M. D. Abramoff, P. J. Magalhaes and S. J. Ram, "Image Processing with ImageJ," *Biophotonics International,* vol. 11, no. 7, pp. 36-42, 2004.

[16] S. Patankar, "Numerical Heat Transfer and Fluid Flow," 1980.

[17] S. Bhattacharjee, M. King and C. and Paolini, "Structure of Downward Spreading Flames: A Comparison of Numerical Simulation, Experimental Results, and a Simplified Parabolic Theory," *Combustion Theory Modeling,* pp. 1-17, 2003.

[18] S. Bhattacharjee, S. Takahashi, K. Wakai and C. P. Paolini, "Correlating Flame Geometry in Opposed Flow Flame Spread over Thin Fuels," *Proceedings of the Combustion Institute,* vol. 33, 2010.

Western States Section of the Combustion Institute – Fall 2017 Meeting
Hosted by the University of Wyoming
October 2-3, 2017

Two-Dimensional Heat Transfer Analysis Within Non-Thermally Thin Poly(Methyl Methacrylate) That is Burned in a Narrow Channel Apparatus

Nicholas A. Lage *Fletcher J. Miller*

Combustion and Solar Energy Laboratory
Department of Mechanical Engineering, San Diego State University, San Diego, CA 92182
Fletcher.miller@sdsu.edu

The San Diego State University (SDSU) Narrow Channel Apparatus (NCA) has been proven through comparison with the previous flame spread studies in the Microgravity Science Glovebox (MSG) experiments conducted on the International Space Station (ISS) to emulate microgravity flow and atmospheric conditions. The NCA approximates microgravity by flowing mixtures of nitrogen and oxygen through a narrow gap in order to suppress buoyancy above the burning sample. The purpose for studying flame behavior in microgravity is to improve fire safety protocol for spacecraft as well as creating alternative measures to study microgravity flames that are much less expensive compared to flame spread experiments conducted on spacecraft.

Using Fire Dynamics Simulator (FDS), version 6.2.0, coupled with Gpyro, a two-dimensional model was developed for non-thermally thin samples of poly(methyl methacrylate) (PMMA) that are burned in the NCA. A fuel length of 30 mm is used as well as a 5 mm gap height and the assumption that the flow at the inlet is laminar and parabolic, which has been verified previously with hot wire anemometer measurements in the experiment. Since the flow is laminar based on Reynold's number as well, direct numerical simulation (DNS) is used to solve the momentum, energy, and species equations for the flow. Gpyro is used to calculate the 2-D temperature field and pyrolysis in the burning solid. A complex pyrolysis model is used to simulate the decomposition of PMMA into methyl methacrylate (MMA) vapor, which consists of a first order, single step Arrhenius reaction rate equation. A second order, single step Arrhenius reaction rate is used to model the combustion reaction of MMA vapor with oxygen in the gas phase. Complete combustion is assumed.

Simulations with varying fuel thicknesses were run to analyze the effect of thickness on flame spread. A comparison between one-dimensional and two-dimensional heat conduction within the solid is also made to show the effect the heat transfer parallel to the direction of flame propagation has on flame spread rate.

Results showed that there are variations between the 2D conduction and the 1D conduction results. The flames for the 2D conduction cases showed higher spread rates compared to the 1D conduction cases. There are also variations in the temperature profiles within the sample. In other words, the isotherms for the 1D conduction cases follow a "V" shaped pattern, while the 2D conduction cases had a more gradual curve. The results show the importance of modeling 2-D conduction. A second result is that the thicker fuel cases spread faster than thinner fuel cases, and work needs to be done to explain or correct this, since it goes against previous experiments and numerical models.

1. Introduction

It is important to implement stringent fire safety protocol to prevent incidents such as the Apollo 1 accident from occurring. To promote improved safety measures, it is imperative to gain a better understanding of flame behavior in microgravity-type environments. However, conducting experiments in true microgravity environments, such as on the International Space Station (ISS), is costly, so the National Aeronautics and Space Administration (NASA) is limited on the number of tests they can do. The Burning and Suppression of Solids II (BASS II) tests were a set of experiments that were conducted on the ISS [1]. NASA is currently funding the Narrow Channel Apparatus experiments, which is the least expensive method for studying microgravity combustion (albeit simulated), to justify the method of testing as a valid ground-based alternative as well as the development of numerical models that simulate the NCA experiments. Through previous studies, it has been proven with thermally thin fuels that the NCA can mimic microgravity-type flames on the ground. In this paper we concern ourselves with non-thermally thin fuels.

The modeling this paper describes parallels the simulated microgravity experiments that have been conducted using the San Diego State University Narrow Channel Apparatus (SDSU NCA). The NCA experiments are loosely based on the NASA flammability test for nonmetallic solids, NASA-STD-6001 Test 1 [2]. For this standard of testing, the sample is oriented in a vertical position and is ignited at the bottom, causing upward spreading flames. It is heavily influenced by buoyancy, and is therefore not sufficient for testing the flammability of materials for use in a microgravity environment. By contrast, the NCA experiments are oriented horizontally, as shown in Figure 1. The figure also illustrates single-sided burning of thermally thick poly(methyl methacrylate) (PMMA), with a small gap height, typically 5 mm, that suppresses vertical flow developing due to buoyancy effects.

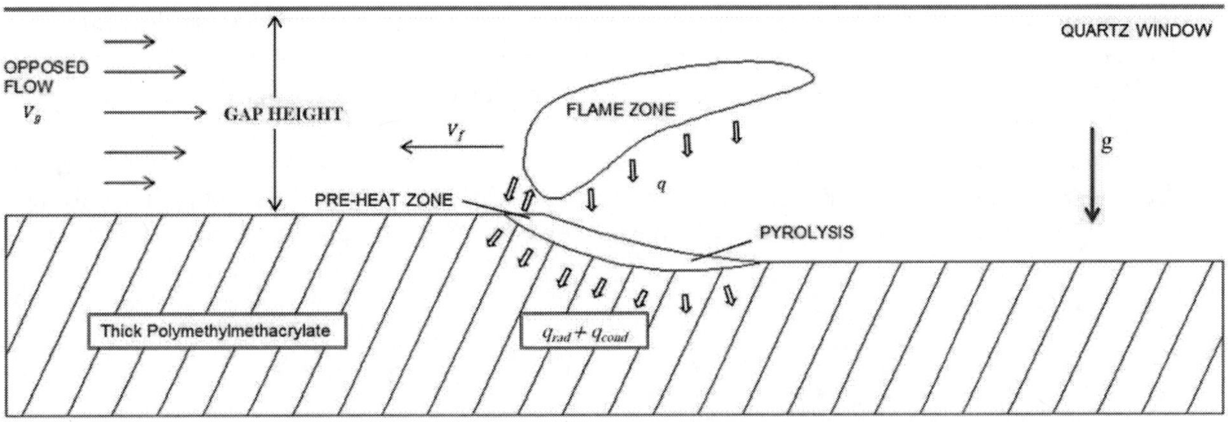

Figure 1: Schematic of thermally thick PMMA burning in the NCA

This paper examines the two-dimensional heat conduction that occurs within thermally intermediate PMMA and compares it to cases with one-dimensional heat conduction. The motivation behind this comparison is that FDS alone can only account for one-dimensional heat conduction normal to the fuel surface, whereas in reality there is conduction parallel to the fuel surface as well. The only way to account for the latter is to couple FDS with Gpyro, which was the approach used here. Our previous work included using only FDS [3], and here we extend that to show the effect of 2-D conduction.

2. Model Setup

A computational model was created using Fire Dynamics Simulator (FDS) 6.2.0 coupled with Gpyro to model the two-dimensional conduction that occurs within the fuel sample. FDS, developed by the National Institute of Standards and Technology (NIST), is a computational fluid dynamics (CFD) software that computes thermally-driven flows [4]. SmokeView, also developed by NIST, is a visualization software that complements FDS, and is used to represent data in contour profiles. Gpyro, developed by Chris Lautenberger, is a software that models condensed-phase heat transfer as well as thermal decomposition due to condensed-phase reactions [5]. The purpose for using the coupled software is to simulate two-dimensional heat conduction within the burning of thermally intermediate samples of PMMA as well as the pyrolysis of PMMA.

A two-dimensional model was created with a sample length of ~3 cm, and a gap height of ~5 mm as well as a channel width of 50.8 mm. A schematic of the model is shown in Figure 2, which represents the FDS domain. The Gpyro domain is superimposed to the PMMA obstruction in the FDS domain, and has half-grid cells along the boundaries that must be accounted for. The uniform control volume size is set to be 0.130 mm in the x-direction and 0.150 mm in the z-direction, and is consistent for both the FDS and Gpyro domain. The domain length, in terms of FDS, is set to be ~4 cm (sufficient to get steady flame spread, but small enough to minimize computational time), and the domain height, h, corresponds to the fuel thickness plus the gap height. Gravity is set in the negative z-direction. The inlet and the ambient temperatures are specified to be 20 °C.

The igniter is a two grid cell by two grid cell heat source that is placed 0.45 mm above the sample and ~2 mm away from the edge of the sample, with a temperature that is ramped from ambient to 1800 °C in half of a second, and is left on through the entirety of the simulation. A laminar, fully-developed flow, with a mean inlet velocity of 15 cm/s is used consisting of an oxygen-nitrogen mixture, with an oxygen concentration of 21%. At the outlet, an open boundary with a dynamic pressure set to 0 is specified. A sheet that is 3.175 mm thick and has the specific heat and thermal conductivity of mica, is set beneath the fuel sample to serve as insulation, and a sheet that is 10 mm thick and has quartz properties is set as the top boundary, which serves as a heat loss and represents the quartz window on the real NCA. Towards the outlet, a block of stainless steel is used to represent the sample holder, and also serves as a heat loss as well as provides enough space for the flame to develop.

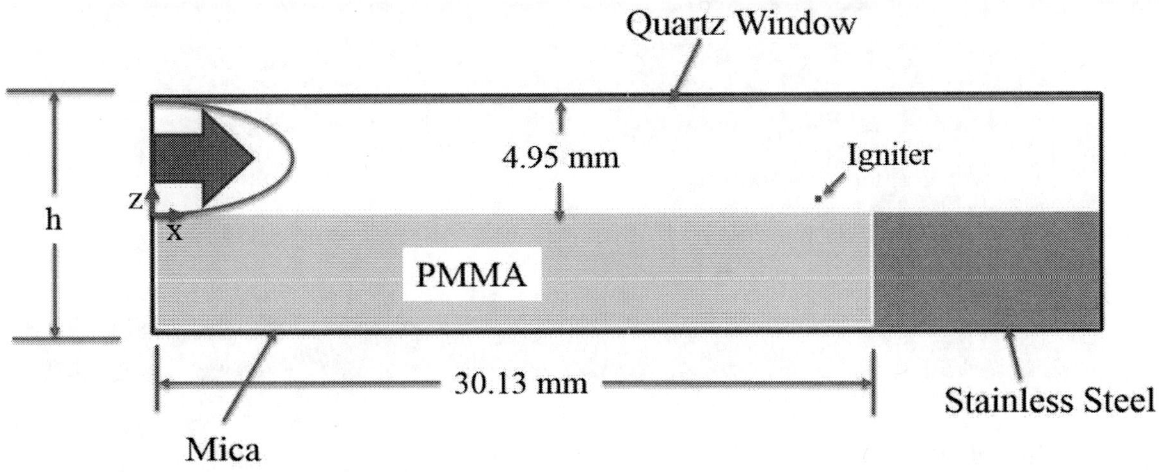

Figure 2: Schematic of the 2D model used to simulate the burning of thermally intermediate PMMA

The combustion was modeled using finite rate kinetics. Equation 1 is the global reaction between methyl methacrylate (MMA) fuel vapor and oxygen. Equation 2 is the solid-phase Arrhenius reaction equation that corresponds to the MMA fuel vapor that is generated. Equation 3 is the gas-phase Arrhenius reaction equation that equates to the amount of fuel that is consumed through combustion.

$$C_5O_2H_8 + 6O_2 \rightarrow 4H_2O + 5CO_2 \tag{1}$$

$$\omega_{di}''' = \bar{\rho}Y_iZexp\left(-\frac{E_s}{RT}\right) \tag{2}$$

$$\frac{dC_5O_2H_8}{dt} = -AC_{C_5O_2H_8}C_{O_2}exp\left(-\frac{E_g}{RT}\right) \tag{3}$$

The material and kinetic properties used are listed in Table 1. The properties were obtained from Garrett Bornand [3]. The only quantity that is different is the density, which was acquired by measuring the dimensions as well as the mass of a PMMA sample, and then using the measurements to calculate it. The original density, 1190 kg/m^3, is set as the solid density. The reason it is different is due to the fact that Gpyro has two different inputs for density, one of them being the solid density, which is the density assuming the solid is a nonporous material, and the other density is that which is based on a reference temperature. The reference temperature is 20 °C, which is equivalent to the ambient temperature. The solid density is 1190 kg/m^3, and the other is the density specified in Table 1.

Table 1: PMMA material and kinetic properties used for the simulation

Gas Phase Kinetic Paramters	Value
Pre Exponential Factor (cm^3/mol·s)	8.93E12
Activation Energy (kJ/kmol)	88.9E3
Heat of Combustion (kJ/kg)	25900
N_S (Reaction Order)	1.0, 1.0
PMMA Material and Pyrolysis Parameters	**Value**
Conductivity (W/m·K)	0.19
Emissivity	0.92
Specific Heat (kJ/kg·K)	1.4
Density (kg/m^3)	1181
Absorption Coefficient (1/m)	2700
Pre Exponential Factor (1/s)	2.82E9
Activation Energy (kJ/kmol)	129.8E3
Heat of Reaction (kJ/kg)	941

With Gpyro, thermal conductivities could differ for each direction. The thermal conductivity in the direction parallel to flame propagation was set to zero in some cases, with such cases defined as 1D conduction cases. The 2D conduction cases had isotropic thermal conductivities defined. Simulations varied by fuel thickness and by 1D and 2D conduction. Other cases were conducted with the igniter shutting off or disappearing from the domain at a specified time, a plane with an adiabatic bottom, etc., but for the sake of brevity, those cases are excluded from this paper and will be reported elsewhere.

For FDS, an open vent is specified for the top boundary of the quartz window to allow for natural convection. The boundary beneath the mica and the steel are set to be exposed to the ambient air. The sides of the inert obstructions that border the domain have a mirrored boundary. The thermal conductivities as well as the specific heats of both quartz and mica are tabulated in the appendix. The boundary conditions for all edges in the Gpyro domain are obtained from the calculated values of the gas-phase cells in FDS that are adjacent to the sample. Thus, it is required for the Gpyro domain to be completely embedded within the FDs domain as well as a one-grid-cell, gas-phase buffer between the PMMA and the inert solid surfaces that surround it within the FDS domain.

3. Results and Discussion

Simulations with one-dimensional and two-dimensional conduction, and with varying fuel thicknesses (1 mm, 3 mm, and 5 mm) are compared. Graphs of temperatures, flame spread rates, heat release rates, and equivalence ratios are presented.

Figure 3 shows the SmokeView profile of the heat release rate per unit volume (hrrpuv) of the 3 mm thick, 2D conduction case at 20 s after the igniter turns on. This quantity (hrrpuv) is a representation of the flame shape. The hrrpuv profiles are used to track the flame position using Spotlight-16. Spotlight is a software that conducts analysis on an image or a sequence of images by tracking an object within that image or image sequence [6].

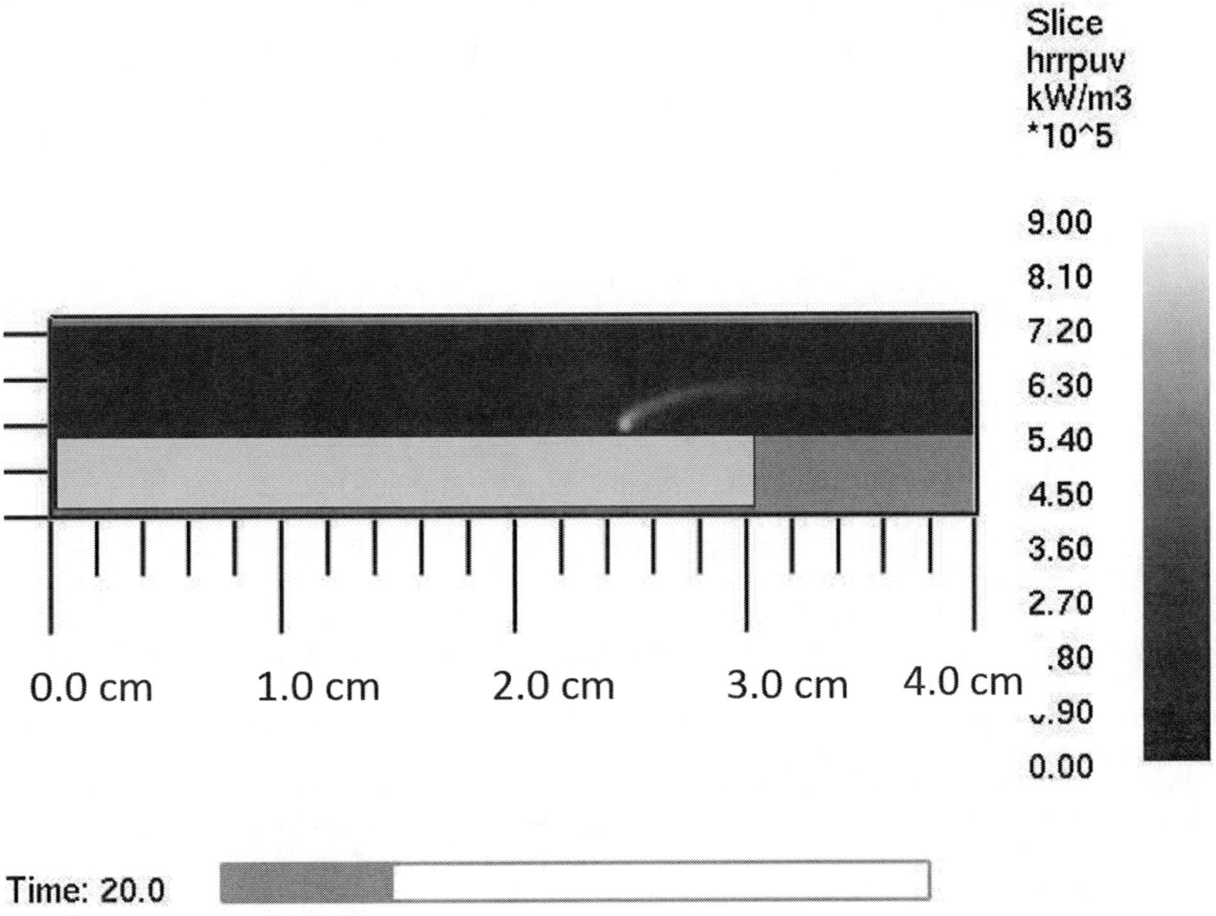

Figure 3: SmokeView profile of the heat release rate per unit volume at 20 seconds for the 3 mm, 2D conduction case

Figure 4 shows the bulk density at 20 seconds for the 3 mm thick, 2D conduction case, which shows that Gpyro does take fuel that is burned away into consideration in terms of quantities such as temperature, but not so in terms of domain or flow.

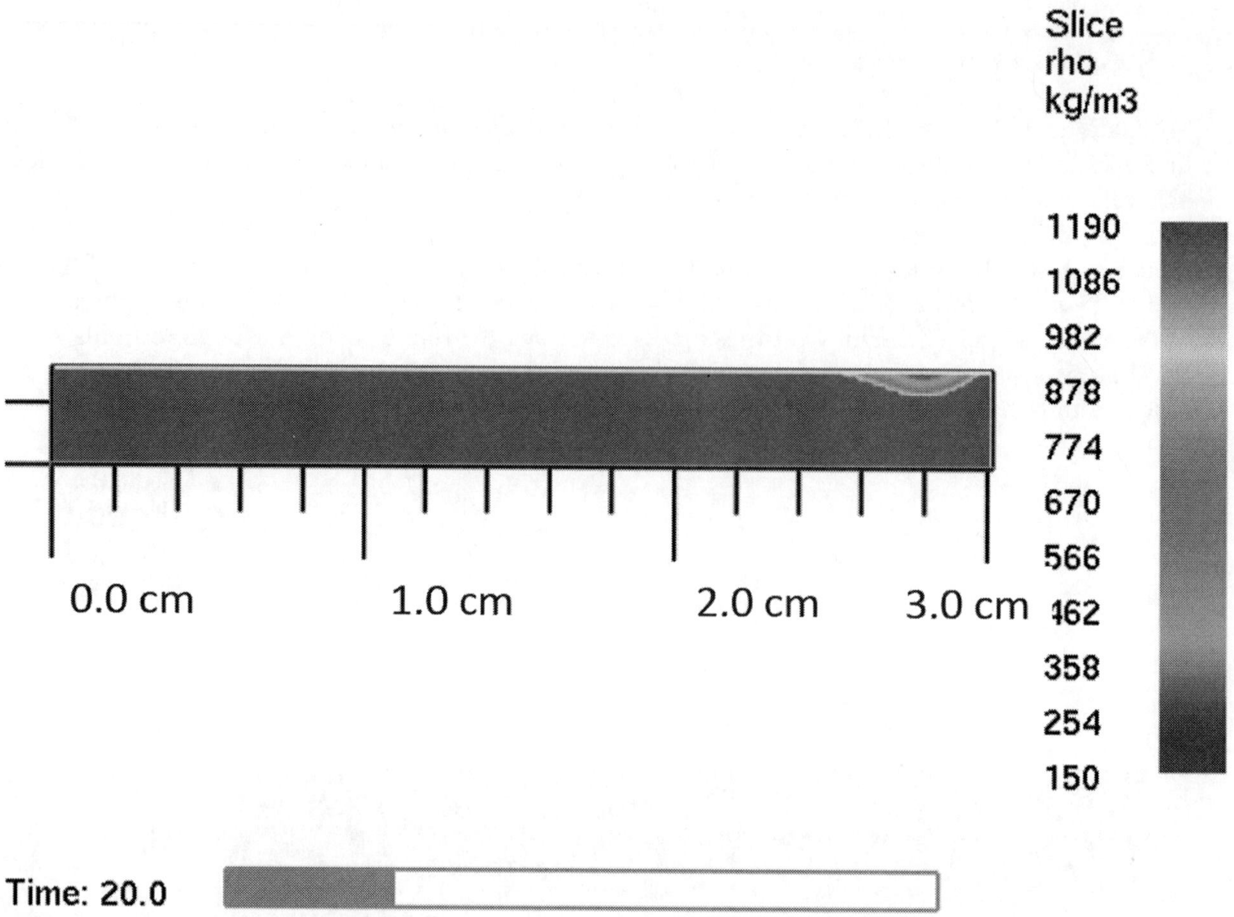

Figure 4: SmokeView profile of the Solid-phase bulk density

The temperature contour graphs for Figures 5 and 6 were plotted using Matlab. A script was created to loop a batch command that reads data from SmokeView profiles and writes it to CSV files for multiple points in time. Another script reads the data from the CSV files and plots it. Figures 5 and 6 consist of temperature contours, where the deepest point of the trough or the valley, is directly underneath the igniter. Figure 5 shows the temperature contours in terms of the flame front, which is determined by tracking the position of the flame, of the 1 mm thick samples at 9 seconds, which is when the 1D conduction case starts to diverge. Divergence occurs because the fuel directly beneath the igniter burns completely through. Neither of the 1 mm cases lasted long, so not enough data could be collected in order to make a good comparison between the other cases. Figure 6 shows the temperature contours with respect to the flame front for both 1D and 2D conduction and for 3 mm and 5 mm thick fuels at 20 seconds. The graphs show that surface regression is taken into consideration in terms of temperature, with the maximum temperature presumed to be the modeled surface and the deepest point in the troughs or valleys being directly underneath the igniter. However, surface regression is not accounted for in terms of domain. For each thickness, the 1D conduction cases tend to burn deeper underneath the igniter as compared to the 2D conduction cases, and the preheat length for the 2D conduction cases tend to be longer compared to the 1D conduction cases. The 1D conduction cases do show a flame front of ~0.5 mm, which means that there is some radiation from the flame that heats up the sample ahead of the flame since conduction is disabled.

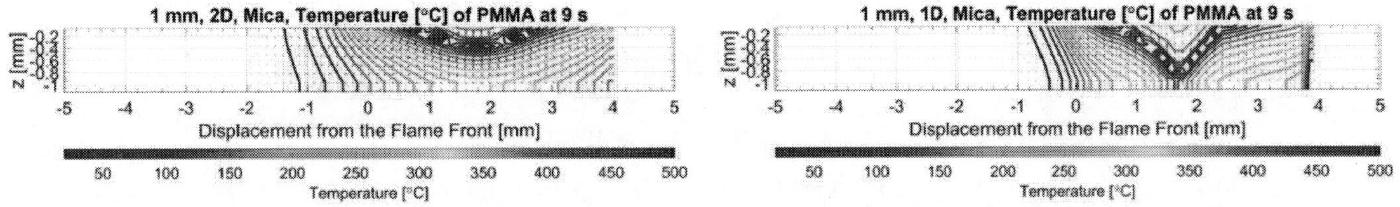

Figure 5: Solid-phase temperature contour plots of the 1 mm thick cases at 9 seconds with (a) 2D conduction set (left) and (b) 1D conduction set (right)

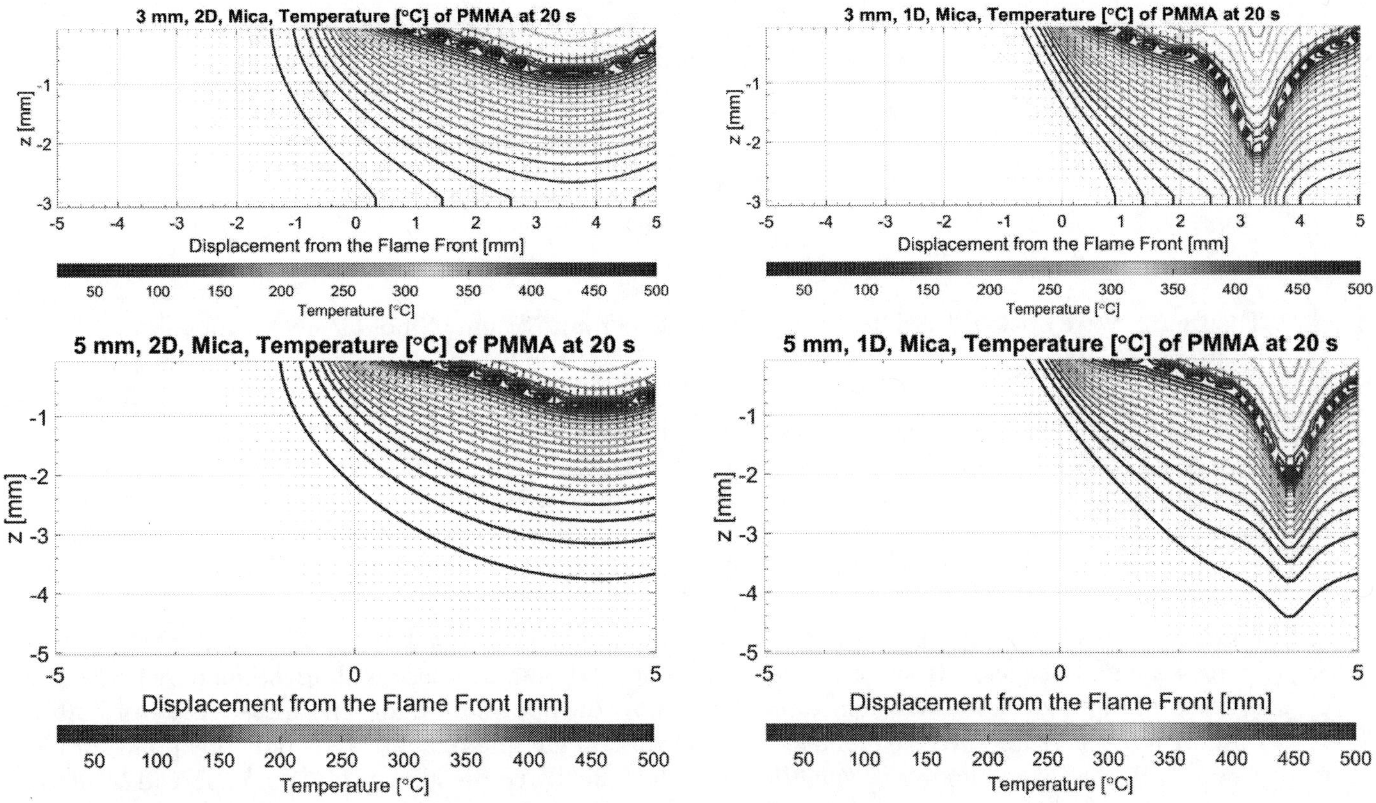

Figure 6: Solid-phase temperature contour plots at 20 seconds for (a) the 3 mm thick, 2D conduction set (top left), (b) 3 mm thick, 1D conduction set (top right), (c) 5 mm thick, 2D conduction set (bottom left) (d) 5 mm thick, 1D conduction set

Figure 7 illustrates the total heat release rate with respect to time. It shows that the 1 mm cases do not last long. After around 10 seconds, the 1 mm thick, 1D conduction case starts to diverge. For the 1 mm thick, 2D conduction case, the heat release rate plummets because of quenching through the mica. Like the 1 mm thick, 1D conduction case, the 1D conduction cases for the 3 mm thick and 5 mm thick cases ended in convergence errors as the flame burned all the way through the sample. It is also apparent that after 10 seconds, the 5 mm thick fuels produce higher heat release rates in relation to the 3 mm thick cases.

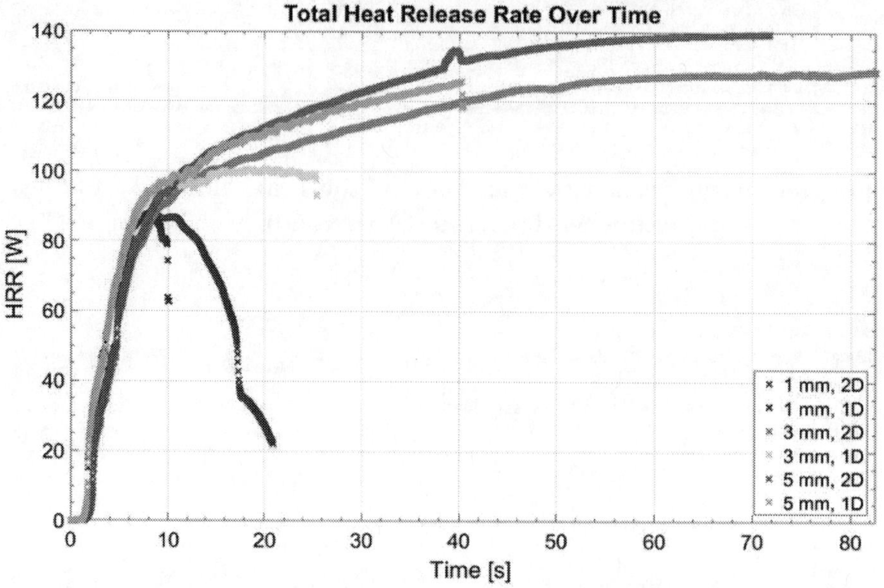

Figure 7: Heat release rate with respect to time

The flames were tracked from the heat release rate per unit volume SmokeView profiles using Spotlight-16. Each case is tracked from 5 seconds, when the flame is developed, and is tracked to the end of the simulation. Each case is tracked based on the scale of 0 to 9 x 10^5 kW/m^3, and with a threshold of 0.6862. The flame positions from tracking are used to define the leading edge for the majority of the figures presented in this paper. It is apparent from Figure 8 that the spread rates are virtually constant, with linear regression values of around 0.99 for the majority of the cases. The only exception was for the 1D conduction, 1 mm thick case, which has an R^2 value of about 0.93, which is likely due to the fact that it was only tracked for about 3 seconds.

Figure 9 shows the flame spread rates for each case. It is apparent that the flame spread rate increases with thickness. Based on experimentation, it should decrease with fuel thickness [7], and it is not yet fully understood why the comptations show the opposite trend. The spread rates for both of the 1 mm cases are virtually identical, so it appears in the graph that there is only one point for the 1 mm thick case, when there are actually two. Also, the 2D conduction cases tend to spread faster as compared to the 1D conduction cases. This is because the 2D conduction cases have a longer preheat length, hence more fuel is fed back into the flame, causing it to spread faster.

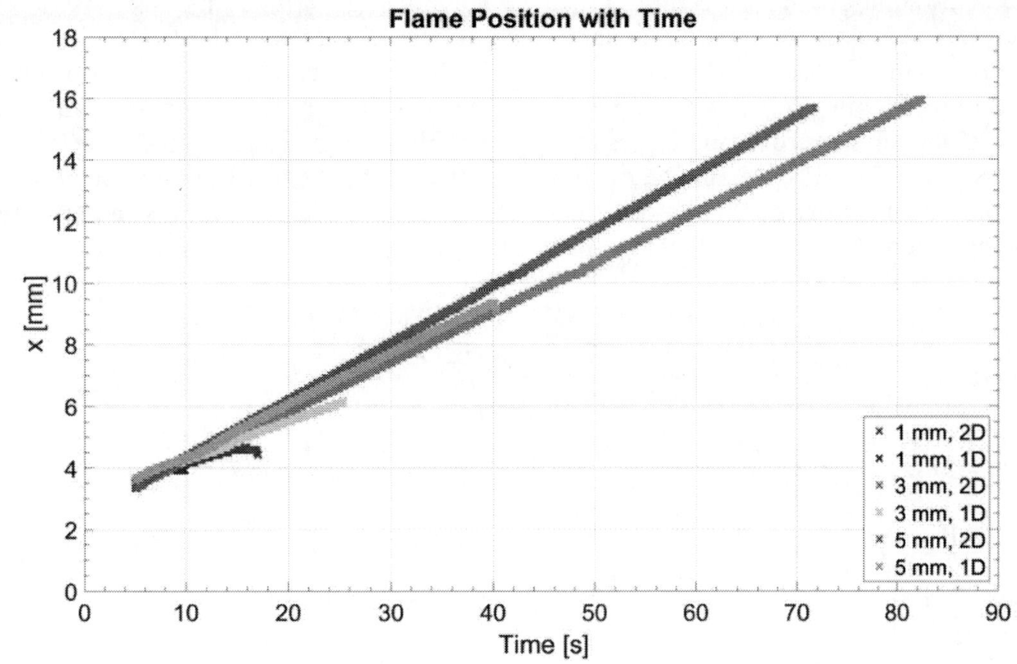

Figure 8: Flame position of each case with time

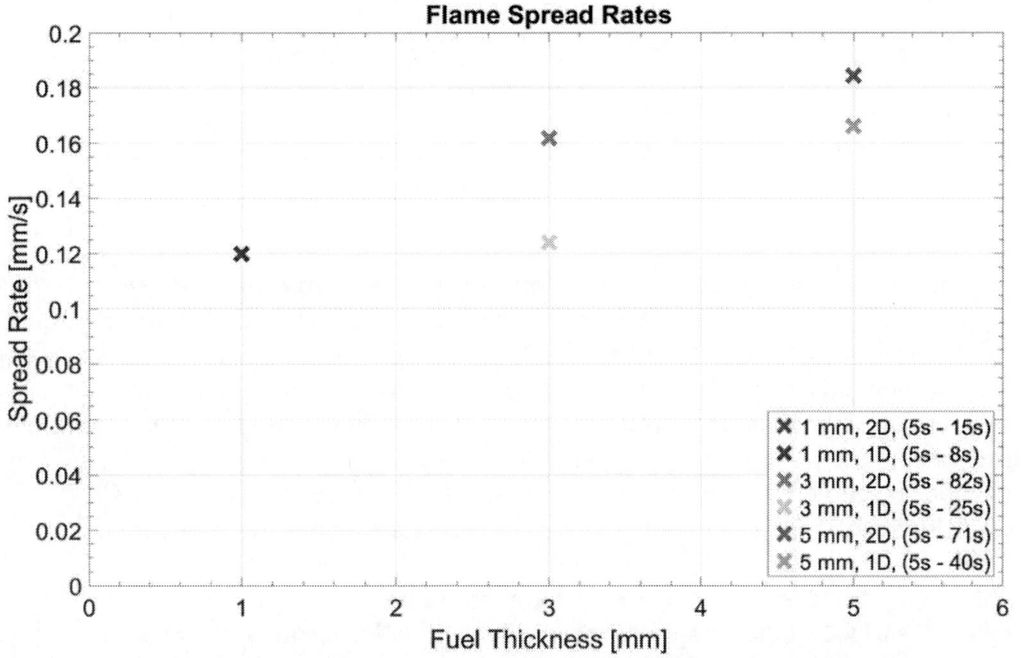

Figure 9: Flame spread rates with respect to fuel thickness

Figure 10 is the solid-phase temperature plot at the leading edge of the flame with respect to depth at 20 seconds. From looking at the plot, it can be seen that the 2D cases tend to have a more gradual curve relative to the 1D cases. This is because the 2D cases have conduction that propagates parallel to the direction of flame propagation, so less heat transferred depth-wise, proportionally. Furthermore, the top surface temperatures ($z = 0$) have high variation between cases. This is because the position of the flame front could be off by one or two grid cells (0.130 mm to 0.260 mm), and there is a high temperature gradient along the sample surface, which could cause a major shift in surface temperature.

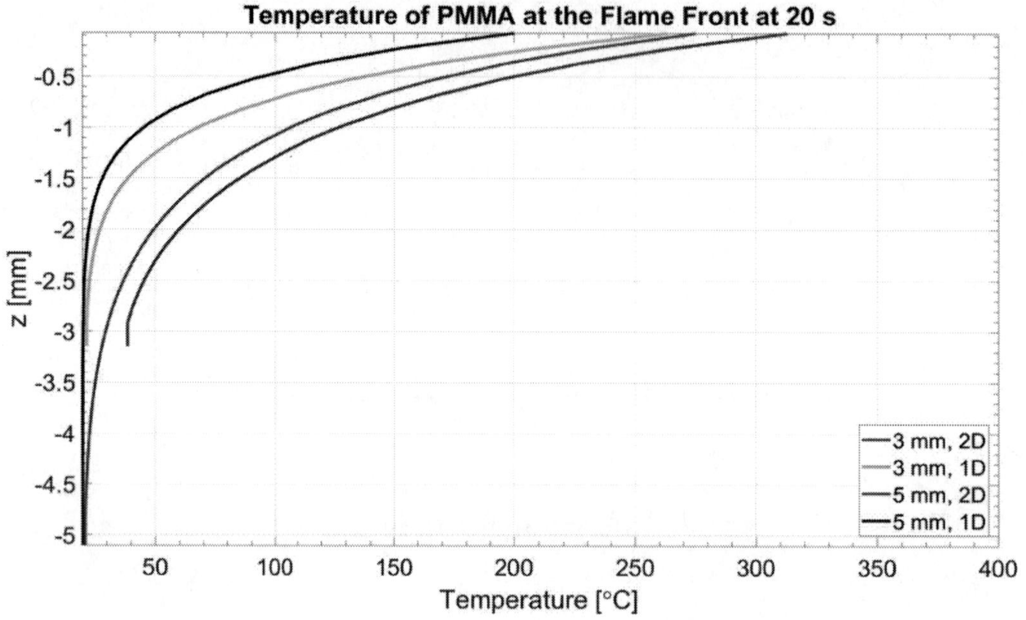

Figure 10: Solid-phase temperature at the flame front with respect to fuel depth at 20 seconds

Figure 11 shows the equivalence ratio at 0.600 mm above the sample and at 20 seconds. The reason for plotting the equivalence ratio at 0.600 mm above the sample is that that is around the height where the flame front was tracked based on first meeting the threshold. Figure 11 shows that at the flame front, the equivalence for all cases are at around 2, meaning that the fuel mixture is rich at that height. The plot also shows that the 5 mm thick cases tend to have a higher equivalence ratio in front of the flame (negative x-direction) as compared to the 3 mm thick cases, which means slightly more MMA fuel vapor is produced in front of the flame for the 5 mm thick cases compared to the 3 mm thick cases. Additionally, the Figure 12 is a plot that illustrates the net heat flux at the top surface of the PMMA. Unlike the equivalence ratio shown in Figure 11, Figure 12 shows that the net heat flux in front of the flame front for all cases are virtually the same. Thus, the flames emit the same amount of heat to the top surface of the sample for each case, which leaves the questions as to why the thicker cases produce slightly more fuel vapor ahead of the leading edge, and the 1-D cases produce more fuel vapor than the 2-D case, so far unanswered.

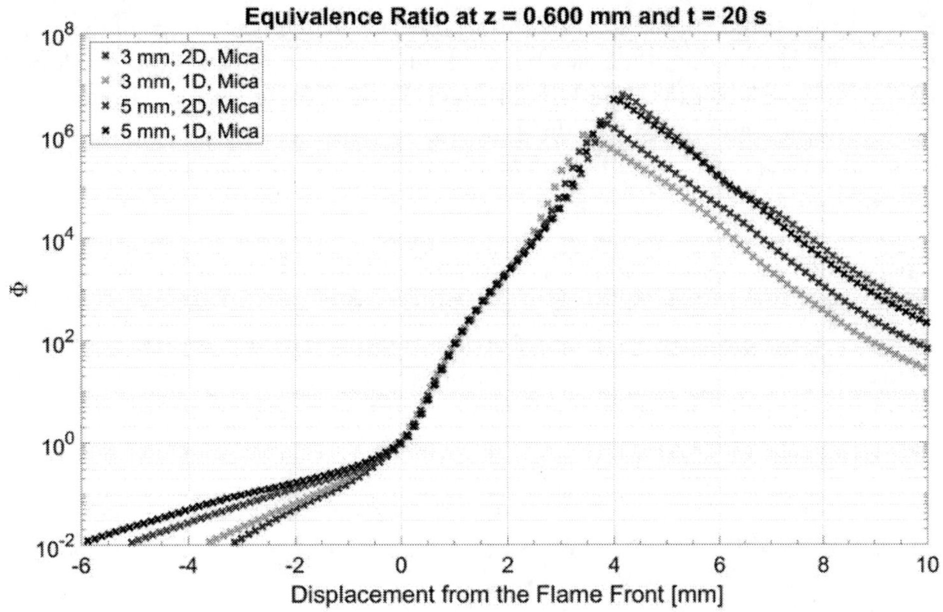

Figure 11: Equivalence ratios at 0.600 mm above the PMMA with respect to the displacement from the leading edge

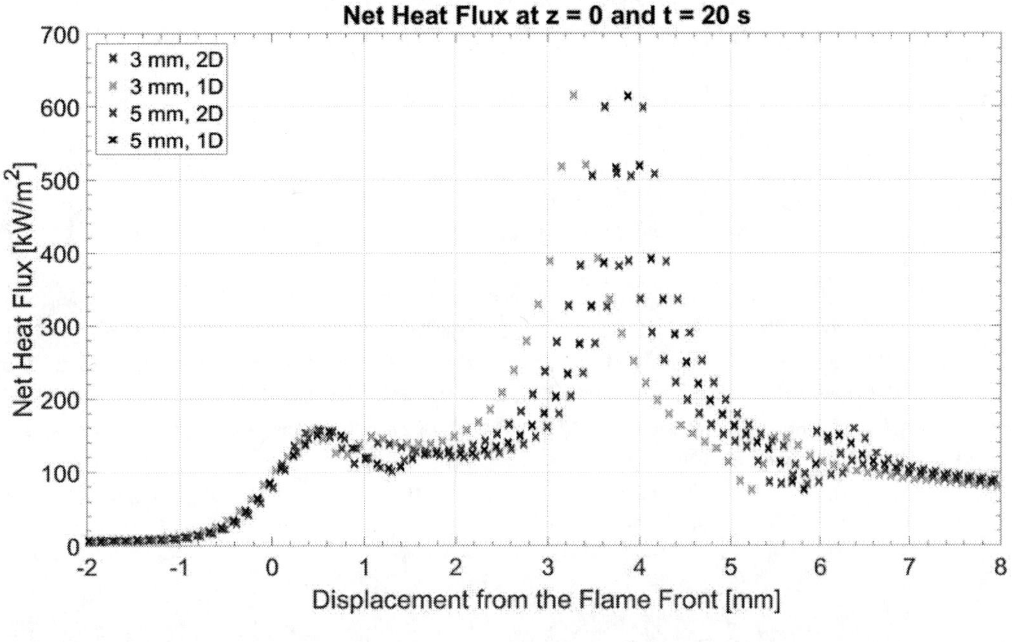

Figure 12: Net heat fluxes at the top surface of the PMMA sample at 20 seconds with respect to the leading edge. The large peak behind the flame is the igniter.

4. Conclusions

It can be concluded that there is a great difference in flame behavior between one-dimensional and two-dimensional conduction, depending on what is compared. The spread rates, at least before the 1-D conduction cases extinguish, are fairly similar, though the 2-D cases are higher. This is likely due to the fact that the 2D conduction cases have a longer preheat length, so more fuel is heated in front of the flame, hence more MMA fuel vapor is fed back into the flame. However, the burn patterns are very different, and indeed all the 1-D conduction cases extinguished, while the 2-D cases did not. This points to the need for modeling heat conduction in 2-D in thermally intermediate or thick fuels.

Inexplicably, it can also be seen that flame spread rate increases with fuel thickness, when it should decrease with thickness due to quenching to the bottom. The current consensus as to why the thicker fuels spread faster in the simulations is likely due to the condensed-phase kinetics. A parametric study for the kinetic parameters must be conducted in order to solve the spread rate discrepancies.

Acknowledgements

This research was funded by NASA under the grant NNX10AD96A. Thanks are due to Dr. Indrek Wichman and Dr. Sandra Olson for their support as well as Dr. Chris Lautenberger for creating Gpyro, and updating/debugging it in his free time.

References

[1] "Space Flight Systems: BASS," NASA, [Online]. Available: https://spaceflightsystems.grc.nasa.gov/sopo/ihho/psrp/msg/bass/. [Accessed 12 July 2017].

[2] NASA, "Flammability, Offgassing, and Compatibility Requirements and Test Procedures," NASA, Washington, DC, 2011.

[3] G. Bornand, "Polymethylmethacrylate Combustion in a Narrow Channel Apparatus Simulating a Microgravity Environment," Master's thesis, San Diego State University, 2014.

[4] K. McGrattan, R. McDermott, C. Weinschenk, K. Overholt, S. Hostikka and J. Floyd, *Fire Dynamics Simulator User's Guide,* 6th ed., Baltimore, Maryland: NIST Special Publication 1019, 2015.

[5] C. Lautenberger, *Gpyro - A Generalized Pyrolysis Model for Combustible Solids Users Guide,* Berkeley: Reax Engineering Inc., 2016.

[6] R. Klimek and T. Wright, "Spotlight: Image Analysis and Object Tracking Software," NASA, 22 July 2008. [Online]. Available: https://spaceflightsystems.grc.nasa.gov/spotlight/. [Accessed 31 August 2017].

[7] F. A. Lastrina, R. S. Magee and R. F. McAlevy, "Flame Spread over Fuel Beds: Solid-Phase Considerations," *Symposium (International) on Combustion,* vol. 13, no. 1, pp. 935-948, 1971.

[8] Y. Touloukian, Thermophysical Properties of Matter: the TPRC Data Series; a Comprehensive Compilation of Data, vol. 2, New York: IFI/Plenum, 1970, p. 826.

[9] Y. Touloukian, Thermophysical Properties of Matter: the TPRC Data Series; a Comprehensive Compilation of Data, vol. 5, New York: IFI/Plenum, 1970, p. 1545.

Appendix

Table 2: Thermal properties of mica used in the model [8] [9]

Mica Properties			
T [°C]	k [W/m-K]	T [°C]	c_p [kJ/kg-K]
25	0.795	25	0.8067
100	0.46	27	0.8109
300	0.23	127	0.9686
500	0.23	227	1.0565
		327	1.1167
		427	1.1644
		527	1.2046
		627	1.2414
		727	1.2753

Table 3: Thermal properties of quartz used in the model [3]

Quartz Properties			
T [°C]	k [W/m-K]	T [°C]	c_p [kJ/kg-K]
25	0.795	25	0.8067
100	0.46	27	0.8109
300	0.23	127	0.9686
500	0.23	227	1.0565
		327	1.1167
		427	1.1644
		527	1.2046
		627	1.2414
		727	1.2753

Western States Section of the Combustion Institute – Fall 2017 Meeting
Hosted by the University of Wyoming
October 2-3, 2017

Temperature Measurement of Glowing Firebrands with Multi-Color Pyrometry

James L Urban[1], Dorsa Shirazi[2], Michela Vicariotto[2], Daniel Murphy[1], Derek Dunn-Rankin[2] and Carlos Fernandez-Pello[1]

[1]Department of Mechanical Engineering, University of California, Berkeley, CA, 94720
[2]Department of Mechanical & Aerospace Engineering, University of California, Irvine, CA, 92697

Abstract

Surface temperature measurements were made on wood glowing embers using Multi-Color Pyrometry (MCP) in a laboratory environment. MCP allows for temperature measurement of an incandescing grey or black body by measuring the intensities of two or more wavelengths of light. Specifically, with MCP the surface temperature of a black/grey body can be directly related to the ratio of the intensities of emitted light at several wavelengths of light. The light can be observed with a DSLR camera. The results show that it is possible to measure the temperature of glowing embers using a digital camera MCP with an accuracy comparable to that of a commercial IR camera. The temperatures of glowing combustion zones on the embers were found to be between 600 and 1100C depending on the intensity of the smolder (glowing) reaction of the ember's surface. The results presented here indicate that it is possible that this technique could remotely measure the temperatures of moving glowing embers. This capability coupled with particle tracking velocity algorithms (PTV) would allow for real-time measurement of ember locations and temperatures. This information could be used for assessment of ember exposure of communities during Wildland-Urban Interface (WUI) fires.

1. Introduction

Each year as Wildland Urban Interface (WUI) fires wreak havoc on nearby communities, the intensity and the costs associated with fighting these fires continue to rise. Recently the number of structures lost each year has increased significantly due to the rapid expansion of communities into wildland areas. WUI fires have been persistent throughout history, but are not chronicled and reported to the same extent of urban fires. One example noted by Caton et al. [1] is the Peshtigo Fire which occurred in the same week as the Great Chicago Fire of 1871. The Peshtigo Fire killed roughly 2000 people while the Great Chicago Fire killed only 300. Further analysis shows that the Peshtigo Fire burned almost 1000 times more land than the Great Chicago Fire [1]. Other wildland and WUI fires that occurred afterward led to substantial work being done by the United States Forest Service (USFS) to control and suppress these fires. Despite these efforts, the frequency and severity of wildland and WUI fires continue to rise [2]. There have been many notable WUI fires in the past quarter century: the 1991 Oakland Hills Fire [3], the 2012 Waldo Canyon Fire, the 2003, 2007, and 2014 San Diego Firestorms [1], the Texas Bastrop county fires [4] and the Valley Fire [5]. There have also been many significant WUI fires which occurred outside of the United States [1] in the Mediterranean Area [6], South America [7], Australia [8] and New Zealand [9]. There does not appear to be a consensus on the fraction of structure ignitions caused by firebrands [1] as it is difficult to determine the ignition source after the fire. However, recent studies attribute the majority of structure ignitions to ember spotting [2,10].

WUI fires can spread through fire radiation and convection, or ember spotting. Ember spotting is the process by which burning pieces of biomass or other flammable materials are moved significant distances downstream by the wind and/or gas dynamics induced by the fire. Recent studies have shown ember spotting to be the primary cause of structure loss in WUI fires [2,10]. Typically, there are three aspects of embers spotting: the production of the embers, their transport, and finally the spot fire ignition process. To date, the transport of the firebrands is the best-understood aspect of the process [11,12]. However, topics such as ember exposure (number, mass, or energy fluxes of embers), and firebrand ignition capability including the structure ignition vulnerabilities have only recently been studied [13].

The ignition of structures can be the result of a single firebrand igniting a part of the structure with a lower fire resistance or by showers of firebrands which can accumulate and ignite more resilient parts of the structure (e.g. siding, mulch, roofs, decks, etc.) [10,14]. Both the National Institute of Standards and Technology (NIST) and Insurance Institute for Business & Home Safety (IBHS) have investigated these vulnerabilities [15–17] with the aid of NIST's Dragon [13]. The Dragon simulates a shower of embers which can be used in laboratory ember ignition experiments [14]. While the ignition of structures by firebrands is important, the firebrands can also ignite vegetation. Ignition of other vegetation can accelerate the spread of a fire beyond what is capable through convective and radiative spread [18] and can even allow the fire to spread to structures that are not yet in contact with the fire front [10].

To help prevent firebrands in WUI fires from igniting structures, an ember exposure scale such as the one proposed by NIST [2] could be used by developers of performance-based building codes of building external surfaces which would provide guidelines for defending structures against firebrand attack. A crucial step towards developing an ember exposure scale is the quantification of relevant properties of ember exposure such as the ember flux (mass or number of embers per unit area) as well as the ember temperatures, burning rates and Heat Release Rates (HRR). This would require, among others, the possibility of real time remote measurement of the temperatures of moving burning or glowing embers, while tracking their location and velocity.

The goal of this work is use Multi-Color Pyrometry (MCP) to measure the surface temperature of glowing firebrands. This method combined with firebrand tracking using techniques such as Particle Tracking Velocimetry (PTV) could be used to quantify the number of glowing firebrands produced by a given fire and show the direction that the firebrands are heading. This would allow for measurement of ember exposure in terms of number of glowing embers, their energy, and their HRR.

Traditional thermography (thermal imaging), which attempts to deduce temperature by measuring the intensity of infrared light emitted by objects, is of limited value in this application because of both fundamental and technical limitations. First, thermography requires knowledge of the emissivity of the material under study to accurately invert the Stephan-Boltzmann equation to recover the temperature. Second, thermal imaging sensors are currently limited in both their resolution and their response time, making them ill-suited to the observation of small, quickly moving firebrands. MCP addresses these concerns admirably, in that it is insensitive to material emissivity, particle size and speed. Further, it can be performed with conventional visible-light cameras, which are small, provide exceptional spatial resolution, and are readily available [19]. MCP using digital cameras has been performed effectively on hot objects in flames such as silicon carbide wires (Thin Filament Pyrometry) [20–22].

2. Multi-Color Pyrometry (MCP)

MCP allows for temperature measurement of a glowing grey or black body by measuring the intensities of two or more wavelengths of light. Specifically, with MCP the surface temperature of a black/grey body can be directly related to the ratio of the intensities of emitted light at two wavelengths of light. Performing MCP with light of discrete, discernable wavelengths requires high spectral resolution and sensitivity which is not easily achieved with a digital camera. Instead a similar method can be used measuring ratios of the combined spectral light intensities that are sensed by the r, g, b pixels on the camera's CMOS detector [23].

Care must be taken to turn off all image enhancements options on the camera and to save the images as RAW image (or NEF for Nikon cameras). RAW images formats do not do color correcting post processes done by other formats. For the experiments shown here a Nikon D90 camera was used, with an aperture setting of f/0 and ISO set to 200. Then the raw image can be converted to a linear 16 bit depth TIFF file with a suitable software such as *DCraw*. Converting to a linear image is important because other methods will post-process the image, making it more like what is seen by the naked eye, but not suitable for MCP. The different intensity values at each pixel from the TIFF file can be used to determine the temperature for that pixel.

This can be done by characterizing the camera's spectral response for each color channel, $\alpha_i(\lambda)$ and the transmissivity of any optics in between the camera and the ember, $\tau(\lambda)$. Then assuming the Planck distribution, one can calculate the expected color intensity ratio that the camera should observe for a surface of a given temperature.

$$I_{i,predicted} = \int \epsilon(\lambda) I_{Plank}(\lambda, T_{surface}) \tau(\lambda) \alpha_i(\lambda) d\lambda \quad for\ i \in \{r, g, b\}$$

Then if we take the ratios of these intensities assuming $\epsilon(\lambda)$ is constant for wavelengths considered (grey/blackbody assumption).

$$\xi_{i,j} = \frac{I_{i,predicted}}{I_{j,predicted}} = \frac{\int I_{Plank}(\lambda, T_{surface}) \tau(\lambda) \alpha_i(\lambda) d\lambda}{\int I_{Plank}(\lambda, T_{surface}) \tau(\lambda) \alpha_j(\lambda) d\lambda} \quad for\ i, j \in \{r, g, b\}\ and\ i \neq j$$

This can be integrated for each color ratio, which will produce an intensity ratio, $\xi_{i,j}$, for a given surface temperature, $T_{surface}$. For values used here, the data was collected for wavelengths between 400 – 720nm (the visible spectrum) with resolution of 1nm and that was the range of integration. The results of this can be turned into a look-up table allowing for the temperature to be predicted based on an observed intensity ratio. A key advantage of this method is that ratios, rather than absolute intensities of color channels, are used. Thus, it is insensitive to the emissivity of the target and other factors.

3. Experimental Method and Results

The objective of this work is to develop and implement MCP to remotely measure the surface temperature of firebrands. The experimental system consists of a stand where the glowing ember is held (shown in Figure 1), calibrated Nikon D90 camera and the software required to post process the images of the camera into a temperature map of the ember as described above.

Representative results from the ember tests are shown in Fig. 3. The sub-figures in the figure are a photograph of the ember as it burns (3a), a temperature image using MCP with a colorbar for the temperature (3b) and a histogram of the measured temperatures (3c). The results show that the temperature of the glowing patches on the ember's surface vary between 600 and 1100°C, which are reasonable values for glowing combustion in an airflow. Comparison of the images 3a and 3b shows that the MCP method can provide a reliable localized distribution of temperatures over the surface of the ember.

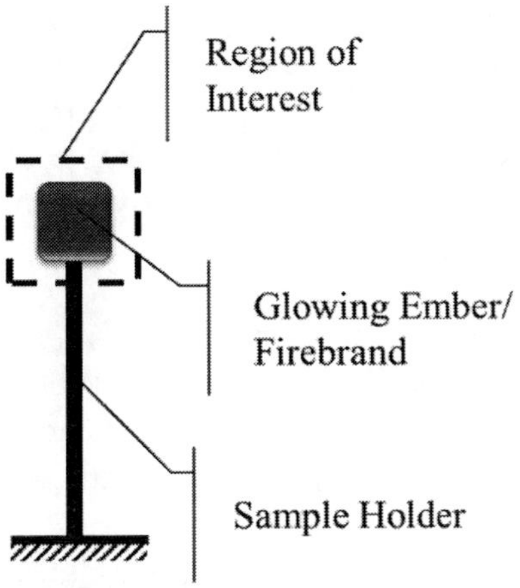

Figure 1: Experimental set up for MCP ember/firebrand experiments

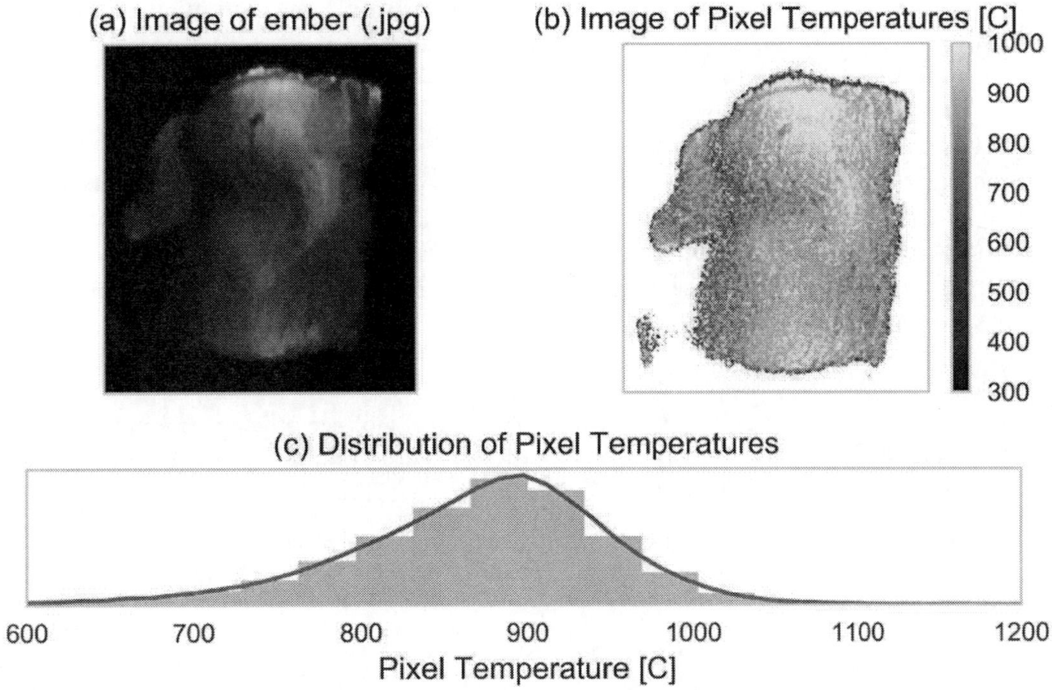

Figure 3: a) Jpeg image of the ember (how ember appears to the naked eye); b), pixel temperature image using MCP, and c) distribution of pixel temperatures (histogram with estimated probability density function).

Independent verification of the temperatures measured by MCP was attempted, but largely not successful with the methods available to the authors. Measurements with a 0.005" (0.127mm) diameter type K thermocouple reported temperatures up to 577°C, but were unreliable because the area around the thermocouple visibly cooled (specifically ceased to glow) when touched by the thermocouple bead. Ember temperatures were also imaged with an FLIR E60 IR (Infra-Red) camera, which measured the temperature of the ember to be over 670°C, the maximum temperature the IR camera could measure. These temperatures were more in accordance with the MCP measurements.

Temperature measurement of the ember surface temperature in the non-glowing patches was not considered with the current hardware configuration. The temperatures of these regions do not produce enough light for the camera to detect and provide a reliable intensity ratio from which to calculate a temperature. Using different filters and more sensitive camera detectors may make temperature measurement of these portions possible.

4. Conclusions

Surface temperatures of glowing combustion on an ember (firebrand) were performed with MCP with a digital camera. It was shown that the MCP method can provide a temperature distribution along the ember surface that agrees at least qualitatively with normal photography of the ember. The measured temperatures were between 600 and 1100°C and vary based on the local intensity of char oxidation reactions and heat losses. The promising results indicate that it is possible that this technique could remotely measure the temperatures of moving glowing embers. This capability coupled with particle tracking velocity algorithms would allow for real-time measurement of ember locations and temperatures. This information could be used for assessment of ember exposure of communities during Wildland-Urban Interface (WUI) fires.

Acknowledgements

This research was made possible with funds from the Almy C. Maynard and Agnes Offield Maynard Endowed Chair fund awarded to the senior author (CFP) and National Defense Science and Engineering Graduate (NDSEG) Fellowship for the first author (JLU). UROP (Undergraduate Research Opportunities Program) for funding, FERCo for the use of their blackbody source and Dr. Yu-Chien Chien for research support.

References

[1] S.E. Caton, R.S.P. Hakes, D.J. Gorham, A. Zhou, M.J. Gollner, Review of Pathways for Building Fire Spread in the Wildland Urban Interface Part I: Exposure Conditions, Fire Technol. (2016) 1–45. doi:10.1007/s10694-016-0589-z.

[2] A. Maranghides, W. Mell, Framework for Addressing the National Wildland Urban Interface Fire Problem--Determining Fire and Ember Exposure Zones Using a WUI Hazard Scale, NIST Tech. Note 1748 Framew. (2013) 1–33. doi:http://dx.doi.org/10.6028/NIST.TN.1748.

[3] P. Pagni, Causes of the 20 October 1991 Oakland Hills Conflagration, Fire Saf. J. 21 (1993) 331–339.

[4] S.G. Badger, 2011 Large Loss Fires in the United States, NFPA J. (2012).

[5] Valley Incident Damage Inspection Report CALNU008670, September 12, 2015, 2015.

[6] C. Lampin-Maillet, A. Mantzavelas, L. Galiana, M. Jappiot, M. Long, G. Herrero, O. Karlsson, A. Iossifina, L. Thalia, P. Thanassis, Wildland urban interfaces, fire behaviour and vulnerability: characterization, mapping and assessment, (n.d.).

[7] P. Reszka, A. Fuentes, The Great Valparaiso Fire and Fire Safety Management in Chile, Fire Technol. 51 (2015) 753–758. doi:10.1007/s10694-014-0427-0.

[8] B. Teague, R. McLeod, S. Pascoe, Victorian Bushfires Royal Commission Final Report, 2010.

[9] N.Z.F. Service, The New Zealand Fire Service Emergency Incident Statistics 2009-2010, Wellington, NZ, 2010.

[10] W.E. Mell, S.L. Manzello, A. Maranghides, D. Butry, R.G. Rehm, The wildland–urban interface fire problem – current approaches and research needs, Int. J. Wildl. Fire. 19 (2010) 238–251. doi:10.1071/WF07131.

[11] S. Tarifa, P.P. Del Notario, F.G. Moreno, On the Flight Paths and Lifetimes of Burning Particles of Wood, Tenth Symp. Combust. (1965) 1021–1037.

[12] J.P. Woycheese, P.J. Pagni, D. Liepmann, Brand Propagation From Large-Scale Fires, J. Fire Prot. Eng. 10 (1999) 32–44. doi:10.1177/104239159901000203.

[13] S.L. Manzello, S. Suzuki, Y. Hayashi, Enabling the study of structure vulnerabilities to ignition from wind driven firebrand showers: A summary of experimental results, in: Fire Saf. J., 2012: pp. 181–196. doi:10.1016/j.firesaf.2012.06.012.

[14] S.L. Manzello, S.H. Park, T.G. Cleary, Investigation on the ability of glowing firebrands deposited within crevices to ignite common building materials, Fire Saf. J. 44 (2009) 894–900. doi:10.1016/j.firesaf.2009.05.001.

[15] A. Maranghides, W. Mell, A Case Study of a Community Affected by the Witch and Guejito Wildland Fires, Fire Technol. 47 (2011) 379–420. doi:10.1007/s10694-010-0164-y.

[16] A. Maranghides, D. McNamara, W. Mell, J. Trook, B. Toman, NIST Technical Note 1796 A, 2013. doi:10.6028/NIST.TN.1796.

[17] S. Quarles, P. Leschak, R. Cowger, K. Worley, R. Brown, C. Iskowitz, Lessons Learned From Waldo Canyon: Fire Adapted Communities Mitigation Assessment Team Findings, 2012.

[18] S.L. Manzello, T.G. Cleary, J.R. Shields, J.C. Yang, Ignition of mulch and grasses by firebrands in wildlandurban interface fires, Int. J. Wildl. Fire. 15 (2006) 427–431. doi:10.1071/WF06031.

[19] B.C. Connelly, S.A. Kaiser, M.D. Smooke, M.B. Long, Two-Dimensional Soot Pyrometry with a Color Digital Camera, Jt. Meet. U.S. Sect. Combust. Insitiute. 2 (2005).

[20] J.D. Maun, P.B. Sunderland, D.L. Urban, Thin-filament pyrometry with a digital still camera., Appl. Opt. 46 (2007) 483–488. doi:10.1364/AO.46.000483.

[21] R.E. Padilla, Structure and Behavior of Water-laden Methane/Air Flames, University of California, Irvine, 2016.

[22] F.H. Wu, R.E. Padilla, D. Dunn-Rankin, G.B. Chen, Y.C. Chao, Thermal structure of methane hydrate fueled flames, Proc. Combust. Inst. 36 (2016) 4391–4398. doi:10.1016/j.proci.2016.06.012.

[23] P.B. Kuhn, B. Ma, B.C. Connelly, M.D. Smooke, M.B. Long, Soot and thin-filament pyrometry using a color digital camera, Proc. Combust. Inst. 33 (2011) 743–750. doi:10.1016/j.proci.2010.05.006.

Western States Section of the Combustion Institute – Fall 2017 Meeting
Hosted by the University of Wyoming
October 2-3, 2017

Pyrolysis of Live Vegetation at Slow Heating Rates

*Elham Amini, Mohammad-Saeed Safdari, Mahmood Rahmati, Joel Howarth, Jonathan DeYoung, Thomas H. Fletcher**

Department of Chemical Engineering, Brigham Young University, Provo, Utah 84602, USA
Corresponding Author Email: tom_fletcher@byu.edu

Wildland fire is an important component of many American ecosystems. Wildland fires can have necessary ecological influences in many North American ecosystems or can dangerously affect life, property and natural resources. These hazards decrease with the proper application of controlled burning. Controlled burning destroys smaller plants and also decreases combustible materials and the potential for ignition and fire propagation in the fuel bed. Fuel bed pyrolysis and ignition determine fire ignition and propagation rates in both wildland fires and prescribed burning, but the details of solid fuel reaction under wildland fire conditions remain poorly understood. The purpose of this study is to provide fundamental information about pyrolysis of live shrub fuels at slow heating rates. A programmable heater apparatus was used to study the pyrolysis products of three different kind of live shrubs from forests of the Southeastern United States. Low and high molecular weight pyrolysis products were collected in a cold trap and then analyzed using GC-MS.

1. Introduction

Solid fuel samples degrade irreversibly at high temperatures through the two main processes of pyrolysis and combustion. Pyrolysis involves irreversible thermal decomposition of organic material at high temperatures and occurs in the absence of oxygen. Pyrolysis occurs early in the combustion process and is affected by the major components of biomass [1]. Cellulose (around 50% on dry basis), hemi-cellulose (20–40% in herbaceous biomass on dry basis and 10–30% in woods) and lignin (10–40% in herbaceous and 20–40% in woods) are the three main groups of natural polymeric materials in biomass. Extractives are generally smaller organic molecules or polymers like salts, protein and acids [2, 3]. Minerals are inorganic components, such as potassium, calcium, sodium, silicon, phosphorus and magnesium and also chlorine [2, 3].

Pyrolysis rates depend on the solid fuel particle temperature which in turn is a function of heat transfer to the unburnt live fuels. Pyrolysis produces volatile gaseous (such as CO, CO_2, H_2O, and H_2), condensable gases (tars), solid residue (char), and mineral ash. The yields of these products depend on heating rate, temperature, fuel type, and fuel properties. The heat transfer process in pyrolysis of live fuels is shown in Figure 1.

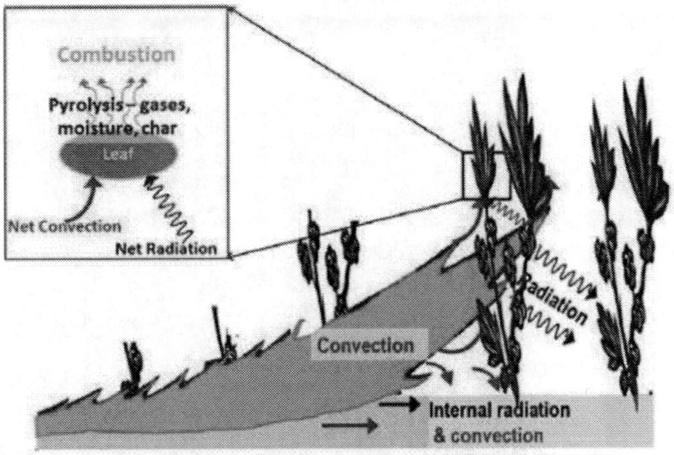

Figure 1. Heat transfer mechanisms in pyrolysis of live fuels

Primary pyrolysis and secondary pyrolysis are considered as two steps of the pyrolysis process. Primary pyrolysis is the process that the biomass degrades into volatile gases and char. CO, CO_2, H_2O, and light and heavy hydrocarbons are the products of primary pyrolysis of biomass and coals [4]. Tars are defined as hydrocarbons which condense at room temperature. Secondary pyrolysis is the process that tars at elevated temperature decompose either to lighter gases or polymerize to form soot [5]. The orange color of flames is because of the radiation of the tiny soot particles in the fuel-rich part of the flame. The secondary pyrolysis products that remain unburned appear as smoke.

Live fuels with a significant amounts of moisture are involved in a considerable number of large fires in the United States [6]. The records show that the large fires occurred in Los Angeles area and Santa Monica [7] when the moisture content of live fuels decreased below 79% and 77%.

The research presented here is related to the research on pyrolysis of live shrubs during rapid heating experiments above a fuel-rich flat-flame burner at 750°C reported by Safdari et al. [8]. The combination of slow-heating experiments and rapid heating experiments will help to determine the extent of secondary pyrolysis that occurs in the rapid pyrolysis experiments.

2. Methods

The foliar leaf species which was analyzed in this study is Swamp Bay (Figure 2). A programmable heater apparatus was used to measure pyrolysis products at slow heating rate (10 K/min) and low temperature (500 °C) to generate gas and tar samples (Figure 3). Approximately 2 grams of sample were positioned in a U-like metal tube made from ¾ in. stainless steel and placed in the heater. A small flow of 10 slpm of N_2 was used to purge the pyrolysis gases and provide an oxygen-free environment for pyrolysis. Gas condensers were constructed by packing fine glass wool into test tubes and using rubber stoppers to close the top. The stoppers had two holes drilled into them through which ¼ in. stainless steel tubing was tightly fitted. The gases entered and passed through the glass wool before exiting. The four condensers were placed in an ice bath filled with dry ice to aid condensation. Tar vapors were condensed in condensers and were analyzed by GC-MS after solvent extraction using GC grade dichloromethane. A Thermo Scientific Trace 1310 gas chromatograph (GC) equipped with a thermal conductivity detector (TCD) was used with a HP 5890 GC combined with a HP 5972 mass spectrometer (MS). Non-condensable gases were collected in gas sampling bags for transfer to gas analysis devices. The reaction was stopped at specified temperature and the apparatus was cooled to prevent further reaction.

Figure 1. Swamp Bay

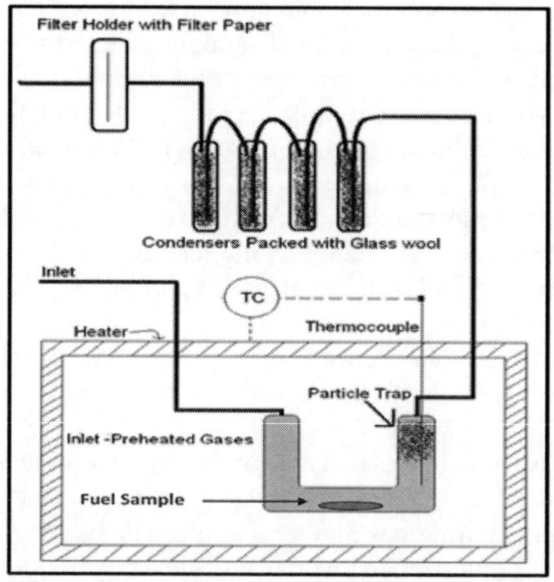

Figure 2 Schematic of programmable heater used for generating tar and gas samples

3. Results and Discussion

3.1. Tar Analysis

Figure 4 shows the peaks and the retention times for pyrolysis resulted by GC-MS analysis. The main identified tar components from GC-MS analysis are presented in Table 1. Tar analysis showed that tars are a complex mixture of condensable hydrocarbons, and for low temperatures as 500 °C, phenols (i.e., benzene rings with OH attachments) are the main chemical compounds in the tar from this experiment.

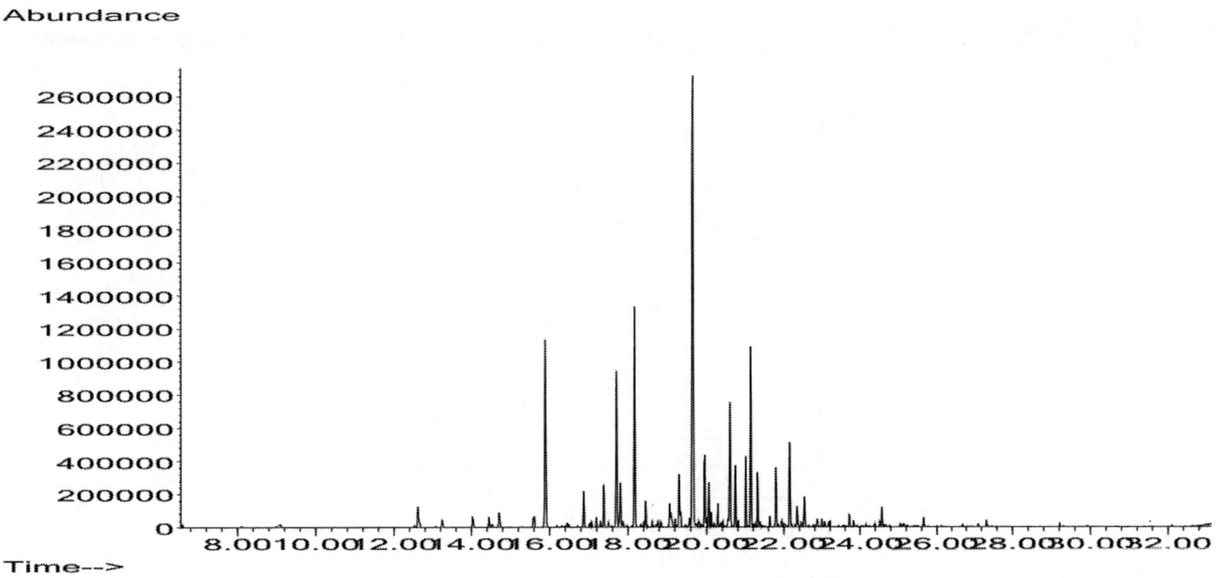

Figure 3. GC spectrum of tars from pyrolysis of Swamp Bay.

Table 1. Retention times and compositions of tar compounds (Swamp Bay)

Peak No.	Peak RT (min)	Component name	Molecular formula	Composition(mole%)
1	12.618	2-Cyclopenten-1-one	C_5H_6O	1.32
2	14.712	2-Cyclopenten-1-one, 2-hydroxy-	$C_5H_6O_2$	0.91
3	15.611	2-Cyclopenten-1-one, 3-methyl-	C_6H_8O	0.83
4	15.885	Phenol	C_6H_6O	10.23
5	16.872	1,2-Cyclopentanedione, 3-methyl-	$C_6H_8O_2$	1.46
6	17.376	Phenol, 2-methyl-	C_7H_8O	1.63
7	17.705	Phenol, 4-methyl-	C_7H_8O	7.60
8	17.804	5-Hepten-3-one, 5-methyl-	$C_8H_{14}O$	1.99
9	18.166	Phenol, 2-methoxy-	$C_7H_8O_2$	9.35
10	18.451	Larixic acid	$C_6H_6O_3$	0.95
11	19.065	Phenol, 3,4-dimethyl-	$C_8H_{10}O$	1.68
12	19.306	Phenol, 4-ethyl-	$C_8H_{10}O$	2.90
13	19.657	1,2-Benzenediol	$C_6H_6O_2$	24.53
4	19.953	Phenol, 2-methoxy-5-methyl-	$C_8H_{10}O_2$	3.09
15	20.051	Benzofuran, 2,3-dihydro-	C_8H_8O	1.49
16	20.599	1,4-Benzenediol	$C_6H_6O_2$	6.53
17	20.731	1,2-Benzenediol, 3-methyl-	$C_7H_8O_2$	2.49
18	21.005	1,2-Benzenediol, 3-methoxy-	$C_7H_8O_3$	2.56
19	21.137	1,2-Benzenediol, 4-methyl-	$C_7H_8O_2$	7.49
20	21.312	Phenol, 4-ethyl-2-methoxy-	$C_9H_{12}O_2$	2.13
21	21.794	1-Hydroxy-2-acetyl-4-methylbenzene	$C_9H_{10}O_2$	2.62
22	22.145	Phenol, 2,6-dimethoxy-	$C_8H_{10}O_3$	3.82
23	22.343	1,2,3-Benzenetriol	$C_6H_6O_3$	0.86
24	22.529	1,3-Benzenediol, 4-ethyl-	$C_8H_{10}O_2$	1.51

3.2. Proximate Analysis

Table 2 shows the proximate analysis for Swamp Bay. Volatile content includes tars and gases. For this species, the tar and gas yields are 23% and 60.15%, respectively. The non-condensable gases and their yields are presented in Table 3.

Table 2. Proximate analysis results

Species	Moisture Content (%, Dry Basis)	Wt.% Moisture (Wet Basis)	Wt. % Volatiles (Moisture free)	Wt. % Fixed Carbon (Moisture Free)	Wt. % Ash (Moisture Free)
Swamp Bay	116.8	53.9	83.1	14.4	2.5

Table 3. Non-condensable gas composition

Species	Mole%
CO	46
CO_2	24
CH_4	17
H_2	13

4. Summary and Conclusions

Live samples of Swamp Bay leaves were pyrolyzed in a programmable heater at 10 K/min with a maximum temperature of 500°C. The sample was purged with N_2 during heating to facilitate collection of pyrolyzed gases. Pyrolyzed tars were condensed in a cold trap using dry ice. Non-condensable pyrolysis gases were collected in a transfer bag. Tars were analyzed using a GC/MS system. The tar analysis showed that the major components are phenolic compounds (i.e., aromatic rings with at least one hydroxyl (OH) attachment). Other live species are currently being analyzed. Results from this study will be compared with results from high heating rate experiments obtained by Safdari et al. [8].

Acknowledgements

This research was funded by DOD/EPA/DOE Strategic Environmental Research and Development Program, Project RC-2640, funded through contract 16-JV-11272167-024, administered by the USDA Forest Service PSW Research Station.

References

[1] Haykiri-Acma, H., S. Yaman, and S. Kucukbayrak, "Comparison of the Thermal Reactivities of Isolated Lignin and Holocellulose During Pyrolysis," *Fuel Processing Technology*, 91(7) (2010) 759-764.

[2] Zabaniotou, A. A., "Pyrolysis of Forestry Biomass by-Products in Greece," *Energy Sources*, 21(5) (1999) 395-403.

[3] Sharma, A., V. Pareek, and D. K. Zhang, "Biomass Pyrolysis-a Review of Modelling, Process Parameters and Catalytic Studies," *Renewable & Sustainable Energy Reviews*, 50 (2015) 1081-1096.

[4] Rothermel, R. C., 1972, A Mathematical Model for Predicting Fire Spread in Wildland Fuels. USDA Forest Service, Intermountain Forest and Range Experiment Station, Ogden, UT, Report No. INT-115.

[5] DiBlasi, C., "Numerical Simulation of Cellulose Pyrolysis," *Biomass and Bioenergy*, 7 (1994) 87-98.

[6] Ferguson, S. C., A. Dahale, B. Shotorban, S. Mahalingam, and D. R. Weise, "The Role of Moisture on Combustion of Pyrolysis Gases in Wildland Fires," *Combustion Science and Technology*, 185(3) (2013) 435-453.

[7] Dennison, P. E. and M. A. Moritz, "Critical Live Fuel Moisture in Chaparral Ecosystems: A Threshold for Fire Activity and Its Relationship to Antecedent Precipitation," *International Journal of Wildland Fire*, 18(8) (2009) 1021-1027.

[8] Safdari, M.-S., M. Rahmati, E. Amini, and T. H. Fletcher, "Analysis of Pyrolysis Products from Live Shrub Fuels," 10th National Combustion Institute Meeting, University of Maryland, April 23-26 (2017).

Western States Section of the Combustion Institute – Fall 2017 Meeting
Hosted by the University of Wyoming
October 2-3, 2017

Effect of Reduced Plume Entrainment on the Burning Rate of Porous Fuel Beds

Sara McAllister

RMRS Missoula Fire Sciences Laboratory, USDA Forest Service, 5775 W US Highway 10, Missoula, MT 59808

Large outdoor fires often exhibit unexpected behavior, particularly compared with much smaller fires. Large fires are often characterized by very tall plumes that likely entrain relatively little, and may act in a similar fashion to a chimney. To explore the possible changes in the steady burning rate due to this chimney effect, experiments were conducted with wood cribs burned with a chimney of variable height. Sixteen crib designs were tested with chimney heights ranging from 0 to 3.7 m. Eleven of the crib designs had burning rates that monotonically increased with chimney height with increases as high as 190%. The increase in flow through the fuel bed due to the chimney not only increases the equivalence ratio of the gas phase (particularly important for ventilation-limited cribs), but also increases the char oxidation rate. Five of the crib designs, however, had burning rates that first increased then decreased with chimney height. The common factors with these cribs are lower non-dimensional heat release rates ($Q^* < 2.5$), which tipped the balance between heat generated and lost. The maximum normalized burning rate can be well correlated to the ratio of the fuel surface area to the crib porosity.

1. Introduction

Large outdoor fires often exhibit unexpected behavior, particularly compared with much smaller fires. One noticeable characteristic of large fires is that they often have a very tall plume which can extend into the stratosphere [1]. This indicates that relatively little ambient air is entrained, particularly as the actively burning area increases. Such a tall column of hot gases creates a local low pressure region above the burning area. The pressure difference between the plume and ambient air induces a flow along the ground. This induced flow potentially produces a feedback mechanism, whereby the burning rate of the porous, charring fuel may further increase, explaining the change in burning behavior when fires get sufficiently large. Unfortunately, not much is known about the change in burning rates of porous fuel beds in such a scenario. Altering plume entrainment by various other means, such as the trench effect (see for example [2] or in room corners (see for example [3]), has been studied and is known to effect the flame height and flame spread rate. The work here is the first known work to examine the change in burning rate with reduced plume entrainment. This is accomplished by burning a wide variety of wood cribs under a chimney with a variable height.

2. Methods

A total of sixteen crib designs were tested (Table 1). These crib designs were chosen to give a range of porosities (as defined by Heskestad [4]), geometric variation (width versus height), footprint size relative to the chimney, and initial mass. Knot-free ponderosa pine wood was used. All cribs were conditioned in a conditioning chamber at 35°C and 3% relative humidity for at least three days so that the calculated equilibrium moisture content was approximately 1%. Due to a change in available equipment, crib design #4 was conditioned at 30°C and 10% relative humidity with a calculated equilibrium moisture content of 2.4%. As a baseline, all cribs were tested without a chimney. Four chimney heights were then tested with each crib design: 0.61 m, 1.22 m, 2.44 m, and 3.66 m. In all cases the chimneys were constructed of thin stainless steel sheets bent into a 30.5 x 30.5 cm square cross section. Each crib design and chimney height combination was tested at least three times and the results averaged.

The testing apparatus consisted of a fixed stand to mount the chimney and a rolling platform on which the cribs were placed (Fig. 1). This rolling platform was on tracks so that it could easily be rolled into the correct position. Due to the variation in crib designs, the height of the rolling platform was adjusted via a scissor lift-type feature so that the top of the crib was always 3 cm from the bottom of the chimney. Three 6-kg capacity load cells on the platform were used to weigh the cribs during the tests. To minimize heat transfer to the load cells, the crib was placed on a thin aluminum sheet (91.4 cm x 91.4 cm) with multiple sheets of ceramic paper. This aluminum sheet was only in contact with the load cells via 1.3 cm ball bearings. As the spacing between the crib and the support platform has been shown to considerably change the burning rate [5], all cribs were placed on spacers that kept the bottom of the crib 7.6 cm from the platform. In some cases, significant amounts of embers were lofted into the chimney. To reduce this effect, a 30.5 x 30.5 cm flattened expanded steel screen (1.3 cm (1/2") diamond) was placed on top of all cribs. Scoping tests indicated that the presence of this screen had no effect on the burning rate.

Table 1. Crib designs tested. Cribs with porosity under 0.05 cm are considered densely packed.

Crib design #	Stick thickness (b, cm)	Stick length (l, cm)	Number of sticks/layer (n, [ ])	Number of layers (N, [ ])	Heskestad porosity (φ, cm)	Chimney heights tested (H_c, m)
1	1.27	12.7	5	10	0.0215	0, 0.61, 1.22, 2.44, 3.66
2	1.27	20.3	6	14	0.0390	0, 0.61, 1.22, 2.44, 3.66
3	1.27	25.4	4	21	0.1202	0, 0.61, 1.22, 3.66
4	1.27	25.4	11	2	0.0625	0, 3.66
5	1.27	30.5	10	5	0.0757	0, 1.22, 3.66
6	0.64	12.7	6	24	0.0205	0, 0.61, 1.22, 2.44, 3.66
7	0.64	20.3	3	45	0.1213	0, 1.22, 3.66
8	0.64	20.3	10	16	0.0270	0, 1.22, 3.66
9	0.64	25.4	6	10	0.2110	0, 1.22, 3.66
10	0.64	25.4	10	6	0.1197	0, 0.61, 1.22, 2.44, 3.66
11	0.64	25.4	10	10	0.7246	0, 0.61, 1.22, 3.66
12	0.64	25.4	14	15	0.0213	0, 0.61, 1.22, 3.66
13	0.64	30.5	8	14	0.1210	0, 0.61, 1.22, 3.66
14	0.64	30.5	20	7	0.0270	0, 1.22, 3.66
15	0.32	25.4	14	30	0.0252	0, 3.66
16	0.32	30.5	6	40	0.1215	0, 3.66

Fig. 1. Crib design #5 with 1.2 m chimney.

To ignite each crib, the platform was wheeled out from underneath the chimney. The cribs were then dipped in 99% pure isopropyl alcohol and allowed to drain. The total mass of alcohol was 10% or less of the mass of the crib. This procedure ensured simultaneous and uniform ignition. To avoid igniting a premixed alcohol-air mixture in the confined area of the chimney, the cribs were first ignited outside the chimney. Once it was noted that the alcohol had burned off and the sticks themselves were burning (on the order of seconds to tens of seconds), the platform with burning crib was pushed under the chimney. In all cases, steady burning was achieved under the chimney. The burning rate results in the following discussion are obtained using a linear regression of the mass data in the steady burning region and the maximum value is reported.

3. Results and Discussion

Though not quite every chimney height has been tested yet for all crib designs, it appears that for eleven of the sixteen crib designs, the burning rate increased with chimney height. Figures 2a and 2b show the burning rate variation for the chimney heights tested for these eleven crib designs. In these figures, the burning rate has been normalized by the no-chimney case. Each data point is the average of at least three tests. The error bars represent one standard deviation of the data for that particular set of experimental variables. The average standard deviation for all tests conducted was 8.4% of the mean values. In general, the most noticeable rise in the burning rate occurred for the smallest chimney height (0.61 m). While still increasing the burning rate, the effect of the chimney seemed to begin leveling off for chimneys taller than 1.22 m. The most shocking increase in the burning rate was seen for crib design #12, where the burning rate increased 190% compared to the no-chimney case.

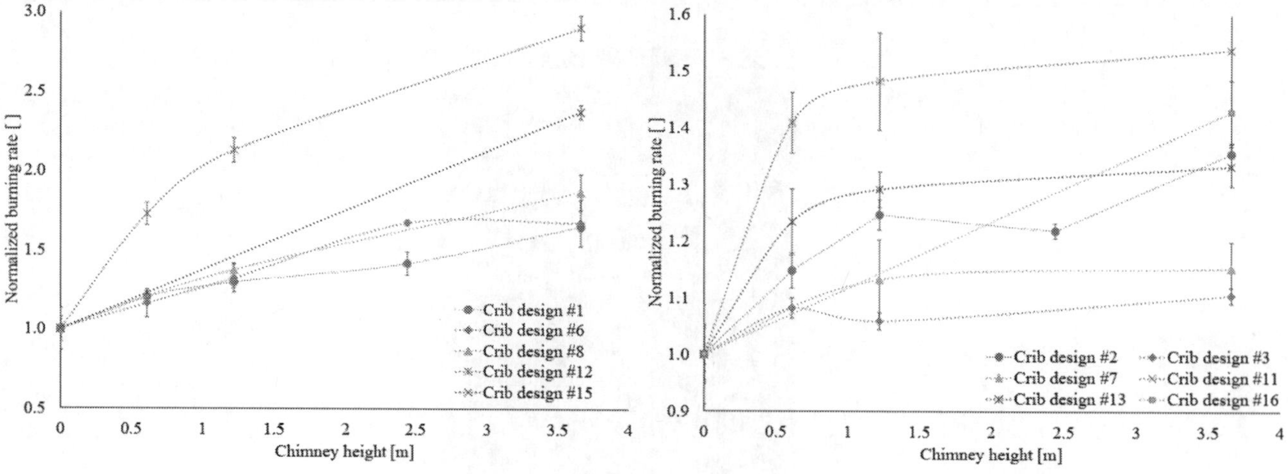

Fig. 2a (left) and 2b (right). Normalized burning rate for different chimney heights for eleven crib designs showing increasing trend.

As seen in Fig. 3, for the other five crib designs tested, the burning rate first increased for chimneys less than about 1.22 m tall, then decreased as the chimney height increased further. Interestingly, the chimney height where the decrease in burning rate is seen seems corresponds to the height where the effectiveness of the chimney appears to generally diminish in Fig. 2. Figures 4 and 5 show photographs comparing the burning behavior of both the no chimney and 3.66 m chimney tests for crib design #9 and #10, respectively. As shown, these crib designs visibly burned less completely with taller chimneys compared to the no-chimney tests. In fact, they frequently burned only in the middle, leaving the perimeter unconsumed and often even uncharred. Figure 6 shows one of the most dramatic cases of this localized burning. Figure 7 is a close up view of crib design #14 under a 3.66 m chimney showing that large pockets of pyrolyzates aren't consumed near the fuel bed.

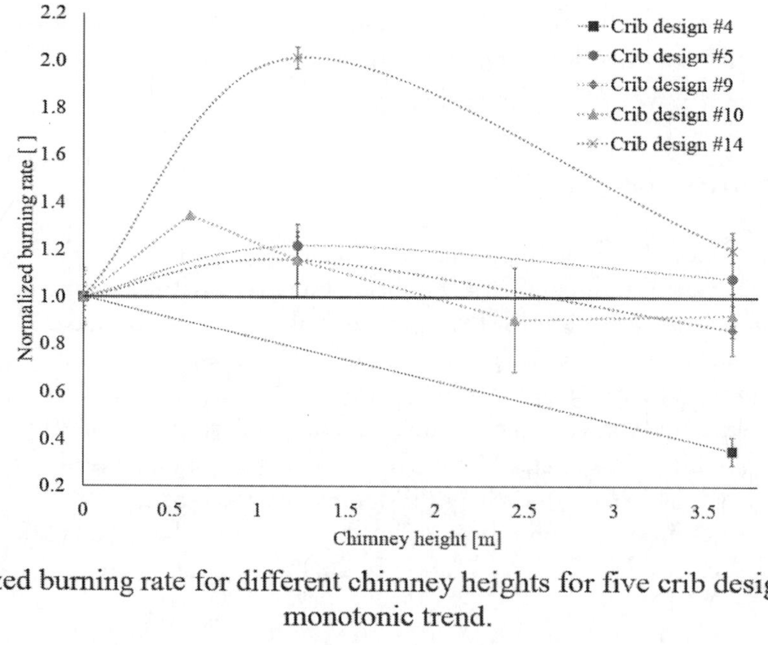

Fig. 3. Normalized burning rate for different chimney heights for five crib designs showing non-monotonic trend.

Fig. 4. Crib design #9 with no chimney (left) and with 3.66 m chimney (right).

Fig. 5. Crib design #10 with no chimney (left) and with 3.66 m chimney (right).

Fig. 6. Highly localized burning of crib design #4 with 3.66 m chimney.

Fig 7. Crib design #14 with 3.66 m chimney.

Figure 8 shows the mass and the mass loss rate (dm/dt) as a function of time for crib designs #12 (left) and #14 (right). The mass loss rate plots were generated by taking the slope of the smooth spline fit of the data with ten degrees of freedom. As shown above, the burning rate crib design #12 monotonically increased with chimney height and #14 did not. In both cases, the chimney had the noticeable effect of shortening the duration of the burning and reducing the residual mass. This reduction in residual mass was observed experimentally as a higher proportion of ash compared to char. Figure 8 also shows that the mass loss rate appears to be more "spiky" with a shorter duration of maximum burning rate when a chimney is used.

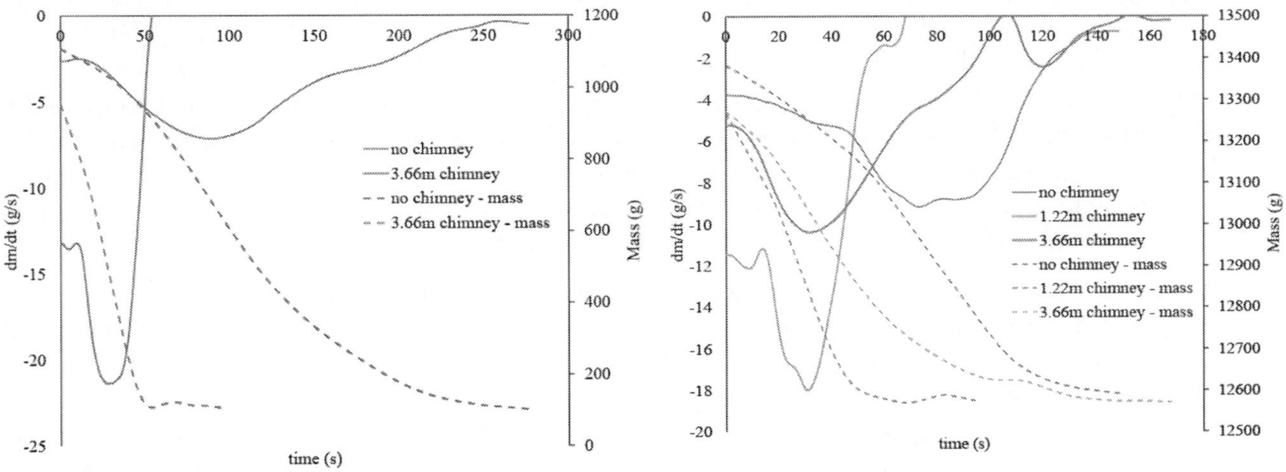

Fig. 8. Mass loss rate (dm/dt) and mass as a function of time for crib designs #12 (left) and #14 (right).

These trends in the burning rate have several causes. As the chimney height is increased, the hydrostatic pressure above the fuel bed decreases. This creates a larger pressure difference that increases the flow at the ground level and through the fuel bed. This increases the ventilation of the fuel bed, which will increase the burning rate for densely packed fuel beds. This increase in flow through the bed increases the burning rate through another mechanism as well. Because the cribs are built with a charring material, this increased flow increases the char oxidation rate. This could have important implications for wildland fires as it could promote flaming combustion over smoldering combustion, further increasing the fire intensity. For the cases where the burning rate decreases, the fuel beds show signs of being "over ventilated" in a way that tips the balance between heat losses and heat generation. In these cases, the outer perimeter shows signs of being effected by convective cooling. Additionally, if the flow velocities through the crib are such that the gaseous fuel and air doesn't mix and burn as well near the fuel bed, the heat feedback from the flame is reduced.

Table 2 shows the numerical values of the normalized burning rates along with the non-dimensional heat release rate (Q^*) for each of the no chimney cases. The non-dimensional heat release rate is one of the primary scaling parameters of the flame height (see for example [6-11]). It is very similar to the square root of the Froude number, and is frequently defined as [6-11]:

$$Q^* = \frac{\dot{Q}}{\rho_\infty c_p T_\infty g^{1/2} D^{5/2}} = \frac{R \Delta h_c}{\rho_\infty c_p T_\infty g^{1/2} D^{5/2}} \quad (1)$$

where $\dot{Q}$ is the heat release rate (kW), ρ_∞ is the density of the ambient air (kg/m^3), c_p is the specific heat of the ambient air (kJ/kg-K), T_∞ is the ambient air temperature (K), g is the acceleration due to gravity (m/s^2), D is the characteristic length (m), R is the burning rate (g/s), and Δh_c is the heat of combustion (kJ/g). For our purposes, we assumed that the heat of combustion was 14.1 kJ/g [12] and the properties of air were evaluated at 300 K. Three burning regimes are often identified based on the value of Q^*. For Q^* larger than 1, the fire is said to be in the "tall flame" regime where the flame height is independent of the fire diameter and only dependent on the heat release rate [6]. This is the regime expected under most spreading wildfire conditions [6]. As seen in Table 2, almost all of our tests fall into this regime. For Q^* smaller than 0.1, one single flame or plume cannot form, and the fire breaks into "distributed flamelets" [6]. This is the regime of true mass fires. In between these two regimes is the intermediate regime, where the flame transitions between these two behaviors. It is possible that the truly large wildland fires that show extreme behavior may have Q^* values lower than are tested here and should be considered in future work.

As seen in Table 2, the crib designs that showed a decrease in the burning rate (designs #4, 5, 9, 10, and 14) are the shorter cribs, but also the cribs with the lowest non-dimensional heat release rate. Though not a perfect criteria, the shorter crib height would mean that more of the fuel elements would be exposed to the cold air flowing through the crib from the bottom. Previous work has shown that a large portion of the airflow through a crib in fact originates from underneath [5]. A lower non-dimensional heat release rate (Q^*) implies that the buoyant advection is becoming comparable to the fuel flow rate. This could explain why it appeared that the gaseous fuel and air was not mixing and combusting as near to the fuel bed for the cases with decreased burning rates. The transition between the two burning regimes seen here appears to happen around a Q^* of 2.5. This is only a rough estimate however, because the heat of combustion used in calculating Q^* was pulled from the literature, not measured.

Table 2. Results of experiments, including burning rate normalized to no-chimney case.

Crib design #	Surface area (A_s, cm^2)	Heskestad porosity (φ, cm)	Crib height (h, cm)	Chimney height (H_c, m)	Normalized burning rate ($\bar{R}$, [ ])	St dev (% of mean)	Q^* ([ ])
1	2661.29	0.0215	12.7	0	1.000	2.936	4.167
	2661.29	0.0215	12.7	0.61	1.204	3.322	
	2661.29	0.0215	12.7	1.22	1.298	3.613	
	2661.29	0.0215	12.7	2.44	1.418	7.269	
	2661.29	0.0215	12.7	3.66	1.651	2.774	
2	7432.243	0.038997	17.78	0	1.000	5.275	4.288
	7432.243	0.038997	17.78	0.61	1.149	3.123	
	7432.243	0.038997	17.78	1.22	1.247	2.647	
	7432.243	0.038997	17.78	2.44	1.219	1.336	
	7432.243	0.038997	17.78	3.66	1.356	1.792	
3	10077.4	0.120172	26.67	0	1.000	1.760	4.102
	10077.4	0.120172	26.67	0.61	1.082	1.793	
	10077.4	0.120172	26.67	1.22	1.059	1.528	
	10077.4	0.120172	26.67	3.66	1.105	1.397	
4	2519.35	0.062478	2.54	0	1.000	10.392	0.985
	2519.36	0.062478	2.54	3.66	0.352	14.452	
5	6612.89	0.075721	6.35	0	1.000	5.793	1.851
	6612.89	0.075721	6.35	1.22	1.216	5.978	

	6612.89	0.075721	6.35	3.66	1.081	19.592	
6	4093.54	0.0205	15.24	0	1.000	1.868	6.412
	4093.54	0.0205	15.24	0.61	1.164	8.810	
	4093.54	0.0205	15.24	1.22	1.322	8.784	
	4093.54	0.0205	15.24	2.44	1.673	1.741	
	4093.54	0.0205	15.24	3.66	1.675	14.599	
7	6757.245	0.121348	28.575	0	1.000	4.539	6.917
	6757.245	0.121348	28.575	1.22	1.133	7.031	
	6757.245	0.121348	28.575	3.66	1.154	4.626	
8	7177.405	0.026995	10.16	0	1.000	7.829	4.571
	7177.405	0.026995	10.16	1.22	1.376	4.086	
	7177.405	0.026995	10.16	3.66	1.869	11.461	
9	3658.057	0.211	6.35	0	1.000	4.110	2.327
	3658.057	0.211	6.35	1.22	1.156	9.896	
	3658.057	0.211	6.35	3.66	0.863	10.579	
10	3516.122	0.119657	3.81	0	1.000	10.832	1.762
	3516.122	0.119657	3.81	0.61	1.344	0.370	
	3516.122	0.119657	3.81	1.22	1.155	15.193	
	3516.122	0.119657	3.81	2.44	0.905	22.031	
	3516.122	0.119657	3.81	3.66	0.926	9.389	
11	5806.44	0.072459	6.35	0	1.000	8.731	2.685
	5806.44	0.072459	6.35	0.61	1.410	5.323	
	5806.44	0.072459	6.35	1.22	1.483	8.579	
	5806.44	0.072459	6.35	3.66	1.538	16.375	
12	11504.82	0.021277	9.525	0	1.000	13.400	2.831
	11504.82	0.021277	9.525	0.61	1.726	6.980	
	11504.82	0.021277	9.525	1.22	2.130	7.682	
	11504.82	0.021277	9.525	3.66	2.905	7.714	
13	8090.306	0.121048	8.89	0	1.000	3.040	3.003
	8090.306	0.121048	8.89	0.61	1.234	5.837	
	8090.306	0.121048	8.89	1.22	1.292	3.172	
	8090.306	0.121048	8.89	3.66	1.334	3.542	
14	9016.111	0.027028	4.445	0	1.000	9.161	2.201
	9016.111	0.027028	4.445	1.22	2.011	4.640	
	9016.111	0.027028	4.445	3.66	1.199	4.926	
15	12487.07	0.025157	9.54	0	1.000	5.677	3.430
	12487.07	0.025157	9.54	3.66	2.375	4.352	
16	9055.627	0.12146	12.72	0	1.000	4.945	4.121
	9055.627	0.12146	12.72	3.66	1.430	5.507	

By considering the physical factors that increase the burning rate, the maximum burning rate can be predicted. Figure 9 shows the correlation of the largest increase in the normalized burning rate regardless of chimney height to a function of the stick surface area (A_s) and the crib porosity (φ). For the eleven crib designs that showed increased burning rate with chimney height, the maximum was measured for the tallest chimney (3.66 m). For the four of the five crib designs that showed decreased burning rate with chimney height, the maximum was measured at chimney heights from 0.61 – 1.22 m. The fifth crib has not yet been tested with these intermediate chimney heights and is work in progress. Future work will also include more cribs with low non-dimensional heat release rate ($Q^* < 1$), but also a wider variety of chimney heights (both larger than 3.66 m and between the range of 0.61 to 1.22 m) to determine the actual maximum burning rate. Preliminary results of a densely packed ($\phi = 0.007435$) crib (b = 0.64 cm, l = 25.4 cm, n = 27, N = 3) with a Q* of 1 show an interesting burning behavior. As shown in Fig. 10, there are two burning regimes – one slow but very steady burning regime and one short peak. The location of the peak burning in relation to the steady burning is different depending on if a chimney is used. Without a chimney, the peak occurs towards the end as the edges are consumed, effectively reducing the porosity of the remaining fuel. With a chimney, the burning rate peaks just after it is placed underneath the chimney, lasts only a moment, then the burning dies down considerably, only burning in one or two small areas. In one test, the flames went out entirely. This burning behavior is quite unique and further testing is required.

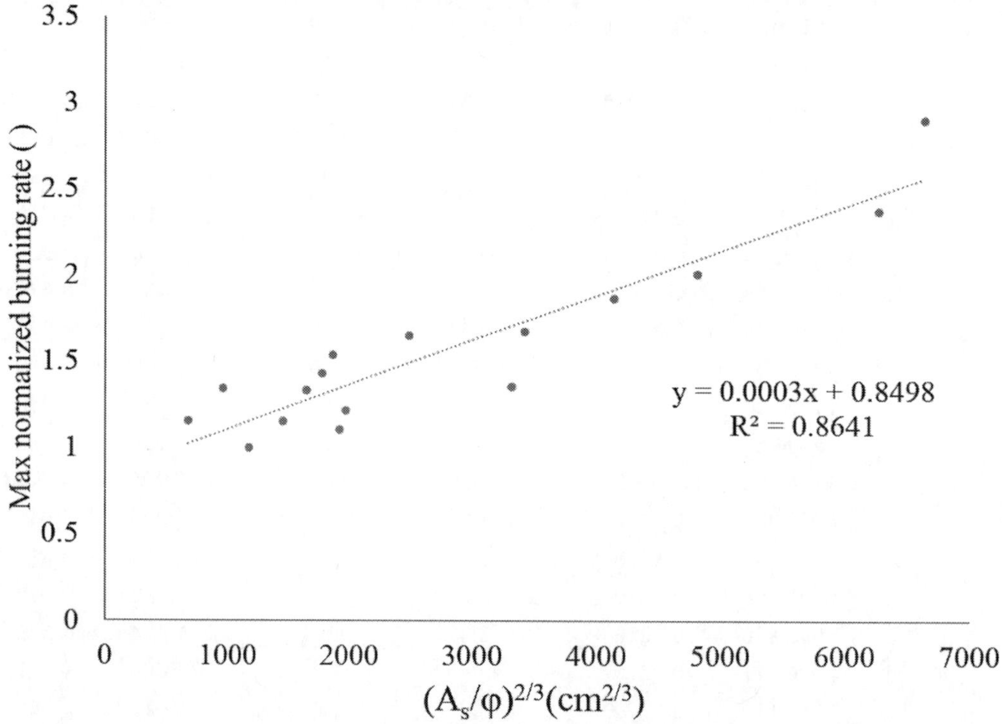

Fig. 9. Maximum normalized burning rate regardless of chimney height as a function of stick surface area (A_s) and crib porosity (φ).

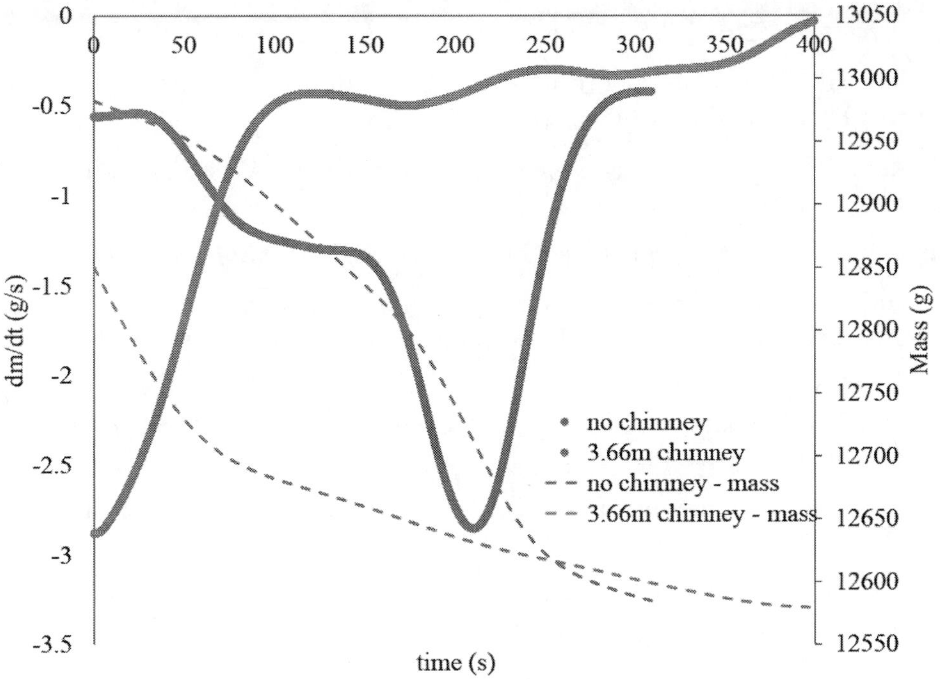

Fig. 10. Mass loss rate (dm/dt) and mass as a function of time for dense crib with Q* = 1.

Acknowledgements

This research was funded by the National Fire Decision Support Center. The author wants to thank Sophia Vernholm and Chelsea Phillips for building the many cribs and help performing the experiments, Mark Finney and Jason Forthofer for many discussions on this idea, Torben Grumstrup for help performing the experiments, Randy Pryhorocki and Josh Deering for building the apparatus, and Cyle Wold for setting up the data acquisition system.

References

[1] Fromm, M., Bevilacqua, R., Servranckx, R., Rosen, J., Thayer, J. P., Herman, J., and Larko, D. "Pyro-cumulonimbus injection of smoke to the stratosphere: Observations and impact of a super blowup in northwestern Canada on 3–4 August 1998," *J. Geophys. Res.-Atmos*. 110:D08205, (2005).

[2] Drysdale, D.D., Macmillan, A.J.R. "Flame spread on inclined surfaces," Fire Safety Journal 18:245-254 (1992).

[3] Takahashi, W., Tanaka, H., Sugawa, O., Ohtake, M. "Flame and Plume behavior in and near a corner of walls," Fire Safety Science – Proceedings of the Fifth International Symposium: 261-271 (1997).

[4] Heskestad, G. "Modeling of enclosure fires," *Symposium (International) on Combustion* 14:1021-1030 (1973).

[5] McAllister, S., Finney, M. "Burning rates of wood cribs with implications for wildland fires," *Fire Technology* 52(6):1755-1777 (2016).

[6] Zukoski, E.E. "Properties of fire plumes," *Combustion Fundamentals of Fire* (ed. G. Cox), Academic Press, London, 1995, p. 101-219.

[7] Zukoski, E.E. "Fluid dynamic aspects of room fires," *Proceedings of First International Symposium on Fire Safety Science*:1-30 (1986).

[8] Quintiere, J.G. "Scaling applications in fire research," *Fire Safety Journal* 15:3-29 (1989).

[9] Heskestad, G. "Fire Plumes, flame height, and air entrainment," *SFPE Handbook of Fire Protection Engineering (4th ed.)*, National Fire Protection Association, Quincy, MA, 2008, p. 2-1.

[10] Quintiere, J.G. *Fundamentals of fire phenomena*, John Wiley & Sons, West Sussex, 2006, p. 318-322.

[11] Drysdale, D. *An introduction to fire dynamics (2nd ed)*, John Wiley & Sons, West Sussex, 1998, p. 107-111, 126-134.

[12] Heskestad, G. "Heat of combustion in spreading wood crib fires with application to ceiling jets," *Fire Safety Journal* 41(5):343-348 (2006).

Western States Section of the Combustion Institute – Fall 2017 Meeting
Hosted by the University of Wyoming
October 2-3, 2017

Comparison of Laminar Pulverized Coal Flame Experiment and Simulation

Clint Dunn[1] Emily Beagle[2] Michael Stoellinger[3] Erica Belmont[4]
[1]Mechanical Engineering, University of Wyoming, 1000 E. University Laramie Wy 82072
[2]Mechanical Engineering, University of Wyoming, 1000 E. University Laramie Wy 82072
[3]Mechanical Engineering, University of Wyoming, 1000 E. University Laramie Wy 82072
[4]Mechanical Engineering, University of Wyoming, 1000 E. University Laramie Wy 82072

Abstract

The accurate simulation of multiphase reactive flows commonly rests upon empirical models. The parameters of these models are determined by comparing computational and experimental results in conditions that, as much as possible, illustrate relevant phenomenon while avoiding complicating factors. We are fortunate at the University of Wyoming to have both the experimental and computational resources available to perform this comparison entirely "in-house".

In this study a Hencken burner with a laminar jet of pulverized coal entrained in air surrounded by a hot methane flame co-flow is paired with an axisymmetric simulation of the same reactor. In particular we are interested in obtaining parameters for coal devolatilization models as well as knowledge regarding the importance and necessary complexity of the chemical reaction models. By using the co-flow Hencken burner we avoid the added complexity of turbulent combustion and can therefore examine the isolated processes of coal particle devolatilization and laminar coal volatile combustion. Any experimental information obtained about these two processes can then be used directly in optimizing computational models for future use in more complex simulations. Our first step, detailed in this paper, is to make a baseline qualitative comparison of the experiment and simulation.

1 Introduction

As computational modeling of complex flows such as those found in pulverized coal reactors becomes more mainstream and commercially viable there is a need to collect and validate parameters for many of the sub-models in these simulations. The present work sets the groundwork for future projects which will enable the validation of several key sub-model parameters.

In order to emphasize and isolate the effects of coal devolatilization rates, combustion kinetics and coal volatile gas composition we are investigating a laminar flow. This permits these processes/features to be evaluated in the absence of complicating turbulent factors. In this work there are two main objectives.

First, illustrate qualitative agreement between experimental and computational results of a laminar pulverized coal jet in a hot methane co-flow. Secondly we investigate the effect of variations in the composition of coal volatile gases. Often in computational fluid dynamic simulations the chemical kinetics of combustion are drastically simplified and equilibrium chemistry is used. The difference in volatile gas composition is then unimportant. Here we use a simple yet finite-rate combustion mechanism that permits the flexibility to see some difference in flame structure depending on the assumed volatile composition. Two simulations with varying volatile compositions are performed. The first is referred to as "heavy" and its volatile gas contains relatively more large hydrocarbons, the second is referred to as "light" and it contains mostly small hydrocarbons in its volatile makeup.

Beyond these two points this paper details both the experimental and computational background for the paired experiment/simulation. Specifically the baseline versions of the computational models that are the subject of future work are detailed.

2 Methods

2.1 Experimental Methods

A Hencken burner modified to handle solid fuel particles was utilized to experimentally create a laminar pulverized coal flame in this study. The burner consisted of a single small tube of diameter 1.6 $[mm]$ in the center surrounded by a honeycomb structure of fuel and oxidizer tubes. The dimensions of the square burner surface were 35 $[mm]$ × 35 $[mm]$. Methane was used as the co-flow fuel at an equivalence ratio of 0.75. Powder River Basin coal was used as the solid fuel. Coal particles were dried at 100° $[C]$ for six days before being sieved to a particle range of $20 - 45$ $[\mu m]$. Coal was fed into the burner using a twin-screw auger feeder (Coperion feeder) at a mass flow rate of 0.04 $[g/s]$ where it was then entrained with air at a flow rate of 4.0 $[SLPM]$ before reaching the burner surface and combusting. Photos of a steady state coal flame were taken using a Nikon D90 camera and analyzed for lift off height using ImageJ image analysis software.

2.2 Numerical Simulation

The open-source package OpenFOAM v4.1 is used for the simulations [1]. The core solvers for the continuous fluid phase and for the Lagrangian coal particle tracking are all taken directly from OpenFOAM without modification. The mass, energy, momentum, and species conservation equations for the continuous fluid phase are solved with the built-in finite volume of the coalChemistryFoam solver from OpenFOAM. The flow is assumed to be entirely laminar and therefore no turbulence model is used.

OpenFOAM also provides a framework for modeling more advanced aspects of the Lagrangian particles and their interaction with the continuous gas phase. This includes mass and energy exchange with the continuous phase as well as models specific to coal particle simulations such as particle devolatilization and char surface reaction. Not only are some basic devolatilization and surface reaction models provided but the code is implemented such that modifications to these sub-models are easy to implement. This is taken advantage of in this work to implement a devolatilization model and a coal particle surface reaction model.

2.3 Continuous Phase Modeling

2.3.1 Laminar Combustion Modeling

One of the reasons that a laminar flame is desirable for comparisons of numerical simulations to experiments is the circumvention of complex turbulence/reaction interactions. Without the need to consider these interactions the combustion model used in this work is relatively simple. The unmodified laminar combustion model available in OpenFOAM is used. The laminar combustion model treats each finite volume cell as a perfectly stirred reactor. The contents of the cell are evolved in time according to the user specified combustion mechanism and the sources of species and energy are calculated and used directly in their respective finite volume equations.

The two (light and heavy) simulations vary in their volatile gas composition but the same modified version of the DRM-22 mechanism is used in both simulations [2]. The modification to the DRM mechanism allows for the use of Benzene as a tar surrogate in this study. A single reaction is taken from Fortsch et al. [3] (reaction C4 in that paper) to model the secondary pyrolysis of large tar molecules into smaller hydrocarbons already contained in the DRM-22 mechanism. Although in its original form the reaction from Forstch permits the breakdown of a generic, but presumably large, hydrocarbon into CO_2 and H_2 we have assumed the only tar molecule to be benzene (C_6H_6) and have therefore fixed the stoichiometric coefficients. The Arrhenius parameters for the reaction rate were also provided in [3] and are used directly in this work.

$$\nu_{tar}C_xH_yO_z + \nu_{O_2}O_2 \rightarrow \nu_{CO}CO + \nu_{H_2}H_2 \tag{1}$$

The addition of Equation 1 to the DRM-22 mechanism provides an opportunity to investigate the effect of variations in volatile gas composition on flame structure. The two compositions, and the process through which they were selected, is detailed in Section 2.5.1.

2.4 Dispersed Coal Particle Phase Modeling

OpenFOAM provides a framework for not only Lagrangian parcel injection and tracking but momentum and energy transfer between those parcels and the continuous gas phase. Additionally OpenFOAM provides several Lagrangian sub-models specific to pulverized coal simulations. These sub-models are written in a way that enables users to readily implement their own devolatilization and surface reaction (char burnout) models. The models implemented in this work are detailed below.

2.4.1 Parcel Velocity and Temperature Equations

The Lagrangian framework represents physical particles with computational parcels. Each parcel can be representative of many real particles but only particles of the same diameter are grouped together into a parcel. This parcel approach significantly reduces the computational cost of tracking and evolving the dispersed phase. Each computational parcel is tracked through the flow field with velocity determined by

$$\frac{d\boldsymbol{U_p}}{dt} = \boldsymbol{F_{drag}} + \boldsymbol{F_{gravity}}. \tag{2}$$

Where

$$\boldsymbol{F_{gravity}} = m_p * \boldsymbol{g} \tag{3}$$

with particle mass m_p, is used for the force of gravity and a simple spherical drag approximation is used to model drag

$$F_{drag} = \boldsymbol{U_{diff}} * \frac{0.75 \cdot m_p \cdot \mu_c \cdot C_d}{\rho {D_p}^2}. \tag{4}$$

Here C_d is a function of the particle surface Reynolds number, defined as

$$Re_s = \frac{\rho_s \cdot \|\boldsymbol{U_{diff}}\| \cdot D_p}{\mu_s} \tag{5}$$

The subscript s on μ_s and ρ_s indicate these properties are evaluated at the particle surface. In OpenFOAM these surface properties are approximated using the "two thirds rule" to determine the particle surface temperature and then assuming linear property variations with small temperature ranges. For example the particle surface density ρ_s is calculated as follows where ρ_c is the gas phase density interpolated to the parcel position, while T_p, T_s and T_c are, respectively the particle, particle surface and interpolated gas phase temperatures.

First T_s and T_{ratio} are determined with the "two thirds rule".

$$T_s = \frac{2T_p + T_C}{3}; \quad T_{ratio} = \frac{T_c}{T_s} \tag{6}$$

then ρ_s can be determined

$$\Rightarrow \rho_s = \rho_c \cdot T_{ratio}. \tag{7}$$

μ_s is calculated similarly. With Re_s known, C_d is then determined

$$C_d = \begin{cases} 0.424(Re_s) & Re_s > 1000 \\ 24 \cdot \left(1 + 6 \cdot {Re_s}^{-\frac{2}{3}}\right) & Re_s \leq 1000. \end{cases}$$

The temperature of the parcels is assumed to be uniform and its evolution is determined by

$$m_p \frac{d\left(c_{p,p} T_p\right)}{dt} = S_{dev} + S_{char} + \dot{q}_{p,rad} + \dot{q}_{p,c}. \tag{8}$$

Here, $c_{p,p}$ is the coal's specific heat capacity, it is determined from the thermodynamic properties of the constituents in the parcel, scaled by their mass fractions, as well as the current parcel temperature and the seen pressure. The heat of devolatilization is neglected, ($S_{dev} = 0$). The enthaply source from char surface reactions is determined as

$$S_{char} = \frac{m_{char}{}^{*} \cdot H_{char}(T) - m_{CO_2}{}^{*} \cdot H_{CO_2}(T)}{dt} \tag{9}$$

where $m_{char}{}^{*}$ and $m_{CO_2}{}^{*}$ are the mass of solid carbon consumed and CO_2 formed during the surface reaction over time step dt. $H_{char}(T)$ and $H_{CO_2}(T)$ are the sensible enthalpies in units of $\left[\frac{J}{kg}\right]$ taken, as all thermodynamic data in this work are, from the JANAF tables provided in OpenFOAM. The mass loss rates associated with parcel surface reaction and parcel devolatilization are detailed in Sections 2.4.3 and 2.4.2.

The third source term in Equation 8, $\dot{q}_{p,rad}$ is the parcel radiation enthaply source, as radiation is neglected in this study, $\dot{q}_{p,rad} = 0$.

The final term in Equation 8

$$\dot{q}_{p,c} = \frac{h\left(T_c - T_p\right)}{A_s}$$

represents the convective heat transfer from the gas phase to the parcel. Here A_s is the particle surface area and the heat transfer coefficient h is determined with the Ranz Marshall model which is implemented in OpenFOAM as

$$h = \frac{\kappa_s \cdot Nu}{D_p}. \tag{10}$$

Where κ_s is the thermal conductivity evaluated at the particle surface temperature, D_p is the coal parcel diameter (all physical coal particles represented in a single parcel share the same diameter) and Nu is the parcel Nusselt number determined as

$$Nu = 2.0 + 0.6\sqrt{Re_s}\,\sqrt[3]{Pr_s} \tag{11}$$

where Pr_s is the Prandtl number determined from properties evaluated at the surface of the parcel

$$Pr_s = \frac{c_p\,\mu_s}{\kappa_s}$$

and Re_s is the Reynolds number of the parcel as discussed above.

Equation 27, which governs the evolution of parcel mass is given later in this section, after the devolatilization and surface reaction models have been described.

2.4.2 Distributed Activation Energies Devolatilization Model

The model for the parcels contains four different mass fractions $Y_{vol}, Y_{char}, Y_{ash}$ and Y_{liquid} which are in this case representative of, respectively, the coal volatiles, solid carbon, ash and moisture within the coal particles. Each of these mass fractions varies throughout the lifetime of a coal parcel. The initial values corresponding to these mass fractions at the beginning of the parcel's lifetime are denoted as $Y_{vol,0}; Y_{char,0}; Y_{ash,0}$ and $Y_{liquid,0}$, see Table 2 for these initial values. The devolatilization model is active whenever the parcel temperature is over 400 [K] and the moisture of the parcel has been evaporated entirely. In this study the coal is modeled as dry, so this latter condition is met for all parcels at injection time. Once these conditions are met the volatile gas fraction is devolatilized in accordance with the model until satisfying the user selected minimum mass fraction threshold $Y_{vol} = 0.001$. At which point the little remaining volatile mass is transferred to the gas phase and parcel is considered to be completely devolatilized, surface reactions may then begin.

The mass loss during devolatilization is assumed to follow the Distributed Activation Energy Model (DAEM) [4, 5, 6]. In this model a large number of first order decomposition reactions occur in parallel. Each reaction has the same pre-exponential factor A_{v0} and activation energies that are distributed according to a Gaussian distribution. The devolatilization rate at time t depends on the complete temperature history of the coal parcel:

$$\dot{m}_{p,vol} = -\left(Y_{vol,0} \cdot m_{p,0}\right) \int_0^\infty \left[\exp\left(-\int_0^t k(E,\tau)d\tau\right) k(E,t)\right] f(E)dE, \tag{12}$$

where $m_{p,0}$ is the initial particle mass and $\dot{m}_{p,vol} = \frac{d}{dt}(Y_{vol} \cdot m_p)$. The reaction rates are given by

$$k(E,t) = A_{v0} \exp\left(-\frac{E}{RT_p(t)}\right), \tag{13}$$

and the Gaussian distribution of the activation energy with mean E_{v0} and standard deviation σ_{v0} is

$$f(E) = \frac{1}{\sigma_{v0}(2\pi)^{1/2}} \exp\left(\frac{-(E-E_{v0})^2}{2\sigma_{v0}^2}\right). \tag{14}$$

The best performance of any devolatilization model is obtained if the model parameters are fitted to experimentally obtained devolatilization curves for the particular coal of interest. Unfortunately, no such data was available and hence we have used parameters suggested by [4] for a bituminous coal: $A_{v0} = 1.67 \cdot 10^{13}\,1/s$, $E_{v0} = 212.6 \cdot 10^3\,J/mol \cdot K$ and $\sigma_{v0} = 29.4 \cdot 10^3\,J/mol \cdot K$.

An efficient method to solve Equation 12 numerically has been proposed by [6]. The method is based on a modified Gauss-Hermite quadrature rule. Here we use three quadrature points leading to the following numerical approximation

$$\dot{m}_{p,vol} \approx -(Y_{vol,0} \cdot \pm_{,0}) \sum_{i=1}^{3} W_i \frac{m}{\sqrt{\pi}} \exp\left(\frac{-(E_i - E_{v0})^2}{2\sigma^2}\right) \cdot \exp\left(-\int_0^t k(E_i,\tau)d\tau\right) k(E_i,t). \tag{15}$$

Details about the scaling factor m, the quadrature points E_i and quadrature weights W_i are given in [6]. The time integral in the above expression is approximated by a simple Euler method

$$\int_0^t k(E_i,\tau)d\tau \approx k(E_i,t_0)dt_0 + k(E_i,t_1)dt_1 + \cdots + k(E_i,t_N)dt_N, \quad t_N = t. \tag{16}$$

2.4.3 Smith Intrinsic Surface Reaction Model

The volatiles are released until $Y_{vol} \leq 0.001$. At this time the char burnout begins, it is modeled according to the intrinsic model of Smith [7]. The model takes the bulk diffusion of the oxygen to the char particle external surface, the chemical reaction and pore diffusion into account. The oxygen surface reaction order is assumed to be unity. The mass rate of change of a particle due to char combustion is given by

$$\dot{m}_{p,char} = -\pi {D_p}^2 p_c Y_{s,O_2} \frac{R_{ox}R_{ch}}{R_{ox} + R_{ch}}, \tag{17}$$

where Y_{s,O_2} is the oxygen mass fraction seen by the coal parcel, $\dot{m}_{p,char} = \frac{d}{dt}(Y_{char} \cdot m_p)$ and p_c is the gas phase pressure interpolated to the parcel position. The bulk molecular mass diffusion rate R_{ox} is

$$R_{ox} = C_{diff} \frac{T_m^{0.75}}{D_p} \tag{18}$$

where T_m is the mean temperature. The chemical rate R_{ch} is expressed by the intrinsic chemical rate and pore diffusion rate

$$R_{ch} = \eta \frac{D_p}{6} \rho_p A_g k_i, \tag{19}$$

where η is the effectiveness factor, A_g is the specific internal surface area and k_i is the intrinsic reactivity calculated from the Arrhenius expression

$$k_i = A_i \exp\left(-\frac{E_i}{RT_p}\right). \tag{20}$$

The effectiveness factor η is calculated as

$$\eta = \frac{3}{\phi^2}\left(\phi \coth\phi - 1\right), \tag{21}$$

where ϕ is the Thiele modulus [8, 9]

$$\phi = \frac{D_p}{2}\left(\frac{S_b \rho_p A_g k_i p_c Y_{s,O_2}}{D_e \rho_c Y_{s,O_2}}\right)^{1/2}. \tag{22}$$

Here, $S_b = 1.33$ is the stoichiometric coefficient assuming that CO is formed by the heterogeneous reaction, D_e is the effective diffusion coefficient in the particle pores and ρ_c is the gas phase density interpolated to the parcel position. Assuming that bulk and Knudsen diffusion proceed in parallel [8, 9],

$$D_e = \frac{\theta}{\zeta^2}\left[\frac{1}{D_{\mathrm{Kn}}} + \frac{1}{D_{ox}}\right]^{-1}, \tag{23}$$

where θ is the porosity of the char, ζ is the tortuosity of the pores, D_{ox} is the oxygen diffusion coefficient and D_{Kn} is Knudsen diffusion coefficient calculated as

$$D_{\mathrm{Kn}} = C_{\mathrm{Kn}} \bar{r}_{po} \sqrt{\frac{T_p}{M_{w,O_2}}}, \tag{24}$$

where C_{Kn} is a constant, $\bar{r}_{po}$ is the mean pore radius and M_{w,O_2} is the molecular weight of oxygen. The oxygen diffusion coefficient is obtained from [10]

$$D_{ox} = D_0 \left(\frac{T_m}{T_0}\right)^{1.75}. \tag{25}$$

The initial porosity and the specific internal surface area of the char are estimated from the relations [11]

$$A_g = 1546.3 \cdot C^2 - 2834.9 \cdot C + 1301.7 \ [m^2/g], \quad \theta = \frac{A_g}{1000\, m^2/g}, \tag{26}$$

where C is the fixed carbon mass fraction (daf) of the parent coal. All parameters of the char reaction model are given in Table 1. It is assumed that the char particles burn at a constant diameter $\dot{D}_{p,char} = 0$.

Table 1: Parameters for the intrinsic char reaction model.

symbol	value, reference	unit
C_{diff}	$5 \cdot 10^{-12}$, [10, 8]	$\frac{kg}{m^2 \cdot s \cdot Pa}$
A_i	0.052, [11]	$\frac{kg}{m^2 \cdot s \cdot Pa}$
E_i	161.5, [11]	$\frac{kJ}{mol}$
S_b	1.33, [8]	-
θ	0.1, [11]	-
ζ	$\sqrt{2}$, [8]	-
C_{Kn}	97.0, [8]	$\frac{kg \cdot m}{K^{0.5} \cdot \mathrm{kmol}^{0.5} \cdot s}$
$\bar{r}_{po}$	$6 \cdot 10^{-8}$, [8]	m
D_0	$3.13 \cdot 10^{-4}$, [10]	$\frac{m^2}{s}$
A_g	104, eq.(26) [11]	$\frac{m^2}{g}$

The char combustion lasts until $m_p = (1 - Y_{vol,0} - Y_{char,0})\, m_{p,0}$. At this stage, the mass of the parcel does not change anymore and is considered to consist of ash ($Y_{ash,0} = 1 - Y_{vol,0} - Y_{char,0}$). The "inert" ash parcels are not removed from the simulation and are still transported through the domain.

The rate of change of the coal parcel mass can be written as

$$\frac{dm_p}{dt} = \dot{m}_{p,char} + \dot{m}_{p,vol}. \tag{27}$$

2.5 Simulation Geometry

The Hencken burner geometry is modeled as an axisymmetric wedge rather than the real rectangular shape. This modeling approach decreases the computational cost considerably. The entire numerical domain is a 30° wedge as pictured in Figure 1. The inflow areas of the coal flow, co-flow and nitrogen shroud are appropriately scaled to represent their experimental counterparts (with a correction for the co-flow and nitrogen shroud to account for the area taken by the honeycomb structures). The axial length of the domain is $0.25[m]$ this length was selected based on earlier experimental results that suggested devolatilization, combustion of volatiles and char burnout all occurred within this range.

A structured mesh is used with significant stretching to accommodate the large shear layers between inlet streams near the entrance. The use of axisymmetric cyclic boundary conditions on the azimuthal sides of the geometry dictates a minimum of three cells in the azimuthal direction, three azimuthal cells are used here.

Currently the simulation extends radially only to the outside of the nitrogen shroud. Potentially a region of quiescent air may be included beyond the nitrogen shroud in future work. This would permit the entrainment of air into the flow, but as it is currently very little nitrogen is entrained which suggests even that the exterior air entrained would have little effect on the simulation.

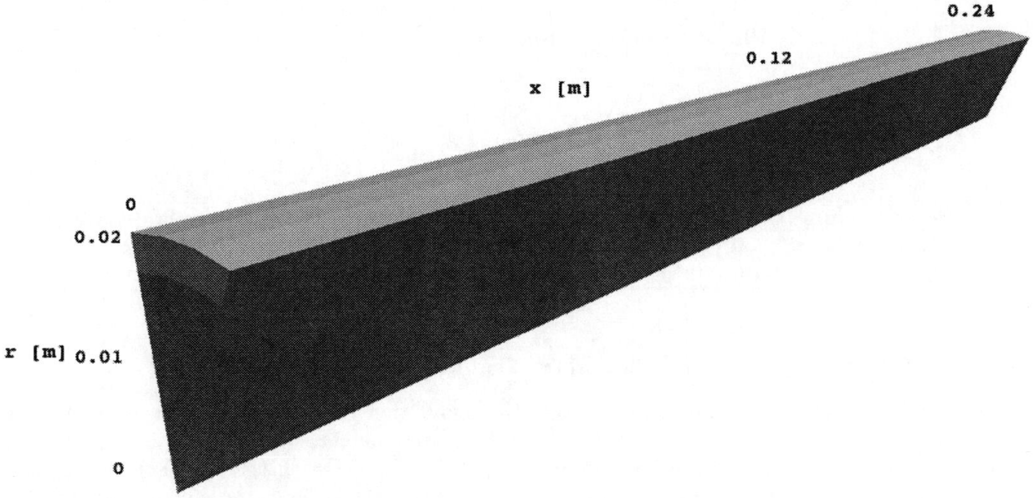

Figure 1: Simulation geometry with co-flow (red), N_2 shroud (green), coal stream (very small tan section below co-flow) inflow boundaries. The teal patch is the front cyclic periodic boundary and the yellow patch is the radial edge of the simulation.

2.5.1 Coal Inlet Boundary Conditions

The coal particles and the entraining air are at room temperature ($T_{air} = 295\,[K]$) upon injection into the domain. The velocity of this inlet, $U_{coalflow} = 27.3\,[m/s]$ in the axial direction, is set to match the corresponding mass flow rate from the experiment. The mass flow rate of coal particles entering the domain is matched to the experimental rates $\dot{m}_{coal} = 0.04\,[g/s]$, and the initial velocity with which the parcels enter the domain is equal to the entraining air.

Table 2 shows the modified proximate analysis of the coal used in this simulation. The phase fractions are modified from the Powder River Basin coal proximate analysis. A "Q-factor" of 1.2 is applied to the proximate analysis values to arrive at the figures in Table 2. The likely conservative value of Q assumed here is intended to account for the very high heating rates experienced by the coal.

In this study two possible coal volatile compositions are considered, denoted "light" and "heavy", both are detailed in Table 3. The two mixtures are selected to present as wide a range in species composition as possible while maintaining energetic and elemental consistency with the coal analysis. Energetic consistency is achieved by ensuring the lower heating value of both mixtures is within 2.5 $\left[\frac{Mj}{kg}\right]$ of the analyzed volatile sample where the lower heating value of the volatiles in the sample is determined as

$$L_{vol} = \frac{L_{coal} - L_{char} \cdot Y_{char}}{Y_{vol}}. \tag{28}$$

Where L_{vol} , L_{coal} and L_{char} are the lower heating values of the volatile gas, dry ash-free coal and char respectively and Y_{vol} and Y_{char} are the volatile and char mass fractions of the dry ash free coal. L_{coal} is provided by the coal ultimate analysis and L_{char} is assumed to be 32.9 $\left[\frac{Mj}{kg}\right]$. Y_{vol} and Y_{char} are determined from the dry ash-free coal proximate analysis (Table 2).

Elemental consistency is achieved by ensuring that the elemental mass fractions of both mixtures, along with the remaining solid carbon in the coal, match the dry ash-free ultimate analysis. The mass fractions of the species in both the heavy and light mixtures in Table 2 satisfy this relationship. An exception has been made for elemental sulfur as the modified DRM-22 used in the simulation does not contain sulfur. The small amount of sulfur remaining in the dry ash-free coal is neglected.

Table 2: Modified Proximate Analysis (Q = 1.2)

Phase	**Gas**	**Liquid**	**Solid**
Phase Mass Fraction	0.539	0.0	0.461

Table 3: Volatile Composition

	Composition [mass fraction]							**LHV** $\left[\frac{Mj}{kg}\right]$
Fuels	N_2	CO_2	CH_4	H_2	CO	C_2H_4	C_6H_6	$LHV_{analysis} = 26.65$
Heavy	0.021	0.330	0.076	0.000	0.000	0.140	0.433	27.96
Light	0.021	0.016	0.106	0.001	0.400	0.154	0.303	28.94

2.5.2 Hot Methane Co-Flow Boundary Conditions

The simulated methane flame co-flow enters the simulation domain mixed and at equilibrium conditions rather than as separate fuel and oxidizer streams through adjacent honeycombs (as in the experiment). The equilibrium conditions are determined with the open-source chemistry solver Cantera [12]. The inputs to Cantera are the equivalence ratio of the methane air mixture, its temperature and its pressure, which are all specified from the experiment. This mixture is then equilibrated adiabatically and at constant pressure with the modified DRM-22 combustion mechanism.

This preliminary chemistry simulation provides much of the necessary inputs for the cfd simulation. The list of species predicted in the Cantera are filtered to include only those species with a mass fraction (following equilibration) of greater than $1e^{-10}$. The resulting co-flow mixture is described in Table 4.

The temperature of the incoming co-flow $T = 1840[K]$ is set as the adiabatic flame temperature of the mixture, which is also obtained from the Cantera simulation. The inflow velocity is determined by assuming the mass flow rate of the unburnt reactants in the experiment equals that of the preburnt products flowing into the simulation. With the equilibrium density calculated by Cantera the velocity can be determined. The velocity of the co-flow is therefore set as 1.456 $\left[\frac{m}{s}\right]$ in the axial direction.

2.5.3 Nitrogen Shroud Boundary Conditions

The nitrogen shroud of the Hencken burner is modeled as the outer annulus around the co-flow and coal/air stream. The area of the axisymmetric section modeled here is consistent with the area of the real shroud on the Hencken burner. Nitrogen is the only gas species injected here ($Y_{N_2} = 1$). The stream is at room temperature and is modeled as 295 $[K]$. The stream axial velocity is matched to the co-flow as in the experiment and is therefore $U_{shroud,axial} = 1.456 \left[\frac{m}{s}\right]$.

Table 4: Co-Flow Inlet Composition

Species	Mass Fraction
N_2	0.74
CO_2	0.107
H_2O	0.088
O_2	0.067
OH	$4.8e^{-4}$
CO	$9.9e^{-5}$
O	$2.8e^{-5}$
H_2	$3.6e^{-6}$
HO_2	$7.0e^{-7}$
H	$1.4e^{-7}$
H_2O_2	$3.7e^{-8}$

3 Results and Discussion

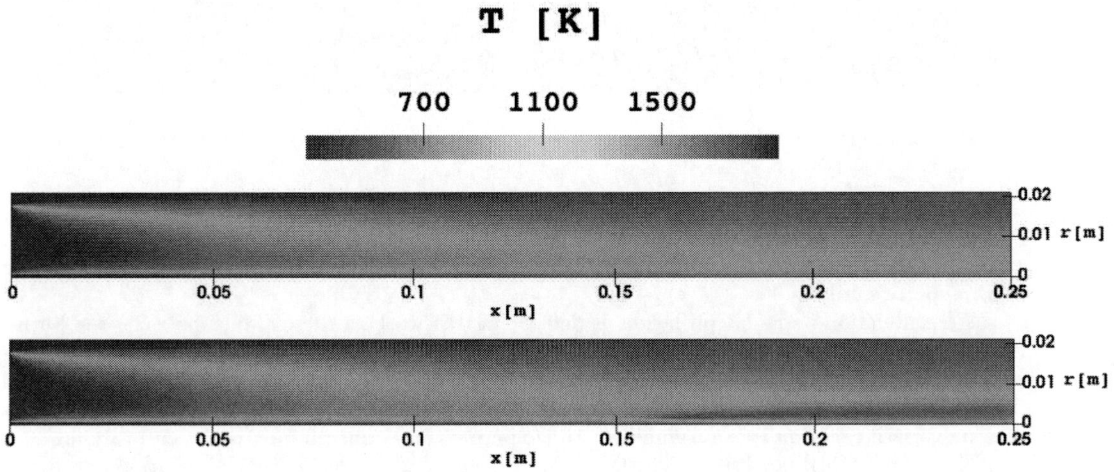

Figure 2: Temperature contour plots of the simulated domain. Only an axisymmetric wedge is modeled and only the center plane of that wedge is illustrated here. The heavy mixture simulation results are presented in the top plot, and the light mixture results in the bottom.

Two comparisons are made in this paper. The first is made between the two simulation cases. The volatile gas composition of the coal is varied between the two simulations. The first case, where the volatile mixture is referred to as "light", used a high percentage of small hydrocarbons in the volatile gases while attempting to minimize the percentage of large molecules like benzene. In the second case, with the volatile composition referred to as "heavy", this trend was reversed, the volatile composition was specified in a manner that maximized the percentage of large molecules.

Even with the simple combustion mechanism there is a noticeable difference in flame structure between the two simulations. Temperature contour plots of both the heavy and light mixtures are presented in Figure 2. Using the temperature as a indicator for the location of the flame we can immediately see a difference in the flame location of the two simulations. The light fuel ignites much more quickly and produces a flame within the domain of the simulation at approximately $0.16\,[m]$. In contrast, while the heavy case devolatilizes it does not form a flame within the space allotted for the simulation.

The second comparison made in this paper is between the experimental data and the simulations. Since both the experimental and simulated results are preliminary only the lift off height is compared and it is found to be in poor agreement. As the heavy simulation does not even contain

Figure 3: Hencken burner running with visible pulverized coal flame and hot methane co-flow.

a flame its comparison to experiment is not useful and is not mentioned here. The light simulation does contain at least the beginning of a flame within the modeling domain so is used to compare with experimental results.

The simulation flame lift off height is defined in this work as the distance between the burner surface and the location of a strong positive gradient in temperature (beyond the hot methane co-flow at the entrance). From the second plot in Figure 2 the simulated flame lift off height is approximately 160 $[mm]$. Preliminary experimental results show a lift off height of 10.8 $[mm]$ for the pulverized coal flame. An example of the experimental setup running is shown in Figure 3.

There are several possible explanations for the poor agreement between simulation and experiment. Likely a significant source of error is the devolatilization model parameters selected in this work. As mentioned in Section 2.4.2 the parameters are taken from a study of bituminous coal while the the Powder River Basin coal used in this work is ranked as sub-bituminous. Changes in coal structure can produce large changes in the devolatilization characteristics. Although volatile species concentration data is not yet available for this experiment it is clear from the flame lift off height results that the observed experimental devolatilization rates are considerably greater than the simulated rates.

Another probable contribution in the discrepancy between the two results involves a mismatch in the coal mass flow rate provided by the feeder and the actual consumption rate of the burner. The coal mass flow rate was measured at the exit of the twin auger feeder system but during transport to the burner some of the mass adhered to the inside of the tubing and did not reach the burner surface. Therefore it is likely that the actual coal entrainment rate was lower than the measured value which was used to specify the corresponding mass flow rates in the simulations.

Earlier iterations of the simulation in this study suggest that a decreased coal particle flow rate ($\dot{m}_{coal}$) tends to decrease the flame lift off height, at least near the current operating point. It seems that when $\dot{m}_{coal}$ is decreased the concomitant decrease in the coal stream's heat capacity allows the devolatilization of the remaining coal parcels to proceed at an increased rate, thus shifting the flame towards the burner surface. The combination of these two effects provides further explanation of the poor agreement in flame lift off heights.

4 Conclusions

In this study simultaneous simulations and experiments of a laminar pulverized coal flame with a methane air co-flow in a Hencken burner were performed. Two qualitative comparisons are made of the results. The first compares two different simulations with varying coal volatile gas compositions. In this comparison it was determined that representing volatile gases with different fuel species had a significant impact on the resultant flame structure.

A second comparison is also made between one of the simulations and the experimental results. Since both the simulation and experiment are rather preliminary only a comparison of the lift off height of the coal flame is made. The simulation is found to be in poor agreement with experiment. Two likely causes for this discrepancy are proposed. First it is likely that the devolatilization model parameters used in the simulation are ill-suited to the type of coal used in the experiment. Second the rate at which coal is entrained in the experiment is inaccurately measured. This flawed rate was taken as the coal parcel injection rate in simulations.

Future work will involve obtaining more accurate devolatilization models and coal entrainment rates. The first of these objectives will be accomplished with the commercial solid fuels modeling software PC Coal Lab© [13]. With this software more accurate devolatilization model parameters can be obtained that are specific to the Powder River Basin coal considered in this work. The second objective, increasing the accuracy with which the coal entrainment rate can be determined will be accomplished through mechanical modifications to the feeder/burner design as well as the transport tube between them.

Acknowledgements

This research was funded by the University of Wyoming's Office for Academic Affairs and School of Energy Resources.

References

[1] OpenCFD. *OpenFOAM - The Open Source CFD Toolbox - User's Guide.* OpenCFD Ltd., United Kingdom, 4.1 edition.

[2] A. Kazakov and M. Frenklach. `http://www.me.berkeley.edu/drm/`.

[3] Dieter Förtsch, Frank Kluger, Uwe Schnell, Hartmut Spliethoff, and Klaus R.G. Hein. A kinetic model for the prediction of no emissions from staged combustion of pulverized coal. *Symposium (International) on Combustion*, 27(2):3037–3044, jan 1998.

[4] D.B. Anthony, J.B. Howard, H.C. Hottel, and H.P. Meissner. Rapid devolatilization of pulverized coal. *Proceedings of the Combustion Institute*, 15:1303 – 1317, 1975. The Combustion Institute.

[5] Peter R. Solomon, Michael A. Serio, and Eric M. Suuberg. Coal pyrolysis: Experiments, kinetic rates and mechanisms. *Progress in Energy and Combustion Science*, 18(2):133 – 220, 1992.

[6] E. Donskoi and D. L. S. McElwain. Optimization of Coal Pyrolysis Modeling. *Combustion and Flame*, 122:359–367, 2000.

[7] I. W. Smith. The Combustion Rates of Coal Chars: A Review. *Proceedings of the Combustion Institute*, 19:1045 – 1065, 1982. The Combustion Institute.

[8] R. I. Backreedy, L. M. Fletcher, L. Ma, M. Pourkashanian, and A. Williams. Modelling pulverised coal combustion using a detailed coal combustion model. *Combustion Science and Technology*, 178(4):763–787, 2006.

[9] N. Schaffel, M. Mancini, A. Szlek, and R. Weber. Mathematical modeling of MILD combustion of pulverized coal. *Combustion and Flame*, 156:1771–1784, 2009.

[10] M. M. Baum and P. J. Street. Predicting the Combustion Behavior of Coal Particles. *Combustion Science and Technology*, 3(5):231–243, 1971.

[11] A. Williams, R. I. Backreedy, R. Habib, J. M. Jones, and M. Pourkashanian. Modelling coal combustion: the current position. *Fuel*, 81:605–618, 2002.

[12] David G. Goodwin, Harry K. Moffat, and Raymond L. Speth. Cantera: An object-oriented software toolkit for chemical kinetics, thermodynamics, and transport processes. `http://www.cantera.org`, 2017. Version 2.3.0.

[13] PC Coal Lab Version 5.0. `http://www.niksaenergy.com/products/pc-coal-lab/`, 1997 - 2017.

Western States Section of the Combustion Institute - Fall 2017 Meeting
Hosted by the University of Wyoming
October 2-3, 2017.

A Fast-Running Simulation Tool for Axisymmetric Oxy-Coal Combustors

Todd Williams[1] *Bradley Adams*[1]

[1]*Mechanical Engineering, Brigham Young University, Provo, Utah*

FENICS (Flame Equilibrium Numeric Iterative Calculation Software) is a code developed to be a fast running tool for use in designing axisymmetric, oxy-coal furnaces. The code calculates gas temperature and composition, wall heat flux, and wall and refractory temperatures using 1D, equilibrium chemistry and a 3D discrete ordinates radiation solver. Chemistry and temperature calculations are made for a discrete number of axial zones wherein the temperature and species composition are assumed to be uniform. Radiation calculations are performed on a finer grid and assume gray gases.
The code was validated by comparing to experimental data obtained from the L1500 Multifuel Furnace. The experimental data were taken from a single oxy-coal combustion experiment with flue gas recirculation run at a firing rate of ~880 kW and an air-fuel ratio of 3.6. FENICS was able to qualitatively match the wall flux and temperature profiles quite well. The major exception to this being that peak flux cannot accurately be predicted due the assumptions utilized 'smearing' the flame. Quantitatively, FENICS showed good agreement with the data. Wall temperatures were accurate to within 150 K (13 %) while total wall radiative heat loss was within 4 %. FENICS was also used to evaluate temperatue and flux profiles for a new pilot-scale oxy-coal reactor.

1 Introduction

Due in part because of concerns with CO_2 emissions, the use of coal as an energy source is projected to decrease over the next 30 years, with the void created thereby expected to be filled with natural gas and so-called renewable energy [1]. Oxy-combustion along with carbon capture and sequestration would allow coal plants to continue operating while greatly reducing carbon emissions, all at much lower cost than replacing said plants [2]. Furthermore, the cost of designing upgrades for current facilities can be reduced through the use of computational design tools. The Flame Equilibrium Numeric Iterative Calculation Software, FENICS, was developed to be a fast running tool to aid in the early stages of the design of oxy-coal systems.

2 Methods

2.1 Experimental Setup

Experimental results were taken from an experiment run with the L1500 Multi-Fuel Furnace, housed at the University of Utah. The L1500 is capable of either air or oxy-firing of both natural gas and/or pulverized coal at up to 1.5 MW. Coal is fed into the reactor pneumatically, rather than being conveyed hydraulically as part of a slurry. The furnace consists to two different sections, a radiative section and a convective section (see Fig. 1). The radiative section is prismatic,

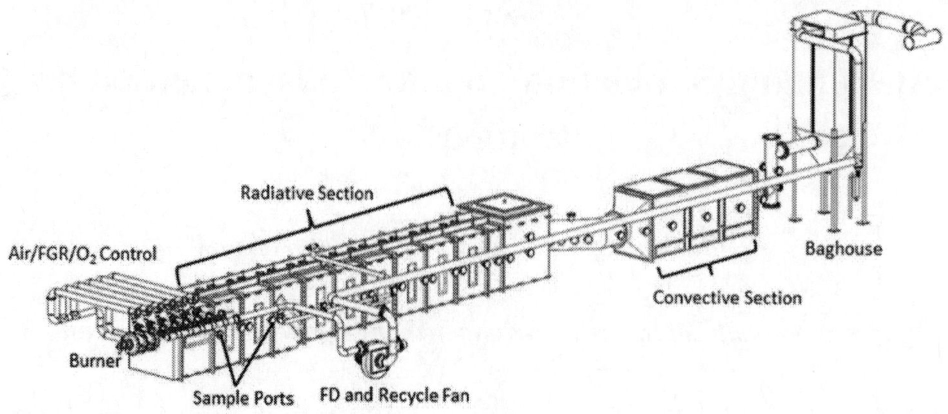

Figure 1: Schematic of the L1500 Multifuel Furnace at the University of Utah

measures 43 ft (13.1 m) long, has a square cross-section measuring 3.3 ft (1 m) on a side, and is divided into 12 sections. The first 10 sections measure 4 ft (1.22 m) each, whereas the last two measure 2 ft and 6 ft (0.61 m and 1.83 m) respectively; each has its own set of instruments, cooling plates, view ports, etc. The data used for the current study came from an experiment where pulverized coal was fired with O_2 and recirculated flue gas at a rate of 879 kW and a air-fuel ratio of 3.6. The composition of the coal burned in the L1500 is found in Table 1.

Table 1: L1500 Coal Composition

Proximate Analysis				
Volatile	Fixed Carbon	Moisture	Ash	
38.60	49.39	3.18	8.83	
Ultimate Analysis				
C	H	O	N	S
70.60	5.05	10.39	1.42	0.53

2.2 Computational Tool

FENICS makes its calculations based on the following assumptions:

- The reactor is axisymmetric
- The reactor is at a uniform pressure
- Temperature, gas composition, and particle dispersion are uniform in the radial direction
- Chemical reactions happen fast enough for the gas to always be in chemical equilibrium
- The only species of interest are H, O, OH, H_2, CO, O_2, CO_2, H_2O, N_2, and SO_2

- The global reaction is fuel lean
- The coal particles are all uniform size
- The coal particles are assumed to be at the gas temperature and have negligible heat capacity
- The volatile gases are released evenly over a specified range of the reactor
- The fixed carbon is oxidized according to an exponential function that decays over the length of the reactor
- The moisture in the coal evaporates prior to entering the reactor

FENICS begins by computing source terms for both volatile gases and char, based on prescribed profiles. The cumulative distribution of volatile gases takes the form

$$S_{vol,cum}(z) = \Theta\left((vol_{strt} - z)(z - vol_{stp})\right) \frac{z - vol_{strt}}{vol_{stp} - vol_{strt}} + \Theta(z - vol_{stp}) \quad (1)$$

where $S_{vol,cum}(z)$ is the cumulative volatile gas source term at axial distance, z, Θ is the Heaviside theta function, vol_{strt} is the axial location where the coal begins to devolatilize and vol_{stp} is where devolatilization finishes. The cumulative char oxidation profile has the form

$$S_{char,cum} = \Theta(z - Z_0)\left[1 - Exp\left(-\frac{Ln(20)\ L}{Z_{95} - Z_0}(z - Z_0\ L)\right)\right] \quad (2)$$

where $S_{char,cum}$ is the cumulative char oxidation source term, Z_0 is the location where the char beings to oxidize, and Z_{95} is the axial location where $95\ \%$ of the char has oxidized. The percentage of volatiles or fixed carbon released in each cell is found by taking the difference between their respective cumulative distribution functions at the faces of each cell.

Source terms for the individual elements are calculated using

$$S_{C,i} = \frac{m_f}{MW_C}\left[m_{FC} S_{char,i} + (\alpha - m_{FC}) S_{vol,i}\right] \quad (3a)$$

$$S_{H,i} = \beta \frac{m_f}{MW_H} S_{vol,i} \quad (3b)$$

$$S_{O,i} = \gamma \frac{m_f}{MW_O} S_{vol,i} \quad (3c)$$

$$S_{N,i} = \delta \frac{m_f}{MW_N} S_{vol,i} \quad (3d)$$

$$S_{S,i} = \varepsilon \frac{m_f}{MW_S} S_{vol,i} \quad (3e)$$

where $S_{C,i}$, $S_{H,i}$, $S_{O,i}$, $S_{N,i}$, $S_{S,i}$ are the sources in cell i of carbon, hydrogen, oxygen, nitrogen, and sulfur respectively; α, β, γ, δ, and ε are the fractions of coal mass that are carbon, hydrogen, oxygen, nitrogen, and sulfur respectively; m_f is the mass of the fuel; MW_C, MW_H, MW_O,

MW_N, and MW_S are the molecular weights of carbon, hydrogen, oxygen, nitrogen, and sulfur respectively; m_{FC} is the fraction of the coal mass that is fixed carbon; and $S_{char,i}$ and $S_{vol,i}$ are the sources of char and volatiles respectively in cell i. From the source terms, the composition, neglecting dissociation, of each cell is then calculated using

$$N^I_{i+1,O_2} = N^I_{i,O_2} + \frac{1}{4}\left(-4\, S_{C,i} - S_{H,i} - 4\, S_{S,i} + 2\, S_{O,i}\right) \tag{4a}$$

$$N^I_{i+1,CO_2} = N^I_{i,CO_2} + S_{C,i} \tag{4b}$$

$$N^I_{H_2O,i+1} = N^I_{H_2O,i} + \frac{1}{2} S_{H,i} \tag{4c}$$

$$N^I_{N_2,i+1} = N^I_{N_2,i} + \frac{1}{2} S_{N,i} \tag{4d}$$

$$N^I_{SO_2,i+1} = N^I_{SO_2,i} + S_{S,i} \tag{4e}$$

where N^I_{i,O_2}, N^I_{i,CO_2}, $N^I_{H_2O,i}$, $N^I_{N_2,i}$, and $N^I_{SO_2,i}$ are the number of moles in cell i resulting from ideal combustion for O_2, CO_2, H_2O, N_2, and SO_2 respectively.

Having calculated the gas composition in each cell, the net enthalpy of the volatile gases (enthalpy of formation minus enthalpy of devolatilezation) can be calculated using

$$H_{vol,net} = m_f\, HHV - \left[\left(\sum N_j\, \bar{h}_{form,j}\right)_{in} - \left(\sum N_j\, \bar{h}_{form,j}\right)_{out}\right] \tag{5}$$

where $H_{vol,net}$ is the net enthalpy of the volatiles, HHV is the higher heating value of the coal, and $\bar{h}_{form,j}$ is the molar specific formation enthalpy of species j.

Next, starting with an initial guess for the radiative heat flux to the walls, FENICS calculates gas temperature and composition for each cell by solving the energy balance

$$Q_{rad,i} + Q_{extra,i} + H_{vol,net}S_{vol,i} + \Delta\left[\sum_j N^E_j(T)\, \bar{h}_{tot,j}(T)\right]_{in-out} = 0 \tag{6}$$

where $Q_{rad,i}$ is the radiative heat loss of cell i, $Q_{extra,i}$ is additional heat losses in cell i[1], $N^E_{j,i}$ is the number of moles of the species j in cell i at chemical equilibrium, $\bar{h}_{tot,j}$ is the molar specific enthalpy of species j, and T is the temperature. $N^E(T)$ is found using the dissociation reactions [3]

[1] Q_{extra} is used to account for potential heat losses due to view ports, cooling plates, etc. not modeled in the radiation routine.

$$\frac{1}{2}H_2 \rightleftharpoons H \tag{R1}$$
$$\frac{1}{2}O_2 \rightleftharpoons O \tag{R2}$$
$$\frac{1}{2}H_2 + \frac{1}{2}O_2 \rightleftharpoons OH \tag{R3}$$
$$H_2 + \frac{1}{2}O_2 \rightleftharpoons H_2O \tag{R4}$$
$$CO + \frac{1}{2}O_2 \rightleftharpoons CO_2 \tag{R5}$$

and solving the system of equations [4]

$$\sum_j \nu_{j,C} N^E_{j,i} = \sum_j \nu_{j,C} N^I_{j,i} \tag{7a}$$
$$\sum_j \nu_{j,H} N^E_{j,i} = \sum_j \nu_{j,H} N^I_{j,i} \tag{7b}$$
$$\sum_j \nu_{j,O} N^E_{j,i} = \sum_j \nu_{j,O} N^I_{j,i} \tag{7c}$$
$$N^E_{N_2,i} = N^I_{N_2,i} \tag{7d}$$
$$N^E_{SO_2,i} = N^I_{SO_2,i} \tag{7e}$$
$$K_1 = \frac{N^E_{H,i}\sqrt{\frac{P}{N^E_{tot,i}}}}{N^E_{H_2,i}} \tag{7f}$$
$$K_2 = \frac{N^E_{O,i}\sqrt{\frac{P}{N^E_{tot,i}}}}{N^E_{O_2,i}} \tag{7g}$$
$$K_3 = \frac{N^E_{OH,i}}{\sqrt{N^E_{H_2,i}\, N^E_{O_2,i}}} \tag{7h}$$
$$K_4 = \frac{N^E_{H_2O,i}}{N^E_{H_2,i}\sqrt{\frac{N^E_{O_2,i}\, P}{N^E_{tot,i}}}} \tag{7i}$$
$$K_5 = \frac{N^E_{CO_2,i}}{N^E_{CO,i}\sqrt{\frac{N^E_{O_2,i}\, P}{N^E_{tot,i}}}} \tag{7j}$$
$$N^E_{tot,i} = \sum_j N^E_{j,i} \tag{7k}$$

where $\nu_{j,k}$ indicates the number of atoms of element k in species j, K_ℓ is the equilibrium constant for reaction ℓ, P is the absolute pressure, and $N^E_{tot,i}$ is the total number of moles at equilibrium in cell i.

After the composition and temperature of each cell are calculated, a radiation subroutine is called that calculates the radiative intensity field, net radiative source terms (i.e., the flux divergence), wall incident and net heat fluxes, and wall temperatures for the reactor and returns radiative heat losses for use in Eq. 6. The radiative heat transfer calculation is performed in three dimensions, although the results are essentially axisymmetric since the medium properties and boundary conditions are axisymmetric. Radiation intensities are caluclated using the discrete ordinate method [5]. The code then iterates between thermochemistry and radiation calculations until T and Q_{rad} converge to a specified tolerance.

2.3 Numerical Simulations

The accuracy of FENICS was first evaluated by modeling the radiative section of the L1500. As mentioned above, the furnace has a square cross-section, however FENICS can only handle prismatic, axisymmetric geometry. Because of this, three different cases were run. In the first, the reactor diameter was set to $1\,\mathrm{m}$ to match the side length of the L1500. In the second, the diameter was set to $1.13\,\mathrm{m}$ so that the volumes would match. Finally, in the third simulation, the diameter was set to $1.27\,\mathrm{m}$ so that the surface areas would match. The values of vol_{strt}, vol_{stp}, Z_0, and Z_{95} were set to $0\,\mathrm{m}$, $2.44\,\mathrm{m}$, $2.44\,\mathrm{m}$, and $6.09\,\mathrm{m}$, corresponding to combustion in the first 5 segments. Values of $75\,\mu\mathrm{m}$ and $1300\,\frac{\mathrm{kg}}{\mathrm{m}^3}$ were used for the coal particle diameter and density, respectively.

Speed and accuracy were also evaluated by modeling the proposed design of Brigham Young University's pressurized oxy-coal (POC) reactor and comparing results from FENICS to results obtained from Reaction Engineering International using their two-phase reacting CFD code, GLACIER. The reactor is cylindrical, with a diameter of $8\,\mathrm{in}$ and a length of $72\,\mathrm{in}$ ($0.2032\,\mathrm{m}$ and $1.829\,\mathrm{m}$ respectively). Similar to the L1500, the POC reactor will utilize a pneumatic feed system. The composition for the coal used for the POC simulations can be found in Table 2. The simulations used a firing rate of $100\,\mathrm{kW}$ and a CO_2 flow rate to coal feed rate ratio of 1.8. The O_2 flow rate was set such that the exit gas would contain $3\,\%$ dry O_2.

Table 2: POC reactor coal composition

Proximate Analysis				
Volatile	Fixed Carbon	Moisture	Ash	
39.2	40.8	10.0	10.0	
Ultimate Analysis				
C	H	O	N	S
63.00	4.60	1.20	10.60	0.58

3 Results and Discussion

3.1 L1500

A comparison between the experimental and numerical temperature plots can be seen in Fig. 2. The equal surface area simulation matches most closely with the experimental data, being accurate to within $150\,\mathrm{K}$ or $13\,\%$. A comparison of the flux profiles is seen in Fig. 3. The data exhibits a spike in the flame region that is absent in all three simulations, however the total radiative heat loss was accurate to within $6\,\%$, $2\,\%$, and $4\,\%$ for cases 1, 2, and 3 respectively.

The discrepancy between the experimental and simulated profiles have two probable causes. First, in the flame region of the actual reactor, the maximum temperature is much higher than the average temperature. This results in a spike in heat flux to the walls. Additionally, the high flux in the flame region transfers more heat out of the gases, cooling them down and in turn lowering the temperature of the downstream walls. Additionally, the wall temperatures span a range of $600\,\mathrm{K}$. The conduction coefficients of the materials used in the L1500 can vary by as much as $5.3\,\%$ in this range. Because of this, it may not be appropriate to assume constant wall conductivity at an average temperature. For this reason, some preliminary work has been done to investigate the effect of temperature dependent properties, though no conclusive work has yet been done.

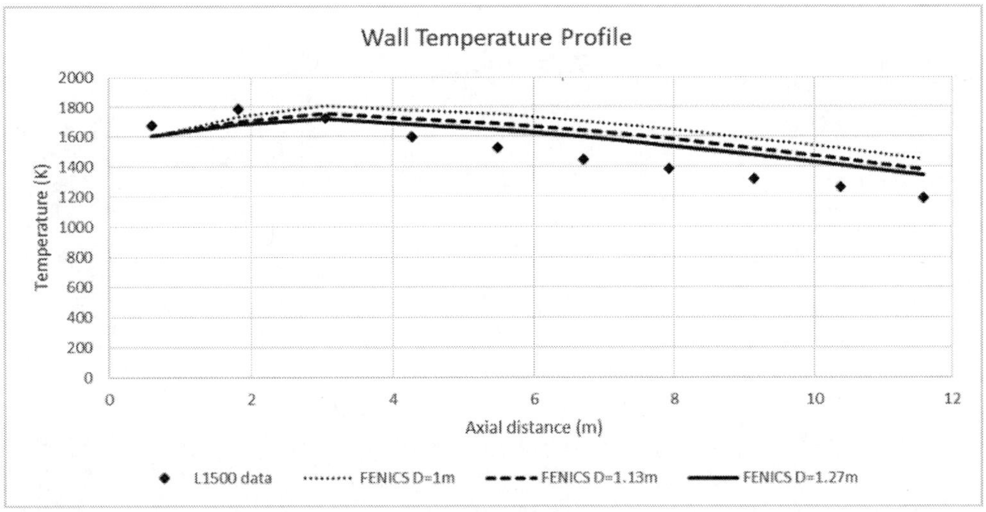

Figure 2: Wall temperature profiles from the L1500 experimental data and from FENICS calculations

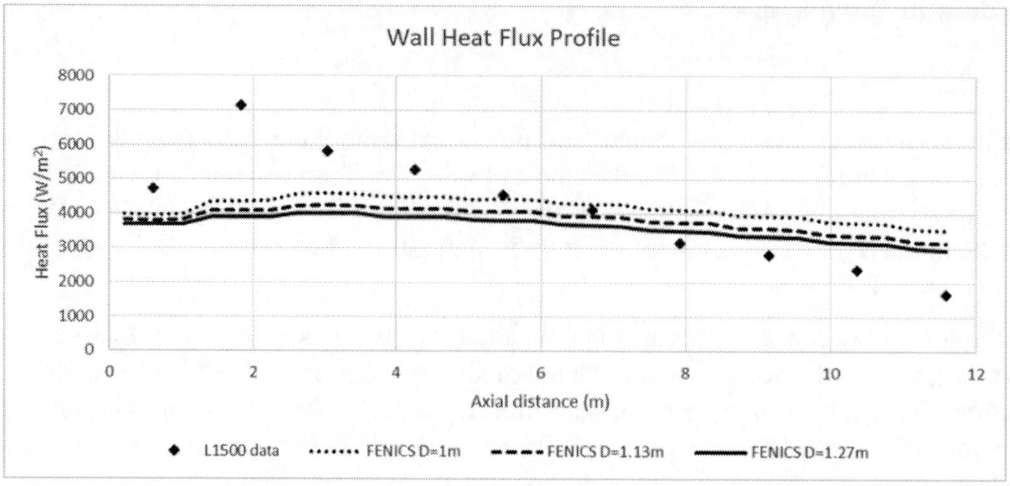

Figure 3: Wall heat flux profiles from the L1500 experimental data and from FENICS calculations

3.2 POC

A comparison of average radial gas temperature as a function of axial position is found in Fig. 4. The results agree to within $4.1 \pm 3.3\,\%$. The FENICS run of the POC reactor was performed with 15 thermochemistry cells and 8460 radiation cells on a single core of an Intel® Core™i5-6500 CPU and ran in 5 min 6 sec. The GLACIER simulation was also performed on a single core, but used 4.7 million cells and took approximately 240 h.

The GLACIER calculations show a higher maximum, radially averaged, gas temperature as well

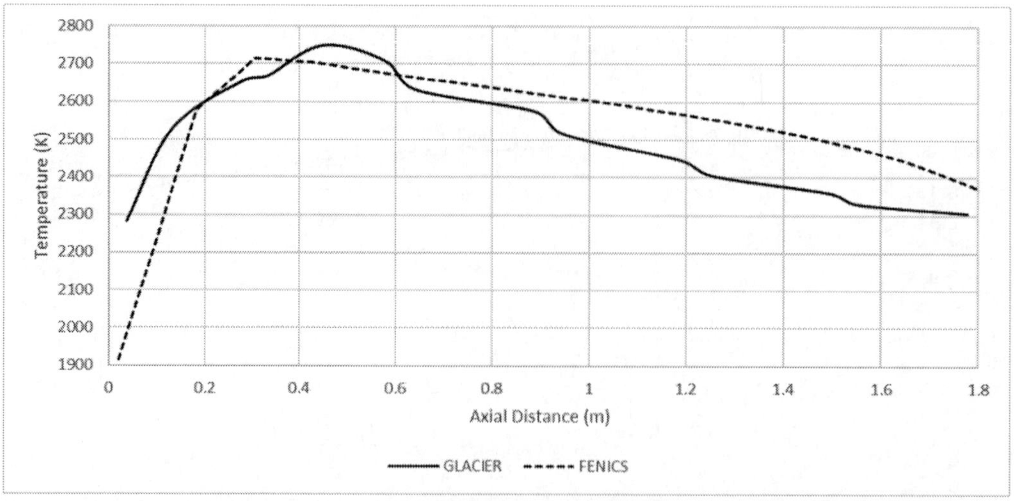

Figure 4: Gas temperature profiles for the POC reactor design from FENICS and GLACIER

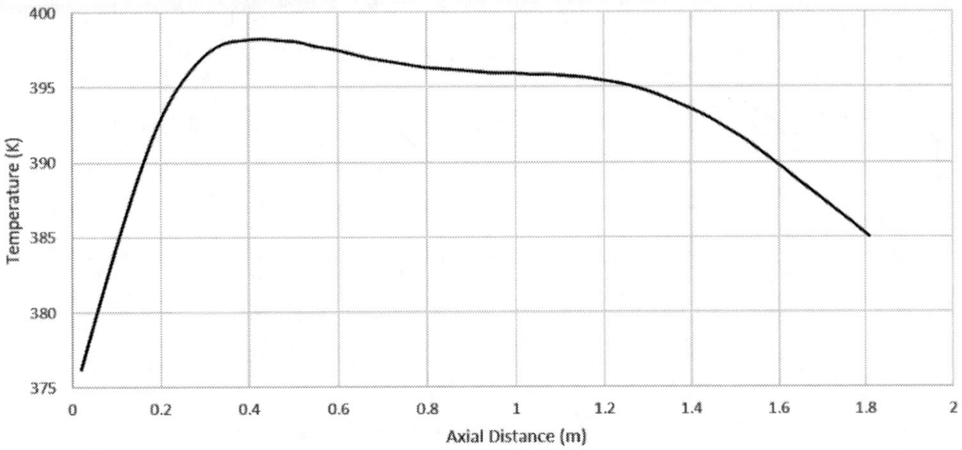

Figure 5: FENICS output of refractory outer shell temperature

as a lower radially averaged gas exit temperature. The discrepancy in maximum gas temperature is most likely due to two sources. First, FENICS does not have any recirculation, whereas GLACIER is able to model recirculation in the flame region. Second, the even radial distribution of particles is believed to increase the optical thickness of the flame region, resulting in less radiative heat transfer from the downstream walls, gases, and particles to the inlet region. The difference in gas exit temperature is likely due to the presence of 15 view ports in the reactor walls. In the REI simulations, these ports were open, resulting in additional heat losses that were not accounted for in the FENICS runs.

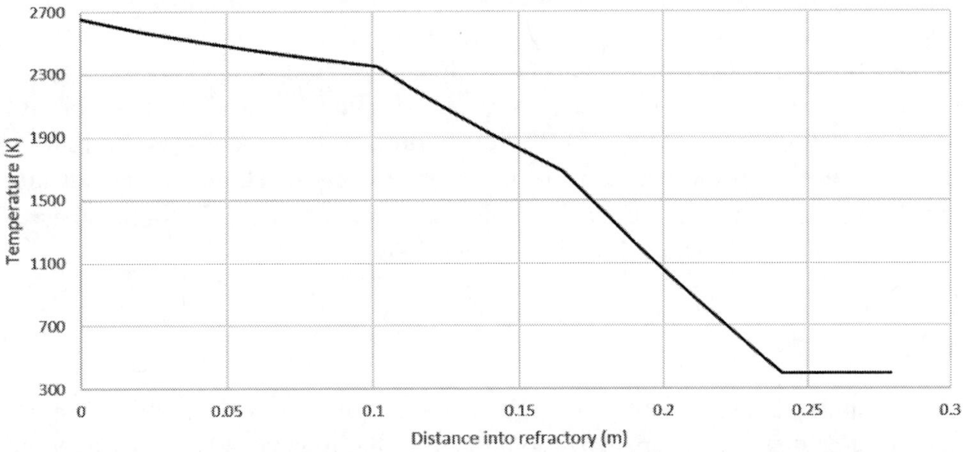

Figure 6: Radial temperature profile at axial distance where wall temperatures are at a maximum

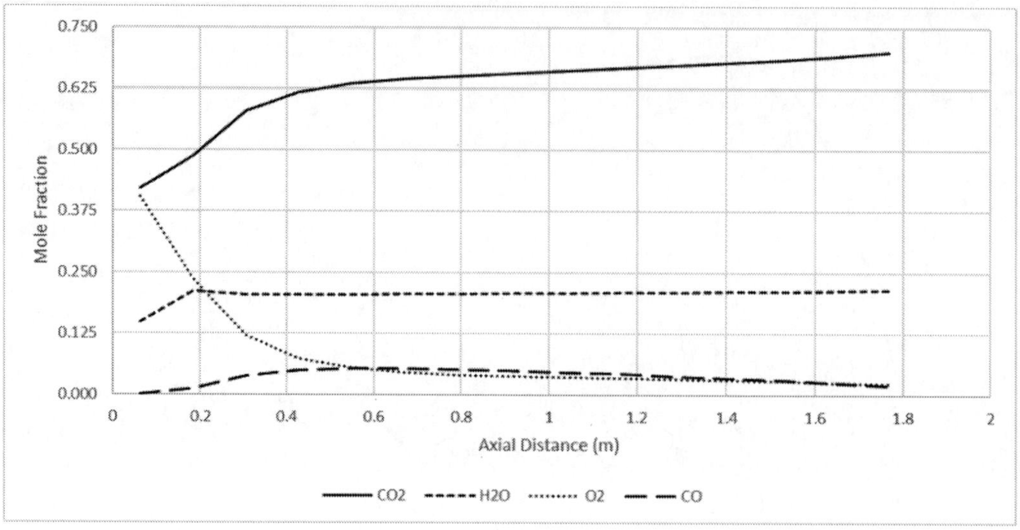

Figure 7: FENICS output of concentrations of selected gases in the POC reactor

As mentioned above, FENICS also calculates incident flux, refractory temperatures, and species concentrations. Figure 5 shows a plot of the temperature of the outer shell of the reactor. The shell temperature profile is an important design constraint in that temperatures less than 450 K allow the use of more economically advantageous materials for the pressure vessel. Figure 6 shows a plot of the radial temperature profile at the axial location where the wall temperatures are at a maximum. Plots such as these have been used as part of the design process in determining the materials and dimensions of the refractory layers, type and placement of thermocouple instrumentation within the reactor walls, and limits on reactor operating conditions.

Species concentrations from the POC run can be seen in Fig. 7. As the coal devolatizes and the char then begins to oxidize, the CO_2 levels raise quickly in the first 20 % of the reactor length. CO_2 production rates then taper off as the char is depleted. H_2O jumps quickly in the first 10 % of the reactor length after which it remains nearly constant, dipping slightly, immediately following the flame region because of dissociation. The O_2 levels drop monotonically as the fuel is consumed, with most of it gone within the first third of the reactor length. The CO concentration increases rapidly, once the gas temperatures surpass 2600 K. As the gas cools, CO diminishes at a lower rate than it was created. This is because there is less availible O_2.

4 Conclusion

FENICS has demonstrated sufficiently accurate results for use in exploring the design space of an axisymmetric oxy-coal furnace, especially in designing the materials and instrumentation used in the refractory walls. Additionally, it shows a significant improvement in computation time over full CFD simulations. The accuracy could be improved, at the expense of run time, by incorporating temperature dependent material properties. Accuracy could also be improved by adding radial discretization for the gas temperature, gas species, and particle property distributions.

Acknowledgments

Acknowledgment: This material is based upon work supported by the Department of Energy under Award Number DE-FE0029157.

References

[1] Energy Information Association. Annual energy outlook 2017: With projections to 2050, 2017.

[2] F. P. Sioshansi. *Generating Electricity in a Carbon-Constrained World.* Academic, Oxford, 2009.

[3] C. R. Ferguson. *Internal combustion engines.* John Wiley & Sons, New York, 2nd ed. edition, 2001.

[4] S. R. Turns. *An introduction to combustion.* McGraw-Hill, New York, 1996.

[5] B. R. Adams and P. J. Smith. *Three-dimensional Discrete-ordinates Modeling of Radiative Transfer in a Geometrically Complex Furnace.* Combustion Science and Technology, Vol. 88, pp. 293-308, 1993.

Western States Section of the Combustion Institute – Fall 2017 Meeting
Hosted by the University of Wyoming
October 2-3, 2017

Modeling Effects of Annealing on Coal Char Reactivity to O_2 and CO_2 Based on Preparation Conditions

Troy Holland[1,2] Thomas H. Fletcher[2]

[1]Chemical Engineering Dept., Brigham Young University, Provo, UT, USA 84602
[2] Los Alamos National Laboratory, Los Alamos, NM, USA

Abstract
Oxy-fired coal combustion is a promising potential carbon capture technology. Predictive CFD simulations are valuable tools in evaluating and deploying oxy-fuel and other carbon capture technologies either as retrofit technologies or for new construction. However, accurate predictive combustor simulations require physically realistic submodels with low computational requirements. A recent sensitivity analysis of a detailed char conversion model (Char Conversion Kinetics (CCK)) found thermal annealing to be an extremely sensitive submodel. In the present work, further analysis of the previous annealing model revealed significant disagreement with numerous data sets from experiments performed after that annealing model was developed. The annealing model was accordingly extended to reflect experimentally observed reactivity loss due to thermal annealing of a variety of coals in diverse char preparation conditions. The model extension was informed by a Bayesian calibration analysis. Additionally, because oxy-fuel conditions include extraordinarily high levels of CO_2, development of a first-ever CO_2 reactivity loss model due to annealing is presented.

Key Words: coal, char oxidation, oxy-fuel, kinetics, sensitivity analysis

Nomenclature

Parameter(s)	Description
a	A parameter for determining the mean of the annealing activation energy distribution
$A_{d,0}$	The initial value of A_d (s^{-1})
A_d	The preexponential factor of the annealing reaction (s^{-1})_
b	A parameter for determining the mean of the annealing activation energy distribution
B_f	A factor used to bifurcate the log normal distribution of the annealing activation energy
B_r	A factor used to bifurcate the log normal distribution of the annealing activation energy
c	A parameter for determining the mean of the annealing activation energy distribution
CBK	Carbon Burnout Kinetics model
CCK	Carbon Conversion Kinetics model
E_d	The activation energy of some thermal deactivation process (kcal/mol)
f	The fraction of carbon converted (burned) from the char, post devolatilization
f_i	The fraction of active sites in bin "i" in the thermal annealing model
HR	The *initial* particle heating rate (K/s)
HTT	The high temperature heat treatment time (s)
MC_R	Measured **relative** coal reactivity (post annealing)
PC_R	Model-predicted **relative** coal reactivity (post annealing)
p_0	An NMR parameter for the fraction of intact bridges in the coal pseudo monomer.
T_{peak}	Peak particle temperature during heating (K)
V_{ASTM}	ASTM volatile yield (wt% daf)
y	The prediction (with uncertainty) of the model+discrepancy+observational error
δ	The discrepancy between a model and reality, typically due to incomplete system knowledge or the ubiquitous, imperfect assumptions used to develop a model
ε	The observational error (noise) in experimental observations
η	The model (or an emulator thereof)
μ_{Ed}	The mean of the annealing activation energy distribution (kcal/mol)
σ_0	The initial standard deviation of the annealing activation energy distribution (kcal/mol)
σ_{Ed}	The standard deviation of the annealing activation energy distribution (kcal/mol)

1. Introduction

A comprehensive single-particle coal combustion model (Carbon Conversion Kinetics or CCK) was recently subjected to a global sensitivity analysis.[1] The results indicated that the parameters of the thermal annealing submodel were overwhelmingly the most sensitive of the parameter sets (relative to char burnout and temperature profile predictions) besides the chemical kinetic parameters. In the present work, thermal annealing is used as an umbrella term to include both the radical changes in char reactivity due to coal pyrolysis and the lesser (but still substantial) reactivity loss induced by thermal treatment of the post-pyrolysis char. These effects may include reactivity loss due to changes in coal morphology (swelling, changes in pore structure, molten ash that physically plugs pores) and a shift in coal chemistry (cross-linking, loss of reactive groups, rearrangement of char carbon structure, loss of inorganic catalytic activity etc.). Such a broad definition of annealing is adopted because, at the temperatures of practical coal combustion, it may well be infeasible to separate the numerous effects since they occur on similar time scales and may be better viewed as continuous rather than discrete events.[2]

2. Literature Observations

The impact of thermally-activated reactivity loss on carbon oxidation was observed decades ago.[3] Since then, numerous researchers have documented char reactivity loss,[4-11] and developed several models to incorporate variable char reactivity in combustion modeling. The available annealing literature spans an eclectic mix of carbon-based materials, preparation conditions, and char structural changes (both chemical and physical). Though the myriad processes of char annealing are not fully understood, the aggregate literature strongly implies that a distributed activation energy annealing model could be appropriate, and that the activation energy distribution should depend on heating rate, peak temperature, and coal precursor. [2, 12-23]

3. Experimental

The annealing model employed in CBK has been frequently reused in comprehensive char conversion models. The CBK annealing model was first calibrated to the relative paucity of experimental data available at the time, i.e., five papers published between 1973 and 1996 [4-7, 11] The older data taken over a large time span resulted in highly diverse experimental methods, making it quite difficult to arrive at a single, consistent comparison. For example, particle sizes varied greatly, the reaction regime may have been zone I or zone II, measures of reactivity were not uniform, treatment temperatures were generally hundreds of Kelvin lower than practical combustion conditions, and some of the precursors were carbon sources other than coal. Also, except for the most recent paper listed,[4] the data were obtained using exceptionally low heating rates (well below 1 K/s) and long heat treatment times (up to 2 hours). These data are potentially useful for the regimes they were taken in, but more recent data show clearly that the most dramatic and dynamic annealing occurs in the first tens or hundreds of milliseconds, implying that annealing models should focus on (or at least include) short timescale data. Fortunately, much data has been collected on the millisecond to second timescale in the two decades since Hurt et al.[24] published the CBK model. Unfortunately, the bulk of these data lack one or more crucial model input or output, such as a reasonably well-defined heating rate, comparable reactivity measurements, or a recorded proximate and ultimate analysis, etc. Nevertheless, the literature contains several times more applicable data now than at the advent of the CBK model, which implies the potential for a more broadly applicable, less uncertain annealing model. The data discussed here were obtained from a detailed search of the literature. However, because the conversion of graphical data to a tabulated format entailed considerable effort and is likely of great value to related model development, the tables are made available elsewhere.[25]

3.1. The Error Factor

To accurately assess model success, a quantity termed the "error factor" is defined. The error factor is the larger of the ratio of model prediction to experimental measurement or the reciprocal of that same ratio. The result is the factor by which the prediction differs from the measurement, and by taking the larger value, under-prediction is treated on equal footing with over-prediction (i.e., a ratio of PC_R/MC_R of 0.1 or 10 are both penalized as being off by a full order of magnitude).

4. O_2 Reactivity Model Development, Results, and Discussion

4.1. Parent Model

Hurt et al.[24] surmised that an appropriate oxidation preexponential factor model could have the functional form of Equation 1. They dealt with the precursor dependence as a separate correlation (Equation 2), which left the annealing model as a function of the time-temperature profile only. The annealing model functional dependence was determined by apportioning the total (unknown) number of active sites N_0 in "i" bins each with a unique activation energy, where relative bin size was determined by the lognormal distribution. The annealing model then reduces to a series of i first order kinetic expressions, where N_i is the number of active sites in bin i, A_d is the preexponential factor of the annealing reaction, and $E_{d,i}$ is the annealing activation energy associated with bin i as seen in Equation 3.

$$A_{ox} = \mathrm{f}[precursor, T(t)] \tag{1}$$

$$\ln(A_0) = 10.96 - 0.07136 * \mathrm{C} \tag{2}$$

$$\frac{dN_i}{dt} = -A_d \exp\left(-E_{d,i}/(RT_p)\right) N_i \tag{3}$$

Again, the actual value of N_i is unknown (i.e., there is no way to count the number of actual active sites), so N_i is normalized by N_0. The model output is the ratio of the number of active sites remaining to the initial number of active sites, N_A/N_0=f[T(t)], where the ratio is known to initially be unity. Experimentally, this requires at least two reactivity data points to compare model output as a ratio of the two experimental measurements, which cancels the unknown N_0 from the equation and allows the model parameters to be fit to data.

N_i/N_0 initially follows a log normal distribution as in Equation 4. In Equation 4, $\ln(E_d)$ is considered as a variable rather than E_d, and thus the normal distribution of $\ln(E_d)$, yields a log normal distribution of E_d, where μ_{Ed} and σ_{Ed} are, respectively, the mean and standard deviation of the distribution. For any given value of activation energy for bin i ($E_{d,i}$), Equation 5 yields the ratio N_i/N_0 at t=0. Equation 6 is a discretized approximation for $i<\infty$, and holds at t=0, but as time proceeds, the series of i ODEs (Equation 7, where $f_i=N_i/N_0$) reduces the value f_i, so the distribution of active sites degrades to some irregular distribution. The sum of f_i at any time t>0 is some value less than unity, and, at fully kinetically-limited conditions, is directly proportional to the relative reactivity of the partially annealed char at a given time t.

$$N\left(\ln(E_d)\,; \mu_{E_d}, \sigma_{E_d}\right) = \frac{1}{\sigma_{E_d}\sqrt{2\pi}} \exp\left[-\frac{\left(\ln(E_d)-\mu_{E_d}\right)^2}{2{\sigma_{E_d}}^2}\right] \tag{4}$$

$$\frac{N_i(E_{d,i})}{N_0} = \frac{1}{\sigma_{E_d}\sqrt{2\pi}} \exp\left[-\frac{\left(\ln(E_{d,i})-\mu_{E_d}\right)^2}{2\sigma_{E_d}{}^2}\right] \quad (5)$$

$$\sum_i \Delta E_d \left[\frac{N_i}{N_0}\right]_{initial} = \sim 1 \quad (6)$$

$$\frac{df_i}{dt} = -A_d \exp\left(-E_{d,i}/(RT_p)\right) f_i \quad (7)$$

4.2. Conceptual Development

Considering the aggregate picture presented in the literature, and the experimental data, it is clear that coal annealing depends heavily on the precursor, heating rate, treatment time, and peak temperature. The literature also clearly highlights several of the most prominent thermal deactivation processes (loss of heteroatoms, carbon structure reordering, ash fusion etc.). Such diverse physical and chemical changes cannot be adequately captured by a single activation energy, at least not for a broad array of preparation conditions and precursors. Unfortunately, neither data nor computational power are available in sufficient quantity to rigorously model the vast diversity of annealing processes, so a distributed activation energy extended from the prior CBK model by Hurt et al.[24] is used. The literature observations in section 2 also imply that an annealing model should only affect the preexponential factor in combustion kinetics, and is ambivalent on the subject of reactivity loss of O_2 vs. CO_2. Thus, the addition of an identical annealing model is possibly necessary to capture CO_2 activity loss (with appropriate parameter values tuned to CO_2 data). While, the CBK annealing submodel is an excellent starting point, it lacks a more subtle point found in recent literature observations. The annealing model as a whole depends on time, peak temperature, and heating rate, with the expectation that precursor differences fall under the purview of an additional submodel, while the actual distribution of activation energies is fixed for all coals and preparation conditions. Both the annealing literature and a basic understanding of coal combustion indicate that this is an oversimplification. The activation energy distribution simply states the relative abundance of various deactivation pathways. The following examples briefly describe why this distribution depends heavily on precursor and preparation conditions:

- In the case of the precursor, coals have widely varying chemical structures, sometimes even when their elemental composition is nearly identical. This variation can be observed not only between different coal seams, but also within the same seam to some degree. A different chemical structure implies a different distribution of annealing pathways.
- In the case of peak temperature, consider a coal with high catalytic ash content. This ash may limit pyrolysis or encourage graphite crystal growth by catalyzing crosslinking or carbon crystal rearrangement.[17-20] In such a case, the ash will fuse, eliminating the catalyzed deactivation pathway, if and only if a sufficient peak temperature is attained. A catalyst is defined, in part, by lowering the activation energy of a reaction pathway, so a loss of a catalyst certainly changes the correct form of the distribution of annealing activation energies.
- In the case of heating rate, the importance of initial heating rate on pyrolysis is well established. As the extent of pyrolysis shifts for various heating rates, the chemistry of the newly formed char necessarily adjusts as well. Again, different char chemistry unavoidably leads to a different activation energy distribution.

Since annealing depends on so many variables, the lognormal activation energy employed in CBK and its many successors is intrinsically flawed. This lack of generality was unavoidable given the data available to Hurt et al.,[24] and while significantly more relevant data are now available, the data are far from sufficient to construct a perfect annealing model. However, given the arguments of more recent literature and data, it is possible to extend the CBK annealing model to include the effects of coal type, heating rate, and peak temperature. As in the Hurt annealing model, Equation 4 represents the log-normal distribution of the activation energy, which is divided into i "bins" with i activation energies (i=100 was found to be adequate). N_i, N_o, and f_i are also computed as discussed previously and shown in Equations 5-7.

Equation 7 is integrated over a time-temperature profile to arrive at a fraction of remaining reactivity, rather than an actual value for post-annealing reactivity. This is because the initial reactivity and N_o are not obtainable experimentally. In contrast to the CBK annealing model, the mean and standard deviation of the activation energy used here, along with the preexponential factor (Equations 8-10) become functions of coal type, heating rate (HR), and peak temperature experienced during heat treatment (T_{peak}).

$$\mu_{E_d} = \mathrm{f}(T_p, \text{Precursor}, \text{HR}, \text{reactive gas}) \quad (8)$$

$$\sigma_{E_d} = \mathrm{f}(T_p, p_0, \text{HR}, \text{reactive gas}) \quad (9)$$

$$A_d = \mathrm{f}(T_p, p_0, \text{HR}, \text{reactive gas}) \quad (10)$$

In the equations above, p_0 represents a coal-type effect based on chemical structure. While p_0 ranges only between 0 and unity, the majority of coals fall in a range with roughly 30% variation; the model form uses scaled parameters so that the 0-1 range of p_0 is not overwhelmed by the ~1000-2000 K range that influences the T_{peak} parameter.

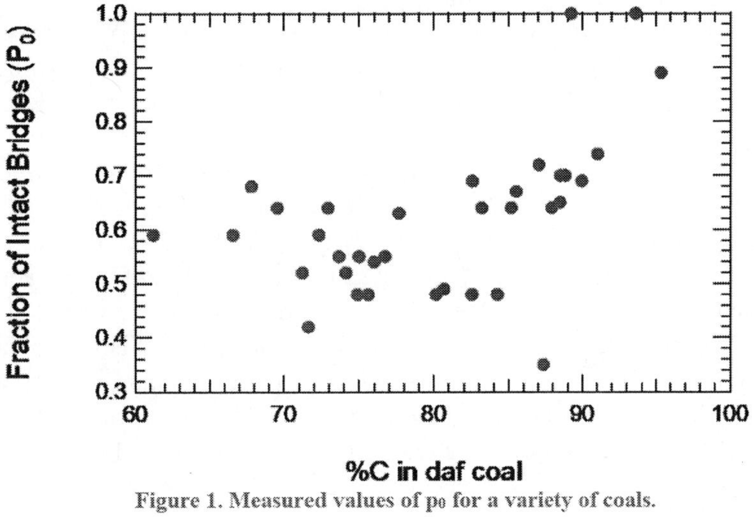

Figure 1. Measured values of p_0 for a variety of coals.

4.3. Model Calibration and Optimization

Having arrived at a theoretical dependency for an extended annealing model, finding the functional form relies on a statistical calibration tool, insight from past annealing experiments, a good deal of trial and error, and an optimization routine. The statistical calibration tool is complex, and numerous publications have been devoted to various iterations, applications, and development of the method. The method is not detailed here, but details are available elsewhere.[26-29] The calibration can be broken into six steps, as demonstrated in Figure 2.

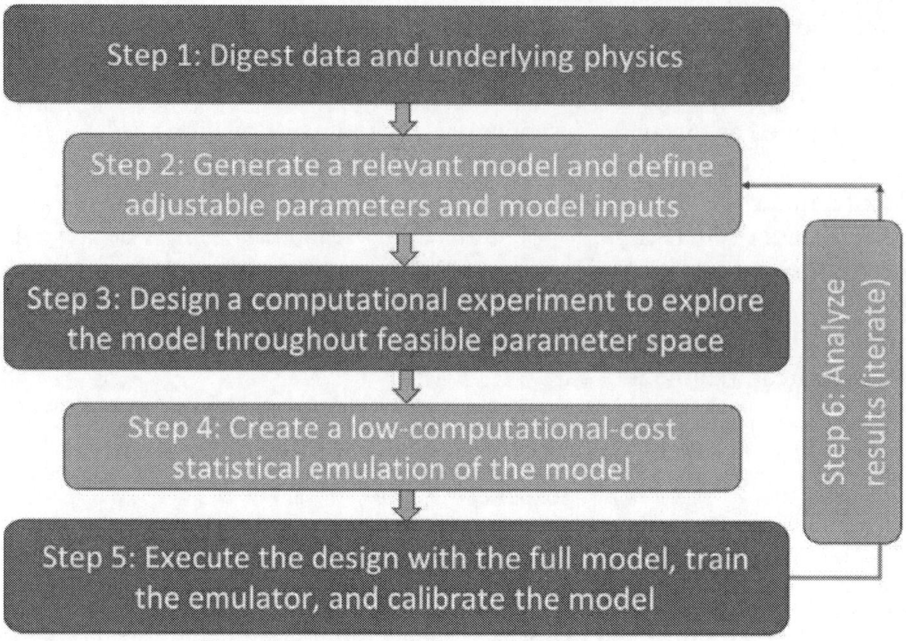

Figure 2. Logic map of the model calibration process

4.3.1. Calibration Results and Analysis

In the interest of brevity, the details of the statistical analysis are not given. To summarize, the analysis found the CBK annealing model to be overwhelmingly uncertain, which is in fact consistent with the statement by Hurt et al.[24] in the publication of the original annealing model. The model uncertainty was substantially reduced by applying a greater wealth of data, a better exploration of parameter space, and a coal structural parameter as a dummy input, but considerable uncertainty remained. The dummy parameter had no impact on the model whatsoever, but it did allow the statistical machinery to attach a discrepancy to the input, and indicate the level of functional dependence an annealing model *should* have. The subsequent analysis indicated that the time-temperature profile and the coal structural parameter were by far the largest sources of discrepancy between the model and the data. This was quantified in an internal statistical parameter for the covariance matrix of the discrepancy (δ) term. The covariance parameter necessary to capture the discrepancy due to precursor structure and the time-temperature profile were each about a factor of five greater than discrepancy in other parameters.

The insights from the statistical calibration tools and the observations of the literature led to a campaign of informed trial and error to uncover the model functional form (step 6 in Figure 2); this ultimately reduced the model uncertainty and average prediction error substantially. The extended annealing model is executed precisely as the prior CBK annealing model, with all changes successfully confined to the annealing activation energy distribution and the annealing preexponential factor. The model was inspired by literature implications for functional dependence on heating rate, T_{peak}, and the precursor chemical structure, but balanced with a need to minimize the total number of parameters that would effectively be relegated to fudge factors in an over-parameterized optimization. Note however, that despite considerable model improvement, annealing is still an umbrella term for numerous processes that include enormous variability. The current model performs well, given the current data, but substantial improvements could be made by both specifying which processes should fall under the auspices of annealing, and conducting sharply focused research on the specified processes.

Equations 13-17 describe the final functional form of the preexponential parameter for annealing (A_d), the mean annealing activation energy (μ_{Ed}), and the standard deviation for the annealing activation energy distribution (σ_{Ed}). Note that the model inputs are p_0, HR, and T_p. The values of HR and T_p are given in K/s and K, respectively, but the actual inputs are divided by their native units so that all inputs are unitless. The input p_0 is naturally unitless, and the parameters have the units of $\ln(\mu_{Ed})$ (i.e., ln(kcal/mol)).

$$if\ HR < 10^4 \qquad A_d = \frac{p_0 * A_{d,0}}{\ln(HR+2.7)} \tag{13}$$

$$if\ HR \geq 10^4 \qquad A_d = \frac{p_0 * A_{d,0}}{\ln(10^4)} \tag{14}$$

$$if\ T_p \leq 1500 \qquad \ln(\mu_{E_d}) = a * p_0 + b + T_p * c/1000 \tag{15}$$

$$if\ T_p > 1500 \qquad \ln(\mu_{E_d}) = a * p_0 + b \tag{16}$$

$$\ln(\sigma_{E_d}) = \frac{\ln(\sigma_0)}{p_0} \tag{17}$$

Equations 13 and 14 express the heating rate dependence on the original number of active sites. Cai et al.[22] found that higher heating rates increased the reactivity of a char considerably with all other preparations conditions held constant. The natural log of the heating rate captured the reactivity increase well up to about 10^4 K/s, after which a plateau was reached for all coals tested. The increase in activity is speculatively attributed to an increase in micropores during increasingly rapid devolatilization. This implies that the annealing model is perhaps not the ideal submodel to include this information, but in general, comprehensive char combustion codes do not include a submodel to address preparation condition-based pore development. Therefore, the umbrella of annealing with built-in preparation condition dependence is the most appropriate submodel available. Any model employing an estimate of porosity based on heating rate would be well served to eliminate the heating rate dependence of the annealing preexponential factor. Note that the heating rate dependence is located in the denominator because A_d describes how rapidly annealing proceeds. If the surface area increases by some factor "F", the number of active sites is expected to increase by the same factor "F" (assuming uniform active site density). This is captured in the annealing submodel by reducing the rate of site destruction by "F," where F=ln(HR+2.7) (note that F reduces to ~1 at very low heating rates). This is not mathematically identical to increasing the number of sites by "F," but the number of active sites was normalized to unity as is appropriate for a lognormal probability density function, and it is undesirable to disrupt the normalization. Since the annealing model is far from mathematically perfect in any case, it was deemed conceptually adequate to adjust the preexponential factor to decrease the rate of reactivity loss instead. Finally, the structural parameter p_0 is included, where p_0 specifies the fraction of intact bridges in the coal pseudo-monomer. The intact bridge fraction is experimentally found via NMR spectroscopy, or derived from a correlation via the proximate and ultimate analysis as described by Genetti et al.[30, 31] There are four NMR parameters commonly used to describe coal structure, all of which appear strongly correlated with each other and were found to be equally suitable for a coal structural parameter. The NMR parameter p_0 was chosen to represent coal structure after only moderate success with the simple ratio of carbon to hydrogen from the ultimate analyses. The C/H ratio does not distinguish between radically different coal structures with a similar elemental composition, but the NMR structural parameter approach has been successfully employed elsewhere.[32, 33]

Equations 15 and 16 are a straightforward linear model to predict the mean (μ_{E_d}) of the log-normal activation energy distribution, given coal structure and peak temperature, with the addition of some constant, "b." The only subtlety is that the peak temperature dependence may be profitably turned off at peak temperatures above 1500 K, where the literature indicates many T_{peak} effects were no longer relevant. In fact, a series of models that employed the T_{peak} term *above* 1500 K (and turned off "c" *below* 1500 K) found that the optimal value of "c" was driven towards zero (i.e., T_{peak} is not an important parameter input for the mean activation energy above 1500 K). Equation 17 predicts the standard deviation of the distribution, and, together with Equations 15 and 16, fully defines the log-normal activation energy distribution. The model standard deviation was found to be relatively constant with respect to T_{peak} and heating rate, with a direct dependence on char precursor. The functional dependence of σ_{EA} on precursor only is reasonable as a given coal should have some distribution of activation energies, but T_{peak} and HR likely shift the center of that distribution more than they adjust the range. Once μ_{Ed} and σ_{Ed} are defined, the log-normal distribution is known. After the log normal distribution is defined, it is split into a bimodal distribution in accordance with numerous observations of two distinct annealing regimes in the literature.[34-36] This is done by the factors B_f and B_r, which are both optimized values, so the actual location of the bimodal trough is also determined by optimization. The factors B_f and B_r split the log normal distribution by isolating the activation energies in the range: $\mu_{Ed} - B_f < \mu_{Ed} < B_f + \mu_{Ed}$. The density of active sites within that range is divided by B_r. In other words, B_f indicates the breadth of a trough in the bimodal distribution (centered on the mean), and B_r determines the depth of the trough. The formerly log-normal distribution is then renormalized, which maintains the total number of initial active sites, but allows for a bimodal distribution with an optimized weight between the two regimes.*

Figure 3 below is a sample log normal distribution that shows the fraction of active sites in any given bin. The exact mean and variance of the distribution depends on heating rate, precursor NMR parameters, and peak temperature. Figure 4 shows an irregular, bimodal distribution after parameters B_r and B_f are applied and the distribution is renormalized. The second figure has two striking features. First, the majority of the low activation energy sites vanish. This is not a problem in either of the two distributions, because the original log normal distribution resulted in a highly exaggerated rate of initial annealing, which was compensated for by using an excessively high initial preexponential value. The reduced initial annealing of the second figure dispenses with the need for an excessively high initial preexponential factor. The second notable feature is that the second peak of the distribution is highly irregular. This indicates that, after the very rapid initial annealing, the remaining activated annealing processes have a large number of

---

* The bimodal distribution factors B_f and B_r are a complex series of operations as shown in the script "match_reactivity" found at https://github.com/tmholland86/CCK-oxy (LA-UR-17-24629).

active sites at a high activation energy, so the remaining annealing processes will be relatively gradual.

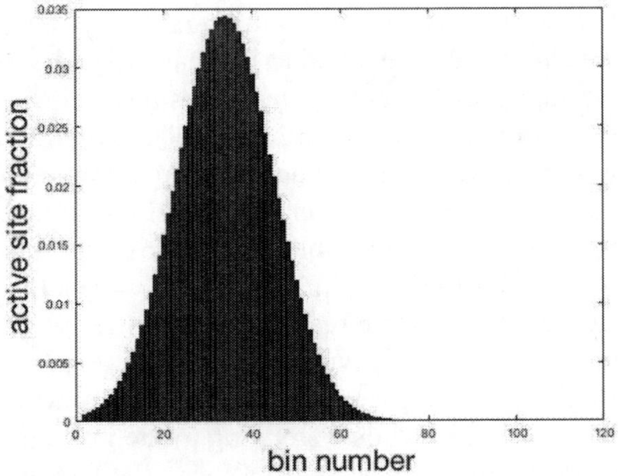

Figure 3. Sample log normal distribution of the fraction of active sites in any given bin.

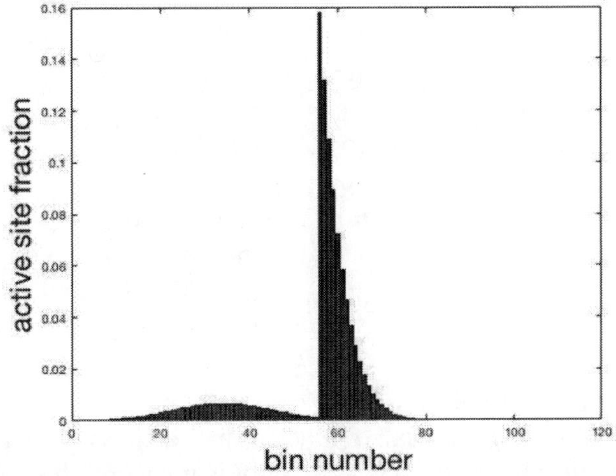

Figure 4. Sample "bimodal" distribution after applying the parameters B_r and B_f and renormalizing the distribution.

4.4. Model Optimization

Once a reasonable model form was developed, the model parameters were optimized. The optimization routine was coded using fmincon in MATLAB, which traversed the bounded parameter space of the model in an attempt to minimize an objective function. Thus, the annealing data are not fit via a standard linear regression. In fact, the model parameters cannot be directly regressed because even the conceptual form of the model is far from linear in its parameters, and even if the model could be linearized, such a mathematical transform would render a regression statistically invalid. Instead, the optimization routine finds a narrow subsection of parameter space containing numerous local minima, many of which lead to approximately the same model output. Any vector of parameter spaces that leads to a reasonable local minimum is valid, and finding the absolute minimum in a high dimensional, non-linear parameter space is essentially impossible. Instead, the optimization is carried out by implementing various parameter bounds and constraints, and initializing the optimization from different locations in parameter space. Each optimization required roughly 12 hours on a 2016 MacBook Air, and, after dozens of optimizations, approximately 1/3 failed (settled on a local minimum that was clearly far from the global minimum) and 2/3 succeeded (located local minima that were all roughly equal, and presumed to be near the global minimum). Table 1 gives the values of the annealing model parameters used in the results section. The optimization routine yields a single local minimum. In producing Table 1, the optimization routine was executed numerous times with different initial values in the parameter space, and in about half of all cases a local minimum in the objective function was found within 5% of the objective function value produced by the parameters in Table 1. In almost all other cases, the objective value was within 50% of the superior local minima, and there were a handful of optimization failures. The particular values in Table 1 were chosen because they were the lowest found, but in this model there are numerous essentially equivalent local minima, and an infinite number of surrounding points in parameter space that are equivalent for any practical purpose. Note that the parameters a and b are on similar scales, while c (relating to the influence of peak particle temperature) is roughly an order of magnitude smaller. Since the temperature and NMR parameters were scaled to be similar, this indicates the relatively low impact of peak particle temperature, which is expected, since most the available data are at relatively high temperature.

Table 1. Annealing model parameter values.

Parameter	Value	Units
$A_{d,0}$	9.71×10^{11}	s^{-1}
B_f	45.55	kcal/mol
B_r	176.66	-
a	0.46	ln(kcal/mol)
b	1.77	ln(kcal/mol)
c	7.32×10^{-2}	ln(kcal/mol)/K
$\ln(\sigma_o)$	0.65	ln(kcal/mol)

5. Model Execution Results and Discussion

The final annealing model predicts the relative reactivity of a char compared to some reference char, as discussed previously. This is because reactivity is measured directly, but the number of active sites (either initial or final) cannot be measured. Instead, by taking the ratio of reactivity in two annealed chars in a zone 1 combustion regime, the ratio of final active sites can be determined (relative to the initial active sites). Therefore, each set of annealing data has a single char designated as a reference, and the data points are the ratio of the reactivity between each char and the designated reference char. The annealing model is considered to be performing well when the measured relative reactivity is close to the predicted relative reactivity. Because the model parameters were optimized, an objective function was required to quantify “close.” Because the annealing data covered a very wide range of coals and conditions, the results cannot be conveniently displayed in a meaningful way. Simply plotting the measured and predicted reactivities on the ordinate axis (with an index on the abscissa) resulted in an unreadable mess due to the variation in scale between experiments. Ultimately, it was decided that several measures of model success should be included. A log-scale parity plot is shown in Figure 5 partially to show all of the O_2 data in one compact location, and partially because the original CBK annealing model results were displayed on a log parity plot. A linear parity plot has the dual difficulties of requiring a very large axis to display the outlier points from the Hurt model, while simultaneously compressing the majority of the points into a tight region not amenable to visual inspection. The log-scale plot on the other hand, can be quite misleading and give the impression that the extended annealing model developed in the present work is only a minor improvement over the prior model. Note that, in accordance with the discussion in the model development section, the **relative** reactivity is used in Figure 5 because the actual degree of activity loss can only be observed as a ratio between two chars that received different heat treatments. Thus, every data set includes a reference char with a measured reactivity, and the ratio of reactivities between the reference char and any other char in the data set is the (unitless) relative reactivity.

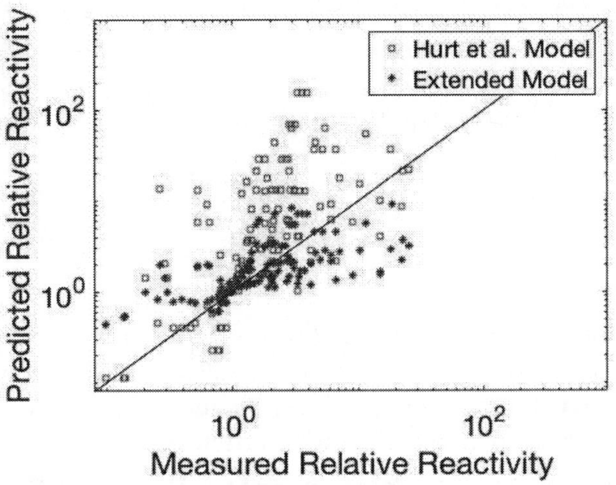

Figure 5. Parity plot of measured and predicted relative reactivities in O_2 for the Hurt et al.[24] mod vs the extended annealing model

In contrast, Table 2 includes the results of the sum-squared error as an equally misleading measure. Superficially, the sum-squared error indicates that the extended model is ~2 orders of magnitude superior to the prior model, but in reality a relatively small handful of points result in enormous error. The best measure of model success, both for an optimization objective function and for reasonable model comparison, is referred to in Table 2 as the error factor (which is a quantity defined and described in detail in the experimental section). Table 2 includes the mean and range of the error factor for entire body of data as well as several subsets. This breakdown shows that the extended model is certainly an improvement, but the improvement is closer to a factor of four rather than a factor of 100. Additionally, the quartile breakdown shows that the extended model performs well across the board with only a small handful of egregious failures, while the CBK annealing model has numerous large errors in all but the best performing quartile. In other words, the prior (Hurt) annealing model is effectively either on or off, and accurate only in a small subset of conditions. Note that, while it would be desirable to compare the original CBK annealing model directly to the extended annealing model, the extended annealing model requires more detailed information that was not available in the referenced literature for the Hurt model, so direct comparison using the original subset of data from the Hurt model is impossible.

Table 2. Measures of Extended Model Success Compared with the Hurt et al.[24] model for the Entire Annealing Data Set

	Hurt et al. Model			Extended Model		
Model Quantification	Mean	Min	Max	Mean	Min	Max
Sum Squared Error	1.45×10^5 †	N/A	N/A	2.43×10^3†	N/A	N/A
Error Factor: All Points	6.08	1.00	51.97	2.24	1.00	9.96
Error Factor: Least Successful Quartile	17.28	7.00	51.97	4.44	2.30	9.96
Error Factor: Most Successful quartile	1.13	1.00	1.25	1.10	1.00	1.20
Error Factor: Central Quartiles	2.78	1.25	6.50	1.63	1.21	2.27

On average, the extended annealing model predictions differ by a factor of about 2.24 from measured values. An average error of 124% is far from an optimal model, but in comparison to past work (with an average error of 508%), it is a significant improvement (a factor of ~4). It is worth noting that carbon sources are notorious for enormous variability in their respective reactivities. In fact, it seems likely that the umbrella of effects referred to as annealing are responsible for such a large range of reactivities. There are insufficient replicates in the data for a detailed statistical analyses of variance within individual data sets, but a brief examination of the few replicates or pseudo-replicate data points is enlightening. Among these data,[18, 37] the variation between replicates ranges from roughly 20-50%. Thus, 124% error is not at all unreasonable, and is in fact a substantial stride towards mitigating the vast uncertainty of coal combustion rate modeling, perhaps even to the extent that a coal-general kinetic correlation may be feasible if derived in conjunction with a comprehensive char conversion model.

5.1. Model Predictivity

The present annealing model was developed with a minimal number of adjustable parameters specifically to avoid a large number of "fudge factors" that would fit virtually any data set while having very low predictive power. However, a model with few parameters is no guarantee of predictive power. As observed above, the annealing model employed in CBK and its successors often predicted annealing quite far from experimental data, probably because the model was calibrated using relatively few data points from a narrow selection of precursors and preparation conditions. Given that a substantial amount of uncertainty remains in the current annealing model, it was entirely possible that the new model would suffer from a similar handicap. Therefore, several data points were initially excluded from the optimization objective function to test the model predictivity. In general, the model predicted the new points with remarkable

† This is a single value, not a mean of multiple values.

success, and when the model parameters were re-optimized with all data points, the new data had a minimal impact on the overall model calibration. One such data set is shown below in Figure 6 (data collected by Jayaraman et al.[38] for a high-ash Indian coal). The ordinate axis is the relative reactivity NOT the error factor. The error factor (calculated as described in Section 3) in this case ranged from 1.00-1.12, with a mean of 1.08. In general, any ratio less than ~2 is considered quite good in the context of reactivity loss. The abscissa is potentially misleading in that up to six input variables influence each data point (the heating rate, treatment time, and peak temperature for the char particle, and the same three variables for the reference char). However, the peak temperature of the char particle is most likely to both change and have a dramatic impact between points, so it was deemed the best variable for the abscissa in the absence of six-dimensional plotting software. Nevertheless, the apparent replicates are not actually identical data points.

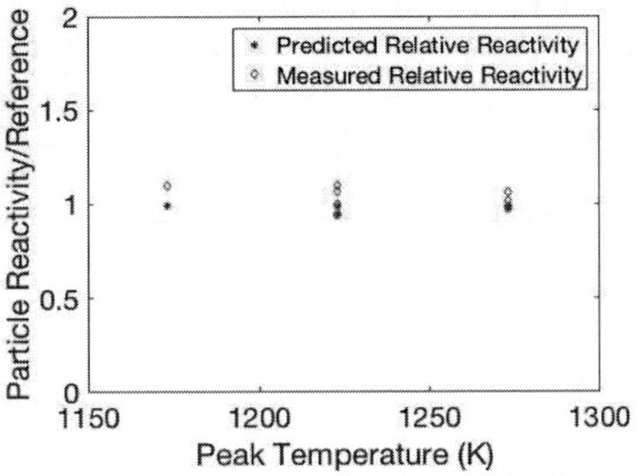

Figure 6. Model predictions compared to experimental measurements of an Indian coal (data were not used to calibrated model).

At first glance, Figure 6 can be misleading. It is natural to intuitively view the annealing values close to unity as uninformative. However, an annealing value near unity does NOT indicate a lack of annealing. Instead, it indicates that the annealing of the data point and the annealing of the reference char were very similar. The agreement in Figure 6 is therefore an exceptional example of model success in that it tracks annealing quite well even at conditions or treatment times that are very similar. The predictions and data are not identical, and the predictions have a slight but consistent negative bias, which indicates that the annealing model is not exact, but it is correctly responding to subtle changes in heat treatment.

Additional comparisons of predictions and measured data (and associated error factors) are shown in Figure 7 and Table 3. In all cases, the ordinate access is the relative reactivity NOT the error factor. The error factor mean and range are reported in Table 3. Figures 7a and b both show data from experiments with Pittsburgh 8 coal, but from two different investigators. Figures 7c through f show a variety of other coals over the range of experimental data. In general, the model predictions are quite good, with obvious outliers at the lowest heat treatment temperatures. The poor predictions at low temperature are much worse than the average predictions, and while the lack of agreement in this region is undesirable, it is not unexpected. The raw coal precursors are quite diverse, and low heat treatment temperatures do not allow the chars to progress as far towards the thermodynamic minimum of a perfect graphite crystal. Thus, the mildly annealed chars retain greater diversity. Also, low-temperature heat treatments likely do not fuse the ash. The chemical identity of the ash is rarely characterized in individual coals, so ash variation between char precursors is not reflected in the annealing model. This creates additional model uncertainty for low-temperature char preparation because ash has fusion-dependent chemical and physical effects on the coal. More importantly, the low-temperature heat-treatment implies a drastically different devolatilization process, which significantly alters the chemistry of the char. Ultimately, the low-temperature predictions are of far less practical importance, and are not cause for significant concern in practical char conversion models.

5.2 Model Failure

Despite the general success of the model in fitting data (and predicting data it had not yet been calibrated to), there were a number of significant model failures (5.1%), defined as model predictions greater than a factor of five from the experimental data. Among these failures, all but a single point of data had several commonalities. First, all of the data points in question came from two authors,[21, 42] and both of these authors formed their char in an flat flame burner. The data sets from both authors appeared to have a very high variance, in that supposedly similar experiments led to widely different results without a definite trend; however, the actual variance cannot be computed, as replicates are not given. Third, both sets of experiments sampled data on a short time scale, and fourth, both authors obtained a number of points that agreed remarkably well with model predictions, both before and after the questionable data was included in the optimization routine. Finally, both data sets included coals that were used by other authors in annealing experiments, with vastly different results.

It is possible that the annealing model is simply unable to predict flat flame burner chars. However, given that many of the chars in the same data sets were predicted very accurately, it seems far more likely that the short time scale and sampling limitations in the flat flame burner led to noisy data. A sample of the worst model failures and adjacent successes to supposedly similar experiments is given in Figure 8 (data from Shim and Hurt[21]).

Table 3. Error Factors Associated with a Diverse Selection of Coals.

Coal	Min Error Factor	Max Error Factor	Mean Error Factor
Pittsburgh 8 [39]	1.01	1.92	1.41
Pittsburgh 8 [22]	1.20	3.84	2.00
HVB [40]	1.06	1.37	1.26
Lindby [22]	1.01	4.50	1.92
South African [41]	1.44	2.73	2.06
Pocahontas [39]	1.00	3.84	1.95

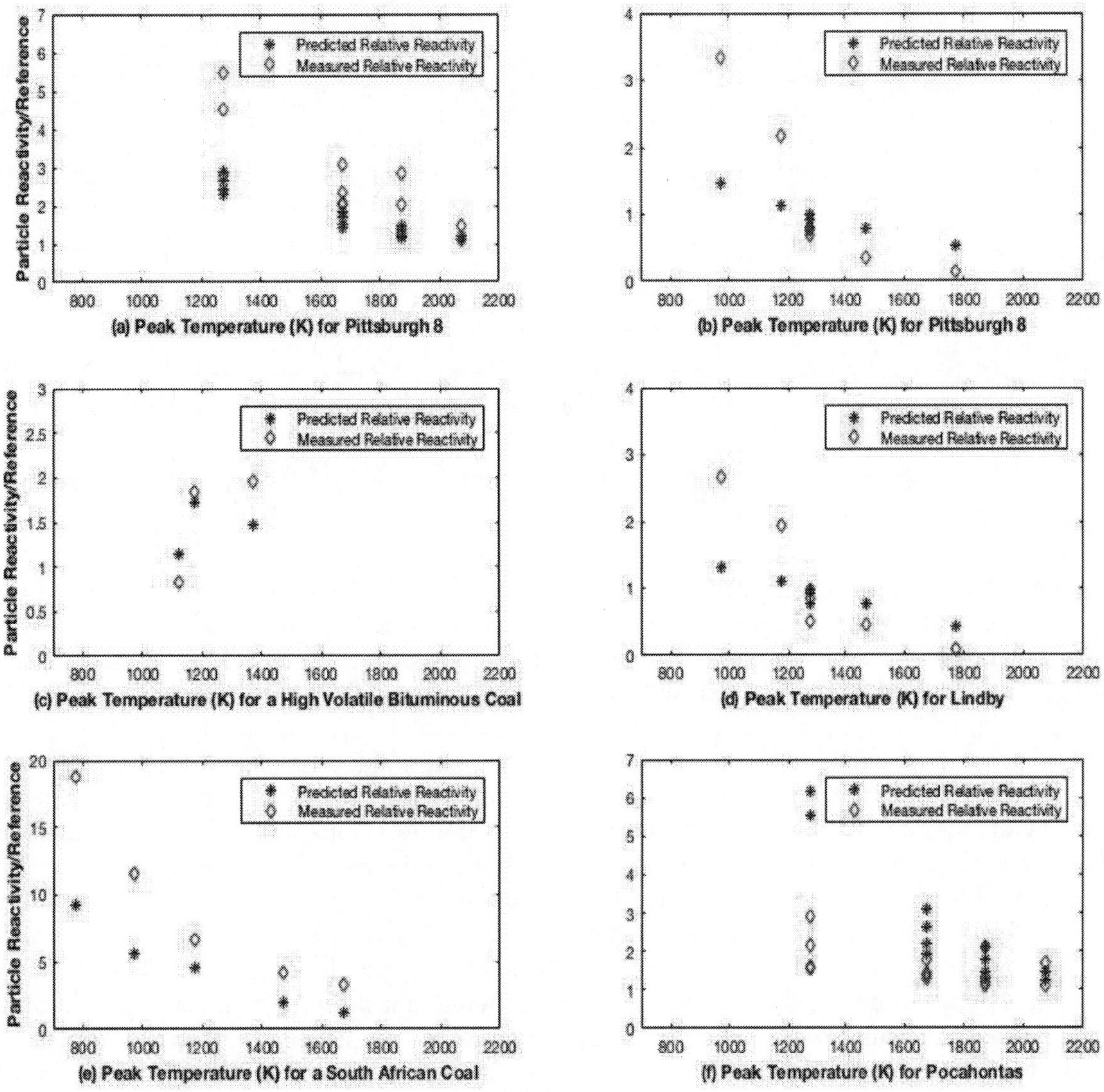

Figure 7. Diverse selection of char relative reactivities (both predicted and measured). See Table 3.

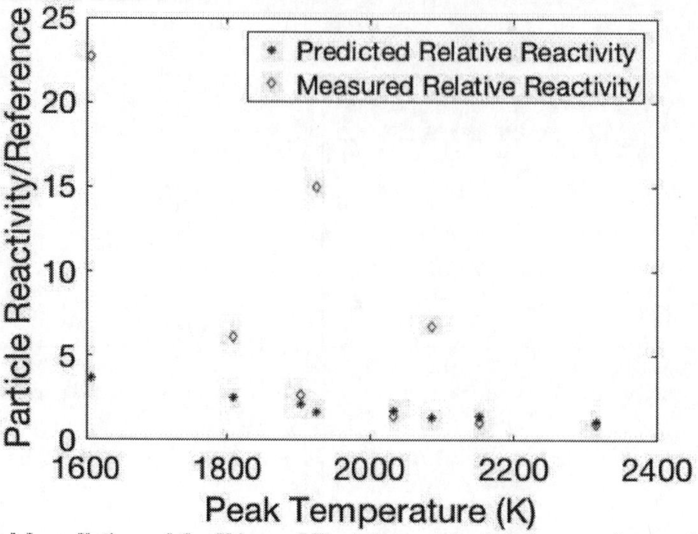

Figure 8. Model predictions of the Shim and Hurt data set, both failures and exceptional successes.

5.2 Sample Model Predictions

While the extended annealing model was broadly successful in predicting char reactivity loss due to preparation and combustion conditions, there was one large data set that consistently failed. The failure was not large in terms of error factor, as with the handful of anomalous, erratic points. Instead, it was a consistent failure to track a subtle trend in char reactivity with treatment time in a Chilean coal.[36] Figure 9 shows that the model does reasonably well predicting the degree of annealing in an absolute sense, but an additional dimension of treatment time shows that the model remains essentially constant with the very small changes in treatment time (shown by the vertical distribution of points at any temperature). This was ultimately determined to be the reason that the statistical calibration tool found substantial model discrepancy with treatment time. Physically, it is caused by the approximation of the initial heating time-temperature profile. Because detailed data were not available to reasonably estimate the time-temperature profile, the initial heating rate was used as a crude substitute. In general, the estimate is sufficient, but when the heating time is a significant portion of the total annealing time, the error can become noticeable. This error was further verified by attempts to fit only the data in Figure 9 without any other experimental data, but the results were not improved, even when several different model functional forms were attempted. Also, when the data were excluded from the objective function, the annealing model predictions did not change notably (and still gave the reasonable predictions seen below). Conversely, when the data of Figure 9 were included with all other data, model predictions were also not significantly influenced; the extended annealing model simply cannot fit the subtle trend of a data set with very short treatment times in the absence of a more accurate heating profile.

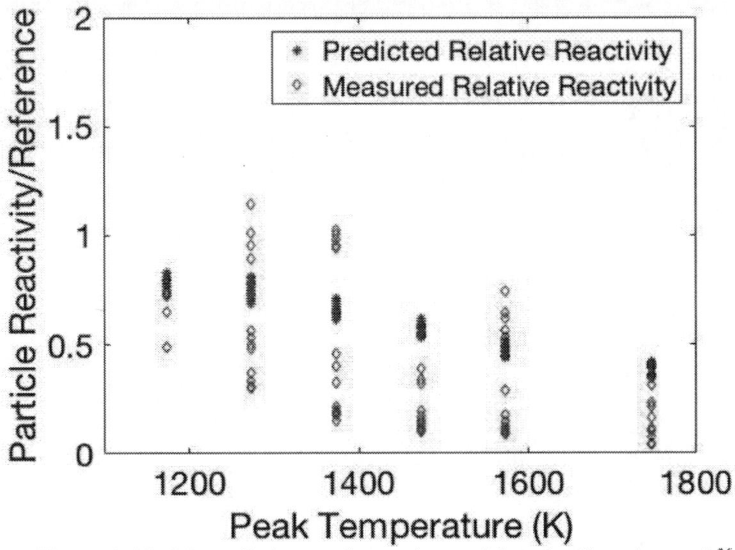

Figure 9. Model predictions and experimental data for Cerrejon coal.[36]

5.3 CO_2 Model Development and Results

The literature implied that activity loss to CO_2 gasification potentially proceeds along different kinetic pathways than O_2 annealing. By extension, steam reactivity reduction may also behave differently than combustion annealing. In general, CO_2 and steam may be neglected as reactants in pulverized coal combustion, but the extended annealing model is intended to function in an oxy-fuel setting as well, where CO_2 and H_2O may have appreciable influence on the combustion process. Therefore, it seemed prudent to tabulate CO_2 annealing data and attempt a similar calibration for an alternative annealing model form, or at least determine alternative parameter values if the same functional form turned out to be adequate. However, as a matter of curiosity, the O_2 calibrated annealing model was applied directly to the CO_2 and steam data, and in general performed astonishingly well considering the lack of calibration. It must be noted that there was relatively little CO_2 data available, and that one large set of data fit extremely well, forcing the mean error factor down. However, the results in Table 4 and Figures 10-11 clearly show that, within the limits of available data, the model is able to predict CO_2 and H_2O annealing adequately even with the parameters calibrated to O_2 annealing. Figure 10 contains data for a South African Coal, which was also used in several of the O_2 data sets, although the CO_2 experiments were from a different lead author and institution.[41] This may well have improved the uncalibrated model prediction in the case of South African coal, but the uncalibrated model was also successful with other coals in the CO_2 environment. The same annealing model form and parameters therefore seem appropriate for both O_2 and CO_2 activity loss. Recall that Figure 10 shows the model predictions, not the error factor, so the model is actually performing even better than the plot implies. Figure 11 represents the sole H_2O data set available.[38] The small sample size limits inference, but the model appears to perform exceptionally well for H_2O annealing as well. Because the model form and parameters captured all available data well (O_2, CO_2, and H_2O reactivity) even when the model was calibrated to O_2 only, the final model and the values reported in Table 1 were calibrated using all available data in all conditions and reactive gases.

Table 4– Uncalibrated CO_2 and H_2O extended model annealing predictions.

Model Quantification	Mean	Minimum	Maximum
Sum Squared Error	268*	N/A	N/A
Error Factor: All Points	1.66	1.00	4.96
Error Factor: Least Successful Quartile	2.84	2.05	4.96
Error Factor: Most Successful quartile	1.04	1.00	1.08
Error Factor: Central Quartiles	1.37	1.09	2.02

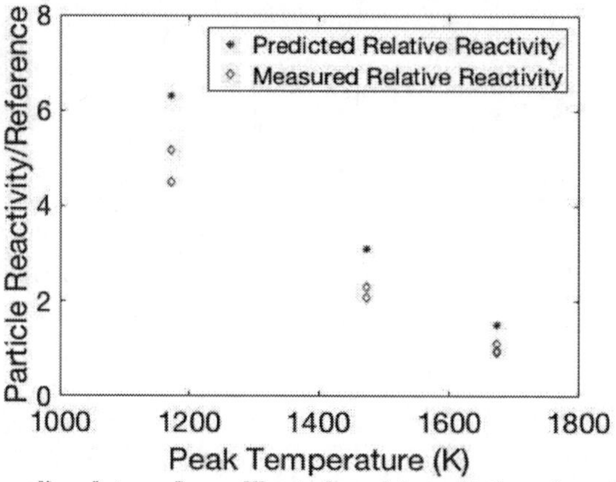

Figure 10. CO_2 Annealing data and uncalibrated model predictions for a South African coal.

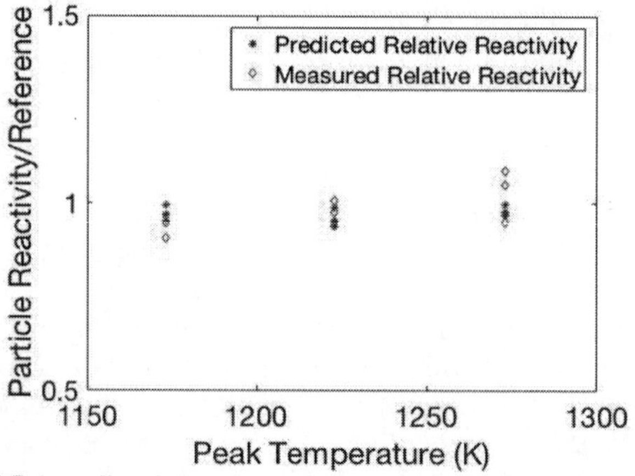

Figure 11. H_2O Annealing data and uncalibrated model predictions for an Indian Coal.

5.4 Model Implications

The literature indicates that the vast bulk of annealing occurs very rapidly, followed by a much more moderate loss of reactivity due to heat-treatment. The model described here has been designed to reflect this trend, which is thought to be due to initial, dramatic annealing during

* This is a single value, not a mean of multiple values.

pyrolysis, and a lesser degree of annealing during subsequent heat-treatment. The annealing model therefore offers flexibility (when employed in a comprehensive coal conversion model) to allow char combustion predictions over a broad array of combustion conditions. However, no effective, coal-general kinetic correlation exists, so data are generally required to tune coal conversion rate constants. If data are available to tune kinetic constants, the initial annealing effects are embedded in the data, while the subsequent annealing effects may or may not be negligible, depending on coal type and combustion conditions. Thus the annealing model may be circumvented entirely in certain conditions, but is of considerable utility in comprehensive coal conversion models. In particular, the model shown here has been successfully employed in just such a comprehensive model to explain data over a broad range of conditions.[33]

Perhaps even more importantly, the literature data used in this work clearly show several orders of magnitude in char reactivity change due to char preparation condition, and both the data and the model show trends (at high T_{peak}) of converging char reactivities for diverse precursors. The historical failure of coal-general kinetic correlations has depended in large part upon many orders of magnitude of variation in the body of literature data between char reactivities, while no truly consistent pattern between proximate and ultimate analysis and char reactivity is readily discernible. The enormous impact of preparation condition on char reactivity (and the great inconsistency in char preparation methods in the literature) certainly account for a portion of char reactivity variability. It is possible that preparation conditions even accounts for the bulk of observed variability, at least at practical (high T_{peak}) combustion conditions. If so, the present annealing model could potentially account for the preparation condition variability, and allow for a coal-general kinetic correlation based on structural parameters and/or the proximate/ultimate analysis.

6. Summary and Conclusions

A thermal annealing model functional form was developed based on observations supported by multiple authors, experimental methods, and detection systems. Qualitatively, coal char annealing is found to depend heavily on the time-temperature profile, the initial particle heating rate, the peak particle temperature, and the parent coal chemical structure as indicated by the NMR parameters. Trends in functional form dependence were quantified by optimizing a number of model forms for the annealing activation energy distribution to fit a broad array of literature data. The resulting model was shown to be a significant improvement; average error decreased to roughly a factor of two rather than a factor of five as compared to the preceding annealing model. The improvement is largely due to a model form that accounts for coal chemical structure, heating rate, and T_{peak}, as well as a much larger data set. Note that both the predicted degree of annealing, and the annealing activation energy distribution depend heavily on heating rate, T_{peak}, and chemical structure. Model functional dependency on HR and T_{peak} were previously included implicitly in the model form, which integrated a reaction rate at a specific temperature for a relevant increment of time. However, a more realistic model must incorporate explicit dependence of the annealing activation energy distribution on HR, T_{peak}, and chemical structure. For example, ash fusion occurs solely as an effect of T_{peak}. Ash fusion in turn affects several annealing and char conversion pathways, such as catalyzing carbon crystallite rearrangement. If the ash fuses, a catalytic annealing pathway is eliminated, and the distribution of activation energies should shift accordingly. Similar arguments can be made for other annealing effects such as devolatilization, coal morphology, and coal precursor.

The annealing model was trained by a selection of 25 different data sets and 167 data points. Some data sets were initially excluded to test model predictivity; the model was found to predict the excluded data approximately as well as the training data. The model shows rapid decreases in char reactivity during heat-up and devolatilization before significant char oxidation occurs, and generally matches the decrease in reactivity best at temperatures above 1200 K (and especially above 1800 K), which was expected. The dependence on coal type was modeled using a chemical structure parameter measured by solid-state NMR spectroscopy.
While the average error factor was greatly reduced, much of the remaining error is due to a relatively few outliers with data at low values of T_{peak}. The weaker model results at low temperature are thought to be due to both greater diversity and poorer char characterization (i.e., annealing is far from complete, and the chars may have unfused ash, catalytic effects, residual volatiles, greater crosslinking etc.). Bearing in mind the disproportionate error in low-temperature experiments, and given that practical combustion occurs are temperatures in the range of ~2000-2300 K, the annealing model presented here works especially well in practical circumstances.

It was shown that annealing data relevant to CO_2 gasification is predicted at an acceptable level by the annealing model trained from O_2 oxidation reactivity data. This result is somewhat surprising as the limited literature data imply that O_2, CO_2, and H_2O active sites may all be different.[43] However, because current knowledge of coal reaction pathways is woefully incomplete, this result is certainly not impossible. The single set of available H_2O gasification data (with sufficient recorded experimental detail for this model), is also predicted quite well by the annealing model trained to O_2 oxidation data.

Despite model improvements, there is still considerable work that could be done to improve the annealing model. Such work includes data for an even broader array of coals, data collected over a large range of temperatures and heating rates with very short treatment times, and data that allow more detailed predictions of the initial time-temperature profile (prior to a fixed temperature soak time). There are also numerous other questions of interest that may prove more difficult to address. These questions include disentangling the coal swelling models and pore development from thermal annealing, a broader study of char ash content on thermal annealing (examining both ash quantity and chemical identity), and further investigation into early and late burnout effects. Late burnout effects are not related to annealing, but tend to obscure char reactivity (for example pore networks that are initially blocked during the metaplast phase, but may become unblocked even at a low level of char conversion).

Finally, it should be emphasized that this annealing model, like most prior literature models, impacts only the conversion rate law preexponential factor by decreasing the number of active sites available. It was implemented directly in a comprehensive coal char conversion model with the parameters derived in this work (with a remarkable level of success).[33] However, the comprehensive model typically requires that key kinetic parameters be fit to data, since it lacks an effective coal-general reactivity correlation. This is a universal weakness in char conversion models, and it should be noted that the annealing model will impact the preexponential factor, so optimization or regression of kinetic coefficients for any char conversion model should take place *after* implementing the char annealing model.

Acknowledgements
This material is based upon work supported in part by the Department of Energy, National Nuclear Security Administration, under Award Number DE-NA0002375. Funding for this work was provided by the Department of Energy through the Carbon Capture Simulation Initiative (CCSI) and the Carbon Capture Simulation for Industry Impact ($CCSI^2$) projects. Los Alamos National Laboratory is operated by LANLS, LLC for the NNSA of the U.S. DoE under contract No. DE – AC52-06 NA 25396.

References

1. Holland, T.; Fletcher, T. H., Global Sensitivity Analysis for a Comprehensive Char Conversion Model in Oxy-fuel Conditions. *Energy & Fuels* **2016**.
2. Senneca, O.; Salatino, P., Loss of gasification reactivity toward O-2 and CO2 upon heat treatment of carbons. *Proceedings of the Combustion Institute* **2002,** 29, 485-493.
3. Nagle, J.; Strickland-Constable, R. F. In *Oxidation of carbon between 1000°C and 2000°C*, Fifth Carbon Conference, Pennsylvania State University, 1962; Macmillan, Ed. Pergamon Press: Pennsylvania State University, 1962; p 154.
4. Beeley, T.; Crelling, J.; Gibbins, J. R.; Hurt, R.; Lunden, M.; Man, C. K.; Williamson, J.; Yang, N. Y. C., Transient high-temperature thermal deactivation of monomaceral-rich coal chars. *Proceedings of the Combustion Institute* **1996**, (26th), 3103-3110.
5. Jenkins, R. G.; Nandi, S. P.; Walker, P. L., Reactivity of Heat-treated Coals in Air at 500 Degrees C. *Fuel* **1973,** 52, (4), 288-293.
6. McCarthy, D. J., Some Effects of Overheating and Carbon Bunr-off on the Performance of Fluidized-bed Combustors. *Fuel* **1982,** 61, (3), 298-302.
7. Radovic, L. R.; Walker, P. L.; Jenkins, R. G., Importance of Carbon Active-sites in the Gasification of Coal Chars. *Fuel* **1983,** 62, (7), 849-856.
8. Suuberg, E. M.; Wojtowicz, M.; Calo, J. M., Some Aspects of the Thermal Annealing Process in a Phenol-formaldehyde Resin Char. *Carbon* **1989,** 27, (3), 431-440.
9. Senneca, O.; Russo, P.; Salatino, P.; Masi, S., The relevance of thermal annealing to the evolution of coal char gasification reactivity. *Carbon* **1997,** 35, (1), 141-151.
10. Sahu, R.; Levendis, Y. A.; Flagan, R. C.; Gavalas, G. R., Physical-Properties and Oxidation Rates of Chars from 3 Bituminous Coals. *Fuel* **1988,** 67, (2), 275-283.

11. Radovic, L. R.; Walker, P. L.; Jenkins, R. G., Effect of Lignite Pyrolysis Conditions on Calcium-oxide Ddispersion and Subsequent Char Reactivity. *Fuel* **1983,** 62, (2), 209-212.
12. Senneca, O.; Salatino, P., Overlapping of heterogeneous and purely thermally activated solid-state processes in the combustion of a bituminous coal. *Combustion and Flame* **2006,** 144, (3), 578-591.
13. Senneca, O.; Salatino, P.; Menghini, D., The influence of thermal annealing on oxygen uptake and combustion rates of a bituminous coal char. *Proceedings of the Combustion Institute* **2007,** 31, 1889-1895.
14. Senneca, O.; Salatino, P.; Masi, S., Heat treatment-induced loss of combustion reactivity of a coal char: the effect of exposure to oxygen. *Experimental Thermal and Fluid Science* **2004,** 28, (7), 735-741.
15. Senneca, O.; Salatino, P.; Masi, S., The influence of char surface oxidation on thermal annealing and loss of combustion reactivity. *Proceedings of the Combustion Institute* **2005,** 30, 2223-2230.
16. Salatino, P.; Senneca, O.; Masi, S., Assessment of thermodeactivation during gasification of a bituminous coal char. *Energy & Fuels* **1999,** 13, (6), 1154-1159.
17. Shibaoka, M.; Ohtsuka, Y.; Wornat, M. J.; Thomas, C. G.; Bennett, A. J. R., Application of Microscopy to the Investigation of Brown-coal Pyrolysis. *Fuel* **1995,** 74, (11), 1648-1653.
18. Senneca, O.; Salatino, P.; Masi, S., Microstructural changes and loss of gasification reactivity of chars upon heat treatment. *Fuel* **1998,** 77, (13), 1483-1493.
19. Feng, B.; Bhatia, S. K.; Barry, J. C., Structural ordering of coal char during heat treatment and its impact on reactivity. *Carbon* **2002,** 40, (4), 481-496.
20. Wang, W.; Thomas, K. M.; Poultney, R. M.; Willmers, R. R., Iron-Catalyzed Graphitization in the Blast-Furnace. *Carbon* **1995,** 33, (11), 1525-1535.
21. Shim, H. S.; Hurt, R. H., Thermal annealing of chars from diverse organic precursors under combustion-like conditions. *Energy & Fuels* **2000,** 14, (2), 340-348.
22. Cai, H. Y.; Guell, A. J.; Chatzakis, I. N.; Lim, J. Y.; Dugwell, D. R.; Kandiyoti, R., Combustion reactivity and morphological change in coal chars: Effect of pyrolysis temperature, heating rate and pressure. *Fuel* **1996,** 75, (1), 15-24.
23. Hurt, R. H.; Gibbins, J. R., Residual Carbon from Pulverized Coal-Fired Boilers .1. Size Distribution and Combustion Reactivity. *Fuel* **1995,** 74, (4), 471-480.
24. Hurt, R.; Sun, J. K.; Lunden, M., A kinetic model of carbon burnout in pulverized coal combustion. *Combustion and Flame* **1998,** 113, 181-197.
25. Holland, T. M. A Comprehensive Coal Conversion Model Extended to Oxy-coal Conditions (in progress). Brigham Young University, Provo, UT, 2017.
26. Sacks J., W., W.J., Mitchell, T.J. and Wynn, H.P. , *Statistical Science* **1989,** 4, 409-423.
27. Welch, W. J.; Buck, R. J.; Sacks, J.; Wynn, H. P.; Mitchell, T. J.; Morris, M. D., Screening, Predicting, and Computer Experiments. *Technometrics* **1992,** 34, (1), 15-25.
28. Kennedy, M. C.; O'Hagan, A., Bayesian calibration of computer models. *Journal of the Royal Statistical Society Series B-Statistical Methodology* **2001,** 63, 425-450.
29. Gattiker, J., Myers, K., Williams, B., Higdon, D., Carzolio, M., & Hoegh, A. , Gaussian Process-Based Sensitivity Analysis and Bayesian Model Calibration with GPMSA. In *Handbook of Uncertainty Quantification*, Ghanem, H., Owadhi, Ed. Springer: 2017; pp 1-41.
30. Genetti, D. An Advanced Model of Coal Devolatilization Based on Chemical Structure. Brigham Young University, Provo, UT, 1999.

31. Genetti, D.; Fletcher, T. H.; Pugmire, R. J., Development and application of a correlation of C-13 NMR chemical structural analyses of coal based on elemental composition and volatile matter content. *Energy & Fuels* **1999,** 13, (1), 60-68.
32. Shurtz, R. C.; Kolste, K. K.; Fletcher, T. H., Coal Swelling Model for High Heating Rate Pyrolysis Applications. *Energy & Fuels* **2011,** 25, (5), 2163-2173.
33. Holland, T.; Fletcher, T. H., Comprehensive Model of Single Particle Pulverized Coal Combustion Extended to Oxy-Coal Conditions. *Energy & Fuels* **2017**.
34. Zolin, A.; Jensen, A.; Jensen, P. A.; Frandsen, F.; Dam-Johansen, K., The influence of inorganic materials on the thermal deactivation of fuel chars. *Energy & Fuels* **2001,** 15, 1110-1122.
35. Zolin, A. Reactivity of Solid Fuels. Technical University of Denmark, Lyngby, Denmark, 2001.
36. Feng, B.; Jensen, A.; Bhatia, S. K.; Dam-Johansen, K., Activation energy distribution of thermal annealing of a bituminous coal. *Energy & Fuels* **2003,** 17, (2), 399-404.
37. Senneca, O.; Cortese, L., Thermal annealing of coal at high temperature and high pressure. Effects on fragmentation and on rate of combustion, gasification and oxy-combustion. *Fuel* **2014,** 116, 221-228.
38. Jayaraman, K.; Gokalp, I.; Bonifaci, E.; Merlo, N., Kinetics of steam and CO2 gasification of high ash coal–char produced under various heating rates. *Fuel* **2015,** 154, 370-379.
39. Russell, N. V.; Gibbins, J. R.; Man, C. K.; Williamson, J., Coal char thermal deactivation under pulverized fuel combustion conditions. *Energy & Fuels* **2000,** 14, (4), 883-888.
40. Naredi, P.; Pisupati, S. V., Interpretation of char reactivity profiles obtained using a thermogravimetric analyzer. *Energy & Fuels* **2008,** 22, (1), 317-320.
41. Bar-Ziv, E.; Zaida, A.; Salatino, P.; Senneca, O., Diagnostics of carbon gasification by Raman microprobe spectroscopy. *Proceedings of the Combustion Institute* **2000,** 28, 2369-2374.
42. Gale, T. K. Effects of Pyrolysis Conditions on Coal Char Properties. Brigham Young University, Provo, UT, 1994.
43. Liu, G. S.; Niksa, S., Coal conversion submodels for design applications at elevated pressures. Part II. Char gasification. *Progress in Energy and Combustion Science* **2004,** 30, 679-717.

Western States Section of the Combustion Institute – Fall 2017 Meeting
Hosted by the University of Wyoming
October 2-3, 2017

Co-gasification of Coal and Biochar in Modified Drop Tube Reactor and TGA

E. Beagle[1], *Y. Wang*[2], *D. Bell*[2], *E. Belmont*[1]
[1]*Department of Mechanical Engineering, University of Wyoming, 1000 E University Ave, Laramie, WY 82071, USA*
[2]*Department of Chemical Engineering, University of Wyoming, 1000 E University Ave, Laramie, WY 82071, USA*

The use of biomass as a fuel to reduce greenhouse gas emissions from the electricity generation sector is gaining attention. Biochar, a byproduct of biomass pyrolysis, has many potential applications including use as a gasification feedstock to produce syngas for power and other useful products. Co-gasification of biochar with coal can address several issues with both fuels, including reducing emissions associated with coal while also alleviating biomass feedstock availability concerns. However, potential interactions between the two fuels as well as differing reaction behaviors must be understood and addressed for optimum results. In this study, co-gasification of two types of biochar (pine and oak) with coal is studied in a modified drop tube reactor (MDTR) and using thermogravimetric analysis (TGA). The MDTR allows for determination of rapid heating kinetics such as would be experienced in an industrial gasifier, while TGA is used to assess the existence of additive or synergistic effects. Results include extents of fuel conversion, conversion rate, and rapid heating kinetic parameters from the MDTR, as well as mass loss profiles and additive analysis results from the TGA.

1. Introduction

With increasing concerns about the negative consequences of anthropogenic climate change, energy sources that reduce greenhouse gas emissions without impacting energy security and reliability are in increasing demand. One attractive alternative to fossil fuels is biomass because it is widely available, doesn't suffer from intermittency issues, and has been demonstrated to be a drop-in ready fuel for some current generation processes. With sustainable management, biomass can be considered a carbon neutral fuel [1] [2] [3]. One factor that currently limits widespread biomass utilization is a lack of understanding of conversion characteristics, such as kinetics, over a variety of fuels and processing conditions.

One pathway of biomass utilization for energy production is gasification. Gasification is the partial oxidation of a fuel which produces primarily a mixture of carbon monoxide (CO) and hydrogen (H_2), commonly known as synthesis gas or syngas. The product mixture can also contain CO_2 and H_2O. Produced syngas can be combusted for power production or upgraded to higher value products [4] [5]. Disadvantages to the widespread use of biomass gasification include uncertainty in feedstock availability and composition. Co-gasification of biomass with coal can address these concerns and meet the goals of reducing fossil fuel use and emissions [6]. However, a thorough understanding of feedstock properties and co-conversion behavior is needed to further advance this power production strategy.

Undesirable traits of raw biomass, including high moisture content and low energy density, can be improved through torrefaction or pyrolysis [7] [8]. Additionally, biomass pyrolysis is being explored as a primary process because it produces valuable bio-oil products. Gasification provides a means by which biochar, a pyrolysis byproduct, can be beneficially used [9]. Several studies have examined the co-gasification of different biomass feedstocks with coal for synergistic effects; however, results have not been consistent [10] [11]. Alkali ash components, such as potassium, have been shown to have both inhibitory and synergistic effects in a study on the co-conversion of switchgrass with fossil fuels [12]. Co-gasification of pine with coal was shown to have increased reactivity and synergy at higher biomass fractions and process temperatures [13] while another study of co-gasification of cedar bark with coal did not show any synergy [11].

Many different reactor types have been used in gasification studies, including thermogravimetric analyzers [10], fixed bed reactors [13], fluidized bed reactors and drop tube furnaces [10] [14]. Drop tube furnaces mimic the rapid heating conditions of commercial plants, allowing for determination of realistic conversion parameters and behavior [10]. The random pore model (RPM), which considers the development of pore surfaces available for reaction, has been found to simulate well the observed gasification behavior of coal and biomass [11] [13]. In the current study, co-gasification of pine and oak biochars with Powder River Basin coal was tested using a modified drop tube reactor (MDTR) and a thermogravimetric analyzer (TGA) to assess the co-conversion behaviors of these fuel blends. The random pore model (RPM) was used in conjunction with results from the MDTR to determine rapid heating kinetics while additive and synergistic effects were explored using the TGA.

2. Methods

a. Sample Preparation

The biochar used in this study was produced by pyrolysis in the Thermochemical Process Development Unit at the National Renewable Energy Laboratory (NREL) in Golden, CO. The two biomass feedstocks considered were oak and pine. Nitrogen was used as the carrier gas to feed the biomass feedstock into the entrained flow reactor. The solid char products produced were separated via cyclonic separators. The collected solids were passivated using low percent air and then provided for this study. For both feedstocks considered, the pyrolysis temperature was 480°C with a residence time of ~3.5 seconds.

The coal used in this study was sub-bituminous coal from the Powder River Basin provided by Idaho National Laboratory. All tested feedstocks (pine biochar, oak biochar and coal) were dried and sieved to a particle size of 106 – 250 μm prior to all gasification tests in both the MDTR and the TGA. Samples were stored in sealed glass jars in a desiccator in between tests to prevent moisture uptake.

Sample blends were created by mixing dried and sieved coal and biochar particles at a 50/50 weight ratio. Blends were well mixed prior to gasification tests to ensure a homogenous distribution of coal and biochar particles in each tested sample. Ultimate and proximate analyses and heating values of the pine and oak biochars and the coal were measured using ASTM standards D3176, D3172 and D5865 respectively and are given in Table 1. The elemental ash analysis for the primary feedstocks are given in Table 2. The properties for the blends are calculated based on the properties of the individual components and the 50/50 blend ratio by weight.

Table 1. Ultimate and proximate and heating values for feedstocks tested

	Pine Biochar	Oak Biochar	Pine Biochar & Coal	Oak Biochar & Coal	Coal
Proximate Analysis (As Rec'd)					
Moisture	<0.01	<0.01	13.40	13.40	26.8
Ash	6.34	3.49	6.26	4.83	6.17
Volatile Matter	16.33	17.64	24.42	25.08	32.51
Fixed Carbon	77.33	78.87	55.93	56.70	34.52
Total	100	100	100	100	100
Ultimate Analysis (Dry)					
Moisture	--	--	--	--	--
Hydrogen	3.36	3.41	3.96	3.99	4.56
Carbon	82.93	84.08	74.66	75.24	66.39
Nitrogen	0.14	0.14	0.47	0.47	0.79
Sulfur	0.03	0.02	0.34	0.33	0.64
Oxygen	7.20	8.86	13.16	13.99	19.13
Ash	6.34	3.49	7.38	5.96	8.43
Total	100	100	100	100	99.95
Heating Value (MJ/kg)	31.12	31.20	26.57	26.61	19.73

Table 2. Elemental ash analysis of feedstocks tested

	Pine Biochar	Oak Biochar	Coal
SiO_2	47.83	4.52	40.41
Al_2O_3	6.39	1.90	16.86
TiO_2	0.04	<0.01	0.93
Fe_2O_3	3.19	0.68	5.56
CaO	13.60	34.80	19.55
MgO	5.76	2.21	3.98
Na_2O	0.84	0.45	1.08
K_2O	11.20	21.10	0.81
P_2O_5	4.33	2.00	0.68
SO_3	0.47	0.31	8.00
Cl	0.04	0.01	-
CO_2	1.92	26.89	-
MnO2	-	-	0.02

BaO	-	-	0.8
SrO	-	-	0.33
Total	*95.6*	*94.9*	*99.0*

b. Experimental Set-up

All five feedstock types (pine biochar, oak biochar, coal, pine biochar and coal blend, and oak biochar and coal blend) were gasified at three different temperatures (833°C, 900°C and 975°C) in a modified drop tube reactor (MDTR) and by thermogravimetric analysis (TGA). The MDTR used to carry out gasification experiments is detailed in Wang et al [14]. The reactor consisted of an outer Inconel 600 tube of dimensions 33 mm ID x 38 mm OD x 1067 mm length and an inner quartz tube with dimensions 28 mm ID x 31 mm OD x 241 mm length. The MDTR was heated to the desired reaction temperature, and once the reactor was at temperature, the sample was dropped into the reactor from above and fell onto a quartz frit located at the bottom of the reactor. This reactor setup allows for results from rapid heating of the sample, as it is initially dropped into the high temperature reactor, as well as longer timeframe results for complete gasification as the sample sits on the quartz frit inside the heated reactor for an extended period of time. An MKS Cirrus 2 quadrupole mass spectrometer (MS) was used to analyze the product gases. All tests were conducted with gas flow rates through the reactor of 1.6 SLPM CO_2 and 0.4 SLPM Ar at ambient pressure (75 kPa in Laramie, WY) with an initial sample mass of 1.0 g. Each test was repeated twice to ensure repeatability.

The TGA used in this study was a TA Instruments Q500. Samples were heated in an inert environment (Ar) at 30°C/min until the desired gasification temperature was reached, at which point the gas environment was switched to mimic the environment of the drop tube reactor (80 mL/min CO_2 purge and 20 mL/min Ar purge). The experiments in the TGA allow for study of the behavior of the blended fuels while all particles are in close proximity throughout the duration of the test.

3. Results and Discussion

a. Modified Drop Tube Reactor

The CO signal from the MS was used to calculate conversion extent with time from the MDTR. The rate of conversion extent (dX/dt) was assumed to be equal to the difference between the reaction CO signal and the baseline CO signal using an introduced proportionality rate signal k, as shown in Equation 1

$$\frac{dX}{dt} = k([CO] - [CO]_{baseline}) \,. \qquad (1)$$

The integral of the conversion rate from the beginning of the reaction t_0 to any given time t was defined as the extent of conversion $X(t)$, as shown in Equation 2

$$X(t) = \int_{t_0}^{t} \frac{dX}{dt} = \int_{t_0}^{t} k([CO] - [CO]_{baseline})dt \,. \qquad (2)$$

At the completion of the reaction the CO signal returned to its baseline level at which point the final conversion extent X_f was equal to one, as shown in Equation 3

$$X_f = 1 = \int_{t_0}^{t_f} \left(\frac{dX}{dt}\right) dt = \int_{t_0}^{t_f} k([CO] - [CO]_{baseline}) dt \; . \tag{3}$$

Using the CO signal data from the MS and *Excel* Solver, the value of k was determined. The determined value of k was then used with Equations 1 and 2 to calculate the rate of conversion (dX/dt) and extent of conversion (X) versus time.

Figure 1 shows the extent of conversion X versus the residence time from the MDTR for the five different feedstocks at the three gasification temperatures tested. At all three temperatures, the coal reached complete conversion ($X=1$) more quickly than the other feedstocks, indicative of coal's higher reactivity. The conversion plots for the two blends fall in between the curves for the primary feedstocks. At 833° C, the pine biochar reacts the slowest, although this trend is reversed at the other two temperatures as oak biochar reacts more slowly.

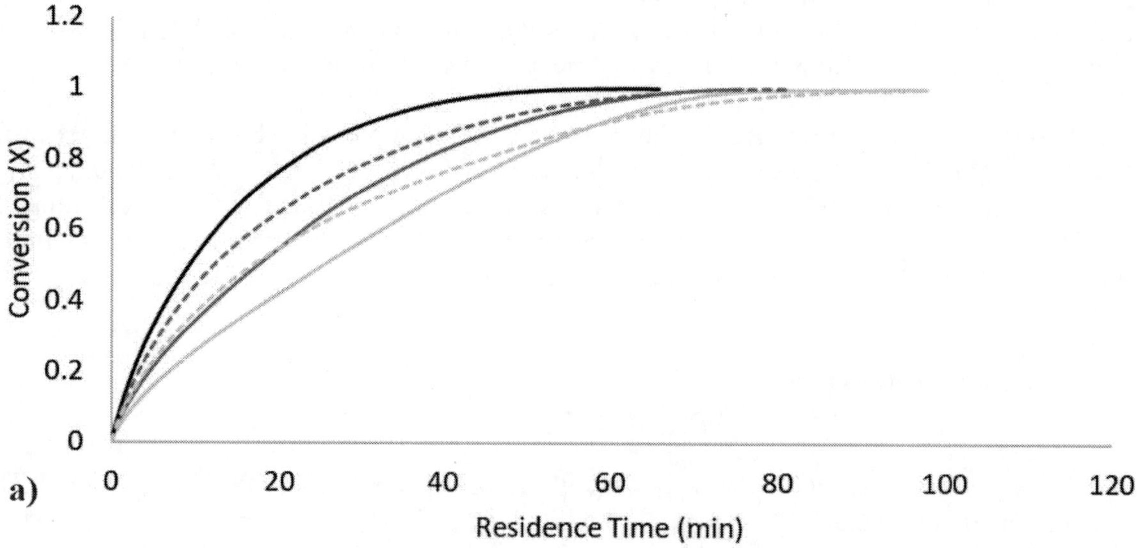

a)

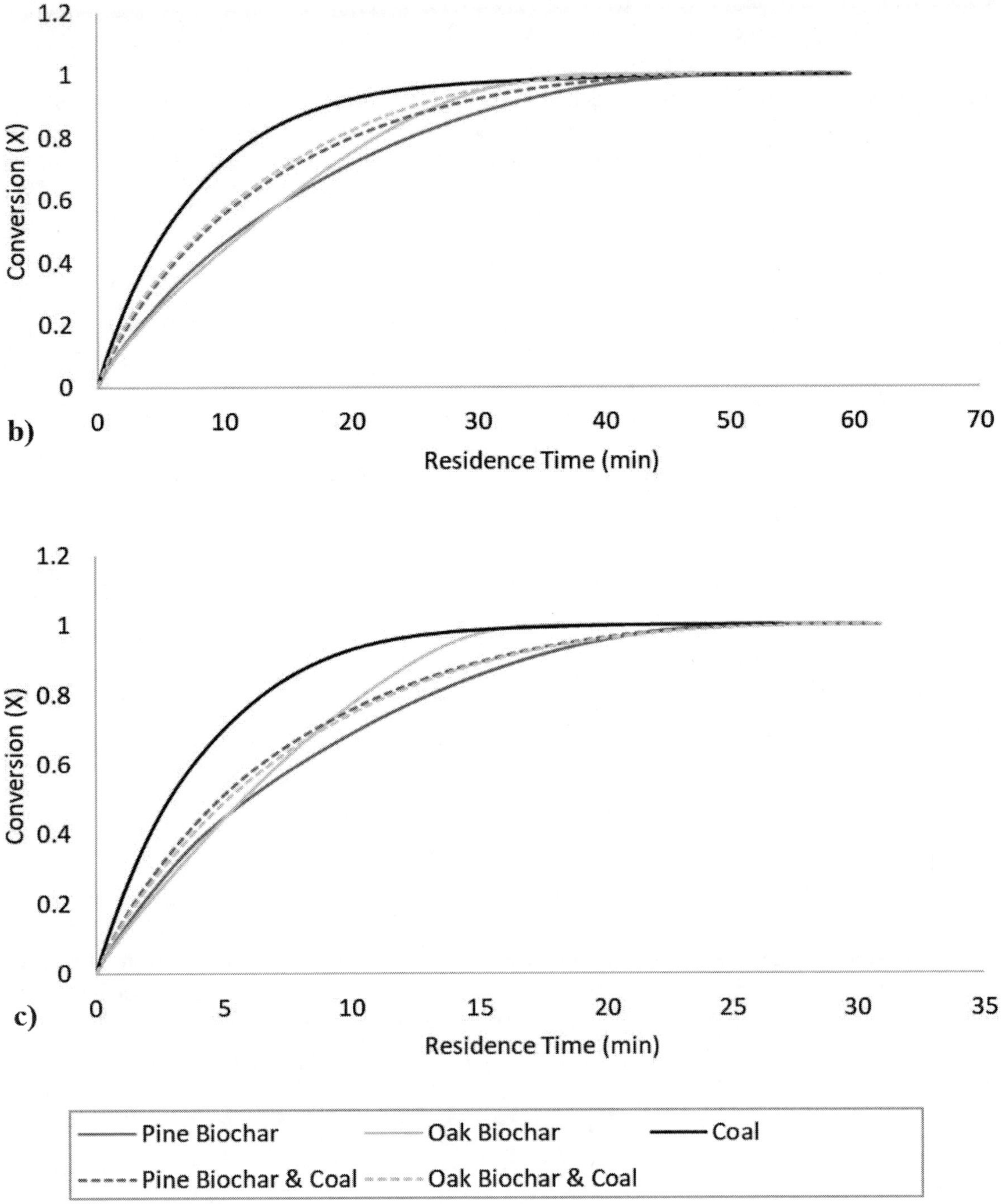

Figure 1. Extent of conversion versus residence time for the five feedstocks tested at a) 833°C, b) 900°C and c) 975°C

The residence time at the end of the reaction, defined as the time when $X = 0.95$, is shown in Table 3. The residence time decreases with increasing reaction temperature for all feedstock types with average residence times of 59 mins, 30 mins and 16 mins for the 833°C, 900°C and 975°C cases respectively. The average uncertainty of residence time with 95% confidence for all test conditions is 2.9 mins. The coal had significantly lower residence times for all three temperatures, with the largest difference being at 833°C. The pine biochar and coal blend residence times fall in between that of the primary constituents, suggestive of additive effects, except at 833°C. However, for the case of the oak biochar and coal blends at 833°C, the residence time of the blends is higher than those of the primary constituents, suggestive of inhibitory reaction behavior. This is not observed for the oak biochar and coal blends at 900°C. At 975°C, the residence time of the blend closely resembles that of the oak biochar.

Table 3. Residence times for the five feedstocks and three gasification temperatures tested

	Residence Times (min)				
	Pine	**Oak**	**Pine & Coal**	**Oak & Coal**	**Coal**
833C	56	63	59	77	38
900C	36	30	33	31	22
975C	20	13	19	18	10

b. Random Pore Model (RPM)

The RPM approach models fluid-solid reactions by assuming the solid is porous and that the gasification reaction rate is determined by the rate of gasification of pore walls within the char particle. The model accounts for the change in char reactivity due to changing pore morphology with reaction progression [15] [16]. This method has been shown to represent experimental gasification behavior well, including in a related study of coal gasification in a MDTR by Wang et al [11] [14]. Equation 4 gives the RPM gasification rate equation

$$\frac{dX}{dt} = k_p(1-X)\sqrt{1-\psi\ln(1-X)} \qquad (4)$$

Where k_p is the reaction rate constant and ψ is a dimensionless structural parameter. Though ψ does have a physical meaning, it is typically used as a data fitting parameter, as is done in this study. K_p is usually defined as the reaction rate at the beginning of the reaction when $X=0$; however, in this case the initial reactions observed were pyrolysis which is not accounted for the in RPM formulation. Therefore, it was assumed that gasification began at $X=0.3$ such that k_p was determined based on Equation 5

$$k_p = \frac{dX}{dt}|_{X=0.3}\,. \qquad (5)$$

The value of ψ was calculated using a sum-of-squared errors approach with *Solver* in Microsoft Excel to minimize the difference between the model results and the experimental data. Table 4 gives the fitted parameters of the RPM results with the uncertainty calculated using a *student-t* test approach and a 95% confidence interval.

Table 4. Fitted parameters from the random pore model with uncertainty calculated over 95% confidence intervals

Feedstock	Temperature	k_p	ψ
Pine	833°C	4.39E-04±3.87E-05	2.41±0.43
	900°C	7.47E-04±3.58E-05	1.81±0.14
	975°C	1.36E-03±1.06E-04	1.72±0.25
Oak	833°C	2.92E-04±3.06E-05	4.28±0.45
	900°C	6.52E-04±2.44E-05	3.94±0.19
	975°C	1.42E-03±1.19E-04	4.10±0.00
Pine Biochar & Coal	833°C	6.08E-04±6.20E-05	0.84±0.21
	900°C	9.48E-04±5.95E-05	1.45±0.03
	975°C	1.64E-03±2.27E-05	1.39±0.29
Oak Biochar & Coal	833°C	4.17E-04±9.06E-05	0.96±0.00
	900°C	8.78E-04±9.09E-05	1.99±0.25
	975°C	1.49E-03±4.42E-05	2.02±0.36
Coal	833°C	8.49E-04±3.93E-05	1.45±0.04
	900°C	1.60E-03±1.49E-04	1.12±0.05
	975°C	2.92E-03±1.52E-04	1.63±0.36

c. MDTR Kinetics

The Arrhenius equation, shown in Equation 6, is commonly used to describe reaction kinetics and allows for the determination of activation energy (E_a) and pre-exponential constant (A_0) for the five feedstocks tested in the present study using the reaction rate constant k_p determined from the RPM

$$\ln(k_p) = \frac{-E_a}{R}\frac{1}{T} + \ln(A_0) \ . \qquad (6)$$

Figure 2 shows the resulting Arrhenius plot for the five tested feedstocks with linear trend line fits for the data indicated by the dashed lines. These trendlines were used to calculate E_a and A_0 for the five feedstocks, results of which are given in Table 5. Of the three primary feedstocks, pine biochar had the lowest activation energy of 91.4 kJ/mol while the oak biochar had the highest of 127.9 kJ/mol. The activation energy of the pine biochar and coal blend was lower than that of either constituent at 80.2 kJ/mol. The activation energy of the oak biochar and coal blend fell between its two constituents at 102.8 kJ/mol, although this value is much closer to the coal activation energy of 100.4 kJ/mol than the oak biochar. There was significant variation in the pre-exponential constant for the five feedstocks, ranging from 3.7 s^{-1} for the pine biochar and coal blend to 319.4 s^{-1} for the oak biochar. Results from other studies have found the activation energy for coal to be in the range of 29-280 kJ/mol and for biochar to be in the range of 88-250 kJ/mol under similar gasification test conditions [17] [18].

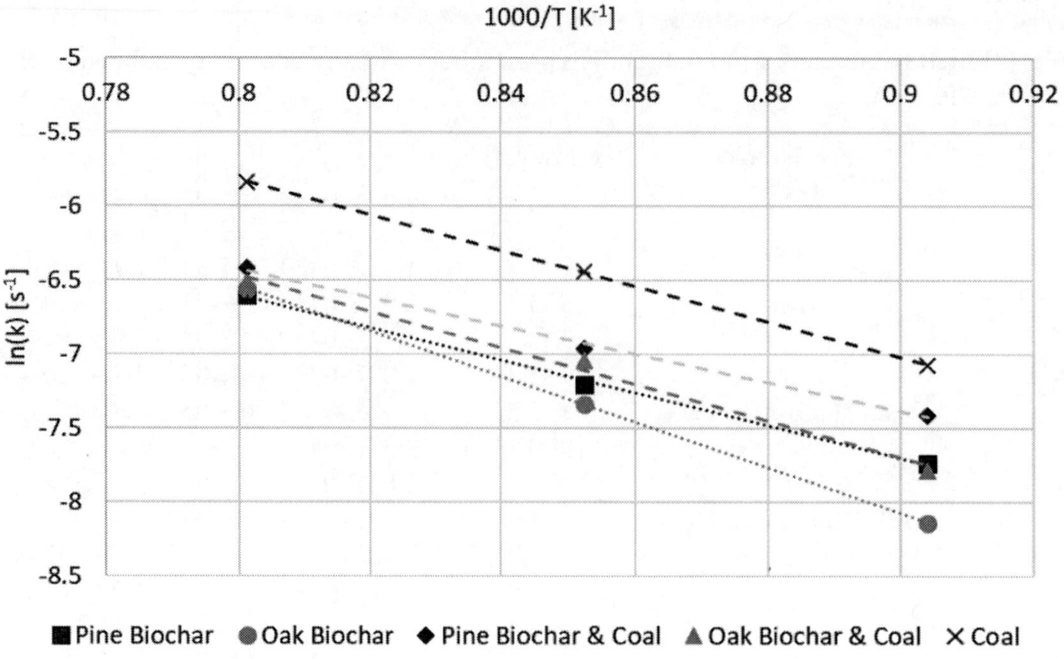

Figure 2. Arrhenius plot for different feedstocks using k_p from the random pore model analysis

Table 5. Calculated Arrhenius parameters (activation energy and pre-exponential constant) from trend line fit of Arrhenius plot (Figure 2)

	Pine	*Oak*	*Pine & Coal*	*Oak & Coal*	*Coal*
Ea (kJ/mol)	91.4	127.9	80.2	102.8	100.4
A_0 (s^{-1})	9.0	319.4	3.7	30.9	44.9

d. Additive Analysis

The existence of synergistic or inhibitive effects during conversion is an important consideration for the co-utilization of biomass and coal resources. The presence of such effects, which may be quite complex, must be accounted for when determining conversion rates, while the absence of these effects implies that conversion behavior can be predicted for a feedstock blend through a simple additive analysis. In this study, an additive analysis was done using results from both the MDTR and the TGA to determine whether synergistic effects occurred during co-gasification of the biochar and coal. Theoretical conversion profiles representative of only additive behavior were calculated using Equation 7

$$X_{theor} = m_c \cdot X_c + m_b \cdot X_b \tag{7}$$

where X_{theor} is the theoretical blend result assuming only additive effects, m_c and m_b are the mass fractions of the coal and biochar samples in the blend respectively, and X_c and X_b are the experimental results for the coal and biochar samples respectively. For the MDTR, the result of interest was the extent of conversion for comparison of sample residence time, and for the TGA the result of interest was the derivative mass loss profile as this provides a more detailed picture of mass loss behavior.

Figure 3 shows the theoretical and experimental blend results based on conversion extent from the MDTR tests. For all reaction conditions considered, the theoretical blend reaches complete conversion slightly more rapidly than the experimental blends, suggestive of inhibitory effects during the experimental co-gasification runs.

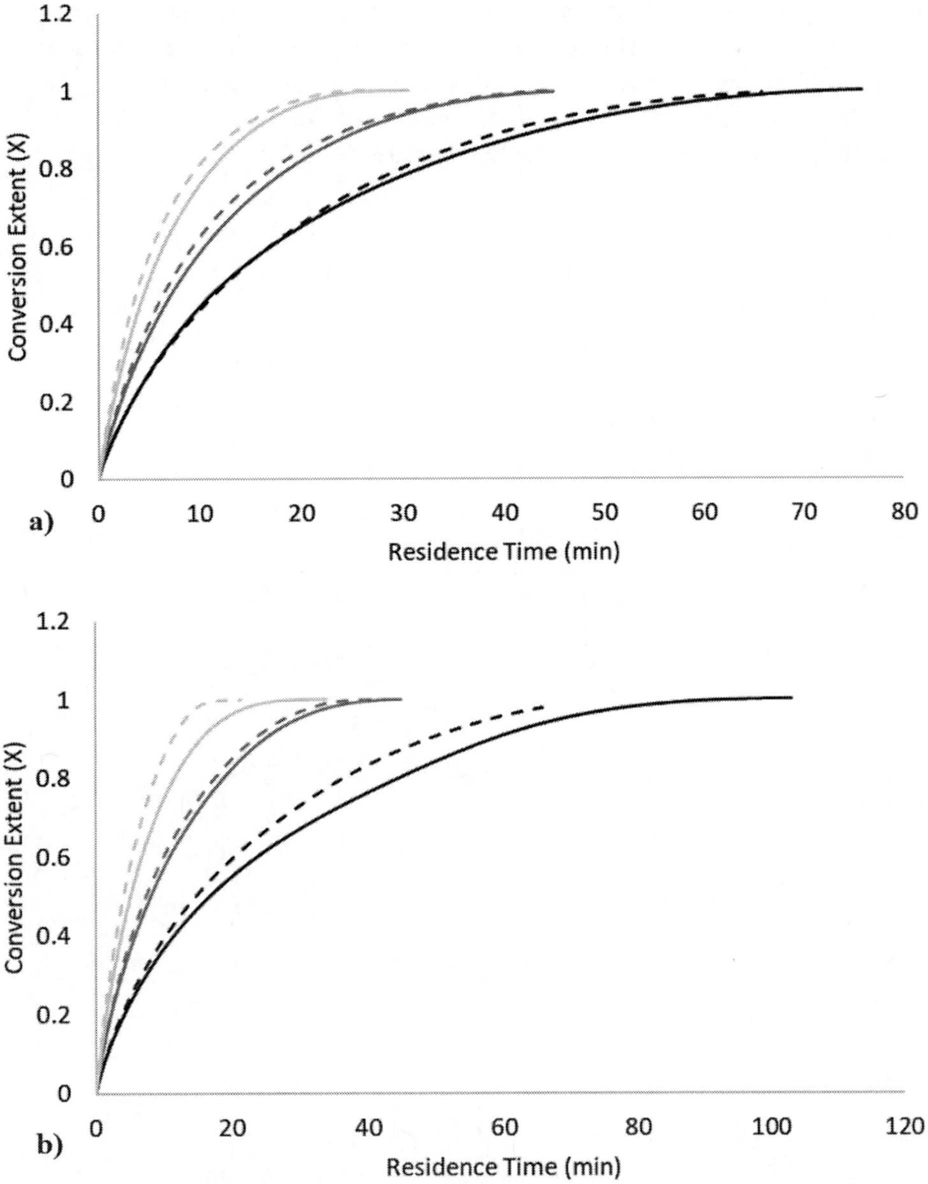

Figure 3. Additive analysis results for conversion extent (*X*) from MDTR results for a) pine biochar and coal blend and b) oak biochar and coal blend

Figure 4 shows the additive analysis results based on experimental TGA tests. For the oak biochar and coal blend (Figure 4-b), the theoretical derivative mass loss profiles were higher, indicating a higher reactivity, than the experimental blend results for all temperatures tested. This suggests a similar inhibitory behavior in the experimental blends as that observed with the MDTR results. For the pine biochar and coal blend (Figure 4-a), at 833°C and 900°C the experimental derivative mass loss is slightly higher than the theoretical derivative mass loss, suggesting mild synergistic behavior at these temperatures. However, at 975°C the theoretical derivative mass loss is slightly higher than the experimental, suggesting mild inhibitory behavior.

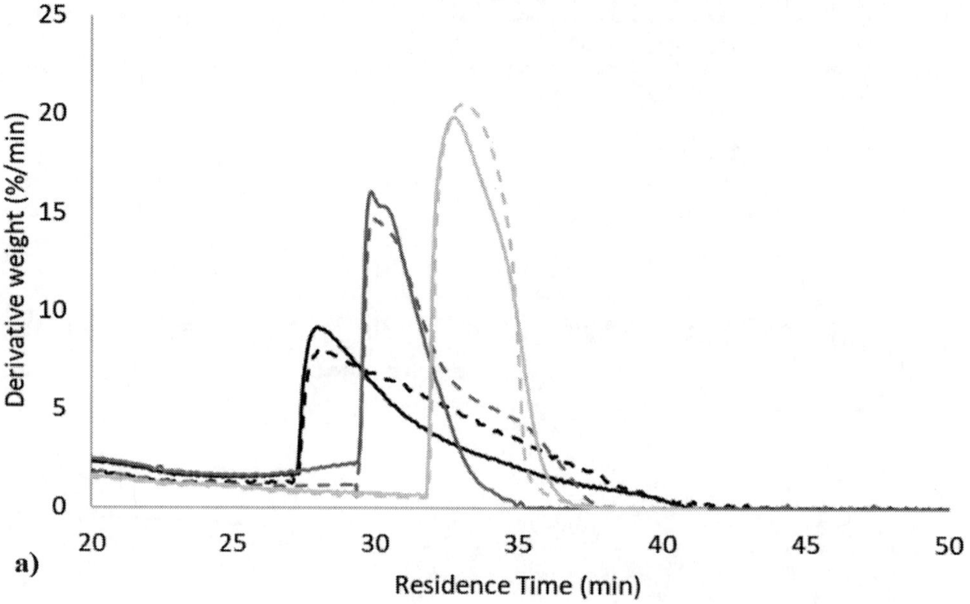

a)

b)

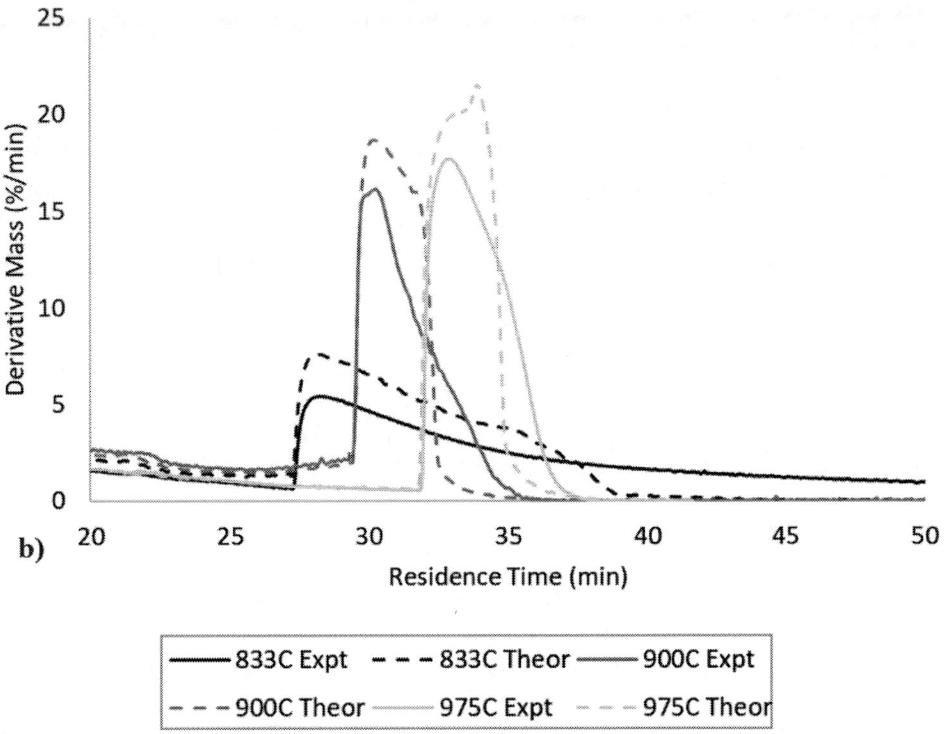

Figure 4. Additive analysis results for derivative mass loss from TGA results for a) pine biochar and coal blend and b) oak biochar and coal blend

The non-additive behavior observed in both the MDTR and TGA results is attributed to the mineral compositions of the pine and oak biochar ashes. Other studies have noted the possible importance of alkali and alkaline metal ash components in biomass on synergistic effects during co-gasification [19]. Potassium and calcium are of particular interest because they have been shown to increase the gasification reactivity of coal char and carbon [20]. These minerals are found in substantial quantities in the biochar feedstocks in this study with K_2O accounting for 11.2% and 21.1% and CaO accounting for 13.6% and 34.8% of the ash for the pine and oak biochars respectively. Additionally, the pine biochar ash has significantly higher amounts of silica and aluminum than the oak biochar ash, which has been shown to form inhibitory compounds with other minerals at elevated temperatures [21] [22]. The interactions between these minerals and the gasification reaction pathways at various temperatures is complex, which could account for the discrepancy of inhibitory and synergistic effects observed in the pine biochar and coal blends at varying reaction temperatures.

4. Conclusions

Co-gasifying biochar with coal provides an effective means of utilizing biomass pyrolysis byproducts and reducing fossil fuel consumption. In order to increase co-gasification implementation, reaction characteristics of fuel blends must be understood. In this study, five feedstocks (pine biochar, oak biochar, coal, pine biochar and coal blend, and oak biochar and coal blend) were gasified in a CO_2 environment in a modified drop tube reactor and in a thermogravimetric analyzer. The modified drop tube reactor mimics rapid particle heating rates that would be experienced by fuels in industrial reactors while the TGA maintains constant contact of blend particles to fully assess synergistic effects.

Experiments were carried out at 833°C, 900°C and 975°C at atmospheric pressure. The higher reaction temperatures increased the reaction rates for all tested fuels. Kinetic parameters for the rapid heating in the MDTR were determined using the Arrhenius equation and the reaction rate constant determined by the random pore model. Additive analysis from both the MDTR and TGA results suggest non-additive behavior, with both synergistic and inhibitory effects observed depending on feedstock type and reaction temperature. Further research is required to fully understand the complex interactions of fuel chemistry and reactivity that lead to the observed blend behaviors.

Acknowledgements

This material is based upon work supported by the National Science Foundation Graduate Research Fellowship under Grant no. 12-52375.
The authors would like to thank Ester Wilcox and Matt Oliver of the National Renewable Energy Laboratory for providing the biochar samples and other support for this study.

References

[1] M. Sami, K. Annamalai and M. Wooldridge, "Co-firing of coal and biomass fuel blends," *Progress in Energy and Combustion Science,* pp. 171-214, 2001.

[2] S. Sahu, N. Chakraborty and P. Sarker, "Coal-biomass co-combustion: an overview," *Renewable and Sustainable Energy Reviews,* pp. 575-586, 2014.

[3] M. J. Prins, K. J. Ptasinski and F. J. Janssen, "More efficient biomass gasification via torrefaction," *Energy,* pp. 3458-3470, 2006.

[4] P. Basu, Biomass gasification, pyrolysis and torrefaction: practical design and theory, Academic Press, 2013.

[5] B. Digman, H. S. Joo and D.-S. Kim, "Recent progress in gasification/pyrolysis technologies for biomass conversion to energy," *Environmental Progress & Sustainable Energy,* vol. 28, no. 1, pp. 47-51, 2009.

[6] L. Taba, M. Irfan, W. Daud and M. Chakrabarti, "The effect of temperature on various parameters in coal, biomass and co-gasification: A review," *Renewable and Sustainable Energy Reviews,* vol. 16, pp. 5584-5596, 2012.

[7] D. Medic, M. Darr, A. Shah, B. Potter and J. Zimmerman, "Effects of torrefaction process parameters on biomass feedstock upgrading," *Fuel,* vol. 91, no. 1, pp. 147-154, 2012.

[8] C. Couhert, S. Salvador and J.-M. Commandre, "Impact of torrefaction on syngas production from wood," *Fuel,* vol. 88, no. 11, pp. 2286-2290, 2009.

[9] C. E. Brewer, K. Schmidt-Rohr, J. A. Satrio and R. C. Brown, "Characterization of Biochar From Fast Pyrolysis and Gasification Systems," *Environmental Progress & Sustainable Energy,* vol. 28, no. 3, pp. 386-396, 2009.

[10] C. Gao, F. Vejahati, H. Katalambula and R. Gupta, "Co-gasification of biomass with coal and oil sand in a drop tube furnace," *Energy & Fuels,* vol. 24, pp. 232-240, 2010.

[11] S. Kajitani, Y. Zhang, S. Umemoto, M. Ashizama and S. Hara, "Co-gasification reactivity of coal and woody biomass in high-temperature gasification," *Energy & Fuels,* vol. 24, pp. 145-151, 2010.

[12] R. Habibi, J. Kopyscinski, M. Masnadi, J. Lam, J. Grace, C. Mims and J. Hill, "Co-gasification of Biomass and Non-biomass Feedstocks: Synergistic and Inhibition Effects of Switchgrass Mixed with Sub-bituminous Coal and Fluid Coke During CO2 Gasification," *Energy & Fuels,* vol. 27, pp. 494-500, 2013.

[13] H. Jeong, S. Park and J. Hwang, "Co-gasification of coal-biomass blended char with CO2 at temperatures of 900-1100C," *Fuel,* vol. 116, pp. 465-470, 2014.

[14] Y. Wang and D. Bell, "Reaction kinetics of Powder River Basin coal gasification in carbon dioxide using a modified drop tube reactor," *Fuel,* vol. 140, pp. 616-625, 2015.

[15] S. K. Bhatia and D. D. Perlmutter, "A random pore model for fluid-solid reactions: I. Isothermal, kinetic control," *AIChE Journal,* vol. 26, no. 3, pp. 379-386, 1980.

[16] J. Fermoso, B. Arias, C. Pevida, M. G. Plaze, F. Rubiera and J. J. Pis, "Kinetic models comparison for steam gasification of different nature fuel chars," *Journal of Thermal Analysis and Calorimetry,* vol. 91, no. 3, pp. 779-785, 2008.

[17] M. F. Irfan, M. R. Usman and K. Kusakabe, "Coal gasification in CO2 atmosphere and its kinetics since 1948: A brief review," *Energy,* vol. 36, pp. 12-40, 2010.

[18] C. D. Blasi, "Combustion and gasification rates of lignocellulosic chars," *Progress in Energy and Combustion Science,* vol. 35, pp. 121-140, 2009.

[19] K. Miura, K. Hashimoto and P. L. Silveston, "Factors affecting the reactivity of coal chars during gasification, and indices representing reactivity," *Fuel,* vol. 68, no. 11, pp. 1461-1475, 1989.

[20] Y. Zhang, S. Hara, S. Kajitani and M. Ashizawa, "Modeling of catalytic gasification kinetics of coal char and carbon," *Fuel,* vol. 89, pp. 152-157, 2010.

[21] H. Risnes, J. Fjellerup, U. Henriksen, A. Moilanen, P. Norby, K. Papdakis, D. Posselt and L. H. Sorensen, "Calcium addition in straw gasification," *Fuel,* vol. 82, pp. 641-651, 2003.

[22] P. Lahijani, Z. A. Zainal, M. Mohammadi and A. R. Mohamed, "Conversion of the greenhouse gas CO2 to the fuel gas CO via the Boudouard reaction: A review," *Renewable and Sustainable Energy Reviews,* vol. 41, pp. 615-632, 2015.

Western States Section of the Combustion Institute – Fall 2017 Meeting
Hosted by the University of Wyoming
October 2-3, 2017

Comparison of raw and torrefied healthy and beetle kill pine in co-combustion with coal

Alexandra Howell Emily Beagle Erica Belmont

Department of Mechanical Engineering, University of Wyoming, 1000 E University Ave, Laramie, WY 82071, USA

The abundance of beetle-kill lodgepole pine in proximity to existing coal-fired power plants in the Rocky Mountain Region of the United States presents a unique opportunity for the integration of fuelwood biomass into electricity generation while simultaneously allowing forest management to reduce the associated safety hazards of the beetle-kill trees. In this study, both healthy and beetle-kill pine were torrefied at 200, 250 and 300 °C to assess the increase in their energy densities and improvement in their grindability for integration into existing infrastructure. Raw and torrefied samples were combusted and co-combusted with sub-bituminous Power River Basin coal using thermogravimetric analysis. Results showed increased grindability and mass loss of the beetle-kill as opposed to the healthy pine of both raw and torrefied pine feedstock, which is attributed to the degraded state of the wood caused by the bark beetles and associated fungi. Additive behavior was observed for the co-combustion of all the beetle-kill and healthy pine samples with coal. The results showed good interchangeability between healthy and beetle-kill pine for future implementation.

1. Introduction

As the electricity generation sector seeks to reduce greenhouse gas (GHG) emissions, the use of biomass as a feedstock alone and blended with coal for co-firing has been increasingly explored and implemented as a viable GHG mitigation option [1, 2]. The Department of Energy recently released its Billion Ton Report, showing that plentiful biomass resources exist, including forest resources, within the United States (US) to help meet energy demands [3]. Local biomass availability and needs must be considered when deciding on renewable energy strategies, as assessments of biomass resource location and demands in the US have shown that supply and intended end use are not necessarily collocated [4]. Recent analyses have shown that that, in order to achieve high levels of emissions reductions at minimal additional cost, transportation of biomass over significant distances should be avoided and local biomass resources should be considered for regional energy needs such as electricity [5, 6].

One biomass resource which is plentiful in the Rocky Mountain region of the US is lodgepole pine that has been killed by bark beetles. Estimates indicate that as much as 19.7 billion cubic feet of standing dead timber could potentially be salvaged from across the western US [7]. Additionally, if left unmanaged, devastating negative effects of large swaths of dead and dry forest on forest fire severity and risk to life and property have been identified [8]. Many different groups, including the National Forest Service, recognize the potential for bioenergy development utilizing bark beetle kill trees and encourage bioenergy projects that utilize beetle kill biomass as a feedstock [9]. Furthermore, a recent technoeconomic assessment found that there are large amounts of standing and fallen dead pine in proximity to power plants in the region, and a recent study that quantified the co-firing potential of beetle kill pine by volume and proximity found it to be a plentiful resource [10]. Therefore, these assessments of beetle kill wood and the interest in increased use of biomass for electricity generation suggest that the investigation of beetle kill pine as a combustion and co-combustion feedstock is warranted.

There are technical challenges in co-firing biomass feedstock with coal in power plants, many of which are due to the differences in properties between coal and various biomasses [11]. Unfavorable properties of raw biomass include high moisture content and hygroscopicity, low energy density and low grindability relative to coal. These properties can be improved for combustion with coal through mild pyrolysis, called torrefaction, in which raw biomass is heated in an inert atmosphere to temperatures between 200-300 °C. Specifically, torrefaction reduces the oxygen and hydrogen contents of biomass, thereby concentrating the carbon content and increasing the heating value [12]. The hemicellulose portion of biomass is substantially degraded and the biomass is rendered hydrophobic during torrefaction due to the loss of OH functional groups, which improves the storability of the biomass [13]. Additionally, torrefaction degrades the fibrous nature of raw biomass, thereby improving its grindability and handling [14, 15]. As a result of these property changes, torrefaction can render biomass properties more similar to coal, improve combustion efficiency, and reduce transportation and storage costs of biomass.

A number of studies have used thermogravimetric analysis (TGA) to examine the combustion of healthy raw and torrefied pine [16, 17, 18] and its co-combustion with coal. Gil et al. [19] studied the combustion of bituminous coal and raw pine sawdust blends using TGA and found no significant synergistic behavior between the coal and biomass, as the blend behaved as an additive sum of the mass fractions and mass loss characteristics of each component. The coal, pine and blends presented one, two and three major mass loss events during combustion, respectively; these reaction steps were attributed predominantly to char oxidation for coal and volatile release followed by char oxidation for pine, with the blends demonstrating behavior of both coal and pine. Wang et al. [20] also studied the combustion of bituminous coal and pine sawdust using TGA and did observe both negative synergistic behavior at low conversion extents, which was attributed to coal restricting the devolatilization of biomass, and positive synergistic behavior which was attributed to the alkali content of the biomass ash. Zhang et al. [21] studied the co-combustion of coal gangue and pine sawdust using TGA and also observed synergistic behavior, which was attributed to a thermal effect in which heat release from pine combustion was lost to coal gangue, thereby slowing pine combustion and accelerating coal gangue combustion.

This study characterizes the torrefaction of beetle kill (BK) pine wood and its co-combustion with Powder River Basin (PRB) coal in order to assess the technical viability of using beetle kill wood as a co-firing feedstock. Healthy pine (HP) is also assessed to elucidate the differences between BK and HP, as any co-firing strategy that purports to use BK wood will need to plan for an alternative feedstock in the future as BK wood is depleted. TGA and a fixed bed reactor are used to torrefy BK and HP wood at 200-300 °C, and the grindability of the raw and torrefied feedstocks are assessed. HP and BK mass loss during torrefaction and the combustion of these feedstocks alone and their co-combustion with PRB coal in a 20/80% by energy blend of wood/coal is assessed by TGA. Differences in the composition, handling, and combustion characteristics of the feedstocks are discussed.

2. Materials and Methods

2.1 Feedstock Preparation

The biomass feedstocks used in this study were a sub-bituminous coal from the Powder River Basin in WY (coal), pine sawdust produced from healthy pine (HP), and pine sawdust produced from standing dead beetle kill pine (BK). The pine wood was collected from the Medicine Bow National Forest following all proper permitting requirements. Ultimate, proximate and heating values for the three tested feedstocks are summarized in Table 1. Biomass feedstock heating value, ultimate, and proximate analyses were performed by Wyoming Analytical Laboratories in Laramie, WY. Coal feedstock analyses were performed by Idaho National Laboratory.

Table 1. Ultimate, proximate and heating value analyses of as-received raw healthy pine (HP), beetle kill pine (BK), and coal are summarized. Results are on a weight % basis unless otherwise specified.

	HP	BK	Coal
Moisture	7.16	6.53	26.8
Volatile Matter	74.6	76.0	32.5
Fixed Carbon	17.3	17.0	34.5
Ash	0.97	0.42	6.17
Carbon	47.2	48.6	48.6
Oxygen	38.9	38.4	14.0
Hydrogen	5.39	5.70	3.34
Nitrogen	0.38	0.28	0.58
Sulfur	0.02	0.02	0.47
HHV (kJ/kg)	18,401	19,006	19,725

HP and BK sawdust in a particle size range of 0.425-3.35 mm were torrefied in a fixed bed reactor in order to produce samples for grindability assessment and combustion. The reactor was constructed from 316 SS pipe and fittings, and the main section of the reactor in which biomass was loaded for torrefaction was approximately 15 cm long with a 20 mm inside diameter. An inlet at the bottom of the reactor allowed inert gas to flow through the sample during torrefaction, and an outlet at the top permitted the venting of inert and evolved gases. A thermocouple was inserted into the sample and the sample temperature was continuously monitored during torrefaction. The reactor was loaded with approximately 5 g of pre-dried sample and loaded into a preheated oven, while nitrogen (N_2) was flowed through the reactor at 0.5 $L \cdot min^{-1}$. HP and BK samples underwent an average heating rate of 20 $^{\circ}C \cdot min^{-1}$ up to a final torrefaction temperature of 200, 250 and 300 °C and were held at torrefaction temperature for one hour. Following torrefaction, samples were cooled in an inert environment to below 100 °C before they were removed from the oven. Samples were then transferred to glass jars and flushed with N_2 for storage. Heating values for HP and BK torrefied at 200 °C (HP_200 and BK_200, respectively), 250 °C (HP_250 and BK_250, respectively) and 300 °C (HP_300 and BK_300, respectively) were measured and are summarized in Table 2.

Table 2. Heating values for torrefied HP and BK, as-received, are shown.

	HHV (kJ/kg)
HP_200	18,908
HP_250	19,682
HP_300	22,551
BK_200	18,966
BK_250	19,757
BK_300	22,341

2.2 Grindability

Grindability was assessed for raw and torrefied HP and BK samples in a particle size range of 0.425-3.35 mm. Prior to milling the samples, raw and torrefied BK and HP were sieved into size ranges of <75 μm, 75-150 μm, 150-425 μm, 0.425-1.18 mm, 1.18-2.00 mm, 2.00-3.35 mm, and >3.35 mm. A Gilson shaker was used to separate samples into the sieves using a shaking period of 2 min. After sieving, mass fractions were weighed and the fraction of total mass in each size range was calculated. After weighing all separated samples, the fractions were recombined and milled in an SPEX Industries ball mill for 1 minute using a tungsten vessel and milling balls. The milled HP and BK samples were then sieved and weighed again. The 75-150 μm fraction of each sample was retained and used for TGA combustion tests. The remainder of the sample was used for ultimate, proximate, and heating value analyses.

2.3 Thermogravimetric Analysis

Torrefaction mass loss and combustion were conducted using a TA Instruments Q500 TGA. Biomass and coal samples were loaded into alumina sample pans of 0.1 mL volume for all tests. N_2 (99.999%) and air (industrial compressed air, dried and filtered) were used for inert and reactive gas environments, respectively, and flowed at 100 $mL \cdot min^{-1}$ for all tests. All tests were conducted at atmospheric pressure, which is 75 kPa in Laramie, WY. Torrefaction mass loss was quantified for HP and BK sawdust at each of the three tested torrefaction temperatures (200, 250 and 300 °C) by loading 5-7 mg of sawdust into a sample pan, and then carrying out torrefaction in the TGA. A temperature profile of heating to 100 °C at 100 $°C \cdot min^{-1}$, then a 10-minute isothermal dehydration step, followed by heating at 20 $°C \cdot min^{-1}$ to the final torrefaction temperature of 200, 250 or 300 °C, and finally an isothermal step at torrefaction temperature for 1 hour was applied. This method mimicked the heating profile used in larger scale torrefaction in the fixed bed reactor. Mass loss during torrefaction was determined on a dry basis following dehydration. Torrefaction was repeated five times to confirm repeatability and conduct statistical analysis on mass loss results.

Combustion, and co-combustion of pulverized HP and BK, with particle sizes between 75-150 μm, were conducted using TA Instruments modulated TGA, in which feedstocks underwent a 2 °C/min linear heating rate and a sinusoidal temperature modulation was imposed upon the linear heating profile [22, 23]. Modulated TGA was utilized in order to facilitate future determination of kinetic parameters. Combustion was conducted in air at temperatures up to 800 °C of nine unblended samples: raw HP and BK; HP and BK each torrefied at 200, 250 and 300 °C; and coal. Co-combustion of blends of HP or BK with coal were prepared by blending 20% by energy of HP or BK with coal, and these blends were combusted in the TGA. For all combustion and co-combustion tests, samples of 5-7 mg were used to avoid mass and heat transfer limitation effects and tests were repeated twice to confirm repeatability.

The presence of additive or synergistic behavior was assessed by calculating theoretical derivative mass loss curves for feedstock blends from the derivative mass loss curves of the individual components using Eqn. 1

$$dm/dt_{model} = x_{coal}(dm/dt)_{coal} + x_{biomass}(dm/dt)_{biomass} \quad (1)$$

where x_{coal} and $x_{biomass}$ are the mass fractions of coal and biomass in the blends, respectively, and $(dm/dt)_{coal}$ and $(dm/dt)_{biomass}$ are the individual mass loss rates of coal and biomass, respectively. TGA is an excellent method for testing for the presence of synergistic or additive behavior between two feedstocks because feedstock particles are maintained in contact throughout the duration of the heating process.

3. Results and Discussion

3.1 Torrefaction Mass Loss

Figure 1 summarizes the mass loss due to torrefaction, as determined by TGA. The average uncertainty, calculated using Student t-distribution with a 95% confidence interval for five replicates, for all data sets is ± 0.9%, and is not shown in Figure 1 due to its small magnitude. Results show that BK mass loss is slightly greater than HP mass loss at 200 °C torrefaction at 10.2% and 9.3%, respectively. The difference in mass loss increases at a torrefaction temperature of 250 °C, where BK loses 28.4% of its initial mass and HP loses 26.4% of its initial mass. The greatest difference in mass loss is observed at 300 °C torrefaction temperature, where BK and HP lose 68.5% and 64.7% of their initial masses, respectively. These results suggest that HP is more recalcitrant to degradation at torrefaction temperatures, which could be attributed to the natural degradation that BK undergoes upon death by infestation of bark beetles and the fungi that the bark beetles introduce to the tree. Specifically, cellulose and lignin degradation onset during torrefaction occurs at lower temperatures for BK than for HP because the recalcitrant compounds in pine wood of lignin and cellulose have undergone some previous deterioration due to beetle-associated fungus [24, 25].

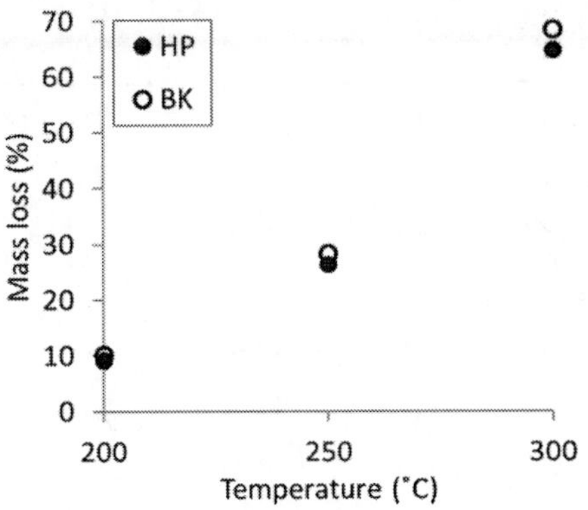

Figure 1. Mass loss of HP and BK due to torrefaction at 200, 250, and 300 ˚C are shown.

3.2 Grindability

Table 3 summarizes the size distributions of raw and torrefied BK and HP before and after milling. The majority of unmilled biomass, including both raw and torrefied samples, is observed to be between 0.425-3.35 mm before milling as expected, although small fractions were measured to be larger and smaller than this range. Significant size reduction occurs after milling for all samples, with the greatest fraction of small particle sizes observed at the highest torrefaction level of 300 ˚C. In addition to the effective size reduction achieved by torrefaction at 300 ˚C, it is noteworthy that torrefaction at this level altered BK and HP to have heating values which exceed that of coal (Table 2). In contrast, BK and HP torrefied at 200 ˚C have heating values below that of coal, and HP and BK torrefied at 250 ˚C have heating values nearly equal to coal. Thus, the benefit of torrefaction in altering biomass to improve its suitability in co-firing with or substituting for coal is seen, with 300 ˚C torrefaction being particularly effective, in terms of size reduction and heating value increase.

Table 3. The results of grindability are shown, with the percentages of particles found in each size range before and after milling given.

	Size range percentages (before milling)						
	< 75 µm	75-150 µm	150-425 µm	0.425-1.18 mm	1.18-2.00 mm	2.00-3.35 mm	> 3.35 mm
BK_raw	0.0	0.2	2.5	40.4	37.1	16.9	2.9
BK_200	0.0	0.2	2.2	42.9	35.2	17.8	1.6
BK_250	0.1	0.4	3.8	46.6	34.2	12.9	2.1
BK_300	0.1	0.6	5.0	45.9	33.1	11.8	3.4
HP_raw	0.0	0.0	1.1	34.4	32.8	30.4	1.3
HP_200	0.0	0.1	1.7	37.2	32.9	25.5	2.6
HP_250	0.0	0.1	2.7	46.3	31.1	18.4	1.3
HP_300	0.1	0.1	2.1	37.1	35.8	22.6	2.1
	Size range percentages (after milling)						

	< 75 µm	75-150 µm	150-425 µm	0.425-1.18 mm	1.18-2.00 mm	2.00-3.35 mm	> 3.35 mm
BK_raw	0.1	2.9	18.0	62.2	14.9	1.9	0.2
BK_200	0.4	6.2	33.9	56.7	2.9	0.0	0.0
BK_250	0.8	9.3	49.1	39.9	0.7	0.1	0.0
BK_300	5.8	22.3	59.5	12.5	0.0	0.0	0.0
HP_raw	0.0	1.6	10.8	47.9	27.8	11.2	0.7
HP_200	1.1	4.5	18.6	54.9	20.6	0.1	0.3
HP_250	0.4	11.4	33.4	48.1	6.3	0.4	0.0
HP_300	14.5	39.8	36.2	9.5	0.1	0.0	0.0

3.3 Combustion and co-combustion

Figure 2 shows the mass loss of coal, raw HP and BK, and 300 °C torrefied HP and BK during combustion. Results are abbreviated to show the region of most significant mass loss between 100 °C and 500 °C, after which the raw and torrefied biomass show negligible mass loss but coal continues to gradually lose mass via oxidation. In contrast to coal, raw and torrefied HP and BK clearly show multiple stages of mass loss. These two regions of mass loss correspond to Stage 1 of biomass combustion, which is devolatilization of biomass by pyrolysis and homogeneous gas phase combustion, and Stage 2 combustion of the devolatilized biomass char, which is a heterogeneous carbon oxidation process. BK and HP generally behave quite similarly, particularly when torrefied at 300 °C, although raw BK shows slightly earlier onset of mass loss as compared to raw HP. Additionally, BK_300 and HP_300 show mass loss during combustion before 300 °C, the fixed bed reactor torrefaction temperature. During severe torrefaction, the pine biomass undergoes extensive devolatilization. This event significantly increases the amount of oxidation sites available within the feedstock. Once the sample is exposed to air during combustion, it is predisposed to oxidation. Hence, the early onset of mass loss of the torrefied HP and BK in Figure 2 is attributed to the onset of reactions at these available oxidation sites. Since HP and BK behave quite similarly, mass loss during combustion of BK, including raw and torrefied at 200, 250 and 300 °C, is shown in Figure 3 to highlight the impact of torrefaction level on combustion behavior, with the understanding that raw and torrefied HP exhibit similar behavior. Mass loss behavior is observed to be quite similar between raw, 200 and 250 °C BK, although the torrefied biomass shows later mass loss onset, likely due to the loss of light volatiles during torrefaction. Notably, raw and 200 °C torrefied BK loses mass monotonically at combustion temperatures above 100 °C, while 250 °C torrefied BK gains 0.1 % mass and 300 °C torrefied BK gains 0.4 % mass before losing mass. This mass gain is attributed to oxygen adsorption, and the observed increase in more highly torrefied BK is attributed to the higher level of carbonization which provides more active sites for oxygen adsorption.

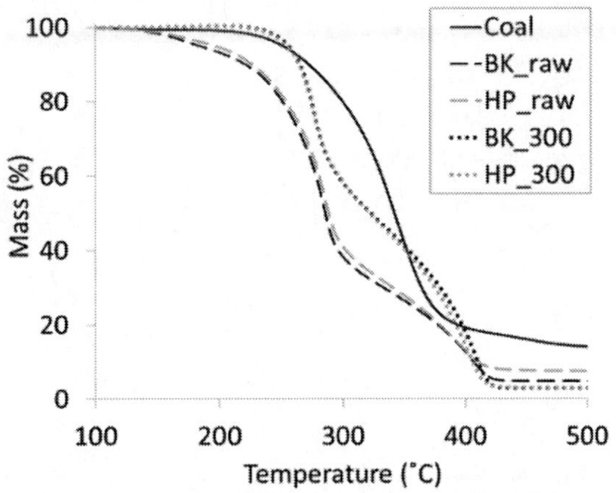

Figure 2. Mass loss profiles of coal, raw BK and HP, and 300 ˚C torrefied HP and BK during combustion are shown.

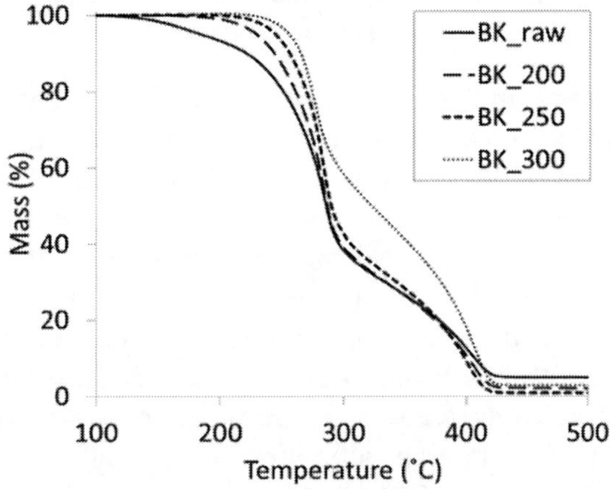

Figure 3. Mass loss profiles of raw and 200, 250 and 300 ˚C torrefied BK during combustion are shown.

Figure 4 shows the derivative mass loss profiles of coal, raw HP and BK, and 300 ˚C torrefied HP and BK. The derivative mass loss profiles for 200 ˚C and 250 ˚C torrefied biomass are not shown for brevity, but exhibit similar dual peak characteristics to the raw and torrefied biomasses. These dual peaks correspond to pyrolysis (Stage 1 of biomass combustion) and char oxidation (Stage 2 of biomass combustion). In contrast, coal shows a single stage of combustion, consisting predominantly of char oxidation.

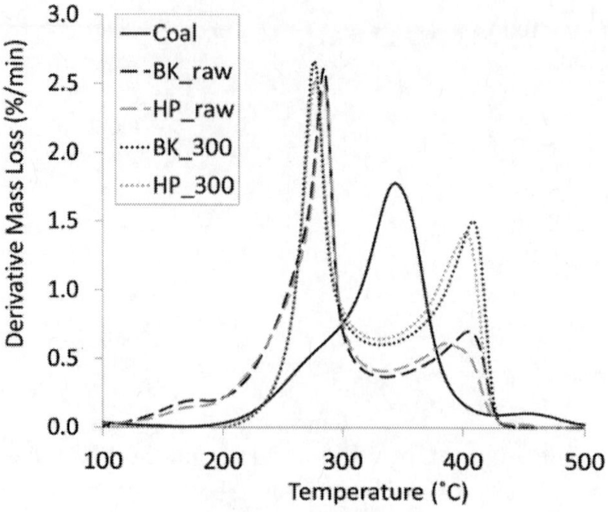

Figure 4. Derivative mass loss profiles for combustion of coal, raw HP and BK, and 300 ˚C HP and BK are shown.

Figures 5 and 6 show the experimentally measured derivative mass losses of raw and 300 ˚C torrefied HP and BK blends with coal, respectively, as well as the calculated model derivative mass loss (Eqn. 1). Results are abbreviated between 100 ˚C and 500 ˚C in order to highlight the region of maximum substantial mass loss, and results are not shown for 200 ˚C and 250 ˚C torrefied BK and HP blended with coal for brevity because results exhibit quite similar trends to those observed for raw and 300 ˚C torrefied BK and HP. Results show generally excellent agreement between experimentally measured and calculated theoretical derivative mass loss, suggesting that the combustion behavior of both raw biomass/coal and torrefied biomass/coal blends is additive. This indicates that there is no significant catalytic behavior resulting from the ash contents of biomass or coal, nor is there any apparent alteration to biomass devolatilization imposed by the presence of coal among the feedstocks tested in this study. Raw HP and BK blends with coal (Fig. 5) both show a maximum rate of Stage 1 (pyrolysis) mass loss at 286 ˚C, and a maximum rate of Stage 2 (char combustion) mass loss at 346 ˚C. The 300 ˚C torrefied biomass/coal blends (Fig. 6) exhibit lower pyrolysis peaks as compared to raw biomass/coal blends, as expected due to the loss of volatiles during torrefaction, with 300 ˚C HP/coal and 300 ˚C BK/coal blends showing maximum Stage 1 mass loss at 278 ˚C. Stage 2 maximum mass loss for 300 ˚C HP and BK blends with coal occur at 346 ˚C, as was observed for raw biomass/coal blends. While not shown for brevity, 200 ˚C and 250 ˚C HP and BK blends with coal also exhibit maximum Stage 2 mass loss at 346 ˚C. Thus, following pyrolysis of the raw or thermally pretreated biomass/coal blends, Stage 2 combustion proceeds in a uniform manner regardless of feedstock blend. Stage 1 peak mass loss for 200 ˚C biomass/coal blends and 250 ˚C biomass/coal blends (not shown) occurs at 285 ˚C and 284 ˚C, respectively. These results suggest that HP and BK combustion behavior is similar enough that a seamless transition from BK to HP could be made in the future when BK resources are depleted.

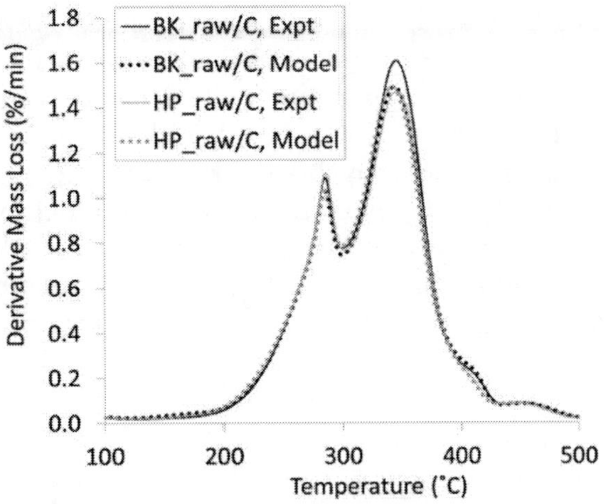

Figure 5. Experimental (Expt) and calculated (Model) derivative mass loss during combustion of raw BK and raw HP blended with coal (BK_raw/C and HP_raw/C, respectively)

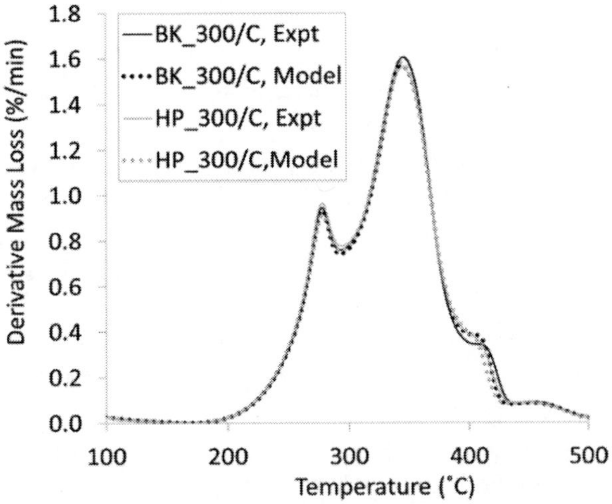

Figure 6. Experimental (Expt) and calculated (Model) derivative mass loss during combustion of 300 ˚C torrefied BK and raw HP blended with coal (BK_300/C and HP_300/C, respectively)

4. Conclusions

This study assessed beetle kill pine as a potential feedstock for combustion and co-combustion with coal, and compared its performance with healthy pine. Healthy and beetle kill pine were torrefied at 200, 250 and 300 °C using thermogravimetric analysis and a fixed bed reactor. Increased mass loss was observed in torrefaction of the beetle kill pine, which was attributed to the natural degradation of the wood that occurs after beetle- and associated fungus-induced death. Torrefaction of both pine feedstocks was found to improve fuel properties by increasing energy density and grindability. Beetle kill pine was found to have a generally higher grindability than its healthy counterpart, which would translate to decreased processing energy required when beetle kill pine is used as a feedstock. Combustion of raw and torrefied healthy and beetle kill pine alone and co-combustion of these feedstocks blended with coal in an amount of 20% by energy were conducted using thermogravimetric analysis. Combustion and co-combustion results, including mass loss and derivative mass loss for all fuels and blends tested, showed similar trends for beetle kill and healthy pine at each torrefaction level. These results suggest the interchangeability of these fuels in combustion and co-combustion applications, which has important implications for the use of both beetle kill and healthy pine as combustion feedstocks, as healthy pine can be readily utilized when beetle kill pine is depleted without substantial combustor performance changes. Additive analysis of the blend results showed no synergistic or inhibiting effects in co-combustion of the coal and biomass. Thus, in summary, the results of this study show a viable path for beetle kill pine utilization in energy applications whereby beetle kill pine has favorable pretreatment behavior but otherwise behaves similarly to healthy pine during conversion.

5. Acknowledgements

This material is based upon work supported by the National Science Foundation Graduate Research Fellowship under Grant No. 12-52375.

References

[1] R. Froese, D. Shonnard, C. Miller, K. Koers and D. Johnson, "An evaluation of greenhouse gas mitigation options for coal-fired power plants in the US Great Lakes States," *Biomass and Bioenergy,* vol. 34, no. 3, pp. 251-262, 2010.

[2] M. Sami, K. Annamalai and M. Wooldridge, "Co-firing of coal and biomass fuel blends," *Progress in Energy and Combustion Science,* vol. 27, pp. 171-214, 2001.

[3] M. Langholtz, B. Stokes and L. Eaton, "2016 Billion-Ton Report: Advancing Domestic Resources for a Thriving Bioeconomy, Volume 1: Economic Availability of Feedstocks," U.S. Department of Energy, 2016.

[4] K. Mullins, A. Venkatesh, A. Nagengast and M. Kocoloski, "Regional Allocation of Biomass to U.S. Energy Demands under a Portfolio of Policy Scenarios," *Environmental Science & Technology,* vol. 48, pp. 2561-2568, 2014.

[5] E. Beagle and E. Belmont, "Comparative Life Cycle Assessment of Biomass Utilization for Electricity Generation in the European Union and the United States," in *Proceedings of the 25th European Biomass Conference and Exhibition*, Stockholm, Sweden, 2017.

[6] P. Basu, Biomass Gasification, Pyrolysis and Torrefaction: Practical Design and Theory, Amsterdam: Academic Press, 2013.

[7] J. Prestemon, K. Abt, K. Potter and F. Koch, "An Economic Assessment of Mountain Pine Beetle Timber Salvage in the West," *Western Journal of Applied Forestry,* vol. 28, no. 4, pp. 143-153, 2013.

[8] USFS, "Western Bark Beetle Strategy: Human Safety, Recovery and Resiliency," USDA, 2011.

[9] USFS, "Requests for Proposals: 2014 Statewide Wood Energy Teams Cooperative Agreement; and 2014 Wood to Energy Grant," Federal Register, 2014.

[10] E. Beagle and E. Belmont, "Technoeconomic assessment of beetle kill biomass co-firing in existing coal fired power plants in the Western United States," *Energy Policy,* vol. 97, pp. 429-438, 2016.

[11] A. Demirbas, "Sustainable cofiring of biomass with coal," *Energy Conversion and Management,* vol. 44, pp. 1465-1479, 2003.

[12] B. Madanayake, S. Gan, C. Eastwick and H. Ng, "Thermochemical and structural changes in Jatropha curcas seed cake during torrefaction for its use as a coal co-firing feedstock," *Energy,* vol. 100, pp. 262-272, 2016.

[13] J. Tumuluru, S. Sokhansanj, J. Hess, C. Wright and R. Boardman, "A review on biomass torrefaction process and product properties for energy applications," *Industrial Biotechnology,* pp. 384-401, 2011.

[14] M. Phanphanich and S. Mani, "Impact of torrefaction on the grindability and fuel characteristics of forest biomass," *Bioresource Technology,* vol. 102, pp. 1246-1253, 2011.

[15] M. Gil, R. Garcia, C. Pevida and F. Rubiera, "Grindability and combustion behavior of coal and torrefied biomass," *Bioresource Technology,* vol. 191, pp. 205-212, 2015.

[16] D. Shen, S. Gu, K. Luo, A. Bridgwater and M. Fang, "Kinetic study on thermal decomposition of woods in oxidative environment," *Fuel,* vol. 88, pp. 1024-1030, 2009.

[17] S. Yorulmaz and A. Atimtay, "Investigation of combustion kinetics of treated and untreated waste wood samples with thermogravimetric analysis," *Fuel Processing Technology,* vol. 90, pp. 939-946, 2009.

[18] B. Mi, Z. Liu, W. Hu, P. Wei, Z. Jiang and B. Fei, "Investigating pyrolysis and combustion characteristics of torrefied bamboo, torrefied wood and their blends," *Bioresource Technology,* vol. 209, pp. 50-55, 2016.

[19] M. Gil, D. Casal, C. Pevida, J. Pis and F. Rubiera, "Thermal behaviour and kinetics of coal/biomass blends during co-combustion," *Bioresource Technology,* vol. 101, pp. 5601-5608, 2010.

[20] G. Wang, J. Zhang, J. Shao, Z. Liu, G. Zhang, T. Xu, J. Guo, H. Wang, R. Xu and H. Lin, "Thermal behavior and kinetic analysis of co-combustion of waste biomass/low rank coal blends," *Energy Conversion and Management,* vol. 124, pp. 414-426, 2016.

[21] Y. Zhang, Z. Zhang, M. Zhu, F. Cheng and D. Zhang, "Interactions of coal gangue and pine sawdust during combustion of their blends studied using differential thermogravimetric analysis," *Bioresource Technology,* vol. 214, pp. 396-403, 2016.

[22] R. Blaine and B. Hahn, "Obtaining Kinetic Parameters by Modulated Thermogravimetry," *Journal of Thermal Analysis,* vol. 54, pp. 695-704, 1998.

[23] K. Cheng, W. Winter and A. Stipanovic, "A modulated-TGA approach to the kinetics of lignocellulosic biomass pyrolysis/combustion," *Polymer Degradation and Stability,* vol. 97, pp. 1606-1615, 2012.

[24] H. Solheim and P. Krokene, "Growth and virulence of mountain pine beetle associated blue-stain fungi, Ophiostoma clavigerum and Ophiostoma montium," *Canadian Journal of Botany,* vol. 76, pp. 561-566, 1998.

[25] S. Lee, J. Kim and C. Breuil, "Pathogenicity of Leptographium longiclavatum associated with Dendroctonus ponderosae to Pinus contorta," *Canadian Journal of Forest Research,* vol. 36, no. 11, pp. 2864-2872, 2006.

Exploring Continuous Monitoring Methods for SO_3 in Flue Gas Conditions

Andrea Biasioli, Yu-Chien Chien, and Derek Dunn-Rankin

Mechanical and Aerospace Engineering Department, University of California, Irvine, California

Abstract

The combustion of coal in power plants gives rise to sulfur compounds, including sulfur trioxide (SO3) and a higher amount of sulfur dioxide (SO_2). In wet flue gas, SO_3 can react to form sulfuric acid (H_2SO_4), a misty acid that can be difficult to capture and that has several adverse effects on both plant operation and the surrounding environment. Despite the importance of these reactive sulfur compounds, currently there are no reliable methods of continuously measuring SO_3/H_2SO_4 in coal-fired flue gas; continuous measurement techniques for SO_3 would improve the operational performance of power plants by optimizing the amount of sorbent needed to neutralize the SO_3 and by extending the lifespan of the power plant components. This work has the objectives of exploring feasible methods for monitoring SO_3 and as part of that objective of verifying that under relevant flue gas conditions the gaseous SO_3 and H_2SO_4 are in a mutual chemical equilibrium, as hypothesized by previous work. If this is the case, the SO_3/H_2SO_4 partitioning could be characterized by just measuring one of the species. The relationship between SO_3 and H_2SO_4 is investigated experimentally in simulated coal-fired power plant conditions, using Differential Optical Absorption Spectroscopy (DOAS) and External Cavity Quantum Cascade Laser (EC-QCL) techniques, assisted by a controlled condensation system that acts as an industry accepted benchmark. The first results are inconclusive among the different techniques, and they highlight the intrinsic complexity of working with these highly reactive and corrosive compounds.

Introduction

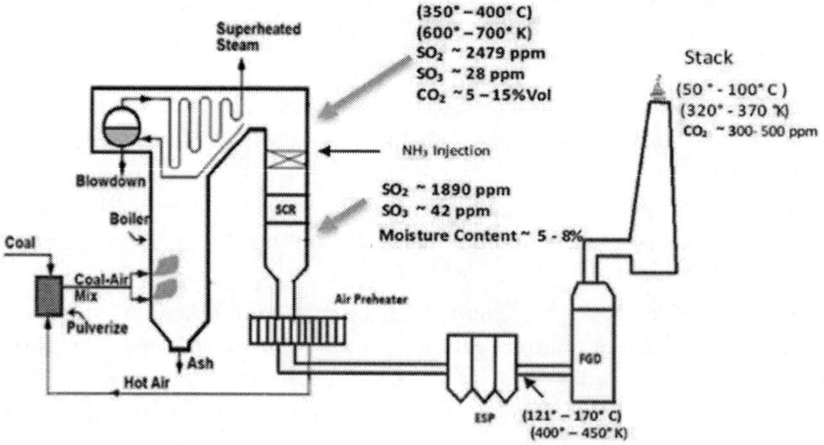

Figure 1. Coal power plant typical simplified diagram with conditions and possible measurement points.

The formation of sulfur compounds is an integral part of the combustion of coal in power plants. Most of the fuel sulfur content is oxidized to sulfur dioxide (SO_2), while a small percentage is further oxidized to sulfur trioxide (SO_3) in the boiler, or across the catalyst when powerplants use selective catalytic reduction (SCR) for NO_x reduction (Moretti, 2012). In wet flue gas, SO_3 can react to form H_2SO_4, a misty acid difficult to capture due to the small size of the acid droplets, with several adverse effects on both plant operation and the surrounding environment, such as corrosion of ductwork surfaces (Halstead and Talbot, 1980), flyash contamination (Bayless et al, 1996), formation of visible plumes (Mueller and Imhoff, 1994), powdered activated carbon (PAC) poisoning (Moretti, 2012), formation of ammonium bisulfate due to ammonia slip from SCR or SNCR (selective non-catalytic reduction) processes (Bionda, 2002), and formation of particulate level PM2.5, a known health hazard (Sanyal and Parker, 1999). Despite its importance, however, challenges associated with its reactivity have prevented any reliable method of continuously monitoring SO_3/H_2SO_4 concentration and transformation in coal-fired flue gas.

Among other opportunities, one example of operational performance improvement of power plants that could come from continuous measurement of SO_3 in flue gas is that the quantity of sorbent used to neutralize the SO_3 could be optimized. This real-time tuning would decrease the operational costs of the power plant as well as extend the lifespan of components downstream from the boiler.

Previous Research

Recent methods proposed for monitoring SO_3 and H_2SO_4 include those from Breen (AbSensor – SO_3), from SICK (MCSO3), ProCeas (SO3/SO2 Analyzer), Thermo Scientific (Arke SO3 System), and the Severn Science Analyser. Several of these approaches rely on the expectation that at essentially all relevant flue gas conditions the gaseous SO_3 and H_2SO_4 are in a mutual chemical equilibrium, while the SO_2 concentration is kinetically frozen (Menasha, 2010). If this is the case, the SO_3/H_2SO_4 partitioning could be characterized by just measuring one of the species. Some techniques take advantage of this assumption by capturing a flue gas sample and then returning it to its in-situ condition but in an external cell where the species are measured by IR absorption (e.g., the SICK MCSO3). The Breen approach is to measure the acid dewpoint temperature and to then relate that value to the concentration (Breen, 2014).

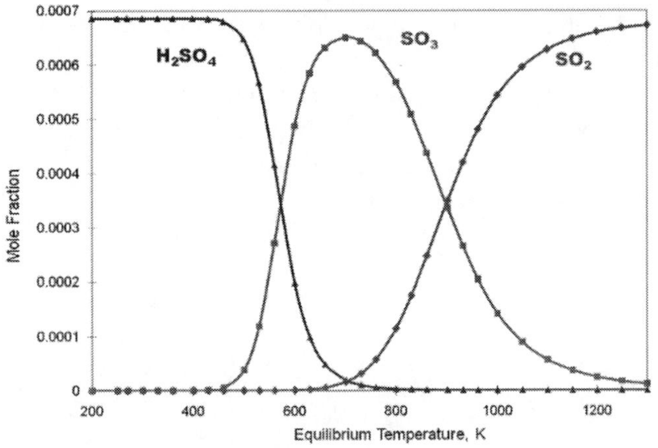

Figure 2. Effect of temperature on the equilibrium (Menasha, 2010).

The current paper concentrates on the development of the experimental apparatus, including a performance evaluation along with some of the important lessons learned when working with the challenging acid gas typical as effluent from coal combustion. The ultimate goal of the work is to experimentally investigate the relationship between SO_3 and H_2SO_4 in simulated coal-fired power plant conditions in order to confirm the hypothesis of inter-conversion equilibrium between SO_3 and H_2SO_4, and to demonstrate the in situ optical measurement of SO_3 using:

- Differential Optical Absorption Spectroscopy (DOAS)
- External Cavity Quantum Cascade Laser (EC-QCL)

Experimental Setup and Methods

The experimental setup includes a high temperature measurement cell on top of a controlled flue gas production apparatus able to replicate appropriate levels of flue gas species (SO_2, SO_3, H_2O, N_2, Air) by flowing a gas mixture through a Haldor Topsoe catalyst (VK-48). The system allows flow ranging between 0 and 12 slpm, with multiple injection points to create a broad range of relevant gas compositions. The temperature of the catalyst section spans from 360 °C to 500 °C ± 1%. Flow piping sections connecting the flue gas generator to the measuring cell benefit from an extended controlled range of temperature (T_{amb} to 500 °C ± 1%).

Each flowing gas component is regulated by a digital mass flow controller or rotameter, and the operational range for the sulfur dioxide line is 0 to 300ppmv ± 3%. The flue-gas simulator can operate in completely dry conditions or up to 15% v/v of water vapor by activating the injection-evaporator system.

After the high temperature measurement cell, a gas sample is extracted from the flowing gas mixture and the sulfur is collected according to the EPA Method 8A (Jain, 1996) into a condensation coil and a train of impingers.

The first tests will not include water vapor addition (dry tests) to allow unambiguous identification of the sulfur trioxide high temperature spectral signature in the UV and the IR regimes in dry conditions. Subsequently, with water vapor inclusion the experiment is aimed to demonstrate that SO_3 and H_2SO_4 are in equilibrium (or at accurately predictable ratios) at typical flue gas conditions. As mentioned above, the results will be verified using a controlled condensation method strictly derived from EPA Method 8A, with the aid of ion chromatography (IC) to evaluate the sulfate concentration collected by the condensation coil and the residual sulfur trapped by bubbling gas through impingers.

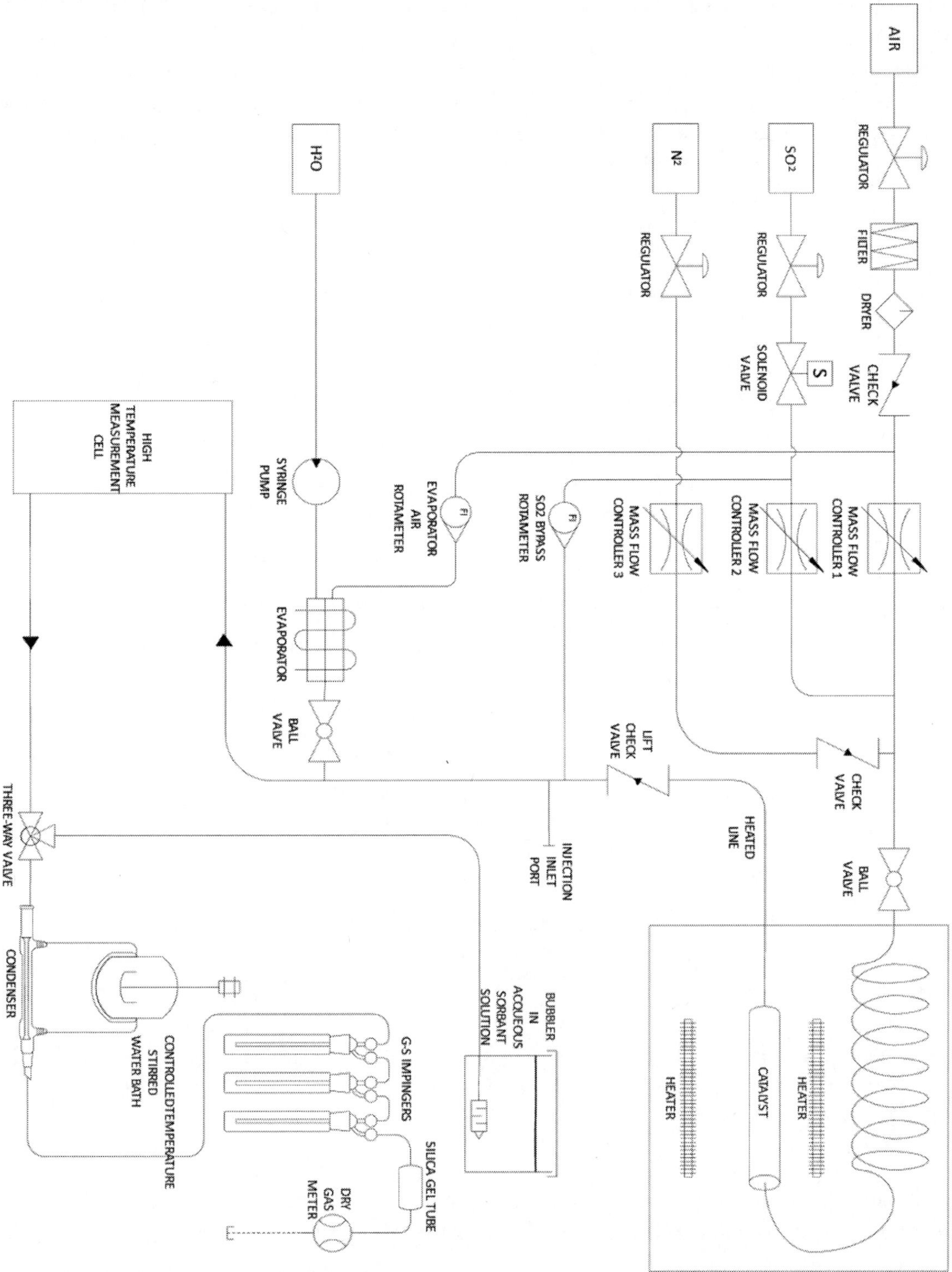

Figure 3: Schematic of the experimental facility.

Dry testing is the first step to verify the conversion rate of the catalyst system, to capture the sulfur trioxide signature in the UV and IR regimes, and to determine the detectability limits for different measuring techniques in controlled conditions. Figure 4 (UV) and 5 (IR) show the expected spectral signature for the sulfur oxides. Figure 4 is experimental in hot conditions and Figure 5 is theoretical for room temperature. Figure 5 also includes water lines to show regions of possible interferences. The figure assumes very low water concentration at room temperature but for higher water content the interference can be more significant.

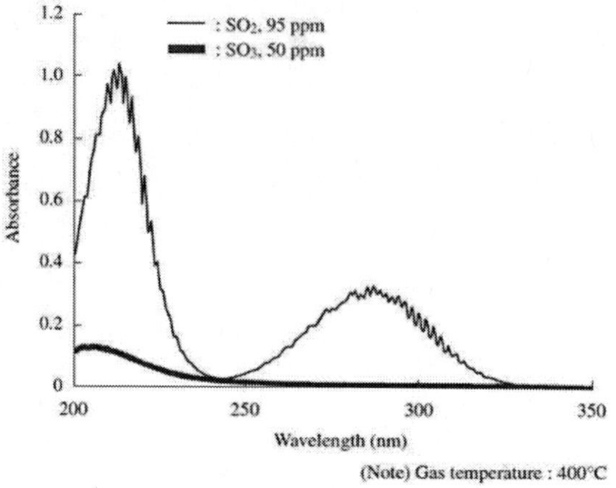

Figure 4. Reference UV spectrum for sulfur dioxide and sulfur trioxide (Kurata, 2003)

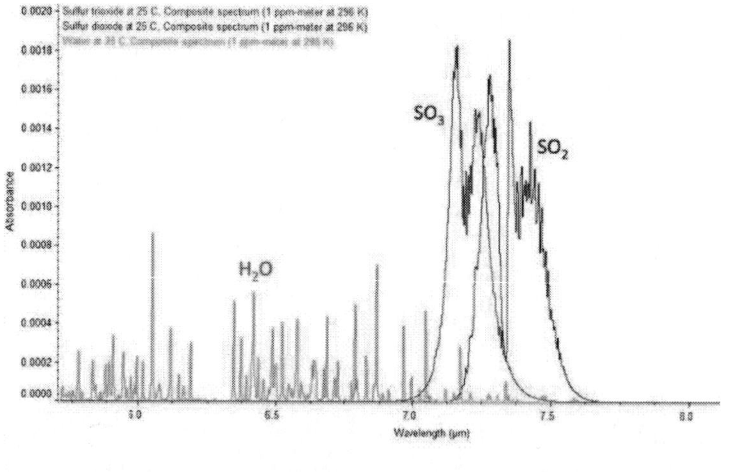

Figure 5. Simulated SO_2, SO_3 and H_2O mid-IR spectra around 7 µm at room temperature. All species are at 1 ppm so this condition is far from typical flue gas conditions but the figure shows the spectral regions of interest.

Wet tests introduce a further layer of complexity, since water at levels on the order of a few percent will introduce interfering features that can dominate the spectral response in the wavelengths where SO_2 and SO_3 are expected, as seen in Figure 5.

Wet tests, apart from being significant for on-site operations, offer the advantage of giving an estimation of the total sulfur content of the flow because controlled condensation can be used to later trap the acid in a quantitative fashion. When the temperature of the gas flow in the condensation coil drops below the dew point temperature of the acid, the sulfuric acid will condense. At those temperatures, sulfur trioxide is completely converted to sulfuric acid, according to chemical equilibrium. The impingers collect any remaining sulfur dioxide. Analyzing with ion chromatography the sulfate ion content of the solutions from the impingers and the condensation coil, the total sulfur content (sulfur dioxide, sulfur trioxide, and sulfuric acid) can be determined.

The conversion from $ppm\ \left[\frac{mg}{L}\right]$ given by the IC technique, to volumetric parts per million of sulfur trioxide in the gas flow, is given by:

$$V_w\ [L] = m_w \cdot 10^{-3}$$

$$m_{SO_4^{2-}}\ [mg] = V_w \cdot c_{SO_4^{2-}}$$

$$n_{SO_4^{2-}}\ [mol] = \frac{10^{-3}}{M_{m,SO_4^{2-}}} m_{SO_4^{2-}}$$

$$V_{SO_3}\ [sL] = 22.414\ n_{SO_4^{2-}}$$

$$\dot{V}_{SO_3}\ [sL/min] = \frac{V_{SO_3}}{t_{test}}$$

$$x_{SO_3}\ [ppmv] = \frac{\dot{V}_{SO_3}}{\dot{V}_O}\ 10^6$$

Where m_w is the mass of water used to collect the sample from the condensation coil, $c_{SO_4^{2-}}$ is the concentration of sulfate in ppm given by the IC machine, $M_{m,SO_4^{2-}}$ is the molar mass of the sulfate ion, 22.414 is the conversion factor from moles to standard liters for an ideal gas, t_{test} is the duration of the sample collection phase, and $\dot{V}_O$ is the overall gas flow rate.

Preliminary Findings

The most important finding is that the experimental apparatus provides excellent thermal and gas composition control over the conditions desired. The system permits a wide range of gas composition, both wet and dry.

The first dry tests with the DOAS identified a significant drop in the concentration of sulfur dioxide after the catalyst, but the instrument did not detect a SO_3 signature in the expected 220 to 250 nm region. The approach of creating SO_3 using a catalyst, while avoiding handling pure SO_3 liquid and the complexity of oxidizing sulfur dioxide with ozone as others have attempted, followed the expected conversion efficiency of the catalyst manufacturer (e.g., conversion increased with decreasing face velocity and increasing catalyst temperature) based on the decrease in SO_2 but independent quantitative confirmation of SO_3 formation was not accomplished.

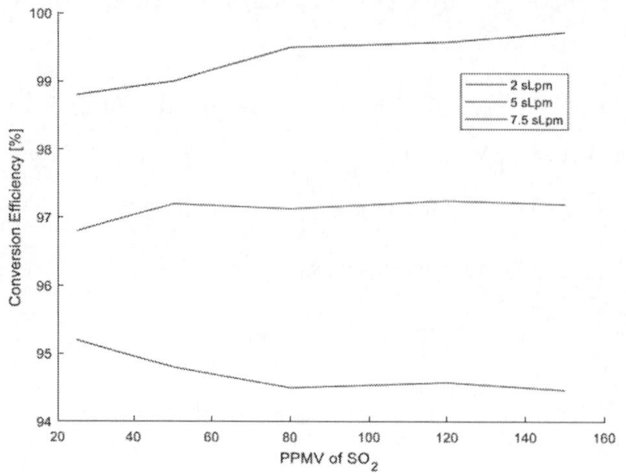

Figure 6. Catalyst conversion efficiency based on DOAS SO_2 drop in dry tests.

Like the DOAS experiments, dry tests with the EC-QCL did not detect a clear signature for SO_3 after conversion in the catalyst. In this case, the optical method indicated lower conversion based on the sulfur dioxide (i.e., its concentration was halved after the catalyst) but there should still have been sufficient sulfur trioxide to be measured. This result was inconsistent with the one given by the DOAS, but both were unable to show SO_3 presence. There are many possible explanations for the lack of sulfur trioxide detection, including loss of sulfur through sulfuric acid condensation in parts of the system before the test section, spatial variation of temperature and species, uncertain absorption strength of sulfur trioxide at high temperature, and interference from sulfur dioxide absorption. While the system does not yet produce unequivocal quantitative concentrations of sulfur species the downstream collection of the test gases confirms that substantial conversion of SO_2 to SO_3 is occurring through the catalyst. That is, tests that incorporated water vapor demonstrated the presence of H_2SO_4 in the gas flow by condensation in the Method 8A coil, although with an estimated concentration well below the expected one.

In addition, wet tests involve sulfuric acid mist, a component that reacts with CaF_2 windows needed for IR measurements with the EC-QCL and can cause corrosion problems below its dew point temperature.

Summary and Future Work

The tests and improvements of this unique flue gas measurement system are ongoing. Further results are forthcoming, but preliminary findings show that the results of different measuring techniques are inconsistent, under the same operating conditions. That is, the DOAS indicates higher conversion of SO_2 than does the EC-QCL measurement.

The controlled condensation method confirms the presence of sulfur trioxide, although at levels below expectations. The presence of sulfuric acid in the measurement cell causes the optical windows surfaces to etch, contributing to sulfur depletion; a further mechanism could be the presence of cold spots inside the measuring cell, where the sulfuric acid can condense and react with the walls.

Future steps will involve different solutions for the optical windows, and the analysis of the composition of the gas stream with the controlled condensation method in different points of the system, including before the measurement cell. The analysis will indicate whether the total sulfur content is conserved across the catalyst. If there are no reactions decreasing the sulfur content along the way to the measurement cell, the sulfur conservation is confirmed. A second effort will involve a different thermal management of the measurement cell; in fact, if the temperature is not uniform across the chamber, due to the shifting equilibrium some of the SO_3 may convert to H_2SO_4, which is not detected in the wavelengths of interest.

References

Moretti, Albert L., Christopher S. Jones, and Power-Gen Asia. "Advanced emissions control technologies for coal-fired power plants." Babcock & Wilox power generation group, Bangkok, Thailand. Technical Paper, BR-1886 (2012).

Halstead, W. D., and J. R. W. Talbot. "Sulphuric acid dewpoint in power station flue gases." Journal of the Institute of Energy 53 (1980): 142-5.

Bayless, D.J., Schroeder, A.R., Olsen, M.G., Johnson, D.C., Peters, J.E., Krier, H., and Buckius, R.O. (1996) "The Effects of Cofiring Natural Gas and Coal on Sulfur Retention in Ash," Combustion and Flame, 106, 3, 231-240.

Parker, K.R. and Sanyal, A. (1999) "Effective Capture of Respirable Sized Particulates Using Electrostatic Precipitator Technology," Proceedings of the American Power Conference, editor A.E. McBride, Vol. 61, 174-179.

Menasha, J. (2010) "Characterization of Ammonium Bisulfate Formation in a Bench-Scale Single-Channel Air Preheater," M.S. Thesis, University of California, Irvine.

Menasha, J., Dunn-Rankin, D., Muzio. L., and Stallings, J. (2011) "Ammonium Bisulfate Formation Temperature in a Bench-Scale Single-Channel Air Preheater," Fuel, 90, 7, 2445-2453.

Mueller, S.F. and Imhoff, R.E. (1994) "Estimates of Particle Formation and Growth in Coal-Fired Boiler Exhaust. 1. Observations," Atmospheric Environment, 28, 4, 595-602.

Bionda, J. (2002) "Flue Gas SO3 Determination – Importance of Accurate Measurements in Light of Recent SCR Market Growth," presented at Conference on SCR and SNCR Reduction NOx Control, Pittsburgh, PA, May 15 -16.

Ashok K. Jain, NCASI Southern Regional Center (1996) "Method 8a - Determination of sulfuric acid vapor or mist and sulfur dioxide emissions from kraft recovery furnaces" present on EPA website.

Breen (2014), "Acid Gas (SO_3 and HCl) Measurement Solutions" present on Breen website; http://breenes.com.

Acknowledgments

This work was supported by Electric Power Research Institute (EPRI) under the guidance of Richard Himes. We appreciate the useful information on QCL laser absorption from Jason Kriesel (OKSI) and on the catalyst behavior from Larry Muzio at FERCo. We appreciate the help of Andre Nodem and Harshil Dadhaniya on ion chromatography.

Western States Section of the Combustion Institute – Fall 2017 Meeting
Hosted by the University of Wyoming
October 2-3, 2017

Strategy for background-free measurements of high-speed OH in turbulent flames using SLIPI

Bo Zhou[1] and Jonathan H. Frank[1]

[1]*Combustion Research Facility, Sandia National Labs, Livermore, California*

Abstract

In present work, the combined high-speed OH/Tomographic PIV measurement was performed in a counterflow flame intended to study the temporal dynamics of the local extinction events, which, however, is prevented by the presence of background interference in high speed OH imaging. It is shown that experimental condition did not favor conventional means for background rejection. Therefore, Structured Laser Illumination Planar Imaging (SLIPI) was implemented, for the first time, to high speed OH imaging to demonstrate a background-free OH detection. Issues with the employment of a high speed detection system as well as the achievable imaging spatial resolution with SLIPI are discussed.

1. Introduction

Recent development of the combined high-speed OH/Tomographic PIV (TPIV) diagnostics allows temporally resolved simultaneous visualization of flame topology and flow field [1-3], which provides unique possibilities to study turbulent combustion phenomena such as flame straining, turbulence mixing, local extinction and re-ignition from the Lagrangian perspective. Although the Lagrangian description of turbulence-flame interaction [4] demands preferably three-dimension measurement at high repetition rates imposing more diagnostic challenging, it is conceptually natural compared to its Eulerian counterpart, and it is also a necessity for understanding of many turbulent combustion phenomena. For example, it is of particular interest to study the temporal dynamics of the local extinction events which exemplify an important scenario where the classic flamelet concept fails [5, 6], and unburnt reactants and burnt mixtures could be mixed by turbulence. This turbulent mixing could significantly alter reactant composition and temperature which could have profound impacts on sustaining combustion. To study the effect of such turbulent mixing, it is critical to unambiguously determine the local extinction/burnt regions to allow synthetic particle tracking using the TPIV data with known particle "properties" (i.e. burnt or unburnt). OH signals are known to exist primarily in the reaction zone and burnt regions, and could potentially be used for identifying the quenched/burnt regions. However, as will be shown later, high speed OH LIF signals can potentially be noticeably contaminated by flame OH^* chemiluminescence which can not be spectrally filtered or sufficiently gated in time under certain circumstances using conventional means for background rejection. Therefore, in order to take the advantage of high speed measurements for turbulent combustion studies, high signal fidelity of the measurement has to be first ensured.

Structured Laser Illumination Planar Imaging (SLIPI) [7] is known as a technique mainly applied for spray visualization [8] and was recently demonstrated for Rayleigh thermometry [9] to suppress multiple scattered light and/or stray light. This technique is based on spatial modulation of laser sheet such that laser-induced signals from the first light-matter interaction events are inherently coded with the modulated pattern while background does not bear the coded signature. Conventional SLIPI (referred as 3p-SLIPI [7]) requires recording of three phase-shifted sub-images, which increases experimental complexity and limits its application to mainly time-averaged measurements. Very recently, one-phase SLIPI (1p-SLIPI) has been demonstrated for instantaneous spray imaging [10] and Rayleigh thermometry [11] using a spatial frequency lock-in algorithm [10] which greatly simplifies the experimental requirement. However, application of 1p-SLIPI to achieving background-free imaging of reactive scalars at high repetition rates has not been reported. It is expected that high speed reactive scalar imaging is more susceptible to the background interference due to weaker signals constrained by the pulse energy available. One prerequisite to implement SLIPI is that the spatial modulation frequency of the laser sheet should be sufficient to separate the modulated SLIPI signals from background in the Fourier domain while the modulation frequency should still be discernible by the detection system. Most of SLIPI applications have been demonstrated using an unintensified CMOS camera. For high speed scalar imaging, the employment of an intensified high speed CMOS leads to poorer optical performance in transferring contrast of spatial variations, which can compromise the spatial resolution, and therefore limits the range of applicable modulation frequencies for SLIPI.

In present work, we performed the combined high speed OH/TPIV measurements in a counterflow turbulent flame [12] to illustrate local extinction events and the existence of background interference. As a preliminary result, we demonstrated high speed SLIPI OH imaging in a lifted turbulent jet flame showing that SLIPI can be implemented to successfully recover pure OH signals even under the presence of significant background interference. This diagnostic development would benefit the employment of high speed measurements to be applied for Lagrangian study of turbulent combustion.

2. Experimental Setup

Figure 1 depicted the combined 10 kHz OH/TPIV system described in details elsewhere [1, 13]. The probe volume for TPIV is 18 × 14 × 1.3 mm^3, containing 178 × 140 × 14 velocity vectors with 100 μm spacing. Excitation of OH was made with a pulse energy of 60 μJ/pulse at 283.3 nm to access the $Q_1(7)$ transition of the A-X (1-0) OH band. The OH laser beam was reshaped into an 18-mm laser sheet and spatially overlapped with the central plane of the TPIV probe volume (i.e. z=0). OH LIF signal was recorded using a high speed intensified CMOS camera (Vision Research, Phanton v7.3, 640 × 480 px^2) with a bandpass filter (FWHM=10nm) centered at 310 nm. The OH camera gate width was set to 1 μs to minimize the detection of stray light but also allow some tolerance for temporal jittering of laser pulses. For SLIPI measurement, a Ronchi ruling mask was inserted into the OH optical path to generate a spatial modulation of the OH laser sheet with a selected frequency.

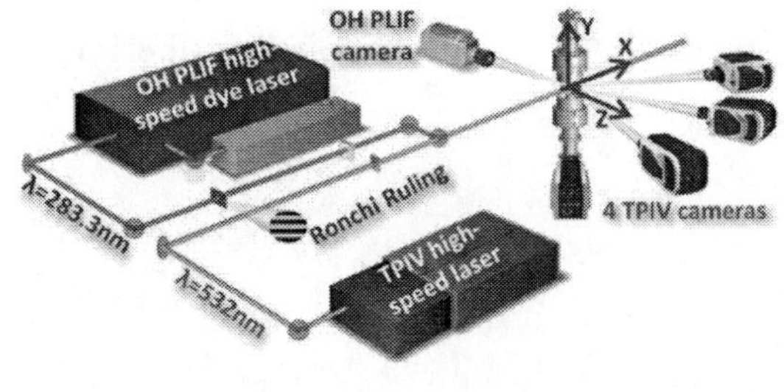

Fig.1 Experimental setup for 10 kHz SLIPI-OH/TPIV system.

The combined OH/TPIV measurement was first performed in a turbulent flame stabilized on a counterflow burner. It has been demonstrated in previous study [12] that the studied flame experiences frequent local extinctions, which makes it a suitable target for studying turbulent mixing of the burnt and unburnt mixtures during local extinction events. Figure 2 shows an example of the temporal dynamics of a local extinction event illustrated by simultaneous OH/TPIV measurement. It is evident that local extinction has occurred as the OH layer is no longer continuous, and turbulence at the regions of local extinction has stirred the unburnt and burnt mixtures. However, the presence of background interference blurred the OH layer, and the regions of OH intensity less than 50% of maximal OH signal (shown as green and blue regions) were lack of a sharp boundary and were completely blended into the wide-spreading background, making it difficult to unambiguously discriminate the regions of local extinction. The weak temporal dynamics of the background interference suggests that OH^* chemiluminescence is the interference source. Considering the finite gate width feasible for high speed scalar imaging, the long line-of-sight of the counterflow turbulent flames, being an unfavorable scenario, strengthens the influence of flame chemiluminescence.

As a preliminary demonstration, the background-free high speed OH measurement using SLIPI was performed in a lifted turbulent jet flame with an object inserted on purpose as a background interference source to the measured OH signals. However, the OH detection system has been kept same as that used for the counterflow flame.

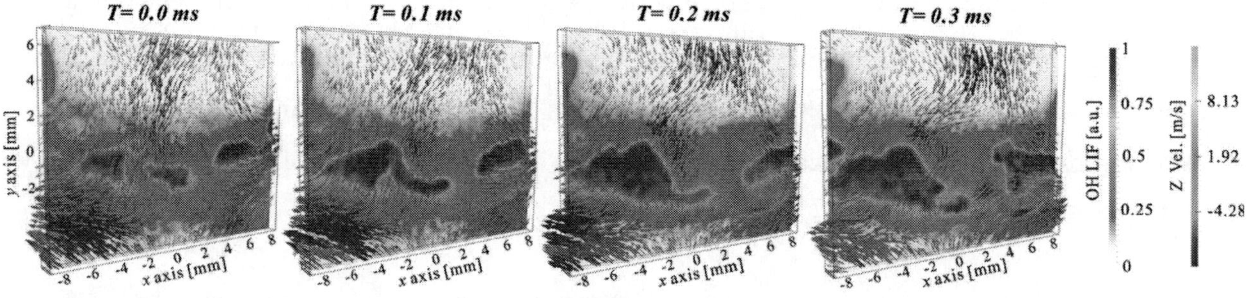

Fig.2 Example of the temporal dynamics of a local extinction event in the counterflow turbulent flame illustrated by simultaneous OH/TPIV at 10 kHz. The presence of background interference in OH signals is obvious.

3. Results and Discussion

3.1. Characterization of high speed OH detection system

The resolution of an optical system can be quantified by the modulation transfer function (MTF) which characterizes the transfer of the contrast at various spatial frequencies [14]. Specifically, the contrast of a sine wave object (C_{object}) at a spatial frequency, f, is determined by the $C_{object}=(I_{max}-I_{min})/(I_{max}+I_{min})$ where I_{max} and I_{min} are the maximum and minimum intensities of the sine wave object, respectively. The sine wave object is imaged by an optical system, giving the contrast of the image,

C_{image}, defined in the same way as C_{object}. The MTF at spatial frequency *f*, is then defined as MTF(*f*)= C_{image}/C_{object}. A standard means of measuring the MTF includes imaging an oblique sharp-edge blade, i.e. the slanted edge method [15, 16], which gives the step-response function (SRF) near the blade edge as shown in Fig. 3(a). The line spread function (LSF) is obtained from the derivative of the SRF, which describes the image blurring of an infinitely thin line. The MTF is simply the Fourier transform of the LSF. As shown in Fig.3 (b), the MTF drops dramatically with increasing spatial frequency. At the Rayleigh limit, the MTF is approximately 30% which corresponds to a limiting spatial frequency of 4 lp/mm, i.e. a resolution of 125 μm that was digitized approximately four pixels. For practical applications, imaging noise can significantly compromise the claimed resolution. Therefore, the modulation frequency should be considerably lower than the limiting frequency to be able to properly resolve the modulation pattern. In the other words, the imaging contrast at the selected frequency should remain reasonable as it directly determines the amplitudes of modulation frequency components in the Fourier domain. In present work, , a spatial modulation frequency of 1.67 lp/mm (i.e. 300 μm pitch width) was selected, giving a MTF of 76%.

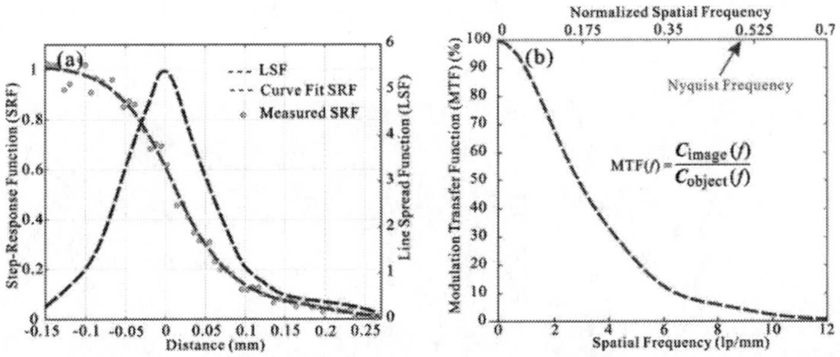

Fig.3 Characterization of the intensified high speed OH detection system. (a) Step-Response Function (SRF, red dashed curve) was recorded by imaging a sharp blade edge, and its derivative defined the Line Spread Function (LSF, blue dashed curve). (b) Modulation Transfer Function.

3.2. High speed SLIPI OH measurements

Figure 4 (a-c) illustrates how background-free OH images are recovered through post processing the SLIPI images. Figure 4 (a1-a2) shows two subsequent high speed SLIPI OH images recorded in a lifted turbulent jet flame. An object as outlined by the red dashed curve in Fig. 4(a1-a2) was inserted slightly behind the OH imaging plane to obscure the OH signal on purpose, resulting a signal-to-background ratio (SBR) of 0.4. The Fourier transform of Fig.4 (a1-a2) was shown in Fig.4 (b1-b2). The center of the Fourier image represents zero frequency. OH signals in the Fourier domain consist of both zeroth and higher orders frequency components which are associated with the modulation frequency, while background interference and imaging noise do not exhibit any characteristic frequency components due to the nature of randomness. As both background and OH signals contain less amount of finer structures in the images which correspond to high frequency components in the Fourier domain, the associated frequency components with noticeable amplitudes are congested near the center, and marched towards higher frequencies but with progressively decreasing amplitudes. Imaging noises are manifested in the Fourier domain as a cloud of scattered frequency components of comparably lower amplitudes around the center. The first order frequency components of the OH signal appear at a shifted frequency, that is, the modulation frequency, providing a separation from the main background frequency components. An elliptical bandpass filter in the Fourier domain (cf. Fig.4 (b1-b2)) was employed to select the modulated signal frequency components which were inverse Fourier transformed to recover background-free OH images as shown in Fig.4 (c1-c2), and the flame temporal dynamics are evident by comparing Fig.4 (c1) and (c2). On the other hand, if the bandpass filter was designed to select the zeroth order frequency components, the original images, i.e. Fig.4 (a1-a2) will be obtained.

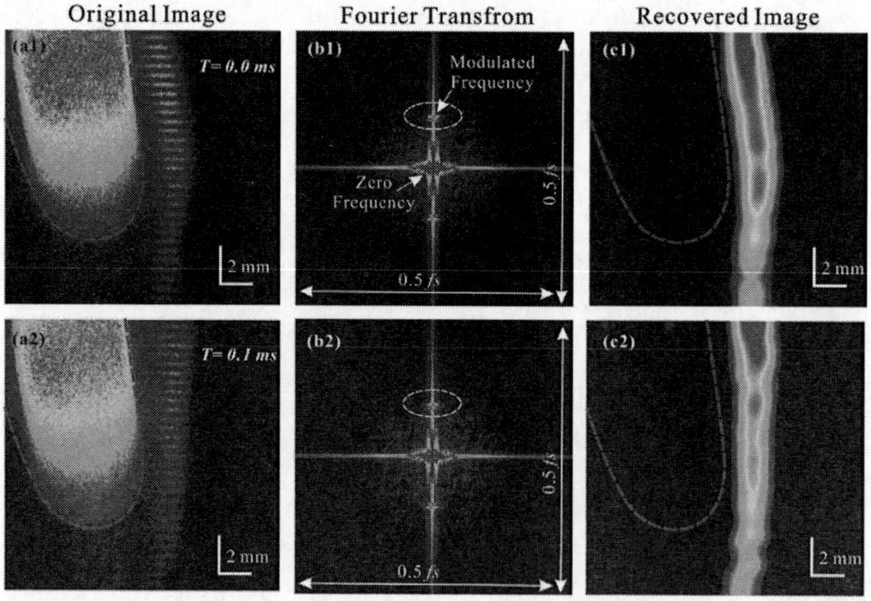

Fig.4 (a1-a2) two subsequent high speed SLIPI OH images recorded in a lifted turbulent jet flame at 10 kHz. Red dashed curve outlines the shape from the object that was inserted slight behind the OH imaging plane to obscure the OH signal on purpose. (b1-b2) Fourier transform of the original SLIPI OH images. Only half of the Fourier domain was shown for clarity. (c1-c2) Recovered OH images showing that the background interference can be successfully removed through inverse Fourier transfer of lock-in frequency components in the Fourier domain.

Note that the removal of background inference comes at the cost of spatial resolution. The spatial resolution of the recovered image was fundamentally limited by the spatial modulation pitch width, which is 300 μm in present work. The employed bandpass filter in the Fourier domain imposes additional limit on the spatial resolution that is determined by the radius of the elliptical filter. Along the horizontal direction, there is no limit to approach the 300 μm resolution, while in the vertical direction, the radius of the elliptical filter is constrained to reject the background frequency components, providing a practical resolution of 900 μm. The achievable resolution depends on the background interference levels; Higher background level leads to higher amplitudes of background frequency components extending towards the modulation frequency, and therefore smaller frequency range has to be selected by the filter applied. Note that the preliminary result shown in Fig.4 was made with significant background, i.e. a SBR of 0.4. As for lower levels of background interference shown in Fig.2 where the SBR is about 2.5, the spatial resolution along y direction can be further improved.

4. Conclusions

Combined high speed OH/TPIV provides the potential to unveil new information from studying the temporal dynamics of turbulent flames. Experimental data with high signal fidelity can be critical for this purpose. In present work, preliminary experimental results have shown that SLIPI technique can be feasible for high speed scalar imaging with high signal fidelity even under the presence of significant background interference. Characterization of the high speed detection system and therefore selection of a proper spatial modulation frequency is important to ensure a successful SLIPI measurement. The demonstrated high tolerance of SLIPI signals to background suggests that high speed scalar imaging with SLIPI could be applicable to even harsher diagnostic environment.

Acknowledgements

The authors gratefully thank Erxiong Huang for his technical support, and acknowledge the support of the US Department of Energy, Office of Basic Energy Sciences, Division of Chemical Sciences, Geosciences, and Biosciences. Sandia National Laboratories is a multimission laboratory managed and operated by National Technology and Engineering Solutions of Sandia, LLC., a wholly owned subsidiary of Honeywell International, Inc., for the U.S. Department of Energy's National Nuclear Security Administration under contract DE-NA-0003525.

References

[1] B. Coriton, A.M. Steinberg, J.H. Frank, High-speed tomographic PIV and OH PLIF measurements in turbulent reactive flows, Exp Fluids 55 (2014).

[2] A.M. Steinberg, B. Coriton, J.H. Frank, Influence of combustion on principal strain-rate transport in turbulent premixed flames, Proceedings of the Combustion Institute 35 (2015) 1287-1294.

[3] B. Coriton, J.H. Frank, Impact of heat release on strain rate field in turbulent premixed Bunsen flames, Proceedings of the Combustion Institute 36 (2017) 1885-1892.

[4] P.K. Yeung, Lagrangian investigations of turbulence, Annu Rev Fluid Mech 34 (2002) 115-142.

[5] N. Peters, Laminar flamelet concepts in turbulent combustion, Proceedings of the Combustion Institute 21 (1988).

[6] B. Zhou, C. Brackmann, Q. Li, Z.K. Wang, P. Petersson, Z.S. Li, M. Alden, X.S. Bai, Distributed reactions in highly turbulent premixed methane/air flames Part I. Flame structure characterization, Combust Flame 162 (2015) 2937-2953.

[7] E. Berrocal, E. Kristensson, M. Richter, M. Linne, M. Alden, Application of structured illumination for multiple scattering suppression in planar laser imaging of dense sprays, Opt Express 16 (2008) 17870-17881.

[8] E. Kristensson, Structured laser illumination planar imaging SLIPI applications for spray diagnostics, PhD thesis, Lund University, 2012.

[9] N.J. Kempema, M.B. Long, Quantitative Rayleigh thermometry for high background scattering applications with structured laser illumination planar imaging, Appl Optics 53 (2014) 6688-6697.

[10] Y.N. Mishra, E. Kristensson, M. Koegl, J. Jonsson, L. Zigan, E. Berrocal, Comparison between two-phase and one-phase SLIPI for instantaneous imaging of transient sprays, Exp Fluids 58:100 (2017).

[11] E. Kristensson, A. Ehn, J. Bood, M. Alden, Advancements in Rayleigh scattering thermometry by means of structured illumination, Proceedings of the Combustion Institute 35 (2015) 3689-3696.

[12] B. Coriton, J.H. Frank, A. Gomez, Interaction of turbulent premixed flames with combustion products: Role of stoichiometry, Combust Flame 170 (2016) 37-52.

[13] B. Coriton, J.H. Frank, Experimental study of vorticity-strain rate interaction in turbulent partially premixed jet flames using tomographic particle image velocimetry, Phys Fluids 28 (2016).

[14] G.H. Wang, N.T. Clemens, Effects of imaging system blur on measurements of flow scalars and scalar gradients, Exp Fluids 37 (2004) 194-205.

[15] International Organization for Standardization (ISO-12233), Resolution and Spatial frequency response, 2014.

[16] S.E. Reichenbach, S.K. Park, R. Narayanswamy, Characterizing Digital Image Acquisition Devices, Opt Eng 30 (1991) 170-177.

Paper # 29DI-0086 **Topic: Laser diagnostics**

Western States Section of the Combustion Institute - Fall 2017 Meeting
Hosted by the University of Wyoming
October 2-3, 2017.

Direct comparison of simulated OH Fluorescence and experimental results in a non-premixed laminar diffusion coflow flame at high pressure conditions

David Escofet-Martin[1] *Alex Torredemer*[2] *Yu-Chien Chien*[1]
Derek Dunn-Rankin[1]

[1]*Mechanical and Aerospace Engineering, University of California, Irvine 92697, USA*

[2]*Industrial Engineering, Polytechnic University of Catalonia , Terrassa 08222, Catalonia*

Modern laser diagnostic tools have been a staple for non-intrusive experimental combustion research to evaluate high temperature reacting flow environments, and to contribute fundamental knowledge or sensors needed for improving combustion systems; they also represent a critical tool for delivering the data to validate computational models. Understanding the behavior of diagnostics when they are to be used in practical high-pressure combustion environments is particularly important to overcome the substantially increased difficulty in the measurements. The current work examines practical challenges of high-pressure fluorescence measurements and aims to provide a semi-quantitative comparison between a numerical simulation and the experiments. An OpenFOAM simulation is used to discuss spatial distribution of the OH molecule, and obtain insights from the fluorescence signal as a function of pressure. For this comparison, accurate evaluation of the absorption overlap integral effects, temperature dependence and quenching are quantitatively discussed at each different pressure. A good agreement is obtained between predicted and experimental fluorescence results. Challenges in obtaining chemistry insights by comparing different mechanisms from a 2D-configuration remain although physical combustion characteristics can still be inferred.

1 Introduction

The majority of combustion systems are operated under high pressure, such as internal combustion engines and gas turbines. At high pressures thermodynamic efficiency and specific energy production both are enhanced. These attributes produce systems with improved fuel efficiency, reduced emissions of combustion generated pollutants, including CO_2, which is a significant contributor to global warming. Despite the modern trend towards higher pressure, the bulk of combustion research is made at atmospheric pressure conditions where experimental conditions can be controlled more easily, and data analysis can be obtained more reliably, although extrapolating atmospheric pressure results for higher pressure conditions could produce prediction errors. Also, high pressure experimental data is very much needed, in particular, to validate models, which due to experimental costs of high pressure experiments are a focus of attention when developing the new high pressure technology. With increased pressure, the reaction layer becomes very thin, and it is increasingly difficult to use sampling or other intrusive techniques. Laser diagnostics offer the unmatched spatial resolution that is necessary in order to resolve spatial gradients that appear at high pressure.

These advantages come with additional challenges such as increased collisional quenching, increased absorption, and pressure shifting, among others. High pressure burners have been studied previously [1, 2, 3, 4, 5, 6, 7, 8], mostly for soot production but there is no previous study for detailed planar laser induced fluorescence measurements for a laminar non-premixed coflow flame as a function of pressure. To complement the experimental fluorescence measurements a simulation in OpenFOAM[9, 10] is completed after matching the experimental conditions. OpenFOAM is an open source CFD software offering a community supported long term reliable CFD code. OpenFOAM has been modified for diffusion or coflow combustion purposes in several works [11, 12], and in this work it is used with a detailed spectroscopic model to simulate the fluorescence signal using three different chemical mechanisms.

2 Methods

2.1 Experimental setup

2.1.1 Chamber, burner and flow delivery

The main apparatus component is the high pressure chamber made of SS304 stainless steel and designed to withstand over 200 $bars$ of pressure. The inner dimensions of the chamber are approximately $\varnothing\, 0.26 \times 0.33\ m$ with a total of 17.5 dm^3 internal volume; the outer dimensions are $\varnothing\, 0.33 \times 0.51\ m$. The chamber is equipped with 4 sapphire windows that allow optical access to the chamber at angles of 90°, the windows are placed in four viewports with a $\varnothing\, 7.6\ cm$. The burner and flow delivery are inspired by [13, 6], the coflow burner has an inner $\varnothing\, 2\ mm$ centered fuel tube for methane flow and a $\varnothing\, 2.16\ cm$ coflow surrounded with air. The burner is operated with the mass flowrate equivalent of 15 cm/s at NTP (293 K and 1 bar) both in the air and in the fuel side. A metallic porous material is placed in both the air coflow and methane annulus in order to homogenize the flow and induce a small pressure drop in the tip of the burner that reduces the disturbances of small pressure oscillations from the chamber to the flow ducts. The flow delivery system can be seen in Figure 1 and consists of 3 different lines, air, methane, and nitrogen, with each line equipped with a regulator, flow meter, and needle valve. A dome loaded back pressure regulator is connected to the top of the chamber in order to maintain the level of desired pressure in the experiment.

2.1.2 Optical setup

The PLIF measurements are conducted by pumping OH at 283.2 nm $Q_1(7)$ in the $A_2\Sigma^+ \leftarrow X\Pi_i\ (1,0)$ band and detecting fluorescence in the $A_2\Sigma^+ \rightarrow X\Pi_i\ (0,0),\ (1,1)$ band. The specific setup used includes a 10 Hz Surelite III Nd:YAG Laser, followed by a dye laser (Vista) and a doubling crystal. A spherical convex and a cylindrical concave lens is used to form a laser sheet with a height of 30 mm. The fluorescence light is collected using an EM-ICCD PIMAX-4 with a filter (Semrock FF01-320/40-25). The fluorescence images are recorded at a sampling rate of 10 Hz using a 30 ns gate, which is sufficient to collect the fully integrated signal of the fluorescence at high pressures while minimizing the effects of background chemiluminescence, natural soot luminescence and soot laser induced incandescence. 300 images are collected for each pressure with an

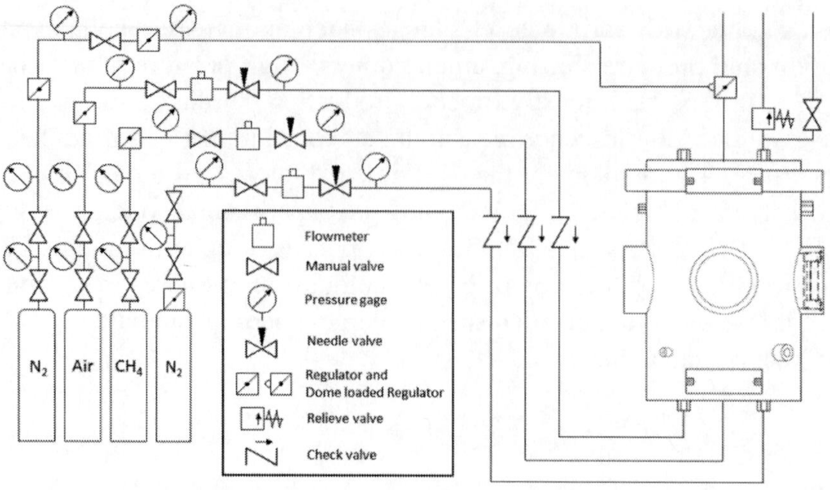

Figure 1: High pressure chamber and experimental flow system schematic.

on CCD accumulation of 20 exposures per image, and the background emission is subtracted from the raw fluorescence images. The soot luminescence is also corrected by off resonant excitation since soot luminescence from LII becomes important as the pressure increases.

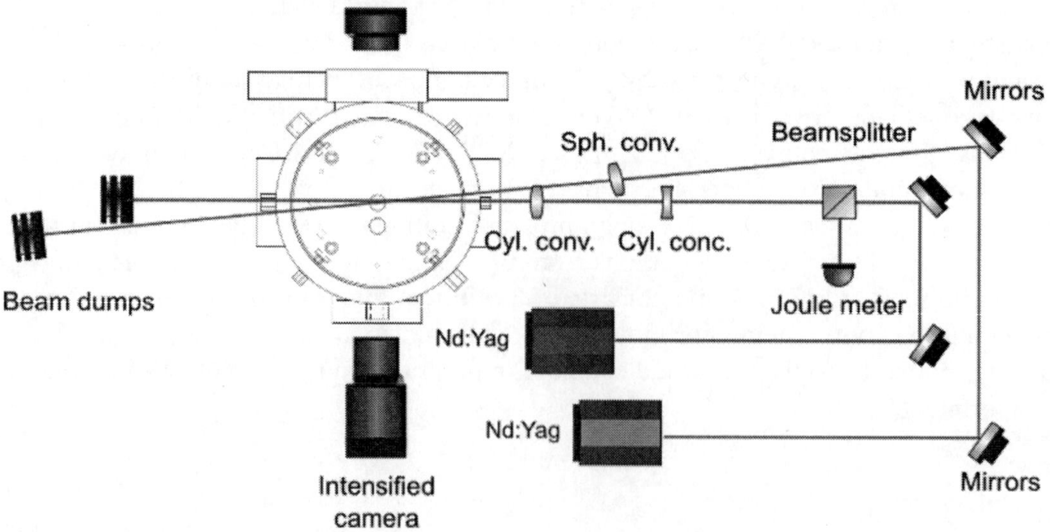

Figure 2: OH PLIF experimental setup.

2.2 Experimental procedure

The experimental procedure for each pressure is described as the following. First, setting up the back pressure regulator to the target pressure, closing air and methane lines and adjusting the nitrogen flow to the desired pressure. Second, once the desired pressure is reached, tunning the air, nitrogen and methane flows to the desired value until pressure is stabilized. Third, igniting the flame with laser ignition waiting again for steady pressure, and then readjust needle valves if needed in order to obtain the desired flow rate. The ignition is performed through laser induced breakdown, with a Quanta ray INDI Nd:YAG doubled to $532\ nm$ and focused using a spherical convex lens with a $fl = 25\ cm$. All the experiments are conducted under an excess flow of nitrogen in the environment to keep the global mixture inside the chamber outside of flammability limits. The current design does not allow a steady flame over $12\ bar$ due to the buoyant instabilities, which produce an oscillatory vertical motion that complicates steady fluorescence measurements.

2.3 OpenFOAM simulation

The Navier-Stokes equation and chemical kinetics are solved with ReactingFOAM in an OpenFOAM environment, the detailed description is out of the scope of this work and a more detailed description can be found in [9]. Transport properties use the Leonard-Jones parameters, to obtain viscosity and thermal conductivity for the different species. The current OpenFOAM 4.0 version with the ReactingFoam solver lacks the implementation to use those values to calculate some properties. The current ReactingFoam uses the Sutherland model to estimate viscosity that is dependent on temperature. A Cantera interface with Python and Matlab is used to fit the different viscosity profiles from the Leonard Jones parameters. A fit to a nonlinear function (lsqcurvefit) is used with the air coefficients as initial conditions ($A_s = 1.67212 \times 10^{-6}\ \frac{kg}{ms\sqrt{K}}$, $T_s = 170.672\ K$), for each species and each chemical mechanism. For the PAH mechanism described later, the transport data was not available, and the transport data from a previous iteration of the chemical model was used [14]. The mixture viscosity properties are calculated by a mass fraction weighted average. The current OpenFOAM version does not allow to use binary diffusion coefficients, and an assumption has to be made to obtain mass diffusivities from the viscosity calculations. Based on the methane fuel used a $Sc = 0.7$ is chosen as a better assumption. The thermal conductivity is calculated with the modified Eucken formulation [15]. The thermodynamic properties are included in each mechanism in the form of NASA polynomials [16].

Figure 3 and Table 1 show the mesh blocks, mesh size, and number of cells; blocks 6, 7, and 8 have nonuniform spacing in the radial direction, blocks 2, 5, 8 and 11 have nonuniform spacing in the axial direction to reduce the computational time for a larger cell size further from the flame sheet. The mesh size in the flame region has been chosen to be in the same order of magnitude as the resolution obtained in the experimental configuration (radial pixel size= 50µ m and radial cell size =33µ m). The overall dimensions are $2 \times 4\ cm$, and the angle of the section is 5°. The time step $\Delta t = 10\ \mu s$ and the mesh size has been selected to produce results that don't change when decreasing the time step to $\Delta t = 5\ \mu s$, or doubling the number of cells in both the radial and axial direction. The procedure for each simulation has three steps.

- Step 1: Using as initial conditions the domain filled with (76% N_2 24% O_2 by mass), and the boundary conditions that are shown in Table 2, except with $T = 800\ K$ for methane flow, methane tube, and air flow to ignite the flame. Each simulation is run until steady state is reached, with a single step mechanism. An adjustable time step is used for this case:

$$\Delta t = 0.1 \frac{1}{\left(\frac{u_x}{\Delta x} + \frac{u_y}{\Delta y}\right)} \tag{1}$$

- Step 2: Using as initial conditions the steady state solution of step 1, and boundary conditions from Table 2, with a single step mechanism and fixed time step $\Delta t = 10\ \mu s$.
- Step 3: Using as initial conditions the steady state solution of step 2 with boundary conditions from Table 2, compute a solution with the desired more complete chemical mechanism with a fixed time step $\Delta t = 10\ \mu s$.

There are three mechanisms used in this high pressure simulation: GRI-Mech 3.0 [17] is the reference mechanism for methane chemistry in combustion, DRM-19 [18] which is a skeletal mechanism based on the GRI-Mech, and what we will refer to as PAH mechanism [19]; this latter mechanism was developed to predict the formation of poly-aromatic hydrocarbons(PAH) and the growth of up to five aromatic rings for fuels such as methane and ethylene. Here we used a modification from it found in [20].

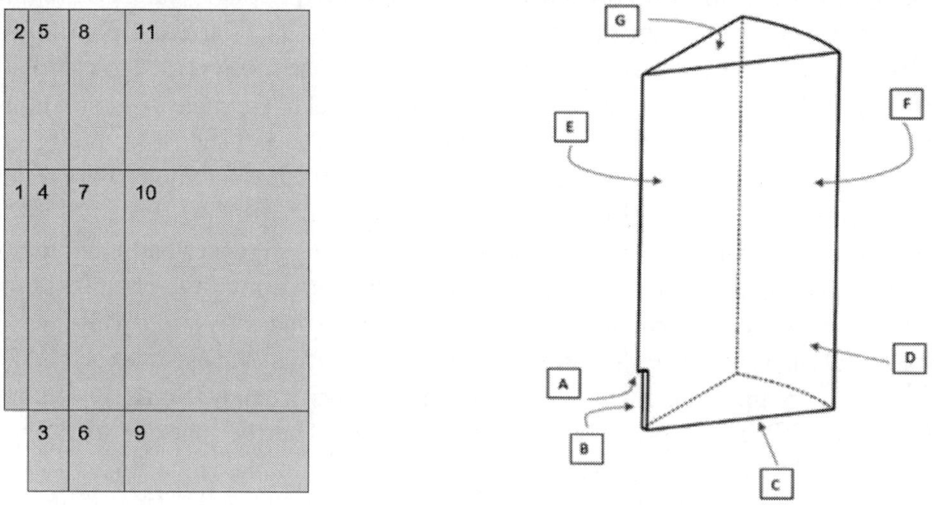

(a) Mesh sections. (b) Face labels.

Figure 3: OpenFOAM mesh.

2.4 Simulated fluorescence corrections

A simplified description of the fluorescence model is given here, as more details can be found elsewhere [21]. The Fluorescence signal F is given by:

$$F = \eta I_l G N_t(T) B_{12} \Phi(T, J') \tag{2}$$

$$\eta = \frac{hv}{c}\frac{\Omega}{4\pi} l A \tag{3}$$

Where h is Plancks constant; v is the frequency of the emitted fluorescence; Ω is the collection solid angle; A is the focal area of the laser beam, l is the axial extent of the beam from where the fluorescence is observed, c is the speed of light, I_l is the laser intensity, G is the overlap integral effects, η

	Block dimensions (mm)	Number of cells	Cell dimensions (μm)
Block 1	1×18	3600 (30×120)	33.3×150
Block 2	1×12	1200 (30×40)	33.3×200 to 33.3×400
Block 3	6×6	7200 (180×40)	33.3×150
Block 4	6×18	21600 (180×120)	33.3×150
Block 5	6×12	7200 (180×40)	33.3×200 to 33.3×400
Block 6	3.8×18	1800 (45×40)	56.1×150 to 112.3×150
Block 7	3.8×18	5400 (45×40)	56.1×150 to 112.3×150
Block 8	3.8×12	1800 (45×40)	56.1×200 to 112.3×400
Block 9	9.2×6	2400 (60×40)	153.3×150
Block 10	9.2×18	7200 (60×120)	153.3×150
Block 11	9.2×12	2400 (60×40)	153.3×200 to 153.3×400

Table 1: Mesh dimensions.

	Mesh face	Velocity	Pessure	Temperature
Methane flow	1A	Fully developed $V_{avg} = 0.15\ \frac{m}{s}$	Zero gradient	Uniform $T = 310\ K$
Methane tube	3B	No-slip	Zero gradient	Uniform $T = 310\ K$
Air flow	3C & 6C	Top-hat $V = 0.15\ \frac{m}{s}$	Zero gradient	Uniform $T = 310\ K$
Burner rim	9C	No-slip	Zero gradient	Zero gradient
Side flow	9F, 10F & 11F	Zero gradient	Zero gradient	Zero gradient
Top flow	2G, 5G, 8G & 11G	Zero gradient	Uniform $P = 1\ bar$	Zero gradient

Table 2: Boundary conditions.

comprises several constants including collection efficiency, $N_t(T)$ is the population of the ground state excited transition, B_{12} is the stimulated emission Einstein coefficient and $\Phi(T, J')$ is the fluorescence quantum yield. Equation 2, is derived from the 2-level model, assuming steady state conditions, linear regime, and generalizing the equation with the concept of $\Phi(T, J')$. Evaluating equation 2 with the experimental assumptions, similar η, I_l and B_{12} for the different pressures, the effects of G, N_t and $\Phi(T, J')$ need to be modeled in detail.

2.4.1 Broadening effects and overlap integral

The overlap integral effects G are important because flames have large spatial temperature gradients, and the line shape and line-width can change, based on pressure (pressure broadening) and temperature (Doppler broadening). Hence the balance between Doppler dominated and pressure broadened transitions can change through the flame. The transitions are modeled following a Voigt profile that combines Gaussian and Lorentzian profiles [21].

The Doppler broadening follows a Gaussian line shape:

$$g_D(\nu) = \frac{c}{\nu_0}\sqrt{\frac{m}{2\pi kT}} \exp\left[-4\ln 2\frac{(\nu-\nu_0)^2}{\Delta\nu_D^2}\right] \tag{4}$$

Where c is the speed of light; ν_0 the transition center frequency; m the molecular mass of the absorbing molecule; k, Boltzmann's constant; $\Delta\nu_D$, the transition width (FWHM); and T, temperature. Where $\Delta\nu_D$ is:

$$\Delta\nu_D^2 = \frac{2\nu_0}{c}\sqrt{\frac{2\ln(2)kT}{m}} \tag{5}$$

The pressure broadening follows a Lorentzian line shape:

$$g_c(\nu) = \frac{\Delta\nu_c}{2\pi}\frac{1}{(\nu-\nu_0)^2 + (\Delta\nu_c/2)^2} \tag{6}$$

The $\Delta\nu_c$ is the transition width (FWHM), both $\Delta\nu_c$ and $\Delta\nu_D$ are calculated from the local composition temperature and pressure from OpenFOAM simulations, of the flame for each mesh cell; the pressure broadening coefficients are obtained from [22] and [23]. There is no literature for oxygen, and for this particular case the same coefficient as nitrogen is assigned; as can be seen in [22] the trends in air and nitrogen are similar. Finally, the combination of both broadening phenomena is given by:

$$g(\nu) = 2\sqrt{\frac{\ln 2}{\pi}}\frac{V(a,x)}{\Delta\nu_D} \tag{7}$$

$$V(a,x) = \frac{a}{\pi}\int_{-\infty}^{\infty}\frac{e^{-y^2}}{a^2+(x+y)^2}dy \tag{8}$$

$$a = \sqrt{\ln 2}\frac{\Delta\nu_c}{\Delta\nu_D} \tag{9}$$

$$x = 2\sqrt{\ln 2}\frac{(\nu-\nu_0)}{\Delta\nu_D} \tag{10}$$

The lineshape is then normalized for each simulation point:

$$\int_{-\infty}^{\infty} g(\nu)d\nu = 1 \tag{11}$$

The laser linewidth is estimated from the excitation scans to be a Gaussian with $(FWHM) = 0.15$ cm^{-1}.

$$\phi(\nu) = \exp\left[-\frac{4\ln 2(\nu-\nu_0)^2}{(FWHM)^2}\right] \tag{12}$$

The overlap integral is calculated as:

$$overlapintegral = \int_{-\infty}^{\infty}\phi(\nu)\cdot g(\nu)d\nu \tag{13}$$

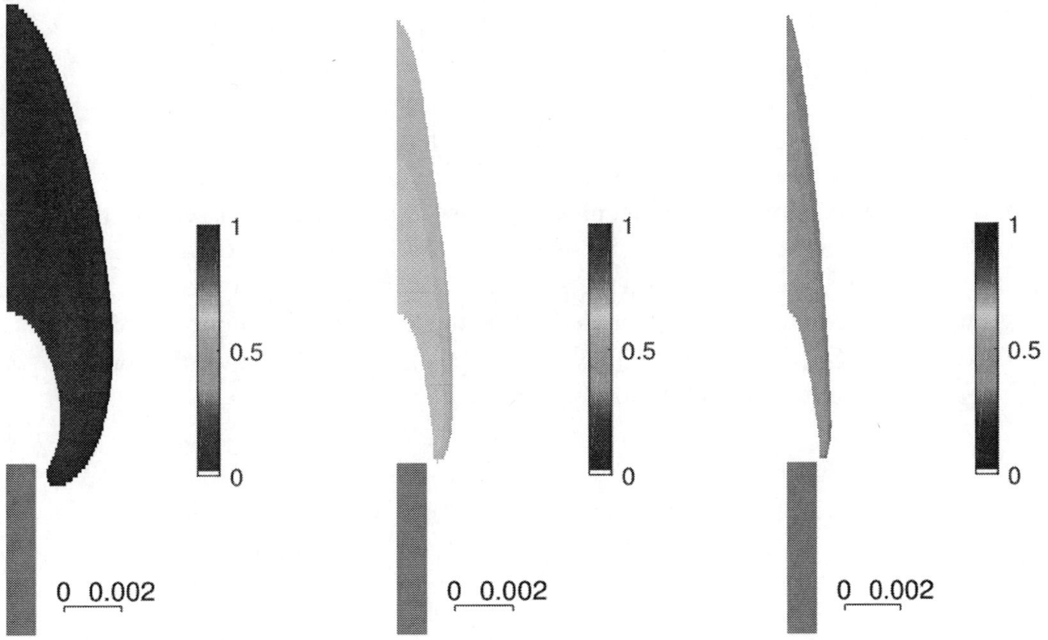

(a) 1.4 bar. (b) 5.7 bar. (c) 11.1 bar.

Figure 4: Simulated overlap integral effects as a function of pressure.

For an easier visualization of the results and relative effect in the fluorescence signal, all the images are normalized to the point with the higher value of the $overlapintegral$ in the condition of $1.4\ bar$. To reduce the computational cost of the overlap integral, only the simulation points with OH mole fraction higher than 1% are considered. The simulated results that are seen in Figure 4 are in qualitative agreement with [4], as the overall values of the overlap integral decay as a function of pressure, due to the effects of pressure broadening. The relative effects at $1.4\ bar$, give higher values of the overlap integral at the cold OH regions and lower values in the hot regions because the shape is dominated by Doppler broadening. On the other hand at $11.1\ bar$ this relative trend is the opposite due to the line shape being dominated by pressure broadening.

2.4.2 Quantum fluorescence yield

The determination of the quantum fluorescence yield $\Phi(T, J')$ is key to evaluate the fluorescence signal in a semi-quantitative manner. The spontaneous decay coefficient is calculated from the natural fluorescence lifetime of OH, which is about two orders of magnitude longer than the typical quenching lifetime at atmospheric pressure. $A_{J'J''} \approx 10^6\ s^{-1}$ and $Q_{J'J''} \approx 10^8\ s^{-1}$ so the correction effects from spontaneous decay are very small. The transient energy exchange in the upper electronic state complicates the modeling. The broadband detection used in this work ($320 \pm 20\ nm$) covers almost all the rotational lines and two vibrational transitions. This width allows the approximation of Q_{eff} as being collision independent of the molecule distribution in the upper excited state [24]. With these assumptions, the quenching becomes only a function of the density, temperature and collisional partners.

$$Q_{J'J''} = N_{tot} \cdot \sum_i \chi_i \sigma_i \upsilon_i \tag{14}$$

$$\upsilon_i = \left(\frac{8k_b T}{\pi \mu_{i,m}} \right) \tag{15}$$

Where N_{tot} is the total number of molecules, σ_i is the quenching cross section of species i, χ_i are the mole fractions, υ_i is the relative velocity between OH and the collisional partner, k_b the Boltzmann constant and $\mu_{i,m}$ the reduced mass of species i and excited molecule m. The quenching cross sections and the coefficients are modeled following [25, 26]. The modeled quenching species include H_2O, CH_4, CO_2, O_2, H_2, and C_2. N_2 effects can be neglected since the relative quenching effect is very minimal compared to that of the molecules included.

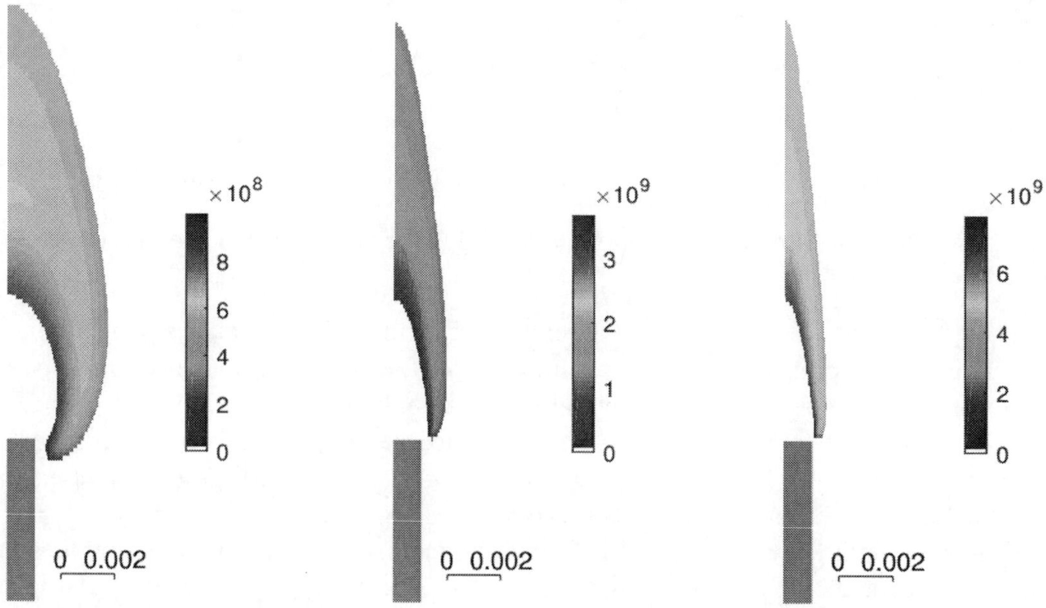

(a) 1.4 bar. (b) 5.7 bar. (c) 11.1 bar.

Figure 5: Simulated quenching in (s^{-1}) as a function of pressure.

The simulations seen in Figure 5 are consistent with the literature [4]. As expected, quenching is smaller in the high temperature zones dues to the decreased $N_t \propto \frac{1}{T}$ despite the increased relative speed between the collisional partners and OH molecules $\upsilon \propto \sqrt{T}$. The methane side of the non-premixed flame quenching is higher than the air side because CH_4 is a good quencher; on the other hand, on the air side O_2 is the dominant quencher, but it is present at a smaller mole fraction compared to CH_4 on the fuel side.

2.4.3 Temperature dependence

The transition selected from the $A_2\Sigma^+ \leftarrow X\Pi_i$ $(1,0)$ is the $Q_1(7)$. This transition has a relatively small temperature sensitivity over the range of temperatures where OH is present [27] ($1000\ K$ to peak flame temperature), and the Q branch is a good choice due to the line strength. Often other weaker branches such as the P branch are preferred to minimize the absorption effect through the flame at high pressures, but with such a small flame (methane tube i.d $2\ mm$) those effects are negligible compared to soot absorption which will become present at higher pressures. At high pressures, the quantum fluorescence yield is smaller, and the overlap integral is also reduced. To study this temperature sensitivity, the temperature dependence is modeled of the ground state of the OH molecule [28]. The images shown evaluate Figure 6 in each point of the OpenFOAM simulation results, only considering cells with an OH mole fraction higher than 1%. Figure 7 shows the normalized temperature effect.

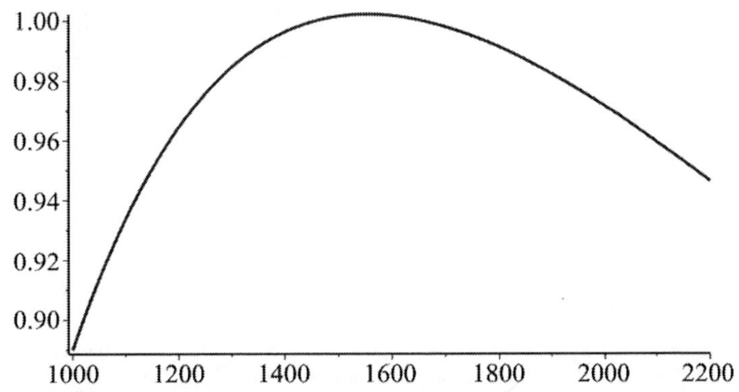

Figure 6: $Q_1(7)$ temperature dependence.

The qualitative trend through all the pictures is very similar. The hottest spots $T > 1900$ have a smaller fluorescence emission; as seen in Figure 6. This position corresponds to the OH cloud closer to the tube (inner gradient); farther from the inner tube the OH cloud cools down, and the relative fluorescence goes through the peak of the temperature dependence and then goes back down as the OH cloud cools down further from the tube. With the above described analysis tools and methods, it will be possible to quantitatively assess OH fluorescence from the OpenFOAM simulations and compare it with the experiments.

3 Results and discussion

Figure 8, 9, and 10 compares the spatial distribution of the fluorescent OH with the simulated fluorescence using the GRI, DRM, and PAH mechanisms. The simulated and the experimental fluorescence have a good overall qualitative agreement. It is important to point out that the flame height is matched very well throughout all the pressures; this is due to constant mass flow rate in all the cases, so while the diffusion coefficient decreases as pressure increase the convective axial velocity decreases at the same rate producing a constant flame height. Despite the general agreement, however, the simulated model fails to reproduce several important details in the experimental results.

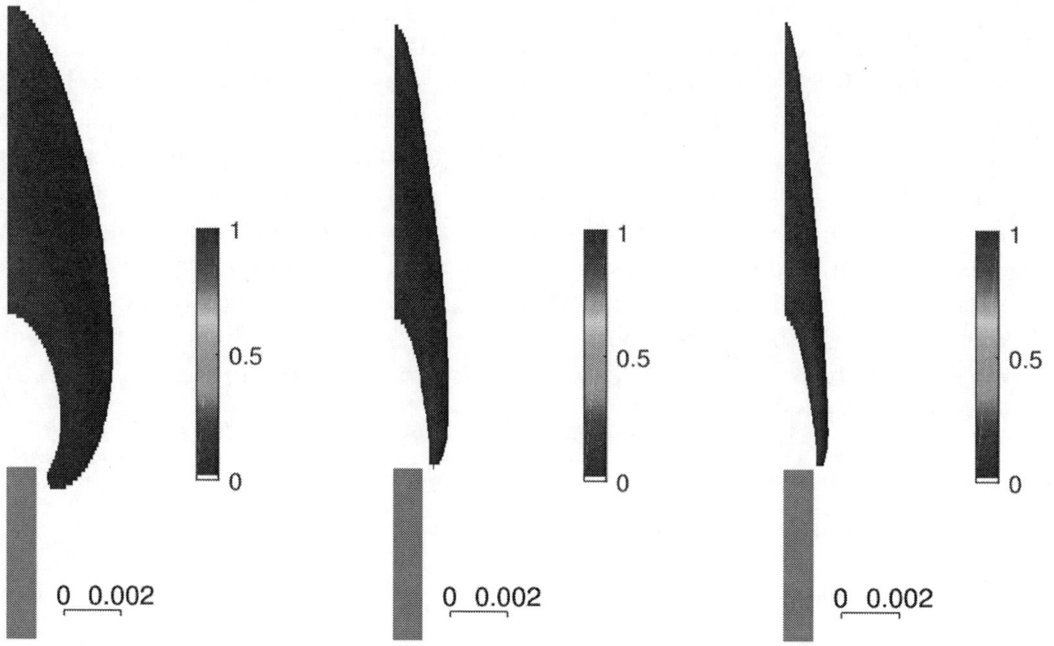

(a) 1.4 bar. (b) 5.7 bar. (c) 11.1 bar.

Figure 7: Simulated normalized temperature dependence.

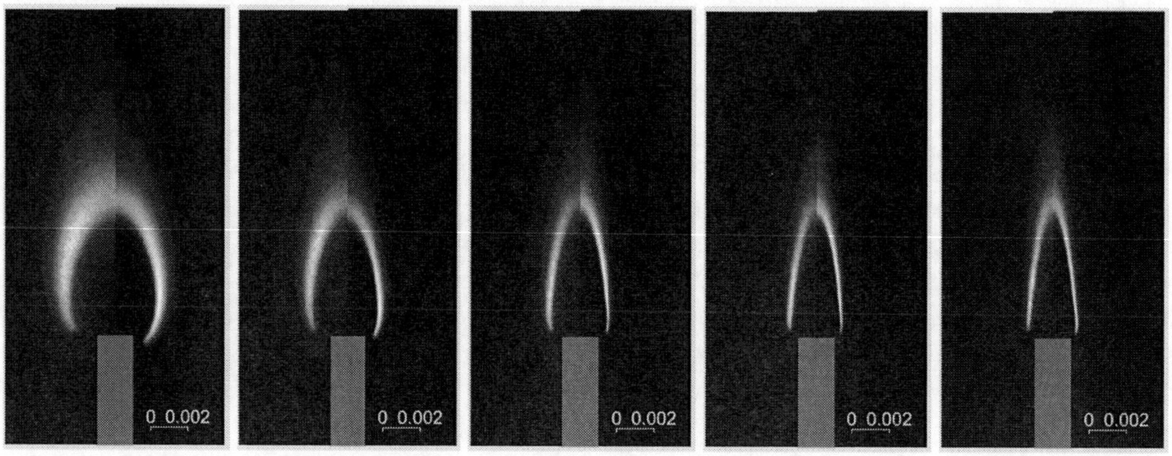

(a) 1.4 bar. (b) 3.3 bar. (c) 5.7 bar. (d) 8.9 bar. (e) 11.1 bar.

Figure 8: Experiment (left side) vs GRI-mech simulated fluorescence (right side).

- The flame base in the 1.4 bar case is not predicted properly; this is probably due to difficulties in the modeling of low temperature chemistry for the flame initiation and near the burner wall (T_{burner} constant at all the pressures). On the other hand, the base of the flame lifts as the pressure increases, which is a phenomenon that is also seen in the experiment.
- The overall OH thickness is underpredicted by the simulation; this can also be seen in

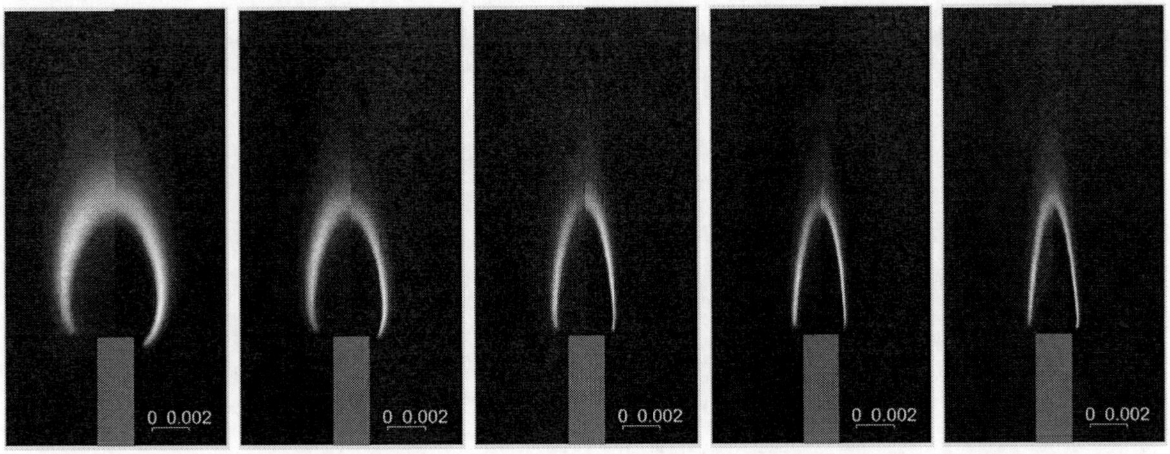

(a) 1.4 bar. (b) 3.3 bar. (c) 5.7 bar. (d) 8.9 bar. (e) 11.1 bar.

Figure 9: Experiment (left side) vs DRM simulated fluorescence (right side).

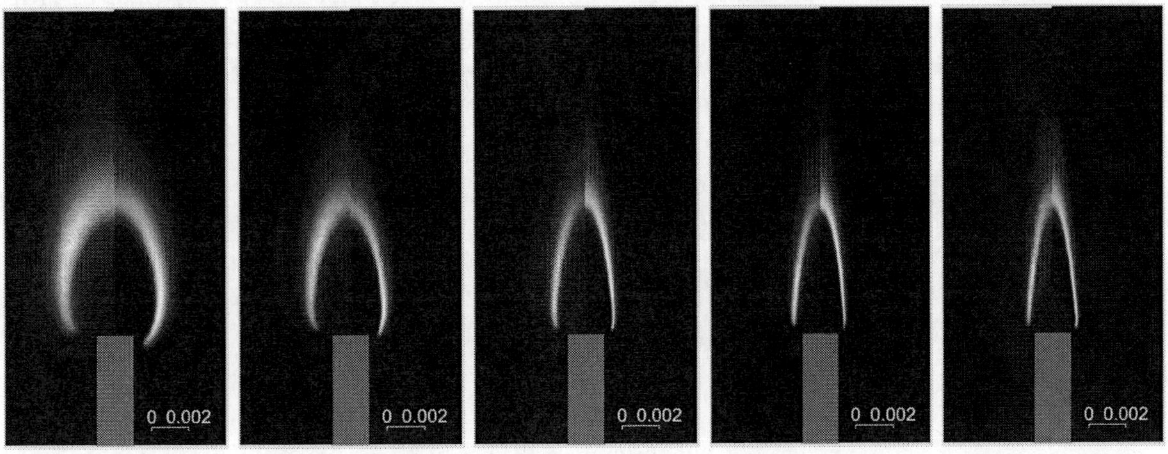

(a) 1.4 bar. (b) 3.3 bar. (c) 5.7 bar. (d) 8.9 bar. (e) 11.1 bar.

Figure 10: Experiment (left side) vs PAH-mech simulated fluorescence (right side).

[29] in a counterflow configuration, which suggests that OH consumption rates might be overestimated. The overall qualitative trend is again consistent, as the overall reaction layer thickness decreases with pressure, both in simulation and experiment. This pattern can be explained due to the smaller thickness of the mixing layer and the reduced OH diffusivity.

The three mechanisms share similar qualitative distributions with very subtle differences. The differences between the DRM and the GRI mechanisms are negligible, but the PAH mechanism has differences that become more significant as the pressure increases, at 11.1 bar the tip of the flame has a relatively higher intensity, and the flame thickness is greater.

To evaluate the contribution of each parameter, including the number of OH molecules interacting with the laser, temperature, quenching, and overlap integral, the following values are calculated for each pressure, based on the simulated fluorescence corrections described.

$$OH\ mol. = \sum_{cells} \frac{P_{cell}}{T_{cell}} \chi_{OH} \tag{16}$$

$$Temperature\ eff. = \frac{1}{OH\ mol.} \sum_{cells} T_{eff} \cdot \frac{P_{cell}}{T_{cell}} \chi_{OH} \tag{17}$$

$$Q_{inteff} = \frac{1}{OH\ mol.} \sum_{cells} Q_{eff} \cdot \frac{P_{cell}}{T_{cell}} \chi_{OH} \tag{18}$$

$$Overlap\ int. = \frac{1}{OH\ mol.} \sum_{cells} Overlap_{eff} \cdot \frac{P_{cell}}{T_{cell}} \chi_{OH} \tag{19}$$

$$Sim\ fluo. = \sum_{cells} \frac{P_{cell}}{T_{cell}} (\chi_{OH} T_{eff} Q_{eff} Overlap_{eff}) \tag{20}$$

$OH\ mol.$ is the number of molecules interacting with the laser sheet, $Temperature\ eff.$ represents the effects of temperature in the simulated fluorescence signal, Q_{inteff} is the quenching effects in the simulated fluorescence signal, $Overlap\ int.$ is the broadening effects in the simulated fluorescence signal, and $Sim\ fluo.$ is the overall fluorescence signal.

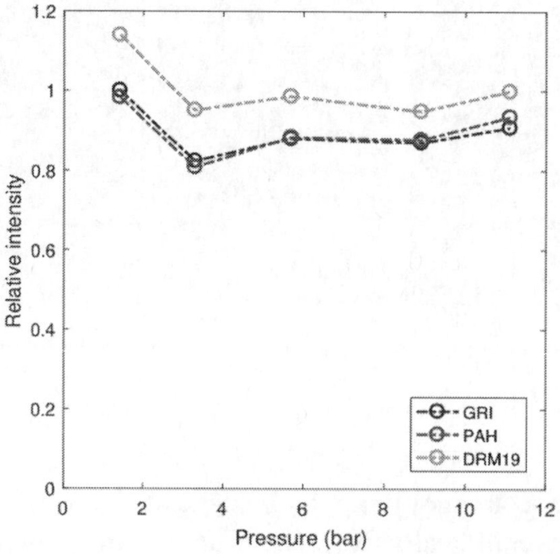

Figure 11: OH mol. mechanism comparison.

Figure 11 shows an equivalent of OH molecules present at a given moment that are interacting with the laser sheet, see Equation 16. To retrieve the total number of OH molecules at a given moment, it would be necessary to add the symmetric convolution which would increase the number of molecules at lower pressures since the reaction zone stands further from the symmetry axis. There is no clear dependence of the OH molecules as a function of pressure. Both detailed chemical mechanisms agree quantitatively with a small difference at $11.1\ bar$ where the DRM19 mechanism has a higher number of molecules. On the other hand, the skeletal mechanism over-predicts both of the detailed mechanisms by 14%-18% throughout all the pressures. The trends of the other mechanisms are also represented in the skeletal mechanism, which suggests that the pressure effects are captured in the skeletal mechanism but misses some key reactions that lead to the absolute disagreement. The OH distribution seems to be independent of the PAH production as the PAH mechanism, and the GRI-mech mimic the results up to $8.9\ bar$, with a small difference at $11.1\ bar$. This small difference suggests that the effects of soot formation on OH are relatively low for the OH chemistry, despite the fact that relatively high carbon to soot conversion rate (5%) have been reported in similar configurations at $10\ bar$ [6].

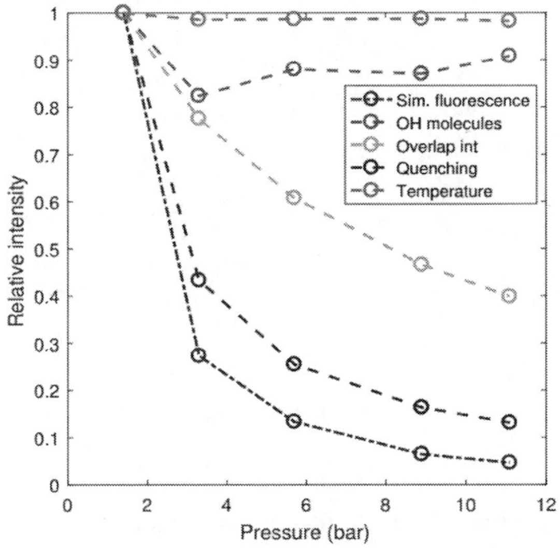

(a) GRI-mech fluorescece parameters.

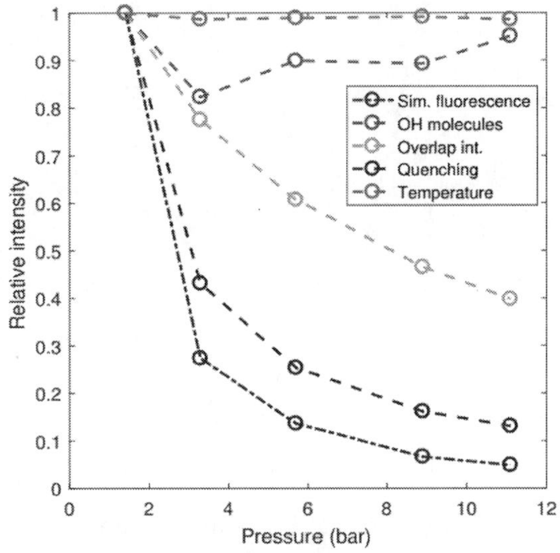

(b) PAH-mech fluorescece parameters.

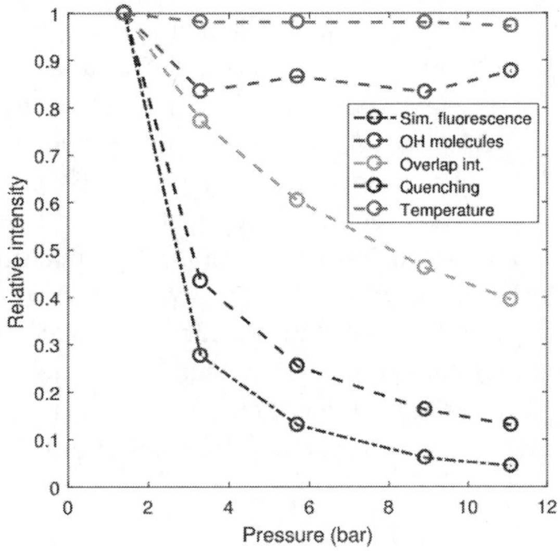

(c) DRM19-mech fluorescece parameters.

Figure 12: Mechanism fluorescence parameters.

Figure 12 shows the effects of the different corrections for each mechanism studied, see Equations (16)-(20) . The highest contribution is accomplished by the quenching correction, which as expected has an approximate dependence as $Q \propto \frac{1}{P}$. The second most important effect is the overlap integral, which has a higher gradient at lower pressures. As the transitions broaden, the dependence on pressure becomes weaker. The temperature correction decreases slightly, as the average temperature where OH exists increases, as high pressure increases dissociation decreases, the energy not consumed for dissociation reactions increases the peak flame temperature.

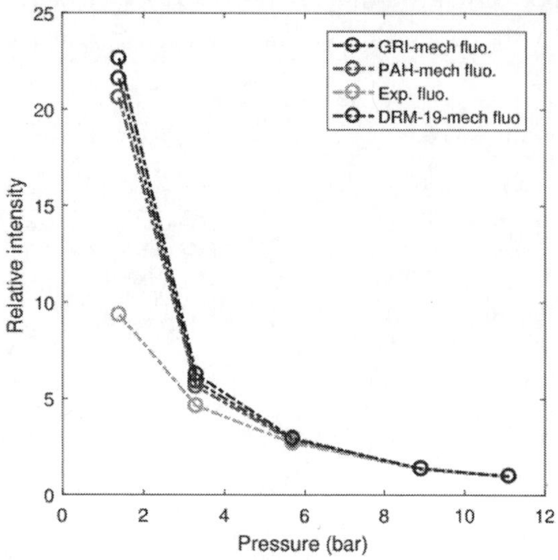

(a) All pressures.

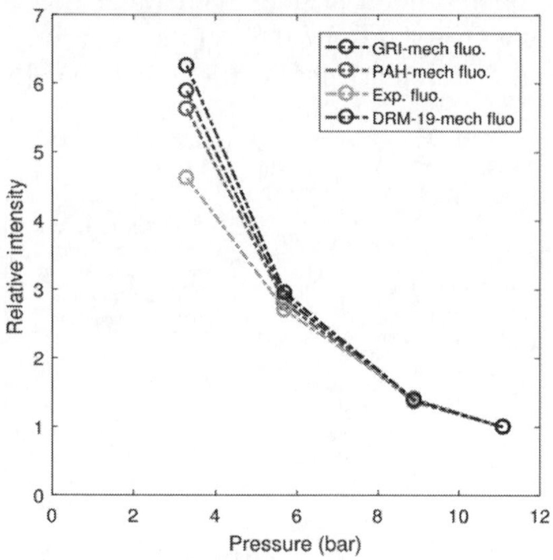

(b) Excluding (1.4 bar).

Figure 13: Integrated fluorescence vs integrated simulation fluorescence.

Figure 13 compares the overall simulated fluorescence signal of the three different mechanisms with the experimental results. The results are normalized to the highest pressure point; the assumptions such as steady state fluorescence are reached faster and are more accurate at higher pressures. The simulation and the experiment agree qualitatively in the decay. The match is good for 3.3 bar, 5.7 bar, 8.9 bar and 11.1 bar. Assuming that the normalization at 11.1 bar is accurate, there is a 100% discrepancy at 1.4 bar; this can be explained by a combination of effects. The choice of laser intensity is complicated due to the dynamic range associated with a decay in the intensity by a factor of 20 throughout the pressures. In order to maximize SNR the laser intensity choice is a compromise to the linear dependence at the lowest pressure, and hence the fluorescence decay time for 1.4 bar is non-negligible in the nanosecond time scale. To maximize the filtering of the LII signal, it is important to minimize the gate window time, and this could affect the fluorescence collection efficiency. The mechanism that is closest to the experiment is the PAH mechanism that stands as the most updated which includes the contribution of PAH formation.

4 Conclusions

This work aims to compare trends of the experimental OH fluorescence signal as pressure increases, with a numerical CFD simulation that match the experimental conditions and allows comparison of different chemical mechanism and transport datasets. The simulated fluorescence and the experimental dataset is compared in 2 aspects.

- The spatial distribution of the OH molecule. The numerical simulation has an excellent agreement matching the OH layer position and the overall flame height; this flame height is mostly driven by the diffusion coefficients. The simulations fail to quantitatively match the thickness of the OH layer, transport, and chemical effects are convoluted, in the flame thickness calculation, on the other side, the overall flame heights are matched accurately, suggesting that the issue could be on the chemical mechanisms. The flame base is not matched, specifically, at low pressures, more accurate modeling of the burner could stretch the differences; also, low temperature chemistry is known to be challenging, especially when mixing is dominated by diffusion.

- The comparison in the trends of the number of OH molecules as pressure increases. All the mechanisms capture the same qualitative trend as the pressure increases, but the skeletal mechanism has a 20 % higher number of OH molecules difference with the detailed mechanisms. This result reflects that the number of OH molecules present in the studied flame does not depend strongly on pressure and furthermore, pressure does not have a significant impact on the overall number of OH molecules that interact with the laser sheet. The comparison between the experimental results and the three simulations is good in the high pressure region, and the agreement is similar to the only similar study in a counterflow flame [4]. The use of a 2D configuration for quantitative evaluation is challenging, in particular for this case where the variation of OH is small to the corrections applied, although excellent agreement overall in the trends is observed primarily in the high pressure regime.

Acknowledgments

This research was supported by the NSF grant - under grant CBET 1605533, with Song-Charng Kong as program manager. The authors acknowledge Sunny Karnani and Royal Welding for the design of the high pressure chamber, and the Balsells foundation for their financial support to first author.

References

[1] Sarah Tedder and Yolanda Hicks. (2012) journal.

[2] Kevin A. Thomson, Ömer L. Gülder, Elizabeth J. Weckman, Roydon A. Fraser, Greg J. Smallwood, and Dave R. Snelling. *Combustion and Flame*, 140 (2005) 222–232.

[3] G. Singla, P. Scouflaire, C. Rolon, and S. Candel. *Combustion and Flame*, 144 (2006) 151–169.

[4] A. Matynia, M. Idir, J. Molet, C. Roche, S. De Persis, and L. Pillier. *Applied Physics B: Lasers and Optics*, 108 (2012) 393–405.

[5] Peter H. Joo, Jinlong Gao, Zhongshan Li, and Marcus Aldén. *Review of Scientific Instruments*, 86 (2015) 1–7.

[6] Hyun I. Joo and Ömer L. Gülder. *Proceedings of the Combustion Institute*, 32 (2009) 769–775.

[7] C D Carter, G B King, and N M Laurendeau. *Applied Optics*, 31 (1992) 1511–22.

[8] Brett E. Battles and Ronald K. Hanson. *Journal of Quantitative Spectroscopy and Radiative Transfer*, 54 (1995) 521–537.

[9] Hrvoje Jasak, Aleksandar Jemcov, and Zeljko Tukovic. *International Workshop on Coupled Methods in Numerical Dynamics*, (2007) journal.

[10] OpenFOAM Foundation. OpenFOAM - The Open Source CFD Toolbox - User Guide. Technical report, 2014.

[11] Hassan I. Kassem, Khalid M. Saqr, Hossam S. Aly, Mohsin M. Sies, and Mazlan Abdul Wahid. *International Communications in Heat and Mass Transfer*, (2011) journal.

[12] A. Cuoci, A. Frassoldati, T. Faravelli, and E. Ranzi. *Combustion and Flame*, (2013) journal.

[13] Nasa Tn, Iruin M Miller, and G M Howard. High-pressure flame system for pollution studies with results for methane-air diffusion flames. Technical Report June, 1977.

[14] Jörg Appel, Henning Bockhorn, and Michael Frenklach. *Combustion and Flame*, (2000) journal.

[15] Marc J Assael, J P Martin Trusler, and Thomas F Tsolakis. *Thermophysical Properties of Fluids*, Vol. 1. Imperial College press, jul 1996.

[16] Bonnie J. McBride, Sanford Gordon, and Martin A. Reno. Coefficients for Calculating Thermodynamic and Transport Properties of Individual Species. Technical report, 1993.

[17] Gregory P. Smith, David M. Golden, Michael Frenklach, Nigel W. Moriarty, Boris Eiteneer, Mikhail Goldenberg, C. Thomas Bowman, Ronald K. Hanson, Soonho Song, William C. Gardiner, Vitali V. Lissianski Jr., Qin, and Zhiwei. GRI-Mech3.0.

[18] A. Kazakov and M. Frenlack. DRM.

[19] N. A. Slavinskaya and P. Frank. *Combustion and Flame*, (2009) journal.

[20] Victor Chernov, Murray J. Thomson, Seth B. Dworkin, Nadezhda A. Slavinskaya, and Uwe Riedel. *Combustion and Flame*, (2014) journal.

[21] Alan C. Eckbreth. *Journal of Chemical Information and Modeling*, (1989) journal.

[22] Soon Muk Hwang, Jun N. Kojima, Quang-Viet Nguyen, and Martin J. Rabinowitz. *Journal of Quantitative Spectroscopy and Radiative Transfer*, 109 (2008) 2715–2724.

[23] E.C. Rea, A.Y. Chang, and R.K. Hanson. *Journal of Quantitative Spectroscopy and Radiative Transfer*, 41 (1989) 29–42.

[24] Katharina Kohse-Höinghaus. *Progress in Energy and Combustion Science*, 20 (1994) 203–279.

[25] Nancy L. Garland and David R. Crosley. *Symposium (International) on Combustion*, 21 (1988) 1693–1702.

[26] P.H. Paul. *Journal of Quantitative Spectroscopy and Radiative Transfer*, 51 (1994) 511–524.

[27] Andreas Brockhinke, Andreas Bülter, Juan C. Rolon, and Katharina Kohse-Höinghaus. *Applied Physics B: Lasers and Optics*, (2001) journal.

[28] G.H. Dieke and H.M. Crosswhite. *Journal of Quantitative Spectroscopy and Radiative Transfer*, 2 (1962) 97–199.

[29] Rosa Padilla. *Structure and Behavior of Water-laden Methane/Air Flames*. PhD thesis, 2016.

Sub Topic: Diagnostics

Western States Section Combustion Institute 2017 Fall Meeting
October 2-3, 2017 • Laramie, Wyoming

Light Extinction Based Image Analysis Technique for Rotating Fluidized Beds

Z. Lavrich, Z. Taie, D. Wagner, C. Hagen*

School of Mechanical, Industrial and Manufacturing Engineering,
Oregon State University - Cascades, Bend, OR, USA
**Corresponding Author Email: taiez@oregonstate.edu*

Abstract: Oregon State University - Cascades has developed an intensified ethane conversion reactor from an internal combustion (IC) engine. The engine reactor includes a small (5 cm diameter) rotating fluidized bed (RFB) that contains the required catalyst. Unlike conventional fluidized beds, for which the pressure drop across the bed gives an indication of the onset of fluidization, RFBs require a different diagnostic. A light extinction-based image analysis technique was developed to investigate the fluidization of catalyst particles within an RFB. An RFB was 3-D printed in a transparent plastic, which was then coated with clear acrylic for increased transparency. The technique utilizes a strong diffuse halogen light source placed behind the RFB, with light intensity captured on the other side with a CMOS sensor. A MATLAB code is employed to analyze the intensity and determine fluidization metrics such as onset of fluidization and fluidization quality. Preliminary results show that static bed heights can be accurately measured with this imaging technique. Fluidization metrics based on light intensity profiles are under construction.
Keywords: Rotating Fluidized Bed, Fluidization, Image Analysis, Ethylene Production

1. Introduction

The consumption of ethane in the United States is projected to double between 2016 and 2018. The primary driver of the increase in ethylene consumption is an increase in demand for plastic, of which ethylene is a major feedstock. New ethylene production plants are scheduled to be constructed in the United States in the coming years. However, the traditional industrial scale cracking plants have long construction times and large upfront capital expenses. In the best case, these plants take years to construct, and cannot be used to quickly meet increasing demand. For this reason, smaller plants that have shorter construction times and relatively insignificant capital expenses are desirable. Smaller plants require significantly more intensified processes, which requires developing new chemical processing technology.

Many industries, ethylene production among them, require chemical reactors which utilize solid phase catalysts to enhance gas phase reactions. The catalyst can take many physical forms depending on, for instance, the necessary reaction conditions, the catalyst chemical composition, and the economic and physical constraints. For example, vehicle exhaust treatment systems use monolithic style catalysts since the exhaust system is constrained to a small size (~0.006 m^3 reactor volume), while large commercial chemical processing plants utilize fluidized bed reactors which can be over 10 meters tall (~22.25 m^3 reactor volume). These large-scale fluidized beds provide many benefits over monolith style catalysts, such as increased mixing of the solid and gas phases which increases heat and mass transfer, as well as a decreased pressure drop at high gas throughput. These attributes are desirable, as they allow for increased gas throughput with less work input, and decrease the relative requirements for catalyst within the reactor.

Though there are distinct advantages to using a fluidized bed reactor as opposed to a monolith, there are inherent challenges associated with the architecture. Fluidized bed reactors require a large capital expense and must be large enough to allow the fluidized particles to be entrained in the gas for a long enough time for the reactions to occur. The increased throughput and rapid mixing can also lead to pockets with a lower concentration of catalyst particles, or even channels of fluid that are not in contact with catalyst particles at all. As the gas throughput is increased, there becomes a point where the velocity of the incoming gas produces a drag force on the particles that is larger than the gravitational force acting on the particles. At this point, the particles are carried upward together as a packed bed and no longer act fluidized. To overcome this velocity limitation, another force is necessary to balance the large drag force. One approach is to use centrifugal force to counter the drag force, which can be accomplished by rotating the particles at high angular velocities. This architecture is termed a rotating fluidized bed (RFB).

Many applications of RFBs can be found in use today, such as chemical processing [1], [2], biomass drying [3] and gasification [4], [5]. There are two techniques to induce particle rotation: by mechanically rotating the device, which in turn induces particle rotation, or by injecting high pressure gas into the bed with angled inlets to induce a swirling flow pattern. The rotating gas then induces particle rotation. The latter option, termed a "static geometry" RFB provides less mechanical complexity. Testing of this architecture is currently underway at Oregon State University - Cascades.

In order to achieve the advantages offered by RFBs, good fluidization quality is necessary. An even distribution of particles rotating within the reactor presents the best gas/solid mixing, and provides the best chance for reactant gas flowing through the RFB to be in contact with solid catalyst. Channeling or slugging within the rotating bed indicates that some of the gas within the RFB is allowed to bypass the catalyst, which has a negative impact on the yield of the desired product. For RFBs used as chemical reactors, a uniform distribution of particles with ample gas-solid contact is expected to achieve better chemical conversion.

There are a number of design parameters that may influence the fluidization quality within the RFB. These include the number, width, and angle of the inlet channels, the diameter and height of the RFB, the chimney design, and the size and amount of catalyst particles. It is difficult to measure the impact of modifying these design parameters without taking intrusive measurements. In standard fluidized beds the pressure drop across the bed provides an indication if the bed is fluidized or not. At the onset of fluidization the pressure drop across the bed remains constant. However, in RFBs there is no such analog with pressure drop. Optical measurements are the default metric used to determine good fluidization in optically accessible beds. However, subjective person-to-person observations are not as consistent as quantitative measurements. Videos can be taken of the RFB in operation, but the lack of depth perception makes video measurements less clear than in-person observation. This work details an approach that was designed to provide a quantitative image processing technique based on light extinction to evaluate the performance of various RFB designs.

2. RFB Test Facility

The Energy Systems Laboratory at Oregon State University – Cascades has constructed a RFB testing apparatus to analyze various RFB designs. The testing apparatus consists of a RFB stand that is situated in between a 500 watt halogen light source and a Cannon EOS50D camera equipped with a CMOS sensor. Unlike incandescent, LED, or CFL lights, the halogen bulbs provided an unmodulated light source, which is important for the measurement technique yet to be outlined. To decrease the amount of incident light that was captured by the camera, all extraneous room lighting was turned off prior to data collection.

The RFBs were 3D printed in Formlabs standard clear resin using stereolithography. To increase their transparency, a clear acrylic was applied to the RFBs after sanding and the outer annulus intake manifold was made of transparent PVC pipe. A picture of the RFB is shown in Figure 1. The RFB used here was designed with 4 inlet slots, each with a width of 0.25 mm. A mass of 7.6 grams of 40-63 micron diameter silica gel from Alfa Aesar was placed into the RFB to be used as a catalyst surrogate. These slot dimensions and particle properties are arbitrary for this study, and are given here for reference only. To stop the particles from escaping through either the inlet or outlet ports, mesh screens of size 400 mesh were glued over the inlets.

The camera used was equipped with a zoom lens and allowed for the International Standards Organization light sensitivity rating (ISO), shutter speed, and aperture to each be manually set. First, all the extraneous lights were switched off in the room and the halogen backlight was turned on. The ISO, aperture, and shutter speed were adjusted until the region of interest was close to being fully saturated. This procedure allowed for the full range of the CMOS sensor to be utilized in the region of interest.

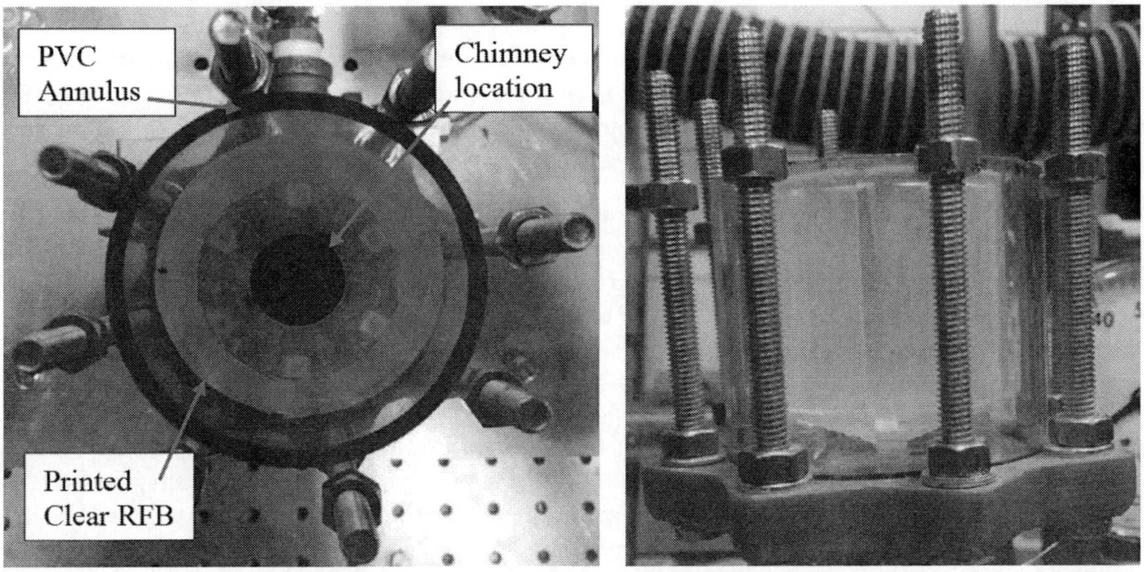

Figure 1. Top view of the RFB (left), and side view of RFB (right).

3. Diagnostic Procedure

There are two steps to consider when using this diagnostic procedure. The first is acquiring the photographs. The second is analyzing the images to reveal information regarding fluidization.

Image collection

After the ISO, shutter speed, and aperture were appropriately adjusted, photographs were taken with the bed completely empty, and then with the bed filled with particles. These two photographs represent the two background images. They can also be thought of as the bounding cases that the CMOS sensor will see: no light obstruction from particles (empty background), and the maximum obstruction that can occur from particles (full background). An example of the empty background image can be seen in Figure 2, left.

Figure 2: Background image of empty RFB (left) and backlit RFB with test amount of catalyst surrogate (right).

With the background images taken, the photographs of interest (or subject photos) are taken. These are the images that will be analyzed using the image processing procedure described below. While taking the subject photos it is important to keep the camera settings consistent between the background and subject photographs. It is also important to keep the alignment of the test apparatus constant during the two sets of images. The background images will be subtracted, pixel by pixel, from the subject photos. As such, the images need to have been taken under similar conditions.

An example of a subject photograph is shown in Figure 2, right. In this image, the RFB was filled with 7.6 grams of 40-63 micron diameter silica gel particles. There is no gas flowing through the bed in this photo; it is simply a static resting bed of particles. A clear distinction can be made between the bottom of the bed which is full of particles and the top of the bed that has no particles. This indicates the camera setting have been correctly chosen.

Image Processing

The images shown above, in above Figure 2, are from a new design that provides better optical access to the internal RFB than older versions. However, test data have not yet been processed for this design. As such, for the remainder of this paper, an old design, shown in Figure 3, will be used to demonstrate the image processing technique. The full and empty background, along with the half-filled images are shown below in Figure 3.

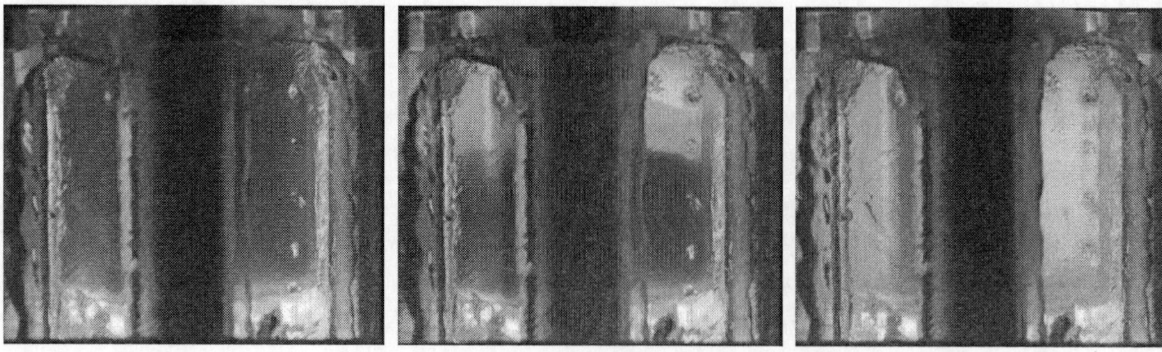

Figure 3. From left to right: Filled background, half filled (subject), empty background images.

As can be seen, the surface finish on the old design, Figure 3, is much rougher. The newer design seen in Figure 2 corrects this, but the older version is sufficient to convey the analysis technique. The subject image in this case is a bed of particles resting in the bottom of the RFB. This subject frame can be any image that is of interest. During fluidization, many images are taken every second for post processing to get an idea of the transient behavior of the RFB.

The first step in the image processing is to select the area of the images to analyze. It is important to choose a region that will sufficiently show if the particles are fluidized, while also having the least influence from surface imperfections. In this design, that is the area just to the right of the chimney (identified with red dashed lines in (Figure 4). Next, the identified areas are then cropped out of the images as is shown in step A of Figure 5. This reduces the images to a vertical 2-dimensional representation of the bed.

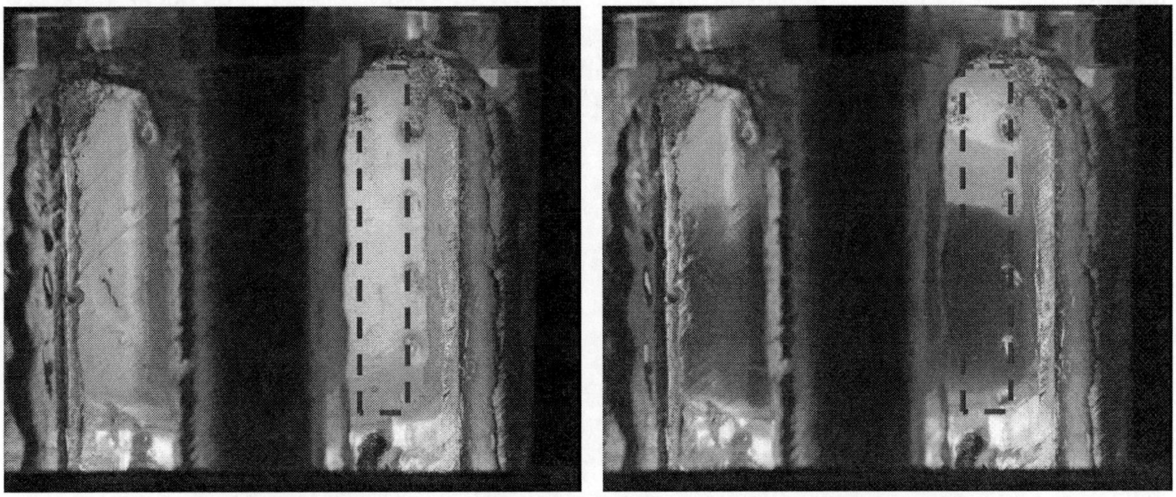

Figure 4. Section that is cropped from the empty background and subject photo.

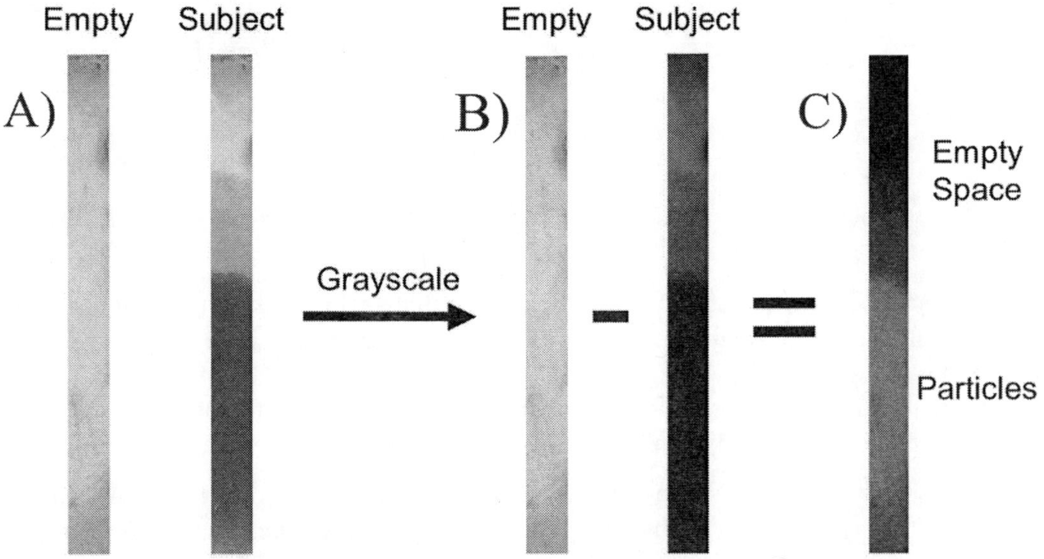

Figure 5. A) Cropping the section of interest out of images, B) converting the cropped sections from RGB to grayscale, C) subtracting the subject frame from the empty background.

The level of the packed bed is clear from inspecting the difference between the background and subject image above. However, during fluidization the difference is not as stark, and a better quantification metric needs to be determined. To do this, the images are converted from RGB format (three matrices representing the red, green, and blue color intensity at each pixel) into grayscale images that represent the light intensity at each pixel by a single number in the range of 0 to 255. This allows for a single matrix to represent the image, and makes manipulating the images through matrix math simple.

The grayscale subject image is then subtracted from the empty background image (shown as step B in Figure 5) to produce a single grayscale image (C in Figure 5). In this image, dark areas represent space which has fewer particles, whereas light areas have a higher density of particles. This is because the lightest portions of the subject and background images have a high light intensity (e.g. closer to 255), while the regions where the light was obstructed by the particles have a low light intensity (e.g. closer to 0). When a light region is subtracted from a light region, the result is a dark area (e.g. 255-255=0), and when a dark region is subtracted from a light region a light area remains (e.g. 255–0=255). The remaining 2-dimensional greyscale matrix can then be plotted as an intensity map, show in Figure 6.

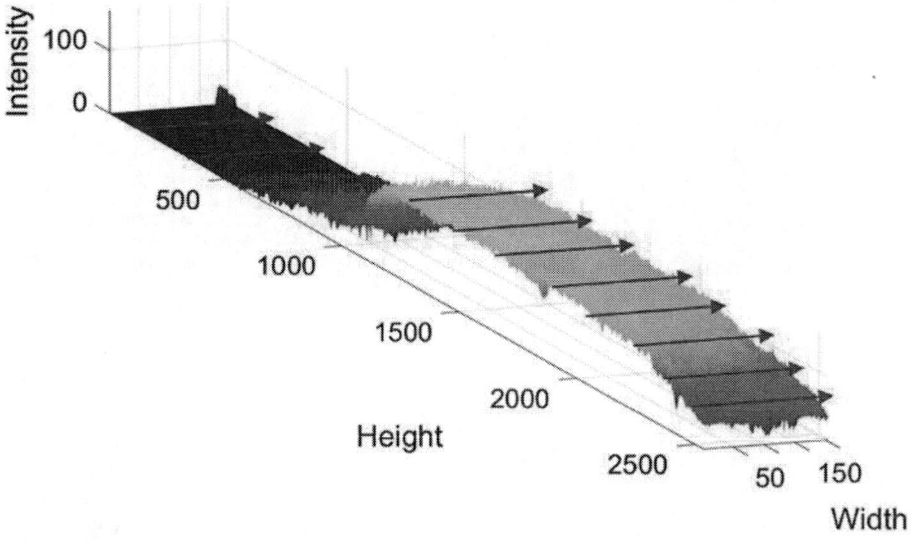

Figure 6. Intensity map of the post-subtraction greyscale image. The red arrows indicate the dimension the averaging was carried over.

In this representation, it is much easier to see the impact that surface impurities and roughness has on the light intensity. These differences in localized intensity necessitate averaging across the horizontal dimension, as is indicated by the red arrows in Figure 6. This average leaves a single horizontal intensity for each vertical pixel row which can be plotted to reveal the average light intensity as a function of height. This plot is shown in Figure 7.

The process that has been outlined above subtracts the subject image from the *empty* background image. However, it is just as reasonable to subtract the *full* background image from the subject image. This produces a final intensity map that is inverse of the method outlined above (i.e. low intensity represents a higher density of particles, and a high intensity represents fewer particles). This approach has also been conducted, and is represented as the blue line in Figure 7.

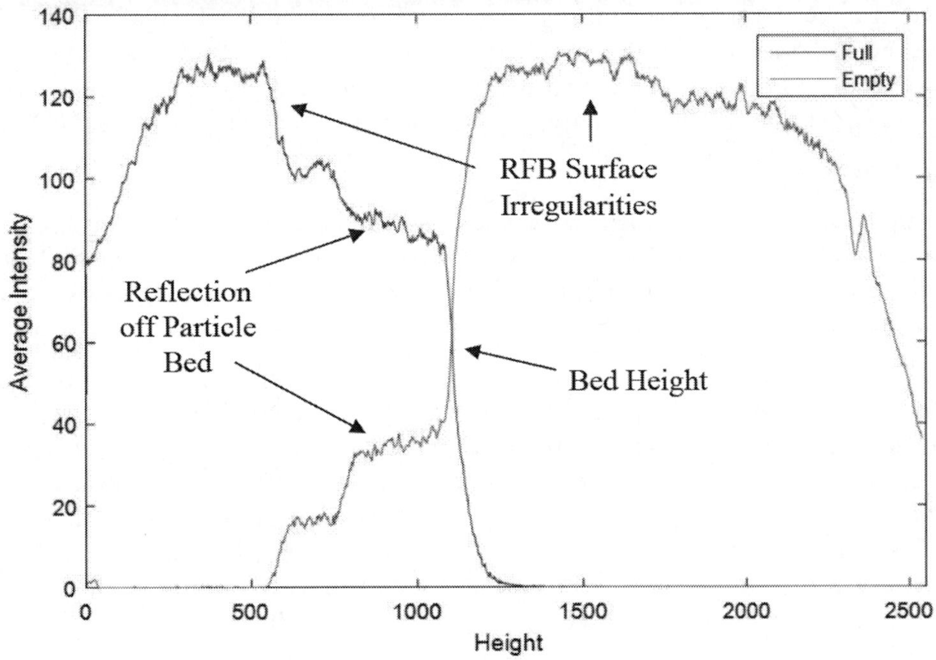

Figure 7. Intensity vs height of the final averaged and subtracted greyscale image. Full denotes using the full background image, empty denotes using the empty background image.

Inspecting the resultant curves in Figure 7 reveals some of the limitations and successful aspects of this technique. The step change in intensity in both curves (around a height of 1200) successfully indicates the height of the static bed. This is encouraging, and shows the method works in the most basic case. However, the non-uniformity in the intensity near the top of each curve shows the effects of the RFB surface irregularities. There is also a small "shelf" that is present in both curves just above the top of the particle bed. It is believed this was caused by light reflecting off the top of the particle bed, and is not observed in fluidized images. However, it does speak to the impact that light reflection can have on the final intensity map.

4. Future Method Additions

Several additions are being be made to this baseline method. As was discussed, this technique lends itself well to analyzing the transient behavior of a RFB by analyzing each frame of a video, of any framerate. These frames can then be averaged temporally to produce a metric for the time averaged fluidization as a function of bed height. The images can also be taken from another plane to provide additional temporal information. For instance, a clear top and bottom can be manufactured for the RFB, and the light and camera can be placed vertically above and below the RFB. This plane provides information on whether the particle concentration is higher near the chimney or walls. Finally, including a length standard in one of the baseline images (such as a ruler or bed dimension) can be used to accurately measure the height of the static bed, or any other area of interest.

5. Conclusion

A noninvasive, quantitative diagnostic for measuring rotating bed fluidization quality was developed. This technique involves capturing images of a transparent RFB which is placed between a light source and a camera. Using the images captured during the operation of the RFB, the difference in light intensity reaching the camera indicates the local concentration of catalyst particles. The method outlined above assigns a quantitative value to this concentration, and can be used to determine the impact of changing variables within the RFB. This technique is nonintrusive, and has shown to be successful at calculating the height of a static bed of particles within the RFB.

6. Acknowledgements

This information, data or work presented herein was funded in part by the Advanced Research Projects Agency-Energy (ARPA-E), U.S. Department of Energy, under Award Number DE-AR0000681. The views and opinions of authors expressed herein do not necessarily state or reflect those of the United States Government or any agency thereof.

Sub Topic: Diagnostics

[1] J. De Wilde, "Gas–solid fluidized beds in vortex chambers," *Chem. Eng. Process. Process Intensif.*, vol. 85, pp. 256–290, Nov. 2014.

[2] J. De Wilde and A. de Broqueville, "Rotating fluidized beds in a static geometry: Experimental proof of concept," *AIChE J.*, vol. 53, no. 4, pp. 793–810, Apr. 2007.

[3] P. Eliaers, J. Ranjan Pati, S. Dutta, and J. De Wilde, "Modeling and simulation of biomass drying in vortex chambers," *Chem. Eng. Sci.*, vol. 123, pp. 648–664, Feb. 2015.

[4] J. De Wilde, G. Richards, and S. Benyahia, "Qualitative numerical study of simultaneous high-G-intensified gas–solids contact, separation and segregation in a bi-disperse rotating fluidized bed in a vortex chamber," *Adv. Powder Technol.*, vol. 27, no. 4, pp. 1453–1463, Jul. 2016.

[5] J. Z. Kovacevic, M. N. Pantzali, K. Niyogi, N. G. Deen, G. J. Heynderickx, and G. B. Marin, "Solids velocity fields in a cold-flow Gas–Solid Vortex Reactor," *Chem. Eng. Sci.*, vol. 123, pp. 220–230, Feb. 2015.

Western States Section of the Combustion Institute – Fall 2017 Meeting

Hosted by the University of Wyoming
October 2-3, 2017

Opposed-flow heterogeneous reactor for partial oxidation of methane

*Ying Lin[1], Xuesong Li[2], Martyn V. Twigg[3], William F. Northrop[1]**

*[1] University of Minnesota, Department of Mechanical Engineering, 111 Church St. SE, Minneapolis, MN 55455, USA; *correspondence to: wnorthro@umn.edu, (612) 625 6854*

[2] Shanghai Jiaotong University, School of Mechanical Engineering, 800 Dongchuan Road, Shanghai 200240, China

[3]Twigg Scientific & Technical Ltd, Caxton, Cambridge CB23 3PQ, UK

**correspondence to: wnorthro@umn.edu, (612) 625 6854*

Abstract

This work experimentally evaluates a novel non-premixed opposed-flow reactor for reforming gaseous fuels. Our reactor is like the canonical opposed-flow reactor commonly used in flame studies but has a catalyst-coated metal mesh located between the air inlet and fuel inlet. Experiments were conducted to evaluate both temperature profile and species profiles of the rector with methane as fuel at different catalytic mesh location. We operated the reactor using wash coat supported platinum on the metal mesh as the active material. A CHEMKIN diffusion opposed-flow model with modified GRI-30 mechanism was created based on reactor boundary condition. This model provided a preliminary estimate of local equivalence ratio based on species diffusion. Our results show that the reactor can convert methane to reforming products and mesh stoichiometry plays a critical role in optimizing the reforming products. Moreover, the data experimentally confirm that the catalytic mesh serves like a flame sheet in an opposed flame reactor by completely reacting the oxygen without crossover.

1. Introduction

Catalytic partial oxidation (CPO) is one of the most common industrial reactions. It is used to convert heavier hydrocarbon fuels into light hydrocarbon fuels or synthesis gas for a variety of applications. CPO systems usually incorporate metal catalysts, such as nickel, platinum, rhodium, etc. [1-3], impregnated on wash-coated porous media to enable enhanced surface reaction rates.

There are different CPO schemes to enable desired reforming reactions; these include plug-flow reactors [4-5], wire gauge reactors [6], and flash volatilization techniques [7-8]. To the authors' knowledge, most CPO applications implement premixed schemes that mix the fuel, oxidizer (usually oxygen), and other species (such as steam) before the mixture reaches the catalyst. Such implementations enjoy benefits including simple structures and flow fields. However, disadvantages of premixed reformers include autoignition caused by preheating the mixture before reactions [1] and catalyst deactivation through carbon formation.

To overcome the disadvantages of premixed CPO, some have attempted non-premixed catalytic reformers. For instance, Aicher et al developed a non-premixed catalytic converter mainly for evaporation liquid diesel fuel by the heat release from non-premixed catalytic reactions and secondary stage premixed reformer [9]. In our previous work [10], the authors developed a novel catalytic opposed flow reformer that directly evaporate liquid diesel fuel to create a gaseous jet, meanwhile the local equivalence ratio was controlled by the location of the catalytic wire mesh.

The concept of diesel fuel direct evaporation and opposed flow reforming has the potential in both fundamental studies, such as reforming kinetics investigations; and practical applications such as enabling single fuel RCCI engine operation [11]. However, utilizing diesel fuels such as n-dodecane or ULSD yields limited information, detailed chemistry and the complex physical phenomenon for such fuels is not well known.

This work investigates a novel opposed flow scheme CPO process with methane as the fuel instead of liquid fuels. Using methane as the fuel leads to steady boundary conditions and flow fields, furthermore, the investigation of catalytic reaction mechanisms for methane is more thorough than heavier fuels and detailed methane reforming mechanisms have been reported by many groups [12-13]. As such, the catalytic reforming process can be accurately modeled with 1D opposed flow models that has been widely utilized in the combustion community. This work introduces the implementation of methane catalytic reforming with platinum as the catalyst. Experimental investigations, such as temperature and species measurements, were performed. The data was then analyzed and discussed, finally preliminary considerations in developing a 1D numerical model with detailed chemistry was discussed.

2. Experimental Setup and Theory

The opposed flow reactor illustrated in Figure 1 was designed to study non-premixed gaseous reforming reactions. This reactor was modified from an opposed flow reactive volatilization reactor in a previous study [10]. The reactor has two concentric tube chambers for feeding gaseous reactants. The top chamber delivers the heated oxidizer inlet to initiate the catalytic reforming process; and the bottom chamber is the unheated gaseous hydrocarbon fuel inlet. A round catalytic mesh with an outside diameter of 3 inches was placed between two chambers. The catalytic mesh is controlled by a linear translation stage. A 1.5 mm thick ring support holds the catalytic mesh in desired reaction position. The gaseous oxidizer (air & N_2 mixture) was preheated to 300 °C by an inline electric heater and a band heater before entering the reaction zone. The unheated gaseous fuel (research grade methane in this research) was fed from the bottom chamber to the reaction zone. The distance between two chambers, which is defined as the reaction zone, was set as 15mm. Ceramic honeycombs were used as flow straighteners, placed in the both chamber to generate a uniform flow to the reaction zone. A nitrogen purge system shields the reaction zone from the effect of the ambient air.

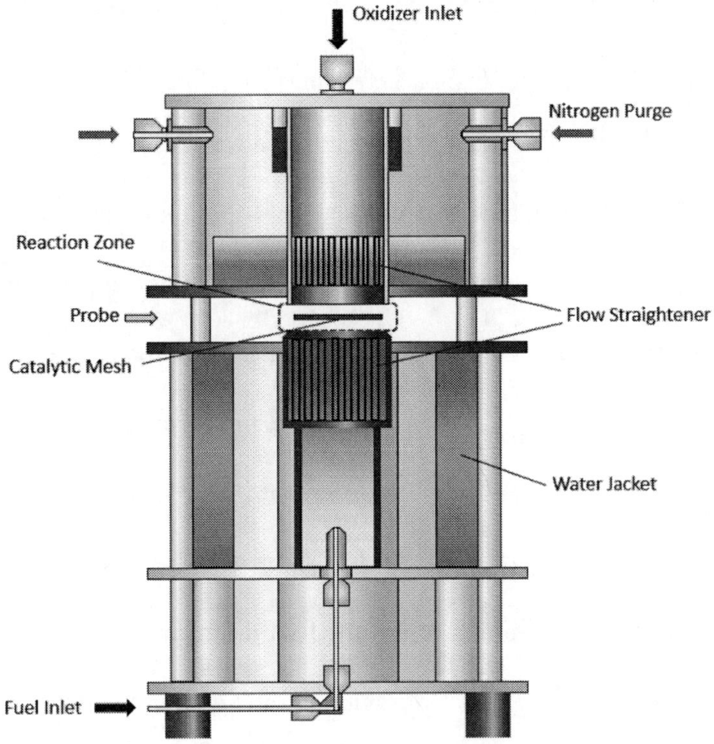

Figure 1. Diagram of the experimental setup.

Centerline temperature profile was measured by a 400 μm OD K-type thermocouple. Centerline species concentration sample was acquired using a 2.0 mm diameter uncooled quartz microprobe with a small orifice inlet. Both measurement probes were control by a linear stage system to adjust their vertical location. This system located the thermocouple or microprobe to the desire centerline axial position in the reaction zone. Reaction productions were detected by a Raman laser gas analyzer (RLGA) system (Atmosphere Recovery Inc., Model RLGA-1800L-BF). RLGA provided detailed speciation of hydrogen, carbon monoxide, carbon dioxide etc. The sampling flow rate was set as 220 mL/min. Other hydrocarbon species were determined experimentally to be negligible, therefore in this research only RLGA was chosen for speciation purposes.

The reactor was started by heating the oxidizer inlet to 450 °C. After the catalytic mesh light-off, the heating element was set to 300 °C, controlled by a temperature controller. The reaction could be operated at a variety of preheat temperatures after the reaction started and heat source could be cut-off; however, the reaction was not stable without an elevated inlet temperature. All reported data in this experiment were run at 300°C oxidizer inlet temperature. Experiments were conducted with 5 mesh positions for one flow condition. The methane fuel flow rate was set to 25 standard liters per minute (SLPM). The oxidizer is a mixture of compressed air diluted with nitrogen that can reduce the mesh temperature, therefore decrease the risk of fuel auto-ignition. The oxidizer flow contained 14% oxygen and 86% nitrogen and overall flow rate was set to 30 SLPM. Condition 1 through condition 5 represent five different mesh locations under the same flow situation. Conditions details, global equivalence ratio ϕ_g and local equivalence ϕ_{local} for this experiment are listed in Table 1.

Table 1. Experimental conditions at different mesh locations.

Condition	Oxidizer flow (SLPM)	Fuel flow (SLPM)	Mesh Pos. (mm)	Local ϕ_{local} *(Pt)*	Global ϕ_g (Pt)
1	30	25	6	58.51	2.33
2	30	25	7	11.45	2.33
3	30	25	8	2.79	2.33
4	30	25	9	0.80	2.33
5	30	25	10	0.26	2.33

3. Results and Discussion

3.1. Speciation Spatial Profile

Mesh location directly affects the mesh stoichiometry in the opposed flow arrangement, and it was found that local equivalence ratio is extremely sensitive to axial position as in our previous study [10]. Mesh stoichiometry plays a critical role in optimizing the reforming products and avoid undesired products such as soot and tar. In this work, we chose maximize carbon monoxide and hydrogen mixture (synthesis gas) production, the key product of catalytic partial oxidization reaction as the desired reforming products. To achieve this goal, product concentration in each condition was measured and studied. Figure 2 depicts the synthesis gas concentration distribution below the catalytic mesh for different mesh location from fuel chamber (mesh location = 0 mm) to oxidizer chamber (mesh location = 15 mm). Gas samples were taken at closed measurement position (1 mm) below the mesh location for each condition.

A CHEMKIN diffusion opposed-flow model was used to provide an estimation for the local equivalence ratio at different conditions. It should be noted, to the authors' knowledge, that a 1D reaction model with detailed gas phase mechanisms and surface catalytic reactions is not readily available and is the focus of our current work. To support the customized 1D reactor model being developed by the authors, it is desirable to establish a simple model in order to estimate local equivalence ratio. For this purpose, modified GRI 30 mechanism without chemical reaction was used in the simple model shown here to provide a preliminary guess of local ϕ based on species transportation/diffusion. Consequently, temperature, speciation concentration and flow profile cannot be precisely calculated with this current model. This model only provided a rough estimate for local equivalence ratio. Table 1 lists the estimated local equivalence ratio (ϕ_{local}) corresponding with mesh location for each test condition, with boundary conditions set based on experiments.

Condition 1 (with the highest ϕ_{local}) and condition 5 (with the lowest ϕ_{local}) had lower production rates of synthesis gas compared to the other conditions. In Figure 2, synthesis gas concentration approached zero at these two conditions since catalytic mesh location was too far away from the stagnation plane. Fuel or oxidizer could not diffuse to the mesh location so that reaction surface didn't have enough fuel or oxidizer to react at these conditions. Furthermore, product concentration at condition 3 (rich $\phi_{local}>1$) rapid dropped to a low level at condition 4 (lean $\phi_{local}<1$). As mentioned in previous discussions, mesh stoichiometry played an important role in the resulting product concentrations. At the flowrates used in this study, stoichiometric ($\phi_{local}=1$) axial position was location between 8 mm and 9 mm above the fuel inlet chamber, so that in a fuel lean conditions such as condition 4 and 5, the mass fraction of methane was too low to maintain an active reforming process to produce high concentration of synthesis gas, rather, the fuel was mostly converted to water vapor and carbon dioxide, indicating a catalytic combustion process. In comparison, syngas production rate approximates the maximum near $\phi_{local}=2.0$ from the ideal partial oxidization reaction $CH_4 + \frac{1}{2}O2 -> 2H_2 + CO$, therefore the rapid drop in syngas production is evident. From this experiment, condition 3 ($\phi_{local}=2.79$) has been found with the optimized synthesis gas production. The authors believe at this point $\phi_{local}\sim 2.0$ and the discrepancy shall be explained by the error introduced by the model. Nevertheless, the conclusion is that the local equivalence ratio can be utilized to estimate the optimized mesh axial location for synthesis gas production.

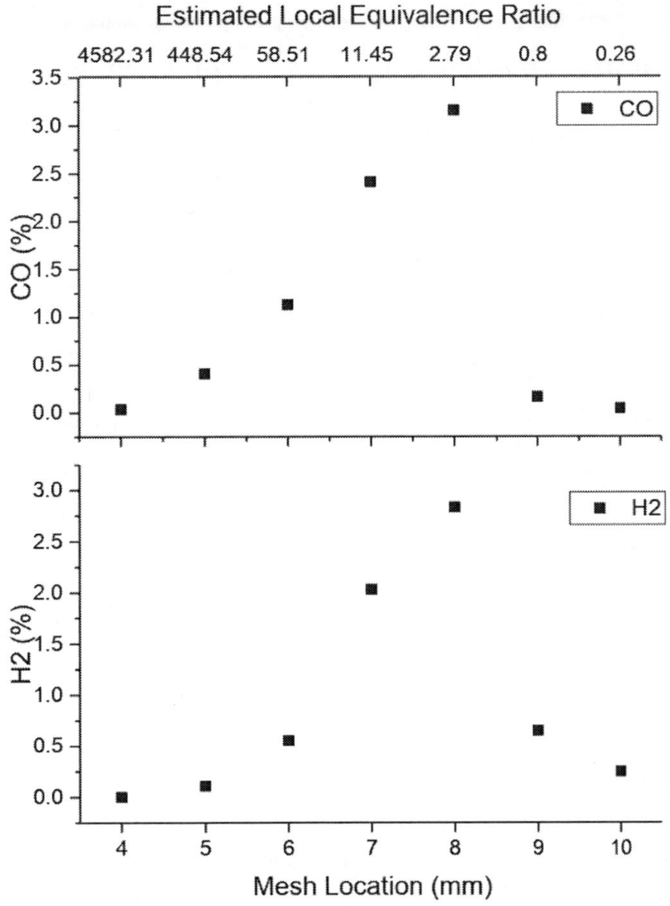

Figure 2. CO and H_2 mole fraction by changing mesh location for platinum catalyst. Speciation concentration for mesh location below 4 mm or above 10 mm approach to zero is not listed in this figure.

To further analyze the syngas production, partial oxidization product concentration profile of condition 3 is shown in Figure 3. In this effort, gas sample was taken at 13 different locations along the center axial with a 1 mm spacing. The hydrocarbon composition in this experiment cannot be clearly identified by RLGA measurement. In some cases, the gas phase reaction produce light hydrocarbon species such as C_2H_6 and C_2H_4 at low temperatures around 650 °C. As mentioned in the work conducted by the Schmidt group [14], the short residence time of reactants increases the rate of the mass transfer-limited to surface reaction so that only limited amount of light hydrocarbon produced in this experiment. In our experiment, most hydrocarbon is considered as unburned methane since opposed flow reactor also have short catalyst residence times, and the production of small amounts of heavier hydrocarbons, is not within the scope of this paper.

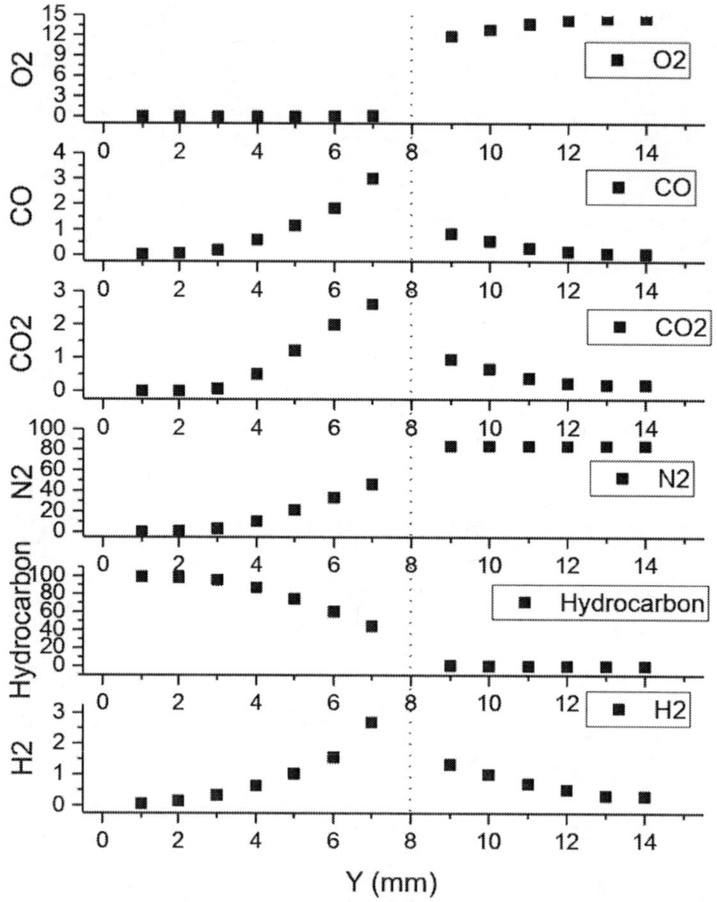

Figure 3. Mole fractions of partial oxidation products as a function of axial distance at the centerline for condition 3.

In the opposed flow reformer, catalyst mesh has a similar behavior as a flame sheet as most reactions occur at or very near the catalytic mesh. From Figure 3, the hydrocarbon and O_2 concentration has a significant decrease while across the catalytic mesh. There is almost no trace of any O2 in the fuel side of the mesh. It presents surface reaction at the mesh dominates in this reactor while the mesh itself serves like a flame sheet in an opposed flame reactor. Moreover, reforming products have their peak concentration next to the catalytic mesh demonstrating that the catalytic reactions on the mesh dominate any gas phase chemistry.

3.2. Temperature profile

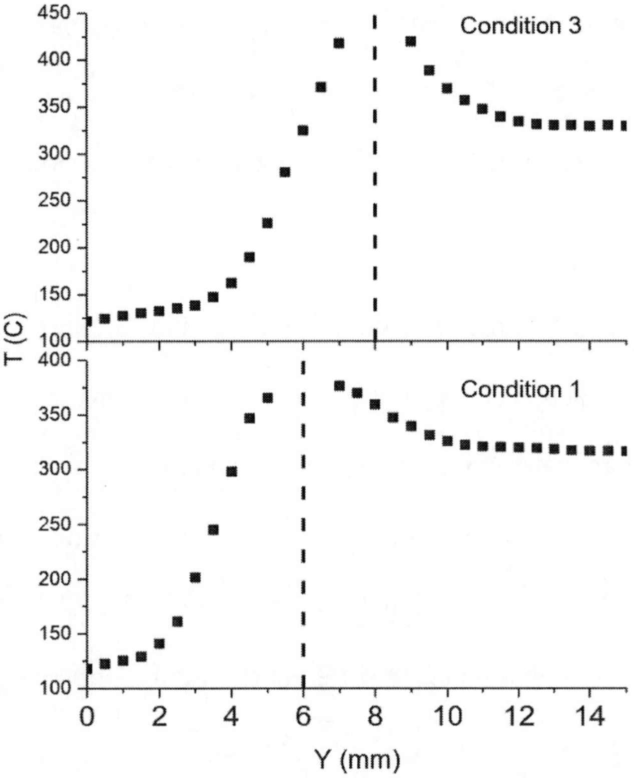

Figure 4. Axial centerline temperature for condition 1 and condition 3.

Temperature data from 28 data points were collected along the center axial of the reaction zone with a 0.5 mm spacing. The oxidizer inlet is preheated to 300 °C and fuel inlet is unheated. Higher temperature tends to produce more synthesis gas during partial oxidization process [14]. In figure 4, two temperature axial profiles are compared. The average peak temperature for condition 3 was

420°C, which was 43°C higher than the average peak temperature of condition 1. As previously discussed, condition 3 had the highest synthesis gas production due to 1) higher reaction temperature than condition 1 and 2) favorable local equivalence ratio at the catalytic mesh.

4. Conclusion

This research investigated a gaseous opposed-flow reactor for reforming gaseous fuel. The reactor was constructed by two gas chambers contain fuel and oxidizer separately. A catalytic mesh was placed between the two inlets. Temperature and speciation data was collected and analyzed. A non-reaction CHEMKIN diffusion opposed-flow model was created to estimate the local equivalence ratio. Optimized mesh location of synthesis gas production was determined by local equivalence ratio ϕ_{local}.

Modeling work to simulate the reactor is ongoing and the authors are establishing a Cantera-based one dimensional opposed flow model, involving the catalytic mesh with surface reactions with a modified Jacobian matrix for the reaction system [15]. Such efforts would greatly improve the current state of knowledge in understanding the catalytic reforming process using the catalytic opposed flow reactor. The model can also determine optimized mesh location, reaction speciation profile and temperature distribution under different reforming scenarios.

Acknowledgements

This research was funded by the National Science Foundation [grant number 1350709].

References:

[1] J.J. Krummenacher, K.N. West, L.D. Schmidt, "Catalytic partial oxidation of higher hydrocarbons at millisecond contact times: decane, hexadecane, and diesel fuel," Journal of Catalysis, 215 (2003) 332-343.

[2] J. R. Rostrup-Nielsen, "New aspects of synthesis gas production and use," Catal. Today, 63, (2000) 159–164.

[3] G. J. Panuccio, B. J. Dreyer, L. D. Schmidt, "A comparison of the catalytic partial oxidation of C1 to C16 normal paraffins," AIChE journal, 53 (2007) 187-195.

[4] A. Prabhu, R. Radhakrishnan, and S. Ted Oyama, "Supported nickel catalysts for carbon dioxide reforming of methane in plug flow and membrane reactors," Appl. Catal. A Gen., 183 (1999) 241–252.

[5] A. M. O'Connor and J. R. H. Ross, "The effect of O2 addition on the carbon dioxide reforming of methane over Pt/ZrO2 catalysts," Catal. Today, 46 (1998) 203–210.

[6] C. R. H. de Smet, M. H. J. M. de Croon, R. J. Berger, G. B. Marin, and J. C. Schouten, "An experimental reactor to study the intrinsic kinetics of catalytic partial oxidation of methane in the presence of heat-transport limitations," Appl. Catal. A Gen., 187 (1999) 33–48.

[7] C. M. Balonek, J. L Colby,, L. D. Schmidt, "Millisecond catalytic reforming of monoaromatics over noble metals," AIChE journal, 56 (2010) 979-988.

[8] J. L Colby, P. J. Dauenhauer, and L. D. Schmidt, "Millisecond autothermal steam reforming of cellulose for synthetic biofuels by reactive flash volatilization," Green Chem., 10, (2008) 773.

[9] T. Aicher and L. Griesser, "Novel process to evaporate liquid fuels and its application to the catalytic partial oxidation of diesel," J. Power Sources, 165 (2007) 210–216.

[10] X. Li, M.V. Twigg, and W. Northrop, "Opposed-Flow Reactive Volatilization Reactor for Reforming Low Volatility Fuels", 36th International Combustion Symposium, Seoul, South Korea, 2016.

[11] J. Hwang, X. Li, and W. Northrop, "Exploration of Dual Fuel Diesel Engine Operation with On-Board Fuel Reforming," (2017).

[12] R. Quiceno, J. Pérez-Ramírez, J. Warnatz, and O. Deutschmann, "Modeling the high-temperature catalytic partial oxidation of methane over platinum gauze: Detailed gas-phase and surface chemistries coupled with 3D flow field simulations," Appl. Catal. A Gen., 303 (2006)166–176.

[13] U. Oemar, Y. Kathiraser, L. Mo, X. K. Ho, and S. Kawi, "CO2 reforming of methane over highly active La-promoted Ni supported on SBA-15 catalysts: mechanism and kinetic modelling," Catal. Sci. Technol., 6 (2015) 1173–1186.

[14] D. A. Hickman and L. D. Schmidt, "Synthesis gas formation by direct oxidation of methane over Pt monoliths," J. Catal., 138 (1992) 267–282.

[15] D. G. Goodwin, "One-Dimensional Flames," Cantera Workshop. (2004).

Western States Section of the Combustion Institute–Fall 2017 Meeting
Hosted by the University of Wyoming
October 2–3, 2017

Preliminary analysis of smoldering combustion in cellulose and hemicellulose mixtures

Tejas Chandrashekhar Mulky[1] *and Kyle E. Niemeyer*[1,*]

[1]*School of Mechanical, Industrial and Manufacturing Engineering, Oregon State University, Corvallis OR, United States*
[*]*Corresponding author:* `kyle.niemeyer@oregonstate.edu`

Abstract: Smoldering combustion contributes significantly to the carbon emissions of wildland forest fires, can persist for long times, and can transition back into flaming combustion, but questions remain about the mechanisms that control the ignition and propagation of smoldering in wood fuels. Wood and forest duff consist of cellulose, hemicellulose, and lignin in various proportions. Here, we use a one-dimensional model for a reactive, porous media developed with the open-source software Gpyro. The boundary condition at the top surface for validation is set by providing the temperatures obtained from experimental measurements, while the bottom surface is kept insulated. We use different kinetic schemes available for cellulose to represent the smoldering chemistry. These include various steps for drying, pyrolysis, and oxidation reactions. The performance of these models are then validated with experimental results. Finally, we examine the effects of physical parameters such as moisture content and density of the fuel on the smoldering propagation speed for these fuels.
Keywords:* *Smoldering, Cellulose, Hemicellulose, Moisture content, Woody fuels

1. Introduction

Smoldering is a flameless, slow, and low-temperature form of combustion. Smoldering combustion can be self-sustained due to heat released when oxygen attacks the solid surface of the fuel [1]. Smoldering combustion is a dominant phenomenon in forest fires. It is considered hazardous because compared with flaming combustion it produces more toxic gases and is difficult to suppress. It also lasts for a long period of time and can initiate flaming combustion even in the presence of a weak heat source [2]. Controlled smoldering combustion has also been used for our benefits. Rein et al. [3] recently demonstrated a technique using controlled smoldering combustion to detect land mines in peat soils.

A self-sustained smoldering combustion occurs in porous fuels which have the tendency to form char when it undergoes thermal degradation [4]. This is because char oxidation is the main source of heat for a self-sustained smoldering [1]. Smoldering combustion has been studied in many fuels like polyurethane foam [5], cellulose [6, 7], peat [8–10], saw dust [11], cigarettes [12], and pine wood [13].

Since smoldering combustion plays a major role in forest fire, study of smoldering combustion in forest duff, peat, and wood becomes important. Cellulose is a major constituent in forest duff, peat, and wood [9, 13, 14]. Hence, it is crucial to understand the smoldering reaction mechanism of cellulose and how well it represents the actual smoldering process. The propagation speed of smoldering depends significantly on reaction rates [7]. Studies like those done by Kashiwagi et al. [15], Di Blasi et al. [7], and Moussa et al. [6] provide various models, and associated reaction parameters, for cellulose smoldering. Other studies have determined reaction parameters for smoldering in peat and pine wood, where the main fuel is represented as cellulose, hemicellulose, and/or lignin [9, 13]. However, a comparison of all these reaction parameters has not been performed to see which represents smoldering in cellulose most accurately. There also have been no computational studies on the effects of moisture content and bulk density of cellulose, on the downward propagation speed of smoldering.

In this paper, we use a one-dimensional model for a reactive, porous media developed with the open-source software Gpyro. We use this model along with four kinetic models available from the literature and compare these against the experimental results. From these comparisons, we select the best-performing model and further examine the sensitivity of smoldering to density and moisture content. In future studies, we will look at the effect of mixing cellulose with hemicellulose on the smoldering propagation speed.

2. Computational model

In the experiment, cellulose was loaded in the container with the dimensions $0.2 \times 0.2 \times 0.0875\,\mathrm{m}^3$; see Cowan et al. [16] for additional details on the experimental configuration. The top surface was open to the atmosphere while the bottom surface was insulated. In this paper, we will be investigating the downward propagation of smoldering. Hence, the computational model developed is one dimensional with a computational domain of 0.0875 m. In this model, the condensed phase and gas phase are assumed to have the same temperature. The shrinkage of the sample during the process is taken into consideration. All the simulations were run with an initial time-step size of 0.02 s and cell size of 10^{-4} m. These values were selected to keep a balance between accuracy and simulation time.

2.1 Governing equations

Smoldering combustion is modeled using the open-source software Gpyro [17]. The one-dimensional governing equations used in this model include condensed-phase mass conservation (1), gas-phase mass conservation (2), condensed-phase species conservation (3), gas-phase species conservation (4), condensed-phase energy equation (5), and gas-phase momentum equation (6), shown below. Lautenberger and Fernandez-Pello [17] provide more information about these equations.

$$\frac{\partial \overline{\rho}}{\partial t} = -\dot{\omega}'''_{fg} \quad (1)$$

$$\frac{\partial (\rho_g \overline{\psi})}{\partial t} + \frac{\partial \dot{m}''}{\partial z} = \dot{\omega}'''_{fg} \quad (2)$$

$$\frac{\partial (\overline{\rho} Y_i)}{\partial t} = \dot{\omega}'''_{fi} - \dot{\omega}'''_{di} \quad (3)$$

$$\frac{\partial (\rho_g \overline{\psi} Y_j)}{\partial t} + \frac{\partial (\dot{m}'' Y_j)}{\partial z} = -\frac{\partial}{\partial z}(\overline{\psi} \rho_g D \frac{\partial Y_j}{\partial z}) + \dot{\omega}'''_{fj} - \dot{\omega}'''_{dj} \quad (4)$$

$$\frac{\partial (\overline{\rho h})}{\partial t} + \frac{\partial (\dot{m}'' h_g)}{\partial z} = \frac{\partial}{\partial z}(\overline{k} \frac{\partial T}{\partial z}) + \sum_{k=1}^{K} (-\dot{\omega}'''_{di,k}) \Delta H_k \quad (5)$$

$$\dot{m}'' = -\frac{\overline{K}}{\nu} \frac{\partial P}{\partial z}, \quad (6)$$

where T is the temperature, t is the time, Y is mass fraction, $\dot{\omega}'''$ is reaction rate, ρ is the bulk density, z is the depth, $\dot{m}''$ is the mass flux, h is the enthalpy, k is thermal conductivity, and subscripts f, d, i, j are formation, destruction, condensed-phase species and gas-phase species, respectively.

2.2 Boundary conditions

The ambient pressure (P_∞) and temperature (T_∞) were set to 1 atm and 293 K, respectively. To validate the model against the experiments, the boundary condition at the top surface ($z = 0$ m) is specified by setting the temperature equal to the experimentally obtained thermocouple values at the top surface, shown in equation (7). After validation, we will look at the effect of density and moisture content on smoldering. For those cases, the boundary condition at top surface is specified by applying a heat flux ($\dot{q}''$) represented by the equation (8). After applying the heat flux a convective-radiative heat balance is applied to the top surface for the remaining simulation time which is represented by equation (9). The pressure at the top surface is set equal to the ambient pressure which is 1 atm. The convective heat transfer coefficient used in the simulation is $h_{c0} = 1.52 \Delta T^{1/3} \approx 10\,\mathrm{W/m^2K}$ [9]. The mass transfer coefficient (h_{m0}) is set to $0.02\,\mathrm{kg/m^2s}$ [8], while oxygen and nitrogen mass fraction are fixed to 0.23 and 0.77, respectively. The emissivity (ε) is set to 0.95. The equations of the boundary conditions at the top surface are shown below.

$$T|_0 = T_{\mathrm{exp}} \quad (7)$$

$$-\overline{k} \frac{\partial T}{\partial z}|_0 = -h_{c0}(T_0 - T_\infty) + \overline{\varepsilon} \dot{q}''_e - \overline{\varepsilon} \sigma (T|_0^4 - T_\infty^4)] \quad (8)$$

$$-\overline{k} \frac{\partial T}{\partial z}|_0 = -h_{c0}(T_0 - T_\infty) - \overline{\varepsilon} \sigma (T|_0^4 - T_\infty^4) \quad (9)$$

$$-(\overline{\psi} \rho_g D \frac{\partial Y_j}{\partial z})|_0 = h_{m0}(Y_{j\infty} - Y_j|_0) \quad (10)$$

$$P|_0 = P_\infty \quad (11)$$

The convective heat transfer coefficient (h_{cL}) at the bottom surface is set to $3\,\mathrm{W/m^2K}$ [8]. This takes into account the small amount of heat transfer across the insulated wall. The mass transfer coefficient (h_{mL}) and mass flux ($\dot{m}''_L$) is set to 0. The equation used for the boundary conditions at the bottom surface is stated below.

$$-\bar{k}\frac{\partial T}{\partial z}|_L = -h_{cL}(T|_L - T_\infty) \tag{12}$$

$$-(\overline{\psi}\rho_g D\frac{\partial Y_j}{\partial z})|_L = h_{mL}(Y_{j\infty} - Y_j|_L) \tag{13}$$

$$\dot{m}''|_L = 0 \tag{14}$$

2.3 Physical properties

The model used in this work considers three condensed-phase species: cellulose, char, and ash. The kinetic model of Huang et al. [9] has two types of char: alpha-char and beta-char. The alpha-char is obtained from cellulose pyrolysis while beta-char is obtained from oxidative degradation of cellulose. However, the properties of alpha-char and beta-char are assumed to be the same [9]. Table 1 provides the physical properties of the condensed-phase species, including bulk density (ρ), solid density (ρ_s), thermal conductivity (k), and heat capacity (c_p). The bulk density of cellulose was measured experimentally [16]. The bulk density of char and ash were calculated using the relations $\rho_{\text{char}} = 0.25 \times \rho_{\text{cellulose}}$ [18] and $\rho_{\text{ash}} = \text{IC}/100 \times 10 \times \rho_{\text{cellulose}}$, where IC stands for inorganic content which is taken as 0.3 [19].

Species	Bulk density (kg/m^3)	Solid density (kg/m^3)	Thermal conductivity (W K/m)	Heat capacity (J K/kg)	Source
Cellulose	188.57	1500	0.356	1674	[20]
Char	47.1425	1300	0.26	1260	[9, 21]
Ash	5.5	2500	1.2	880	[9, 21]

Table 1: Thermophysical properties of condensed phase species

Porosity (ψ_i) and effective thermal conductivity (k_i) are calculated as $\psi_i = 1 - \frac{\rho_i}{\rho_{s,i}}$ and $k_i = k_{s,i}(1-\psi_i) + \gamma_i \sigma T^3$, respectively, where γ is the parameter controlling the radiation heat transfer across pores [9, 17]. The pore diameter (d_p), permeability (K), and the parameter controlling the radiation heat transfer across pores (γ) are calculated using the equations (15), (16), and (17), respectively, which were obtained from [9, 22, 23]:

$$d_{p,i} = \frac{1}{S_i \times \rho} \tag{15}$$

$$K \approx 10^{-3} \times d_{p,i}^2 \tag{16}$$

$$\gamma_i \approx 3 \times d_{p,i}\,, \tag{17}$$

where S is the particle surface area. The value of $S_{\text{cellulose}} = S_{\text{char}} = 0.024$ m^2/g and $S_{\text{ash}} = 0.096$ m^2/g [9, 24, 25]. Table 2 lists the calculated values obtained for d_p, K, and γ.

Condensed-phase species	$d_p \times 10^{-4}$ (m)	$K \times 10^{-11}$ (m^2)	$\gamma \times 10^{-4}$ (m)
Cellulose	2.2	4.8	6.6
Char	8.8	77	26
Ash	19	359	57

Table 2: Pore diameter, permeability, and the parameter controlling the radiation heat transfer across pores of condensed phase species.

2.4 Chemical kinetics

The smoldering reaction mechanism, in general, involves three-step global reactions which include pyrolytic (endothermic) and oxidative (exothermic) degradation of fuel, both of which give char as a product, and oxidation of that char (exothermic) which gives ash. The smoldering reaction mechanism also includes a drying step if there is a significant amount of moisture content in it. However, in our case, the experiments were performed after oven-drying the sample [16]. Hence, we can assume that the moisture content was negligible in the sample. Chemical kinetic parameters, which include pre-exponential factor (Z), activation energy (E), heat of reactions (ΔH), order of reaction (n), and (n_{O_2}), were all obtained from the literature.

We considered four kinetic schemes here, and Table 3 lists all the relevant parameters. The values used for stoichiometric coefficient (υ) were taken from that of Kashiwagi and Nambu [15], where υ_{char} obtained from pyrolysis reaction of cellulose is 0.24 kg/kg, υ_{char} obtained from oxidative degradation of cellulose is 0.21 kg/kg, and υ_{ash} obtained from char oxidation is 0.03 kg/kg. Value for $\upsilon_{O_2,k}$ consumed is calculated using the relation $\upsilon_{O_2,k} = \Delta H/(-13.1)$ MJ/kg [9, 26].

3. Results and discussion

Simulations were run using the four sets of reaction parameters, which are listed in Table 3. We validated the model by comparing the temperature profile and smoldering wave propagation speed against the experimentally obtained values. The propagation speed was calculated by taking the derivative of the location of peak temperature with respect to time. Since the experiment recorded temperature profile at discrete depths 0, 2.5, 5, and 7.5 cm, there are three propagation speeds (v_1, v_2, v_3) available to compare. After the validation, the best model is selected and the sensitivity of smoldering to density and moisture content is further examined.

3.1 Comparison against experimental results

Figure 1 shows the temperature profiles obtained from the model for the above-mentioned depths using the four kinetic schemes of Table 3, along with the experimentally obtained profiles. From Figure 1, it can be seen that the temperature profiles obtained from the experiments is qualitatively similar to the temperature profile obtained from the models which used reaction parameters from Di Blasi [7], Moussa et al. [6], and Huang and Rein [9] compared with the temperature profiles obtained from the model which used reaction parameters from Kashiwagi and Nambu [15].

Table 4 lists the values of the smoldering wave propagation speeds obtained from the models,

Kashiwagi and Nambu [15]					
Reaction	$\log(Z)(\log(\min^{-1}))$	E (kJ/mol)	ΔH (MJ/kg)	n	n_{O_2}
cellulose pyrolysis	19.08	220	0.57	1.8	–
cellulose oxidation	14.18	160	-5.7	1.3	0.5
char oxidation	11.53	160	-25	1	0.78
Di Blasi [7]					
Reaction	$\log(Z)(\log(s^{-1}))$	E (kJ/mol)	ΔH (MJ/kg)	n	n_{O_2}
cellulose pyrolysis	17.3	220	0.68	1	–
cellulose oxidation	15.4	170.4	-5.7	1	1
char oxidation	12.7	125.7	-25	1	1
Moussa et al. [6, 27]					
Reaction	$\log(Z)(\log(s^{-1}))$	E (kJ/mol)	ΔH (MJ/kg)	n	n_{O_2}
cellulose pyrolysis	6.8	108.8	0.35	1	–
cellulose oxidation	16.28	226	-1.8	1	1
char oxidation	12.38	184	-32.2	1	1
Huang and Rein [9]					
Reaction	$\log(Z)(\log(s^{-1}))$	E (kJ/mol)	ΔH (MJ/kg)	n	n_{O_2}
cellulose pyrolysis	11.7	156	0.5	1	–
cellulose oxidation	24.2	278	-28.2	1.73	0.74
α-char oxidation	7.64	120	-28.8	1.25	0.89
β-char oxidation	12.2	177	-27.8	0.93	0.52

Table 3: Reaction parameters from literature

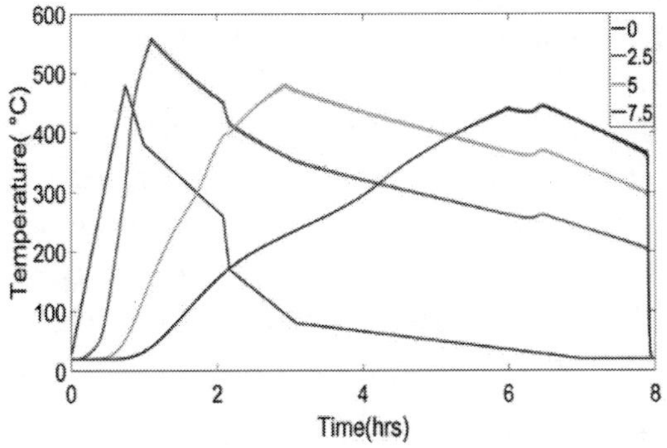

(a) Di Blasi parameters [7]

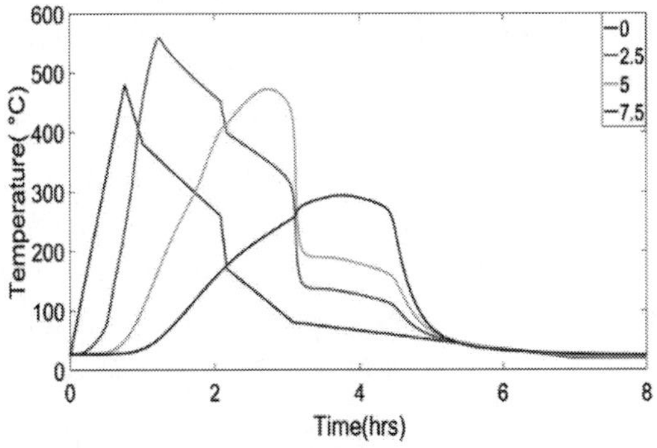

(b) Kashiwagi and Nambu parameters [15]

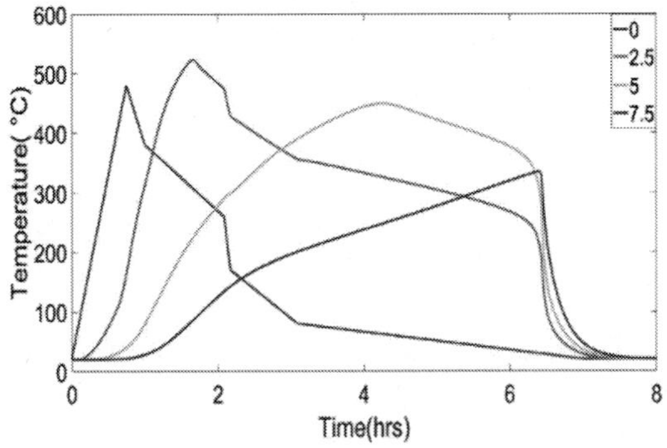

(c) Moussa et al. parameters [6]

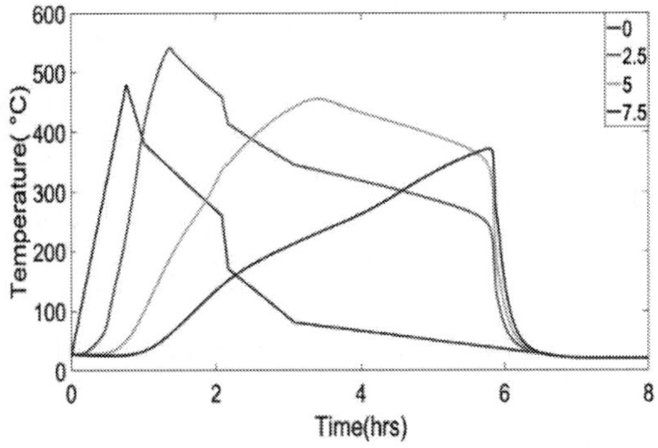

(d) Huang and Rein parameters [9]

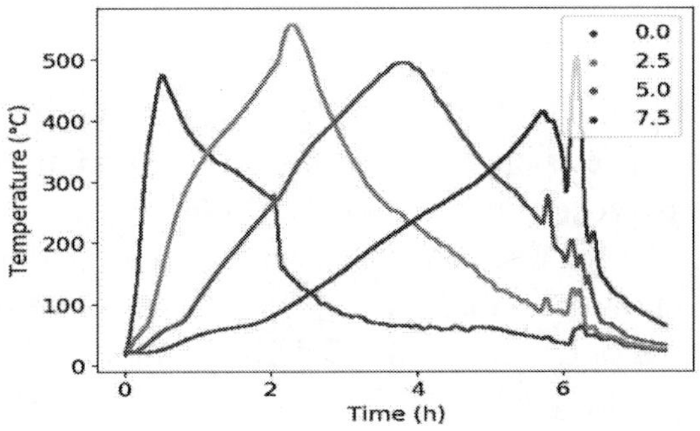

(e) Experimental

Figure 1: Temperature profile at the depth 0, 2.5, 5, and 7.5 cm from $z = 0$ using reaction parameters from various references and experiment.

along with the experimental values. From Table 4 we can see that the values obtained for v_3 from the model which used reaction parameters from Moussa et al. [6], and Huang and Rein [9] closely match the experimental values. But, the values obtained for v_1 from the model are higher compared with the experimental value, which was expected since the peak at the depth 2.5 cm for all the reaction scheme appeared earlier than experiment as we can see in Figure 1. The reason for this could be the presence of moisture content or some other impurities in the sample which are not accounted in the model. There also could be uncertainty in the measurement of temperature at z=0 since, measurement were not taken directly below the ignition source. On comparing the average values propagation speed, among all the reaction parameters used in this paper, the model using the reaction parameters from Moussa et al. [6] closely represents the experiments, whereas the model using the reaction parameters from Kashiwagi and Nambu [15] is the least accurate. Hence, the model using the reaction parameters from Moussa et al. [6] is next used to examine the effect of bulk density and moisture content on smoldering in cellulose.

Reference	Propagation speed (cm/s)			
	v_1	v_2	v_3	v_{avg}
[15]	0.0014	0.00046	0.00068	0.00086
[7]	0.00195	0.00038	0.0002	0.00084
[6]	0.00077	0.00027	0.00032	0.00045
[9]	0.0011	0.00034	0.00029	0.00059
Experiment	0.00039	0.0004	0.00029	0.00037

Table 4: Propagation speed obtained from different reaction scheme along with experimental values.

3.2 Sensitivity to fuel density

The effect of bulk density was examined, by systematically increasing the bulk density of cellulose from 188.57 kg/m^3 to 400, 600, and 800 kg/m^3, to see its effect on the temperature profile, propagation speed, and maximum temperature. For this set of simulations, the boundary condition at the top surface was set by applying a heat flux of 15 kW/m^2 for the first 30 min after which the heat flux was linearly reduced to 0, in the span of the next 10 min, and then a convective-radiative balance was applied until the end of simulation. Figure 2 shows the effect of bulk density on the temperature profile at the depths of 0, 2.5, 5, and 7.5 cm. When the bulk density of cellulose is increased, the peak temperature at all depths increases except at $z = 0$. The peak temperature at the top surface was unaffected by the density of the fuel.

Table 6 shows that, as the bulk density of the fuel increases, the maximum temperature attained by smoldering combustion increases while propagation speed decreases.

3.3 Sensitivity to moisture content

Moisture content (MC) is the ratio of the mass of water to the mass of cellulose [8]. It is assumed that when water is mixed with cellulose, water only occupies the pores, thus the volume

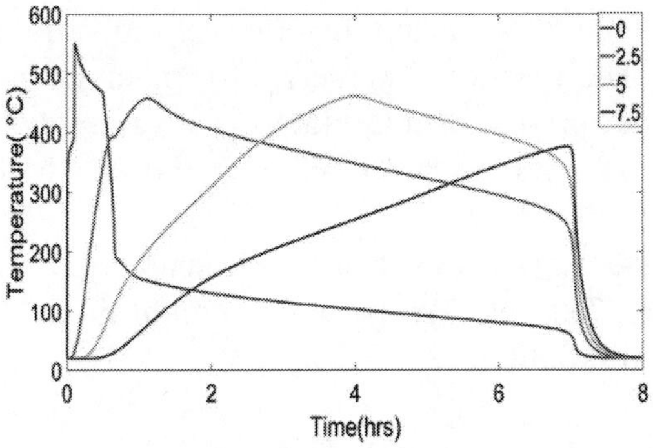

(a) Bulk density of 188.57 kg/m^3

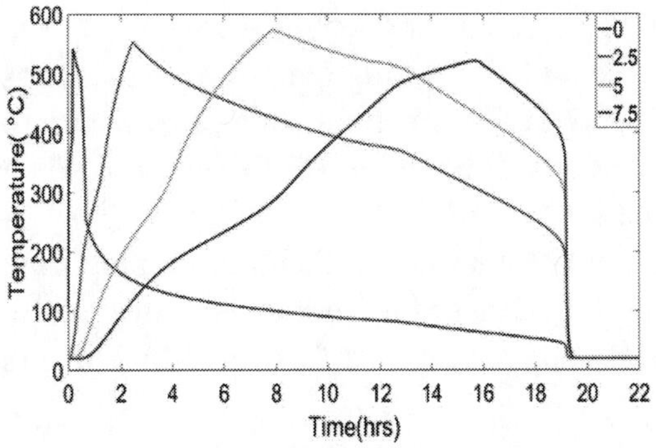

(b) Bulk density of 400 kg/m^3

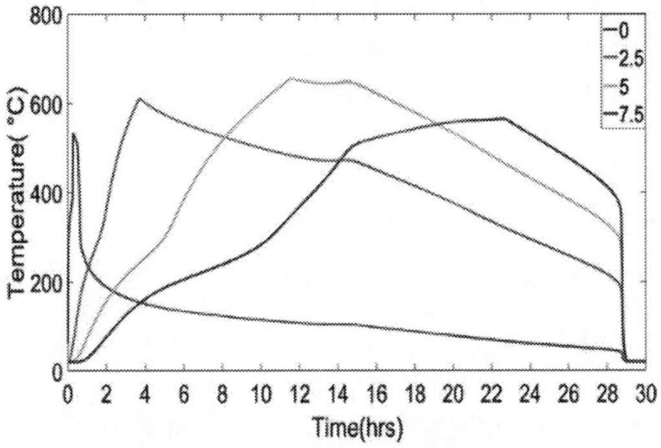

(c) Bulk density of 600 kg/m^3

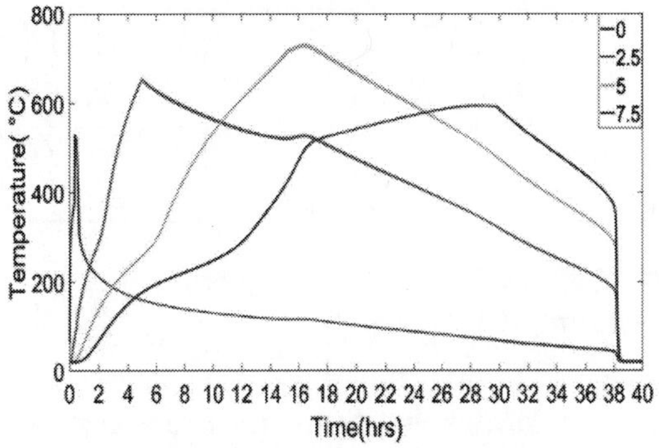

(d) Bulk density of 800 kg/m^3

Figure 2: Temperature profile at the depth 0, 2.5, 5, and 7.5 cm from $z = 0$ for increasing values of bulk density.

Density (kg/m^3)	Propagation speed ($\times 10^{-3}$ cm/s)				T_{max} (°C)
	v_1	v_2	v_3	$v_{average}$	
188.57	0.7	0.24	0.23	0.39	551
400	0.3	0.13	0.089	0.173	573
600	0.2	0.088	0.063	0.117	657
800	0.15	0.061	0.057	0.089	733

Table 5: Effect of bulk density on propagation speed and maximum temperature for the reaction scheme [6].

of the mixture does not change [28]. The bulk density of the mixture of water and cellulose is $\rho_{mix} = \rho_{cellulose}(1+MC)$ [28]. The values of thermal conductivity (k), specific heat capacity (c_p), solid density (ρ_s), and the bulk density of water (ρ) are taken as 0.6 W K/m, 4186 J K/kg, 1000 kg/m^3, and 1000 kg/m^3, respectively [9, 21]. To take into account the effect of moisture content, we added a drying step to our reaction scheme. The reaction parameters for the drying step were obtained from Huang et al. [8], where the values of pre-exponential factor, activation energy, order of reaction, and heat of reaction (ΔH) are 8.12 log(s^{-1}), 67.8 kJ/mol, 2.37, and 2.26 MJ/kg, respectively.

Simulations were run with increasing moisture content from 0–60 % in increments of 20 % and fixing the value of the bulk density of cellulose to 188.57 kg/m^3. The boundary condition at $z=0$ is set by applying a heat flux of 15 kW/m^2 during the simulation.

MC (%)	Propagation speed ($\times 10^{-3}$ cm/s)				T_{max} (°C)
	v_1	v_2	v_3	$v_{average}$	
0	0.70	0.31	0.22	0.41	607
20	0.65	0.29	0.22	0.39	589
40	0.61	0.16	0.44	0.40	579
60	0.58	0.14	0.50	0.41	577

Table 6: Effect of moisture content on propagation speed and maximum temperature for the reaction scheme [6]

Table 6 shows that as the moisture content increases, the propagation speeds v_1 and v_2 decrease while v_3 increases. However, the average value of the propagation speed is almost constant, which means that overall propagation speed is unaffected by changing the moisture content from 0–60 %. Further investigation is required to determine the reason for different trends in the local velocity. One reason could be the coarseness of the data points (i.e., temperature at four depths were used), selected to match the experimental data.

4. Conclusions and future work

In this paper, a one-dimensional model was developed using the open-source software Gpyro. Using this model, along with the different kinetic parameters available for smoldering reaction mechanisms, temperature profiles were obtained at depths of 2, 5, and 7.5 cm, which were then used to calculate the smoldering propagation speed. Validation of the model was performed by comparing the propagation speed against the experimental values. When the model was used with reaction parameters obtained from Moussa et al. [6], the average propagation speed more closely matched the experimental values, compared with the other reaction schemes from the literature. The kinetic model of Moussa et al. [6] was then used to study the sensitivity of smoldering combustion to bulk density and moisture content of the fuel. Downward propagation speed of smoldering decreases and the maximum temperature increases with increasing bulk density of fuel. When the moisture content increases from 0–60 %, the maximum temperature drops but moisture content had no significant effect on the average propagation speed.

Future work will involve looking at the effects of fuel composition (cellulose, hemicellulose, and lignin) on the smoldering process, and we will more closely investigate the effects of moisture content.

Acknowledgements

This research was funded by the Strategic Environmental Research and Development Program (SERDP) award RC-2651 under contract number W912HQ-16-C-0045. I would like to thank Dr. David Blunck, Benjamin Smucker and Daniel Cowan from Oregon State University for providing their experimental temperature measurements data.

References

[1] T. J. Ohlemiller, "Smoldering combustion", in: SFPE Handbook of Fire Protection Engineering, Third Edition, ed. by P. J. DiNenno, D. Drysdale, C. L. Beyler, and W. D. Walton, NIST, 2002, chap. 2, pp. 2/200–210, URL: http://fire.nist.gov/bfrlpubs/fire02/art074.html.

[2] G. Rein, "Smoldering combustion", in: SFPE Handbook of Fire Protection Engineering, Fifth Edition, Springer, 2016, pp. 581–603, DOI: 10.1007/978-1-4939-2565-0_19.

[3] G. Rein, X. Huang, F. Restuccia, and T. McArdle, Detection of landmines in peat soils by controlled smouldering combustion: Experimental proof of concept of *O-Revealer*, Experimental Thermal and Fluid Science 88 (2017) 632–638. DOI: 10.1016/j.expthermflusci.2017.07.016.

[4] D. Drysdale, An Introduction to Fire Dynamics, 3rd ed., Wiley, UK, 2011.

[5] G. Rein, A. C. Fernandez-Pello, and D. L. Urban, Computational model of forward and opposed smoldering combustion in microgravity, Proceedings of the Combustion Institute 31 (2007) 2677–2684. DOI: 10.1016/j.proci.2006.08.047.

[6] N. A. Moussa, T. Y. Toong, and C. A. Garris, Mechanism of smoldering of cellulosic materials, Symposium (International) on Combustion 16 (1977) 1447–1457. DOI: 10.1016/S0082-0784(77)80427-X.

[7] C. Di Blasi, Mechanisms of Two-Dimensional Smoldering Propagation Through Packed Fuel Beds, Combustion Science and Technology 106 (1995) 103–124. DOI: 10.1080/00102209508907769.

[8] X. Huang, G. Rein, and H. Chen, Computational smoldering combustion: Predicting the roles of moisture and inert contents in peat wildfires, Proceedings of the Combustion Institute 35 (2015) 2673–2681. DOI: 10.1016/j.proci.2014.05.048.

[9] X. Huang and G. Rein, Thermochemical conversion of biomass in smouldering combustion across scales: The roles of heterogeneous kinetics, oxygen and transport phenomena, Bioresource Technology 207 (2016) 409–421.

[10] R. M. Hadden, G. Rein, and C. M. Belcher, Study of the competing chemical reactions in the initiation and spread of smouldering combustion in peat, Proceedings of the Combustion Institute 34 (2013) 2547–2553.

[11] K. Palmer, Smouldering combustion in dusts and fibrous materials, Combustion and Flame 1 (1957) 129–154.

[12] A. Rostami, J. Murthy, and M. Hajaligol, Modeling of a smoldering cigarette, Journal of Analytical and Applied Pyrolysis 66 (2003) 281–301.

[13] A. Anca-Couce, N. Zobel, A. Berger, and F. Behrendt, Smouldering of pine wood: Kinetics and reaction heats, Combustion and Flame 159 (2012) 1708–1719. DOI: 10.1016/j.combustflame.2011.11.015.

[14] K. Miyanishi, "Duff Composition", in: Forest Fires: Behavior and Ecological Effects, Academic Press, 2001, pp. 437–475.

[15] T. Kashiwagi and H. Nambu, Global kinetic constants for thermal oxidative degradation of a cellulosic paper, Combustion and Flame 88 (1992) 345–368. DOI: 10.1016/0010-2180(92)90039-R.

[16] D. A. Cowan, B. D. Smucker, and D. L. Blunck, Sensitivity of Smoldering Combustion to Cellulose and Hemicellulose Content, 10th US Combustion meeting (2017) 1–8.

[17] C. Lautenberger and C. Fernandez-Pello, Generalized pyrolysis model for combustible solids, Fire Safety Journal 44 (2009) 819–839. DOI: 10.1016/j.firesaf.2009.03.011.

[18] R. E. Mark, Handbook of physical and mechanical testing of paper and paperboard, 2 vols., Dekker, 1983.

[19] X. Huang, Personal communication, 2017.

[20] C-Therm Technology, C-Therm Technology, http://ctherm.com/products/tci_thermal_conductivity/helpful_links_tools/thermal_physical_properties_conductivity_effusivity_heat_capacity_density2/, Accessed: 2017-08-29.

[21] R. T. Jacobsen, E. W. Lemmon, S. G. Penoncello, Z. Shan, and N. T. Wright, "Thermophysical Properties of Fluids and Materials", in: Heat Transfer Handbook, ed. by A. Bejan and A. D. Kraus, vol. 1, John Wiley & Sons, 2003, chap. 2.

[22] F. Yu, G. Wei, X. Zhang, and K. Chen, Two effective thermal conductivity models for porous media with hollow spherical agglomerates, International Journal of Thermophysics 27 (2006) 293–303.

[23] B. Punmia and A. K. Jain, Soil mechanics and foundations, Laxmi Publication Pvt Limited, 2005.

[24] Sigma-Aldrich, Sigma-Aldrich, http://www.sigmaaldrich.com/catalog/product/sigma/c8002?lang=en&region=US, Accessed: 2017-08-29.

[25] M. Schure, P. A. Soltys, D. F. Natusch, and T. Mauney, Surface area and porosity of coal fly ash, Environmental science & technology 19 (1985) 82–86.

[26] C. Huggett, Estimation of rate of heat release by means of oxygen consumption measurements, Fire and Materials 4 (1980) 61–65.

[27] N. Moussa, Mechanism of Smouldering Combustion in Cellulosic Materials, PhD thesis, MIT, 1976.

[28] X. Huang and G. Rein, Computational study of critical moisture and depth of burn in peat fires, International Journal of Wildland Fire 24 (2015) 798–808.

Topic: Solid, Heterogeneous

Western States Section of the Combustion Institute–Fall 2017 Meeting
Hosted by the University of Wyoming
October 2–3, 2017

Effects of Fuel Composition and Packing Density on Horizontal Smoldering Propagation

Benjamin D. Smucker, Daniel A. Cowan, and David L. Blunck

School of Mechanical, Industrial,and Manufacturing Engineering, Oregon State University, Corvallis, OR 97331, USA

Abstract: Smoldering combustion of natural fuels can be hazardous for both ecosystems and humans by destroying plant litter, transitioning to flaming combustion, and releasing significant quantities of pollution. Understanding the parameters that control smoldering is necessary to help mitigate these hazards. Smoldering has been shown to be sensitive to different fuel compositions at different densities. In this paper, the packing density and fuel composition were systematically varied to experimentally identify their combined effects on horizontal smoldering propagation. The horizontal smoldering propagation velocity was determined at constant packing densities for fuels with a cellulose content ranging from 50% to 100% by mass (with the remainder being hemicellulose). Additionally, fuels with cellulose contents of 50%, 75%, and 100% were burned with densities ranging from 200 to 400 kg/m^3 The fuel was burned in an insulated reactor box. An infrared (IR) camera was used to measure the propagation velocity. At constant packing density, the propagation velocity increased as cellulose content decreased. The largest sensitivities in propagation velocity to cellulose content occurred at higher cellulose contents. The propagation velocity decreased with respect to packing density when the fuel content was constant. The propagation velocity varied more as the density changed for fuels with more cellulose. These trends are attributed to changes in oxygen availability, heat loss, and activation energy.
Keywords: ***Wildland Fires, Smoldering, Solid Combustion***

1. Introduction

Smoldering combustion is the slow, flameless, low-oxygen concentration oxidation of fuel. Smoldering fires in plant litter poses several hazards that are different than those of flaming combustion [1]. Smoldering fires can release larger amounts of smoke, haze, and pollutants than those of flaming combustion due to the lower temperatures and incomplete combustion. In particular, smoldering releases larger amounts of carbon monoxide (CO) and volatile organic compounds (VOCs). [2–6] Smoldering can also produce large amounts of greenhouse gases, such as in 1997 the carbon emissions from smoldering peatland fires in Indonesia were equivalent to at least 13% of the annual global carbon emissions [7]. Smoldering can persist for long periods of time, burning the plant litter layer and destroying vegetation[8]. While smoldering combustion persists there is a chance for it to transition to flaming combustion [1].

Most smoldering combustion research of plant litter has been performed with peat. Frandsen and Huang found that combustion limits of peat are dependent on both moisture content and inorganic content [9–11]. Increased moisture and increased packing density both resulted in a decrease in the smoldering propagation velocity. However, sensitivities to density changed as moisture content changed [12], indicating that a coupling can occur between parameters that affect the propagation velocity.

Smoldering characteristics can change significantly with the origin of plant litter [9, 13, 14]. This can make it challenging to forecast smoldering characteristics if the specific fuel and the specific conditions have not been studied. For example, samples of plant litter were collected from across North America and tested for the moisture content that limits sustainable smoldering. The limiting moisture content varied from 40% to higher than 100% depending on the plant litter [9]. Additionally, the smoldering propagation in cotton at 100 kg/m^3 was about 1.5 mm/min, but dry peat at a similar density had propagation velocities that were about half of that [12, 15]. One notable difference within plant litter is the chemical composition. The three primary organic compounds in plant litter are cellulose, hemicellulose, and lignin. The ratio of each component in the litter changes with the physical location and amount of time it has decomposed [16]. Numerous studies have performed thermograviemetric analysis on these constituents in a N_2 environment. Of the three, lignin has the slowest pyrolysis rate and reacts over the largest temperature range[17, 18]. Hemicellulose and cellulose have similar in pyrolysis rates and temperature ranges, but hemicellulose has a lower activation energy and begins pyrolysis at a lower temperature. [19, 20]. Despite the pyrolysis differences, little research has been performed to identify the smoldering characteristics of these fuel constituents. Such studies are needed to help develop a more universal understanding of smoldering characteristics which can be applied to a broader range of plant litter. Surrogates for natural fuels can be developed, just as surrogates have been used for other fuels such as gasoline and jet fuels [21, 22].

Motivated by the significance of smoldering combustion and need to understand the smoldering characteristics of fuel constituents, Cowan and coworkers [23] burned binary samples of cellulose and hemicellulose. The fuel content was varied from 100% to 0% cellulose. The propagation velocity increased roughly linearly by about a factor of three from 0% to 100% cellulose. It is noted that the fuel was loosely packed. As a consequence, the density of the hemicellulose test was about four times that of the cellulose test. Thus, changes in the propagation velocity, attributed to changes in fuel chemistry, could be caused by changes in density. Increasing packing density caused a decrease in the smoldering propagation in cotton and peat [12, 15, 24].

With this background and motivation, the objective of this study is to identify the effects of packing density and fuel composition on the propagation velocity of smoldering combustion. This knowledge can be used to help validate numerical models and provide insights into the smoldering sensitivity to density and fuel constituents.

2. Methods/Experimental

The experimental arrangement (shown in Figure 1) consisted of a reactor box and an infrared (IR) camera with a mirror. The reactor box contained the powderized fuel (cellulose/hemicellulose mixtures). After the fuel was ignited the IR camera (FLIR SC6700) captured images once every ten seconds with an integration time of 0.00048 ms. The images were reflected off of a stainless steel mirror angled at an angle of 45° placed 75 cm above the reactor box, with the camera being at a fixed location approximately 100 cm from the mirror.

Figure 1: Experimental arrangement for measuring the smoldering propagation velocities during smoldering combustion.

The fuel samples were prepared with the desired fuel composition and packing density. The cellulose was α-cellulose (Sigma-Aldrich; CAS#: 9004-34-6). Powderized glucomannan extracted from konjac root represented the hemicellulose (Nutricost; CAS#: 37220-17-0). The fuel content was determined by weighing the mass of each fuel component. The fuel was mixed using a tumbler for at least five minutes to obtain a uniform fuel distribution throughout the sample. Initially, the reactor box was loosely filled with the desired mass of fuel. Then the fuel was compressed until it was level with the sides of the (20 x 20 x 5 cm deep) reactor box so that the density of the fuel was known.

To limit density effects and isolate chemistry effects, the density was held constant for samples with fuel contents ranging from 50% to 100% cellulose by mass. A second set of testing identified density effects. Fuels with 50%, 75%, and 100% were tested over a range of densities.

The fuel was ignited in the center of the reactor using a 20 Watt, 0.64 cm diameter cartridge heater. The cartridge heater was powered via a temperature controller. The heater was placed into the fuel once it reached 400° C and removed once the fuel was ignited. Any fuel that stuck to the heater was scraped into the center if the box. The fuel was allowed to burn until it reached the edges of the reactor box.

The photon count data from the IR camera was used to determine the smoldering propagation velocities. In an image, the fuel was considered ignited at a particular location if the photon count of a pixel surpassed 1800. This value was selected because it is the approximate value where the smoldering front would self-propagate, as lower values would extinguish. The area of the burn was determined for each image based on the number of pixels above the threshold and the spatial calibration obtained using an IR image of a heated ruler placed on the fuel sample. Once a pixel was ignited, it was considered ignited for the rest of the burn to account for areas that cooled after burning. The area burned was fit to be circular, which is reasonable since the shape of the burned area was generally close to a circle. The radius was determined from the area of the burn. Figure 2 shows how the radius of the burned area changes with time for three different densities. The time axis started when the igniter was removed. The radius of the burned area increased nearly linearly with respect to time after the initial start of the burn. This linear section is used to determine the horizontal smoldering propagation velocity using a linear approximation in MATLAB. This limited the impact of the variations in the ignition since the burns did not have an equal burned area when the igniter was removed, as shown in Figure 2.

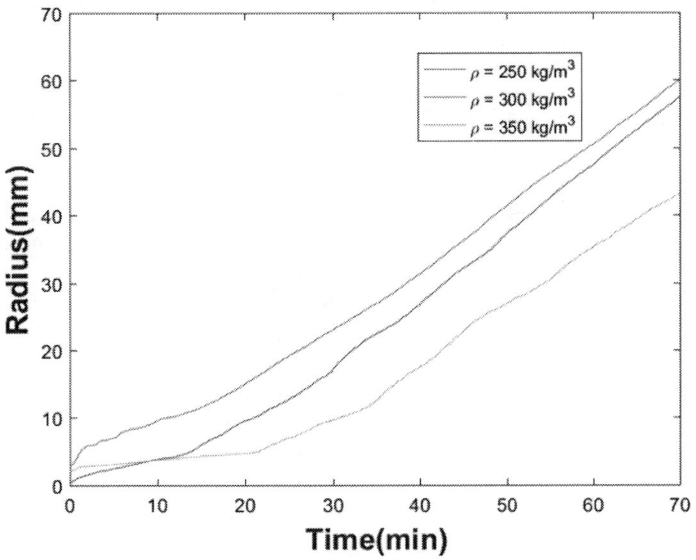

Figure 2: The radius of the burned area with respect to time for three burns of 75% cellulose fuel with different densities. Note that all three have nearly linear increases in radius with time 20 minutes after the ignition source is removed.

3. Results and Discussion

Figure 3 shows the sensitivity of the propagation velocity to fuel composition. The fuels were tested at 300, 310, and 375 kg/m^3. The three densities had similar trends, as fuels with higher cellulose contents had slower propagation velocities. The fuel content had a much larger effect on the propagation velocities for samples with contents between 75 and 100% cellulose than those between 50 and 75% cellulose, particularly from 90 to 95%. In addition, fuels with higher densities generally had lower propagation velocities.

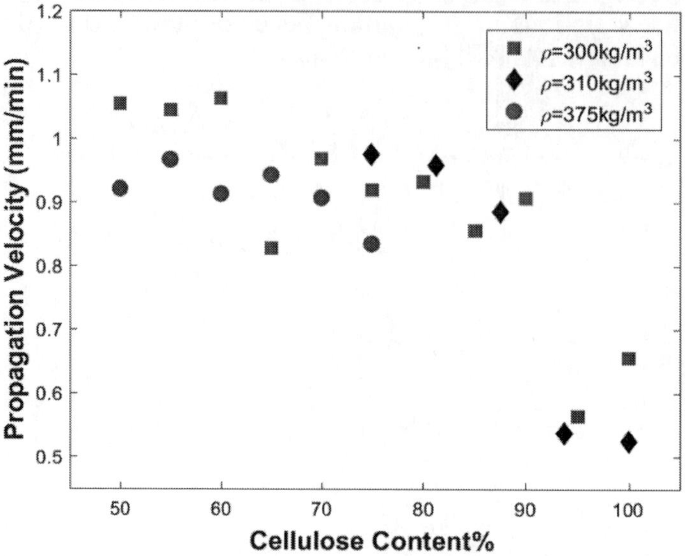

Figure 3: Smoldering propagation velocities at two different densities for cellulose content ranging from 50% and 100%.

Figure 4 shows the propagation velocities for three different fuel mixtures at varying densities. For all three mixtures, the smoldering propagation velocity decreased with increasing density. However, the 100% cellulose case was more much more sensitive to density changes than the other two. For example, the propagation velocity decreased by about 40% while the density increased by 50%. The propagation velocity decreased by about 20% for the 50% and 75% cellulose cases over a similar range of densities.

The trends shown in Figure 3 and Figure 4 are attributed to the differences in physical structure and chemical properties between cellulose and hemicellulose. Various studies have shown that smoldering is often controlled by oxygen availability and heat losses [25–28]. Some of the sensitivities of smoldering propagation to density in cotton and pine needle beds have been attributed to the large changes in permeability. At lower densities, air can flow relatively easily through the porous fuel. As the fuel is compressed, the space between fuel particles becomes smaller, which impacts oxygen availability below the surface [29, 30]. This results in a shallower smoldering front, and more heat is lost in the vertical direction because of larger gradients and decreased contact resistance. As a result, the horizontal propagation velocity is slower. With regard to fuel sensitivity, the cellulose is more fibrous than the hemicellulose, and as a result may have a larger change in permeability as the density increased. Another reason is that hemicellulose has a lower activation energy [19, 20]. Consequently, increasing packing density increases the amount of energy necessary to sustain smoldering[12, 25], but the energy difference is not as large with hemicellulose. A computational model is currently being made to help evaluate the explanations reported here.

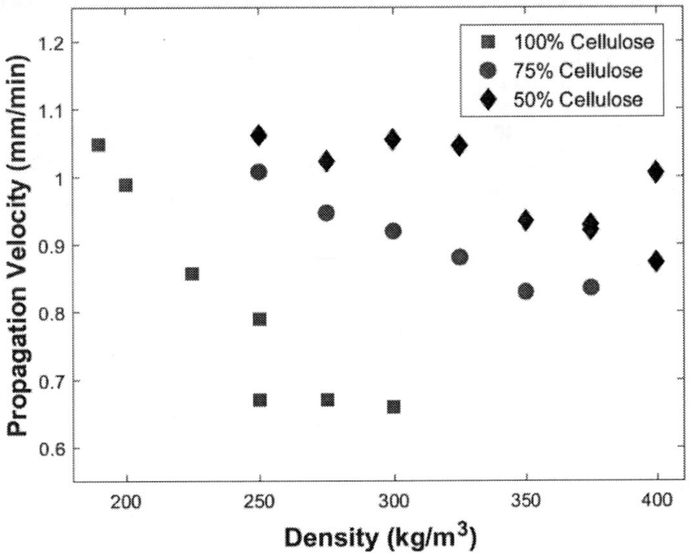

Figure 4: Propagation velocity for three mixtures of cellulose and hemicellulose at varying densities. The fuels were tested at different density ranges because of physical limitations.

4. Summary and Future Work

The smoldering propagation velocity was determined for binary mixtures of cellulose and hemicellulose with systematically varied densities and fuel compositions. The smoldering propagation velocity in samples with higher cellulose content decreased more with increased density than those with lower cellulose content and more hemicellulose. At the same densities, the results showed that increasing cellulose content decreases smoldering propagation velocity. Additional research will include adding lignin to the surrogate fuels to further identify the roles of fuel constituents. Collaboraters are developing a computational model that will to help analyze the results.

5. Acknowledgements

This research was funded by the Strategic Environmental Research and Development Program (SERDP) under contract number W912HQ-16-C-0045). The collaboration with Drs. Jim Reardon, Bret Butler, and Wei Min Hao at the US Forest Service is gratefully acknowledged. Additionally the generous assistance of Kaiden Moore and Jorge Dominguez-Baza in setting up and performing experiments is acknowledged.

5.1 Disclaimer

The views, opinions, and/or findings contained in this report are those of the authors and should not be construed as an official Department of Defense position of decision unless so designated by other official documentation. [31] [32] [33]

References

[1] G. Rein, Smouldering Fires and Natural Fuels, in: Fire Phenomena and the Earth System: An Interdisciplinary Guide to Fire Science, 2013, DOI: 10.1002/9781118529539.ch2.

[2] T. Lee, A. P. Sullivan, L. Mack, J. L. Jimenez, S. M. Kreidenweis, T. B. Onasch, D. R. Worsnop, W. Malm, C. E. Wold, W. Min Hao, J. L. Collett Jr, and J. L. Collett, Chemical Smoke Marker Emissions During Flaming and Smoldering Phases of Laboratory Open Burning of Wildland Fuels, Aerosol Science and Technology 449 (2010).

[3] R. Koppmann, K. Von Czapiewski, and J. S. Reid, A review of biomass burning emissions, part I: gaseous emissions of carbon monoxide, methane, volatile organic compounds, and nitrogen containing compounds, Atmospheric Chemistry and Physics Discussions 5 (2005) 10455–10516.

[4] R. J. Yokelson, R. Susott, D. E. Ward, J. Reardon, and D. W. T. Griffith, Emissions from smoldering combustion of biomass measured by open-path Fourier transform infrared spectroscopy, Journal of Geophysical Research 102 (1997) 18865–18877. DOI: 10.1029/97JD00852.

[5] C. Liu, C. Zhang, Y. Mu, J. Liu, and Y. Zhang, Emission of volatile organic compounds from domestic coal stove with the actual alternation of flaming and smoldering combustion processes, Environmental Pollution 221 (2017) 385–391.

[6] L. M. Mckenzie, W. M. Hao, G. N. Richards, and D. E. Ward, Measurement and Modeling of Air Toxins from Smoldering Combustion of Biomass, Environmental Science and Technology 29 (1995) 2047–2054. DOI: 10.1021/es00008a025.

[7] S. E. Page, F. Siegert, J. O. Riley, H.-D. V. Boehm, A. Jaya, and S. Limin, The amount of carbon released from peat and forest fires in Indonesia during 1997, Nature 420 (2002) 61–65.

[8] G. Rein, N. Cleaver, C. Ashton, P. Pironi, and J. L. Torero, The severity of smouldering peat fires and damage to the forest soil, Catena 74 (2008) 304–309. DOI: 10.1016/j.catena.2008.05.008.

[9] W. H. Frandsen, Ignition probability of organic soils, Canadian Journal of Forest Research 27 (1997) 1471–1477. DOI: 10.1139/x97-106.

[10] X. Huang and G. Rein, Thermochemical conversion of biomass in smouldering combustion across scales: The roles of heterogeneous kinetics, oxygen and transport phenomena, Bioresource Technology 207 (2016) 409–421. DOI: 10.1016/j.biortech.2016.01.027.

[11] X. Huang and G. Rein, Smouldering combustion of peat in wildfires: Inverse modelling of the drying and the thermal and oxidative decomposition kinetics, Combustion and Flame 161 (2014) 1633–1644. DOI: 10.1016/j.combustflame.2013.12.013.

[12] N. Prat-Guitart, G. Rein, R. M. Hadden, C. M. Belcher, and J. M. Yearsley, Propagation probability and spread rates of self-sustained smouldering fires under controlled moisture content and bulk density conditions, International Journal of Wildland Fire 25 (2016) 456–465. DOI: 10.1071/WF15103.

[13] J. Reardon, G. Curcio, and R. Bartlette, Soil moisture dynamics and smoldering combustion limits of pocosin soils in North Carolina, USA, International Journal of Wildland Fire 18 (2009) 326–335. DOI: 10.1071/WF08085.

[14] W. H. Frandsen, The influence of moisture and mineral soil on the combustion limits of smoldering forest duff, Canadian Journal of Forest Research 17 (1987) 1540–1544.

[15] B. C. Hagen, V. Frette, G. Kleppe, and B. J. Arntzen, Effects of heat flux scenarios on smoldering in cotton, Fire Safety Journal 61 (2013) 144–159. DOI: 10.1016/j.firesaf.2013.08.001.

[16] S. a. Waksman, Humus Origin, Chemical Composition, and Importance in Nature, Soil Science 41 (1936) 395. DOI: 10.1097/00010694-193605000-00010.

[17] A. Gani and I. Naruse, Effect of cellulose and lignin content on pyrolysis and combustion characteristics for several types of biomass, Renewable Energy 32 (2007) 649–661. DOI: 10.1016/j.renene.2006.02.017.

[18] C. Quan, N. Gao, and Q. Song, Pyrolysis of biomass components in a TGA and a fixed-bed reactor: Thermochemical behaviors, kinetics, and product characterization, Journal of Analytical and Applied Pyrolysis 121 (2016) 84–92.

[19] H. Yang, R. Yan, H. Chen, D. H. Lee, and C. Zheng, Characteristics of hemicellulose, cellulose and lignin pyrolysis, Fuel 86 (2007) 1781–1788. DOI: 10.1016/j.fuel.2006.12.013.

[20] Z. Chen, M. Hu, X. Zhu, D. Guo, S. Liu, Z. Hu, B. Xiao, J. Wang, and M. Laghari, Characteristics and kinetic study on pyrolysis of five lignocellulosic biomass via thermogravimetric analysis, Bioresource Technology 192 (2015) 441–450.

[21] M. Mehl, W. J. Pitz, C. K. Westbrook, and H. J. Curran, Kinetic modeling of gasoline surrogate components and mixtures under engine conditions, Proceedings of the Combustion Institute 33 (2011) 193–200.

[22] F. L. Dryer, S. Jahangirian, S. Dooley, S. H. Won, J. Heyne, V. R. Iyer, T. a. Litzinger, and R. J. Santoro, Emulating the combustion behavior of real jet aviation fuels by surrogate mixtures of hydrocarbon fluid blends: Implications for science and engineering, Energy and Fuels 28 (2014) 3474–3485. DOI: 10.1021/ef500284x.

[23] D. A. Cowan, B. D. Smucker, and D. L. Blunck, Sensitivity of Smoldering Combustion to Cellulose and Hemicellulose Content, 10th US Combustion meeting (2017) 1–8.

[24] B. C. Hagen, V. Frette, G. Kleppe, and B. J. Arntzen, Transition from smoldering to flaming fire in short cotton samples with asymmetrical boundary conditions, Fire Safety Journal 71 (2015) 69–78. DOI: 10.1016/j.firesaf.2014.11.004.

[25] X. Huang, F. Restuccia, M. Gramola, and G. Rein, Experimental study of the formation and collapse of an overhang in the lateral spread of smouldering peat fires, Combustion and Flame 168 (2016) 393–402. DOI: 10.1016/j.combustflame.2016.01.017.

[26] J. C. Thomas, R. M. Hadden, and A. Simeoni, Experimental investigation of the impact of oxygen fl ux on the burning dynamics of forest fuel beds, Fire Safety Journal (2017) 1–9.

[27] X. Huang and G. Rein, Smouldering Combustion of Soil Organic Matter: Inverse Modelling of the Thermal and Oxidative Degradation Kinetics, EGU General Assembly Conference ... (2013), URL: http://adsabs.harvard.edu/abs/2013EGUGA..1513754H.

[28] J. C. Valette, V. Gomendy, J. Maréchal, C. Houssard, and D. Gihon, Heat transfer in the soil during very low-intensity experimental fires: The role of duff and soil moisture content, International Journal of Wildland Fire 4 (1994) 225–237. DOI: 10.1071/WF9940225.

[29] B. C. Hagen, V. Frette, G. Kleppe, and B. J. Arntzen, Onset of smoldering in cotton: Effects of density, Fire Safety Journal 46 (2011) 73–80.

[30] M. El Houssami, J. C. Thomas, a. Lamorlette, D. Morvan, M. Chaos, R. Hadden, and a. Simeoni, Experimental and numerical studies characterizing the burning dynamics of wildland fuels, Combustion and Flame 168 (2016) 113–126. DOI: 10.1016/j.combustflame.2016.04.004.

[31] S. P. Urbanski, Combustion efficiency and emission factors for wildfire-season fires in mixed conifer forests of the northern Rocky Mountains , US, Atmospheric Chemistry and Physics 13 (2013) 7241–7262. DOI: 10.5194/acp-13-7241-2013.

[32] J. R. Hall, The Smoking-Material Fire Problem, National Fire Protection Association, Fire Analysis and Research Division (2013).

[33] D. M. S. Delicato, Physical-Chemical Properties and Sorption Characteristics of Peat, Ph. D. Dublin City University, 1996.

Western States Section of the Combustion Institute – Fall 2017 Meeting
Hosted by the University of Wyoming
October 2-3, 2017

Flame temperature effect on the transition between soot and graphitic carbon products in premixed stagnation flames

*Joaquin Camacho**

Mechanical Engineering Department, San Diego State University, USA
**Corresponding Author Email: jcamacho@sdsu.edu*

Abstract: Particulate carbon is known to exist in a wide range of forms, some of which exhibit properties that are desirable for emerging technologies. For example, graphitic carbon is widely integrated into energy storage devices due to its charge storage capacity and inherent electrical conductivity. The effect of flame temperature on the structure of carbon products formed in premixed stagnation flames is assessed in this work. Namely, the progression of carbon products from amorphous soot to more ordered carbon structures is examined by tracking the evolution of Raman spectra as a function of flame temperature and time. The stagnation flames are modeled in a pseudo-1D formulation to precisely quantify the temperature-time relationship in the carbon growth region. Raman spectra for carbon formed in all flames consisted of Raman Shifts representing the D and G band close to 1350 and 1600 cm-1, respectively. The relative intensity of the two bands changed as a function of flame temperature and the trend can be interpreted as indicating larger graphitic regions on carbon particles formed at higher flame temperatures. The observations made here provide insight into carbon formation in flames that is useful for soot theory / modeling and flame synthesis processes.

Keywords: Flame Synthesis, Carbon, Nanoparticle, Soot

1. Introduction

Besides being a problematic by-product of combustion processes, carbon, in its many forms, can also be the desired end-product from flames. There are similarities between nuisance soot and more useful graphititc forms of carbon. These similarities have been invoked to describe soot reactivity by referring to experiments [1] on graphite. In addition, soot growth and ageing is regularly described by the degree of graphitization in the carbon structure. Dobbins and co-workers [2, 3] were the first to draw upon classical works on carbonization (see e.g., [4, 5]) to describe the evolution of so-called precursor particles (also known as nascent soot in more recent works [6-8]) to mature soot. For producing useful carbon materials, relatively high flame temperatures are important for achieving carbonization and uniform growth conditions are necessary to obtain precisely controlled properties. The essential commodity known as carbon black is produced industrially using flames operating at the aforementioned conditions.

In this work, the range of flame conditions leading to less-desirable soot formation and more valuable graphitic carbon will be demonstrated. Stretch-stabilized premixed flames will be used because the stabilization mechanism allows for a pseudo one-dimensional flame without significant heat loss [9]. These flame characteristics will enable particle formation under uniform growth conditions with temperatures approaching 2500K. Raman Spectroscopy will serve as a diagnostic for the presence of sp^2 bonded carbon. The raman spectra of amorphous, crystalline and nano-scale carbon have been extensively studied. Soot structural details have also been investigated in raman spectroscopy studies [7, 10-14]. Boehman and co-workers incorporated analysis of the Raman spectra for laboratory [13] and Diesel engine [14] soot to examine soot reactivity and carbon structure. Russo and Ciajolo carried out a systematic Raman study on soot from premixed flames and parsed out subtle differences in soot structure resulting from fuel structure and flame conditions [12].

Empirical [15] and theoretical [16] works showing the relationship between XRD derived crystallite size and the ratio of prominent bands in the Raman spectra marked the beginning of numerous XRD and Raman studies as complimentary carbon diagnostics. XRD peaks attributed to (002), (10) and (110) have been reported for both soot and carbon black [14]. Graphite and turbostratic carbon are known to have one major XRD peak attributed to (002) [17]. In this work, Raman analysis will be carried out to isolate the effect of flame temperature on the carbon products formed in premixed flames. The relatively high-temperature that is accessible in premixed stretch-stabilized flames is expected to allow for observation of higher ordered carbon structure relative to carbon products obtained at lower flame temperatures.

2. Experimental

The experimental setup, summarized in Fig. 1, consists of a burner with an aerodynamically shaped nozzle and a stagnation / deposition surface. The stabilization mechanism of the premixed flames studied herein requires flow stagnation to set the kinematic balance between the local flame speed and normal flow velocity immediately upstream of the flame surface [9]. Microscope slides were placed flush with the surface for deposition of carbon aerosol formed in the flame. The aerodynamic shape of the burner nozzle body is designed to achieve plug flow at the burner exit (1.43 cm nozzle exit diameter). The standing distance between the flame and stagnation / deposition surface, L_s, may be varied by changing the unburned gas flow rate. The gas temperature at the nozzle exit, T_n, was measured with an uncoated fine-wire (125 μm wire diameter) Pt-Rh thermocouple placed near the centerline of the flow. The temperature at the stagnation surface, T_s, was measured by a type-K thermocouple (0.2 cm wire diameter). The gas temperature at the nozzle exit was determined to be T_n = 340 K; and the stagnation surface temperature was T_s = 385 K. Numerical sensitivity calculations show that within the uncertainty values quoted the flame structure exhibits little sensitivity towards the boundary temperature variations.

The carbon structure of the flame products was analyzed by Raman spectroscopy for a series of premixed ethylene flames. The flame diluent was alternated between nitrogen and argon to expand the range of acceptable flame temperatures. The unburned gas composition and cold gas velocity are summarized in Table 1 for each flame. The examined flames range from relatively high velocity flames approaching the limit where stretch-induced flame extinction processes occur to low velocity flames approaching the flashback limit.

A modified OPPDIF computation [18] based on USC Mech II [19] was used to calculate the temperature, velocity and species profiles. The boundary conditions were given by the measured temperatures at the nozzle exit and the stagnation surface/sampling probe. The OPPDIF code invokes the pseudo one-dimensional formulation introduced by Kee, Smooke and co-workers [20, 21]. Plug flow is designated at the burner boundary and non-slip conditions are assumed at the stagnation surface boundary. For particles formed in the flame, the finite residence time is determined by considering the thermophoretic velocity which is driven by the significant temperature gradient, $\partial T/\partial x$, at the stagnation plate. The thermophoretic velocity was calculated numerically and added to the total velocity. The residence time of the soot particles is defined as the time interval for the particle (or precursors) to traverse from the calculated location of the peak flame temperature to the location of the stagnation probe. The carbon flame products were analyzed by Raman Spectroscopy. A Thermo Nicolet DXR Raman Microscope (Thermo Scientific) with a 532 nm laser was used. The laser power was 5 mW and changes to the spectra and sample occurred over the sampling period.

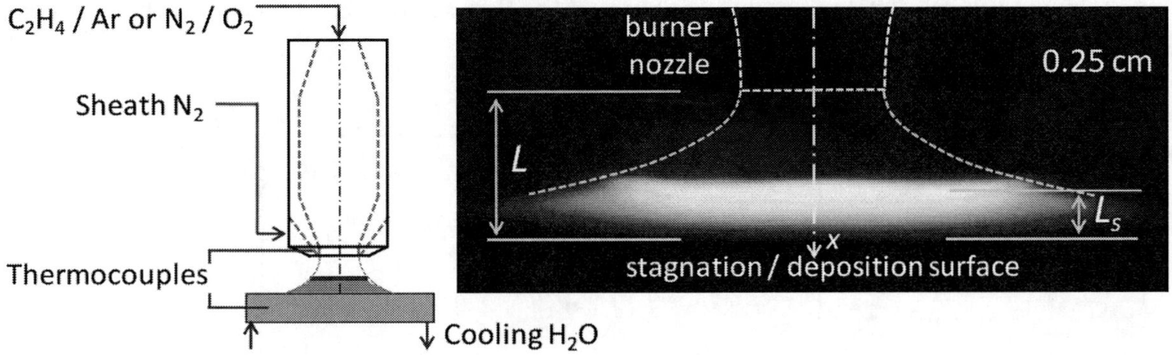

Figure 1: Experimental setup for carbon synthesis and deposition in premixed stagnation flames.

Table 1. Flame parameters[a]

	time, t_p (ms)	$T_{f,max}$ (K)	Velocity, v_0 (cm/s)	Standing distance, L_s (cm)	Separation distance, L (cm)
			15.2 % C_2H_4 / 18.5 % O_2 / 66.2 % N_2		
1	18	1790	20.6	0.48	1.0
2	25	1750	12.0	0.65	1.0
			21.1 % C_2H_4 / 25.3 % O_2 / 53.5 % N_2		
3	6	2125	43.6	0.71	1.0
4	23	2070	43.6	1.85	2.6
			14.9 % C_2H_4 / 17.9 % O_2 / 67.1 % Ar		
5	28	2070	40.4	1.75	2.6
			16.4 % C_2H_4 / 19.7 % O_2 / 63.8 % Ar		
6	19	2150	61.9	1.56	2.6
			19.2 % C_2H_4 / 23.0 % O_2 / 57.8 % Ar		
7	14	2270	71.1	1.62	2.6

3. Results and Discussion

The Raman spectra for the flame-formed carbon showed peaks in the vicinity of known Raman Shifts for the D and G bands. The intensities of the broad peaks close to 1350 and 1600 cm^{-1} were compared for each flame condition. The relative intensity of these bands has been attributed to the change in aromatic layer size [15] and the area of aromatic domains [22]. For the 1775 K maximum flame temperature condition (shown in Fig. 2), there was very little effect of growth time on the Raman spectra. The spectra for flame-formed carbon at the 2100 K is shown in Fig. 3 for two growth times. The intensity of the non-graphite Raman shift at 1300 cm-1 is the same as the graphite peak at 6 ms and gets becomes larger than the graphite peak at 23ms growth time. The spectra for flame-formed carbon for a series of flame temperatures at comparable growth times is shown in Fig. 4. The relative intensity of the non-graphite peak also becomes larger than the graphite peak as the flame temperature approaches 2300 K. Ferrari and Roberston [22] showed that the relative intensity of the D and G bands is proportional to the area of graphitic regions on the carbon surface if the crystallite size is smaller than 2 nm. The estimated characteristic length of the graphitic crystallite region, La, based on the formation of Ferrari and Robertson is shown in Fig. 5. Under this regime, the crystallite region of the flame-formed carbon approaches 1.9 nm for particles formed at 2270 K maximum flame temperature.

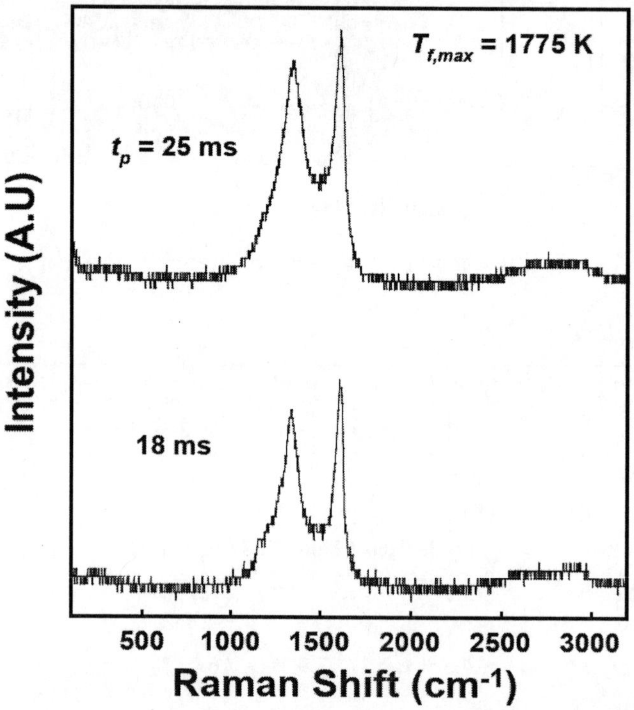

Figure 2: Raman spectra of highly ordered pyrolytic graphite (left) and as-deposited carbon films

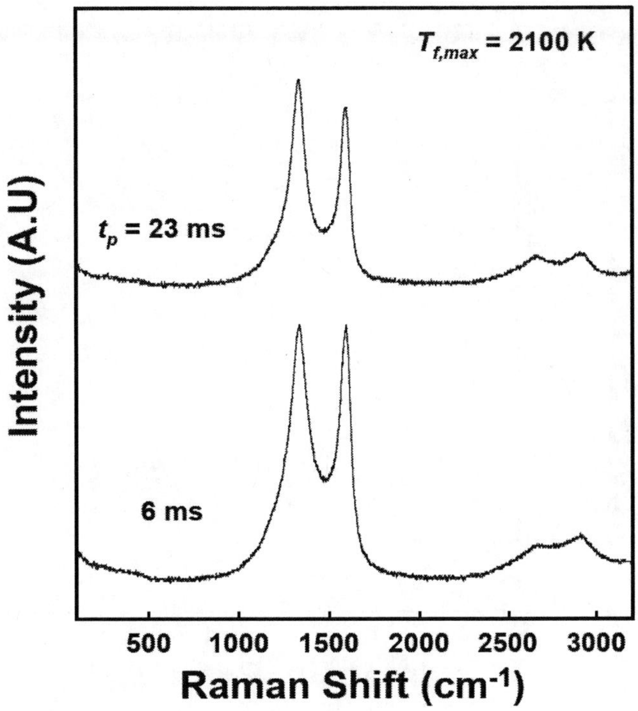

Figure 3: Raman spectra of highly ordered pyrolytic graphite (left) and as-deposited carbon films

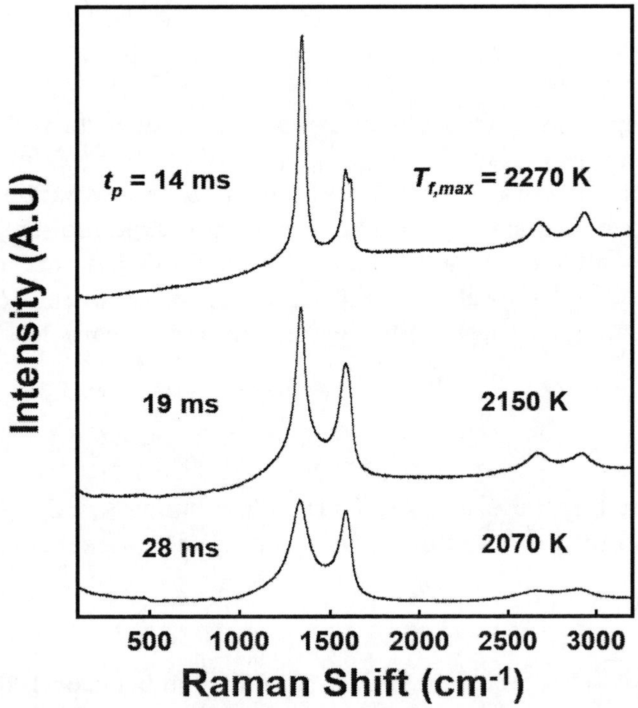

Figure 4: Raman spectra of highly ordered pyrolytic graphite (left) and as-deposited carbon films

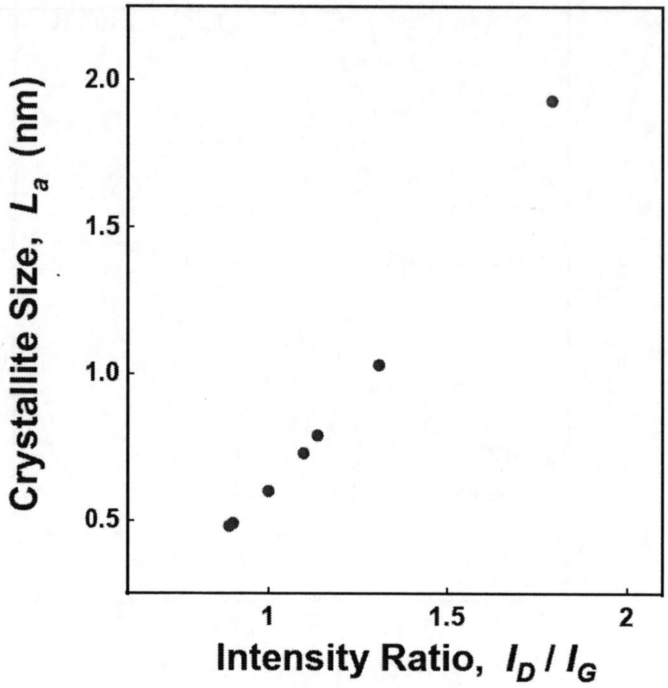

Figure 5: Raman spectra of highly ordered pyrolytic graphite (left) and as-deposited carbon films

4. Conclusion

The effect of flame temperature on the structure of carbon products formed in premixed stagnation flames is assessed in this work. Namely, the progression of carbon products from amorphous soot to more ordered carbon structures is examined by tracking the evolution of Raman spectra as a function of flame temperature and time. Raman spectra for carbon formed in all flames consisted of Raman Shifts representing the D and G band close to 1350 and 1600 cm-1, respectively. The relative intensity of the two bands changed as a function of flame temperature and the trend can be interpreted as indicating larger graphitic regions on carbon particles formed at higher flame temperatures.

5. Acknowledgements

The work was supported by the University Grants Program at San Diego State University. The authors are thankful to Prof. David Pullman for generous access and help with Raman.

6. References

[1] Walls JR, Strickland-Constable RF. Oxidation of carbon between 1000 and 2400 °C. Carbon 1 (1964) 333-334.
[2] Dobbins RA, Govatzidakis GJ, Lu W, Schwartzman AF, Fletcher RA. Carbonization Rate of Soot Precursor Particles. Combust. Sci Tech. 121 (1996) 103.

[3] Dobbins RA. Soot inception temperature and the carbonization rate of precursor particles. Combust. Flame 130 (2002) 204.
[4] Lewis IC. Chemistry of carbonization. Carbon 20 (1982) 519.
[5] Franklin R. The structure of graphitic carbons. Acta Crystall. 4 (1951) 253.
[6] Wang H. Formation of nascent soot and other condensed-phase materials in flames. Proc. Combust. Inst. 33 (2011) 41.
[7] Herdman JD, Connelly BC, Smooke MD, Long MB, Miller JH. A comparison of Raman signatures and laser-induced incandescence with direct numerical simulation of soot growth in non-premixed ethylene/air flames. Carbon 49 (2011) 5298.
[8] Kholghy MR, Veshkini A, Thomson MJ. The core-shell internal nanostructure of soot: A criterion to model soot maturity. Carbon 100 (2016) 508.
[9] Law CK. Combustion Physics. Cambridge: Cambrige Press, 2006.
[10] Lapuerta M, Oliva F, Agudelo JR, Stitt JP. Optimization of Raman Spectroscopy Parameters for Characterizing Soot from Different Diesel Fuels. Combust. Sci Tech. 183 (2011) 1203.
[11] Sadezky A, Muckenhuber H, Grothe H, Niessner R, Paschl U. Raman microspectroscopy of soot and related carbonaceous materials: Spectral analysis and structural information. Carbon 43 (2005) 1731.
[12] Russo C, Ciajolo A. Effect of the flame environment on soot nanostructure inferred by Raman spectroscopy at different excitation wavelengths. Combust. Flame 162 (2015) 2431.
[13] Seong HJ, Boehman AL. Studies of soot oxidative reactivity using a diffusion flame burner. Combust. Flame 159 (2012) 1864.
[14] Seong HJ, Boehman AL. Evaluation of Raman Parameters Using Visible Raman Microscopy for Soot Oxidative Reactivity. Energy Fuel 27 (2013) 1613.
[15] Tuinstra F, Koenig JL. Raman Spectrum of Graphite. J. Chem. Phys. 53 (1970) 1126.
[16] Lespade P, Al-Jishi R, Dresselhaus MS. Model for Raman scattering from incompletely graphitized carbons. Carbon 20 (1982) 427.
[17] Li ZQ, Lu CJ, Xia ZP, Zhou Y, Luo Z. X-ray diffraction patterns of graphite and turbostratic carbon. Carbon 45 (2007) 1686.
[18] Abid AD, Camacho J, Sheen DA, Wang H. Quantitative measurement of soot particle size distribution in premixed flames: The burner-stabilized stagnation flame approach. Combust. Flame 156 (2009) 1862.
[19] Wang H, You X, Joshi AV, Davis SG, Laskin A, Egolfopoulos F, Law CK. USC Mech II. High-Temperature Combustion Reaction Model of H2/CO/C1-C4 Compounds. http://ignis.usc.edu/USC_Mech_II.htm 2007.
[20] Kee RJ, Miller JA, Evans GH, Dixon-Lewis G. A computational model of the structure and extinction of strained, opposed flow, premixed methane-air flames. Symp. (Int.) Combust. 22 (1989) 1479.
[21] Lutz AE, Kee RJ, Grcar JF, Rupley FM. OPPDIF: A Fortan program for computing opposed-flow diffusion flames. Livermore: Sandia National Labs, 1997.
[22] Ferrari AC, Robertson J. Interpretation of Raman spectra of disordered and amorphous carbon. Phys. Rev. B 61 (2000) 14095.

Western States Section of the Combustion Institute – Fall 2017 Meeting
Hosted by the University of Wyoming
October 2-3, 2017

Modeling Soot Formation from Solid Complex Fuels

Alexander J. Josephson[1,2,*], *Emily Hopkins*[2], *Rodman R. Linn*[2], *and David O. Lignell*[1]

[1]*Department of Chemical Engineering, Brigham Young University, Provo, UT*
[2]*Earth and Environmental Sciences Division, Los Alamos National Laboratory, Los Alamos, NM*
[*]*Corresponding author:* `alexanderj@lanl.gov`

Abstract: While the phenomena of soot formation in gaseous flames is well studied and understood, extension of the fundamental mechanisms to systems containing complex solid fuels can introduce large uncertainties and inaccuracies. In this study, we have developed a detailed physics-based model for predicting the evolution of soot particles formed in systems containing complex-solid fuels such as wood or coal. This detailed physics-based model includes two particle-size distributions: that for soot particles and for soot precursor molecules. Sub-models for precursor creation, growth, and thermal cracking are included along with soot particle nucleation, coagulation, surface growth, aggregation, oxidation, and gasification. Validation work is presented comparing experimental results for a coal flame and biomass gasifier against model predicted values for soot concentrations and size distributions. The full detailed model can be computationally expensive when incorporated into CFD simulations; therefore, model simplifications are explored and presented in this work along with some preliminary work of applying particle formation physics to wildfire simulations.
***Keywords:* soot formation, method of moments**

1. Introduction

The presence of soot particles in combustion processes has been observed for centuries, but only in the last several decades have the effects of these particles been studied, evaluated, and understood. Soot particles have significant effects on the thermal radiation emitted by a flame. This thermal radiation impacts both a flame's radiative heat transfer and heat loss: increasing radiative heat flux while at the same time decreasing local temperatures, which in turns affects flame chemistry. In addition, it is known that if soot particles break through the flame's oxidation layer they form air-bound aerosols which are both detrimental to human health and have negative environmental consequences.

Because of the aforementioned effects, it is important for researchers and modelers to understand mechanisms that govern the formation and behavior of soot particles in combusting systems. For gaseous-fuel flames, mechanisms of soot formation have been well-researched and detailed, but not as much for solid complex fuels such as wood or coal.

In general, it has been found that soot formation follows a series of well researched mechanisms described here. Soot precursors are polyaromatic hydrocarbons (PAHs) formed in fuel-rich areas, these precursors nucleate into soot particles. In gaseous fuels, the formation of PAHs is usually the rate-limiting step in soot formation; in solid fuels, however, PAHs are usually released during the initial pyrolysis of the fuel, bypassing this rate-limiting step [1]. After soot particle nucleation occurs, particles grows through kinetic interactions between particle surfaces and the surrounding gas. At the same time, these particles combine together through coagulation. Once the particles reach a critical size, they begin to aggregate, forming chains of spherical particles. Concurrent to this formation process, there are periods of consumption, either through oxidation, or, in special circumstances, through gasification.

In this study, a detailed model has been developed for the formation of soot in solid fuel systems and that model is validated against two different experiments. This model is computationally expensive and thus is not appropriate for large-scale simulations. As a result, subsequent efforts have been made to simplify this model and some of these efforts will be presented here as well.

2. Detailed Model Development

The proposed detailed soot model describes two particle size distributions (PSD): that for soot precursors and that for soot particles. The precursor PSD is represented by a sectional approach and the soot PSD represented by the method of moments. This detailed model contains sub-models for precursor mechanisms (formation, growth, and thermal cracking) as well as soot particle mechanisms (nucleation, surface growth, agglomeration, and consumption).

2.1 Precursor Dynamics

Precursor formation is accomplished in two ways: the release of precursors during the primary pyrolysis of the parent fuel and through the build-up of PAHs from gas-phase mechanisms. In the following validation studies, the coal percolation devolatilization (CPD) [2] model and its biomass adaptation (CPD-bio) [3] were used to predict the release of tar during primary pyrolysis. These models were modified slightly to output a time-evolved sectional distribution of tar instead of an overall tar yield. Tar, which is defined as the volatiles that would condense if cooled to room temperature, are largely aromatic and act as a primary soot precursor in these systems.

The build-up of PAHs from gas-phase mechanisms is modeled using a temperature equilibriated ABF mechanism [4]. This mechanism contains 99 species and 544 reactions and reflects the building of aromatic rings up to pyrene from basic gas-phase components. The formation of pyrene adds to the section of the precursor distribution associated with its molecular weight.

Once precursors are formed, they may grow through the hydrogen-abstraction-carbon-addition (HACA) mechanism well established in the literature [4] or be consumed through oxidation or gasification. Either mechanism effectively moves molecules among different sections of the distribution as molecules either grow or shrink.

In addition, precursors may thermally crack, either shrinking in size or completely cracking to light gases. This process is modeled using a scheme developed by Marias, et al. [5] and adapted to this detailed model. In this submodel, precursors are classified into four different types, dependent on their chemical composition: phenol, toluene, naphthalene, or benzene. From these four types, the scheme depicted in Fig. 1 is applied. Each type cracks at a rate given by the model, when that type cracks it gives off some or all of its weight as light gases and the remaining converts into another type. As it is undesirable to resolve four types for each precursor section, the fraction of each type is kept constant throughout a simulation with the fraction amounts determined through a pre-simulation evaluation which estimates what the time-averaged fractions would be from inception to consumption of all precursors.

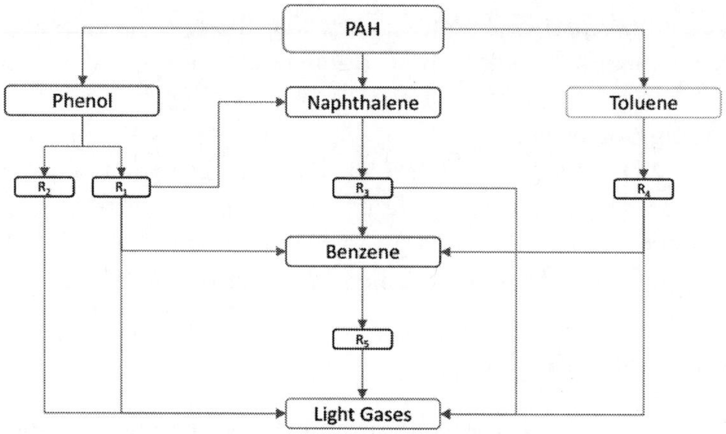

Figure 1: Thermal cracking scheme as applied to soot precursors. [5].

Precursors are also consumed through the nucleation process which converts precursors into primary soot particles.

2.2 Soot Particle Dynamics

Nucleation is accomplished by the coagulation of two precursor molecules. This nucleation is modeled by calculating a frequency of collision between precursors using the kinetic theory of gases. As they coagulate, the moments of the soot PSD are increased according to the PSD sections of the coagulating precursors.

Like the soot precursors, soot particles undergo interactions with the surrounding gases: surface growth via HACA, or consumption via oxidation/gasification. These surface interactions affect the soot PSD moments in the following way [6]:

$$\frac{dM_r}{dt} = \pi \left(\frac{6}{\pi\rho}\right)^{2/3} \frac{k_s m_0^{2/3-d}}{\Delta m} \sum_{k=0}^{r-1} \Delta m^{r-k} M_{k+d}, \tag{1}$$

Where k_s is the rate of reaction in $kg\ m^{-2} s^{-1}$, Δm represents the mass change due to a single reaction, m_0 is the mass of the primary soot particle ($\sim$400 kg/kmole), and d is a shape factor for the particle as introduced by Balthasar and Frenklach [7] to accommodate particle aggregation behavior. Fractal moments are resolved using interpolative closure between resolved whole moments.

Coagulation between particles is modeled according the coagulation scheme developed by Frenklach [8]; in which, coagulation is computed based on frequency of collision between particles in the continuum regime and free-molecular regime, then a weighted average is taken between the two based on a local particle Knudsen number.

Aggregate behavior is accounted for by introducing a 'surface moment,' M_d, as described by Balthasar and Frenklach [7]. This surface moment is used to commpute the shape factor, d, mentioned above and is resolved using submodels present in Balthasar's paper.

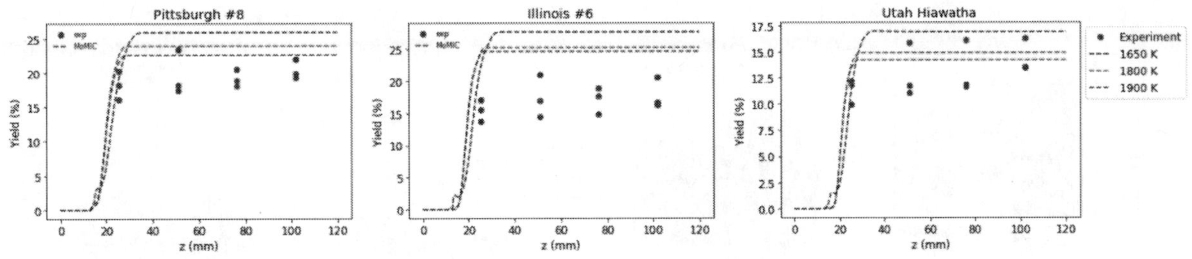

Figure 2: Comparison of model predicted total soot yields versus those measured soot yields for the Pittsburgh #8 coal at experimental temperature of 1650 K.

3. Validation

This model was validated against two different experiments: a coal-fired flat flame burner and a biomass gasifier.

3.1 Coal-System Validation

The coal-system experiment consists of data collected by Ma, et al. [9]. In the experiment, coal was fed into a flat-flame burner and particles were collected from the flame using a water-cooled nitrogen-quenched suction probe at four locations along the flame center-line. Collected particles were sent through a particle separation system which consists of a virtual impactor, a cyclone, and two soot filters. Large soot aggregates (5+ μm in diameter) were separated from char particles using a sieve with 38 μm openings. Six different coal types were tested at three temperatures each.

Results produced by the proposed soot model are shown in Fig. 2 depicts the total soot yield as a percentage of the parent coal mass for three of the coal types. The dotted line represents the model predictions while the points represent the actual measured data. There is reasonable agreement between the model predictions and experimentally measured data as both experiment and model level out to similar concentrations after 30 mm. The model does tend to over-predict yields compared to experimental data, but trends remain consistent and within reasonable error between simulations and experiments.

Due to the configuration of the experiment, there are certain aspects of the proposed soot model that are validated very well by comparisons with this study; however, there are other aspects of the model which are not validated. In the experiment, soot concentrations are dominated by the nucleation of tar molecules to form soot particles, and because of this, the tar formation, PAH cracking, and PAH coagulation portions of the nucleation are well validated. The pyrene formation and surface growth of PAH molecules are not as well tested because the nature of the pre-mixed flame to which the coal particles were added discouraged the formation of pyrene and acetylene, meaning that the effect of these particular two mechanisms were minimal in this set-up.

Just as in the PAH phase, surface growth in the soot phase was also minimal for the same reason stated above. Also, since particles were collected in fuel-rich locations, the effects of soot oxidation/gasification were also minimal. This is consistent with the leveling of the curve shown in Figure 2. Soot coagulation, on the other hand, has been validated.

By measuring particles caught in the two different soot filters, the experimenters were able to

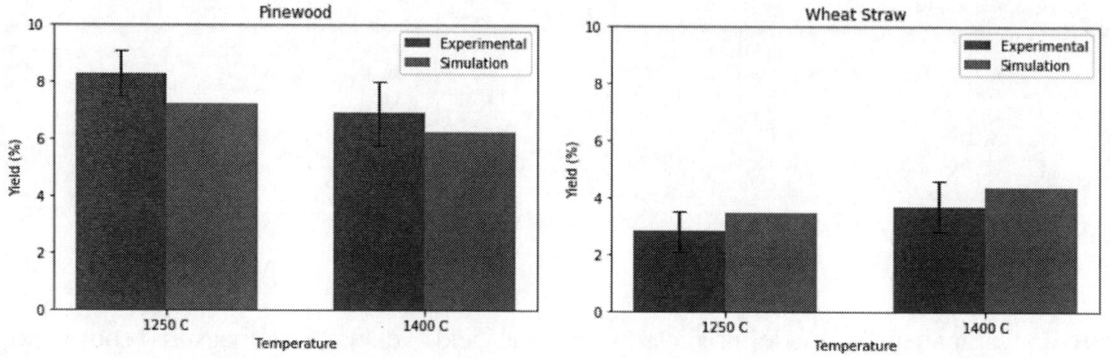

Figure 3: Comparison of model predicted total soot yields versus those measured soot yields for the three different biomass types at temperatures 1250 C and 1400 C.

separate soot particles that were <5 μm in diameter. This allows the validation of the model's coagulation portion due to the shrinking concentration of smaller particles while the overall mass remains constant, indicating coagulation.

3.2 Biomass-System Validation

The wood-system experiment consists of data collected by Trubetskaya, et al. [10]. In this study, biomass was fed into a fast pyrolysis drop tube reactor. This reactor was held under gasification conditions, with no oxygen fed into the reactor. Soot particles were collected from the effluent gas of the reactor and a yield was determined for three different biomass types at two temperatures each.

Figure 3 compares simulations against experimental soot yield for two of the experiments. There is good agreement between yields measured and those predicted by the detailed soot model, with all predicted yields lying within or close to the experimental error bounds reported in the literature.

Like the coal-system, certain aspects of the model were better validated than others. The configuration of the gasifier reactor allowed for pyrene and acetylene to be produced thus validating mechanisms involving these species. However, residence time in the reactor was short, thus only 5-15% of the total soot yield was due to mechanisms involving pyrene or acetylene; whereas the rest was mostly due to nucleation of tars like in the coal experiment. Also like the coal experiment, effects of oxidation and gasification were minimal.

4. Simplified Model Development

As the detailed soot model is computationally expensive, recent research has centered on methods to simplify this detailed-model to more economical models for use in simulations.

Biomass Component	Total Tar Yield (%)	Average Tar Molecular Weight (kg/kmole)
Cellulose	47.9	460
Softwood Hemicellulose	30.5	400
Hardwood Hemicellulose	30.2	396
Softwood Lignin	62.4	528
Hardwood Lignin	53.5	546

Table 1: Total precursor yield along with average molecular weights for different biomass components after primary pyrolysis.

4.1 Sooting Potential

Work has been done to simplify the prediction of soot precursors released from biomass during primary pyrolysis. In the above validation cases, the CPD and CPD-bio maodels were modified to output a sectional distribution of tar released during devolatilization. This is computationally expensive for simulations, so a 'sooting potential' parameter was developed to reflect the total tar released by different components of biomass from primary pyrolysis.

It is known that biomass is made up of cellulose, hemicellulose, and lignin (extractives were excluded from this study as there contribution to biomass is minimal.) Fractional composition varies between species, but the components are the same, throughout all biomass. The exception to that rule being a distinction between hardwood and softwood for hemicellulose and lignin.

Using the CPD-bio model [3], the total tar released by the different components was modeled for a range heating rates (100-1000 K/s), pressures (0.6-1.2 atm), and maximum temperatures (800-3000 K). It was found that variations in these parameters had minimal impact on the total tar yield of the fuel and results for each component can be seen in Table 1. Finding the total tar yield of biomass is simply a matter of summing the total tar yield according to its fractional composition.

4.2 Mono-Dispersed Distributions

One option for model simplification is to assume that each PSD is mono-dispersed rather than a full distribution. This requires a reworking of submodels but is computationally inexpensive as only 3 parameters need to be resolved at each computational node rather than 14+.

The mono-dispersed distributions assume that the molecular weight of soot precursors is set and does not fluctuate enough to significantly affect results. In the case of biomass, that molecular weight is found by taking a weighted average of the components' molecular weights found in Table 1. This assumption means that only one parameter has to be resolved for the precursor PSD: the number density of the molecules.

The soot particle distribution, on the other hand, contains two terms to be resolved: the average weight of particles and the number density. The computational savings of transporting only three parameters as well the simplification of submodels allows considerable applicability of this simplified model, but accuracy decreases. The extent to which accuracy decreases is system dependent and a matter of ongoing research.

A cost-reducing strategy to be mentioned here but has to be done on a case-by-case basis involves a minor sensitivity analysis. Knowing the sensitivity of soot mechanisms to the combustion environment allows researchers to save on computational costs by omitting certain mechanisms that could be considered negligible for their circumstances. For example: it is known that soot gasification can often be considered negligible [1], except in cases of high concentrations of H_2O or CO_2 (ie oxy-fuel combustion). It would be economical for researchers to remove this mechanism from simulations for air-fired combustion.

4.3 Surrogate Modeling

At times, the size of a computational node is too large to capture the physics of soot formation; such is the case for wild-fire simulations. Ongoing research is focused on creating surrogate models which can predict soot formation in such systems. These surrogate models are being creating by first implementing the detailed-physics model described above to basic circumstances encountered in these systems (ie a forest floor, canopy, or grassland). By running fine-detailed simulations of these basic domains, we are able to obtain detailed data of soot yields.

Using these soot yields, we are able to tune simple surrogate models which contain environmental parameters as inputs (fuel density, temperature distribution, moisture content, etc.) Each of these surrogate models would need to be individually tuned to different basic domains, but in cases of repeatable domains, such as a forest, these surrogate models can be powerful in predicting soot yields without the computational cost of detailed physics resolution.

5. Conclusions

A physics-based detailed model for soot formation from solid complex fuels has been developed and validated against two separate experimental cases found in the literature. Simplifications to this model to reduce computational expenses are discussed here and are under further development.

6. Acknowledgements

The work was supported by the Department of Energy, National Nuclear Security Administration under Award DE-NA0002375.

References

[1] T. H. Fletcher, J. L. Ma, J. R. Rigby, A. L. Brown, and B. W. Webb, Soot in coal combustion systems, Progress in Energy and Combustion Science 23 (1997) 283–301.

[2] T. H. Fletcher, A. R. Kerstein, R. J. Pugmire, M. S. Solum, and D. M. Grant, Chemical percolation model for devolatilization. 3. Direct use of carbon-13 NMR data to predict effects of coal type, Energy & Fuels 6 (1992) 414–431.

[3] A. D. Lewis and T. H. Fletcher, Prediction of Sawdust Pyrolysis Yields from a Flat-Flame Burner Using the CPD Model, Energy & Fuels 27 (2013) 942–953.

[4] J. Appel, H. Bockhorn, and M. Frenklach, Kinetic modeling of soot formation with detailed chemistry and physics: Laminar premixed flames of C-2 hydrocarbons, Combustion and Flame 121 (2000) 122–136.

[5] F. Marias, R. Demarthon, A. Bloas, and J. P. Robert-Arnouil, Modeling of tar thermal cracking in a plasma reactor, Fuel Processing Technology 149 (2016) 139–152.

[6] M. Frenklach, Method of moments with interpolative closure, Chemical Engineering Science 57 (2002) 2229–2239.

[7] M. Balthasar and M. Frenklach, Detailed kinetic modeling of soot aggregate formation in laminar premixed flames, Combustion and Flame 140 (2005) 130–145.

[8] M. Frenklach, D. W. Clary, T. Yuan, W. C. Gardiner, and S. E. Stein, Mechanism of Soot Formation in Acetylene-Oxygen Mixtures, Combustion Science and Technology 50 (1986) 79–115.

[9] J. Ma, T. H. Fletcher, and B. W. Webb, Conversion of coal tar to soot during coal pyrolysis in a post-flame environment, Twenty-Sixth Symposium (International) on Combustion, Vols 1 and 2 26 (1996) 3161–3167.

[10] A. Trubetskaya, P. A. Jensen, A. D. Jensen, A. D. G. Llamas, K. Umeki, D. Gardini, J. Kling, R. B. Bates, and P. Glarborg, Effects of several types of biomass fuels on the yield, nanostructure and reactivity of soot from fast pyrolysis at high temperatures, Applied Energy 171 (2016) 468–482.

2017 WSSCI Fall Technical Meeting
Hosted by University of Wyoming
October 2-3, 2017

Investigation of non-Ideal Gasoline-Ethanol Vaporization Behavior Using an Ethanol, Cyclohexane, and Benzene Surrogate

Stephen Burke[1], Matthew Ratcliff[2], Robert McCormick[2], Danielle Bartholet[3], Bret Windom[1]
[1]Mechanical Engineering, Colorado State University, Fort Collins, CO 80523
[2]National Renewable Energy Laboratory, Golden, CO 80401
[3]Bioengineering, University of Florida, Gainesville, FL 32611

Some previous studies have shown a correlation between ethanol content and soot formation in gasoline direct injection engines, while others show a decrease as ethanol displaces particulate matter forming species. Our previous research has shown that this may be due to aromatic enrichment by non-ideal vapor liquid equilibrium (VLE) in the presence of ethanol, and to the high enthalpy of vaporization of ethanol, which cools the combustion chamber and can elongate droplet lifetimes. This aromatic enrichment may simply be due to the high vapor pressure of aromatic species, as their vaporization largely occurs after the ethanol content of the mixture is exhausted for E10-E30 blends. To eliminate vapor pressure differences, two species of different structures and the same boiling point as ethanol, cyclohexane and benzene, were identified. As the volatility of these species at their boiling temperature is nearly identical, composition changes during distillation of this ternary blend is solely a function of the non-ideal VLE, which is affected primarily by ethanol content. Distillations of various compositions were performed, and gas chromatography was used to quantify vapor composition. These compositions, as well as distillation curves, were compared to a UNIFAC-based simulation to examine behavior as a function of initial composition.

1. Introduction

Particulate matter (PM), a combustion product that can present serious health and environmental risks, is formed in fuel-rich zones when fuel and air are not fully premixed, and in particular in a DISI engine when the fuel spray impinges on the cylinder wall or piston [1, 2]. Research has shown that increasing the ethanol percentage in gasoline produces, in most cases, a reduction in PM [1, 3, 4] but in other cases elevated PM levels [5, 6]. General observations suggest that for fuels containing low ethanol concentrations (e.g. E0 – E15) PM emissions are insensitive to ethanol content [2]. For moderate ethanol concentrations (e.g. E20-E40), increased PM emissions relative to E0 have been observed [3, 6]. PM is reduced at high ethanol concentrations, such as E85, because it is likely that blending with ethanol dilutes PM precursors such as aromatic compounds in the gasoline [1].

The fundamental cause of the PM increase seen for moderate ethanol concentrations is not well understood. Current theories suggest that the high HOV of ethanol can cool the cylinder environment slowing the evaporation rate and preventing the highest boiling compounds in gasoline from fully vaporizing and mixing with air in the combustion chamber. The formation of this non-homogenous mixture would be even more likely if there was impingement of the fuel spray on the cylinder wall or piston. The pool or film of fuel can burn in a diffusion flame where these high boiling species are more concentrated in the rich zone leading to hydrocarbon molecular weight growth and increased PM production [5, 6]. Additionally, ethanol as well as other potential biofuels can form azeotropes when blended with gasoline. These non-ideal interactions can alter the boiling temperatures of the fluid as well as the time dependent vapor and liquid composition that can dictate the in-cylinder reactant composition distribution. For example, these interactions can enrich the heavy fractions of the liquid droplet with certain species such as aromatics which have shorter kinetic pathways and greater potential to form PM [3, 7].

Particulate Matter Index (PMI), a prevalent method for predicting emissions of gasolines, does not include effects of non-ideal vapor-liquid equilibrium or HOV, which may affect its suitability for biofuel blends. While there are sooting indices other than PMI, they also do not include HOV and non-ideal vaporization. There are some group-based approximations of threshold sooting index (TSI), but TSI is not intended for oxygenated blends [8]. Oxygen

Extended Sooting Index (OESI) is based on TSI and does take oxygen chemistry into account, but does not incorporate the impact of an oxygenate on relevant phase change and thermophysical properties (e.g. VLE and HOV) [9].

A numerical model was previously created to simulate the droplet evaporation dynamics of ethanol/gasoline fuel blends.[10] Results from these simulations indicate an impact of aromatic enrichment by non-ideal vaporization and an extension of droplet lifetimes by the heat of vaporization present in ethanol. This work provides insight into this enrichment effect from non-ideal VLE by presenting a novel mixture of hydrocarbons that was chosen to eliminate preferential vaporization that occurs because of differences in volatility. In this case, temperature and composition changes during distillation are due solely to the non-ideal interactions within the mixture.

2. Method

2.1 Ethanol Cyclohexane and Benzene Blends

The surrogate blend components were chosen to have the same vapor pressure at their boiling point, roughly 351.5 K. This implies that in the case of ideal vaporization, boiling temperature and vapor composition would remain nearly constant throughout distillation, which was confirmed by the ideal distillation model. In contrast to our previous work [10], this allows us to examine the effect of ethanol on aromatic enrichment based only on the effect of non-ideal vaporization. The test matrix of compositions is shown in Table 1.

2.2 Distillation

Distillations were conducted using the advanced distillation curve method (ADC) [11-15]. A volume of 200 mL of each mixture was prepared for distillation. During distillation, temperature was recorded and condensate sampling was performed at specified intervals based on volume fraction distilled. Distillate sampling was carried out through a rubber septum on the side of the receiver adaptor directly above the glass hammock. A 100 μL syringe with a gas lock valve was used to withdraw the sample. Each sample was then mixed in a vial with 1.5-2 mL of dichloromethane before being analyzed with gas chromatography – flame ionization detection (GC-FID) using a custom isothermal (40 °C) method which was calibrated to allow automated calculation of species mass fractions.

Table 1: Composition of each blend by percent volume of cyclohexane, benzene, and ethanol.

Blend	Ethanol	Cyclohexane	Benzene
1	10	10	80
2	10	40	50
3	10	60	30
4	10	80	10
5	20	30	50
6	40	10	50
7	40	50	10
8	60	10	30
9	60	30	10
10	80	10	10

2.3 Distillation Model

A brief and general description of the model is provided below. Details and equations utilized in the multicomponent fuel droplet evaporation model were described by Burke et al. [10]. Ideally, vapor-liquid equilibrium (VLE) can be determined from Raoult's Law and Dalton's Law. The mole fraction of each component in the liquid is defined as x_i and the mole fraction within the vapor as y_i. The ratio of these values defines the equilibrium constant (K_i) which can be found knowing the ratio of the species' vapor pressure (p_i^0) to the total pressure of the system (p_{tot}), multiplied by an interaction term (γ_i) to account for non-ideal vaporization behaviors (typical of biofuels). If molecular interactions are neglected, the interaction term can be ideally assumed to be unity. To model a droplet of a complex fuel, one needs to be able to predict the change of composition of the fluid as it evaporates. This change in the liquid composition can be described as a function of the instantaneous vapor composition [16], and simplified using the equilibrium ratio to an ordinary differential equation.

2.4 Particulate Matter Index

The PMI of each blend across the domain of possible ratios of ethanol, cyclohexane, and benzene was calculated. PMI will vary between initial blend compositions primarily based on double bond equivalency of each species, (DBE) and non-ideal interaction with ethanol, as the components were chosen for their similar volatility. The masses of ethanol, benzene, and cyclohexane were determined based on volume fraction and density at 273 K. The resulting compositions were used to calculate the PMI of each fuel, as shown in Eq. (1). DBE refers to the double bond equivalency, Wt refers to the percent weight, and $P_v(443K)$ refers to the vapor pressure in kPa of each species at a temperature of 443 K [17].

$$PMI = \sum_i \left[\frac{DBE_i + 1}{P_v(443K)_i} Wt_i\right] \quad (1)$$

3. Results

3.1 Distillations

Ideal and non-ideal vaporization distillation simulations were performed for all possible mixtures of ethanol, cyclohexane and benzene by varying the concentration of each species between 0 and 99 percent of the mixture by volume in one percent intervals. This included binary mixtures of each component calculation (not shown). An example of these distillation curves along with experimental results is shown in Figure 1 for an arbitrary blend. By design, temperature would be constant in an ideal case. Therefore, temperature changes are driven only by non-ideal vaporization. In the example shown in Figure 1, the low saturation temperature between 0 and 70 percent distilled is a result of the strong non-ideal interactions with ethanol. The final temperature remains lower than the ideal case due to the weaker interaction between cyclohexane and benzene. The accuracy of the model is highly sensitive to error in the empirical coefficients used in the calculation of the activity coefficient of each species. However, the UNIFAC method was found to provide an accurate prediction of both composition and temperature.

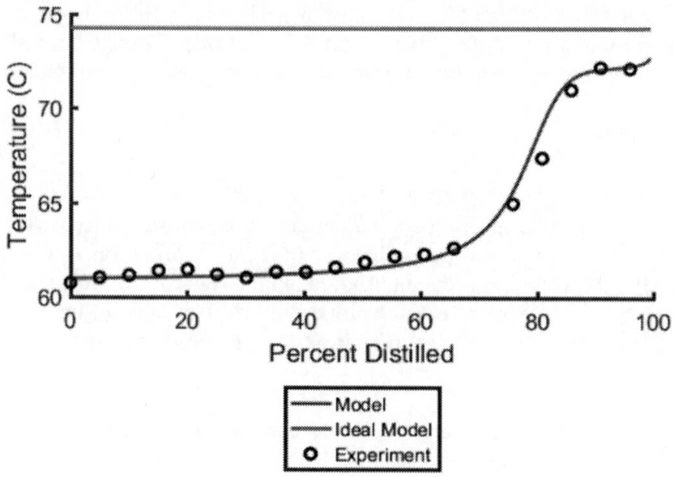

Figure 1: A comparison of the experimental boiling tempartures, the non-ideal model, and the ideal distillation model for distillation of a mixture of 20 percent ethanol. 50 percent benzene, and 30 percent cyclohexane by volume.

Three regions within the initial liquid composition domain were identified based on changes to the liquid composition during distillation/. The first region corresponds to ethanol enrichment, which occurred when the neat mixture was greater than roughly 30 percent volume ethanol. The cyclohexane enrichment region applied to mixtures with less than 30 percent ethanol and greater than 50 percent cyclohexane by volume. The final region, which describes benzene enrichment, occurs when the ethanol content is less than 30 percent and the benzene content at least 50 percent.

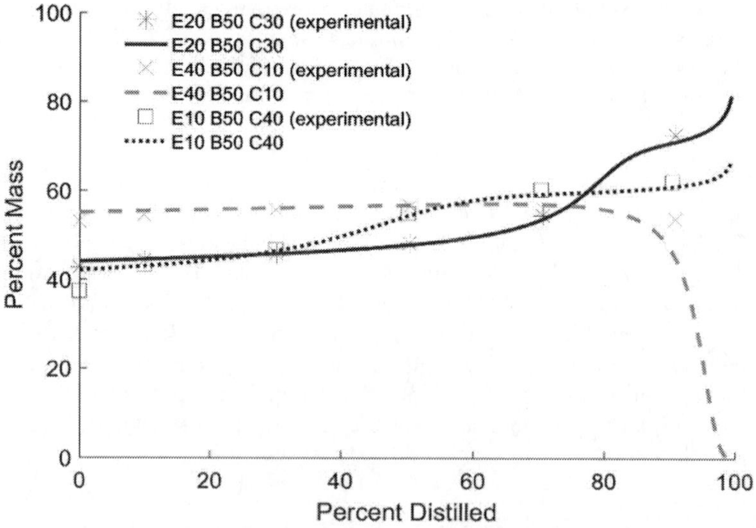

Figure 2: Measured and predicted vapor concentrations of benzene for three mixtures initially containing 50 percent benzene by volume, which vary in ethanol and cyclohexane content.

Based on these regions, 10 blends were selected so that comparisons could be made between blends with the same initial volume of a given species. These volumes were 10, 30, and 50 percent benzene, 10 percent ethanol, and 10 percent cyclohexane. In Figure 2, the vapor benzene concentrations during distillations of the 50 percent benzene blends are shown. These were measure via GC-FID and predicted by the UNIFAC-based non-ideal distillation model. While containing equal amounts of benzene initially, the vapor concentrations differ at the initial boiling point based on the ratio of cyclohexane to ethanol. For the E10 and E20 cases, benzene enrichment occurs at different inflection points, occurring later for E20 as complete vaporization of the ethanol in the blend occurs later in distillation.

Because of this, the final concentration of liquid and vapor benzene increase with ethanol content up to roughly 30 percent ethanol, above which ethanol enrichment instead occurs, as demonstrated in the case of 40 percent ethanol in Figure 2. In the 40 percent ethanol blend, benzene (and to a lesser extent cyclohexane) vaporization is enhanced, resulting in pure ethanol for the final liquid and vapor composition.

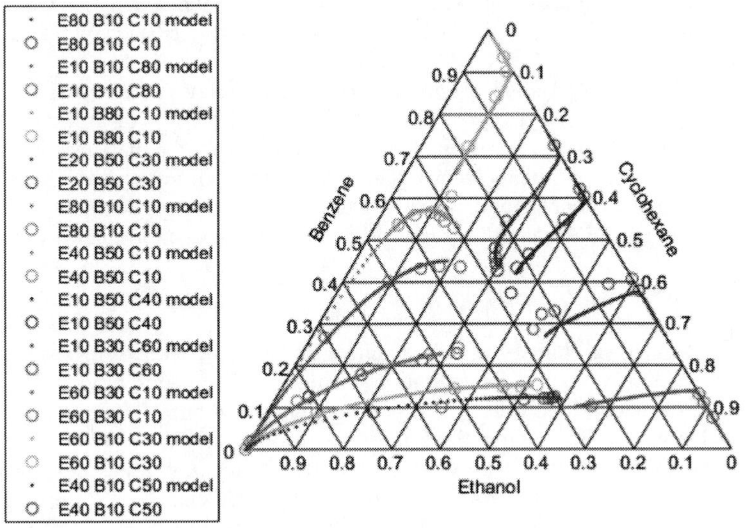

Figure 3: Ternary plot of predicted and measured vapor compositions by mass fraction of each species as it changes throughout distillation for each fuel blend in the test matrix

The enrichment regions were identified based on models of the liquid composition, which was not directly measured, but they are demonstrated in the measured vapor compositions during distillation, as shown in Figure 3. In the ethanol enrichment region, cyclohexane appears to vaporize more rapidly than benzene. As the cyclohexane concentration decreases, there can be an increase in benzene concentration as seen in the 40 percent ethanol 110 percent benzene mixture, though ethanol becomes dominant as distillation progresses. This implies that blends with certain concentrations of ethanol may be particularly sensitive to the aromatic enrichment effect of ethanol described in our previous study of FACE B ethanol blends [10].

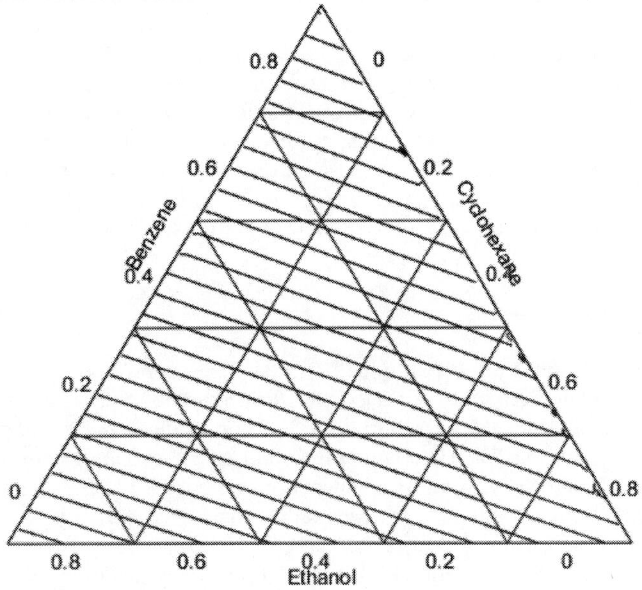

Figure 4: Contour plot showing contours of constant PMI as a function of initial mixture composition by volume fraction.

The inverse may be true when the ethanol in an E10 blend with low aromatic content has fully vaporized, as in the 10 percent ethanol 10 percent benzene mixture. In that case, benzene vaporization is enhanced by cyclohexane when it is no longer in the presence of ethanol. In a gasoline with significant cycloalkanes, and perhaps heavier alkanes and isoalkanes, these species could enhance aromatic vaporization once the ethanol is exhausted, mitigating some of the aromatic enrichment that would have occurred while ethanol vaporized.

3.2 PMI

The particulate matter index was calculated for each mixture in the ternary blend domain in one percent intervals. Contours of PMI were identified such that blend compositions falling along a given contour have equal PMI, as in Figure 4. Many of these contours pass through all three enrichment regimes as previously described. Mixtures along such a contour can thus be chosen so that they would have identical predicted sooting behavior in the absence of preferential vaporization, which PMI does not account for. However, the final composition can be rich in either benzene, cycloalkane, or ethanol in varying concentrations based on the initial fuel composition. Aromatic content of impinged and pooled fuel could be influenced independently of PMI, allowing more direct observation of the effect of the previously described aromatic vaporization on particulate emissions in direct injection engines. Similarly, these surrogates can be used in spray flames to evaluate the effect of preferential vaporization by influencing the location of aromatic vaporization within the flame. Contours of equal premixed ignition delays were also calculated for mixtures of benzene, ethanol, and cyclohexane, but await further validation of kinetic mechanisms before a fuel matrix can be designed (not shown). Such a matrix would have applications to investigation of preferential vaporization on knock and ignition behavior by controlling the order in which species vaporize and mix.

4. Conclusion

A novel ternary gasoline surrogate was developed to measure the non-ideal vaporization effects of oxygenates on aromatic concentration during gasoline vaporization. To control for ideal vaporization effects, two species were identified with similar boiling points to ethanol; benzene and cyclohexane. A UNIFAC-based distillation model was used to identify three regimes of vaporization behavior which determine the final vapor composition. Various test blends were selected for ADC distillation experiments to provide evidence for these regimes. Saturation temperature and vapor mass concentration measurements during these distillations confirm the existence of three regions of non-ideal behavior, as well as the suitability of UNIFAC as a method for estimating the boundaries of these regions.

The enrichment behaviors identified were dependent on the ratio of the ternary components, providing insights into observations made in prior works. For ethanol concentrations beyond 30 percent, there is an inherent tendency to increase the volatility of cyclohexane and thus cause enrichment in aromatics as vaporization progresses. This may partly explain the inconsistent findings on PM effects of intermediate ethanol blends such as E30. At lower ethanol concentrations, cyclohexane and benzene were enriched equally until ethanol was fully vaporized, at which point one of the remaining species was enriched by non-ideal behavior depending on their ratios. It is possible that this may mitigate some of the initial aromatic enrichment effect for blends low in ethanol and aromatics.

Further applications of these surrogates have been explored, including a new methodology for measuring effects of preferential vaporization on pool fire particulate matter emissions. Work is ongoing to adapt these blends into a matrix of surrogate fuel suitable for studying knock and ignition behavior as influenced by preferential vaporization.

References

[1] M. Fatouraie, M.S. Wooldridge, B.R. Petersen, S.T. Wooldridge, Effects of Ethanol on In-Cylinder and Exhaust Gas Particulate Emissions of a Gasoline Direct Injection Spark Ignition Engine, Energy & Fuels 29 (2015) 3399-3412.
[2] M. Fatouraie, M. Wooldridge, S. Wooldridge, In-Cylinder Particulate Matter and Spray Imaging of Ethanol/Gasoline Blends in a Direct Injection Spark Ignition Engine, SAE Int J. Fuels Lubr. (2013).
[3] M.A. Ratcliff, J. Burton, P. Sindler, E. Christensen, L. Fouts, G.M. Chupka, R.L. McCormick, Knock Resistance and Fine Particle Emissions for Several Biomass-Derived Oxygenates in a Direct-Injection Spark-Ignition Engine, SAE Int J. Fuels Lubr. (2016).
[4] M.M. Maricq, J.J. Szente, K. Jahr, The Impact of Ethanol Fuel Blends on PM Emissions from a Light-Duty GDI Vehicle Aerosol Science and Technology 46 (2012) 576-583.
[5] A.D. Butler, R.A. Sobotowski, G.J. Hoffman, P. Machiele, Influence of Fuel PM Index and Ethanol Content on Particulate Emissions from Light-Duty Gasoline Vehicles, SAE International, 2015.
[6] M. Storch, M. Koegl, M. Altenhoff, S. Will, L. Zigan, Investigation of soot formation of spark-ignited ethanol-blended gasoline sprays with single- and multi-component base fuels, Applied Energy 181 (2016) 278-287.
[7] K. Aikawa, T. Sakurai, J.J. Jetter, Development of a Predictive Model for Gasoline Vehicle Particulate Matter Emissions, SAE Int J. Fuels Lubr. (2010).

[8] P. Pepiot-Desjardins, H. Pitsch, R. Malhotra, S.R. Kirby, A.L. Boehman, Structural group analysis for soot reduction tendency of oxygenated fuels, Combustion and Flame 154 (2008) 191-205.
[9] E.J. Barrientos, M. Lapuerta, A.L. Boehman, Group additivity in soot formation for the example of C-5 oxygenated hydrocarbon fuels, Combustion and Flame 160 (2013) 1484-1498.
[10] S. Burke, M. Ratcliff, R. McCormick, R. Rhoads, B. Windom, Distillation-based Droplet Modeling of Non-Ideal Oxygenated Gasoline Blends: Investigating the Role of Droplet Evaporation on PM Emissions, SAE Int. J. Fuels Lubr. 10 (2017).
[11] B.L. Smith, T.J. Bruno, Advanced Distillation Curve Measurement with a Model Predictive Temperature Controller, International Journal of Thermophysics 27 (2006) 1419-1434.
[12] T.J. Bruno, Improvements in the Measurement of Distillation Curves. 1. A Composition-Explicit Approach, Industrial & Engineering Chemistry Research 45 (2006) 4371-4380.
[13] T.J. Bruno, B.L. Smith, Improvements in the Measurement of Distillation Curves. 2. Application to Aerospace/Aviation Fuels RP-1 and S-8, Industrial & Engineering Chemistry Research 45 (2006) 4381-4388.
[14] B.C. Windom, T.M. Lovestead, T.J. Bruno, Application of the Advanced Distillation Curve Method to the Development of Unleaded Aviation Gasoline, Energy & Fuels 24 (2010) 3275-3284.
[15] B.C. Windom, T.J. Bruno, Improvements in the Measurement of Distillation Curves. 5. Reduced Pressure Advanced Distillation Curve Method, Industrial & Engineering Chemistry Research 50 (2011) 1115-1126.
[16] J. Backhaus, Design methodology of bio-derived gasoline fuels, Mechanical Engineering, University of Wisconsin, Madison, WI, 2013.
[17] K. Aikawa, T. Sakurai, J.J. Jetter, Development of a Predictive Model for Gasoline Vehicle Particulate Matter Emissions, (2010).

Western States Section of the Combustion Institute – Fall 2017 Meeting
Hosted by the University of Wyoming
October 2-3, 2017

Performance of a Compression Ignition Engine Fueled with Renewable Diesel Blends Produced from Hydrothermal Liquefaction, Fast Pyrolysis, and Conversion of Ethanol to Diesel

Jessica Tryner[1] *Karl Albrecht*[2] *Justin Billing*[2] *Richard T. Hallen*[2] *Anthony J. Marchese*[1]

[1]*Department of Mechanical Engineering, Colorado State University, Fort Collins, CO*
[2]*Chemical and Biological Process Development Group, Pacific Northwest National Laboratory, Richland, WA*

Hydrothermal liquefaction (HTL) is a promising technology for converting wet biomass to renewable fuel blendstocks. HTL can be combined with catalytic hydrotreating and distillation to produce high-quality hydrocarbon cuts that can be used as blendstocks for diesel fuel, aviation fuel, or gasoline. In this study, six different renewable diesel fuels were blended with certification ultra-low sulfur diesel (ULSD) and tested in a 4-cylinder, 4.5 L, turbocharged, Tier 3, off-highway compression-ignition engine operating at rated speed (2400 rpm) and ~75% load. The renewable blendstocks included fuels derived from: (1) HTL of wood, (2) HTL of microalgae, (3) HTL of a blend of wood and microalgae, (4) HTL of sewage sludge, (5) fast pyrolysis of wood, and (6) conversion of ethanol to diesel. The HTL fuels were blended at 5% by volume with certification ULSD. The fast pyrolysis fuel was blended at 5% and 15% by volume, and the ethanol-to-diesel fuel was blended at 15% by volume. Engine performance was characterized in terms of fuel consumption, thermal efficiency, and in-cylinder heat release rate, as well as emissions of carbon monoxide (CO), oxides of nitrogen (NO_x), total hydrocarbons, and particulate matter. Performance of the engine operating on the renewable fuel blends was compared to performance of the engine operating on conventional petroleum diesel. Results indicated that none of the renewable fuel blends had an adverse impact on the measured performance metrics.

1. Introduction

Hydrothermal liquefaction (HTL) is a thermochemical conversion process used to convert wet biomass from a number of feedstocks—such as wood, algae, and sewage sludge—into biocrude [1,2]. Biocrude, which is a viscous mixture of oxygenated hydrocarbons, can be hydrotreated to produce renewable liquid hydrocarbon fuels [3,4]. In an HTL process, a slurry of biomass particles, typically at 10—20 wt% solids in water, is pressurized to 4—22 MPa and heated to 250—374 °C [2]. HTL may be a particularly advantageous route for converting wet feedstocks, such as microalgae and sewage sludge, into biofuels to avoid drying before conversion [2]. The National Alliance for Advanced Biofuels and Bio-products (NAABB), a research consortium of 39 institutional partners that spent three years working to advance technology for algal biofuels production across the entire value chain, identified HTL combined with catalytic hydrothermal gasification (CHG) as the conversion process that resulted in the highest oil yields, as well as the best process economics and lifecycle assessment [5]. Using HTL, the lipid, carbohydrate, and protein fractions of microalgae cells can all be converted to biocrude and, as a result, cells with a high lipid content are not required [5]. Elliott et al. [3] reported that the yield of liquid hydrocarbon fuel from HTL of *Nannochloropsis* was 40% of the original mass of algae.

To displace petroleum fuels, renewable fuels produced via HTL must meet all standards for fuel properties (e.g., ASTM D 975 and EN 590) and must not have any adverse impacts on engine performance. However, no data are available on how fuels containing renewable fuel blendstocks derived from HTL affect engine performance in terms of fuel efficiency and emissions. In this study, six different renewable diesel fuels—including fuels derived from HTL of wood, a blend of wood and algae, algae, and sewage sludge; a fuel derived from fast pyrolysis of wood, and a fuel produced by converting ethanol to diesel—were blended with certification ultra-low sulfur diesel and tested in a compression-ignition engine. The HTL fuels were blended at 5% by volume, the pyrolysis fuel was blended at 5% and 15% by volume, and the ethanol-to-diesel fuel was blended at 15% by volume. Performance of the engine operating on these renewable fuel blends was compared to performance of the engine operating on conventional petroleum diesel. Engine performance was evaluated in terms of fuel consumption, thermal efficiency, and in-cylinder heat release rate, as well as emissions of carbon monoxide (CO), oxides of nitrogen (NO_X), total hydrocarbons (THC), and particulate matter (PM).

2. Methods

2.1 Fuels tested

Six renewable fuels were tested. These fuels were produced from: (1) HTL of wood, (2) HTL of a mixture of wood and algae, (3) HTL of algae, (4) HTL of sewage sludge, (5) fast pyrolysis of wood, and (6) conversion of ethanol to diesel via the process described in US patent 9,663,416. Properties of these fuels—including density, molar H:C ratio, higher heating value (HHV), lower heating value (LHV), and derived cetane number (DCN)—are shown in **Table 1**. Density was measured using an Anton-Parr DSA 5000 M Density and Sound Velocity Meter. Higher heating value was measured using an IKA C200 Calorimeter. Derived centane number was measured using a fuel ignition tester.

Table 1. Properties of the certification diesel, the renewable fuels, and the renewable fuel blends

Fuel	Volume of renewable fuel (%)	Density (g/mL)	Molar H:C ratio	HHV (MJ/kg)	LHV (MJ/kg)	Test date	No. tests	DCN
Certification diesel	0	0.848	1.8	45.5	42.8	12/2016	10	44.5
Certification diesel	0	0.848	1.8	45.4	42.7	08/2017	10	44.9
Ethanol-to-diesel	100	0.791	2.1	47.0	43.8	-	-	66.7
Wood HTL	100	0.894	1.6	44.5	42.0	-	-	29.7
Wood/algae HTL	100	0.858	1.8	45.1	42.3	-	-	-
Algae HTL	100	0.803	1.9	46.9	44.0	-	-	-
Sewage HTL	100	0.808	1.8	46.5	43.7	-	-	-
Wood pyrolysis	100	0.903	1.6	44.0	41.6	-	-	-
Ethanol-to-diesel blend	15	0.840	1.8	45.7	42.9	12/2016	2	46.1
Wood HTL blend	5	0.851	1.8	45.5	42.7	12/2016	1	43.8
Wood/algae HTL blend	5	0.849	1.8	45.5	42.8	12/2016	2	-
Algae HTL blend	5	0.845	1.8	45.5	42.7	08/2017	3	-
Sewage HTL blend	5	0.845	1.8	45.5	42.7	08/2017	1	-
Wood pyrolysis blend #1	5	0.850	1.8	45.3	42.6	08/2017	2	-
Wood pyrolysis blend #2	15	0.856	1.7	45.2	42.5	08/2017	3	-

The fuels tested in the engine were blends of certification diesel and the renewable fuels listed above. Testing was completed over the course of two days in December 2016 and two days in August 2017, and a different barrel of certification diesel was used for each set of tests. The properties of the certification diesel are also listed in **Table 1**. The density and HHV of the certification diesel were measured. The molar H:C ratio was provided by the distributor.

The properties of the blended fuels are also listed in **Table 1**. These values were calculated based on the properties of the renewable fuel blends, the properties of the certification diesel, and the volume fractions of each fuel in each blend. Each renewable fuel was blended with certification diesel at 5—15% by volume and one to three test replicates were completed with each blend, depending on fuel availability. Certification diesel tests were completed at the start and end of each test day, and in between each test with a blended fuel. Overall, ten certification diesel tests were completed in December 2016, and ten were completed in August 2017.

2.2 Engine

The fuels were tested in a John Deere 4045 PowerTech Plus diesel engine (see **Figure 1**). This engine has been used previously to evaluate the performance of triglyceride blends, biodiesel, and renewable diesel [6], as well as surrogates for fuels derived from upgraded pyrolysis oil [7]. The engine has also been used to investigate potential health impacts of, and study secondary formation of organic aerosol and isocyanic acid from, diesel and biodiesel exhaust [8–10]. The 4-cylinder, 4.5 L engine was equipped with electronically-controlled high-pressure common rail fuel injection, a variable geometry turbocharger, and an intercooler. The engine utilized exhaust gas recirculation and was designed to meet Tier 3/Stage IIIA emissions specification. The engine was not equipped with a diesel particulate filter, diesel oxidation catalyst, or a selective catalytic reduction system. A pneumatic valve installed on a cooling water loop and connected to a PID controller was used to maintain a constant air temperature (60 °C) at the outlet of the intercooler. The engine was operated at 2400 rpm (rated speed) and 338 N·m. The power output at these conditions was 85 kW. Constant speed and load were maintained using an eddy current dynamometer (Midwest Inductor Dynamometer 1014A) and a dynamometer controller (Dyne Systems Dyn-Loc IV).

Figure 1. The 4-cylinder, 4.5 L compression-ignition engine in which the fuels were tested.

2.4 Test Procedure

In between test points, the engine was fueled with dyed-red off-road diesel, which was stored in the main fuel tank for the engine. The certification fuel used for the baseline tests and the fuel blends were stored in five-gallon auxiliary fuel tanks. Three way valves were installed on the fuel and return lines to enable switching between off-road diesel and the test fuels.

Before the start of each test, the tank containing the test fuel was placed on the balance, and a fuel line with a check valve on the end was placed in the tank. The three-way valve at the inlet to the fuel filter was switched so that the fuel pump would draw fuel from the auxiliary tank into the engine. Simultaneously, the three-way valve on the return line was switched so that the return flow would empty into a waste bucket. The engine was operated in this manner, with test fuel being drawn from the auxiliary tank on the balance and return fuel draining to the waste bucket, until no red fuel was observed in the return line (~2 minutes). Then, the return line was placed in the auxiliary tank and the engine operated on the test fuel for five to six minutes while the dilution chamber equilibrated. Once the equilibration period was over, the ten-minute test was started. One-hundred in-cylinder pressure traces were collected half-way through the test.

2.3 Measurement of Fuel Consumption and Efficiency

Fuel consumption was measured by placing the auxiliary fuel tank containing the test fuel on a balance with a capacity of 32.2 kg and a readability of 1 g (Mettler Toledo MS32000L). The output from the balance was logged continuously at 1-2 Hz.

Brake-specific fuel consumption (BSFC) and thermal efficiency (η) were calculated as shown in Equations (1 and (2, respectively:

$$BSFC = \frac{\dot{m}_f}{W_b} \tag{1}$$

$$\eta = \frac{W_b}{\dot{m}_f LHV} \tag{2}$$

where $\dot{m}_f$ is the fuel consumption rate in $g \cdot h^{-1}$, W_b is the brake power output in kW, and LHV is the lower heating value of the fuel in $kJ \cdot g^{-1}$.

2.4 Measurement of Gaseous Emissions

The mole fractions of carbon monoxide (CO) and carbon dioxide (CO_2) in the exhaust were measured using a non-dispersive infrared (NDIR) sensor (Siemens ULTRAMAT 6). Oxygen (O_2) was measured using a paramagnetic detector (Siemens OXYMAT 6), oxides of nitrogen (NO_x) were measured using a chemiluminescence detector (Siemens NOXMAT 600), and total hydrocarbons (THC) were measured using a flame ionization detector (Siemens FIDAMAT 6). Each analyzer was zeroed and spanned every morning using pure N_2 and calibration gases. The flame ionization detector was spanned using methane (CH_4).

The mole fractions of CO, CO_2, O_2, NO_x, and THC in the exhaust gas were logged continuously at 1-2 Hz. Exhaust gases were transported to the analyzers by a heated sample line, and all measurements were made on a dry basis.

The mass flow rate CO, NO_x, and THC in the exhaust was calculated by performing a carbon balance on Reaction 1 [11].

$$C_\alpha H_\beta + \frac{a_s}{\phi}(O_2 + 3.76N_2) \rightarrow n_1 CO_2 + n_2 H_2O + n_3 N_2 + n_4 O_2 + n_5 CO + n_6 H_2 + n_7 CH_z + n_8 NO_x \tag{R1}$$

where a_s is the number of moles of air required for stoichiometric combustion, and ϕ is the equivalence ratio.

The mass flow rate of species i in the exhaust was calculated as shown in Equation 3.

$$\dot{m}_i = \frac{\dot{m}_f y_i^o M_i}{[M_C + (\beta/\alpha)M_H](y_1^o\alpha_1 + y_5^o\alpha_5 + y_7^o\alpha_7)} \quad (3)$$

where $\dot{m}_f$ is the fuel consumption rate in g·h^{-1}, y_i^0 is the dry mass fraction of species i in the exhaust, M_i is the molecular weight of species i in g·mol^{-1}, and α_i is the carbon number of species i.

Brake-specific emission rates (g·kW^{-1}·h^{-1}) for each species were calculated as:

$$BSE_i = \frac{\dot{m}_i}{W_b} \quad (4)$$

2.5 Measurement of Particulate Emissions

Exhaust was sampled into a 200-L dilution chamber to characterize particulate emissions from the engine. The exhaust was diluted with filtered compressed air using a Dekati Diluter L7. The mass flow rate of dilution air was maintained at 50 SLPM using a mass flow controller (Alicat Scientific MC-50SLPM-D-DB15/10M). Diluted exhaust was sampled out of the dilution chamber into a filter sample line, a bypass filter sample line, a scanning mobility particle sizer (SMPS), and an NDIR CO_2 analyzer. An exhaust line with a particulate filter on the outlet maintained a constant pressure inside the chamber.

Diluted exhaust flowed through the filter sample line to two parallel filter trains. One filter train consisted of a 2.5 μm cutpoint cyclone (URG Corp. 16.7 LPM, 2.5 μm) followed by a 47-mm-diameter PTFE filter (Whatman 7592-104) and a pre-baked quartz filter (Pallflex Tissuquartz, 2500 QAT-UP). The second filter train consisted of a 2.5 μm cyclone followed by two quartz filters. The PTFE filter sample was used to determine the total mass of $PM_{2.5}$ emitted. The front quartz filter in the second train was used to determine the mass fractions of elemental and organic carbon in the collected particulate matter. The quartz filter behind the PTFE filter was used to correct the organic carbon fraction for positive artifact resulting from adsorption of gas-phase organic compounds [12].

When a test was not in progress, a total flow rate of 33.4 L·min^{-1} was maintained through a separate set of parallel filter trains. This "bypass" filter sample line was used to keep the total flow rate out of the dilution chamber constant.

The PTFE filters were pre- and post-weighed using a balance with a resolution of 1 μg (Mettler-Toledo MX-5) after equilibrating in the microbalance room for 24 hours. The total mass of $PM_{2.5}$ emitted was calculated based on the mass of $PM_{2.5}$ collected on the PTFE filter, the known mass flow rate of diluted exhaust through the filter, the dilution ratio, and the total mass flow rate of the exhaust. The dilution ratio was calculated based on the mole fractions of CO_2 in the exhaust and in the dilution chamber.

The quartz filters were analyzed according to NIOSH method 5040 using a Sunset Laboratory Inc. Lab OC-EC Aerosol Analyzer. Quartz filters were stored at -25 °C for 5 to 14 days prior to analysis.

The size distribution of the particles emitted from the engine was characterized by sampling diluted exhaust into a scanning mobility particle sizer (GRIMM SMPS+C). The SMPS sized particles between 20.70 nm and 473.58 nm in mobility diameter into 45 channels. A new scan began every 298 s, and two full scans were collected during each test.

2.6 Measurement of In-Cylinder Heat Release

The pressure in cylinder 1 of the engine was measured using a Kistler 6056A high-speed pressure transducer. Crank angle position was measured using an incremental encoder. The data was post-processed using a program written in LabView. The crank angle position and in-cylinder pressure were used to calculate the net rate of heat release (J·CA°$^{-1}$) during the combustion process as shown in Equation 5:

$$\frac{dQ}{d\theta} = \frac{\gamma}{\gamma - 1} P \frac{dV}{d\theta} + \frac{1}{\gamma - 1} V \frac{dP}{d\theta} \quad (5)$$

where Q is the net heat release in J, γ is the ratio of specific heats, P is the in-cylinder pressure in Pa, V is the cylinder volume in m^3, and θ is the crank angle in degrees. High-frequency noise was removed from the heat release profile using low-pass filter. The data was filtered in both the forward and reverse directions to negate the phase shaft.

3. Results and Discussion

The baseline performance of the engine fueled with conventional petroleum diesel (2007 Certification Diesel) is described by the metrics listed in **Table 2**. The brake-specific emissions listed in **Table 2** are below the limits listed for Tier 3 exhaust emissions standards for nonroad compression ignition engines [13]. Results were relatively similar for the December 2016 and August 2017 tests.

Table 2. Brake-specific fuel consumption, thermal efficiency, brake-specific emissions, and ratio of elemental to organic carbon in the particulate matter for the engine fueled with 2007 Certification Diesel.

	December 2016			August 2017		
	Minimum	Mean	Maximum	Minimum	Mean	Maximum
BSFC (g·KW^{-1}·hr^{-1})	212	214	217	229	231	232
Efficiency (%)	38.9	39.4	39.8	36.3	36.6	36.8
CO (g·KW^{-1}·hr^{-1})	0.64	0.66	0.69	0.66	0.70	0.73
NO_x (g·KW^{-1}·hr^{-1})	3.07	3.31	3.48	3.48	3.56	3.63
THC (g·KW^{-1}·hr^{-1})	0.091	0.097	0.109	0.083	0.086	0.089
$PM_{2.5}$ (g·KW^{-1}·hr^{-1})	0.095	0.102	0.114	0.107	0.111	0.114
EC/TC (%)	63	81	91	81	86	88

Changes in brake-specific fuel consumption (BSFC) and thermal efficiency measured with the renewable fuel blends, relative to the certification diesel baseline, are shown in **Figure 2**. The ranges overlap for the 15% ethanol-to-diesel blend, the 5% wood/algae HTL blend, the 5% algae HTL blend, and both of the upgraded pyrolysis oil blends. The results for the 5% wood HTL blend and the 5% sewage sludge HTL blend fall outside of the range measured for the 10 certification diesel tests. These apparent decreases in BSFC (and corresponding increases in thermal efficiency) may be a result of the lack of replicate tests conducted for those blends, since the LHVs of the blends were all within 0.5% of that of the certification diesel.

Changes in brake-specific CO and NO_x emissions measured with the renewable fuel blends, relative to the certification diesel baseline, are shown in **Figure 3**. No changes in CO emissions were observed with any of the fuel blends; the average values and the total ranges for all of the blends overlap the ranges measured for certification diesel. Similarly, no changes in NO_x emissions were observed; the average values for all of the blends overlap the ranges measured for certification diesel. The slightly lower CO emissions and slightly higher NO_x emissions observed for the test with the 5% wood HTL blend suggest that in-cylinder temperatures were higher during that test.

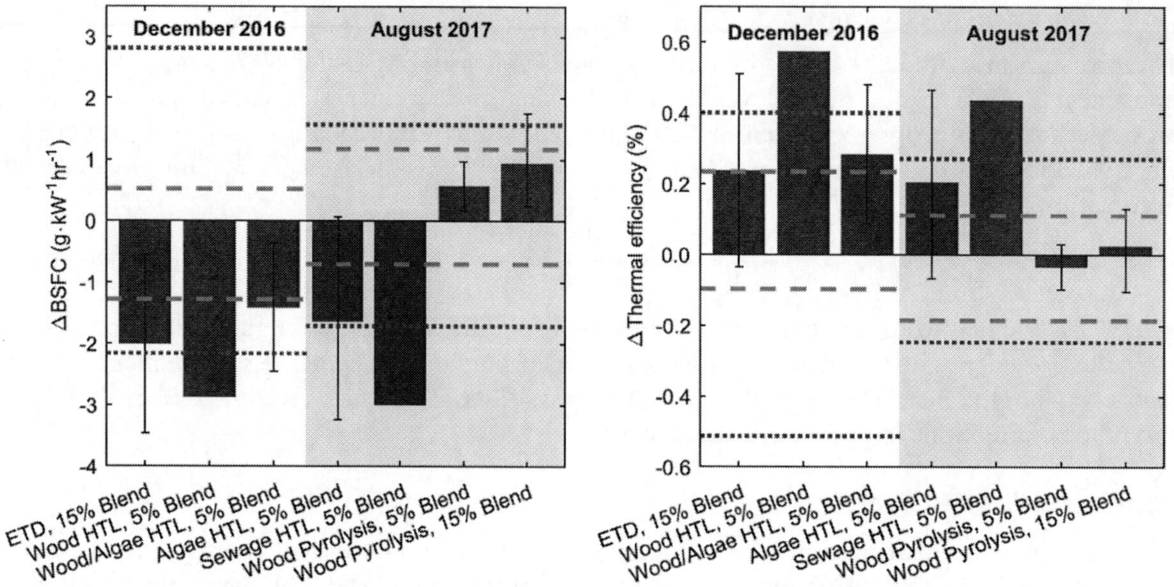

Figure 2. Change in brake-specific fuel consumption (left) and thermal efficiency (right), relative to the baseline certification diesel tests, for the seven renewable fuel blends. The height of the bars represents the difference between the mean for the fuel blend and the mean for the 10 certification diesel tests. The error bars represent the total range of the measurements. No error bars are shown for the 5% wood HTL blend or the 5% sewage HTL blend because only one data point was collected for each of those fuels. The orange dashed line represents the interquartile range for the 10 certification diesel tests, and the gray dotted lines represent the total range of the 10 certification diesel tests.

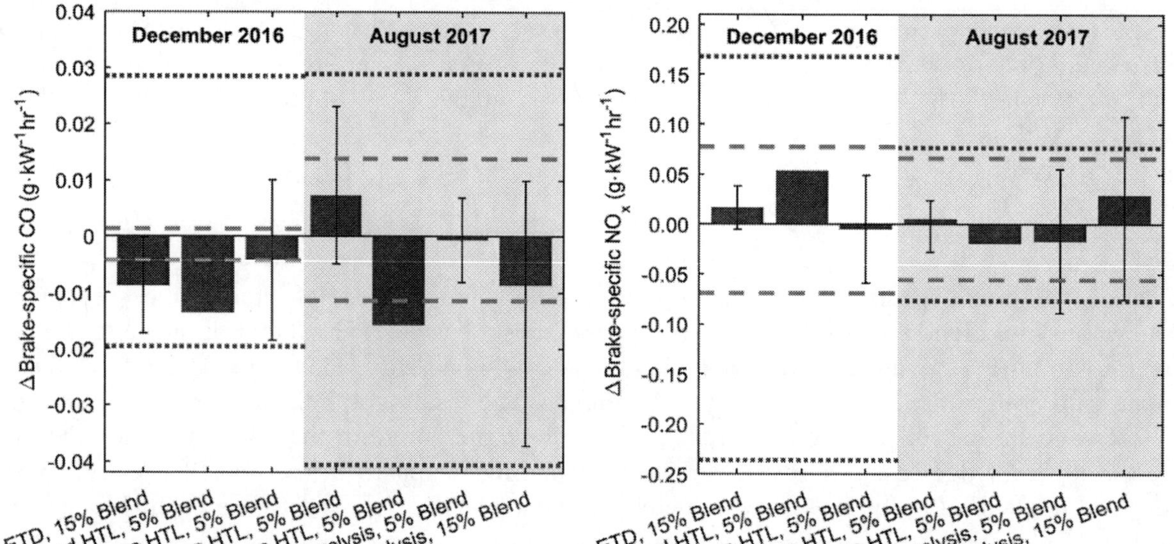

Figure 3. Change in brake-specific CO emissions (left) and brake-specific NO_x emissions (right), relative to the baseline certification diesel tests, for the seven renewable fuel blends. The height of the bars represents the difference between the mean for the fuel blend and the mean for the 10 certification diesel tests. The error bars represent the total range of the measurements. No error bars are shown for the 5% wood HTL blend or the 5% sewage HTL blend because only one data point was collected for each of those fuels. The orange dashed line represents the interquartile range for the 10 certification diesel tests, and the gray dotted lines represent the total range of the 10 certification diesel tests.

Changes in brake-specific total hydrocarbon (THC) emissions are shown in **Figure 4**. No changes were observed for the HTL or upgraded pyrolysis oil blends. The THC emissions measured during the tests with the 15% ethanol-to-diesel blend, on the other hand, were lower than those measured during any of the certification diesel tests.

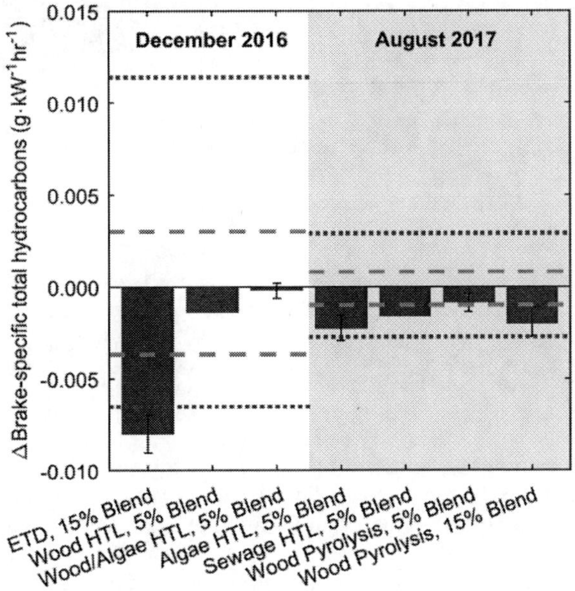

Figure 4. Change in brake-specific total hydrocarbon (THC) emissions, relative to the baseline certification diesel tests, for the seven renewable fuel blends. The height of the bars represents the difference between the mean for the fuel blend and the mean for the 10 certification diesel tests. The error bars represent the total range of the measurements. No error bars are shown for the 5% wood HTL blend or the 5% sewage HTL blend because only one data point was collected for each of those fuels. The orange dashed line represents the interquartile range for the 10 certification diesel tests, and the gray dotted lines represent the total range of the 10 certification diesel tests.

Changes in brake-specific $PM_{2.5}$ emissions, and the fraction of the total carbon collected on the quartz filters that was elemental carbon, are shown on the left side of **Figure 5**. The $PM_{2.5}$ emissions measured during the single test with the 5% Wood HTL blend was at the lower edge of the range of values measured during the certification diesel tests. This result was somewhat unexpected, given that the Wood HTL fuel contained more cyclic an aromatic molecules than the wood/algae HTL fuel or the ethanol-to-diesel fuel. However, only one data point was collected, and the result is consistent with the lower CO and higher NO_x emissions observed for this test. As noted above, lower CO and higher NO_x emissions are likely indicative of higher in-cylinder temperatures, which would be expected to also result in more complete oxidation of particles. Relatively high $PM_{2.5}$ emissions were measured during the first test with the 5% blend of upgraded pyrolysis oil. Emissions measured during the second test with the same fuel were similar to the average baseline emissions. No change in brake-specific $PM_{2.5}$ emissions, relative to the baseline case, was observed with the 15% blend of upgraded pyrolysis oil, or with any of the other renewable fuel blends.

The wide range of EC/TC values measured for certification diesel in December 2016 were the result of a low measurement of 63% and a high measurement of 91%. The other eight measurements from December 2016 fell between 79% and 84%. The range of EC/TC values measured for certification diesel in August 2017 was narrower (81—88%). For the 5% blend of algae HTL fuel and the 15% blend of upgraded pyrolysis oil, the average EC/TC ratio was slightly higher than the maximum ratio measured during the baseline tests.

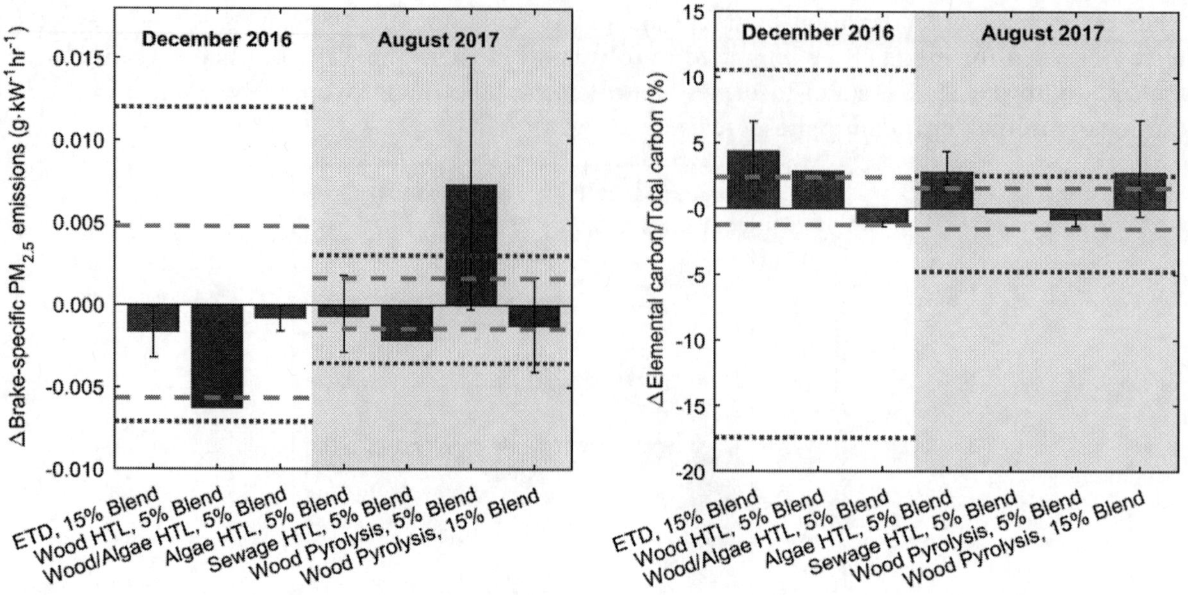

Figure 5. Change in brake-specific $PM_{2.5}$ emissions (left) and the fraction of total carbon in the particulate matter that is elemental carbon (right), relative to the baseline certification diesel tests, for the seven renewable fuel blends. The height of the bars represents the difference between the mean for the fuel blend and the mean for the 10 certification diesel tests. The error bars represent the total range of the measurements. No error bars are shown for the 5% wood HTL blend or the 5% sewage HTL blend because only one data point was collected for each of those fuels. The orange dashed line represents the interquartile range for the 10 certification diesel tests, and the gray dotted lines represent the total range of the 10 certification diesel tests.

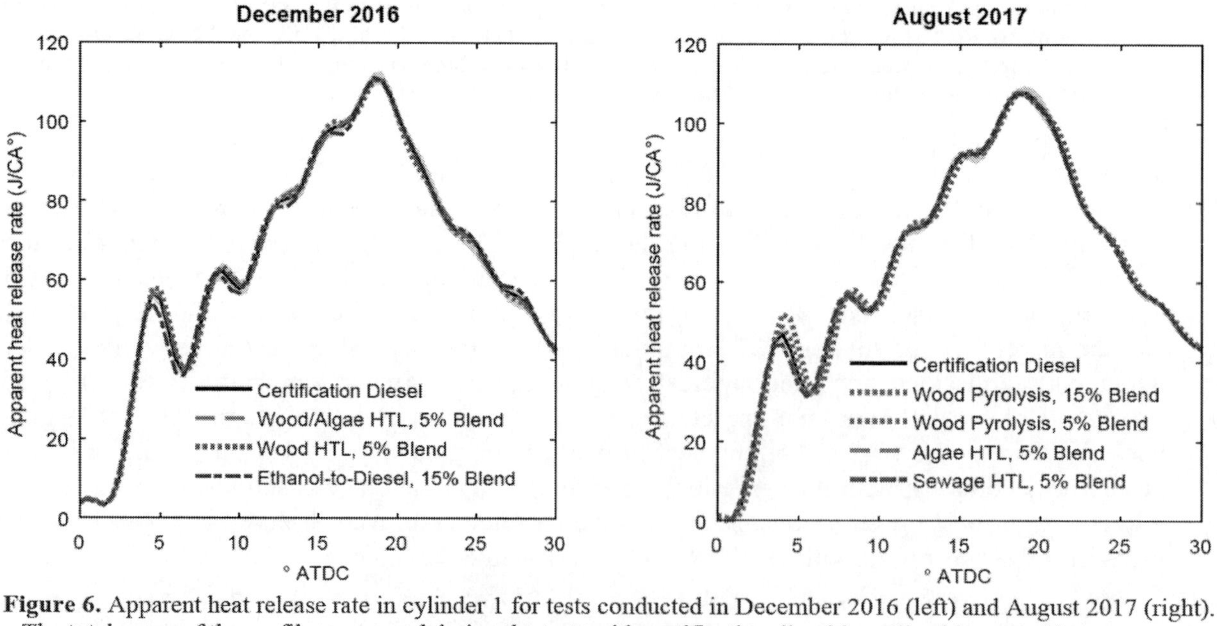

Figure 6. Apparent heat release rate in cylinder 1 for tests conducted in December 2016 (left) and August 2017 (right). The total range of the profiles measured during the tests with certification diesel is outlined in gray. Lines represent average heat release rates.

The average apparent heat release rate calculated for each fuel is shown in **Figure 6**. For the December 2016 tests, the profiles for the 5% HTL fuel blends were almost identical to the profiles for certification diesel. For the 15% ethanol-to-diesel blend, heat release occurred slightly earlier.

The ethanol-to-diesel fuel consisted of almost 100% isoalkanes and had a high derived cetane number (66.7). For the August 2017 results, the heat release profiles for the 5% HTL fuel blends were very similar to each other and to the profile for the baseline certification diesel tests. For the upgraded pyrolysis oil blends, heat release occurred slightly later. A greater difference was observed between the heat release profile for the baseline tests and the heat release profile for the 15% upgraded pyrolysis oil blend than for the 5% upgraded pyrolysis oil blend.

4. Conclusions

Compared to the baseline case of the engine running on 100% certification petroleum diesel, no substantial changes in brake specific fuel consumption, thermal efficiency, heat release rate, or the masses of CO, NO_x, total hydrocarbons, or $PM_{2.5}$ emitted were observed when the engine was fueled with certification diesel blended with renewable fuels derived from hydrothermal liquefaction of wood, HTL of a blend of wood and microalgae, HTL of microalgae, or HTL of sewage sludge at 5% by volume. Similarly, no substantial change in these performance metrics was observed when the engine was fueled with upgraded pyrolysis oil blended with certification diesel at 5% or 15% by volume. Heat release occurred slightly later with the 15% upgraded pyrolysis oil blend. When the engine was fueled with ethanol-to-diesel fuel blended with certification diesel at 15% by volume, heat release occurred slightly earlier and a small reduction in brake-specific total hydrocarbon emissions was observed. These results suggest that compression-ignition engines can be operated on fuel blends that contain 5% HTL-derived fuels by volume with no adverse impacts on fuel consumption, thermal efficiency, or the masses of CO, NO_x, total hydrocarbons, or $PM_{2.5}$ emitted.

Acknowledgements

Pacific Northwest National Laboratory (PNNL) is operated by Battelle Memorial Institute for the U.S. Department of Energy (DOE) under Contract No. DE-AC05-76RL01830. This work was supported by the DOE Office of Energy Efficiency and Renewable Energy through the Bioenergy Technologies Office.

The authors thank James Tillotson, Mark James, Kirk Evans, Cody Vanderheyden, and Brenna King for their help with data collection. The authors thank Shantanu Jathar for use of his dilution chamber and PM sampling equipment.

References

[1] S.S. Toor, L. Rosendahl, A. Rudolf, Hydrothermal liquefaction of biomass: A review of subcritical water technologies, Energy. 36 (2011) 2328–2342.

[2] D.C. Elliott, P. Biller, A.B. Ross, A.J. Schmidt, S.B. Jones, Hydrothermal liquefaction of biomass: Developments from batch to continuous process, Bioresour. Technol. 178 (2015) 147–156.

[3] D.C. Elliott, T.R. Hart, A.J. Schmidt, G.G. Neuenschwander, L.J. Rotness, M.V. Olarte, A.H. Zacher, K.O. Albrecht, R.T. Hallen, J.E. Holladay, Process development for hydrothermal liquefaction of algae feedstocks in a continuous-flow reactor, Algal Res. 2 (2013) 445–454.

[4] J.M. Jarvis, J.M. Billing, R.T. Hallen, A.J. Schmidt, T.M. Schaub, Hydrothermal Liquefaction Biocrude Compositions Compared to Petroleum Crude and Shale Oil, Energy Fuels. 31 (2017) 2896–2906.

[5] J.A. Olivares, I. Baxter, J. Brown, M. Carleton, R.A. Cattolico, D. Taraka, J.C. Detter, T.P. Devarenne, S.K. Dutcher, D.T. Fox, et al., National Alliance for Advanced Biofuels and Bio-Products Final Technical Report, Donald Danforth Plant Science Center, St. Louis, MO, 2014.

[6] A.C. Drenth, D.B. Olsen, P.E. Cabot, J.J. Johnson, Compression ignition engine performance and emission evaluation of industrial oilseed biofuel feedstocks camelina, carinata, and pennycress across three fuel pathways, Fuel. 136 (2014) 143–155.

[7] M.E. Baumgardner, T.L. Vaughn, A. Lakshminarayanan, D. Olsen, M.A. Ratcliff, R.L. McCormick, A.J. Marchese, Combustion of Lignocellulosic Biomass Based Oxygenated Components in a Compression Ignition Engine, Energy Fuels. 29 (2015) 7317–7326.

[8] B. Hawley, C. L'Orange, D.B. Olsen, A.J. Marchese, J. Volckens, Oxidative Stress and Aromatic Hydrocarbon Response of Human Bronchial Epithelial Cells Exposed to Petro- or Biodiesel Exhaust Treated with a Diesel Particulate Filter, Toxicol. Sci. 141 (2014) 505–514.

[9] M.F. Link, B. Friedman, R. Fulgham, P. Brophy, A. Galang, S.H. Jathar, P. Veres, J.M. Roberts, D.K. Farmer, Photochemical processing of diesel fuel emissions as a large secondary source of isocyanic acid (HNCO): Photochemical Source of Isocyanic Acid, Geophys. Res. Lett. 43 (2016) 4033–4041.

[10] S.H. Jathar, B. Friedman, A.A. Galang, M.F. Link, P. Brophy, J. Volckens, S. Eluri, D.K. Farmer, Linking Load, Fuel, and Emission Controls to Photochemical Production of Secondary Organic Aerosol from a Diesel Engine, Environ. Sci. Technol. (2017) , doi:10.1021/acs.est.6b04602.

[11] A.T. Kirkpatrick, C.R. Ferguson, Internal Combustion Engines: Applied Thermosciences, 3rd ed., John Wiley & Sons, Inc., Chichester, West Sussex, 2016.

[12] R. Subramanian, A.Y. Khlystov, J.C. Cabada, A.L. Robinson, Positive and Negative Artifacts in Particulate Organic Carbon Measurements with Denuded and Undenuded Sampler Configurations Special Issue of *Aerosol Science and Technology* on Findings from the Fine Particulate Matter Supersites Program, Aerosol Sci. Technol. 38 (2004) 27–48.

[13] US EPA, Nonroad Compression-Ignition Engines: Exhaust Emission Standards, EPA-420-B-16-022, (2016). https://nepis.epa.gov/Exe/ZyPDF.cgi?Dockey=P100OA05.pdf (accessed August 26, 2017).

Western States Section of the Combustion Institute – Fall 2017 Meeting
Hosted by the University of Wyoming
October 2-3, 2017

Crankcase Blow-by Gas Particulate Matter Characterization and Filtration on a John Deere 4.5 l Diesel Engine

Benjamin D. Rice, James Tillotson, and Daniel B. Olsen

Department of Mechanical Engineering, Colorado State University

Fort Collins, Colorado 80523

ABSTRACT

The EPA fully promulgated Title 40 CFR part 1039 Tier 4 nonroad diesel engine emissions regulations in 2007. Engine manufacturers found it necessary to reduce particulates in the engine crankcase blow-by gases to meet Tier 4 particulate matter limits. Current filtration systems used to reduce PM emissions from the crankcase vent are costly and have a limited life. This article presents research characterizing PM and gaseous emissions from the engine crankcase of a John Deere 4045 4.5 l Tier 3 diesel engine. The study explores different filtration methods, including the stock John Deere filter used on the 4.5 l Tier 4 diesel engine, prototype filtration methods, and crank case vent locations on the valve cover, to reduce PM emissions. Measurements of CO, CO2, O2, total hydrocarbons, NOx, and gravimetric PM with elemental carbon and organic carbon (EC/OC) fractions in the emissions from the crank case vent were collected. The data indicate that the particulates contained in crankcase gases are primarily organic carbon particles, indicative of fine oil droplets formed when the blow-by gases mixed with atomized oil in the crankcase, not black carbon from the combustion process. The evaluation of the 9 valve cover port locations suggests the geometry of the valve cover affects the PM emission rate by approximately 45%. Filtration methods evaluated successfully reduced the oil particulates in the crankcase gases. The most effective, prototype method evaluated filtered the crankcase gas through diatomaceous earth, reducing particulate matter by approximately 84%. Data collected in this study may inform the development and analysis of new cost-effective and efficacious strategies to reduce emissions and extend filter life.

1 INTRODUCTION

The removal of blow-by crankcase gases from an internal combustion (IC) engine is important in maintaining the health and performance of the engine, preventing failures that may occur from internal pressure buildup [1]. The gasses that accumulate in the crankcase result from gases escaping past the piston rings during the combustion cycle. Dusane et al. [1] provides a developed explanation on the formation of blow-by gases and their relevance in crankcase pressure buildup. These gases, however, are a combination of clean air from the compression stroke, exhaust gases from the power stroke, and atomized oil particulates contained inside the crankcase. The necessary removal of these blow-by gases contributes to the overall emissions of an engine, predominantly its particulate matter (PM) emissions.

PM is generally described as submicron liquid or solid particles suspended in the air. The PM exhausted into the atmosphere by diesel engines is typically smaller than 2.5 microns [2]. These particles are easily inhaled and can pose a health risk to those around the operating engine. Exposure to these particles can exacerbate existing ailments in those with lung disease, asthma, and bronchitis. Long-term exposure to PM has been linked to pulmonary and cardiovascular disease (i.e. reduced lung function) and recurring bronchitis [2, 3].

The EPA has established emissions standards all IC engines must meet in Title 40 CFR part 1039 to remain in operation. This area deals with the standards set for all non-road diesel engines. The progression of these standards, deemed Tier 1 through Tier 4 by the EPA, have continuously reduced the allowable emissions from IC engines. The evolution of these standards took close to a 20-year period to move from Tier 0 (engines with unregulated emissions) to final Tier 4. The first Tier 1 engines were introduced in 1996, and the final Tier 4 implementation began in 2015 [4]. The final Tier 4 PM emissions standard for non-road diesel engines in the $100 \leq hp < 175$ range is 0.02 g/kWh [5].

Due to the improvements made on post combustion exhaust treatments, the PM from IC engine blow-by gas is now significant. Research suggests that the crankcase gases emitted to the atmosphere now consist of up to 70% of the total PM emissions from engines [6, 7]. The crankcase gases are typically removed through an exhaust port located on the valve cover. The specific location for this port differs depending on the engine manufacturer and the model of the engine. Engine manufacturers and secondary companies have developed systems to remove the PM from the crankcase gases that are highly efficient yet costly. A typical Tier 4 filtration system utilizes a coalescing filter that the gas is forced through to remove the PM it contains.

The objective of the experimental work done was to characterize the blow-by crankcase gases, investigate alternative crankcase exhaust locations (relative to the current locations on the engine used) on the valve cover to reduce the overall PM emissions from non-road diesel engines, and develop new and lower cost means to effectively remove the PM from the crankcase gases. In this paper, the results from 5-gas emissions testing, PM sampling and elemental and organic carbon (EC/OC) composition analysis are presented from the independent sampling of crankcase gases on a John Deere 4045 4.5L Tier 3 engine at Colorado State University's Engines and Energy Conversion Laboratory (EECL) seen in Figure 1. The results are analyzed using new filtration methods and systems for PM removal from crankcase gases.

Figure 1 – John Deere 4045 4.5L engine test cell set-up used for experimentation at the EECL.

2 METHODS

2.1 TEST CELL SETUP

Engine performance and emission assessments were conducted using a 4-cylinder, 16 valve, turbocharged and intercooled, 4.5 liter, 175 hp, John Deere 4045 PowerTech Plus test engine. The test engine, shown in Figure 1, is configured with a variable geometry turbocharger (VGT), exhaust gas recirculation (EGR), and electronically controlled high-pressure common rail (HPCR) fuel injection and meets Tier 3/Stage IIIA emissions specifications. The test engine is connected to an eddy current dynamometer (Midwest Inductor Dynamometer 1014A). The dynamometer and dynamometer controller (Dynesystems Dyn-LocIV) were used to load the engine and maintain a constant engine speed and load for each baseline, filtration device, and visualization test. The engine's standard fuel tank is filled with dyed off-road petroleum diesel used for engine warm-up and cool-down. Pressure, temperature, and blow-by volumetric flow values can be independently controlled and values logged via National Instrument's data acquisition hardware (DAQ) and LabVIEW virtual instrument (VI) software. Engine control unit (ECU) data was also recorded [8].

2.2 TEST FIXTURE SETUP

Key to the experimentation done for the collection of independent blow-by samples was the fixture constructed to sample the crankcase gases free of the engine's overall exhaust. The system was designed to route the road draft tube from the crankcase to a fixture attached to square piping secured to the engine test cell floor beam support. The gas was then run through a stainless-steel tube that took temperature and pressure measurements, a flow meter, then into a larger two-inch diameter stainless steel pipe. This pipe had two ports in it where samples of the 5-gas and the PM were taken. This design allowed the blow-by samples to come directly from the crankcase gases without routing it into the exhaust of the engine. This allowed for a specialized brake specific particulate matter (BSPM) rate measurement and compositional analysis of the gases exiting the crankcase.

Figure 2 shows the test set-up used for the sampling of the crankcase gases from the engine. Item #1 is the original draft tube used for the crankcase ventilation on the engine. It is set up for the efficient implementation of filtration devices. A thermocouple denoted #2 is used to measure the temperature of the crankcase gas before the sample is taken and a Swagelok fitting denoted by #3 with a pressure gauge and transducer to take measurements at this point. The flowmeter used, JTEC VF563B, is denoted by 4 to measure the volumetric flowrate of the gas for particulate emissions calculations. Item #5 denotes a pitot tube used to direct a sample of the gas to the dilution tunnel and collection cart for PM and EC/OC sampling. Item #6 is a 5-gas probe used to measure $CO_2, O_2, NO_x, CO, and\ THC$. Item #7 and #8, like #2 and #3, are an assembly of a thermocouple and Swagelok fitting with a pressure gauge and transducer to measure the gas temperature and pressure after the sample is taken.

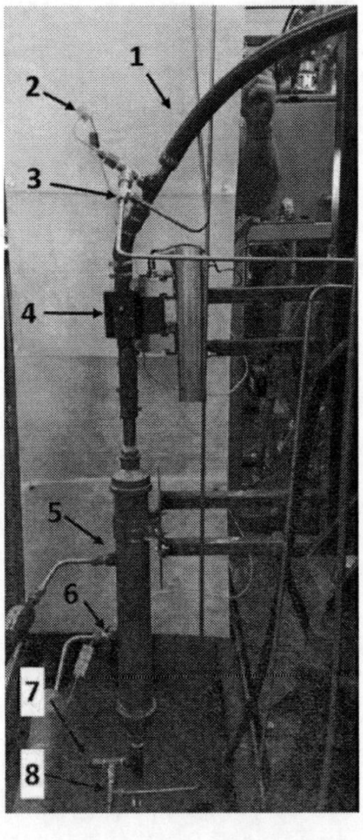

Figure 2 – Sampling fixture fabricated for the independent sampling from the overall exhaust from the engine of the crankcase gases PM and composition.

2.3 DATA COLLECTION

Gaseous and particulate matter (PM) emissions were measured from the crank case vent for some test cases. The gaseous emissions measured were Carbon Monoxide (CO) Carbon Dioxide (CO_2) Nitrous Oxides (NO_x), Oxygen (O_2) Paramagnetic and Total Hydrocarbons (THCs). Gaseous Carbon Dioxide (CO_2) and Carbon Monoxide (CO) emissions were measured using non-dispersive infrared (NDIR), O_2 was measured using Paramagnetic, THCs were measured using flame ionization detection (FID,) and NO_x was measured using Chemiluminescence. All analyzers were Siemens Ultramat 6 except the NOx which was a Siemens NOXMAT 600 (ULTRAMAT 6 and NOXMAT 600, Siemens AG, Munich, Germany).

For PM measurements of the blow-by flow from the crankcase vent were made using a 20 to 1 diluted mixture of exhaust and filtered room air. A fraction of the diluted sample passed into a residence chamber where the sample gas had a resonance time of 5 minutes. Next, the sample flowed through a PM_{10} cyclone which removed particles greater than 10 microns before moving to the filter train. The filter train consists of two parallel pathways. The sample split into two equal parts, controlled by mass flow controllers downstream of the filters. One path contains a single quartz fiber filter, and the other path comprises a Teflon coated glass fiber filter in front of a quartz fiber filter. The quartz filter absorb all elemental and organic particle phase material (particulates) and organic vapors in the sample while the Teflon coated glass fiber filters capture particulates only, while allowing organic vapors to pass through. The split sample train allows the capture of all particulates on the path with the quartz filter, and a differentiation of particulates and organic vapors on the pathway with the Teflon filter in front of the quartz filter [9]

Once collected, the quartz and Teflon coated glass fiber filters were preserved in sealed containers and kept at -80C until processing in an OCEC instrument (Model 5L, Sunset Laboratory Inc., Tigard, OR). To perform the EC/OC

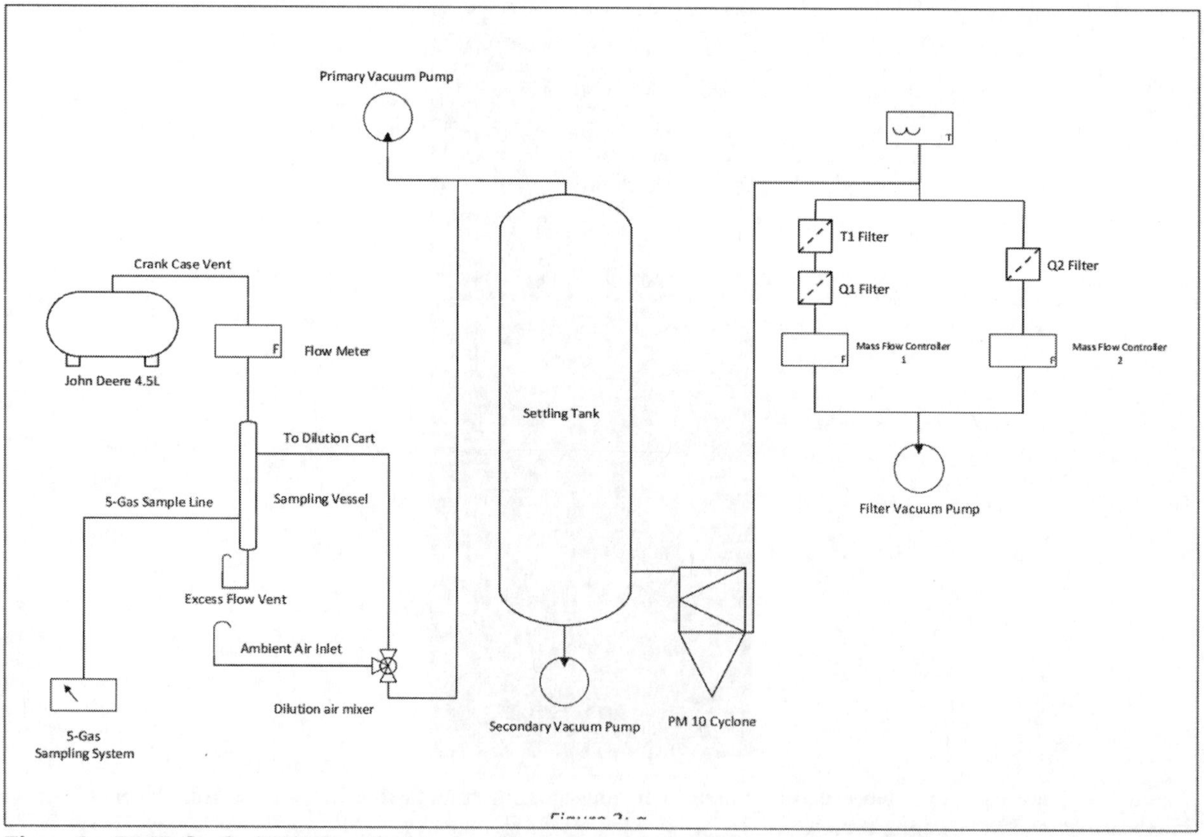

Figure 3 – P&ID for the EC/OC particulate sampling train used to collect emissions related data

analysis a punch from the quartz fiber filter is placed in the OCEC instrument. The sample is heated in a helium purged atmosphere oven which is then ramped in a step profile to 870 °C. This process thermally desorbs the organic compounds collected by the filter. Once desorbed, the products are passed into an oxidizing oven and any constituents which contained Carbon are converted to CO_2 where it is then catalyzed with hydrogen gas in a nickel catalyst know as a methanator and converted to methane CH_4 which is measured by a FID quantifying the carbon atoms which are liberated during the desorption phase of the OCEC measurement. Once the oven reaches 870 °C the desorption cycle is complete and temperature is reduced to 550 °C and the purge gas is switched from pure Helium to Helium and Oxygen. The elemental carbon which remained on the filter during the previous temperature ramp is now oxidized and measured using the same method applied to the products of the organic carbon elements. In the case of the oxidation phase, CO_2 is formed directly as the elemental carbon is oxidized, eliminating the need for the methanator. During both measurements, a He-Ne laser is passed through the filter. The organic carbon present on the filter absorbs the light of the laser and the absorption of light is used to quantify the formation of additional elemental carbon during the desorption process. All the carbon measured from the filter while between the start of the oxidizing phase of the analysis, and the point when the laser absorption level matches the initial absorption value made prior to the analysis is assumed to have been created by the desorption phase.

For a typical exhaust sample, the laser transmittance decreases as the temperature increases during the organic desorption phase as some of the organic carbon is reduced and forms elemental carbon. Once the oxidation phase begins carbon collected during the sample as well as carbon created by the organic desorption phase is oxidized and measured.

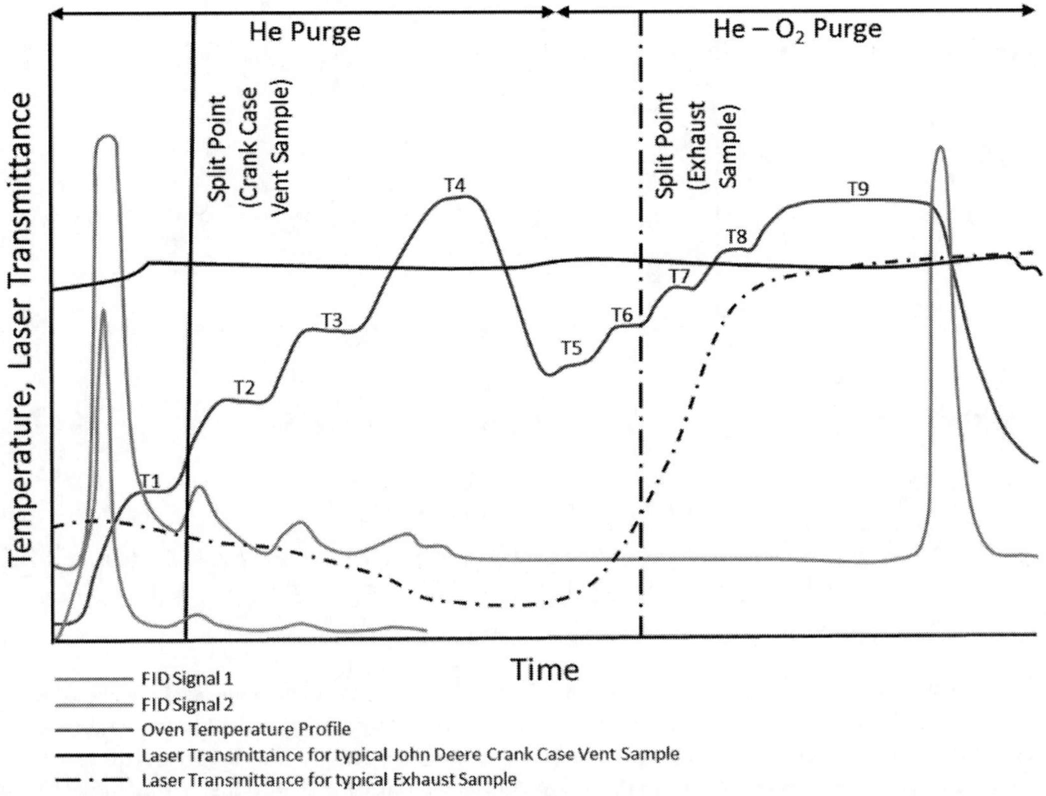

Figure 4 – Sample thermal gram, FID signals, and Laser Transmittance over time during the analysis of a quartz fiber filter in the Model 5L, Sunset Laboratory OCEC Instrument

2.4 BRAKE SPECIFIC PARTICULATE MATTER CALCULATION

To calculate the brake specific particulate matter (BSPM) rate from the crankcase blow-by gas, several calculations were used from a derivation pertaining to the dilution tunnel and collection cart. The calculation began by taking known values acquired from the integrated flowmeter and thermocouples to find the total mass flow rate from the crankcase, $\dot{m}_{ccv,t} = \rho_{act} * \dot{V}_{act}$, using the actual density of air, $\rho_{act} = \frac{P_{act}}{R_{air}T_{act}}$. From the collection cart, the mass flow rate to the filter, $\dot{m}_f = \rho_{std} * \dot{V}_{std}$, was calculated using the standard density of air, $\rho_{std} = \frac{P_{std}}{R_{air}T_{std}}$. Next, the rate of the PM through the filter, $\dot{m}_{PM,f} = \frac{m_{PM}}{\Delta t}$, and the mass fraction of the PM in the gas flowing through the filter, $X_f = \frac{\dot{m}_{PM,f}}{\dot{m}_f}$, were calculated. To equate the information obtained from the collection cart through the dilution tunnel, the mass fraction of the sample, $X_{ccv,s} = X_f * DR$, was calculated using the dilution ratio, $DR = \frac{\dot{N}_t}{\dot{N}_s} = \frac{\dot{m}_t}{\dot{m}_{ccv,s}}$. Because the mass fraction of the sample equals the mass fraction of the total blow-by gas from the crankcase, $X_{ccv,t} = X_{ccv,s}$, the mass fraction of the sample was used in calculating the total mass flow rate of PM from the crankcase, $\dot{m}_{ccv,t,PM} = X_{ccv,t} * \dot{m}_{ccv,t}$. The BSPM was then calculated, $BSPM = \frac{\dot{m}_{ccv,t}}{W_b}$, to present the PM emissions in g/kWh. Figure 6 depicts a diagram of the complete sampling cycle used with the equations relating to their position in the sampling.

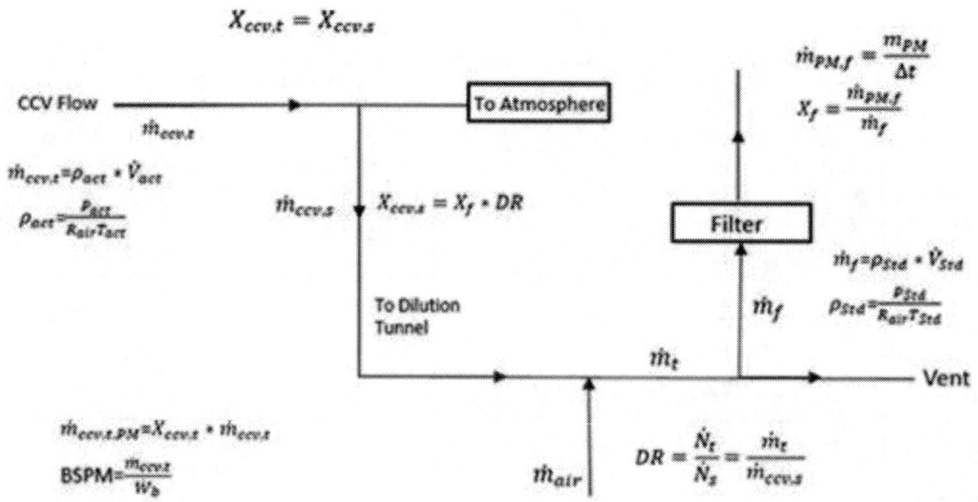

Figure 6 – Diagram of sampling cycle and derivation of PM sampling used during experimentation through the dilution tunnel and sampling cart for the collection of PM on Teflon filters for BSPM calculation.

2.5 VISUALIZATION FIXTURE

Another key experiment performed was the visualization of the oil splash at certain locations on the valve cover. To do this, a new valve cover was machined with six additional ventilation ports and fitted with aluminum tubing on the top of these ports and aluminum mesh at the bottom. The original ventilation port was also fitted with this tubing and mesh. Aluminum was used in this design to match the valve cover material, which was cast aluminum. The final fabricated fixture can be seen in Figure 7. The fixture also used rubber stops that fit into the tubing to prevent oil splash and airflow through the sample areas. Preweighed cotton balls were placed in each port location for a period of time and reweighed after their exposure to the oil splash within the crankcase to quantify the amount of oil splash in each location.

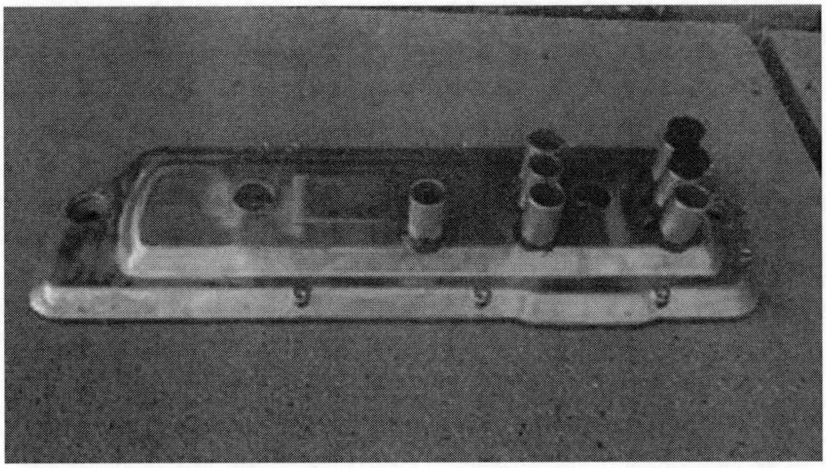

Figure 7 – John Deere 4045 4.5L engine valve cover fabricated as a test fixture for crankcase oil splash visualization testing.

3 RESULTS AND DISCUSSION

3.1 Quantification and Characterization of Crankcase Specific BSPM

The amount of PM emitted by the engine directly depends on the engine operating conditions. This idea typically follows the correlation that higher engine power output creates more PM emitted to the atmosphere. This idea was confirmed through the baseline data collected at two different power outputs. Calculating the BSPM rate provides a normalized metric to relate the PM data collected from an engine at different operating conditions. Baseline particulate data was obtained from the engine with no filter to characterize the blow-by gases and provide data for comparison for the different filtration methods and blow-by characterization. The baseline data collected from the engine through the Tier 3 valve cover was run at two different engine set-points. The first engine speed and load was 1500 rpm at 480 Nm of torque (set-point 1). This data provided a qualitative BSPM rate from the engine crankcase of 0.0043 g/kWh. The second set-point on the engine using this valve cover was 2400 rpm at 385 Nm of torque (set-point 2). This set-point yielded a BSPM of 0.0039 g/kWh. The two set-points used for testing were chosen based on the torque curve provided by John Deere for the engine. Set-point 1 is at rated engine speed and 75% of the maximum allowable torque on the engine due to facility limitations. Set-point 2 is the maximum allowable engine speed via the torque curve and 75% torque.

A secondary round of baseline tests were completed using the Tier 4 valve cover currently on the 4045 4.5L John Deere engine models that have Tier 4 emissions ratings. They have a different ventilation port location approximately an inch and a half to the right of the Tier 3 location. The reason for the second round of baseline testing was due to the design and safeguards for the filtration devices. The Tier 4 valve cover contained a check valve that would open when the engine crankcase pressure reached 5 kPa. Because the filtration methods to be tested had not been tested before, the backpressure they may cause was unknown. To avoid over-pressurization of the crankcase, the design of the tests and filtration devices was developed around the Tier 4 valve cover. The same engine set-points were tested for baseline data collection. Set-point 1 provided a BSPM rate of 0.0078 g/kWh while set-point 2 provided a BSPM rate of 0.0061 g/kWh. Table 1 shows the test port location, engine speed and load, and calculated BSPM values found for the baseline tests. The BSPM rates were calculated using the derivation seen in Figure 6 (presented earlier) and are the BSPM rate of blow-by gas from the engine crankcase independent of the total exhaust of the engine.

The particulate data collected from the different set-points and the different port locations showed that the higher BSPM correlates directly with the load applied to the engine. Subsequent tests to evaluate efficiencies of the Tier 4 filter used by John Deere and the filtration prototypes used the maximum BSPM set-point (1500rpm, 480 Nm of torque) to eliminate variables and expose the filtration systems to a high rate of particulates. The baseline data shows that the blow-by gasses have a relatively high BSPM rate for the relatively low volumetric flow rate of the crankcase ventilation flow (approximately 60 to 70 L/min). The EPA Tier 3 and Tier 4 PM emission limits are 0.3 and

Test Number	Valve Cover	Engine Speed (rpm)	Applied Load (Nm)	BSPM (g/kWh)
1	Tier 3	1500	480	0.0043
2	Tier 3	2400	385	0.0039
3	Tier 4	1500	480	0.0078
4	Tier 4	2400	385	0.0061

Table 1 – Baseline BSPM test results run on the John Deere 4045 4.5L engine with no crankcase ventilation filter at two different port locations and two separate test points.

0.02 g/kWh, respectively, for this engine [5]. The PM emissions from the crankcase ventilation port are a significant fraction of the Tier 4 limit (up to 40%). Thus, it is important to address crankcase ventilation port emissions for Tier 4 engines. Each port location had similarities with higher BSPM rates at set-point 1 than set-point 2, but the Tier 3 valve cover and port location BSPM rates were much lower than Tier 4 valve cover and port locations BSPM rates. The noticeable difference between the two sets of bassline testing led to the expansion of the scope of engine testing to include port location as well as new filtration methods.

3.2 Crankcase Ventilation Location

The difference in BSPM from the Tier 4 valve cover port location and the Tier 3 valve cover port location for set-point 1 was 0.0035 g/kWh (see Figure 9). To scale, this is a substantial difference without any method of after treatment filtration. The BSPM rate difference between the two locations is 45%. This higher rate of particulate matter increases the engine overall emissions without a filtration device and reduces the life and efficiency of the current filters being used.

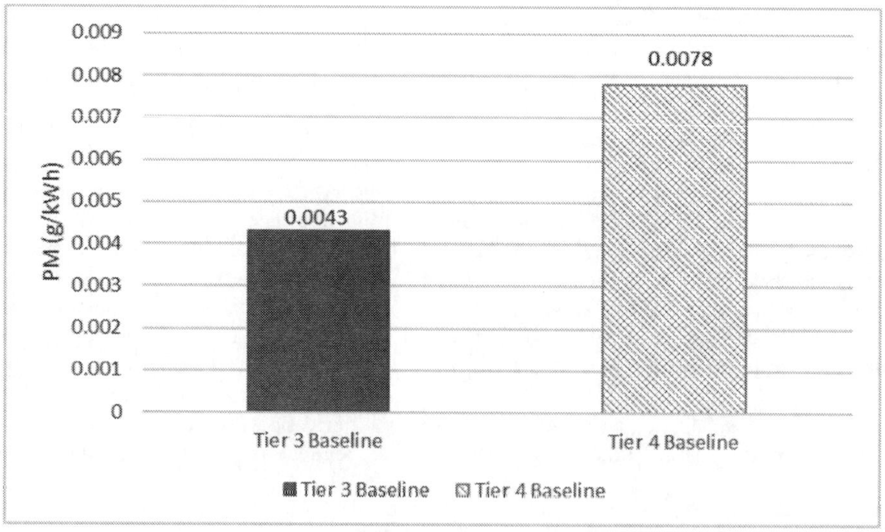

Figure 9 – Calculated baseline BSPM rates for the two valve covers and port locations at set-point 1.

To further explore PM emissions from the two different port locations, six tests were run using the current coalescing filter used by John Deere on their 4045 4.5L Tier 4 engines. This filter when new removes PM at essentially 100% efficiency for the first ~50 hours on the engine until the filter is saturated and the efficiency decreases to a lower value. To successfully obtain data for the BSPM of the engine at both locations through the filter, it needed to be conditioned. The timeline and scope of this project did not allow for the filter to be conditioned on the engine. A method was developed to condition the filter with a similar total PM delivered to the inlet as it would have when conditioned on an engine. The EC/OC analysis (presented in a subsequent section) showed that the measured PM from the crankcase vent was primarily oil droplets. To condition the filter, a shop vac nozzle was attached to the outlet of the filter, pulling air through the inlet. At the inlet, a fine paint sprayer filled with 10W-40 motor oil sprayed into the inlet as the shop vac pulled the fine oil mist through the filter. This process used 200g of oil to fully condition the filter. The visual saturation gradient of the filter was compared to saturated filter images provided by John Deere to arrive at the required mass of oil. Two tests were run at the Tier 3 port location and four tests were run at the Tier 4 location, measuring the PM emission downstream of the Tier 4 filter. Figure 10 shows the difference in BSPM rate from the crankcase from each test and the average difference.

The average BSPM through the conditioned Tier 4 filter from the Tier 3 port location was 0.00047 g/kWh while the average BSPM through the conditioned Tier 4 filter through the Tier 4 port location was 0.00083 g/kWh. These tests show the PM from the Tier 3 port location being 43% less than the Tier 4 port location. This difference is similar to the difference comparing direct port emissions (upstream of filter) of 45%, described earlier. The difference between the largest value at the Tier 3 location and the smallest value at the Tier 4 location is 0.00017 g/kWh, which is still a difference of 22%.

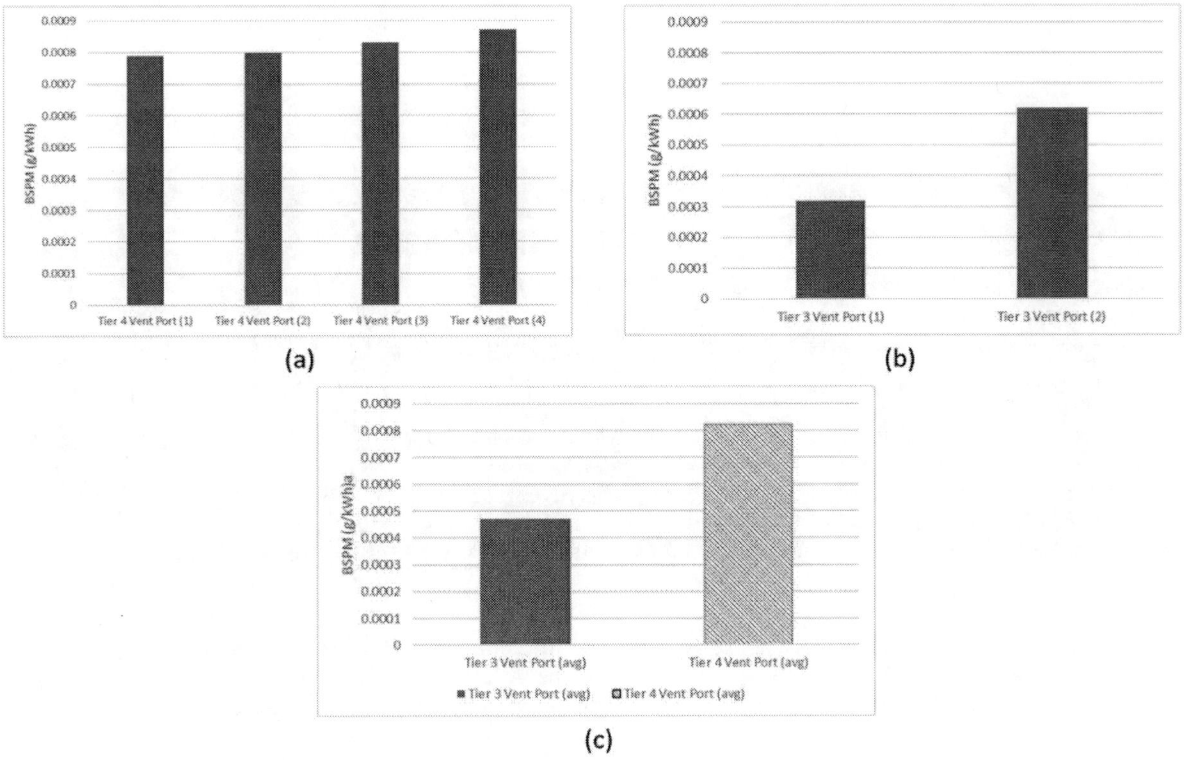

Figure 10 – Graphs of the BSPM rates collected from the John Deere 4045 4.5L engine from the Tier 3 and Tier 4 port locations on the valve covers through the conditioned blow-by coalescing filter used on John Deere Tier 4 engines. The four tests run on the Tier 4 location can be seen in (a) and are very similar in value. The two tests run on the Tier 3 location are seen in (b). The average values from each location and the difference between them can be seen in (c).

An oil splash visualization test was also carried out to research new port locations for the crankcase gases to be removed based on oil splash. This was also used to correlae BSPM from Tier 3 and Tier 4 port locations and the amount of oil splash seen in those areas. This test was done using the fabricated visualization fixture and cotton balls to collect the oil splash while running the engine at set-point 1. The baffle was removed from the valve cover to eliminate its effect. Three preconditioned and preweighed cotton balls were placed in each port location for one minute and reweighed after their exposure to the oil splash within the crankcase. Figure 11 shows the average value in g/min of oil splash inside the crankcase at each location overlaid on the engineering drawing and test locations.

The visualization fixture provided qualitative data regarding the amount of oil splash at multiple port locations. The port location with the highest average weight of oil absorbed was at the far end of the valve cover towards the middle at 0.618g, and the lowest weight of oil absorbed was at the current Tier 3 port location at 0.031g. The geometry of the valve cover and multiple walls of protection are the reasons the splash from the Tier 3 port location is so low. The port tested that is closest to the Tier 4 valve cover port location measured oil splash at 0.143g. The largest amount of oil splash was seen at the side of the valve cover near the crankcase walls. There is very little shielding geometrically in the crankcase and on the valve cover in this area. Figure 12 provides a visual image of the underside of the Tier 3 and Tier 4 valve covers and ventilation ports. The port tested resembling the Tier 4 port location the closest had the highest oil splash observed relative to the other ports near it. The ventilation port location on the Tier 3 cover has more shielding from the surrounding area in the crankcase due to the geometry of the valve cover surrounding its location. From this test and the baseline test on the engine, it can be concluded that the oil behavior and amount of oil splash surrounding a port location, even with a baffle, provides evidence that higher concentrations of oil splash near ventilation port locations directly contribute to the rate of particulate matter emitted from the crankcase.

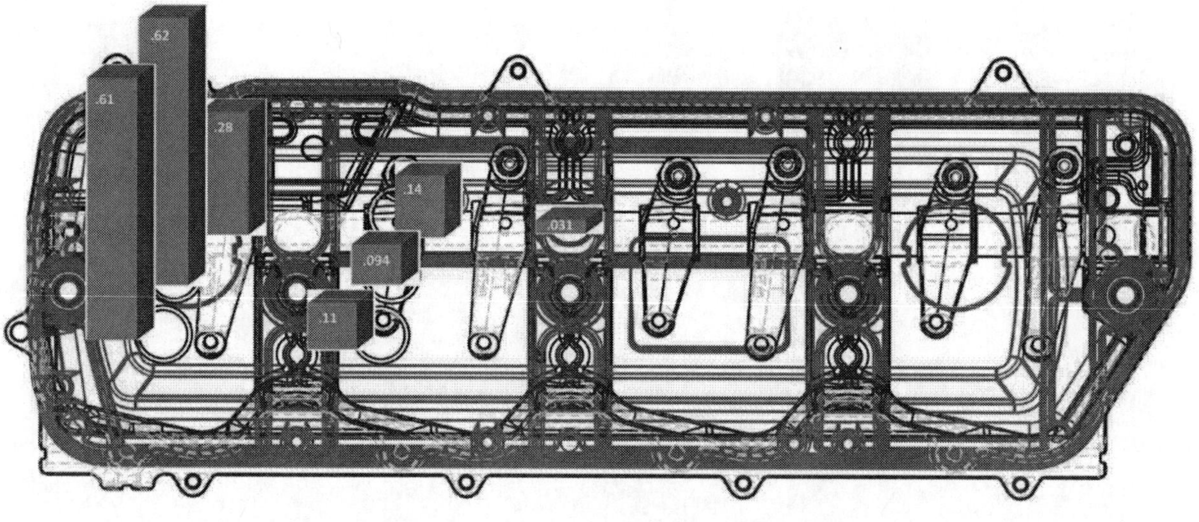

Figure 11 – An engineering drawing of the valve cover and the components inside the crankcase with their location relative to the different port locations tested for oil splash. The average value of oil collected from the splash inside the crankcase during these tests are overlaid on the locations tested and are measured in g/min.

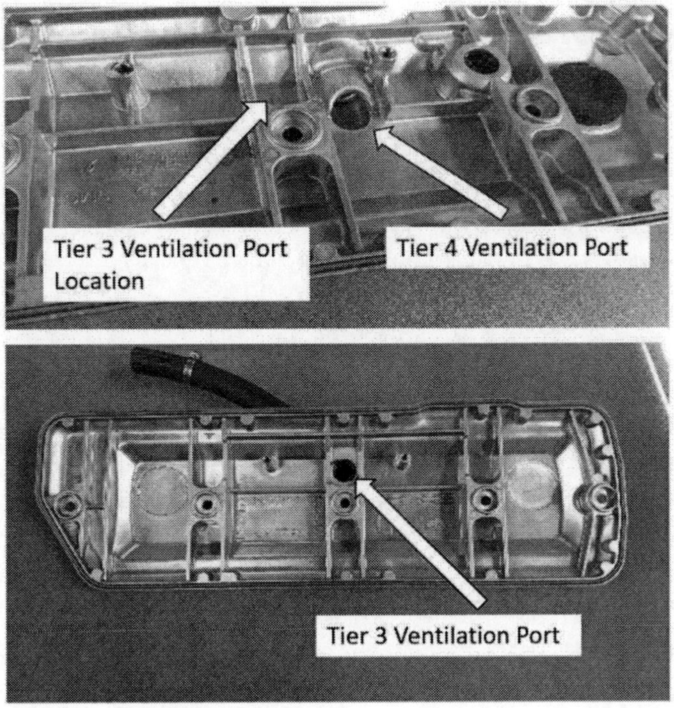

Figure 12 – John Deere 4045 4.5L engine Tier 3 and Tier 4 valve covers showing the different location of the blow-by ventilation ports and underside geometry of the valve covers.

3.3 Carbon Particulate Characterization

In addition to the particulate data collected during the second baseline testing using the Tier 4 valve cover, EC/OC data was collected at both set-points to characterize the carbon contained in the crankcase gases. The data collected regarding the carbon composition can be seen in Table 2. For both set-points on the engine, the composition of carbon in the crankcase gases yielded minute values of elemental carbon (black soot). These values were so diminutive they can be neglected as a contributor to the particulate emissions from the blow-by gas. The carbon particulates contained in these gases are organic carbon (heavy, condensable hydrocarbons). Based on the EC/OC results and the physical appearance of the filters the particulate matter is most likely fine oil droplets. This information is key in understanding the best filtration methods for particulate matter contained in blow-by gases from the crankcase to reduce overall engine emissions.

Sample	OC (ug/sq cm)	EC (ug/sq cm)
20170317 Q1	5.80	0
20170317 Q2	190	2.3
20170317 Q3	190	0.38
20170317 Q4	8.8	0

Table 2 – Organic carbon and elemental carbon masses collected on quartz filters during the two EC/OC tests run at the two set-points of the engine. Samples Q1 and Q2 were run at set-point 1 and samples Q3 and Q4 were run at set-point 2. These values were determined using samples from the original filters using EC/OC analysis techniques.

3.4 Characterization of Crankcase Gases

The blow-by gases were sampled with an averaging probe and analyzed by the 5-gas analyzer. The data showed the gases contained in the crankcase were characteristic diluted engine exhaust gas. The data is presented in Figure 8. The oxygen percent was approximately 19%,dry. The exhaust O_2% for these conditions is approximately 10%. Thus, the O_2% in the crankcase blow-by gases is a mixture of air and combustion products. Comparing the 5-gas data collected at the two different set-points shows a substantial decrease in CO from set-point 1 to set-point 2 and a small increase in NOx levels from set-point 1 to set-point 2. This slight change in NOx may explain the reason why set-point 2 (with a higher level of NOx) has a lower BSPM rate, since NOx and PM are typically inversely correlated.

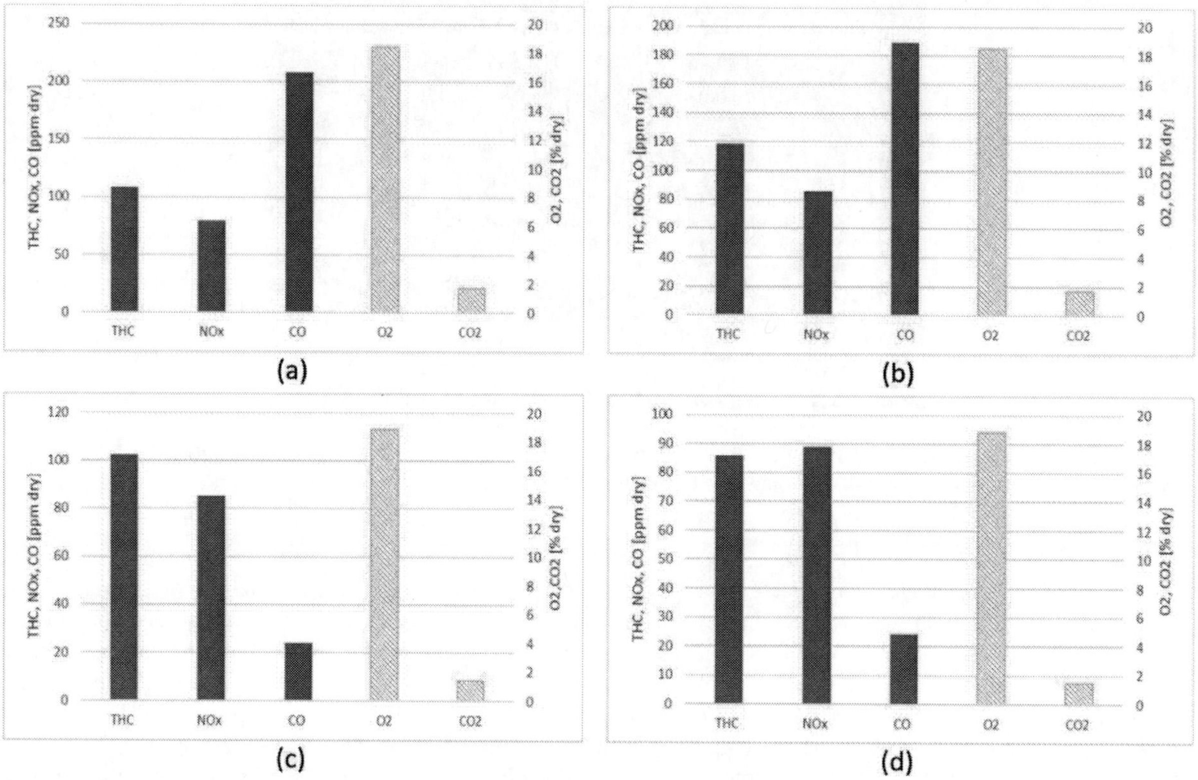

Figure 8 – Crankcase blow-by 5-gas data collected from the baseline engine tests at the two different set points. The tests run using the Tier 3 valve cover and port location at set-point 1 is denoted by (a) and at set-point 2 is denoted by (c). The tests run using the Tier 4 valve cover and port location at set-point 1 is denoted by (b) and at set-point 2 is denoted by (d)

3.4 Crankcase PM Filtration

New filtration systems were designed around solely removing atomized oil particulates from the blow-by gas, based on the EC/OC test results. Additionally, the back pressure in the crankcase was limited to 5kPa. The external design was to be dimensionally similar to the current filtration system housing used by John Deere and easily integrated with the crankcase gas sampling system. To do this, a water filter housing was used to house all the filtration designs. Each method utilized a similar approach to directing the gas through the filtration medium as seen in Figure 13. The gas flows down to the bottom of the housing and is forced up through the oil removal medium toward the exit port at the top. The two prototype methods used were an oil impinger, which bubbles blow-by gas through an oil bath, and oil absorbent desiccant, which forces blow-by flow through an absorbent solid.

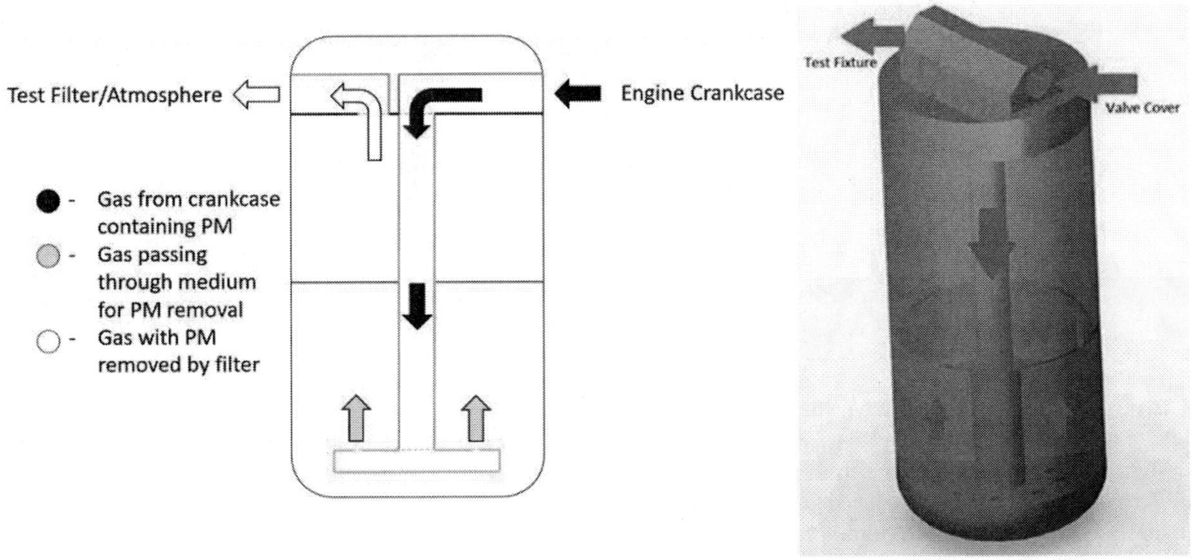

Figure 13 – Cross-sectional view and three-dimensional CAD model of the filtration system and housing used to test the different media of PM removal from the blow-by gas.

An impinger is a technique in which suspended particles, typically in air, are removed by impinging, forcing particles through a medium creating sharp collisions, the suspended particulate load through a liquid medium [10, 11]. In this prototype, the liquid medium used was oil. The idea to utilize a filtration prototype impinger method for removing particulates from the blow-by gas was supported by tests performed by Technikon, LLC using the same principle to collect PM emissions form metal casting operations in foundries [12]. The impinger filtration method required the most iterations and bench tests. A flat bubbler plate evolved into a domed shape as further researched and testing proved it an advantageous concept. The dome shape allowed for the air to pass into the oil at different frequencies, preventing bubbles from coalescing together to create large bubbles, preventing the suspended particles in the gas from contacting the oil bath with an optimal surface area. Oil was selected as the medium for this filtration device because the particulates being removed were primarily oil, allowing the oil to be drained, filtered, and refilled automatically. This would provide a low amount of maintenance with no need for replacement parts. Deflector plates and grade 000 steel wool were placed above the oil level to reduce violent oil splash and atomization of new oil particles. The steel wool also acted as a labyrinth system to further filter the PM from the blow-by gases. Various fluids were evaluated and 10W-40 motor oil was shown to be the most effective.

The desiccants selected for testing were chosen for their ability to efficiently absorb oil. For the desiccant prototypes, the oil from the impinge filtration method was placed in the prototype housing with oil absorbent desiccants. The first was a silica free desiccant (Pig Lite-dri) that was tested dry and saturated. This medium was chosen due to its silica free material and common use for oil spill cleanup. The other oil absorbing solid used was diatomaceous earth. This desiccant was selected due to its microscopic structure and general use as an oil spill cleanup material. Diatomaceous earth yielded better results in reducing the blow-by PM than the Pig Lite-dri. The microstructure of diatomaceous earth is unique because it contains thousands of holes ranging in diameter from a few microns to submicron sizes [13]. Because the diameter of the particulates in the blow-by gas are typically less than a micron, the microstructure of the diatomaceous earth provides an effective, natural filter for this particle size reduction. These prototypes were compared to the John Deere current filtration method used as well as the baseline data collected. Table 3 provides the BSPM for each filtration method.

Test Number	Prototype Test Type	BSPM (g/kWh)
1	Oil Impinger with Steel Wool (grade 000)	0.0021
2	Diatomaceous Earth (test 1)	0.0021
3	200g Oil Conditioned Pig Lite-dri	0.0017
4	Diatomaceous Earth (test 2)	0.0013

Table 3 – BSPM values downstream of the filters for different filtration media using the filtration system design and housing. Each of these tests was performed at set-point 1. Numerous iterations and tests were done on these designs both on and off the engine. The results in this table provide the final BSPM rates of the final designs tested.

The final iterations of the prototype designs were successful in reducing the particulates in the crankcase gases. The final designs of the filtration systems used for particulate testing on the engine can be seen in Figure 14. The impinger design reduced the particulates from the Tier 4 valve cover baseline by 73%, the final Pig Lite-dri conditioned desiccant filtration system reduced the particulates from the Tier 4 valve cover baseline by 79%, and the final diatomaceous earth desiccant filtration system reduced the particulates from the Tier 4 valve cover baseline by 84%. These particulate reductions were well below the new Tier 4 EPA PM emissions standard of 0.02 g/kWh . The conditioned Tier 4 filter reduced the PM by 89%, which is slightly higher than the best prototype design., The diatomaceous earth was only 5.5% lower in the PM reduction than the Tier 4 filter. The advantage of diatomaceous earth is cost; it is more cost effective than the filter currently used by John Deere on their new Tier 4 engines. The R540710 coalescing filter used has a life of 1500 hours and is its recommended replacement rate is every two years. The typical price for a replacement is $69.95. The price of diatomaceous earth, however, is much less. A 25 lb bag can be purchased for $9.99 and the amount used to fill a similar volume as the filter weighed approximately 0.5 lb. The cost to replace the diatomaceous earth filter would be $0.20. This is a 99.7 % decrease from the cost to replace the coalescing filter currently used.

The engine tests performed provided adequate data concluding that the new and cost-effective prototypes were successful in the reduction of the particulates from the crankcase according to Tier 4 EPA emissions standards. The visualization test fixture and tests performed on the different port locations through the Tier 4 filter concluded that the original Tier 3 port location was more efficient in reducing the particulate matter from the crankcase than the Tier 4 port location.

Figure 14 – Final prototype filtration system designs using a water filter housing to test different media for PM removal from the blow-by crankcase gas. The oil bath impinger with steel wool is seen in (a), the Pig Lite-dri desiccant is seen in (b), and the diatomaceous earth desiccant is seen in (c).

4 CONCLUSIONS

The reduction of PM contained in blow-by gases expelled to the atmosphere is key in the reduction of diesel engine overall emissions. This paper presents results from 21 tests performed on a John Deere 4045 4.5L Tier 3 diesel engine with a focus on engine crankcase particulates. The analysis of the blow-by gases characterized the particulates expelled in the atmosphere, the effect of valve cover internal geometry on particulate rates, and new effective and inexpensive methods for particulate removal. Specific conclusions are:

- The PM contained in blow-by gases from the engine crankcase is primarily organic carbon (atomized oil droplets), containing almost no elemental carbon (black soot) from the combustion process.
- The PM emissions emitted directly from the crankcase ventilation port are a significant fraction of the Tier 4 limit (up to 40%).
- The BSPM rate of the engine is reduced by approximately 45% without any filtration device from the Tier 4 baseline by relocating the ventilation port to the Tier 3 port location. The internal geometry of the valve cover surrounding the port location and its reduction of oil splatter directly affect this reduction.
- When paired with the coalescing filter used currently on John Deere's Tier 4 models, the Tier 3 valve cover port location is approximately 94% efficient at reducing blow-by PM. This is approximately 5% more efficient than the Tier 4 valve cover port location.
- The three filtration prototypes successfully reduce blow-by PM by 73% to 84%. The most effective filtration medium is diatomaceous earth, which is only approximately 5% less efficient than the current coalescing filters used on Tier 4 models.

NOTATION

$X_{ccv,t}$	Mass fraction of PM in total
$X_{ccv,s}$	Mass fraction of PM in sample
$\dot{m}_{ccv,s}$	Sample mass flow rate from crankcase
ρ_{act}	Actual density of air
$\dot{V}_{act}$	Actual volumetric flow rate
P_{act}	Actual pressure of air
R_{air}	Actual specific gas constant of air
X_f	Mass fraction of PM in filter
T_{act}	Actual temperature of air
DR	Dilution ratio of dilution tunnel
$\dot{m}_{ccv,t,PM}$	Total mass flow rate of PM
$\dot{m}_{ccv,t}$	Total mass flow rate from crankcase
$BSPM$	Brake specific particulate matter
$\dot{W}_b$	Brake specific work
$\dot{m}_t$	Total mass flow rate to dilution tunnel
$\dot{m}_{PM,f}$	Mass flow rate of PM through filter

Δt	Total time of collection
m_{PM}	Mass of PM on filter
$\dot{m}_f$	Mass flow rate to the filter
ρ_{std}	Standard density of air
$\dot{V}_{std}$	Standard volumetric flow rate
P_{std}	Standard pressure of air
R_{air}	Specific gas constant of air
T_{std}	Standard temperature of air
$\dot{V}_{act}$	Actual volumetric flow rate

ACKNOWLEDGMENTS

This research was funded by the Power Systems group from John Deere located in Waterloo, Iowa. This funding was provided as a donation for the 2016 – 2017 senior mechanical engineering student's senior design project at Colorado State University. A special thanks to Dr. John Petro (lead senior design professor), Tyler Hupp (engineering contact at John Deere), Kirk Evans, Christian L'Orange, Dr. Shantanu Jathar, and the John Deere senior design team (Tyler Whitmore, Charley Thomas, Casey Parks, and Jon Miller).

REFERENCES

[1] Dusane, M. and Nandgaonkar, D. (2014). Development of Oil Mist Separator and Evaluation with Oil Droplet Diameter Using Multiphase Simulation for Diesel Engine. International Journal of Research in Engineering & Advanced Technology, [online] 2(5). Available at: http://www.ijreat.org [Accessed 15 Sep. 2016].

[2] Manufacturers of Emission Controls Association (2007). Emission Control Technologies for Diesel-Powered Vehicles. [online] Washington, D.C. Available at: http://www.meca.org/galleries/files/MECA_Diesel_White_Paper_12-07-07_final.pdf [Accessed 14 Sep. 2016].

[3] Valentini, D. and Cerio, R. (2014). Diesel Emission Reduction in Construction Equipment. [ebook] The University of Rhode Island Transportation Center and Outreach Center, pp.1-3. Available at: https://ntl.bts.gov/lib/51000/51500/51514/S000118.pdf [Accessed 14 Apr. 2017].

[4] Ohta, H., Nagasaka, S. and Katou, T. (2014). Compliance with Tier 4 Final Emissions Regulations Development of 125 mm and 140 mm Bore Engines. Komatsu Technical Report, [online] 60(167). Available at: http://www.komatsu.com/CompanyInfo/profile/report/pdf/167-E02.pdf [Accessed 9 Aug. 2017].

[5] Environmental Protection Agency (2017). 40 CFR Parts 9, 69 et al. Control of Emissions of Air Pollution from Nonroad Diesel Engines and Fuel; Final Rule. Vol. 69 No 124

[6] Donaldson, "Spiralce TM Crankcase Filtration Technology for Open (OCV) and Closed (CCV) Systems" Brochure No. F111118 (9/08), 2008.

[7] Johnson, B.T., Hargrave, G.K., Reid, B.A., and Page, V.J. 2011. Crankcase sampling of PM from a fired and motored compression ignition engine. SAE International Journal of Engines, 4 (2), pp. 2498-2509.

[8] Drenth, A., Olsen, D., Cabot, P. and Johnson, J. (2014). Compression ignition engine performance and emission evaluation of industrial oilseed biofuel feedstocks camelina, carinata, and pennycress across three fuel pathways. *Fuel*, 136, pp.143-155.

[9] Subramanian, R., Khlystov, A., Cabada, J. and Robinson, A. (2004). Positive and Negative Artifacts in Particulate Organic Carbon Measurements with Denuded and Undenuded Sampler Configurations Special Issue ofAerosol Science and Technologyon Findings from the Fine Particulate Matter Supersites Program. *Aerosol Science and Technology*, 38(sup1), pp.27-48.

[10]impinger. (2017). In: Merriam-Webster. [online] Available at: https://www.merriam-webster.com/dictionary/impinger [Accessed 9 Aug. 2017].

[11]impinge. (2017). In: Merriam-Webster. [online] Available at: https://www.merriam-webster.com/dictionary/impinging [Accessed 9 Aug. 2017].

[12] Shey, PhD, S., Glowacki, CIH, C., Prabhu, PhD, A., Chang, PhD, M., Chow, PhD, J. and Watson, Phd, J. (2005). Method Comparison Between a Cyclone/Impinger and a Dilution Tunnel for the Characterization of Particulate Matter from a Ferrous Metal Foundry. [AWMA Paper] Casting Emissions Reduction Program, Technikon # 1410-3116.

[13] Zima, G. (2013). Diatomaceous Earth - A Versatile Raw Material. Inside ACAT, 13(17), pp.12 - 13.

Western States Section of the Combustion Institute – Fall 2017 Meeting
Hosted by the University of Wyoming
October 2-3, 2017

Evaluation of Fuel Savings for a Ship System with Variable Frequency Drives and Energy Storage using a Cummins QSK50 Tier 4 Diesel Engine

Josh Moothart, Daniel M. Wise, and Daniel B. Olsen

Department of Mechanical Engineering, Colorado State University, Fort Collins, CO

New innovations in marine vessel control systems allow the diesel engine generator system to be optimized for minimum fuel consumption. Variable frequency AC power immediately rectified and distributed as DC power permits traditionally fixed speed diesel generators to run at a variable speed for optimal brake specific fuel consumption. Stored energy systems (e.g. batteries) can be directly tied to distributed DC power to provide peak load shaving and generator bridging (providing power during generator transition and startup). The purpose of this study is to analyze the differences in efficiency between a conventional fixed speed, balanced diesel generator system and a variable speed diesel generator system with energy storage. Fuel consumption data was collected by running power sweeps at constant speed and optimal variable speeds on a Cummins QSK50 50 liter Tier 4 diesel engine. Typically, for a single engine at part load, the minimum brake specific fuel consumption is below rated speed, at the minimum speed where the desired power can be met with stable operation. A Matlab simulation program was written and used to extrapolate the single engine data to model a four engine ship system. Weights based on operating profiles are applied to calculate a single value for the brake specific fuel consumption. The findings show that the variable speed system with energy storage is 5-7% more fuel efficient than the conventional constant speed, balanced system.

1. Introduction

Common Diesel Electric propulsion applications require constant speed operation of the engines in order to maintain constant generator output frequency. New innovations in marine vessel control systems allow the diesel engine generator system to be optimized for minimum fuel consumption. Variable frequency AC power immediately rectified and distributed as DC power permits traditionally fixed speed diesel generators to run at a variable speed for optimal brake specific fuel consumption. Stored energy systems (e.g. batteries) can be directly tied to distributed DC power to provide peak load shaving and generator bridging (providing power during generator transition and startup). The described system enables variable speed operation of Diesel engines powering generators to directly match loading. Analysis of the system using operating characteristics provided by engine manufacturers indicates that the variable speed system will result in fuel savings, and potentially reduced engine maintenance costs. It is desired to conduct performance testing to validate the analysis through comparison of data collected from actual engine operation.

2. Methods

The objective was to collect fuel consumption data at engine operating conditions of speed and load simulating conventional constant speed diesel electric propulsion and then repeated under conditions simulating variable speed propulsion. The data sets were then compared and analyzed to quantify the differences in performance between the test conditions.

- Engine selection: Operation of the engine at variable speeds is a firm requirement for conduct of the testing. The engine selected was the Cummins QSK50 diesel engine (Figure 1), which produces approximately 1700 kW depending on the calibration and hardware configuration [1]. For testing, the engine was configured for variable speed operation and Tier 4 emissions calibration with a selective catalytic reduction (SCR) catalyst system. This engine is a laboratory engine and not a consumer Tier 4 certified engine.
- Testing facility: The engine was operated under specified conditions of load and speed while the fuel consumption was recorded. Fuel flow was measured using a Coriolis flow meter.

Figure 1: Cummins QSK50 50 liter diesel engine loaded with an eddy current dynamometer

The main comparison approach follows ISO 8178 [2] where weighted averages were used to compare fuel consumption for different engine configurations operating on a specific load profile. The process for the comparison testing is summarized below:

1. Performed constant speed variable load testing at each load –10, 25, 50, 65, 75, 90% torque at 1800 rpm. Applied analysis using weighting factors to produce one fuel consumption value.
2. Speed sweeps were conducted to determine the optimal speed for minimum fuel consumption at each power level in step 1.
3. Performance data was acquired at each load at the optimized speed. Weighting factors were applied to data to calculate single values for fuel consumption. Variable speed was then compared to constant speed results.

3. Results and Discussion

Single Engine

The maximum power at 1800 rpm was 1703kW. This was established during an initial maximum torque vs. speed map carried out with a non-SCR calibration. Power operating points (i.e. 10%, 25%, etc.) were set relative to this value. A load profile for a typical marine application is shown in Table 1.

To simulate variable speed generation an optimum speed was selected at each power level that minimized fuel consumption. This occurred at the lowest speed where the power target was achieved. Thus, the optimal speed was selected at the lowest speed where the power target was achieved and operation was stable. This is illustrated in Figure 2 for 65% power. Initial optimization testing was performed with a different calibration (SCR Cal 1) with no reagent dosing for the SCR catalyst system. Subsequent testing with full emissions was performed using a slightly different calibration (SCR Cal 2) that matched the SCR catalyst configuration installed. This primarily impacted the SCR catalyst system performance and NOx emissions. The fuel consumption was slightly different between the two calibrations in some cases. SCR Cal 2 was used throughout the data analysis to ensure consistency. At 90 and 100% loads the optimal speed was ~1800 rpm so the data points at these loads for variable speed and constant speed generation were the same.

Assumptions were made about the 10% load and the 100% load at 1800rpm for the ISO8178-4 calculations. The 10% load could not be reached at 1800 rpm on the testing chassis, and the engine could not reach a full 100% load for long enough to collect data when the SCR calibration was employed. A point was taken from the initial maximum torque vs. speed map for the 100% load fuel consumption data point. The 10% load had to be increased to an average power of 180 kW which makes it about 10.5% of the max load.

The 25 and 50% load data were used for the typical marine application with corresponding weights to aid in testing. The towing and transit weightings were also combined since the loading was the same. The simplified marine application load profile is shown in Table 2. Choosing common load points between the two test cycles saved test time. The 90% loads for the standard generation and variable speed were duplicated because 1800 rpm was the minimum to provide 90% load on this motor. Table 3 shows load profile weights per ISO 8178 [2].

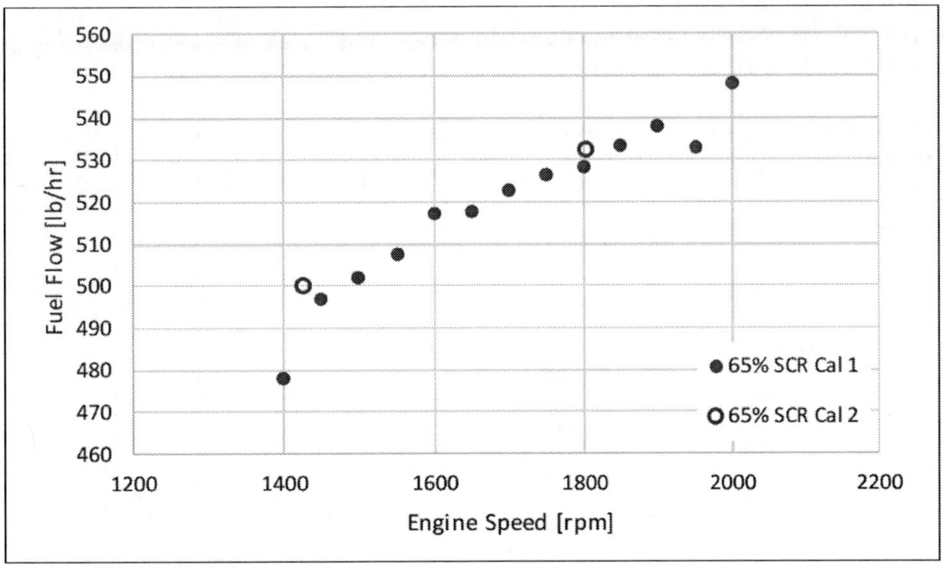

Figure 2: Speed Sweep of 65% Load

Table 1: Example of breakdown of underway time for typical marine application

	T-ATS(x)
0-05 knots Salvage and Station Keeping (Load- 10- 40% installed power)	50%
05-08 knots- Transit (Load =40- 50% installed power)	5%
05-08 knots -Towing (Load =70- 80% installed power)	15%
10-12 knots Transit (Load =60- 70% installed power)	10%
10-12 knots Towing (Load =85- 95% installed power)	15%
Top Speed- Transit (Load =85- 95% installed power)	5%

Table 2: Simplified load and weight values from Table 1

Loads %	25	50	65	90
Weights	.5	.05	.1	.2

Table 3: Test cycles from ISO8178-4, showing test cycle D2 recommended for various generator applications, including on board ships

Mode number (cycle B)	1	2	3	4	5	6	7	8	9	10	11
Mode number (cycle D1)	**1**	**2**	**3**								
Speed 1)	Rated speed					Intermediate speed					Low-idle speed
Torque 1), %	100	75	50								
Weighting factor	0,3	0,5	0,2								
Mode number (cycle D2)	**1**	**2**	**3**	**4**	**5**						
Speed 1)	Rated speed					Intermediate speed					Low-idle speed
Torque 1), %	100	75	50	25	10						
Weighting factor	0,05	0,25	0,3	0,3	0,1						
1) See ISO 8178-1:1996, 11.5, and clauses 3.5, 3.6, 5 and 6 of this part of ISO 8178.											

Equations 1-3 illustrate how weights are used to compute a single brake specific fuel consumption (BSFC) value for a load profile.

Equation 1: Brake Specific Fuel Consumption

$$BSFC = \frac{\sum \dot{m}_i w_i}{\sum \dot{W}_i w_i}$$

Equation 2: Calculation for ISO8178-4 (subscripts denote load percentage)

$$BSFC = \frac{(\dot{m}_{10} w_{10}) + (\dot{m}_{25} w_{25}) + (\dot{m}_{50} w_{50}) + (\dot{m}_{75} w_{75}) + (\dot{m}_{100} w_{100})}{(\dot{W}_{10} w_{10}) + (\dot{W}_{25} w_{25}) + (\dot{W}_{50} w_{50}) + (\dot{W}_{75} w_{75}) + (\dot{W}_{100} w_{100})}$$

Equation 3: Calculation for Typical Marine Application (subscripts denote load percentage)

$$BSFC = \frac{(\dot{m}_{25} w_{25}) + (\dot{m}_{50} w_{50}) + (\dot{m}_{65} w_{65}) + (\dot{m}_{75} w_{75}) + (\dot{m}_{90} w_{90})}{(\dot{W}_{25} w_{25}) + (\dot{W}_{50} w_{50}) + (\dot{W}_{65} w_{65}) + (\dot{W}_{75} w_{75}) + (\dot{W}_{90} w_{90})}$$

where
$\dot{m}_i$ = Fuel Flow Rate at Load % [g/hr]
w_i = Weight at Load %
$\dot{W}_i$ = Power [kW]

Fuel Consumption Results

Figures 3 and 4 show the fuel consumption differences at each power level. Generally the magnitude of the variable speed generation BSFC reduction [(constant speed) – (variable speed)] increases as power is reduced. Table 4 shows the weighted average fuel consumption results and the percent savings for the variable speed generation. Fuel savings for a single engine were calculated to be between 5.46 and 7.15% for typical marine and ISO load profiles, respectively.

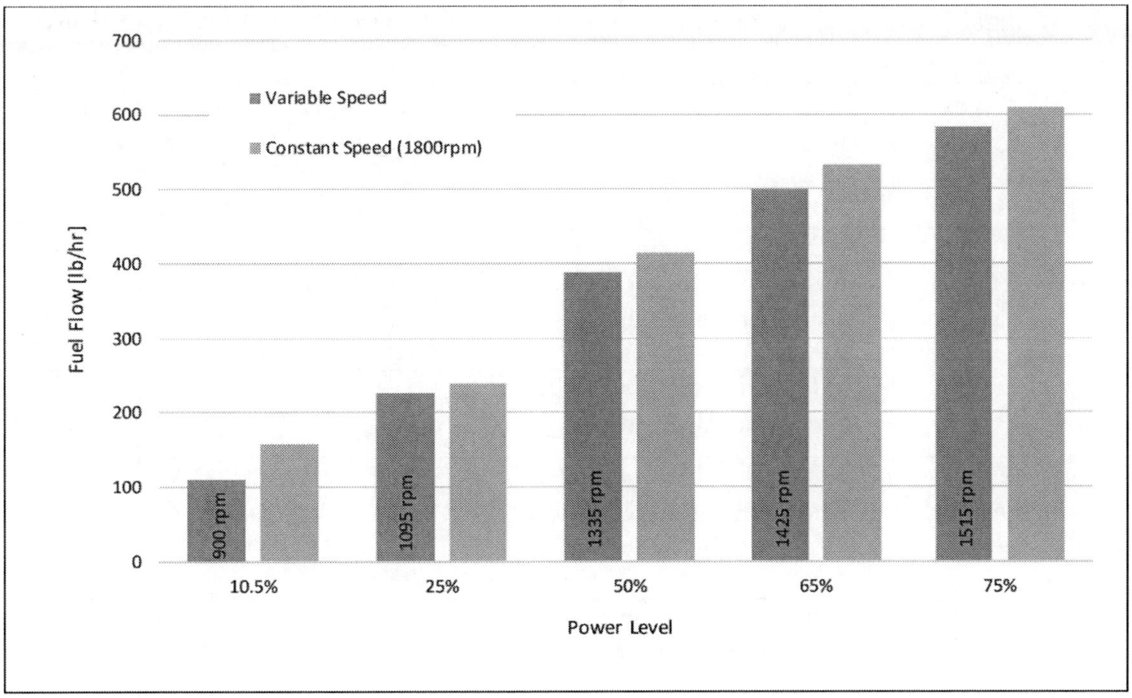

Figure 3: Fuel mass flowrate at different power levels

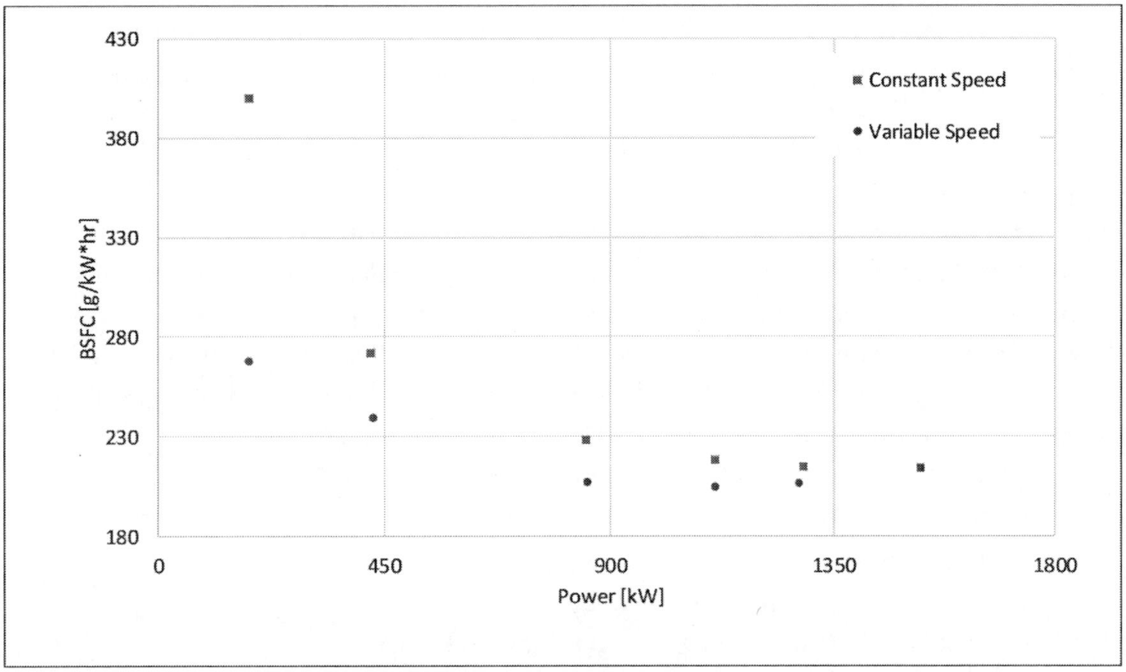

Figure 4: Brake Specific Fuel Consumption Averaged Points

Table 4: Final Brake Specific Fuel Consumption Using Weighted Values, % Fuel Savings

		Brake Specific Fuel Consumption [g/bkW*hr]	Variable Speed Fuel Reduction (%)
Constant Speed 2, or 4 Engines	ISO8178-4 Recommended Test Cycles for Generator Applications	231	N/A
	Underway Time for Typical Marine Applications	230	N/A
Variable Speed Using Optimal Speed	ISO8178-4 Recommended Test Cycles for Generator Applications	214	7.15
	Underway Time for Typical Marine Applications	217	5.46

Ship System

In the previous section results were presented for single engine operation. For use with marine applications a ship system was considered. The ship system was assumed to have four engines. In the multiple engine system fuel consumption is effected by how many engines are running concurrently at each value of total power for the ship system.

Assumptions/Rules

Rules were constructed in conjunction with marine specialists to simulate the ship system. The variable speed system is assumed to contain stored power, which permits a closed bus and allows the system to run with an odd number of engines. The ship system with constant speed generators is not allowed to run an odd number of engines. We assume that the ship system must be balanced in the number of engines per side and the power each engine produces (i.e. each engine produces the same power).This is due to the lack of stored energy and an open bus requirement.

- 340 kW used to power ship service electronics (5% of total)

Constant Speed:

- Either two or four engines
- Total power / # of engines (split equally)
- 2 engines until 3060 kW, then 4 engines until 6800 kW
- 1800 rpm required

Variable Speed:

- Rpm can vary between engines
- Total power can be split in whichever way is optimal
- Keep under 90% load if possible
- No minimum engine number because of stored energy and closed bus

Simulation

A four engine ship system was considered with the rules above. The data in the single engine section was used to optimize the four engine system. A cubic spline curve fit was applied to the single engine data (Figure 5) to determine fuel consumption at power levels in between data points. Y axis points at zero power were added to the data to show the fuel consumption that would theoretically be necessary to overcome the internal friction and other parasitic losses.

A Matlab code was run to optimize the fuel consumption in [g/kW] throughout the ship system power sweep while complying with the rules above. The power sweep was from 340kW to 6800kW. 340kW is 5% of 6800kW and is considered the minimum because of the ship's base load. Figure 6 shows a clear representation of how many engines are running to produce the optimal fuel consumption. The constant speed system was allowed either two or four engines with evenly distributed power.

Figure 7 shows power distribution over the entire power sweep for the four engines. Note there are many points where the power level is split evenly across multiple engines.

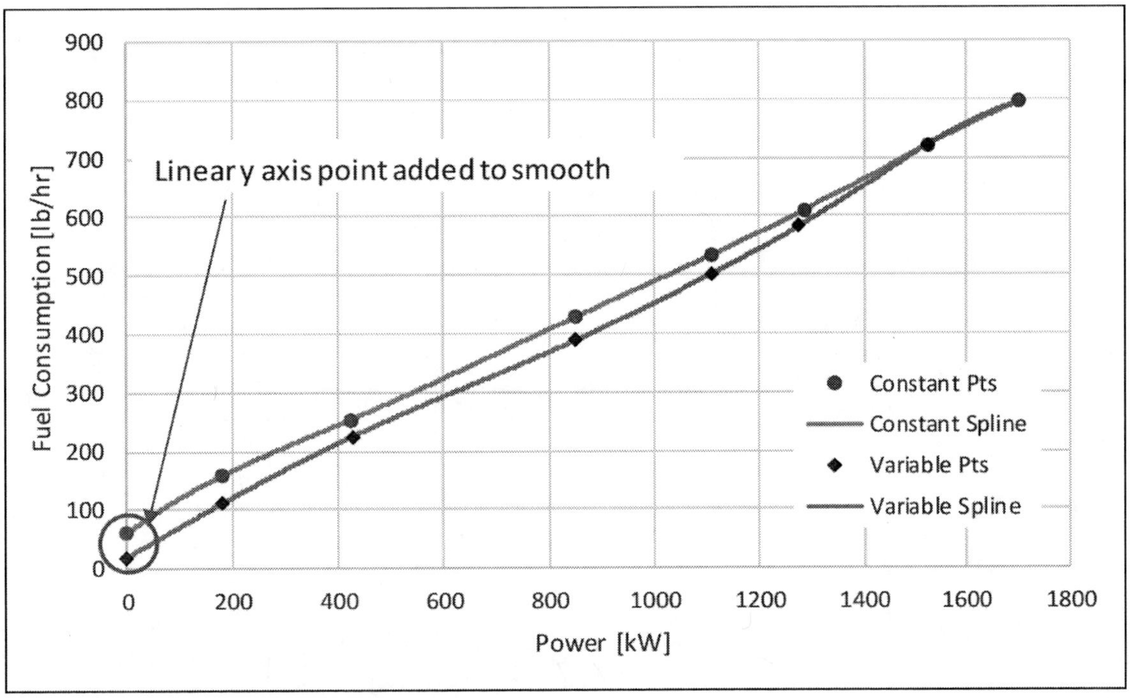

Figure 5: Single Engine Cubic Spline with Averaged Data Points

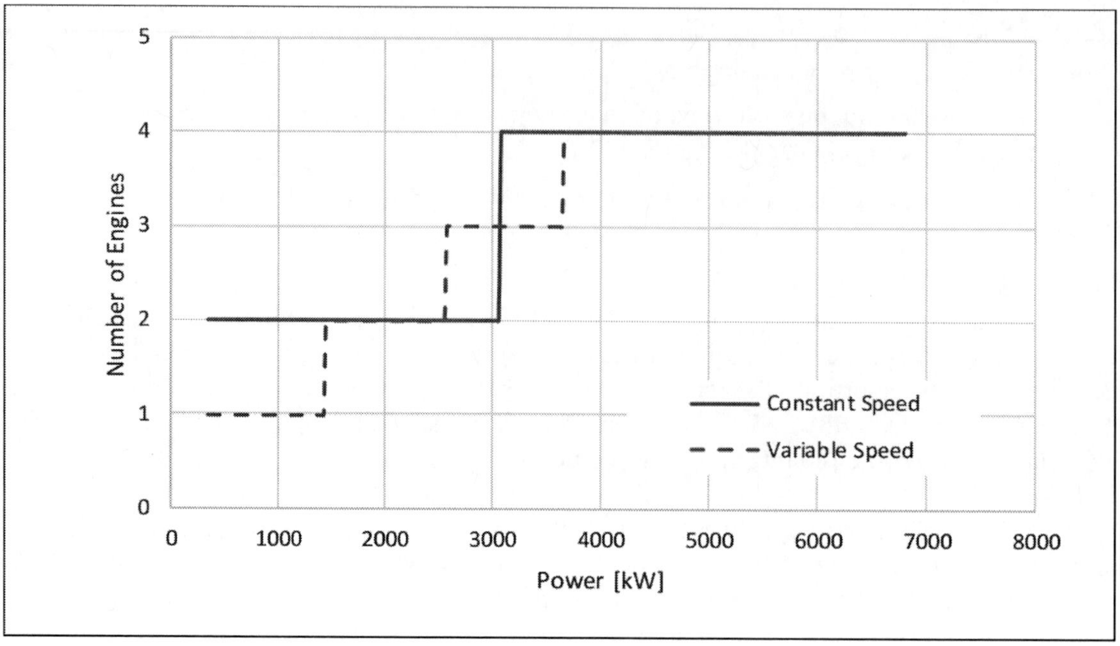

Figure 6: Engine Number over System Power Level

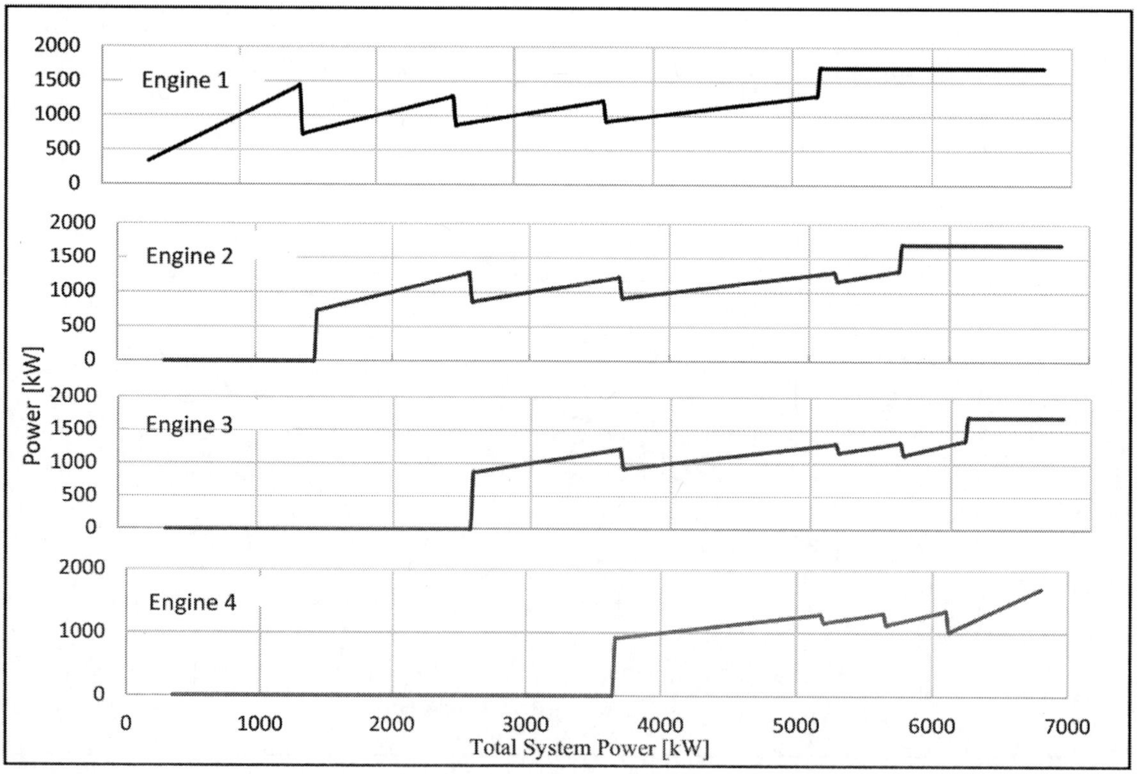

Figure 7: Engine Power Levels in Variable Optimization over System Power

Tables 5 and 6 show the optimization at the power levels used to find the minimum BSFC. The values for power and rpm are shown for the variable speed. The engines in the constant speed system are, by definition, run at 1800 rpm. The constant speed system also required the engines to run at equal power levels. The optimization shows which engines would be engaged and at what power level to be the most fuel efficient.

Figure 8 shows the BSFC for constant and variable speed across the load range. It highlights the optimum power levels (minimum BSFC) for variable speed. The minimum BSFC occurs between 800 and 1300 kW. This is why the variable speed engine attempts to keep as many engines as possible running in this range. There are often multiple engines running at this power level.

Table 5: Variable Speed Optimization Breakdown at System Power Levels

Variable Speed	% Total Load	10%	25%	50%	65%	75%	90%	100%
	Power [kW]	680	1700	3400	4420	5100	6120	6800
		Rpm, kW						
Gen 1	1700	1222, 680	1334, 850	1434, 1135	1424, 1105	1512, 1275	1800, 1700	1800, 1700
Gen 2	1700	0	1334, 850	1434, 1135	1424, 1105	1512, 1275	1800, 1700	1800, 1700
Gen 3	1700	0	0	1430, 1130	1424/, 1105	1512, 1275	1800, 1700	1800, 1700
Gen4	1700	0	0		1424, 1105	1512, 1275	1399, 1020	1800, 1700

Table 6: Constant Speed Breakdown at System Power Levels (1800 rpm)

Constant Speed	% Total Load	10%	25%	50%	65%	75%	90%	100%
	Power [kW]	680	1700	3400	4420	5100	6120	6800
		kW						
Gen 1	1700	340	850	850	1105	1275	1530	1700
Gen 2	1700	340	850	850	1105	1275	1530	1700
Gen 3	1700	0	0	850	1105	1275	1530	1700
Gen4	1700	0	0	850	1105	1275	1530	1700

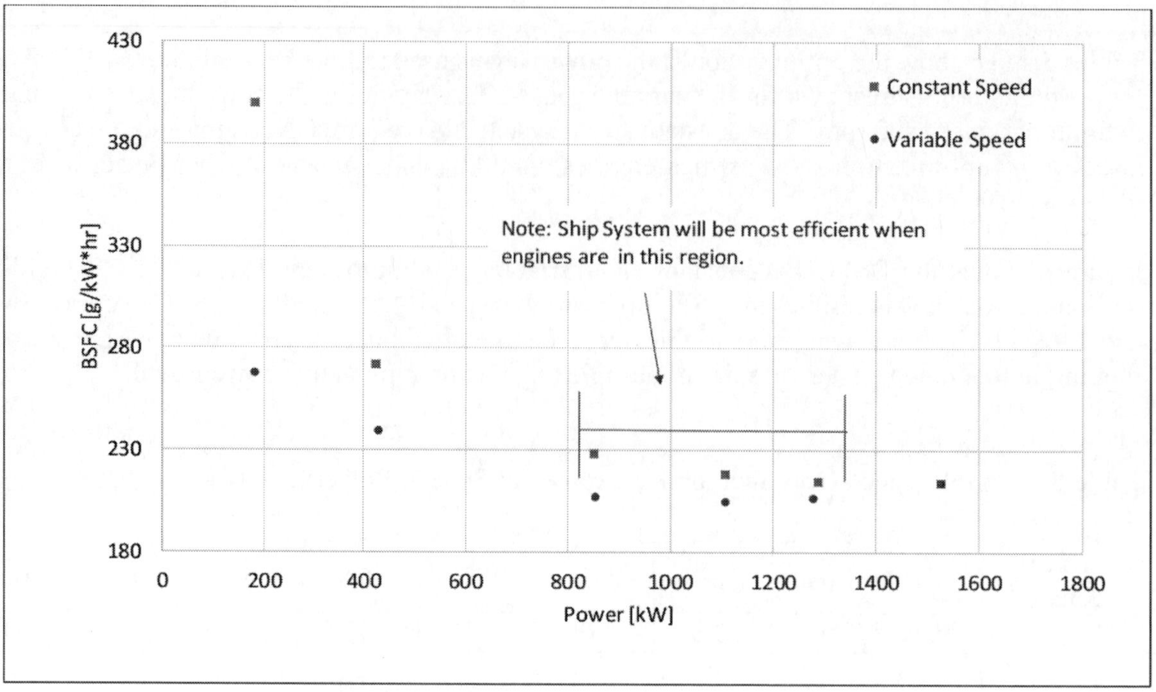

Figure 8: Minimum in Brake Specific Fuel Consumption for Single Engine

Fuel Consumption Values

The total fuel consumption for each ship system was calculated. Figure 9 shows the ship fuel consumption over the entire power sweep for constant and variable speed. The fuel savings in percent was also calculated using the fuel consumption data and is shown in Figure 10. It was discovered that the highest percent savings is at low ship power, and the middle range of power when the constant speed system requires four engines.

The weighted average brake specific fuel consumption values for the ship system are shown in Table 7. The percent savings are similar to the single engine. The fuel savings are 5.15 and 7.09% for typical marine and ISO load profiles, respectively. For low power operations however, the fuel savings are much higher (up to nearly 40%).

Using the data from the ship system simulation the fuel consumption for low power salvage and station keeping operation was calculated. Table 8 shows the load profile for the new weight system. It was assumed that the weights for the operation were distributed evenly among the seven loads. The fuel savings, shown in Table 9, increase substantially to around 10%. This demonstrates the advantage for low power operations with the variable speed system.

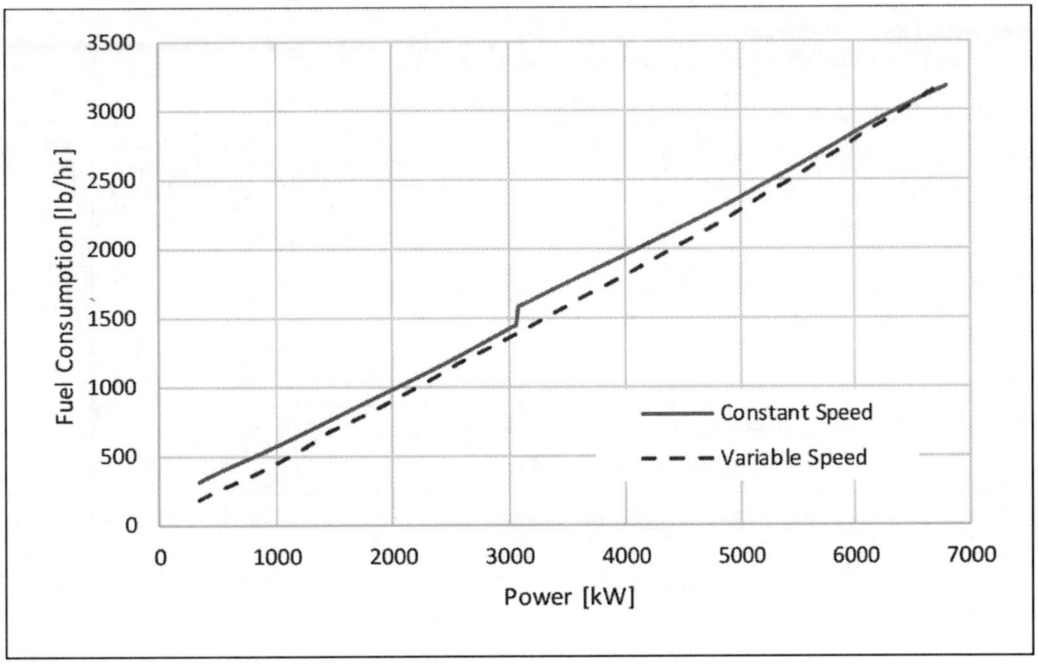

Figure 9: Total Fuel Consumption over System Power

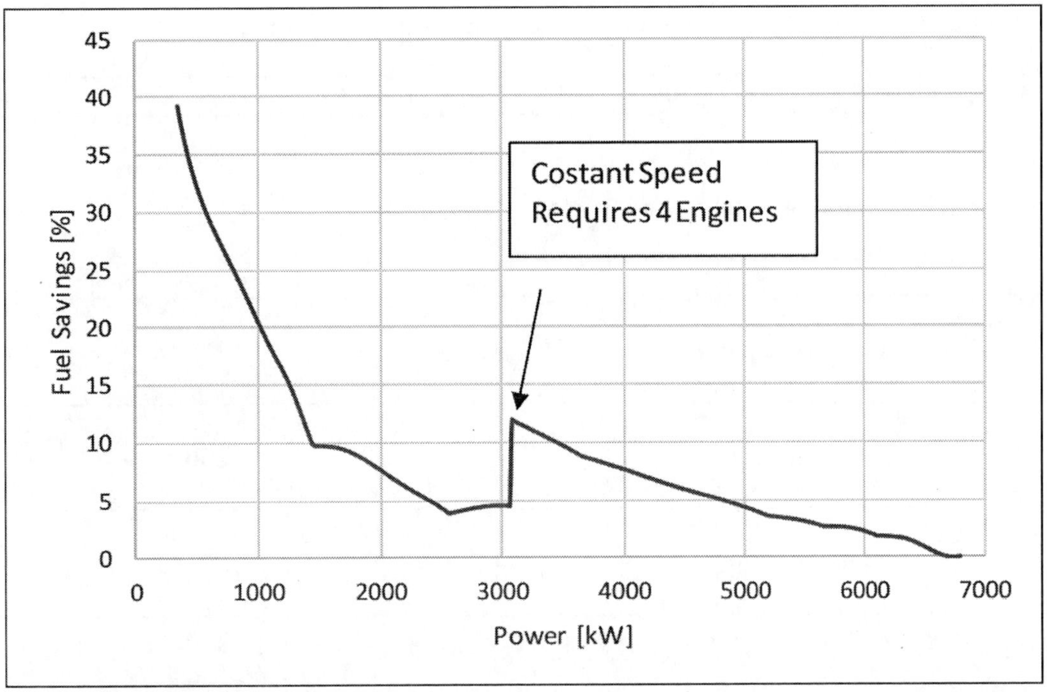

Figure 10: Fuel Savings of Variable Speed over System Power

Table 7: Fuel Savings using Weighted Averages for Ship System

		Brake Specific Fuel Consumption [g/bkW*hr]	Variable Speed Fuel Reduction (%)
Constant Speed 2, or 4 Engines	ISO8178-4 Recommended Test Cycles for Generator Applications	222.6	N/A
	Underway Time for Typical Marine Applications	219.0	N/A
Variable Speed Using Optimal Speed	ISO8178-4 Recommended Test Cycles for Generator Applications	206.8	7.09
	Underway Time for Typical Marine Applications	207.8	5.15

Table 8: Loads for Salvage and Station Keeping, Weights Distributed Evenly

10%	15%	20%	25%	30%	35%	40%

Table 9: Fuel Savings for Salvage and Station Keeping

	BSFC [g/kW*hr]
Constant Speed	228.8
Variable Speed	206.5
Fuel Savings	9.76%

4. Conclusions

Testing was carried out on a Cummins QSK50 1.7 MW Tier 4 diesel engine to evaluate the fuel savings associated with implementation of variable speed ship engine control system. At each engine power level of interest a speed sweep was conducted to determine the optimum speed to be used for simulation of the variable speed control system. Assessments of single engine and a 4 engine ship system were carried out. A Matlab simulation model was written to facilitate the ship system evaluation. Specific conclusions from the work are provided below:

- Variable speed engine testing showed that at each power level tested the optimum speed is the minimum speed for which the engine can meet the required power level and can achieve stable operation.
- Fuel savings for a single engine are calculated to be between 5.5 and 7.2% for typical marine and ISO load profiles, respectively.
- The fuel savings for the 4 engine ship system are predicted to be 5.2 and 7.1% for typical marine and ISO load profiles, respectively.
- For ship operations at single, low power values, however, the fuel savings are much higher (up to nearly 40%).

Acknowledgements

This research was funded by Siemens Corporation. The authors would like to thank the staff at the Colorado State University Engines & Energy Conversion Laboratory that operated the engine, Kirk Evans, Mark James, and James Tillotson.

References

[1] Engine Specifications at https://cumminsengines.com/showcase-item.aspx?id=260&title=QSK50+for+Mining+(Tier+4+Final)&Filters=3%3ATier+4+Final+%2F+Stage+IV#overview

[2] ISO 8178-4, Geneva, Switzerland: International Organization for Standardization; 2007

Western States Section of the Combustion Institute – Fall 2017 Meeting
Hosted by the University of Wyoming
October 2-3, 2017

Lagrangian analysis of premixed autoignition in compressible turbulence

Colin A. Z. Towery[1,*], *Ryan Darragh*[2], *Alexei Y. Poludnenko*[3], *and Peter E. Hamlington*[1]

[1]*Mechanical Engineering, University of Colorado, Boulder, CO, 80309, USA*
[2]*Aerospace Engineering Sciences, University of Colorado, Boulder, CO, 80309, USA*
[3]*Aerospace Engineering, Texas A&M University, College Station, TX, 77843, USA*
[*]*Corresponding author:* colin.towery@colorado.edu

Abstract: Autoignition plays a fundamental role in high-speed combustion systems, such as scramjet engines, where combustion occurs at large Karlovitz numbers and the turbulence can be highly compressible, with turbulence Mach numbers $Ma_t > 0.1$. The effects of compressible turbulence on the ignition delay time and intermittency of autoignition in high-speed combustion are investigated here using direct numerical simulations of three-dimensional, reactive, homogeneous isotropic turbulence with single-step Arrhenius reaction kinetics. Probability distributions of the fuel mass-fraction and reaction rate reveal large increases in intermittency and small-scale structure between the linear and non-linear compressibility regimes, with eddy shocklets appearing at $Ma_t > 0.4$. Detailed time histories of fluid parcel pathlines are analyzed in the Lagrangian reference frame for $Ma_t = 0.2$, 0.4, and 0.6. Results indicate that at the turbulence Mach numbers expected in scramjet engines, compressibility significantly shortens ignition delay time, and both subsonic and supersonic spontaneous autoignition waves can be formed simultaneously in both the linear and non-linear compressibility regimes.

1 Introduction

Autoignition of turbulent reacting flows can play a significant role in the reliability and sustainability of high-speed combustion systems, including in supersonic combustion ramjet (scramjet) engines [1, 2, 3, 4]. In high-speed combustion at very large Karlovitz numbers (i.e., the ratio of the shortest turbulence timescale, the Kolmogorov timescale, with the laminar flame self-crossing time [5]), the reactants are often partially or fully premixed [4] and the turbulence can be non-linearly compressible, wherein turbulent velocity fluctuations directly generate localized shock waves known as eddy shocklets [6]. Recent research [7], in particular, has indicated that spontaneous transition from deflagration to detonation modes of burning may occur for turbulent Mach numbers, Ma_t, as low as 0.12, and it is of substantial interest to study the exact mode of combustion after ignition occurs [8].

The study of even non-reactive compressible turbulence is made challenging by dynamical dependencies on the speed of sound, specific heat capacities, and thermal diffusion, all of which depend on the local fluid temperature. At the same time, the divergence-free velocity constraint that is central to the structure and evolution of incompressible turbulence no longer applies. As Ma_t increases, a flow will experience several distinct dynamical regimes marked by the increasing impact of the dilatational velocity mode on the thermodynamics. These regimes are (*i*) the quasi-isentropic regime, $Ma_t \lesssim 0.1$, where the low-Mach, or Boussinesq, approximation is valid, (*ii*) the linear thermodynamic regime, $0.1 \lesssim Ma_t < 0.3$, where turbulence and acoustic modes begin interacting but pressure fluctutations are approximately linearly-related to temperature and density fluctuations, (*iii*) the subsonic non-linear regime, $0.3 \lesssim Ma_t < 1.0$, where the Mach number is less than unity but locally supersonic flow gives rise to small-scale eddy shocklets and non-linear thermodynamic fluctuations, and (*iv*) the supersonic non-linear regime, $Ma_t \geq 1.0$, where the Mach number is approximately unity or greater, dilatational fluctuations dominate solenoidal fluctuations, and kinetic energy is rapidly converted to heat by viscous dissipation [6, 9, 10]. The exact Ma_t boundaries of these regimes are currently not well understood and, therefore, it is necessary to properly isolate and quantify the impact of turbulence compressibility on autoignition separate from other physical processes in high-speed combustion systems [8].

In order to quantitatively investigate small-scale compressible turbulence-chemistry interactions leading to autoignition, new three-dimensional direct numerical simulations (DNS) of reacting homogeneous isotropic turbulence (HIT) are performed, where the simulations fully resolve the smallest active length and time scales of the turbulence and the dynamical modes of reaction[23] are investigated in the Lagrangian frames of reference, which provides detailed time histories of igniting fluid parcels. The assumption of small-scale homogeneity and premixed combustion allows for highly detailed analyses of turbulence-chemistry interactions, such as in previous analyses of multi-scale and cross-scale interactions in the transport of turbulence kinetic energy [11, 12, 13]. The present set of simulations and analytical approach directly continues the Eulerian- and Lagrangian-frame analyses presented in Towery, et al (2017) [14].

Particular attention is paid in the following to the subsonic non-linear compressibility regime, as it is in this regime that the turbulence becomes highly thermodynamic (i.e., pressure, temperature, and density fluctuations become strongly coupled). Additionally, high-speed combustion systems can be expected to operate at least some of the time in the presence of non-linearly compressible turbulence, in addition to bulk-flow shock waves and high-frequency oscillating flow structures that shed acoustic waves [1, 15, 4]. For instance, the air entering the inlet of a hypersonic vehicle flying at $Ma_\infty \geq 5$ will pass-through a complicated shock-train flow structure in the scramjet engine isolator [2] before decelerating to a supersonic bulk velocity $1 < Ma \lesssim 4$. These shock-trains and the subsequent supersonic channel flow in the isolator section can easily generate combustor inflow turbulence with an intensity on the order of $5 - 10\%$ ($0.05 \lesssim Ma_t \lesssim 0.4$), even at low friction Reynolds numbers [16]. All scramjets also intentionally generate additional turbulence at the sites of fuel injection – for instance by injecting fuel with high-pressure jets directly into supersonic crossflow [17] or by injecting fuel into subsonic shear layers produced by bluff bodies, struts [18], or cavities [15] – to ensure efficient mixing in the face of extremely short combustor residence times. These injection configurations directly interact with and amplify the upstream turbulence intensity, and therefore one can reasonably expect post-injection turbulence compressibility in both the linear thermodynamic and subsonic non-linear compressibility regimes.

In the following, the physical configuration and analytical methodology of the computational simulations are described first. Results from a series of simulations are then presented and discussed, with a focus on differences in ignition intermittency and combustion mode between compressibility regimes. Finally, key observations in the data are revisited and a discussion of necessary future work are provided in the conclusions.

2 Methods

2.1 Direct numerical simulations

The physical and numerical configuration of the present simulations is identical to our previous work [14], but with slightly different values of several parameters, including the reference viscosities, thermal conductivity, and mass diffusivity. Additionally, we imposed a smaller Courant-Freidrichs-Lewy condition on the simulations in the nonlinear compressibility regime to improve the computational accuracy of both the grid-resolved eddy shocklets and shock-captured detonation waves.

As in our previous work, the physical dynamics of the simulations are described by the reacting flow equations for a calorically-perfect ideal gas reactant undergoing a single-step irreversible reaction whereby the fuel-mass-fraction, Y, is consumed at a rate

$$\dot{w} = -\rho Y B \exp\left(-\frac{Q}{RT}\right), \tag{1}$$

where B is the pre-exponential factor and Q is the activation energy. The corresponding chemical energy release is denoted q. While the single-step reaction used here is not tuned to model a specific fuel, it provides a sufficiently stiff reaction rate to approximate autoignition while still being fast enough to completely consume all reactant in the domain in an amount of time comparable to the turbulence eddy-turnover time at all turbulence Mach numbers. This agnostic modeling approach has been used successfully in the past for studying fundamental dynamics [19].

The simulations for each case are peformed as follows. We initialize four simulations at $t = 0$ with the reaction term turned off, uniform temperature and pressure, and a random velocity field with a prescribed isotropic energy spectrum and turbulence Mach number Ma_t significantly lower than the target turbulence Mach numbers at the initiation of reactions, Ma_init, to ensure the dynamical inconsistency in the initial conditions does not impact the simulations at later times. [20] We let the turbulent flow field develop for at least five eddy-turnover times (as defined in [14]) until the (ensemble-averaged) $a_\mathrm{t} \approx Ma_\mathrm{init}$, and reactions are turned on. For each set of simulations, t_init is the (uniform) time at which reactions are initiated.

We define the onset of autoignition as the time, denoted t_ign, at which the probability of finding completely-burned fluid is non-zero; that is, when $P[Y \leq 0.01] > 0$. In order to further improve the ensemble size used to compute statistics, we restart each of the four simulations non-deterministically from a restart checkpoint at least $4\tau_\ell$ before t_ign with a new random seed value for the turbulence forcing mechanism. Data are collected from all eight resulting simulations at a relatively high rate, but only for a short time before and after an estimated t_ign. From the new data, a final determination of t_ign is made, and the analysis outputs closest to this time from each simulation are presented in §3.1. The output rate of analysis fields from the current set of simulations is significantly higher than that in [14], resulting in data which is much closer to the actual ignition time.

For the Lagrangian analysis, two of the original four simulations for $Ma_\mathrm{init} = 0.2$, 0.4, and 0.6 are restarted deterministically from t_init, $t_\mathrm{init} + \tau_\ell$, and $t_\mathrm{init} + \tau_\ell$, respectively, with very small data output intervals $\Delta t = \tau_\ell/378.2$, $\tau_\ell/975.5$, and $\tau_\ell/975.5$, respectively. This output rate is continued for the entire course of combustion for the purposes of offline Lagrangian fluid parcel tracking. Pathlines are traced forward in time from 1728 uniformly-distributed initial positions in each data set, resulting in 3456 fluid parcel pathlines per Ma_t.

T_r	293.0 K	reference temperature
p_r	1.01325×10^6 erg/cm^3 = 1.0 atm	reference pressure
ρ_r	1.11467×10^{-3} g/cm^3	reference density
M	26.8 g/mol	molecular weight
B	1.0×10^{13} cm^3/(g s)	pre-exponential factor
Q	$46.1\,RT_\mathrm{r}$	activation energy
q	$25.6RT_\mathrm{r}/M$	reaction heat release
$\nu_\mathrm{r} = \alpha_\mathrm{r} = \lambda_\mathrm{r}$	0.163084 cm^2/s	reference transport diffusivities
$N_x \times N_y \times N_z$	$512 \times 512 \times 512$	uniform simulation grid dimensions
$(\eta_\mathrm{K}/\Delta x)_\mathrm{init}$	2.0	target η_K resolution at t_init
T_init	1000 K	target $\{T\}$ at t_init
Ma_init	$\{0.1, 0.2, 0.3, 0.4, 0.5, 0.6, 0.8, 1.0\}$	target Ma_t at t_init
Re_init	≈ 90	target Re_t at t_init
τ_ℓ	$\{234, 59.0, 26.2, 14.7, 9.43, 6.40, 3.49, 2.18\} \times 10^{-7}$ s	eddy turnover time at t_init

Table 1: Physical model parameters for the direct numerical simulations.

In total, eight simulations are performed for each of eight different values of Ma_t spanning $Ma_\mathrm{t} = 0.1$ in the quasi-isentropic regime to $Ma_\mathrm{t} = 1$ in the supersonic nonlinear regime, Values of relevant parameters used in the present simulations are provided in Table 1.

Figures 1 and 2 show fields of dilatation and temperature for three simulations of non-reacting HIT performed previously, corresponding to $Ma_\mathrm{t} \in [0.2, 0.6, 1.0]$ and $Re_\mathrm{t} \approx 135$, as visual examples of compressible HIT dynamics. These figures show the emergence of eddy shocklets in the non-linear subsonic compressibility regime and the increasingly strong and localized fluctuations in temperature.

2.2 Reaction mode identification

Fundamentally, there are only two kinds of combustion possible in this configuration: the slow, conduction- and diffusion-driven reaction fronts of a classical deflagration, and explosion waves which propagate 'spontaneously', driven by approximately-adiabatic heat release and without the need for thermal or mass transport between reactive fluid parcels. These explosive waves are generally referred to as 'spontaneous propagation' (SP) or autoignition waves when the wave front propagates subsonically in the wave frame, and as detonation waves when propagating supersoni-

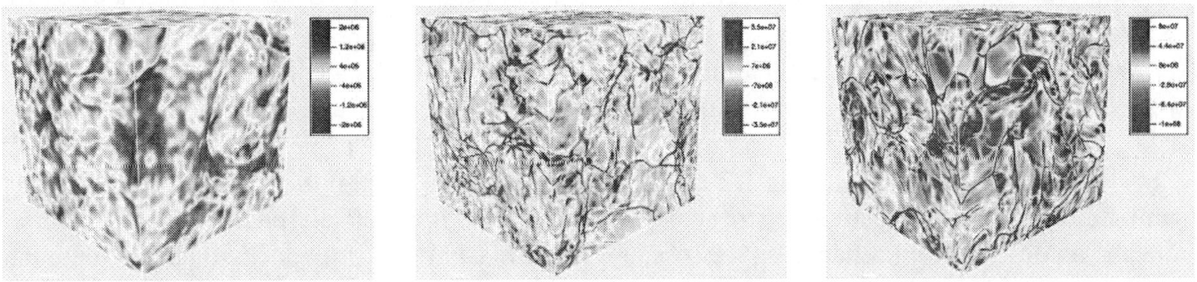

Figure 1: Pseudocolor fields of dilatation for non-reacting simulations of HIT at $Ma_\mathrm{t} = 0.2$ (left), 0.6 (center), and 1.0 (right) and $Re_\mathrm{t} \approx 135$.

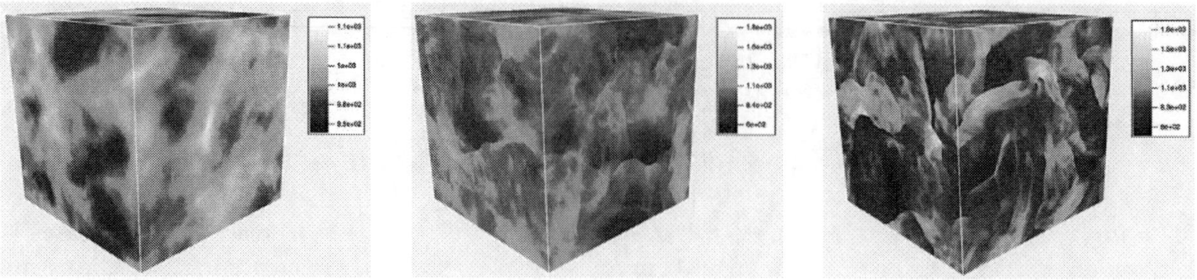

Figure 2: Pseudocolor fields of temperature for the same simulations as in Figure 1.

cally with a shock wave [21].

In principle, different types of propagating reaction waves (i.e., spontaneous explosion, deflagration, and detonation waves) could be identified in the Eulerian frame of reference by calculating the propagation speed and Rankine-Hugoniot jump conditions across a reaction front. These properties could then be compared, for example, with steady-state laminar solutions for deflagrations and with Zeldovich-Neumann-Döring (ZND) detonation profiles [21, 22] in order to determine the type of reaction wave. However, due to the very strong turbulent fluctuations both upstream and downstream of eddy shocklets and emerging reaction fronts, choosing the proper pre- and post-jump states is not straightforward. Moreover, the fluid flow is highly three-dimensional and reaction fronts can have high curvatures; by contrast, classical combustion theories rely on the dual assumptions of planar one-dimensional flow with a steady mass flux across the front, both of which are violated in radially-expanding fronts.

Although reaction waves can be difficult to identify in the Eulerian frame, the Lagrangian frame provides an ideal qualitative method for reaction mode identification based on the ratio of the diffusive flux and reaction rate, $\mathcal{D}/\dot{w}$, and the degree of monotonicity in the thermodynamic state during peak reaction rate along fluid parcel pathlines. Three different chemical regimes can be identified from the diffusion-reaction ratio when $\dot{w} < -\varepsilon$, where ε is some small-but-significant reaction rate which indicates that the fluid is actively reacting: (*i*) diffusion-driven reaction, $\mathcal{D}/\dot{w} > 1$, (*ii*) quasi-adiabatic reaction (i.e., autoignition), $|\mathcal{D}/\dot{w}| < 1$, and (*iii*) diffusion-driven extinction, $\mathcal{D}/\dot{w} < -1$ [23]. In all cases, reactions and diffusion are assessed locally and instantaneously.

It should be noted that the chemical mode of diffusion-driven extinction indicates only that fuel is being diffused into a fluid parcel faster than reactions can convert that fuel to product. For instance, a laminar deflagration wave experiences all three chemical modes in sequence, and yet at no point is the flame extinguished. Instead, a necessary consequence of mass and heat conservation is $\mathcal{D}/\dot{w} < -1$ in the reaction zone, where $\mathcal{D}/\dot{w} > 1$ in the flamelet pre-heat zone.

With respect to autoignition, although a deflagration wave does contain a very thin region where $|\mathcal{D}/\dot{w}| < 1$ in the region of transition between the pre-heat and reaction zones (i.e., for $0.4 \lesssim Y \lesssim 0.5$), $|\mathcal{D}/\dot{w}| > 1$ for almost all values of Y. In contrast, any fluid parcel for which $|\mathcal{D}/\dot{w}| < 1$ in regions where $Y \not\approx 0.45$ can be said to undergo autoignition. Furthermore, if that same fluid parcel also experiences strong negative dilatation immediately preceding peak reaction rate and a non-monotonic change in pressure, as occurs in Chapman-Jouguet detonations, then that fluid parcel can be said to undergo detonation.

3 Results and discussion

3.1 Global dynamics and Eulerian analysis

Time series of turbulence Mach number, Ma_t, mass-averaged temperature, $\{T\}$, and mass-averaged scalar reaction rate, $\{\dot{w}\}$, are shown in Figure 3. Time series are additionally ensemble-averaged over the four simulations started from initial conditions for each Mach number.

In near-perfect consistency with previous results [14], Figure 3(a) shows that at the ignition point, each of the observed values of Ma_t were close to the target values, denoted Ma_{init}. Only for the lowest values of Ma_{init} were the observed values of Ma_t relatively constant for all times. This change in Ma_t was due, at least in part, to adiabatic heating of the closed domain by kinetic energy dissipation, which is also evident in Figure 3(b) by the linear increase in temperature with time before the onset of autoignition.

The increasing width of the active reaction period in Figure 3(c) is a consequence of the increasingly smaller τ_ℓ as Mach number increases, while the average reaction timescale remains relatively constant. The reaction rate is shown in Figure 3(c) normalized by τ_ℓ so that the areas under each curve are approximately equal. By contrast, the reaction rate is shown against absolute time in Figure 3(d), indicating that the mean effect of increasing Mach number is to shorten the ignition delay time, with roughly an order of magnitude difference in the ignition delay time between Mach 0.1 and 1.

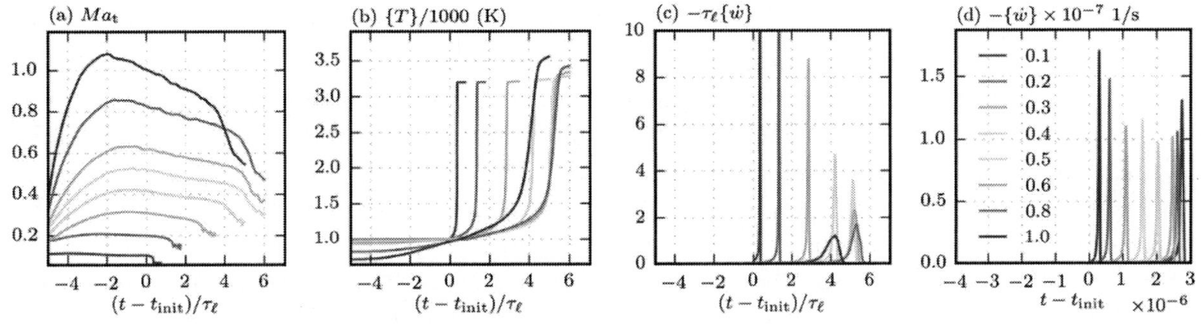

Figure 3: Time series of global quantities ensemble-averaged over four independent simulations for each Ma_{init} (Table 1 summarizes each case). Results are shown for a range of target Mach numbers Ma_{init}, as indicated by the legend in panel (d).

Figure 4 shows the structure of the turbulence fluctuations at t_{init} and the structure of the reaction fluctuations at t_{ign}. We found previously [14] that as the turbulence Mach number increases, the ratios of maximum to minimum pressures in the HIT increase by over two orders of magnitude from $Ma_t = 0.2$ to 1.0, and the maximum to minimum temperature ratio increases by over a factor of 3. These increases were accompanied by increasingly broad probability distributions of temperature, dilatation, fuel mass-fraction, and reaction rate as Ma_t increases, indicating that intermittency of quantities relevant to reaction ignition increase as the degree of turbulence compressibility increases. The new set of simulations confirm these general trends with one unexpected difference: the probability density functions (PDFs) of fuel mass reaction rate, $\rho\dot{w}$, shown in Figure 4(d), display significantly higher intermittency with increasing Mach number, with maximum reaction rates up to four times higher than the previous data set. We attribute this difference to both the increased sample size and the increased temporal resolution of this data set compared to the previous set.

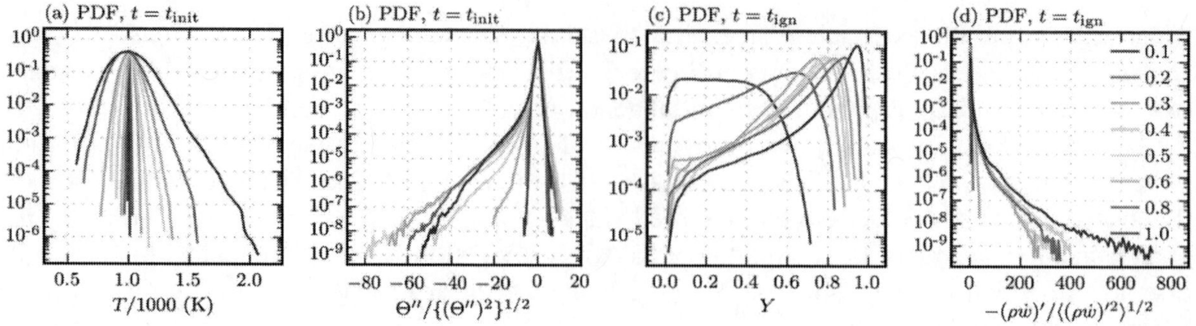

Figure 4: Probability density functions of (a) temperature T and (b) dilatation Θ at initiation of reactions, t_{init}, and of (c) fuel mass-fraction Y and (d) reaction rate $\rho\dot{w}$ at onset of autoignition, t_{ign}. Results are shown for a range of target Mach numbers Ma_{init}, as indicated by the legend in panel (d).

3.2 Lagrangian analysis

In order to compare individual fluid parcel time histories, each temporal trajectory is phase-shifted to the time at which the fluid parcel reaches a fuel mass-fraction of $Y_{\text{pr}} = 0.265$, which corresponds to the point of maximum reaction rate magnitude for a steady-state laminar deflagration wave with the given reaction model and initial conditions. The time at which each fluid parcel state reaches this mass fraction is denoted as t_{pr}.

Detailed time histories averaged over 35 fluid parcel trajectories are shown in Figure 5. The trajectories included in the averaging correspond to those trajectories with maximum pressures in the zeroth and ninety-eighth percentiles, as computed based on the full set of 3456 total parcels in each case. Here the peak pressure experienced by each fluid parcel is taken to be a surrogate for reaction propagation speed, with zeroth percentile trajectories assumed to pass through slower reaction fronts than the ninety-eighth percentile trajectories. The ninety-eighth percentile is used rather than the ninety-ninth percentile to control for statistical outliers in the limited data.

For all three values of Ma_{init}, the average time history of the ninety-eighth percentile shows distinct differences from the zeroth percentile, including an order of magnitude difference or more in peak reaction rate, a non-monotonic spike in temperature, and corresponding shock and expansion structure in the dilatation field, all of which strongly suggest at least some of these pathlines passed through a detonation wave at all three Mach numbers.

In order to confirm that detonations have occurred for all three cases, as well as to identify the presence of SP and deflagration waves, Lagrangian joint probability distributions (pdfs) of pressure P versus specific volume $1/\rho$ and diffusion-reaction ratio $D/\dot{w}$ versus fuel mass-fraction Y are constructed from all 3456 pathlines in each Mach case. Figure 6 shows the P-$1/\rho$ state-space for each Mach case side-by-side for comparison.

In each panel of Figure 6, the joint pdf has an envelope corresponding to the two Hugoniot curves for pure fuel ($Y = 1$, cyan lines) and pure product ($Y = 0$, magenta lines). Each panel also

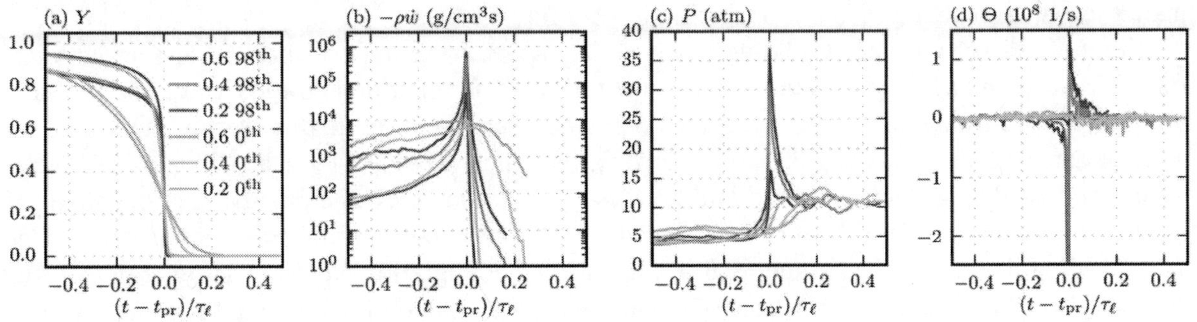

Figure 5: Fluid parcel phase-averaged time histories of (a) reaction rate, (b) temperature, (c) fuel mass-fraction, and (d) dilatation for parcels in the zeroth (0th) percentile and the ninety-eighth (98th) percentile of peak reaction rate magnitude and for $Ma_{init} = 0.2$, 0.4, and 0.6. Results are shown for a range of target Mach numbers Ma_{init}, as indicated by the legend in panel (d).

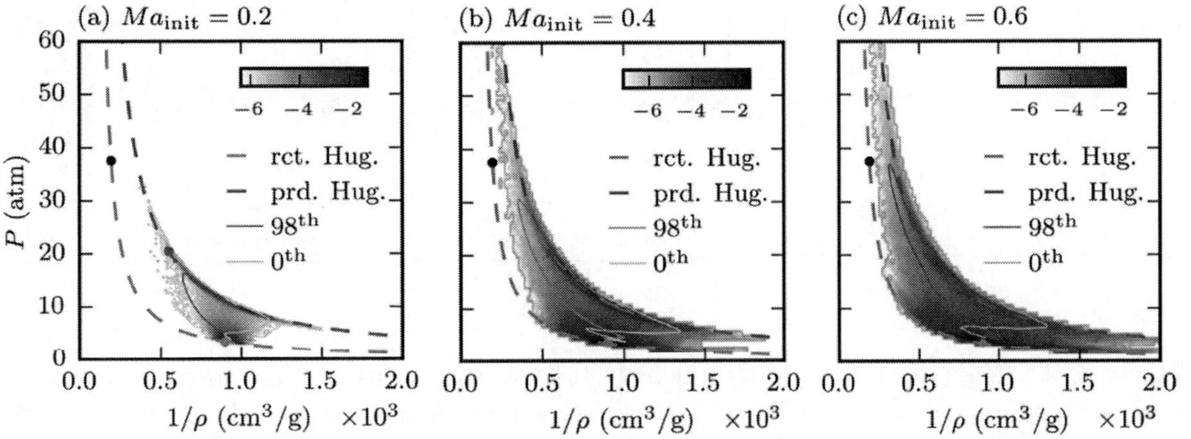

Figure 6: Lagrangian joint probability density functions of pressure versus specific volume for (a) $Ma_{init} = 0.2$, (b) $Ma_{init} = 0.4$, and (c) $Ma_{init} = 0.6$. Hugoniot curves for pre- and post-combustion are plotted as thick dashed colored lines (cyan and megenta, respectively), and the thin solid colored lines correspond to the phase-averaged Lagrangian time histories shown in Figure 5. Also shown are the average initial fluid condition at t_{init} (cyan dots), von Neumann post-shock state (black dots) and Chapman-Jouguet sonic point (magenta dots) for a laminar ZND detonation wave.

highlights the average fluid condition at t_{init}, the von Neumann post-shock peak, and Chapman-Jouguet sonic points for a laminar ZND detonation [22] propagating through a homogeneous fluid at the average initial conditions. Phase-averaged time histories from Figure 5 are also shown, indicating the path in thermodynamic phase space followed by reacting fluid parcels.

The joint pdfs in Figure 6 indicate that the ninety-eighth percentile trajectories correspond to weak detonations at $Ma_{init} = 0.2$ and 0.4 and nearly ideal ZND detonations at $Ma_{init} = 0.6$. The main difference in the latter case is that the reaction rate is non-negligible inside the leading shock wave, while ZND theory assumes that reactions occur only after the shock. This finding reinforces previously-published work [7] detailing the deflagration-to-detonation transition of a statistically-planar turbulent methane flame propagating in cold fuel with $Ma_t = 0.12$.

Figure 7 shows the corresponding $\mathcal{D}/\dot{w}$-Y state-space joint pdfs of the Lagrangian pathlines for each Ma_{init}, along with the state-space trajectory for a laminar deflagration wave and the same phase-averaged time histories from Figures 5 and 6. It can be seen that the zeroth percentile trajectories are in fact subsonic SP autoignition waves, since $|\mathcal{D}/\dot{w}| < 1$ for all Y, in stark contrast to the laminar deflagration pathline. Unlike $Ma_{init} = 0.2$ and 0.4, large positive and negative values of $\mathcal{D}/\dot{w}$ greater than the laminar deflagration solution occur at $Ma_{init} = 0.6$, as seen in panel 7(c). While this does indicate the possible existence of deflagration waves in some regions of the $Ma_{init} = 0.6$ simulations, the presence of equally strong extinction regions indicates that these deflagration fronts may also be highly unstable or propagating in the distributed or broken reaction zones regimes of premixed turbulent combustion [5].

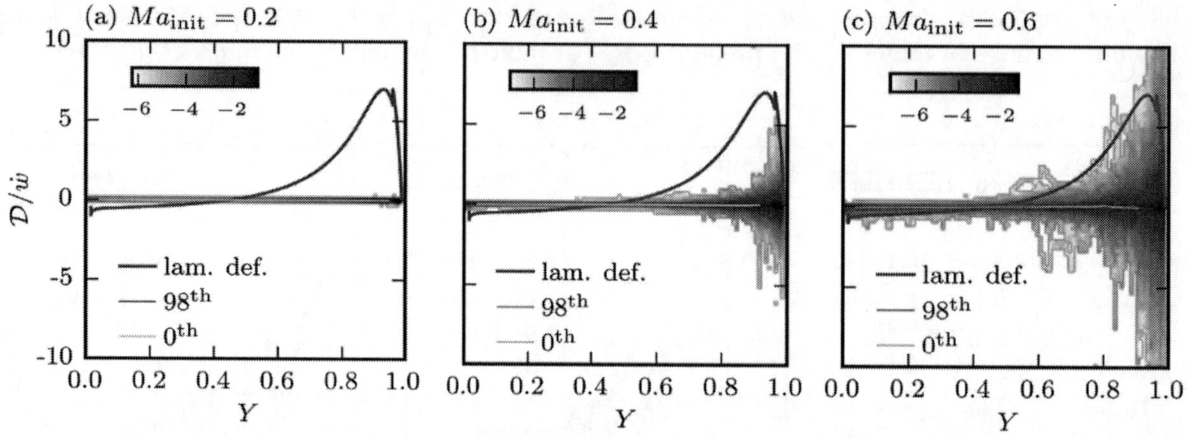

Figure 7: Lagrangian joint probability density functions of diffusion-reaction ratio versus fuel mass-fraction for (a) $Ma_{init} = 0.2$, (b) $Ma_{init} = 0.4$, and (c) $Ma_{init} = 0.6$. The laminar deflegration solution is shown as a thick solid blue line and the thin solid colored lines correspond to the phase-averaged Lagrangian time histories shown in Figure 5.

4 Conclusion

Autoignition in the presence of highly compressible turbulence is a problem of fundamental interest for the design and characterization of high-speed propulsion systems such as scramjets. In the present study, we have investigated the role of turbulence compressibility on autoignition using fully resolved direct numerical simulations (DNS) of reactive homogeneous isotropic turbulence (HIT) with a single-step chemical kinetics model. We performed an ensemble of simulations at eight target turbulence Mach numbers spanning a wide range of turbulence compressibility regimes.

Detailed time histories of Lagrangian fluid parcels computed for the turbulence Mach number 0.2, 0.4, and 0.6 cases revealed that both linear and non-linear compressibility can induce fast autoignition reaction fronts, which appear to be detonations. It is emphasized that detonations appear to be present even at relatively low Mach numbers traditionally considered to be in the linear thermodynamic or non-linear transition regimes. This finding is consistent with the earlier observation

[7] that deflagration-to-detonation transition can occur for a premixed flame with compressibility levels as low as $Ma_t = 0.12$ in cold reactants.

These results have important implications for the understanding of combustion in high-speed propulsion systems, since we find that turbulence compressibility can induce explosive modes of combustion and, in particular, that the resulting reaction waves are frequently detonations. This finding suggests that further experimental studies are necessary to understand the probability and impact of such detonations, and potentially to adjust scramjet designs based on the improved understanding of ignition causes and modes provided here.

5 Acknowledgements

CAZT and PEH were supported in part by AFOSR awards FA9550-17-1-0144, and by a DoD HPCMP Frontier Project award. RD was supported by a National Defense Science and Engineering Graduate Fellowship. AYP was supported in part by AFOSR award F4FGA05260G001. Computing resources were provided by DoD HPCMP under the Frontier Project award.

References

[1] A. Ben-Yakar and R. K. Hanson, *Journal of Propulsion and Power*, 17:4 (2001) 869–877. doi:10.2514/2.5818

[2] D. J. Micka and J. F. Driscoll, *Proceedings of the Combustion Institute*, 32:2 (2009) 2397–2404. doi:10.1016/j.proci.2008.06.192

[3] M. Berglund, E. Fedina, C. Fureby, J. Tegnér, and V. Sabel'Nikov, *AIAA Journal*, 48:3 (2010) 540–550. doi:10.2514/1.43746

[4] Z. Wang, H. Wang, and M. Sun, *Proceedings of the Institution of Mechanical Engineers, Part G: Journal of Aerospace Engineering*, 228:1 (2014) 2718–2735. doi:10.1177/0954410014521172

[5] N. Peters, Turbulent Combustion, Cambridge University Press, Cambridge, U.K., 2000.

[6] P. Sagaut and C. Cambon, Homogeneous Turbulence Dynamics, Cambridge University Press, Cambridge, U.K., 2008.

[7] A. Y. Poludnenko, T. A. Gardiner, and E. S. Oran, *Physical Review Letters*, 107:5 (2011) 1–4. doi:10.1103/PhysRevLett.107.054501

[8] A. Ingenito and C. Bruno, *AIAA Journal*, 48:3 (2010) 515–525. doi:10.2514/1.43652

[9] D. A. Donzis and S. Jagannathan, *Journal of Fluid Mechanics*, 733, (2013) 221–244. doi:10.1017/jfm.2013.445

[10] S. Jagannathan and D. A. Donzis, *Journal of Fluid Mechanics*, 789, (2016) 669–707. doi:10.1017/jfm.2015.754

[11] H. Kolla, E. R. Hawkes, A. R. Kerstein, N. Swaminathan, and J. H. Chen, *Journal of Fluid Mechanics*, 754, (2014) 456–487. doi:10.1017/jfm.2014.392

[12] C. A. Z. Towery, A. Y. Poludnenko, J. Urzay, J. O'Brien, M. Ihme, and P. E. Hamlington, *Physical Review E*, 93:5 2016, p. 053115. doi:10.1103/PhysRevE.93.053115

[13] J. O'Brien, C. A. Z. Towery, P. E. Hamlington, M. Ihme, A. Y. Poludnenko, and J. Urzay, *Proceedings of the Combustion Institute*, 36:2 (2017) 967–1975. doi:10.1016/j.proci.2016.05.005

[14] C. A. Z. Towery, R. Darragh, A. Y. Poludnenko, and P. E. Hamlington, *AIAA Paper 2017-4286*, (June 2017). doi:10.2514/6.2017-4286

[15] N. Zhuang, F. S. Alvi, M. B. Alkislar, and C. Shih, *AIAA Journal*, 44:9 (2006) 2118–2128. doi:10.2514/1.14879

[16] S. Heinz, *AIAA Journal*, 44:1 (2006) 3040–3050. doi:10.2514/1.22259

[17] W. M. Van Lerberghe, J. G. Santiago, J. C. Dutton, and R. P. Lucht, *AIAA Journal*, 38:3 (2000) 470–479. doi:10.2514/2.984

[18] F. Génin and S. Menon, *AIAA Journal*, 48:3 (2010) 526–539. doi:10.2514/1.43647

[19] E. Mastorakos, T. A. Baritaud, and T. J. Poinsot, *Combustion and Flame*, 109:1 (1997) 198–223. doi:10.1016/S0010-2180(96)00149-6

[20] R. Samtaney, D. I. Pullin, and B. Kosović, *Physics of Fluids*, 13:5 (2001) 1415–1430. doi:10.1063/1.1355682

[21] Y. B. Zeldovich, *Combustion and Flame*, 39:2 (1980) 211–214. doi:10.1016/0010-2180(80)90017-6

[22] K. K. Kuo, Principles of Combustion, 2nd ed., John Wiley & Sons, New York, (1986) 354–436.

[23] Z. Luo, C. S. Yoo, E. S. Richardson, J. H. Chen, C. K. Law, and T. Lu, *Combustion and Flame*, 159:1 (2012) 265–274. doi:10.1016/j.combustflame.2011.05.023

Western States Section of the Combustion Institute – Fall 2017 Meeting
Hosted by the University of Wyoming
October 2-3, 2017

The Effect of Spray Burner Outlet Configuration, Nozzle Position and Nozzle Type on the Flame Behavior and Stability of Liquid Fuels

Radi A. Alsulami[1] and Bret C. Windom[1]

[1]Department of Mechanical Engineering, Colorado State University, Fort Collins, CO 80523

Abstract: A spray burner was designed and manufactured to investigate the relationship between the evaporation and flame/combustion behaviors (spray combustion) of real liquid fuels. The spray burner was designed with the flexibility to explore multiple configurations including spray nozzle types (e.g. hollow or solid cone spray shape). In addition, it is capable of delivering high fuel and air co-flow rates. This work focuses on studying the effect of the spray burner configuration/characteristics on the flame/combustion behavior and stability by measuring lift off heights and lean blow out limits for liquid fuels including n-heptane and n-dodecane. Results indicate a strong influence of air and/or fuel flow rates as well as fuel volatility and other physical properties of the liquid fuel on the flame stability. The solid cone spray nozzle generates a more uniform atomized spray than the hollow cone nozzle at lower fuel flow rates and thus produces more stable and higher lifted flames. This initial characterization of the burner is an important step towards future investigations which will explore the role of preferential vaporization on the combustion of complex fuels.

Keywords: Spray combustion, Flame stabilization, Liftoff height, Blowout limits.

1. Introduction

The energy conversion process inside many heat engines, such as gas turbines, begins with the injection of the liquid fuel through a spray nozzle, followed by atomization and droplet evaporation before it combusts. The phase change process and the turbulent environment, as well as, the injector characteristics can influence the combustion of the liquid fuel inside the engine. As a result, understanding the coupling between these processes which can affect the spray combustion of a liquid fuel, *e.g.* flame stability, flame structure, and sooting potential, is desirable. This understanding will ultimately lead to the co-design of an engine and the fuel, with optimal combustion and physical properties, to maximize energy conversion efficiency and minimize associated emissions.

In general, most of the previous works studied the stability of diffusion flame for gaseous or pre-vaporized liquid fuels [1,2]. However, a more limited number of numerical and experimental studies have investigated the stability and structure of spray flames [3-6]. The effect of two different nozzle positions on the flame stability and structure was studied for n-heptane fuel by Mansour et al. [3]. The flame structure and fuel droplet properties of a spray flame was studied by Verdier et al. [4] using laser-based techniques. They used a hollow cone nozzle and measured the droplet size and temperature at different vertical positions. Experimental and simulation characterization of turbulent ethanol spray was done by Düwel et al. [5] by measuring droplet size, droplet temperature, liquid phase temperature and gas phase temperature. Liu et al. [7] investigated the effect of the spray characteristics on flame stability, *e.g.* nozzle configuration and position. The lean blowout behavior for spray flame at different air preheated temperature was experimentally predicted by Grohmann et al. [8,9] for multiple single component fuels including n-hexane, n-dodecane and iso-octane. The differences in physical and chemical properties of each fuel were responsible for differences in the observed lean blowout limits. Iso-octane was noticed to blowout faster than the two n-alkane fuels and the responsible of that is the chemical structure and kinetic of iso-octane. The mean reasons for differences in lean blowout between n-hexane and n-dodecane are the physical properties, such as density, viscosity and surface tension, which effect the atomization and mixing processes. In addition, they concluded that the processes of atomization, mixing, turbulence, evaporation and flame chemistry can all influence the flame blowout limits. O'Loughlin and Masri [10] studied the liftoff height of single component fuels including alcohol and n-alkane fuels. They discussed the effect of different physical and chemical properties of the fuel, such as boiling temperature, enthalpy of vaporization and ignition temperature on the liftoff height.

Recently a spray burner was constructed to study the relationship between the physical and chemical properties of real liquid fuels and their flame/combustion behaviors. The present experimental study focuses on investigating the effect of the spray burner configuration on the flame dynamics and stability by measuring flame liftoff heights and lean blow out limits for n-alkane liquid fuels including n-heptane and n-dodecane without using bluff-body or pilot flame stabilization. A number of the burner components are adjusted to investigate their influence on the flame stability including, (1) the effect of the fuel spray shape by using two different nozzle types, a hollow and a solid cone nozzle, (2) the effect of air co-flow entrainment and (3) the effect of varying the fuel flow rate. This initial characterization of the burner is an important step in future investigations linking the role of fuel properties on the combustion of complex multi-component liquid fuels

2. Experiment

A spray burner was constructed with the ability to accurately deliver large flow rates of fuel and air to achieve stable flame. The cross-section schematic of the spray burner is shown in **Fig. 1**. It consists of a central axisymmetric spray injector (DELAVAN 0.40 GPH 80°) surrounded by a heated annular air co-flow. Two nozzle types are used in this study, a hollow and a solid cone nozzle, with the same

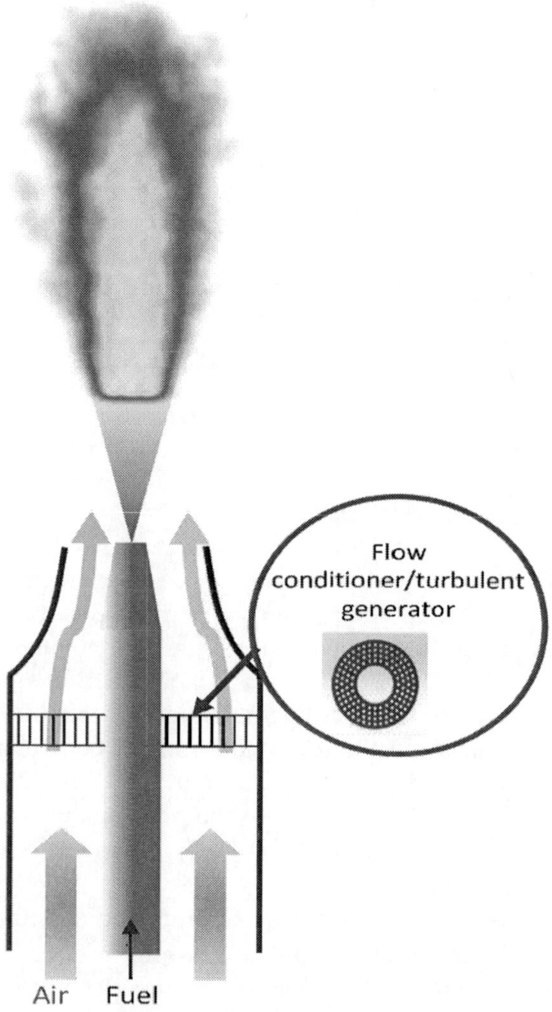

Figure 1. Schematic of the turbulent spray burner.

specifications for nozzle orifice size and spray angle. Concentricity is maintained by a turbulent ring generator/flow straightener, which has 165 circular slots evenly distributed over the ring's area. The exit area of the co-flow air is kept constant throughout this study at 685.5 mm^2. The apparatus consists of the spray burner flow controller system that has the ability to deliver high air flow rates, up to 3000 L/min (Alicat, MCR-3000SLPM-D-PAR) with a stated accuracy of ±0.8% of reading, and 0.2% of full-scale. The fuel is supplied by a high-pressure syringe pump (ISCO, Model 260D) capable of delivering up to 107 ml/min of liquid fuel. Experiments were carried out in the atmospheric environment of Fort Collins, CO, at T = 298 ± 5 K and P = 0.85 atm. The inlet flow rate conditions of

Table 1. Experimental flow conditions at the burner exit for flame liftoff measurments with n-heptane and n-dodecane (outlet area of 685.5 mm^2).

Fuel Flow Rate [mL/min]	Air Flow Rate [L/min]	Air Velocity @ the burner exit [m/s]	Reynolds Number
40 & 45	0	-	-
40 & 45	175	4.26	4670
40 & 45	250	6.1	6671
40 & 45	300	7.3	8006
40 & 45	550	13.4	14678

Table 2. Physical properties of selected fuels investegated in this work. All properties evaluated at T=298 K and P=0.85 atm.

Fuel	Boiling Temperature [K]	Vapor Pressure [kPa]	Density [kg/m3]	Dynamic Viscosity [mPa.s]	Surface Tension [mN/m] @20°C
n-heptane	371.53	6.02	679.71	0.38906	20.14
n-dodecane	489.3	0.0176	745.83	1.3622	25.35

the liquid fuel (n-heptane or n-dodecane) and co-flow air are varied as illustrated in Table 1. The fuel and the air flow rates were carefully controlled to maintain desired global equivalence ratio. The two fuels were chosen to be from the same hydrocarbon moiety (n-alkane) to eliminate the chemical kinetic effect on the flame stability and rather focus on the influence of physical properties, highlighted by those listed in **Table 2**.

The effect of the spray burner configuration/characteristics on the flame/combustion behavior and stability is investigated by measuring flame lift off heights and lean blow out limits. The flame liftoff height is measured by averaging 10 images of flames generated for each fuel at the specified fuel and air flow rates (and the Reynolds number indicated previously in **Table 1**. The lean blowout limits are evaluated by ramping up the air flow rate while keeping the fuel flow rate constant until the flame becomes destabilized and extinguishes. The reported data is the average of at least two measurements for each of the four fuel flow rates, 40, 45, 50, and 55 ml/min. The repeatability of the blowout limit measurements was acceptable with relative standard deviation from the mean value between 0-2.5% for the solid cone and between 0.8-3.8% for the hollow cone. The air flow rates and global equivalence ratios at the time of blowout are recorder for all the conditions as listed in **Tables 3 & 4**.

3. Results and Discussion

The liftoff heights and the blowout limits were measured for n-heptane and n-dodecane for different spray, co-flow and fuel flow rate conditions. Consistent with previous works [3,6], a double flame structure appears in all flames. Blue color (partially premixed) flame in the upstream near the burner tip followed by soot rich yellow flame. **Figure 2** shows the flames of n-heptane at different air flow rates, 0, 175, 250, 300 and 550 mL/min, respectively. As the co-flow velocity increases the flame becomes unstable until it extinguishes. At 0 ml/min air flow, the flame is rich causing yellow soot.

Table 3. Blowout conditions of solid cone nozzle.

Fuel Flow Rate [mL/min]	nc7 - Air Flow Rate at Blowout [L/min]	nc7 - $Ø_{BO}$	nc12 - Air Flow Rate at Blowout [L/min]	nc12 - $Ø_{BO}$
40	696.67	0.61	937.5	0.49
45	643.33	0.74	1055	0.49
50	570	0.93	1082.5	0.53
55	490	1.19	1055	0.6

Table 4. Blowout conditions of hollow cone nozzle.

Fuel Flow Rate [mL/min]	nc7 - Air Flow Rate at Blowout [L/min]	nc7 - $Ø_{BO}$	nc12 - Air Flow Rate at Blowout [L/min]	nc12 - $Ø_{BO}$
40	782	0.54	775	0.59
45	837.5	0.57	840	0.62
50	855	0.62	945	0.61
55	883.33	0.66	987.5	0.6

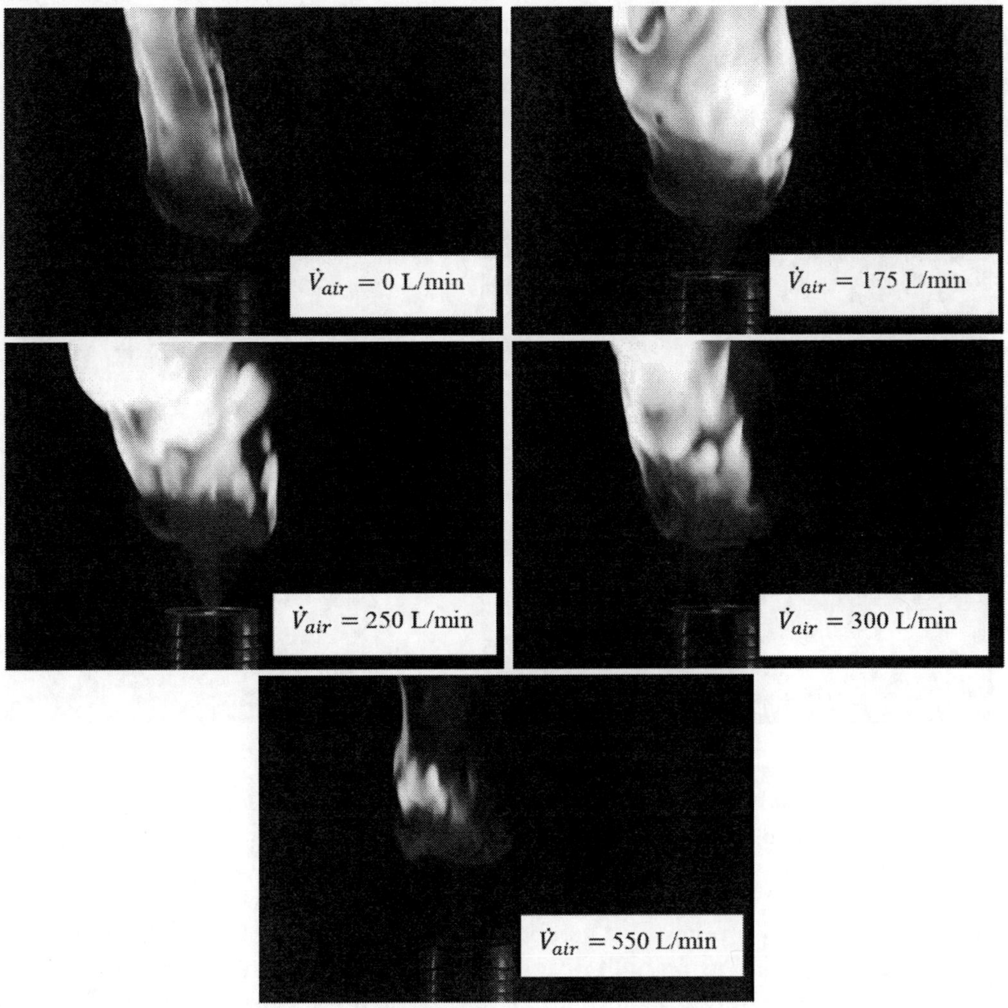

Figure 2. Photographs of n-heptane flames at fuel flow rate of 40 mL/min and different air flow rates.

Figure 3 shows a n-dodecane flame of both spray types at 40 mL/min fuel flow rate and 250 L/min air flow rate. The hollow cone spray at these lower fuel flow rates is not as completely atomized compared to the solid cone nozzle producing larger droplets spray which penetrate further into the spray. n-dodecane generates bigger flame than n-heptane as shown in **Fig 4**. For heavier fuel, *e.g.,* n-dodecane, unburned droplets were seen passing through the flame as illustrated in **Fig. 4b.**

Figure 5a, illustrates the liftoff height for n-heptane flames at two different fuel flow rates, 40 & 45 mL/min, for both a solid (-S) and hollow (-H) cone nozzle. In general, increasing the fuel and/or the co-flow air flow rates increase the liftoff height of the flame. For the hollow cone nozzle, at low fuel flow rates, flame liftoff was below the observed liftoff for the solid cone nozzle. It was noticed that at low air flow rates, the fuel is not properly atomized, resulting in the clear observation of larger droplets, which travel slower and result in a lower stabilized flame as shown in **Fig 3b**. As the air flow rate is increased, the shear force caused by the air flow helps atomize the fuel droplets producing liftoff heights which are almost identical for the two nozzle types at higher air flow rate for n-heptane.

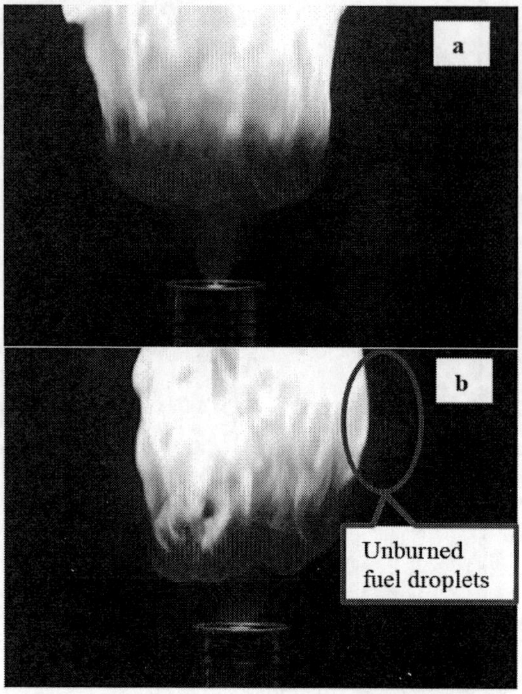

Figure 3. Flame photographs of n-dodecane for a) solid and b) hollow cone nozzles at 40 mL/min and 250 L/min fuel and air flow rates, respectively.

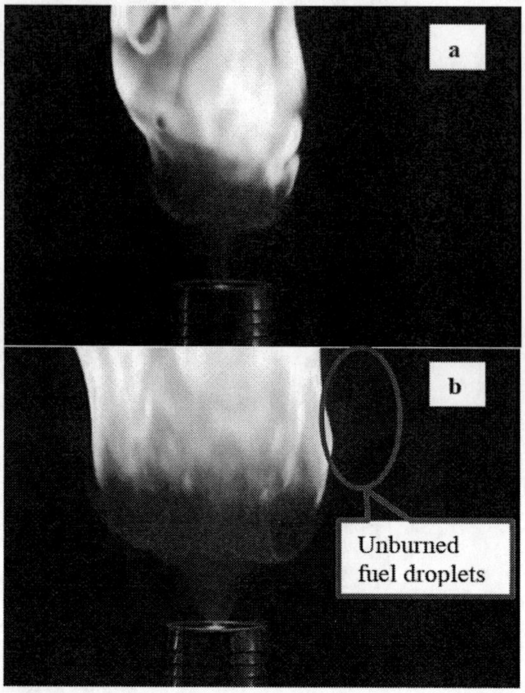

Figure 4. Flame photographs of a) n-heptane and b) n-dodecane at 40 mL/min and 175 L/min fuel and air flow rates, respectively.

The role of volatility on flame lift can be observed in **Figure 5b**, which compares flame liftoff for n-heptane and n-dodecane using both nozzle types at a constant fuel flow rate of 40 ml/min. For the solid cone, the liftoff of n-dodecane flame is larger than that of the n-heptane as a result of slower evaporation. An opposite trend is observed for the hollow cone spray with the liftoff height of n-dodecane to be much lower than that of n-heptane, which is assumed to be caused by the poor atomization exacerbated by the lower volatility of the n-dodecane, as depicted in **Fig 3b**.

Figure 6 shows the blowout limits of the two fuels for both the solid and hollow cone nozzles. For the solid cone, n-heptane blowout was observed at higher equivalence ratio ($Ø_{BO}$) than n-dodecane. A recent study which noted a similar trend attributed the difference in blowout limits of two n-alkane fuels with very different molecular weights to the local liquid fuel loading and the fuel air mixing. The higher volatility of n-heptane results in a faster transition into the gas phase improving mixing resulting in a leaner local equivalence ratio closer to the nozzle exit than for n-dodecane given the same flow conditions. On the other hand, n-dodecane has a lower vapor pressure and higher viscosity, density and surface tension, which means slower vaporization and bigger droplet size [9]. As a result, more regions have higher local equivalence ratio compared to n-heptane. As such, less air is required per mass of fuel to create the lean conditions responsible for localized flame extinction for n-heptane as compared to n-dodecane resulting in larger (i.e. more rich) lean blowout limits. The opposite trend is observed at less than 50 mL/min fuel flow rate when the hollow cone nozzle is used, again, resulting from the poor atomization of the hollow cone sprayer. In general, larger n-alkanes require higher air flow rates (or higher Reynolds numbers) to achieve blowout generating more turbulence which can enhance the atomization and mixing and stabilize a flame. On the other hand, the increased velocity can shorten the resident time of the droplets inside the flame preheat zone which can impact the droplet evaporation. It is believed that the increasing Ø_{BO} with increasing fuel flow rates for the solid cone n-heptane case is caused from the decreased droplet residence time in the flame preheat zone as droplet velocities are increased.

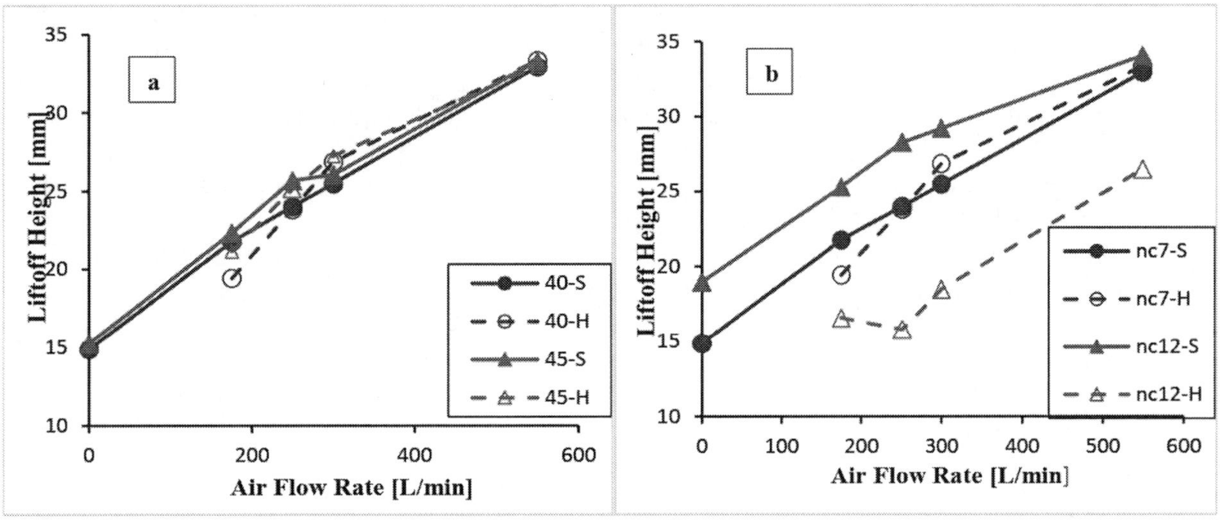

Figure 5. a) Liftoff heights for n-heptane for solid and hollow cone spray shapes at different air and fuel flow rates, and b) Liftoff heights for n-heptane and n-dodecane for solid and hollow cone spray shapes at 40 L/min fuel flow rate and varying air flow rates.

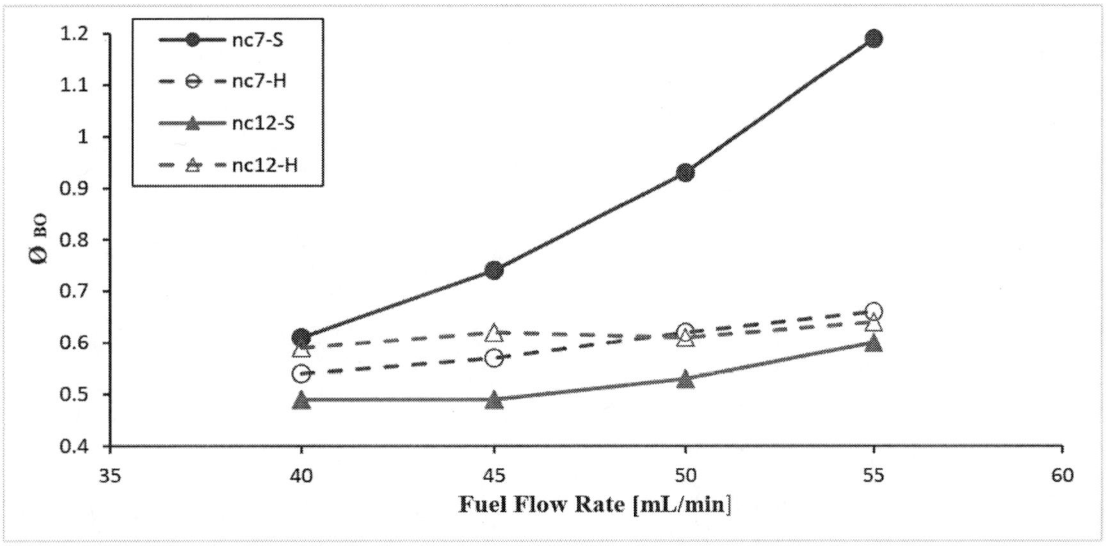

Figure 6. Blowout limits for n-heptane and n-dodecane for solid and hollow spray shapes at different fuel flow rates.

4. Conclusions and Future Work

A turbulent spray burner was constructed to investigate the role of fuel physical properties on the flame/combustion of complex fuels. Initial characterization of the burner is conducted in the current work focusing on investigating the effect of the spray burner configuration on the flame/combustion behavior and stability by measuring lift off heights and lean blow out limits for n-alkane liquid fuels including n-heptane and n-dodecane. The effect of the fuel spray shape by using two different nozzle types, hollow and solid cone nozzles, the effect of air co-flow entrainment and the effect of varying the fuel flow rate on the stability of the flame were examined. It was noticed that increasing the fuel and/or the co-flow air flow rates increase the liftoff height of the flame. The hollow cone spray, especially at low fuel flow rates, was not fully atomized generating uncertainty in the results and lower flame liftoff heights. It was noticed that n-heptane flames blowoff at richer equivalence ratios ($Ø_{BO}$) than n-dodecane, when solid cone nozzle is used. However, the opposite occurs at fuel flow rates less than 50 mL/min when the hollow cone nozzle is utilized. It was concluded that the processes of atomization, mixing, turbulence, evaporation and flame chemistry influence the flame stability (*e.g.* liftoff height and blowout limit).

More burner components will be adjusted in the future to investigate their influence on the flame stability, such as (1) a contoured versus straight co-flow outlet configuration to understand the influence of the air co-flow aerodynamics on fuel spray development and flame stability and (2) the vertical nozzle location relative to the outlet of the co-flow to determine its influence on the flame dynamics. In addition, heated air co-flow will be supplied to the burner to further investigate the role of vaporization on the combustion of complex fuels

References

[1] Windom, Bret, et al. "Study of ignition chemistry on turbulent premixed flames of n-heptane/air by using a reactor assisted turbulent slot burner." Combustion and Flame 169 (2016): 19-29.

[2] Won, Sang Hee, et al. "The role of low temperature fuel chemistry on turbulent flame propagation." Combustion and Flame 161.2 (2014): 475-483.

[3] Mansour, Morkous S., Issam Alkhesho, and Suk Ho Chung. "Stabilization and structure of n-heptane flame on CWJ-spray burner with kHZ SPIV and OH-PLIF." Experimental Thermal and Fluid Science 73 (2016): 18-26.

[4] Verdier, Antoine, et al. "Experimental study of local flame structures and fuel droplet properties of a spray jet flame." Proceedings of the Combustion Institute 36.2 (2017): 2595-2602.

[5] Düwel, I., et al. "Experimental and numerical characterization of a turbulent spray flame." Proceedings of the Combustion Institute 31.2 (2007): 2247-2255.

[6] Marley, S. K., et al. "Effects of leading edge entrainment on the double flame structure in lifted ethanol spray flames." Experimental thermal and fluid science 29.1 (2004): 23-31.

[7] Liu, Cunxi, et al. "Investigations of the effects of spray characteristics on the flame pattern and combustion stability of a swirl-cup combustor." Fuel 139 (2015): 529-536.

[8] Grohmann, J., et al. "Influence of Single-Component Fuels on Gas-Turbine Model Combustor Lean Blowout." Journal of Propulsion and Power (2017).

[9] Grohmann, Jasper, et al. "Comparison of the Combustion Characteristics of Liquid Single-Component Fuels in a Gas Turbine Model Combustor." ASME Turbo Expo 2016: Turbomachinery Technical Conference and

Exposition. American Society of Mechanical Engineers, 2016.

[10] O'Loughlin, W., and A. R. Masri. "A new burner for studying auto-ignition in turbulent dilute sprays." Combustion and Flame 158.8 (2011): 1577-1590.

Western States Section of the Combustion Institute – Fall 2017 Meeting
Hosted by the University of Wyoming
October 2-3, 2017

Nitrogen Oxide Emissions from Reacting Jets in Vitiated Crossflow

Matthew D. Sirignano[1] Vedanth Nair[1] Benjamin Emerson[1] Jerry Seitzman[1] Timothy C. Lieuwen[1]

[1]School of Aerospace Engineering, Georgia Institute of Technology, 270 Ferst Drive, Atlanta, GA 30332, USA

Nitrogen oxides (NOx) are an undesirable byproduct of hydro-carbon combustion in air. In lean, premixed combustion, NOx levels are exponential functions of temperature and linear functions of residence time. Historically, lean premixed technologies have enabled local flame temperatures below the value where substantial NOx production occurs; consequently, most historical combustor technology development has focused on fuel/air mixing technologies and dealt with other operational challenges of premixed combustion (e.g., blowoff, flashback, etc). However, improved cooling technologies and improved high temperature materials have enabled steadily increasing combustor exit temperatures (>1800K), to the extent that this strategy is no longer capable of low NOx levels. While reducing residence times is one potential approach for counterbalancing the increasing flame temperatures, such approaches significantly limit the turndown capability of the system, due to excessive CO emissions at lower powers/temperatures. Axial staging of combustion has been identified as a means to combat these issues at high combustor temperatures. This paradigm shift motivates research into the canonical problem of a reacting jet-in-crossflow (RJICF). The NOx emissions of RJICF are influenced by the degree of premixing of the fuel jets before injection and jet/crossflow mixing before combustion. In turn, jet/crossflow mixing is controlled by the hydrodynamic stability of the jet, as well as degree of flame lifting. The presented work investigates the NOx emissions and flame stabilization characteristics of rich premixed methane/air jets injected into a high temperature (1650K – 1800K) vitiated crossflow of lean ($0.5 \leq \Phi \leq 0.65$) natural gas/air combustion products. The study establishes the existence of a combustor firing temperature threshold, above which a NOx emissions benefit to axial staging is realized. Reductions in NOx production, for set levels of axial stage heat release, are shown to correspond with reacting jets of lower momentum flux ratio and higher jet equivalence ratio. Increasingly rich jet equivalence ratios are observed to lead to significant lifting of the associated flame with a correlated impact on the NOx emissions. This relationship is, however, confounded due to the sensitivities of lift-off distance to jet parameters that are shown to directly impact the NOx production of the reacting jet.

1. Introduction

Over the past several decades, the gas turbine industry has experienced significant increases in combined cycle efficiency while simultaneous reducing undesirable emissions such as nitrogen oxides (NOx). Gains in efficiency have been achieved primarily through steady improvements in turbine cooling technologies, enabling ever higher turbine inlet temperatures. In parallel, advances in injector and post-flame mixing technologies have reduced NOx emissions through the elimination of locally high temperature combustion. These advances have led to the current state of the art in gas turbine Dry Low NOx (DLN) combustor architectures. In order to achieve further improvements in combined cycle efficiency, however, higher turbine inlet temperatures are required. DLN combustors will be unable to meet current emission regulations while simultaneously delivering the necessary elevated turbine inlet temperatures due to the nature of NO production.

NO is formed through a variety of mechanisms both within and after a flame reaction zone. In lean, high temperature environments this production is dominated by thermal NO production, often referred to as "Zeldovich NOx". Thermal NO is created from reactions of N_2 and O_2 at high temperatures. An approximate formula (Equation 1) developed by Bowman [1] highlights the key contributing factors to NO production.

$$[NO] \propto [O][N_2]e^{-38{,}379/T}\tau_{res} \quad (1)$$

As indicated by Equation 1, NO production is extremely sensitive to temperature and is linearly dependent on the combustor residence time (τ_{res}) and the concentration of atomic oxygen. These sensitivities suggest several potential methodologies to reduce combustor NOx. Reducing τ_{res} at high temperatures is one potential approach to alleviating NOx emission, but the extent to which this strategy can be implemented for single point injection is limited by turndown considerations; i.e., longer residence times are required for CO oxidation at part load conditions. In contrast, axial staging enables a reduction in residence time for a portion of the reacting mixture.

Axial combustion staging has the benefit of addressing all three sensitives of NO production simultaneous. By delaying the injection and thus combustion of a portion of the fuel, the time spent at the final turbine inlet temperature is reduced. In addition, burning the remaining fuel in the vitiated crossflow created by the primary burner further reduces NO by creating a reduced oxygen environment [2]. The theoretical benefits of axial staging of fuel have been established by previous work; e.g., see Ahrens *et al.* [3] and Goh *et al.* [4]. However, without sufficient mixing with the crossflow, axial staging can also significantly increase NO formation, as local non-premixed combustion at stoichiometric fuel/air ratios generates significant NO levels.

A reacting jet in crossflow (RJICF) is a natural choice for implementing axial staged concepts, as it does not require insertion of hardware (e.g., bluff bodies, swirl vanes) into the high temperature gas stream. Due to numerous industrial applications and its importance as a canonical fluid mechanics problem, jets in crossflow (JICF) have been the subject of extensive research [5]. As a result the fundamental form of time averaged velocity and scalar fields for subsonic, non-reacting, momentum driven JICF is well established, and has been shown to be primarily governed by the jet momentum flux ratio given by Equation 2:

$$J = \frac{\rho_j u_j^2}{\rho_\infty u_\infty^2} \quad (2)$$

with ρ and u denoting the density and velocity of the jet and crossflow with subscripts j and ∞ respectively. A common resultant jet trajectory correlation is given by Equation 3:

$$\frac{z}{d_j\sqrt{J}} = A\left(\frac{x}{d_j\sqrt{J}}\right)^B \quad (3)$$

with x and z corresponding to the trajectory location in direction of the crossflow flow and transverse to it respectively. The jet diameter is denoted by d_j and the coefficients A and B are parameters that account for various effects such as boundary layer thickness, jet exit velocity profile, and jet trajectory definition (i.e., maximum velocity, scalar concentration, or other important fluid-mechanic features).

The impact of reacting flows on JICF behavior is less well understood and is an ongoing subject of research, with work divided into two areas of focus: non-premixed RJICF [6, 7] and premixed RJICF[8-11]. The focus of previous premixed work has primarily been on lean or near stoichiometric jets. This work investigates the behavior of rich premixed jets with equivalence ratios well above the rich flammability limit. Particular attention is paid to the NOx production of these reacting jets.

Bandaru and Turns [12] investigated NOx emissions from a reacting jet in a crossflow of air over a range of jet compositions and crossflow velocities and observed a lack of sensitivity of the NOx levels to these parameters, with the exception of fuels noted for high soot formation such as ethylene. The heat loss to the soot particles causes a reduction in product temperature and thus a noticeable decrease in thermal NO formation. Lyle *et al.* [13] investigated the impact of partial premixing in jet flame emissions by injecting rich premixed methane/air jets with a co-flow of air into quiescent air. No significant deviation from NOx emissions produced by fully non-premixed jet flames was seen for jet equivalence ratios (ϕ_{Jet}) greater than 5.0. For $5.0 \geq \phi_{Jet} \geq 1.5$ a 25% reduction in NOx emission was observed compared to the non-premixed case. For values of ϕ_{Jet} even closer to stoichiometric condition a significant increase in NOx emission was observed. This trend is heavy supported by equivalence ratio's impact on flame temperature and thermal NOx production's high degree of sensitivity to temperature as the jet flames evolved from high temperature diffusion flames, to low temperature rich premixed flames, and on to high temperature near stoichiometric premixed flames. Prathap *et al.* [14] showed that, for a given temperature rise across a RJICF, jet trajectory exhibited strong sensitivity to jet parameters such as momentum flux ratio and Reynolds number, but NOx production experienced little variation. Roa *et al.* [15] demonstrated that NOx production from a RJICF was a strong function of the temperature rise of the flow due to the jet flame (ΔT). These results show that NOx production by a RJICF is primarily driven by the associated ΔT, but did not explore NO sensitivities to parameters such as jet fuel/air ratio or J for a given ΔT. Ahrens *et al.* [3] showed, for lean premixed jets in a vitiated crossflow, that local NOx production is strongly correlated to local mixture fraction, suggesting rapid mixing prior to combustion as a key element in delivering the potential NO emissions benefits at a given ΔT.

A better fundamental understanding of the relevant phenomena governing a reacting jet in cross flow and in particular their impact with the associated production of pollutants is key to addressing current energy challenges. The work presented in this paper seeks to assist in building this understanding by investigating the sensitivities of NOx formation of rich premixed jets reacting in a hot vitiated crossflow to varying jet parameters at several ΔT values. The work was performed over a range of jet fuel/air ratios, cross flow temperatures, and J values, which also produced the indirect impact of varying degrees of flame attachment/lifting to the jet exit. These studies enable us to analyze the effects of these parameters for a given ΔT value and study the coupling of fluid mechanic and kinetic processes in influencing NO emissions.

2. Methods

2.1. Experimental Facility

The experimental facility was designed to provide lean, high temperature products of natural gas/air combustion in order to auto-ignite premixed jets of methane and air at atmospheric pressure. A schematic of the facility is provided by Figure 1. The primary components of the facility include the main burner, flow conditioning section, test section with secondary injection, and chemical quench section.

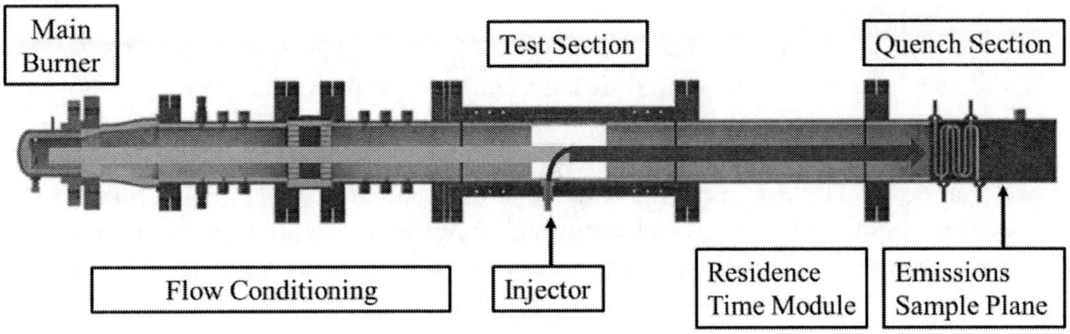

Figure 1: Schematic of experimental facility.

A premixed natural gas/air mixture preheated to 591K (±5K) is tangentially injected through 4 identical injection ports into the main burner to produce a stable high swirl primary flame. Flow to the main burner was metered using sub-critical orifices. The main burner demonstrated stable operation across the entire range of test conditions. The flow conditioning section consists of a circular to rectangular cross-section transition piece followed by a 30 cm (3.7 channel hydraulic diameters) settling chamber, two ceramic honeycomb flow straighteners, and a second 30 cm settling chamber in sequence. The honeycomb flow straighteners are Mullite Versagrid units comprised of 6.35 mm by 6.35 mm passages. The test section includes optical access (sides and top) and the secondary injection jet. The test section cross-section is 63.5 mm wide by 114 mm tall. This cross-section initiates immediately after the second flow straightener 48cm upstream of the jet exit plane. The jet is injected via a 12mm inner diameter stainless steel tube flush with the ceramic lining of the test section floor. The jet delivered preheated and premixed methane/air mixtures with preheat temperatures ranging from 420 – 460K. Flow to the jet was metered using critical orifices. The injection tube is straight for 40 diameters prior to the jet exit to ensure fully developed exit flow profile, and premixing occurring approximately 100 diameters prior to the jet exit.

After the test section and residence time module, the hot gases enter the NOx measurement section. Because of the highly non-uniform NOx and flow velocity profiles (which implies that simply mapping out spatial NOx values is insufficient, as NOx fluxes can also vary spatially), the flow is cooled to freeze NOx formation, then mixed, then measured. This quench section consists of a 15 cm ceramic lined entrance section followed by 15cm of pressurized water filled radiator tube banks. The tube banks are made up of two sets of six tubes shaped as indicated in Figure 1 and set opposite each other. Each of the two banks is fed by a manifold that is itself supplied at the sides and top. The waste water is expelled into a similar manifold. The quench section was verified to reduce the gas flow temperature to less than 700K. The NOx emissions sampling and measurement system is described in the next section.

Ultimately the facility delivered a vitiated crossflow of natural gas/air combustion products at equivalence ratios ranging from 0.5 to 0.65 at a velocity of 17.3 m/s. Corresponding test section temperatures ranged from 1650 – 1810K and cross flow Reynolds numbers from 4375 - 4800 based on a hydraulic diameter of 8.16 cm. The jet Reynolds number ranged from 3900 - 7000 with a wide range of equivalence ratios and mass flow rates.

2.2. Diagnostics

Emissions sampling was conducted using a slotted probe located 1.25 cm after the section of radiator tube banks. The spatial variation of NO was mapped out in a 4 by 3 grid with single point measurements taken at the center of each 3 cm wide by 3.5 cm tall box for a stoichiometric jet injected with $J \approx 12$ into the vitiated crossflow of the main burner with $\phi_{HE} = 0.55$. The spatial variation of NO measurements in the horizontal and vertical direction was less than 1 ppm (corresponding to less 15% of average), respectively. Spatial variations in the vertical direction were further assessed by mapping the center column of the 4 by 3 grid for following jets injected into a similar vitiated cross flow as previously: $J = 1.3$, $\phi_{Jet} = 4.0$ & $J = 6.4$, $\phi_{Jet} = 2.5$. In these instances NO measurements spatially varied vertically up to 14 ppm (corresponding to approximately 100% of average). Consequently, all NO measurements reported here were obtained from a slot probe that intakes gas through a 1.6 mm wide and 12 cm tall slot in order to simultaneously sample the entire vertical span. Gas collected by the probe is funneled into a 7.75 inner diameter tube which connects the probe to the gas analyzer. Gas samples were analyzed by a Horiba PG-350 gas analyzer which uses a cross-flow modulation chemiluminescence detection method to measure NOx (NO and NO_2 in this instance) with a repeatability of ±1% and a linearity of ±2% at a rate of 1 Hz. After a steady state was achieved for each reacting jet condition, a two-minute average was obtained.

Side-on CH* chemiluminescence imaging was conducted at 2000 fps using a Photron Fastcam SA-1 high speed camera and a Lambert HiCATT high-speed intensifier coupled to a single-band bandpass filter with a center wavelength of 434 nm and a minimum bandwidth of 17nm paired with a long wave pass IR filter with an edge wavelength of 364 nm. 909 images were taken for each test condition.

2.3. Test Matrix

The test matrix was designed to look at the NOx sensitivity to different fractions of heat release in the axial stage versus the main burner. It was also developed to investigate the differences in NOx between achieving a given ΔT across the axial stage by using richer, low J jets or leaner, high J jets which can provide some insight into the most beneficial manner, from an emissions standpoint, in providing premixed fuel in an axial staged combustor. These studies were performed at three different post-axial stage temperatures (also referred to as turbine inlet temperature) of nominally 1873K, 1915K and 1956K. The rest of this section describes the test matrix development approach.

Based on the average preheat conditions of the main burner these target temperatures were achieved with global equivalence ratios of 0.6, 0.625, and 0.65 respectively. The equivalence ratio of the main burner (ϕ_{HE}) was varied from $\phi_{HE} = 0.5$ to 0.65 in increments of 0.025. In order to fix the crossflow conditions, the mass flow rate was reduced with increasing equivalence ratio to account for higher flow temperatures; ensuring the crossflow velocity ($u_\infty = 17.3$ m/s) was constant between all the cases. The mass flow reduction was based on test section temperature measurements via spectroscopic measurement of the irradiance of a ceramic shielded type R thermocouple in place of the secondary jet.

Once the crossflow conditions were fixed and the turbine inlet temperature was chosen, the jet conditions were then iterated by changing J and the jet equivalence ratio (ϕ_{Jet}). The center panel of Figure 2 shows the test matrix data points corresponding to a given turbine inlet temperature of 1873K. The J ratios were chosen in the range of 2 to 5 to ensure that the jet stays within the confines of the viewable region of the test section. The trend lines on the plots show the computed target ϕ_{Jet} for jets of specific J values to reach the ϕ_{Global}. Two series of test points were created: the "adiabatic series," which assumes the crossflow temperature is equal to the adiabatic flame temperature at ϕ_{HE}, and the "non-adiabatic series," which utilizes measured test section temperatures and thus has higher associated ϕ_{Jet} values. The left panel of Figure 2 illustrates the interdependent nature of J and ϕ_{Jet}, by plotting the measured values for all test points. Visible are the sweeps of constant $\Delta\phi_{Global}$, that drive ϕ_{Jet} down as J increases. Figure 2 also indicates the execution of the target parameters by plotting measured J (left), and the measured ϕ_{Global} against the target turbine inlet temperature (right). Measured J grouped based on its target have mean values of 1.5, 2.2, 2.9, and 3.6 with variations from that value up to 14%, 8%, 8%, and 23% respectively for each grouping of similar J.

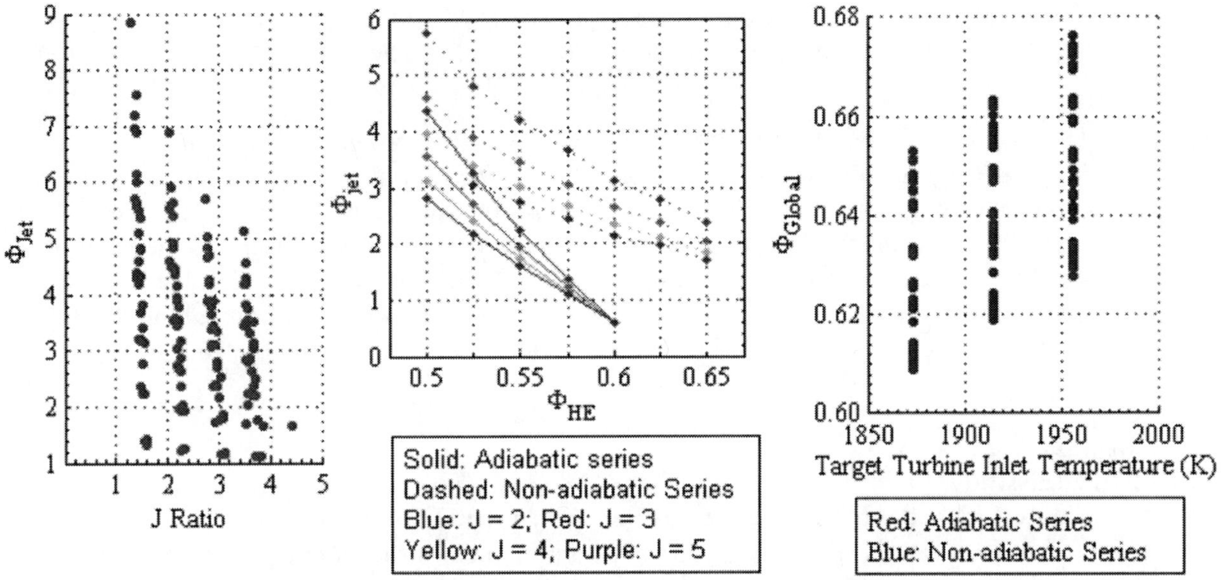

Figure 2: Study test parameters. Jet equivalence ratio vs momentum flux ratio (left), test matrix points for target turbine inlet temperature of 1873K (center) and ϕ_{Global} vs target turbine inlet temperature (right)

2.4. Image Processing

The instantaneous and time averaged CH* chemiluminescence images were analyzed to determine the spatial flame structure dependence on operating point. Of particular interest was the degree of flame lifting, which strongly influences jet/free stream premixing before combustion. The raw images were normalized and filtered using a guided image filter [16] to remove most of the noise. An example instantaneous raw image for a given case is shown on the left of Figure 3. The filtered image is then binarized by using a threshold based on Otsu's method [17]; the resulting binarized image for this example is shown in the center pane of Figure 3. The jet trajectory marked on all the images corresponds to the trajectory correlation by Smith & Mungal [18] for the jet diameter, jet velocity profile, and momentum flux ratios from this study.

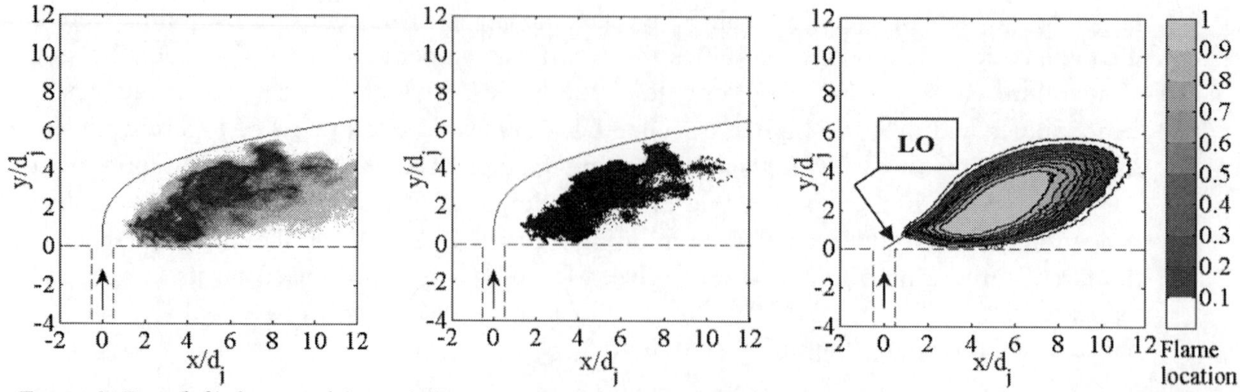

Figure 3: Raw (left), binarized (center) flame images with jet trajectory marked, and probability density field of flame position (right) with 0.3 contour corresponding to LO highlighted in red. Images from test case with $\phi_{HE} = 0.604$, $\phi_{Jet} = 1.402$, and $J = 1.56$.

The averaged flame characteristics were analyzed by looking at the time-averaged intensity field by averaging all the instantaneous CH* chemiluminescence images. The binarized instantaneous images were combined to create a probability density field, an example of which is shown in the right panel of Figure 3, where the value at each pixel describes the fraction of time (referred to henceforth as the probability) that the flame was present at that location through the image set for a given case. The lift-off distance (LO) was defined as the point along the 0.3 contour, highlighted in red, in this probability density field which was the closest to the center of the jet exit.

3. Results and Discussion

3.1. Emissions

Figure 4 plots the NOx generated by the experimental facility as a function of the equivalence ratio of the system (ϕ_{Global}). ϕ_{Global} directly corresponds to the final temperature achieved by the flow. As expected, NOx increases monotonically with increasing ϕ_{Global} and, thus, final temperature. In addition, the change in equivalence ratio across the axial stage ($\Delta\phi_{Global}$) normalized by ϕ_{Global}, which is an indicator of the amount of heat release added by the axial stage relative to the main burner and is henceforth referred to as the stage ratio, is indicated for each data point in Figure 4 via color scale. In addition, lines showing fitted NOx levels for the stage ratios of 15%, 19%, and 22% (corresponding to first, second, and third quartiles) are included. If axial staging is of benefit to reducing NOx emission, higher $\Delta\phi_{Global}$ should correspond to lower measured NOx levels. In the region below $\phi_{Global} \approx 0.70$, there is no clear stratification of NOx emission due to $\Delta\phi_{Global}$. In many instances in this region, increased utilization of the axial stage increases NOx emission, presumably due to local burning at near-stoichiometric equivalence ratios rather than the average value. Utilizing the quartile lines, the first quartile is associated with the lowest NOx levels and the third quartile the highest. However, for $\phi_{Global} > 0.70$, a stratification appears with high values of $\Delta\phi_{Global}$ producing lower NOx. This is confirmed by the quartile lines crossing each other and the third quartile line becoming associated with the lowest NOx values. For these high flow temperatures, the production rate of NO is so high that it becomes critical to reduce the residence time at elevated temperatures. Axial staging provides this reduction. In short, there exists a temperature threshold above which the residence time reduction afforded by axial staging wins out over any imperfect mixing effects associated with injection of a rich premixed jet into the crossflow

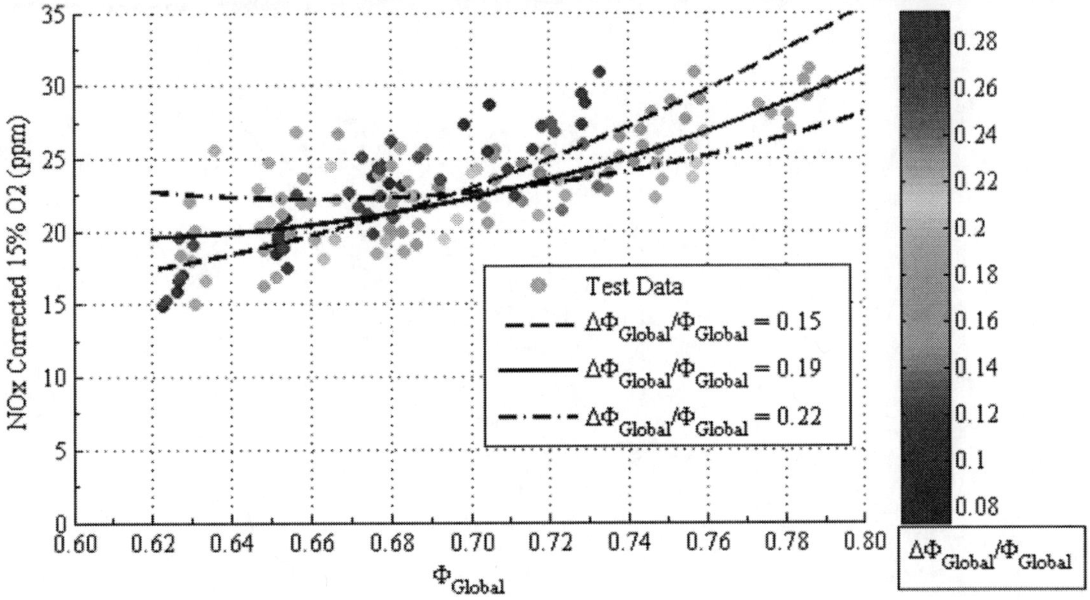

Figure 4: Dependence of NOx emissions upon the global equivalence ratio of the system. Color indicates the change in equivalence ratio due to the axial stage normalized by global equivalence ratio.

To investigate the sensitivity of NOx production when achieving a given ΔT, NOx contribution from the axial stage (ΔNOx) is utilized. ΔNOx is calculated by subtracting an estimated main burner NOx emission from the measured NOx level. The main burner NOx contribution was established during characterization of the main burner performance.

The axial stage NOx contribution is plotted against the associated rise in equivalence ratio in Figure 5. A clear monotonic relationship between ΔNOx and $\Delta\phi_{Global}$ is observed. The more temperature rise the reacting jet produces, the more NOx it creates. In addition, Figure 5 highlights the impact of the jet parameters by grouping data points of similar momentum flux ratio (an exponential fit line of each group is also included). Doing so reveals a trend where the lower J, and therefore richer jets (Figure 2), produce less NOx than their less rich and higher J counterparts at similar $\Delta\phi_{Global}$. Figure 6 further illustrates this behavior.

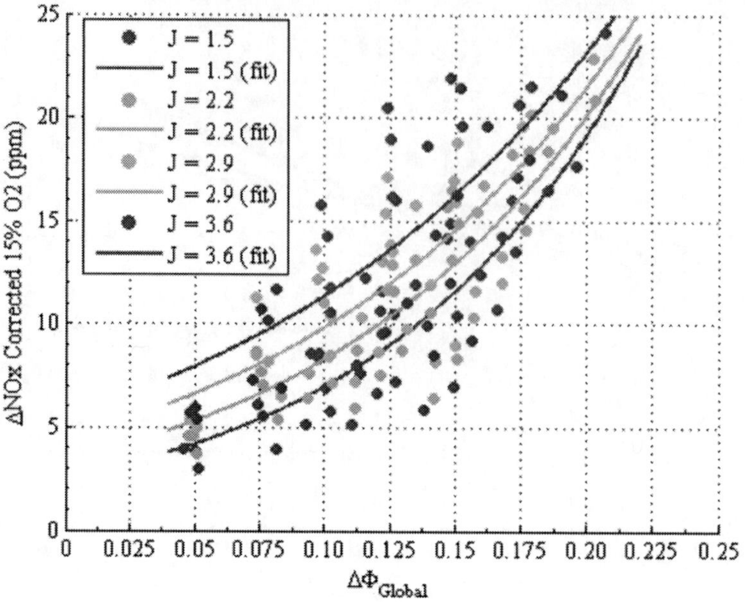

Figure 5: Dependence of axial stage contribution of NOx emissions upon the rise in flow equivalence ratio due to the axial stage. Color indicates groupings of similar values for J.

Figure 6 replots the same data, this time showing the axial stage NOx contribution dependence upon the equivalence ratio of the reacting jet (ϕ_{Jet}), with $\Delta\phi_{Global}$ correspondingly increasing to the top right. As in Figure 5, the data subsets of similar *J* are indicated via color with an associated exponential fit line. In Figure 6 it is clearly evident by the bands of differing momentum flux ratio that the richer, lower *J* jets are producing less NOx than their less rich, higher *J* counterparts.

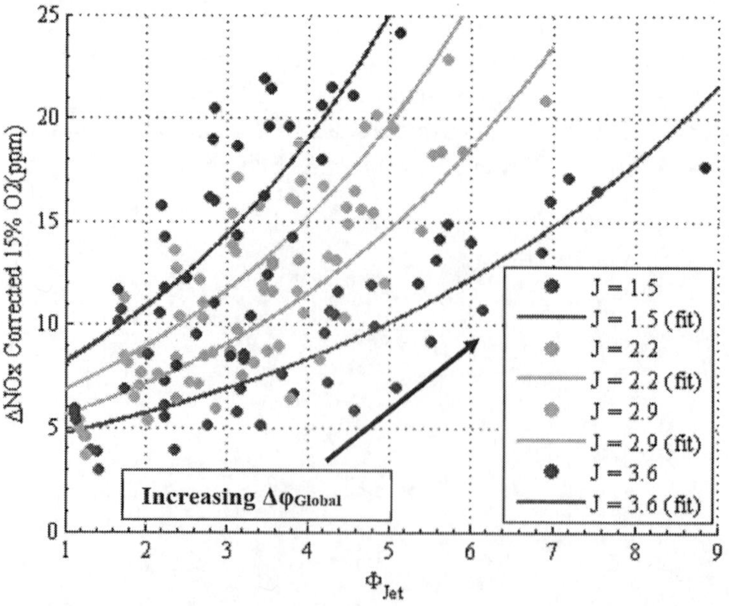

Figure 6: Dependence of axial stage contribution of NOx emissions upon the reacting jet equivalence ratio. Color indicates groupings of similar values for J.

Also varying across these experiments is the flame lifting. The flame stabilization mechanism which is dominant in this experiment is unclear as it is not possible to determine whether it is auto-ignition stabilized or turbulent flame propagation stabilized. A number of prior studies have considered flame stabilization mechanisms and liftoff [8, 9], showing that they can be both flame propagation and autoignition controlled, depending upon crossflow temperature. The chemiluminescence imaging done by both the above mentioned studies observed that there are leeward as well as windward regions where the flame attaches in some cases. This study, on the other hand, demonstrates that the flame is located to leeward of the jet, with almost no burning to the windward side, as indicated by Figure 7, which shows the time averaged flame images and estimated jet trajectory of several jet conditions. A primary distinction between the studied test conditions is that both Wagner [8] and Schmitt [9] use lean premixed jets while this study uses only jets with rich equivalence ratios. Appendix A provides further analysis of flame lift-off distance, LO, extracted from these data, and the correlation of LO with different parameters.

These flame images suggest that the jet is reacting in a partially premixed mode. As previously described in Section 2, for a given desired axial stage temperature rise, higher J cases are less rich than lower J cases. The trend indicated in Figure 6 can thus be accounted for by the following: if the high ϕ_{Jet} cases are burning in the vicinity of the rich flammability limit, then the reduction in ϕ_{Jet}, potentially combined with enhanced mixing due to a larger momentum flux ratio, drives the equivalence ratio of combustion closer to stoichiometric than for its low J, high ϕ_{Jet} counterpart. Reducing equivalence ratio of combustion towards stoichiometric increases the flame temperature leading to higher NOx local production from the reacting jet.

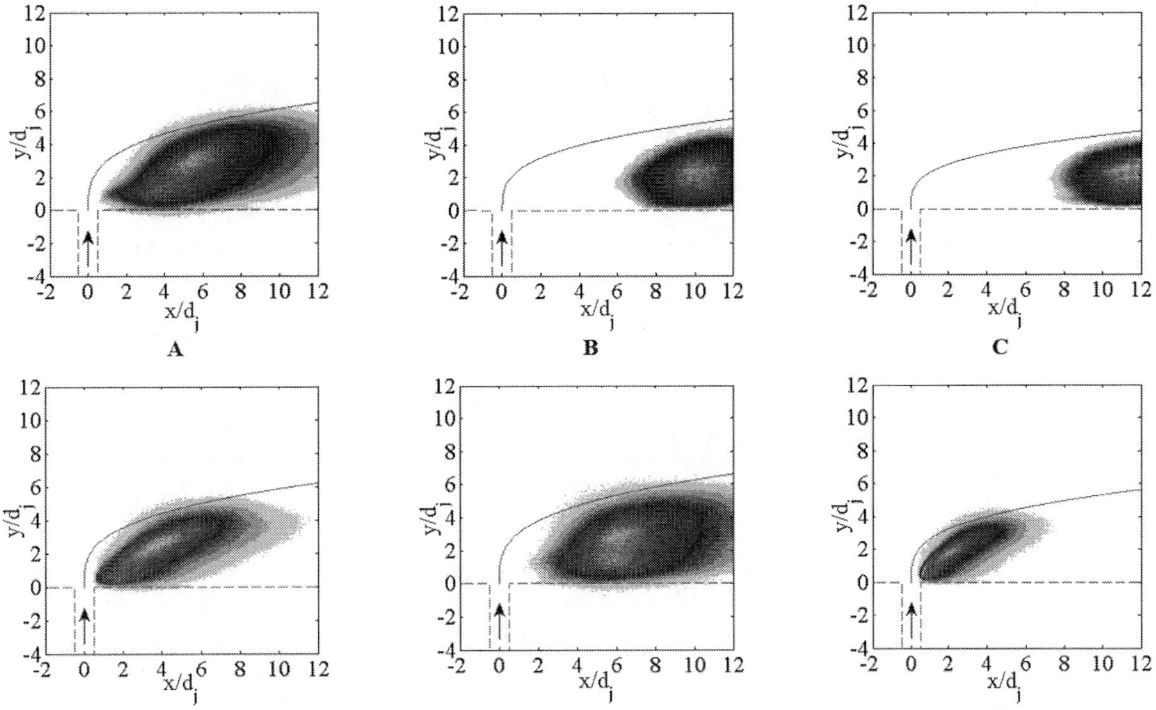

Figure 7: Characteristic time averaged flame image. Images from following test cases: a) $\phi_{HE} = 0.51$, $\phi_{Jet} = 3.49$, $J = 3.47$, b) $\phi_{HE} = 0.535$, $\phi_{Jet} = 3.879$, $J = 2.17$, c) $\phi_{HE} = 0.535$, $\phi_{Jet} = 5.57$, $J = 1.38$, d) $\phi_{HE} = 0.55$, $\phi_{Jet} = 1.73$, $J = 3.02$, e) $\phi_{HE} = 0.575$, $\phi_{Jet} = 3.13$, $J = 3.67$, f) $\phi_{HE} = 0.575$, $\phi_{Jet} = 1.22$, $J = 2.27$.

Figure 8 plots ΔNOx against lift-off distance (LO). The pre-flame mixing between the cross-flow and a lean jet is essential for NOx reduction demonstrated by Kolb *et al.* [10]. No strong correlation is indicated between lifting and NOx production when considering the data set as a whole. However Figure 8 also groups the data by similar values of momentum flux ratio via color with an exponential fit line, which shows an increase in ΔNOx with increasing LO. This observation is seemingly contrary to prior observations, such as that presented by Kolb *et al.* [10]. However, $\Delta\phi_{Global}$, which was shown to directly impact ΔNOx in Figure 5, increases to the top right of Figure 8. When focusing on series of near constant stage ratio ($\Delta\phi_{Global}/\phi_{Global}$), several examples of which are indicated in Figure 8, ΔNOx is shown to decrease with increasing LO. This increase in LO is aligned with transitioning from high J, low ϕ_{Jet} jets to low J, high ϕ_{Jet} jets, a transition which was shown to directly impact ΔNOx in Figure 6. While flame lifting certainly has an impact on NOx formation, LO is itself sensitive to parameters such as ϕ_{Jet}, J, and crossflow temperature that directly impact NOx contribution of the reacting jet. More work is therefore necessary to specifically isolate and assess the impact of lifting the flame on NOx production. Lifting would indicate increased mixing with cross flow prior to ignition and in the cases observed as part of this study could be responsible for keeping highly rich ($\phi_{Jet} > 4$) jets reacting in an apparent partially premixed manner as opposed to a more diffusion based modality. Lifting also likely promotes mixing of rich jets with the cross-flow prior to burning, thus reducing their equivalence ratio closer to stoichiometric and cancelling some of the other, above-noted benefits of lifting on NOx emissions.

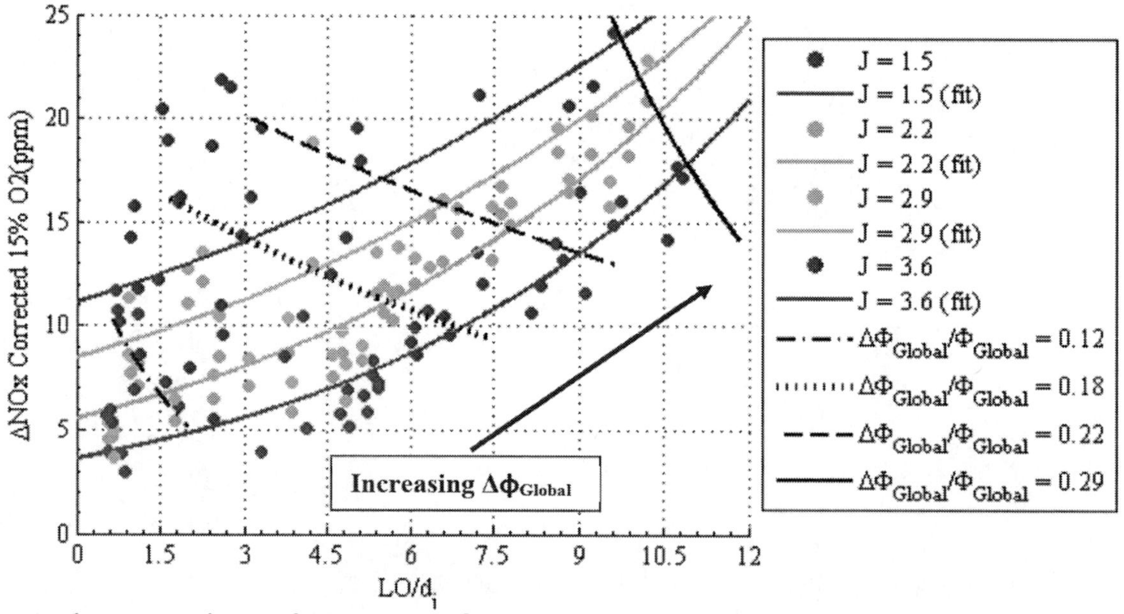

Figure 8: Axial stage contribution of NOx emissions for an axially staged combustor versus the flame lift-off distance. Color indicates groupings of similar values for J.

4. Conclusions

The work presented investigates the emissions characteristics and flame behavior of rich premixed methane/air jets into a high temperature vitiated crossflow.

As anticipated, NOx production of the axial stage has been shown to be monotonically increasing with increasing $\Delta\phi_{Global}$. For axial stages of similar $\Delta\phi_{Global}$, NOx contribution was shown to decrease with increasing jet equivalence ratio. Due to the interdependency between ϕ_{Jet} and J in this study, this trend translates to increases in NOx production with increasing momentum flux ratio. The reacting jets were observed to be reacting in a partially premixed modality and therefore this trend can be explained by combustion occurring at values in closer proximity to stoichiometric proportions for the higher J and lower ϕ_{Jet} cases due to both the jet initial condition and increased entrainment of the crossflow.

At jet equivalence ratios above values of 4.0 significant lifting of the flame was observed. Liftoff was observed to be highly sensitive to parameters, such as ϕ_{Jet}, that highly impact NOx production directly. Lift-off distance was observed to monotonically increase with increasing jet equivalence ratio. For constant $\Delta\phi_{Global}$, NOx levels were shown to decrease with increasing lift-off distance. All jets were observed to be reacting in a partially premixed modality as opposed to diffusion based, the high degree of lifting is most likely responsible for this in the highly rich ($\phi_{Jet} > 4$) jets that were tested.

Further investigations are underway to expand to parameter space of this work including a wide range of jet momentum and equivalence ratios. Of particular interest is to observe the impact on NOx production of jets of constant $\Delta\phi_{Global}$ over a broader range of fuel/air ratios that encompass both rich and lean jets. As mentioned, the impact of liftoff proved difficult to isolate in this study. Additional investigations are underway to specifically investigate this phenomena and more accurately ascertain its impact on NOx production.

Acknowledgements

This research was partially supported by the University Turbine Systems Research (contract #DE-FE0025344), contract monitor Dr. Mark Freeman.

References

[1] C. T. Bowman, "Control of combustion-generated nitrogen oxide emissions: technology driven by regulation," in *Symposium (International) on Combustion*, 1992, vol. 24, no. 1, pp. 859-878: Elsevier.

[2] T. C. Lieuwen and V. Yang, *Gas turbine emissions*. Cambridge University Press, 2013.

[3] D. Ahrens, M. Kolb, C. Hirsch, and T. Sattelmayer, "NOx Formation in a Reacting Premixed Jet in Hot Cross Flow," in *ASME Turbo Expo*, Düsseldorf, Germany, 2014, no. 45691, p. V04BT04A016, Volume 4B: Combustion, Fuels and Emissions.

[4] E. Goh, M. Sirignano, V. Nair, B. Emerson, T. Lieuwen, and J. Seitzman, "Modeling of Minimum NOx in Staged-Combustion Architectures at Elevated Temperatures," in *ASME Turbo Expo*, Charlotte, NC, USA, 2017, no. 50848, p. V04AT04A048, Volume 4A: Combustion, Fuels and Emissions.

[5] A. R. Karagozian, "Transverse jets and their control," *Progress in Energy and Combustion Science*, vol. 36, no. 5, pp. 531-553, 2010.

[6] A. M. Steinberg, R. Sadanandan, C. Dem, P. Kutne, and W. Meier, "Structure and stabilization of hydrogen jet flames in cross-flows," *Proceedings of the Combustion Institute*, vol. 34, no. 1, pp. 1499-1507, 2013.

[7] R. Sullivan, B. Wilde, D. R. Noble, J. M. Seitzman, and T. C. Lieuwen, "Time-averaged characteristics of a reacting fuel jet in vitiated cross-flow," *Combustion and Flame*, vol. 161, no. 7, pp. 1792-1803, 2014.

[8] J. A. Wagner, S. W. Grib, M. W. Renfro, and B. M. Cetegen, "Flowfield measurements and flame stabilization of a premixed reacting jet in vitiated crossflow," *Combustion and Flame*, vol. 162, no. 10, pp. 3711-3727, 2015.

[9] D. Schmitt, M. Kolb, J. Weinzierl, C. Hirsch, and T. Sattelmayer, "Ignition and Flame Stabilization of a Premixed Jet in Hot Cross Flow," no. 55102, p. V01AT04A053, 2013.

[10] M. Kolb, D. Ahrens, C. Hirsch, and T. Sattelmayer, "A Model for Predicting the Lift-Off Height of Premixed Jets in Vitiated Cross Flow," *Journal of Engineering for Gas Turbines and Power,* vol. 138, no. 8, pp. 081901-081901-9, 2016.

[11] J. A. Wagner, M. W. Renfro, and B. M. Cetegen, "Premixed jet flame behavior in a hot vitiated crossflow of lean combustion products," *Combustion and Flame,* vol. 176, pp. 521-533, 2017.

[12] R. V. Bandaru and S. R. Turns, "Turbulent jet flames in a crossflow: effects of some jet, crossflow, and pilot-flame parameters on emissions," *Combustion and Flame,* vol. 121, no. 1, pp. 137-151, 2000/04/01/ 2000.

[13] K. H. Lyle, L. K. Tseng, J. P. Gore, and N. M. Laurendeau, "A study of pollutant emission characteristics of partially premixed turbulent jet flames," *Combustion and Flame,* vol. 116, no. 4, pp. 627-639, 1999/03/01/ 1999.

[14] C. Prathap *et al.*, "Analysis of NOX Formation in an Axially Staged Combustion System at Elevated Pressure Conditions," *Journal of Engineering for Gas Turbines and Power,* vol. 134, no. 3, pp. 031507-031507-8, 2012.

[15] M. Roa, W. G. Lamont, S. E. Meyer, P. Szedlacsek, and R. P. Lucht, "Emission Measurements and OH-PLIF of Reacting Hydrogen Jets in Vitiated Crossflow for Stationary Gas Turbines," no. 44687, pp. 491-498, 2012.

[16] K. He, J. Sun, and X. Tang, "Guided image filtering," *IEEE transactions on pattern analysis and machine intelligence,* vol. 35, no. 6, pp. 1397-1409, 2013.

[17] N. Otsu, "A threshold selection method from gray-level histograms," *IEEE transactions on systems, man, and cybernetics,* vol. 9, no. 1, pp. 62-66, 1979.

[18] S. H. Smith and M. G. Mungal, "Mixing, structure and scaling of the jet in crossflow," *Journal of Fluid Mechanics,* vol. 357, pp. 83-122, 1998.

[19] L. Gevorkyan, T. Shoji, D. R. Getsinger, O. I. Smith, and A. R. Karagozian, "Transverse jet mixing characteristics," *Journal of Fluid Mechanics,* vol. 790, pp. 237-274, 2016.

Appendix A: Liftoff Analysis

The term "attached flame" has been used in reacting JICF literature to describe jets whose ignition point is right above the lip of the jet exit. In this study, degree of anchor point movement is characterized by looking at the standard deviation (σ_{LO}) in the lift-off heights from the binarized instantaneous images. Figure 9 demonstrates that the jets which ignite closer to the jet exit have a lower variation in instantaneous lift-off distance (LO). This would suggest that in addition to autoignition, there are other effects which contribute to the stability of the flame at the lip of the jet exit. The cluster of points within the box near the bottom left of the graph will be categorized as "attached" while the remaining flames will be categorized as "lifted". Lifted flames do not have any fixed attachment point and have a higher variance of lift-off height, suggesting that these flames may be highly sensitive to autoignition chemistry.

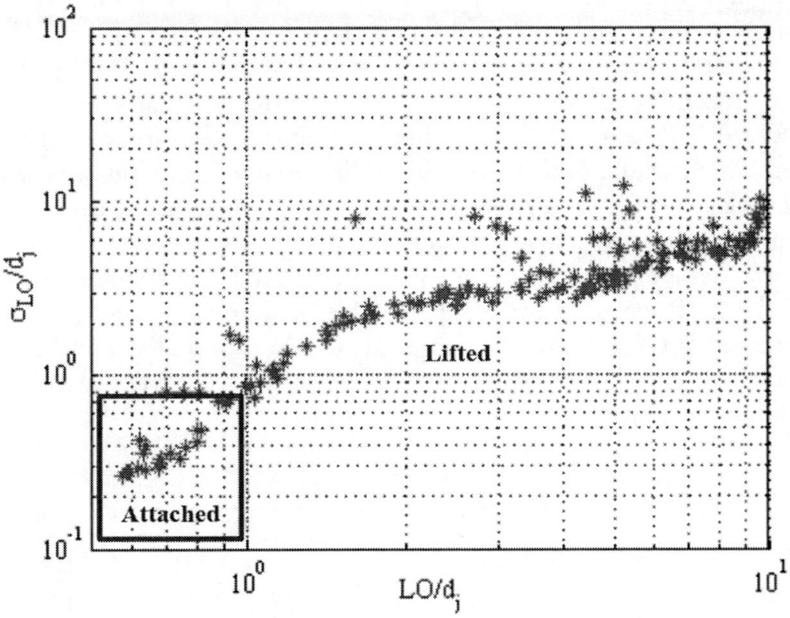

Figure 9: Lift-off distance vs std deviation of flame position.

Flame lift-off height/distance is an import metric to characterize the flame behavior of the system. Previous RJICF work has demonstrated models for characterizing lift-off height in non-premixed [6, 7] as well as premixed [10] RJICF. The flame lift-off height is a strong function of the crossflow temperature, jet temperature, equivalence ratio as well as the momentum flux ratio (J) as demonstrated in Kolb *et al.* [10]. The left panel of Figure 10 shows us that LO is correlated with respect to the jet equivalence ratio when the crossflow conditions, including the main burner equivalence ratio (ϕ_{HE}), are fixed. Note that the jet momentum flux ratio is also changing because the experiment was performed with each condition trying to reach a fixed turbine inlet temperature. Thus, the effect of increasing the equivalence ratio of the jet is confounded by the effect of reducing the jet momentum flux ratio. This may contribute to the lack of a clear relationship between LO and J as shown in the right panel of Figure 10.

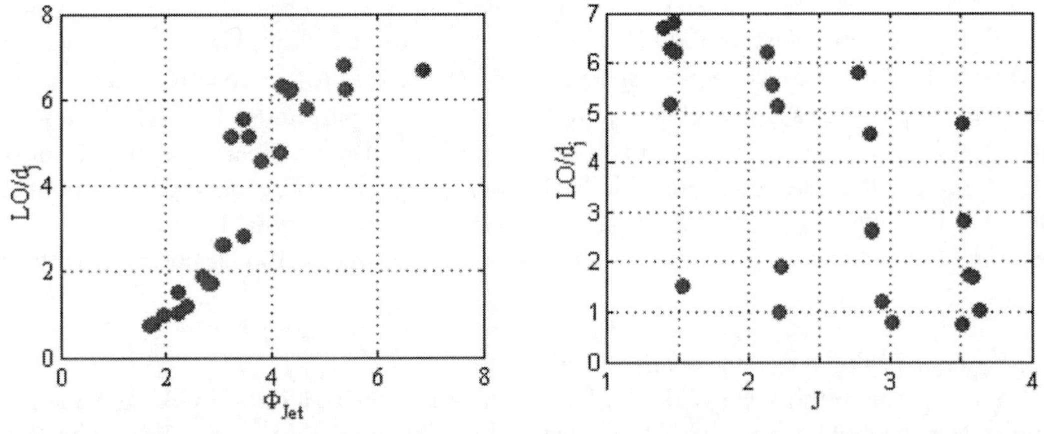

Figure 10: Influence of jet equivalence ratio (left) and momentum flux ratio (right) upon lift-off distance

In an attempt to isolate the impact of jet momentum flux ratio on lift-off distance, we refer to the results presented in Kolb *et al.* [10] which suggests that the lift-off height increases with the momentum flux ratio because of increasing jet velocity. The impact of the jet momentum flux ratio can also be analyzed by looking at the mixing performance of jets of different momentum flux ratio's [19]. Gevorkyan *et al.* [19] noted that at a given location the 'time' of exposure of high J jets is greater, which would in this case aid in the mixing of the hot crossflow products to reduce the chemical time scale of the mixture. As these are two contrasting effects, in this study, the effect of increase in jet velocity appears to be countered to a larger degree by the changing equivalence ratio of the jet as demonstrated by Figure 10. This would suggest that in the case of rich jets above the flammability limit, pre-flame mixing has a higher impact on ignition position than increasing the convective velocity from the jet. Jet equivalence ratios closer to stoichiometric conditions would have shorter ignition delay times and thus would have lower lift-off heights if the flame was purely auto-ignition stabilized.

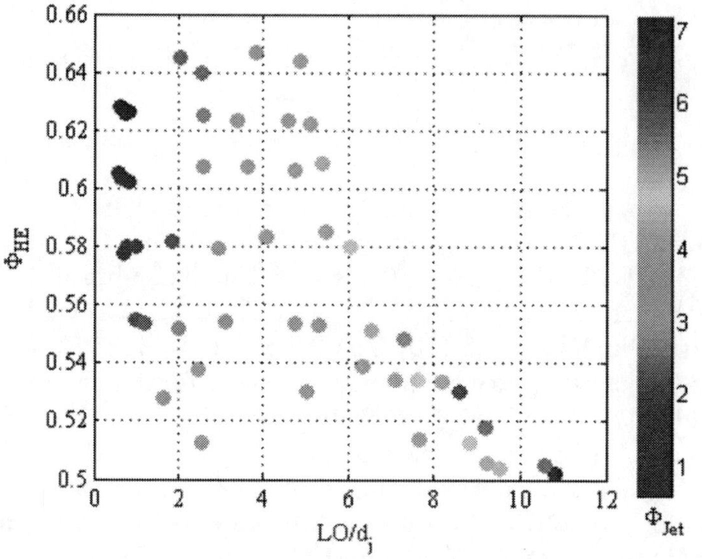

Figure 11: Comparison of Phi HE vs LO for TIT 1956

Figure 11 demonstrates the effect of the crossflow conditions on LO. In this study, ϕ_{HE} is inversely proportional to change in equivalence ratio due to the reacting jet and as a result ϕ_{Jet} generally increases as ϕ_{HE} decreases, could contribute in part to the shape of the above graph. The effect of increasing ϕ_{HE} corresponds to increasing crossflow temperature, which has a significant effect on the lift-off height as noted in Kolb *et al.* [10]. This effect observable amongst data points with similar values of ϕ_{Jet}, where LO is shown to increase with decreasing ϕ_{HE}. This result is intuitive as increasing the crossflow temperature has a major impact on increasing the reactivity of the jet which needs to mix with the crossflow before burning. This increased reactivity occurs for hotter (but lower oxygen concentration) crossflows because of the exponential temperature dependence of the autoignition chemistry.

To summarize the sensitivities of lift-off distance, the reduction in chemical time through the entrainment of hot crossflow products results in a decrease in LO. The changes in the momentum flux ratio as well as the jet equivalence ratio both impact the chemical time through changing the amount of hot crossflow required to obtain a highly reactive mixture.

Western States Section of the Combustion Institute – Fall 2017 Meeting
Hosted by the University of Wyoming
October 2-3, 2017

Effect of stoichiometric mixture fraction on hydrogen edge-flames in a counter-flow burner

Zhenghong Zhou, Paul D. Ronney

Department of Aerospace and Mechanical Engineering
University of Southern California, Los Angeles, CA 90089-1453, USA
Email: zhenghoz@usc.edu, Ronney@usc.edu

Abstract: The influences of Lewis number and fuel chemistry effect are investigated by measuring the propagation speeds (U_{edge}) of nonpremixed hydrogen edge-flames as a function of stoichiometric mixture fraction (Z_{st}) and global strain rate (σ), for which Le_F is much less than unity and the fuel chemistry is far different from hydrocarbons. Unlike hydrocarbons, in hydrogen-oxygen-inert mixtures only monotonic trends were observed, with U_{edge} increasing with Z_{st}. At sufficiently high Z_{st}, where a shift from oxygen-deficient to fuel-deficient reaction occurs, cellular structures characteristic of low Le were observed which led to combustion under conditions that apparently cannot sustain reaction in planar structures. For more-dilute mixtures this transition Z_{st} is lower. It was found that different structure transformation modes exist and the cellular structures can transform from a propagating flame to isolated, stationary cells as the dilution level increases or as Z_{st} decreases at high strain rate. It is concluded that the combined influences of Lewis number and chemical effects are also present in hydrogen-oxygen-inert mixtures, however, the differences in these effects compared to hydrocarbon-oxygen-inert mixtures leads to different responses to σ, Z_{st} and dilution level.

Keywords: edge-flames, counter-flow, mixture fraction

1. Introduction

Edge-flames are the transition regions between burning and non-burning regions of a flame sheet. Edge-flame behaviors have been studied to understand many non-uniform flame phenomena including flames in highly turbulent flows such as "holes" in flame sheets inside reciprocating-piston internal combustion engines; flames stabilized near a cold wall or splitter plate; or leading edges of flames spreading across condensed-phase fuel surfaces [1].

The most significant property of an edge-flame is its propagation speed (U_{edge}), defined as the speed of the edge that moves relative to the unburned gases in the direction parallel to the flame sheet. Previous theoretical studies in premixed [2] [3] and nonpremixed [4] [5] configurations predict that edge-flames may propagate from the burning region into the non-burning region, forming a "propagating flame" with $U_{edge} > 0$ or retreat from the burning region into the burned region, forming a "retreating flame" with $U_{edge} < 0$. U_{edge} is affected by mixture strength (characterized by the adiabatic flame temperature, T_{ad}), global strain rate (σ), Lewis numbers (Le, ratio of mixture thermal diffusivity to reactant mass diffusivity) of fuel and oxidant (Le_F and Le_O, respectively), heat losses, and for nonpremixed edge-flames the stoichiometric mixture fraction $Z_{st}=1/(1+\upsilon X_f/X_o)$. In the equation above, υ is the stoichiometric oxygen-to-fuel mass ratio and X_f and X_o are the mass fractions of fuel and oxygen in the unburned mixture streams.

While traditional nonpremixed combustion generally employs pure fuel mixing and burning with highly-diluted oxygen (air in most cases) corresponds to low Z_{st} (typically 0.06), new fuels and combustion technologies (biofuels, oxyfuel combustion, massive Exhaust Gas Recirculation, etc.) expands to much broader Z_{st} ranges – up to 0.8 for pure oxygen burning with highly-diluted hydrocarbon fuel and even higher than 0.93 for highly-diluted hydrogen. In a counter-flow burner, increasing Z_{st} moves the flame location from the oxidizer side toward the fuel side, which results in significant differences in the reactant temperature/composition/time history. For hydrogen flames, at high Z_{st} or weak mixture strength, cellular flame structures characteristic of flames with low Le (~0.3) were observed, which enabled combustion under conditions that apparently cannot sustain reaction in planar (flat) flame structures [6].

Accordingly, this work focuses on Z_{st} effects on hydrogen edge-flame properties for varying σ (by changing flow rate) and varying mixture strengths (by changing the amount of diluent gas), and to draw the flame behavior regime maps regarding Z_{st}. Following prior work [7] [8] a counter-flow slot-jet apparatus is employed because slot-jets provide extensional strain orthogonal to the slot plane yet little convection along the slot length, thus edge-flame propagation speeds in the laboratory frame are basically the propagation speed relative to the cold unburned gas far ahead of the edge-flame (or behind, for retreating edge-flames). So the slot-jet counter-flow burner simplifies the interpretation of experimental data.

2. Methods

The counter-flow slot-jet apparatus and procedures were similar to those employed in the previous work [7] [8]. Mass flow controllers regulated jet exit velocities (U_f, U_o for fuel and oxidizer respectively) to achieve specified strain rate $\sigma = (U_o+U_f)/d$ (d is the jet spacing, 7.5 mm) and mixture composition (H_2+diluent and O_2+diluent) at required X_o and X_f to obtain specific stoichiometric mixture fraction Z_{st}. The apparatus was enclosed in a ventilation hood to suppress influences of room draft. A Helium-Neon laser shadowgraph system was applied to "visualize" the invisible H_2 flame by shadowgraph technique. The change of H_2 edge-flame propagation speed and structure were recorded by high-speed camera. Figure 1 and Figure 2 Schematic illustrates the apparatus and shadowgraph system.

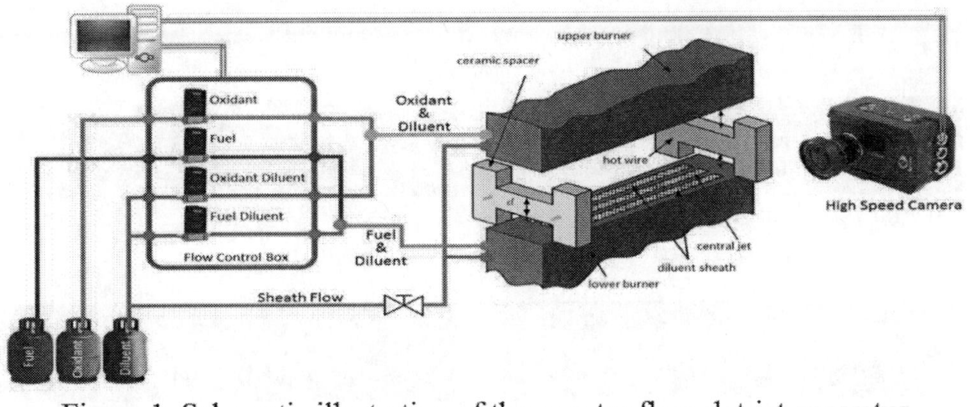

Figure 1: Schematic illustration of the counter-flow slot-jet apparatus

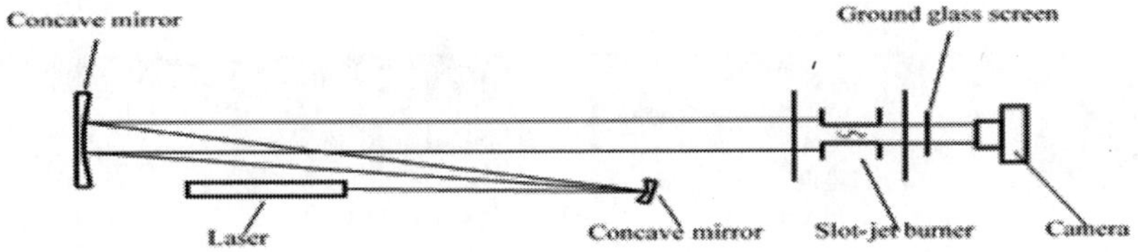

Figure 2: Schematic illustration of Shadowgraph system

In the cases that $U_{edge} > 0$, a small N_2 jet was used to "erase" the flame from right slot end to nearly the left end and was then retracted, letting the edge-flame propagate. In the cases that $U_{edge} < 0$, a mixture with $U_{edge} > 0$ was first introduced, and then with heating wires electrically heated at both slot ends, X_o and X_f were slowly reduced to the required values to achieve $U_{edge} < 0$ conditions, and then the N_2 jet was used to tap on one flame end to observe retreating flames. The heating wires enhanced local flame temperature, and thus increased reaction rates at the flame ends, anchoring these ends under conditions which they would retreat without localized heating, but had no influence on the edge-flame temperature or the edge-flame propagation speed [7].

In the cases of stationary edge-flames with cellular structures, a mixture with cellular propagating flame was first introduced, and then X_o and X_f were slowly reduced to the required values to induce stationary cellular edge-flames.

Videos were analyzed to obtain U_{edge}. Because the slot-jet aspect ratio was finite, there was a slight extensional flow along the slot length which slightly affected U_{edge} in the laboratory frame; as in prior work [7], this bias was nullified by interpolating U_{edge} vs. relative position along the slot to the jet's vertical centerline.

3. Results and Discussion

The propagation speed U_{edge} of $H_2/O_2/N_2$=2/1/N, N=15, 16, 17, 18, 19, 20 were measured. Figure 3 shows the shadowgraph profile of a typical propagating flame and a retreating flame. Similar to the prior work on hydrocarbon fuels, the tips of propagating flames are round while the tips (or "tails") of retreating flames are sharp. Figure 4 shows the shadowgraph profile of a typical propagating cellular flame and stationary cells. The cellular structures in propagating cellular flames always split and formed immediately from the moving front, not after the formation of a flat flame. The stationary cells were isolated and if one was "erased" then there would not be recovery of the cell.

Figure 3: Flat propagating flame at σ=80/s, Z_{st}=0.70, N=17(left) and flat retreating flame at σ=80/s, Z_{st}=0.20, N=17 (right)

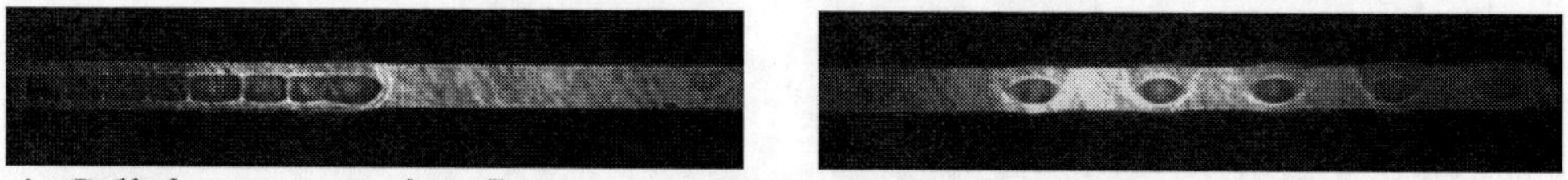

Figure 4: Cellular propagating flame at σ=120/s, Z_{st}=0.85, N=17 (left) and stationary cellular flame at σ=80/s, Z_{st}=0.85, N=19 (right)

Figure 5: Cellular propagating flame at σ = 280/s, Z_{st} = 0.93, N=17 (left) and turbulent cellular flame at σ = 370/s, Z_{st} = 0.93, N=17 (right)

Scaled propagation speed plots of mixture strength $H_2/O_2/N_2$=2/1/17 and $H_2/O_2/N_2$=2/1/18 regarding Z_{st} and global strain rate σ were plotted as shown in Figure 6. And it should be noted that every point in each plot in Figure 6 corresponds to exactly the same mixture (same ratio of $H_2/O_2/N_2$, T_{ad}, and thus S_L). These dimensionless propagation speeds were scaled by the numerical flame speed of premixed flame and density of burnt and unburnt gas to eliminate the influence of thermal expansion. Non-filled markers represent cellular edge-flames.

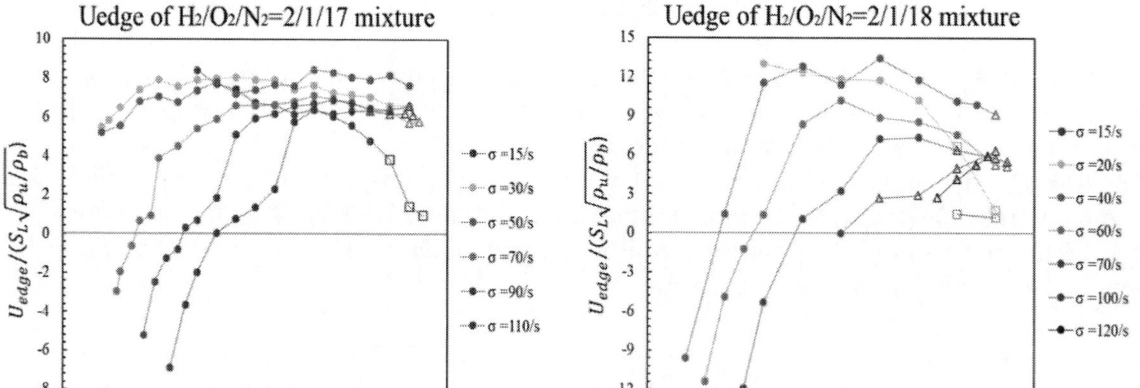

Figure 6: U_{edge} of mixture strength $H_2/O_2/N_2$=2/1/17 and $H_2/O_2/N_2$=2/1/18

From low Z_{st} to high Z_{st}, U_{edge} increases monotonically for strain rates higher than 20/s, in which the heat loss to the ambient had smaller influence on U_{edge} compared to other factors. Retreating flames occur in cases with either large global strain rate or low Z_{st}. And for a weaker mixture the threshold σ for retreating flame to occur become smaller and Z_{st} become larger.

At higher Z_{st}, cellular structures start to appear. There is no clear trend of how cellular structures would affect the U_{edge}, and from the two plots in figure 6 it seems that the cellular structures do not have large influence on U_{edge}. And also in both plots no retreating cellular edge-flames are observed.

In the attempt to identify (if possible) an easily-determined 1D flame parameter characterizing the effects of Z_{st}, the adiabatic extinction strain rate σ_{ext} vs. Z_{st} in a 1D counter-flow was computed for N=17 and N=18 mixture that was studied numerically same apparatus dimension settings as in the experiments in CHEMKIN. Figure 7 shows the computed effect of Z_{st} on the simulation for extinction strain rate (σ_{ext}) and σ_{ext} in the experiments.

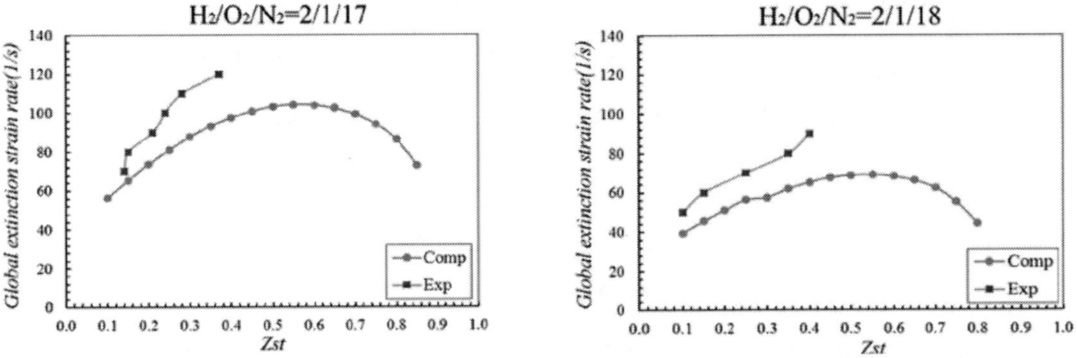

Figure 7 computed σ_{ext} and σ_{ext} in the experiments

Table 1: Comparisons of measured and computed extinction strain rates (σ_{ext} unit s^{-1}).

Mixture	Z_{st}	σ_{ext} (exp)	Z_{st}	σ_{ext} (comp)
	0.10	56.41	0.14	70
	0.15	65.52	0.15	80
	0.20	73.75	0.21	90
	0.25	81.10	0.24	100
	0.30	87.78	0.28	110
	0.35	93.19	0.37	120
$H_2/O_2/N_2$	0.40	97.67		
=	0.45	100.94		
2/1/17	0.50	103.36		
	0.55	104.23		
	0.60	104.00		
	0.65	102.44		
	0.70	99.26		
	0.75	94.20		
	0.80	86.15		
	0.85	73.07		

Computation showed that the σ_{ext} would first increase with growing Z_{st} and then drop. For lower Z_{st} the σ_{ext} from computation and experiments matched, however when it went to high Z_{st}, the cellular structures sustain the flame where it would extinguish without the mechanism, So in experiments no extinction was observed at high Z_{st}.

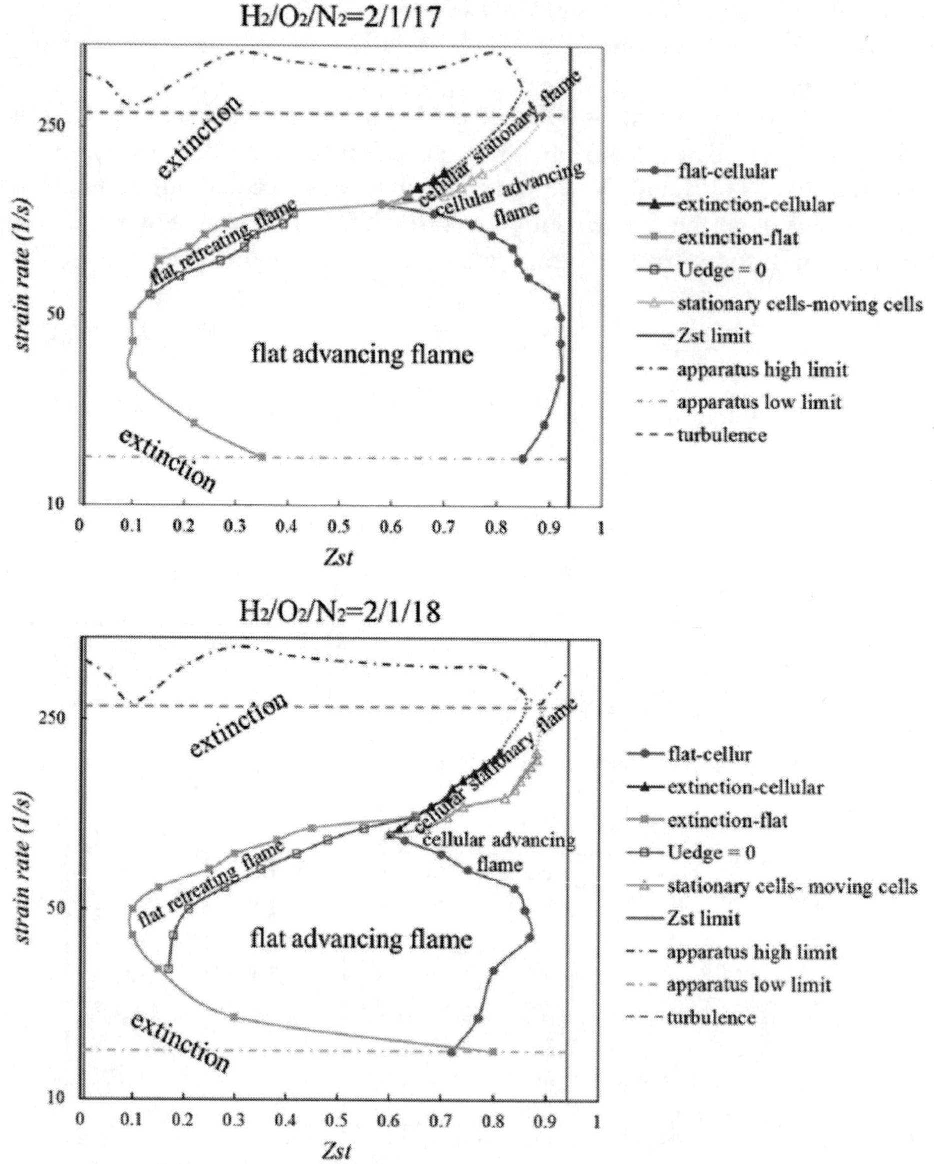

Figure 8: Flame behavior regime map for $H_2/O_2/N_2$=2/1/17 and $H_2/O_2/N_2$=2/1/18

Figure 8 shows the flame behavior regime maps. The two straight solid lines are the Z_{st} limits lines and the horizontal dashed line is the limit strain rate line of the counter-flow apparatus. Extinction more easily occurs when strain rate is high or low. Only with intermediate strain rate would extinction happen at very low Z_{st}. Retreating flames are flat and sit in where Z_{st} is low and strain rate is relatively high. When Z_{st} is large, a transition from flat advancing flame to cellular advancing flame would occur. And when strain rate keeps increasing, the cellular advancing flame would become stationary and separated. And also it could be told from Figure 8 that weaker mixture leads to larger chance for retreating flames and cellular flames to exist, and the flat advancing flame region would shrink.

4. Conclusions

This work investigated the effect of stoichiometric mixture fraction (Z_{st}) on the propagation speeds of nonpremixed edge-flames (U_{edge}) in a variety of hydrogen-diluent vs. O_2-diluent combinations, in order to understand the effects of heat and mass transport. Interpretation of results was facilitated by the use of values of fuel and oxidizer mass fractions such that, all stoichiometric combinations of hydrogen and oxidizer streams result in the same mixture strength (thus same T_{ad}, density ratio and S_L) even though Z_{st} varies widely depending on how the diluent is distributed between these streams.

First of all with this approach to varying Z_{st}, it was found that U_{edge} would increase monotonically with increasing Z_{st} in most of the cases, Secondly, at high Z_{st} no extinction strain rate was found as a simple computation would suggest because of the effect of Lewis number that leads to cellular sructures, although observed U_{edge} vs. Z_{st} trends are consistent with computed values of extinction strain rate (σ_{ext}) at low Z_{st} in a 1D counter-flow. Thirdly, under different strain rate, the transition of structures could change from one mode to another.

These results indicated that the behavior of H_2 edge-flame depends critically on stoichiometric mixture fraction Z_{st} (degree of dilution of both fuel and oxidizer), and strain rate.

5. Acknowledgements

This research was funded by the U.S. National Science Foundation, grant CBET-1236892

6. References

[1] J. D. Buckmaster, Edge-flames, *Progress in Energy and Combustion Sci.* 28.5 (2002): 435-475.

[2] J. D. Buckmaster, *Prog. Energy Combust. Sci.* 28 (2002) 435-475.

[3] T. G. Vedarajan, J.D. Buckmaster, P. D. Ronney, *Proc. Combust. Inst.* 27 (1998) 537-544.

[4] J. Daou, A. Liñán, *Combust. Flame* 118 (1999) 479-488.

[5] R. Daou, J. Daou, J. Dold, *Proc. Combust. Inst.* 29 (2002) 1559–1564.

[6] C. Kaise, J.B. Liu, and P.D. Ronney. Diffusive-thermal instability of counterflow flames at low Lewis number. *38th Aerospace Sciences Meeting and Exhibit* (2000), p. 576

[7] M. S. Cha, P.D. Ronney, *Combust. Flame* 146 (2006) 312-328.

[8] H. Song*, P. Wang, R. Boles, D. Matinyan, H. Prahanphap, J.Piotrowicz, P. D. Ronney. Effects of Mixture Fraction on Edge-flame Propagation Speeds. *Proceedings of the Combustion Institute 36* (2017) 1403–1409

Western States Section of the Combustion Institute – Fall 2017 Meeting
Hosted by the University of Wyoming
October 2-3, 2017

Transported and Presumed Joint PDF Modeling of Turbulent Jet Flames

Vasu Jaganath[1] *Michael Stoellinger*[2]

[1,2] *Department of Mechanical Engineering, University of Wyoming, Laramie, WY 82071,USA*

This paper presents a framework for comparative numerical modeling study of turbulent jet flames. Two Reynolds Averaged Navier-Stokes simulation (RANS) models are used to study jet flames, in one the shape of the joint composition probability density function (PDF) is presumed and in another the joint composition PDF is obtained by solving the transport equation for it. In both these models chemistry is modeled by the Flamelet Generated Manifold (FGM). The two models are implemented in the open-source CFD framework OpenFOAM. The transported PDF model uses a Monte Carlo solver and consistency of the implementation of the Monet Carlo method is shown for a simple planar jet mixing case and for a planar jet flame. Furthermore, a comparison between the different models for the turbulence-chemistry interaction are presented for a simple mixing planar flame.

1. Introduction

Modeling of Turbulent Combustion is one of the open problems in the field of reactive flow simulations. However a direct approach where the entire combustion phenomena is computed at runtime, is prohibitively expensive for flows of practical interest. Often simplifying assumptions are made to solve those problems, most notably instead of a detailed chemical mechanism, a tabulated chemistry method is used. In this study we employ a tabulated chemistry method called FGM [2].

FGM model builds on the flamelet concept [3] and combines it with a manifold approach [2].
The multi-dimensional turbulent flame is considered as a set of 1D flamelets. A low dimensional manifold now describes the mixing, reaction and other properties of the turbulent flame which is constructed based on the laminar flamelets calculation. In the turbulent flame simulation, only the transport equations of the controlling variables, instead of transport equations for each species are solved in addition to continuity and momentum. In this way the costly chemical reaction process is decoupled from the flow simulation.

When FGM is applied, the turbulence-chemistry interaction (TCI) is accounted for by the probability density function (PDF) of the controlling parameters, typically a mixture fraction and progress variable. The PDF can be obtained by solving a modeled transport equation for the joint PDF or it can be assumed that the PDFs obey a certain shape which can be determined by the first and second moments of the controlling parameters. The flow simulation method used here is Unsteady Reynolds Averaged Navier Stokes (URANS).

All the simulations are carried out using open source CFD package OpenFOAM and all the solvers are implemented within this framework. The FGM solver is based on the implementation of Likun Ma et al [4] and the transported composition PDF solver is developed by the authors.

The goal of this work is to develop a methodology to investigate these two (transported and presumed PDF) different TCI models in conjunction with FGM for the target flame Sandia D [5] and assess the typical assumptions involved in the presumed PDF model. However, in this paper we will only consider the simpler case of a plane-jet configuration and show some first results for a pure mixing in a round jet.

2. Mathematical Modeling

Flamelet Generated Manifolds

FGM is the tabulated chemistry method used in this study. Neglecting the effect of radiative heat transfer, the scalars such as temperature, species mass fractions, density, source terms are stored in the lookup table as function of mixture fraction and progress variable.

Mixture fraction used here according to the Bilger's formula

$$Z = \frac{b - b_o}{b_f - b_o}$$

Where $b = 2\frac{Y_C}{W_C} + 0.5\frac{Y_H}{W_H} - \frac{Y_O}{W_O}$ for hydrocarbon fuels. The subscripts o and f refer to the oxidizer and fuel streams. Subscripts C, O and H correspond to the elements Carbon, Hydrogen and Oxygen respectively. Mixture quantifies the extent of mixing between fuel and oxidizer streams.

A progress variable C is a parameter that characterizes the evolution of global chemical reaction status from "pure mixing" (C = 0) to "fully burnt" (C = 1).

It is generally defined as a linear combination of species mass fraction, in our case

$$Y_c = a\frac{Y_{H_2O}}{W_{H_2O}} + b\frac{Y_{CO_2}}{W_{CO_2}} + c\frac{Y_{H_2}}{W_{H_2}} + d\frac{Y_{CO}}{W_{CO}}$$

Where a = 4, b = 2, c = ½ and d = 1 are the values arrived after a parametric study by Likun Ma et al [4] and progress variable in normalized form C

$$C = \frac{Y_c - Y_c^u}{Y_c^b - Y_c^u}$$

Where subscripts b and u respectively represent the burnt and unburnt status. From here on, Y_c will be referred to as Progress variable and C as scaled progress variable. The two dimensional manifold (FGM table) is parametrized by Z and C, $\Phi = \Phi(Z,C)$, where Φ is any themro-chemical variable such as density or temperature.

Presumed PDF method

The effect of turbulent fluctuations on the flamelets is accounted through the joint PDF of the independent variables, i.e. Z and C. The Favre-averaged scalars are calculated as

$$\tilde{\phi} = \iint_0^1 \phi(Z,C)\, \tilde{P}(Z,C) dZdC$$

The Reynolds-averaged mean of scalar $\overline{\phi}$ can be calculated as follows

$$\overline{\phi} = \overline{\rho} \iint_0^1 \frac{\phi(Z,C)}{\rho(Z,C)}\, \tilde{P}(Z,C) dZdC$$

and thus for density we obtain

$$\overline{\rho} = \left[\iint_0^1 \frac{\tilde{P}(Z,C)}{\rho(Z,C)}\, dZdC \right]^{-1}$$

where $\tilde{P}(Z,C)$ is the joint Favre PDF of mixture fraction and scaled progress variable.

In the presumed shape method, the PDFs are assumed to obey a certain shape which can be determined by first and second order moments of the controlling parameters. It is a common practice to further assume that Z and C are statistically independent of each other.

$$\tilde{P}(Z,C) = \tilde{P}(Z)\, \tilde{P}(C)$$

For our study, we also assume that *Z* and *C* are statistically independent and we use a β- and δ-function as the representation of PDF of mixture fraction *Z*. The β-function PDF is defined as

$$\tilde{P}(\phi) = \frac{\tilde{\phi}^{\alpha-1}\left(1-\tilde{\phi}\right)^{\beta-1}\,\Gamma(\alpha+\beta)}{\Gamma(\alpha)\Gamma(\beta)},$$

and the shape parameters are related to the moments of the scalars as follows

$$\alpha = \tilde{\phi}\left[\frac{\tilde{\phi}\,(1-\tilde{\phi}\,)}{\widetilde{\phi''^2}} - 1\right]$$

$$\beta = (1-\tilde{\phi}\,)\frac{\alpha}{\tilde{\phi}}.$$

Fluctuations for the progress variable are neglected (i.e a δ-function is used for the PDF of progress variable). To describe the presumed β-function PDF, the mean and variance of mixture fraction and mean progress variable are required. The means are modeled by

$$\frac{\partial\bar{\rho}\tilde{Z}}{\partial t} + \frac{\partial\bar{\rho}\tilde{u}_j\tilde{Z}}{\partial x_j} = \frac{\partial}{\partial x_j}\left(\bar{\rho}\,\tilde{D}\frac{\partial\tilde{Z}}{\partial x_j} - \Phi_{Z,j}\right),$$

$$\frac{\partial\bar{\rho}\tilde{Y}_c}{\partial t} + \frac{\partial\bar{\rho}\tilde{u}_j\tilde{Y}_c}{\partial x_j} = \frac{\partial}{\partial x_j}\left(\bar{\rho}\,\tilde{D}\frac{\partial\tilde{Y}_c}{\partial x_j} - \Phi_{Y_c,j}\right) + \bar{\tilde{\omega}}_{Y_C},$$

where $\tilde{Y}_c$ can be converted to $\tilde{C}$ by

$$\tilde{C} = \frac{\tilde{Y}_c - \tilde{Y}_c^u}{\tilde{Y}_c^b - \tilde{Y}_c^u}.$$

The Reynolds fluxes of mixture fraction and progress variable are modeled with standard gradient diffusion assumption

$$\Phi_{Z,j} = -\bar{\rho}\,D_t\frac{\partial\tilde{Z}}{\partial x_j}$$

$$\Phi_{Y_c,j} = -\bar{\rho}\,D_t\frac{\partial\tilde{Y}_c}{\partial x_j},$$

where $D_t = \frac{\nu_t}{\mathrm{Sc}_t}$ is the turbulent diffusivity and $\mathrm{Sc_t}$ is the turbulent Schmidt number. The transport equation for mixture fraction variance is

$$\frac{\partial \bar{\rho}\widetilde{Z''^2}}{\partial t} + \frac{\partial \bar{\rho}\tilde{u}_j\widetilde{Z''^2}}{\partial x_j} = \frac{\partial}{\partial x_j}\left(\bar{\rho}\,\tilde{D}\frac{\partial \tilde{Z}}{\partial x_j} - \Phi_{Z_v,j}\right) - 2\bar{\rho}\tilde{s}_{\chi Z} - 2\Phi_{Z,j}\frac{\partial \tilde{Z}}{\partial x_j}$$

and variance fluxes are again modeled with a gradient diffusion model

$$\Phi_{Z_v,j} = -\bar{\rho}\,D_t\frac{\partial \widetilde{Z''^2}}{\partial x_j}.$$

Transported PDF method

A detailed review of transported PDF methods can be found in [1]. In this study composition PDF method is employed to obtain the joint PDF of the controlling parameters, Z and C.

In our implementation we work with the mass density function (MDF) $F_\phi = \rho f_\phi$ such that

$$\bar{\rho} = \int F_\phi\, d\varphi$$

$$\phi = (Z, C)$$

The transport equation for the MDF is

$$\frac{\partial F_\phi}{\partial t} + \frac{\partial \tilde{u}_j F_\phi}{\partial x_j} + \frac{\partial S_\alpha F_\phi}{\partial \psi_\alpha} = -\frac{\partial}{\partial x_j}\left[\langle u''_j|\Psi\rangle F_\phi\right] + \frac{\partial}{\partial \psi_\alpha}\left[\frac{1}{\rho}\left\langle\frac{\partial J_j^\alpha}{\partial x_j}\middle|\Psi\right\rangle F_\phi\right]$$

The unclosed terms on the RHS are modeled as following:

Turbulent transport term is modeled with gradient diffusion.

$$\langle u''_j|\Psi\rangle = \frac{-\left[D_t\frac{\partial (F_\phi/\bar{\rho})}{\partial x_j}\right]}{F_\phi}$$

where $D_t = \frac{\nu_t}{\mathrm{Sc}_t}$ is the turbulent diffusivity and $\mathrm{Sc_t}$ is the turbulent Schmidt number. The micro-mixing term is modeled with Interaction by exchange with mean (IEM) mixing model.

$$\left\langle\frac{\partial J_j^\alpha}{\partial x_j}\middle|\Psi\right\rangle = 0.5C_\phi\rho\,\frac{\varepsilon}{k}\,(\psi_\alpha - \tilde{\phi}_\alpha)$$

where ε and k are turbulent dissipation and turbulent kinetic energy.

As the MDF transport equation is untenable to solve using traditional finite volume or finite difference methods, we solve for the PDF using a Monte Carlo methods based on an equivalent system of Lagrangian stochastic ordinary differential equation (SDE):

$$dx_i^* = \widetilde{u_i^*}dt + \left(\frac{1}{\bar{\rho}}\frac{\partial D_t}{\partial x_j}\right)^* + \left(\frac{2}{\bar{\rho}}\,D_t\right)^{*1/2} dW_i\,,$$

$$d\phi_\alpha^* = S_\alpha\,(\phi_\alpha^*) - 0.5C_\phi\omega(\phi_\alpha^* - \tilde{\phi}_\alpha)dt\,,$$

where * denotes a particle, i denotes the discrete time step and α is the index of the composition vector, in our composition vector we only have mixture fraction *Z* and progress variable *C*.

In addition to solving the SDE using the Monte-Carlo method, one needs to implement a positional correction algorithm to ensure consistency between elemental (Finite Volume cell) mass and Lagrangian particle mass within that element. A detailed formulation for one such algorithm is given in Zhang and Haworth [6] and is briefly summarized here.

A three stage procedure is adopted to maintain consistency between the FV mass density and PDF mass density through establishing a correction velocity field for particle advection:

Stage 1: Establish a continuous approximation to FV mean velocity field using discrete values at mesh vertices and interpolation functions

Stage 2: Compute a correction velocity to enforce consistency between FV cell-face mass flow rates and cell-face mass flow rates implied by Stage 1. (cell = finite volume mesh element)

Stage 3: Monitor the deviation between FV element-level mass density (denoted by q) and the mean particle mass density(denoted by ρ) in each element, and compute second correction velocity to drive down this mass residual (typically using another PISO iteration)

3. Results and Discussion

Plane jet pure mixing results

We will first demonstrate the consistency of the implementation of the transported PDF solver regarding the mass density by considering a pure mixing problem of a hot and a cold stream. A plane-jet is studied having three inlets, one for hot the jet and two for cold coflows. The hot jet has velocity of 20 m/s and a temperature of 1000K and the coflows have a velocity of 5m/s and 0.1m/s and a temperature of 300K. Figure 1 shows a visualization of the mixture fraction represented by Monte Carlo particles.

Next, we compare the particle mass density q with the mean density $\bar{\rho}$. The particle mass density is simply given by the mass of the particle over the cell volume. The mean density is the average of the density of the particles in the cell. If the position control algorithm is not employed ρ differs significantly from q as shown in figure 2.

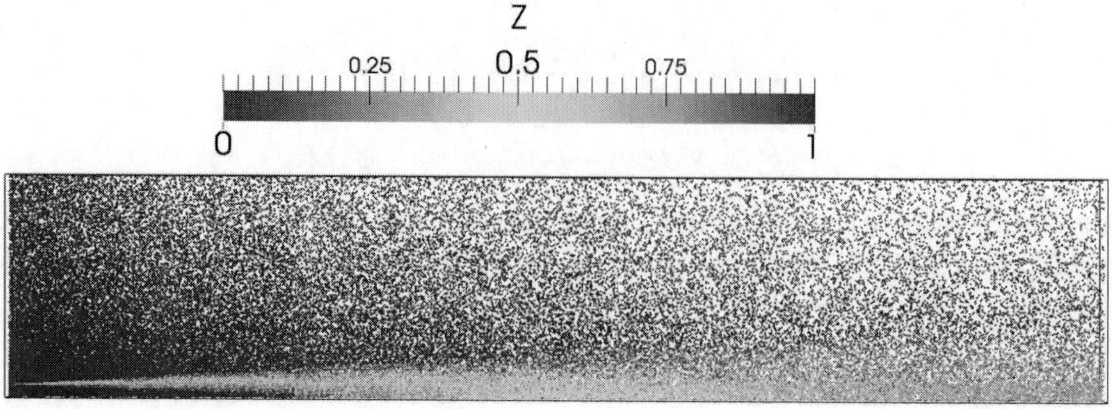

Figure 1: Visualization of the Monte Carlo particle representation of the mixture fraction PDF.

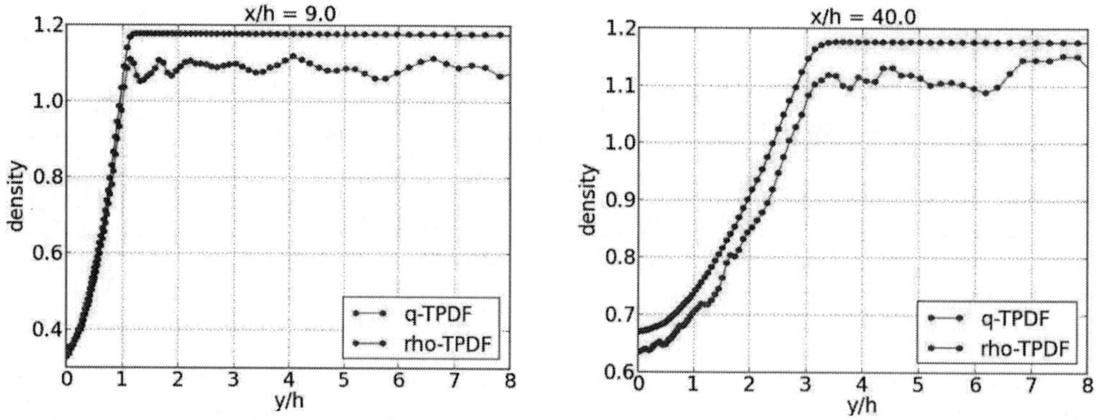

Figure 2: Plots of the mean particle density q and the mean density ρ obtained from the particles for a planar jet mixing problem without using the mass consistency algorithm.

Figure 3 shows the same comparison but now using all three stages of the mass constancy algorithm and it can be seen that the consistency between the two densities is established.

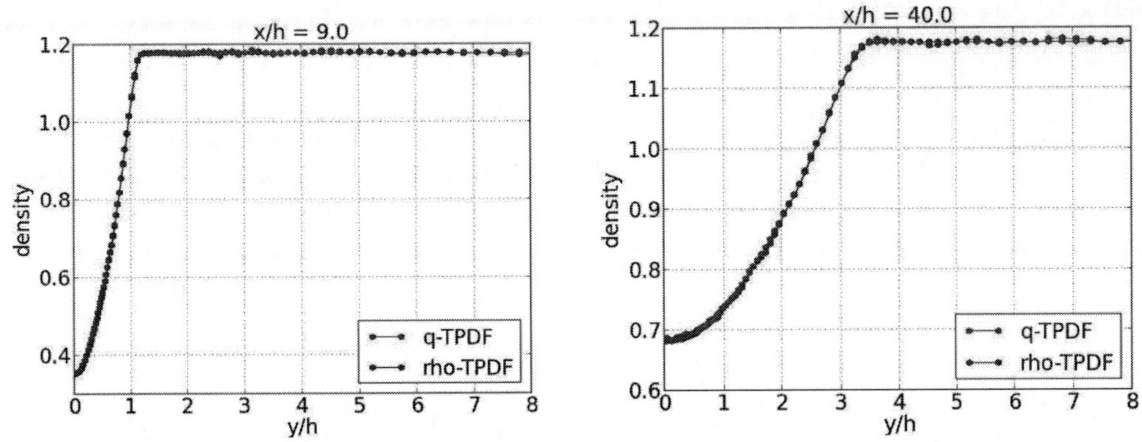

Figure 3: Plots of the mean particle density q and the mean density ρ obtained from the particles for a planar jet mixing problem using the full mass consistency algorithm.

For the pure mixing case, the PDF model implies the same modeled mean and variance equation as presented above. Thus, a second consistency check is given by solving the mean and variance equation for the mixture fraction Z and comparing these values to mean and variance obtained from the Monte Carlo particle system. This comparison is shown in figure 4 and it can be seen that the means agree quite well but there is some deviation in the variance. This difference can be explained by the fact that the Monte Carlo solution is essentially numerical diffusion free.

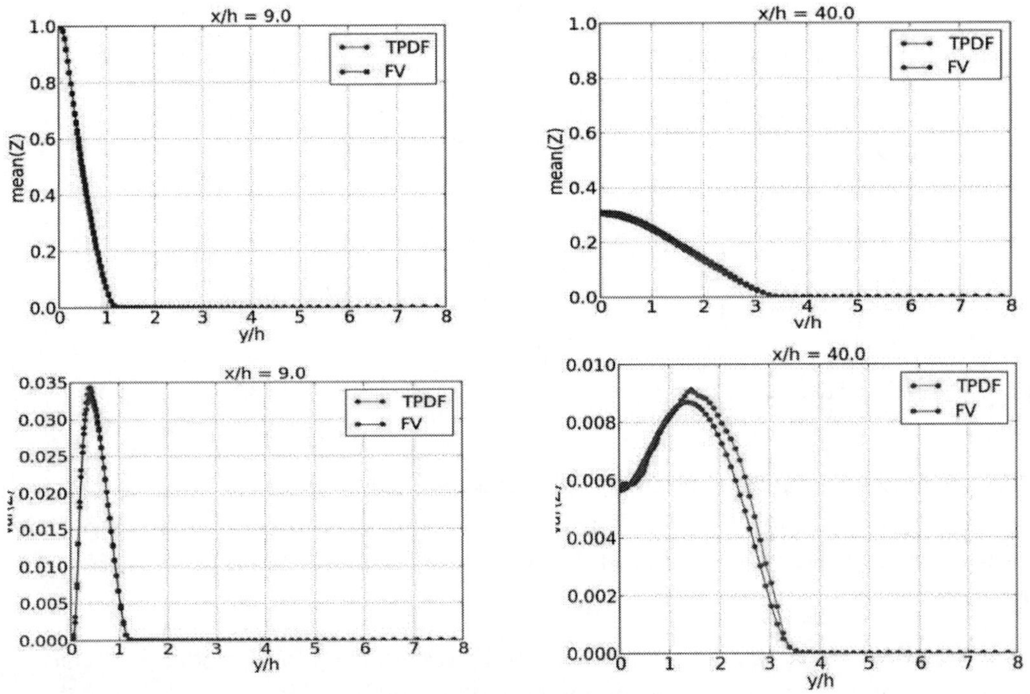

Figure 4: Plots of the mean mixture fraction (top) and mixture fraction variance (bottom) obtained from the particles (green line) and from solving the modeled transport equations (blue line).

Planar partially premixed flame results

Next we consider planar flame results using the transported PDF (TPDF) Monte Carlo solver and a FGM based on 10 steady state flamelets. The FGM table is discretized into 200x200 points in mixture fraction and progress variable space. A partially premixed methane/air jet is ignited by a hot coflow. The case is run using the full three-stage mass consistency algorithm. Figure 5 (top) shows the consistency in the mass density for this flame case and the bottom plots show the consistency of the mean mixture fraction.

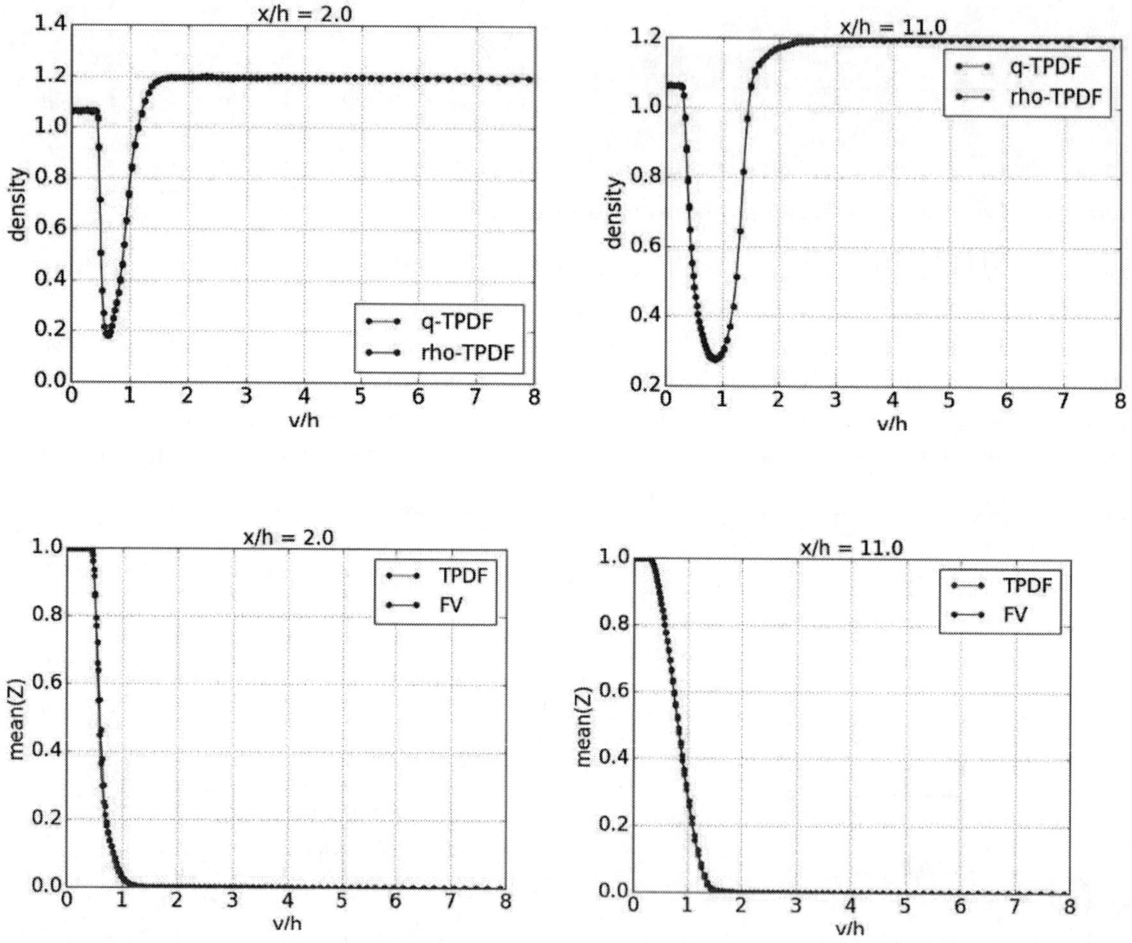

Figure 5: Plots of the mean particle density q and the mean density ρ (top) for a planar jet flame with using the mass consistency algorithm. The bottom plots show a comparison between the mean mixture fraction obtained from the particles and from the solution of the transport equation.

Figure 6 shows contour plots for mean temperature obtained with the TPDF model (left) and with the presumed PDF model using delta-functions (i.e. completely neglecting TCI). Much higher temperatures are observed when neglecting TCI as in the TPDF results which is in line with findings from other studies.

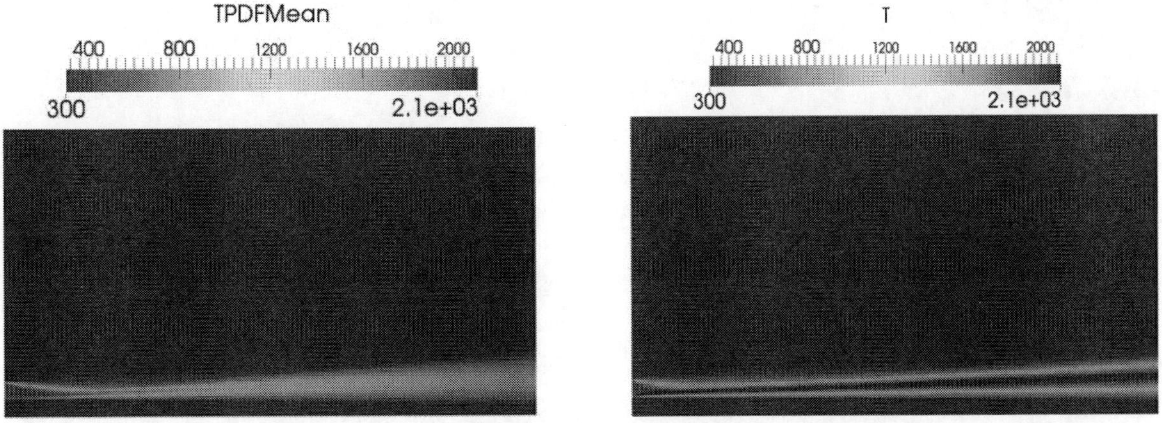

Figure 6: Contour plots of the mean temperature obtained using the transported PDF model (left) and neglecting turbulence chemistry interaction i.e. using delta-PDF's (right).

Round jet pure mixing results

OpenFoam uses special boundary conditions for reduced dimension cases in lieu of special 2-d and 2-d axisymmetric solver implementations. However, implementing the Monte Carlo solver for the transported PDF equation in a 2-d axisymmetric would require special treatment of certain aspects of the solver. To avoid such code specialization, we adopt a simplified 3-d solution approach by considering a small wedge of the axisymmetric problem which is three cells thick in the azimuthal direction and employs periodic boundary conditions on the sides. The wedge domain is visible in figure 7 which shows a contour plot of the mean mixture fraction obtained from the Monte Carlo particles.

Figure 7: Contour plots of the mean mixture fraction in the round jet pure mixing case using a three-cell thick wedge with period boundary conditions.

Figure 8 shows a comparison of the mean mixture fraction obtained from the Monte Carlo particles and from the solution of the corresponding transport equation. At x/h=2 (close to the inlet) the two results match as they should but farther downstream at x/h=11 we observe a rather large discrepancy especially close to the axis. The reason for this discrepancy is currently under investigation.

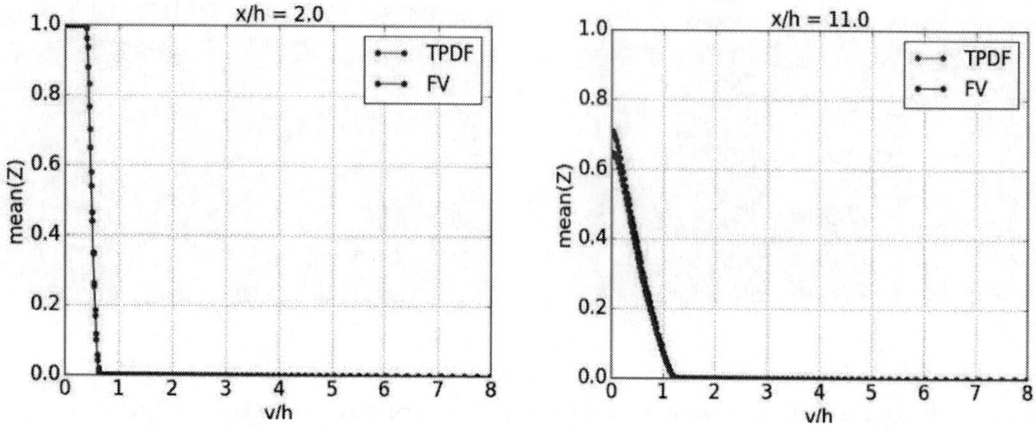

Figure 8: Plots of the mean mixture obtained from the particles (green line) and from solving the modeled transport equations (blue line) in the round jet mixing case.

4. Conclusions

Both transported PDF and presumed PDF solvers are tested on plane-jet non-reacting and reacting cases, for the transported PDF solver the position correction algorithm is validated first with non-reacting plane-jet case and then with reacting plane-jet case. For the both planar configurations the consistency of Monte Carlo solver for the transported PDF could be established. For the round jet problem (modeled as a three cell wedge with cyclic boundary conditions) the consistency could not be established. This issue is currently investigated. Once the problem is found, we can proceed to simulations of the Sandia D flame and provide an in-depth analysis of the main assumptions made in presumed-shape PDF approaches using the FGM chemistry model by comparison with transported PDF result.

References

[1] D.C Haworth and M.W. Jones. *Progress in Energy and Combustion Science*, 36, issue 2 (2014) 168–259

[2] J.A. van Oijen and L.P.H. de Goey, *Combustion Science and Technology* 161 (2000) 113

[3] N.Peters, *Progress in Energy and Combustion Science* 10 (1984) 319

[4] L.Ma and D.J.E.M Roekaerts *Combsution and Flame* 165 (2016) 402-423

[5] R.S. Barlow and J.H. Frank, *Proceedings of Combustion Institute* 27 (1998) 1087

[6] Y.Z. Zhang and D.C. Haworth, *Journal of Computational Physics 194* (2004) 156-193

Western States Section of the Combustion Institute – Fall 2017 Meeting
Hosted by the University of Wyoming
October 2-3, 2017

Sub-model Fidelity in Pulverized Coal Combustion Simulations

Clint Dunn[1], Michael Stoellinger[2]
[1]Mechanical Engineering, University of Wyoming, 1000 E. University Laramie Wy 82072
[2]Mechanical Engineering, University of Wyoming, 1000 E. University Laramie Wy 82072

Abstract

Many assumptions become necessary to produce a computationally tractable simulation of a multiphase reactive flow. In this study we model a pulverized coal reactor in which previous studies have shown poor agreement with experiment for some operating conditions. We would like to perturb some of the assumptions made in these previous simulations in an effort to determine the assumption's role in the disagreement with experiment.

Specifically we are interested in changing the models for combustion kinetics and devolatilization as well as the chemical composition of the coal volatiles. Often it is assumed in complex and computationally expensive simulations like these that the chemical time scales are so much smaller than those of turbulent mixing that chemical equilibrium can be accurately assumed. By incorporating finite-rate chemical models of varying complexity we can investigate this assumption by comparing results to experiment. Devolatilization models often assume rates are species independent, the validity of this assumption too will be investigated by using more general devolatilization models capable of species dependent rates. Finally by including complex molecules that are usually neglected the suitability of the assumed, yet likely overly simplistic, set of volatile gas species can be investigated.

1 Introduction

As interest in numerical modeling of complex multiphase reactive flows with computational fluid dynamics (cfd) grows, so too does the need to evaluate the primary limitations of these approaches. Of particular importance, especially in pulverized coal flows, is the many sub-models required in these simulations. These models are often highly empirical and complex, requiring drastic simplifications be made for the sake of decreasing computationally expense. It is therefore of considerable interest to elucidate which of these sub-models is insufficient and needs to be reconsidered. Of particular interest in this study, and future work, are the models for coal particle devolatilization, coal particle surface reaction, coal particle dispersion, coal volatile composition, subgrid turbulent combustion model and combustion chemistry mechanism.

The goal of this study is to first produce a flame simulation qualitatively similar to the result in the experimental work of Michel, Payne [1] with a baseline set of sub-models. With a similar (if somewhat erroneous) flame simulation we can begin to probe the effect of increasing the fidelity of sub-models, one at a time, on the overall flame structure and it's agreement with experiment. In this paper first the benchmark flame is simulated with the the baseline set of sub-models. Then the combustion model is changed from the very simply laminar combustion model to a Partially Stirred Reactor (PaSR) turbulent combustion model.

2 Methods

2.1 Reactor Geometry and Mesh

A thorough description of the B1 flame is given in [1, 2]. Here a brief description is provided. Figure 1 illustrates the axisymmetric geometry considered in this work. The primary coal inlet in which air with entrained coal particles enters is surrounded by a hot co-flow of air. The rest of

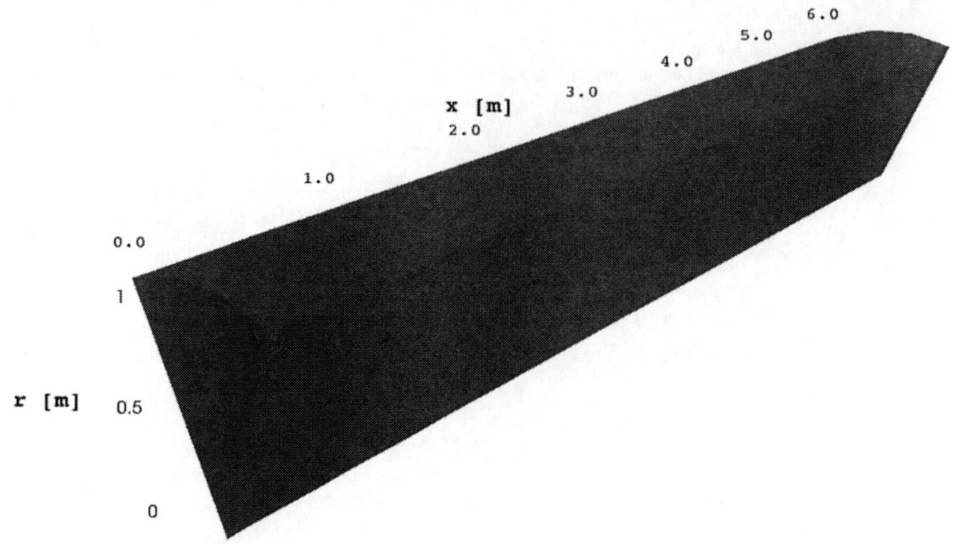

Figure 1: The 30° simulation geometry with air(red), coal stream (small green triangle under the red) inflow patches. The blue patch is the outside reactor wall and the purple patch represents the front cyclic periodic boundary.

the geometry is surrounded with solid adiabatic walls. The flow mixes throughout the reactor and then exits the reactor through a single outlet.

A structured mesh is used to model the domain with considerable stretching near the shear layers in the flow, particularly at the entrance between the co-flow and the primary coal inlet. The mesh contains approximately 180,000 hexagonal cells. Due to limitations of OpenFOAM particle tracking axisymmetric cyclic boundary conditions were applied rather than wedge. This choice necessitates a minimum of three cells in the azimuthal direction. Three cells were used in this work. Each of these azimuthal cells is a wedge of 10° leading to a total axisymmetric geometry of 30°.

2.1.1 Primary Coal Inlet Boundary Conditions

The primary coal inlet injects air entrained with coal parcels into the simulation domain. The air and parcel velocities are $(40.7, 0, 0)$ $[m/s]$. The air and parcel temperatures are $463[K]$. Since this is a reactive flow the composition of the air is important and the typical simplified mixture of oxygen and nitrogen is used, $Y_{O_2} = 0.23$ and $Y_{N_2} = 0.77$ where $Y_{[]}$ signifies a mass fraction.

The the species and phase composition of the coal parcels in mass fractions are given below in Table 2.1.1. The distribution of coal particle diameters is specified by a Rosin-Rammler distribution with $d = 6.3e^{-5}$; $n = 1.1$; $min = 5e^{-6}$; and $max = 0.0005$ where d is the fineness parameter, n is the spread parameter and min/max are the minimum and maximum diameters respectively.

The turbulent kinetic energy k and its dissipation rate ε are calculated based on the incoming turbulence intensity specified by [2].

Table 1: Coal Properties

Phase	Gas			Liquid	Solid	
Species	CH_4	H2	CO_2	H_2O	ash	C
Species Mass Fraction	0.604	0.099	0.297	1.0	0.155	0.845
Phase Mass Fraction	0.506			0.0	0.494	

2.1.2 Secondary Inlet (Co-Flow)

The secondary inlet is where the hot air co-flow enters the domain. The air has the same species mass fractions as the primary air (i.e. $Y_{O_2} = 0.23$ and $Y_{N_2} = 0.77$). The secondary inlet air temperature is $773[K]$. The turbulent kinetic energy k and its dissipation rate ε are calculated based on the incoming properties specified by [2] and then assuming the flow in the annulus to have a fully developed boundary layer prior to reaching the reactor inlet.

2.2 Numerical Modeling

The open-source package OpenFOAM v4.1 and a modified version of the coalChemistryFoam solver is used here. The modifications to the solver allowed for the implementation of some custom Lagrangian phase sub-models which are detailed below in this section. OpenFOAM provides many of the tools necessary for a turbulent reactive flow simulation.

- Numerical tools required for the finite-volume solution of the Reynolds Averaged Navier Stokes (RANS) equations including separate equations for reactive species
- A large collection of RANS turbulence models
- Combustion models for RANS simulations (including some with turbulence/chemistry interactions that are templated upon specific RANS turbulence models)
- Lagrangian parcel tracking and including injection and mass/energy exchange with the continuous phase
- Templated classes for Lagrangian parcel devolatilization and surface reaction models specific to coal particle combustion

2.3 Continuous Gas Phase Modeling

We use the coalChemistryFoam solver with no modifications regarding the continuous phase solution. The pressure velocity coupling is handled with the pimple algorithm. Then the variable density Favre averaged Navier Stokes equations for mass, momentum and energy; the transport equations for reactive species and the transport equations for the turbulent quantities k (turbulent kinetic energy) and ε (dissipation of turbulent kinetic energy) are solved with OpenFOAM built-in finite volume solvers.

2.3.1 Combustion Modeling

This work focuses on the importance of accounting for turbulent mixing in combustion. Therefore two combustion models that each treat the chemical-turbulence interaction differently are investigated, both are used as implemented in OpenFOAM without modification. The first and simplest is referred to as the laminar combustion model. This model simply takes the mean cell species concentrations and temperature and numerically integrates a stiff system of coupled ordinary differential equations (corresponding to a selected combustion mechanism) over the computational timestep. This approach entirely discounts the effect of turbulent mixing on combustion and therefore tends to overestimate the reaction rates. It assumes either that the flow is laminar (and therefore $\omega \approx \overline{\omega}$) or that turbulent fluctuations are irrelevant, e.g.

$$\overline{\omega\left(\Phi\right)} = \omega\left(\overline{\Phi}\right). \tag{1}$$

where ω is a species production rate and Φ is a temperature/species concentration state vector.

The Partially Stirred Reactor (PaSR) [3, 4], is the second model utilized in this study. It represents an improvement on the laminar combustion model by incorporating the effect of a turbulent mixing time scale τ_{mix} in addition to the chemical reaction time scale τ_{chem}. A thorough development of the theory supporting the PaSR model is provided in [3]. Here the PaSR as implemented in OpenFOAM is presented.

The PaSR model builds upon the laminar model discussed above. It makes the assumption that only in the special case when τ_{chem} is significantly larger than τ_{mix} (where τ_{chem} and τ_{mix} are respectively the characteristic time scales of chemical reaction and turbulent mixing) the reaction rate will approach the equivalent laminar model rate. Otherwise the rate predicted by PaSR can be much lower than the laminar one under the same conditions. The laminar reaction rate R_{lam} is scaled by a factor $\kappa \in [0, 1]$ to produce the turbulent reaction rate R_{turb}.

$$R_{turb} = \kappa \cdot R_{lam} \tag{2}$$

The factor κ is dependent on the relationship between τ_{chem} and τ_{mix}

$$\kappa = \frac{\tau_{chem}}{\tau_{chem} + \tau_{mix}}. \tag{3}$$

where

$$\tau_{mix} = C_{mix} \cdot \sqrt{\frac{\mu_{eff}}{\rho\varepsilon}} \tag{4}$$

for constant model parameter C_{mix}, turbulent kinetic energy dissipation rate ε, fluid density ρ and the effective viscosity μ_{eff} (equal to the sum of the fluid viscosity μ and the turbulent viscosity μ_t). $\tau_{chem,k}$, the chemical timescale for a particular finite volume cell is determined in OpenFOAM as

$$\tau_{chem,k} = \frac{N_R \cdot C_{sum,k}}{t_k} \tag{5}$$

where N_R and C_{sum} are respectively the number of reactions in the combustion mechanism and the sum of species concentrations in the kth cell. The factor t_k for the kth cell is calculated

$$t_i = \sum_{i}^{reactions} \left\{ \sum_{j}^{r.h.s.} (s_j \cdot p_{f,i} \cdot c_{f,i}) \right\}. \tag{6}$$

Where *reactions* on the exterior sum indicates a summation over all reactions in the combustion mechanism and *r.h.s.* on the interior sum indicates summation over all of the species on the right hand side of the current (i.e. i^{th} reaction). Then s_j, and $c_{f,j}$ are respectively the stoichiometric coefficient of the jth element and the reference species concentration from the left hand side of the ith reaction. The reference species is specified as the species with the lowest concentration and thus the first to attain an unphysical negative concentration when it's reaction is integrated over an inappropriate time step. $p_{f,i}$ is the forward production rate of the ith reaction

$$p_{f,i} = k_{f,i} \cdot \prod_{l}^{l.h.s} \{c_l\}. \tag{7}$$

Where again the *l.h.s.* on the product symbol implies multiplication over the species on the left hand side of the ith equation. Then $k_{f,i}$ is the forward Arrhenius reaction rate of the ith equation and c_l is the species concentration of the lth species on the left hand side of the ith reaction.

As mentioned in [3] the parameter C_{mix} can be estimated by using an alternative definition for τ_{mix} and comparing the two formulations. By equating the two estimations of τ_{mix}

$$\tau_{mix} = \sqrt{\frac{k}{\varepsilon}} \cdot \sqrt{\frac{\nu}{\varepsilon}} = C_{mix} \cdot \sqrt{\frac{\mu_{eff}}{\rho\varepsilon}} \tag{8}$$

and making the assumptions

$$\frac{\mu_t}{\rho} = C_\mu \cdot \frac{k}{\varepsilon};\ Sc_t = 1;\ \text{and}\ Re_t = \frac{k^2}{\varepsilon\nu} \tag{9}$$

where $C_\mu = 0.09$ is the standard constant from the k-epsilon turbulence model; ν is the fluid kinematic viscosity; and Re_t and Sc_t are respectively the turbulent Reynolds and Schmidt numbers, an equation for C_{mix} emerges

$$C_{mix} = \sqrt{\frac{1}{1 + C_\mu Re_t}} \tag{10}$$

The model is implemented such that a single global value of C_{mix} must be selected for the simulation, in this work we take a commonly accepted value for moderate turbulence levels from [3] and set $C_{mix} = 0.1$

2.3.2 Radiation Modeling

Is this work the P1 radiation model is used. The OpenFOAM implementation is used unmodified and is described here. The effect of radiation on the simulation is represented by a source term in the enthaply equation

$$S_h = (aG - E) - \left(4e\sigma T^4\right). \tag{11}$$

Where $a = 0.5$ is the absorption coefficient of the gas phase, $e = 0.5$ is the emission coefficient of the gas phase, $E = 0$ is the omission contribution (source), T is the local temperature, $\sigma = 5.67e^{-8}\left[W/\left(m^2K^4\right)\right]$ is the Stefan Boltzmann constant, and G is the incident radiation in units of $\left[W/m^2\right]$ determined by solving the following transport equation

$$\nabla \cdot \Gamma \nabla G - aG + 4e\sigma T^4 = E. \tag{12}$$

Where

$$\Gamma = \frac{1}{3a + \sigma_{eff}} \tag{13}$$

and

$$\sigma_{eff} = \frac{A_p\,(1-f)\,(1-\epsilon)}{V \cdot dt} \tag{14}$$

is the particulate scattering factor and it is highly dependent on the dispersed coal solid particle phase. In Equation 14, $\epsilon = 0.9$ is the emissivity of the coal particles, $f = 0.5$ is the coal particle scattering factor, V is the current cell volume in units of $[m^3]$ and dt is the overall timestep in $[s]$. A_p is the sum of projected parcel areas in $[m^2]$ and is defined as a sum over the parcels in the current cell.

$$A_p = \sum_{parcels\,in\,cell} dt_L \cdot n_p \cdot a_p \tag{15}$$

Here dt_L is the Lagrangian parcel time step in units of $[s]$, n_p is the number of real coal particles represented by the parcel (see section 2.3 for an explanation of relationship between real particles and their computational representatives parcels) and a_p is the projected area of a single physical particle $a_p = (0.25)\,\pi D_p^2\,[m^2]$ for parcel diameter D_p.

2.4 Lagrangian Coal Particle Modeling

OpenFOAM provides a framework for not only Lagrangian parcel injection and tracking but momentum and energy transfer between those parcels and the continuous gas phase. OpenFOAM also provides several Lagrangian sub-models for particle devolatilization and surface reaction specific to pulverized coal simulations. These sub-models are written in a way that enables users to readily implement their own devolatilization and surface reaction functionality. A devolatilization model as well as a surface reaction model are implemented in this work and are detailed below.

2.4.1 Parcel Velocity and Temperature Equations

The Lagrangian framework represents physical particles with computational parcels. Each parcel can be representative of many real particles but only particles of the same diameter are grouped together into a parcel. This parcel approach significantly reduces the computational cost of tracking and evolving the dispersed phase. Each computational parcel is tracked through the flow field with velocity determined by

$$\frac{d\boldsymbol{U_p}}{dt} = \boldsymbol{F_{drag}} + \boldsymbol{F_{gravity}}. \tag{16}$$

Where

$$\boldsymbol{F_{gravity}} = m_p * \boldsymbol{g} \tag{17}$$

with particle mass m_p, is used for the force of gravity and a simple spherical drag approximation is used to model drag

$$F_{drag} = \boldsymbol{U_{diff}} * \frac{0.75 \cdot m_p \cdot \mu_c \cdot C_d}{\rho {D_p}^2}. \tag{18}$$

Here C_d is a function of the particle surface Reynolds number, defined as

$$Re_s = \frac{\rho_s \cdot \|\boldsymbol{U_{diff}}\| \cdot D_p}{\mu_s} \tag{19}$$

The subscript s on μ_s and ρ_s indicate these properties are evaluated at the particle surface. In OpenFOAM these surface properties are approximated using the "two thirds rule" to determine the particle surface temperature and then assuming linear property variations with small temperature ranges. For example the particle surface density ρ_s is calculated as follows where ρ_c is the gas phase density interpolated to the parcel position, while T_p, T_s and T_c are, respectively the particle, particle surface and interpolated gas phase temperatures.

First T_s and T_{ratio} are determined with the "two thirds rule".

$$T_s = \frac{2T_p + T_C}{3}; \quad T_{ratio} = \frac{T_c}{T_s} \tag{20}$$

then ρ_s can be determined

$$\Rightarrow \rho_s = \rho_c \cdot T_{ratio}. \tag{21}$$

μ_s is calculated similarly. With Re_s known, C_d is then determined

$$C_d = \begin{cases} 0.424(Re_s) & Re_s > 1000 \\ 24 \cdot \left(1 + 6 \cdot {Re_s}^{-\frac{2}{3}}\right) & Re_s \leq 1000. \end{cases}$$

The temperature of the parcels is assumed to be uniform and its evolution is determined by

$$m_p \frac{d\left(c_{p,p} T_p\right)}{dt} = S_{dev} + S_{char} + \dot{q}_{p,rad} + \dot{q}_{p,c}. \tag{22}$$

Here, $c_{p,p}$ is the coal's specific heat capacity, it is determined from the thermodynamic properties of the constituents in the parcel, scaled by their mass fractions, as well as the current parcel temperature and the seen pressure. The heat of devolatilization is neglected, ($S_{dev} = 0$). The enthaply source from char surface reactions is determined as

$$S_{char} = \frac{{m_{char}}^* \cdot H_{char}(T) - {m_{CO_2}}^* \cdot H_{CO_2}(T)}{dt} \tag{23}$$

where ${m_{char}}^*$ and ${m_{CO_2}}^*$ are the mass of solid carbon consumed and CO_2 formed during by the surface reaction over time step dt. $H_{char}(T)$ and $H_{CO_2}(T)$ are the sensible enthalpies in units of $[J/kg]$ taken, as all thermodynamic data in this work are, from the JANAF tables provided in OpenFOAM. The mass loss rates associated both parcel surface reaction and parcel devolatilization are detailed in Sections 2.3.3 and 2.3.2.

The third source term in Equation 22, $\dot{q}_{p,rad}$ is the parcel radiation enthaply source and is determined as

$$\dot{q}_{p,rad} = A_s \cdot \epsilon \left(\frac{G_c}{4} - \sigma T_p^4\right).$$

Here G_c is the incident radiation interpolated to the parcel position, A_s the surface area of the spherical coal parcel and the other variables are consistent with their definition in Section 2.2.2.

The final term in Equation 22

$$\dot{q}_{p,c} = \frac{h\left(T_c - T_p\right)}{A_s}$$

represents the convective heat transfer from the gas phase to the parcel. Here A_s is the particle surface area and the heat transfer coefficient h is determined with the Ranz Marshall model which is implemented in OpenFOAM as

$$h = \frac{\kappa_s \cdot Nu}{D_p}. \tag{24}$$

Where κ_s is the thermal conductivity evaluated at the particle surface temperature, D_p is the coal parcel diameter (all physical coal particles represented in a single parcel share the same diameter) and Nu is the parcel Nusselt number determined as

$$Nu = 2.0 + 0.6\,\sqrt{Re_s}\,\sqrt[3]{Pr_s} \tag{25}$$

where Pr_s is the Prandtl number determined from properties evaluated at the surface of the parcel

$$Pr_s = \frac{c_p\,\mu_s}{\kappa_s}$$

and Re_s is the Reynolds number of the parcel as discussed above.

Equation 41, which governs the evolution of parcel mass is given later in this section, after the devolatilization and surface reaction models have been described.

2.4.2 DAEM Devolatilization Model

The model for the parcels contains four different mass fractions $Y_{vol}, Y_{char}, Y_{ash}$ and Y_{liquid} which are in this case representative of, respectively, the coal volatiles, solid carbon, ash and moisture within the coal particles. Each of these mass fractions varies throughout the lifetime of a coal parcel. The initial values corresponding to these mass fractions at the beginning of the parcel's lifetime are denoted as $Y_{vol,0}; Y_{char,0}; Y_{ash,0}$ and $Y_{liquid,0}$, see Table 2.1.1 for these initial values. The devolatilization model is active whenever the parcel temperature is over 400 [K] and the moisture of the parcel has been evaporated entirely. In this study the coal is modeled as dry, so this latter condition is met for all parcels at injection time. Once these conditions are met the volatile gas fraction is devolatilized in accordance with the model until satisfying the user selected minimum mass fraction threshold $Y_{vol} = 0.001$. At which point the little remaining volatile mass is transferred to the gas phase and parcel is considered to be completely devolatilized, surface reactions may then begin.

The mass loss during devolatilization is assumed to follow the Distributed Activation Energy Model (DAEM) [5, 6, 7]. In this model a large number of first order decomposition reactions occur in parallel. Each reaction has the same pre-exponential factor A_{v0} and activation energies that are distributed according to a Gaussian distribution. The devolatilization rate at time t depends on the complete temperature history of the coal parcel:

$$\dot{m}_{p,vol} = -\left(Y_{vol,0}\cdot m_{p,0}\right)\int_0^\infty \left[\exp\left(-\int_0^t k(E,\tau)d\tau\right)k(E,t)\right] f(E)dE, \tag{26}$$

where $m_{p,0}$ is the initial particle mass and $\dot{m}_{p,vol} = \frac{d}{dt}\left(Y_{vol}\cdot m_p\right)$. The reaction rates are given by

$$k(E,t) = A_{v0}\,\exp\left(-\frac{E}{RT_p(t)}\right), \tag{27}$$

and the Gaussian distribution of the activation energy with mean E_{v0} and standard deviation σ_{v0} is

$$f(E) = \frac{1}{\sigma_{v0}(2\pi)^{1/2}}\exp\left(\frac{-(E-E_{v0})^2}{2\sigma_{v0}^2}\right). \tag{28}$$

The best performance of any devolatilization models is obtained if the model parameters are fitted to experimentally obtained devolatilization curves for the particular coal of interest. Unfortunately, no such data was available and hence we have used parameters suggested by [5] for a bituminous coal: $A_{v0} = 1.67\cdot 10^{13}\,1/s$, $E_{v0} = 212.6\cdot 10^3\,J/mol\cdot K$ and $\sigma_{v0} = 29.4\cdot 10^3\,J/mol\cdot K$.

An efficient method to solve Equation 26 numerically has been proposed by [7]. The method is based on a modified Gauss-Hermite quadrature rule. Here we use three quadrature points leading to the following numerical approximation

$$\dot{m}_{p,vol} \approx -\left(Y_{vol,0}\cdot m^*_{p,0}\right)\sum_{i=1}^{3} W_i\frac{m}{\sqrt{\pi}}\exp\left(\frac{-(E_i-E_{v0})^2}{2\sigma^2}\right) \cdot\exp\left(-\int_0^t k(E_i,\tau)d\tau\right)k(E_i,t). \tag{29}$$

Details about the scaling factor m, the quadrature points E_i and quadrature weights W_i are given in [7]. The time integral in the above expression is approximated by a simple Euler method

$$\int_0^t k(E_i,\tau)d\tau \approx k(E_i,t_0)dt_0 + k(E_i,t_1)dt_1 + \cdots + k(E_i,t_N)dt_N, \ t_N = t. \tag{30}$$

2.4.3 Smith Intrinsic Surface Reactions Model

The volatiles are released until $Y_{vol} \leq 0.001$. At this time the char burnout begins, it is modeled according to the intrinsic model of Smith [8]. The model takes the bulk diffusion of the oxygen to the char particle external surface, the chemical reaction and pore diffusion into account. The oxygen surface reaction order is assumed to be unity. The mass rate of change of a particle due to char combustion is given by

$$\dot{m}_{p,char} = -\pi {D_p}^2 p_c Y_{s,O_2} \frac{R_{ox} R_{ch}}{R_{ox} + R_{ch}}, \tag{31}$$

where Y_{s,O_2} is the oxygen mass fraction seen by the coal parcel, $\dot{m}_{p,char} = \frac{d}{dt}\left(Y_{char} \cdot m_p\right)$ and p_c is the gas phase pressure interpolated to the parcel position. The bulk molecular mass diffusion rate R_{ox} is

$$R_{ox} = C_{diff} \frac{T_m^{0.75}}{D_p} \tag{32}$$

where T_m is the mean temperature. The chemical rate R_{ch} is expressed by the intrinsic chemical rate and pore diffusion rate

$$R_{ch} = \eta \frac{D_p}{6} \rho_p A_g k_i, \tag{33}$$

where η is the effectiveness factor, A_g is the specific internal surface area and k_i is the intrinsic reactivity calculated from the Arrhenius expression

$$k_i = A_i \exp\left(-\frac{E_i}{RT_p}\right). \tag{34}$$

The effectiveness factor η is calculated as

$$\eta = \frac{3}{\phi^2}\left(\phi \coth\phi - 1\right), \tag{35}$$

where ϕ is the Thiele modulus [9, 10]

$$\phi = \frac{D_p}{2}\left(\frac{S_b \rho_p A_g k_i p_c Y_{s,O_2}}{D_e \rho_c Y_{s,O_2}}\right)^{1/2}. \tag{36}$$

Here, $S_b = 1.33$ is the stoichiometric coefficient assuming that CO is formed by the heterogeneous reaction, D_e is the effective diffusion coefficient in the particle pores and ρ_c is the gas phase density interpolated to the parcel position. Assuming that bulk and Knudsen diffusion proceed in parallel [9, 10],

$$D_e = \frac{\theta}{\zeta^2}\left[\frac{1}{D_{\text{Kn}}} + \frac{1}{D_{ox}}\right]^{-1}, \tag{37}$$

where θ is the porosity of the char, ζ is the tortuosity of the pores, D_{ox} is the oxygen diffusion coefficient and D_{Kn} is Knudsen diffusion coefficient calculated as

$$D_{\text{Kn}} = C_{\text{Kn}} \bar{r}_{po} \sqrt{\frac{T_p}{M_{w,O_2}}}, \tag{38}$$

where C_{Kn} is a constant, $\bar{r}_{po}$ is the mean pore radius and M_{w,O_2} is the molecular weight of oxygen. The oxygen diffusion coefficient is obtained from [11]

$$D_{ox} = D_0 \left(\frac{T_m}{T_0}\right)^{1.75}. \tag{39}$$

The initial porosity and the specific internal surface area of the char are estimated from the relations [12]

$$A_g = 1546.3 \cdot C^2 - 2834.9 \cdot C + 1301.7 \ [m^2/g], \quad \theta = \frac{A_g}{1000\, m^2/g}, \tag{40}$$

where C is the fixed carbon mass fraction (daf) of the parent coal. All parameters of the char reaction model are given in Table 2. It is assumed that the char particles burn at a constant diameter $\dot{D}_{p,char} = 0$.

Table 2: Parameters for the intrinsic char reaction model.

symbol	value, reference	unit
C_{diff}	$5 \cdot 10^{-12}$, [11, 9]	$\frac{kg}{m^2 \cdot s \cdot Pa}$
A_i	0.052, [12]	$\frac{kg}{m^2 \cdot s \cdot Pa}$
E_i	161.5, [12]	$\frac{kJ}{mol}$
S_b	1.33, [9]	-
θ	0.1, [12]	-
ζ	$\sqrt{2}$, [9]	-
C_{Kn}	97.0, [9]	$\frac{kg \cdot m}{K^{0.5} \cdot \mathrm{kmol}^{0.5} \cdot s}$
$\bar{r}_{po}$	$6 \cdot 10^{-8}$, [9]	m
D_0	$3.13 \cdot 10^{-4}$, [11]	$\frac{m^2}{s}$
A_g	104, eq.(40) [12]	$\frac{m^2}{g}$

The char combustion lasts until $m_p = (1 - Y_{vol,0} - Y_{char,0})\, m_{p,0}$. At this stage, the mass of the parcel does not change anymore and is considered to consist of ash ($Y_{ash,0} = 1 - Y_{vol,0} - Y_{char,0}$). The "inert" ash parcels are not removed from the simulation and are still transported through the domain.

The rate of change of the coal parcel mass can be written as

$$\frac{dm_p}{dt} = \dot{m}_{p,char} + \dot{m}_{p,vol}, \tag{41}$$

2.4.4 Particle Dispersion Model

The OpenFOAM Lagrangian particle dispersion model "gradientDispersionModel" is used here. It accounts for turbulence through reference to the turbulent kinetic energy k and its dissipation ε fields interpolated to the location of the parcel. As mentioned in the OpenFOAM documentation this model perturbs the particle velocity in the direction of $-\nabla(k)$. The model also injects some randomness into the particle dispersion. It is important to note that the purpose of this sub-model is to compute the modified fluid velocity U_C^* in the cell, not the new parcel velocity. The parcel velocity is calculated later with a particle drag model (described in Section 2.3.1) which uses U_C^* rather than the unmodified cell mean fluid velocity U_C.

The model as implemented in OpenFOAM is now described. First a relative velocity magnitude and in turn a turbulence timescale are defined (U_{rel} and τ_{turb} respectively).

$$U_{rel} = \|\mathbf{U_p} - (\mathbf{U_c} + \mathbf{U_{turb}})\| \text{ and } \tau_{turb} = min\left(\frac{k}{\varepsilon}; \frac{0.16432 \cdot k^{1.5}}{\varepsilon\, U_{rel}}\right) \tag{42}$$

Next the parcel time step dt is compared to τ_{turb}. If $dt > \tau_{turb}$ then the cell's mean fluid velocity is returned, unmodified (i.e. $U_C^* = U_c$). If $dt < \tau_{turb}$ then

$$U_C^* = U_C + U_{turb} = U_C + \left(\sqrt{\frac{2k}{\varepsilon}} \cdot f \cdot \frac{-\nabla k}{\|\nabla k\|}\right) \tag{43}$$

where f is a random variable drawn from $N(0,1)$.

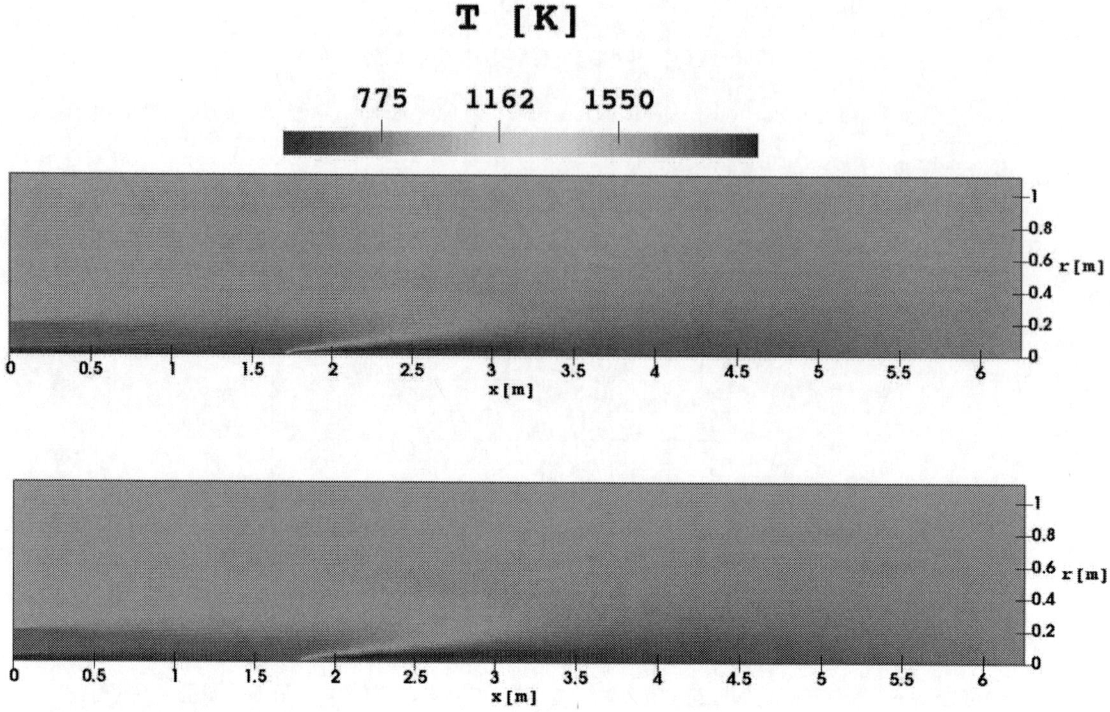

Figure 2: Top: Mean temperature contour plot of the laminar combustion model simulation results. Bottom: Mean temperature contour plot of the PaSR combustion model simulation results. In both plots only the top half of the reactor is shown.

3 Results and Discussion

Two simulations are performed and compared in this work to determine the effect of the combustion model, further both simulated results are compared with experimental results from Michel et al. [1]. The first simulation uses the simple laminar combustion model in which reactions proceed based on the mean thermophysical state. This model ignores the effects of turbulence. The second simulation uses the Partially Stirred Reactor (PaSR) combustion model. The PaSR builds upon the laminar model by incorporating a turbulent mixing timescale (τ_{mix}) that permits the attenuation of reaction rates depending on the relationship between τ_{mix} and the chemical reaction time scale τ_{chem}.

In Figure 2 the mean temperature fields are presented for the top half of the symmetric geometry. The bottom of each of these figures is located at the central axis of the reactor and the geometry extends radially outwards from there to the reactor wall. A qualitative comparison of these two contour plots immediately suggests a trend that is reinforced throughout this section, the results for both simulations are very similar. There is a very slight difference in the shape of the flame front, but its location is nearly identical in both simulations.

Figures 3 and 4 show the mean axial velocity, mean temperature and mean CO_2 and O_2 molar fraction trends along the central axis of the reactor for both simulations and the experiment. The temperature plot in Figure 3 shows a poor prediction for the axial location of the flame front. This analysis is corroborated by Figure 4. Comparison with experiment suggests that the simulated flames sit nearly a meter too far downstream. One likely cause for this error is the use of an inaccurate devolatilization model that results in low rates of devolatilization. This explanation is contradicted when similar work from Stoellinger et al.[2] with the same devolatilization model and parameters is considered. There the flame front location was found to be in much better agreement with experiment. It is therefore unlikely that the devolatilization model is solely responsible for

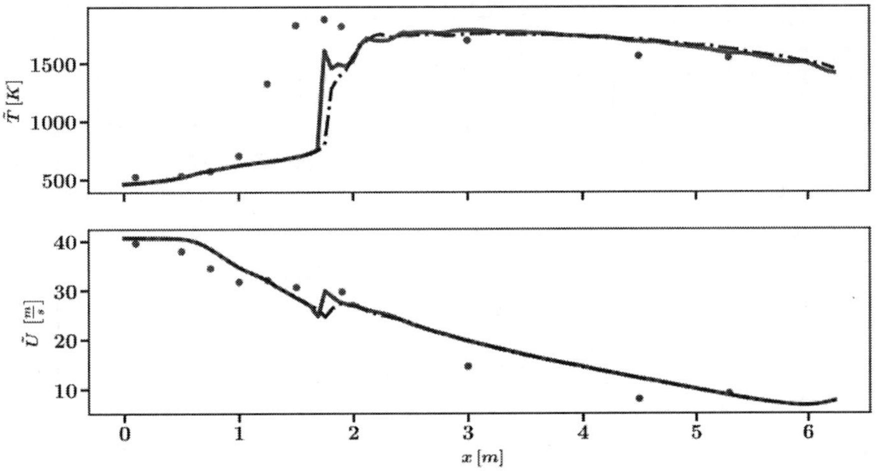

Figure 3: Centerline temperature and axial velocity component profile of measured (circle), PaSR combustion model simulation (solid line) and laminar combustion model simulation (broken line).

the error.

An alternative explanation is that the Lagrangian particle dispersion model is inaccurate. If the simulated dispersion is insufficient then the particle heating rates are unphysically low, this results in delayed devolatilization and therefore a displaced flame front. Unfortunately increasing the complexity and accuracy of the dispersion model will likely necessitate a concomitant increase in the turbulence model complexity as well, beyond the simple k-ε model used here.

The axial velocity plot in Figure 3 shows that the axial velocity component along the centerline is in general over predicted. In the region near the flame front (both experimental and from the simulation) the velocity is under predicted. This can be partially explained by the significant offset of the simulation and experimental flames. But the flame location offset does not explain the severity of the axial velocity gradients.

These sharp gradients are attributable to insufficient mesh resolution in the near the flame. The parcels here heat rapidly and therefore rapidly are devolatilized. As the volatile gases leave the parcels and are injected into the continuous gas phase they produce a momentum source that creates these velocity gradients. The mesh is poorly resolved in this area because the location of simulated flame front was not known prior to meshing. In future work the mesh will be resolved more appropriately. Increased parcel dispersion would also mitigate these sharp jumps since the devolatilization would become less localized.

Figures 5 and 6 illustrate the simulated and experimental radial trends of the temperature and axial velocity at a variety of axial locations. These plots again show how similar the two simulation results are. Particularly the radial axial velocity profile is nearly identical at all axial locations examined.

As should be expected from the axial results presented above the predictions of the simulations are in poor agreement with experimental data near the axial location of the flame (i.e. at $x = 1.9$ and $3.0\,[m]$) for both velocity and temperature. Considering the temperature and velocity profiles in comparison with experiment at $x = 0.1$ and $1.9\,[m]$ shows that the trends are at least similar but the magnitudes, particularly at larger values of r (i.e. nearer the wall) are in disagreement. This difference is the result of increased re-circulation in the experiment compared to the simulations. This trend is most prominently illustrated in Figure 6 at $x = 3.0\,[m]$. There the experimental velocity near the wall shows considerable backflow while both of the simulations show near stagnation.

Similarly Figures 8 and 7 show that the species concentration results support the aforementioned trends. They show that there is a low concentration of combustion products (represented by CO_2 here) near the wall in the simulations while the experimental results show much higher

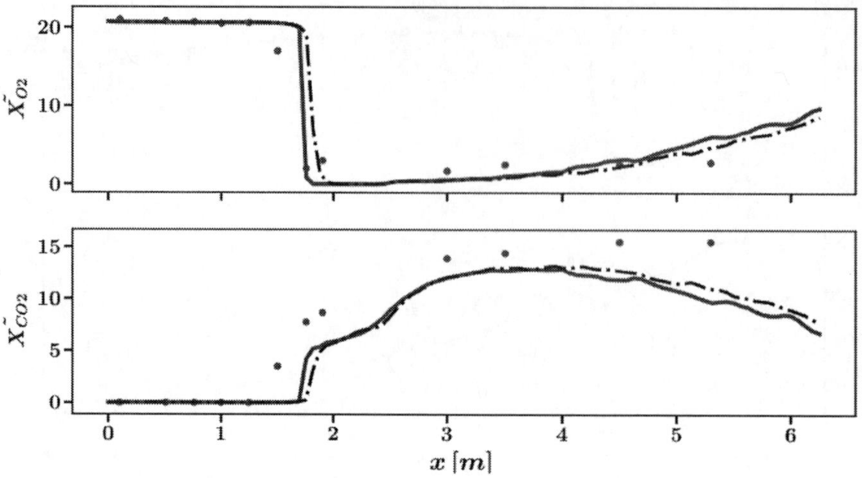

Figure 4: Centerline O_2 and CO_2 molar fraction profile of measured (circle), PaSR combustion model simulation (solid line) and laminar combustion model simulation (broken line).

concentrations of CO_2 in these regions. In the experiment where there is a powerful backflow these product gases are transported from the rear of the reactor to the front. In the simulations no such backflow is present and the concentrations remain similar to their initial state (air) in these regions. The cause of this discrepancy between simulation and experiment is currently unclear and will be a major focus of future work.

4 Conclusion

In this study the IFRF B1 pulverized coal flame is simulated using OpenFOAM with two different combustion models and compared to experimental results from Michel et al.[1]. The two combustion models vary in their treatment of the turbulence/chemistry interactions. The laminar combustion model is the simpler of the two, it completely ignores the subgrid turbulence and predicts the combustion reactions according to mean properties. The Partially Stirred Reactor (PaSR) combustion model represents an incremental improvement on the laminar model. With PaSR a turbulent mixing time scale is considered in addition to the chemical reaction time scale. The PaSR model then scales the reaction rates generated with the laminar model depending on the subgrid turbulence levels. Both simulations produced very similar results, regardless of the combustion model selected.

There is considerable disagreement between the two simulations and the experimental results of the same flame from Michel et al. [1]. In the experiment the flame is much closer to the inlet side of the reactor; further the species, temperature and velocity profiles indicate that the backflow predicted in the simulations is much less significant than that observed in experiments. It is proposed that the Lagrangian parcel dispersion model leads to insufficient dispersion. This is suspected to be a cause for the slow devolatilization and combustion rates seen in simulations and will be the immediate focus of future work. The cause of the backflow rate discrepancy between the simulations and experiment is currently unclear and will be the subject of future work following parcel dispersion model adjustments.

Acknowledgements

This research was funded by the University of Wyoming's Office for Academic Affairs and School of Energy Resources.

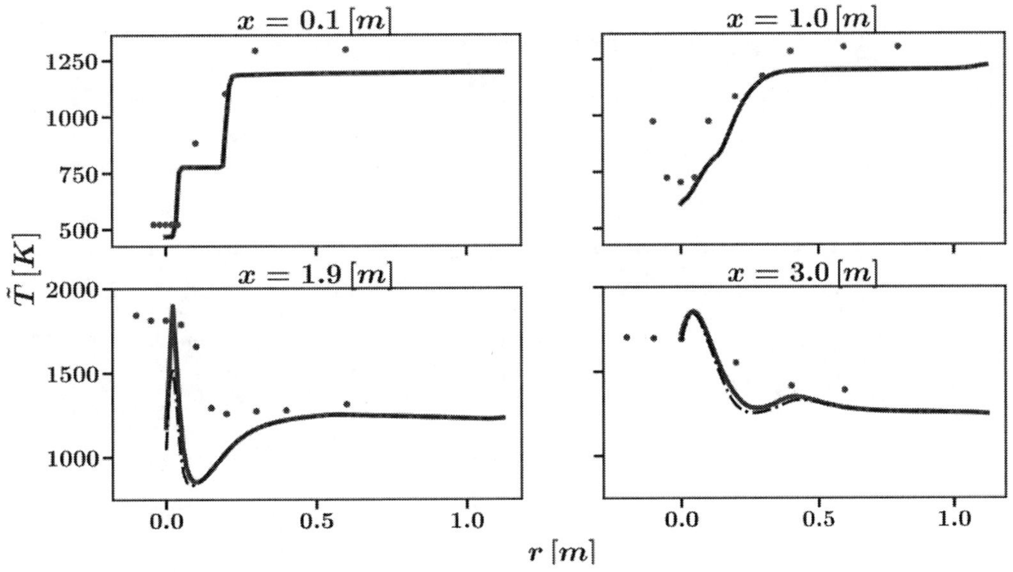

Figure 5: Radial temperature distribution of measured (circle), PaSR combustion model simulation (solid line) and laminar combustion model simulation (broken line).

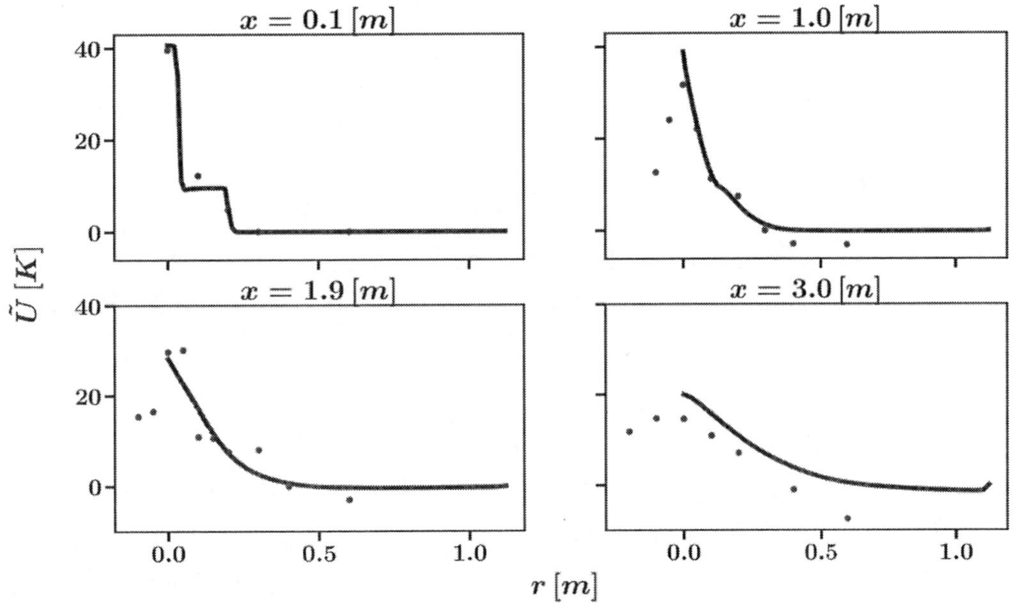

Figure 6: Radial distribution of the axial velocity component of measured (circle), PaSR combustion model simulation (solid line) and laminar combustion model simulation (broken line).

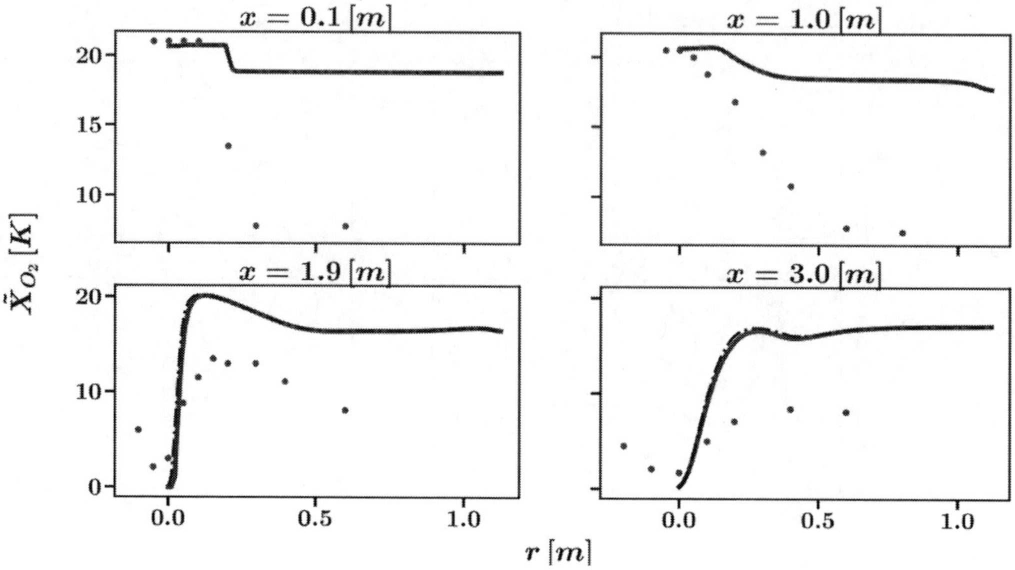

Figure 7: Radial distribution of the molar fraction of oxygen for measured (circle), PaSR combustion model simulation (solid line) and laminar combustion model simulation (broken line).

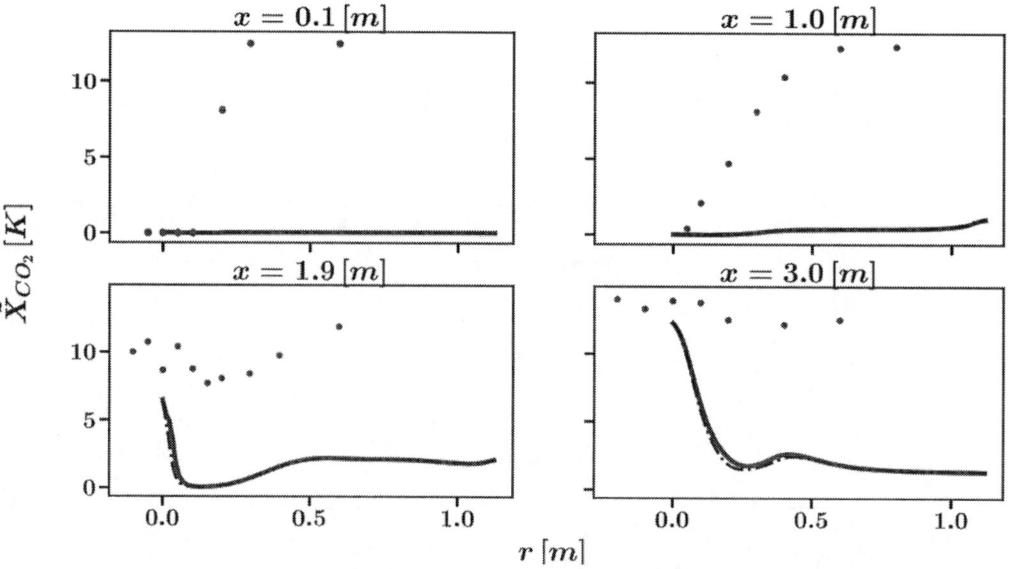

Figure 8: Radial distribution of the molar fraction of carbon dioxide for measured (circle), PaSR combustion model simulation (solid line) and laminar combustion model simulation (broken line).

References

[1] J. Michel and R. Payne. Detailed Measurement of Long Pulverized Coal Flames for the Characterization of Pollutant Formulation. Technical report, International Flame Research Foundation, 1980.

[2] Michael Stöllinger, Bertrand Naud, Dirk Roekaerts, Nijso Beishuizen, and Stefan Heinz. PDF modeling and simulations of pulverized coal combustion – Part 2: Application. *Combustion and Flame*, 160(2):396–410, 2013.

[3] Lorenzo Bartolucci, Naser Hamedi, and Håkan Nilsson. CFD with OpenSource software EngineFoam: implementation of a different combustion model and the new Janaf thermo equations. 2014.

[4] Fabian Peng Kärrholm. Numerical Modelling of Diesel Spray Injection, Turbulence Interaction and Combustion. 2008.

[5] D.B. Anthony, J.B. Howard, H.C. Hottel, and H.P. Meissner. Rapid devolatilization of pulverized coal. *Proceedings of the Combustion Institute*, 15:1303 – 1317, 1975. The Combustion Institute.

[6] Peter R. Solomon, Michael A. Serio, and Eric M. Suuberg. Coal pyrolysis: Experiments, kinetic rates and mechanisms. *Progress in Energy and Combustion Science*, 18(2):133 – 220, 1992.

[7] E. Donskoi and D. L. S. McElwain. Optimization of Coal Pyrolysis Modeling. *Combustion and Flame*, 122:359–367, 2000.

[8] I. W. Smith. The Combustion Rates of Coal Chars: A Review. *Proceedings of the Combustion Institute*, 19:1045 – 1065, 1982. The Combustion Institute.

[9] R. I. Backreedy, L. M. Fletcher, L. Ma, M. Pourkashanian, and A. Williams. Modelling pulverised coal combustion using a detailed coal combustion model. *Combustion Science and Technology*, 178(4):763–787, 2006.

[10] N. Schaffel, M. Mancini, A. Szlek, and R. Weber. Mathematical modeling of MILD combustion of pulverized coal. *Combustion and Flame*, 156:1771–1784, 2009.

[11] M. M. Baum and P. J. Street. Predicting the Combustion Behavior of Coal Particles. *Combustion Science and Technology*, 3(5):231–243, 1971.

[12] A. Williams, R. I. Backreedy, R. Habib, J. M. Jones, and M. Pourkashanian. Modelling coal combustion: the current position. *Fuel*, 81:605–618, 2002.

Western States Section of the Combustion Institute – Fall 2017 Meeting
Hosted by the University of Wyoming
October 2-3, 2017

Smoldering behavior of compacted sawdust

Patryk P. Radyjowski *Robert H. Bush* *Janet L Ellzey*

Department of Mechanical Engineering
University of Texas at Austin
Austin, TX 78712

Biomass is an important fuel used around the world for heating and cooking in both developing and developed countries. In most practical devices, wood or wood by-products are burned through smoldering rather than flaming combustion. In this study, smoldering of sawdust is studied experimentally with a focus on understanding the conversion process. Temperature and emission measurements were taken in a radial, steel, combustion chamber which showed the evolution of the combustion process: an initial conversion of the raw sawdust to char followed by the conversion of sawdust to ash. In general, the char front propagated symmetrically in the radial direction while the ash front was not symmetric, and typically followed paths where oxygen was most readily available. Experiments with vitiated air, in which the oxygen content of air was diluted by adding nitrogen, were conducted to determine a limit at which combustion was no longer self-sustaining. Experiments showed that vitiated air with 7% oxygen in the supply air did not support self-sustaining combustion while those with 11% or more of oxygen did support smoldering. Mass loss rate and emissions for two cases with different oxygen concentrations are also discussed.

1. Introduction

Globally, over two billion homes use some form of biomass such as dung, agro crop residue, or wood as a source of cooking or heating fuel [1]. Biomass, as a source of fuel for individual and residential use, has also seen a rise in the developed countries of central and northern Europe [2], with wood being a commonly selected fuel [3]. Utilizing this resource as a fuel source is seen as mostly sustainable as it is renewable and is considered carbon neutral.

Much of the work on wood combustion has arisen from an interest in fire spread and suppression and has focused on ignition and propagation rates [4, 5, 6, 7]. Data on emissions from biomass combustion are often related to the performance of practical cook stoves [8, 9, 10, 11] and on wildland fires [12, 13] rather than on the fundamental combustion processes. The biomass type and cook stove design have significant effects on the performance of the stove as shown by Zhang [10] who compared 56 fuel/stove combinations in terms of the total products of incomplete combustion. The stoves' performances differed by more than a factor of 10 depending on which fuel/stove combination was considered, highlighting the importance of understanding the fundamental processes more thoroughly.

Sawdust is a particularly interesting form of biomass in that it is a waste product from the wood processing industry. Estimates of waste, which includes sawdust, trims and shavings, from a sawmill are in excess of 50% of the original wood mass [14]. This by-product is expected to continue in importance as the demand for wood products grows [15]. Sawdust is currently used as a fuel in cook stoves in many parts of the world and, as with other biomass, research has focused on the performance and design of cookstoves using this fuel [16, 17].

In most practical devices using wood or a wood by-product, the combustion process is dominated by smoldering rather than flaming combustion due to the longer burn times during this lower temperature combustion mode. The understanding of smoldering has progressed through studies on polyurethane or cellulosic foams [18, 19, 20, 21] and much of this work can inform the study of smoldering of biomass. Leach [20] showed that for reverse smolder initially an increased inlet velocity resulted in faster propagation due to the increase in oxygen supply, but eventually the cooling effect of higher inlet flows resulted in a slower propagation. Controlled studies of smoldering of logs [22] confirmed that oxygen supply and heat loss were the main factors determining whether combustion was sustained.

The purpose of the current study is to study smoldering of sawdust with the intent of understanding the combustion process more fully. An experimental apparatus was constructed to allow the visual observation of the smoldering process as well as measurement of the temperature in the fuel bed and emissions from the exhaust as the smoldering front evolved. These data provide insights into the characteristics of the smoldering process.

2. Methods

The combustion chamber is a steel cylinder, filled with sawdust, around a mesh inner cylinder creating an annulus for airflow. The final apparatus used in the experiments is shown in Figure 1. The air enters the chamber through a 6.35 cm diameter stainless steel mesh. A 7.62 cm diameter mesh cylinder, extends vertically through the chamber. This mesh structure stabilizes the fuel bed and allows removal of the ignition coil. The ignition coil, used to initiate smoldering, is a 624 watt heating

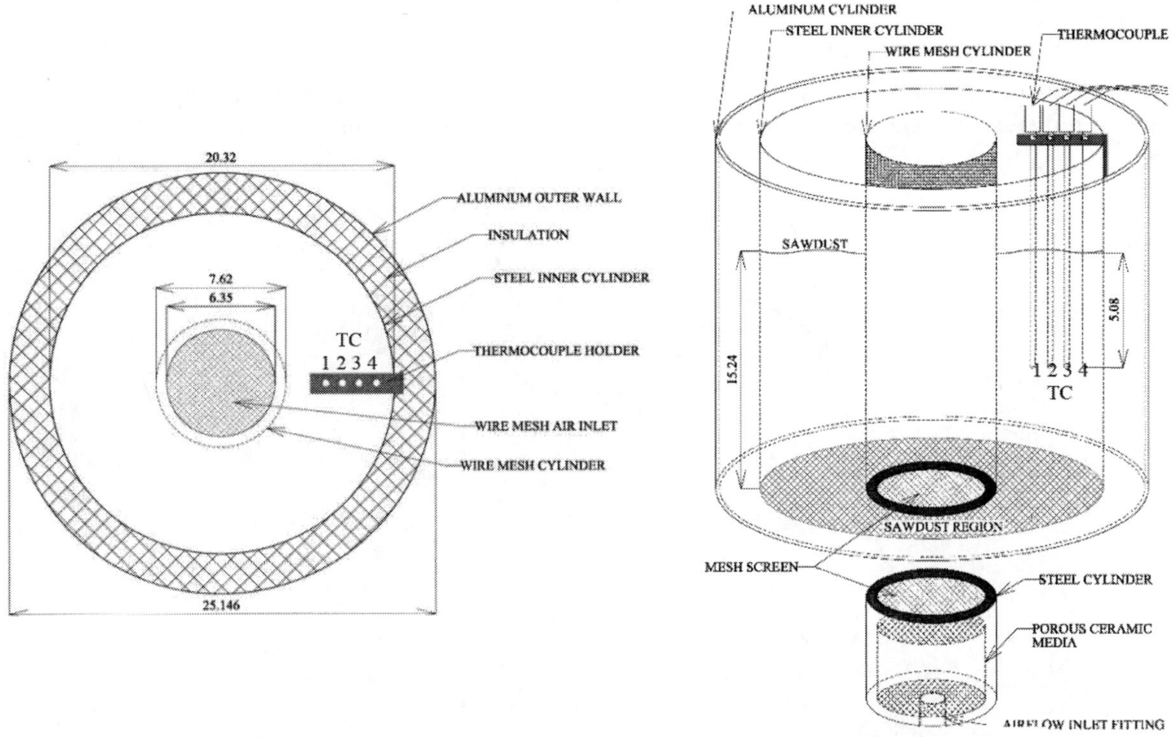

top view isometric view

Figure 1: Experimental apparatus; all dimensions in centimeters.

tape. The chamber was filled and compacted up to 15.2 cm height with well mixed sawdust. A chimney was attached to the top of the combustion chamber to mitigate the effects of external air flow as well as an aid in sample procurement. CO and HC samples were obtained from the chimney exit in 10-minute interval and analyzed by TPI 707 and Rosemount 400A analyzers respectively.

The results presented in this paper were all acquired with poplar sawdust. Earlier experimentation utilized mesquite sawdust, which has been studied in greater detail by Lo [23]. All sawdust was obtained from local lumberyards or woodworking shops. Samples of poplar sawdust were sent to Galbraith Labs (Knoxville, TN) for proximate and ultimate analysis to characterize the constituents and these results are shown in Table 1.

Ultimate Analysis				**Proximate Analysis**			
Species				**Volatile Matter**	**Fixed Carbon**	**Ash**	**Limit of Detection**
H	**C**	**N**	**O**				
8.0%	58.9%	0.5%	34.1%	88.3%	10.9%	0.6%	2.8%

Table 1: Ultimate and proximate analysis for constituents of studied sawdust specie

3. Results and Discussion

Numerous experiments were conducted and they are discussed below in terms of overall smoldering characteristics, emissions behavior, and effect of oxygen concentration.

3.1 Overall Smoldering Characteristics

The evolution of the temperature profile through the bed is shown in Figure 2 for four thermocouples each at a different radial depth as described above along with the mass loss curve. Multiple tests (not shown) at a base condition of 3 lpm airflow revealed behavior similar to that of thermocouple 1 in

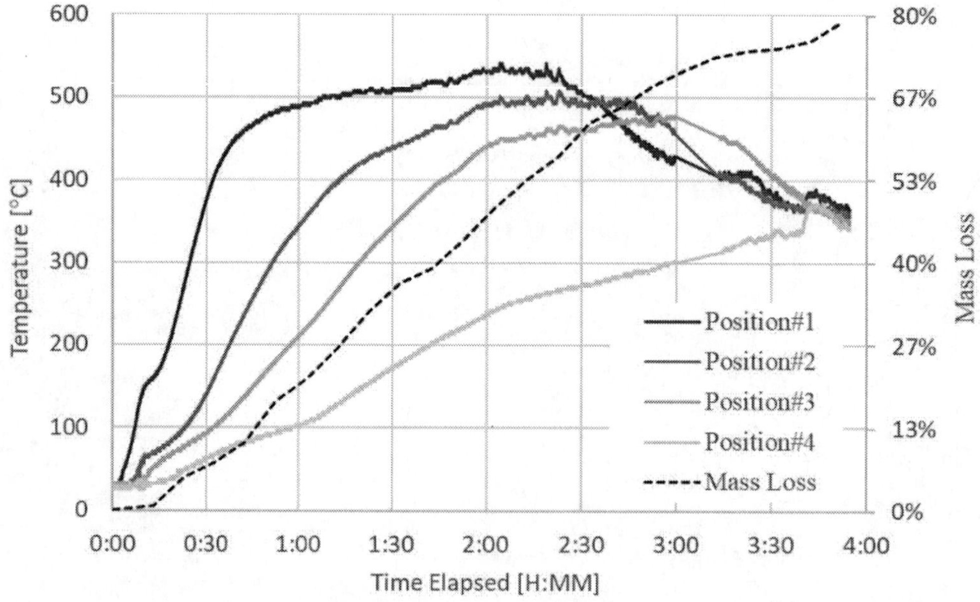

Figure 2: Base Case with Mass Loss

Figure 2: an initial sharp increase in temperature of the innermost thermocouple until ~10 minutes associated with the heating due to the coil, continuing strong increase in temperature until ~30-45 minutes, followed by a decrease in the slope or heating rate. The decrease in temperature observed after 2.0 hours is associated with the changing structure of the bed: as mass is consumed and the bed becomes more porous and shrinks, the thermocouples lose contact with the solid and the temperature is influenced by the surrounding air which is heated by the combustion process. Thermocouples at positions 2, 3, and 4 show the propagation of the front in time. The mass loss rate follows a fairly linear path from the onset of self-sustained smolder at ~15 minutes until 3 hours at which time the 70% of the solid mass has been combusted. Ultimately, about 95% of the mass is burned and this final stage occurs after the solid has collapsed into the bottom of the chamber.

Although the chemistry of biomass combustion is complicated, progress has been made in understanding the key chemical processes [24]. To describe the overall bulk conversion process, however, Ohlemiller [6] suggests that three reactions are sufficient:

Endothermic Pyrolysis: $Fuel \rightarrow Char + Gases$ (1)

Exothermic Thermal Oxidation: $Fuel + O_2 \rightarrow Char + Gases'$ (2)

Exothermic Char Oxidation: $Char + O_2 \rightarrow Ash + Gases''$ (3)

The propagating front is represented in Figure 3(a) for 1-D forward smolder in a porous solid fuel. The endothermic pyrolytic reaction (1) competes for fuel with an exothermic oxidative reaction (2) required for self-sustained smolder followed by oxidation of char.

In practice, of course, the process is not one-dimensional as shown in Figure 3(a). Videos of the smoldering sawdust in the current experiments, however, revealed these distinct processes as illustrated in Figure 3(b). In the first phase, heat propagates radially outward and the unburned sawdust

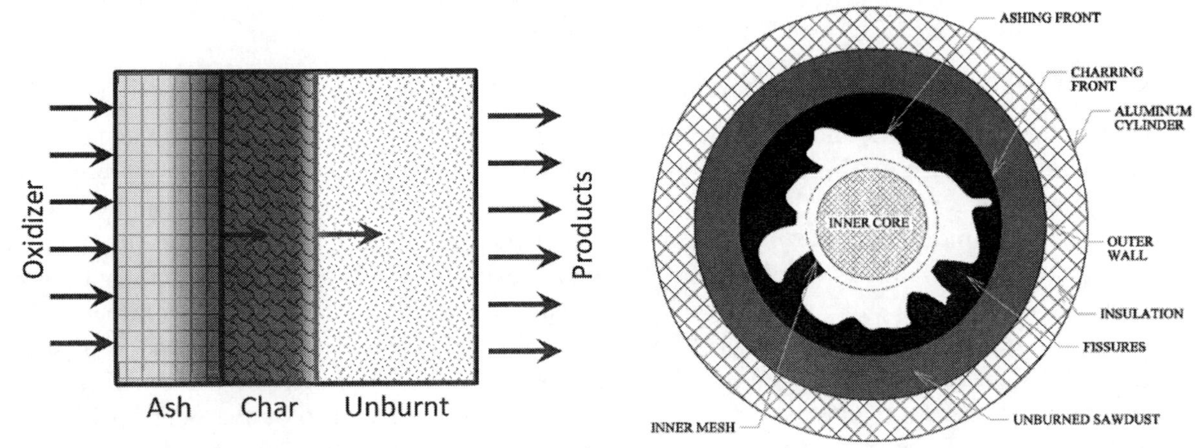

(a) (b)

Figure 3: Forward smolder front in porous solid fuel (a) 1-D representation (b) actual process observed in experiment

is converted to char. In the second phase, small fissures form in the char allowing the influx of air which promotes the ashing process. Correlation of thermocouple recordings and visual observation indicated that charring occurs at approximately 200°C followed by ashing when temperature exceeds 500°C.

3.2 Emissions

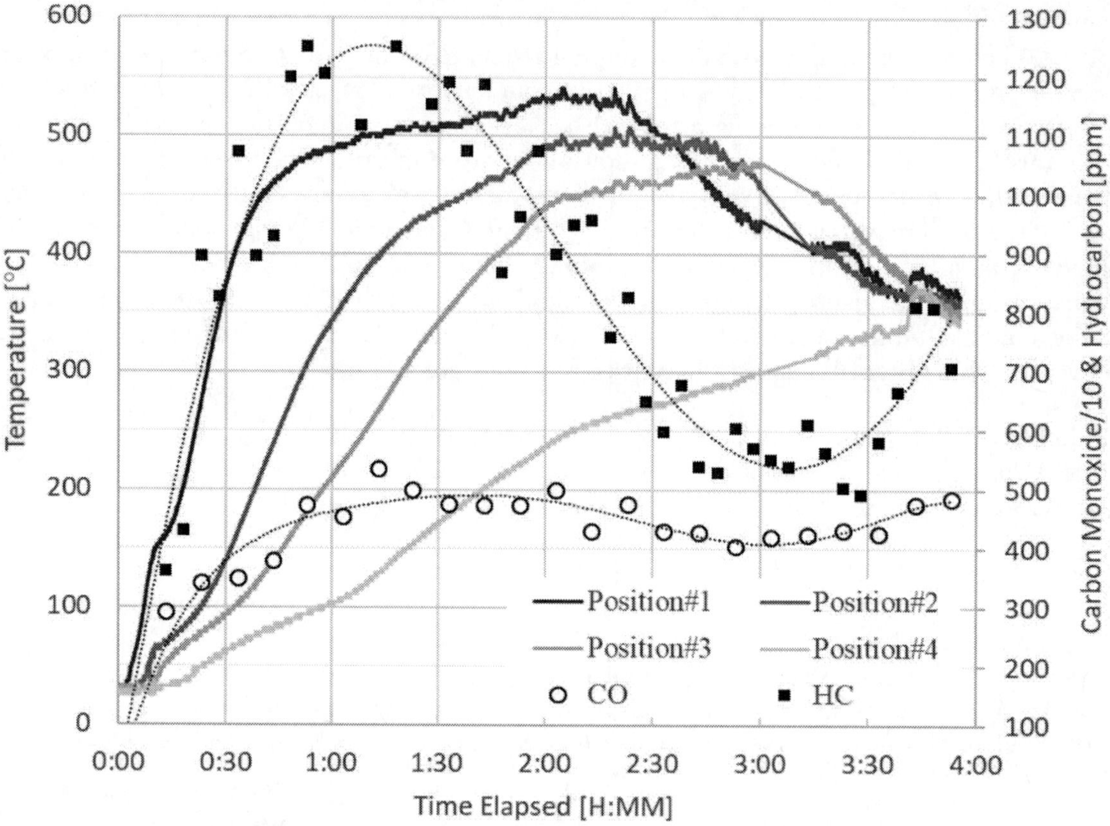

Figure 4: Emissions from smoldering of poplar with 3 lpm airflow

The measurements of CO and unburned HC for the base case are overlaid on the temperature measurements in Figure 4. The progression of combustion is evident as the peak temperature occurs at subsequently longer times for the thermocouples further from the annular surface. The hydrocarbons (HC) emissions rapidly increase in the initial part of the experiment alongside the propagation of charring zone. This is related to radial geometry, which causes the reaction front (circular) to increase with growing radius. Once the charring zone arrives at the edge of container at approximately 80min it reaches the maximum in excess of 1250 ppm. As the solid is consumed the bed becomes more porous and the availability of oxygen increases, thus promoting the conversion of HCs. This, together with change in chemical reactions from charring to ashing, is believed to be the source of lower HC emissions in the later part of the experiment. Meanwhile the CO production shows much less sensitivity to a particular process. The initial growth can be accounted for the geometry configuration discussed above, but unlike HC it stays relatively constant at approximately 4000ppm level once the initial reaction front propagates through the fuel bed indicating there is insufficient oxygen to promote conversion of CO to CO_2.

3.3. Effect of Oxygen Concentration and Airflow Rate

The chemistry outlined above suggests that the availability of oxygen is critical to the smoldering process and will play an important role in the conversion rate. Four different cases examining this effect are shown in Figure 5. Presented temperatures are for the first thermocouple, which captures

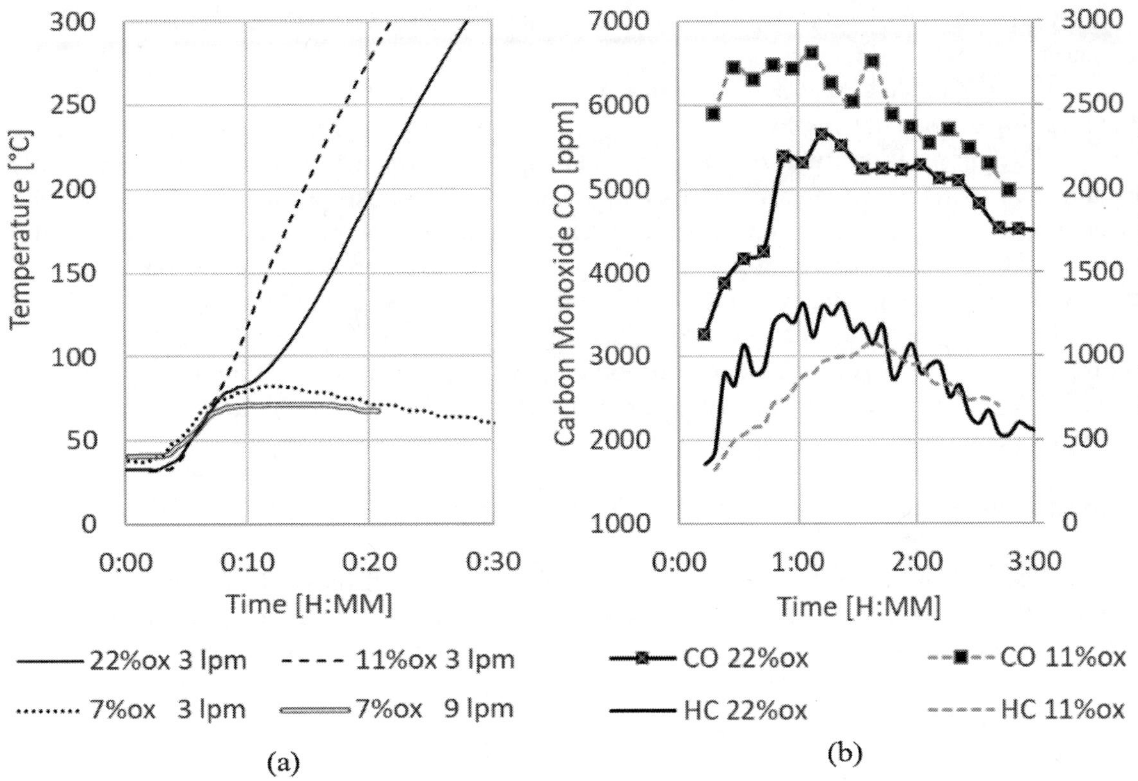

Figure 5: Oxygen concentration influence on combustion (a) Temperature (b) CO and HC emissions

the initiation of self-sustained smolder. Both the base case (O_2=22%) and the case with reduced oxygen of 11% show self-sustained smolder at ~10 minutes. When the oxygen concentration is reduced to 7% while maintaining the same inlet velocity, self-sustained smolder is not attained and the front dies out at ~20 minutes. This is due to the endothermic energy consumption by pyrolysis, and a lack of exothermic oxidation to support further smoldering combustion. Noteworthy at the same concentration of 7%, a self-sustained front is not obtained at a triple flowrate of the base case indicating that higher flow did not succeed in supplying sufficiently more oxygen to the front. Based on these results, the critical concentration of oxygen for self-sustained smolder appears to be between 7-11%. The mass loss data for smoldering poplar sawdust show that at 2.5 hours, about 55 to 60% of the total mass has been consumed and the rate is roughly linear in both cases. The emissions data in Figure 5b provides more insight into the smoldering process for the two cases discussed above. Emissions of CO are higher when the oxygen level is lower which may be due to insufficient oxygen to further conversion of CO to products of complete combustion. The emissions of HCs are initially higher for the case with 22% oxygen which may be associated with the higher fuel consumption rate.

4. Conclusions

This study examined smoldering combustion of a bed of poplar sawdust. A cylindrical apparatus was constructed and an experimental procedure established that provided repeatable results.

Temperatures at four discrete locations were recorded together with exhaust concentrations of CO and HC and mass loss.

Visual records and temperature data indicated a clear, symmetrical charring front developing at the leading edge of the smolder front. It is followed by a highly irregular ashing front, that completes the wood decomposition to gases and ash. Charring was observed to have a very small influence on reactant bed structure. Meanwhile the ashing process is responsible for significant changes such as fissures and bed collapse as the bulk of charred wood is consumed. The emissions data revealed increased HC production during the charring process, with sharp decrease once all virgin wood was fully converted to char. Meanwhile CO emissions were relatively steady thorough the whole process.

Reducing the oxygen content to 11% of the incoming air led to a slower propagation rate and a complete failure to ignite at a concentration of 7% O_2. Furthermore the increase in overall flowrate in the 7% O_2 case did not lead to ignition, meaning that ignition failure is tied to balance of endo- and exothermic reactions.

Acknowledgements

The authors would like to thank Dr. Amara Holder from EPA Office of Research and Development at National Risk Management Research Laboratory for her guidance and comments on this research. The authors also acknowledge the sawdust donation by Precision Craft Woodworks in Austin, TX.

References

[1] E. Rehfuess, Fuel for Life: Household Energy and Health, Geneva, 2006.

[2] D. Haluza, A. Kaiser, H. Moshammer, C. Flandorfer, M. Kundi and M. Neuberger, "Estimated health impact of a shift from light fuel to residential wood-burning in Upper Austria," *J. Expo. Sci. Environ. Epidemiol.*, vol. 22, no. 4, p. 339–43, 2012.

[3] E. Hedberg, A. Kristensson, M. Ohlsson, C. Johansson, P. Å. Johansson, E. Swietlicki, V. Vesely, U. Wideqvist and R. Westerholm, "Chemical and physical characterization of emissions from birch wood combustion in a wood stove," *Atmos. Environ.*, vol. 36, no. 30, p. 4823–4837, 2002.

[4] R. Bilbao, J. Mastral, M. Aldea, J. Ceamanos, M. Betrán and J. Lana, "Experimental and Theoretical Study of the Ignition and Smoldering of Wood Including Convective Effects," *Combust. Flame*, vol. 126, no. 1-2, pp. 1363-1372, 2001.

[5] M. Spearpoint and J. Quintiere, "Predicting the Burning of Wood Using an Integral Model," *Combust. Flame*, vol. 123, no. 3, pp. 308-325, 2000.

[6] T. Ohlemiller, "Smoldering Combustion Propagation on Solid Wood," in *Fire Safety Science—Proceedings of the Third International Symposium*, 1991.

[7] A. A. H.R. Baum, "A Model of Transport of Fuel Gases in a Charring Solid and Its Application to Opposed-Flow Flame Spread," *Proc. Combust. Inst.*, vol. 31, no. 2, pp. 2633-2641, 2007.

[8] H. Lamberg, K. Nuutinen, J. Tissari, J. Ruusunen, P. Yli-Pirilä, O. Sippula, M. Tapanainen, P. Jalava, U. Makkonen, K. Teinilä, K. Saarnio, R. Hillamo, M.-R. Hirvonen and J. Jokiniemi, "Physicochemical Characterization of Fine Particles from Small-Scale Wood Combustion," *Atmos. Environ.*, vol. 45, no. 40, pp. 7635-7643, 2011.

[9] J. Osán, B. Alföldy, S. Török and R. Van Grieken, "Characterization of Wood Combustion Particles Using Electron Probe Microanalysis," *Atmos. Environ.*, vol. 36, no. 13, pp. 2207-2214, 2002.

[10] J. Zhang, K. Smith, Y. Ma, S. Ye, J. Jiang, W. Qi, P. Liu, M. Khalil, R. Rasmussen and S. Thorneloe, "Greenhouse Gases and Other Airborne Pollutants from Household Stoves in China: a Database for Emission Factors," *Atmos. Environ.*, vol. 34, no. 26, pp. 4537-4549, 2000.

[11] J. Jetter and P. Kariher, "Solid-fuel household cook stoves: Characterization of performance and emissions," *Biomass Bioenergy,* vol. 33, no. 2, p. 294–305, 2009.

[12] M. Schulz, A. Heil, J. Hoelzemann, A. Spessa, K. Thonicke, J. Goldammer, A. Held, J. Pereira and M. Van Het Bolsher, "Global Wildland Fire Emissions from 1960 to 2000," *Global Biogeochem. Cycles,* vol. 22, p. GB2002, 2008.

[13] B. Porterie, J. Consalvi, A. Kaiss and J. Loraud, "Predicting Wildland Fire Behavior and Emissions Using a Fine-Scale Physical Model," *Numer. Heat Transfer, Part A,* vol. 47, pp. 571-591, 2005.

[14] S. Zafar, "A Glance at Woody Biomass Resources," BioEnergy Consult, 15 October 2015. [Online]. Available: http://www.bioenergyconsult.com/tag/wood-wastes/.

[15] E. Pepke, "Global Wood Markets: Consumption, Production and Trade," 18 May 2010. [Online]. Available: https://www.unece.org/fileadmin/DAM/timber/mis/presentations/PepkeGlobalWoodMkts050510.pdf.

[16] H. Mukunda, S. Dasappa, B. Swati and U. Shrinivasa, "Studies on a Stove for Powdery Biomass," *Int. J. of Energy Res.,* vol. 17, p. 281–291, 1993.

[17] C. Bhaskar Dixit, P. Paul and H. Mukunda, "Part I: Experimental studies on a pulverized fuel stove," *Biomass Bioenergy,* vol. 30, pp. 673-683, 2006.

[18] S. Tse, A. Fernandez-Pello and K. Miyasaka, "Controlling Mechanisms in the Transition from Smoldering to Flaming of Flexible Polyurethane Foam," *Symp. (Int.) Combust., [Proc.],* vol. 26, no. 1, pp. 1505-1513, 1996.

[19] J. Torero and A. Fernandez-Pello, "Forward Smolder of Polyurethane Foam in a Forced Air Flow," *Combust. Flame,* vol. 106, pp. 89-109, 1996.

[20] S. Leach, J. Ellzey and O. Ezekoye, "Convection, Pyrolysis, and Damkohler Effects on Extinction of Reverse Smoldering Combustion," *Symp. (Int.) Combust., [Proc.],* vol. 27, pp. 2873-2880, 1999.

[21] S. Leach, G. Rein, J. Ellzey, O. Ezekoye and J. Torero, "Kinetic and Fuel Property Effects on Forward Smoldering Combustion," *Combust. Flame,* vol. 120, pp. 346-358, 2000.

[22] E. Carvalho, C. G. Veras and J. C. Jr., "Experimental Investigation of Smouldering in Biomass," *Biomass Bioenergy,* vol. 22, no. 4, pp. 283-294, 2002.

[23] C. Lo, *Characteristics of Smoldering Combustion of Sawdust,* University of Texas at Austin, 2013.

[24] C. Di Blasi, "Modeling chemical and physical processes of wood and biomass pyrolysis," *Prog. Energy Combust. Sci.,* vol. 34, pp. 47-90, 2008.

[25] M. Ortega, *Improved combustion in wood stoves,* Norwegian University of Science and Technology, 2008.

[26] K. Ragland, D. Aerts and A. Baker, "Properties of wood for combustion," *Bioresour. Technol.,* vol. 37, no. 2, p. 161–168, 1991.

Western States Section of the Combustion Institute – Fall 2017 Meeting
Hosted by the University of Wyoming
October 2-3, 2017

Impact of Particle Properties on Radiative Heat Flux in an Oxy-Coal Reactor

Ty Hosler[1] *Bradley Adams*[1]

[1]*Dept of Mechanical Engineering, Brigham Young University, Provo UT 84602*

A sensitivity study testing the significance of particle size, scattering phase functions, and distribution in oxy-coal combustion was conducted. Literature review suggests that particle size and number density play a more critical role in determination of radiative properties than the complex index of refraction. Radiative data was produced from numerical experiments using one of two particle distribution profiles (narrow or wide). A comparison of highly forward scattering cases for two particle sizes (25μm vs. 75μm) revealed a significant difference in incident heat flux maximum (≈ 25%). This difference was observed in both profiles examined. It was also generally observed that as particle size increases the difference between incident heat flux for forward and backward scattering decreases. No significant difference was observed between the large diffuse sphere and no scattering cases, suggesting that accurate modelling of larger particle size distributions can be achieved while neglecting scattering and phase functions. This result agrees with current practice in industry for air – coal mixtures. However, data suggests that phase function/scattering coefficient influence becomes more and more important with increasingly fine particle size.

1. Introduction

In many combustion applications, especially industrial furnaces, radiative heat transfer plays a dominant role accounting for upwards of 90% of the total heat transfer in the combustion chamber [1]. This dominance is due to heat transfer rates being proportionate to temperature changes to the fourth degree. Therefore, it is essential that this phenomenon be correctly understood and modelled to best take advantage of the energy available in a furnace setting. Considerable experimentation and research has been conducted in this field by a variety of experts. Goodwin and Mitchner as well as others [2-5] have done extensive research on coal slag and fly ash properties. Additionally, Menguc et al. [1,6,7] have shown that particle number density, temperature, and concentration distributions play a larger role in radiative heat transfer than the complex index of refraction. Menguc et al. also reported measured effective values for extinction and scattering coefficients. This study looks to build upon and supplement such previous work by examining the impact of properties on radiative heat flux through the pursuit of the following objectives:

- What effect does particle distribution have on incident heat flux and radiative flux divergence (narrow vs. wide)?
- How does particle size affect heat flux?
- How do particle scattering phase functions impact radiative heat flux?
- What is the estimated effect on computational time when accounting for particle scattering phase functions?

2. Methods

RADIATION SOLUTION

The oxy-coal reactor was modeled by dividing the reactor into 20 axial zones. Gas temperature, gas composition, and particle composition were prescribed and constant throughout each of the 20 zones. Radiative heat transfer in the reactor was calculated based on the radiative heat transfer equation (**Eq. 1**) for a gray absorbing-emitting-scattering medium.

$$\nabla \cdot \vec{\Omega} I = -\kappa_{ext} I + e_{bg} + e_{bp} + \frac{\kappa_S}{4\pi} \int_{\Omega''=4\pi} \Phi(\Omega'', \Omega) I'' d\Omega'' \qquad \textbf{Eq. 1}$$

The left side of **Eq. 1** represents the divergence of intensity in the direction of propagation, whereas the right side represents, respectively, the attenuation of directional intensity due to gas and particle absorption and particle out-scattering, the augmentation of directional intensity due to gas and particle emission, and the contribution to directional intensity due to particle in-scattering from the surrounding medium. Calculations here assumed gray walls, gases and particles.

Whereas the gas and particle properties were prescribed zone by zone in one-dimension, the radiative heat transfer calculation was performed in three dimensions, although the results were essentially axisymmetric since the medium properties and boundary conditions were axisymmetric. Radiation intensity, surface heat fluxes and net radiative source terms (i.e., the flux divergence) were calculated using the discrete ordinate method [8]. Particle scattering and absorption coefficients were calculated from the following relations,

$$\kappa_S = \frac{\pi}{4}(Q_s n d^2) \qquad \textbf{Eq. 2}$$

$$\kappa_a = \frac{\pi}{4}(Q_a n d^2) \qquad \textbf{Eq. 3}$$

where Q_s and Q_a are particle scattering and absorption efficiencies, respectively, n is the particle number density, and d is the particle diameter. Efficiency calculations were completed using Mie theory based on particle size and complex index of refraction.

SYSTEM DESCRIPTION

The reactor dimensions used (1.829 m high x 0.18 m x 0.18 m) were based on a pilot-scale oxy-coal combustor that is currently being designed and built at Brigham Young University. **Table A** lists theoretical temperatures and ash mass fractions for each axial zone. The reactor coal flow rate was 0.0037 kg/s and the combined oxygen-CO_2 flow rate was 0.03 kg/s. The gas composition was assumed to be constant throughout the reactor as 65% CO_2, 27% H_2O, 3% O_2, 5% N_2. Particle refractive indices used were n=1.85 and κ=0.25 for coal and n=1.5 and κ=0.02 for ash.

Two prescribed particle distribution profiles were used in experiments (narrow and wide). The narrow profile concentrated all particles within a center square of size 0.09 x 0.09 m, with no particles in the outer areas near the wall. Contrastingly, the wide profile used the same incoming mass of particles, but distributed the particles across the full 0.18 x 0.18 m cross-section of the reactor. Particle concentrations (i.e., number densities) were the same at all axial locations; only the particle composition changed with axial position.

Table A. Prescribed Reactor Gas and Particle Properties

Axial Position (m)	Temperature (K)	Ash Mass Fraction (%)
0.0457	2100	.10
0.1372	2400	.20
0.2286	2300	.30
0.3201	2200	.40
0.4115	2100	.50
0.5030	2050	.60
0.5944	2000	.70
0.6859	1950	.75
0.7773	1900	.80
0.8688	1850	.85
0.9602	1800	.90
1.0517	1750	.91
1.1431	1700	.92
1.2346	1650	.93
1.3260	1600	.94
1.4175	1550	.95
1.5089	1500	.96
1.6004	1450	.97
1.6918	1425	.98
1.7833	1400	.99

ASSUMPTIONS

The following assumptions were made for this study:

- Gas temperature, gas composition and particle composition (ash mass fraction) profiles were prescribed over the 20 zones in the reactor. The profiles were the same for all cases studied.
- Particle temperatures were assumed to be the same as gas temperatures.
- Mono-size spherical particle distributions were used (25μm, 75μm).
- Each phase function evaluated was assumed constant throughout the reactor.

Four types of phase functions were evaluated – isotropic, linear anisotropic, Henyey-Greenstein, and large diffuse sphere. It is generally accepted in literature and practice that coal and fly ash particles are mostly forward scattering. When modeling highly forward scattering media, Menguc and Viskanta recommend using a two term delta-Eddington approximation for accuracy and computational efficiency [1,9]. However, without careful consideration of proper constraints the delta-Eddington function can return negative values. This breakdown can be overcome with minimal losses in accuracy by using the Henyey – Greenstein approximation, as was done in this study. For backward scattering, Modest shows that the phase function for large diffuse spheres has a backward-scattering peak [10]. Therefore, its phase function was used to model backward scattering along with Henyey – Greenstein in this study. A linear anisotropic phase function was also implemented to model mild forward scattering. Isotropic scattering was a subset of the anisotropic model. Ultimately, selection of the proper phase function depends on a variety of factors and interactions, most of which are beyond the scope of this study.

3. Results and Discussion

PARTICLE DISTRIBUTION EFFECTS ON INCIDENT HEAT FLUX AND DIVERGENCE

Incident heat flux at the wall of the reactor was observed to be higher for the wide particle distribution in all cases. This was due to the particles local to the reactor wall providing a much closer radiation source than those in the narrow profile. However, the difference in maximum incident heat flux between narrow and wide profiles of the same particle size were seen to decrease as particle diameter increased (see **Figure 1**). This suggests that the impact of particle distribution on incident heat flux diminished with increasing particle size. Additionally, all of the heat flux values, regardless of their particle distribution, dropped off rapidly after the flame region of the reactor. This rapid drop in value was strongly linked to the prescribed temperature profile.

In addition to higher incident heat flux values, the wide particle distribution likewise contributed to a marked difference in flux divergence at the wall of the reactor (see **Figure 2**). Flux divergence was much more sensitive to temperature effects for the wide distribution although all cases were temperature dependent. This is consistent with the higher particle absorption and scattering coefficients near the wall for the wide distribution cases. As with heat flux, the flux divergence distribution was more sensitive to the smaller 25μm particle size, reaching a peak approximately six times greater than the 75μm particles. This again suggests that the influence of particle distribution diminished with increased particle size. **Figure 2** also reveals that particle size does play a significant role in wide distributions, but narrow distributions show only minor differences for the 25μm and 75μm profiles. This is due to particle emissive power being much higher than gas emissive power (particle absorption and scattering coefficients are 20-50 times larger than gas absorption coefficients). Once the larger emissive power is generated in the center of the combustor (whether by 25μm or 75μm particles) and propagated out towards the walls, the lack of particles near the walls for the narrow distribution means there is little impact on intensity absorption or emission in this region.

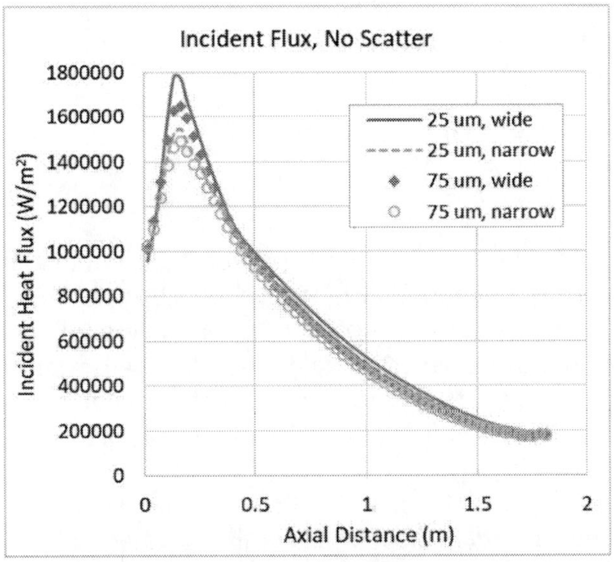

Figure 1: A comparison of narrow and wide particle distribution effects on heat flux between 25μm and 75μm particles with no scattering.

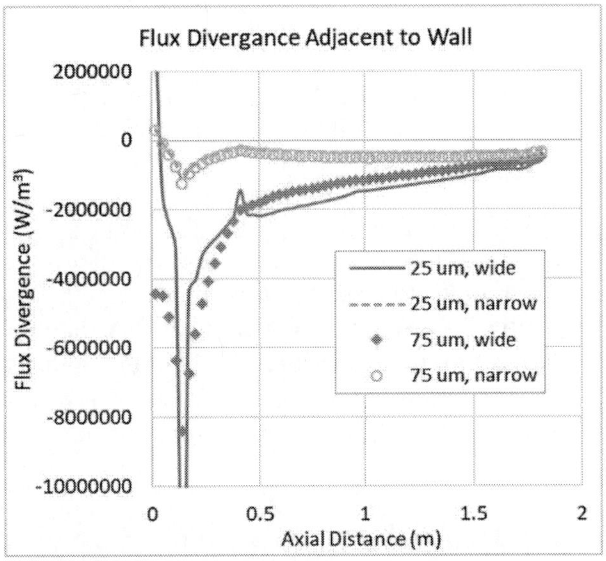

Figure 2: A comparison of narrow and wide particle distribution effects on heat flux between 25μm and 75μm particles with no scattering.

Figure 3 compares the radiative flux divergence at the combustor centerline and near wall as a function of axial distance for 75μm particle in narrow and wide distributions. Results showed that the flux divergence was sensitive to particle distribution. The differences between centerline and near wall flux divergence for the narrow distributions were significant. In the center where particle number densities and particle absorption and scattering coefficients are large, flux divergence was very sensitive to temperature changes. Near the wall where there are no particles, the flux divergence was much less sensitive to temperature changes. For the wide particle distributions, the flux divergence values were much closer. This was because the temperatures and particle properties were the same at the centerline and near wall for these distributions.

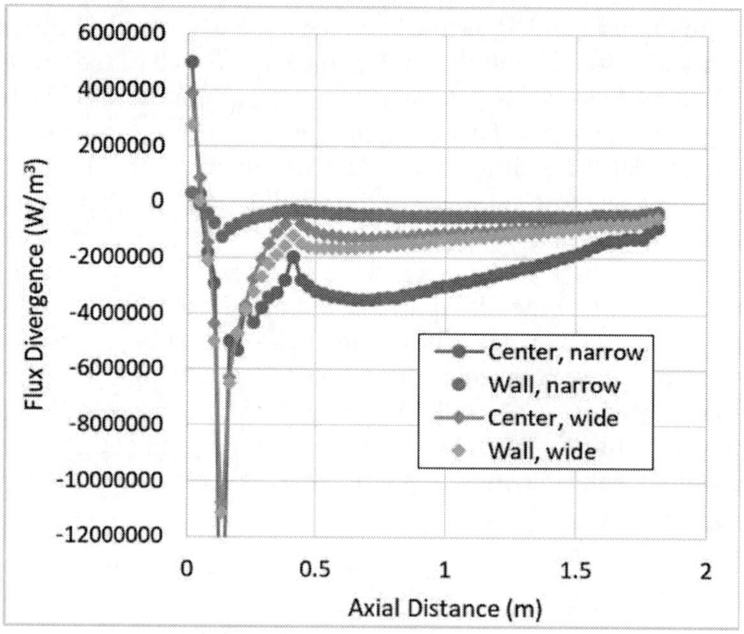

Figure 3: Flux divergence comparison of 75μm particles at combustor centerline and near wall for narrow and wide distributions. All cases had no scattering.

PARTICLE SIZE EFFECT ON HEAT FLUX:

It has already been observed that particle size plays a role in wide particle distributions, but its singular role in affecting incident heat flux was also examined. **Figure 4** shows that the 25μm profile has a higher incident heat flux than the 75μm profiles. This pattern holds true regardless of phase function or particle distribution, suggesting that finer particle size distributions will result in higher heat flux values. This result can be attributed to the larger number of smaller diameter particles (assuming the same mass of fuel), which increases number density in turn raising the value of the absorption and scattering coefficients as shown in **Eq. 2** and **3**. The effect of increasing the number density outweighs the effect of decreasing the particle diameter. From a radiative heat transfer perspective, finer particle size distributions or rather distributions with a smaller mean diameter are desirable. However, particle size also affects the combustion characteristics of a flame and, in practice, obtaining finer particle size distributions requires additional milling and increases operational costs.

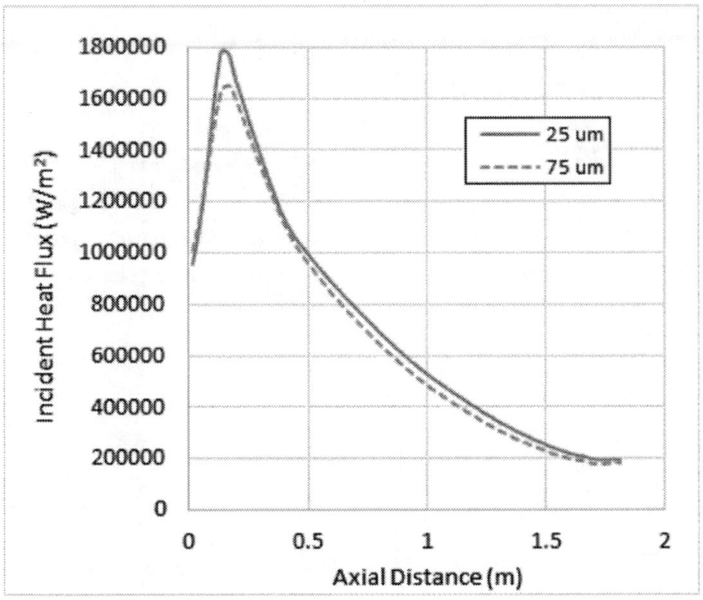

Figure 4: Incident heat flux comparison of different size particles in a strongly forward scattering case implementing the Henyey-Greenstein phase function (g = 0.9) with wide particle distribution.

PARTICLE SCATTERING PHASE FUNCTION IMPACTS:

Often in practice, scattering phase function contributions in **Eq. 1** are neglected in favor of simplicity and time. The objective of this study was to examine the impacts of phase functions specifically for the oxy-coal reactor outlined previously. The phase functions evaluated were the isotropic, linearly anisotropic with an asymmetry factor of 0.9, Henyey-Greenstein with an asymmetry factor of g=0.9, and large diffuse sphere. **Figure 5** plots different phase functions as a function of scattering angle and asymmetry factors. Forward scattering phase functions have higher magnitude near zero degrees, whereas backward scattering functions have higher magnitude near 180 degrees. All phase functions other than the Henyey-Greenstein function have relatively small differences. The Henyey-Greenstein functions have a very strongly forward scattering behavior as the asymmetry factor, g, approaches 0.9.

Heat flux results based on the four phase functions used in this study as well as a no-scattering case are plotted in **Figure 6**. The full range of values is shown on the left and a zoomed range of values at the flux peaks is shown on the right. All the profiles were quite similar to each other. The maximum difference in fluxes at the peak was approximately seven percent. The greatest difference was between the no scattering and isotropic scattering. Interestingly, the large diffuse sphere (moderate backscattering) and anisotropic (moderately forward scattering) phase functions produced flux results within one percent. The Henyey-Greenstein results (strongly forward scattering) fell directly in line with the no scattering results. This was expected as the no scattering condition was the same as assuming all scattering occurred in the forward direction.

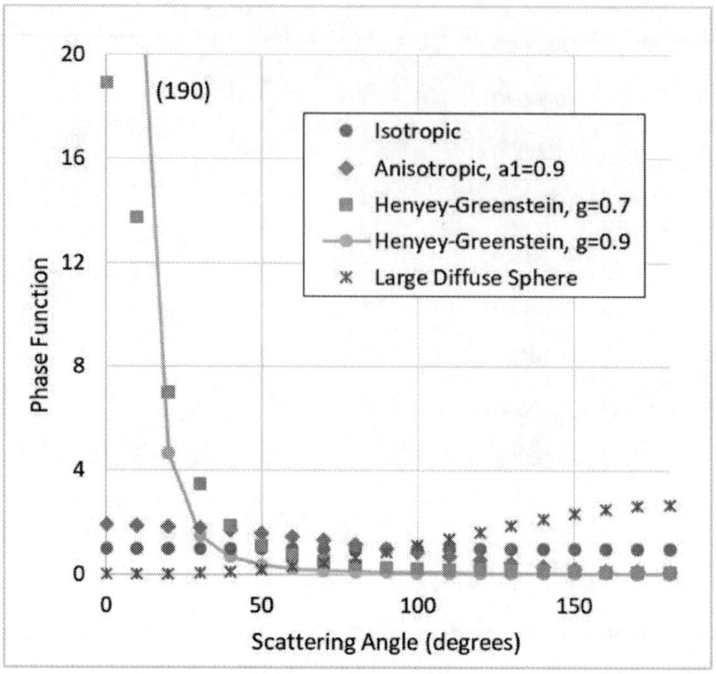

Figure 5: Phase function magnitude as a function of scattering angle for different phase functions and asymmetry factors.

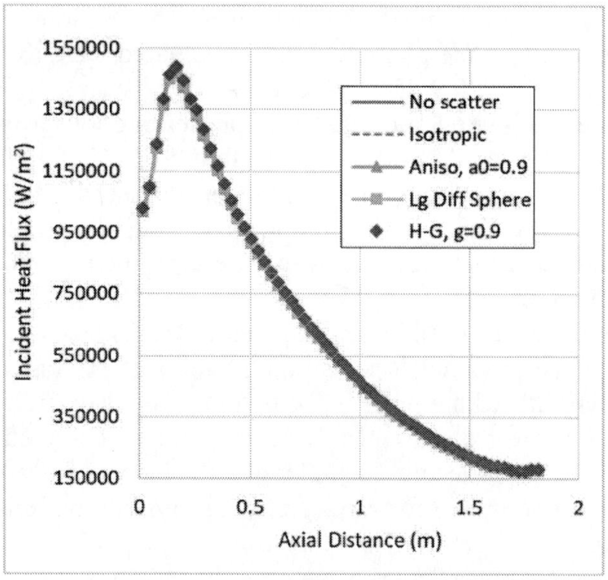

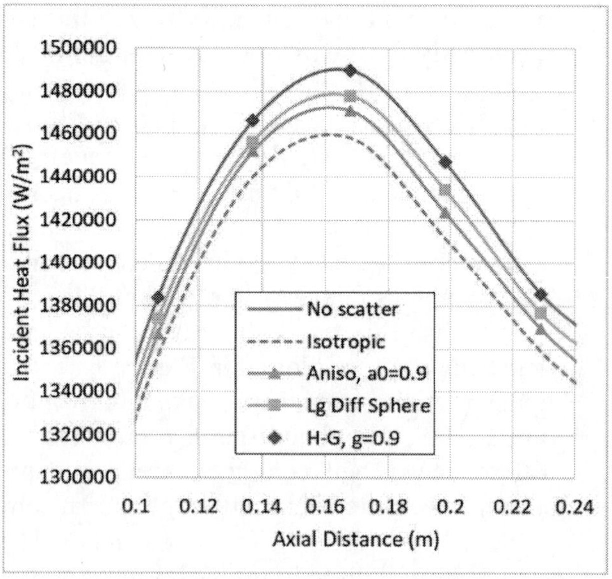

Figure 6: A phase function comparison of incident heat flux for no scattering, isotropic scattering, moderately forward scattering (linear anisotropic), backward scattering (large diffuse sphere) and strongly forward scattering (Henyey-Greenstein) for the 75μm particles with narrow particle distribution. Full range (left), zoomed range at flux peak (right).

The similarities in predicted fluxes for different phase functions was also observed for the wide distribution case, with two exceptions. First, the flux value magnitudes were consistent with the 75μm wide case shown in **Figure 1** and second, the relative magnitude of the fluxes for isotropic scattering and strongly forward scattering were reversed (see **Figure 7**), suggesting a minor impact from particles scattering differently near the combustor walls. Overall, the results showed that differences between phase functions were negligible for the geometry and narrow distribution studied. Even differences between backward and forward scattering were negligible. The results for the 25μm particle narrow and wide distributions showed similar behavior, i.e., negligible differences in predicted incident heat fluxes for different scattering phase functions.

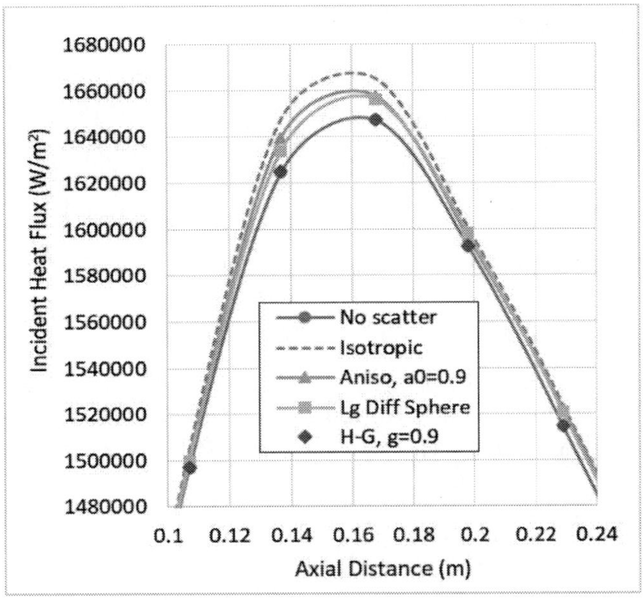

Figure 7: A phase function comparison of heat flux for no scattering, isotropic scattering, moderately forward scattering (linear anisotropic), backward scattering (large diffuse sphere) and strongly forward scattering (Henyey-Greenstein) for the 75μm particles with wide particle distribution. Zoomed range at flux peak.

Figure 8 compares the radiative flux divergence at the combustor centerline for different phase functions with the 75μm particle and wide distribution. The profiles were similar except for the region from the temperature and incident flux peak (maximum radiative emission) at approximately 0.17m to approximately 0.4m. Similar to the incident heat flux, the strong forward scattering and isotropic scattering had the largest differences, and the moderately forward scattering and backward scattering functions were similar. Beyond 0.4m, all profiles were very similar.

Similar plots are shown in **Figure 9** for the narrow distribution cases. For these cases, the higher particle absorption and scattering coefficients near the combustor centerline resulted in greater differences between the flux divergence profiles. The no scattering and Henyey-Greenstein functions were still aligned, and were most different from isotropic scattering. The plot also appears to show oscillations near the peak emission point where the flux divergence was greatest. These oscillations are a function of the high absorption and scattering coefficients in the center of the combustor and are somewhat unexpected given that the temperature profile in the reactor is fixed.

This type of oscillation could present numerical challenges when the flux divergence terms appear in the energy equation in a full combustion simulation where temperature and radiation are coupled.

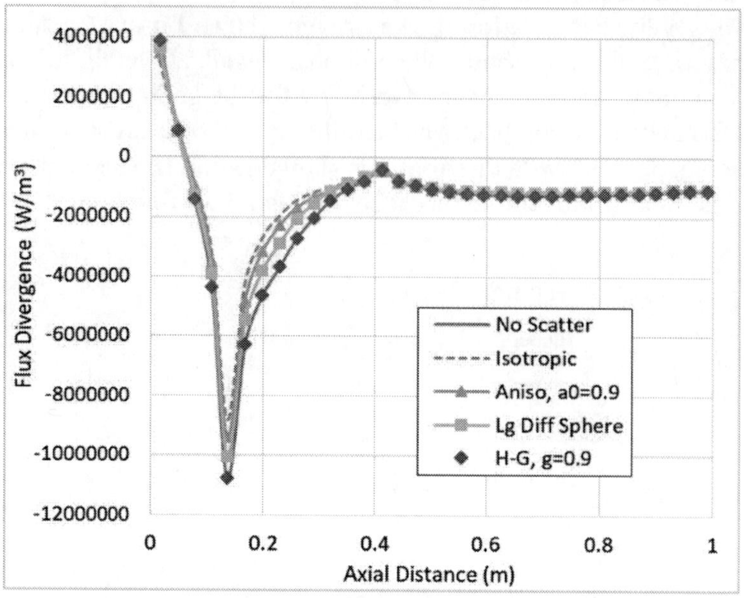

Figure 8: A phase function comparison of flux divergence with no scattering, isotropic scattering, moderately forward scattering (linear anisotropic), backward scattering (large diffuse sphere) and strongly forward scattering (Henyey-Greenstein) for the 75μm particles with wide particle distribution.

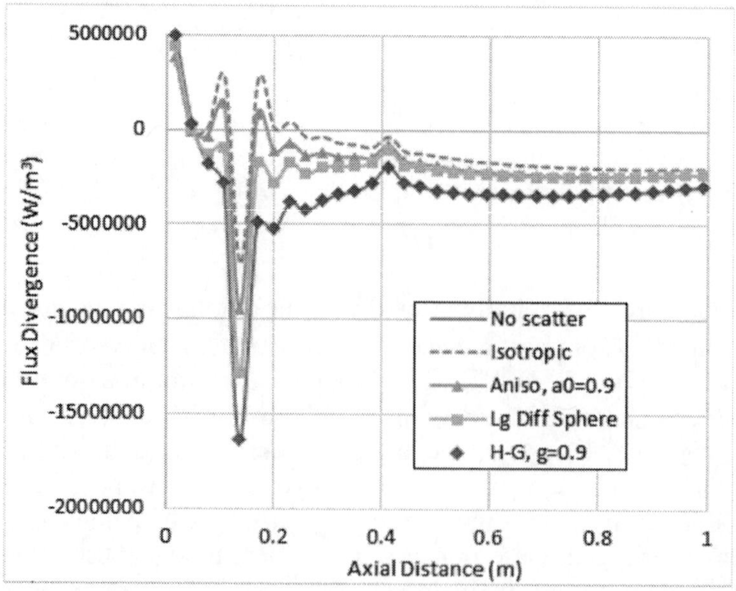

Figure 9: A phase function comparison of flux divergence with no scattering, isotropic scattering, moderately forward scattering (linear anisotropic), backward scattering (large diffuse sphere) and strongly forward scattering (Henyey-Greenstein) for the 75μm particles with narrow particle distribution.

The key outcome from the incident flux and flux divergence results is that accurate heat flux predictions can be obtained without including scattering phase functions, that is, results predicted with scattering provide no noticeable improvement over results without scattering. However, for regions where particles are highly concentrated, that is, where particle number densities are large and particle absorption and scattering coefficients are large, the flux divergence does differ for different phase functions in regions of high emissive power. This could affect numerical convergence when radiation calculations are coupled with the energy equation.

COMPUTATIONAL TIME EFFECT WHEN ACCOUNTING FOR PHASE FUNCTIONS:

Finding the right balance between accuracy and computational time is always a factor with computational methods. It is for this very reason why many choose to neglect addressing the contribution of a particle scattering phase function in the first place. For this study, a general estimate was acquired by recording the time the program took to compute surface heat fluxes and net flux source terms with and without scattering.

Table B. Approximate Run Time Comparisons

Size (um)	Distribution	Phase Function	Runtime (s)
75	wide	No scatter	2
75	wide	Isotropic	6
75	wide	Anisotropic, a0=0.9	7
75	wide	Large Diffuse Sphere	5
75	wide	H-G, g=0.9	4
75	narrow	No scatter	2
75	narrow	Isotropic	8
75	narrow	Anisotropic, a0=0.9	8
75	narrow	Large Diffuse Sphere	5
75	narrow	H-G, g=0.9	5
25	wide	No scatter	2
25	wide	Isotropic	9
25	wide	Anisotropic, a0=0.9	8
25	wide	Large Diffuse Sphere	6
25	wide	H-G, g=0.9	6
25	narrow	No scatter	2
25	narrow	Isotropic	11
25	narrow	Anisotropic, a0=0.9	11
25	narrow	Large Diffuse Sphere	9
25	narrow	H-G, g=0.9	9

When accounting for scattering, the program run time was approximately 3-5 times longer than without. The isotropic and linearly anisotropic cases had the longest run times. The large diffuse sphere and Henyey-Greenstein run times were comparable. The longest run times (and most difficult convergence) was with the 25μm particles with narrow distribution. This was consistent with observations that particle scattering had the greatest impacts for this condition. For this study, that time made little difference, but if that relationship held for a large system it could mean hours or days of difference in run time. Based off this result, the decision to include scattering phase function calculations in a program should be based on required model fidelity and resources/computational time available.

4. Conclusions

Study results for the small oxy-coal reactor considered showed the following:

1. Wider particle distributions (of the same mass and particle count) contribute to higher incident heat fluxes and larger magnitude flux divergence than narrow distribution. Particle size is also a factor in distributions, decreasing their effect as particle diameter increases.
2. Smaller particle sizes contribute to higher heat flux values (compared to larger sizes with the same mass). This is mostly due to increased number density values, resulting in larger scattering coefficients in the Radiative Transport Equation.
3. The effects of phase functions used for this geometry and particle properties are minimal and scattering can be neglected for incident heat flux predictions. However, for regions where particles are highly concentrated, that is, where particle number densities are large and particle absorption and scattering coefficients are large, the flux divergence does differ with phase function in regions of high emissive power. This could affect numerical convergence when radiation calculations are coupled with the energy equation.
4. Accounting for the impacts of particle scattering with phase functions in the Radiative Transport Equation increased run times approximately 3-5 times more than neglecting them. Smaller particles (25μm) with higher concentration (narrow distribution) had the longest run times.

Additional research is recommended to assess the conclusions in a commercial-scale system, evaluate poly-dispersed size distributions, and couple radiation results with energy balance to update gas and particle temperatures for different radiative conditions.

Acknowledgements

This material is based upon work supported by the Department of Energy under Award Number DE-FE0029157. Arun Bose Program Manager.

References

[1] Manickavasagam, S., and Menguc, M. P., "Effective Optical Properties of Pulverized Coal Particles Determined from FT-IR Spectrometer Experiments," Energy & Fuels, **7**(6) (1993) 860-869.

[2] Goodwin, D. G., Mitchner, M., "Infrared optical Constants of Coal Slags: Dependence on Chemical Composition," Journal of Thermophysics, 3(1) 1989 53-60

[3] Goodwin, D. G., and Mitchner, M., "Flyash Radiative Properties and Effects on Radiative Heat Transfer in Coal-Fired Systems," International Journal of Heat and Mass Transfer, **32**(4) (1989) 627-638.

[4] Gupta, R. P., and Wall, T. F., "The Optical Properties of Fly Ash in Coal Fired Furnaces," Combustion and Flame, **61**(2) (1985) 145-151.

[5] Aleksandar Saljnikov, Biljana Vucicevic, Mirko Komatina, Milan Gojak, Darko Goricanec, and Zoran Stevanoic, "Spectroscopic research on infrared emittance of coal ash deposits," Experimental Thermal and Fluid Science, 33 (2009) 1133-1141.

[6] M. P. Menguc, Viskanta, R., "A sensitivity analysis for radiative heat transfer in pulverized coal-fired furnaces," HTD, 45 (1985) 221-229.

[7] M. P. Menguc, S. Manickavasagam, and D. A. D'Sa, "Determination of radiative properties of pulverized coal particles from experiments," 73(4) (1994)

[8] Adams, B.R. and Smith, P.J., "Three-dimensional Discrete-ordinates Modeling of Radiative Transfer in a Geometrically Complex Furnace," *Combust. Sci. Technol.,* 88 (1993) 293-308.

[9] J.H. Joseph, W.J. Wiscombe, J.A. Weinman "The Delta Eddington Approximation for Radiative Flux Transfer (1976).

[10] Modest, Michael F., "Radiative Heat Transfer," McGraw-Hill series in mechanical engineering, 1st edition. (1993) 402-409.

Western States Section of the Combustion Institute – Fall 2017 Meeting
Hosted by the University of Wyoming
October 2-3, 2017

CFD Simulations of Hydropyrolysis

Gideon D. Baldwin, Clint W. Dunn, Michael Stoellinger

Mechanical Engineering, College of Engineering and Applied Science,
1000 E. University Ave. Laramie, WY 82071

1 Abstract

The purpose of this project is to develop a supporting computational fluid dynamics (CFD) model for a flash hydropyrolysis process using Powder River Basin (PRB) coal. A flash hydropyrolysis processes reacts coal volatiles and tars at elevated temperature and hydrogen pressure. Long residence times yield gaseous products, however, shortening residence times via quenching yields more marketable oils in the BTX range. Literature and previous simulations have predicted an approximate 11 percent coal carbon conversion into this BTX range.

To simulate the hydropyrolysis process, CFD software OpenFoam will be coupled with PC Coal Lab devolatilization software. Complex reaction kinetics can predict reactor yields but are computationally impractical to fully implement. Mechanism reduction will have to be implemented. The goal is to find a compromise between simulation accuracy and feasibility such that an experimental reactor's dimensions and operating conditions can be determined.

2 Introduction

In recent years, increasing environmental pressure, sources of renewable energy, and low natural gas prices have reduced the demand for the direct combustion of coal. This in turn has been detrimental to the Wyoming economy [7]. While there are several clean coal technologies under development, coal hydropyrolysis process is a unique method to open up a new market sector for Powder River Basin coal. The process can be thought of as a "coal refinery". The goal is to input coal and a syngas mixture of H_2 and CO to produce light oils rich in the benzene, toluene, and xylene (BTX). In flash hydropyrolysis, coal is exposed to hot hydrogen gas in an oxygen free environment for a short residence time. An increased hydrogen partial pressure is favorable for light oil production. The hydrogen shifts the competing reactions of polymerization and hydrocracking of coal tars to produce more light compounds [2].

The concept of hydropyrolysis, and flash hydropyrolysis, is not a new technology. A large amount of research can be found from the mid 1970s to the 2000s. The end goal of this research is to implement and modify existing models into a CFD simulation to optimize the process of hydropyrolysis. This final goal requires the understanding of several different topics. As a starting point, it was decided to investigate the chemical kinetics and devolatilization. A CFD geometry is not needed for a basic understanding, thus a zero dimensional reactor was implemented.

3 Methods

3.1 Devolatilization

The first component in modeling flash hydropyrolysis is the devolatilization software. For this simulation, coal devolatilization was modeled via C3M software produced by the Multiphase Flow Science group and PC Coal Lab produced by Niksa Energy Associates. C3M utilized the CPD (Chemical Percolation Devolatilization) method to estimate the devolatilization and mass loss of the coal and will henceforth be referred to as C3M-CPD. PC Coal Lab (PCCL) simulates coal devolatilization via FLASHCHAIN. The licensing in this simulation included the hydropyrolysis and tar conversion into gases and polycyclic aromatic hydrocarbons modules. Input conditions for both devolatilization softwares can be found in Table 1.

Table 1: Experimental Conditions For Simulated Experiments

	Zibin Zhu et al.	Tatterson et. al	
Initial Temperature	100	25	°C
Final Temperatures	500, 600, 650 700, 750, 800	621, 816, 843 871, 932	°C
Syngas Fraction	1	1	$\frac{kg_{H2}}{kg_{H2}+kg_{CO2}}$
Final Pressure	60	4.1	Bar
Mixture Fraction, Z	0.635, 0.451, 0.361 0.301, 0.262, 0.237	$\geq 90\%$	$\frac{kg_{H2}}{kg_{H2}+kg_{Volatiles}}$
Coal Residence Time	1 to 2	2.0e-03 to 3.0e-03	sec
Volatile Residence Time	42	2.0e-03 to 3.0e-03	sec
Heating Rate (assumed)	10,000	600,000	$\frac{K}{sec}$
Coal Feed Rate	0.10	324 to 1200	$\frac{kg}{h}$
Reactor length	2	0.46	m

C3M-CPD assumes that coal consists of only two portions, a fixed carbon portion that is pure carbon and a volatile matter portion. The coal used in this experiment was not ash free. To account for the ash, the fixed carbon and volatile matter fractions were normalized with the proximate analysis percentage of ash. It was also assumed that no carbon was contained in the ash fraction.

C3M-CPD only predicts the gaseous products CO, CO_2, H_2O, CH_4 and tar yields. PCCL predicts CO, CO_2, H_2O, CH_4, alkanes and alkenes up to propane, H_2, H_2S, HCN, and tar yields. The tar chemical composition was modeled from the results presented by the European Commission [2]. Tar composition mass fractions are found in Table 2. While the experimental conditions used to generate this hydropyrolysis tar are unknown, the result gives a starting point for an approximate model. The tar model also compromised between a more realistic tar chemistry and consistent elemental composition. The addition of compounds led to the creation or reduction of elements in the coal. For example, the Zibin Zhu trial at 800°C had a carbon, hydrogen and oxygen content that was increased by 15.6 percent, increased by 1.5 percent, and decreased by 32.8 percent respectively from the ultimate analysis of Dongshen coal.

Table 2: EU Feasibility Study Tar Model Species in Percent Mass Fraction

Benzene	Toluene	Xylene	Phenol	Cresol	Indene	Naphthalene	Biphenyl	Fluorene	C14H10	Pyrene
3.56	5.52	4.12	7.07	8.23	5.82	1.57	10.50	5.60	8.90	39.10

To investigate the accuracy of C3M-CPD and PCCL devolatilization, an experimental procedure with flash pyrolysis at ultrashort residence times was needed. Tatterson et al. conducted such an experiment and the conditions can be observed in Table 1 [3]. In these conditions, chemical kinetics did not have the residence time needed to convert the coal devolatiles into different products. This was supported by a zero dimensional reaction using the Creck Complete mechanism. In the highest temperature experiment analyzed by this report (932°C), the maximum change in a species throughout the reaction was 43ppm with an average species conversion of 4.6ppm per species. Tatterson et al. also reported heating rates sufficient to heat the coal to the final temperature in 1ms. This was confirmed by both empirical calculations and photographic analysis. Heating rates of up to 820,000 $\frac{K}{s}$ would be needed to accomplish this. PCCL encountered solution convergence errors above heating rates of 700,000 $\frac{K}{s}$. Thus, the heating rate was limited to 600,000$\frac{K}{s}$ to ensure convergence. The properties of Illinois No. 6 coal used by Tatterson et at. can be found in Table 3.

3.2 Chemical Kinetics

To compare and verify the simulated chemical kinematic model, the article *Flash hydropyrolysis of northern Chinese coal* [9] was selected. This research was selected on the basis that Zibin Zhu et al. provided the convergence in BTX and PCX (phenol, cresol, and xylenol) to one species. Reproduction of detailed product yields in BTX and PCX would give validity to a mechanism if the same trends could be modeled. Furthermore, the gas long residence time of 42 seconds allowed for more detail to be observed in the chemical kinetics instead of the devolatilization of the coal.

However, the article gave an ambiguous heating rate and did not define all of the reactor's geometry or the feed rate of the hydrogen gas. The feed rate of the hydrogen gas could be calculated from residence time and given geometry. Missing geometry details were insignificant as long as a fixed heating rate was assumed. Gaseous yields and fixed carbon fractions were not significantly affected by a change in heating rate. For a temperature of 800°C, specific product and fixed carbon yields only changed by a maximum of 1.6 and 1.1 percent respectively over a heating rate range of 10,000 to 790,000 $\frac{K}{s}$ in C3M-CPD simulations. Specific experimental conditions can be seen in Table 1 and the Northern Chinese Dongshen Bituminous coal properties are in Table 3. Six different temperature dependent trials were run so six different mixture fractions had to be calculated.

Table 3: Coal Analysis of Dongshen Bituminous and Illinois No. 6

Dongshen Bituminous Coal				Illinois No. 6 Coal			
Ultimate Analysis		Proximate Analysis		Ultimate Analysis		Proximate Analysis	
C	66.6	Ash	8.8	C	78.09	Ash	10.85
H	4.5	VM	33.6	H	5.29	VM	37.75
N	1.6	FC	57.6	N	11.26	FC	51.40
O (Diff.)	18.1			O (Diff.)	1.49		
S	0.8			S	3.85		
Ash	8.8			Ash	10.85		

3.3 Reactor and Software

The simulation model was created in Python3.5.3 with the combustion package Cantera2.3.0. An overall process of the experimental model can be seen in Figure 2. Experimental conditions were input into C3M-CPD which predicted the mass fractions of the fixed carbon (assumed to be pure carbon) and the volatile mass fraction. These two fractions were then normalized with the ash content of coal. Due to the fact that Cantera reports gas mixtures as mass fractions, an arbitrary mass of 10 grams was chosen as simulated coal mass. This 10 grams was then multiplied by the normalized fractions of the volatiles, fixed carbon, and ash content. The tar mass was further divided up via the hydropyrolysis tar given in the EU feasibility study. A syngas mixture fraction of H_2 and CO was known from experimental conditions. In the case of Zibin Zhu et al, the mass of the syngas added was calculated from the residence time, reactor geometry, temperature, and pressure of the reactor. With Tatterson et al. a feed loading ratio was given which was assumed to be the ratio of hydrogen mass to coal mass. Syngas was then added to the reactive mixture and reacted in the zero dimensional reactor. In this procedure, the coal was modeled as being instantly devolatilized. This assumption is valid for these simulated experiments. The mass loss time scale of Zibin Zhu et al. is two orders of magnitude smaller than the residence time (Figure 1). Tatterson et al. verified that their coal was fully devolatilized via empirical calculations and photographic analysis. The nonreative mixture bypassed the reactor, but, the mass of the carbon inside of this mixture was needed to calculate carbon conversion results.

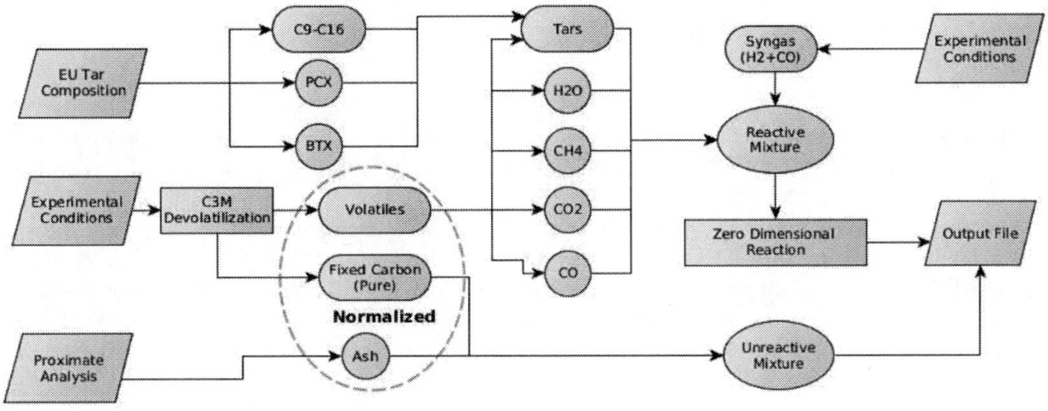

Figure 1: Reactor Simulation Flow Chart

The reactor was modeled with Cantera's zero dimensional constant pressure reactor. The reactor is a good representation of hydropyrolysis because the real process will be well mixed with hot syngas jets and maintained at a constant pressure. Three conservation equations are implemented in this reactor and are as follows:

Conservation of Mass:

$$\frac{dm}{dt} = \sum_{in} \dot{m}_{in} - \sum_{out} \dot{m}_{out} + \dot{m}_{wall} \tag{1}$$

Conservation of Species:

$$\frac{d(mY_k)}{dt} = \sum_{in} \dot{m}_{in} Y_{k,in} - \sum_{out} \dot{m}_{out} Y_k + \dot{m}_{k,gen} \tag{2}$$

Conservation of Energy:

$$mc_p \frac{dT}{dt} = -\dot{Q} - \sum_{k} h_k \dot{m}_{k,gen} + \sum_{in} \dot{m}_{in}(h_{in} - \sum_{k} h_k Y_{k,in}) \tag{3}$$

3.4 Mechanism

As stated in the abstract, a goal of this study was to compare mechanisms to find a viable reduced mechanism for CFD research. All of the mechanisms chosen for this study came from the Creck research group and are as follows: Creck Complete with NOx, Creck Complete, Creck Primary Reference Fuels [PRA] with Alcohols [ALC] and Ethers [ETH], and Creck Primary Fuels and Polycyclic Aromatic Hydrocarbons [PAH] [4]. These mechanism were chosen due to their availability and claimed applicability to the simulation. The complete mechanism is described by Creck as a "detailed and lumped mechanism of pyrolysis, partial oxidation and combustion of hydrocarbon[s]". All of mechanisms are the low temperature variation. They also had to have the species CSOLID and the one reaction pertaining to it deleted due to its nonlinear nature. The Complete mechanism with NO_x formation was established as a bench mark. The NOx mechanism has been verified for cool flame creation down to 700 Kelvin [8]. A comparison of each mechanism and the number of reactions and species can be seen in Table 6.

Table 4: Creck Mechanism Reaction and Species Comparison

	Complete NOx	Complete	PRA with ALC and ETH	PRA and PAH
Reactions	19341	17848	7645	11790
Species	484	451	317	300

4 Results and Discussion

4.1 Investigation of Devolatilization Yields

To investigate the accuracy of C3M-CPD and PCCL devolatilization, yields were compared to Tatterson et al. It became apparent that PCCL predicted much slower weight loss rates from the coal than C3M-CPD and experimental results(Figure 2). In the short residence time of 2.5ms, PCCL never predicted a weight loss greater than 10 percent. Such low predictions made it impractical to directly compare PCCL devolatilizations yields and C3M-CPD and experimental data in same, short residence times. However, it was observed that PCCL weight loss was comparable to C3M-CPD at short residence times if the coal was assumed fully devolatilized. Thus, a long residence time of 15 seconds was implemented in PCCL simulations when comparing to the short residence time of 2.5 milliseconds of Tatterson et. al experimental data and C3M-CPD simulations.

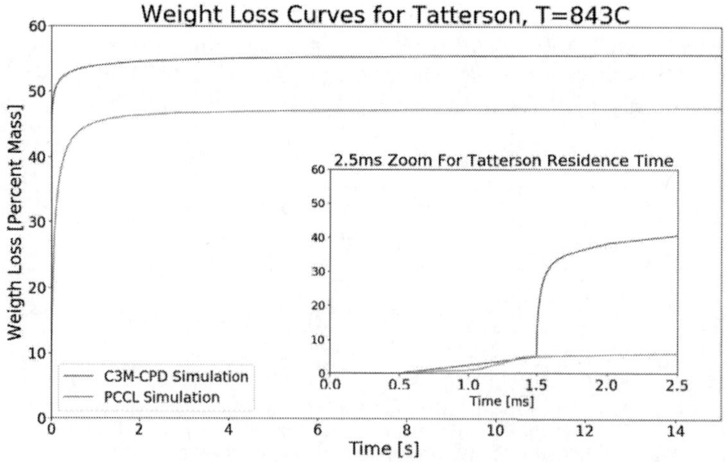

Figure 2: C3M-CPD and PCCL Weight Loss Curves

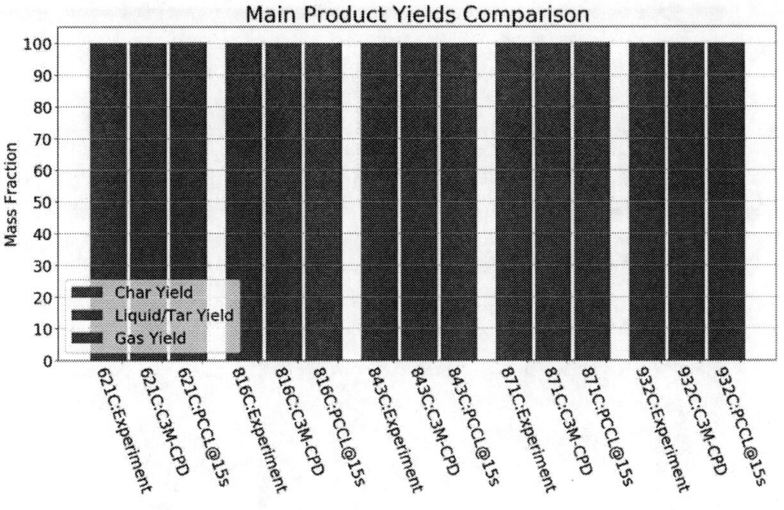

Figure 3: Tatterson, C3M-CPD, and PCCL Main Product Devolatilization Comparison

Main product volatiles were compared with the following assumptions. First, the liquid yields reported by Tatterson et al. were assume to be the tar yields of both C3M-CPD and PCCL. Second, chemical kinetics did not have the residence time to convert the coal volatilizes into different species. Low chemical reactions was supported by a zero dimensional reaction using the Creck complete mechanism. With those assumptions, char, gaseous and liquid yields were compared for five experimental trials (Figure 3). Both C3M-CPD and completely devolatilized PCCL simulations are in good agreement with experimental data. The agreement is improved as temperatures approach those of a hydropyrolysis plant (800°C-950°C).

As discussed in Section 3.2, the tar yield will be approximated by the EU tar approximation. Gaseous yields are given by the simulations themselves and were compared in Figure 4. To begin, Tatterson et al. reported no water. C3M, however, alway predicted a water mass fraction of approximately 50 percent of the total gaseous yield. PCCL predicts a small consumption of water, which better matches experimental data. Both simulations under predict CO. Methane, a main product of hydropyrolysis, is better predicted by PCCL. Furthermore, PCCL has the ability to alkanes and alkenes up to propane. PCCL generally over predicts these larger gases. In the Creck simulation to estimate chemical reactions, gaseous products were not included. It is hypothesized that some of these larger gaseous products would be reduced to methane.

By observing Figures 3 and 4, C3M-CPD is superior to PCCL at predicting the total yields of gaseous, liquid (tar), and solid char in the given time scales. However, the under prediction of methane and the vast over prediction of water makes the C3M-CPD method unable to simulate gaseous yields. To obtain accurate simulations the following procedure is presented. First, C3M-CPD simulations will be utilized to predict weight loss rates. Then, it will be verified main product yields from PCCL at a long residence time are similar to C3M-CPD results. Lastly, PCCL gaseous yields and EU tar predictions will be used to obtain species yields.

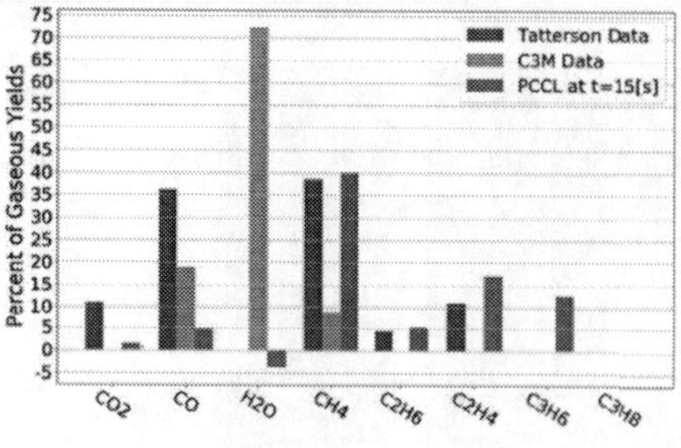

(a) 621°C Reactor Temperature

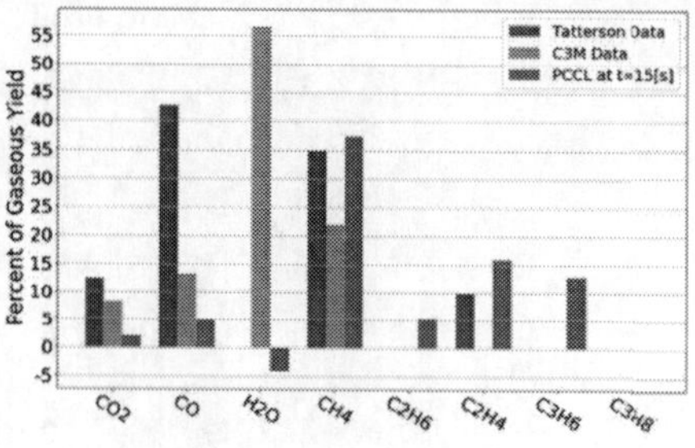

(b) 816°C Reactor Temperature

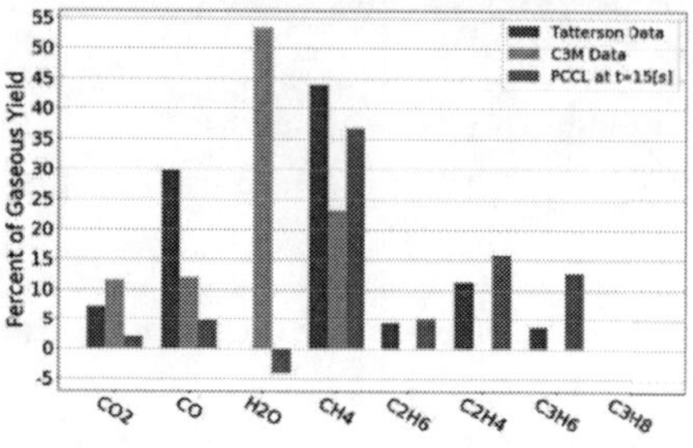

(c) 843°C Reactor Temperature

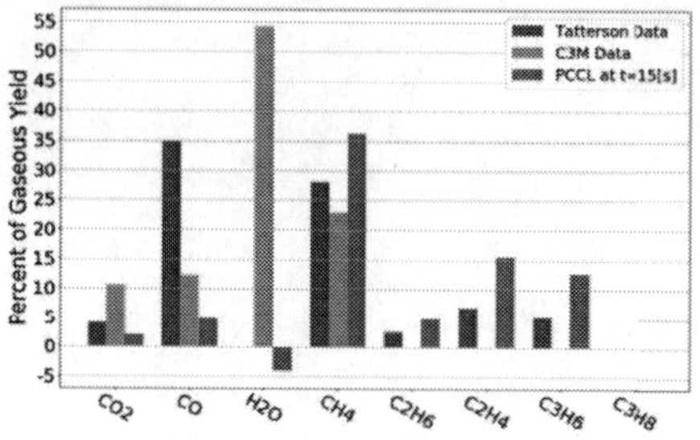

(d) 871°C Reactor Temperature

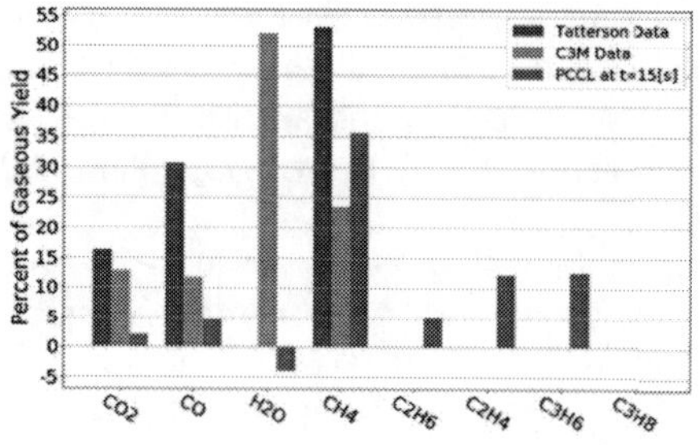

(e) 932°C Reactor Temperature

Figure 4: Tatterson et. al Experimental Resuluts in Comparison to C3M-CPD and PCCL

4.2 Investigation of Chemical Kinetics

The simulation was then directly compared against the experimental yields presented by Zibin Zhu et. al in Figure 5. The following conclusions can be drawn. Due to time constraints, only C3M-CPD predictions were used for the following simulation.

1: The mechanism is able to simulate trends observed in liquid yields. As temperature is increased, more of the tar in the coal is converted into BTX. Also as temperature is increased, more methyl groups are removed from toluene and xylene making the BTX yield primarily benzene. PCX yields also follow trends seen in the experimentation. The yields of PCX are significantly lower because the mechanism used contains no xylenol. Xylenol is the main component seen in PCX as observed by Zibin Zhu et. al.
2: The mechanism struggles to accurately simulate gaseous yields. Methane, which should be a main carbon conversion yields, is not present in significant quantities in the simulated yields. These simulations, however, only used the results of C3M-CPD. Of the several factors that contribute to formation of methane, errors in yields can be narrowed down to devolatilization, gasification, and mechanism kinetics.

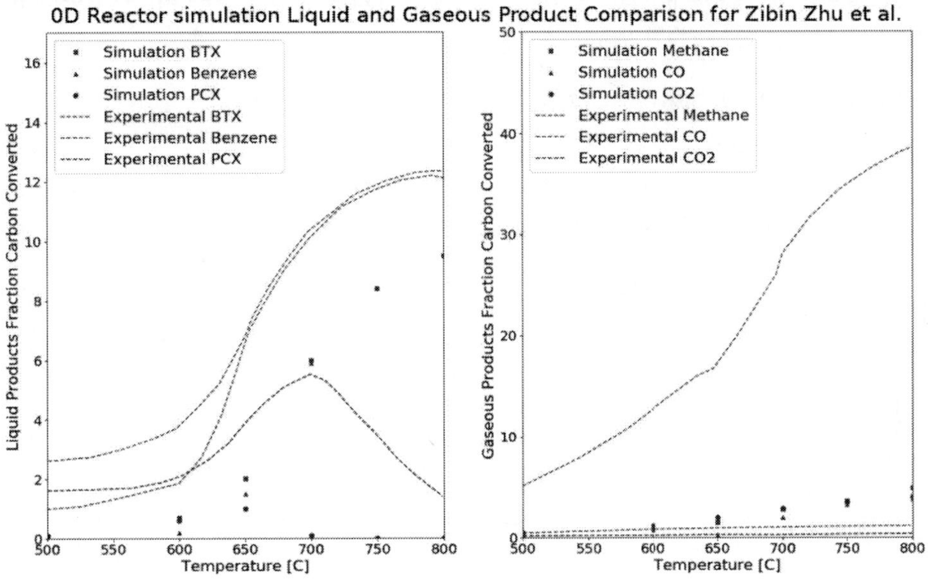

Figure 5: Simulation Vs. Experimental Data for Liquid and Gaseous Yields

Already discussed was that C3M-CPD contains errors in gaseous product predictions. From these errors, average percent errors were calculated for CO, CO_2 and CH_4. Correction factors were then applied to the Zibin Zhu et al. C3M-CPD devolatilization products. After these corrections the methane production was increased from 3.2 to 5.5 percent carbon conversion for a temperature of 800 Celsius. BTX yields percent carbon conversion also decreased to 4.8 percent. Utilization of the PCCL simulations would help the discrepency in gaseous products.

4.3 Investigation into Char Gasification

From literature it is well know that with enough temperature and residence time, solid carbon will react with oxygen, carbon dioxide, and hydrogen to form carbon monoxide, hydrogen, and methane [1]. It is also supported that in a hydropyrolysis process, methane production is directly proportional to hydrogen consumption [6]. A lack of methane formation in this simulation of Zibin Zhu et al. led to an investigation if the nonreactive solid carbon that bypassed the reactor should be reacted with gaseous products. The model adapted to this model was presented while trying to model char gasification in a oxy-coal combustion simulations [5]. The three reactions and governing equation of the process are:

$$Reaction(I) : 2C_{(s)} + O_2 \rightarrow 2CO \tag{4}$$

$$Reaction(II) : C_{(s)} + H_2O \rightarrow CO + H_2 \tag{5}$$

$$Reaction(III) : C_{(s)} + 2CO_2 \rightarrow 2CO \tag{6}$$

$$\dot{m}_{C,i} = 4\pi r^2 P_{i,g} \frac{k_{diff,i} k_{kin,i}}{k_{diff,i} + k_{kin,i}} \tag{7}$$

where $\dot{m}_{C,i}$ is the mass of carbon reacted in each reaction, r is the radius of the coal particle, $P_{i,g}$ is the partial pressure of the gas that is being reacted in the ith reaction, and $k_{diff,i}$ and $k_{kin,i}$ are the calculated diffusion and kinematic coefficients respectively. The conversion of char to gaseous products is given in kilograms per second. An assumption made to adopt this model was that all gaseous products needed for the above reactions were always available. It was then calculated that for 42 seconds of residence time, 19.7

micromols of carbon would be produced. This in turn corresponds to 0.00033 percent of total carbon in the coal. With such a low percentage of carbon able to be converted from carbon to methane, it seems implausible that this is the reason for a lack of methane.

4.4 Investigation into Mechanism

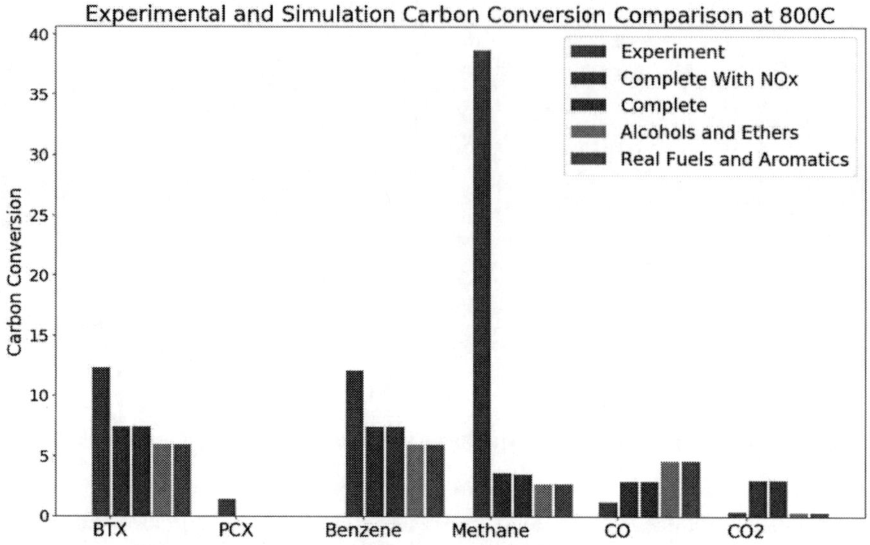

Figure 6: Simulations with Various Mechanisms Vs. Zibin Zhu Yields

The difference in yields for four different Creck mechanisms can be seen in Figure 6. It can be observed that the Complete mechanism without NO_x produces the same results as the mechanism with NO_x. The results obtained from the three hundred species mechanisms were also the same up to one-hundredth of a percent. As species and reactions are removed for to produce the next mechanism, BTX yields decrease.

It is worth noting that all of the above reactions contain a C20H10 and C20H16 species named BIN1B and BIN1A respectively. These bins are precursors to soot. It was hypothesized that because the Creck mechanisms were developed mainly for combustion processes, they over predict soot formation in hydropyrolysis. Soot formation is not a topic discussed in hydropyrolysis papers. Also, if the carbon that was being transformed into soot precursors was instead hydrocracked to methane, the methane yields would match those of Zibin Zhu et al. To test this theory, all BIN reactions were removed from the Complete mechanism. Removal of these 300 reactions led to extremely minimal reaction rates occurring during the 42 second residence time. This would present the theory that during hydropyrolysis simulations all modeled tar species are combine into these large molecules and then hydrocracked into smaller species. It is known that the reactions of polymerization and hydrocraking compete during hydropyrolysis. However, this mechanism appears to treat these two reactions in series instead of parallel competing reactions. A visual reaction pathway generation was attempted via Cantera software, however, the complexity of the Complete mechanism made this task impractical with current resources.

5 Conclusion

A hydropyrolysis plant would be able to help revitalize the Wyoming economy by re-purposing the use of Powder River Basin coal. Modeling of this process is needed to establish parameters to design a contentious closed loop pilot plant. Current modeling of the tar and chemical kinetics predict BTX yields and trends. However, the carbon conversion yields for methane were much low than experimental values. Through a systematic investigation the following results were obtained. One, C3M-CPD is able to predict gaseous, liquid, and char yields, however the composition of the gas yields is questionable. Coupling this software with PCCL will increase gaseous yield accuracy. Two, gasification of char to methane is an unlikely due to the time needed for these reactions. Three, the mechanism used in this research is appropriate for cool flames, however, hydropyrolysis may be out of its usability. Instead of competing between hydrocracking and polymerization, the mechanism polymerizes everything and then hydrocracks the larger molecule.

References

[1] Maarten van der Burgt Christopher Higman. *Gasification*. Gulf Professional Publishing, Burlington, MA, USA, second edition, 2008.

[2] COMMISION OF THE EUROPEAN COMMUNITIES. Feasibility study of the hydropyrolysis of coal. Report, Luxembourg, 1992.

[3] Terry L. Marker David F. Tatterson, Ken K Robinson. Coal flash pyrolysis in a free-jet reactor. *Industrial and Engineering Chemical Research*, 27(9):1606–1613, apr 1988.

[4] The CRECK Modeling Group. Kinetic mechanisms, dec 2014.

[5] Ahmed F. Ghoniem Lei Chen, Sze Zheng Yong. Oxy-fuel combustion of pulverized coal: Characterization, fundamentals, stabilization and cfd modeling. *Progress in Energy and Combustion Science*, 38:156–214, dec 2012.

[6] Peter S. Maa Martin L. Gorbaty. A critical temperature thrshold for coal hydropyrolysis. *Fuel Processing Technology*, 15:91–100, 1987.

[7] M. Soellinger. Numerical modeling of flash pyrolysis of prb coal. Proposal, may 2017.

[8] E. Ranzi T. Faravelli, A. Frassoldati. Kinetic modeling of the interactions between no and hydrocarbons in the oxidation of hydrocarbons at low temperatures. *Combustion and Flame*, 132:188–207, jul 2003.

[9] Chengfang Zhang Haihua Jin Zibin Zhu, Zhihua Ma and Xinrong Wang. Flash hydropyrolysis of northern chinese coal. *Fuel*, 75(12):1429–1433, jan 1996.

Western States Section of the Combustion Institute – Fall 2017 Meeting
Hosted by the University of Wyoming
October 2-3, 2017

Impact of acoustic-induced pressure fluctuations on the dynamics of laminar premixed flames

G. Beardsell[1,*] and G. Blanquart[1]

[1] *Department of Mechanical Engineering*
California Institute of Technology, Pasadena, USA
[*] *Corresponding author, e-mail: gbeardse@caltech.edu*

Thermo-acoustic instabilities are problematic in the design of continuous-combustion propulsion systems such as rocket motors. They are the result of a feedback loop in which acoustic waves are in sync with heat release oscillations and can thus become self-excited. Generally, the coupling between acoustics and flame dynamics involve multiple mechanisms including flame area variation, oscillations of the flame front and pulsating flame speed. The latter is the focus of the present work and is caused by the pressure fluctuations that accompany an acoustic wave. Since results extracted from laminar flame simulations are the building blocks of many models used to simulate complex geometries such as combustion chambers, a thorough understanding of the impact of acoustic waves on the global properties of laminar flames is essential. Using the low Mach number solver NGA with detailed chemistry, we investigate the response of one-dimensional, flat, hydrocarbon/air premixed flames subjected to pressure oscillations. The amplitude and phase responses of global flame quantities such as heat release and fuel consumption rate are extracted for different frequencies.

1 Introduction

Despite the remarkable body of work that has been achieved since Lord Rayleigh formulated his famous eponymous criterion in 1878 [1], thermo-acoustic instabilities remain problematic in the design of continuous-combustion propulsion systems such as rocket motors, jet engine afterburners and ramjets [2]. These instabilities are characterized by large fluctuations of the flow field in the combustor. Low-frequency oscillations can cause significant mechanical vibrations in the system, which in turn increase fatigue and noise radiation. In some cases, it can even lead to catastrophic failure [3]. In other instances, the oscillations can cause partial or total flame blow-off. They result from a feedback loop in which acoustic waves are in sync with heat release oscillations and become self-excited. A review of the different flame-acoustic interaction mechanisms in premixed combustion is given by Lieuwen *et al.*[2]. Fluctuations in heat release can conceptually be divided into two broad categories: changes due to the flow (flame position and surface area) and changes due to modifications of the flame structure.

In this work, we are concerned with the changes to the internal structure of a flat flame caused by the density and temperature fluctuations that accompany an incident acoustic wave. We consider fully premixed configurations and sound waves that can be treated with linear (isentropic) acoustic theory, which has been claimed to correctly describe waves with amplitudes up to 170 dB spl ($A_p \sim 6$ kPa) [4].

A significant amount of theoretical work has been performed on the subject [5–9]. Many of the results are summarized by McIntosh [9], who emphasizes the role played by the relative time and length scales of the flame and pressure perturbation, i.e., $\omega\tau_F$ and l_F/λ. Here, $\omega = 2\pi f$ is the angular frequency of the acoustic perturbation, τ_F is the flame time scale, $l_F = (T_b - T_u)/\max|\partial T/\partial x|$ is the flame thickness and λ is the acoustic wavelength. The subscripts $(\,)_b$ and $(\,)_u$ refer to quantities evaluated in the burnt and unburnt region, respectively. Although the flame steady-state response to small pressure changes typical of acoustic waves is fairly weak, the unsteady response can cause the local heat release to exceed its quasi-steady value by an order of magnitude [2]. Also, the acoustic pressure and unsteady heat release are always in phase [10]. Therefore, the Rayleigh criterion [1]

$$\iiint_V \int_t^{t+\tau_a} p_a \dot{Q}' dV dt > 0\,, \tag{1}$$

is satisfied and the flame-acoustics system is always unstable when acoustic damping is neglected. In Equation (1), V is the volume of the system, τ_a is the acoustic period, $\dot{Q}'$ is the fluctuating part of the heat release, and p_a is the acoustic pressure. This has consequences reaching beyond the mostly academic study of flat, laminar flames interacting with an incident acoustic wave. Large Eddy Simulations (LES) using tabulation approaches have emerged as a powerful tool to investigate turbulent combustion [11]. A popular tabulation approach is the laminar flame model, in which one populates a look-up table with global flame properties obtained from simulations of one-dimensional flames with detailed kinetics. For the study of configurations prone to thermo-acoustic instabilities, these laminar flame libraries need to capture the relevant flame-acoustics dynamics. Despite the abundant theoretical work on the response of a laminar flame subjected to acoustic waves, the experimental [12] and numerical [10, 13] literature is scarce, and focuses on small hydrocarbons (methane, propane) and hydrogen. There is thus a need for a better understanding of the flame-acoustics interactions involving large hydrocarbon species, which are commonly used in practical combustion devices.

This is the aim of this work. We perform direct numerical simulations of the flame-acoustics interactions for the combustion of a large hydrocarbon, n-heptane, in air at standard conditions. We also consider a methane/air configuration for comparison and validation purposes. For a wide range of acoustic frequencies, we extract the phase and amplitude of the unsteady heat release response, which are directly related to the Rayleigh criterion and thus the stability of the system.

This paper is organized as follows. In Section 2, we present the flow configuration investigated in this work, as well as the cases considered and the numerical methodology used to simulate them. The results are presented in Section 3 and compared with existing experimental and numerical data. We then highlight some of the differences observed between the responses of large (n-heptane) and small (methane) hydrocarbon fuels.

2 Methodology

2.1 Flow configuration

Figure 1 shows a schematic of the flow configuration considered in this work. Fully premixed reactants (blue) enter the domain on the left with velocity u_{in}, and burnt products (orange) exit on the right. A laminar planar flame with speed S_L is subjected to a standing acoustic wave with frequency $f = c_u/\lambda_u$, where c is the sound speed. To isolate the effects of pressure fluctuations on the combustion process, the flame position x_F is chosen to coincide with a velocity node.

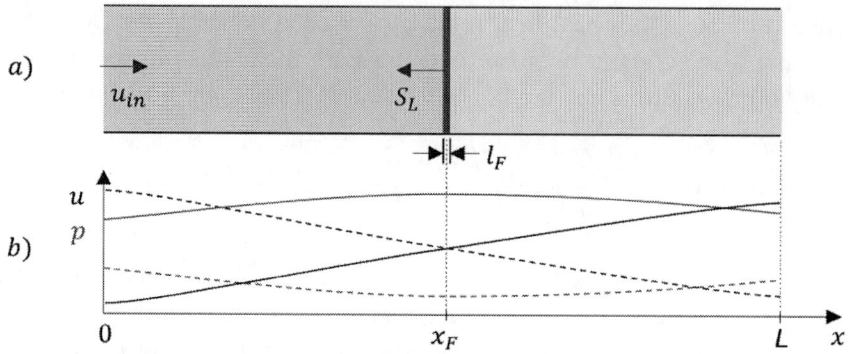

Figure 1: a) Schematic of the flow configuration investigated in this work. b) Example of pressure (red) and velocity (black) signals at different times (solid and dashed lines). The flame position $x = x_F$ is chosen to be a velocity node.

We consider two fuels, methane (CH_4) and n-heptane (C_7H_{16}), at standard pressure and temperature. The simulation parameters are given in Table 1. The methane cases allow us to compare our results with the experimental data of Wangher *et al.*[12] and the numerical simulations of Schmidt and Jiménez [13]. The n-heptane cases are performed to explore possible differences in the response of small and large hydrocarbons. The laminar flame speed is computed as

$$S_L = -\frac{1}{\rho_{u,0} Y_{F,u}} \int_{x=0}^{L} \rho\, \dot{\omega}_F \, dx\,, \tag{2}$$

where $\dot{\omega}_F$ is the fuel consumption rate ($\dot{\omega}_F < 0$). The subscript $(\,)_0$ refers to the unperturbed, steady-state flame.

Table 1: Parameters of the simulations: $p_{t,0}$ is the mean thermodynamic pressure, $T_{u,0}$ is the temperature of the fresh mixture at $p_{t,0}$, ϕ is the equivalence ratio, L is the length of the simulation domain, l_F is the laminar flame thickness, $S_{L,0}$ is the steady-state laminar flame speed, f is the acoustic frequency, $A_p = \max|p_a|$ is the acoustic pressure amplitude, and $\tau_F = l_F/S_{L,0}$ is the flame time scale.

Fuel	T_u [K]	$p_{t,0}$ [atm]	ϕ	# of grid points	L [mm]	l_F [mm]	$S_{L,0}$ [m/s]	f [Hz]	A_p [Pa]	τ_F [ms]
CH_4	300	1	0.625	900	12	0.87	0.13	50 to 10,000	500	1.3
C_7H_{16}			0.9			0.39	0.35		1000	0.17

2.2 Numerical method

For the range of frequencies considered in this work ($f = 50 - 10,000$ Hz), the ratio of the laminar flame thickness to the acoustic wavelength in the unburnt region is small, i.e., $l_F/\lambda_u \ll 1$. In particular, for the highest frequency, $l_F/\lambda_u < 3\%$. This allows us to use an approach similar to the one presented and validated by Schmidt and Jiménez [13], which is briefly summarized here.

In the limit $l_F/\lambda_u \ll 1$, the acoustic pressure gradients vanish in the vicinity of the flame, and the flame only experiences a time-varying acoustic pressure $p_a(t)$. Because the Mach number of the flow is small ($M = S_L/c \sim 10^{-3}$), we can employ a low Mach number approach [14]. We set the background pressure to $p_t(t) = p_{t,0} + p_a(t)$, with $p_a = A_p \sin(\omega t)$. The governing equations for mass conservation, species mass fractions, momentum and temperature are

$$\frac{\partial \rho}{\partial t} + \nabla \cdot (\rho \boldsymbol{u}) = 0\,, \tag{3}$$

$$\frac{\partial \rho Y_i}{\partial t} + \nabla \cdot (\rho \boldsymbol{u} Y_i) = -\nabla \cdot \boldsymbol{j}_i + \rho \dot{\omega}_i\,, \qquad i = 1, 2, ..., n \tag{4}$$

$$\frac{\partial \rho \boldsymbol{u}}{\partial t} + \nabla \cdot (\rho \boldsymbol{u}\boldsymbol{u}) = -\nabla p_1 + \nabla \cdot \boldsymbol{\tau}\,, \tag{5}$$

$$\frac{\partial \rho T}{\partial t} + \nabla \cdot (\rho \boldsymbol{u} T) = \frac{1}{c_p}\frac{dp_t}{dt} + \nabla \cdot (\rho c_p \alpha \nabla T) - \frac{1}{c_p}\sum_{i=1}^{n} c_{p,i}\boldsymbol{j}_i \cdot \nabla T + \rho \dot{\omega}_T\,. \tag{6}$$

In Equation (4), the subscript $(\,)_i$ denotes the ith species, Y_i are the mass fractions, $\dot{\omega}_i$ are the production terms, and n is the total number of species transported. The diffusion mass fluxes $\boldsymbol{j}_i$ are computed as

$$\boldsymbol{j}_i = -\rho Y_i \left(D_i \frac{\nabla X_i}{X_i} + \boldsymbol{u}_c \right)\,, \tag{7}$$

where $\boldsymbol{u}_c$ is the correction velocity required to enforce a zero net diffusion flux, X_i are the mole fractions, and D_i are the species diffusivities. In Equation (5), p_1 is the hydrodynamic pressure and $\boldsymbol{\tau}$ is the viscous stress tensor. In Equation (6), α is the mixture diffusivity and c_p is the mixture heat capacity at constant pressure. The temperature production term $\dot{\omega}_T$ is given by

$$\dot{\omega}_T = -\frac{1}{c_p}\sum_{i=1}^{n} h_i \dot{\omega}_i\,, \tag{8}$$

where h_i are the species enthalpies. The heat release is computed as

$$\dot{Q} = \int_{x=0}^{L} \rho c_p\, \dot{\omega}_T\, dx\,. \tag{9}$$

We use the equation of state for a perfect gas

$$p_t = \rho RT \sum_{i=1}^{n} \frac{Y_i}{W_i}, \tag{10}$$

where W_i are the species molar masses and R is the universal gas constant. The inlet temperature T_{in} is continuously adjusted according to the isentropic relation

$$T_{in} = T_{u,0} \left(\frac{p_t}{p_{t,0}} \right)^{\frac{\gamma_u - 1}{\gamma_u}}, \tag{11}$$

where γ is the heat capacity ratio. Note that while Equation (11) assumes γ to be constant, this is not the case when solving the governing equations. Equations (3) to (10) are solved using the energy-conservative, finite-difference low Mach number solver NGA [15]. We use second-order accurate temporal and spatial discretizations. For all simulations, $L = 12$ mm with 900 uniformly spaced grid points. This corresponds to 65 and 29 grid points per flame thickness for methane and heptane, respectively. We use a mixture-averaged formulation to compute the diffusion mass fluxes [16]. The species viscosities μ_i are obtained from standard gas kinetic theory [17] and the mixture-averaged viscosity μ is calculated using a modified form of Wilke's formula [18, 19]. The species thermal diffusivities λ_i are evaluated using a modified version of Eucken's formula [20]. The mixture-averaged thermal conductivity λ is computed following Mathur *et al.* [21].

To isolate and quantify the impact of acoustic pressure fluctuations on the flame response, care must be taken to limit the impact of the acoustic velocity field on the flame. To this end, we perform one-dimensional simulations. This ensures that the flame remains planar and thus prevents multidimensional effects from developing. For example, flame area variation could result from a Rayleigh-Taylor instability triggered by the pulsating acoustic velocity field [22].

Even in 1D, accelerations due to the acoustic velocity field generate body force terms that could influence the flame response. Figure 2a shows the velocity profile at different times, for the n-heptane configuration at $f = 5000$ Hz, with a fixed inlet velocity ($u_{in} = S_{L,0}$). Using this boundary condition, the flame is not exactly positioned at a velocity node (the inlet is a velocity node). The flame thus experiences periodic accelerations caused by the acoustic velocity field. By modulating u_{in}, we can choose the velocity node to roughly coincide with the flame location x_F. Assuming a constant unburnt density ρ_u, i.e., $\rho = \rho_u$ for $x \leq x_F$, we perform a simple control volume analysis and get

$$u(x_F, t) = u_{in} - \frac{1}{\rho_u} \frac{d\rho_u}{dt} x_F = u_{in} - \frac{1}{\gamma_u} \frac{1}{p_t} \frac{dp_t}{dt} x_F \,. \tag{12}$$

The last equality in Equation. (12) is obtained by linearizing the isentropic relation between density and pressure. We then set

$$u_{in} = S_{L,0} + \frac{1}{\gamma_u} \frac{1}{p_0} \frac{dp_0}{dt} x_{F,0} \,, \tag{13}$$

where

$$x_{F,0} = \frac{1}{\rho_{u,0} Y_{F,u}} \left[\int_{x=0}^{L} \rho\, Y_F\, dx \right]_{t=0} \tag{14}$$

is the initial flame position. Figure 2b shows the velocity profile at different times, when the inlet velocity is modulated according to Equation (12). As desired, the flame now sits at a velocity node.

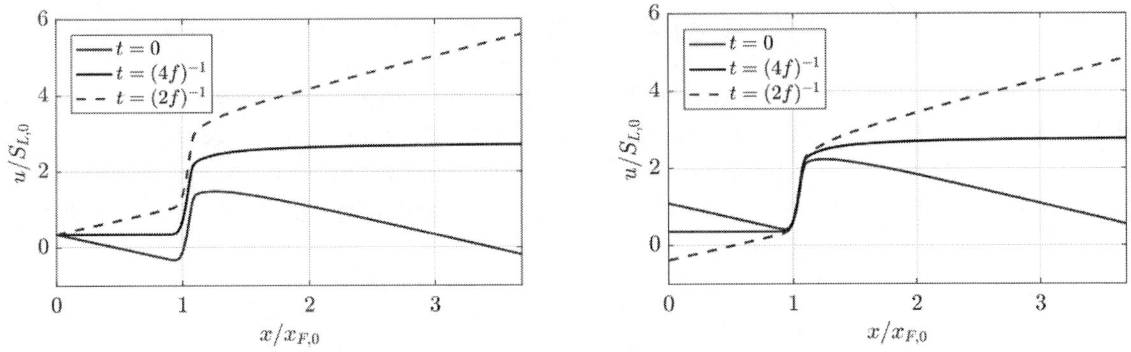

(a) Fixed inlet velocity ($u_{in} = S_{L,0}$). (b) Inlet velocity modulated according to Eq. (13).

Figure 2: Velocity profile at different times during the acoustic cycle for the n-heptane configuration at $f = 10,000$ Hz. The flame is located at $x_{F,0}/\lambda_u = 0.057$.

2.3 Chemical models

The chemical model used for methane is the well-established GRI-Mech 3.0 mechanism [23], which includes 52 species and 634 reactions. For n-heptane, we use a reduced version of the mechanism of Bisetti et al. [24], from which the aromatic species have been removed. This is justified by the slightly lean conditions used in the n-heptane cases. The reduced mechanism contains 35 species and 217 reactions.

2.4 Spatial and temporal convergence

The time steps used in this study were systematically verified by repeating each simulation with a time step 2 to 5 times smaller. Differences up to 3 % and 0.01 rad were deemed acceptable for the heat release fluctuation amplitude and phase shift, respectively. Unsurprisingly, the required time step becomes smaller as the frequency increases. To verify that 900 grid points were enough, additional simulations were performed for each fuel with twice as many points, for $f = 1000$ Hz. The differences in amplitude and phase were found to be within the tolerances used for the temporal convergence, and the coarser grid was used for all subsequent simulations.

2.5 Extraction of the amplitude and phase responses

The imposed pressure signal is chosen to be a sine wave, but the flame response, e.g, $\dot{Q}$, might not be sinusoidal. Here, we only assume that the response reaches a periodic steady-state with frequency f. We determine the amplitude of the response using

$$A_\psi = \frac{1}{2N} \sum_{i=N_0+1}^{N} \left[\psi_{max,i} - \psi_{min,i}\right], \tag{15}$$

where $\psi_{min,i}$ and $\psi_{max,i}$ are the extrema of a flame metric ψ (e.g., $\psi = \dot{Q}$), on the interval $t \in [f(i-1), f\,i]$. Here, N is the total number of cycles simulated and N_0 is the number of cycles rejected because the system had not reached steady state yet. We consider that the system reaches a steady state when the differences in amplitude and phase differ by less than 1% between all subsequent cycles. For all the cases considered here, $N_0 \leq 1$, and the simulations were carried for $N = 3$ to $N = 50$ cycles, depending on the frequency. Low frequencies require more computational time per cycle and less cycles were simulated. To compute the phase angle (in radians), we use

$$\phi = \frac{\pi f}{N} \sum_{i=1}^{N} \left[\left(t_{max,i} - \left(i - \frac{3}{4} \right) f \right) + \left(t_{min,i} - \left(i - \frac{1}{4} \right) f \right) \right], \tag{16}$$

where $t_{min,i}$ and $t_{max,i}$ are the time at which $\psi_{min,i}$ and $\psi_{max,i}$ occur, respectively. In Figure 3, we plot a portion of the normalized time response of $\dot{Q}'$ and S_L' for the methane configuration at $f = 5$ kHz. The heat release phase shift ϕ is shown. Primed quantities denote the fluctuating part of a signal, i.e., $\psi' = \psi - \overline{\psi}$.

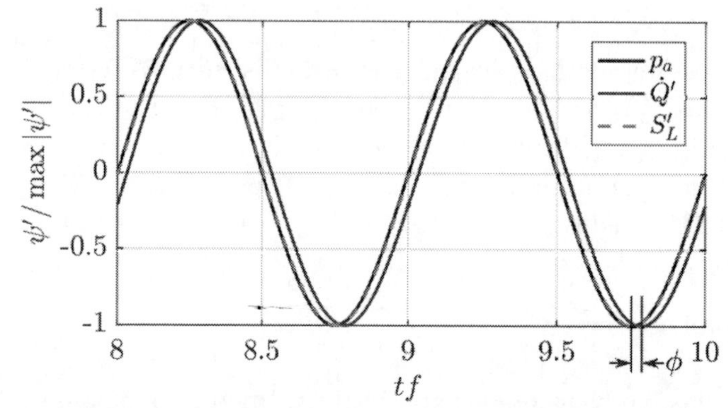

Figure 3: Simulated time response of $\dot{Q}'$ and S_L', for the methane configuration at $f = 5$ kHz. All quantities are normalized with their maximum value. Two cycles are shown, the 9th and the 10th after the acoustic perturbation was initiated.

3 Results and discussion

In this section, we first compare our results for methane with the simulations of Jiménez *et al.*[10] and the experimental data of Wangher *et al.*[12]. Next, the results obtained for n-heptane are presented and compared with the methane cases.

3.1 Comparison with other simulation results

In Figure 4, the normalized amplitude and phase shift of $\dot{Q}'$ are shown. Our results are plotted in red, while the data of Jiménez *et al.* is plotted in blue. The agreement between our simulations and the ones of Jiménez is poor for $A_{\dot{Q}'}$, especially given that the flow conditions are the same ($T_{u,0}$, $p_{t,0}$, ϕ, A_p), and that we both use the GRI-Mech 3.0 mechanism and a low Mach number solver. The results should therefore be identical. These discrepancies require further investigation.

3.2 Comparison with experimental measurements

To the author's best knowledge, the results reported by Wangher *et al.* [12] form the only experimental data set providing measurements of the direct impact of acoustic pressure variations on planar premixed flames. They maintained a flame at a velocity node in an open-ended tube while one of the resonant frequencies of the tube was excited. They then measured the OH* emission intensity. In Figure 4, their results are plotted in black and directly compared with the numerical results for heat release. The error bars associated with their measurements have not been included here but are available in their paper. We see a reasonable agreement between the simulated and the experimental results. The amplitudes are of the same order of magnitude, and the phase angle drops as the frequency is increased. Jiménez *et al.* [10] discuss the limitations associated with the direct use of OH* chemiluminescence as a proxy for $\dot{Q}'$. In particular, for the methane configuration, they show that OH* chemiluminescence over-predicts ϕ at low frequencies, which is what is observed in Figure 4b.

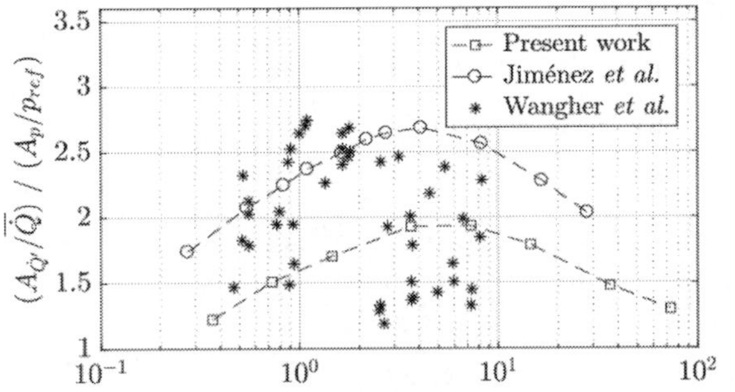

(a) Normalized amplitude

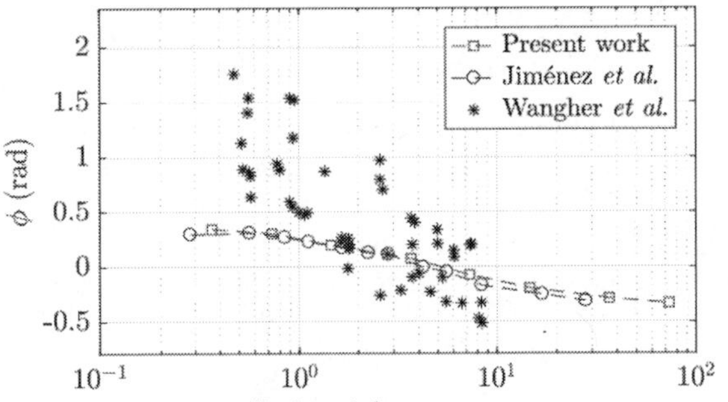

(b) Phase angle

Figure 4: Amplitude and phase angle of the heat release fluctuations (simulations, blue and red) and OH* emission intensity (experiments, black) as a function of the reduced acoustic frequency $\omega\tau_F$, for the methane configuration. The red squares are from the present study, the blue circles are numerical results from Jiménez *et al.*[10], and the black stars are experimental results from Wangher *et al.*[12].

3.3 Differences in the response of small and large hydrocarbons

In Figure 5, we plot the fluctuating heat release response of the methane and n-heptane configurations, for $50 \leq f \leq 10{,}000$ Hz. This corresponds to $0.41 \leq \omega\tau_F \leq 82$ for methane and $0.055 \leq \omega\tau_F \leq 11$ for n-heptane. The shape of the normalized heat release amplitude responses $A_{\dot{Q}}(\omega\tau_F)$ are similar for the two configurations investigated, with a peak response around $\omega\tau_F \sim \mathcal{O}(1)$. In the limit $f \rightarrow 0$, the system is in quasi-steady state. This enables us to compute the amplitude response by solving for the steady-state solutions at $p_t = p_{t,0} \pm A_p$, with the inlet temperature specified by Equation (11). The flame responses for $f \rightarrow 0$ are denoted by horizontal dashed lines in Figures 5a and 6. Compared with their respective quasi-steady-state solutions, the peak unsteady response for $A_{\dot{Q}'}$ is seen to be larger for methane than for n-heptane. Figure 5b shows the phase shift between the heat release and pressure signals. A positive phase ($\phi > 0$) means that $\dot{Q}'$ is in advance with respect to p_a. The two phase responses are similar.

Figure 6 shows the amplitude of the fluctuating flame speed $A_{S_L'}(\omega\tau_F)$. Similarly to $\max(A_{\dot{Q}'})$, $\max(A_{S_L'})$ is larger for methane than for n-heptane, when compared with their respective quasi-steady-state solutions.

In all cases, the mean heat release and flame speed change by less than 0.1% with respect to the steady, non-perturbed values, i.e., $\overline{S_L} \approx S_{L,0}$ and $\overline{\dot{Q}} \approx \dot{Q}_0$.

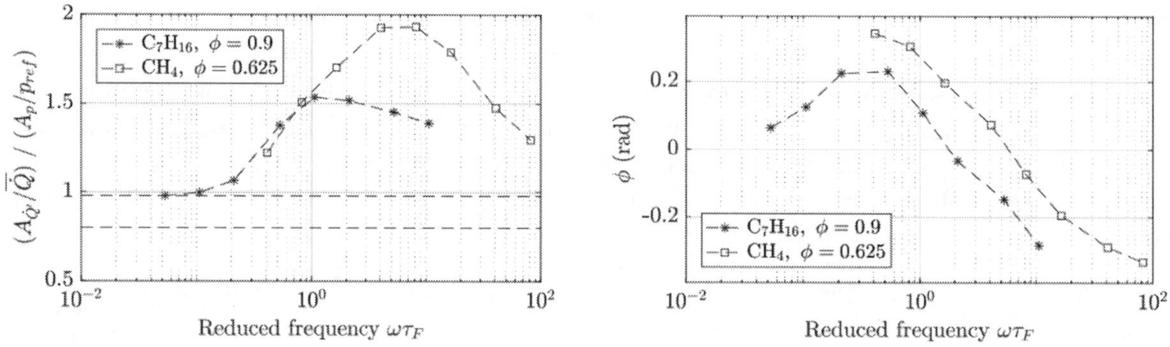

(a) Normalized amplitude of the heat release fluctuations. The horizontal lines denote the solutions for $\omega \rightarrow 0$.

(b) Phase angle of the heat release fluctuations.

Figure 5: Amplitude and phase angle of the heat release fluctuations as a function of the reduced acoustic frequency $\omega\tau_F$, for the methane (red squares) and n-heptane (blue stars) configurations.

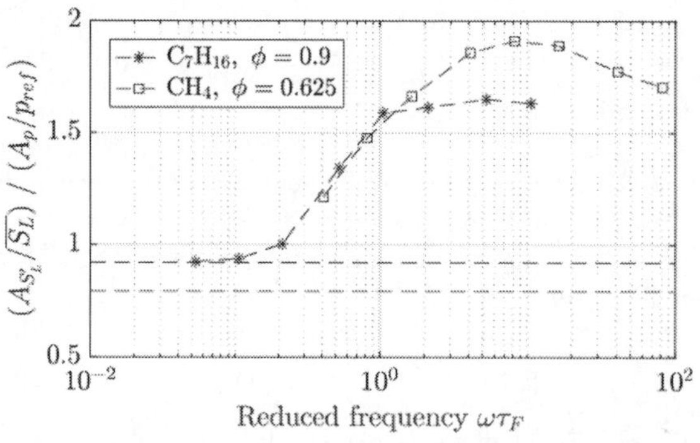

Figure 6: Amplitude of the flame speed fluctuations $A_{S_L'}$ as a function of the reduced acoustic frequency $\omega\tau_F$, for the methane (red squares) and n-heptane (blue stars) configurations. The horizontal lines denote the solutions for $\omega \rightarrow 0$.

4 Conclusion

In this work, the direct impact of acoustic pressure fluctuations on the response of laminar, flat, hydrocarbon/air premixed flames was investigated. A better understanding of the relevant physics at play in such flow configurations is important since it could help to develop more accurate LES frameworks using laminar flame libraries, especially for simulations aimed at capturing thermo-acoustic instabilities. In particular, the impact of acoustic waves on the unsteady heat release response is important as it is directly involved in the Rayleigh criterion. Using the low Mach number solver NGA with detailed chemistry, we extracted the amplitude and phase responses of global flame quantities such as heat release and flame speed, for a wide range of frequencies. For the methane configuration, we compared our results with the ones of Jiménez *et al.*[10]. Discrepancies in the calculated heat release fluctuation amplitudes were observed, despite the fact that the same numerical approach was used and that the simulations were performed at the same flow conditions. The results should therefore be identical, and further investigation is required. We then compared the responses of two hydrocarbon fuels, CH_4 and C_7H_{16}. Despite having different molecular weights and chemistry, the two responses were found to be similar when the acoustic frequency is normalized by the flame time scale. It would be interesting to gather data for other fuels and at different flow conditions. Are all the responses similar when properly normalized?

References

[1] J. W. S. B. Rayleigh. Nature, 18 (1878) 319–321.

[2] T. C. Lieuwen. Journal of Propulsion and Power, 19 (2003) 765–781.

[3] S. M. Candel. Symposium (International) on Combustion, 24 (1992) 1277–1296.

[4] P. Davies and K. R. Holland. Journal of Sound and Vibration, 271 (2004) 849–861.

[5] A. C. McIntosh. Combustion Science and Technology, 75 (1991) 287–309.

[6] G. Ledder and A. K. Kapha. Combustion Science and Technology, 76 (1991) 21–44.

[7] A. C. McIntosh. Combustion Science and Technology, 91 (1993) 329–346.

[8] G. A. Batley, A. C. McIntosh, and J. Brindley. Combustion Science and Technology, 92 (1993) 367–388.

[9] A. C. McIntosh. Philosophical Transactions of the Royal Society of London A: Mathematical, Physical and Engineering Sciences, 357 (1999) 3523–3538.

[10] C. Jiménez, J. Quinard, J. Graña-Otero, H. Schmidt, and G. Searby. Combustion and Flame, 159 (2012) 1894–1908.

[11] H. Pitsch. Annual Review of Fluid Mechanics, 38 (2006) 453–482.

[12] A. Wangher, G. Searby, and J. Quinard. Combustion and flame, 154 (2008) 310–318.

[13] H. Schmidt and C. Jiménez. Combustion and Flame, 157 (2010) 1610–1619.

[14] B. Müller. Low Mach Number Asymptotics of the Navier-Stokes Equations and Numerical Implications, 1999.

[15] O. Desjardins, G. Blanquart, G. Balarac, and H. Pitsch. J. Comput. Phys., 227 (2008) 7125–7159.

[16] R. B. Bird, W. E. Stewart, and E. N. Lightfoot. Transport phenomena. John Wiley & Sons, 2007.

[17] R. B. Bird, J. O. Hirschfelder, and C. F. Curtiss. Molecular theory of gases and liquids. John Wiley, 1954.

[18] C. R. Wilke. The journal of chemical physics, 18 (1950) 517–519.

[19] S. Lapointe, B. Savard, and G. Blanquart. Combust. Flame, 162 (2015) 3341–3355.

[20] J. H. Ferziger and H. G. Kaper. American Journal of Physics, 41 (1973) 601–603.

[21] S. Mathur, P. K. Tondon, and S. C. Saxena. Molecular physics, 12 (1967) 569–579.

[22] A. McIntosh. Pressure-driven disturbances in fluid dynamic interactions with flames. In J. Buckmaster and T. Takeno, editors, Modeling in Combustion Science, Vol. 449 of Lecture Notes in Physics, Berlin Springer Verlag, pages 176–192, 1995.

[23] G. P. Smith, D. M. Golden, M. Frenklach, N. W. Moriarty, B. Eiteneer, M. Goldenberg, C. T. Bowman, R. K. Hanson, S. Song, W. C. Gardiner, V. V. Lissianski, and Z. Qin. GRI-Mech 3.0, http://www.me.berkeley.edu/gri_mech/.

[24] F. Bisetti, G. Blanquart, M. E. Mueller, and H. Pitsch. Combust. Flame, 159 (2012) 317–335.

Sub Topic: Laminar flames

Western States Section of the Combustion Institute – Fall 2017 Meeting
Hosted by the University of Wyoming
October 2-3, 2017

Impact of non-ideal corrections to equation of state, thermodynamics, and chemical kinetics on one-dimensional freely propagating flame structure

Shashank Yellapantula[1,], John B. Bell[2], Emmanuel Motheau[2], Marc S. Day[2], and Ray W. Grout[1]*

[1]*National Renewable Energy Laboratory (NREL), Golden, CO, USA*
[2]*Lawrence Berkeley National Laboratory (LBNL), Berkeley, CA, USA*
[*]*Corresponding author:* shashank.yellapantula@nrel.gov

Abstract: In the current generation industrial combustion devices for propulsion and energy generation, a common trend of increase in operating pressure can be observed. Primary motivation behind this increase in operating pressure is improvement in efficiency and a compact combustion chamber. At these high-pressure conditions observed in combustion chambers, an ideal gas equation of state is no longer valid. Moreover, at operating conditions similar and above the critical point of the fuel-air mixture the discrete interface between gas and liquid phase is no longer observed and instead a more diffused interface is seen. Thermodynamic and transport properties undergo a sharp but continuous change around this diffused interface, resulting in a large deviation from the ideal gas predictions.

In the present study, the impact of departure from the ideal gas is studied on a canonical freely propagating one-dimensional premixed flame. A cubic equation of state (EOS) namely the Soave-Redlich-Kwong is used as the constitutive relationship and is used to compute other thermodynamic information such as specific heat at constant pressure, Cp. Chemical kinetics are also modified to accommodate chemical activity coefficients instead of molar concentrations. Impact of all these non-ideal modifications on chemical kinetics, thermodynamics, on 1-D premixed flame structure will be investigated in this study. Additionally, flame structure corresponding to different fuels namely hydrogen and methane are also highlighted in the context of non-ideal modifications to thermodynamics.
***Keywords:* EOS, Laminar Flame, Soave-Redlich-Kwong**

1. Introduction

Mixing and combustion characteristics at pressure and temperature beyond the critical point of gases is an active area of research and has large implications on design and operation of industrial combustion chambers in propulsion. One of the commonly used design philosophy to increase combustor efficiency is to increase the operating pressure in a combustion chamber. For example, in the next generation aircraft engine for the Boeing 777, the GE9x, a pressure ratio of 60:1 is being targeted [1]. Some of the other application areas where improved understanding of mixing and combustion at high pressure are essential for safe and efficient operation are rocket motor design, diesel engine combustion and super-critical CO_2 based power generation turbines.

[1]https://www.geaviation.com/commercial/engines/ge9x-commercial-aircraft-engine

At conditions close to and beyond the critical point, gases exhibit strong departures from ideal gas behavior. Instead of a sharp interface between phases a diffused interface is typically observed [1]. Thermodynamic and transport properties undergo a sharp but continuous change around this diffused interface. Modeling of this departure from ideal gas behavior is of significant importance for an accurate simulation of mixing and combustion. In this study a thermodynamically consistent formulation for non-ideal corrections to equation of state (EOS), thermodynamic properties and chemical kinetics is used to study combustion characteristics at high pressure conditions. Soave-Redlich-Kwong cubic EOS is used as the constitutive relation. This form of the cubic equation EOS is further used to derive non-ideal corrections to thermodynamic quantities such as enthalpy, Gibbs free energy, internal energy, specific heat capacities and entropy. Additionally, chemical kinetics are modified to include the non-ideal corrections in the form of chemical activity coefficients. Impact of all these non-ideal modifications are then studied on freely propagating 1-D premixed flames with $CH_4/O_2/He$ and $H_2/O_2/N_2$ as fuel-oxidizer mixtures. Results from these simulations demonstrate the improvement in accuracy of flamespeed predictions at higher pressures due to the non-ideal EOS and corrections to thermodynamics and chemical kinetics.

2. Mathematical Formulation

Fully compressible and conservative form of the continuity, species, momentum and energy equations are considered in this study.

$$\frac{\partial}{\partial t}(\rho)+\nabla\cdot(\rho u)=0, \tag{1}$$

$$\frac{\partial}{\partial t}(\rho Y)+\nabla\cdot(\rho u Y)=\nabla\cdot\mathscr{F}+\omega \tag{2}$$

$$\frac{\partial}{\partial t}(\rho u)+\nabla\cdot(\rho u\otimes u)+\nabla p=\nabla\cdot\Pi, \tag{3}$$

$$\frac{\partial}{\partial t}(\rho E)+\nabla\cdot(\rho u E+pu)=\nabla\cdot\mathscr{Q}+\nabla\cdot(\Pi u) \tag{4}$$

where ρ, u and p denote the density, velocity vector and pressure, respectively, of the mixture. The mass fraction of the k-th species is given by Y_k, and is denoted in vector form as Y. The different species in the gaseous mixture are assumed to be in thermal equilibrium, that is, at a common temperature, T. The chemical species reaction source term, in the vector form, is denoted by ω. In the species transport equation, the Soret and Dufour effects are neglected.

In the momentum equation, the tensor Π is the viscous stress and, under the Newtonian assumption, is given by:

$$\Pi=\eta(\nabla u+(\nabla u)^T)+(\kappa-\frac{2}{3}\eta)\mathbf{I}\nabla\cdot u \tag{5}$$

2.1 Equation of State

The Soave-Redlich-Kwong (SRK) equation of state (EOS), in the following form, is considered in this study.

$$p=RT\sum\frac{Y_k}{W_k}\frac{1}{\tau-b_m}-\frac{a_m}{\tau(\tau+b_m)} \tag{6}$$

where T is temperature, τ is the specific volume, Y_k are mass fractions and R is the universal gas constant and where b_m and a_m are mixture attractive and repulsive terms, respectively.

Specification of the SRK EOS for a mixture of species requires mixing rules for the attractive and repulsive terms in the equation of state. The terms, $a_m = a_m(T, Y_k)$ and $b_m = b_m(Y_k)$ for the mixture are specified with standard mixing rules

$$a_m = \sum_{ij} Y_i Y_j \alpha_i \alpha_j \quad b_m = \sum_k Y_k b_k \tag{7}$$

Typically, a_i and b_i are defined using the critical quantities and temperature dependent correction factors [2]:

$$\begin{aligned} \alpha_i &= \sqrt{a_i} \\ a_i(T) &= a_i^* \left(T_{c,i}, P_{c,i}\right) \bar{a}_i(T/T_{c,i}) \\ &= 0.42748 \frac{\left(RT_{c,i}\right)^2}{W_i^2 p_{c,i}} \bar{a}_i \left(T/T_{c,i}\right) \\ b_i &= 0.08664 \frac{RT_{c,i}}{W_i p_{c,i}} \end{aligned} \tag{8}$$

where

$$\bar{a}_i(T/T_{c,i}) = \left(1 + f\left(\omega_i\right)\left(1 - \sqrt{T/T_{c,i}}\right)\right)^2$$

where ω_i are the accentric factor and

$$f\left(\omega_i\right) = 0.48508 + 1.5517\omega_i - 0.151613\omega_i^2$$

For chemically unstable species such as radicals, critical temperatures and pressures are not available. For species where critical properties are not available, the procedure discussed in [3, 4] is used to estimate the attractive and repulsive factors using the Lennard-Jones potential for the relevant species.

To be precise, for the ith species we define

$$T_{c,i} = 1.316 \frac{\varepsilon_i}{k_b} \quad a_i(T_{c,i}) = 5.55 \frac{\varepsilon_i \sigma_i^3}{m_i^2} \quad \text{and} \quad b_i = 0.855 \frac{\sigma_i^3}{m_i}$$

where σ_i, ε_i are the Lennard-Jones potential molecular diameter and well-depth, respectively, m_i the molecular mass, and k_b is Boltzmann constant.

For the SRK equation of state, the integrals for internal energy and entropy can be evaluated to obtain

$$e_m = \sum_k Y_k e_k^{id} + \left(T \frac{\partial a_m}{\partial T} - a_m\right) \frac{1}{b_m} ln \left(1 + \frac{b_m}{\tau}\right)$$

and

$$s_m = \sum_k Y_k s_k^{id} - \sum_k \frac{Y_k R}{W_k} ln \left(\frac{\tau}{\tau - b_m}\right) + \frac{\partial a_m}{\partial T} \frac{1}{b_m} ln \left(1 + \frac{b_m}{\tau}\right) \tag{9}$$

Enthalpy defined as $h_m = e + p\tau$, yields

$$h_m = \sum_k Y_k h_k^{id} + \left(T\frac{\partial a_m}{\partial T} - a_m\right)\frac{1}{b_m}\ln\left(1+\frac{b_m}{\tau}\right) + RT\sum \frac{Y_k}{W_k}\frac{b_m}{\tau - b_m} - \frac{a_m}{\tau + b_m}$$

Additionally chemical potentials, defined as the derivatives of the Gibbs free energy with respect to composition (Eq. 10), are used to define chemical kinetics rate laws.

$$\mu_k = \left(\frac{\partial f}{\partial Y_k}\right)_{\tau,T,Y_j} \tag{10}$$

For SRK EOS, the species chemical potentials are given by

$$\begin{aligned}\mu_k = \mu_k^{id} + \frac{RT}{\overline{W}}\frac{1}{\tau - b_m}\frac{\partial b_m}{\partial Y_k} + \frac{RT}{W_k} ln\left(\frac{Y_k}{W_k(\tau - b_m)}\right) \\ - \frac{1}{b_m} ln\left(1+\frac{b_m}{\tau}\right)\frac{\partial a_m}{\partial Y_k} + \frac{a_m}{b_m^2} ln\left(1+\frac{b_m}{\tau}\right)\frac{\partial b_m}{\partial Y_k} - \frac{a_m}{b_m}\left(\frac{1}{\tau+b_m}\right)\frac{\partial b_m}{\partial Y_k}\end{aligned} \tag{11}$$

2.2 Kinetics

In order to be thermodynamically consistent, reactions rates are specified in terms of the chemical potentials. For chemical kinetics rescaled chemical potentials given by the following expression are used

$$\hat{\mu} = \frac{1}{k_B T}\mathscr{W}\mu$$

where $\mathscr{W}$ is a diagonal matrix of molecular weights. In terms of the rescaled chemical potentials, the molar production rates are given by

$$\Omega_i = \sum_j (\nu_{ij}^b - \nu_{ij}^f)\mathscr{R}_j$$

where

$$\mathscr{R}_j = k_j^s\left[exp(\sum_k \nu_{jk}^f \hat{\mu}_k) - exp(\sum_k \nu_{jk}^b \hat{\mu}_k)\right] \tag{12}$$

where k_j^s is the symmetric reaction rate. In order to reduce asymptotically to the ideal gas limit, chemical activity coefficients, A_k, are defined in the following manner

$$A_k = exp[\hat{\mu}_k - \hat{\mu}_k^{u,id}] \tag{13}$$

In the ideal gas limit these activity coefficients reduce to molar concentrations $[X_k]$. In the case of SRK EOS, these chemical activities become

$$\begin{aligned}A_k = exp(\frac{W_k}{RT}[\frac{RT}{\overline{W}}\frac{1}{\tau - b_m}\frac{\partial b_m}{\partial Y_k} + \frac{RT}{W_k} ln\left(\frac{Y_k}{W_k(\tau - b_m)}\right) \\ - \frac{1}{b_m} ln\left(1+\frac{b_m}{\tau}\right)\frac{\partial a_m}{\partial Y_k} + \frac{a_m}{b_m^2} ln\left(1+\frac{b_m}{\tau}\right)\frac{\partial b_m}{\partial Y_k} - \frac{a_m}{b_m}\frac{1}{\tau+b_m}\frac{\partial b_m}{\partial Y_k}])\end{aligned} \tag{14}$$

Using this framework, the reaction rate laws can be defined in the traditional ideal gas framework, allowing the sharing of significant amount of code for reaction rate evaluation with ideal gas chemistry. Specifically, simply replacing molar concentration $[X_k]$ with chemical activity A_k is the only change needed to specify kinetics that are thermodynamically consistent with the non-ideal system. To summarize, the chemical kinetic rate laws are expressed as

$$\mathscr{R}_j = k_j^f \prod_k (A_k)^{\nu_{jk}^f} - k_j^b \prod_k (A_k)^{\nu_{jk}^b} \tag{15}$$

In this form the forward and reverse rates are related:

$$k_j^f = k_j^b K_j^{eq} \tag{16}$$

where

$$K_j^{eq} = exp\left(\left(\nu_{jk}^f - \nu_{jk}^b\right) \hat{\mu}_k^{id,u}\right) \tag{17}$$

The equilibrium rate constant here is the same as is used for ideal gases, all of the non-ideal behavior is captured in the definition of the A_k. By using k_j^f in traditional Arrhenius form with this definition of equilibrium rate constant ensures thermodynamic consistency.

3. Numerical formulation

Numerical simulations described in this study are performed using a compressible flow solver, `PeleC`, developed as a part of the Exascale computing project on combustion[2].`PeleC` uses an explicitly integrated compressible formulation and is built on the AMReX framework [3] . AMReX provides the infrastructure necessary to build highly scalable block-structured adaptive mesh refinement applications. AMReX is openly available and used for a wide variety of applications in the combustion, astrophysics, multi-phase flow, and other communities. Within the AMReX paradigm, AMReX manages mechanics of the data decomposition, distributed memory communication, load balancing and supports parallel I/O. It also provides support for hyperbolic, parabolic and elliptic solvers on a multi-level adaptive grid structure and support for particles. The application, in this case PeleC, implements physics on a patch – a regular block of cell, edge, or nodal data where AMReX provides the size of the data block as well as a region that the application code should update. PeleC treats the hyperbolic term with an unspilt piecewise parabolic method (PPM) Godunov scheme that is widely used in the astrophysics community and described in detail in the CASTRO users guide[4]. Diffusion is advanced semi-implicitly, and the reactions are integrated with an implicit ODE solver (CVODE)[5]. Within PeleC, the processes are coupled together using a spectral deferred correction based approach analogous to that described in [5]. In this approach two iterations are sufficient for formal second-order accuracy in time and space. Evaluation of chemical reaction rates and transport coefficients is done by generating unrolled C code from chemkin mechanism files, suitable for compiler optimization using an in-house preprocessor and interface code that is API compatible with EGLib[6].

[2] `https://exascaleproject.org/`

[3] `https://amrex-codes.github.io`

[4] `http://bender.astro.sunysb.edu/Castro/staging/Castro/Docs/CastroUserGuide.pdf`

[5] `http://computation.llnl.gov/projects/sundials/sundials-software`

[6] `http://blanche.polytechnique.fr/www.eglib/`

4. Results

The impact of non-ideal gas EOS and thermodynamics are presented in this study in the context of a one-dimensional freely propagating premixed flame. The initial conditions for these simulations are obtained from the 1-D flame simulations, using ideal gas EOS, performed using Cantera [6]. As mentioned in the previous section, fully compressible unsteady Navier-Stokes equations on a one-dimensional domain are solved till a steady state solution is reached. Inflow and outflow boundary conditions are applied to the left and the right most end of the domain. The inflow boundary condition values for velocity, internal energy and species are computed from the Cantera ideal gas solution. A flame tracer based on mass fractions of OH in the case of hydrogen and CH_2O for methane flame are used to determine the adaptive mesh refinement. All the simulations described in this study were run with a limiting CFL criterion of 0.7. Flamespeed computed using the following formula was used as the criterion for steady state convergence.

$$S_l = -\frac{\int_{-\infty}^{\infty} \omega_F dx}{\rho_{ub} Y_{F,ub}} \tag{18}$$

where ρ_{ub} and $Y_{F,ub}$ are the density of the unburned fuel-air mixture and the unburnt massfraction of fuel, respectively. This expression can be derived by integrating the steady state transport equation for fuel massfraction, with zero-gradient boundary condition applied at the physical boundaries. In all the results presented in the following sections, flames starting from an ideal gas solution were integrated till a steady evolution of the flame speed was obtained. Figure 1 shows a typical evolution of flamespeed from one of the simulations described in the next section.

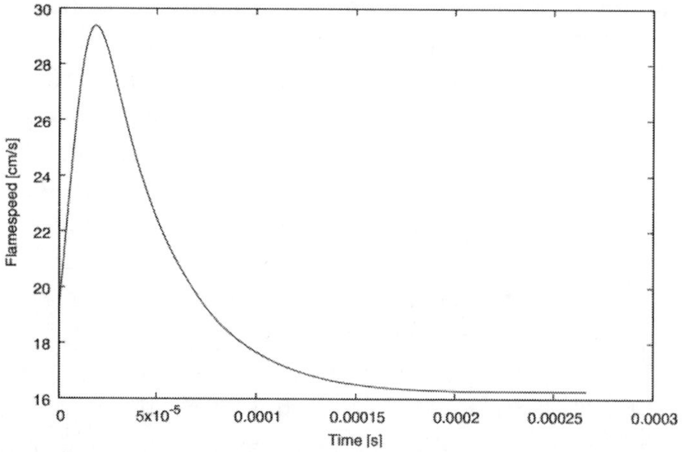

Figure 1: Transient evolution of the flamespeed as function of physical simulation time obtained from the computation of a 1-D premixed flame with stoichiometric proportions of CH4/O2/He. SRK EOS was used in the simulation.

Before the results are presented from simulations it is informative to look at the critical temperature and pressure for all the fuel and oxidizer species considered in this study. The following table, Tab. 1, lists out the critical conditions for the fuel-oxidizer species considered in this study.

Species	T_c [K]	P_c [bar]
CH_4	190.564	45.992
H_2	33.145	12.964
O_2	154.581	50.430
N_2	126.192	33.958
He	5.1953	2.2746

Table 1: Critical pressure and temperature for all the fuel and oxidizer species considered in this study

4.1 Methane flame

Studies presenting experimental measurement of laminar flame speeds at very high pressure conditions are rare. One of the only studies that made an attempt to measure the flamespeeds is the work by Rozenchan[7]. In that study the authors imaged and measured flame speed of spark-ignited expanding spherical methane/oxygen/helium flames up to 60 *atm* pressure and 300*K* temperature, using Schlieren cinematography. As can be observed from Table 1, the unburnt methane/oxygen/helium mixture conditions in the experiments are higher than the critical temperature and pressure of the species in the mixture and thus are in supercritical state. The goal of this study is to investigate the behavior of ideal gas and SRK EOS in reproducing the major combustion characteristics in the form of flame speed.

Figure 2 shows the comparison between experimental measurements of flame speed and from simulations using ideal gas and SRK EOS. For these simulations `DRM 19` chemical mechanism [7], reduced from the detailed GRI mechanism 1.2, with 19 species and 84 reactions is used for both ideal gas and SRK EOS simulations. As can be observed from the Fig. 2, simulations with SRK EOS predict the decrease in flamespeed with pressure, accurately when compared against the experimental measurements and simulations with ideal gas EOS. The slowing down of flamespeed

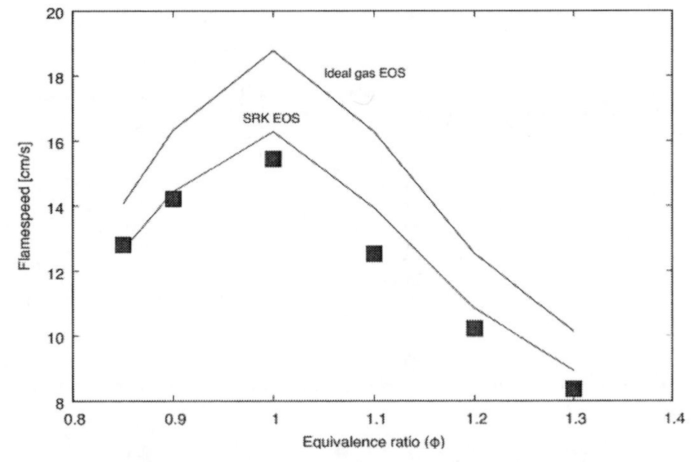

Figure 2: Comparison of flamespeed obtained from experimental measurements [7] and simulations performed using ideal gas and SRK EOS. The filled symbols are experimental measurements, red line is the result from simulations with SRK EOS and the blue line is from simulations with ideal gas EOS

[7] A. Kazakov and M. Frenklach, `http://www.me.berkeley.edu/drm/`

in the case of SRK EOS can also be inferred from the plot of reaction rate of CH_4 corresponding to the flame with stoichiometric fuel-oxidizer ratio, shown in Fig. 3. Figure 4 compares density and

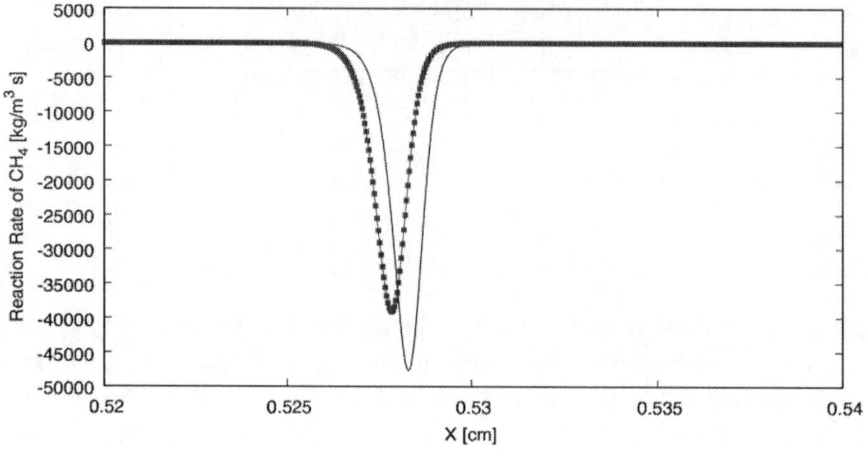

Figure 3: Comparison of reaction rate of CH_4 obtained from simulations performed using ideal gas and SRK EOS. The red line with filled symbols is from simulations with SRK EOS and the blue line is from simulations with ideal gas EOS

temperature profiles from 1-D premixed flame computed at stoichiometric equivalence ratio using ideal gas and SRK EOS. At these supercritical conditions, ideal gas EOS computes higher unburnt mixture density values compared to SRK. However in the burnt stream at higher temperatures where compressibility factor is close to 1, both ideal gas and SRK EOS predict the same density field. In terms of temperature profiles, both ideal gas and SRK EOS simulation results match except a peak in the post flame region. This peak in temperature corresponds to a higher heat release associated with increase in net production of OH radical in the post flame region, as seen in Fig. 5. This increased formation of OH species is consistent with other studies in the literature

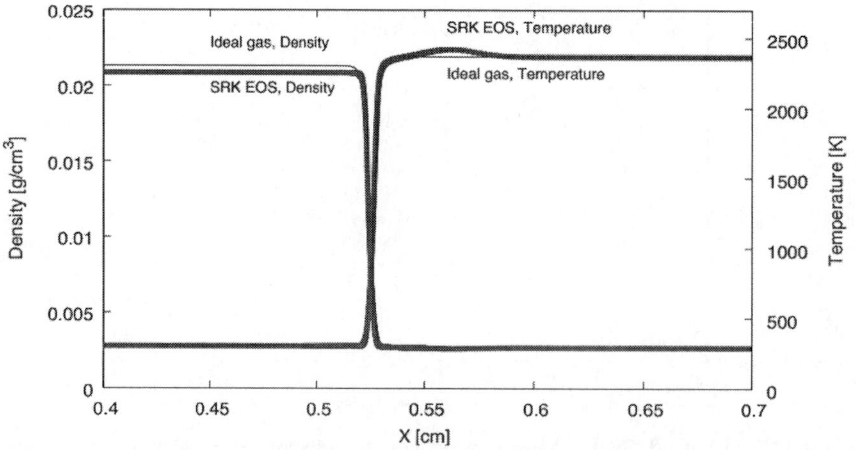

Figure 4: Density and temperature profile from the 1-D premixed freely propagating flame with CH_4/O_2/He mixture at equivalence ratio = 1.0. Red line with symbols is from a simulation with ideal gas EOS and the blue line is from the SRK EOS simulation.

including the study by Rozenchan et. al [7], where the authors observed that the following chain branching reaction becomes very active and plays a significant role in decreasing the flame speed with increasing pressure.

$$HO_2 + CH_3 \rightarrow OH + CH_3O \tag{19}$$

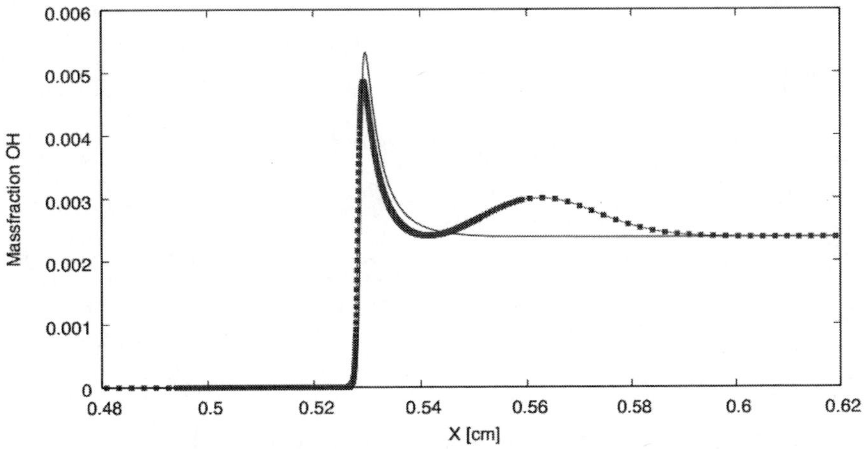

Figure 5: OH massfraction profile from the 1-D premixed freely propagating flame with fuel-air mixture at equivalence ratio = 1.0. Blue line is from a simulation with ideal gas EOS and the red line with symbols is from the SRK EOS simulation.

4.2 Hydrogen flame

To further study the impact of non-ideal corrections to EOS and thermodynamics, a 1-D premixed hydrogen-air flame was also considered. Operating pressure of 100 atm was considered for these flames. Critical temperature of hydrogen is 33.145 K and therefore at room temperature non-ideal effects are relatively small in a hydrogen-air mixture. To bring out significant non-ideal effects the unburnt temperature for hydrogen-air mixture was set 100 K and 1-D premixed flame simulations were performed with SRK EOS. A H_2-Air chemical mechanism consisting of 9 species [8] was chosen to describe the chemical kinetics. For the non-ideal case a laminar flame speed of 1.9 [cm/sec] was obtained, which is in close agreement with the steady non-ideal simulation results presented by Giovangigli *et al.* [9]. The trend, similar to what was observed in methane flames, of reduction of flame speed with increasing pressure is also observed in the case of hydrogen-air flames. Studies by Rozenchan et. al [7] attribute this reduction of flame speed in high pressure hydrogen-flames to the following chain branching mechanisms.

$$\begin{aligned} 2HO_2 &\rightarrow H_2O_2 + O_2 \\ H_2O_2 + M &\rightarrow 2OH + M \\ HO_2 + H &\rightarrow 2OH \end{aligned} \tag{20}$$

From Fig. 6, it can be observed that contrary to the case with methane as fuel, density profiles from the hydrogen-air premixed flames with SRK EOS demonstrate a strong departure when compared against the profiles from ideal gas EOS. However density predicted by SRK and ideal gas

EOS match very accurately in the burnt product stream. These differences in prediction of unburnt density fields from ideal gas and SRK EOS can be attributed to the lower unburnt hydrogen-air mixture. Figure 6 also show the temperature profiles obtained from SRK and ideal gas simulations. No appreciable difference between SRK and ideal gas predicted temperature profiles can be observed. In future studies, multiple simulations varying the operating pressure and fuel-air mixture equivalence ratio will be studied and difference between ideal gas and SRK EOS will be elucidated.

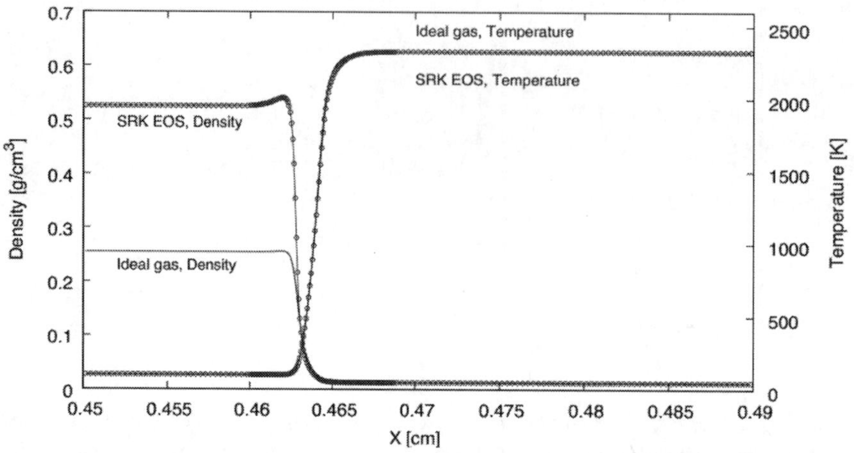

Figure 6: Density and temperature profile from the 1-D premixed freely propagating flame with H_2-Air mixture at equivalence ratio = 1.0. Blue line is from a simulation with ideal gas EOS and the blue line with symbols is from the SRK EOS simulation.

5. Conclusions

In the present study, non-ideal gas corrections to equation of state (EOS), thermodynamics and chemical kinetics were implemented and their impact on 1-D premixed flame simulations were investigated. Soave-Redlich-Kwong cubic equation of state was used as the constitutive relation and also to construct departure functions for various thermodynamic quantities. Chemical kinetics were also modified to accommodate chemical activity coefficients instead of molar concentrations. Freely propagating 1-D premixed flames with CH_4/O_2/He and $H_2/O_2/N_2$ as fuel-oxidizer mixtures were simulated. Results from these simulations were used to demonstrate the improvement in accuracy of flamespeed predictions at higher pressures.

6. Acknowledgements

This research was supported by the Exascale Computing Project (17-SC-20-SC), a collaborative effort of two U.S. Department of Energy organizations (Office of Science and the National Nuclear Security Administration) responsible for the planning and preparation of a capable exascale ecosystem, including software, applications, hardware, advanced system engineering, and early testbed platforms, in support of the nation's exascale computing imperative.
Helpful technical discussions with Dr. Hariswaran Sitaraman are also gratefully acknowledged.

Sub Topic: Laminar flames

References

[1] W. O. Mayer, A. H. Schik, B. Vielle, C. Chauveau, I. Gokalp, D. G. Talley, and R. D. Woodward, Atomization and breakup of cryogenic propellants under high-pressure subcritical and supercritical conditions, Journal of Propulsion and Power 14 (1998) 835–842.

[2] B. E. Poling, J. M. Prausnitz, and J. P. O 'connell, The properties of gases and liquids, vol. 5, Mcgraw-hill New York, 2001.

[3] W. A. Canas-Marn, U. E. Guerrero-Aconcha, and J. D. Ortiz-Arango, Comparison of different cubic equations of state and combination rules for predicting residual chemical potential of binary and ternary Lennard-Jones mixtures: Solid-supercritical fluid phase equilibria, Fluid Phase Equilibria 234 (2005) 42–50.

[4] W. A. Canas-Marn, J. D. Ortiz-Arango, U. E. Guerrero-Aconcha, and C. P. Soto-Tavera, Thermodynamic derivative properties and densities for hyperbaric gas condensates: SRK equation of state predictions versus Monte Carlo data, Fluid Phase Equilibria 253 (2007) 147–154.

[5] A. Nonaka, J. Bell, M. Day, C. Gilet, A. Almgren, and M. Minion, A deferred correction coupling strategy for low Mach number flow with complex chemistry, Combustion Theory and Modelling 16 (2012) 1053–1088.

[6] D. G. Goodwin, H. K. Moffat, and R. L. Speth, *Cantera: An Object-oriented Software Toolkit for Chemical Kinetics, Thermodynamics, and Transport Processes*, Version 2.3.0, 2017, URL: `http://www.cantera.org`.

[7] G. Rozenchan, D. Zhu, C. Law, and S. Tse, Outward propagation, burning velocities, and chemical effects of methane flames up to 60 ATM, Proceedings of the Combustion Institute 29 (2002) 1461–1470.

[8] M. P. Burke, M. Chaos, Y. Ju, F. L. Dryer, and S. J. Klippenstein, Comprehensive H2/O2 kinetic model for high-pressure combustion, International Journal of Chemical Kinetics 44 (2012) 444–474.

[9] V. Giovangigli, L. Matuszewski, and F. Dupoirieux, Detailed modeling of planar transcritical H2-O2-N2 flames, Combustion Theory and Modelling 15 (2011) 141–182.

Western States Section of the Combustion Institute - Fall 2017 Meeting
Hosted by the University of Wyoming
October 2-3, 2017.

Numerical Simulations of a Co-Flow Methane/Air Flame including Ions and Excited Species under Different Gravity Conditions

Claudia-F. Lopez-Camara *Derek Dunn-Rankin*

Environmental Engineering Program, University of California Irvine, 2101 Engineering Gateway, Irvine, California

Buoyancy forces due to gravity affect flames by changing reaction zone properties and flame geometry. These changes have a direct impact on the production and location of species within the flame. Looking toward the possibility of current and future space missions to Mars and in other environments where gravity conditions differ from those on Earth, a prediction of flame behavior has been performed under Earth gravity, microgravity, partial gravity and supergravity conditions. These predictions include the quantification and localization of major species present in the flame -such as fuel, oxygen, or nitrogen- and of minor species -such as electronically excited chemiluminescent species and naturally produced ions in the flame. In addition, the gravity body force has similarities to the electric body force so the work will also be relevant to future studies of electric field forcing of flames.

Simulations of a methane/air co-flow laminar flames are conducted for a 2D axisymmetric geometry using the open-source software OpenFOAM®. The chemical kinetics used in these simulations are from the literature (for neutral species and ionic species) with an addition of a submechanism for excited chemiluminescent species. The simulations are compared with results from experiments of co-flow flames under the influence of an applied electric field to approximate different gravity conditions. Moreover, simulations for methane/air flames are also performed in a zero gravity environment in order to study the ion production without the complication of buoyancy, and to prepare for future electric field microgravity experiments on the ISS as part of NASAs ACME program.

1 Introduction

For more than half a century, the combustion community has been trying to understand the contribution of both ionization chemistry and ion-driven convection in flames. While there has been significant experimental evidence demonstrating that ions and electrons in flames can influence combustion, the effects are often subtle, making their interpretation difficult. However, with the development of comprehensive chemical kinetic models and powerful computational tools, there is chance now to understand these complex phenomena. For example, previous work trying to decouple these effects has helped describe the flame stability limits in non-buoyancy conditions [1].

Understanding the nature of flame behavior should allow us to develop applications that will not only help existing combustion related devices, but also create new ones, as suggested in previous studies [2,3]. Furthermore, understanding how buoyancy effects modify flame behavior could provide interesting insights should we choose to use combustion processes in current and future missions to Mars and other environments where gravity conditions differ from those on Earth.

In addition, recent studies have suggested the use of electric fields as a body force that could be able to interact with the reacting zone of the flame as buoyancy forces do [2,4]

Hence, in this paper, simulations for methane/air flames are performed in a zero gravity environment without the complication of buoyancy and to prepare for future electric field microgravity experiments on the ISS as part of NASAs ACME program. Moreover, simulations under Earth gravity, partial gravity, and supergravity have been carried out in order to see how different buoyancy forces affect the same methane/air diffusion flame.

2 Results

2.1 Chemical Kinetic Model

During recent years, significant progress has been made to improve the molecular transport coefficients and chemical kinetic models to predict methane flame behavior [5,6]. This current work presents a new combination chemical kinetic model (i.e., a selected collection of reactions from published models) for the prediction of major species and minor species such as ions and excited species (table 2 in the Annex). Reactions 1 to 280 come from the model proposed by Prager et al. [6], except that the parameters have been reviewed and checked with the references proposed by [6]. A submodel proposed by Walsh [7] that predicts excited species (OH* and CH*) has been added to the combined chemical kinetic model proposed (table 2, reactions 281 to 299).

Species transport coefficients were predicted by using Cantera® software using the same approach as previous work from Tinajero [4]. The excited species transport coefficients were presumed to be the same as their respective ground state specie. The ionic species transport was similarly presumed to be the same as was used for air as first approximation. Refinements to the model will include using transport coefficients for ions that match their neutral counterpart but for the current work this was not necessary because the concentration of these species is so small.

2.2 Geometry

The axisymmetric geometry models a co-flow methane flame, with the inlet tube (from which fuel is flowing through) extruded from the burner. The left on figure 1 shows the geometry of the burner studied and the right shows the mesh used to represent this burner configuration.

2.3 Boundary Conditions

The boundary condition for the velocity is that the normal component is zero and tangential components are zeroGradient; the top wall (43mm higher than the bottom part of the mesh) has a zeroGradient condition.

For the temperature, a condition of zeroGradient -i.e. normal gradient of T is zero- has been

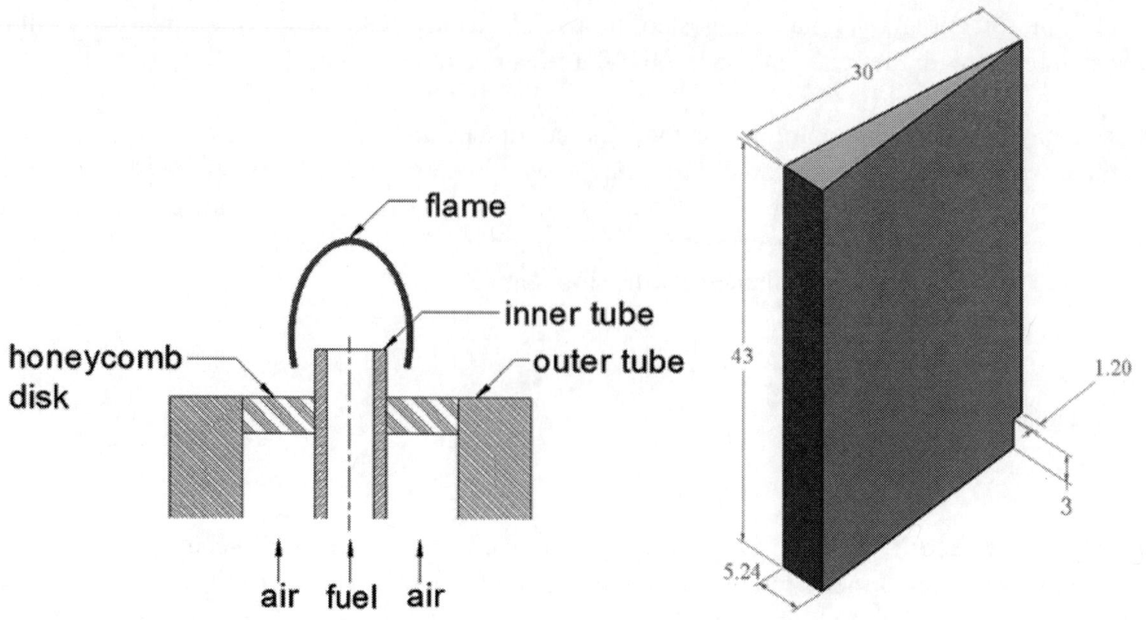

Figure 1: ***Left: Co-Flow Extruded burner. Right: Axial-symmetric Mesh (main shape, [mm])***

set on all boundary walls. Since only steady state is studied for all cases, the simulations were run from ignited simplified cases in which air and fuel were entering to the system at 800K. Once the simplified computation reached steady state, these temperatures were changed to 300K and the simulation cases were re-run until they reached steady state again.

The air was considered as molar fractions of 0.76, 0.2395 and 0.0005 for nitrogen (N_2), oxygen (O_2) and carbon dioxide (CO_2) respectively. The initial pressure is atmospheric for all cases and the pressure rise due to ignition is minimal. For all cases, a non-uniform parabolic velocity profile of the inlet methane was calculated using the approach from [4].

The results were checked to assure that they were mesh independent. The final mesh used had 144234 squared cells formed by 2896417 points, with distribution finer in the flame reaction zone and tip of the burner and gradually expanded(i.e. bigger cells) further from the reaction zone.

2.4 Numerical Model

The solver used was a modified version of the reactingFoam solver in OpenFOAM® . The mass transport equation was modified by imposing an Schmidt number equal to 0.7 condition. Previous studies showed that this approximation was suitable for combustion processes [8,9,10]. Our research group also compared simulations using different approximations (Le=1 and Sc=0.7) and the results showed that there was a better correlation in location of the flame while using the Sc = 0.7 approach by comparing these simulations with experimental results.

2.5 Simulations

Table 1 shows the NASA values for the ratio to Earth gravity of the planets in the Solar System and the Moon. The simulations of this study predict flames behavior under microgravity (0G), partial gravity (0.5G), gravity (1G) and supergravity (1.5G, 2G and 3G). Hence, the gravity conditions used in the simulations are in all the range of the gravity values for the Solar System.

Table 1: Gravity of Planets and Moon (ratio to Earth values)

	Mercury	Venus	Earth	Moon	Mars	Jupiter	Saturn	Uranus	Neptune	Pluto
Gravity (G)	0.378	0.907	1	0.166	0.377	2.36	0.916	0.889	1.12	0.071

Predicted temperature and minor species profiles (H_3O^+, HCO^+ and CH*) are shown in figure 2 under different gravity conditions. The discussion of these results follows.

3 Discussion

Figure 2 shows the main flame characteristics that will be discussed. The profiles shown are the temperature in the flame as well as the CH* (which acts as the visible flame marker) profile and the profile of the two main ions produced in the flame, H_3O^+ and HCO^+ (which highlights the source of electric body forces if a field were to be applied).

3.1 Temperature profile

Due to the co-flow air that is surrounding the inside fuel jet, the completely spherical shape expected from a 0G flame is slightly modified in the wings since air is coming from the sides. As gravity increases, buoyancy forces pull the flame upward and make it narrow. The effects observed in the temperature profiles were not unexpected in response to changing gravitational acceleration. New experimental data is expected to be available from the ACME project in order to validate the 0G profile.

3.2 CH* profile

Chemiluminescence associated with the relaxation of CH* to CH is directly correlated with the luminosity of the flame and how the flame looks to the eye. As shown in figure 2 by the 1G flame, the flame height correlates with that found in the literature [4] and the location of the CH* is well predicted. Moreover, the concentration of CH* is in the range to be the expected by comparing it with the literature [7]. Also, the flame height was observed to be the same while applying different buoyancy forces, and a brighter flame was observed under supergravity conditions.

Detaching from the tip of the burner is predicted, which potentially occurs due to the numerical challenge of accurately predicting edge flame behavior close to these sharp burner boundaries. Hence, more work is needed to predict the bottom of the flame properly.

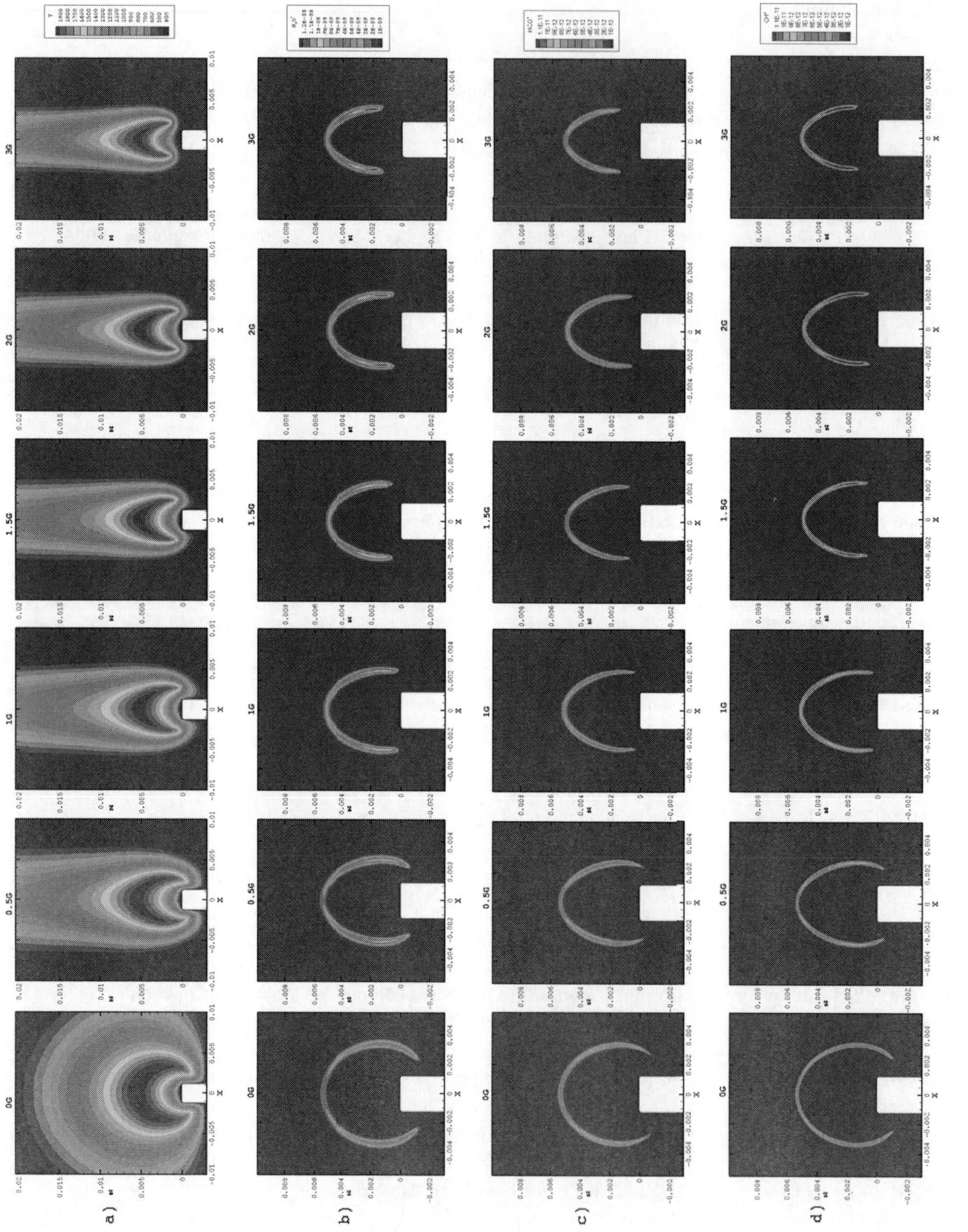

Figure 2: a) Temperature profiles, b) H_3O^+ profiles, c) HCO^+ profiles, d) CH^* profiles

3.3 H_3O^+ and HCO^+ profiles

The main ions naturally produced by the flame, H_3O^+ and HCO^+, are shown to be in similar or higher concentration than the CH*. This fact strengthens the hypothesis of modification of flame behaviors by using electric fields and interacting with these species, since it has been observed that electric fields change the luminosity and flame shape of laminar diffusion flames [1,2,4]. Although detaching from the burner tip is also observed in these cases, the location of the ions correlate with previous Langmuir probes measurements [11]. In fact, H_3O^+ has higher production under stronger gravity environments, and in all cases, it is transported far away from the peak generation region, which supports three hypotheses from the literature: i) HCO^+ is produced but is very short lived and reacts very quickly into H_3O^+; ii) H_3O^+ has a much longer life time, which explains how it is able to travel further outside the flame and have much higher concentrations than other ions; iii) the Langmuir probe may in fact provide a reliable measurement of local ion concentration.

On the other hand, as previously mentioned for CH* profiles, the height of the flame does not seem to be affected by the change in buoyancy forces but the flame width does shrink to create a thinner flame. This might be the explanation of the higher production of H_3O^+ species at super-gravity, since the flame shrinking more tightly packs the species.

4 Conclusion

Simulations including excited species and ionic species reproduced successfully many of the main characteristics of the flame, such as flame height, flame temperature, as well as major and minor species naturally produced by the flame.

Flame height was shown to not be affected by a change in gravity field while flame width was affected by producing a thinner plume as it was expected. However, detachment with the burner tip was observed numerically though it was not seen experimentally, and it is attributed to the numerical challenge related to the prediction of these boundary conditions.
Quantitative and qualitative predictions were made for minor species such as CH*, H_3O^+ and HCO^+. CH* profiles showed a brighter flame while going to higher gravity environments (super gravity conditions). Ionic profiles agreed with previous literature by showing that there is a higher concentration of H_3O^+ produced and H_3O^+ is transported far away from the flame. Also, higher concentration of naturally produced ions has been observed under supergravity conditions, which supports the hypothesis of flame modification by using electric fields. Due to the concentration of ions, the electric field effect could be increased under strong gravity environments. Previous studies showed that flame luminosity is affected by the application of an electric field, and since this is directly correlated with CH*, which is in the same order of magnitude (or lower) than H_3O^+ and HCO^+, it is reasonable to consider that ionic species such as H_3O^+ and HCO^+ will also be affected by electric fields.

Concluding, more work has to be done to accurately predict the bottom of the flame and also the interaction of the flame with electric fields. Nevertheless, the new chemical kinetic model proposed containing excited species and ions has shown to predict important characteristics of the non-premixed methane laminar jet flame.

Acknowledgments

The authors would like to acknowledge important discussions with Jesse A. Tinajero and the support from NASAs ISS Research Project via agreement NNX07AB55A, with Dennis Stocker as contract monitor.

References

[1] S.V. Karnani. PhD Dissertation. Electric Field-Driven Flame Dynamics. University of California at Irvine (2011).

[2] J. Lawton and F.J. Weinberg. Electrical aspects of combustion. Oxford University Press (1970).

[3] A.B. Fialkov. Investigations on ions in flames. Prog. Energy Combustion. 399 (1997) 23.

[4] J.A. Tinajero. PhD Dissertation. Dynamics of chemi-ion driven flows in an applied electric field. University of California at Irvine (2017).

[5] T. Pedersen and R.C. Brown. Simulation of electric field effects in premixed methane flames. Combustion and Flame. 433 (1993) 94.

[6] J. Prager, U. Riedel and J. Warnatz. Modeling ion chemistry and charged species diffusion in lean methane-oxygen flames. Proc. Combust. Inst.1129 (2007) 31.

[7] K. T. Walsh. PhD Dissertation. Quantitative Characterizations of Coflow Laminar Diffusion Flames in a Normal Gravity and Microgravity Environment. Yale University (2002).

[8] J.O. Hinze. Turbulence. The McGraw-Hill Book Company Inc., NewYork. 372 (1987).

[9] N.R. Panchapakesan and J.L. Lumley. Turbulence measurements in axi-symmetric jets of air and helium. Journal of Fluid Mechanics. 225 (1993) 246.

[10] I. Yimer, I. Campbell and L.Y. Jiang. Estimation of the turbulent Schmidtnumber from experimental proles of axial velocity and concentration forhigh-Reynolds-number jet ows. Canadian Aeronautics and Space Journal. 195 (2002) 3.

[11] A. Aguilera Roman. Masters thesis. Relating CH* chemiluminescence with charged species in a non-premixed methane flame. Universitat Politcnica de Catalunya (2015).

Annex

Table 2: Chemical kinetic model. A in [$cm^3\,mole^{-1}\,s^{-1}$)], Ea in [$cal\,mol^{-1}$]. Index n and p in some species denotes negative ions (n) and positive ions(p). HCOp is defined as CHOp in these table.

#	Chemical kinetic model	A	b	E_a
1	O2+H $\rightleftharpoons$ OH+O	9.700E+13	0.0	14832.330
2	H2+O $\rightleftharpoons$ OH+H	5.060E+04	2.670	6281.647
3	H2+OH $\rightleftharpoons$ H2O+H	1.000E+08	1.600	3296.073
4	OH+OH $\rightleftharpoons$ H2O+O	1.500E+09	1.140	100.315
5	H+H+M $\rightleftharpoons$ H2+M	1.800E+18	-1.000	0.000
	H2 /1.0 /H2O /6.5 /O2 /0.4 /AR /0.35 /N2 /0.4 /CO /0.75 /CO2 /1.5 /CH4 /3.0 /			
6	O+O+M $\rightleftharpoons$ O2+M	5.400E+13	0.000	-1767.460
	H2 /1.0 /H2O /6.5 /O2 /0.4 /AR /0.35 /N2 /0.4 /CO /0.75 /CO2 /1.5 /CH4 /3.0 /			
7	H+OH+M $\rightleftharpoons$ H2O+M	5.530E+22	-2.000	0.000
	H2 /1.0 /H2O /2.55 /O2 /0.4 /AR /0.35 /N2 /0.4 /CO /0.75 /CO2 /1.5 /CH4 /3.0 /			
8	H+O2+M $\rightleftharpoons$ HO2+M	2.100E+18	-0.800	0.0
	H2 /1.0 /H2O /6.5 /O2 /0.4 /AR /0.35 /N2 /0.67 /CO /0.75 /CO2 /1.5 /CH4 /3.0 /			
9	HO2+H $\rightleftharpoons$ OH+OH	1.500E+14	0.0	907.614
10	HO2+H $\rightleftharpoons$ H2+O2	3.000E+13	0.0	955.384
11	HO2+H $\rightleftharpoons$ H2O+O	3.000E+13	0.0	1719.69
12	HO2+O $\rightleftharpoons$ OH+O2	2.000E+13	0.0	0.0
13	HO2+OH $\rightleftharpoons$ H2O+O2	6.000E+13	0.0	0.0
14	HO2+HO2 $\rightleftharpoons$ H2O2+O2	4.220E+14	0.0	11975.733
	DUPLICATE			
15	HO2+HO2 $\rightleftharpoons$ H2O2+O2	1.320E+11	0.0	-1628.929
	DUPLICATE			
16	OH+OH(+M) $\rightleftharpoons$ H2O2(+M)	7.230E+13	-0.370	0.0
	LOW / 5.530E+19 -0.760 0.0 /			
	TROE / 0.500 1.0E-30 1.0E+30 /			
	H2 /1.0 /H2O /6.5 /O2 /0.4 /AR /0.35 /N2 /0.4 /CO /0.75 /CO2 /1.5 /CH4 /3.0 /			
17	H2O2+H $\rightleftharpoons$ H2+HO2	1.700E+12	0.0	3749.881

Table 2: Chemical kinetic model. A in [$cm^3\,mole^{-1}\,s^{-1}$)], Ea in [$cal\,mol^{-1}$]. Index n and p in some species denotes negative ions (n) and positive ions(p). HCOp is defined as CHOp in these table.

#	Chemical kinetic model	A	b	E_a
18	H2O2+H $\rightleftharpoons$ H2O+OH	1.000E+13	0.0	3582.689
19	H2O2+O $\rightleftharpoons$ OH+HO2	9.630E+06	2.0	3972.007
20	H2O2+OH $\rightleftharpoons$ H2O+HO2	5.400E+12	0.0	1003.153
21	CO+OH $\rightleftharpoons$ CO2+H	4.760E+07	1.230	69.265
22	CO+HO2 $\rightleftharpoons$ CO2+OH	1.500E+14	0.0	23574.09
23	CO+O+M $\rightleftharpoons$ CO2+M	1.540E+15	0.0	2999.905
	H2 /1.0 /H2O /6.5 /O2 /0.4 /AR /0.35 /N2 /0.4 /CO /0.75 /CO2 /1.5 /CH4 /3.0 /			
24	CO+O2 $\rightleftharpoons$ CO2+O	2.500E+12	0.0	47769.180
25	CH+O $\rightleftharpoons$ CO+H	4.000E+13	0.0	0.0
26	CH+O2 $\rightleftharpoons$ CHO+O	0.843E+13	0.0	0.0
27	CH+CO2 $\rightleftharpoons$ CHO+CO	3.400E+12	0.0	692.653
28	CH+H2O $\rightleftharpoons$ CH2O+H	4.560E+12	0.0	-764.307
29	CH+H2O $\rightleftharpoons$ Th-CH2+OH	1.140E+12	0.0	-764.307
30	CH+OH $\rightleftharpoons$ CHO+H	3.000E+13	0.0	0.0
31	CHO+M $\rightleftharpoons$ CO+H+M	4.500E+14	0.0	15763.829
	H2 /1.0 /H2O /6.5 /O2 /0.4 /AR /0.35 /N2 /0.4 /CO /0.75 /CO2 /1.5 /CH4 /3.0 /			
32	CHO+H $\rightleftharpoons$ CO+H2	9.000E+13	0.0	0.0
33	CHO+O $\rightleftharpoons$ CO+OH	3.000E+13	0.0	0.0
34	CHO+O $\rightleftharpoons$ CO2+H	3.000E+13	0.0	0.0
35	CHO+OH $\rightleftharpoons$ CO+H2O	1.000E+14	0.0	0.0
36	CHO+O2 $\rightleftharpoons$ CO+HO2	3.000E+12	0.0	0.0
37	CHO+CHO $\rightleftharpoons$ CH2O+CO	3.000E+13	0.0	0.0
38	Th-CH2+H $\rightleftharpoons$ CH+H2	6.000E+12	0.0	-1791.344
39	Th-CH2+O $\rightarrow$ CO+H+H	7.600E+13	0.0	0.0
40	Th-CH2+Th-CH2 $\rightleftharpoons$ C2H2+H2	1.200E+13	0.0	812.076
41	Th-CH2+Th-CH2 $\rightleftharpoons$ C2H2+H+H	1.100E+14	0.0	812.076
42	Th-CH2+CH3 $\rightleftharpoons$ C2H4+H	4.200E+13	0.0	0.0

Table 2: Chemical kinetic model. A in [$cm^3\, mole^{-1}\, s^{-1}$)], Ea in [$cal\, mol^{-1}$]. Index n and p in some species denotes negative ions (n) and positive ions(p). HCOp is defined as CHOp in these table.

#	Chemical kinetic model	A	b	E_a
43	Th-CH2+O2 $\rightleftharpoons$ CO+OH+H	1.300E+13	0.0	1480.845
44	Th-CH2+O2 $\rightleftharpoons$ CO2+H2	1.200E+13	0.0	1480.845
45	On-CH2+M $\rightleftharpoons$ Th-CH2+M	1.200E+13	0.0	0.0
	H2 /1.0 /H2O /6.5 /O2 /0.4 /AR /0.35 /N2 /0.4 /CO /0.75 /CO2 /1.5 /CH4 /3.0 /			
46	On-CH2+O2 $\rightleftharpoons$ CO+OH+H	3.100E+13	0.0	0.0
47	CH3+H $\rightleftharpoons$ On-CH2+H2	6.000E+13	0.0	15095.061
48	Th-CH2+O $\rightleftharpoons$ CO+H2	4.400E+13	0.0	0.0
49	CH2O+M $\rightleftharpoons$ CHO+H+M	1.400E+36	-5.54	96632.274
	H2 /1.0 /H2O /6.5 /O2 /0.4 /AR /0.35 /N2 /0.4 /CO /0.75 /CO2 /1.5 /CH4 /3.0 /			
50	CH2O+H $\rightleftharpoons$ CHO+H2	1.270E+08	1.62	2149.613
51	CH2O+O $\rightleftharpoons$ CHO+OH	4.150E+11	0.57	2770.612
52	CH2O+OH $\rightleftharpoons$ CHO+H2O	3.400E+09	1.18	-446.642
53	CH2O+HO2 $\rightleftharpoons$ CHO+H2O2	3.000E+12	0.0	13064.871
54	CH2O+CH3 $\rightleftharpoons$ CHO+CH4	7.830E-08	6.1	1958.536
55	CH2O+O2 $\rightleftharpoons$ CHO+HO2	6.000E+13	0.0	40770.995
56	CH2O+M $\rightleftharpoons$ H2+CO+M	3.250E+36	-5.54	96632.274
	H2 /1.0 /H2O /6.5 /O2 /0.4 /AR /0.35 /N2 /0.4 /CO /0.75 /CO2 /1.5 /CH4 /3.0 /			
57	CH3+M $\rightleftharpoons$ Th-CH2+H+M	1.000E+16	0.0	90522.596
	H2 /1.0 /H2O /6.5 /O2 /0.4 /AR /0.35 /N2 /0.4 /CO /0.75 /CO2 /1.5 /CH4 /3.0 /			
58	CH3+M $\rightleftharpoons$ CH+H2+M	6.900E+14	0.0	82409.001
	H2 /1.0 /H2O /6.5 /O2 /0.4 /AR /0.35 /N2 /0.4 /CO /0.75 /CO2 /1.5 /CH4 /3.0 /			
59	On-CH2+CH3 $\rightleftharpoons$ C2H4+H	1.600E+13	0.00	-568.453
60	CH3+O $\rightleftharpoons$ CH2O+H	8.430E+13	0.0	0.0
61	CH4(+M) $\rightleftharpoons$ CH3+H(+M)	2.400E+16	0.000	104853.35
	LOW / 1.290E+18 0.000 379.97 /			
	TROE / 0.000 1350.0 1.0 7830.0 /			
	H2 /1.0 /H2O /6.5 /O2 /0.4 /AR /0.35 /N2 /0.4 /CO /0.75 /CO2 /1.5 /CH4 /0.66 /			

Table 2: Chemical kinetic model. A in [$cm^3\, mole^{-1}\, s^{-1}$)], Ea in [$cal\, mol^{-1}$]. Index n and p in some species denotes negative ions (n) and positive ions(p). HCOp is defined as CHOp in these table.

#	Chemical kinetic model	A	b	E_a
62	CH3+OH → CH3O+H	4.000E+14	0.0	15477.214
63	CH3O+H → CH3+OH	4.750E+16	-0.13	21018.439
64	CH3+OH ⇌ On-CH2+H2O	3.000E+13	0.0	2780.166
65	CH3+O2 ⇌ CH2O+OH	1.700E+11	0.0	8932.837
66	CH3+HO2 ⇌ CH3O+OH	1.600E+13	0.0	0.0
67	CH4+O2 ⇌ CH3+HO2	3.900E+13	0.0	56845.324
68	CH3+CH3(+M) ⇌ C2H6(+M)	3.610E+13	0.000	0.00
	LOW / 3.630E+41 -7.000 11.60 /			
	TROE / 0.620 73.0 1180.0 0.0 /			
	H2 /1.0 /H2O /6.5 /O2 /0.4 /AR /0.35 /N2 /0.4 /CO /0.75 /CO2 /1.5 /CH4 /3.0 /			
69	CH3O+M ⇌ CH2O+H+M	5.000E+13	0.0	25078.82
	H2 /1.0 /H2O /6.5 /O2 /0.4 /AR /0.35 /N2 /0.4 /CO /0.75 /CO2 /1.5 /CH4 /3.0 /			
70	CH3O+H ⇌ CH2O+H2	1.800E+13	0.0	0.0
71	CH3O+O2 ⇌ CH2O+HO2	3.000E+10	0.0	2125.729
72	CH2O+CH3O ⇌ CH3OH+CHO	1.150E+11	0.0	1241.999
73	CH3O+O → O2+CH3	1.320E+13	0.0	0.0
74	CH3+O2 → O+CH3O	0.600E+14	0.0	31288.813
75	CH3O+O ⇌ OH+CH2O	1.800E+12	0.0	0.0
76	CH3O+OH ⇌ CH2O+H2O	1.810E+13	0.0	0.0
77	CH2OH+M ⇌ CH2O+H+M	5.000E+13	0.0	25078.82
	H2 /1.0 /H2O /6.5 /O2 /0.4 /AR /0.35 /N2 /0.4 /CO /0.75 /CO2 /1.5 /CH4 /3.0 /			
78	CH2OH+H ⇌ CH2O+H2	3.000E+13	0.0	0.0
79	CH2OH+O2 ⇌ CH2O+HO2	1.570E+15	-1.0	0.0
	DUPLICATE			
80	CH2OH+O2 ⇌ CH2O+HO2	7.230E+13	0.0	3575.523
	DUPLICATE			
81	CH2OH+H ⇌ CH3+OH	1.000E+13	0.0	0.0

Table 2: Chemical kinetic model. A in [$cm^3\,mole^{-1}\,s^{-1}$)], Ea in [$cal\,mol^{-1}$]. Index n and p in some species denotes negative ions (n) and positive ions(p). HCOp is defined as CHOp in these table.

#	Chemical kinetic model	A	b	E_a
82	CH3+O2(+M) $\rightleftharpoons$ CH3O2(+M)	7.800E+08	1.2	0.0
	LOW / 1.650E+26 -3.300 0.0 /			
	TROE / 0.495 2325.5 10.0 0.0 /			
	H2 /1.0 /H2O /6.5 /O2 /0.4 /AR /0.35 /N2 /0.4 /CO /0.75 /CO2 /1.5 /CH4 /3.0 /			
83	CH3O2+CH2O $\rightleftharpoons$ CH3O2H+CHO	0.130E+12	0.0	9004.49
84	CH3O2+CH3 $\rightleftharpoons$ CH3O+CH3O	1.500E+13	0.0	-1194.23
85	CH3O2+HO2 $\rightleftharpoons$ CH3O2H+O2	2.400E+11	0.0	-1576.383
86	CH3O2+CH3O2 $\rightarrow$ CH2O+CH3OH+O2	2.800E+10	0.0	-788.191
87	CH3O2+CH3O2 $\rightarrow$ CH3O+CH3O+O2	2.700E+10	0.0	-3788.191
88	CH4+H $\rightleftharpoons$ H2+CH3	1.330E+04	3.000	8032.388
89	CH4+O $\rightleftharpoons$ OH+CH3	6.900E+08	1.560	8479.029
90	CH4+OH $\rightleftharpoons$ H2O+CH3	1.540E+07	1.830	2770.612
91	CH4+HO2 $\rightleftharpoons$ H2O2+CH3	1.000E+13	0.0	24625.012
92	CH4+CH $\rightleftharpoons$ C2H4+H	3.000E+13	0.0	-406.038
93	CH4+Th-CH2 $\rightleftharpoons$ CH3+CH3	4.300E+12	0.0	10031.528
94	CH3+OH(+M) $\rightleftharpoons$ CH3OH(+M)	6.000E+13	0.000	0.00
	LOW / 4.560E+44 -8.200 0.00 /			
	TROE / 0.820 200.0 1438.0 0.0 /			
	H2 /1.0 /H2O /6.5 /O2 /0.4 /AR /0.35 /N2 /0.4 /CO /0.75 /CO2 /1.5 /CH4 /3.0 /			
95	CH3OH+H $\rightleftharpoons$ CH2OH+H2	3.100E+13	0.0	6090.57
96	CH3OH+H $\rightleftharpoons$ CH3O+H2	0.900E+13	0.0	6090.57
97	CH3OH+O $\rightleftharpoons$ CH2OH+OH	3.880E+05	2.5	3078.724
98	CH3OH+O $\rightleftharpoons$ CH3O+OH	1.300E+05	2.5	4996.656
99	CH3OH+OH $\rightleftharpoons$ CH2OH+H2O	3.000E+04	2.65	-883.730
100	CH3OH+OH $\rightleftharpoons$ CH3O+H2O	5.300E+03	2.65	-883.730
101	CH3OH+HO2 $\rightleftharpoons$ CH2OH+H2O2	0.620E+13	0.0	19370.402
102	CH3OH+CH3 $\rightleftharpoons$ CH4+CH2OH	3.190E+01	3.17	7165.377

Table 2: Chemical kinetic model. A in [$cm^3\,mole^{-1}\,s^{-1}$)], Ea in [$cal\,mol^{-1}$]. Index n and p in some species denotes negative ions (n) and positive ions(p). HCOp is defined as CHOp in these table.

#	Chemical kinetic model	A	b	E_a
103	CH3OH+CH3 $\rightleftharpoons$ CH4+CH3O	1.410E+01	3.1	6926.531
104	CH3O+CH3OH $\rightleftharpoons$ CH2OH+CH3OH	1.500E+12	0.0	6998.185
105	CH3OH+CH2O $\rightleftharpoons$ CH3O+CH3O	0.153E+13	0.0	79583.454
106	CH3OH+O2 $\rightleftharpoons$ HO2+CH2OH	2.050E+13	0.0	45165.76
107	CH3O2H $\rightleftharpoons$ CH3O+OH	6.000E+14	0.0	42299.609
108	OH+CH3O2H $\rightleftharpoons$ H2O+CH3O2	7.300E+11	0.0	-238.846
109	O+CH3O2H $\rightleftharpoons$ OH+CH3O2	1.990E+13	0.0	4753.033
110	CH3O2+H2O2 $\rightleftharpoons$ CH3O2H+HO2	2.400E+12	0.0	9983.759
111	CH3O2+CH4 $\rightleftharpoons$ CH3O2H+CH3	1.810E+11	0.0	18582.211
112	CH3O2+CH3OH $\rightleftharpoons$ CH3O2H+CH2OH	1.810E+11	0.0	13781.408
113	C2H+O $\rightleftharpoons$ CO+CH	1.000E+13	0.0	0.0
114	C2H+O2 $\rightleftharpoons$ HCCO+O	3.000E+12	0.0	0.0
115	C2H+O2 $\rightleftharpoons$ CO2+CH	9.000E+12	0.0	0.0
116	C2H+CH4 $\rightleftharpoons$ C2H2+CH3	1.810E+12	0.0	0.0
117	HCCO+H $\rightleftharpoons$ Th-CH2+CO	1.400E+14	0.0	0.0
118	HCCO+O $\rightarrow$ CO+CO+H	9.640E+13	0.0	0.0
119	HCCO+Th-CH2 $\rightleftharpoons$ C2H3+CO	3.000E+13	0.0	0.0
120	CH+CO $\rightleftharpoons$ HCCO	2.800E+11	0.0	-1695.806
121	C2H2+M $\rightleftharpoons$ C2H+H+M	3.600E+16	0.0	106525.271
	H2 /1.0 /H2O /6.5 /O2 /0.4 /AR /0.35 /N2 /0.4 /CO /0.75 /CO2 /1.5 /CH4 /3.0 /			
122	C2H2+O2 $\rightleftharpoons$ HCCO+OH	2.000E+08	1.5	30094.583
123	C2H2+H $\rightleftharpoons$ C2H+H2	6.023E+13	0.0	27801.663
124	C2H2+O $\rightleftharpoons$ Th-CH2+CO	2.200E+06	2.1	1569.218
125	C2H2+O $\rightleftharpoons$r HCCO+H	5.100E+06	2.1	1569.218
126	C2H2+OH $\rightleftharpoons$ H2O+C2H	6.200E+13	0.0	12945.448
127	CH2CO+M $\rightleftharpoons$ Th-CH2+CO+M	1.000E+16	0.0	59233.783
	H2 /1.0 /H2O /6.5 /O2 /0.4 /AR /0.35 /N2 /0.4 /CO /0.75 /CO2 /1.5 /CH4 /3.0 /			

Table 2: Chemical kinetic model. A in [$cm^3\, mole^{-1}\, s^{-1}$)], Ea in [$cal\, mol^{-1}$]. Index n and p in some species denotes negative ions (n) and positive ions(p). HCOp is defined as CHOp in these table.

#	Chemical kinetic model	A	b	E_a
128	CH2CO+H $\rightleftharpoons$ CH3+CO	3.000E+13	0.0	3367.727
129	CH2CO+O $\rightleftharpoons$ CH2O+CO	0.750E+12	0.0	1361.422
130	CH2CO+O $\rightarrow$ CHO+H+CO	0.750E+12	0.0	1361.422
131	CH2CO+O $\rightleftharpoons$ CHO+CHO	0.750E+12	0.0	1361.422
132	CH2CO+OH $\rightleftharpoons$ CH3+CO2	2.520E+12	0.0	0.0
133	CH2CO+OH $\rightleftharpoons$ CH2O+CHO	4.680E+12	0.0	0.0
134	C2H3(+M) $\rightleftharpoons$ C2H2+H(+M)	1.900E+14	0.000	39720.073
	LOW / 1.000E+42 -7.500 190.4 /			
	TROE / 0.350 1.0E-30 1.0E+30 /			
	H2 /1.0 /H2O /6.5 /O2 /0.4 /AR /0.35 /N2 /0.4 /CO /0.75 /CO2 /1.5 /CH4 /3.0 /			
135	C2H3+OH $\rightleftharpoons$ C2H2+H2O	5.000E+12	0.0	0.0
136	C2H3+H $\rightleftharpoons$ C2H2+H2	1.200E+13	0.0	0.0
137	C2H3+O $\rightleftharpoons$ C2H2+OH	1.000E+13	0.0	0.0
138	C2H3+O $\rightleftharpoons$ CH3+CO	1.000E+13	0.0	0.0
139	C2H3+O $\rightleftharpoons$ CHO+Th-CH2	1.000E+13	0.0	0.0
140	C2H3+O2 $\rightleftharpoons$ CH2O+CHO	5.420E+12	0.0	0.0
	DUPLICATE			
141	C2H3+O2 $\rightleftharpoons$ CH2O+CHO	7.000E+14	-0.6	3128.881
	DUPLICATE			
142	C2H3+O2 $\rightleftharpoons$ CH2CHO+O	2.460E+15	-0.78	3133.658
143	C2H3+O2 $\rightleftharpoons$ C2H2+HO2	1.340E+06	1.61	-384.542
144	CH3+CO(+M) $\rightleftharpoons$ CH3CO(+M)	5.058E+11	0.000	6871.597
	LOW / 3.109E+14 0.000 15.88 /			
	TROE / 0.500 1.0E-30 1.0E+30 /			
	H2 /1.0 /H2O /6.5 /O2 /0.4 /AR /0.35 /N2 /0.4 /CO /0.75 /CO2 /1.5 /CH4 /3.0 /			
145	CH3CO+H $\rightleftharpoons$ CH2CO+H2	2.000E+13	0.0	0.0
146	CH2CHO+H $\rightleftharpoons$ CH2CO+H2	2.000E+13	0.0	0.0

Table 2: Chemical kinetic model. A in [$cm^3\, mole^{-1}\, s^{-1}$)], Ea in [$cal\, mol^{-1}$]. Index n and p in some species denotes negative ions (n) and positive ions(p). HCOp is defined as CHOp in these table.

#	Chemical kinetic model	A	b	E_a
147	C2H4+M ⇌ C2H2+H2+M	1.000E+17	0.0	71653.77
	H2 /1.0 /H2O /6.5 /O2 /0.4 /AR /0.35 /N2 /0.4 /CO /0.75 /CO2 /1.5 /CH4 /3.0 /			
148	C2H4+M ⇌ C2H3+H+M	7.400E+17	0.0	96493.744
	H2 /1.0 /H2O /6.5 /O2 /0.4 /AR /0.35 /N2 /0.4 /CO /0.75 /CO2 /1.5 /CH4 /3.0 /			
149	C2H4+H ⇌ C2H3+H2	5.200E+14	0.0	14903.984
150	C2H4+O ⇌ CH2CHO+H	4.800E+06	1.88	179.134
151	C2H4+O ⇌ CHO+CH3	8.100E+06	1.88	179.134
152	C2H4+O ⇌ CH2CO+H2	6.800E+05	1.88	179.134
153	C2H4+OH ⇌ C2H3+H2O	2.050E+13	0.0	5947.263
154	C2H4+CH3 ⇌ C2H3+CH4	3.500E+12	0.0	11130.219
155	CH3CHO(+M) ⇌ CH3+CHO(+M)	2.200E+15	0.0	81876.375
	LOW / 5.100E+12 0.000 131.4 /			
	TROE / 0.500 1.0E-30 1.0E+30 /			
	H2 /1.0 /H2O /6.5 /O2 /0.4 /AR /0.35 /N2 /0.4 /CO /0.75 /CO2 /1.5 /CH4 /3.0 /			
156	CH3CHO+H ⇌ CH3CO+H2	2.150E+09	1.16	2,388.459
157	CH3CHO+H ⇌ CH2CHO+H2	1.850E+09	1.16	2,388.459
158	CH3CHO+O ⇌ CH3CO+OH	5.600E+12	0.0	1815.229
159	CH3CHO+O ⇌ CH2CHO+OH	2.400E+11	0.0	1815.229
160	CH3CHO+O2 ⇌ CH3CO+HO2	4.000E+13	0.0	39242.381
161	CH3CHO+OH ⇌ CH3CO+H2O	2.300E+10	0.73	-1098.691
162	CH3CHO+HO2 ⇌ CH3CO+H2O2	3.100E+12	0.0	11942.295
163	CH3CHO+Th-CH2 ⇌ CH3CO+CH3	2.500E+12	0.0	3797.65
164	CH3CHO+CH3 ⇌ CH3CO+CH4	2.050E-06	5.60	2460.113
165	CH3CHO+OH ⇌ CH2CHO+H2O	0.050E+10	0.73	-1098.691
166	C2H4+H(+M) → C2H5(+M)	3.970E+09	1.280	1289.768
	LOW / 6.980E+18 0.000 3.20 /			
	TROE / 0.760 40.0 1025.0 0.0 /			

Table 2: Chemical kinetic model. A in [$cm^3\,mole^{-1}\,s^{-1}$)], Ea in [$cal\,mol^{-1}$]. Index n and p in some species denotes negative ions (n) and positive ions(p). HCOp is defined as CHOp in these table.

#	Chemical kinetic model	A	b	E_a
	H2 /1.0 /H2O /6.5 /O2 /0.4 /AR /0.35 /N2 /0.4 /CO /0.75 /CO2 /1.5 /CH4 /3.0 /			
167	C2H5(+M) → C2H4+H(+M)	8.200E+13	0.000	39839.496
	LOW / 3.400E+17 0.000 139.60 /			
	TROE / 0.750 97.0 1379.0 0.0 /			
	H2 /1.0 /H2O /6.5 /O2 /0.4 /AR /0.35 /N2 /0.4 /CO /0.75 /CO2 /1.5 /CH4 /3.0 /			
168	C2H5+H ⇌ CH3+CH3	6.000E+13	0.0	0.0
169	C2H5+O ⇌ CH3CHO+H	5.320E+13	0.0	0.0
170	C2H5+O ⇌ CH2O+CH3	1.300E+13	0.0	0.0
171	C2H5+O2 ⇌ C2H4+HO2	1.000E+10	0.0	-2173.498
172	C2H5+CH3 ⇌ C2H4+CH4	1.140E+12	0.0	0.0
173	C2H5+C2H5 ⇌ C2H4+C2H6	1.400E+12	0.0	0.0
174	C2H5O ⇌ CH3CHO+H	2.000E+14	0.0	23168.052
175	C2H5O ⇌ CH2O+CH3	8.000E+13	0.0	21496.131
176	C2H5O+O2 ⇌ CH3CHO+HO2	6.000E+10	0.0	1671.921
177	C2H5O+OH ⇌ CH3CHO+H2O	1.000E+14	0.0	0.0
178	C2H5O+H ⇌ CH3CHO+H2	1.000E+14	0.0	0.0
179	C2H5O+O ⇌ CH3CHO+OH	1.210E+14	0.0	0.0
180	CH3CHOH ⇌ CH3CHO+H	1.000E+14	0.0	25078.82
181	CH3CHOH+H ⇌ CH3CHO+H2	3.000E+13	0.0	0.0
182	CH3CHOH+OH ⇌ CH3CHO+H2O	1.510E+13	0.0	0.0
183	CH3CHOH+O ⇌ CH3CHO+OH	1.200E+14	0.0	0.0
184	CH3CHOH+O2 ⇌ CH3CHO+HO2	1.200E+13	0.0	0.0
185	CH2CH2OH ⇌ C2H4+OH	1.000E+14	0.0	33438.426
186	CH2CH2OH+H ⇌ CH3CHO+H2	5.000E+13	0.0	0.0
187	C2H5OH ⇌ CH3+CH2OH	3.100E+15	0.0	80538.837
188	C2H5OH ⇌ C2H5+OH	5.000E+16	0.0	91143.595
189	C2H5OH ⇌ C2H4+H2O	1.000E+14	0.0	76645.649

Table 2: Chemical kinetic model. A in [$cm^3\, mole^{-1}\, s^{-1}$)], Ea in [$cal\, mol^{-1}$]. Index n and p in some species denotes negative ions (n) and positive ions(p). HCOp is defined as CHOp in these table.

#	Chemical kinetic model	A	b	E_a
190	C2H5OH+OH ⇌ CH3CHOH+H2O	5.250E+06	2.0	453.807
191	C2H5OH+OH ⇌ C2H5O+H2O	1.150E+06	2.0	907.614
192	C2H5OH+OH ⇌ CH2CH2OH+H2O	8.130E+06	2.0	597.115
193	C2H5OH+O ⇌ CH3CHOH+OH	7.940E+12	0.0	3248.304
194	C2H5OH+O ⇌ C2H5O+OH	4.790E+13	0.0	6854.877
195	C2H5OH+O ⇌ CH2CH2OH+OH	1.000E+14	0.0	7475.877
196	C2H5OH+H ⇌ CH3CHOH +H2	4.400E+12	0.0	4561.957
197	C2H5OH+H ⇌ C2H5+H2O	5.900E+11	0.0	3439.381
198	C2H5OH+HO2 ⇌ CH3CHOH+H2O2	6.300E+12	0.0	19370.402
199	C2H5OH+CH3 ⇌ CH3CHOH+CH4	2.040E+11	0.0	8693.991
200	C2H5OH+CH3 ⇌ CH2CH2OH+CH4	2.040E+11	0.0	8693.991
201	C2H5OH+CH3 ⇌ C2H5O+CH4	7.490E+10	0.0	9386.644
202	C2H5OH+CH3O ⇌ CH3CHOH+CH3OH	2.000E+11	0.0	6998.185
203	C2H5OH+CH2O ⇌ C2H5O+CH3O	1.530E+12	0.0	79583.454
204	C2H5OH+C2H5O ⇌ C2H5OH+CH3CHOH	2.000E+11	0.0	6998.185
205	C2H6+H ⇌ C2H5+H2	1.550E+09	1.5	7428.107
206	C2H6+O ⇌ C2H5+OH	1.000E+09	1.5	5827.84
207	C2H6+OH ⇌ C2H5+H2O	7.200E+06	2.0	859.845
208	C2H6+HO2 ⇌ C2H5+H2O2	1.330E+13	0.0	20516.863
209	C2H6+O2 ⇌ C2H5+HO2	6.000E+13	0.0	51829.56
210	C2H6+Th-CH2 ⇌ C2H5+CH3	2.200E+13	0.0	8670.106
211	C2H6+CH3 ⇌ C2H5+CH4	1.500E-07	6.0	6066.686
212	C2H6+CH ⇌ C2H4+CH3	1.080E+14	0.0	-262.730
213	CH+H ⇌ C+H2	1.200E+14	0.00	0.0
214	C+O2 ⇌ CO+O	5.000E+13	0.00	0.0
215	CH+O ⇌ CHOp+E	2.512E+11	0.0	1700.105
216	CHOp+E ⇌ CO+H	7.400E+18	-0.68	0.000

Table 2: Chemical kinetic model. A in [$cm^3\,mole^{-1}\,s^{-1}$)], Ea in [$cal\,mol^{-1}$]. Index n and p in some species denotes negative ions (n) and positive ions(p). HCOp is defined as CHOp in these table.

#	Chemical kinetic model	A	b	E_a
217	CHOp+H2O $\rightleftharpoons$ H3Op+CO	1.506E+15	0.00	0.000
218	CHOp+C2H5OH $\rightleftharpoons$ H3Op+CO+C2H4	6.000E+14	0.00	0.000
219	H3Op+E $\rightleftharpoons$ H2O+H	2.291E+18	-0.50	0.000
220	H3Op+E $\rightleftharpoons$ OH+H+H	7.949E+21	-1.37	0.000
221	H3Op+E $\rightleftharpoons$ H2+OH	1.253E+19	-0.50	0.000
222	H3Op+E $\rightleftharpoons$ O+H2+H	6.000E+17	-0.30	0.000
223	H3Op+C $\rightleftharpoons$ CHOp+H2	6.022E+12	0.00	0.000
224	CHOp+CH2CO $\rightleftharpoons$ C2H3Op+CO	1.259E+15	-0.048	0.000
225	CHOp+CH3 $\rightleftharpoons$ C2H3Op+H	7.763E+14	-0.006	0.000
226	C2H3Op+E $\rightleftharpoons$ CH2CO+H	2.291E+18	-0.500	0.000
227	H3Op+CH2CO $\rightleftharpoons$ C2H3Op+H2O	1.204E+15	0.000	0.000
228	C2H3Op+E $\rightleftharpoons$ CO+CH3	2.403E+17	-0.050	0.000
229	C2H3Op+O $\rightleftharpoons$ CHOp+CH2O	2.000E+14	0.000	0.000
230	CHOp+CH3OH $\rightleftharpoons$ CH5Op+CO	8.710E+14	-0.056	0.000
231	H3Op+CH3OH $\rightleftharpoons$CH5Op+H2O	1.506E+15	0.000	0.000
232	CH5Op+E $\rightleftharpoons$ CH3OH+H	2.403E+17	-0.050	0.000
233	CH5Op+CH2CO $\rightleftharpoons$ C2H3Op+CH3OH	1.486E+15	-0.077	-82.880
234	O2+E+O $\rightleftharpoons$ O2n+O	3.627E+16	0.00	0.000
235	O2+E+O2 $\rightleftharpoons$ O2n+O2	1.523E+21	-1.00	1191.602
236	O2+E+H2O $\rightleftharpoons$ O2n+H2O	5.077E+18	0.00	0.000
237	O2n+H2 $\rightleftharpoons$ H2O2+E	6.022E+14	0.00	0.000
238	O2n+H $\rightleftharpoons$ HO2+E	7.226E+14	0.00	0.000
239	O2n+C2H3Op $\rightleftharpoons$ O2+CH3CO	2.090E+18	-0.50	0.000
240	O2n+OH $\rightleftharpoons$ OHn+O2	6.022E+13	0.00	0.000
241	O2n+H $\rightleftharpoons$ OHn+O	1.084E+15	0.00	0.000
242	OHn+O $\rightleftharpoons$ HO2+E	1.204E+14	0.00	0.000
243	OHn+H $\rightleftharpoons$ H2O+E	1.084E+15	0.00	0.000

Table 2: Chemical kinetic model. A in [$cm^3\,mole^{-1}\,s^{-1}$)], Ea in [$cal\,mol^{-1}$]. Index n and p in some species denotes negative ions (n) and positive ions(p). HCOp is defined as CHOp in these table.

#	Chemical kinetic model	A	b	E_a
244	E+OH+M ⇌ OHn+M	1.088E+17	0.00	0.000
	H2 /1.0 /H2O /6.5 /O2 /0.4 /AR /0.35 /N2 /0.4 /CO /0.75 /CO2 /1.5 /CH4 /3.0 /			
245	OHn+C ⇌ CHO+E	3.001E+14	0.00	0.000
246	OHn+CH ⇌ CH2O+E	3.001E+14	0.00	0.000
247	OHn+CH3 ⇌ CH3OH+E	6.022E+14	0.00	0.000
248	OHn+CO2+O2 ⇌ CHO3n+O2	2.760E+20	0.00	0.000
249	OHn+CO2+H2O ⇌ CHO3n+H2O	1.104E+21	0.00	0.000
250	CO3n+H ⇌ OHn+CO2	1.020E+14	0.00	0.000
251	CO3n+O ⇌ O2n+CO2	4.800E+13	0.00	0.000
252	CHO2n+H ⇌ CO2+H2+E	1.159E+14	0.00	0.000
253	OHn+CHO ⇌ CHO2n+H	2.959E+15	-0.14	-105.331
254	E+O+O2 ⇌ On+O2	3.627E+16	0.00	0.000
255	E+O+O ⇌ On+O	3.021E+17	0.00	0.000
256	On+C ⇌ CO+E	3.011E+14	0.00	0.000
257	On+H2 ⇌ OHn+H	1.987E+13	0.00	0.000
258	On+CH4 ⇌ OHn+CH3	6.022E+13	0.00	0.000
259	On+H2O ⇌ OHn+OH	8.431E+14	0.00	0.000
260	On+CH2O ⇌ OHn+CHO	5.601E+14	0.00	0.000
261	On+CH2O ⇌ CHO2n+H	1.307E+15	0.00	0.000
262	On+C2H6 ⇌ C2H5+OHn	6.130E+15	-0.50	0.000
263	On+H ⇌ OH+E	3.011E+14	0.00	0.000
264	On+H2 ⇌ H2O+E	4.215E+14	0.00	0.000
265	On+CH ⇌ CHO+E	3.011E+14	0.00	0.000
266	On+On-CH2 ⇌ CH2O+E	3.011E+14	0.00	0.000
267	On+CO ⇌ CO2+E	3.914E+14	0.00	0.000
268	On+O ⇌ O2+E	8.431E+13	0.00	0.000
269	On+C2H2 ⇌ CH2CO+E	7.226E+14	0.00	0.000

Table 2: Chemical kinetic model. A in [$cm^3\, mole^{-1}\, s^{-1}$)], Ea in [$cal\, mol^{-1}$]. Index n and p in some species denotes negative ions (n) and positive ions(p). HCOp is defined as CHOp in these table.

#	Chemical kinetic model	A	b	E_a
270	On+H2O $\rightleftharpoons$ H2O2+E	3.613E+11	0.00	0.000
271	O2n+O $\rightleftharpoons$ On+O2	1.987E+14	0.00	0.000
272	On+CO2+O2 $\rightleftharpoons$ CO3n+O2	1.123E+20	0.00	0.000
273	On+C2H3Op $\rightleftharpoons$ O+CH3CO	2.090E+18	-0.50	0.000
274	CHO3n+C2H3Op $\rightleftharpoons$ CH3CO+CO2+OH	2.000E+18	0.00	0.000
275	CHO3n+CH5Op $\rightleftharpoons$ CH3OH+H2O+CO2	2.000E+18	0.00	0.000
276	On+C2H3Op $\rightleftharpoons$ O+CH2CHO	1.000E+18	0.00	0.000
277	O2n+C2H3Op $\rightleftharpoons$ O2+CH2CO+H	1.000E+18	0.00	0.000
278	On+C2H3Op $\rightleftharpoons$ O+CH2CO+H	1.000E+18	0.00	0.000
279	O2n+CH5Op $\rightleftharpoons$ O2+CH3+H2O	1.000E+18	0.00	0.000
280	On+CH5Op $\rightleftharpoons$ O+CH3+H2O	1.000E+18	0.00	0.000
281	C2H+O $\rightleftharpoons$ CH*+CO	1.08E+13	0.00	0.000
282	C2H+O2 $\rightleftharpoons$ CH*+CO2	2.17E+10	0.00	0.000
283	CH* $\rightarrow$ CH+hν	1.85E+06	0.00	0.000
284	CH*+N2 $\rightleftharpoons$ CH+N2	3.03E+02	3.40	-381
285	CH*+O2 $\rightleftharpoons$ CH+O2	2.48E+06	2.14	-1720
286	CH*+H2O $\rightleftharpoons$ CH+H2O	5.30E+13	0.00	0.000
287	CH*+H2 $\rightleftharpoons$ CH+H2	1.47E+14	0.00	1361
288	CH*+CO2 $\rightleftharpoons$ CH+CO2	2.40E-01	4.30	-1694
289	CH*+CO $\rightleftharpoons$ CH+CO	2.44E+12	0.50	0.000
290	CH*+CH4 $\rightleftharpoons$ CH+CH4	1.73E+13	0.00	167
291	CH+O2 $\rightleftharpoons$ OH*+CO	3.25E+13	0.00	0.000
292	OH*= $\rightarrow$ OH	1.45E+06	0.00	0.000
293	OH*+N2 $\rightleftharpoons$ OH+N2	1.08E+11	0.50	-1238
294	OH*+O2 $\rightleftharpoons$ OH+O2	2.10E+12	0.50	-482
295	OH*+H2O $\rightleftharpoons$ OH+H2O	5.92E+12	0.50	-861
296	OH*+H2 $\rightleftharpoons$ OH+H2	2.95E+12	0.50	-444

Table 2: Chemical kinetic model. A in [$cm^3\,mole^{-1}\,s^{-1}$)], Ea in [$cal\,mol^{-1}$]. Index n and p in some species denotes negative ions (n) and positive ions(p). HCOp is defined as CHOp in these table.

#	Chemical kinetic model	A	b	E_a
297	OH*+CO2 ⇌ OH+CO2	2.75E+12	0.50	-968
298	OH*+CO ⇌ OH+CO	3.23E+12	0.50	-787
299	OH*+CH4 ⇌ OH+CH4	3.36E+12	0.50	-635

Western States Section of the Combustion Institute - Fall 2017 Meeting
Hosted by University of Wyoming
October 2-3, 2017

Influences of stoichiometry on steadily propagating triple flames in counterflows

P. Rajamanickam *W. Coenen* *A. L. Sánchez* *F. A. Williams*

Department of Mechanical and Aerospace Engineering, University of California San Diego, La Jolla, CA 92093–0411, USA

Most studies of triple flames in counterflowing streams of fuel and oxidizer have been focused on the symmetric problem in which the stoichiometric mixture fraction is $1/2$. There then exist lean and rich premixed flames of roughly equal strengths, with a diffusion flame trailing behind from the stoichiometric point at which they meet. In the majority of realistic situations, however, the stoichiometric mixture fraction departs appreciably from unity, typically being quite small. With the objective of clarifying the influences of these effects of stoichiometry, attention is focused on one of the simplest possible models, addressed here mainly by numerical integration. A leading point is identified along one of the premixed flames, which controls the propagation velocity and is found to be dominant to such an extent that the diffusion flame and the other premixed flame are very weak by comparison, when the stoichiometric mixture fraction departs appreciably from $1/2$. These curved partially premixed flames then are expected to be more representative of flames in realistic counterflow triple-flame configurations.

1 Introduction

Triple flames, first identified by Phillips over fifty years ago [1], play a fundamental role in many practical combustion systems. Since they have been observed to move along mixing layers that are strained in laminar and turbulent jet flows, for example, there is interest in investigating their response to the imposed strain rate. The counterflow mixing layer separating two opposed planar jets of fuel and oxidizer, used in previously in theoretical [2, 3] and experimental [4, 5] studies, provides an attractive canonical problem for analyzing these effects. An appreciable literature has now been developed on the subject, as may be seen from a recent review [6]. The present contribution is intended to offer some additional observations concerning such steadily propagating triple flames, by building on a simplification of the formulation of Daou and Liñán [3], who emphasized effects of Lewis numbers by parametrically studying, both numerically and analytically, these triple flames in mixtures with unequal diffusivities. Since underlying influences of stoichiometry may tend to be obscured by varying Lewis numbers, the present considerations are restricted to equi-diffusional systems in which all Lewis numbers are unity. Under that restriction, implications are considered here for systems with stoichiometries that are quite likely to be encountered in practice. The simplifications that will be introduced in the formulation will be identical to those in this reference, which also are employed in a number of other publications, thereby facilitating comparisons.

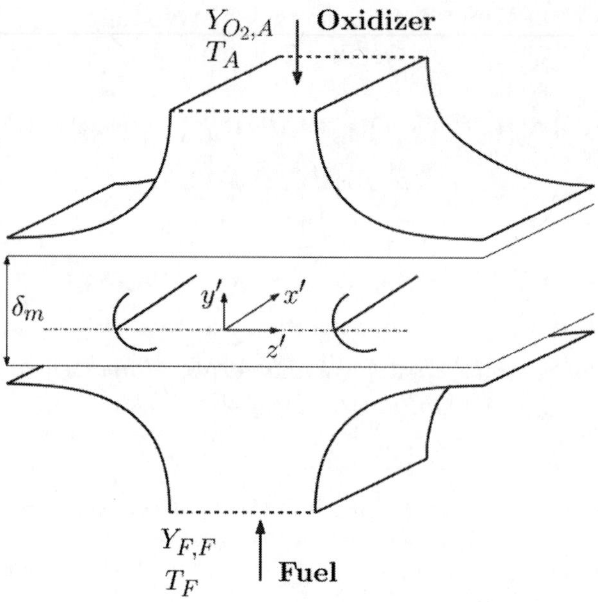

Figure 1: Schematic view of counterflow mixing layer.

2 Problem formulation

The analysis adopts a one-step irreversible reaction for the chemistry description, with the unit mass of fuel reacting with s units of mass of oxygen to generate products according to

$$\text{Fuel} + s\text{O}_2 \rightarrow (1+s)\text{Products} + q, \tag{1}$$

with q denoting the amount of energy released in the process. The number of moles of fuel burned per unit volume per unit time

$$\omega = B\left(\frac{\rho Y_F}{W_F}\right)\left(\frac{\rho Y_{O_2}}{W_{O_2}}\right)\mathrm{e}^{-E_a/RT} \tag{2}$$

is assumed to have a first-order dependence on each reactant concentration and an exponential dependence on the temperature T through an Arrhenius constant involving a pre-exponential factor B and an activation energy E_a. Here, ρ and T are the density and temperature of the gas mixture and R is the universal gas constant. Mass fractions and molecular weights of species i are represented by Y_i and W_i, respectively.

Following [3], we investigate triple-flame propagation in a strained mixing layer separating two-dimensional counterflowing streams of fuel and oxidizer, with the front propagating at a constant speed U in the direction perpendicular to the plane of counterflow motion. To render the problem steady, a reference frame moving with the front will be used in the description, as indicated in figure 1, with the counterflowing streams approaching from $y' = \pm\infty$ and the front propagating in the negative x' direction. Figure 2 further illustrates the situation by showing a representative curved flame region in the plane of symmetry for an advancing triple flame, with streamlines appropriate to the approximations that are to be introduced, for a stoichiometric mixture fraction less than 1/2. This figure shows the leading point along the flame as an enlarged dot, where the rate of advancement is effectively determined. The illustrated leading point is the point on the propagating front where the local flow velocity with respect to the front is normal to the front, the component of the flow velocity tangential to the front being in opposite directions on opposite sides of the leading point. The conditions at the leading point determine the steady propagation velocity of the flame pattern in the sense that, irrespective of the manner in which the propagation velocity varies with the equivalence ratio, the flame shape can readily change to adjust to the variations of the incoming flow velocity that are encountered away from the leading point.

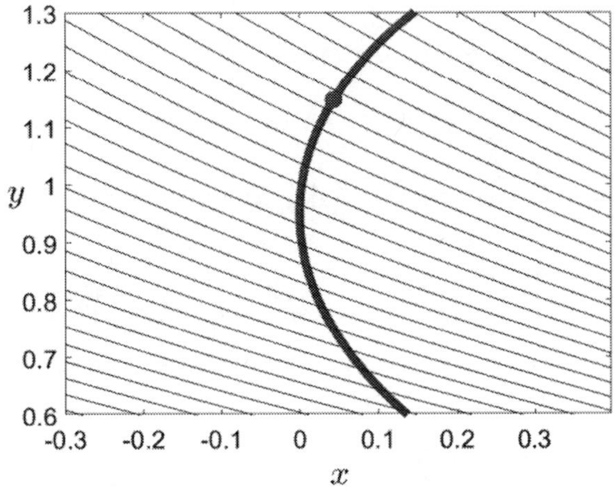

Figure 2: A typical propagating flame front for $S \gg 1$. Leading point is denoted by the red dot, where the front has only normal component of local burning velocity.

In the analysis, the thermo-diffusive approximation (i.e. constant density and constant transport properties) is adopted, following the earlier work [2, 3]. Then the counterflow velocity field reduces to the familiar stagnation-point solution $(v, w) = (-Ay', Az')$ in terms of the strain rate A, which defines, together with the thermal diffusivity D_T, the characteristic mixing-layer thickness $\delta_m = (D_T/A)^{1/2}$. With the dimensionless variables

$$x = \frac{x'}{\delta_m},\ y = \frac{y'}{\delta_m},\ \hat{Y}_F = \frac{Y_F}{Y_{F,F}},\ \hat{Y}_O = \frac{Y_{O_2}}{Y_{O_2,A}},\ \hat{T} = \frac{T - T_A}{\gamma T_A}, \tag{3}$$

and the dilution-adjusted stoichiometric ratio S (i.e. the amount of air needed to burn the unit mass of the fuel stream), non-dimensional heat release γ, and non-dimensional pre-exponential factor $\hat{B}$,

$$S = \frac{sY_{F,F}}{Y_{O_2,A}},\ \gamma = \frac{qY_{F,F}}{c_p T_A(1+S)},\ \hat{B} = \frac{\rho B Y_{O_2,A}}{W_{O_2}} \tag{4}$$

(where c_p is the specific heat at constant pressure), the relevant conservation equations become

$$\frac{U}{(D_T A)^{1/2}}\frac{\partial \hat{Y}_F}{\partial x} - y\frac{\partial \hat{Y}_F}{\partial y} = \frac{\partial^2 \hat{Y}_F}{\partial x^2} + \frac{\partial^2 \hat{Y}_F}{\partial y^2} - \frac{\hat{B}}{A}\hat{Y}_F\hat{Y}_O \mathrm{e}^{-E_a/RT}, \tag{5}$$

$$\frac{U}{(D_T A)^{1/2}}\frac{\partial \hat{Y}_O}{\partial x} - y\frac{\partial \hat{Y}_O}{\partial y} = \frac{\partial^2 \hat{Y}_O}{\partial x^2} + \frac{\partial^2 \hat{Y}_O}{\partial y^2} - S\frac{\hat{B}}{A}\hat{Y}_F\hat{Y}_O \mathrm{e}^{-E_a/RT}, \tag{6}$$

$$\frac{U}{(D_T A)^{1/2}}\frac{\partial \hat{T}}{\partial x} - y\frac{\partial \hat{T}}{\partial y} = \frac{\partial^2 \hat{T}}{\partial x^2} + \frac{\partial^2 \hat{T}}{\partial y^2} + (1+S)\frac{\hat{B}}{A}\hat{Y}_F\hat{Y}_O \mathrm{e}^{-E_a/RT}. \tag{7}$$

The temperature and composition of the feed streams provide the boundary conditions

$$\hat{T} - \hat{T}_F = \hat{Y}_F - 1 = \hat{Y}_O = 0 \text{ as } y \to -\infty \quad \text{and} \quad \hat{T} = \hat{Y}_F = \hat{Y}_O - 1 = 0 \text{ as } y \to \infty. \tag{8}$$

Upstream from the flame the mixture is chemically frozen, corresponding to the boundary distributions

$$\hat{T}/\hat{T}_F = \hat{Y}_F = 1 - \hat{Y}_O = \frac{1}{2}\mathrm{erfc}\left(\frac{y}{\sqrt{2}}\right) \quad \text{as } x \to -\infty, \tag{9}$$

whereas the trailing diffusion flame that emerges downstream leads to the boundary condition

$$\frac{\partial \hat{T}}{\partial x} = \frac{\partial \hat{Y}_F}{\partial x} = \frac{\partial \hat{Y}_O}{\partial x} = 0 \text{ as } x \to \infty. \tag{10}$$

The two-dimensional boundary–value problem defined by (5)–(10) constitutes an eigenvalue problem that determines the propagation velocity U as an eigenvalue.

The above problem can be simplified by noting that, when Lewis numbers are unity, species mass fractions and the temperature are related to the mixture fraction

$$Z = \frac{S\hat{Y}_F - \hat{Y}_O + 1}{S+1} = \frac{\hat{T} + (1+S)\hat{Y}_F}{\hat{T}_F + 1 + S}, \tag{11}$$

where $\hat{T}_F = (T_F - T_A)/(\gamma T_A)$ measures the difference in temperature between fuel and oxidizer streams. The mixture fraction Z is unity in the fuel feed stream and zero in the oxidizer feed stream; it also satisfies a convective-diffusive balance, so that, under the present assumptions, its solution is given simply by

$$Z = \frac{1}{2}\mathrm{erfc}\left(\frac{y}{\sqrt{2}}\right). \tag{12}$$

Local stoichiometric conditions, defined by the equation $S\hat{Y}_F = \hat{Y}_O$, are reached on the planar surface $y = y_s$, where

$$Z_s = \frac{1}{S+1} = \frac{1}{2}\mathrm{erfc}\left(\frac{y_s}{\sqrt{2}}\right), \tag{13}$$

as follows from (11) and (12). The problem then reduces to that of integrating (7) supplemented with (11) and (12).

In looking for an appropriate scale for U, one may note that for weakly strained systems, the front becomes thinner than the mixing layer, and its propagation velocity U approaches the propagation velocity of the stoichiometric planar deflagration $S_{L\infty}$, which depends on the values of the stoichiometric flame temperature

$$T_s = (1+\gamma)T_A + (T_F - T_A)Z_s \quad (14)$$

and stoichiometric Zel'dovich number

$$\beta_s = \frac{E_a}{RT_s}\frac{T_s - T_A}{T_S}. \quad (15)$$

The value of $S_{L\infty}$ and the associated flame residence time $D_T/S_{L\infty}^2$ are to be used as scales in defining the dimensionless front velocity $\tilde{U}$ and the dimensionless strain rate $\tilde{A}$ (or Karlovitz number) as

$$\tilde{U} = \frac{U}{S_{L\infty}}, \quad \tilde{A} = \frac{D_T A}{S_{L\infty}^2}, \quad (16)$$

thereby reducing (7) to

$$\frac{\tilde{U}}{\sqrt{\tilde{A}}}\frac{\partial \hat{T}}{\partial x} - y\frac{\partial \hat{T}}{\partial y} = \frac{\partial^2 \hat{T}}{\partial x^2} + \frac{\partial^2 \hat{T}}{\partial y^2} + \frac{\hat{\omega}}{Z_s \tilde{A}}, \quad (17)$$

where

$$\hat{\omega} = \frac{\beta_s^3 \hat{Y}_F \hat{Y}_O}{4(1-Z_s)} \exp\left[-\frac{\beta_s(\hat{T}_s - \hat{T})}{1-\sigma_s(\hat{T}_s - \hat{T})}\right] \quad (18)$$

is the dimensionless reaction rate, with $\sigma_s = \gamma T_A/T_s$ and $\hat{T}_s = (T_s - T_A)/(\gamma T_A) = 1 + \hat{T}_F Z_s$. The problem reduces to that of integrating (17) subject to the boundary conditions

$$y \to -\infty: \quad \hat{T} = \hat{T}_F, \quad y \to \infty: \quad \hat{T} = 0, \quad (19)$$

$$x \to -\infty: \quad \hat{T} = \hat{T}_F Z, \quad x \to \infty: \quad \frac{\partial \hat{T}}{\partial x} = 0. \quad (20)$$

The reaction rate (18) must be evaluated with use of $\hat{Y}_F = Z\hat{T}_s - Z_s\hat{T}$ and $\hat{Y}_O = (1-Z) + (1-Z_s)(Z\hat{T}_F - \hat{T})$, obtained from (11), with $Z(y)$ given in (12). Note that, since the extinction strain rate A_E for the one-dimensional diffusion flame that develops behind the triple flame is of order $S_{L\infty}^2/D_T$, the solution for the propagating triple flame can be anticipated to exist only for values of $\tilde{A}$ in the range $0 < \tilde{A} < \tilde{A}_E$ with $\tilde{A}_E = A_E/(S_{L\infty}^2/D_T) \sim 1$.

3 Numerical Results

A pseudo-transient finite-element scheme with adaptive grid is employed in the numerical integration, which is marched in time until a steady solution is reached. The integration domain is a rectangular box defined to be sufficiently large to satisfy the boundary conditions smoothly. The results shown below correspond to $x_{min} = -20$, $x_{max} = 20$, $y_{min} = -8$, $y_{max} = 8$. Since the problem defined above exhibits invariance under translations in the x direction, to anchor the flame the additional condition $\hat{T} = 0.3$ is imposed at $x = 0$ along the stoichiometric line $y = y_s$. The

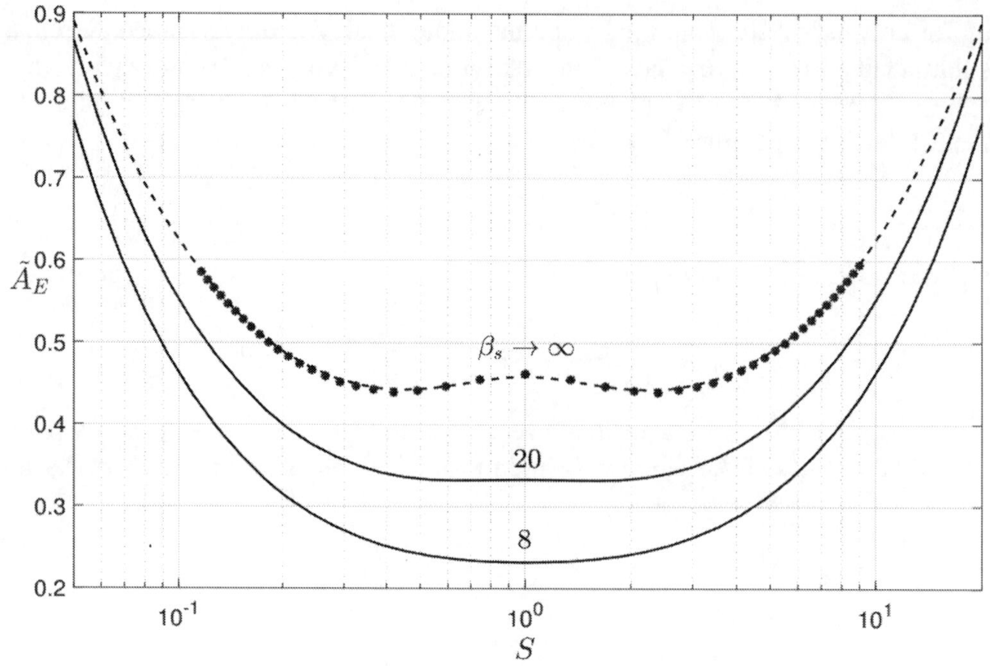

Figure 3: Extinction strain rate $\tilde{A}_E$ as a function of stoichiometric ratio S for $\beta_s = 8,\ 20,\ \sigma_s = 0.85$ and $\hat{T}_F = 0$. The solid lines are obtained from numerical integration of (21). The dashed line is from [7] and the points are from [8].

parametric values $\beta_s = 8$ & 20, $\sigma_s = 0.85$ and $\hat{T}_F = 0$ (equal feed temperatures) are used in the integrations for three different values of stoichiometric ratio $S = (1, 4, 17.2)$, with $S = 4$ and $S = 17.2$ selected as representative of the conditions found in methane-oxygen and methane-air combustion, respectively.

As previously mentioned, the steady triple flame cannot exist for strain rates above the extinction value $\tilde{A}_E$ of the one-dimensional diffusion flame that emerges for $x \to \infty$, described by integration of

$$\frac{d^2\hat{T}}{dy^2} + y\frac{d\hat{T}}{dy} + \frac{\hat{\omega}}{Z_s\tilde{A}} = 0, \quad \hat{T}(-\infty) = \hat{T}_F,\ \hat{T}(\infty) = 0, \tag{21}$$

obtained by neglecting x-derivative terms in (17). The value of $\tilde{A}_E$, obtained as the maximum value of $\tilde{A}$ for which there exists a solution to (21), is shown in figure 3 as a function of S for the two values of the $\beta_s = 8$ & 20 considered here. For completeness, the numerical results are compared with the asymptotic predictions for $\beta_s \gg 1$ [7, 8].

The influence of the stoichiometry of fuel stream on the structure of the propagating flame is investigated in figure 4 for $\beta_s = 8$ using contours of reaction rates $\hat{\omega}$ defined in (18). The front shapes for $\beta_s = 20$, not shown here, were found to be quite similar to those for $\beta_s = 8$, except for an overall reduction in the spatial extent of the reaction region, consistent with the stronger temperature sensitivity associated with the increase in activation energy.

To better identify the relative position of the flame, the stagnation plane $y = 0$ and the stoichiometric plane $y = y_s$ are represented in each plot by a dot-dashed line and a solid line, respectively.

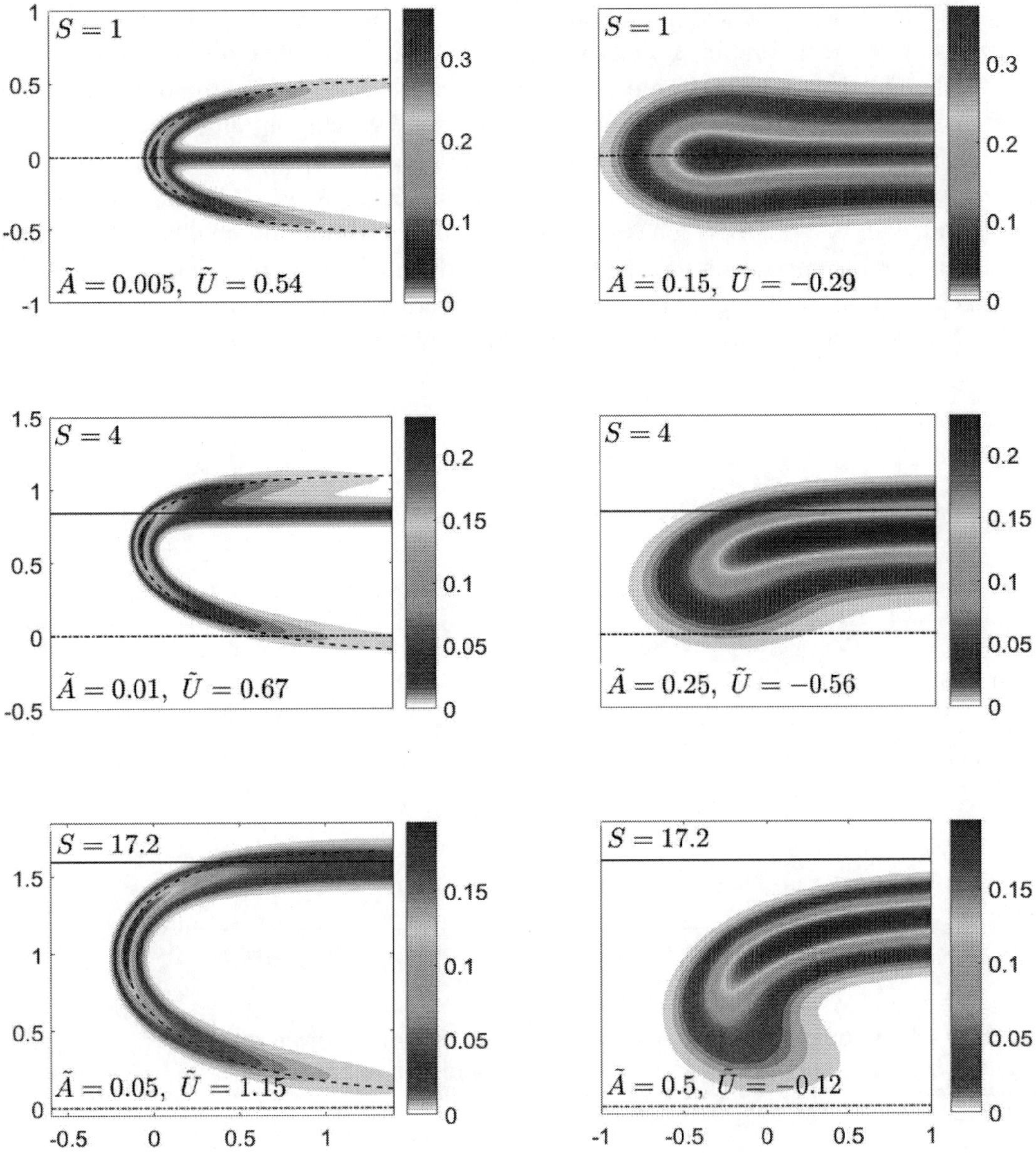

Figure 4: Reaction-rate $\hat{\omega}$ contours for fixed values of $\beta_s = 8$, $\sigma_s = 0.85$ and $\hat{T}_F = 0$ and for three different values of S at different strain rates $\tilde{A}$. Left plots correspond to advancing fronts and right plots show retreating fronts. Solid lines represent stoichiometric locations, and the dash-dotted line is the stagnation plane. The dashed curves are solutions of (24).

Two values of the strain rate are selected in the figure for each value of S, with the smaller value corresponding to an advancing front with $\tilde{U} > 0$ and the higher value corresponding to a retreating front with $\tilde{U} < 0$. The symmetric solutions for $S = 1$, similar to those reported earlier [2, 3], result in a triple-flame structure for low strain rates and an edge-flame structure for near-extinction strain rates. As can be seen, the symmetric character is lost for $S = 4$, with the flame migrating to the oxidizer side of the mixing layer and the associated lean flame that develops for $y > y_s$ becoming very weak. This lean branch disappears altogether for $S = 17.2$, and the propagating front adopts a C shape, with one of the wings of the premixed front evolving into the trailing diffusion flame as $x \to \infty$. Also of interest is that at the lower strain rate, the front for $S = 17.2$ is found to propagate at a velocity $\tilde{U} = 1.15 > 1$ higher than its stoichiometric value. This behavior, already reported for the range $\tilde{A} \ll \beta_s^{-2}$ by Daou & Liñán [3] through asymptotic analysis, can be explained by investigating the composition dependence of one-dimensional planar flames, as will be discussed in greater detail in a different publication. The retreating fronts are seen in the figure for $S \gg 1$ to tend to become hook-shaped when the advancing fronts are C-shaped.

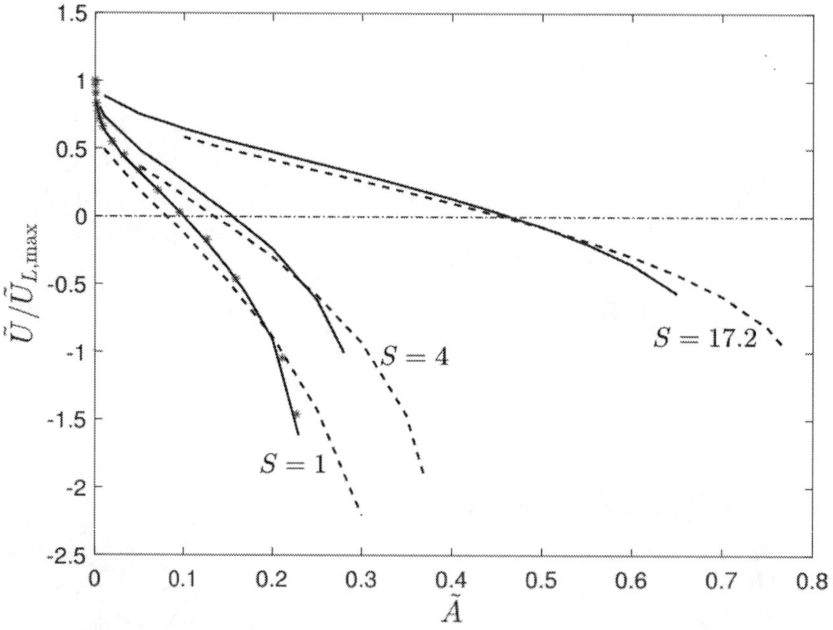

Figure 5: Triple-flame propagation velocity $\tilde{U}$ as a function of strain rate $\tilde{A}$ for $S = 1,\ 4,\ 17.2$, $\sigma_s = 0.85$ and $\hat{T}_F = 0$. Solid curves correspond to $\beta_s = 8$ and dashed curves to $\beta_s = 20$. The asterisk marks are from [3].

The computed dependence of the flame velocity on the strain rate is shown in figure 5 for the same three different values of S. Since for a given activation energy β_s, the two-dimensional flame propagation velocity cannot exceed its one-dimensional planar maximum speed for any strain rate $\tilde{A}$ between ignition and extinction i.e, $\tilde{U} < \tilde{U}_{L,\max}$, it is appropriate to plot the two-dimensional flame velocity normalized by its one-dimensional maximum velocity. As can be seen, for each value of S there exists an intermediate value of $\tilde{A}$ at which $\tilde{U} = 0$, thereby defining the boundary between advancing fronts with $\tilde{U} > 0$ and retreating fronts with $\tilde{U} < 0$. The negative value of the propagation velocity of the retreating front diverges as the strain rate approaches the extinction value $\tilde{A}_E$ shown in figure 3, seen, consistent with that figure, to be different for the two different activation energies considered here.

4 Kinematics of thin fronts

The initial analytical description of flame-front structures and propagation velocities for thin fronts, corresponding to low-strain feed streams, is due to Dold *et al.* [9, 10, 11], later extended to fuels with non-unity Lewis numbers by Daou & Liñán [2, 3], both works invoking activation-energy asymptotics in the description. The following analysis can be considered to be a generalization of those previous analyses to apply also for more complex chemistry.

The non-dimensional laminar flame thickness for a low strain rate can be written as $\delta_L(y)/\delta_m = \sqrt{\tilde{A}}/\tilde{U}_L$. In the limit $\sqrt{\tilde{A}} \ll 1$, a premixed wing of a triple flame can be considered to be a thin front, leading to a simple description $x = f(y)$ for the front shape. For fronts with small curvature, the local burning velocity is given in dimensional form [12, 13, 14] by

$$U_f = U_L - U_L \mathcal{M} \delta_L \kappa + \mathcal{M} \delta_L \boldsymbol{n} \cdot \nabla \boldsymbol{v} \cdot \boldsymbol{n}, \tag{22}$$

where the components of the velocity vector in the present application are $(u, v) = (U', -Ay')$. Here the Markstein number, in the thermo-diffusive approximation, reduces to $\mathcal{M} = 1$ [3, 15], $\boldsymbol{n}$ is the unit normal pointing towards unburnt gas, and $\kappa = \nabla \cdot \boldsymbol{n}$ is the front curvature, $-\boldsymbol{n} \cdot \nabla \boldsymbol{v} \cdot \boldsymbol{n}$ being the imposed strain rate associated with the velocity gradients.With $\boldsymbol{n} = -\nabla G/|\nabla G|$, the equation for the front ptopagation can be written as,

$$\boldsymbol{v} \cdot \nabla G = U_f |\nabla G|, \tag{23}$$

and it is convenient to put $G(x, y) = x - f(y)$.

Treating $\epsilon = \sqrt{\tilde{A}}$ as small parameter and defining front slope as $p = df/dy$, the non-dimensional front-propagation equation reduces to

$$\tilde{U} + \epsilon y p = \tilde{U}_L \sqrt{1 + p^2} - \epsilon \frac{1}{1 + p^2} \frac{dp}{dy} - \epsilon^2 \frac{p^2}{\hat{U}_L \sqrt{1 + p^2}}, \tag{24}$$

which becomes a nonlinear eigenvalue problem because this equation needs to be integrated until the slope on either side of the front becomes large (i.e., $|p| \to \infty$), with a small curvature (at least $p^{-2} dp/dy \to 0$). At the boundaries these conditions yield from (24)

$$p \to \frac{\tilde{U}}{\pm(\tilde{U}_L - \epsilon^2/\tilde{U}_L) - \epsilon y}. \tag{25}$$

It is clear from this equation that $|p|$ peaks at a finite y, placing constraints on the domain of integration; the locations at which p becomes large are obtained by solving for the roots of the denominator of the equation. These two roots corresponding to the lower and upper bounds of the y domain, are denoted by $y_{-\infty}$ and y_∞, respectively. Beyond these limits, the equation carries no physical meaning.

Once $p(y)$ and $\tilde{U}$ are determined, the leading point, as defined earlier, can be identified by seeking for the location at which the tangential velocity goes to zero, tantamount to finding the root of

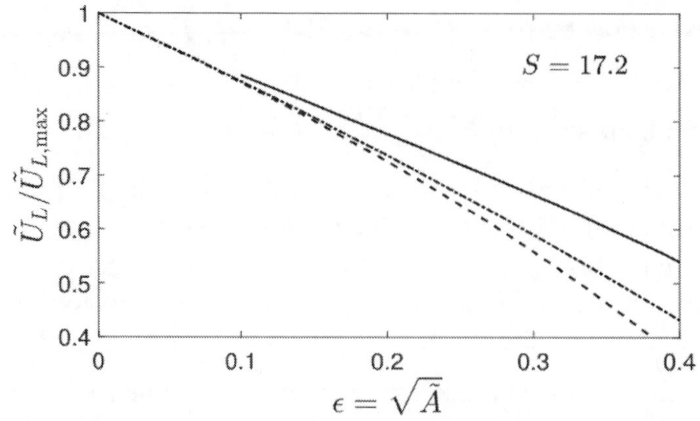

Figure 6: Dashed lines are obtained from the integration of (24), and dashed-dotted curves are the results of the perturbation analysis given in (26). Solid curves are the actual two-dimensional computational results.

$\tilde{U}p - \epsilon y = 0$. A singular-perturbation analysis, based on the small parameter ϵ, similar to that of [9], predicts the eigenvalue as

$$\tilde{U} = U_m - \epsilon\sqrt{-U''_m} + \frac{\epsilon^2}{U_m}\left(\frac{\gamma_1^2}{2} - \gamma_2 - y_m\gamma_1\right), \tag{26}$$

where

$$U_m = \tilde{U}_{L,\max},\ y_m = y_{\max},\ \gamma_1 = y_m - \frac{U'''_m}{3(-U''_m)},\ \gamma_2 = 1 - \frac{7}{72}\frac{U'''^2_m}{U''^2_m} - \frac{1}{8}\frac{U^{iv}_m}{(-U''_m)}, \tag{27}$$

and $U''_m, U'''_m, \ldots$ are derivatives of $\tilde{U}_L(y)$ evaluated at the maximum location. This formula for the eigenvalue implicitly assumes that all of the derivatives of $\hat{U}_L$ are continuous and exist. An additional correction of order $\epsilon^{3/2}$ is required if a discontinuity occurs within the range $|y - y_m| \sim \epsilon^{1/2}$. The result provides the dashed curves in the left-hand plots of figure 4, exhibiting good agreement with the previous analysis. Figure 6 shows how well the asymptotic results agrees with results of numerical integrations.

5 Conclusions

The canonical problem formulated above provides an attractive framework for investigations of strain effects and stoichiometry effects on flame propagation in counterflow mixing layers. The numerical computation presented here reveals some interesting observations of concerning propagating flames in counterflows, for a simple model. The shapes of propagating flame which would occur for representative realistic values of the stoichiometric ratio S, which increase with increasing dilution of the oxidizer stream, is found to be quite different from the expected tri-brachial structure. A simple kinematic balance, based on the thin-front approximation, is seen to provide accurate predictions of flame shapes and associated propagation velocities for weakly strained flames. While the present study employs a simple model to focus more directly on effects of stoichiometry, more complicated descriptions of triple flames, accounting for differential diffusion, thermal expansion, and more realistic chemistry, are worth investigating in future work.

Acknowledgements

This work was supported by the US AFOSR Grant No. FA9550-16-1-0443.

References

[1] H. Phillips. *Sym. (Int.) Combust.*, 10 (1965) 1277–1283.

[2] J. Daou and A. Liñán. *Sym. (Int.) Combust.*, 27 (1998) 667–6743.

[3] J. Daou and A. Liñán. *Combust. Th. Model.*, 2 (1998) 449–477.

[4] M.S. Cha and P.D. Ronney. *Combust. Flame*, 146 (2006) 312–328.

[5] H. Song, P. Wang, R.S. Boles, D. Matinyan, H. Prahanphap, J. Piotrowicz, and P.D. Ronney. *Proc. Combust. Inst.*, 36 (2017) 1403–1409.

[6] A.J. Juanós and W.A. Sirignano. *Combust. Th. Model.*, 18 (2014) 454–473.

[7] A. Liñán. *Acta Astronautica*, 1 (1974) 1007–1039.

[8] A. Liñán, D. Martínaz-Ruiz, M. Vera, and A.L. Sánchez. *Combust. Flame*, 175 (2017) 91–106.

[9] J.W. Dold. *Combust. Flame*, 76 (1989) 71–88.

[10] L.J. Hartley and J.W. Dold. *Combust. Sci. Tech.*, 80 (1991) 23–46.

[11] J.W. Dold, L.J. Hartley, and D. Green. Dynamics of laminar triple-flamelet structures in non-premixed turbulent combustion. In *Dynamical issues in combustion theory*, pages 83–105. Springer, 1991.

[12] N. Peters. *Turbulent combustion*. Cambridge university press, 2000.

[13] P. Clavin and G. Searby. *Combustion Waves and Fronts in Flows: Flames, Shocks, Detonations, Ablation Fronts and Explosion of Stars*. Cambridge University Press, 2016.

[14] A. Liñán and F.A. Williams. *Fundamental Aspects of Combustion*. Oxford University Press, New York, 1993.

[15] P. Clavin. *Proc. Energy Combust. Sci.*, 11 (1985) 1–59.

Western States Section of the Combustion Institute – Fall 2017 Meeting
Hosted by the University of Wyoming
October 2-3, 2017

A new reduced thermal diffusion model for H and H_2

J. Schlup[1,*] and G. Blanquart[2]

[1] *Graduate Aerospace Laboratories*
[2] *Mechanical Engineering*
California Institute of Technology, Pasadena, USA
[*] *Corresponding author, e-mail: jschlup@caltech.edu*

Species and mixture transport property evaluation in numerical simulations often incurs large computational costs. Detailed multicomponent models typically scale with the number of species squared (or cubed, for non-iterative methods), which can become expensive in large three-dimensional domains typical of turbulent flame configurations. Reductions of the multicomponent transport models have been proposed, including mixture-averaged viscosity, conductivity, and mass diffusion, as well as a few models for thermal diffusion (Soret and Dufour effects). In this work, a new thermal diffusion model (utilizing thermal diffusion coefficients of only atomic and molecular hydrogen) is developed, based on an existing mixture-averaged thermal diffusion model by Chapman and Cowling (1970). Multicomponent thermal diffusion is used to validate the reduced thermal diffusion model with a variety of fuels, diluents, and operating conditions for a flat one-dimensional flame. Numerous hydrogen-air flames with a range of equivalence ratios are also examined using the same flat flame configuration. Finally, the reduced model is validated with a curved flame at a single equivalence ratio. Flame speeds and thermal diffusion fluxes from the reduced model show excellent agreement to the multicomponent model.

1 Introduction

The increase in computing power available to researchers has many implications, from larger problem sizes to more detailed models of relevant physics and chemistry. For combustion applications, this advancement in supercomputer platforms has similar impacts, e.g., using detailed kinetic modeling with large chemical models, small grids such that turbulent features are resolved, and inclusion of transport effects that have often been neglected. One such oft-neglected transport property, especially in turbulent reacting flows, is thermal diffusion.

The importance of including thermal diffusion in reacting flow simulations (especially of lean hydrogen-air mixtures) has been known for many years. Early theoretical work by García-Ybarra et al. [1] showed that thermal diffusion has a non-negligible impact on flat, and to a greater extent, curved hydrogen-air flames. Subsequent research using numerical simulations have shown similar results in hydrogen-air flames with various levels of transport property and chemistry modeling [2–7]. Additional work using fuels such as methane [3, 4] and "heavy" hydrocarbons [8] have also shown the impact of thermal diffusion in flame structure and extinction strain rate. More recently, thermal diffusion has been included in a few studies regarding lean premixed hydrogen flames of varying complexity [6, 7, 9–12]. For these recent works, thermal diffusion was included using either a multicomponent approach [6, 7, 9, 10] or a mixture-averaged model [10–12], with the exception of Hall and Pitz [10] who also proposed a fourth-order polynomial fit of multicomponent thermal diffusion coefficients for H and H_2, as a function of temperature.

One well-known reduction in computational cost for transport property evaluation is the iterative method of Ern and Giovangigli [13]. The iterative multicomponent method has a computational complexity of $\mathcal{O}(n^2)$, where n is the number of species in the chosen chemical model; detailed descriptions of multicomponent calculation timings can be found elsewhere (e.g., Ern and Giovangigli [14] and Ambikasaran and Narayanaswamy [15]). This cost is even higher if direct inversion of the multicomponent transport linear systems is performed [16]. The mixture-averaged thermal diffusion method of Chapman and Cowling [17] also has a computational complexity of $\mathcal{O}(n^2)$. As a result of these scalings, and its perceived trivial impact, thermal diffusion (and multicomponent transport property evaluation, in general) is often neglected in simulations with detailed chemistry and large numbers of grid points (such as Direct Numerical Simulations (DNS) of turbulent flames). Thus, a simplified model of thermal diffusion that scales as $\mathcal{O}(n)$ (or less) is highly desirable. One such model was detailed very recently by Ambikasaran and Narayanaswamy [15]; however, this multicomponent mass and thermal diffusion model is still quite complex and requires several matrix-vector multiplications and summations.

In this work, a new thermal diffusion model is developed, which requires thermal diffusion coefficients of only two species to be computed (without the need for multicomponent mass diffusion). This model is validated using the multicomponent method for a variety of fuel, oxidizer, and diluent combinations under various operating conditions for one-dimensional unstretched flames. Multiple chemical models are considered for the unstretched validation cases. The model is also tested across a range of equivalence ratios for a given fuel and chemistry model. Finally, a stretched flame is considered, and the capability of the reduced model to capture flame curvature effects is identified.

2 Review of mixture-averaged model

We begin with a brief review of the mixture-averaged thermal diffusion model proposed by Chapman and Cowling [17]. The mixture-averaged (MA) thermal diffusion coefficients for species i, $D^T_{i,MA}$, are given by Paul and Warnatz [18] as

$$D^T_{i,MA} = \rho \frac{W_i}{W} D_{i,m} k_{Ti}, \tag{1}$$

where ρ is the mixture density, W_i is the molecular weight of species i, W is the mixture molecular weight, $D_{i,m}$ is the mixture-averaged mass diffusion coefficient of species i defined as

$$D_{i,m} = \frac{1 - Y_i}{\sum\limits_{j \neq i}^{n} X_j / \mathcal{D}_{ij}}, \tag{2}$$

and k_{Ti} is the thermal diffusion ratio of species i, given as [17]

$$k_{Ti} = \frac{W^2}{R\rho} \sum_{j=1}^{n} \frac{1.2C^*_{ij} - 1}{\mathcal{D}_{ij}} \frac{Y_i a_j - Y_j a_i}{W_i + W_j}. \tag{3}$$

In Eq. (2), Y_i is the mass fraction of species i, X_j is the mole fraction of species j, and $\mathcal{D}_{ij}$ is the binary diffusion coefficient of species i diffusing into species j. For Eq. (3), R is the ideal gas constant, C^*_{ij} is a ratio of collision integrals [19],

$$C^*_{ij} = \frac{1}{3}\frac{\Omega^{(1,2)}_{ij}}{\Omega^{(1,1)}_{ij}}, \tag{4}$$

and a_i is based on the form from Oran and Boris [20], given as

$$a_i = \frac{\lambda_{i,mon} X_i}{\Delta_i}. \tag{5}$$

Δ_i is given by

$$\Delta_i = X_i + \sum_{j \neq i}^{n} X_j \Phi_{ij}, \tag{6}$$

where Φ_{ij} is a term appearing in mixture-averaged viscosity formulations (examined later) and $\lambda_{i,mon}$ is the monatomic species conductivity

$$\lambda_{i,mon} = \frac{15}{4}\frac{R\mu_i}{W_i}. \tag{7}$$

Finally, C^*_{ij} is computed using a rational function fit detailed in Appendix A.

Φ_{ij} in Eq. (6) is a function of species molecular weights and species viscosities, μ_i, and takes the form [20]

$$\Phi_{ij} = \frac{1.065}{2\sqrt{2}} \frac{\left[1 + \left(\frac{\mu_i}{\mu_j}\right)^{1/2} \left(\frac{W_j}{W_i}\right)^{1/4}\right]^2}{\left[1 + \left(\frac{W_i}{W_j}\right)\right]^{1/2}}, \tag{8}$$

where μ_i is defined using kinetic theory [21]

$$\mu_i = \frac{5}{16}\frac{\sqrt{\pi m_i k_B T}}{\pi \sigma_i^2 \Omega_i^{(2,2)*}}. \tag{9}$$

In Eq. (9), m_i is the mass of species i, k_B is the Boltzmann constant, T is the temperature, $\Omega_i^{(2,2)*}$ is a collision integral normalized by its rigid sphere value (see Appendix A), and σ_i is the collision diameter of species i.

In a reacting flow code where the species transport equations are solved, one must satisfy mass conservation. This includes the constraint that there is zero net diffusion flux, $\sum_i \mathbf{j}_i = 0$, where $\mathbf{j}_i$ is the diffusion flux of species i. In order to ensure this condition is met for mixture-averaged diffusion, a correction velocity, $\mathbf{u}_c$, is added to the species diffusion flux, leading to

$$\mathbf{j}_i = -\rho Y_i D_{i,m}\frac{\nabla X_i}{X_i} - D_i^T \frac{\nabla T}{T} + \rho Y_i \mathbf{u}_c. \tag{10}$$

The correction velocity contains correction terms due to mixture-averaged mass diffusion ($\mathbf{u}_c^D$) and thermal diffusion ($\mathbf{u}_c^T$),

$$\mathbf{u}_c = \mathbf{u}_c^D + \mathbf{u}_c^T. \tag{11}$$

The two correction terms are thus given as

$$\mathbf{u}_c^D = \frac{\nabla W}{W} \sum_i D_{i,m} Y_i + \sum_i D_{i,m} \nabla Y_i \tag{12}$$

and

$$\mathbf{u}_c^T = \frac{1}{\rho} \frac{\nabla T}{T} \sum_i D_i^T. \tag{13}$$

3 Development of reduced model

We now consider three approximations for various terms comprising the mixture-averaged thermal diffusion model. In order to reduce the computational cost of Eq. (1), we desire to remove the summations over species. The first approximation is to write Φ_{ij} in terms of species i only. We first consider reducing Δ_i (Eq. (6)), which we approximate to have the form

$$\Delta_i \approx \Delta_i' = \sum_{j=1}^{n} X_j \Phi_{ij}. \tag{14}$$

Upon substitution of Eq. (9) into Eq. (8), Φ_{ij} becomes

$$\Phi_{ij} = \frac{1.065}{2\sqrt{2}} \frac{\left[1 + \left(\frac{\sigma_j}{\sigma_i}\right)\left(\frac{\Omega_j^{(2,2)^*}}{\Omega_i^{(2,2)^*}}\right)^{1/2}\right]^2}{\left[1 + \left(\frac{W_i}{W_j}\right)\right]^{1/2}}. \tag{15}$$

A fit of σ_i as a function of W_i is given in Appendix A. Additionally, Rosner [22] demonstrated that $\Omega_i^{(2,2)^*}$ can be described by a power law fit as a function of Lennard-Jones reduced temperature, $T_i^* = Tk_B/\epsilon_i$, of the form $\Omega_i^{(2,2)^*} \approx 1.22(T_i^*)^{-0.16}$ over $T_i^* \in [3, 300]$, where ϵ_i is the potential well depth of species i. Then, substitution of a fit for k_B/ϵ_i (also given in Appendix A) yields $\Omega_i^{(2,2)^*} \approx T^{-0.16} W_i^{0.0928}$. With these fits for σ_i and $\Omega_i^{(2,2)^*}$, Φ_{ij} can be approximately written as a function of molecular weights only,

$$\Phi_{ij} \approx \Phi_{ij}' = \frac{1.065}{2\sqrt{2}} \frac{\left[1 + \left(\frac{W_j}{W_i}\right)^{0.376}\right]^2}{\left[1 + \left(\frac{W_i}{W_j}\right)\right]^{1/2}}. \tag{16}$$

Figure 1 shows Φ_{ij}' evaluated for $W_i/W_j \in [0.01, 100]$ and a power law fit as a function of W_i/W_j, resulting in a relationship of the form

$$\Phi_{ij}' \approx 1.022 \left(\frac{W_j}{W_i}\right)^{0.602}. \tag{17}$$

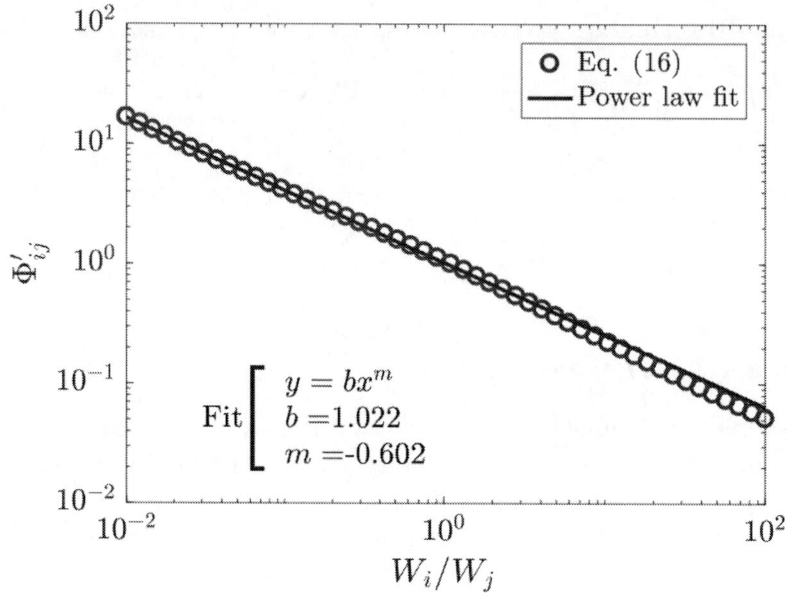

Figure 1: Φ_{ij}, computed using Eq. (16), as a function of $W_i/W_j \in [0.01,\ 100]$, approximating the range of W_i/W_j in CaltechMech [23].

Since the power law dependence on W_j is of order one, it is possible to write the mixture-averaged Φ'_{ij} (i.e., Δ'_i) by replacing $\sum_j X_j W_j$ with W, namely

$$\Delta'_i \approx \sum_{j=1}^{n} X_j \Phi'_{ij} \approx 1.022 \left(\frac{W}{W_i}\right)^{0.602}. \tag{18}$$

Alternatively, we propose to retain the form of Φ_{ij} and instead replace properties of species j with properties of the mixture, leading to

$$\Delta'_i \approx \Phi_{i,m} = \frac{1.065}{2\sqrt{2}} \frac{\left[1 + \left(\frac{\mu_i}{\mu}\right)^{1/2} \left(\frac{W}{W_i}\right)^{1/4}\right]^2}{\left[1 + \left(\frac{W_i}{W}\right)\right]^{1/2}}, \tag{19}$$

where μ is the mixture viscosity.

We now approximate a_i (Eq. (5)) using $\Phi_{i,m}$ and substitute this form of a_i into Eq. (1). Our first approximation thus yields approximate mixture-averaged thermal diffusion coefficients, $D^T_{i,m}$, defined as

$$D^T_{i,m} = \frac{15}{4} W_i^2 X_i D_{i,m} \sum_{j \neq i}^{n} \left[\frac{1.2C^*_{ij} - 1}{\mathcal{D}_{ij}} X_j \left(\frac{W_j}{W_i + W_j}\right) \left(\frac{\mu_j}{W_j^2 \Phi_{j,m}} - \frac{\mu_i}{W_i^2 \Phi_{i,m}}\right)\right]. \tag{20}$$

We have now reduced the double summation (due to Δ_i) over all species in Eq. (1) ($\mathcal{O}(n^2)$) to a single summation over all species ($\mathcal{O}(n)$).

The second approximation is that $W_i \ll W_j$ (where i is H or H_2). This approximation yields two results: $\mu_j/W_j^2\Phi_{j,m} \ll \mu_i/W_i^2\Phi_{i,m}$ and $W_i + W_j \approx W_j$. The first relation is a direct consequence of the molecular weight scalings for μ_i (Eq. (9)), i.e.,

$$\mu_i \propto \frac{W_i^{1/2}}{\left(W_i^{0.33}\right)^2 \left(W_i\right)^{0.0928}} = W_i^{-0.253}. \tag{21}$$

Using Eq. (17), we see that $\mu_i/W_i^2\Phi_{i,m} \propto W_i^{-1.65}W^{-0.602}$ and thus $\mu_j/W_j^2\Phi_{j,m} \ll \mu_i/W_i^2\Phi_{i,m}$. Using the two relations of this second approximation, Eq. (20) becomes

$$D_{i,m}^T \approx -\frac{15}{4}\frac{\mu_i X_i}{\Phi_{i,m}} D_{i,m} \sum_{j \neq i}^{n} \left(\frac{1.2C_{ij}^* - 1}{\mathcal{D}_{ij}} X_j \right). \tag{22}$$

The third approximation is to replace C_{ij}^* using properties of species i and the mixture. Appendix A shows that C_{ij}^* is largely independent of T_{ij}^* for a range of $T_{ij}^* \in [1, 100]$, representative of a reacting flow. Here, $T_{ij}^* = Tk_B/\epsilon_{ij}$ is computed using the interaction well depth $\epsilon_{ij} = \sqrt{\epsilon_i \epsilon_j}$. A deviation of only 5% is found around a nominal value $C_{ij}^* \approx 0.93$ for $T_{ij}^* > 2$. Because C_{ij}^* is insensitive to T_{ij}^*, we desire a mixture-averaged $C_{i,m}^*$, which would be a function of k_B/ϵ_i and some mixture k_B/ϵ_m. This is accomplished by utilizing the power law fit for k_B/ϵ_i and a similar fit for the mixture, $k_B/\epsilon_m \approx [37.15W^{0.58}]^{-1}$. Then,

$$T_{i,m}^* = \frac{1}{37.15} T \left(W_i W\right)^{-0.29}. \tag{23}$$

Now, $C_{i,m}^* = f(T_{i,m}^*)$ is independent of species j.

Combining the three assumptions made above, Eq. (20) then reduces to an expression with no summations,

$$D_{i,UF}^T = -\frac{15}{4}\frac{X_i \mu_i}{\Phi_{i,m}} \left(1.2C_{i,m}^* - 1\right)\left(1 - Y_i\right), \tag{24}$$

where $\Phi_{i,m}$ simplifies from Eq. (19) to

$$\Phi_{i,m} = \frac{1.065}{2\sqrt{2}} \left[1 + \left(\frac{\mu_i}{\mu}\right)^{1/2} \left(\frac{W}{W_i}\right)^{1/4} \right]^2 \tag{25}$$

under the assumption that $W_i \ll W$. The cost to compute thermal diffusion coefficients for H and H_2 is now independent of the size of the chemistry model, and implementing this model requires only computing the correction velocity (see Section 4.4). This model is called the "unfitted reduced model" (UF). Note that a scaling factor may be introduced to Eq. (24) to account for systematic errors arising from the model assumptions. This is explored in Section 4.1, and the resulting model is called the "reduced model" (RM).

4 Reduced model validation

The reduced thermal diffusion model is validated using both unstretched (Sections 4.1 and 4.2) and stretched (Section 4.3) flame conditions. First, the model is validated across a range of unstretched flame properties and conditions. Subsequently, the model is validated across a range of equivalence ratios for H_2-air mixtures. Finally, stretch is introduced such that the reduced model can be validated under the influence of flame curvature.

Table 1: One-dimensional flame cases used for mixture-averaged thermal diffusion model validation.

Case	Fuel	Chem.	Diluent	ϕ	T_u [K]	p [atm]	L^2_{H} [%]	$L^2_{\mathrm{H_2}}$ [%]	$\epsilon^{RM}_{S_L}$ [%]
1	H_2	[25]	N_2	0.4	298	1	2.89	2.69	0.26
2	H_2	[25]	N_2	0.4	298	5	2.06	2.79	0.38
3	H_2	[25]	N_2	1.0	298	1	6.43	12.00	0.70
4	H_2	[25]	N_2	0.4	500	1	1.16	3.65	0.27
5	H_2	[25]	—	0.4	298	1	12.31	21.82	1.40
6	H_2	[28]	Ar	0.4	298	1	2.29	3.91	0.90
7	CH_4	[28]	N_2	1.0	298	1	0.89	1.19	0.15
8	NC_7	[29]	N_2	1.0	298	1	0.94	1.46	0.18

4.1 Effects of flame conditions

Table 1 summarizes the one-dimensional unstretched laminar flame conditions tested. For this validation, we change the fuel, oxidizer, diluent, equivalence ratio (ϕ), unburnt temperature (T_u), and pressure (p). Each case uses multicomponent (MC) thermal diffusion and mixture-averaged mass diffusion for species transport. The cases are run to steady state using the variable density, low Mach number, reacting flow code NGA [24]. The inlet velocity boundary condition matches the laminar flame speed of each case. Each case uses a different grid, with at least 20 points through the laminar flame thickness defined using the maximum temperature gradient, $l_F = (T_{max} - T_{min}) / |\nabla T|_{max}$. At steady state, $D^T_{i,UF}$ from Eq. (24) are computed. Three chemical models are used in this validation, including a hydrogen-air chemistry model by Hong et al. with updated reaction rates from the same group [25–27], GRI-Mech 3.0 [28], and a reduced n-heptane model (without aromatic species) from Bisetti et al. [29].

First, the thermal diffusion coefficients using the multicomponent model ($D^T_{i,MC}$) and the unfitted reduced model ($D^T_{i,UF}$) are shown in Fig. 2. The excellent collapse of all data on a single line with unity slope indicates agreement between $D^T_{i,MC}$ and $D^T_{i,UF}$ is strong over a wide range of flame operating conditions. Small deviations from the unity line are present, particularly at low values of D^T_i; however, the important parameter for the solution of the species transport equations is the thermal diffusion flux, $\mathbf{j}^T_i = D^T_i \nabla T / T$.

A least squares regression (LSR) of the thermal diffusion fluxes is performed to identify a possible fitting parameter α_i such that

$$\mathbf{j}^T_{i,MC} \approx \mathbf{j}^T_{i,RM} = \alpha_i \left(D^T_{i,UF} \frac{\nabla T}{T} \right). \tag{26}$$

While a LSR using data from all cases is preferable, this was found to result in large relative L^2_i errors of the diffusion fluxes, defined as

$$L^2_i = \sqrt{\frac{\sum \left(\mathbf{j}^T_{i,MC} - \mathbf{j}^T_{i,RM}\right)^2}{\sum \left(\mathbf{j}^T_{i,MC}\right)^2}}. \tag{27}$$

Significantly lower L^2_i are found by performing the LSR without Cases 3 and 5. The optimal correction factors neglecting Cases 3 and 5 are $\alpha_{\mathrm{H}} = 0.895$ and $\alpha_{\mathrm{H_2}} = 0.910$. These fitting

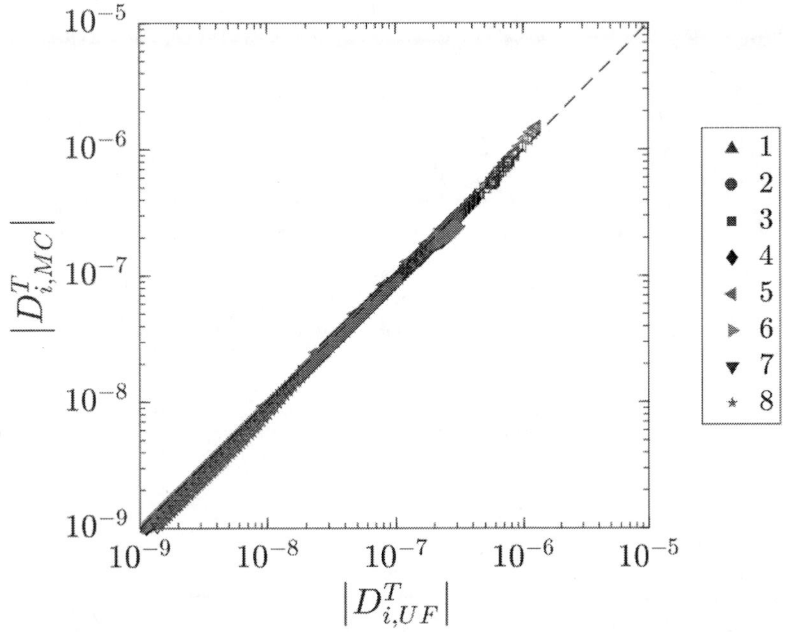

Figure 2: Thermal diffusion coefficients for H (filled symbols) and H_2 (open symbols) for all cases using the multicomponent model ($D^T_{i,MC}$) and the unfitted reduced model ($D^T_{i,UF}$) given by Eq. (24). The symbols indicate each considered test case.

parameters are used to calculate the RM thermal diffusion coefficients $D^T_{i,RM} = \alpha_i D^T_{i,UF}$. The relative L^2_i error norms are computed for all cases and reported in Table 1. For a majority of the cases, the relative errors are below 4%. The maximum L^2_i errors occur for Cases 3 and 5; this is unsurprising as the α_i values were computed neglecting these cases. The error norms indicate that the thermal diffusion fluxes can be captured accurately for many cases using the RM model.

Next, a plot of the thermal diffusion fluxes through the flame for Case 1 is given in Fig. 3. All four thermal diffusion models (i.e., MC, MA, UF, and RM) are included for H and H_2. There is a slight discrepancy between the MC and MA models, where the MA model overpredicts the magnitude of H and H_2 thermal diffusion coefficients. This overprediction is made lager using the UF model. After applying α_i, the coefficients using RM agree extremely well with the MC model, as expected based on the value of L^2_i.

Each of the cases in Table 1 are then run to steady state using the RM model. The unstretched laminar flame speeds, S^o_L, are calculated for the RM and MC transport cases, using the equation

$$S^o_L = -\frac{\int \rho \dot{\omega}_{\mathrm{F}} dx}{\rho_u Y_{\mathrm{F},u}}, \tag{28}$$

where $\dot{\omega}_{\mathrm{F}}$ is the fuel production rate, ρ_u is the unburnt gas density, and $Y_{\mathrm{F},u}$ is the unburnt fuel mass fraction. The superscript o indicates unstretched flame speeds. The relative error in S^o_L is calculated using

$$\epsilon^{RM}_{S_L} = \left|\frac{S^{o,\mathrm{MC}}_L - S^{o,RM}_L}{S^{o,\mathrm{MC}}_L}\right|. \tag{29}$$

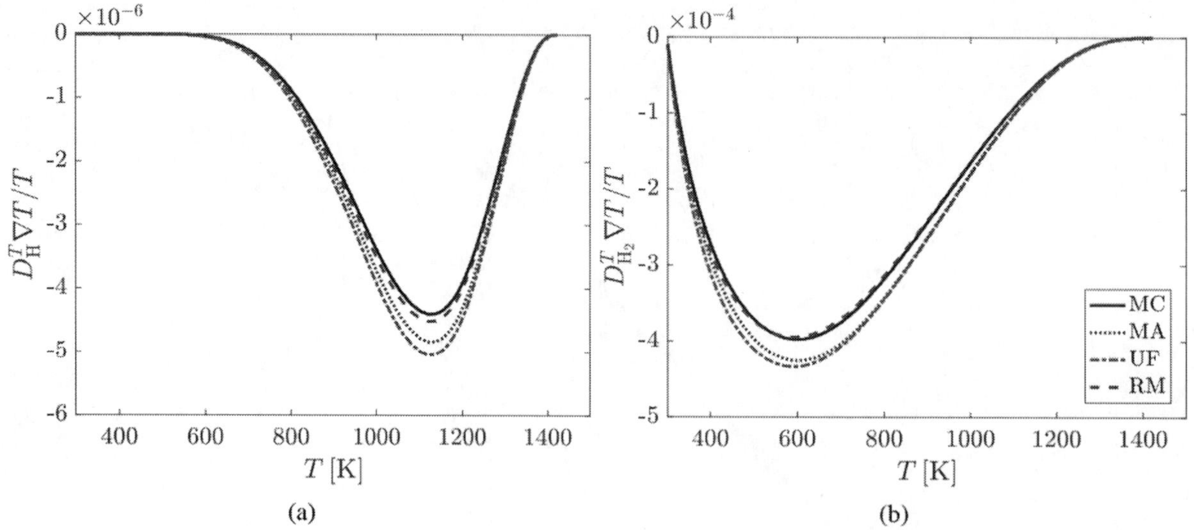

Figure 3: Thermal diffusion fluxes for (a) H and (b) H_2 as a function of T for Case 1 using the MC (solid black), MA (dotted blue), unfitted reduced (dot-dashed green), and RM (dashed red) models. Note the MC and RM cases nearly overlap.

The relative errors for each case are listed in Table 1. The unstretched laminar flame speeds agree extremely well across all cases, including cases with higher relative L^2 norms for H and H_2 thermal diffusion fluxes (i.e., Cases 3 and 5).

4.2 Effects of equivalence ratio

A range of equivalence ratios are considered ($\phi \in [0.35, 1.0]$) to determine the effect of thermal diffusion on laminar flame speeds, as well as the accuracy of the RM model over a range of ϕ. The inlet mixture has an unburnt temperature of $T_u = 298$ K and pressure of $p_o = 1$ atm. The inlet velocity boundary condition, u_{BC}, matches the laminar flame speed. Similar to Section 4.1, each equivalence ratio uses a different grid, containing at least 20 points through the laminar flame thickness. An identical grid is used for all diffusion models for a given equivalence ratio.

Figure 4 shows the percent deviation (from the MC laminar flame speed) of the flame speeds calculated using the various thermal diffusion models using Eq. (28). This percent deviation is defined using Eq. (29) with the superscript d specifying the diffusion model, i.e., MA, RM, or no thermal diffusion (xx). As can be seen, $S_L^{o,\mathrm{MA}}$ and $S_L^{o,\mathrm{RM}}$ closely agree with $S_L^{o,\mathrm{MC}}$ for all equivalence ratios. Additionally, the error incurred by the thermal diffusion models are an order of magnitude lower than the error incurred by neglecting thermal diffusion. Further, at $\phi = 0.4$, all models indicate that S_L^o decreases by approximately 5% if thermal diffusion is included. This reduction in flame speed is consistent with previous studies of thermal diffusion in one-dimensional flat flames [5]. Finally, while there is a clear difference between flame speeds with and without thermal diffusion, the choice between the MA and RM thermal diffusion models has little impact on the results.

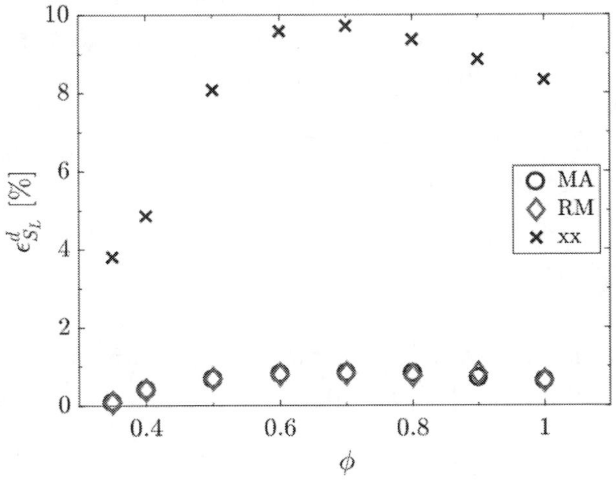

Figure 4: Laminar flame speed comparison of the MA (red circles), RM (green diamonds), and xx (blue crosses) models with the MC model. The MA and RM cases overlap at many equivalence ratios.

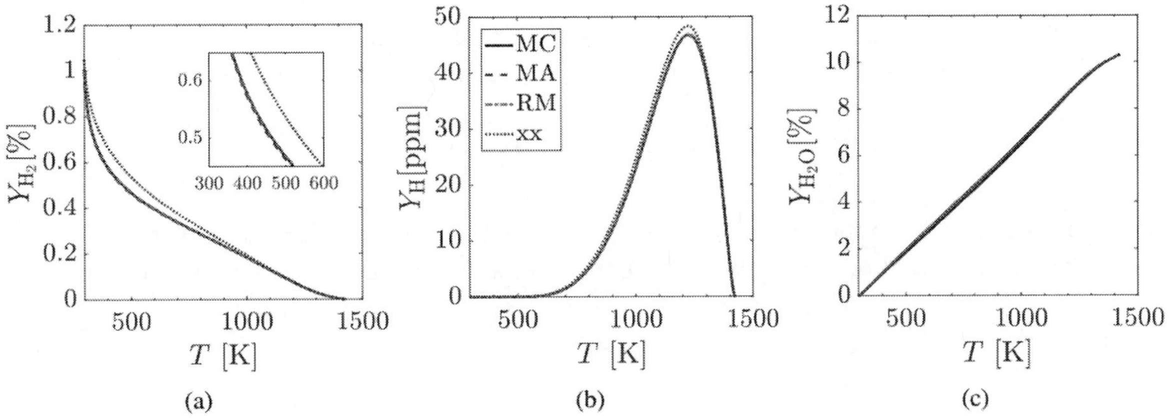

Figure 5: Comparison of species mass fractions as functions of temperature for the different diffusion models. MC (black), MA (dashed red), RM [30] (dot-dashed green), and no thermal diffusion (dotted blue) results are shown. The cases including any thermal diffusion model overlap in these figures.

While the laminar flame speeds provide insight into the global effect of thermal diffusion, local comparisons between the multicomponent and mixture-averaged models can be made by examining species mass fraction profiles. Figure 5 shows a comparison of the mass fractions of two major chemical species (H_2 and H_2O) and H radicals for all four cases. For the light species (H and H_2), including thermal diffusion causes a deviation of the mass fraction profiles from those found by neglecting thermal diffusion. However, the inclusion of thermal diffusion has little impact on the profiles of heavier chemical species. These local and global results show that, for a one-dimensional flame, using either the MA or RM models accurately predicts the laminar flame speeds and species profiles of the MC case.

4.3 Effects of curvature

As the next logical step in flame complexity, this section examines a one-dimensional stretched flame. The flame considered is a one-dimensional, outwardly propagating, cylindrical flame such that the flame is always convex to the unburnt mixture (positive curvature). The unburnt mixture has an equivalence ratio of $\phi = 0.4$ with $T_u = 298$ K and $p_o = 1$ atm. The computational domain has a maximum radius $R = 0.2$ m, with a grid spacing of $\Delta r = l_F/40$. The boundary at $r = R$ is set as an outflow boundary condition; there is no inflow at the origin.

The flame is initialized with a one-dimensional flat solution with an initial flame position normalized by the flame thickness of $\kappa l_F = 0.5$. The temperature corresponding to the peak source term of H_2O in a one-dimensional flat flame (denoted by T_{peak}) is used to locate the flame. The curvature, κ, is defined as the reciprocal of the flame radius, $1/r_F$. The simulation is run until the flame radius reaches $\kappa l_F = 0.1$. Then, this solution is re-interpolated to $\kappa l_F = 0.5$ (original starting flame radius), and the simulation is restarted. This two-step procedure permits a slightly better initial guess of the flame profile at high curvatures. The simulation then runs until $\kappa l_F = 0.01$, where the flame is only slightly curved. Measurements of the flame speed and flame position are considered during this second simulation run. Figure 6 shows the normalized flame speed as a function of normalized curvature, with the flame speed being computed as

$$S_L = \frac{\rho_b}{\rho_u}\frac{dr_F}{dt} \tag{30}$$

where ρ_b is the burnt mixture density. Because the burnt gases are at rest inside the outwardly propagating flame, the time derivative of the flame radius represents the flame speed relative to the burnt gases. Utilizing the density ratio between the burnt and unburnt gases produces the more common 'unburnt' flame speed. The flame speed is then averaged over ten time steps, to smooth out the differentiation operation.

An initial transient exists due to the initial flame profile being an approximate stretched solution; thus, Fig. 6 shows only the portion after the transient has been removed. This figure shows that, regardless of the inclusion of thermal diffusion, the flame speed approaches the unstretched laminar flame speed as the curvature decreases. This behavior is expected, and in the limit of an infinitely large flame radius, $S_L/S_L^o \to 1$. Additionally, Fig. 6 shows that the inclusion of thermal diffusion, using either the MA or RM thermal diffusion models, increases the flame speed at a given positive curvature. For the entire range of flame curvature presented, the increase of S_L

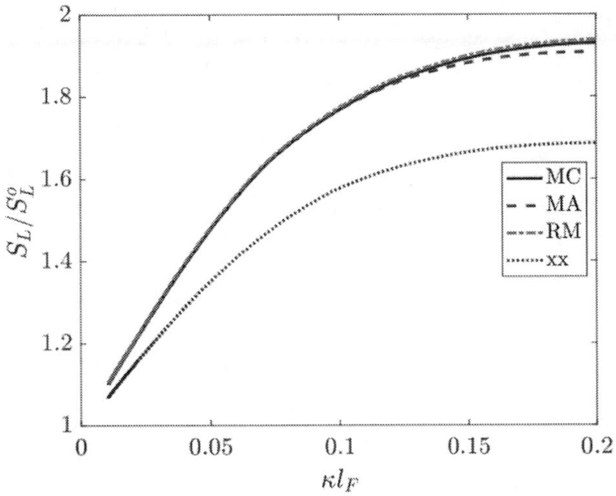

Figure 6: Normalized flame speed as a function of normalized flame curvature with $\phi = 0.4$.

over S_L^o, i.e., $S_L/S_L^o - 1$, is consistently 33% larger for the thermal diffusion cases than the case neglecting thermal diffusion. This result is also expected as thermal diffusion acts to increase the local equivalence ratio in regions of positive curvature [6], leading to a strengthening of the flame speed. Central to our investigation, the RM model reproduces accurately (to within 1% of S_L/S_L^o) the effects of curvature on the flame speed computed using multicomponent thermal diffusion.

4.4 Computational cost

As mentioned in Section 3, implementing $D^T_{i,RM}$ significantly reduces the computational cost of including thermal diffusion by eliminating summations or iterations over all species in the chemical model. The computational cost to include thermal diffusion using the RM method can thus be split into two contributions: the first cost is computing $D^T_{i,RM}$, while the second cost is computing $\mathbf{u}_c^T$. From Section 3, it is evident that the cost of computing $D^T_{i,RM}$ should be a constant value, regardless of the number of species contained in the chemical model, provided the mixture properties are known. The cost to compute $\mathbf{u}_c^T$, however, scales linearly with the number of species.

In order to investigate the cost of these calculations more carefully, a new problem set-up was used. Four new flames were simulated, using four separate chemical models. First, a lean hydrogen-air flame ($\phi = 0.4$) is simulated using the nine-species mechanism of Hong *et al.* [25] described in Section 4.1. Then, a methane-air flame with $\phi = 0.9$ is performed using the GRIMech3.0 [28] chemical model, which includes 53 species and was utilized in Section 4.1. Finally, two n-C_7H_{16} flames are considered, using the full CaltechMech2.4 [23], containing 171 species, and the 35-species reduced mechanism of Bisetti *et al.* [29], also utilized in Section 4.1. These four flames are simulated in a three-dimensional domain with 1024 points in the streamwise direction and 32 points in the two periodic directions. The inlet velocity boundary condition still matches the laminar flame speeds. The flames were initialized using unstretched laminar flame solutions from FlameMaster [31] for each of the chemistry models. The flames are then run for 20 iterations on TACC's Stampede cluster using the Extreme Science and Engineering Discovery Environment

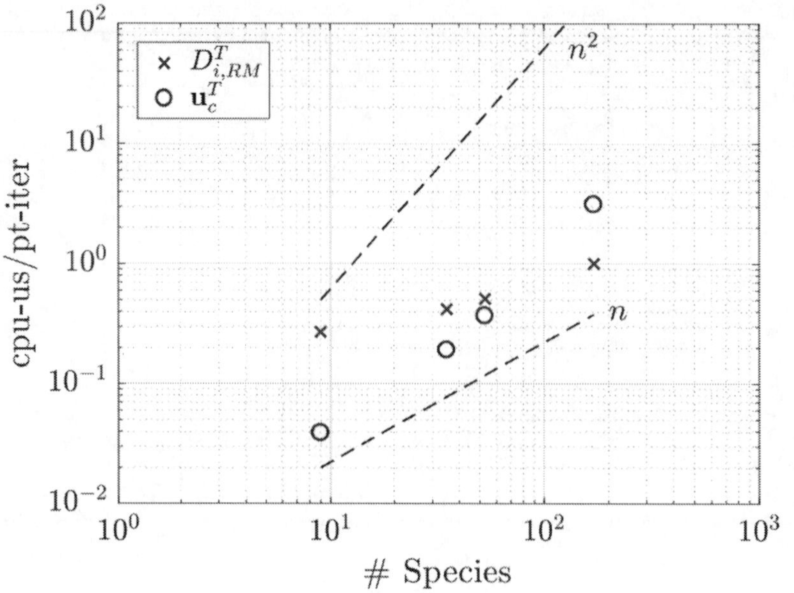

Figure 7: Computational cost for $D^T_{i,RM}$ (red crosses), including the computation of W, and $\mathbf{u}^T_i$ (blue circles).

(XSEDE) [32] in order to record computational timings of various parts of NGA.

Figure 7 shows the time to compute both $D^T_{i,RM}$ and $\mathbf{u}^T_c$, multiplied by the number of processors and divided by the number of points in the computational domain. It should be noted that in the computation of $D^T_{i,RM}$, the mixture viscosity μ has been calculated prior to timing $D^T_{i,RM}$, but the mixture molecular weight is computed during the timing.

As can be seen, the cost of $D^T_{i,RM}$ increases slightly with n, although at a slower-than-linear rate. This is primarily because the computation of W is included in the cost. The cost of $\mathbf{u}^T_c$ also increases with n, with a rate slightly above n. As the number of species increases, the comparative cost of $D^T_{i,RM}$ to $\mathbf{u}^T_c$ decreases. Finally, the scaling of $D^T_{i,RM}$ being less than n is a significant improvement over $\mathcal{O}(n^2)$ methods, such as iterative multicomponent methods [13] and the original mixture-averaged method [17].

5 Conclusions

In this work, a thermal diffusion model based solely on thermal diffusion of H and H_2 was developed, starting from the mixture-averaged thermal diffusion model of Chapman and Cowling [17]. Timings of this model indicate computational scalings of $\mathcal{O}(n)$ or less. This model was then fitted to the rigorous multicomponent thermal diffusion model. For eight test cases spanning various fuels, diluents, equivalence ratios, unburnt temperatures, and pressures, a single set of fitting parameters was identified that accurately captures both the thermal diffusion fluxes and laminar flame speeds of one-dimensional unstretched laminar flames. The model was further validated across a range of equivalence ratios for lean hydrogen-air flames, where thermal diffusion is known to be important. Results indicate that the reduced model calculates laminar flame speeds to the same accuracy as the mixture-averaged thermal diffusion model. Finally, the reduced model was validated in a one-dimensional stretched flame. Again, the reduced model predicted flame speeds over

Table 2: Coefficients of the rational functions fitting $\Omega_i^{(2,2)^*}$ and C_{ij}^*.

Variable	Coefficient	$m=0$	$m=1$	$m=2$	$m=3$	$m=4$
$\Omega_i^{(2,2)^*}$	c_m	3.35306	2.53272	2.90242	0.11186	—
	d_m	0.86623	1.39139	3.15849	0.18973	0.00018
C_{ij}^*	c_m	0.73680	-0.16258	0.70953	1.39801	0.06796
	d_m	0.81695	-0.06989	1.01867	1.45230	0.07200

a wide range of flame curvature.

Acknowledgments

The authors gratefully acknowledge funding from the U.S. Department of Energy-Basic Energy Sciences (DE-SC0006591) under the supervision of Dr. Wade Sisk.

A Empirical fits for transport property evaluation

In Section 2 and 3, several variables are computed using fits of tabulated data. The values of $\Omega_i^{(2,2)^*}$ and C_{ij}^* are computed using rational function fits to tabulated data presented by Monchick and Mason [33], which take the form of

$$\Omega_i^{(2,2)^*} = \frac{\sum\limits_m c_m \left(T_i^*\right)^m}{\sum\limits_m d_m \left(T_i^*\right)^m} \tag{31}$$

and

$$C_{ij}^* = \frac{\sum\limits_i c_m \left(T_{ij}^*\right)^m}{\sum\limits_m d_m \left(T_{ij}^*\right)^m}. \tag{32}$$

Table 2 presents the coefficients for Eqs. (31) and (32) [31]. Figure 8 shows the tabulated data and rational function fits of $\Omega_i^{(2,2)^*}$ and C_{ij}^* over a relevant range of reduced temperatures for reacting flow simulations.

The values of σ_i and ϵ_i/k_B are fit using powers of the species molecular weight [31]. The collision diameter uses an exponential of the form

$$\sigma_i = 1.234 W_i^{0.33}. \tag{33}$$

Additionally, ϵ_i, the potential well depth of species i, is well approximated as a function of molecular weights,

$$\frac{\epsilon_i}{k_B} = 37.15 W_i^{0.58}. \tag{34}$$

Figure 9 shows Eqs. (33) and (34) with aggregated transport property data used in CaltechMech2.4 [23].

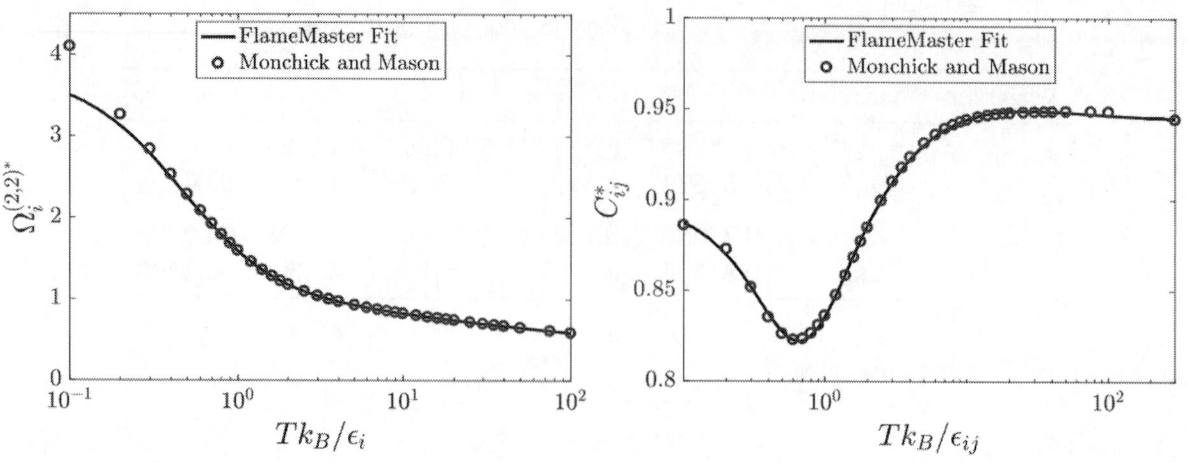

Figure 8: $\Omega_i^{(2,2)^*}$ (left) and C_{ij}^* (right), with rational function fits (black lines), using coefficients from FlameMaster [31], and tabulated data (red circles) from Monchick and Mason [33].

References

[1] P. Garcia-Ybarra, C. Nicoli, and P. Clavin. Combust. Sci. Technol., 42 (1984) 87–109.

[2] J. Warnatz. Influence of Transport Models and Boundary Conditions on Flame Structure, pages 87–111. Vieweg+Teubner Verlag, Wiesbaden, 1982.

[3] A. Ern and V. Giovangigli. Combust. Theor. Model., 2 (1998) 349–372.

[4] A. Ern and V. Giovangigli. Combust. Sci. Technol., 149 (1999) 157–181.

[5] H. Bongers and L. P. H. De Goey. Combust. Sci. Technol., 175 (2003) 1915–1928.

[6] J. F. Grcar, J. B. Bell, and M. S. Day. Proc. Combust. Inst., 32 (2009) 1173–1180.

[7] F. Yang, C. K. Law, C. J. Sung, and H. Q. Zhang. Combust. Flame, 157 (2010) 192–200.

[8] D. E. Rosner, R. S. Israel, and B. La Mantia. Combust. Flame, 123 (2000) 547 – 560.

[9] J. De Charentenay and A. Ern. Combust. Theor. Model., 6 (2002) 439–462.

[10] C. A. Hall and R. W. Pitz. Combust. Theor. Model., 20 (2016) 328–348.

[11] K. K. J. Ranga Dinesh, H. Shalaby, K. H. Luo, J. A. van Oijen, and D. Thévenin. Int. J. Hydrog. Energ., 41 (2016) 18231–18249.

[12] J. Schlup and G. Blanquart. Combust. Theor. Model., (2017, Under Review) xx–xx.

[13] A. Ern and V. Giovangigli. Multicomponent transport algorithms, Vol. 24. Springer-Verlag, 1994.

[14] A. Ern and V. Giovangigli. J. Comput. Phys., 120 (1995) 105–116.

[15] S. Ambikasaran and K. Narayanaswamy. Proc. Combust. Inst., (2016) journal.

[16] R. J. Kee, F. M. Rupley, and J. A. Miller. Chemkin-II: A Fortran chemical kinetics package for the analysis of gas-phase chemical kinetics. Technical report, Sandia National Labs., Livermore, CA (USA), 1989.

[17] S. Chapman and T. G. Cowling. The mathematical theory of non-uniform gases: An account of the kinetic theory of viscosity, thermal conduction and diffusion in gases. Cambridge University Press, 1970.

[18] P. Paul and J. Warnatz. Symp. (Int.) Combust., 27 (1998) 495 – 504.

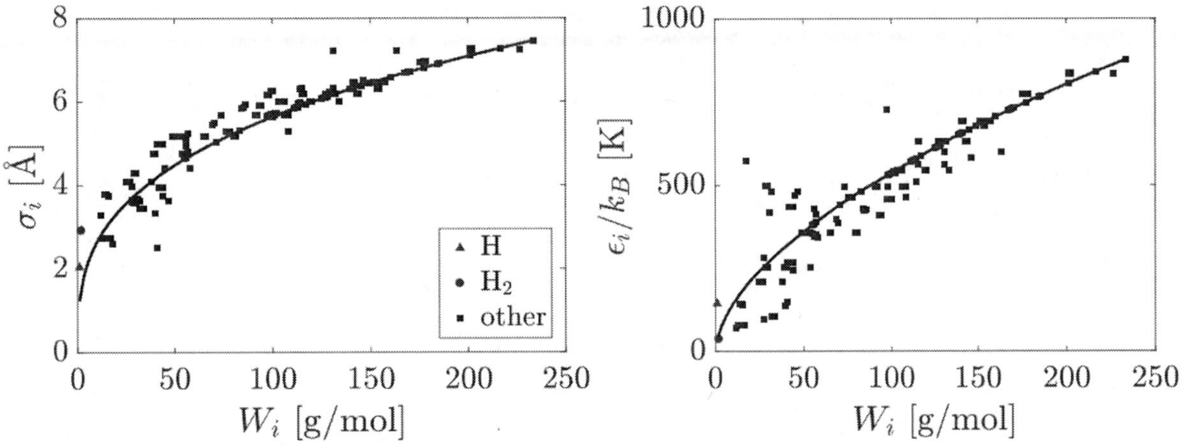

Figure 9: Power law fits (black lines) of σ_i using Eq. (33) and ϵ_i/k_B using Eq. (34) as functions of W_i. The fit coefficients are from FlameMaster [31] and the transport property data (symbols) are from multiple references [23]. H is the red triangle and H_2 is the blue circle.

[19] G. Dixon-Lewis. Flame structure and flame reaction kinetics. ii. Transport phenomena in multicomponent systems. In Proc. R. Soc. London, Ser. A, Vol. 307, pages 111–135. The Royal Society, 1968.

[20] E. S. Oran and J. P. Boris. Prog. Energy Combust. Sci., 7 (1981) 1–72.

[21] J. O. Hirschfelder, C. F. Curtiss, and R. B. Bird. Molecular theory of gases and liquids. Wiley, New York, 1964.

[22] D. E. Rosner. Transport processes in chemically reacting flow systems. Butterworth Publishers, Stoneham, MA, 1986.

[23] G. Blanquart. CaltechMech, version 2.4. http://www.theforce.caltech.edu/CaltechMech, 2017.

[24] O. Desjardins, G. Blanquart, G. Balarac, and H. Pitsch. J. Comput. Phys., 227 (2008) 7125 – 7159.

[25] Z. Hong, D. F. Davidson, and R. K. Hanson. Combust. Flame, 158 (2011) 633 – 644.

[26] Z. Hong, K.-Y. Lam, R. Sur, S. Wang, D. F. Davidson, and R. K. Hanson. Proc. Combust. Inst., 34 (2013) 565 – 571.

[27] K.-Y. Lam, D. F. Davidson, and R. K. Hanson. Int. J. Chem. Kinet., 45 (2013) 363–373.

[28] G. P. Smith, D. M. Golden, M. Frenklach, N. W. Moriarty, B. Eiteneer, M. Goldenberg, C. T. Bowman, R. K. Hanson, S. Song, Jr. Gardiner, W. C., V. V. Lissianski, and Z. Qin. GRI-Mech 3.0. http://combustion.berkeley.edu/grimech/, 1999.

[29] F. Bisetti, G. Blanquart, M. E. Mueller, and H. Pitsch. Combust. Flame, 159 (2012) 317 – 335.

[30] J. Schlup and G. Blanquart. Combust. Flame, (In preparation) xx–xx.

[31] H. Pitsch. A C++ computer program for 0D combustion and 1D laminar flame calculations. Technical report, RWTH Aachen University, Aachen, Germany, 1998.

[32] J. Towns, T. Cockerill, M. Dahan, I. Foster, K. Gaither, A. Grimshaw, V. Hazlewood, S. Lathrop, D. Lifka, G. D. Peterson, R. Roskies, J. R. Scott, and N. Wilkins-Diehr. Comput. Sci. Eng., 16 (2014) 62–74.

[33] L. Monchick and E. A. Mason. J. Chem. Phys., 35 (1961) 1676–1697.

Western States Section of the Combustion Institute - Fall 2017 Meeting
Hosted by the University of Wyoming
October 2-3, 2017.

Thermal explosions in spherical vessels at large Rayleigh numbers

D. Moreno[1] *I. Iglesias*[2] *A.L. Sánchez*[1] *A. Liñán*[3] *F.A. Williams*[1]

[1] *University of California San Diego. La Jolla, CA 92093-0411, USA.*
[2] *Universidad Carlos III de Madrid. 28911 Leganés, Spain.*
[3] *Universidad Politécnica de Madrid. 28006 Madrid, Spain.*

This paper investigates effects of buoyancy-driven motion on the "slowly reacting" mode of combustion, and its thermal-explosion limits, of an initially cold gaseous mixture enclosed in a spherical vessel with a constant wall temperature. As in Frank-Kamenetskii's seminal analysis, the strong temperature dependence of the effective overall reaction is modeled with a single irreversible reaction with an Arrhenius rate having a large activation energy. Besides the classical Damköhler number Da, measuring the ratio of the heat-release rate by chemical reaction evaluated at the wall temperature to the rate of heat removal by heat conduction to the wall, the solution is seen to depend on the Rayleigh number Ra, measuring the effect of buoyancy-induced motion on the heat-transport rate. For values of Da below a critical value Da_c the system evolves in a slowly reacting mode where the heat losses to the wall limit the temperature increase associated with the chemical reaction, whereas for $Da > Da_c$ the initial stage of slow reaction ends abruptly at a well-defined ignition time, at which a thermal runaway occurs. Transient numerical integrations of the initial stage of slow reaction, formulated in the distinguished limit $Da \sim 1$ and $Ra \sim 1$ with account taken of the effects of the temporal pressure variation, are used to investigate influences of natural convection on thermal-explosion development, including changes in ignition times for $Da > Da_c$ and modified explosion curves. Our analysis reveals that Frank-Kamenetskii's criterion for the determination of critical explosion conditions, based on the investigation of existence of steady solutions, provides values of $Da_c(Ra)$ that are identical to those extracted from the transient computations. Specific consideration is given to the structure of the steady solution in the asymptotic limit $Ra \gg 1$ in which the flow includes a thin chemically frozen near-wall boundary layer of downward moving cold gas bounding a central inviscid region of slowly rising reacting flow driven by the boundary-layer entrainment, with the critical explosion conditions predicted to occur for $Da_c \sim Ra^{1/4}$. The problem is solved with use made of an approximate integral method. The resulting asymptotic prediction for the critical Damköhler number $Da_c = 0.655Ra^{1/4}$ is found to be in excellent agreement with the results obtained by integration of the complete conservation equations for $Ra \gg 1$.

1 Introduction

In developing the excellent stationary theory of thermal explosions [1], Frank-Kamenetskii (FK) investigated the role of the strong temperature sensitivity of the reaction rate in the combustion of enclosed reactive mixtures. Explosion limits were defined as the critical conditions for existence of a steady, slowly reacting mode of combustion in which the heat-conduction losses to the wall limit the temperature rise in such a way that the reaction rate does not exceed in order of magnitude its near-wall value. With the constant-density approximation used in the early work, heat transfer to the wall occurred only by conduction, its effect being measured by a Damköhler number Da, defined below in (3) as the ratio of the time of heat conduction to the wall to the homogeneous thermal-explosion time. The curve giving the variation of the peak temperature with Da exhibits a first turning point at a critical value $Da = Da_c$ of order unity (e.g., $Da_c = 3.32$ for spherical containers), which identifies the explosion limit in FK theory. Using the value of Da_c together with the definition of the Damkhöhler number leads to explicit expressions for critical explosion sizes, giving results in close agreement with experimental observations, a remarkable achievement of the early theory, given the many different simplifying assumptions involved in its derivation [2]. This success has motivated recent extensions of the early theory incorporating realistic chemistry in descriptions of hydrogen-oxygen systems that have been shown to predict explosion conditions in spherical vessels in excellent agreement with experiments [3], including critical pressures along the so-called third-explosion limit [4]. The FK problem was recently revisited in [5] to investigate influences on ignition times of initial conditions and of temporal pressure variations in spherical vessels with fixed walls.

For gaseous mixtures under normal gravity the small temperature differences characterizing the slowly reacting mode of combustion are known to generate a convective motion that promotes heat transfer to the wall, measured by a relevant Rayleigh number Ra, defined below in (4). Influences of buoyancy on the resulting critical Damköhler numbers, discussed in early references [2], have been investigated numerically for simplified vessel geometries, as referenced below, including the infinite slab, the infinite horizontal cylinder, and the sphere, the last one being the configuration considered in the present paper.

The problem for a slab configuration, with a reactive gas bounded by two isothermal infinite horizontal walls, has been addressed in numerous theoretical analyses [6, 7, 8, 9, 10]. As in the closely related case of the Rayleigh-Bénard problem of convection produced by wall-temperature differences, a motionless quasi-steady combustion mode may exist for values of the Rayleigh number below a critical value, at which a bifurcation occurs to a convective state, as described by linear stability analyses [6, 7]. A noteworthy feature of the slab geometry is that, contrary to widespread expectations, convection does not always hinder the development of the thermal explosion [8]. While sufficiently small eddies with widths on the order of or smaller than the slab height raise the explosion threshold by enhancing heat removal from the reaction region, long-wave convection rolls favor explosions by promoting the formation of hot spots [8], an aspect of the problem further investigated in [9] with use made of a prescribed velocity field. The closely related problem of a horizontal cylindrical container of square section with adiabatic vertical walls has been also considered by different researchers [11, 12, 13]. For this geometry, the critical Damkhöhler number Da_c was found to increase in the presence of buoyancy-induced motion [11]. The rich nonlinear dynamics of weakly reactive solutions for subcritical values of the Damkhöhler number $Da < Da_c$ has now been investigated in more recent bifurcation analyses [12, 13].

Unlike the horizontal infinite slab, for reactive gases in cylindrical and spherical vessels any gravity force at all establishes buoyant motion because the horizontal variations of the density that are present in these configurations generate forces that cannot be counteracted by a vertical pressure gradient. The case of a horizontal circular cylinder has been investigated both experimentally [11] and numerically [14]. This last work includes transient computations for Rayleigh numbers up to 10^4, for different initial temperatures both above and below the wall value. Influences of buoyancy on thermal explosions in spherical vessels have also been considered in the past. In connection with the present work, the most closely related investigations are the numerical computations of Hayhurst and coworkers [15, 16, 17]. To enable comparisons with previous experimental results [18] to be made, the thermal decomposition of azomethane vapor was considered in [15] when selecting the specific chemical-rate parameters, for values of the Rayleigh numbers up to 21900. Resulting values of Da_c were found to differ by only a few percent from those of the buoyancy-free predictions for values of Ra as large as a few hundred, in agreement with earlier results [10, 11, 14]. Larger values of the Rayleigh numbers were considered in subsequent direct numerical simulations [16, 17] addressing transient problems with the initial gas temperature equal to the vessel wall temperature. The flow was found to remain laminar for values of $Ra < 10^6$. More recent numerical work by this same research group includes investigation of combined natural and forced convection on thermal explosions [19] as well as consideration of non-isothermal walls by use of a Robin condition including a heat-transfer coefficient [20].

The remarkably small influence of buoyancy on the critical ignition Damkhöhler number Da_c for moderate values of Ra was studied recently by asymptotic methods [21]. At leading order in the limit $Ra \ll 1$, the analysis revealed that the spherically symmetrical temperature field corresponding to the buoyancy-free FK solution is accompanied by a toroidal vortex, which may be called the Frank-Kamenetskii vortex, resulting from the balance between buoyancy and viscous forces. The modifications to the explosion curve, determined by extending the perturbation analysis to a higher order, were found to scale with Ra^2, but the numerical factors in the expansion for Da_c are extremely small, so that the corrections associated with buoyancy remain smaller than 10% even at $Ra = 300$. While this previous analysis showed that effects of weak buoyancy associated with values of Ra on the order of a few hundred can be suitably addressed by asymptotic methods, the description of more pronounced buoyancy effects for larger values of Ra requires numerical integrations. Results of relevant integrations are reported below for values of the Rayleigh number up to $Ra = 7 \times 10^6$, thereby covering the entire range of flow conditions under which the flow may be expected to remain laminar. Besides numerical integrations, perturbation methods are used to investigate the flow structure that develops in the asymptotic limit $Ra \gg 1$, which is shown here to include a thin chemically frozen boundary layer driven by buoyancy, surrounding the vessel surface, and an inner chemically reacting inviscid region. The analysis at leading order provides the formulation of a boundary-value problem that determines the rescaled FK explosion curve, including the critical explosion conditions. The solution is obtained by an approximate method involving an integral form of the boundary-layer equations, which is found to produce predictions for the critical Damköhler number that compare favorably with the results of numerical integrations.

The presentation begins in section 2 with the complete set of equations for the slowly reacting mode of combustion with natural convection, with account taken of the nonnegligible effect of the temporal pressure variations in the energy equation. Results of steady and transient computations are presented in section 3, including the associated FK predictions for the thermal explosion limits in the presence of buoyancy-driven flow. The asymptotic limit $Ra \gg 1$ is considered next, in section 4. The paper ends in section 5 with concluding remarks.

2 Formulation for the transient slowly reacting mode of combustion in the presence of buoyancy

The analysis considers a spherical vessel of radius a filled with a reactant ideal-gas mixture at a strictly uniform temperature T_I, assumed to be sufficiently low for the chemical reaction to be initially negligible. In order to promote reaction, the wall temperature is rapidly increased at the initial instant of time to a fixed constant value $T_o > T_I$. This temperature rise is assumed to occur in a time that is short compared with the heat-conduction time across the vessel but still much larger than the characteristic acoustic time, so that spatial pressure variations remain very small. The initial reactant mass fraction Y_I is uniform, and any changes in the average molecular weight of the gas are considered to be negligible. The initial average gas density ρ_I is the ratio of the mass of gas present in the chamber to the chamber volume, and the initial average pressure p_I is determined from T_I and ρ_I by application of the ideal gas law. Initial pressure variations of the gas, at rest in hydrostatic equilibrium, are extremely small for chambers of sizes of interest in the Earth's gravitational field and give rise to correspondingly small density variations in the isothermal gas. The average pressure p_o is defined by employing ρ_I and T_o in the ideal gas law.

As in Frank-Kamenetskii's work [1], an overall Arrhenius reaction is adopted here, with the mass of reactant consumed per unit volume per unit time $\dot{m}$ given by

$$\dot{m}/\rho = k(T)Y = YB\exp[-E/(RT)], \tag{1}$$

where ρ is the gas density, T its temperature, and Y is the reactant mass fraction. The temperature-dependent reaction-rate constant $k = B\exp[-E/(RT)]$ includes the frequency factor B, the activation energy E, and the universal gas constant R. In this simple chemistry description, the heat-release rate per unit volume is $q\dot{m}$, where q denotes the amount of heat released per unit mass of reactant consumed. The ratio of the heat released per unit mass of gas mixture qY_I to the gas enthalpy evaluated at the wall temperature c_pT_o, with c_p denoting the specific heat at constant pressure, defines the exothermicity parameter $\alpha = (qY_I)/(c_pT_o)$, of order unity or larger in applications.

The formulation of the transient thermal-explosion problem follows our earlier work [21]. The nondimensional activation energy $\beta = E/(RT_o)$ based on the wall temperature T_o is assumed to be large. As a result, small temperature increments $T - T_o$, of the order of the FK temperature T_o/β, suffice to increase the reaction rate (1) by a factor of order unity. The associated chemical time needed to increase the enthalpy by an amount of order c_pT_o/β is given by

$$t_e = \frac{1}{\alpha\beta B\exp[-E/(RT_o)]}. \tag{2}$$

This time is compared with the characteristic heat-conduction time across the vessel $t_c = a^2/D_T$ in defining the FK Damköhler number,

$$Da = \frac{t_c}{t_e} = \frac{a^2}{D_T}\frac{qY_I}{c_pT_o}\frac{E}{RT_o}B\exp[-E/(RT_o)], \tag{3}$$

where D_T is the thermal diffusivity. The other important parameter in the description is the Rayleigh number (based on the ordering of the temperature difference),

$$Ra = \frac{\beta^{-1}ga^3}{\nu D_T}, \tag{4}$$

where g is the acceleration of gravity and ν is the kinematic viscosity. This Rayleigh number will be seen to measure the convective transport in the energy equation, while the Grashof number, $Gr = Ra/Pr = \beta^{-1}ga^3/\nu^2$, involving the Prandtl number $Pr = \nu/D_T$, similarly measures convective transport in the momentum equation.

In the distinguished regime $Da \sim 1$ with $\beta \gg 1$ and $\alpha \sim 1$ considered here, variations of the temperature of order $T - T_o \sim T_o/\beta$ are expected to develop everywhere across the vessel after a transient period, scaling with the heat-conduction time t_c, during which the gas temperature increases from T_I to T_o. For values of Da larger than a critical value $Da_c(Ra)$ this transient stage ends with the development of a thermal explosion, characterized by a localized rapid increase of the temperature $T - T_o \sim \alpha T_o$ at a well-defined ignition time, followed by propagation of a premixed flame from the ignition kernel. By way of contrast, for subcritical values $Da < Da_c$ the temperature evolves to approach a quasi-steady distribution, with $T - T_o \sim T_o/\beta$, resulting from a balance between the heat released by the chemical reaction in the vessel interior and the heat losses to the wall.

In the following non-dimensional formulation lengths are scaled with a, time t with $t_c = a^2/D_T$, and velocity $\boldsymbol{v}$ with $v_g = \beta^{-1} g a^2/\nu$, the characteristic value resulting from a balance between the viscous and buoyancy forces, associated with density variations of order ρ_I/β. Dimensionless variables $\phi = \beta(T - T_o)/T_o$ and $\hat{\rho} = \beta(\rho - \rho_\mathrm{I})/\rho_\mathrm{I}$ are introduced for the temperature and density, as is appropriate for studying the slowly reacting mode of combustion. As noted in [5, 21], in rigid confined vessels the changes of the density are restricted to ensure that the mean spatial value of the density remains equal to ρ_I, resulting in temporal variations of the pressure of order p_o/β, much larger than the spatial variations of the pressure associated with the low-Mach-number buoyancy-induced gas motion. These temporal changes of the pressure, which had been overlooked in previous numerical studies [15, 16, 17], are taken into account here when writing the energy balance and the equation of state with use made of the order-unity variable $\hat{p}(t) = \beta(p - p_o)/p_o$.

Only laminar solutions symmetric with respect to the vertical axis are considered in the following analysis. Cylindrical dimensionless coordinates are employed in the description, with r and z denoting the distance to the vertical axis and the distance to the equatorial plane, respectively, and v_r and v_z being the corresponding radial and vertical velocity components. The amount of reactant consumed during the transient stage considered here, corresponding to times of order $t_c \sim t_e$, is $(Y_\mathrm{I} - Y)/Y_\mathrm{I} \sim (\alpha\beta)^{-1} \ll 1$, so that reactant consumption can be neglected in the first approximation when computing ignition times t_i and critical Damköhler numbers Da_c. Correspondingly, the equations describing the evolution of temperature and velocity reduce to

$$\nabla \cdot \boldsymbol{v} = 0, \tag{5}$$

$$\frac{1}{Pr}\left(\frac{\partial \boldsymbol{v}}{\partial t} + Ra\, \boldsymbol{v} \cdot \nabla \boldsymbol{v}\right) = \nabla^2 \boldsymbol{v} - \nabla p' + \phi\, \boldsymbol{e}_z, \tag{6}$$

$$\frac{\partial \phi}{\partial t} - \frac{\gamma - 1}{\gamma}\frac{\mathrm{d}\hat{p}}{\mathrm{d}t} + Ra\, \boldsymbol{v} \cdot \nabla \phi = \nabla^2 \phi + Da\, \mathrm{e}^{\phi}, \tag{7}$$

where the FK linearization $\mathrm{e}^{\phi/(1+\phi/\beta)} \simeq \mathrm{e}^{\phi}$ has been adopted for the Arrhenius exponent of the reaction–rate constant, as is appropriate in the limit $\beta \gg 1$. Additional symbols that appear in (6) and (7) are defined in the following paragraph.

In this near-explosion regime, associated with small density differences of order ρ_I/β, the contribution of the temporal density variation to mass conservation gives a higher-order correction, of order $(\beta\, Ra)^{-1}$, which has been neglected in (5). Also, leaving out terms of order $1/\beta$, variations of transport coefficients and specific heat from their values evaluated for $T = T_o$ and $\rho = \rho_\mathrm{I}$ have been omitted in (6) and (7). In the momentum equation (6), $\boldsymbol{e}_z$ is a unit vector pointing upwards (against gravity), and p' represents the pressure differences from the hydrostatic value that would apply when $\phi = 0$, scaled with $\rho_I g a/\beta$; these are essential for establishing the fluid motion, even though the magnitudes of the associated pressure changes are small compared with the uniform temporal variations of the pressure. The problem thus defined depends on the gas properties through the Prandtl number Pr appearing in (6) and the specific-heat ratio γ, which is present in the term arising from the time derivative of the pressure in the energy-conservation equation (7). The values $Pr = 0.7$ and $\gamma = 1.4$ are adopted for all computations performed below, an appropriate selection for most fuel-air combustion mixtures.

The above equations are supplemented by the equation of state written in the linearized form

$$\hat{\rho} + \phi = \hat{p}(t). \tag{8}$$

To ensure that the spatially mean density remains ρ_I, as it must, the time-dependent pressure perturbation $\hat{p}$ must obey the relation

$$\hat{p} = \bar{\phi} = \int_{-1}^{+1}\left(\int_0^{\sqrt{1-z^2}} \tfrac{3}{2}\phi\, r\, \mathrm{d}r\right)\mathrm{d}z \tag{9}$$

involving the non-zero mean value $\bar{\phi}$ of the non-dimensional temperature increase.

Inside the spherical vessel of constant wall temperature, the slowly reacting mode of combustion accounting for buoyancy-induced motion is expected to have analytical solutions of (5)–(8), bounded and symmetric with respect to

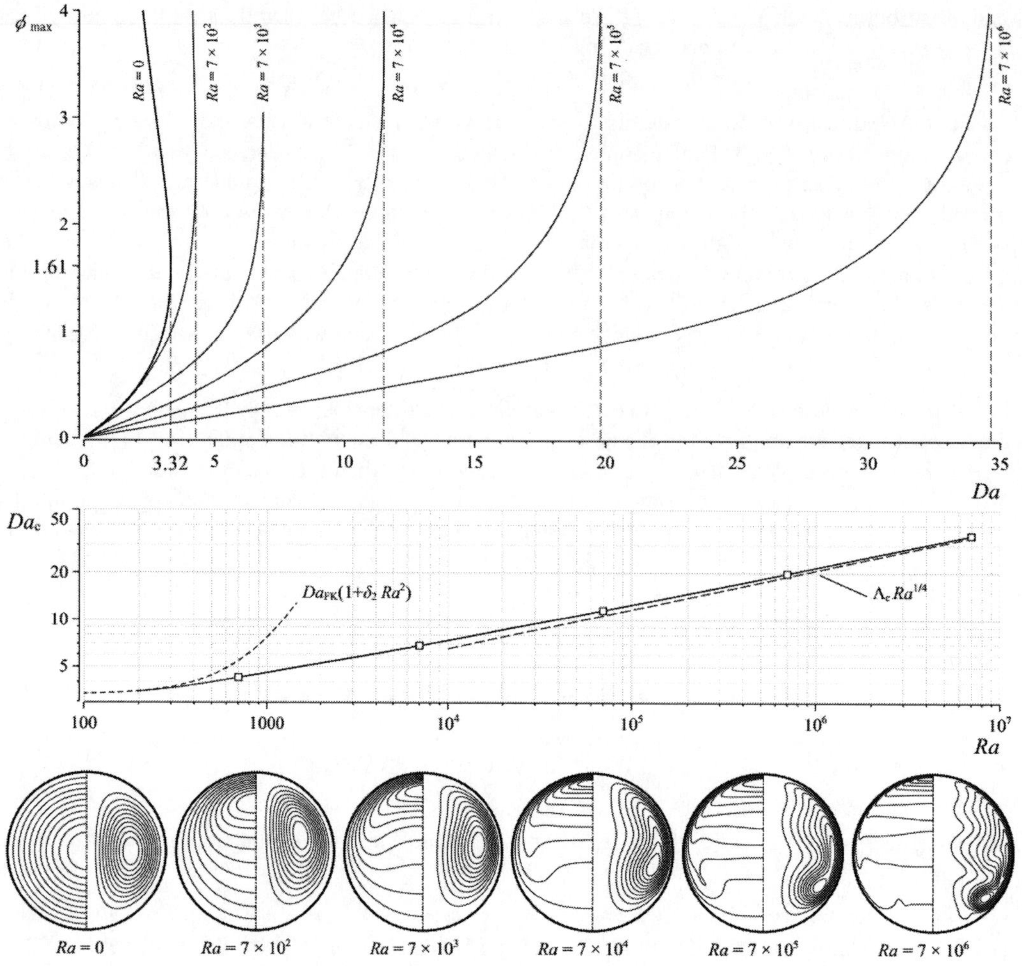

Figure 1: Results of numerical integrations of steady reactive flows for different values of *Ra* and *Da*. The upper plot shows the variation with *Da* of the peak temperature in the vessel for six different values of $Ra = 7 \times (0, 10^2, 10^3, 10^4, 10^5, 10^6)$. The middle plot shows the variation with *Ra* of the critical Damköhler number, including as dashed curves the asymptotic predictions $Da_c = Da_{FK}(1 + \delta_2 Ra^2)$, derived in [21] for $Ra \ll 1$, as well as the new result $Da_c = \Lambda_c Ra^{1/4}$ for $Ra \gg 1$, to be derived later in section 4. The bottom plots show ten isotherms (left half) and ten streamlines (right half) for the conditions corresponding to the six turning points of the upper plot.

the vertical axis $r = 0$, satisfying the boundary conditions

$$v = \phi = 0 \tag{10}$$

at the vessel wall $r^2 + z^2 = 1$. Computation of the transient evolution requires specification of the initial conditions. In the following, we consider an initially cold stagnant gas mixture at temperature T_I such that $T_o - T_I \sim T_o/\beta$, yielding the initial conditions

$$v = \phi - \phi_I = 0 \quad \text{at} \quad t = 0, \tag{11}$$

with $\phi_I = \beta(T_I - T_o)/T_o < 0$. As discussed in [5], the analysis becomes somewhat more complicated when the wall-temperature rise is larger, such that $T_o - T_I \sim T_o$, in which case the heating of the gas by heat conduction from the wall, which proceeds initially with negligible chemical reaction, involves relative variations of the density, temperature, and pressure of order unity.

3 Selected numerical results

As previously mentioned, the evolution of the reactive flow in the vessel depends fundamentally on the Damkhöhler number. If Da is smaller than a critical value Da_c, larger for larger Ra, the temperature and velocity evolve to approach a steady distribution for moderately large values of t, whereas for $Da > Da_c$ the transient stage ends with a thermal runaway at a finite ignition time. The determination of Da_c for a given Ra based on transient computations requires consideration of increasing values of Da until critical conditions are achieved, as was done previously in [15, 16, 17]. Alternatively, the explosion limit for a given Ra can be obtained directly by investigating the existence of steady solutions for the slowly reacting mode of combustion, with the value of Da_c corresponding to the turning point of the stationary explosion curve giving the variation of the peak temperature with Da, that being the criterion proposed in Frank-Kamenetskii's original theory. We shall see below that both computational approaches lead to predictions of explosion limits that are virtually identical.

3.1 FK analysis of thermal-explosion limits

We begin by considering steady FK solutions, described by integrating the steady version of (5)–(7) supplemented by (9) with the boundary conditions (10) at the vessel wall. A standard iterative root–finding technique (Newton–Raphson) is employed in the integrations. The linearized system of equations is solved using the software package FreeFem++ [22], which employs finite elements for the spatial discretization. A summary of the steady solutions obtained for different values of Da and Ra is given in Fig. 1. The upper plot in Fig. 1 shows explosion diagrams (i.e., curves representing the peak temperature $\phi_{\max}$ as a function of the Damköhler number Da) for six different values of the Rayleigh number, including the classical FK results corresponding to $Ra = 0$, for which the turning point occurs at $Da_{\rm FK} \simeq 3.32$ with $\phi_{\max} \simeq 1.61$ [2]. Because of the enhanced heat transfer associated with the convective motion, the turning points of the resulting curves occur at larger values of Da_c for increasing Ra. This variation is represented in the intermediate plot of Figure 1. As expected from previous analyses, the departures $Da_c - Da_{\rm FK}$ of the critical Damköhler number from the buoyancy-free value $Da_{\rm FK}$ remain fairly small for values of Ra on the order of a few hundred. For these conditions of weak convective motion the asymptotic prediction $Da_c = Da_{\rm FK}(1 + \delta_2 Ra^2)$, derived previously in the limit $Ra \ll 1$ [21], agrees well with the numerical results. The quantitative effect of buoyancy on the explosion boundary becomes clearly noticeable for $Ra \gtrsim 500$, for which the departures $Da_c - Da_{\rm FK}$ become of order unity. The monotonic growth of Da_c with Ra continues for increasing Ra. As seen in the plot, for sufficiently large values of $Ra \gtrsim 10^4$, the curve $Da_c(Ra)$ displays the asymptotic behavior $Da_c \propto Ra^{1/4}$, corresponding to a straight line with slope 1/4 in the log-log scale used in the figure. The dashed curve $Da_c = \Lambda_c Ra^{1/4}$ in the plot is the asymptotic prediction for $Ra \gg 1$, to be derived later in Sec. 4. The motion induced by buoyancy and its associated effects on the temperature field are investigated on the bottom plots of Figure 1 by representing the isotherms (left half of each circular plot) and streamlines (right half) corresponding to the solutions at the turning points of the six different explosion curves. A total of ten isocurves of temperature and stream function are shown in each subplot. The isotherms are equally spaced between the value $\phi = 0$ at the wall and the peak value $\phi_{\max}$. Similarly, the selection of streamlines correspond to equally spaced values of the stream function ψ, defined such that $rv_r = \partial\psi/\partial z$ and $rv_z = -\partial\psi/\partial r$ with $\psi = 0$ at the wall, and the peak values being $\psi_{\max} = (1.6822 \times 10^{-3}, 1.5575 \times 10^{-3}, 5.728 \times 10^{-4}, 1.361 \times 10^{-4}, 2.7973 \times 10^{-5}, 5.6811 \times 10^{-6})$ for $Ra = 7 \times (0, 10^2, 10^3, 10^4, 10^5, 10^6)$. It can be seen that the Frank-Kamenetskii vortex induced by buoyancy, symmetric about the equatorial plane for $Ra \ll 1$, evolves in a non-monotonic manner for increasing Ra, with the accompanying velocity, directed upwards in the central hot region, causing a vertical displacement of the peak temperature, which readily destroys the spherical symmetry of the temperature distribution, as noticed in early experimental observations [23, 24]. The streamline pattern indicates that the center of the vortex, which initially migrates upwards, moves eventually downwards to occupy a location near the wall, where a high-velocity boundary-layer flow develops for $Ra \gg 1$, as inferred by the close spacing between adjacent streamlines. The structure of the flow in that limit, which includes a central region of stratified temperature with a hot kernel near the "north pole" of the vessel, is to be considered in detail in section 4.

Unlike numerical schemes based on pseudo-transient methods, the root–finding technique employed here not only is able to describe the lower branch of stable solutions, extending between the origin of the explosion diagram $\phi_{\max} - Da$ and the turning point of the bending bifurcation, but it can also be employed in principle to describe the unstable solutions lying beyond the turning point, corresponding to larger values of $\phi_{\max}$. For illustrative purposes, a detailed view of the turning point of the curve corresponding to $Ra = 700$ is shown in Fig. 2, along with the isotherms and

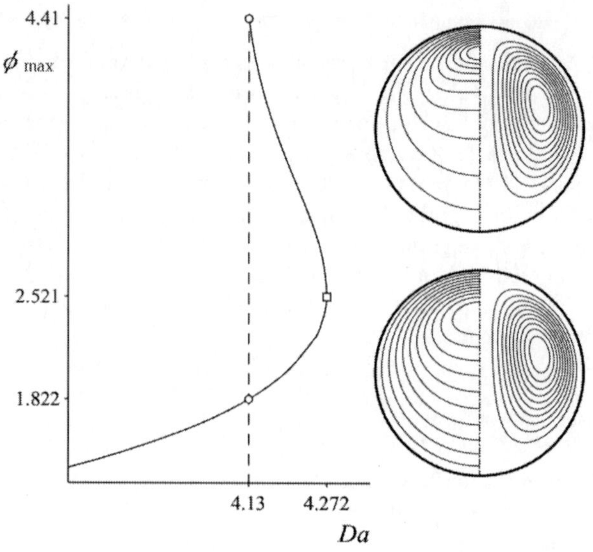

Figure 2: Detailed view of the explosion curve near the turning point for $Ra = 700$. The circular plots show isotherms (left half) and streamlines (right half) corresponding to $Da = 4.13$ for the solutions along the upper (unstable) branch and the lower (stable) branch. A total of 10 equally spaced isocurves is shown in each plot, with peak values of temperature and stream function given by $\phi_{max} = 4.410$ and $\psi_{max} = 1.594 \times 10^{-3}$ for the upper plot and $\phi_{max} = 1.822$ and $\psi_{max} = 1.380 \times 10^{-3}$ for the lower plot.

streamlines corresponding to two different solutions for the same $Da < Da_c$. The description of the solutions along the upper branch becomes more difficult for larger Ra, because the curvature of the associated explosion curves at the turning point becomes smaller, leading to rapid variations of ϕ_{max} for $Da_c - Da \ll 1$ that hinder the convergence of the root-finding scheme. For these larger values of Ra, the value of Da_c reported here is defined as the maximum value of Da for which a converged solution is found. Numerical methods based on continuation techniques might be better suited for the description of the complete multi-branch explosion curve developing beyond the first turning point, but this alternative approach is not further pursued here because of the limited practical interest in these unstable solutions.

3.2 Transient computations

To complement the above steady solutions, transient histories corresponding to the initial conditions (11) were computed by marching in time the equations (5)–(7) supplemented with (9), with the boundary conditions given in (10). Besides the parameters Ra and Da, the Prandtl number $Pr = 0.7$, and the specific-heat ratio $\gamma = 1.4$, the results of the transient computations depend also on ϕ_I, the rescaled initial gas temperature in the container. As previously discussed, to ensure that the chemical reaction is negligible before the wall temperature is suddenly increased at $t = 0$, this negative temperature must satisfy $-\phi_I \gg 1$. The value $\phi_I = -10$ is employed in the computations reported below. As in the steady computations, finite elements were used for the spatial discretization. The time-marching technique employed a characteristics-Galerkin method [22, 25] with a fixed time step Δt. The resulting system of linearized equations was solved at each time step using FreeFem++ [22]. The temporal evolution, including the associated ignition time for supercritical cases, was found to be independent of the choice of Δt provided that a sufficiently small value is employed. The solutions shown below in Fig. 3 correspond to $\Delta t = 10^{-5}$. As expected, the transient computations reproduce the FK explosion boundary $Da_c(Ra)$ shown in the middle plot of Fig. 1 in that, regardless of the value of Ra, the unsteady flow for all subcritical values of the Damköhler number $Da < Da_c(Ra)$ was seen to evolve to reach the final steady–state solutions shown in Fig. 1 with indistinguishable differences, whereas for $Da > Da_c(Ra)$ a localized thermal runaway occurs at a finite time $t = t_i$, larger for smaller values of $Da - Da_c$.

The variation of t_i with Da is shown in Fig. 3 for $Ra = 7 \times 10^4$. In the plot, the ignition time is defined as the last time for which a bounded solution is found in the numerical integration. Because of the precipitous character of the temperature increase, other definitions of the ignition time, based for instance on a given threshold value for

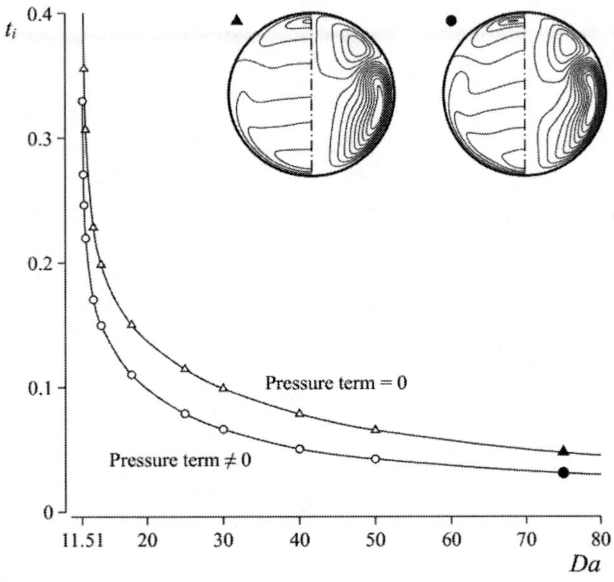

Figure 3: Ignition time for $Ra = 7 \times 10^4$ as a function of $Da > Da_c(Ra) \simeq 11.51$. The inset shows isotherms and streamlines for $Da = 75$ at $t \approx t_i$.

the local rate of temperature increase, would produce almost identical results. The plot includes the curve $t_i(Da)$ obtained by neglecting in the energy equation (7) the term involving the temporal pressure increase. In view of these results, it can be concluded that computations of thermal explosions omitting this term, as was done in many previous studies [16, 17], may result in significant overpredictions of ignition times in chambers with rigid walls. The plot also indicates that, as expected, the computation of the critical value Da_c at which $t_i \to \infty$ is not affected by the presence of this unsteady term. To illustrate changes in the morphology of the ignition kernel, the insets in the figure show snapshots of isotherms and streamlines immediately before ignition for $Da = 75$. For this particular case, the ignition kernel is centered at a point along the axis when the term $\mathrm{d}\hat{p}/\mathrm{d}t$ is removed, but ignition occurs in a thin reacting annulus, slightly detached from the container axis, when this term is retained in the computation, as may be seen at the top of the insets.

As previously stated, the results in Fig. 3 correspond to $\phi_I = -10$. Since the duration of the heating period needed to increase the gas temperature in the container to values close to the wall value depends fundamentally on ϕ_I, lower values of ϕ_I result in larger ignition times, as was verified in additional computations. Nevertheless, the explosion limits, measured by the critical Damköhler number Da_c, were found to be independent of the selection of ϕ_I for all $\phi_I \leq 0$, in agreement with our previous results for buoyancy-free systems [5]. The role of buoyancy in the development of the thermal explosion is investigated in Fig. 4 for $Ra = 7 \times 10^4$ and $\phi_I = -10$ when the Damköhler number is $Da = 11.75$, only slightly above the critical value $Da_c = 11.51$ predicted by the turning-point FK criterion. The upper left panel shows the temperature at the center of the vessel ϕ_o and the peak temperature $\phi_{\max}$, the latter displaying a precipitous growth at $t = t_i \simeq 0.28$. The twelve panels at the bottom show isotherms and streamlines at representative instants of time, marked with circles in the curve $\phi_o(t)$. The shading is used in the panels to indicate positive temperatures (above the wall value) and positive values of the stream function, corresponding to counterclockwise motion in the right-hand-side panels. The heating of the gas from the hot wall affects initially a thin boundary layer that moves upwards under the action of gravity, entraining the surrounding cold gas and inducing a counterclockwise motion on the right inside the vessel. Because of the spherical geometry of the container, the boundary layer collides at the top of the sphere, ejecting downwards along the axis the gas that has been heated while ascending next to the wall. A secondary vortex with clockwise motion is seen to emerge from the collision region, thereby further complicating the dynamics. This convective transport rapidly heats the upper half of the vessel, while the bottom, nearly stagnant, remains initially cold.

The chemical reaction, initially negligible, begins to occur near the top of the sphere, causing the temperature to increase there above the wall value (i.e. positive values of ϕ) for $t \gtrsim 0.017$. The shaded region representing the hot

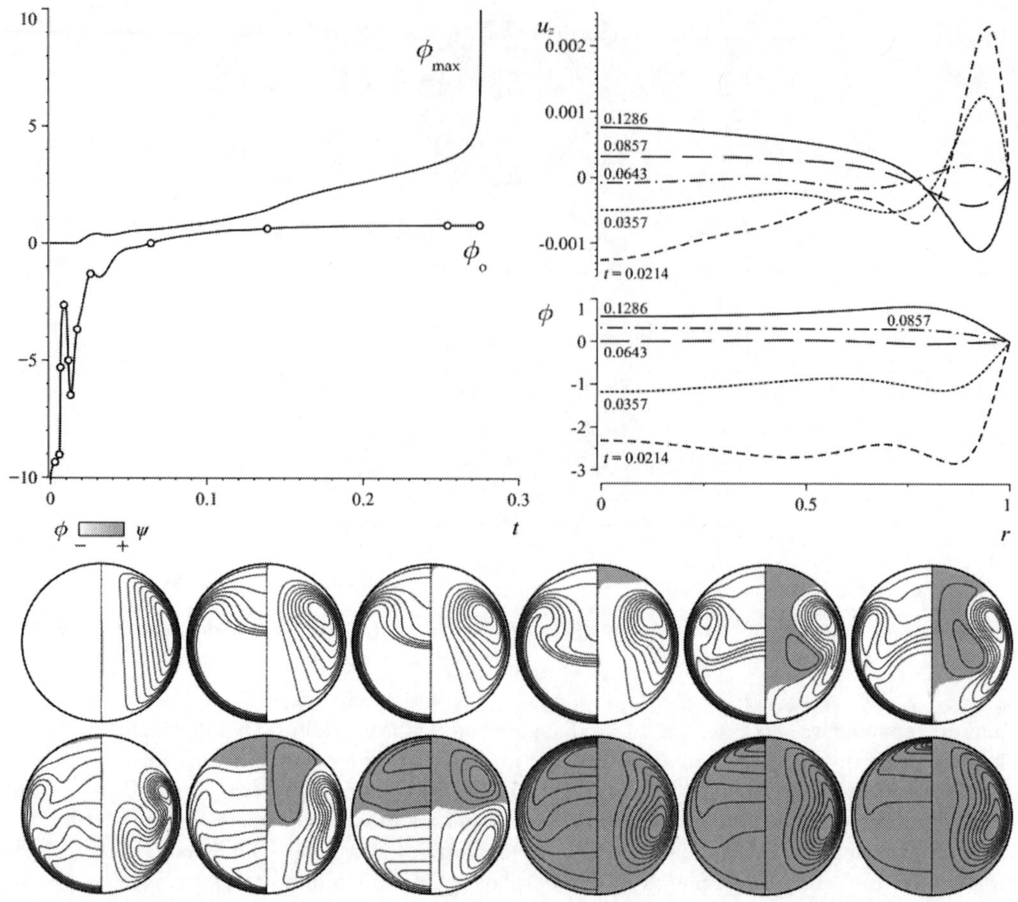

Figure 4: Results of transient numerical integrations of (5)–(7) with initial temperature $\phi_I = -10$, $Da = 11.75$, $Ra = 7 \times 10^4$. The temporal evolution of the temperature at the center ϕ_o and of the peak temperature $\phi_{\max}$ is shown in the top left panel, whereas sample equatorial profiles of vertical velocity u_z and temperature ϕ are plotted in the top right panel for times indicated in the figure. As in Figure 1, isotherms (left) and streamlines (right) are shown in the bottom panel for times marked with circles in the curve $\phi_o(t)$. The values of the stream function and temperature corresponding to each isocurve are selected to be equally spaced between the zero wall values and the peak values. The shade indicate positive values of the temperature (left half) and stream function (right half).

gas extends rapidly downstream, changing in a fundamental way the gas motion. This can be seen in the snapshot corresponding to $t = 0.0643$ (the third from the left on the bottom row). In the upper half of the vessel $\phi > 0$, so that in the corresponding near-wall boundary layer the gas temperature is lower than that found outside. Consequently, buoyancy pulls the fluid downwards near the wall, supporting clockwise motion in this region. The opposite effect is found on the lower half of the vessel, where the temperature still remains negative. As a result, the flow at this intermediate time is characterized by the existence of two annular counterrotating vortices. For longer times, the hot gas occupies the entire vessel, and the resulting motion is given by a single vortex with clockwise motion, as in the steady cases displayed in Fig. 1. The change in sign of the heat flux from the wall and the corresponding flow reversal along the boundary layer are illustrated in the upper right plot of Fig. 4 by plotting the radial distributions of vertical velocity and temperature at the equatorial plane $z = 0$ at different instants of time. As seen in the last three snapshots in Fig. 4, corresponding to $t = (0.1386, 0.2543, 0.2750)$, for this near-critical case the transient ignition history shows a relatively long stage of slow evolution in which the streamline pattern at all times is very similar to that of the FK solution at the turning point (see the plot for $Ra = 7 \times 10^4$ in Fig. 1). This quasi-steady motion is accompanied by a slow temperature evolution in which the values found in the central region remain almost constant, as seen in the evolution of ϕ_o, while the temperature distribution near the top develops an increasingly sharp peak that eventually transitions into a rapid runaway.

4 Slowly reacting flow for $Ra \gg 1$

4.1 Flow structure

The computations shown in Fig. 1 reveal that the structure of the flow in the limit $Ra \gg 1$ includes a central region of slowly moving hot gas bounded by a high-velocity near-wall boundary layer. The scales and the dominant balances applying in each region can be identified by an order-of-magnitude analysis. Thus, since the temperature increment must be $\phi \sim 1$ for the thermal explosion to develop, the buoyancy force in (6) can be anticipated to be of order unity when $Da \sim Da_c$, inducing streamwise velocities of order $Ra^{-1/2}$ in a near-wall boundary layer of characteristic thickness $Ra^{-1/4}$, as inferred from the convection-diffusion balance in (6). The flow in the central region, outside the boundary layer, is induced by the boundary-layer entrainment, with characteristic velocities of order $Ra^{-3/4}$. This slow motion does not induce an appreciable pressure disturbance inside the container, where (6) reduces in the first approximation to $-\nabla p' + \phi\, e_z = 0$. This hydrostatic balance requires that the pressure p' and the temperature ϕ be a function of z only. This is consistent with the results shown in Fig. 1, where the isotherms for $Ra \gg 1$ are nearly horizontal outside the boundary layer. Since the velocities induced by boundary-layer entrainment are of order $Ra^{-3/4}$, the effective Peclet number, measuring in the central region the relative importance of convective and conductive rates, is of order $Ra^{1/4} \gg 1$, so that the effect of heat conduction can be neglected in the first approximation when writing (7) there. The resulting convection-reaction balance $Ra\, \boldsymbol{v} \cdot \nabla\phi = Da\, e^{\phi}$, with $\boldsymbol{v} \sim Ra^{-3/4}$, indicates that for $Da \sim 1$ the temperature increase associated with the chemical reaction is limited to small values of order $Ra^{-1/4}$, insufficient to produce a significant increase in the reaction rate from its near-wall value, and that temperature increments of order unity, needed to trigger the thermal explosion, require values of the Damköhler number of order $Ra^{1/4}$. In this near-explosion regime, $Da \sim Ra^{1/4}$, the chemical reaction occurs mainly in the central region, while the boundary layer remains chemically frozen in the first approximation, because the transport rates there are larger than the reaction term by a factor of order $Ra^{1/4}$, as can be seen from (7).

Before proceeding with the analysis, it is worth pointing out that the flow structure identified here is fundamentally similar to that emerging in fully developed flow in a curved pipe at large values of the Dean number [26]. In both cases, the inviscid internal flow is coupled with a thin boundary layer that develops between two stagnation-point self-similar regions. Also, as we shall see below, the difficulties in the numerical solution stem in both cases from the elliptic character of the boundary-layer problem, associated with the existence of recirculating flow.

4.2 The boundary-layer problem

The leading-order form of the simplified equations in the chemically frozen boundary layer and in the central inviscid region are written now in terms of appropriately rescaled variables. The boundary layer developing at the wall is described in terms of the local distance to the wall $y = Ra^{1/4}[1-(r^2+z^2)^{1/2}]$ and the angle $\theta = \cos^{-1} z$, which measures the distance along the wall from the vessel top, giving the equations

$$\frac{\partial}{\partial\theta}(U\sin\theta) + \frac{\partial}{\partial y}(V\sin\theta) = 0 \tag{12}$$

$$\frac{1}{Pr}\left(U\frac{\partial U}{\partial\theta} + V\frac{\partial U}{\partial y}\right) = \frac{\partial^2 U}{\partial y^2} + (\phi_c - \phi)\sin(\theta) \tag{13}$$

$$U\frac{\partial\phi}{\partial\theta} + V\frac{\partial\phi}{\partial y} = \frac{\partial^2\phi}{\partial y^2} \tag{14}$$

where $U = Ra^{1/2}(zv_r - rv_z)$ and $V = -Ra^{3/4}(rv_r + zv_z)$ are the rescaled streamwise and transverse velocity components, respectively. The boundary conditions for $0 < \theta < \pi$ are

$$y = 0: \quad U = V = \phi = 0 \tag{15}$$

$$y \to \infty: \quad U = \phi - \phi_c(\theta) = 0. \tag{16}$$

Here $\phi_c(\theta)$ is the temperature outside the boundary layer, in the central inviscid region, given by the convection-reaction balance

$$\bar{v}_z \frac{d\phi_c}{dz} = \Lambda e^{\phi_c} \tag{17}$$

where $\Lambda = Ra^{-1/4}Da$ is a rescaled Damköhler number and $\bar{v}_z = Ra^{3/4}v_z$ is the rescaled vertical velocity, also a function of z, as required for consistency.

The conduction-free equation (17) describes the increase of the temperature of the ascending gas due to chemical reaction, from the value $\phi_c = 0$, found near the bottom $z = -1$, to reach the peak value $\phi_{\max}$ as $z \to 1$. The vertical velocity appearing as a factor in (17) can be obtained at a given height z by equating the upward mass flux $\pi(1-z^2)\bar{v}_z$ in the container interior to the downward mass flux $2\pi(1-z^2)^{1/2}\int_0^\infty U \mathrm{d}y$ across the boundary layer to give

$$\bar{v}_z = \frac{2\int_0^\infty U \mathrm{d}y}{(1-z^2)^{1/2}}, \tag{18}$$

and the accompanying radial velocity $\bar{v}_r = Ra^{3/4}v_r$ is given by $\bar{v}_r = -(r/2)\mathrm{d}\bar{v}_z/\mathrm{d}z$, as follows from continuity. The geometrical relations $\sin\theta = (1-z^2)^{1/2}$ and $\sin(\theta)\mathrm{d}\theta = -\mathrm{d}z$ can be used together with (18) to express (17) in the form

$$\frac{\mathrm{d}\phi_c}{\mathrm{d}\theta} = -\frac{\Lambda \sin^2(\theta) e^{\phi_c}}{2\int_0^\infty U \mathrm{d}y}, \tag{19}$$

yielding the solution

$$\phi_c(\theta) = \ln\left[e^{-\phi_{\max}} + \frac{\Lambda}{2}\int_0^\theta \frac{\sin^2(\tilde{\theta})}{\int_0^\infty U \mathrm{d}y}\mathrm{d}\tilde{\theta}\right]^{-1}, \tag{20}$$

upon integration with the initial condition $\phi_c(0) = \phi_{\max}$.

The integration of (12)–(14) must be initiated at $\theta \ll 1$, where there exists a self-similar solution. At the leading order described here, the solution is identical to that encountered near the attachment point of the buoyancy-driven boundary layer that develops over a hot sphere in a stagnant atmosphere when the Grashof number is large [27]. The local flow, first described for the hot sphere in [28], can be analyzed by introducing the normalized coordinate $\zeta = \phi_{\max}^{1/4} y$ together with the stream function $\psi = \phi_{\max}^{1/4}\theta^2 F(\zeta)$ and the reduced temperature $G = (\phi_{\max} - \phi)/\phi_{\max}$ to give

$$F_{\zeta\zeta\zeta} + (2FF_{\zeta\zeta} - F_\zeta^2)/Pr + G = 0, \tag{21}$$

$$G_{\zeta\zeta} + 2FG_\zeta = 0, \tag{22}$$

with boundary conditions

$$F(0) = F_\zeta(0) = G(0) - 1 = 0 \tag{23}$$

$$F_\zeta(\infty) = G(\infty) = 0, \tag{24}$$

where the subscript ζ denotes differentiation with respect to this coordinate. The integration of (21)–(24) for $Pr = 0.7$ provides the profiles shown in Fig. 5 along with the value of $F(\infty) = 0.63702$, which can be used to write the relationship $\bar{v}_z(z=1) = 2\phi_{\max}^{1/4}F(\infty)$ between the boundary temperature and the uniform entrainment velocity at $\theta = 0$. The uniformity of the boundary-layer entrainment rate near the top is consistent with the negligible radial variation of the vertical velocity $\bar{v}_z(z)$ in the reactive core region. The functions $F(\zeta)$ and $G(\zeta)$ provide the profiles of streamwise velocity and temperature

$$U = \phi_{\max}^{1/2}\theta F_\zeta \quad \text{and} \quad \phi = \phi_{\max}(1-G) \qquad \text{at} \quad \theta \ll 1, \tag{25}$$

to be used as initial condition for integration of (12)–(14).

The apparent parabolic nature of (13) and (14) suggests a marching procedure in which (12)–(14) supplemented with (20) are integrated for increasing θ subject to the boundary conditions (15). The integration must be initiated at $\theta \ll 1$ using the self-similar profiles (25). In this scheme, the value of Λ corresponding to a given maximum temperature $\phi_{\max}$ is obtained from (20) as an eigenvalue by imposing the condition that $\phi_c \to 0$ as $\theta \to \pi$. This eigenvalue problem bears similarity with that encountered in the analysis of fully developed flow in curved pipes at large Dean number [26]. As in that case, the time-marching integration strategy delineated above fails because of the appearance of recirculating flow. This is observed to emerge at an intermediate location between $\theta = \pi/2$ and $\theta = \pi$, with the velocity profile U exhibiting regions of negative U. This behavior seems to be consistent with the streamline patterns depicted in Fig. 1 for $Ra \gg 1$.

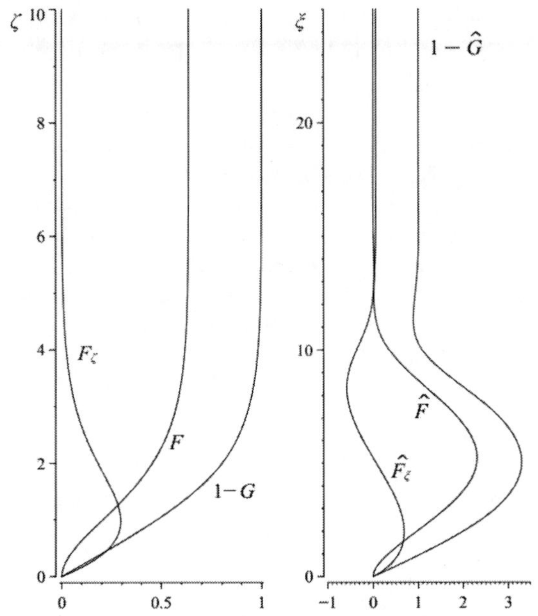

Figure 5: Results of the integration of (21)–(24) (left panel) and of (26)–(29) (right panel), showing the functions $F(\zeta)$, $F'(\zeta)$, $1-G(\zeta)$ and $\hat{F}(\xi)$, $\hat{F}'(\xi)$, $1-\hat{G}(\xi)$, respectively. The integrations indicate that $F(\infty)=0.63702$ and $\hat{F}(\infty)=0.05903$.

Because of the presence of backflow, the character of the resulting boundary-layer problem changes from parabolic to elliptic. The associated boundary-value problem requires consideration of the specific conditions found near the bottom $\theta=\pi$. There, the flow exhibits a self-similar solution, with the core temperature vanishing on approaching $\theta=\pi$ according to the power law $\phi_c=A(\pi-\theta)^\lambda$, consistent with the local scalings $U\propto(\pi-\theta)^{1+\lambda/2}$ and $y\propto(\pi-s)^{-\lambda/4}$, where the value of the exponent $\lambda=8/5$ is determined from the convective-reactive balance (19). Writing (12)–(14) in terms of the local coordinate $\xi=A^{1/4}(\pi-\theta)^{2/5}y$, stream function $\psi=A^{1/4}(\pi-\theta)^{12/5}\hat{F}(\xi)$, and temperature increment $\phi_c-\phi=A(\pi-\theta)^{8/5}\hat{G}(\xi)$ yields

$$\hat{F}_{\xi\xi\xi}+\left(\tfrac{9}{5}\hat{F}_\xi^2-\tfrac{12}{5}\hat{F}\hat{F}_{\xi\xi}\right)/Pr+\hat{G}=0 \tag{26}$$

$$\hat{G}_{\xi\xi}-\tfrac{12}{5}\hat{F}\hat{G}_\xi-\tfrac{8}{5}\hat{F}_\xi(1-\hat{G})=0 \tag{27}$$

with boundary conditions

$$\hat{F}(0)=\hat{F}_\xi(0)=\hat{G}(0)-1=0 \tag{28}$$

$$\hat{F}_\xi(\infty)=\hat{G}(\infty)=0, \tag{29}$$

where the subscript ξ denotes differentiation with respect to this coordinate. The profiles $\hat{F}$, $\hat{F}_\xi$, and $\hat{G}$ are shown in Fig. 5 for $Pr=0.7$. In connection with this self-similar solution it is worth noting that the boundary-layer thickness diverges as $(\pi-\theta)^{-2/5}$ as $\theta\to\pi$, indicative of a boundary layer that empties near the bottom, a flow feature consistent with the streamline patterns observed in the bottom plots of Fig. 1 for $Ra\gg1$. This type of local behavior is fundamentally different from that emerging in free convection from a heated sphere, where the eruption of the fluid into the plume above the sphere is the result of the collision of the boundary layer at the upper stagnation point, as described by Potter and Riley [27]. Also of interest is that the rescaled streamwise velocity $\hat{F}_\xi$ shown in Fig. 5 displays an oscillatory profile with multiple zeros. An emptying boundary layer with oscillatory velocity profile is also encountered near the inside bend of curved pipes at large Dean number [29], thereby further highlighting the similarities between both flows. The self-similar profiles can be used to write the velocity and temperature distributions as

$$U=A^{1/2}(\pi-\theta)^{9/5}\hat{F}_\xi,\quad \phi_c-\phi=A(\pi-\theta)^{8/5}\hat{G}\ \text{at}\ \pi-\theta\ll1, \tag{30}$$

where the constant factor $A=[(5\Lambda)/(16\hat{F}(\infty))]^{4/5}$, obtained from (19), can be evaluated with use made of of the boundary value $\hat{F}(\infty)=0.05903$ determined from the integration of (26)–(29).

Equations (12)–(14) must be integrated for $0 < \theta < \pi$ together with (20) subject to the boundary conditions (15) and to the local boundary distributions (25) and (30) at $\theta \ll 1$ and $\pi - \theta \ll 1$, respectively. In the solution, the value of Λ corresponding to a given $\phi_{\max}$ is an eigenvalue of the problem. The numerical solution of the above boundary–value problem was attempted using different methods, involving either a pseudo-transient scheme for advancing (12)–(14) or the simultaneous solution for U, V, and ϕ by means of a Newton-Raphson method, with the eigenvalue Λ updated after each iteration using (20). Despite significant efforts, difficulties associated with the strong sensitivity of the results to small changes in Λ precluded convergence of the method, so that a numerical solution to the problem could not be achieved. It is worth pointing out that similar difficulties have been encountered in connection with the boundary-value problem arising in curved pipes at large Dean number [26], whose numerical solution also remains elusive to date. For this latter problem, an approximate solution based on the integral form of the boundary-layer equations was successfully employed by Ito [30]. The same strategy is followed below to obtain the reduced explosion curve $\phi_{\max}$ as a function of Λ and the associated critical value Λ_c.

4.3 Von-Karman integral formulation

The approximate solution begins by integrating (13) and (14) across the boundary layer to give the two ordinary differential equations

$$\begin{aligned}\frac{1}{Pr}\frac{\mathrm{d}}{\mathrm{d}\theta}\left(\sin\theta\int_0^\infty U^2\mathrm{d}y\right) &= -\sin\theta\left.\frac{\partial U}{\partial y}\right|_{y=0} \\ &\quad +\sin^2\theta\int_0^\infty(\phi_c-\phi)\mathrm{d}y, \end{aligned} \tag{31}$$

$$\begin{aligned}\frac{\mathrm{d}}{\mathrm{d}\theta}\left(\sin\theta\int_0^\infty U(\phi_c-\phi)\mathrm{d}y\right) - \sin\theta\frac{\mathrm{d}\phi_c}{\mathrm{d}\theta}\int_0^\infty U\mathrm{d}y \\ = \sin\theta\left.\frac{\partial\phi}{\partial y}\right|_{y=0}. \end{aligned} \tag{32}$$

Guided by the method employed in [30], it is further assumed that the boundary layer has a finite thickness $\delta(\theta)$ and that the temperature and streamwise velocity across the boundary layer can be expressed in the form

$$\phi = \phi_c k(\eta) \quad \text{and} \quad U = \phi_c \sin(\theta)\,\delta^2 f(\eta) + U_o h(\eta), \tag{33}$$

where the description for U incorporates a velocity correction $U_o(\theta)$ in addition to the direct effect of the gravitational acceleration. The functions $k(\eta)$, $f(\eta)$, and $h(\eta)$ of the normalized transverse coordinate $\eta = y/\delta$ are introduced to define the shape of the profiles. These shape functions are defined such that $k-1=f=h=0$ for $\eta \geq 1$, with polynomial expressions employed below for describing $k(\eta)$, $f(\eta)$, and $h(\eta)$ in the range $0 \leq \eta \leq 1$. In principle, a second-order polynomial for k and f and a third-order polynomial for h suffice to satisfy the boundary conditions

$$\begin{cases} k=f=h=0 \text{ and } \frac{\mathrm{d}^2k}{\mathrm{d}\eta^2} = \frac{\mathrm{d}^2f}{\mathrm{d}\eta^2}+1 = \frac{\mathrm{d}^2h}{\mathrm{d}\eta^2} = 0 \text{ at } \eta=0 \\ k-1=f=h=0 \text{ at } \eta=1, \end{cases} \tag{34}$$

stemming from (15). Smoother profiles with vanishing derivatives at $\eta = 1$ can be obtained by increasing the order of the polynomials. In particular, the results shown below correspond to

$$k=\eta(2-2\eta^2+\eta^3), f=\tfrac{1}{6}\eta(1-\eta)^3, h=\eta(1+3\eta)(1-\eta)^3 \tag{35}$$

which satisfy the additional boundary conditions $k_\eta = f_\eta = h_\eta = k_{\eta\eta} = f_{\eta\eta} = h_{\eta\eta} = 0$ at $\eta = 1$.

The temperature in the interior of the vessel $\phi_c(\theta)$, the velocity correction $U_o(\theta)$, and the boundary-layer thickness $\delta(\theta)$ are to be computed by integration of (19) written in the form

$$\frac{\mathrm{d}\phi_c}{\mathrm{d}\theta} = -\frac{\Lambda\sin^2(\theta)e^{\phi_c}}{2\delta\left(\phi_c\sin(\theta)\delta^2\int_0^1 f\mathrm{d}\eta + U_o\int_0^1 h\mathrm{d}\eta\right)}, \tag{36}$$

together with the momentum and energy equations (31) and (32) written with use made of the expressions

$$\int_0^\infty U^2 \mathrm{d}y = \phi_c^2 \sin^2(\theta)\,\delta^5 \int_0^1 f^2 \mathrm{d}\eta + 2U_o\phi_c \sin(\theta)\,\delta^3 \int_0^1 fh \mathrm{d}\eta + U_o^2 \delta \int_0^1 h^2 \mathrm{d}\eta, \tag{37}$$

$$\int_0^\infty U(\phi_c - \phi)\mathrm{d}y = \phi_c^2 \sin(\theta)\,\delta^3 \int_0^1 f(1-k)\mathrm{d}\eta + U_o\phi_c\,\delta \int_0^1 h(1-k)\mathrm{d}\eta, \tag{38}$$

$$\int_0^\infty (\phi_c - \phi)\mathrm{d}y = \phi_c\,\delta \int_0^1 (1-k)\mathrm{d}\eta, \tag{39}$$

$$\int_0^\infty U \mathrm{d}y = \phi_c \sin(\theta)\,\delta^3 \int_0^1 f \mathrm{d}\eta + U_o\,\delta \int_0^1 h \mathrm{d}\eta, \tag{40}$$

$$\left.\frac{\partial U}{\partial y}\right|_{y=0} = \phi_c \sin(\theta)\,\delta \left.\frac{\mathrm{d}f}{\mathrm{d}\eta}\right|_{\eta=0} + \frac{U_o}{\delta}\left.\frac{\mathrm{d}h}{\mathrm{d}\eta}\right|_{\eta=0}, \tag{41}$$

$$\left.\frac{\partial \phi}{\partial y}\right|_{y=0} = \frac{\phi_c}{\delta}\left.\frac{\mathrm{d}k}{\mathrm{d}\eta}\right|_{\eta=0}. \tag{42}$$

The equations were integrated by marching in θ. The integration is started with the initial values

$$\phi = \phi_{\max}, \quad U_o = C_U\,\phi_{\max}^{1/2}\theta, \quad \delta = C_\delta\,\phi_{\max}^{-1/4} \quad \text{at} \quad \theta \ll 1, \tag{43}$$

consistent with the previously identified self-similar solution (25). The constant factors C_U and C_δ are determined from the system of algebraic equations obtained by substituting (43) into (31) and (32), yielding the values $C_U = 0.021683$ and $C_\delta = 4.05827$ for the shape functions (35). The condition $\phi_c(\pi) = 0$ is used to determine the value of Λ corresponding to a given value of $\phi_{\max}$, producing the results shown in Fig. 6.

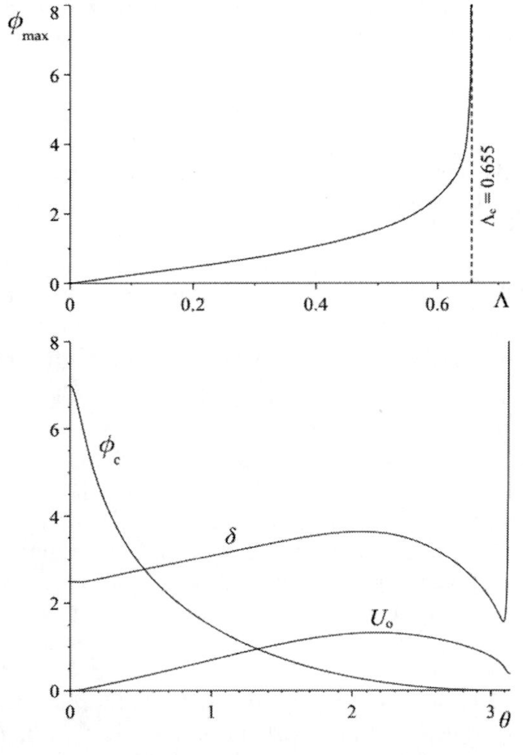

Figure 6: Results of numerical integrations of (31), (32) and (36) for the shape functions given in (35).

The lower plot in Fig. 6 represents typical profiles of δ, U_o, and ϕ_c for a large value of $\phi_{\max}$ (e.g., $\phi_{\max} = 7$ in the computation shown in the figure). Although the model (33) allows for the presence of backflow in the boundary layer, as would occur if U_o reaches sufficiently large negative values, such recirculating flow conditions were not achieved for any $\phi_{\max}$; instead, U_o was seen to remain always positive for $0 \leq \theta \leq \pi$. The nonzero value of U_o as $\theta \to \pi$, indicative of boundary-layer collision, can be attributed to the inherent limitations of this approximate model. As in Ito's analysis [30], the thickness of the mixing layer $\delta(\theta)$ shows a non-monotonic behavior for increasing θ and eventually diverges as $\theta \to \pi$, the latter boundary-layer thickening being consistent with the self-similar scaling $\delta \sim (\pi - \theta)^{-2/5}$ identified earlier for $\pi - \theta \ll 1$.

The upper plot in Fig. 6 shows the reduced explosion curve (i.e., the variation of the reduced Damköhler number Λ with $\phi_{\max}$). The integral method is not capable of describing the bending bifurcation; it yields instead a value of Λ that approaches a limiting value $\Lambda = \Lambda_c$ as $\phi_{\max} \to \infty$, seen in the upper plot. The vertical asymptote, rather than the turning point, identifies in this case the explosion limit, thereby providing the leading-order prediction

$$Da_c = \Lambda_c Ra^{1/4}. \tag{44}$$

For the polynomials (35) used in Fig. 6 the critical Damköhler number is found to correspond to $\Lambda_c \simeq 0.655$. Although a different selection of shape functions $k(\eta)$, $f(\eta)$, and $h(\eta)$ results in a different value of Λ_c, the associated changes were verified in additional computations to be only moderate. For instance, use of third-order polynomials for k and f and a fourth-order polynomial for h (satisfying (34) and having vanishing first-order derivatives at $\eta = 1$) yields instead the value $\Lambda_c = 0.616$. Since (44) can be expected to have relative errors of order $Ra^{-1/4}$ (that being the order of magnitude of the terms left out in the leading-order asymptotic description), the differences in Λ_c, on the order of 5%, are not very important for practical purposes, in that up to large values of $Ra \sim 10^6$ they lead to differences in predictions of Da_c that are comparable to or smaller than the inherent errors of the leading-order result (44).

The asymptotic prediction (44) evaluated with $\Lambda_c = 0.655$ is compared in the middle plot of Fig. 1 with the critical Damköhler numbers computed numerically from integration of (5)–(7). As can be seen, the accuracy of the prediction is truly remarkable. The relative departures between the two curves remain of order $Ra^{-1/4}$, consistent with the relative errors present in the leading-order development leading to (44). Achievement of this degree of success in numerical accuracy with a von-Karman integral method is perhaps somewhat surprising.

5 Concluding remarks

Simplified equations, derived for the slowly reacting mode of combustion and accounting for buoyancy-induced motion as well as for the time variations of the pressure in the confined environment of the vessel, provide a useful basis for developing insights into fluid dynamics and heat-transfer mechanisms associated with this ignition process. Numerical integrations of the transient problem successfully describe the development of the thermal explosion. The integrations indicate that, despite the complex dynamics arising during the transient stage, the predictions of explosion limits derived from Frank-Kamenetskii's criterion (i.e., based on the existence of steady solutions with negligible fuel consumption) remain accurate over the whole range of Rayleigh numbers explored in the paper.

In the limit $Ra \gg 1$ appropriate scalings can be identified for deriving predictions of critical explosion conditions when the flow is dominated by buoyancy. A characteristic structure emerges, including an inviscid convective-reactive core of slowly rising hot fluid, surrounded by a descending near-wall boundary-layer flow driven by the negative buoyancy associated with the cooler walls. The presence of backflow in the boundary layer gives this boundary-layer problem an elliptic character, resulting in an appreciably complicated boundary-value problem which needs results of analyses of special, simpler boundary layers at the top and bottom of the vessel for determining the appropriate boundary condition for the computation of the boundary layer along the curved walls. An approximate solution method, employing an integral form of the boundary-layer equations, combined with presumed shapes for the temperature and velocity profiles, circumvents computational difficulties that are encountered in attempts to integrate numerically the elliptic boundary-value problem that arises for the boundary layer on the curved walls. The most significant result of the asymptotic analysis for $Ra \gg 1$ is the leading-order prediction (44) of the critical Damköhler number for explosion as a function of the Rayleigh number. The value of the proportionality factor $\Lambda_c \simeq 0.655$, obtained from the approximate analysis appears to be surprisingly accurate, as may be seen in the comparisons of critical explosion conditions shown in the middle plot of Fig. 1.

This work serves to extend the Frank-Kamentskii theory of thermal explosions of ideal gas mixtures in rigid spherical vessels throughout the entire range of conditions over which buoyancy produces interior flows that remain laminar.

The associated fluid-flow influences on explosion conditions become important only for Rayleigh numbers above a few hundred, and transitions to turbulent flow occur at Rayleigh numbers of a few million, whence the current results may be useful over about four orders of magnitude of the Rayleigh number. Both numerical and conceptual challenges were encountered, but the correct scalings for laminar flows at large Rayleigh numbers were established, along with the determination of reasonably accurate values for associated constants that appear in the scaling relations. Future studies for large Rayleigh numbers may address peculiar geometrical effects that are known to be encountered in vessels of different shapes (as indicated in the introduction), or turbulent flows at higher Rayleigh numbers.

Acknowledgments

We would like to thank Norman Riley for his numerous enlightening comments regarding our work. In particular, he pointed out the similarities between the boundary-layer problem identified here in the limit $Ra \gg 1$ and that emerging in fully developed flow in curved pipes at large values of the Dean number, and he also suggested the use of the integral formulation to obtain an approximate solution. The work of FAW is supported by the US National Science Foundation through award #CBET-1404026.

References

[1] DA Frank-Kamenetskii. *Zh. Fiz. Khim.*, 13 (1939) 738–755.

[2] David Al'bertovich Frank-Kamenetskii. *Diffusion and heat exchange in chemical kinetics*. Princeton University Press, 1955.

[3] Antonio L Sánchez and Forman A Williams. *Prog. Energy Combust. Sci.*, 41 (2014) 1–55.

[4] Antonio L Sánchez, Eduardo Fernández-Tarrazo, and Forman A Williams. *Combust. Flame*, 161 (2014) 111–117.

[5] Amable Liñán, Daniel Moreno-Boza, Immaculada Iglesias, Antonio L Sánchez, and Forman A Williams. *Combust. Theory Modelling*, 20 (2016) 1010–1028.

[6] DR Jones. *Int. J. Heat Mass Transfer*, 16 (1973) 157–167.

[7] AK Kolesnikov. *Int. J. Heat Mass Transfer*, 35 (1992) 1091–1102.

[8] G Joulin, A.B. Mikishev, and G.I. Sivashinsky. A Semenov–Rayleigh–Benard problem. Unpublished, 1996.

[9] L Kagan, H Berestycki, G Joulin, and G Sivashinsky. *Combust. Theory Modelling*, 1 (1997) 97–111.

[10] EA Shtessel', KV Pribytkova, and AG Merzhanov. *Combust. Explos. Shock Waves*, 7 (1971) 137–146.

[11] AG Merzhanov and EA Shtessel'. *Acta Astronaut.*, 18 (1973) 191–199.

[12] Marc Massot, Stephane Génieys, Thierry Dumont, and VA Volpert. *SIAM J. Appl. Math.*, 63 (2002) 351–372.

[13] Michaël Belk and Vitaly Volpert. *Chaos: An Interdisciplinary Journal of Nonlinear Science*, 14 (2004) 263–273.

[14] DR Jones. *Int. J. Heat Mass Transfer*, 17 (1974) 11–21.

[15] AN Campbell, SSS Cardoso, and AN Hayhurst. *Chem. Eng. Sci.*, 62 (2007) 3068–3082.

[16] Ting-Yueh Liu, Alasdair N Campbell, Silvana SS Cardoso, and Allan N Hayhurst. *Phys. Chem. Chem. Phys.*, 10 (2008) 5521–5530.

[17] Ting-Yueh Liu, Alasdair N Campbell, Allan N Hayhurst, and Silvana SS Cardoso. *Combust. Flame*, 157 (2010) 230–239.

[18] WH Archer. *Heat transfer mechanisms in exothermic reactions*. PhD thesis, The University of Manchester, 1977.

[19] Ting-Yueh Liu and Silvana SS Cardoso. *Combust. Flame*, 160 (2013) 191–203.

[20] AN Campbell. *Phys. Chem. Chem. Phys.*, 17 (2015) 16894–16906.

[21] Antonio L Sánchez, Immaculada Iglesias, Daniel Moreno-Boza, Amable Liñán, and Forman A Williams. *Combust. Theory Modelling*, 20 (2016) 1029–1045.

[22] F. Hecht. *J. Numer. Math.*, 20 (2012) 251–265.

[23] B.J. Tyler. *Combust. Flame*, 10 (1966) 90–91.

[24] P G Ashmore, B J Tyler, and T A B Wesley. *Proc. Combust. Inst.*, 11 (1967) 1133–1140.

[25] J. Carpio and J. L. Prieto. *Comput. Methods in Appl. Mech. Eng.*, 273 (2014) 77–99.

[26] SCR Dennis and N Riley. *Proc. Roy. Soc. Lond. A*, 434 (1991) 473–478.

[27] Janet M Potter and N Riley. *J. Fluid Mech.*, 100 (1980) 769–783.

[28] T Chiang, A Ossin, and CL Tien. *J. Heat Transfer*, 86 (1964) 537–541.

[29] FT Smith. *J. Fluid Mech.*, 71 (1975) 15–42.

[30] H Ito. *Z. Angew. Math. Mech.*, 49 (1969) 653–663.

Western States Section/Combustion Institute
P.O. Box 969 M S 9052
Livermore, California 94551-0969
USA

ISBN 978-1-5108-5203-7

9 781510 852037

CURRAN ASSOCIATES INC.
proceedings.com